The Handbook of Science and Technology Studies

Handbook Advisory Board

The Handbook of Science and Technology Studies

Third Edition

Edited by

Edward J. Hackett
Olga Amsterdamska
Michael Lynch
Judy Wajcman

Published in cooperation with the Society for Social Studies of Science

The MIT Press
Cambridge, Massachusetts
London, England

For information about special quantity discounts, please email special_sales@mitpress.mit.edu

This book was set in Stone Serif and Stone Sans by SNP Best-set Typesetter Ltd., Hong Kong.

Printed and bound in the United States of America.

Library of Congress Cataloging-in-Publication Data
The handbook of science and technology studies / Edward J. Hackett . . . [et al.], editors.—3rd ed.
    p. cm.
  Includes bibliographical references and index.
  ISBN 978-0-262-08364-5 (hardcover : alk. paper)
  1. Science. 2. Technology. I. Hackett, Edward J., 1951– II. Society for Social Studies of Science.
Q158.5.N48 2007
303.48′3—dc22

                                                                                2007000959

10 9 8 7 6 5 4 3 2 1

In memory of David Edge

Contents

Preface

The Handbook of Science and Technology Studies, Third Edition, testifies to a thriving field of research in social studies of science, technology, and their interactions with society. The editors of the third Handbook have done a tremendous job by mapping a multifaceted but now clearly maturing field. This volume shows the richness of current empirical and theoretical research. The volume also displays—indirectly, in the notes, references, and bibliographies—the institutional strengths of the field in terms of journals, book series, research institutes, and graduate and undergraduate programs. And it significantly highlights that research in science and technology studies is increasingly engaging with the outside world. This engagement is partly directed toward other academic disciplines and practices and partly toward addressing questions of policy and governance in public and political institutions.

This Handbook was produced under the aegis of the Society for Social Studies of Science (4S). The Society selected the proposal by the editorial team and constituted the Handbook Advisory Board to monitor and assist in the process. Most importantly, the editors drew on the wealth of scholarship produced by the 4S membership. During 4S annual meetings, consecutive steps for developing the Handbook were presented by the editors and discussed with 4S members. The Handbook thus bears witness to the richness within the STS scholarly community, encompassing different generations of researchers, different research agendas, and different styles of engagement. It is, then, with conviction and pride that 4S grants its imprimatur to this Handbook.

4S extends its gratitude to all who have contributed to the realization of this grand project: the contributing authors, the members of the Advisory Board, and the staff at MIT Press. First and foremost, however, the Society is indebted to the editors Ed Hackett, Olga Amsterdamska, Mike Lynch, and Judy Wajcman. They succeeded in producing a truly exciting handbook that maps the current state of the field while also offering new challenges and innovative perspectives for future research.

Wiebe E. Bijker
Michel Callon

Acknowledgments

An effort of this magnitude and duration incurs many debts, and it is a pleasure for us to thank the many who helped us with the Handbook. We do so knowing, with some embarrassment, that we will surely overlook others who helped us. We thank them here first, adding our apologies for forgetting to thank them by name.

We are very grateful to the Handbook Advisory Board, led by Wiebe Bijker and Michel Callon, in their capacities as 4S president and Advisory Board chair, respectively, and to Jessie Saul, who coordinated the board's activities. Wiebe, Michel, and Jessie, accompanied at times by Wes Shrum, worked openly with us and quietly behind the scenes on the strategy, substance, and logistics of the Handbook.

Chapters were reviewed by Alison Adam, Anne Balsamo, Isabelle Baszanger, Anne Beaulieu, Stuart Blume, Bob Bolin, Geoff Bowker, Suzanne Brainard, Steve Brint, Phil Brown, Larry Busch, Alberto Cambrosio, Monica Casper, Claudia Casteneda, Daryl Chubin, Adele Clarke, Simon Cole, Dave Conz, Elizabeth Corley, Jennifer Croissant, Norman Denzin, Jose van Dijk, Gilli Drori, Joseph Dumit, Ron Eglash, Wendy Faulkner, Jennifer Fishman, John Forrester, Michael Fortun, Scott Frickel, Joan Fujimura, Jan Golinski, David Gooding, Michael Gorman, Alan Gross, Hugh Gusterson, David Guston, Rob Hagendijk, Martin Hajer, Patrick Hamlett, David Healey, Joseph Hermanowicz, Kathryn Henderson, Stephen Hilgartner, Eric von Hippel, Christine Hine, Rachelle Hollander, Alan Irwin, Paul Jones, Sarah Kember, Nick King, Daniel Kleinman, Jack Kloppenberg, Martin Kusch, David Livingstone, Scott Long, Ilana Löwy, William Lynch, Harry Marks, Brian Martin, Joan McGregor, Martina Merz, Carolyn Miller, Philip Mirowski, Thomas Misa, Chandra Mukerji, Greg Myers, Thomas Nickles, Paul Nightingale, Jason Owen-Smith, Andrew Pickering, Ted Porter, Lawrence Prelli, Paschal Preston, Rayna Rapp, Nicholas Rasmussen, Judith Reppy, Alan Richardson, Toni Robertson, Susan Rosser, Joseph Rouse, Dan Sarewitz, Simon Shackley, Sara Shostak, Susan Silbey, Sergio Sismondo, Sheila Slaughter, Radhamany Sooryamoorthy, Knut Sørensen, Susan Leigh Star, Nico Stehr, Jane Summerton, Judy Sutz, Karen-Sue Taussig, Paul Thagard, Stefan Timmermans, Sherry Turkle, Frederick Turner, Stephen Turner, Gerard de Vries, Clare Waterton, Robin Williams, Ned Woodhouse, Sally Wyatt, Steven Yearley, Petri Yikoski, Steve Zavestoski, and Steve Zehr.

In our editorial work we were ably assisted on our campuses by Julian Robert, Ceridwen Roncelli, Kiersten Catlett, Stephanie Meredith, and Nicole Heppner. We also thank the Consortium for Science Policy and Outcomes at Arizona State University, which served as genial host for the secret handbook Web site.

The staff of MIT Press, particularly Marguerite Avery and Sara Meirowitz, and Peggy Gordon, production editor, expertly carried the volume into print.

Edward J. Hackett
Olga Amsterdamska
Michael Lynch
Judy Wajcman

Introduction

Edward J. Hackett, Olga Amsterdamska, Michael Lynch, and Judy Wajcman

In the mid-1970s, Ina Spiegel-Rösing and Derek J. de Solla Price organized and edited *The Handbook of Science, Technology, and Society* because they felt "a strong need for some sort of cross-disciplinary mode of access to this entire spectrum of scholarship" and also wanted to "contribute to the intellectual integration" of the emergent field. Spiegel-Rösing and Price were visionary in setting themselves the task and spectacularly successful in seeing it through: a field of scholarship was born and took flight. Some 18 years later *The Handbook of Science and Technology Studies* (note the title change) was published, providing "a map of a half-seen world" characterized by "excitement and unpredictability" (Jasanoff et al., 1995: xi). Introducing the third edition of this episodic series challenges us to find the right metaphor for activity in our field today. If the 1970s was an era of disciplinary juxtaposition and integration and the 1990s a time for mapping a half-seen world of shifting continents and emerging countries, then in our time the field of science and technology studies (STS) may be characterized by its engagement with various publics and decision makers, its influence on intellectual directions in cognate fields, its ambivalence about conceptual categories and dichotomies, and its attention to places, practices, and things.

STS has become an interdisciplinary field that is creating an integrative understanding of the origins, dynamics, and consequences of science and technology. The field is not a narrowly academic endeavor: STS scholars engage activists, scientists, doctors, decision makers, engineers, and other stakeholders on matters of equity, policy, politics, social change, national development, and economic transformation. We do so with some hesitation and considerable self-reflection because we seek academic respectability and institutionalization and their accompanying resources (professorships, departments, degrees, and research grants), yet also strive for change in the service of justice, equity, and freedom. Establishing and holding the right balance will be challenging, with the risk of irrelevance and disengagement on the one side and cooptation and loss of prestige and resources on the other. Through three decades of interdisciplinary interaction and integration, shifting intellectual continents and cataclysmic conceptual shocks, perseverance and imagination, STS has become institutionalized and intellectually influential, and STS scholars have become engaged in various arenas of activism and policy. A decade ago, STS was mired in the "science wars"; today, STS scholars are invited (and supported) to engage in a spectrum of studies with implications for science and technology policies and practices.

Place, time, and editorial process have figured prominently in Handbook introductions, perhaps reflecting a professional commitment to situating knowledge and disclosing institutional circumstances and influences. The 1977 volume was conceived within the International Council for Science Policy Studies, which was established at the International Congress of the History of Science in Moscow in 1971. The editors selected "a team of authors from all the different disciplines and fields we felt had to be incorporated, [and then] the contents and boundaries of the sections had to be specified and negotiated with prospective authors" (Spiegel-Rösing & Price, 1977: 2). Over a four-year period, authors and editors met and worked in Moscow, Schloss Reiseburg (Germany), Amsterdam, Delhi, and Paris. Despite such peregrinations, the editorial process was linear, centralized, and dispassionate.

The second Handbook, and the first to bear the imprimatur of the Society for Social Studies of Science (4S, founded a year before the first Handbook was published), was produced in circumstances of greater passion and less certainty than its predecessor. The opening paragraph of the introduction conveys uncertainties in language that is candid, inviting, and explicitly spatial. The field is "still emerging," so do not expect "the traditional, treatiselike handbook that would clinically describe the field of STS . . . for it had not yet achieved the hoary respectability that merits such dispassionate, and unimaginative, treatment" (Jasanoff et al., 1995: xi). Instead, the Handbook offers "scholarly assessments of the field . . . definitive roadmaps of the terrain . . . that project the field's broad interdisciplinary and international outlook . . . [and] capture for readers who come fresh to STS a little of the excitement and unpredictability that have drawn scholars . . . to claim STS as their primary intellectual home . . . an unconventional but arresting atlas of the field at a particular moment in its history" (Jasanoff et al., 1995: xi–xii). To elicit and organize this mass of diverse material the editors first marked the "meridians and parallels" of STS, then opened their mailbox, and "proposals came flooding in" for topics on the map and topics not anticipated. "The field, it seemed, was intent on defining itself in ways not initially contemplated. We decided to accept this movement toward self-definition. Rather than search for authors to occupy every vacant slot in the proposal, we decided to redraw the boundaries to include more of the topics that authors did wish to address . . . [which yielded] a more interesting and comprehensive, if not always more coherent, guide to the field" (Jasanoff et al., 1995: xiii). In contrast to its predecessor, the second volume was intimately co-produced (or co-edited) by the community of authors; it exhibits irony and passion, contingency and agility.

Our birth story is more mundane. We did not enjoy the exotic travels of the inaugural Handbook: technologies allowed us to work from a distance but did not speed the pace of production: the 1977 volume took six years to complete, the 1995 took seven, and the present volume (the largest of the three) has taken eight years from conception to publication.

Working under the aegis of 4S, we divided the intellectual terrain of STS into four parts: theory and methods, reciprocal relations with other fields, engagement with the public sphere, and enduring themes and new directions. In fall 2003 we issued a call for chapter proposals to the entire membership of 4S, posted it on relevant bulletin boards of other societies, and listed it in appropriate newsletters. We aimed for a handbook that would consolidate the field's accomplishments, welcome new scholars to enter STS, and indicate promising research pathways into the future. Twenty of the chapters in this book were written in response to that call. For the balance, with guidance from our advisory board, we identified topics that were essential but overlooked and solicited manuscripts to address them.

In reconciling the top down and bottom up editorial processes, our meridians and parallels shifted, producing unforeseen alignments. Theory chapters focused on problems such as technological determinism or social worlds rather than on competing schools and systems of ideas. There were no chapters explicitly devoted to methods. Invitations to consider relations with other fields revealed some new connections (e.g., with communication studies or cognitive sciences) but passed over the traditional interdisciplinary engagements of STS with anthropology, medical sociology, and history. Perhaps such connections are so deep and integral that they escape notice. What emerged instead is a multifaceted interest in the changing practices of knowledge production, concern with connections among science, technology, and various social institutions (the state, medicine, law, industry, and economics more generally), and urgent attention to issues of public participation, power, democracy, governance, and the evaluation of scientific knowledge, technology, and expertise.

These topics are approached with theoretical eclecticism: rather than defending pure positions, authors risked strategic crossovers and melded ideas from different intellectual domains. Normativity, relativism, and evaluation of expertise and scientific knowledge endure from previous volumes but in new ways: no longer just problems for philosophical reflection, such concerns are now posed in terms that seek collective political or social resolution.

Politics, democracy, and participation in scientific and technological decision making are pervasive. Politics is no longer just science policy and is not limited to guidance in various substantive realms (environment, health, information technologies, and so forth), but instead takes the form of a general concern about *which* political systems, institutions, and understandings; *which* participants with *what* qualifications, roles, and responsibilities; and *which* kinds of civil society would be most democratic while preserving the benefits of scientific and technical expertise. And these are not posed as abstract intellectual puzzles but as problems of concrete technologies, practices, and institutions in specific places and circumstances with particular challenges and limitations.

If the first Handbook (1977) characterized STS as a nascent field borrowing disciplinary ideas and theories to explain science and technology, and the second Handbook (1995) as an adolescent field coalescing and establishing its identity, this Handbook presents STS as a maturing field that generates ideas and findings used to address fundamental problems in other disciplines.

How have differences in the intellectual state of the field influenced the organization of STS Handbooks in three eras? The 1977 STS Handbook placed its 15 chapters into three sections: normative and professional contexts, disciplinary perspectives on science studies, and interdisciplinary perspectives on science policy. Then, as now, science studies and science policy occupied separate spheres, with too little discourse between them, and the structure of the 1977 book reflects this segregation. Even in the section on science studies there are independent chapters for the sociology, history, economics, philosophy, and psychology of science.

In the 1995 Handbook, the number of chapters has increased to 28, grouped into seven sections (including an opening section containing a single chapter), and the titles are revealing. "Technology" has become more than a full partner: among section headings the term (or its cognate) is mentioned everywhere that the term "science" appears, and once ("Constructing Technology") by itself. Chapters titled "the [fill in a discipline] of science" have been replaced by chapters concerned with finer social processes (e.g., laboratory studies, boundaries), emergent phenomena (e.g., "machine intelligence," globalization), communication (and other representations of the public), and controversy and politics (which are virtually everywhere).

The Handbook in your hands has 38 chapters in five main sections. The first offers framing ideas and perspectives on STS, sketching the conceptual and historical foundations of the field. The second section is concerned with the people, places, and practices of research, continuing the field's abiding concern for the circumstances of knowledge production. The third considers the diverse publics and politics of science and technology, collecting ideas and empirical studies that demonstrate and extend the relevance of STS scholarship for policy and social change. The fourth section examines the institutions and economics of science and technology, filling a void noted in the 1995 volume. The fifth and final section collects chapters concerned with emergent technologies and sciences, pointing the way for new research.

A handful of powerful themes cut across sections and chapters. First among these is an emphasis on social action and activity: science and technology are as they do, so attention is directed toward arrangements and practices that produce knowledge, meaning, and impact. Second, sharp identities and distinctions are replaced by hybrids and ambiguities, tensions and ambivalences. Sets are fuzzy; categories are blurred; singulars become plurals (sciences, not science; publics, not public, for example); and linear causality, even reciprocal causality, is replaced by processes of co-production that imply deeply integrated action. Third, context, history, and place matter more than ever, and not only at the level of individual action—the individual scientist in organizational and historical context—but also at the larger scales of institutional structure and change. Despite the challenges posed by these increasingly sophisticated conceptualizations, the explanatory objective remains a precise, empirical, multilevel account of the processes of production, influence, and change. But that analytic goal alone no longer suffices: it must be wedded to an agenda of social change, grounded in the bedrock of ethical principles and explicit values (equality, democracy, equity, freedom, and others). Where once it may have seemed adequate to choose to write in either an analytic or normative mode, or to attach normative implications to an analytic argument, the emerging challenge is to integrate or synthesize those modes of thought.

Absent from this Handbook, to our regret, is much systematic treatment of research methods, not only quantitative methods—survey tools, network techniques, bibliometry, experiments, and analytic models—but also qualitative, observational, and text-based techniques that are evolving rapidly across the social sciences. Decades ago some in our field may have been "against method," but since then empirical attention to the practices of science and technology and to the sentient and tacit knowledge embodied within those practices surely argues for making explicit our own methods and epistemic assumptions, if only so others can build upon our successes and learn from our mistakes.

A division between studies of science and studies of science policy has endured for 30 years, to their mutual impoverishment. While the divide remains, the present Handbook offers new opportunities for dialog. The first Handbook attempted, with some success, to bring those worlds into conversation. But in the second Handbook, David Edge noted this persistent divide with displeasure and posed this challenge: "Given that critical STS scholarship paints a distinctive and fresh picture of science—a new 'is'—what are the policy implications (if any)—the new 'ought'— that follow?" (Edge, 1995: 16). In this Handbook the beginnings of an STS answer may be discerned in discussions of governance and democracy, in the consistent attention given to activism, politics, social movements, and user engagement, and in concern for empowerment and egalitarianism. An STS scholar today would substitute plural "oughts" for the singular, recognizing that different groups and their interests would be served by different courses of action, and would examine dynamic, interactive processes that are shaped by circumstances of science and technology, society and history. But the core challenge remains: how to bring the distinctive insights and sensibilities of STS into the analysis of policy and the process of social change.

By pursuing a change agenda, by addressing matters of policy and politics, and by engaging the various parties to such discussions, STS scholars have opened themselves to criticism from various quarters. Scientists, politicians, business interests, religious groups, activists, and others have engaged one another—not always fruitfully—on climate change, human evolution, the inception and end of life, the ethics of stem cell research, the cause and treatment of HIV infection, and much else. What can STS contribute to resolving these conflicts? Many things: strategies for arranging and evaluating evidence, patterns of reasoning from principles of right conduct or democratic process, ways to bridge the logics of law, politics, policy, religion, and common sense, and empirical insights that might explain a causal process or break an impasse. By providing historical perspective, by taking a variety of social points of view, and by revealing the social, political, and ethical implications of technology and science, STS deflates hyperbolic rhetoric about technoscientific miracles and shapes developmental pathways.

David Edge found a hopeful sign of the institutionalization of STS: "In the United States, the National Science Foundation has, for many years, maintained a program of support for activities in the field of 'ethics and values' [of science and technology]. This has survived many metamorphoses but has recently come to embrace aspects of policy research and has joined forces with the history and philosophy of science" (Edge, 1995: 16). We would reassure David that such hopeful signs have grown: not only are the NSF programs flourishing, but there are also vibrant, well-funded programs committed to the ethics, history, philosophy, and social study of science and technology to be found within the research councils and science foundations of many governments.

Looking backward is a way of assessing the distance traveled and taking bearings to guide the way forward. In her opening chapter of the 1977 Handbook, Ina Spiegel-Rösing identified five "cardinal tendencies" of STS (1977: 20–26). The field, she observed, tends to be *humanistic* in its focus on real, acting human beings; *relativistic* in its systematic attention to place, time, and history; *reflexive* in its critical self-awareness of the potential influence of research on the object studied; *de-simplifying* in its commitment to "un-blackboxing" phenomena, understanding mechanisms, and delineating reciprocal influences; and *normative* in its commitment to understanding the ethics and values implicit in science and technology and to using that understanding to guide the transformative powers of science and technology in ways that are more generally beneficial and less potentially harmful. We believe these cardinal tendencies are perhaps the most admirable and wise qualities of STS research and invite the reader to use these to take bearings and chart progress after three decades of travel.

For balance, it is sobering to remember that Spiegel-Rösing also listed four "major and fairly obvious deficiencies" in STS research, which endure in varying degrees: *rhetoric pathos*, or the unfortunate trait of posing a problem without making much progress toward its solution; *fragmentation*, or divisions between disciplines in their studies of science and technology, and between STS and policy-relevant research; a *lack of comparative research* across disciplines and nations; and a bias toward studying the *bigger and harder sciences* (1977: 27–30). Thirty years later, STS has acquired intellectual and institutional integrity, though centrifugal forces swirl beneath its surface; there is a growing amount of research concerned with science and technology in comparative and global perspectives, performed by an increasingly global community of scholars; analytic attention has shifted from "bigger and harder sciences" toward a spectrum of fields, with special concern for their distinctive qualities; and as for rhetoric pathos, we leave that for the reader to judge.

Much has changed and much remains obdurate, but what matters is how the foundations and dynamics summarized in this body of work will shape the next decade of scholarship.

Finally, please note that this edition of the Handbook is dedicated to the memory of David Edge, who died in January 2003. David was the first director of the Science Studies Unit at the University of Edinburgh, co-founding editor of *Social Studies of Science*, and President of the Society for Social Studies of Science. Above and beyond these leadership roles, he was and remains a guiding spirit for the field. As we reflect on the central themes of this Handbook, we are reminded of David's belief that the substance and insights of STS scholarship are "of central concern to humankind. STS analysis points to all the 'higher' aspects of human endeavor—truth and power and justice and equity and democracy—and asks how these can be conserved and consolidated in modern society, so that the immense possibilities of scientific knowledge and technological innovation can be harnessed (in Bacon's words) 'for the relief of man's estate'" (Edge 1995: 19).

References

Edge, David (1995) "Reinventing the Wheel," in Sheila Jasanoff, Gerald E. Markle, James C. Petersen, & Trevor Pinch (eds), *Handbook of Science and Technology Studies* (Thousand Oaks, London, New Delhi: Sage): 3–23.

Jasanoff, Sheila, Gerald E. Markle, James C. Peterson, & Trevor Pinch (eds) (1995) *Handbook of Science and Technology Studies* (Thousand Oaks, London, New Delhi: Sage).

Spiegel-Rösing, Ina & Derek J. de Solla Price (eds) (1977) *Handbook of Science, Technology, and Society* (London, Thousand Oaks: Sage).

I Ideas and Perspectives

Michael Lynch

Section I of the Handbook includes chapters that present ideas and perspectives that apply broadly to science and technology studies (STS). We should keep in mind, however, that broadly does not mean universally. A signal feature of much current STS research is an aversion to universalistic claims about science, knowledge, or STS itself. This aversion does not arise from defensiveness or timidity, but from an acute recognition that disciplinary histories and characterizations of states of knowledge are both topics and resources for STS. Not only are accounts of ideas and perspectives themselves perspectival, they often perform transparent, or not-so-transparent, political work. Such political work sometimes expresses narrow self-promotional agendas, or the ambitions of a "school" or "program" often consisting of a mere handful of people located at one or two academic institutions. More often these days, as indicated by recent STS conference themes and trends in the literature, scholars in our field aim beyond struggles for academic recognition and express ambitions to instigate changes in the world at large (ambitions whose articulations can in themselves become vehicles of academic recognition). In current STS discussions, terms such as "normativity," "activism," "intervention," and "engagement" signal a desire (or sometimes a wish) to critically address extant versions of science and technology, and by so doing to effect changes and redress inequalities in the way scientific, technical, and clinical knowledges are presented and deployed in particular cultural and institutional circumstances (the neologistic pluralization of the word "knowledge" itself signals a refusal to go along with a conception of knowledge as singular and universal).

The tension between simply presenting a history of STS and denying the very possibility of doing so is nicely expressed by Sergio Sismondo when he says in his chapter: "STS in one lesson? Not really." But then he goes on to present "one easy lesson" (but not *the only* lesson) to be drawn from recent trends in the field. Like many other contributors to this and later sections of the Handbook, Sismondo addresses efforts to engage with politics in STS research programs, but he also points to the complications and dilemmas we face when trying to politically mobilize research in a field notorious for its relativism. Instead of five-year plans for reforming science and technology, are we to contemplate five-year programs in situated knowledges? Perhaps so, but it can be baffling to consider what such programs would look like.

Concerns about the politics of science and technology are far from new, as Stephen Turner informs us in his chapter on the pre-Mertonian intellectual origins of science studies. But, as numerous chapters in this Handbook illustrate, STS interest in politics has never been more pervasive than at the present time—a time characterized by paradoxical developments calling for nuanced treatments. Lucy Suchman observes in her chapter that a heightened awareness of the politics of STS is far from a straightforward matter of consolidating our knowledge into normative principles that can then be applied in the political domain: "intervention presupposes forms of engagement, both extensive and intensive, that involve their own often contradictory positionings." Engagement in controversies about science and technology forces us to confront robust conceptions of science and technology that might otherwise seem to have been buried in the past by the philosophical arguments and case studies we present to our students. For example, as Sally Wyatt demonstrates in her chapter, although technological determinism has been reduced to the status of a straw position in technology studies, it is alive and well in business and policy circles and remains so pervasive that it even tends to dwell within our own, unreconstructed patterns of thought.

Some of the chapters in this section revisit familiar themes and perspectives, but rather than accepting established disciplinary agendas and lines of demarcation, they attempt to show how STS research challenges set ways of thinking about science, knowledge, and politics. For example, Adele Clarke and Susan Leigh Star turn their discussion of the "social worlds" perspective (a line of research ostensibly derived from symbolic interactionist sociology) back on sociological theory itself, to challenge general conceptions of theory and method that continue to pervade sociology. Similarly, when Charles Thorpe addresses the place of political theory in STS research, instead of showing how such research derives from one or another political philosophy, he argues for a conception of STS *as* political theory. He points out that the theoretical and political implications of lines of STS research do not predictably follow from the ideological positions on which they are supposedly founded. Considered in this way, STS is not a substantive field of theoretical "application," but is instead a source of critical insight into the conceptual underpinnings of modern social and political theory.

Several of the chapters in this section exemplify this turning from substantive engagement to critical theoretical insight—and not just insight *guided* by critical theory but insight into the very ideas of "theory," "the social," and what it means to be "critical." Warwick Anderson and Vincanne Adams identify postcolonial studies as a point of leverage for challenging univocal notions of technological progress with a pluralized conception of modernity. Suchman's chapter on the "sciences of the artificial" shows how feminist and other lines of STS research propose to redistribute the intellectual, cultural, and economic configurations implied by distinctions between humans and machines, and designers and users. Turner's (pre)history of STS points to the contingencies of history and of the tenuous links between particular political philosophies and versions of science (also see Thorpe's nuanced treatment of the relationship between post-Kuhnian STS and conservative thought). Finally, and characteristically, Bruno Latour eschews the "debunking urge" so often ascribed to STS, and proposes to merge histories of scientists with histories of the worldly things with which scientists (and the rest of us) are concerned. Politics is no longer confined to smoke-filled rooms; political action becomes embedded in the very substance of smoke itself and its effects on, for example, lung tissue and global climate. Consequently, the *politics* of science and technology—the subject and agenda of so many of the chapters in this Handbook—becomes at once mundane and mysterious. It is mundane—about worldly things and accounts of their details—and mysterious—involving hidden agendas compounded by hidden contingencies. Consequently, as the chapters in this section argue and exemplify, "engagement" is as much a condition for doing original STS research as a matter of following through on its lessons in real-worldly settings.

1   Science and Technology Studies and an Engaged Program

Sergio Sismondo

There is the part of Science and Technology Studies (STS) that addresses and often challenges traditional perspectives in philosophy, sociology, and history of science and technology; it has developed increasingly sophisticated understandings of scientific and technical knowledge, and of the processes and resources that contribute to that knowledge. There is also the part of STS that focuses on reform or activism, critically addressing policy, governance, and funding issues, as well as individual pieces of publicly relevant science and technology; it tries to reform science and technology in the name of equality, welfare, and environment. The two parts, which Steve Fuller (1993) has called the "High Church" and "Low Church" of STS, differ simultaneously in goals, attention, and style, and as a result the division between them is often seen as the largest one in the field.

However, this image of division ignores the numerous bridges between the Churches, so numerous that they form another terrain in which the politics of science and technology are explored. There we find theorists increasingly concerned with practical politics of science, articulating positions with respect to questions about the place of expertise in a democracy, or engaging in studies that directly bear on questions of reform and activism. In particular, constructivist STS has created a space for theoretically sophisticated analyses of science and technology in explicitly political contexts. By way of a scandalously short history of STS, this chapter describes that space.

SCIENCE AND TECHNOLOGY STUDIES IN ONE EASY LESSON

STS in one lesson? Not really. However, one important feature of the field can be gained from one lesson: STS looks to how the things it studies are constructed. The history of STS is in part a history of increasing scope—starting with scientific knowledge, and expanding to artifacts, methods, materials, observations, phenomena, classifications, institutions, interests, histories, and cultures. With those increases in scope have come increases in sophistication, as its analyses assume fewer and fewer fixed points and draw on more and more resources to understand technoscientific constructions. A standard history of STS (as in Bucchi, 2004; Sismondo, 2004; or Yearley, 2005) shows how this has played out.

The metaphor of "construction," or "social construction," was so ubiquitous in the 1980s and 1990s that now authors in STS bend over backward to avoid using the term: other terms, like "framing," "constitution," "organization," "production," and "manufacture," fill similar roles, attached to parts of the construction of facts and artifacts. The construction metaphor has been applied in a wide variety of ways in STS; attention to that variety shows us that the majority of these applications are reasonable or unobjectionable (Sismondo, 1993). We may also, though, pay attention to the central implications of the metaphor, the ones that allow it to be used in so many different ways and about so many different subject matters. Social constructivism provides three important assumptions about science and technology, which can be extended to other realms. First, science and technology are importantly *social*. Second, they are *active*—the construction metaphor suggests activity. And third, they do not provide a direct route from nature to ideas about nature; the products of science and technology are *not themselves natural* (for a different analysis, see Hacking, 1999).

A standard history of STS might start with Thomas Kuhn's *Structure of Scientific Revolutions* (1962), which emphasized the communal basis of the solidity of scientific knowledge, the perspectival nature of that knowledge, and the hands-on work needed to create it. More importantly, the popularity of Kuhn's book and iconoclastic readings of it opened up novel possibilities for looking at science as a social activity.

In this way, Kuhn's work helped make space for another starting point in the field, David Bloor's (1976) and Barry Barnes's (1974) articulation of the "strong program" in the sociology of knowledge. The strong program starts from a commitment to naturalist explanations of scientific and mathematical knowledge, to investigating the causes of knowledge. Much traditional history and philosophy of science retained non-naturalist patterns of explanation by explaining beliefs deemed true (or rational) and false (or irrational) asymmetrically, in so doing importing an assumption that truth and rationality have an attractive force, drawing disinterested science toward them. Such asymmetric treatments of science assume that, *ceteris paribus*, researchers will be led to the true and the rational, and therefore there can be no sociology of scientific knowledge but only a sociology of error. The strong program, then, provides a theoretical backdrop for studying the construction of scientific knowledge and not just error.

The strong program was most immediately worked out in terms of interests: interests affect the positions people adopt and shape the claims that count as scientific knowledge (e.g., MacKenzie, 1981; Shapin, 1975). A current body of work in STS largely compatible with interest-based explanations is feminist work revealing the sexism or sexist origins of particular scientific claims, usually ones that themselves contribute to the construction of gender (e.g., Fausto-Sterling, 1985; Martin, 1991; Schiebinger, 1993). This strand of feminist STS shows how ideology, as starting and ending points, contributes to the construction of scientific knowledge.

The empirical program of relativism (EPOR), mostly due to Harry Collins's work in the 1970s, bears much similarity to the strong program (e.g., Collins, 1985). Symmetry is achieved, as it is for many strong program studies, by focusing on controversies, during which knowledge is undetermined. Controversies display interpretive flexibility: materials, data, methods, and ideas can be given a range of interpretations compatible with the competing positions. For this reason, Collins's methodological relativism asserts that the natures of materials play no role in the resolution of controversies. EPOR goes on to show that there is always a regress in scientific and technical controversies. Judgments of interpretations and of the claims they support depend on each other, as participants in a controversy typically see the work and arguments to support a claim as sound to the extent that they see the claim itself as sound. Case studies support a picture of controversies being resolved through actions that define one position as the right and reasonable one for members of an expert community to hold. Thus, the constitution of scientific knowledge contains an ineliminable reference to particular social configurations.

While they are (literally) crucial components of the construction of scientific and technical knowledge, controversies are also only episodes in that construction, episodes in which groups of experts make decisions on contentious issues. To fully understand controversies, we must study how they have been shaped by cultures and events. In the 1970s a number of researchers—most prominently Harry Collins, Karin Knorr Cetina, Bruno Latour, Michael Lynch, and Sharon Traweek—simultaneously adopted a novel approach of studying cultures of science, moving into laboratories to watch and participate in the work of experimentation, the collection and analysis of data, and the refinement of claims. Early laboratory ethnographies drew attention to the skills involved in even the most straightforward laboratory manipulation and observation (Latour & Woolgar, 1979; Collins, 1985; Zenzen & Restivo, 1982). In the context of such skill-bound action, scientists negotiated the nature of data and other results in conversation with each other (Knorr Cetina, 1981; Lynch, 1985), working toward results and arguments that could be published. Attention to such details is consonant with the ethnomethodological study of science advocated by Lynch, which makes epistemology a topic of detailed empirical study (Lynch 1985, 1993); for ethnomethodology, the order of science is made at the level of ordinary actions in laboratories and elsewhere. In all of this, cultures play an enormous role, setting out what can be valued work and acceptable style (Traweek, 1988). The construction of data, then, is heavily marked by skills and cultures and by routine negotiation in the laboratory.

Not only data but phenomena themselves are constructed in laboratories—*labor*atories are places of work, and what is found in them is not nature but rather the product of much human effort. Inputs are extracted and refined, or are invented for particular purposes, shielded from outside influences, and placed in innovative contexts (Latour & Woolgar, 1979; Knorr Cetina, 1981; Hacking, 1983). Experimental systems are tinkered with until stabilized, able to behave consistently (Rheinberger, 1997). Laboratory phenomena, then, are not in themselves natural but are made to stand in for nature; in their purity and artificiality they are typically seen as more fundamental and revealing of nature than the natural world itself can be.

In the seventeenth century, this constructedness of experimental phenomena was a focus of debates over the legitimacy of experimental philosophy. The debates were, as we know, resolved in favor of experiment but not because experiment is self-evidently a transparent window onto nature. They were resolved by an articulation of the proper bounds and styles of discourse within a community of gentlemanly natural philosophers (Shapin & Schaffer, 1985) and by analogy to mathematical construction (Dear, 1995). In the analysis of these and other important developments, STS has opened up new approaches to historical epistemology, studying how and why particular styles of scientific work have arisen (Hacking, 1992); the histories and dynamics of key scientific concepts and ideals, like objectivity (Daston, 1992; Porter, 1995); and the rhetoric and politics of method (Schuster & Yeo, 1986). From the construction of scientific knowledge developed an interest in the construction of scientific methods and epistemologies.

Trevor Pinch and Wiebe Bijker's (1987) transfer of concepts from the study of science to the study of technology, under the title "social construction of technology" (SCOT), argued that the success of a technology depends on the strength and size of the groups that take it up and promote it. Even a technology's definition is a result of its interpretation by "relevant social groups": artifacts may be interpreted flexibly, because what they do and how well they perform are the results of competing goals or competing senses of what they should do. Thus, SCOT points to contingencies in the histories and meanings of technologies, contingencies on actions and interpretations by different social groups.

The symbolic interactionist approach treats science and technology as work, taking place in particular locales using particular materials (e.g., Fujimura, 1988). Moreover, objects serve as symbols that enable work and, through it, the creation of scientific knowledge and technical results (Star & Griesemer, 1989). Attention to the work of science, technology, and medicine alerts symbolic interactionists to the contributions of people not normally recognized as researchers or innovators (e.g., Moore, 1997).

Actor-network theory (ANT) further broadens that picture by representing the work of technoscience as the attempted creation of larger and stronger networks (Callon, 1986; Latour, 1987; Law, 1987). Actors, or more properly "actants," attempt to build networks we call *machines* when their components are made to act together to achieve a consistent effect, or *facts* when their components are made to act as if they are in agreement. Distinctive to ANT is that the networks are heterogeneous, including diverse components that span materials, equipment, components, people, and institutions. In ANT's networks bacteria may rub shoulders with microscopes and public health agencies, and experimental batteries may be pulled apart by car drivers and oil companies. All these components are actants and are treated as simultaneously semiotic and material; ANT might be seen to combine the interpretive frameworks of EPOR and SCOT with the materialism of laboratory studies. Scientific facts and technological artifacts are the result of work by scientists and engineers to translate the interests of a wide group of actors so that they work together or in agreement. ANT's step in the history of constructivist STS is to integrate human and nonhuman actors in analyses of the construction of knowledge and things—controversially, because it may reproduce asymmetries (Collins & Yearley, 1992; Bloor, 1999).

For scientific knowledge and technological artifacts to be successful, they must be made to fit their environments or their environments must be made to fit them. The process of adjusting pieces of technoscience and their environments to each other, or of simultaneously creating both knowledge and institutions, is a process of co-production (Jasanoff, 2004) or co-construction (Taylor, 1995) of the natural, technical, and social orders. Drugs are made to address illnesses that come into being because of the availability of drugs (e.g., Fishman, 2004), classifications of diseases afford diagnoses that reinforce those classifications (Bowker & Star, 1999), and climate science has created both knowledge and institutions that help validate and address that knowledge (Miller, 2004). Part of the work of successful technoscience, then, is the construction not only of facts and artifacts but also of the societies that accept, use, and validate them.

There have been many more extensions of constructivist approaches. Observing that interests had been generally taken as fixed causes of scientific and technological actions, even while interests are also flexible and occasioned (Woolgar, 1981; Callon & Law, 1982), some researchers have taken up the challenge of reflexivity, explaining sociology of knowledge using its own tools (Mulkay, 1985; Woolgar, 1988; Ashmore, 1989). Studies of scientific and technical rhetoric follow the discursive causes of facts and artifacts into questions of genre and styles of persuasion (e.g., Gilbert & Mulkay, 1984; Myers, 1990). The study of boundary work displays the construction and reconstruction of the edges of disciplines, methods, and other social divisions (Gieryn, 1999). Meanwhile, researchers have examined some of the legal, regulatory, and ethical work of science and technology: How are safety procedures integrated with other laboratory practices (Sims, 2005)? How is informed consent defined (Reardon, 2001)? How are patents constructed out of scientific results (e.g., Packer & Webster, 1996; Owen-Smith 2005)? In these and many other ways, the constructivist project continues to find new tools of analysis and new objects to analyze.

The metaphor of construction, in its generic form, thus ties together much of STS: Kuhn's historiography of science; the strong program's rejection of non-naturalist explanations; ethnographic interest in the stabilization of materials and knowledges; EPOR's insistence on the muteness of the objects of study; historical epistemology's exploration of even the most apparently basic concepts, methods, and ideals; SCOT's observation of the interpretive flexibility of even the most straightforward of technologies; ANT's mandate to distribute the agency of technoscience widely; and the co-productionist attention to simultaneous work on technical and social orders. Of course, these programs are not unified, as different uses and interpretations of the constructivist metaphor allow for and give rise to substantial theoretical and methodological disagreements. Yet the metaphor has enough substance to help distinguish STS from more general history of science and technology, from the rationalist project of philosophy of science, from the phenomenological tradition of philosophy of technology, and from the constraints of institutional sociology of science.

THE PROBLEM WITH THE NARRATIVE SO FAR

Unfortunately, the narrative so far is entirely a High Church one, to adopt Fuller's useful analogy.[1] This High Church STS has been focused on the interpretation of science and technology and has been successful in developing sophisticated conceptual tools for exploring the development and stabilization of knowledge and artifacts. While its hermeneutics of science and technology are often explicitly framed in opposition to the more rationalist projects of traditional philosophy and history of science, the High Church occupies a similar terrain.

But there is also a Low Church, less concerned with understanding science and technology in and of themselves, and more with making science and technology accountable to public interests. The Low Church has its most important origins in the work of scientists concerned with ties among science, technology, the military, and industry. For them, the goal is to challenge the structures that allowed nuclear physics to contribute to the development of atomic weapons, that allowed chemistry to be harnessed to various environmentally disastrous projects, or that gave biology a key place in the industrialization of agriculture. Activist movements in the 1940s and 1950s produced the *Bulletin of Atomic Scientists* and organizations like Pugwash, in which progressively minded scientists and other scholars discussed nuclear weapons and other global threats. Put differently, science and technology often contribute to projects the benefits, costs, and risks of which are very unevenly distributed. In recognition of this fact, and in the context of a critique of the idea of progress (Cutliffe, 2000), 1960s activists created organizations like the Union of Concerned Scientists and Science for the People.

Especially in the academy, the Low Church became "Science, Technology, and Society," a diverse grouping united by its combination of progressive goals and orientation to science and technology as social institutions. In fact these two have been connected: For researchers on Science, Technology, and Society, the project of understanding the social nature of science has generally been seen as continuous with the project of promoting a socially responsible science (e.g., Ravetz, 1971; Spiegel-Rösing & Price, 1977; Cutliffe, 2000). This establishes a link between Low and High Churches and a justification for treating them as parts of a single field, rather than as two completely separate denominations. So the second of the elements that distinguish STS from other disciplines that study science and technology is an activist interest.

For the Low Church, key questions are tied to reform, to promoting science and technology that benefit the widest populations. How can sound technical decisions be made through genuinely democratic processes (Laird, 1993)? Can innovation be democratically controlled (Sclove, 1995)? How should technologies best be regulated (e.g., Morone & Woodhouse, 1989)? To what extent, and how, can technologies be treated as political entities (Winner, 1986)? What are the dynamics of public technical controversies, and how do sides attempt to control definition of the issues and the relevant participants (Nelkin, 1979)? As problems of science and technology have changed, so have critical studies of them. Military funding as the central focus has given way to a constellation of issues centered on the privatization of university research; in a world in which researchers, knowledge, and tools flow back and forth between academia and industry, how can we safeguard pure science (Dickson, 1988; Slaughter & Leslie, 1997)?

An assumption behind, and also a result of, research on Science, Technology, and Society is that more public participation in technical decision-making, or at least more than has been traditional, improves the public value and quality of science and technology. So, for example, in a comparison of two parallel processes of designing chemical weapons disposal programs, a participatory model was a vast improvement over a "decide, announce, defend" model; the latter took enormous amounts of time, alienated the public, and produced uniform recommendations (Futrell, 2003). In evaluations of public participation exercises it is argued that these are more successful to the extent that participants represent the population, are independent, are involved early in the decision-making process, have real influence, are engaged in a transparent process, have access to resources, have defined tasks, and engage in structured decision-making (Rowe et al., 2004).

The democratization of science and technology has taken many forms. In the 1980s, the Danish Board of Technology created the consensus conference, a panel of citizens charged with reporting and making (nonbinding) recommendations to the Danish parliament on a specific technical topic of concern (Sclove, 2000). Experts and stakeholders have opportunities to present information to the panel, but the lay group has full control over its report. The consensus conference process has been deemed a success for its ability to democratize technical decision-making without obviously sacrificing clarity and rationality, and it has been extended to other parts of Europe, Japan, and the United States (Sclove, 2000).

Looking at an earlier stage in research processes, in the 1970s the Netherlands pioneered the idea of "science shops," which provide technical advice to citizens, associations, and nonprofit organizations (Farkas, 1999). The science shop is typically a small-scale organization that conducts scientific research in response to needs articulated by individuals or organizations lacking the resources to conduct research on their own. This idea, instantiated in many different ways, has been modestly successful, being exported to countries across Europe and to Canada, Israel, South Africa, and the United States, though its popularity has waxed and waned (Fischer et al., 2004). Thus, the project of Science, Technology, and Society has had some impressive achievements that are not part of the constructivist project, at least as represented in this chapter's earlier narrative of the history of STS. Nonetheless, these two projects have been better linked together than the two Churches analogy would suggest.

A RECONSTRUCTION OF THE DISTINCTION

This chapter does not attempt a religious reconciliation. Easier is to argue that the religious metaphors are out of place. There is undoubtedly considerable distance between the more "theoretical" and the more "activist" sides to STS, but there are plenty of

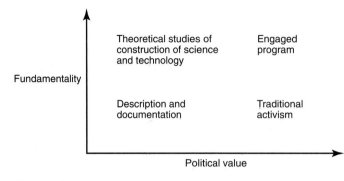

Figure 1.1

overlaps between theory and activism (Woodhouse et al., 2002). There are any number of engaged analyses drawing on constructivist methods and insights, constructivist analyses engaging with policy or politics, and abstract discussions of the connections between theory and the democratization of science and technology. In particular, we can see valuable extensions of constructivist STS to study technoscientific politics, extensions that bridge normative and theoretical concerns.

We might better view the distinction between High Church purely academic work and Low Church political or advocacy work in terms of a double distinction. (There are other revisions of it, around positive and negative attitudes toward science and technology, as well as three-way contrasts among theory, activism, or public policy—see Cutliffe, 2000; Woodhoue et al, 2002; Bijker, 2003.) Let us ask two questions of different pieces of STS scholarship. First, do they aim at results of theoretical or fundamental or wide importance for understanding the construction of science and technology? Second, do they aim at results of political or practical value for promoting democratic control of and participation in science and technology? If we ask these two questions simultaneously, the result is a space defined by two axes: high and low levels of "fundamentality" and high or low levels of "political value." While these axes do not tell a full story of STS, they both distinguish STS from other ways of studying science and technology and capture important dimensions of the field (see figure 1.1). At the lower left of the figure are studies that describe and document. Such studies are not by themselves relevant to either the theoretical or activist projects of STS, though perhaps they may be made so by the right translations. They would typically be left out of the standard characterizations of the field (Cutliffe, 2000; Bucchi, 2004; Sismondo, 2004; Yearley, 2005). At the lower right are studies that aim to contribute primarily to one or another activist project. At the upper left are studies that aim to contribute to theoretical understanding of the construction of science and technology, typically focusing on high-status sciences and technologies and often focusing on their internal dynamics. At the upper right are studies that aim to contribute both to some version of activist projects and to general theoretical perspectives. For ease of reference, this region of intellectual space needs a name: the "engaged program" of STS.

The modest move of the engaged program is to address topics of clear political importance: nuclear energy rather than condensed matter physics, agricultural biotechnology rather than evolutionary systematics. But in so doing the engaged program makes a more sophisticated move by placing relations among science, technology, and public interests at the center of the research program. The engaged program studies science and technology when they are or should be engaged, and as a result, interactions among science, technology, politics, and public interests have become topics for STS and not just contexts of study. Politics has become a site of study rather than a mode of analysis.

The two-dimensional framework allows us to see not a conflict between the goals of theoretical interest and activism but a potential overlap. That overlap is well represented, and increasingly so, in the STS literature. Some of the recent chapters in the history of STS involve the extension of the constructivist program to public sites, with a focus on interactions at the interface of science, technology, law, and government. Without programmatic announcements or even fanfare, the center of gravity of STS has moved markedly toward the terrain of the engaged program. Much of the Low Church has always been there, since many of its representatives intend to contribute to general analyses of the politics of science and technology, treating their subject matters as important case studies. Some strands of feminist STS have also always been there, wherever feminist research met constructivist concerns. So has much symbolic interactionist research, which has been often articulated with attention to issues of power (e.g., Cussins, 1996; Casper & Clarke, 1998). But recently it has become almost the norm for constructivist STS to study cases of public interest, and it has become common to study the interactions of science, technology, and public interests. Consequently, the nature of the politics of science and technology appears to be at the very center of the field. Recent issues of *Social Studies of Science,* certainly one of the highest of High Church central journals in STS, contain any number of articles on a wide variety of topics clearly located in the engaged program.[2] Books on science and technology in an explicitly political context attract attention and win prizes.[3] Indeed, the natures of democracy and politics in a technoscientific world, and the political orders of technoscience, are among the central topics of STS. That movement makes the distinction between two Churches increasingly irrelevant.

CONSTRUCTIVISM AND THE POLITICS OF EXPERTISE

We can see the engaged program converging on the democratization of technoscience. Approaching the problem from the direction of liberal democratic theory, Stephen Turner (2001, 2003a) argues that there is a genuine conflict between expertise and democracy because expertise creates inequalities that undermine citizen rule. As knowledge societies have developed, decisions are increasingly made by or directly responsive to experts and expert commissions. Turner is cautiously optimistic about this new version of democracy, "Liberal Democracy 3.0," arguing that some forms of expertise are effectively democratically accepted, that judgments of expertise are conferred contingently and are always open to challenge, and that therefore the importance of expertise in modern liberal societies is in principle compatible with democracy (Turner, 2001). How best to manage the conflict remains an open theoretical and political project, though.

One set of implications of the (social) constructedness of scientific knowledge is that there is always a way of cashing out knowledge in social terms: that its meaning always includes a social component, and that assumptions about the social world that produced it are embedded in knowledge. When scientific knowledge enters the public arena, those embedded assumptions can come under scrutiny. An interested public may be in an excellent position to see and challenge assumptions about such things as the residence of expertise, the relative values of different interests, and the importance of risks; Steven Yearley (1999) identifies this as one of the key findings in studies of science meeting the public. Constructivism, then, also provides grounds for increasing public participation in science and technology.

Laypeople can develop and possess technical expertise in many ways. Steven Epstein's (1996) study of AIDS activism and its effects on research provides a striking example. Activists were able to recognize that the standard protocols for clinical trials assumed, for example, that research subjects should be expected not to supplement experimental treatments with alternatives or not to share drugs with other research subjects. The protocols effectively valued clean results over the lives and hopes of people living with AIDS, and thus activists were able to challenge both the artificiality of and the ethics embedded in clinical trials. Moreover, it is clear that there are many forms of expertise and that scientists and engineers may lack relevant forms of expertise when their work takes them into public realms. In a somewhat different situation, French muscular dystrophy patients have contributed to research on their disease by organizing the research effort, engaging in their own studies, participating in accredited researchers' studies, and evaluating results (Rabeharisoa & Callon,1999). Because of its considerable resources, l'Association Française contre les Myopathies has become exemplary of a kind of cooperative research between laypeople and scientists (Callon, 1999). Brian Wynne's (1996) study of Cumbrian sheep farmers potentially affected by the 1986 Chernobyl accident is one of the most-discussed pieces of research in STS, precisely because it is about the fate of expertise in a public domain. The farmers were easily able to see that Chernobyl was not the only potential source of irradiation, as the British nuclear power plant Sellafield was already viewed with suspicion, and were also able to see lacunae in government scientists' knowledge, especially about sheep-farming. Thus, they developed a profound skepticism about the government advice.

Outsiders may challenge the seamlessness of scientific and technical expertise. There are competing epistemes in science and law, and when science is brought into the courtroom the value of its forms of knowledge is not straightforwardly accepted (Jasanoff, 1995). Lawyers and judges often understand that scientific expertise contains its own local and particular features. As a result, science can be challenged by routine legal maneuvers, and it may or may not be translated into forms in which it can survive those challenges. Similarly, science typically does not provide the definitive cases for particular policies that both scientists and policymakers hope for, because the internal mechanisms by which science normally achieves closure often fail in the context of contentious policymaking (Collingridge & Reeve, 1986).

THREE PROGRAMMATIC STATEMENTS

Through studies like the above, STS, and particularly that part of the field that we can see as working within the broad constructivist metaphor and as having a High Church history, has turned the politics of science and technology into a topic, indeed, *the* topic. This is not simply to analyze technoscience politically but to analyze technoscientific politics. What follows are three articulations of core substantive issues and normative responses. We can see each of these articulations as attending to the construction of political orders of science and technology and following paths begun in the history of constructivism.

A Normative Theory of Expertise
In a widely discussed paper, H. M. Collins and Robert Evans (2002) identify what they call a "problem of extension": Who should legitimately participate in technical decision-making? That is, given constructivist STS's successful challenge to claims that science has privileged access to the truth, how open should technical decision-making be? In expansive terms, Collins and Evans claim that a version of the problem of extension is "the pressing intellectual problem of the age" (2002: 236).

They offer a normative theory of *expertise* as a framework for a solution to this problem. Experts, they argue, are the right decision-makers because (by definition) they possess relevant knowledge that nonexperts lack. STS has shown, and Collins's work (e.g., 1985) is most prominent in showing, that the solution to scientific and technical controversies rests on judgments by experts and judgments of the location of expertise rather than on any formal scientific method; science and technology are activities performed by humans, not machines. Collins and Evans assume, moreover, that expertise is real and that it represents genuine knowledge within its domains. STS has also shown that legitimate expertise extends much further than merely to accredited scientists and engineers, at least wherever science and technology touches the public domain (e.g., Epstein 1996; Wynne 1996; Yearley 1999). In addition, there are different forms of expertise: contributory expertise allows for meaningful participation in the substance of technoscientific controversies, interactional expertise allows for meaningful interaction with, and often between, contributing experts, and referred expertise allows for the assessment of contributory expertise (Collins & Evans, 2002). Thus, the normative theory of expertise would increase opportunities for participation and would promote an egalitarianism based on ability to participate meaningfully. The problem of extension is to identify how far these different forms of expertise legitimately extend.

Technical decisions are the focus of Collins and Evans's position, the key intersection of science, technology, and politics. This leaves their view open to charges of a "decisionism" (Wynne, 2003; Habermas, 1975) that ignores such matters as the framing of issues, the constitution of expertise, and the dissemination of knowledge (Jasanoff, 2003). We might see a parallel issue in the movement in current political philosophy to value deliberative democracy and active citizenship over aggregative democracy and participation through voting. Thus, we might think that Collins and Evans have construed the topic of the engaged program narrowly, leaving aside terrains where science, technology, and politics intersect.

Civic Epistemologies
Problems with decisionism serve as a point of departure for quite different explorations of science and technology in the public domain. Sheila Jasanoff, in a comparative study of biotechnology in the United States, Britain, and Germany, shows how there are distinct national cultures of technoscientific politics (Jasanoff, 2005). Just as controversies are key moments, but only moments, in the construction of scientific and technical knowledge, decisions are key moments in technoscientific politics. The governments of each of these countries have developed strategies to incubate biotechnological research and industry, even to the extent of being aspects of nation-building. Each has subjected that research and industry to democratic scrutiny and control. Yet the results have been strikingly different: the industries are different, their relations with academia are different, and the regulations dealing with them and their products are different. This is the result of national "civic epistemologies" that shape the democratic practice of science and technology (Jasanoff, 2005: 255).

As Jasanoff describes civic epistemologies, they contain these dimensions: styles of knowledge-making in the public sphere; approaches to, and levels of, accountability and trust; practices of demonstration of knowledge; types of objectivity that are valued; foundations of expertise; and assumptions about the visibility and accessibility of expert bodies (2005: 259). In the United States the level of trust of experts is low, their accountability is grounded in legal or legalistic processes, and neatly congruent with this, the most valued basis of objectivity is formal. In Germany, on the other hand, the level of trust in experts is higher, when they occupy recognized roles, and the basis of objective results is reasoned negotiations among representatives of interested groups. It should be no surprise, then, that the politics of biotechnology are different in the United States and Germany.

The above list of dimensions, which might be expandable, suggests programs of research for all kinds of civic epistemologies and not just national ones. Meanwhile, such a historically grounded and locally situated understanding of technoscientific politics demands historically grounded and locally situated normative approaches. No single template will improve democratic accountability in diverse settings and contexts. And similarly, no single template of active technoscientific citizenship will be adequate to these different settings. If the engaged program foregrounds civic epistemologies, its normative work is multiplied.

Bringing the Sciences into Democracy

For Bruno Latour, the modern world sees nature and politics as two separate domains, their only connection being that nature is taken to provide constraints on politics (Latour, 2004). It has been a central achievement of STS to show that this modern picture is mistaken: what is here being called constructivist STS exposes the work of establishing facts of nature, thus showing that the modernist separation of nature from the social world is a piece of *a priori* metaphysics. Latour aims to bring the sciences into democracy by "blurring the distinction between nature and society durably" (2004: 36). In its place, he proposes the instauration of a collective (or many collectives) that deliberates and decides on its membership. This collective will be a republic of things, human and nonhuman. Just as ANT integrated humans and nonhumans into analyses of technoscience, its contribution to the engaged program should be to integrate humans and nonhumans into technoscientific democracy.

Latour argues that representing nonhumans is no more difficult than representing humans, that there is only one problem of representation, which sometimes appears as a problem of political representation and sometimes as a problem of scientific representation (2004: 55). In both cases, we rely on spokespeople, of whom we must be simultaneously skeptical and respectful. Nonetheless, political philosophy has had enough difficulty dealing with *human* multiculturalisms, with their apparent conflicts over universal rights and national projects, and we might suspect that such conflicts would be more difficult to address if nonhumans were also given consideration.

Perhaps for this reason, Latour's collective would be focused on propositions, determining which propositions belong in a well-ordered common world or cosmos. He divides it into two houses, with separate powers and responsibilities; these houses cut across science and politics, reconceptualizing epistemic processes so that all parties can participate at all stages. An upper house has the power to "take into account," the lower house has the power to "put in order," and both together have the power "to follow up." To effect such powers requires tasks that scientists and politicians would undertake: to be attentive to propositions that might be added to the common world, even if they might challenge members; to determine how to assess propositions that might be included; to arrange propositions in a homogeneous order; to reason toward closure of debates; and so on.

It would be unfair to Latour, given the amount of detail in his descriptions, to say that the organization of the collective is unclear. Nonetheless, Latour's divisions and unities are not described to help us site a new parliament building and populate it with representatives. Rather, he aims to show what a society should do if it took epistemics as central. In effect, Latour's preferred politics of nature is reminiscent of Karl Popper's epistemic liberalism but responsive to research in STS, and specifically to research in ANT. Such a society would not allow propositions to become established without being subjected to the right kinds of scrutiny. It would attempt to institutionalize propositions rationally, yet it would be constantly open to the possibility of revision of its established cosmos. It would not adopt any *a priori* metaphysics, such as that which neatly divides nature and society. And it would be so constituted that these checks and balances would remain in place. So while Latour's politics of nature is intensely normative, it does not make recognizably concrete recommendations.

STS AND THE STUDY OF TECHNOSCIENTIFIC POLITICS

In the above three programmatic statements we can see parallels to programs in the history of constructivist STS. There remain plenty of opportunities to explore further extensions of the field into the terrain of technoscientific politics: constructions of phenomena of interest, natures of interests themselves, histories and uses of civic epistemologies and not only their forms, contingencies of particular understandings of the politics of science and technology, boundary work in political domains, and rhetorical action. As before, these programs need not be unified, as different uses and interpretations of the constructivist metaphor allow for substantial disagreements. Yet the metaphor has enough substance to help guide research in interesting and valuable directions.

Moreover, as the above programmatic statements show, there are opportunities for contributions to a political philosophy that recognize the centrality of science and technology to the modern world. Because it does not separate epistemic and political processes, STS can genuinely study knowledge societies and technological societies rather than treat knowledge and technology as externalities to political processes. This theoretical project is structured so that it already contributes to STS's normative project, providing a broad set of ways of bringing them together.

Notes

I would like to thank Ed Woodhouse and three anonymous reviewers for their excellent comments on an earlier version of this chapter, and Michael Lynch for both his thoughtful suggestions and his keen editor's eye.

1. Fuller's analogy is to two waves of secularization. STS's Low Church resembles the sixteenth and seventeenth century Protestant Reformation, whereas the High Church resembles nineteenth century radical hermeneutical criticism of the Bible (Fuller, 2000: 409).

2. There are, for example, articles on the rhetoric of commentary on tobacco regulation (Roth et al., 2003), environmental management of small islands (Hercock, 2003), race and scientific credit (Timmermans, 2003), social and ethical consequences of pharmacogenetics (Hedgecoe & Martin, 2003), a debate about the nature of engagement in STS (Jasanoff, 2003; Wynne, 2003; Rip, 2003; Collins & Evans, 2003), and a discussion of the politics of expertise (Turner, 2003b).

3. The Society for Social Studies of Science awards two book prizes each year, the Ludwik Fleck and Rachel Carson prizes. The latter, created only in 1996, is explicitly for a book of political or social relevance, but the former is for a book of general interest in STS. Nonetheless, among the Fleck winners are such books as Helen Verran's *Science and an African Logic* (2001), a book that puts relativism in a multicultural context; Adele Clark's *Disciplining Reproduction* (1998), on twentieth century sciences of human reproduction; Donna Haraway's *Modest_Witness@Second-Millennium* (1997), on feminism and technoscience; Theodore Porter's *Trust in Numbers* (1995), which discusses how the ideal of objectivity arises from the democratization of expertise; and Londa Schiebinger's *Nature's Body* (1993), on gender in Enlightenment biology and anthropology.

References

Ashmore, Malcolm (1989) *The Reflexive Thesis: Wrighting Sociology of Scientific Knowledge* (Chicago: University of Chicago Press).

Barnes, Barry (1974) *Scientific Knowledge and Sociological Theory* (London: Routledge & Kegan Paul).

Bijker, Wiebe E. (2003) "The Need for Public Intellectuals: A Space for STS," *Science, Technology & Human Values* 28: 443–50.

Bloor, David (1976) *Knowledge and Social Imagery* (London: Routledge & Kegan Paul).

Bloor, David (1999) "Anti-Latour," *Studies in History and Philosophy of Science* 30: 81–112.

Bowker, Geof & Susan Leigh Star (1999) *Sorting Things Out: Classification and Its Consequences* (Cambridge, MA: MIT Press).

Bucchi, Massimiano (2004) *Science in Society: An Introduction to Social Studies of Science* (London: Routledge).

Callon, Michel (1986) "Some Elements of a Sociology of Translation: Domestication of the Scallops and the Fishermen of St. Brieuc Bay," in John Law (ed), *Power, Action and Belief* (London: Routledge & Kegan Paul): 196–233.

Callon, Michel (1999) "The Role of Lay People in the Production and Dissemination of Scientific Knowledge," *Science, Technology & Society* 4: 81–94.

Callon, Michel & John Law (1982) "On Interests and Their Transformation: Enrollment and Counter-Enrollment," *Social Studies of Science* 12: 615–25.

Casper, Monica & Adele Clark (1998) "Making the Pap Smear into the 'Right Tool' for the Job: Cervical Cancer Screening in the USA, circa 1940–95," *Social Studies of Science* 28: 255–90.

Clarke, Adele (1998) *Disciplining Reproduction: Modernity, American Life Sciences, and "the Problems of Sex"* (Berkeley: University of California Press).

Collingridge, David & Colin Reeve (1986) *Science Speaks to Power: The Role of Experts in Policy Making* (London: Frances Pinter).

Collins, H. M. (1985) *Changing Order: Replication and Induction in Scientific Practice* (London: Sage).

Collins, H. M. & Robert Evans (2002) "The Third Wave of Science Studies: Studies of Expertise and Experience," *Social Studies of Science* 32: 235–96.

Collins, H. M. & Robert Evans (2003) "King Canute Meets the Beach Boys: Responses to the Third Wave," *Social Studies of Science* 33: 435–52.

Collins, H. M. & Steven Yearley (1992) "Epistemological Chicken," in Andrew Pickering (ed), *Science as Practice and Culture* (Chicago: University of Chicago Press): 301–26.

Cussins, Charis (1996) "Ontological Choreography: Agency Through Objectification in Infertility Clinics," *Social Studies of Science* 26: 575–610.

Cutliffe, Stephen H. (2000) *Ideas, Machines, and Values: An Introduction to Science, Technology, and Society Studies* (Lanham, MD: Rowman & Littlefield).

Daston, Lorraine (1992) "Objectivity and the Escape from Perspective," *Social Studies of Science* 22: 597–618.

Dear, Peter (1995) *Discipline and Experience: The Mathematical Way in the Scientific Revolution* (Chicago: University of Chicago Press).

Dickson, David (1988) *The New Politics of Science* (Chicago: University of Chicago Press).

Epstein, Steven (1996) *Impure Science: AIDS, Activism, and the Politics of Knowledge* (Berkeley: University of California Press).

Farkas, Nicole (1999) "Dutch Science Shops: Matching Community Needs with University R&D," *Science Studies* 12: 33–47.

Fausto-Sterling, Anne (1985) *Myths of Gender: Biological Theories About Women and Men* (New York: Basic Books).

Fischer, Corinna, Loet Leydesdorff, & Malte Schophaus (2004) "Science Shops in Europe: The Public as Stakeholder," *Science & Public Policy* 31: 199–211.

Fishman, Jennifer (2004) "Manufacturing Desire: The Commodification of Female Sexual Dysfunction," *Social Studies of Science* 34: 187–218.

Fujimura, Joan (1988) "The Molecular Biological Bandwagon in Cancer Research: Where Social Worlds Meet," *Social Problems* 35: 261–83.

Fuller, Steve (1993) *Philosophy, Rhetoric, and the End of Knowledge: The Coming of Science and Technology Studies* (Madison: University of Wisconsin Press).

Fuller, Steve (2000) *Thomas Kuhn: A Philosophical History for Our Times* (Chicago: University of Chicago Press).

Futrell, Robert (2003) "Technical Adversarialism and Participatory Collaboration in the U.S. Chemical Weapons Disposal Program," *Science, Technology & Human Values* 28: 451–82.

Gieryn, Thomas F. (1999) *Cultural Boundaries of Science: Credibility on the Line* (Chicago: University of Chicago Press).

Gilbert, Nigel & Michael Mulkay (1984) *Opening Pandora's Box: A Sociological Analysis of Scientists' Discourse* (Cambridge: Cambridge University Press).

Habermas, Jürgen (1975) *Legitimation Crisis,* trans. T. McCarthy (Boston: Beacon Press).

Hacking, Ian (1983) *Representing and Intervening: Introductory Topics in the Philosophy of Natural Science* (Cambridge: Cambridge University Press).

Hacking, Ian (1992) "'Style' for Historians and Philosophers," *Studies in History and Philosophy of Science* 23: 1–20.

Hacking, Ian (1999) *The Social Construction of What?* (Cambridge, MA: Harvard University Press).

Haraway, Donna (1997) *Modest_Witness@Second_Millennium. FemaleMan_Meets_OncoMouse: Feminism and Technoscience* (London: Routledge).

Hedgecoe, Adam & Paul Martin (2003) "The Drugs Don't Work: Expectations and the Shaping of Pharmacogenetics," *Social Studies of Science* 33: 327–64.

Hercock, Marion (2003) "Masters and Servants: The Contrasting Roles of Scientists in Island Management," *Social Studies of Science* 33: 117–36.

Jasanoff, Sheila (1995) *Science at the Bar: Law, Science, and Technology in America* (Cambridge, MA: Harvard University Press).

Jasanoff, Sheila (2003) "Breaking the Waves in Science Studies: Comment on H. M. Collins and Robert Evans, 'The Third Wave of Science Studies'," *Social Studies of Science* 33: 389–400.

Jasanoff, Sheila (ed) (2004) *States of Knowledge: The Co-Production of Science and Social Order* (London: Routledge).

Jasanoff, Sheila (2005) *Designs on Nature: Science and Democracy in Europe and the United States* (Princeton, NJ: Princeton University Press).

Knorr Cetina, Karin D. (1981) *The Manufacture of Knowledge: An Essay on the Constructivist and Contextual Nature of Science* (Oxford: Pergamon Press).

Kuhn, Thomas S. (1962) *The Structure of Scientific Revolutions* (Chicago: University of Chicago Press).

Laird, Frank (1993) "Participatory Analysis, Democracy, and Technological Decision Making," *Science, Technology & Human Values* 18: 341–61.

Latour, Bruno (1987) *Science in Action: How to Follow Scientists and Engineers Through Society* (Cambridge, MA: Harvard University Press).

Latour, Bruno (2004) *Politics of Nature: How to Bring the Sciences into Democracy* (Cambridge, MA: Harvard University Press).

Latour, Bruno & Steve Woolgar (1979) *Laboratory Life: The Social Construction of Scientific Facts* (London: Sage).

Law, John (1987) "Technology and Heterogeneous Engineering: The Case of Portuguese Expansion," in Wiebe E. Bijker, Thomas P. Hughes, & Trevor Pinch (eds), *The Social Construction of Technological Systems: New Directions in the Sociology and History of Technology* (Cambridge, MA: MIT Press): 111–34.

Lynch, Michael (1985) *Art and Artifact in Laboratory Science: A Study of Shop Work and Shop Talk in a Research Laboratory* (London: Routledge & Kegan Paul).

Lynch, Michael (1993) *Scientific Practice and Ordinary Action: Ethnomethodology and Social Studies of Science* (Cambridge: Cambridge University Press).

MacKenzie, Donald A. (1981) *Statistics in Britain, 1865–1930: The Social Construction of Scientific Knowledge* (Edinburgh: Edinburgh University Press).

Martin, Emily (1991) "The Egg and the Sperm: How Science Has Constructed a Romance Based on Stereotypical Male-Female Roles," *Signs* 16: 485–501.

Miller, Clark (2004) "Climate Science and the Making of a Global Political Order," in Sheila Jasanoff (ed), *States of Knowledge: The Co-production of Science and Social Order* (London: Routledge): 46–66.

Moore, Lisa Jean (1997) "'It's Like You Use Pots and Pans to Cook. It's the Tool.' The Technologies of Safer Sex," *Science, Technology & Human Values* 22: 434–71.

Morone, Joseph & Edward J. Woodhouse (1989) *The Demise of Nuclear Energy?: Lessons for Democratic Control of Technology* (New Haven, CT: Yale University Press).

Mulkay, Michael (1985) *Word and the World: Explorations in the Form of Sociological Analysis* (London: Allen & Unwin).

Myers, Greg (1990) *Writing Biology: Texts in the Social Construction of Scientific Knowledge* (Madison: University of Wisconsin Press).

Nelkin, Dorothy (1979) *Controversy: Politics of Technical Decisions* (Beverly Hills, CA: Sage).

Owen-Smith, Jason (2005) "Dockets, Deals, and Sagas: Commensuration and the Rationalization of Experience in University Licensing," *Social Studies of Science* 35: 69–98.

Packer, Kathryn & Andrew Webster (1996) "Patenting Culture in Science: Reinventing the Scientific Wheel of Credibility," *Science, Technology & Human Values* 21: 427–53.

Pinch, Trevor J. & Wiebe E. Bijker (1987) "The Social Construction of Facts and Artifacts: Or How the Sociology of Science and the Sociology of Technology Might Benefit Each Other," in Wiebe E. Bijker,

Thomas P. Hughes & Trevor Pinch (eds), *The Social Construction of Technological Systems: New Directions in the Sociology and History of Technology* (Cambridge, MA: MIT Press): 17–50.

Porter, Theodore (1995) *Trust in Numbers: The Pursuit of Objectivity in Science and Public Life* (Princeton, NJ: Princeton University Press).

Rabeharisoa, Vololona & Michel Callon (1999) *Le Pouvoir des malades: L'association française contre les myopathies et la Recherche* (Paris: Les Presses de l'École des Mines).

Ravetz, Jerome R. (1971) *Scientific Knowledge and Its Social Problems* (Oxford: Clarendon).

Reardon, Jenny (2001) "The Human Genome Diversity Project: A Case Study in Coproduction," *Social Studies of Science* 31: 357–88.

Rheinberger, Hans-Jörg (1997) *Toward a History of Epistemic Things: Synthesizing Proteins in the Test Tube* (Stanford, CA: Stanford University Press).

Rip, Arie (2003) "Constructing Expertise: In a Third Wave of Science Studies," *Social Studies of Science* 33: 419–34.

Roth, Andrew L., Joshua Dunsby & Lisa A. Bero (2003) "Framing Processes in Public Commentary on U.S. Federal Tobacco Control Regulation," *Social Studies of Science* 33: 7–44.

Rowe, Gene, Roy Marsh, & Lynn J. Frewer (2004) "Evaluation of a Deliberative Conference," *Science, Technology & Human Values* 29: 88–121.

Schiebinger, Londa (1993) *Nature's Body: Gender in the Making of Modern Science* (Boston: Beacon Press).

Schuster, J. A., & R. R. Yeo (eds) (1986) *The Politics and Rhetoric of Scientific Method* (Dordrecht, the Netherlands: D. Reidel).

Sclove, Richard (1995) *Democracy and Technology* (New York: Guilford Press).

Sclove, Richard (2000) "Town Meetings on Technology: Consensus Conferences as Democratic Participation," in Daniel Lee Kleinman (ed), *Science, Technology and Democracy* (Albany: State University of New York Press): 33–48.

Shapin, Steven (1975) "Phrenological Knowledge and the Social Structure of Early Nineteenth-Century Edinburgh," *Annals of Science* 32: 219–43.

Shapin, Steven & Simon Schaffer (1985) *Leviathan and the Air-Pump: Hobbes, Boyle, and the Experimental Life* (Princeton, NJ: Princeton University Press).

Sims, Benjamin (2005) "Safe Science: Material and Social Order in Laboratory Work," *Social Studies of Science* 35: 333–66.

Sismondo, Sergio (1993) "Some Social Constructions," *Social Studies of Science* 23: 515–53.

Sismondo, Sergio (2004) *An Introduction to Science and Technology Studies* (Oxford: Blackwell).

Slaughter, Sheila & Larry L. Leslie (1997) *Academic Capitalism: Politics, Policies, and the Entrepreneurial University* (Baltimore, MD: Johns Hopkins University Press).

Spiegel-Rösing, Ina & Derek de Solla Price (Eds.) (1977) *Science, Technology and Society: A Cross-Disciplinary Perspective* (London: Sage).

Star, Susan Leigh & James R. Griesemer (1989) "Institutional Ecology, 'Translations' and Boundary Objects: Amateurs and Professionals in Berkeley's Museum of Vertebrate Zoology, 1907–39," *Social Studies of Science* 19: 387–420.

Taylor, Peter (1995) "Building on Construction: An Exploration of Heterogeneous Constructionism, Using an Analogy from Psychology and a Sketch from Socioeconomic Modeling," *Perspectives on Science* 3: 66–98.

Timmermans, Stefan (2003) "A Black Technician and Blue Babies," *Social Studies of Science* 33: 197–230.

Traweek, Sharon (1988) *Beamtimes and Lifetimes: The World of High Energy Physicists* (Cambridge, MA: Harvard University Press)

Turner, Stephen (2001) "What Is the Problem with Experts?" *Social Studies of Science* 31: 123–49.

Turner, Stephen (2003a) *Liberal Democracy 3.0: Civil Society in an Age of Experts* (London: Sage).

Turner, Stephen (2003b) "The Third Science War," *Social Studies of Science* 33: 581–612.

Verran, Helen (2001) *Science and an African Logic* (Chicago: University of Chicago Press).

Winner, Langdon (1986) *The Whale and the Reactor: A Search for Limits in an Age of High Technology* (Chicago: University of Chicago Press).

Woodhouse, Edward, David Hess, Steve Breyman, & Brian Martin (2002) "Science Studies and Activism: Possibilities and Problems for Reconstructivist Agendas," *Social Studies of Science* 32: 297–319.

Woolgar, Steve (1981) "Interests and Explanation in the Social Study of Science" *Social Studies of Science* 11: 365–94.

Woolgar, Steve (ed) (1988) *Knowledge and Reflexivity: New Frontiers in the Sociology of Knowledge* (London: Sage).

Wynne, Brian (1996) "May the Sheep Safely Graze? A Reflexive View of the Expert-Lay Knowledge Divide," in S. Lash, B. Szerszynski, & B. Wynne (eds), *Risk, Environment and Modernity: Towards a New Ecology* (London: Sage): 44–83.

Wynne, Brian (2003) "Seasick on the Third Wave? Subverting the Hegemony of Propositionalism: Response to Collins and Evans (2002)," *Social Studies of Science* 33: 401–18.

Yearley, Steven (1999) "Computer Models and the Public's Understanding of Science: A Case-Study Analysis," *Social Studies of Science* 29: 845–66.

Yearley, Steven (2005) *Making Sense of Science: Understanding the Social Study of Science* (London: Sage).

Zenzen, Michael & Sal Restivo (1982) "The Mysterious Morphology of Immiscible Liquids: A Study of Scientific Practice," *Social Science Information* 21: 447–73.

2 The Social Study of Science before Kuhn

Stephen Turner

The controversy over Thomas Kuhn's astonishingly successful *Structure of Scientific Revolutions* ([1962]1996), which denied the possibility of a rational account of conceptual revolutions and characterized them in the language of collective psychology, created the conditions for producing the field that became "science studies." The book was the immediate product of an existing tradition of writing about science, exemplified by the works of James Bryant Conant and Michael Polanyi, and the distal product of a literature on the social character of science that reaches back centuries. This literature was closely connected to practical problems of the organization of science and also to social theory debates on the political meaning of science. The basic story line is simple: a conflict between two views of science, one of which treats science as distinguished by a method that can be extended to social and political life, and a responding view that treats science as a distinctive form of activity with its own special problems and does not provide a model for social and political life. Interlaced with this story is a puzzle over the relationship between science and culture that flourished especially in the twenties and thirties. In this chapter I briefly reconstruct this history.

BACON, CONDORCET, AND THE BEGINNINGS OF AN EXPLANATORY INTEREST IN SCIENCE

The *fons et origo* of this discussion is Francis Bacon's vision of a political order in which the class of scientists is given power by an enlightened ruler in his House of Solomon in "The New Atlantis" ([1627]1860–62, vol. 5: 347–413). This vision had a practical effect on the attempts by the Royal Society in London to distinguish itself by its methodological practices and internal governance as a type of political body in relation to the Crown (Sprat, [1667]1958: 321–438; Lynch, 2001: 177–96; Shapin, 1994) and to do the same with parallel institutions elsewhere in Europe (Hahn, 1971: 1–34; Gillispie, 2004). The Victorians assured that Bacon would be best known for his ideas about induction as a method (cf. Peltonen, 1996: 321–24) and, as his major German expositor put it, "how his whole nature was, in every way, instinctively opposed to verbal discussions" (Fischer, 1857: 307). But Bacon's extensive body of writings included not only writings on method but also on "counselors" to the Crown, or experts, on the merits of republics, on the nature of political authority, on the proper internal organization of science, on funding and authority over science, and on collective research.

The fundamental issues of science studies can be teased out of these works, but only with difficulty, because of the intentional absurdity of Renaissance style. The main "political" argument, for example, is presented as a fiction, and like other political works by ambitious office seekers, Bacon's message is shrouded in ambiguities. The basic and most influential claim (though he was far more subtle than this) (cf. Whewell, 1984: 218–47; Fischer, 1857) was that scientific truth can be produced by following a technique of assembling facts, generalizing about them, and ascending to higher level generalizations from them; that following this method precluded contestation and controversy, which were the great evils of "the schools"; that the technique is open to all, or public and democratic, because it "places all wits . . . nearly on a level" (quoted in Peltonen, 1996: 323); that it can and should be pursued collectively or cooperatively; that it requires that the mind be freed of prejudices or assumptions (and perhaps of theories); that something like social science or "civil knowledge" was also possible and necessary; and that kings would be better able to accept counsel on the basis of merit than on the basis of trust of obedient favorites. This now familiar picture of science and its extension to the social world was then novel and radical. Bacon's politics fit with his hostility to contestation, and although his recent admirers (e.g., Peltonen, 1996) have argued that he was not the stereotypic proponent of royal absolutism and unfettered state power that he was once thought to be, Bacon's primary role in the history of political thought has been as the archenemy of Edward Coke, the judge who, as defender of the common law and the rule of law, was a key progenitor of modern liberalism (cf. Coke, 2003). It sharpens and assumes new forms.

The Baconian picture is recast in a recognizably modern form in Condorcet's "Fragment on Bacon's *New Atlantis*" ([1793]1976), and in chapter X of Condorcet's *Outlines of an Historical View of the Progress of the Human Mind* ([1795]1955), which promoted the idea that science was the engine of human progress. Condorcet deals with such issues as scientific rivalry, which he regards as a normal product of the passions of scientists for their work but which can take pathological institutional forms; with the failure to utilize talent, which he regards as a major flaw of the old Regime; with concerns about financing and the forms of scientific association and internal governance, which he resolves with an argument for science's need for autonomy, or freedom from political control; and with the need for scientific knowledge of the social and political world. The argument for autonomy is grounded on the consideration that only scientists have the capacity to govern scientific activity. Though Condorcet believed in the benefits of science and the diffusion of scientific knowledge, he, characteristically, also grasped the contrary idea that there was nothing automatic about the benefits of advances in science, and he concluded that the production of these benefits required state action.

Condorcet's preferred method of extending the benefits of science was education, by which he meant the kind of education that was useful for "citizens" and would enable them to think on their own ([1795]1955): 182).[1] But he also recognized that no educational program would make scientists and citizens epistemic equals.[2] Moreover, education was politically ambiguous: not only did it require the exercise of state power, there was a sense, which he shared with other *Philosophes*, that progress resulted from collective submission to reason and science, each understood as authoritative in its own right. Condorcet attempted to put a nonauthoritarian face on this submission to science and scientists: he expressed a "hope" that the citizens thus instructed would acknowledge the "superiority of enlightenment" of their intellectual betters in choosing leaders ([1793]1976: 283). But this would necessarily amount to a regime of expert rule, with democratic consent.

SAINT-SIMON AND COMTE: SCIENCE REPLACES POLITICS

The implication that social knowledge allowed for the replacement of politics was always the most problematic element of this picture, for it placed science and politics in direct competition. In 1803, in the aftermath of the restoration of French politics and as part of the return to normalcy after the revolution, the section on social and political science of the French Academy was suppressed (Columbia Electronic Encyclopedia, 2001–2004).[3] This action served to draw a line between acceptable science and dangerous science and to reject the extension of science to politics. One consequence was that social and political speculation, and in particular speculation on science and politics, now fell to thinkers outside the academy and on the margins of science, notably Henri de Saint-Simon. Saint-Simon's faith in scientists as the saviors of society (a faith which diminished in the course of his life) was similar to Condorcet's, and he carried forward and generalized similar concerns especially with the problem of the full utilization of talent, making this theme central to his social theory, as expressed by the slogan "Each according to his capacity, to each capacity according to its works" on the masthead of the Saint-Simonian newspaper *Le Globe* (Manuel; 1995: 163).[4]

But Saint-Simon radicalized Condorcet's invasion of the political. His explicitly antipolitical and implicitly antiliberal idea that in the future the rule of man over man would be replaced by "the administration of things" proved to have a long future in the hands of Marxist-Leninism.[5] Politics would vanish, he argued, because social antagonism would disappear in a society in which capacities were fitted to tasks. The theory of "capacities" assumed that capacities were transparent. His model was science. Within science, in Saint-Simon's view, scientific merit was sufficiently transparent that scientists would naturally recognize and defer to greatness in others, allowing for the fulfillment and utilization of talent, and creating within science a natural hierarchy. This was in turn a model for the natural hierarchy of the new scientific and industrial order he envisioned.[6] Saint-Simon's young secretary, Auguste Comte, revised and extended his sketchy but illuminating ideas into a complete intellectual system,

Positivism, which provided both a philosophy of science and a model for the relations of science and society, and was also an explicit repudiation of liberalism, which Comte, like most of the advanced continental thinkers of the time, regarded as a transitory historical phenomenon doomed by its overwhelming defects (Comte, [1830–42]1864, [1877]1957).

Saint-Simon was not a methodological thinker, but Comte was. His newly christened science of "sociology" which represented the fulfillment of the dream of extending science to society and politics, required him to reflect extensively on what science was, to classify the sciences, and to give an account of method. His central "discovery," the law of the three stages, which he took to be the core finding of sociology, was a law about the internal development of scientific disciplines: the first stage was the theological or fictitious, the second the metaphysical or abstract, and the third the scientific or positive, in which such metaphysical notions as causation were supposed to disappear, leaving only predictive law (Comte, [1830–42]1858: 25–26). The principle was reflexive, indeed self-exemplifying: sociology was to be the last science to reach the positive stage, and the law predicted that it would do so. Comte never strayed far from the lessons of science as a model. Indeed, the history of science, specifically Joseph-Louis de Lagrange's history of rational mechanics (which explored the filiation and descent of ideas), was the model for the specific "historical" method that he claimed was appropriate for sociology ([1830–42]1858: 496).[7]

The laws themselves were "objective." But in the end, according to Comte's later account, when sociology had reached the positive stage, all the sciences would become subordinate to it, and the relation of all knowledge to the subject, man, would be revealed.[8] At this point the sciences would be the servants of man, by analogy to medicine. Moreover, a fully developed sociology that related all knowledge to the subject would teach the critical anti-individualist lesson of the dependence of each person on others. Sociology would be both policy science and state ideology.

Comte's account of politics was similar to Saint-Simon's but with an even more strident hostility to liberal discussion.[9] Comte ([1830–42]1864, IV: 50ff) expressed his disgust for the idea that everyone should be permitted to have their opinion heard, that the ignorant and expert should be equally empowered, and to "conscience."

[T]here is no liberty of conscience in astronomy, in physics, or even in physiology, that is to say everyone would find it absurd not to have confidence in the principles established by the men of these sciences. (Comte [1830–42]1864, IV: 44n, trans. in Ranulf, 1939: 22)

Science, in particular the science of sociology, was consequently both model and means for overcoming the "anarchy of opinions" by providing consensus. In contrast, liberal politics and free discussion, from the point of view of the prospect of such knowledge, was no more than the politics of ignorance and pointless dissension.

Comte, to put it simply, had assembled all the elements of a powerful argument to resolve the ambiguities of Condorcet by eliminating its liberal squeamishness about authority and consent. For Comte, the issue was this: if science is correct, and science includes knowledge of the social world and politics, why shouldn't scientists rule over the ignorant, or rule through their control of education? And is not the rule, *de facto* if not *de jure,* of scientists the condition of progress? Is the public's failure to consent to such rule anything other than a failure of scientific education? And if the understanding and recognition of the authority of science are the central condition of progress, shouldn't science be imposed on the ignorant, just as the dogmas of Catholicism had been so effectively imposed in the past (Comte [1830–42]1864, IV: 22, 480; V: 231)? Given his premises, the conclusion was difficult to avoid, and even John Stuart Mill, his admirer who rejected his later work, admitted that as a matter of logic, Comte was correct ([1865]1969: 302).

THE LIBERAL CHALLENGE

Although each of these premises, and the related picture of science they depended on, would be rejected by Comte's critics, a fully coherent response, with an alternative image of science, was slow to develop. The main obstacle to constructing an alternative was the notion of scientific method itself. Although Mill was a paragon of liberalism, he was trapped between his father's faith in free discussion, which he expounded in his famous *On Liberty* ([1859]1978), and his own methodological views, which were centered on the idea that the canons of induction lead to proven knowledge. The canons produced consensus apart from discussion, by the following of rules—even in the social sciences, where their value was limited by the problem of causal complexity. Moreover, Mill was a utilitarian, who believed that moral and political questions resolved into questions of the greatest good for the greatest number. So he was compelled, in the conclusion of Book VI of *A System of Logic*, to say that questions of politics were a matter of practical science, subordinate to the principle of utility (1974). To the extent that this is true, there is much for indoctrination to be about, and little if anything for democratic discussion to be about, a point not lost on his critics (cf. Cowling, 1963).[10]

Mill did not resolve the conflict between science and free discussion. In *On Liberty*, science is simply omitted. In his address to the University of St. Andrews that discusses science, freedom of speech is commended, but for schools of theology, and although science education is discussed at length in this text, it is not mentioned in connection to free discussion. Mill's critique of the later Comte expresses concerns about the practical implementation of the authority of science. He notes that Comte's position relies on the consensus of scientists but that this authority "entrusted to any organized body, would involve a spiritual despotism" (Mill, [1865]1969: 314). But he does not question the notion of consensus itself. Mill's conflict is nevertheless deep: if science is distinguished by the possession of a consensus-producing method, its reliance on human institutions is incidental or inessential and the authority of science overrides free discussion.

There were other important, and less ambivalent, though also less direct, responses. When William Whewell wrote the history of core intellectual advances in science, he also wrote about the difficulties that major ideas had in becoming accepted, which undermined the idea that within science truth was readily recognized and acknowledged (Whewell, 1857; cf. 117–20, 130–33, 150–53, 177–79, 184–88). A section of Buckle's *History of Civilization in England*, one of the most influential works of the nineteenth century, argued that state patronage of knowledge in France had diminished French intellectual life ([1857]1924: 490–516). The idea that science was a product of routinizable methods itself became the subject of an intense debate, much of it critical of Mill. This debate set the stage for a new formulation of the basic Baconian picture of science.

PEARSON AND MACH

Although there are questions about the nature of Comte's influence on the next stage of the discussion, Ernst Mach and Karl Pearson in their writings come into focus as transitional figures between two widely separated bodies of thought: Comte's positivism and the Communist theorists of science of the 1930s. One of the latter, Lancelot. T. Hogben, recalled that his generation had "been suckled on the *Grammar of Science*," Pearson's major text on science (Hogben, 1957: 326, quoted in Porter, 2004: 7). Mach developed and popularized a philosophy of science that was congenial to certain subsequent developments, notably Logical Positivism, and served as a carrier for some key ideas of Comte's (Blackmore, 1972: 164–69).[11] Both had a view of science as "economical" or oriented to "efficiency." Pearson connected this to contemporary ideas of national efficiency, Mach to a movement of scientists led by Wilhelm Ostwald called "energeticism" that opposed the atomic theory and extended the law of conservation of total energy to a normative notion of the economy of energy in social life. This idea also influenced their ideas about the relation of theory to data. Because they thought of theory as economical expressions of data, they were hostile to realistic interpretations of theoretical entities that went beyond the data. The standard view placed them together: "just as Mach opposed the atomic theory, so Pearson fought Mendelism" (Blackmore, 1972: 125; cf. Porter, 2004: 269–70 for a more nuanced view).

The Grammar of Science ([1892]1937) began in this vein, with a discussion of the purpose of science, which Pearson claimed was the same as that of any other human activity: to promote the welfare of human society, to increase social happiness, and to strengthen social stability. Stability was strongly associated with consensus, and as in Comte, science was a model for the achievement of consensus. Yet Pearson appeared to be of two minds about the problem of consensus, as indicated in his phrase "unforced consensus," which reflected both his idea, shared with Mach, that the age of force had ended, and his insistence on consensus as a condition of social stability and that social stability was the ultimate goal of science. The conflict lies in the relation of the two ideal elements of the ideal of unforced consensus. One is that "force" in the form of a scientific hierarchy persecuting scientific heresy would be fatal to progress. The other is that consensus is the primary good that science provides. And for Pearson the scientific method assures consensus without force.

Citizens must, of course, accept the consensus produced by science, and this is where education and popularization come in. Pearson was concerned with the right way to inculcate the scientific, unbiased cast of mind. Merely reading about science did not lead to this result: what did so was the close scientific study of some small area ([1892]1937: 15–16). And one could expect such experience to transfer to the role of the citizen. This would produce consensual politics without coercion ([1892]1937: 11–14).

Although he was a socialist, Pearson was no egalitarian with respect to the hierarchy of scientific talent. The role of the semi-educated citizen was still primarily one of respect for the "Priests" of science.[12] But he also believed, in the phrase of nineteenth century Catholicism, in "no rights for the wrong." Lack of conformity to the canons of legitimate inference, Pearson says, is "antisocial" if it involves believing "in a sphere in which we cannot reason," and there is no "right" to holding false beliefs that lead to negative consequences in matters that are "of vital importance to others" ([1892]1937: 54–55). And he argued that "the abnormal perceptive faculty [i.e., the kind that failed to arrive at the consensual conclusion assumed to be more or less automatically produced by persons with normally evolved perceptual powers], whether that of the madman or the mystic, must ever be a danger to human society, for it undermines the efficiency of the reason as a guide to conduct" ([1892]1937: 120).

Pearson's optimism about the efficacy of the scientific method as a source of consensus was grounded in his philosophy of science. The facts of science for him are perceptual successions, and so the idea of arriving at an unforced consensus on them is plausible. What is controversial is the idea that *political* questions can be resolved into issues of perceptual succession. Pearson's examples of how this should work included Poor Law reform, where "the blind social instinct and the individual bias at present form extremely strong factors of our judgment" (Pearson, [1892]1937: 29), preventing their objective solutions through considerations of national efficiency.[13]

THE PROBLEM OF CULTURE

The thinkers we have considered here, in the line from Bacon to Pearson, had an "extensive" conception of science, one in which science, understood for example as a method, could be applied to something beyond its normal subject matter. Science could be conceived "extensively" in a variety of ways: as incorporating technology and engineering, as including "social" and "mental" sciences, as including the policy sciences, and even as a foundation for ethics, a popular theme in the post-Darwinist period. The nature of science came to be discussed in terms of the essence which carried over. It was in response to this that a "liberal" view of science finally emerged. Pearson, and later the heterodox economist Thorstein Veblen, talked about science and engineering as a cast of mind that carries over from one activity or topic to others, and the theme was deeply embedded in the culture of the time (Jordan, 1994). There was also a strong current of sociological thinking that developed a variant of this thesis. William F. Ogburn's *Social Change* ([1922]1966), one of the most influential works of sociology of the interwar years, which introduced the term "cultural lag," into the language, was akin to technological determinism.[14]

The "cultural" significance of science soon became a hotly contested issue. In the German speaking world, the issue took the form of a discussion of the idea of a scientific *Weltanschauung*. Mach and his successors, including the Logical Positivists and especially Otto Neurath, were interpreted, and sometimes interpreted themselves, as providing a scientific alternative to retrograde *Weltanschauungen* (cf. Richardson, 2003), and Neurath used the term *Wissenschaftliche Weltauffassung*, or scientific conception of the world, to distinguish the scientific alternative from mere "world views" (Richardson, 2003: 68–69). This quest for a scientific conception of the world played a role in German thought analogous to the role that the problem of the replacement of traditional religion had played in British and French thought.

The problem of whether science could provide a *Weltanschauung*-substitute in turn produced an issue about the cultural status and character of science that was highly consequential for what followed, first in the German-speaking world (Lassman and Velody, 1989), and ultimately, as Logical Positivism was imported, in the Anglo-American world. But the discussion also led indirectly to a body of explicitly "sociological" thought about the nature of world views and the causal relations between science and civilization, and ultimately to the "classical" sociology of knowledge of Mannheim and to the development of Marxist accounts of science.

The carry-over thesis answered the question of causal direction in the science-society relation by making science the prime mover. But the question could also be put as follows: Did advances in science, or indeed the phenomenon of modern science itself, depend on cultural conditions? Philosophers, such as Alfred North Whitehead ([1925]1967), and civilizational sociologists, such as Sorokin ([1937]1962) and Max Weber ([1904–05]1949: 110; [1920]1958: 13–31), ran the direction of causality in this other way, from culture to science, seeing features of modern western culture as conditions for the growth of science and the scientific mentality.[15]

The idea of the scientific resolution of policy questions, already formulated by Mill, also played a significant role in this period, in a variety of forms. Fabian socialism in Britain and a huge array of reform movements in the United States, as well as bodies such as the German *Verein für Sozialpolitik*, promoted scientific or engineering solutions to social and policy problems and an "efficiency" movement. The Russian Revolution proclaimed itself to be "scientific" in that it was based on the scientific materialism of Marx and Engels: this was a realization in practice of the extension of science to absorb and obliterate politics. The experience of "War Socialism" in Germany during WWI persuaded many thinkers, notably Otto Neurath, of the practicality and desirability of a planned economy (cf. O'Neill, 1995; Steele, 1981). The issue of the efficacy of planning was to become central to the later literature.

The idea of experiment also served as a political model. John Dewey, in such works as *Human Nature and Conduct*, pronounced the experimental method to be the greatest of human achievements, and he promoted the idea of its application to human affairs, replacing "custom" and attachment to traditions, such as constitutional traditions, as a basis for political action (1922). Yet Dewey distinguished the techniques of science from the spirit: he wanted the spirit, and its creativity, in politics, but not the techniques or the experts that employed them, or the experts themselves, whom he dismissed as specialists and technicians whose work needed to be "humanized" (Dewey, [1937]1946: 33). This reasoning, and the movement it represented, was not attractive to scientists themselves (Kuznick, 1987: 215).

In connection with science, the model of "conceptual schemes," under the influence of L. J. Henderson, became a Harvard commonplace (Henderson, [1941–42]1970). The reception of this way of thinking about science was aided by developments in science and mathematics, such as the discovery of non-Euclidean geometries and the broader recognition that what appeared as physical truth was dependent on nonempirical choices of mathematical structures. This was a thesis developed by Poincaré, but quickly absorbed and underlined by other thinkers, notably the Vienna Circle, and in the extended discussion of the theory of relativity that followed (Howard, 1990: 374–375). The broader relativistic implications of this idea were recognized at the time. When Neurath wrote that the choice of mathematical structures for a theory was not an empirical matter, Max Horkheimer cited the passage as evidence that he embraced hyperrelativism ([1947]1972: 165). This assimilation of scientific premises to "culture" took many other forms as well, for example, in such influential texts as Alfred North Whitehead's *Science in the Modern World* ([1925]1967) and *Process and Reality* ([1929]1978), and even more explicitly in E. A. Burtt's *Metaphysical Foundations of Modern Physical Science: A Historical and Critical Essay* (1927). This was part of a larger and pervasive climate of opinion,[16] shared by Mannheim's sociology of knowledge (though Mannheim specifically exempted science from the subject matter of his "sociology of knowledge") but also by Ludwig Fleck, who used the notion of *Denkgemeinschaft* to account for the problem of the reception of scientific ideas ([1935]1979), an issue that was soon to become central.

This general approach was paralleled in France in a series of historical studies broadly influenced by the French neo-Kantian tradition and phenomenology, which focused on conceptual change and difference, and in particular on conceptual breaks and ruptures. Pierre Duhem was one of the pioneers of this approach, especially for his studies of medieval physics, which he showed to be methodologically sophisticated and coherent, and his holism, which led him to reject the idea of crucial experiments. Later French historians of science, such as Alexandre Koyré, who focused on the scientific revolution, stressed the radical nature of change between the conceptual systems it involved (1957). This austerely presuppositional approach, influenced in his case by Husserl, largely ignored experiment and data as relevant to scientific change. His contemporary, Gaston Bachelard, performed a similar analysis of the transformation represented by Einstein's special theory of relativity (1984). His concept of the "epistemological break" was a means of expressing the interconnected or holistic aspect of such transformations, including their relations to general philosophical outlooks. Georges Canguilhem extended this notion of epistemological breaks in relation to the creation of fields of knowledge, especially, in the life sciences, through concepts of normality (1978). Canguilhem was the reporter for Michel Foucault's dissertation on psychiatry. Foucault extended this reasoning to new topics and new disciplinary fields and to the phenomenon of disciplining itself, thus completing the extension of explanations of the history of science in terms of breaks to the explanation of the history of culture. By focusing relentlessly on theory rather than experiment, technology, and instrumentation, and by its concern with rupture, the French discussion (which of course influenced English-language history of science, particularly with respect to the scientific revolution) simply bypassed the issues that arose in the English and German language discussions of science, not only becoming Kuhnians *avant la lêttre* but using this new understanding of science as a model for the understanding of intellectually organized social life generally.

The English and German discussions arrived at a similar point through a much more tortured route, and the reasons are relevant to the subsequent history. During the early twentieth century neo-Kantianism was in "dissolution," but the dissolution took various forms. Both Heidegger and Positivism provided different approaches to the problem of *a priori* truth, and each undermined the "presuppositions" model (cf. Friedman, 1999, 2000, 2001), as did the later Wittgenstein ([1953]1958, para. 179–80). These criticisms pointed in the direction of a notion of practice or tacit knowledge. Karl Popper attacked the presuppositions model by arguing that presuppositions changed every time a theory changed, and he attacked Mannheim for his idea that identifying presuppositions placed one in a position to "critique." The discussion of conceptual schemes, frameworks, and the like persisted in the history of science during this period, but it was not until the fifties, with N. R. Hanson's Wittgenstein-influenced *Patterns of Discovery* (1958), which undermined the notion of raw observational data, that it came into its own in philosophy proper.

WEBER'S "SCIENCE AS A VOCATION"

The German postwar discussion of the idea of science as a *Weltanschauung* produced an especially important response that did not directly figure in the historical and philosophical literature, but later appeared in the influential "sociological" approach to science developed by Merton. The idea that *Wissenschaft* had a cultural and political task of providing a worldview gained significance as a result of the cultural crisis produced by military defeat. This idea was to receive its classic critique in two speeches by Max Weber: one on "Politics as a Vocation" ([1919]1946a), the other on "Science as a Vocation" ([1919]1946b).[17] In "Science as a Vocation" Weber provided a history of motivations for science from Plato through Schwammerdam's proof of God's providence in the anatomy of a louse, dismissing them all and concluding the list with the brutal comment on "the naive optimism in which science, that is the technique of mastering life which rests on science—has been celebrated as the way to happiness. Who believes in this?—aside from a few big children in university chairs" (Weber, [1919]1946b: 143). These big children would have included Pearson, Mach, and Ostwald, whom he took the trouble to denounce in a separate article, particularly for the utilitarian theory of knowledge Ostwald shared with Mach, who spoke of theories as economizations (Weber, [1909]1973: 414). The brunt of his emphasis in the speech, aside from its anti-utilitarian view of science,[18] was on specialization as a condition for genuine achievement. This also undermined the "extensive" conception of science: the achievements of the specialist do not generalize into lessons about the mastery of life.

The message in the speech on politics was also explicit: "the qualities that make a man an excellent scholar and academic teacher are not the qualities that make him a leader . . . specifically in politics" ([1919]1946b: 150). The aspiring political leader was constrained by the realities of modern party politics and the demands of creating a following, as well as the intrinsic demands of the pursuit of power, demands so onerous that very few people had the personal qualities for such a career. This account of the political sphere—with its emphasis on the necessity of power for the achievement of any meaningful end, as well as its relentless reminders that the means specific to the state is violence and that to engage in politics is to contract with diabolical powers—served to place the sphere of the political beyond the prospect of transformation by intellectuals. And Weber made a particular point about the limitations of the bureaucratic mentality in the face of the demands of politics, thus undermining any thought that politics could be replaced by the administration of things.[19]

HESSEN AND THE TRANSFORMATION OF THE DEBATE

In 1931 the discussion of science was transformed by the emergence of a fully developed Marxian account of science, sponsored at the highest level of the Soviet ideological apparatus by Nikolai Bukharin. Bukharin's own main theoretical work was entitled *Historical Materialism* and opened with these sentences: "Bourgeois scholars speak of any branch of learning with mysterious awe, as if it were a thing produced in heaven, not on earth. But as a matter of fact any science, whatever it be, grows out of the demands of society or its classes" (Bukharin, [1925]1965). A volume of articles applying these ideas to the history of science was produced for an international congress of historians of science in London, and it had a profound, galvanizing effect, especially in Britain (Delegates of the U.S.S.R., 1931).[20] The thesis they presented was in fact a dramatic one that had the effect of incorporating "premises" talk into the Marxian theory of base and superstructure. The major point of this text was to show in detailed case studies that science was also the product of the demands of the time for technological results, that the demands were specific to particular social formations and historical situations, and that "theory" was ultimately driven by technological practice, so that the idea of an autonomous realm of pure science was a sham and an ideological construction (Hessen, 1931).

The British discussion of science had evolved differently than the German one. At the 1927 meeting of the British Association for the Advancement of Science, E. A. Burroughs (1927: 32), Bishop of Ripon, suggested a moratorium on science for a decade to allow for a reconsideration of its social consequences. Josiah Stamp pursued this theme in his Presidential Address at the 1936 meeting of the same association when he called on scientists to consider the social responsibility of scientists ([1936]1937). In this context of social concern and deepening economic and political crisis, a message about science from the Soviet Union, already idealized by British Fabian socialists such as the Webbs, was bound to have an impact.

One of Marx's central ideas was that the revolutionary moment occurs when the conflict between the forces of production and the capitalist class structure and system of economic relations is at its height. One of the central ideas of both the fascists and the Soviets was that of rational planning in the economy and other spheres of life. These ideas had a strong grip on the public and on policy makers during the Depression. In the case of science, a large literature developed on "the frustration of science," the idea that capitalists, incompetent bureaucrats, and politicians stood in the way of the kinds of scientific developments that could overcome the failures of capitalism.[21]

These ideas became the core of a Left view of science, which focused on conditions outside of science, such as the demands of the economy for particular kinds of technology, which either propel or retard relevant scientific development.[22] In line with the Marxist theory of history, the explanations of scientific development were implicitly teleological. But the detailed explanations themselves were novel and quite different from other histories of science, especially when they showed how the development of particular ideas was closely entwined with the technology of the time. A particular favorite was the argument that the availability of slaves in the ancient world and the consequent contempt for "work that could be carried out by slaves" led to Aristotle's failures to recognize relevant facts, such as the fact, known to ancient craftsmen but whose significance was not grasped until Galileo, that water could not be raised more than thirty feet by pumps (Hogben, 1938: 367–68).

The leading Marxist commentators on science argued that the Soviet Union was the one country in which science had obtained its "proper function," as its most important figure J. D. Bernal put it.[23] They viewed the Soviet system as benign and also argued that neutrality was impossible for the scientist, especially in face of the antiscientific drive of Fascism. They argued further that money for science would flow freely in a rationally organized planning regime rather than a market economy and that "any subject is capable of being examined by the scientific method" (Huxley, 1935: 31) including the economic system and society. They held that history was presently in a transitional phase moving toward a state in which science, understood extensively as implying "a unified, coordinated, and above all conscious control of the whole of social life" (Bernal, 1939a: 409), would abolish the dependence of man on the material world. Its rightful role was to become the conscious guiding force of material civilization, to permeate all other spheres of culture.[24] This claim allowed Bernal to say, echoing Pearson on unforced consensus, that science already is Communism, since it is performing the task of human society, and in the Communist way, in which "men collaborate not because they are forced by superior authority or because they blindly follow some chosen leader, but because they realize that only in this willing collaboration can each man find his goals. Not orders, but advice, determines action" (1939a: 415–16).[25] In practice, as Bernal envisioned it, scientists would be organized into trade-unions which would cooperate with other trade-unions in producing and carrying out the five-year plans.[26]

THE CRITIQUE OF EXTENSIVENESS

Bernal and his comrades understood that the issue that made their position unpersuasive to other scientists was the notion that planning would be applied to science itself. This raised the question of what sort of freedom of inquiry would exist under planning. These were not issues that could be confined to the Soviet Union. Nazi science was not only planned, it was "extensive" in a problematic sense that was also relevant to Lenin's notion that no cultural organization in the Soviet regimes should be autonomous from the party. Under the Nazis, science was expected, though in practice this often meant little, to conform to Nazi ideology. Scientists who were Jews were expelled, and a loud campaign was mounted against the "Jewish influence" in science. A paper by a German scientist, Johannes Stark, originally published in a Nazi journal, was translated and published in *Nature* (Stark, 1938). Stark's paper focused the anxiety of scientists and the Left about Nazism and prompted a huge response (Lowenstein, 2006). The response in the United States, however, was cast in terms of "freedom" and assertions about the link between scientific freedom and democracy, leading to manifestoes and resolutions in defense of science and democracy (Boas, 1938; Merton, 1942: 115; Turner, 2007).

This discussion provided the initial spur to a renewed debate on the autonomy of science. Bernal, mindful of the successes of German planned science, defined the issues in terms of a conflict between freedom and efficiency, a conflict which he thought could be resolved within the framework of planning. But the issue of freedom under planning was to be a theme in a larger and more wide-ranging political discussion.[27] The issue of planning and the problem of the autonomy of science, which were originally distinct, now converged. Robert Merton, who had emerged as a respected figure for his study of religion and the Royal Society, wrote two papers, "Science and the Social Order"([1938]1973) and "A Note on Science and Democracy" (1942), both about autonomy and written with an eye to Nazi science, which extended Weber's cryptic account of science in "Science as a Vocation." Merton described four norms of science: universalism, organized scepticism, "communism" or sharing of scientific results, and disinterestedness. In 1938 Merton noted that this was a "liberal" argument, for, as he put it, in a liberal society integration derives primarily from the body of cultural norms ([1938]1973: 265). Merton's norms were not rooted in, nor even consistent with, the attitudes of the public, which could be expected to resent them. It was for this very reason that science was vulnerable to fascism, which trades on popular antirationalism and places centralized control on science. But conflicts occurred in democracies as well, especially when the findings of science invalidated dogmas (cf. 1942: 118–19). Thus, science and democracy are not compatible unless there is a recognition of the autonomy of science, and such recognition was always under threat by the normal extension of science into new topics, such as social science investigations of areas considered sacred (1942: 126).

These were writings in "sociology" and reflected one of the dominant research concerns of sociology in the period: the professions. Merton stayed away from issues of scientific content and was careful to avoid taking sides between Left and liberal views of science. For the most part, his argument preceded the bitter debate over planning that broke out in the 1940s between Bernal's Social Relations of Science movement and the antiplanning Society for Freedom in Science (McGucken, 1984: 265–300).

The leading intellect of the anti-Bernal group was Michael Polanyi. Polanyi provided, where Merton did not, an argument for the autonomy of science based on the claim that science had no need for political governance in the form of planning because it was already "governed" sufficiently by its own traditions and because the nature of scientific discovery itself could not be rationalized in the fashion assumed by the planners (an argument that turned into an assault on the notion of scientific method itself). Polanyi, like Conant, who made the issue of reception a centerpiece of his view of science, denied that science proceeded by overthrowing theories on the basis of new observations, noting that it often required the assimilation of significant changes in unarticulated background knowledge (e.g., 1946: 29–31). Science was, Polanyi argued, a community as distinct from the sort of "corporate" bureaucratic order that was subject to planning.[28] Planning would destroy the feature of community life that made possible the growth of ideas, which was, for Polanyi, the ability of scientists to freely choose which ideas to pursue.[29] He based his claims about science on an elaborately developed account of the ultimately inarticulable cognitive processes of scientific discovery and the way in which discovery is dependent on local traditions and a special level of community life that honors "scientific conscience" and the use of scientific judgement (1946: 52–66). This was an attack on any mechanical or "logical" account of science.[30]

Polanyi's argument addresses the problem of science and democracy in a novel way that contrasts with Merton's. If science, understood as nonmechanical activity of discovery dependent on inarticulable knowledge, is subject to democratic control it will not flourish. But science, Polanyi says, is not an anomaly for democracy. It is similar in character to other communities, such as the church and the legal profession, which are granted autonomy on the basis of their strongly traditional, self-governing character. Democracy itself, Polanyi argued, is strongly traditional and moreover depends on a tradition "of free discussion" and decisions based on "conscience" (1946: 67) like that of science. So the relation between science and democracy should be one of mutual recognition and respect, from one traditional community to another, consistent with the recognition that the fruits of science can best be gained by granting autonomy to the scientific community (cf. Polanyi 1939, 1941–43, 1943–45, 1946, [1951]1980).[31]

These were abstract considerations. There was also a practical battleground for the Left view of science: education and public understanding. From Condorcet on, the Left view of science education was that the workers should be made to think scientifically through some sort of basic training in science itself. Nor was this merely a pious hope: many British scientists participated in workingmen's educational projects that realized this goal, and the idea is reflected in the titles of the texts written by the key Left thinkers about science in the 1930s, such as *Mathematics for the Million* (Hogben, [1937]1940; see also Hogben, 1938; Levy, 1933, 1938, 1939; Crowther, 1931, 1932; Haldane, [1933]1971, [1940]1975). The critics of this view included James Bryant Conant (1947: 111–12n), who dismissed as a failure the fifty years of applying Pearson's idea that elementary instruction in science would make for better citizens. He reformed science education at Harvard accordingly, with the idea that, instead of engaging in rudimentary exercises, it was better for students to get some knowledge of the nature of science by working through case studies of major changes in "conceptual schemes"—the favored Harvard language—such as the Copernican and chemical revolutions.[32] The set of case studies that was produced for this course (Conant, 1957), which Conant at first taught and which ran for nearly a decade (Fuller, 2000: 183), became the background for Kuhn's *Structure of Scientific Revolutions*. Kuhn himself, who was recruited by Conant as an instructor for the course (Kuhn, 2000: 275–76), wrote the case study of the Copernican revolution, which became his first book (Kuhn, 1957). Conant was equally aggressive in attempting to reform recruitment into scientific careers, which he hoped to make more open and meritocratic, a goal consistent with his "opportunity" liberalism (1940).

Although there are some differences in emphasis between Conant, Merton, and Polanyi, to a remarkable extent they overlap, and Conant and Polanyi are particularly close. Both Conant and Polanyi had a Liberal approach to science in the following sense: they thought it was best to govern science indirectly, by facilitating competition among scientists.[33] But Conant, acknowledging the realities of "big science," thought it was necessary to have a set of major elite universities with massive resources, analogous to major corporations, in order to make this competition meaningful. The argument for extensiveness depends on a reductive account of science, identifying transportable features, such as a "method" with unique intellectual authority. Conant objected to the notion that there was a universal method of science and to the "wide use of the word science" (i.e., what I have been calling extensiveness).[34] Almost any account of science that characterized the activity of science as continuous with nonscientific forms of reasoning, psychology, perception, and forms of organization, and accounted for it as a complex but distinctive amalgam of these features, made science less transportable. Moreover, this style of explanation inevitably conflicted with the more expansive claims of science to intellectual authority.[35]

POSTWAR SCIENCE STUDIES: THE ERA OF DISCIPLINES AND ITS CONSEQUENCES

The response of physicists to the Bomb, the coming of the Cold War, the betrayal of atomic secrets by scientists, the Oppenheimer case, the Lysenko affair[36] (which finally discredited the Soviet model of science), and the rise of an aggressively anti-Stalinist Left[37] transformed this debate. Scientists on the Left turned to the nuclear disarmament movement. The rapid growth of universities in the postwar period also led to a greater focus on disciplinary discourse and consequently narrowing of interest in topics "belonging" to other disciplines.[38] The previously marginal field of philosophy of science became the most prestigious and powerful subfield of philosophy while shedding its past interests in Left wing politics.[39] Much of its energy was taken up with consolidating the standard view of the logical structure of scientific theories.[40] Sociology of science, however, declined precipitously.

A bibliography by Barber and Merton in 1952 defines its literature: an amalgam of Left commentary on science with studies of technology, including Ogburn's *Social Effects of Aviation* (1946), works by scientists and historians with a "social contexts" component, government documents, Polanyi and Conant, and studies of Soviet Science. The sociology of knowledge and Mannheim were intentionally omitted (Barber & Merton, 1952: 143n); Fleck had yet to be discovered. Ogburn was at the end of a long career, Stern was to die in the fifties. In American sociology only three major scholars, Merton, Barber, and Edward Shils, continued to write on science, and Barber, a follower of Talcott Parsons, was the only one of these to do so systematically and to teach the subject. Merton left the field. Aside from the bibliography with Barber (Barber & Merton, 1952) and the introduction to Barber's book (1952), Merton published only one paper on science, on the importance of claims of priority, between 1942 and 1961. Shils became involved with the atomic scientists' movement, became close to Leo Szilard, sponsored the hiring of Polanyi at the University of Chicago, and was involved, along with Polanyi, in the Congress of Cultural Freedom and its Hamburg conference on Science. What he wrote on science was largely restricted to the scientists' movement (1972: 196–203).[41] This interest did lead to a minor classic, *The Torment of Secrecy* (1956), on the inherent conflict between science and security in liberal democracies. His basic formulation of the autonomy of science split the difference between Merton and Polanyi: like Polanyi, he argued that "there is an inner affinity between science and the pluralistic society" (1956: 176), and that the "tradition of the free community of science" grew up independently of modern individualistic liberalism; like Merton, he was concerned with "populistic hostility to science" (1956: 181) which exacerbated the intrinsic problems of political supervision of science.

Parsons, the inescapable "theorist" of this era in sociology, wrote a great deal on universities as institutions, but little on science.[42] Parsons saw science through the lens of his own view of the professions as essential building blocks of modernity, especially by virtue of their embodiment of the normative commitments of modernity, and thus as sharing in the central values of the society (cf. Parsons, 1986). The same thinking informed Barber's *Science and the Social Order* (1952), the first text that was recognizable as a theoretical and empirical overview of the sociology of science. In his 1990 collection of essays on science, Barber argued for "the special congruence of science with several characteristic subsystems of modern 'liberal' societies'" (1990: 40) as well as "the independent rationality of science." The emphasis on the place of science in the social system was, as Barber commented, "Parsonian all the way" (1990: 39).

Despite Barber's *Science and the Social Order*, which was one of the earliest in a long series of works by Parsons's students that were designed to colonize and bring theoretical order to the study of different societal "subsystems," the study of science did not flourish. Merton became a major figure in sociology but not for his writings on science. Like the Parsonians, he wrote on the professions, engineering, nursing, technologists, and medical students. When he returned to writing on science in the1960s, "ambivalence" replaced the conflict between science and society of his 1938 essay, and the model of ambivalence was the reluctance of patients to accept the authority of physicians' advice ([1963b]1976: 26).

In the 1960s and '70s, Merton and his students became associated with the argument that science functioned meritocratically, which was a version of the argument that the autonomy of science ought to be honored, but it was a characteristically depoliticized argument and avoided issues involving the intellectual substance of science in favor of external indicators, such as Nobel prizes and citations, which could be correlated with one another (e.g., Cole and Cole, 1973).[43] Merton barely acknowledged such thinkers as Polanyi.[44] Although Merton himself was partial to the history of science and not narrowly "sociological" in his writing about science, the abstractly quantitative approach of the Mertonian "program" made it largely irrelevant to the discussion of science that Kuhn's *Structure of Scientific Revolutions* was opening up, which was dominated by issues relating to the collapse of the theory-observation distinction that had been central to the standard model of scientific theory of the Logical Positivists.

Kuhn was the intellectual heir of Conant (though also influenced by Polanyi and the Quinean critique of Carnap), but he was Conant with the politics left out. He was nevertheless a genuinely interdisciplinary thinker who had been especially ensnared by the disciplinary divisions of the 1950s. But this situation was quickly changing. Departments of history and philosophy of science were established at London (1949) and Melbourne (1946), and Indiana (1960) and Pittsburgh (1971), and others were to follow. Kuhn was appointed to a comparable position at Princeton.[45] *Minerva* was established in 1962. At Edinburgh, the interdisciplinary unit of Science Studies was established in 1964. A department of history and sociology of science was established at the University of Pennsylvania (1971). The continuities with the older discussion were highly visible. Polanyi's concerns and those raised by the atomic scientists' movement guided *Minerva*. The historian of the Bernal circle, Gary Werskey, was appointed at Edinburgh, a program motivated in part by concerns about explaining how science actually worked, a project parallel to Conant's but this time pursued by veterans of the Social Responsibility of Science movement, the heir to the Social Relations of Science movement (MacKenzie, 2003).

The institutional stage was thus set for the developments that produced "Science Studies." Ironically, among the central intellectual conditions for the rise of science studies was the separation between the disciplines that occurred in the 1950s. Now it represented an opportunity for debate. The rational reconstructions given by philosophers of science and the Popperian model of falsification became targets for sociologists of science, and the agonistic relation that emerged (Zammito, 2004) was to provide the motive force for the revival of science studies as an interdisciplinary field. The conflict between science as an authoritative technique and science as a form of life was to take a new form: initially defined in disciplinary terms as a conflict between philosophy and sociology of science, and eventually in political terms as a dispute over the authority of science and of experts that Bacon himself would have recognized.

Notes

1. "by a suitable choice of a syllabus and methods of education, we can teach the citizen everything he needs to know in order to be able to manage his household, administer his affairs and employ his labour and his faculties in freedom . . . not to be in a state of blind dependence on those to whom he must entrust his affairs or the exercise of his rights; to be in a proper condition to choose and supervise them; . . . to defend himself from prejudice by the strength of his reason alone; and, finally, to escape the deceits of charlatans . . . " ([1795]1955: 182).

2. "When it comes to the institutions of public instruction, and the incentives that it would be their duty to provide to those who cultivate the sciences, they can have only a single guide: the opinion of men enlightened on these questions, which are necessarily foreign to the greatest number. Now it is necessary to be endowed with a superior reason, and to have acquired much knowledge oneself, to be able to listen to this opinion or to understand it well." ([1793]1976: 286).

3. Originally there were three classes of the Academy (physical and mathematical sciences, moral and political sciences, literature and fine arts), but in 1803 a decree of Napoleon I changed the division to four (physical and mathematical sciences, French language and literature, history and ancient literature, and fine arts), suppressing the second class (moral and political sciences) as subversive to the state.

4. This was later modified into the more famous Marxist version, "to each according to his needs" (Manuel, 1966: 84; Manuel, 1976: 65).

5. A notion greatly expanded by Lenin's account of the withering away of the state in "The State and Revolution" ([1918]1961).

6. Manuel gives a useful account of Saint-Simon's shifting view of the role of the scientist, which was gradually reduced and subordinated to the industrialist (1960), in part as a reflection of his disappointment at the reluctance of scientists—whom he tellingly denounced for their "anarchism" (1960: 348) for failing to submit to the authority of the general theory he proposed—to join his cause.

7. The idea that the history of mathematics might be the key to the understanding of intellectual progress already appears in Saint-Simon (Manuel, 1960: 345).

8. This was Comte's theory of the subjective synthesis (Acton, 1951: 309).

9. As Manuel explains his reasoning,

> Since men of a class would seek to excel in their natural aptitudes, there could be only rivalry in good works, not a struggle for power. When class chiefs owed their prestige to their control of men, they could fight over one another's 'governed', but since there would be no governors and no subjects, from what source would class antagonism be derived? Within a class men of the same capacity would be striving to excel one another with creations whose merits all members of the class would be able to evaluate. Between classes there could be only mutual aid. There was no basis for hostility, no occasion for invading one another's territory. (Manuel, [1962]1965: 134–35).

10. There is a bitter dispute in Mill scholarship over the question of whether Mill believed in free discussion as such or believed in it only as a means of advancing the opinions he preferred (an interpretation that calls attention to his views on education). Maurice Cowling, in a classic polemic, collects the most damning quotations in support of the view that Mill's notion of "the intelligent deference of those who know much to those who know more" (quoted in Cowling, 1963: 34–35) amounted to a plea for deference to an intellectual and cultural elite. Chin-Liew Ten provides a strong refutation, based on an interpretation of Mill's ethical views (Ten, 1980). These ethical views, however, look very different in the light of the consideration, usually ignored, that Mill regarded politics as applied social science. The argument over whether Mill intended to apply the arguments in *On Liberty* to science is summarized in Jacobs, who denies that he did (Jacobs, 2003).

11. Mach's biographer discusses the commonplace that he is a transitional figure between Comte and Logical Positivism. Mach did not use the term positivist, nor did he acknowledge Comte as a source. But the similarities, especially with respect to their campaigns against metaphysics and religion, their insistence that they are not providing a philosophy but rather a science, and their belief in the unity of the sciences, are telling.

12. Pearson's authoritarianism differed from Comte's. Pearson hoped for the establishment of "poets, philosophers, and scientists" as "high priests" (Pearson, 1888: 20), and for the elevation of "reason, doubt, and the enthusiasm of the study" above the "froth and enthusiasm of the marketplace" (Pearson, 1888: 130–131, 133–134). But he labeled this "free thought" and was careful to describe science itself in terms of freedom. He comments at one point that there is no pope in science ([1901]1905: 60) and observes that doubt is integral to science and part of science's mystery ([1892]1937: 50–51). Skepticism of a certain kind (presumably toward religion) is enjoined ethically: "Where it is impossible to apply man's reason, that is to criticize and investigate at all, there it is not only unprofitable but antisocial to believe" ([1892]1937: 55). Later, out of fear of "scientific hierarchy" based on past achievements, he was to say "science has and can have no high priests" (1919: 75).

13. The main finding of science that most concerned him was the basic law of heredity, that "like produces like," which he applied to the problem of government and national strength by proposing eugenic control over population quality emphasizing the elimination of the unfit, the racially inferior, and stressing the necessity for the reproduction of the best minds. This was consistent with socialism, for him, because he believed that socialism required superior persons ([1901]1905: 57–84).

14. Ogburn's thinking in this book was rooted in an argument that Giddings had derived from Pearson's genetics in which the stabilization of intellectual objects was conceived on analogy to the stabilization of species. The reasoning was used in Ogburn's dissertation on the evolution of labor legislation in American states, which showed how the original diversity of legislation was supplanted by close similarity (1912).

15. Ogburn argued a parallel point in his essay on multiple discoveries and inventions (Ogburn & Thomas, 1922), deriving the argument from Alfred Kroeber (1917): variations in talent were likely to be small and could not account for variations in discovery; what could account for these variations, and at the same time account for the striking phenomenon of multiples, was the variation of cultural conditions. Ogburn's examples of "cultural conditions" were all drawn from material culture, an issue that Merton was to attack him over (1936).

16. Ironically, the term climate of opinion was popularized by the historian Carl Becker in *The Heavenly City of Eighteenth-Century Philosophers* to account for the Enlightenment, and it remains one of the best formulations of the idea that intellectual constructions are the product of tacit assumptions of a particular time (1932: 1–32).

17. The text contains a famous farewell to religion as an intellectually serious alternative, on the grounds that at this point in history religion required an "intellectual sacrifice" (Weber, 1946: 155).

18. An important American critique of Pearson's "social aims" interpretation of science was given by C. S. Peirce, writing as a public scientific intellectual (1901).

19. The term *Wissenschaft* is broader than the English "science" and means any field of organized inquiry. Weber's speeches led to a huge controversy, which in turn contributed to what came to be known as the crisis of the sciences. This became the German language form of a set of issues that was to unfold in a quite different way in the English-speaking world. For an introduction to the German dispute that immediately followed "Science as a Vocation," see Peter Lassman and Irving Velody (1989). Weber's account of the career of the scientist, here and in other texts, emphasized the notion of institutional roles and the inappropriateness of promoting values from within this role. It is worth noting that he had earlier attacked the monarchical socialist economists of the generation before his for wrongly thinking that there were scientific solutions to policy questions. They had, he argued, in a bitter controversy in the prewar period, surreptitiously inserted their own values into the advice, deluding themselves into thinking that they got the policy results from the facts alone. Thus, the effect of the paired accounts was to separate the sphere of science from that of politics, giving the scientist nothing to say to the politician except to instruct him on means, and to strip the scientist of the role of cultural leader. This way of thinking about professions eventually led, through Talcott Parsons's and Merton's emphasis on professionalization, to the American "sociology of science" of Barber and Merton, which was depoliticized. It also led, however, to an explicitly political critique of science in the work of Hans J. Morgenthau, the influential international relations theorist (1946; 1972).

20. The story of these events and their impact is well told in Werskey (1978; on Hessen see Graham, 1985).

21. The texts based on this thesis include Hall (1935) and in the United States the writings of "Red" Bernhard Stern (National Resources Committee, 1937: 39–66), a pioneering medical sociologist and precursor to the sociology of science (Merton, 1957), as well as Ogburn.

22. Much has been written on the key members of this group, J. D. Bernal, J. B. S. Haldane, Hyman Levy, Julian Huxley, Joseph Needham, Lancelot Hogben, and others related to them, such as F. Soddy and P. M. S. Blackett. A brief introduction is Filner (1976).

23. All that the phrase "the dictatorship of the proletariat" meant, according to Julian Huxley, writing in the period of the project of the collectivization of agriculture that killed millions through famine and violence, was that things were administered for the benefit of all (1932: 3).

24. Although Bernal was respectful of "dialectics" (cf. 1939b), the respect was superficial. It would be more accurate to say that he owed his picture of the authority of scientist-experts to Pearson and Fabianism, his picture of the ideal organization of science to the "guild socialism" of Fabian apostate G. D. H. Cole, and, as noted above, his picture of the ideal scientist to the image of the altruistic "new socialist man."

25. Bernal, as Veblen had earlier in his discussion of the likelihood of the creation of a Soviet of Engineers (cf. [1921]1963: 131–51), acknowledged that present scientists were too individualistic and competitive to fully realize the communism inherent in science and pointed to a future, transformed scientist, much in the manner of the New Socialist Man sought by the Soviet Union (1939a: 415).

26. Recall that Condorcet, Comte, and Pearson faced the problem that the condition for their account of science to contribute to progress was that citizens become scientifically educated, if only in a limited way, which required scientists to be, in effect, ideologists whose ideology was authoritative for the rest of society. This was tantamount to rule by scientists, which even enthusiasts such as Bernal regarded as impractical, though preferable, raising the question of whether Bernalism was Pearsonism in the vestments of Communism. Communism was not rule by scientists, though it involved an authoritative ideology that regarded itself as scientific.

27. Friedrich Hayek, in *The Road to Serfdom* (1944), emphasized the inherent conflicts between planning and freedom. Interestingly, in *The Counterrevolution of Science* he traced the issues to Saint-Simon's and Comte's ideas ([1941; 1942–44]1955).

28. The term "community of science" came into standard usage in the 1940s primarily through the writings of Michael Polanyi (Jacobs 2002). Until 1968 Merton did not use the term to describe science, except ironically, noting that "For the most part, this has remained an apt metaphor rather than becoming a productive concept" (1963a: 375).

29. And to benefit from the results of the free choices of other scientists.

30. The Logical Positivists responded by dismissing "the mystical interpretation" of discovery and treating their own views as concerned only with the (logical) context of justification as distraction from the (psychological and social) context of discovery (cf. Reichenbach, 1951: 230–31). This immunizing tactic set them up for the intellectual catastrophe, hastened by Kuhn, that in the 1960s destroyed Logical Positivism—the narrow "logical" view of scientific theory could not account for either conceptual change or the role of necessarily theory-laden "data" (cf. Shapere, 1974; Hanson, 1970).

31. In a widely quoted paper, Hollinger treats the argument for autonomy, for what he calls "laissez-faire communitarianism," essentially as an ideological means of extracting government money without government control (1996). It is notable that Bernal's Left approach to science, though it was self-consciously an attempt to balance freedom and efficiency, involved an even broader scope for self-governance.

32. This reform is extensively discussed in Fuller (2000).

33. This characterization raises the question of their relation to Karl Popper, whose connection with Hayek and defense of liberalism made him an outlier in this group. Popper had a brief flirtation with Polanyi, though the two found themselves uncongenial. In any case they were different kinds of liberals. Polanyi was a competent economist. Both were more interested in the liberal model of discussion and in the analogous "free criticism" of scientific discussion (Popper, [1945]1962: 218; Polanyi, 1946). Neither developed the analogy between liberal discourse and scientific discourse, and if they had done so the relations might have been more evident (cf. Jarvie, 2001). Both are limited forms of discourse, governed by a shared sense of boundaries. Popper's way of bounding science, the use of falsification as a demarcation criterion, may have led him to think that there was no need to locate a supporting ethos or tradition. The difference between the verificationist theory of meaning and falsification distinguishes him from the Logical Positivists. Verification faces out, so to speak, to those forms of purported knowledge that science might hope to supplant or discredit. It is directed at the larger community. Falsification looks in, to the process of scientific discussion that it regulates, and in so regulating makes it into a variant of liberal discussion.

34. One exemplary instance of his critique of extensiveness is his dismissal of the notion of the special virtue of the scientist—argued also in Merton's "Note on Science and Democracy" (1942). Conant puts the point succinctly by asking whether "it be too much to say that in the natural sciences today the given social environment has made it very easy for even the emotionally unstable person to be exact and impartial in his laboratory?" His answer is this. It is not any distinctive personal virtue like objectivity but "the traditions he inherits, his instruments, the high degree of specialization, the crowd of witnesses that surrounds him, so to speak (if he publishes his results)—these all exert pressures that make impartiality on matters of his science almost automatic" (1947: 7). These mechanisms, however, exist only for science proper—not for its extensions into politics, where the scientist has no special claim.

35. Both rejected the idea that scientists had relevant authority over such topics as religion and morals (Conant, 1967: 320–28; Polanyi, 1958: 279–86).

36. The Lysenko episode was a test of the credulity of the Bernal circle and of their willingness to compromise science for politics. The actual background to the events is even more bizarre than could have been known at the time, and Stalin's actions were not, as was often assumed, based on ideas about the class basis of science. Bukharin, discussed earlier, had been convicted in the famous show trial and confession that was the basis for Koestler's *Darkness at Noon* (1941). He appears to have been protecting his family rather than performing, as the expression of the time put it, one last service for the party. Hessen apparently died in the gulag around 1940. Lysenko, who was placed in charge of Soviet agricultural science, had tried to justify his opposition to the gene theory by stigmatizing it as bourgeois science, discrediting the British Communists' defense of Soviet science. Stalin, as it happens, was having none of this and personally crossed out such phrases in the official report Lysenko was to deliver on this issue, and "where Lysenko had claimed that 'any science is based on class,' Stalin wrote 'Ha, ha ha, . . . and mathematics? And Darwin?'" (Medvedev and Medvedev, 2004: 195).

37. Sponsored to a significant extent by the CIA (cf. Saunders, 1999: 1, 167–68).

38. Polanyi's career illustrates this: he continued to be a major participant in discussions of science and continued to write seriously about science in an interdisciplinary sense, publishing in 1958 his magnum opus, *Personal Knowledge: Towards a Post-Critical Philosophy*, a text that combined philosophy with psychology and a vivid picture of the social character of science. But the book was largely ignored by philosophers, who took the view that Polanyi, who of course was a scientist by training, was not a real philosopher.

39. Howard (2000). Jonathan Rée tells this story for Britain: "[T]he *Times Literary Supplement* [1957] tried to reassure its readers that [A. J.] Ayer's hostility to metaphysics was part of the now withered 'Leftist tendencies' of the thirties. In those days, thanks to *Language, Truth, and Logic* [(1936)1952] 'logical positivism successfully carried the red flag into the citadel of Oxford philosophy. . . . But now, at last, philosophy has been purged of any taint of leftism'" (Rée, 1993: 7). For the United States, see McCumber's *Time in the Ditch* (2001). Whether a political explanation of the narrowing of Logical Positivism is needed is open to question.

40. A comment by Einstein on Carnap, written in the course of declining an offer to contribute to a volume on Carnap, puts it thus:

> Between you and me, I think that the old positivistic horse, which originally appeared so fresh and frisky, has become a pitiful skeleton following the refinements that it has perforce gone through, and that it has dedicated itself to a rather arid hair-splitting. In its youthful days it nourished itself on the weaknesses of its opponents. Now it has grown respectable and is in the difficult position of having to prolong its existence under its own power, poor thing. (1953, quoted in Howard, 1990: 373–74).

41. Shils was a translator of Karl Mannheim's *Ideology and Utopia* ([1929]1936). Mannheim, who died in 1947, dedicated himself to a non-Marxist yet antiliberal project of social reconstruction that included the "planning of values," of which Shils thought little. His personal attitude toward Mannheim is revealed in Shils's memoir of Mannheim (1995). With respect to science Shils relied on Polanyi, with whom he was close, and made no attempt to produce an original "sociology of science." Shils did, however, write extensively on intellectuals and on the puzzle of why intellectuals—of which the Left scientists of the 1930s were an example—were so often opposed to the societies that supported them, and he developed a fascination for Indian intellectuals. He also took an interest in the university as an institution. The work of his student Joseph Ben-David, who in the 1960s was to write an introductory text on science for a series of works for students in sociology (1971), reflected this interest in institutional structures and disciplines as institutional forms, and he provided an influential periodization of dominant scientific institutional forms, which was at the same time a history of scientific autonomy and how it was achieved and threatened in each period.

42. Two exceptions are a text written to obtain funding for the social sciences that was known to Parsons' students but not published until much later (Parsons, 1986) and "Some Aspects of the Relations between Social Science and Ethics" (Parsons, 1947), a text that compares to Merton's "Note on Science and Democracy" (1942).

43. The Mertonian reflexive history of the school is given in Cole and Zuckerman (1975).

44. When the idea of the scientific community was revived by Derek Price under the heading "Invisible Colleges" (1963), Merton accepted it enthusiastically, leaving it to his student Diana Crane to concede that the data vindicated Polanyi's earlier insistence on the idea (cf. Crane, 1972).

45. M. Taylor Pyne Professor of Philosophy and History of Science, 1964.

References

Acton, H. B. (1951) "Comte's Positivism and the Science of Society," *Philosophy: Journal of the Royal Institute of Philosophy* 26(99): 291–310.

Ayer, A. J. ([1936]1952) *Language, Truth, and Logic* (New York: Dover Publications).

Bachelard, Gaston (1984) *The New Scientific Spirit*, trans. Arthur Goldhammer (Boston: Beacon Press).

Bacon, Francis ([1627]1860–62) "The New Atlantis," in J. Spedding, R. Ellis, & D. Heath (eds), *The Works of Francis Bacon*, vol. 5 (Boston: Houghton, Mifflin): 347–413.

Barber, Bernard (1952) *Science and the Social Order* (Glencoe, IL: The Free Press).

Barber, Bernard (1990) *Social Studies of Science* (New Brunswick, NJ: Transaction Publishers).

Barber, Bernard & Robert Merton (1952) "Brief Bibliography for the Sociology of Science," *Proceedings of the American Academy of Arts and Sciences* 80(2): 140–54.

Becker, Carl (1932) *The Heavenly City of Eighteenth-Century Philosophers* (New Haven, CT: Yale University Press).

Ben-David, Joseph (1971) *The Scientist's Role in Society: A Comparative Study* (Englewood Cliffs, NJ: Prentice Hall).

Bernal, J. D. (1939a) *The Social Function of Science* (New York: Macmillan).

Bernal, J. D. (1939b) "Science in a Changing World," Review of Hyman Levy (1939) Modern Science: *A Study of Physical Science in the World To-Day* (London: Hamish Hamilton, Ltd.), *Nature* 144: 3–4.

Blackmore, John T. (1972) *Ernst Mach: His Work, Life, and Influence* (Berkeley: University of California Press).

Boas, Franz (1938) *Manifesto on Freedom of Science* (The Committee of the American Academy for the Advancement of Science).

Buckle, Henry T. ([1857]1924) *History of Civilization in England,* 2nd ed., vol. I (New York: D. Appleton & Company).

Bukharin, Nikolai I. ([1925]1965) *Historical Materialism: A System of Sociology*, authorized translation from the 3rd Russian ed. (New York: Russell & Russell).

Burroughs, E. A., Bishop of Ripon (1927) "Is Scientific Advance Impeding Human Welfare?" *Literary Digest* 95: 32.

Burtt, Edwin A. (1927) *The Metaphysical Foundations of Modern Physical Science: A Historical and Critical Essay* (London: K. Paul, Trench, Trubner & Co., Ltd.).

Canguilhem, Georges (1978) *The Normal and the Pathological*, trans. Carolyn R. Fawcett (New York: Zone Books).

Coke, Edward (2003) *The Selected Writings and Speeches of Sir Edward Coke* (Indianapolis, IN: Liberty Fund).

Cole, Jonathan R. & Stephen Cole (1973) *Social Stratification in Science* (Chicago: University of Chicago Press).

Cole, Jonathan R. & Harriet Zuckerman (1975) "The Emergence of a Scientific Speciality: The Self-Exemplifying Case of the Sociology of Science," in Lewis A. Coser (ed), *The Idea of Social Structure: Papers in Honor of Robert K. Merton* (New York: Harcourt, Brace, Jovanovich): 139–74.

Columbia Encyclopedia (2001–2004) "Institut de France," The Columbia Encyclopedia, 6th ed. (New York: Columbia University Press).

Comte, Auguste ([1830–42]1864) *Cours de philosophie positive,* 2nd ed. (Paris: J. B. Baillière).

Comte, Auguste ([1830–42]1858) *The Positive Philosophy of Auguste Comte*, trans. Harriet Martineau (New York: Calvin Blanchard).

Comte, Auguste ([1851–54]1957) *Système de politique positive* (Paris: L. Mathias).

Conant, James B. (1940) "Education in a Classless Society," Charter Day Address, University of California, Berkeley, in *Atlantic Monthly* 165 (May): 593–602.

Conant, James B. (1947) *On Understanding Science: An Historical Approach* (New Haven, CT: Yale University Press).

Conant, James B. (ed) (1957) *Harvard Case Histories in Experimental Science* (Cambridge, MA: Harvard University Press).

Conant, James B. (1967) "Scientific Principles and Moral Conduct: The Arthur Stanley Eddington Memorial Lecture for 1966," *American Scientist* 55(3): 311–28.

Condorcet, Marie Jean Antoine Nicolas Cariat, marquis de ([1793]1976) "Fragment on the New Atlantis, or Combined Efforts of the Human Species for the Advancement of Science," in Keith M. Baker (ed), *Condorcet: Selected Writings* (Indianapolis, IN: Bobbs-Merrill): 283–300.

Condorcet, Marie Jean Antoine Nicolas Cariat, marquis de ([1795]1955) *Sketch for a Historical Picture of the Progress of the Human Mind: Being a Posthumous Work of the late M. de Condorcet*, trans. June Barraclough (London: Weidenfeld and Nicolson).

Cowling, Maurice (1963) *Mill and Liberalism* (Cambridge: Cambridge University Press).

Crane, Diana (1972) *Invisible Colleges: Diffusion of Knowledge in Scientific Communities* (Chicago: University of Chicago Press).

Crowther, J. G. (1931) *An Outline of the Universe* (London: K. Paul, Trench, Trubner & Co.).

Crowther, J. G. (1932) *The ABC of Chemistry* (London: K. Paul, Trench, Trubner & Co.).

Delegates of the U.S.S.R. (1931) *Science at the Crossroads: Papers Presented at the International Congress on Science and Technology* (London: Bush House, Kniga Ltd.)

Dewey, John (1922) *Human Nature and Conduct: An Introduction to Social Psychology* (New York: Holt).

Dewey, John ([1937]1946) *The Problems of Men* (New York: Philosophical Library).

Filner, Robert (1976) "The Roots of Political Activism in British Science," *Bulletin of the Atomic Scientists* (January): 25–9.

Fischer, Kuno (1857) *Francis Bacon of Verulam: Realistic Philosophy and Its Age*, trans. John Oxenford (London: Longman, Brown, Green, Longmans, & Roberts).

Fleck, Ludwik ([1935]1979) *Genesis and Development of a Scientific Fact* (Chicago: University of Chicago Press).

Friedman, Michael (1999) *Reconsidering Logical Positivism* (Cambridge: Cambridge University Press).

Friedman, Michael (2000) *A Parting of the Ways: Carnap, Cassirer, and Heidegger* (Chicago: Open Court).

Friedman, Michael (2001) *Dynamics of Reason* (Stanford, CA: CSLI Publications).

Fuller, Steve (2000) *Thomas Kuhn: A Philosophical History for Our Times* (Chicago: University of Chicago Press).

Gillispie, Charles Coulston (2004) *Science and Polity in France: The Revolutionary and Napoleonic Years* (Princeton, NJ: Princeton University Press).

Graham, Loren (1985) "The Socio-Political Roots of Boris Hessen: Soviet Marxism and the History of Science," *Social Studies of Science* 15: 705–22.

Hahn, Roger (1971) *The Anatomy of a Scientific Institution: The Paris Academy of Sciences, 1666–1803* (Berkeley: University of California Press).

Haldane, J. B. S. ([1933]1971) *Science and Human Life* (Freeport, NY: Books for Libraries Press).

Haldane, J. B. S. ([1940]1975) *Science and Everyday Life* (New York: Arno Press).

Hall, Daniel (Ed) (1935) *The Frustration of Science* (Freeport, NY: Books for Libraries Press).

Hanson, N. R. (1958) *Patterns of Discovery: An Inquiry into the Conceptual Foundations of Science* (Cambridge: Cambridge University Press).

Hanson, N. R. (1970) "A Picture Theory of Theory Meaning," in Michael Radner & Stephen Winokur (eds) *Analyses of Theories and Methods of Physics and Psychology*, Minnesota Studies in the Philosophy of Science, vol. IV (Minneapolis: University of Minnesota Press): 131–41.

Hayek, F. A. (1944) *The Road to Serfdom* (Chicago: University of Chicago Press).

Henderson, Lawrence J. ([1941–42]1970) "Sociology 23 Lectures, 1941–42," in Bernard Barber (ed) *L. J. Henderson on the Social System* (Chicago: University of Chicago Press): 57–148.

Hessen, Boris (1931) "Social and Economic Roots of Newton's 'Principia'," in *Science at the Crossroads* (London: Bush House, Kniga Ltd.): 151–212.

Hogben, Lancelot (1957) *Statistical Theory: The Relationship of Probability, Credibility, and Error; an Examination of the Contemporary Crisis in Statistical Theory from a Behaviourist Viewpoint* (London: Allen & Unwin).

Hogben, Lancelot ([1937]1940) *Mathematics for the Million*, rev. and enl. ed. (New York: W.W. Norton).

Hogben, Lancelot (1938) *Science for the Citizen: A Self-Educator Based on The Social Background of Scientific Discovery* (London: George Allen & Unwin).

Hollinger, David (1996) *Science, Jews, and Secular Culture: Studies in Mid-Twentieth-Century American Intellectual History* (Princeton, NJ: Princeton University Press).

Horkheimer, Max ([1947]1974) *Eclipse of Reason* (New York: The Seabury Press).

Howard, Don (1990) "Einstein and Duhem," *Synthese* 83: 363–84.

Howard, Don (2000) "Two Left Turns Make a Right: On the Curious Political Career of North American Philosophy of Science at Midcentury," in Gary Hardcastle & Alan Richardson (eds), *Logical Empiricism in North America* (Mineapolis: University of Minnesota Press): 25–93.

Huxley, Julian (1932) *A Scientist among the Soviets* (New York: Harper & Brothers).

Huxley, Julian (1935) *Science and Social Needs* (New York: Harper & Brothers).

Jacobs, Struan (2002) "Polanyi's Presagement of the Incommensurability Concept," *Studies in History and Philosophy of Science,* part A (33): 105–20.

Jacobs, Struan (2003) "Misunderstanding John Stuart Mill on Science: Paul Feyerabend's Bad Influence," *Social Science Journal* 40: 201–12.

Jarvie, I. C. (2001) "Science in a Democratic Republic," *Philosophy of Science* 68(4): 545–64.

Jordan, John M. (1994) *Machine-Age Ideology: Social Engineering and American Liberalism* (Chapel Hill: University of North Carolina Press).

Klausner, Samuel Z. & Victor M. Lidz (eds) (1986) *The Nationalization of the Social Sciences* (Philadelphia: University of Pennsylvania Press).

Koestler, Arthur (1941) *Darkness at Noon* (New York: Macmillan).

Koyré, Alexandre (1957) *From the Closed World to the Infinite Universe* (Baltimore, MD: Johns Hopkins Press).

Kroeber, A. L. (1917) "The Superorganic," *American Anthropologist* 19(2): 163–214.

Kuhn, Thomas (1957) *The Copernican Revolution: Planetary Astronomy in the Development of Western Thought* (Cambridge, MA: Harvard University Press).

Kuhn, Thomas ([1962]1996) *The Structure of Scientific Revolutions,* 3rd ed. (Chicago: University of Chicago Press).

Kuhn, Thomas (2000) *The Road Since Structure: Philosophical Essays, 1970–1993, with an autobiographical interview* (Chicago: University of Chicago Press).

Kuznick, Peter (1987) *Beyond the Laboratory: Scientists as Political Activists in 1930s America* (Chicago: University of Chicago Press).

Lassman, Peter & Irving Velody (eds) (1989) *Max Weber's "Science as a Vocation"* (London: Unwin Hyman).

Lenin, V. I. ([1918]1961) "The State and Revolution," in Arthur P. Mendel (ed), *Essential Works of Marxism* (New York: Bantam Books): 103–98.

Levy, Hyman (1933) *The Universe of Science* (New York: The Century Co).

Levy, Hyman (1938) *A Philosophy for a Modern Man* (New York: Knopf).

Levy, Hyman (1939) *Modern Science: A Study of Physical Science in the World Today* (New York: Knopf).

Lowenstein, Aharon. "German Speaking Jews in Sciences in Modern Times." Unpublished manuscript.

Lynch, William T. (2001) *Solomon's Child: Method in the Early Royal Society of London* (Stanford, CA: Stanford University Press).

MacKenzie, Donald (2003) "Tribute to David Edge," in Donald MacKenzie, Barry Barnes, Sheila Jasanoff, & Michael Lynch, *Life's Work, Love's Work. Four Tributes to David Edge (1932–2003), EASST Review* 22(1/2) (The European Association for the Study of Science and Technology). Available at: http://www.easst.net/review/march2003/edge.

Mannheim, Karl ([1929]1936) *Ideology and Utopia: An Introduction to the Sociology of Knowledge*, trans. L. Wirth & E. Shils (New York: Harcourt, Brace & World).

Manuel, Frank E. (1960) "The Role of the Scientist in Saint-Simon," *Revue internationale de Philosophie* 53–54(3–4): 344–56.

Manuel, Frank E. ([1962]1965) *The Prophets of Paris* (New York: Harper & Row).

Manuel, Frank E. (Ed) (1966) "Toward a Psychological History of Utopias," in *Utopias and Utopian Thought* (Boston: Houghton Mifflin): 69–100.

Manuel, Frank E. (1976) "In Memoriam: Critique of the Gotha Program, 1875–1975," *Daedalus,* Winter: 59–77.

Manuel, Frank E. (1995) *A Requiem for Karl Marx* (Cambridge, MA: Harvard University Press).

McCumber, John (2001) *Time in the Ditch: American Philosophy and the McCarthy Era* (Evanston, IL: Northwestern University Press).

McGucken, William (1984) *Scientists, Society, and the State: The Social Relations of Science Movement in Great Britain, 1931–1947* (Columbus: Ohio State University Press).

Medvedev, Roy & Zhores Medvedev (2004) *The Unknown Stalin: His Life, Death, and Legacy*, trans. Ellen Dahrendorf (Woodstock, NY: The Overlook Press).

Merton, Robert K. (1936) "Civilization and Culture," *Sociology and Social Research,* Nov/Dec: 103–13.

Merton, Robert (1938[1973]) "Science and the Social Order," in *The Sociology of Science: Theoretical and Empirical Investigations* (Chicago: University of Chicago Press): 254–66.

Merton, Robert (1942) "A Note on Science and Democracy," *Journal of Legal and Political Sociology* 1: 115–26.

Merton, Robert (1957) "In Memory of Bernard Stern," *Science and Society* 21(1): 7–9.

Merton, Robert ([1963a]1973) "Multiple Discoveries as Strategic Research Site," in *The Sociology of Science: Theoretical and Empirical Investigations* (Chicago: University of Chicago Press): 371–82.

Merton, Robert ([1963b]1976) "Sociological Ambivalence" [with Elinor Barber], in *Sociological Ambivalence and Other Essays* (New York: The Free Press): 3–31.

Mill, John Stuart ([1859]1978) *On Liberty,* Elizabeth Rapaport (ed) (Indianapolis, IN: Hackett Publishing Co.).

Mill, John S. ([1865]1969) *August Comte and Positivism*, Vol. X: *Collected Works* (Toronto: University of Toronto Press).

Mill, John Stuart (1974) *A System of Logic Ratiocinative and Inductive: being a connected view of the principles of evidence and the methods of scientific investigation*, book VI, J. M. Robson (ed) (Toronto: University of Toronto Press).

Morgenthau, Hans (1946) *Scientific Man vs. Power Politics* (Chicago: University of Chicago Press).

Morgenthau, Hans (1972) *Science: Servant or Master?* (New York: New American Library).

National Resources Committee (1937) *Technological Trends and National Policy: Including the Social Implications of New Inventions*, Report of the Subcommittee on Technology (Washington, DC: United States House of Representatives).

Ogburn, William F. (1912) *Progress and Uniformity in Child-Labor Legislation: A Study in Statistical Measurement* (Columbia University, Ph.D. diss.).

Ogburn, William F. ([1922]1966) *Social Change with Respect to Culture and Original Nature* (New York: Dell).

Ogburn, William F. (1946) *The Social Effects of Aviation* (Boston: Houghton Mifflin).

Ogburn, William F. & Dorothy Thomas (1922) "Are Inventions Inevitable? A Note on Social Evolution," *Political Science Quarterly* 37(1): 83–98.

O'Neill, John (1995) "In Partial Praise of a Positivist: The Work of Otto Neurath," *Radical Philosophy* 74: 29–38.

Parsons, Talcott (1947) "Some Aspects of the Relation between Social Science and Ethics," *Social Science* 22(3): 213–17.

Parsons, Talcott (1986) "Social Science: A Basic National Resource," in Samuel Klausner and Victor Lidz (eds), *The Nationalization of the Social Sciences* (Philadelphia: University of Pennsylvania Press): 41–120.

Pearson, Karl ([1892]1937) *The Grammar of Science* (London: J. M. Dent & Sons Ltd.).

Pearson, Karl (1888) *The Ethic of Freethought.* (London: T. Fisher Unwin).

Pearson, Karl ([1901]1905) *National Life from a Standpoint of Science: An Address Delivered at Newcastle, November 19, 1900,* 2nd ed. (London: Adam and Charles Black).

Pearson, Karl (1919) *The Function of Science in the Modern State,* 2nd ed. (Cambridge: Cambridge University Press).

Peirce, C. S. (1901) "Pearson's *Grammar of Science*: Annotations on the First Three Chapters," *Popular Science Monthly* 58 (January): 296–306.

Peltonen, Markku (1996) "Bacon's Political Philosophy," in Markku Peltonen (ed), *The Cambridge Companion to Bacon* (Cambridge: Cambridge University Press): 283–310.

Polanyi, Michael (1939) "Rights and Duties of Science," *Manchester School of Economic and Social Studies* 10(2): 175–93.

Polanyi, Michael (1941–43) "The Autonomy of Science," *Memoirs and Proceedings of the Manchester Literary & Philosophical Society* 85(2): 19–38.

Polanyi, Michael (1943–45) "Science, the Universities, and the Modern Crisis," *Memoirs and Proceedings of the Manchester Literary & Philosophical Society* 86(6): 109–63.

Polanyi, Michael (1946) *Science, Faith, and Society* (Chicago: University of Chicago Press).

Polanyi, Michael ([1951]1980) *Logic of Liberty: Reflections and Rejoinders* (Chicago: University of Chicago Press).

Polanyi, Michael (1958) *Personal Knowledge: Towards a Post-Critical Philosophy* (Chicago: University of Chicago Press).

Popper, Karl ([1945]1962) *The Open Society and Its Enemies,* vol. 2 (New York: Harper & Row).

Porter, Theodore M. (2004) *Karl Pearson: The Scientific Life in a Statistical Age* (Princeton, NJ: Princeton University Press).

Price, Derek (1963) *Little Science, Big Science* (New York: Columbia University Press).

Ranulf, Svend (1939) "Scholarly Forerunners of Fascism," *Ethics* 50(1): 16–34.

Rée, Jonathan (1993) "English Philosophy in the Fifties," *Radical Philosophy* 65 (Autumn): 3–21.

Reichenbach, Hans (1951) *The Rise of Scientific Philosophy* (Berkeley: University of California Press).

Richardson, Alan (2003) "Tolerance, Internationalism, and Scientific Community in Philosophy: Political Themes in Philosophy of Science, Past and Present," in Michael Heidelberger & Friedrich Stadler (eds) *Wissenschaftsphilosophie und politik; Philosophy of Science and Politics* (Vienna: Springer-Verlag): 65–90.

Saunders, Frances S. (1999) *Who Paid the Piper? The CIA and the Cultural Cold War* (London: Granta Books).

Shapere, Dudley (1974) "Scientific Theories and Their Domains," in Frederick Suppe (ed), *The Structure of Scientific Theories* (Urbana, IL: University of Illinois Press): 518–65.

Shapin, Steven (1994) *A Social History of Truth: Civility and Science in Seventeenth-Century England* (Chicago: University of Chicago Press).

Shils, Edward (1956) *The Torment of Secrecy: The Background and Consequences of American Security Policies* (Glencoe, IL: The Free Press).

Shils, Edward (1972) *The Intellectuals and the Powers and Other Essays* (Chicago: University of Chicago Press).

Shils, Edward (1995) "Karl Mannheim," *American Scholar* (Spring): 221–35.

Sorokin, Pitirim A. ([1937]1962) *Social and Cultural Dynamics* (New York: Bedminster Press).

Sprat, Thomas (1667]1958) *History of the Royal Society* (St. Louis, MO: Washington University Press).

Stamp, Josiah ([1936]1937) "The Impact of Science on Society," in *The Science of Social Adjustment* (London: Macmillan & Company Ltd): 1–104.

Stark, Johannis (1938) "The Pragmatic and the Dogmatic Spirit in Physics," *Nature* 141(April 30): 770–71.

Steele, David R. (1981) "Posing the Problem: The Impossibility of Economic Calculation under Socialism," *Journal of Libertarian Studies* V(1): 7–22.

Ten, Chin-Liew (1980) *Mill on Liberty* (Oxford: Claredon Press).

Turner, Stephen (2007) "Merton's 'Norms' in Political and Intellectual Context," *Journal of Classical Sociology*. 7(2): 161–78.

Veblen, Thorstein ([1921]1963) *The Engineers and the Price System* (New York: Harcourt, Brace, & World).

Weber, Max ([1904]1949) "'Objectivity' in Social Science and Social Policy," in Edward Shils & Henry Finch (trans. & eds) *The Methodology of the Social Sciences* (New York: The Free Press): 49–112.

Weber, Max ([1904–5]1958) *The Protestant Ethic and the Spirit of Capitalism*, trans. Talcott Parsons (New York: Scribner's).

Weber, Max ([1909]1973) "'Energetische' Kulturtheorie," in *Gesammelte Aufsätze zur Wissenschaftslehre von Max Weber* (Tübingen: J.C.B. Mohr [Paul Siebeck]): 400–426.

Weber, Max ([1919]1946a) "Politics as a Vocation," in H. H. Gerth & C. W. Mills (trans. & eds) *From Max Weber: Essays in Sociology* (New York: Oxford University Press): 77–128.

Weber, Max ([1919]1946b) "Science as a Vocation," in H. H. Gerth & C. W. Mills (trans. & eds) *From Max Weber: Essays in Sociology* (New York: Oxford University Press): 129–56.

Weber, Max ([1920]1958) "Author's Introduction," in *The Protestant Ethic and the Spirit of Capitalism*, trans. Talcott Parsons (New York: Scribner's): 13–31.

Weber, Max (1946) *From Max Weber: Essays in Sociology*, H. H. Gerth & C. W. Mills (trans. & eds) (New York: Oxford University Press).

Werskey, Gary (1978) *The Visible College: The Collective Biography of British Scientific Socialists of the 1930s* (New York: Holt, Rinehart and Winston).

Whewell, William (1857) *History of the Inductive Sciences From the Earliest to the Present Time* (London: John W. Parker and Son).

Whewell, William (1984) *Selected Writings on the History of Science*, Yehuda Elkana (ed) (Chicago: The University of Chicago Press).

Whitehead, Alfred North ([1925]1967) *Science and the Modern World* (New York: The Free Press).

Whitehead, Alfred North ([1929]1978) *Process and Reality: An Essay in Cosmology*, corrected edition, David Ray Griffin & Donald W. Sherburne (eds) (New York: The Free Press).

Wittgenstein, Ludwig ([1953]1958) *The Philosophical Investigations*, trans. G. E. M. Anscombe (New York: Prentice Hall).

Zammito, John H. (2004) *A Nice Derangement of Epistemes: Post-Positivism in the Study of Science from Quine to Latour* (Chicago: University of Chicago Press).

3   Political Theory in Science and Technology Studies

Charles Thorpe

Writing in the late 1970s, the moral philosopher Alisdair MacIntyre argued that the preoccupations of modern philosophy of science merely recapitulated classic debates in ethics and political thought. So we find "Kuhn's reincarnation of Kierkegaard, and Feyerabend's revival of Emerson—not to mention . . . [Michael] Polanyi's version of Burke" (MacIntyre, 1978: 23). Questions of political theory have been important, but often encoded and implicit, within the fields of the philosophy, history, and sociology of science throughout their twentieth century development. Today, the interdisciplinary field of Science and Technology Studies (STS) is increasingly explicitly concerned with political questions: the nature of governmentality and accountability in the modern state, democratic decision-making rights and problems of participation versus representation, and the structure of the public sphere and civil society. This theorization of politics within STS has particular relevance and urgency today as both the polity of science and the structure of the broader polity are being refashioned in the context of globalization.

The political concerns of STS have pivoted around the formulation and criticism of liberalism. Liberal values of individualism, instrumentalism, meliorism, universalism, and conceptions of accountability and legitimacy have been closely related to understandings of scientific rationality, empiricism, and scientific and technological progress. The "Great Traditions" in the philosophy, history, and sociology of science—represented, for example, by the Vienna Circle and Karl Popper in philosophy, George Sarton in history, and Robert K. Merton in sociology—were all in different ways engaged in formulating accounts of science as exemplifying and upholding liberal political ideals and values. The work of Polanyi and Kuhn, which has been taken to challenge the universalistic ambitions of the "Great Tradition," had a strongly communitarian and conservative flavor. I argue that we can read the development of STS in terms of critiques of liberal assumptions, from such diverse perspectives as communitarian and conservative philosophy, Marxism and critical theory, feminism and multiculturalism. In addition, we can see the recent preoccupation in STS with questions of public participation and engagement in science as suggesting a turn toward participatory democratic and republican ideals of active citizenship.

It is no accident that a heightened concern with participation should be alive in the field at a time when neoliberal economic regimes and globalization are restricting the terms and scope of political discourse and presenting a sense of restricted political possibility. At the same time, working in an opposite direction, new social movements are mapping out fresh arenas of political struggle, repoliticizing technicized domains (risk, advanced technologies such as genetically modified organisms [GMOs]), and may be seen as presenting a model for new forms of democratic mobilization. Rethinking the politics of science is central for coming to grips with the implications of globalization for democracy.

In tracing the linkages between STS debates and political thought, I aim to present a case for STS as an arena for questioning and debating what kind of polity of science (Fuller, 2000a; Kitcher, 2001; Turner, 2003a), "technical constitution" (Winner, 1986), or "parliament of things" (Feenberg, 1991), is warranted by democratic ideals. STS can play a key role in clarifying questions about which values and goals we want to inscribe in our scientific and technological constitutions. STS *as* political theory offers a set of intellectual resources and models on the basis of which competing normative political visions of science and technology can be clarified, analyzed, and criticized.

SCIENCE, TECHNOLOGY, AND LIBERAL DEMOCRATIC ORDER

Questions of political theory have been foregrounded in the sociology of scientific knowledge (SSK) by Steven Shapin and Simon Schaffer's *Leviathan and the Air Pump* (1985). In recovering Hobbes's critique of Boyle's experimental method, Shapin and Schaffer provide a symmetrical reading of Hobbes and Boyle both as political theorists. They rediscover the epistemology and natural philosophy of Hobbes and highlight the implicit political philosophy in Boyle's experimental program. This was a debate over the constitution of the "polity of science" and the way in which the product of that polity would operate as "an element in political activity in the state" (Shapin & Schaffer, 1985: 332).

The paradox that Shapin and Schaffer note is that the polity of science established by Boyle was one that denied its political character, and that paradox underlay its success. Boyle suggested that the experimental apparatus separated the constitution of knowledge from the constitution of power. Experiment allowed cognitive agreement to be based on the transparent testimony of nature rather than human authority (Shapin & Schaffer 1985: esp. 339). There is a strong isomorphism between Boyle's polity of science and the political ideals of liberalism emerging in the period—the ideal of a community based on ordered "free action" in which "mastery was constitutionally restricted" (Shapin & Schaffer, 1985: 339; see also 343). Liberalism in particular has tended to draw legitimacy by claiming a relationship between its political ideals and an idealized polity of science. The notion of a liberal society as "the natural habitat of science" has been a key legitimation for liberal democratic politics into the twentieth century (Shapin & Schaffer, 1985: 343).

The polity of science has been adept at masking its political character. Similarly, a key accomplishment of the modern liberal state has been to present itself as neutral with respect to competing group interests. Arguably, the sociotechnical norms embodied in Boyle's experimental practice provided a basis on which to achieve this image of political neutrality. Yaron Ezrahi (1990) has drawn on Shapin and Schaffer's study in presenting a political theory of the long-standing relationship between science and liberal democratic political culture in the West. Science provided a solution to key problems inherent in liberal democratic political order: how to depoliticize routine official or administrative actions, how to present official action as being in the public interest, how to hold public action accountable, how to reconcile individual freedom with social order. Ezrahi suggests that in solving these problems, the liberal polity drew on the norms of the polity of science: instrumentalism, impersonality or depersonalization, ordered free agency, transparency. Presenting state action as merely the technical solution to problems allowed that action to be presented as objective, based on the empirical facts, and therefore separate from the subjective desires or prejudices of the government official. In other words, science provided a model for liberal-democratic legal-rational authority. Ezrahi suggests that liberalism modeled political accountability on the "visual culture" of experimental science, which aimed to "attest, record, account, analyze, confirm, disconfirm, explain, or demonstrate by showing and observing examples in a world of public facts" (Ezrahi, 1990: 74). The attestive public gaze prevents politicians and officials from pursuing private interests or hidden agendas under the guise of public authority. In these ways science has had "latent political functions in the modern liberal-democratic state" (Ezrahi, 1990: 96).

Liberalism tends to technologize the political order. Political scientist Wilson Carey McWilliams has called America a "technological republic" (1993). Ezrahi points to America as the ideal-type model of the interrelationship between science and modern liberalism (Ezrahi, 1990: 105–8, 128–66). Americans have gone further than other nations in insisting on the instrumentality and impersonality of administrative action although charismatic authority operates at the political level, for example, the Presidency (Porter, 1995: esp. 148–89; Jasanoff, 2003a: 227–28). Indeed, the constitutional separation of powers models the polity after a machine with checks and balances providing an engineered equilibrium. And the image of the machine as a model for order has been a key motif in American political culture. But the technologization of the polity has been in conflict with Jeffersonian republican aspirations for virtuous civic engagement. Today's America is faced with depoliticization and disengagement as technological rationality and instrumentalism have overwhelmed democratic politics (McWilliams, 1993: 107–8). The liberal embrace of science and technology has often ended in moral disenchantment—the sense that science and technology have become substantive values pushing aside the humanistic value-attachments of liberalism. Ezrahi traces how the machine has moved from being a model of balance and equilibrium to being "an icon of excess" because of its association with dehumanizing bureaucracy and environmental degradation. As technical rationality is experienced as undermining human values, science loses its utility as a source of political legitimacy (Ezrahi, 1990: 242–43).

This turn away from the scientific model for politics provides the context for the emergence of contemporary Science and Technology Studies. STS is a discourse constructed in relation, and largely in opposition, to traditions of philosophy, history, and sociology of science that sought to codify and uphold science as an ideal model for liberal political order. STS as a project has been driven by doubts about the validity of the image of science (univeralism, neutrality, impersonality, etc.) that underlay the liberal model. Following a sustained intellectual attack on the epistemological, sociological, and historical underpinnings of the liberal model of science, attention within STS is increasingly focused on the political implications of this critique and on what sort of political model is suggested by STS's reformulations of the image of science.

TWENTIETH CENTURY SCIENTIFIC LIBERALISM

The hope that liberal democratic politics could be founded on cognitively firm first principles was a development of the Enlightenment project of seeking rational bases for cognitive and social order. Its clearest expression is perhaps Jefferson's assertion in "The Declaration of Independence": "We hold these truths to be self-evident . . ." MacIntyre (1978) suggests that the problem of the collapse of this self-evidence of philosophical foundations was faced in political thought before it became a problem for professional philosophers of science. The problem for political philosophy since the Enlightenment has been how to find secular grounds for political equality, justice, respect, and rights in the face of value-pluralism and fundamental conflicts of worldview. Political philosophy has long been confronted with the inescapable humanness of the practices it seeks to justify and the declining persuasiveness of appeals to transcendental standards, whether God (divine right, the soul), Reason (the categorical imperative), or Nature (natural law).

In the twentieth century, skepticism about the possibility of founding liberal principles on transcendent foundations fed into attempts to tie liberalism to empirical science. In the pragmatist philosophy of John Dewey, for example, we find a rejection of the search for transcendent foundations for democracy and science. Both science and democracy, for Dewey, are practical activities, sets of habits rather than abstract principles. Dewey saw these habits as intertwined: democracy depends on the extension and diffusion of scientific method and habit through the polity (Dewey [1916]1966: 81–99). This provided his answer also to Walter Lippmann's "realist" argument for the inherent limits on democracy in an age of experts and his elitist vision of technocratic administration by experts. Dewey suggested that the spread of social scientific knowledge through the popular press and education would render expertise compatible with democracy, negating Lippmann's technocratic visions (Lippmann, [1922]1965; Dewey, [1927]1991; Westbrook, 1991: 308–18).

In contrast with Enlightenment confidence in rationality and progress, twentieth century democratic theory proceeded more hesitantly. Paradoxically, even though there is a strong twentieth century tendency to try to present democracy as allied with science (Jewett, 2003), expert knowledge at the same time starts to seem a fickle ally. So Dewey's attempt to link science to the banner of democracy barely outmaneuvers Lippmann's recognition of the antidemocratic elitist tendency toward expert monopoly of knowledge. Liberalism in the twentieth century has been increasingly subsumed and subordinated by technical expertise (Turner, 2003a: 129–43).

The attempt to link liberal democracy with science became particularly marked in the context of the crisis of liberalism in the 1930s and 1940s, with the Great Depression, the rise of fascism and communism, and the descent into world war. During this period, we can see all three major ideologies—liberalism, fascism, communism—in different ways seeking to claim the mantle of science and technology. All three placed faith in technological gigantism, and all could be seen legitimizing their ideology in the name of science. Mid-twentieth century liberalism's assimilation of science to individualism and democratic dialogue represented, in part, an attempt to extract and liberate science from the ideological snares of the Nazis' "racial science" and the Soviets' claims to scientific socialism. But liberalism's uses of science were nonetheless themselves ideological. Sociologist Shiv Visvanathan has suggested that the turn toward an explicitly scientific basis for liberal principles reflected liberalism's embattled status in the period and the exhaustion of other repertoires of legitimation (Visvanathan, 1988: 113; see also Hollinger, 1996: 80–120, 155–74).

Karl Popper and Robert Merton provided what are most often taken in STS to be the classic formulations of the relationship between science and liberalism, and both did so in explicit confrontation with the threats to liberalism from totalitarianism. Steve Fuller has recently sought to rescue the democratic and critical Popper from the caricature one often encounters within STS of Popper as a dogmatic defender of the scientific status quo. Popper's philosophy of science, Fuller argues, embodied a radical republican ideal of a free and open polity, standing in marked contrast with the closed disciplinary communities of modern science (Fuller, 2003).

There is ambivalence in the notion of science as exemplifying the liberal ideal about whether this meant real science as practiced or science as it ought to be. In an era when scientists had lent their expertise to Nazi racial ideology and to technologies of death and destruction, this gap between ideal and reality was hard to avoid. Popper's account also appears ambivalent in comparison with Ezrahi's portrait of the cultural image of science underpinning liberalism. On the one hand, Popper's notions of the testability of scientific knowledge, the ideal openness of scientific discourse to criticism, and the impersonality of objective knowledge, appear to correspond closely to the cultural image described by Ezrahi. However, Popper can also be seen as occupying a pivotal place in relation to Ezrahi's story of the collapse of faith in the ability of science to ground and legitimize liberal democratic practices. Whereas the American Revolution asserted the basis of democracy in "self-evident truth," Popper, from the perspective of the twentieth century, seeks to distinguish science from what he regarded as the violence of ideological certainty. The liberal principles of free and open dialogue asserted by John Stuart Mill are best guaranteed by the search to expose error rather than to uphold certainty. In one sense, Popper's conception of scientific method was a version of what Ezrahi calls "democratic instrumentalism" (Ezrahi, 1990: 226). But Popper's fallibilism could be seen as posing the danger that skepticism might erode the common-sense underpinnings of democratic public life. Ezrahi argues that Popper's critique of knowledge not only attacks the intellectual foundations of totalitarianism but "undermines the premises of meliorist democratic politics as well" (Ezrahi, 1990: 260).

As assertive as mid-twentieth century liberal statements such as *The Open Society and its Enemies* (Popper, 1945) were in associating the values of science with those of liberal democracy, this was in the context of liberalism under mortal threat in a global context. It is not surprising, therefore, to find notes of tentativeness even in these defenses. This is the case also for Merton's classic sociological defense of science as central to the culture of democracy (Merton [1942]1973). Merton famously delineated norms of science that link it with the values of liberalism, including universalism, free exchange of knowledge, and so on. It is, again, a classic statement of what Ezrahi calls "democratic instrumentalism." At the same time, however, Merton's sociological approach introduces tensions and perhaps an unintended tentativeness into the formulation of democratic values. There is a tension in his analysis as to what extent the norms of science are socially contingent and to what extent they derive from some foundational character of scientific knowledge as knowledge. Merton comes close to suggesting that science's universalism is a community norm, and in that sense (paradoxically) local and contingent. And if "organized skepticism" also has the status of a "norm" it would appear that skepticism is limited at the point where this basic normative framework begins: the norm, accepted as part of socialization into a community, is kept exempt from radical skepticism. Whereas the earlier uses by liberals of science as a legitimatory metaphor were aimed at presenting liberal political values as being universal—as universal as science—in the mid-twentieth century we start to have the sense that *both* liberalism and science are culturally located practices. The cultural location of both science and liberalism is further suggested by Merton's application to science of Max Weber's theory of the influence of Protestantism on capitalist modernity.[1]

Merton's liberalism was an embattled one, holding out against Nazism and Communism, but also embattled in the context of the United States, as historian David Hollinger has argued, by Christian attacks on secular culture and the secular university (Hollinger, 1996: 80–96, 155–74). Merton's argument was that science and democracy are interwoven cultural values, that the combination defines a particular kind of social and political community: if you want to think of yourself as this sort of community you need to uphold these sorts of norms. There is no universal imperative here. In Merton, we can see the liberal defense of science begin to take a distinctively communitarian flavor.

There is a structurally similar and related contrast between liberal and communitarian approaches in both political theory and the philosophy and sociology of science. Generally, liberals and communitarians both subscribe to and seek to defend *broadly* liberal-democratic political values (although there are substantive differences between the liberal valuation of individual rights and choice and the communitarian emphasis on collective morality). But they disagree fundamentally over how social, political, and epistemic values can be justified: what meta-standards, if any, can be appealed to. For the communitarian, democratic values and the norms of science are local, contingent, and immanent and can only be defended as such.

COMMUNITARIANISM, CONSERVATISM, AND THE SOCIOLOGY OF SCIENCE

The view that the defense of liberalism required the abandonment of liberalism's attachment to modernist epistemology and philosophy of science was put forward most strongly by Michael Polanyi. In direct opposition to Popper and Merton's equation of science with skepticism, Polanyi argued that both science and liberal democracy depended on trust and authority. His writing peppered with quotations from St. Augustine, Polanyi insisted that science was rooted in faith and the scientific community was a community of believers rather than skeptics. Modern skepticism was corrosive of the sense of social belonging and tradition that maintained scientific authority and liberal democratic political order. In an argument similar to Julien Benda's critique of "la trahison des clercs" (Benda, [1928]1969), Polanyi argued that skeptical and materialist modern philosophies had resulted in totalitarianism. The preservation both of science and democracy meant maintaining a tradition, the most important elements of which were tacit and taken on faith. So, he argued, a free society was not only liberal but "profoundly conservative" (Polanyi, [1958]1974: 244).

Polanyi's conception of the scientific community as a model polity was, in part, an argument against the proposals for the planning of science put forward in the 1930s and 1940s in Britain by J. D. Bernal and other socialist scientists. Despite the apparent tension with his own conservative valorization of tradition, Polanyi insisted that the social order of science was isomorphic with the capitalist free market (Mirowski, 2004: 54–71; Fuller, 2000a: 139–49).

The American counterpart of Polanyi was J. B. Conant. Both politically and philosophically, there are striking parallels between Conant and Polanyi's programs. Philosophically, both reject analytical philosophy of science's emphasis on abstract propositions and their logical relationships and instead treat science as a set of skilled practices, organized in communities of practitioners. The political thrust of their work was also similar. Where Polanyi's philosophy was targeted explicitly against Bernal, Conant was aligned against proposals in the spirit of the New Deal to prioritize and target research toward social welfare (Fuller, 2000a: 150–78, 210–23; Mirowski, 2004: 53–84). Conant's portrait of science fit into a wider discourse of "laissez-faire communitarianism" current among mid-twentieth century American scientists and liberal intellectuals (Hollinger, 1996: 97–120).

Conant was concerned to harness government support for science and to make science useful for the Cold War military-industrial complex while at the same time maintaining the elite autonomy of the academic scientific community. Steve Fuller has emphasized that, in so doing, Conant upheld the twin pillars of the Cold War compact between science and the American state (Fuller, 2000a: 150–78; Mirowski, 2004: 85–96).

Fuller argues that understanding the Cold War background to Conant's thought is crucial for understanding the intellectual development of STS. This is because of the iconic place Thomas Kuhn's work has assumed in the development of the field. Teaching in Harvard's history of science program, Kuhn was in many ways a Conant protégé and was mentored by the Harvard President. Kuhn was also an inheritor of the "laissez-faire communitarian" conception of science (Hollinger, 1996: 112–13, 161–63, 169–71; Fuller, 2000a: esp. 179–221, 381–83). Further, Fuller suggests that Kuhn's conception of "normal science" as mere puzzle solving legitimated an approach to natural and social science that was noncritical and politically acquiescent. The branch of social science most powerfully influenced by Kuhn is, of course, the sociology of scientific knowledge (SSK), and it is a key implication of Fuller's argument that this field has incorporated a conservative orientation via Kuhn (Fuller, 2000a: 318–78).

It is important, however, to distinguish conservative politics from what Karl Mannheim pointed to as a conservative style of thought. Conservative thought-styles do not necessarily entail conservative politics. In contrast to the Enlightenment search for trans-historical, rational, and universal foundations for epistemic, political, and social practices, the conservative style of thought privileges the local over the universal, practice over theory, and the concrete over the abstract. It denies meliorism, instead emphasizing the moral and cognitive imperfectability of human beings (Mannheim, [1936]1985; Oakeshott, [1962]1991; Muller, 1997). In that sense, SSK clearly follows Polanyi, Conant, and Kuhn in adopting a conservative thought-style, and the Edinburgh school philosopher David Bloor is explicit about this (Bloor, [1976]1991: 55–74; Bloor, 1997; see also Barnes, 1994). But whether that has conservative political implications, as Fuller alleges, and in what sense, is questionable. The project of sociology itself has been deeply informed by the conservative tradition (Nisbet, 1952), but that does not make sociology necessarily a politically conservative project.

SSK combined disparate traditions of philosophy and social thought—from the Marxist critique of ideology via Mannheim's notion of total ideology, to anthropological conceptions of cultural knowledge from Durkheim and Mary Douglas, to Polanyite notions of "tacit knowledge" and trust, as well as Kuhn's concepts of paradigms and incommensurability. In what sense the product is a "conservative" theory is debatable as, even more, is the extent to which it is influenced by Kuhn's political orientation. The very fact that Kuhn rejected the relativistic development of his concepts and ideas by SSK seems to point to the way in which ideas can be recontextualized and separated from their originator's intentions. This would suggest that we do not have to see Kuhn's own political orientation as being implicated in post-Kuhnian developments of the sociology of knowledge. In addition, the designation "conservative" is complicated in the context of late modernity. The Burkean valorization of tradition can today, for example, be a basis on which to challenge the radical change wrought by neoliberal economic policies (Giddens, 1995; Gray, 1995). A Polanyite orientation could warrant criticism of the "audit explosion" associated with British neoliberalism, arguably an extreme version of liberal scientism (Power, 1994; Shapin, 1994: 409–17; Shapin, 2004).

However, conservative and communitarian theories of science and politics do seem to beg the questions "whose tradition?" and "which community?" Appeals to communal values and traditions seem less satisfactory if you find yourself in a subordinated or marginalized position within that community (Harding, 1991; Frazer & Lacey, 1993: 155). Further, while Polanyi treated the epistemic standards of science as internal to a form of life, he still wanted science to be socially privileged and to carry special authority. In contrast to Paul Feyerabend's anarchistic "anything goes" (Feyerabend, 1978; 1993), Polanyi's conservative conclusion was essentially that anything the scientific community does, goes. It does seem that Polanyi's communitarianism led him to ignore the potential for conflict between worldviews and to paper-over social difference in favor of a model of society as a whole united around its core values, which for Polanyi meant science.

CRITICAL THEORY, MULTICULTURALISM, AND FEMINISM

As it followed from the Marxist critique of ideology via Mannheim, SSK could be seen as a critical theory in relation to the dominant liberal ideology of science—exposing the class, professional, and institutional interests that were elided and masked by liberal notions of the universality and neutrality of scientific knowledge (e.g., Mulkay, 1976). In that respect, SSK meshes with branches of STS derived from Frankfurt School Marxism that aim to unmask the social biases built into apparently neutral "instrumental reason." Whereas earlier Marxists such as Bernal tended to see science as an ideologically neutral force of production, Marxist science studies since the 1960s have been oriented toward the critique of "neutrality" and, as Habermas put it, of "technology and science as ideology" (Habermas, 1971). The most important example of Marxist-influenced STS today is Andrew Feenberg's critical theory of technology, which develops Marcuse's analysis of one-dimensional thought and culture into a nuanced critique of technology. Feenberg's critical theory aims to expose how biases enter into technological design and how liberatory and democratic interests can instead be engineered into the technical code (Feenberg, 1999).

Feenberg argues parallel to post-Kuhnian sociology of science in distinguishing his critical theory from competing "instrumental" and "substantive" theories of technology (Feenberg, 1991: 5–14). The instrumental conception of technology follows the liberal ideology of science, presenting technique as a neutral means toward given ends. The substantive conception of technology also conceives of technique as neutral. But thinkers such as Heidegger, Ellul, Albert Borgmann (and, arguably, Habermas) regard this neutral technique as increasingly systematically dominating society to the extent that technology becomes a substantive culture in itself, pushing out spiritual and moral values. Feenberg reflects SSK and other sociological critiques of scientific neutrality in arguing against both the bland positivity of the instrumentalists and the fatalism of the substantive theories. Where post-Kuhnian sociological analyses demonstrate the way in which science and technology incorporate and embed particular interests and values, critical theory aims *both* to expose dominatory values and to suggest the possibility of inscribing new values in technological design. In contrast to thinkers such as Heidegger and Ellul, then, the problem is not technology *per se* but rather bias in the dominant technical codes. And the solution is not to push back technology to make way for the charismatic return to the world of moral and religious values. Instead, the way forward consists in finding ways to decide democratically what kinds of values we want our technologies to embody and fulfill.

Langdon Winner arrived at similar conclusions in his key works, *Autonomous Technology* (1978) and *The Whale and the Reactor* (1986). While strongly influenced by Ellul's notion that technology has become an autonomous system, Winner, like Feenberg, rejects Ellul's pessimistic antitechnological stance. Instead, he argues that, just as societies have a political constitution, they also have a technological constitution and the framing of both are matters of human decision—hence the need for the democratization of technological decision-making.

While SSK and critical theories of technology have in common the influence of Marxism, the Polanyite communitarian aspects of SSK pose problems from a critical theory perspective. Just as critical theorists have sought to expose imbalances of power underlying seemingly neutral technical codes, they would also want to question notions of community consensus and shared standards—to ask whether such consensus is real, or whether it is underwritten by power and distorted communication. In contrast to the communitarian or the pragmatist, the critical theorist is unwilling to stop with communal norms or established practices but would suggest that it should always be possible to evaluate and deliberate over which norms and practices to pursue.

Such questions arise in particular in feminist and multicultural approaches. SSK and feminist epistemology have in common the constructivist critique of liberal notions of universality and neutrality, and a "conservative" emphasis on the local over the universal. The latter can be seen, in particular, in Donna Haraway's notion of "situated knowledge" (1991) and Helen Longino's idea of a "local epistemology" (2002: esp. 184–89). Similarly, it has been argued that the feminist critique of liberalism shares much with the communitarian critique. Both are skeptical of liberal claims to universalistic rationality (of notions of rights, justice, etc.), of the liberal conception of the unattached and disembodied individual subject, and of liberalism's attempt to separate political principles from emotion and subjectivity (Frazer & Lacey, 1993: 117–24; see also Baier, 1994). At the same time, feminists also have reason to distrust appeals to communal solidarity (Frazer & Lacey, 1993: 130–62). So, for example, the guild relation of master and apprentice in science, celebrated by Polanyi, is precisely the sort of patriarchal structure that is problematic on feminist grounds.

Communitarian appeals to solidarity and tradition have a similarly complicated relationship to multiculturalism. Pointing out that scientific knowledge is local rather than universal is a key step for multicultural critiques of western cultural dominance (Harding, 1998; Hess, 1995; Nandy, 1988; Visvanathan, 2006). Kuhn's notions of incommensurability and of the plurality of paradigms have become emblematic for feminist and multicultural approaches. Longino writes that "Knowledge is plural" and that standards of truth depend "on the cognitive goals and particular cognitive resources of a given context" (Longino, 2002: 207). This has critical implications anathema to Kuhn's own sensibilities: the notion that knowledge is disunified and plural provides a basis on which to make claims for the cultural integrity of marginalized or suppressed traditions, and to challenge western technoscientific hegemony. In that sense, the localist sensibilities of STS, derived from communitarianism, have developed toward a "politics of difference" (Young, 1990).

LIBERALISM AFTER LIBERALISM?

Ezrahi concludes *The Descent of Icarus* by suggesting that the scientistic legitimation of liberal democratic politics has broken down in the West, probably irretrievably (Ezrahi, 1990: 263–90). Images of neutrality, universality, and objectivity have lost support among intellectuals and increasingly call forth public distrust. The rise of communitarianism and what he calls "conservative anarchism"[2] in both political thought and theories of science is an element of the broader shift away from the cultural repertoires that previously supported liberal democratic governance (Ezrahi, 1990: 285; 347 n.4). Liberalism and democracy today have to look to other repertoires.

In political theory, liberalism was given a new lease on life by John Rawls's *Theory of Justice* (1971). Rawls's thought-experiment of the original position maintained liberalism's conception of the disembodied subject and the search for neutral principles. But in his theory, justice is reduced to the merely procedural notion of fairness. Additionally, the question of the potential universalism of the standards defined by the original position has been at the core of the consequent "liberal-communitarian debate." Rawls's later *Political Liberalism* attenuated any claims to universality and has been seen as offering considerable concessions to communitarianism (Mulhall & Swift, 1993: esp. 198–205). Ezrahi sees Rawls's work as suggestive of the "recent upsurge of skepticism toward generalized ideas of the polity or toward political instrumentalism" (Ezrahi, 1990: 245). Nevertheless, Rawls's re-founding of liberal ideals can be seen as providing a model for attempts within science studies to salvage liberal theory from relativistic communitarian and multicultural critiques.

Philip Kitcher is influenced by Rawls's thought-experiment of the "original position" in setting out his model for a "well-ordered science" in *Science, Truth, and Democracy* (2001: esp. 211). Kitcher's proposals can be seen, in part, as an attempt to rescue the Ezrahian connection between science and liberal democracy in the wake of the post-Kuhnian breakdown of these legitimations. Just as Rawls proposes a procedural solution to the problem of justice, Kitcher proposes a procedural model of ideal deliberation whereby deliberators, with the aid of expert advice, develop "tutored preferences" (Kitcher, 2001: 117–35; see also Turner, 2003a: 599–600). The possibility of unbiased neutral expertise and of neutral standards on which to choose between worldviews is assumed as a background condition for his deliberative ideal (Brown, 2004: 81). Like Rawls's original position, this is a thought-experiment, but the question arises to what degree it smuggles in substantive normative assumptions, for example, market individualism (Mirowski, 2004: 21–24, 97–115). The critiques that social constructivists make of Kitcher's ideal deliberators precisely parallel those which communitarians have made of Rawls's original position (cf. Mulhall and Swift, 1993).

Despite these criticisms, Stephen Turner has argued that the crucial departure of Kitcher's model from Rawls's original position or Habermas's "ideal speech situation" is (because of the role granted to experts as "tutors") in recognizing that the civic model of the perfectly equal "public" is an impossibility in an expertise-dependent age. To the degree that decision-making requires reliance on special expertise, the ideal of a completely free and equal forum is untenable (Turner, 2003b: 608; Turner, 2003a: 18–45). This forms the core issue for Turner's *Liberal Democracy 3.0: Civil Society in an Age of Experts* (2003a). The key problem for contemporary democracy, he argues, is the problem of the ineliminable dependence on expert knowledge.

Turner attempts the redefinition of liberalism in an age of experts, via a rehabilitation of Conant. The lineage from Conant via Kuhn to post-Kuhnian sociology of science is drawn on by Turner to argue that the liberal political philosophy of science most consonant with constructivist sociology was already established by Conant himself (Turner, 2003a: ix–x). Crucially, Conant shares with contemporary sociologists of science, such as Harry Collins and Trevor Pinch, the emphasis on science as a practical activity characterized by a high degree of uncertainty. Conant and Collins and Pinch have in common a perspective on general science education (in Conant's *On Understanding Science* [1951] and in Collins and Pinch's *The Golem* [1996]), which suggests that public understanding of science should be oriented not to knowing scientific facts but rather to understanding how science operates as a practical activity and its practical limitations. This latter kind of knowledge is necessary for the public to be in a position to make decisions about science policy—from assigning research priorities to handling expert opinion and advice. In a sense, they are suggesting that what Kitcher's "ideal deliberators" most require is sociological "tutoring" about the character of science as a form of social activity and practice. While Conant's program was conservative (as Fuller argues) in that he was strongly against any far-reaching democratization, nevertheless Turner suggests that Conant pointed to the way in which expertise can be indirectly brought to serve the values of a liberal democratic society. Liberalizing expertise means "to force expert claims to be subjected to the discipline of contentious discussion that would reveal their flaws, and do so by forcing the experts to make arguments to be assessed by people outside the corporate body of experts in the field." This liberalization of expertise "was [to be] a check on expert group-think, on the 'consensus of scientists'" (Turner, 2003a: 122). Rather than subject expertise to democratic control, Turner, following Conant, advocates a liberal regime in which diverse expert opinions are publicly matched against each other. Where there is a complex division of labor and plural sources of expertise, this complexity will act as a check on expert dominance. The recognition that expertise, while necessary, is fallible, allows some protection against sheer technocracy.

Collins and Pinch similarly suggest that public understanding of the sociology of science would demystify expertise, allowing it to be seen as completely secular and mundane: the use of experts would not differ in principle from the use of plumbers (Collins & Pinch, 1996: 144–45). Their expertise is recognized, but it is recognized as imperfect and subject to the choice of those who would employ the expert for whatever task. Both Turner and Collins and Pinch suggest that the Ezrahian goal of instrumental knowledge at the service of democracy can be preserved by doing away with the rationalist myth of certain knowledge on which understandings of instrumental rationality have often been based. When science is recognized as mundane practice, and as fallible, it can genuinely be instrumentalized (Turner, 2001), but as a set of skills rather than rules.

However, it is unclear how far this model can preserve anything but the semblance of liberal democracy. Turner's book leaves the reader unsure whether "liberal democracy 3.0" is a form of democracy at all, and Turner asks, "is liberal democracy increasingly a constitutional fiction?" (Turner, 2003a: 141). Ian Welsh has argued that the plumber model of expertise is a poor analogy for modern technoscience. The plumber's relatively routine and well-defined set of tasks are very different from "the indeterminate quality of 'post-normal science'." Further, "the trustworthiness of a particular plumber may be determined by a phone call to a previous client" (Welsh, 2000: 215–16). The trustworthiness of, for example, nuclear scientists, operating within secretive bureaucratic institutions, is far harder for citizens to ascertain. If citizens were to be able to treat nuclear experts in the same way as plumbers, this would mean a radical reorganization of institutional and political life in western democracies— overcoming not just the epistemic myths but also the bureaucratic and technocratic institutions that maintain undemocratic expert power. Without such a political-institutional leveling, the plumbing analogy is highly limited.

SCIENCE, TECHNOLOGY, AND PARTICIPATORY DEMOCRACY

The declining efficacy of liberal instrumentalist legitimations of public action can be seen as part of a broader developing crisis of liberal democratic structures of representation (Hardt & Negri, 2005: 272–73). New social movements (NSMs), such as the antinuclear and environmental movements, have played a crucial role in politicizing technical domains that liberal discourse had formerly isolated from the scope of politics (Welsh, 2000; Habermas, 1981; Melucci, 1989).

NSM protest poses a challenge also for the discipline of science policy. This discipline has tended to be oriented toward the technocratic imperatives of state policy.

Science policy academics have tended to treat economic growth and technological development as unproblematic goals and to regard the purpose of the discipline as being to advise policy-makers and to assist the management of the scientific-technological complex in terms of values of growth and instrumental efficacy. In challenging modernist imperatives of growth and economic-instrumental rationality, NSMs also therefore pose a challenge to this orientation of science policy (Martin, 1994). Increasingly, science policy has to address the goals of science and technology as contested rather than given and to regard "policy" as a democratic problem of the public rather than as a merely bureaucratic problem for elites.

The shift in the orientation of STS and science policy studies is indicated by the primacy in contemporary discussions in these fields of the idea of "participation." Demands for participation can be seen as following from what Ezrahi calls the "dein-strumentalization of public actions" (Ezrahi, 1990: 286) or, rather, from the increasingly widespread perception of instrumental justifications of public action as ideological and inadequate. The impersonal instrumental techniques, which Porter (1995) and Ezrahi both argue previously allowed liberal democracies to depoliticize public action in the face of potentially skeptical publics, have themselves become the objects of public distrust (Welsh, 2000).

It is significant that the refrain of STS that the technical is political reflects the new politics of technology that has emerged in antinuclear, antipsychiatric, patients-rights, environmental, anti-GMO, and other movements. In that sense, the STS claim that the technical is political is not only a theoretical claim about epistemology but also a description of the new politics that characterizes the risk society (Beck, 1995; Welsh, 2000: 23–33; Fischer, 2000). However, dominant political, bureaucratic, and scientific institutions have been either slow, or just unable, to adapt to this new politicization of the technical. Possibilities for realizing this new politics through mainstream institutions of representation remain extremely limited. Despite their declining legitimacy, bureaucratic and technocratic mentalities hold sway in mainstream representative and political executive institutions. The importance of nonviolent direct action for NSMs is, in part, due to recognition of the impossibility of pursuing the values of the life-world through the representative and bureaucratic means provided by official culture (Welsh, 2000: 150–205; Hardt & Negri, 2005; Ginsberg, 1982).

STS today is increasingly concerned with how to theorize and make practicable structures of public participation in scientific and technological decision-making and design (Kleinman, 2000). In theoretical terms, the concern has been how to conceptualize the role of democratic agency and "participant interests" in technological design (Feenberg, 1999). There is a growing body of empirical literature on examples of lay participation in decision-making in science, technology, and medicine. Steven Epstein's study of the role of AIDS activists in challenging the norms and procedures of clinical trials remains a crucial point of reference (Epstein, 1996; Feenberg, 1995: 96–120; Doppelt, 2001: 171–74; Hardt & Negri, 2005: 189). A key concern in recent STS work has been how can lay citizen participation become established and institutionalized as part of the process of technological decision-making without the need for protest driven by initial exclusion. Ideas include town meetings, citizen juries, consensus conferences, and the model of the "citizen scientist" (Sclove, 1995; Fischer, 2000; Irwin, 1995; Kleinman, 2000). This literature has also recently spurred debate about the coherence of the category of the "expert," whether the notion of "lay expertise" (Epstein, 1995) goes too far in extending the category (Collins & Evans, 2002). One the other hand, it is argued that the attempt to come up with a neutral demarcation of the expert in terms of social-cognitive capacities ignores the value- or "frame" dependence of knowledge and smuggles back in the assumptions of expert neutrality that constructivist approaches have been aimed at criticizing (Wynne, 2003; Jasanoff, 2003b).

Arguably, however, the STS critique of the institutional contexts of science and technology has remained limited. Discussions within STS have tended to assume that democratizing expertise simply involves tacking new institutional devices (such as citizen juries) onto existing political and institutional structures. But it should be asked whether the STS critique can remain within these bounds or whether it has more radical implications. These implications can be seen in particular when STS engages with the place of technology in the workplace (Noble, 1986; Feenberg, 1991: 23–61). Stephen Turner has noted that the sociological conception of science as practice challenges the distinction between knowledge and skill on which Taylorist conceptions of work-organization (and, one could argue, modern managerial authority) are based (Turner, 2003a: 137). STS arguments that technological decisions are political raise long-standing issues about the relationship of democracy to the workplace and arguably provide renewed justification for worker democracy (Pateman, 1970; Feenberg, 1991).

STS scholarship has implications not only for democratic participation in decisions about the use of GMOs in food production, the location of nuclear power stations, the use and testing of medicines, but also for the structure of authority in the workplace (Edwards & Wajcman, 2005). Tackling technology and the workplace potentially draws STS into engagement with the long-standing tradition of participatory democracy and radical democratic theory (Pateman, 1970). And in that case, as Gerald Doppelt has pointed out, arguments for the democratization of technology need to centrally address the question of the legitimacy of Lockean private property rights. Whereas STS has tended to treat expert authority as a product of technocratic ideology, Doppelt points out that "in the common case where technology is private property, the rights and authority of the designers/experts really rests on the fact that they are . . . representatives of capital," and therefore ultimately on "the Lockean moral code of ownership and free-market exchange" (Doppelt, 2001: 162). STS has been somewhat shy of directly addressing the issue of private property. One exception has been Steve Fuller, who notes that Lockean property rights have been central to liberal thinking about science and criticizes the way in which the liberal regime has allowed economic imperatives to undermine the character of science as an "open society." The critique of science-as-private-property is central to Fuller's "republican" conception of science as depending on the "right to be wrong," a right that, he argues, should be democratically extended beyond credentialed experts (Fuller, 2000b: esp. 19–27, 151–56; see also Mirowski, 2004).

Although the workplace remains of crucial importance for the politics of technology, STS also appreciates how people's relationship with technology is of a much broader scope—taking in people's roles as consumers, patients, residents of communities, and so on. The notion that technical decisions that affect people's lives should be participatory decisions is one that calls into question the very structure of the democratic polity—calling for the radical extension of democracy through everyday life—for democracy to be as pervasive as technology. This means an emphasis on local democracy—in the workplace, community, education, and medical settings. It also means democracy on a global level (Beck, 1995; Hardt & Negri, 2000, 2005).

In the context of globalization, mediating structures of representation and the delegation of authority to experts are increasingly perceived as removing real power from citizens and populaces. Hardt and Negri have recently argued that we are faced with a generalized "crisis of democratic representation" and, they write, "In the era of globalization it is becoming increasingly clear that the historical moment of liberalism has passed" (Hardt & Negri, 2005: 273). This thesis is echoed, with different emphases, by Turner, who notes that "A good deal of the phenomena of globalization is the replacement of national democratic control with control by experts" (Turner, 2003a: 131). This crisis of representation is the context in which questions of the democratization of science and technology come to the fore.

THE LANGUAGE OF STS AND THE LANGUAGE OF POLICY

The broad context of the crisis of representation, and the question of whether institutional reforms can be tacked on to existing structures, gain importance because of the way in which scientific and political elites are beginning to appropriate the language of "participation," at least in the watered-down form of "engagement." It is ironic that the unelected House of Lords in Britain has issued one of the most frequently referred to reports calling for increased public "engagement" in science and technology (House of Lords, 2000). The British government's Office of Science and Innovation, part of the Department of Trade and Industry (DTI), emphasizes the shift from the older PUS (Public Understanding of Science) model to a new PEST (Public Engagement with Science and Technology) approach.

There is reason, beyond the occasionally revealing acronyms, to treat this rhetoric of "engagement" with caution when considering the place of science and technology in the broader policy agenda of agencies such as the DTI. The key question to ask is whether, as the government pursues science and technology policy as a primarily economic strategy in the context of globalization (Jessop, 2002; Fuller, 2000b: 127–30), it is possible to reconcile these strategies with genuine public participation. Official calls for public engagement appear as part of an attempt to co-opt skeptical publics. The rhetoric fits into an elite response to the successful public opposition in Europe to GM foods, as well as to earlier "civic dislocations" (Jasanoff, 1997). Hardt and Negri have written of the loss of legitimacy by dominant political institutions as indicated by the "evacuation of the places of power" (Hardt & Negri, 2000: 212). Elite calls for "engagement" understandably arise from the threat that public dis-engagement (or, what Hardt and Negri call "desertion") poses to dominant institutions' claims to legitimacy. We might ask whether democratization is most genuine when it arises organically from grassroots collective action or when it is conducted via institutional reform from above. The development of STS scholarship *as* political theory is particularly important if the notion of participation is to be given sufficient political and analytical substance to preserve its meaning from the diluting and falsely reassuring language of official policy.

Notes

1. On Merton, see also Stephen Turner's chapter in this volume.
2. Ezrahi mentions Robert Nozick and Richard Rorty.

References

Baier, Annette (1994) *Moral Prejudices: Essays on Ethics* (Cambridge, MA: Harvard University Press).

Barnes, Barry (1994) "Cultural Change: The Thought-Styles of Mannheim

and Kuhn," *Common Knowledge* 3: 65–78.

Beck, Ulrich (1995) *Ecological Politics in an Age of Risk* (Cambridge: Polity Press).

Benda, Julien ([1928]1969) *The Treason of the Intellectuals* (New York: W. W. Norton).

Bloor, David ([1976]1991) *Knowledge and Social Imagery* (Chicago: University of Chicago Press).

Bloor, David (1997) "The Conservative Constructivist," *History of the Human Sciences* 10: 123–25.

Brown, Mark B. (2004) "The Political Philosophy of Science Policy," *Minerva* 42: 77–95.

Collins, H. M. & R. J. Evans (2002) "The Third Wave of Science Studies: Studies of Expertise and Experience," *Social Studies of Science* 32(2): 235–96.

Collins, H. M. & Trevor Pinch (1996) *The Golem: What Everyone Should Know About Science* (Cambridge: Canto).

Conant, James B. (1951) *On Understanding Science: An Historical Approach* (New York: New American Library).

Dewey, John ([1916]1966) *Democracy and Education* (New York: Free Press).

Dewey, John ([1927]1991) *The Public and Its Problems* (Athens, OH: Swallow Press).

Doppelt, Gerald (2001) "What Sort of Ethics Does Technology Require?" *Journal of Ethics* 5: 155–75.

Edwards, Paul & Judy Wajcman (2005) *The Politics of Working Life* (Oxford: Oxford University Press).

Epstein, Steven (1995) "The Construction of Lay Expertise: AIDS Activism and the Forging of Credibility in the Reform of Clinical Trials," *Science, Technology & Human Values* 20: 408–37.

Epstein, Steven (1996) *Impure Science: AIDS, Activism, and the Politics of Knowledge* (Berkeley: University of California Press).

Ezrahi, Yaron (1990) *The Descent of Icarus: Science and the Transformation of Contemporary Democracy* (Cambridge, MA: Harvard University Press).

Feenberg, Andrew (1991) *Critical Theory of Technology* (New York: Oxford University Press).

Feenberg, Andrew (1995) *Alternative Modernity: The Technical Turn in Philosophy and Social Theory* (Berkeley: University of California Press).

Feenberg, Andrew (1999) *Questioning Technology* (London: Routledge).

Feyerabend, Paul (1978) *Science in a Free Society* (London: New Left Books).

Feyerabend, Paul (1993) *Against Method* (London: Verso).

Fischer, Frank (2000) *Citizens, Experts, and the Environment: The Politics of Local Knowledge* (Durham, NC: Duke University Press).

Frazer, Elizabeth & Nicola Lacey (1993) *The Politics of Community: A Feminist Critique of the Liberal-Communitarian Debate* (New York: Harvester Wheatsheaf).

Fuller, Steve (2000a) *Thomas Kuhn: A Philosophical History for Our Times* (Chicago: University of Chicago Press).

Fuller, Steve (2000b) *The Governance of Science: Ideology and the Future of the Open Society* (Buckingham: Open University Press).

Fuller, Steve (2003) *Kuhn vs. Popper: The Struggle for the Soul of Science* (Cambridge: Icon Books).

Giddens, Anthony (1995) *Beyond Left and Right: The Future of Radical Politics* (Stanford, CA: Stanford University Press).

Ginsberg, Benjamin (1982) *The Consequences of Consent: Elections, Citizen Control and Popular Acquiescence* (Reading, MA: Addison-Wesley).

Gray, John (1995) *Enlightenment's Wake: Politics and Culture at the Close of the Modern Age* (London: Routledge).

Habermas, Jürgen (1971) *Toward a Rational Society: Student Protest, Science, and Politics* (London: Heinemann).

Habermas, Jürgen (1981) "New Social Movements," *Telos* 49: 33–37.

Haraway, Donna (1991) *Simians, Cyborgs and Women: The Reinvention of Nature* (London: Free Association Books).

Harding, Sandra (1991) *Whose Science? Whose Knowledge? Thinking from Women's Lives* (Ithaca, NY: Cornell University Press).

Harding, Sandra (1998) *Is Science Multicultural? Postcolonialisms, Feminisms, and Epistemologies* (Bloomington: Indiana University Press).

Hardt, Michael & Antonio Negri (2000) *Empire* (Cambridge, MA: Harvard University Press).

Hardt, Michael & Antonio Negri (2005) *Multitude: War and Democracy in the Age of Empire* (London: Hamish Hamilton).

Hess, David (1995) *Science and Technology in a Multicultural World: The Cultural Politics of Facts and Artifacts* (New York: Columbia University Press).

Hollinger, David (1996) *Science, Jews, and Secular Culture: Studies in Mid–Twentieth-Century American Intellectual History* (Princeton, NJ: Princeton University Press).

House of Lords (2000) *Science and Society* (London: Stationary Office).

Irwin, Alan (1995) *Citizen Science: A Study of People, Expertise and Sustainable Development* (London: Routledge).

Jasanoff, Sheila (1997) "Civilization and Madness: The Great BSE Scare of 1996," *Public Understanding of Science* 6: 221–32.

Jasanoff, Sheila (2003a) "(No?) Accounting for Expertise," *Science and Public Policy* 30(3): 157–62.

Jasanoff, Sheila (2003b) "Breaking the Waves in Science Studies: Comment on H. M. Collins and Robert Evans, 'The Third Wave of Science Studies'," *Social Studies of Science* 33(3): 389–400.

Jessop, Bob (2002) *The Future of the Capitalist State* (Cambridge: Polity Press).

Jewett, Andrew (2003) "Science and the Promise of Democracy in America," *Daedalus* Fall: 64–70.

Kitcher, Philip (2001) *Science, Truth, and Democracy* (Oxford: Oxford University Press).

Kleinman, Daniel Lee (ed) (2000) *Science, Technology and Democracy* (Albany: State University of New York Press).

Lippmann, Walter ([1922]1965) *Public Opinion* (New York: Free Press).

Longino, Helen (2002) *The Fate of Knowledge* (Princeton, NJ: Princeton University Press).

MacIntyre, Alisdair (1978) "Objectivity in Morality and Objectivity in Science," in H. Tristram Engelhardt, Jr. & Daniel Callahan (eds), *Morals, Science and Sociality* (New York: Institute of Society, Ethics and the Life Sciences): 21–39.

Mannheim, Karl ([1936]1985) *Ideology and Utopia: An Introduction to the Sociology of Knowledge* (New York: Harcourt Brace).

Martin, Brian (1994) "Anarchist Science Policy," *Raven* 7(2): 136–53.

McWilliams, Wilson Carey (1993) "Science and Freedom: America as the Technological Republic," in Arthur M. Melzer, Jerry Weinberger, & M. Richard Zinman (eds), *Technology in the Western Political Tradition* (Ithaca, NY: Cornell University Press): 85–108.

Melucci, Alberto (1989) *Nomads of the Present* (London: Hutchinson).

Merton, Robert K. ([1942]1973) "The Normative Structure of Science," in Robert K. Merton & Norman W. Storer (eds), *The Sociology of Science: Theoretical and Empirical Investigations* (Chicago: University of Chicago Press): 267–78.

Mirowski, Philip (2004) *The Effortless Economy of Science?* (Durham, NC: Duke University Press).

Mulhall, Stephen & Adam Swift (1993) *Liberals and Communitarians* (Oxford: Blackwell).

Mulkay, Michael (1976) "Norms and Ideology in Science," *Social Science Information* 15(4–5): 637–56.

Muller, Jerry Z. (1997) "What Is Conservative Social and Political Thought?" in J. Z. Muller (ed), *Conservatism: An Anthology of Social and Political Thought from David Hume to the Present* (Princeton, NJ: Princeton University Press): 3–31.

Nandy, Ashis (ed) (1988) *Science, Hegemony, and Violence: A Requiem for Modernity* (New Delhi: Oxford University Press).

Nisbet, Robert A. (1952) "Conservatism and Sociology," *American Journal of Sociology* 58: 167–75.

Noble, David (1986) *Forces of Production: A Social History of Industrial Automation* (Oxford: Oxford University Press).

Oakeshott, Michael ([1962]1991) *Rationalism in Politics and Other Essays* (Indianapolis, IN: Liberty Press).

Pateman, Carole (1970) *Participation and Democratic Theory* (Cambridge: Cambridge University Press).

Polanyi, Michael ([1958]1974) *Personal Knowledge: Towards a Post-Critical Philosophy* (Chicago: University of Chicago Press).

Popper, Karl (1945) *The Open Society and Its Enemies*, 2 vols. (London: Routledge).

Porter, Theodore M. (1995) *Trust in Numbers: The Pursuit of Objectivity in Science and Public Life* (Princeton, NJ: Princeton University Press).

Power, Michael (1994) *The Audit Explosion* (London: Demos).

Rawls, John (1971) *A Theory of Justice* (Cambridge, MA: Harvard University Press).

Sclove, Richard E. (1995) *Democracy and Technology* (New York: Guilford Press).

Shapin, Steven (1994) *A Social History of Truth: Civility and Science in Seventeenth-Century England* (Chicago: University of Chicago Press).

Shapin, Steven (2004) "The Way We Trust Now: The Authority of Science and the Character of the Scientist," in Pervez Hoodbhoy, Daniel Glaser, & Steven Shapin (eds), *Trust Me, I'm a Scientist* (London: British Council): 42–63.

Shapin, Steven & Simon Schaffer (1985) *Leviathan and the Air Pump: Hobbes, Boyle, and the Experimental Life* (Princeton, NJ: Princeton University Press).

Turner, Stephen (2001) "What Is the Problem with Experts?" *Social Studies of Science* 31(1): 123–49.

Turner, Stephen (2003a) *Liberal Democracy 3.0: Civil Society in an Age of Experts* (London: Sage).

Turner, Stephen (2003b) "The Third Science War," *Social Studies of Science* 33(4): 581–611.

Visvanathan, Shiv (1988) "Atomic Physics: The Career of an Imagination," in Ashis Nandy (ed), *Science, Hegemony and Violence* (New Delhi: Oxford University Press): 113–66.

Visvanathan, Shiv (2006) *A Carnival for Science: Essays on Science, Technology and Development* (Oxford: Oxford University Press).

Welsh, Ian (2000) *Mobilising Modernity: The Nuclear Moment* (London: Routledge).

Westbrook, Robert B. (1991) *John Dewey and American Democracy* (Ithaca, NY: Cornell University Press).

Winner, Langdon (1978) *Autonomous Technology: Technics-out-of-Control as a Theme in Political Thought* (Cambridge, MA: MIT Press).

Winner, Langdon (1986) *The Whale and the Reactor: A Search for Limits in an Age of High Technology* (Chicago: University of Chicago Press).

Wynne, Brian (2003) "Seasick on the Third Wave? Subverting the Hegemony of Propositionalism," *Social Studies of Science* 33(3): 401–17.

Young, Iris Marion (1990) *Justice and the Politics of Difference* (Princeton, NJ: Princeton University Press).

4   A Textbook Case Revisited—Knowledge as a Mode of Existence

Bruno Latour

Would it not be possible to manage entirely without something fixed? Both thinking and facts are changeable, if only because changes in thinking manifest themselves in changed facts.

Ludwick Fleck, [1935]1981: 50

Knowledge and science, as a work of art, like any other work of art, confers upon things traits and potentialities which did not *previously* belong to them. Objections from the side of alleged realism to this statement springs from a confusion of tenses. Knowledge is not a distortion or a perversion which confers upon *its* subject-matter traits which *do* not belong to it, but is an act which confers upon non-cognitive materials traits which *did* not belong to it.

John Dewey, 1958: 381–82

Costello—"I do not know how much longer I can support my present mode of existence." Paul—"What mode of existence are you referring to?" Costello—"Life in public."

Coetzee, 2005: 135

I was struck by the huge label: "A Textbook Case Revisited." Every time I visit New York, I spend some time at the Natural History Museum, on the top floor, to visit the fossil exhibit. This specific time, however, it was not the dinosaur section that attracted my attention but the new presentation of the horse fossil history. Why should anyone revisit textbooks? What happened was that in a marvelous presentation the curators had presented in two parallel rows two successive versions of our *knowledge* of the horse fossils. You did not simply follow the successive fossils of the present horse evolving in time, you could also see the successive versions of our understanding of this evolution evolving in time. Thus, not only one but two sets of parallel lineages were artfully superimposed: the progressive transformation of horses and the progressive transformation of our interpretations of their transformations. To the branching history of life was now added the branching history of the science of life, making for an excellent occasion to revisit another textbook case: this one about what exactly is meant in our field by the affirmation that "scientific objects have a history."

In this chapter, I will tackle three different tasks: (1) I will reformulate with the use of this example the double historicity of science and of its subject matter, (2) I will

Figure 4.1
The two genealogical lines of horses in the Natural History Museum. (Photo by Verena Paravel.)

remind the reader of an alternative tradition in philosophy and science studies that might help refocus the question, and finally (3) I will offer what I believe is a fresh solution to the definition of knowledge acquisition pathways.

KNOWLEDGE IS A VECTOR

An Interesting Experiment in Staging the Collective Process of Science

The reason I was so struck by this parallel between the evolution of horses and the evolution of the science of horse evolution is that I have always found puzzling a certain asymmetry in our reactions to science studies. If you tell an audience that scientists have entertained in the course of time shifting representations of the world, you will get nothing in answer but a yawn of acceptance. If you tell your audience that those transformations were not necessarily linear and did not necessarily converge regularly in an orderly fashion toward the right and definitive fact of the matter, you might trigger some uneasiness and you might even get the occasional worry: "Is this leading to relativism by any chance?" But if you now propose to say that the objects of science *themselves* had a history, that they have changed over time, too, or that Newton has "happened" to gravity and Pasteur has "happened" to microbes, then everyone is up in arms, and the accusation of indulging in "philosophy" or worse in "metaphysics" is soon hurled across the lecture hall. It is taken for granted that "history of science" means the history of our knowledge about the world, *not* of the world itself. For the first lineage, time is of the essence, not for the second.[1] Hence, for me, the teasing originality of this Natural History Museum exhibit.

But first, let us read some of the labels: "This collection represents one of the most famous evolutionary stories of the world." Why is it so famous? Because, says the caption, "Horses are one of the best studied and most frequently found groups of fossils." But why "revisit" it instead of just present it "as we now know it"?

The horses in this exhibit are arranged to contrast two versions of horse evolution. Those along the front curve show the classic "straight-line" concept, that over time, horses became larger, with fewer toes, and taller teeth. We now know, however, that horse evolution has been much more complex, more like a branching bush than a tree with a single main trunk. The horses in the back row show just how diverse this family of mammals has actually been.

To be sure, practicing scientists know perfectly well that their research more often takes the form of a "branching bush" than that of a "straight line," but the nice innovation of this exhibit is that those intertwined pathways are rarely shown to the public and even more rarely shown to parallel the hesitating movement of the objects of study themselves. Each of the two rows is further commented on by the following captions:

The story of horses: the classic version:

In the nineteenth and early twentieth century, scientists arranged the first known horse fossils in chronological order. They formed a simple evolutionary sequence: from small to large bodies, from many to fewer toes and from short to tall teeth. This made evolution seem like a single straight line progression from the earliest known horse *Hyracotherium* to *Equus*, the horse we know today.

This is contrasted with what you can see in the second row:

The story of horses: the revised version:

During the twentieth century, many more fossils were discovered and the evolutionary story became more complicated. Some later horses such as *Calippus* were smaller, not larger than their ancestors. Many others, like *Neohipparion* still had three toes, not one.

If you look at the horses in the back row of this exhibit, you will see examples that don't fit into the "straight line" version.

In addition, so as not to discourage the visitor, the curators added this nice bit of history and philosophy of science:

In fact, in any epoch some horses fit into the "straight line" and others didn't. Scientists concluded that there was no single line of evolution but many lines, resulting in diverse groups of animals each "successful" in different ways at different times. This doesn't mean that the original story was entirely wrong. Horses have tended to become bigger, with fewer toes and longer teeth. It's just that this overall trend is only one part of a much more complex evolutionary tale.

You could of course object that nothing much has changed, since "in the end" "we now know" that you should consider evolution as a "bushy" pathway and not as a goal-oriented trajectory. Thus, you could say that even if it goes from a straight-line conception of evolution to a meandering one, the history of science is still moving forward along a *straight* path. But the curators are much more advanced than that: they push the parallel much further and the whole floor is punctuated by videos of scientists at work, little biographies of famous fossil-hunters at war with one another, with even different reconstructions of skeletons to prove to the public that "we don't know for sure"—a frequent label in the show. If the evolution of horses is no longer "Whiggish," neither is the history of science promoted by the curators. The only Whiggishness that remains, the only "overall trend" (and who in science studies will complain about that one?) is that the more recent conception of science has led us from a rigid exhibition of the final fact of paleontology to a more complex, interesting, and heterogeneous one. From the "classic" version, we have moved to the what? "Romantic"? "Postmodern"? "Reflexive"? "Constructivist"? Whatever the word, we have moved on, and this is what interests me here: objects and knowledge of objects are similarly thrown into the *same* Heraclitean flux. In addition to the type of trajectory they both elicit, they are rendered comparable by the process of time to which they both submit.

The great virtue of the innovative directors and designers of the gallery, on the top floor of the Museum, is to have made possible for the visitors to detect a parallel, a common thrust or pattern, between the slow, hesitant, and bushy movement of the various sorts of horses struggling for life in the course of their evolution, and the slow, hesitant, and bushy process by which *scientists* have reconstructed the evolution of the horses in the course of the *history* of paleontology. Instead of papering over the vastly controversial history of paleontology and offering the present knowledge as an indisputable state of affairs, the curators decided to run the risk—it is a risk, no doubt about that, especially in Bushist times[2]—of presenting the succession of interpretations of horse evolution as a set of plausible and revisable reconstructions of the past. "Contrast," "version," "tale"—those are pretty tough words for innocent visitors—not to mention the skeptical scare quotes around the adjective "successful," which is a sure way to attack the over-optimistic gloss neo-Darwinism has tended to impose on evolution.[3]

What fascinates me every time I visit this marvelous exhibit is that everything is moving in parallel: the horses in their evolution and the interpretations of horses in the paleontologists' time, even though the scale and rhythm is different—millions of years in one line, hundreds of years in the other. Ignoring the successive versions of horse evolution that have been substituted for one another would be, in the end, as if, on the fossil side, you had eliminated all the bones to retain only one skeleton, arbitrarily chosen as the representative of the ideal and *final* Horse. And yet what I find most interesting as a visitor and a science student—admittedly biased—is that even though science had to go through different "versions," even though bones could be displayed and reconstructed in different ways, that does not seem to diminish the respect I have for the scientists any more than the multiplicity of past horses would preclude me from admiring and mounting a *present*-day horse. In spite of the words "contrast," "version," and "revision," this is not a "revisionist" exhibit that would make visitors so doubtful and scornful of science and of scientists that it would be as if they were requested, at the entry of the show, to "abandon all hopes to know something objectively."[4] Quite the opposite.

Such is the source of this present paper. While we take the successive skeletons of the fossil horses not only gratefully, but accept it as a major discovery—evolution being the most important one in the history of biology—why do we find troubling, superfluous, irrelevant, the displaying of the successive versions of the science of evolution? Why do we take evolution of animals as a *substantial* phenomenon in its own right while we don't take the history of science as an equally substantial phenomenon, not at least as something that defines the *substance* of knowledge? When a biologist studies the evolution of a species, he or she hopes to detect the vital characteristics that explain its present form in all its details, and the inquiry is carried out in the same buildings and in the same departments as the other branches of science; but when a historian or a science student accounts for the evolution of science, this is done in another building, away from science, and is taken as a luxury, a peripheral undertaking, at best a salutary and amusing *caveat* to warn hubristic scientists, and not as what makes up the finest details of *what* is known. In other words, why is it difficult to have a history *of science*? Not a history of our representation but of the things known as well, of epistemic things? While we take as immensely relevant for the existence of the present-day horse each of the successive instances of the horse line, we are tempted to throw out and consider as irrelevant all of the successive versions that the history and reconstruction of the horse line by paleontologists have taken. Why is it so difficult to consider each of the successive interpretations as an *organism* for its own sake with its own capacious activity and reproductive risks? Why is it so difficult to take knowledge as a vector of transformation and not as a shifting set aiming toward something that remains immobile and "has" no history? What I want to do here is to de-epistemologize and to re-ontologize knowledge activity: time is of the essence in both.

Revisiting the Textbook Case of Epistemology

What is so nice in the labels of the museum is that they are plain and common sense. They are not coming (as far as I know) from any debunking urge, from some iconoclastic drive by the curators to destroy the prestige of science. They display, if I can say this, a plain, healthy, and innocent *relativism*—by which I mean neither the indifference to others' points of view nor an absolute privilege given to one's own point of view, but rather the honorable scientific, artistic, and moral activity of being able to *shift* one's point of view by establishing relations between frames of reference through the laying down of some instrumentation.[5] And it is this plainness that makes a lot of sense, because, such is my claim in this first part of the paper, in principle the acquisition and rectification of knowledge should have been the easiest thing in the world: we try to say something, we err often, we rectify or we are rectified by others. If, to any uncertain statement, you allow for the addition of *time, instrument, colleagues, and institutions*, you come to certainty. Nothing is more common sense. Nothing *should* have been more common sense than to recognize that the process by which we know objectively is devoid of any mysterious epistemological difficulty.

Provided, that is, that *we don't jump*. William James made a lot of fun of those who wanted to jump through some vertiginous *salto mortale* from several shifting and fragile representations to one unchanging and unhistorical reality. To position the problem of knowledge in this fashion, James said, was the surest way to render it utterly obscure. His solution, unaided by science studies or history of science, was to underline again the simple and plain way in which we rectify our grasp of what we mean by establishing a *continuous* connection between the various versions of what we have to say about some state of affairs. His solution is so well known—but not always well understood—that I can rehearse it very fast, by insisting simply on a point rarely highlighted in the disputes around the so-called "pragmatist theory of truth." Since James was a philosopher, his examples were not taken from paleontology but, quite simply, from moving through the Harvard campus! How do we know, he asks, that my mental idea of a specific building—Memorial Hall—does "correspond" to a state of affairs?

To recur to the Memorial Hall example lately used, it is only when our idea of the Hall has actually terminated in the percept that we know "for certain" that from the beginning it was truly cognitive of that. Until established by the end of the process, its quality of knowing that, or indeed of knowing anything, could still be doubted; and yet the knowing really was there, as the result now shows. We were virtual knowers of the Hall long before we were certified to have been its actual knowers, by the percept's retroactive validating power. (James, [1907]1996: 68)

All the important features of what should have been a common sense interpretation of knowledge-making trajectories are there in one single paragraph. And first, the crucial element: knowledge is a *trajectory*, or, to use a more abstract term, a *vector* that projects "retroactively" its "validating power." In other words, we don't know *yet*, but we *will* know, or rather, we will know whether we *had known* earlier or not. Retroactive certification, what Gaston Bachelard, the French philosopher of science, called "rectification," is of the essence of knowledge. Knowledge becomes a mystery if you imagine it as a jump between something that has a history and something that does not move and has no history; it becomes plainly accessible if you allow it to become a continuous vector where *time* is of the essence. Take any knowledge at any time: you don't know if it is good or not, accurate or not, real or virtual, true or false. Allow for a successive, continuous path to be drawn between several versions of the knowledge claims and you will be able to decide fairly well. At time t it cannot be decided, at time $t + 1$, $t + 2$, $t + n$, it has *become* decidable provided of course you engage along the path leading to a "chain of experiences." What is this chain made of? Of "leads" and of substitutions, as James makes clear by another example, not about horses or buildings, this time, but about his dog. The question remains the same: how do we render comparable my "idea" of my dog and this "furry creature" over there?

To call my present idea of my dog, for example, cognitive of the real dog means that, as the actual tissue of experience is constituted, the idea is capable of leading into a chain of experiences on my part that go from next to next and terminate at last in the vivid sense-perceptions of a jumping, barking, hairy body. (James, [1907]1996: 198)

This plain, healthy, and common sense relativism requires a good grounding in the "actual tissue of experience," a grasp of "ideas," "chains of experiences," a movement "next to next" without interruption, and a "termination" that is defined by a change in the cognitive materials from "idea of the dog" to "the jumping, barking, hairy body" of a dog now seized by "vivid sense perceptions."

There is thus no breach in humanistic [a synonym for radical empiricism] epistemology. Whether knowledge be taken as ideally perfected, or only as true enough to pass muster for practice, it is hung on one continuous scheme. Reality, howsoever remote, is always defined as a terminus within the general possibilities of experience; and what knows it is defined as an experience that "represents" it, in the sense of being substitutable for it in our thinking because it leads to the same associates, or in the sense of "pointing to it" through a chain of other experiences that either intervene or may intervene. (James, [1907]1996: 201)

Contrary to Spinoza's famous motto "the word 'dog' *does* bark" but only *at the end* of a process which is oriented as a vector, which has to be continuous, which has to trigger a chain of experiences, and which generates as a result a "thing known" and an accurate "representation of the thing," but only retroactively. The point of James—totally lost in the rather sad dispute around the 'cash value' of truth—is that knowledge is not to be understood as what relates the idea of a dog and the real dog through some *teleportation* but rather as a chain of experiences woven into the tissue of life in such a way that when time is taken into account and when there is no interruption in the chain, then one can provide (1) a retrospective account of what triggered the scheme, (2) a knowing subject—validated as actual and not only virtual, and finally (3) an object known—validated as actual and not only virtual.

The crucial discovery of James is that those two characters—object and subject—are *not the adequate points of departure* for any discussion about knowledge acquisition; they are not the *anchor* to which you should tie the vertiginous bridge thrown above the abyss of words and world, but rather they are *generated* as a byproduct—and a pretty inconsequential one at that—of the knowledge making pathways themselves. "Object" and "subject" are not ingredients of the world, they are successive *stations* along the paths through which knowledge is rectified. As James said, "there is no breach"; it is a "continuous scheme." But if you interrupt the chain, you remain undecided about the quality of the knowledge claims, exactly as if the *lineage* of one horse species were interrupted due to a lack of offspring. The key feature for our discussion here is not to ask from any statement, "Does it correspond or not to a given state of affairs?" but rather, "Does it lead to a continuous chain of experience where the former question can be settled retroactively?" This paper is entirely about uncovering the difference between the "continuous scheme" and what I will call "the teleportation scheme."[6]

But the problem with James (apart from his use of the unfortunate "cash value" metaphor) is that he took examples of buildings and dogs for drawing his continuous scheme, of entities that were much too mundane to prove its common sense point. It is actually the problem with most classical philosophers: they take as their favorite examples mugs and pots, rugs and mats, without realizing that those are the worst possible cases for proving any point about how we come to know because they are *already much too well known* to prove anything about how we come to know. With them, we never feel the *difficulty* of the knowledge-making pathways, and we take the result of the byproduct of the path—a knowing mind and an object known—as the only two real important components of any given state of affairs. With those all too familiar termini, it seems easy to stage the situation in which I ask: "Where is the cat?" and then without any long, difficult, tortuous pathway, to point out and say: "Here on the mat." This lazy way of taking it would be innocuous enough except when, after having based your theory of knowledge acquisition on such mundane, banal, and utterly familiar objects, you feel sure that what really counts are the subject and the object (the name "dog" on the one hand and the "barking dog" on the other). Then you will tend to think that knowledge *in general* is made of one big jump from one of those components to the other. You are replaying Act I Scene 1 of first empiricism.[7] Of course, it is perfectly true that, once we have become familiar with the pathway, we can most of the time safely ignore the intermediary steps and take the two termini as representative of what knowledge is. But this forgetting is an artifact of familiarity.

Even worse is that we try to use the model of knowledge acquisition adapted to the mundane, familiar object, to raise "The Big Question" of knowledge acquisition about new, unknown, difficult to focus upon, and sophisticated objects such as planets, microbes, leptons, or horse fossils, for which there is not yet any pathway or for which the pathway has not become familiar enough to be represented by its two end points. We tend to treat new entities for which it is absolutely crucial to maintain the continuous scheme as if they had become familiar objects already. And yet, for any new objects, the whole framework that had been defined on mundane objects breaks down entirely, as the last three centuries of epistemology have shown, because there is no way you can use the object/subject tool to grasp any *new* entity. The teleportation scheme based on mundane and habitual states of affairs gives not the slightest clue on how to lay down the continuous path that might provide objectivity on new states of affairs.[8]

Breaking the Habits of Thought Due to the Use of Mundane Artifacts

To realize how much in line with common sense James's basic point is, we have to part company with him and consider cases in which the "chain of experience" and the successive versions leading "next to next" to certainty, should be easily documentable, visible, and studiable. This is what science studies and history of science has shown in the last thirty years. To the too familiar James's dog example, we have to substitute, for instance, the difficulty of paleontologists to make sense of dispersed and hard to interpret fossils. As soon as we do so, it will become obvious to all that we never witness a solitary mind equipped with "ideas" of horse evolution trying to jump in one step to the "Horse Evolution" out there. Not because there is no "out" and no "there," but because the "out" and the "there" are not *facing* the mind: "out" and "there" are designating nothing more than stations along the chain of experience leading through successive and continuous rectifications to other revised versions ("termini" in James's parlance, but there are always more than two). If there is one thing that has made philosophy of science so lame, it is to have used mats and cats, mugs and dogs, in order to discover the right frame of mind to decide how we know with accuracy objects such as black holes and fossils, quarks and neutrinos. It is only by studying controversial matters of fact *before* they can be treated matter of factually that we can witness the obvious phenomenon of the pathways—what I call networks[9]—in plain light before they disappear and leave the two byproducts of object and subject to play their roles as if they had *caused* the knowledge of which they are only the provisional *results*.

No one has seen this better than Ludwick Fleck, whose interpretation of "thought collective" is very close to that of the chain of experiences outlined by James. In spite of the expression "thought" in "thought collective," what Fleck has clearly in mind is the sort of heterogeneous practices laboratory studies have since rendered familiar to us. It is interesting to notice here that Fleck's theory itself has been misrepresented by the idea of "paradigm" thrown onto him by Thomas Kuhn's foreword to the English translation of his book (Fleck, 1981). "Paradigm" is typically the sort of term that has meaning only in the abyss-bridging scheme. It reintroduces the knowing subject (now pluralized) as one of the two anchors of the activity of knowledge together with the supposed "thing in itself." The two are facing one another, and the whole question is where we situate any statement along this bridge: nearer the mind's categories or closer to the thing to be known? This is exactly the position of the problem out of which Fleck (who had to invent sociology of science from scratch) had to extract himself.

When you take the example not of dogs and cats but for instance of the pioneering efforts of syphilis specialists to stabilize the Wasserman reaction (the main example in the book), then the whole situation of knowledge acquisition is modified. With Fleck, as with James, we are at once thrown into the Heraclitean flow of time. The wording might still be ambiguous but not the direction taken:

> To give an accurate historical account of a scientific discipline is impossible. . . . It is as if we wanted to record in writing the natural course of an excited conversation among several persons all speaking simultaneously among themselves and each clamoring to make himself heard, yet which nevertheless permitted a consensus to crystallize. (Fleck, [1935]1981: 15)

Notice that the metaphor of crystallization is not opposed to but follows from that of the flow of experience in an "excited conversation." Because of Kuhn's framing of

Fleck's problem in the foreword to the English translation, readers have often forgotten that the subtitle of the book was even more explicitly historical than James's argument: the "genesis" of the scientific "fact." No more than James, Fleck is talking here about the emergence of our *representations* of a state of affairs: it is the *fact itself* that he is interested in following up through its emergence. He wants to tackle facts much like paleontologists want to reconstruct the horse line, not the ideas we entertained of the horse line. Only a Kantian can confuse the phantoms of ideas with the flesh of facts.

This is how a fact arises. At first there is a signal of resistance in the chaotic initial thinking, then a definite thought constraint, and finally a form to be directly perceived. A fact always occurs in the context of the history of thought and is always the result of a definite thought style. (Fleck, [1935]1981: 95)

What's the difference, one could object, with the notion of a paradigm projecting one's category onto a world that is subjected to an inquiry? The difference lies in the philosophical posture; it comes from what time does to all the ingredients of what is here called "thought style." Fleck does not say that we have a mind zooming toward a fixed—but inaccessible—target. It is the fact that "occurs," that emerges, and that, so to speak, offers you a (partially) new mind endowed with a (partially) new objectivity. Witness the musical metaphor used to register the process of coordination that will account for the stabilization of the phenomenon:

It is also clear that from these confused notes Wassermann heard the tune that hummed in his mind but was not audible to those not involved. He and his coworkers listened and "tuned" their "sets" until these became selective. The melody could then be heard even by unbiased persons who were not involved. (Fleck, [1935]1981: 86)

Fleck adds, "something very correct developed from them, although the experiments themselves could not be called correct."

Fleck's originality here is in breaking away from the visual metaphor (always associated with the bridge-crossing version) and in replacing it by the progressive shift from an uncoordinated to a coordinated movement. I wish the dancing together to a melody to which we become better and better attuned could replace the worn-out metaphor of an "asymptotic access" to the truth of the matter. Fleck derides the visual metaphor by calling it the *veni, vidi, vici* definition of science!

Observation without assumption which psychologically is non-sense and logically a game, can therefore be dismissed. But two types of observation, with variations along a transitional scale appears definitely worth investigating: (1) *the vague initial visual perception,* and (2), *the developed direct visual perception of a form.* (Fleck, [1935]1981: 92)

We find here the same direction of the argument as in James: knowledge flows in the same direction as what is known. It is a "transitional scale." But the scale does not go from mind to object with only two possible anchors, it goes from vague perception to direct—that is, directed!—perception through an indefinite number of intermediary stations, not just two. That is the big difference in posture. Notice the daring and quite counterintuitive reversal of metaphors: it is only once the perception is "developed," that is, equipped, collected, attuned, coordinated, artificial, that it is also "direct," whereas the initial perception appears retrospectively to have been simply "vague." Hence this magnificent definition of what it is to be skilled and learned into perception, what it is to graduate into the coherence of fact genesis:

Direct perception of form (*Gestaltsehen*) requires being experienced in the relevant field of thought. The ability directly to perceive meaning, form and self-contained unity is acquired only after much experience, perhaps with preliminary training. At the same time, of course, we lose the ability to see something that contradicts the form. But it is just this readiness for directed perception of form that is the main constituent of thought style. Visual perception of form becomes a definitive function of thought style. The concept of being experienced, with its hidden irrationality, acquires fundamental epistemological importance. (Fleck, [1935]1981: 92)

Fleck does not say, as in the usual Kantian-Kuhnian paradigm metaphor, that "we see only what we know beforehand," or that we "filter" perceptions through the "biases" of our "presupposition." Such a gap-bridging idea is on the contrary what he fights against because then time could not be part of the substance of fact genesis. This is why he reverses the argument and fuses the notion of "direct" grasp of meaning, with being "directed" and "experienced." It is not a subtle hair-splitting nuance, it is a radical departure, as radical in science studies as what James had done to philosophy. Because, if "direct" and "directed" go together, then we are finally through with all this non-sense about being *obliged to choose* between having categories (or paradigms) or grasping the facts of the matter "as they are." It is because of his shift in philosophical posture that Fleck is able for the first time (and maybe for the last one in science studies!) to take the social, collective, practical elements *positively* and not negatively or critically.[10]

Every epistemological theory is trivial that does not take this sociological dependence of all cognition into account in a fundamental and detailed manner. But those who consider social dependence a necessary evil and unfortunate human inadequacy which ought to be overcome fail to realize that without social conditioning no cognition is even possible. Indeed, the very word "cognition" acquires meaning only in connection with a thought collective. (Fleck, [1935]1981: 43)

Trivial after thirty years of science studies? Not at all! Radical, revolutionary, still very far in the future.[11] Why? Because if we read carefully the way in which he engages the social metaphors in the process of discovery, they are in no way a *substitute* for the knowing subject. Fleck, apparently connected to James or at least to pragmatism, has picked up the general tenor of pragmatism in a unique way.[12] "Social" and "collective" are not there to serve as an expansion or a qualification on Kant's epistemology at all. They are mobilized to ruin the idea that there is a mind facing an object above the abyss of words and world. When he deals with the collective, social, and progressive "aspects" of science, it is not because he has abandoned the idea of grasping reality but just for the opposite reason, because he wants at last a *social ontology*, not a social epistemology.

Truth is not 'relative' and certainly not 'subjective' in the popular sense of the word . . . Truth is not a convention but rather (1) in historical perspective, an event in the history of thought, (2) in its contemporary context, stylized thought constraints. (Fleck, [1935]1981: 100)

"Truth is an *event*," and so is the emergence of the horse in nature, and so is the emergence of the knowledge of the horse lineage. So for Fleck as for James, the key features to be outlined are that (1) knowledge is a vector; (2) ideas are there and have to be taken seriously but only as the beginning of a "chain of experience" ("experimentations" for Fleck); (3) successive rectification and revision are not peripheral but are the substantial part of the knowledge acquisition pathways; (4) rectification by colleagues is essential; (5) so is institutionalization—becoming familiar, black-boxing novelty in instruments, tuning, standardizing, getting used to a state of affairs, and so on; (6) direct perception is the end and not the beginning of the process of fact genesis. Fact is the provisional end of the vector and all the questions of correspondence between statements and states of affairs can indeed be raised but cannot be answered except retrospectively and provided the *Dankollektiv* is kept in place without interruption.

KNOWLEDGE RAISES NO EPISTEMOLOGICAL QUESTIONS

Two Orthogonal Positions for Knowledge-making Pathways

Those comments on James, Fleck, and science studies are simply to remind us that, as John Searle (personal commmunication, 2000) quipped, "science raises no epistemological question." I agree with him entirely, and James would have agreed with him also—no matter how incommensurable their various metaphysics. If by "epistemology" we name the discipline that tries to understand how we manage to bridge the gap between representations and reality, the only conclusion to be drawn is that this discipline has no subject matter whatsoever, because we *never* bridge such a gap—not, mind you, because we don't know anything objectively, but because *there is never such a gap*. The gap is an artifact due to the wrong positioning of the knowledge acquisition pathway. We imagine a bridge over an abyss, when the whole activity consists of a drift through a chain of experience where there are many successive event-like termini and many substitutions of heterogeneous media. In other words, scientific activity raises no especially puzzling epistemological questions. All its interesting questions concern *what* is known by science and *how* we can live with those entities but certainly not *whether* it knows objectively or not—sorry for those who have scratched their head about this last one for so long. Skepticism, in other words, does not require much of an answer.

If we had to summarize what I have called here the healthy, common sense relativism expressed in the labels of the Evolution gallery, in James's radical empiricism, in Fleck's trajectories, or in many good (that is, non-debunking) histories of controversies in science, we could end up with a portrayal of a knowledge path, freed from epistemological questions. Yes, we err often, but not always because, fortunately, (1)

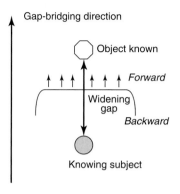

Figure 4.2
The teleportation scheme.

we have time; (2) *we are equipped;* (3) *we are many;* (4) *we have institutions.* A pair of diagrams could summarize the shift in emphasis necessary to absorb the next much more difficult point about the ontology implied by such a common sense description.

In the "teleportation scheme," the great problem of knowledge is to bridge the gap between two distinct domains totally unrelated to one another, mind and nature. Thus, what counts most is to place the cursor along the gradient going from one limit—the knowing subject—to the other—the object known. In this positioning of the problem of knowledge, the key question is to decide whether we move forward—toward the unmoving target of the object to be known—or backward—in which case we are thrown back to the prison of our prejudices, paradigms, or presuppositions.

But the situation is entirely different in the "continuous scheme" invoked by James, Fleck, and much of science studies.[13] Here, the main problem is not to decide whether a statement goes backward or forward along the subject/object pathway (vertically in figure 4.2) but whether it goes backward or forward *in time* (orthogonally in figure 4.3).[14] Now the main problem of knowledge is to deploy the continuous chain of experience to multiply the crossing points at which it will be possible to *retroactively* decide whether we *had been* right or wrong about a given state of affairs. Going "forward" now means that we become more and more "experienced," "cognizant," "attuned" to the quality of the collective, coordinated, instituted knowledge. There is no gap to be bridged, and no mysterious "correspondence" either, but there is a huge difference in going from few crossing points to *many*.

It is rather funny to consider that so much saliva (including mine) has been spent for or against a "correspondence theory of truth" by which proponents and critiques of the theory have always meant a jump between object and subject without ever inquiring about the *type of correspondence.* Trains and subways would have offered a better metaphor for defining what we mean by a correspondence: you don't shift from one subway line to the next without a continuous platform and corridors laid out

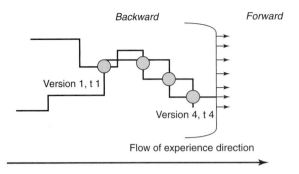

Figure 4.3
The continuous scheme.

allowing you to *correspond* on schedule. So James and Fleck are certainly proponents of a "correspondence theory of truth"—if you keep in mind the train metaphor—whereas they would strongly object to the "*salto mortale* theory of truth." If you accept renewing the metaphor, then you move forward when you go from a simple, isolated, poorly equipped, and badly maintained straight line to a complex network of well-kept-up stations allowing for many correspondences to be established. So "forward" means going from a bad to a good network.[15] Anyone living in a big city with or without a good public transportation network will grasp the difference.

I said earlier that those time-dependent paths could be visible only if we choose to consider, as science studies has done, newer and more complex objects than mugs and rugs. But it is interesting to come back briefly to the mundane cases on which the discontinuous scheme has been honed, once we have tried to follow objects that are less familiar and where it is easier to document the pathways. A lot of energy has been devoted in the course of time to answer skeptics about the so-called "errors of the senses." The classic *topos*, visited over and over again in the course of philosophy, is that I might not be certain, for instance, whether a tower seen from afar is a cylinder or a cube. But what does that prove against the quality of our knowledge? It is perfectly true to say that, at first, I might have misread its shape. But so what? I simply have to walk *closer*, I *then* see that I *was* wrong—or else I take my binoculars or someone else, a friend, a local inhabitant, someone with a better eyesight, to correct me. What could be simpler than this retort? Horse fossils at first seemed to align themselves in a straight line going always in the same direction. Then more fossils were collected, many more paleontologists entered the discipline, the straight line had to be rectified and revised. How could this feed skepticism? To be sure, those rectifications raise interesting questions: why do we err at first—but not always? How come that the equipment is often deficient—and yet quickly upgraded? How come checks and balances of other colleagues often work—but sometimes fail to do so? However, not one of those interesting historical and cognitive science questions should invite us to skepticism. When Descartes asks us to take seriously the question whether or not the

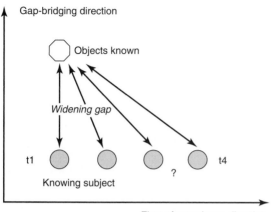

Figure 4.4
The flow of experience scheme.

people walking in the street might not be automats, the only sensible answer should have been: "But René why don't you go down in the street and check for yourself? Or at least ask your valet to go check it for you?" *Ego cogito* might be open to question, but why don't you try *cogitamus*?

The claim that there lies a Big Epistemological Question, so big that if it is not answered it threatens for good the quality of our science and then of our civilization, comes simply from a defect in the first scheme: *there is no place in it for time,* nor for instrument, nor for people, nor for rectification, nor for institution.[16] Or rather there is some place for the successive versions on the *subject* pole side but none for what happens to *the object itself* (Figure 4.4). More exactly, it is because there is no room for the parallel movement in time of the facts themselves that the object becomes isolated "in itself" and "for itself." To use figure 4.2 again, when we add the history of our representations, we register such a distortion that a widening gap is now yawning. It was not there in figure 4.3. Then, but only then, skeptics have a field day. If we have changed our "representations" of the object so often, while the goal, the target, has not changed at all, this could only mean that our mind is weak, and that "we will never know for sure." We will remain forever inside our representations.

Does this prove that skepticism is right? No, it simply proves that epistemology has been silly in proposing such a target for knowledge. It is as if it had offered its throat to be sliced: the temptation to cut it was too great to be resisted. If we think of it, never has any statement been verified by following the vertical dimension of the diagram. Even to check whether a cat is on the mat, we have to engage ourselves in the second dimension—the horizontal one in the figures—and only *retroactively* can we then say: "I *was* right in saying that my sentence 'the cat is on the mat' corresponded to a state of affairs." Contrary to the bad reputation pragmatism often gave its own argument, the time dimension it has so clearly detected in knowledge production is not an *inferior* way of knowing that should be substituted for the higher and more absolute one "because this one, alas, remains inaccessible." The continuous scheme is not an ersatz for the only legitimate realist way to know; on the contrary, it is the teleportation scheme that is a complete artifact. The only way to obtain objective knowledge is to engage, orthogonally, into one of those trajectories, to go with the flow of experience.[17] From the dawn of time, no one has ever managed to jump from a statement to a corresponding state of affairs without taking time into account and without laying down a set of successive versions connected by a continuous path. To be sure, a statement might have led, "next to next," as James said, to a chain of experiences heading toward a provisional terminus allowing, through a substitution of sensory data, for a retrospective judgment about what it was "virtually" earlier. But no statement has ever been judged by its truth content "if and only if" some state of affairs corresponded to it.[18]

Thus, the puzzle for me is not, "How can we decide that a statement about states of affairs is true or false," but rather, "How come we have been asked to take seriously an attempt to transform knowledge production into an impossible mystery, a jump above the abyss?" The true scandal is not to ask, "How come there are bloody relativists attacking the sanctity of science by denying that the gap between representations and objectivity can be bridged?" but instead to ask, "How come a trench has been dug into the paths whose continuity is necessary for any knowledge acquisition?"

If there is no sense in qualifying knowledge out of time, why then does time have to be taken out? Why do we consider that adding, time, rectification, instruments, people, and institutions could be a threat to the sanctity and truth conditions of science when they are its *very stuff*, when they are the only way that exists to lay down the continuous path allowing for ideas to become loaded with enough intersections to decide retroactively if they had been correct or not? In the case of the history museum, does it distract visitors to know that there were paleontologists fighting one another, that fossils had a market value, that reconstitutions have been modified so often, that we "don't know for sure," or, as another label states, "While it's intriguing to speculate about the physiology of long extinct animals we cannot test these ideas conclusively?" The more fossils there are, we feel that the more interesting, lively, sturdy, realistic, and provable are our representations of them; how come we would feel less certain, less sturdy, less realistic about those same representations when they multiply? When their equipment is visible? When the assembly of paleontologists is made visible?

The puzzle I now want to address is not, "Are we able to know objectively with certainty?" but rather the following: "How have we come to doubt that we are able to know objectively, to the point of seeing as proofs of skepticism and relativism the obvious features that allow truth conditions to be met?" I am turning the tables here, against those who have so often accused science students of immorality! After having meekly or provocatively answered those charges for so long, it is time to counter-attack and to doubt the moral high ground they have occupied with no title whatsoever.

KNOWLEDGE IS A MODE OF EXISTENCE

A Real Difficulty in the Knowledge Acquisition Pathways

One possible answer is that we have been asking from objective science something it cannot possibly deliver and should not even try to deliver, thereby opening a large hole into which skepticism could penetrate. And that epistemologists, instead of confessing, "OK, we were wrong to ask this from science," have kept thinking that their main duty was to *fight* against skepticism instead of fulfilling their only duties: to make sure that the truth conditions of science be met, by allowing for time rectification, for the improvement of instruments, for the multiplication of check and balances by colleagues and people, and generally by strengthening the institutions necessary for certainty to be kept up.

What is this added difficulty? Why was this extra baggage added to the burden of science production? One of the answers probably has to do with a denial of the formative quality of time. In the same way as before Darwin individual horses had to be considered as mere *tokens* of the ideal Horse *type*, it has seemed difficult to accept that you could gain certainty by the humble means of rectification, instrumentation, colleagues, and institutions. Actually the parallel goes deeper: in the same way that Darwin's revolutionary insights have really never been swallowed by our intellectual mores and have been instantly replaced by an enterprise to re-rationalize them, it happens that epistemology has never considered that it was enough to let the succession of ideas, plus instruments, plus colleagues, proceed at their own pace in order to obtain a sturdy enough certainty. To the lineages of tokens, they still want to add the *type*. Although there is no God leading the evolution of horses any more, there seems to be still a God, at least an Epistemological Providence, leading the *knowledge* of the horse lineage.

But another reason might have to do with the sheer difficulty of accounting for knowledge formation. It has been noted very often that, although science itself as an activity is a time-dependent, human-made, humble practice, the *result* of its activity—after a while, that is—offers a time-independent, not human made, quite exhilarating objectivity. After all, facts are generated. This is the main conclusion of the constructivist schools in science studies: at some point in the course of the fabrication, facts emerge that are no longer enlightened by the revelation that they have been fabricated or have to be carefully maintained. The double nature of facts—as fabricated and as unfabricated—has become a cliché of history of science and of science studies.

The limit of constructivism is that we have *trouble focusing on the two aspects* with equal emphasis: either we insist too much on the messy, mundane, human, practical, contingent aspects, or too much on the final, extramundane, nonhuman, necessary, irrefutable elements. Quite apart from the temptation to use the results of science to make a mess of politics,[19] it is perfectly true to say that objectivity as a practice is simply difficult to understand and square with our common metaphysics and our common ontology—by "common" I mean what has been made to be common by the first empiricism.

Remember that the puzzle I am trying to understand is not, "How come we manage to know objectively some distant state of affairs"—we do, no question about that—but rather, "How come, in spite of the obvious quality of our knowledge-acquisition pathways, we have engaged objectivity production into an impasse where knowledge becomes a mystery?" The reformulation I am now proposing is the following: "There must be a strange feature in objectivity production that has provided the temptation to engage this innocent, healthy, and rather common sense activity into an impasse that seemed productive for reasons utterly unrelated to objectivity *per se*" (one of them being politics, but this is not the object of this chapter). What then is this strange feature?

We have to admit that something happens to a state of affairs when it is engaged into knowledge acquisition. The dog of James's example, the horse fossils of paleontology, Pasteur's microbes, all undergo a transformation; they enter into a new path, and they circulate along different "chains of experiences" once they are known. This transformation is coded by epistemology—wrongly, as I have proposed earlier—as a grasping by a knowing subject. And we now understand why: the vertical dimension of the gap-bridging scheme in figure 4.4 is unable to detect any important transformation in the object known. Instead, it simply registers retroactively what happens *once* we know for sure: object and subject "correspond" to one another well; they are, as Fleck would have said, coordinated to the same tune and are "directly perceived." We have become able to detect the source of the artifact created by such a view: it takes the consequence, a knowing subject, for the anchor of a mysterious bridge leading to something that is *already* an object waiting to be known objectively. This is the reason that, while the knowing subject appears to have a history, a movement, a series of revisions and rectification, the object itself—the future "thing in itself"—does not move (see figure 4.4). Hence, the opening of the "breach" that volumes of epistemology have tried to fill: *one terminus moves and not the other*. Skepticism engulfs the open space. Yet if the genesis of fact is an event, this eventfulness should be equally shared with the discoverers as well as with the discovered.

Reparative Surgery: Distinguishing Pathways

To grasp this difference in a way that does not make again the same "mistake" as epistemology, it is important to consider first how the object moved *before* being grasped by the knowledge pathways. How was the dog jumping and barking before James tried to make sure his "idea of the dog" "co-responded" to the dog? To phrase it in my rather infamous way: "What was the way of life for microbes *before* Pasteur engaged them into the pathways of nineteenth century microbiology?"[20] If we answer, "Well, they were sitting there, *an sich*, waiting to be known," we at once reopen the gap, the breach, the cleft that no amount of ingenuity will fill in. On the other hand, if we answer: "They date from the moment when the philosophers or the scientists designate them," we open the can of worms of relativism—in the papal pejorative meaning of the word—and soon risk settling upon one of the various idealist positions, no matter how sophisticated we try to be. And yet, in the continuous scheme, something must *have happened to* the tissue of experience in which the various entities we are considering now move in the same direction. What was absurd according to the scenography sketched in figure 4.2 (knowing and known were on two different metaphysical sides of a gap), becomes almost common sense in the scenography of figure 4.3: knowing and known share at least a common "general trend"—and this is why we end up knowing so objectively.[21] To reuse James's metaphor, we should now ask, "What is the fabric of the common tissue?"

It is clear that one character at least is common to all the threads: they are made of vectors that are all aligned, so to speak, in the same struggle for existence. All the horses, at the time when they were alive, were struggling to subsist in a delicate and changing ecology and racing along reproductive paths. For them, too, no doubt about it, there was a difference between going forward or backward! It was the difference between surviving as a horse or becoming extinct. Whatever definition of knowledge we choose, we could agree that such a path must have a different bent, a different movement forward, that it must be made of different segments from what happens to the very few fossilized bones unearthed, transported into crates, cleaned up, labeled, classified, reconstructed, mounted, published in journals, and so on, once paleontologists have crossed path with the ancient horses.

Whatever your metaphysics, you would agree that there must be a nuance between being a horse and having a tiny fraction of the horse existence made visible in the Natural History Museum. The least provocative version of this crossing point is to say that horses benefited from a *mode of existence* while they were alive, a mode that aimed at reproducing and "enjoying" themselves—enjoyment is Alfred North Whitehead's expression—and that, at the intersection with paleontologists, some of their bones, hundreds of thousands of years later, happened to enter into *another mode of existence* once fragments of their former selves had been shunted, so to speak, into paleontological pathways. Let's call the first mode *subsistence* and the second mode *reference* (and let's not forget that there might be many more than two modes).[22]

I am not saying anything odd here: everyone will accept that an organism striving for life does not carry on exactly in the same way as a bone being unearthed, cleaned up, collectively scrutinized, and published about. And yet I have to be careful here to avoid two misrepresentations of this expression, "not being exactly the same."

First, I hope it is clear that I am not trying to revive the romantic cliché of "rich life" versus "dead knowledge"—even though romanticism might have seized rightly on one aspect of this difference. Because for a bone to be carried along the paleontologists' networks, this is a life just as rich, interesting, complex, and risky as for the horse to roam through the great plains. I am just saying that it is not *exactly the same* sort of life. I am not opposing life and death, or object and knowledge of the object. I am simply *contrasting two vectors* running along the same flow of time, and I am trying to characterize both of them by their different mode of existence. What I am doing is simply refusing to grant existence to the object while knowledge itself would be floating around without being grounded anywhere. Knowledge is not the voice-over of a nature film on the Discovery channel.

The second misrepresentation would be to forget that knowledge acquisition is also a pathway, just as much a continuous chain of risky transformations as the subsistence of horses. Except that the latter goes from one horse to the next through the *reproduction* of lineages, while the other one goes from a sandy pit to the History Museum through many segments and transformations, in order to maintain "immutable mobiles," drawing what I have called a *chain of reference*.[23] In other words, my argument makes sense only if we fill in the line going through all the transformations characterizing this second mode of existence without limiting the move at its two putative termini. We know what happens when we forget this long chain of intermediaries: we lose the reference, and we are no longer able to decide whether a statement is true or false. In the same way, when the horse fails to accomplish the reproductive feat, its lineage just dies away. One is a vector that can stop if there is a discontinuity along the path, but *so is the other!* The difference does not come, in other words, from the *vector character* of those two types of entities but from the *stuff* out of which the successive segments of the two vectors are made. The tissue of experience is the same but not the thread from which it is woven. That is the difference I try to convey by the notion of mode of existence.

A few philosophers have learned from Whitehead that it might become possible again, after James's redescription of knowledge, to distinguish those two modes of existence instead of confusing them. Whitehead has called this confusion of the way a horse survives and the way a bone is transported through the paleontologists' knowledge acquisition pathways, "the bifurcation of nature." His argument is that we have been confusing how we know something with how this something is carried over in time and space. This is why he concluded that there is no question which would be clarified by adding that it is known by a subject—a big challenge for science students who pride themselves in doing just that!

There is now reigning in philosophy and in science an apathetic acquiescence in the conclusion that no coherent account can be given of nature as it is disclosed to us in sense-awareness, without dragging in its relations to mind. (Whitehead, 1920: 26)

What he was against was in no way that we know objectively—like Searle, like James, like myself, like all practicing scientists, he would not be interested for a minute in opening to doubt the certainty-acquisition networks. What Whitehead does is to give an even more forceful rendering of the slogan that "science does not raise any interesting *epistemological* questions." Precisely for this reason, Whitehead did not want to *confuse* the procedures, the pathways necessary for the mode of existence called knowledge, with the modes of existence that he calls organisms.

Thus what is a *mere procedure of mind* in the translation of sense-awareness into discursive knowledge has been transmuted into a fundamental character of nature. In this way matter has emerged as being the metaphysical substratum of its properties, and the course of nature is interpreted as the history of matter. (Whitehead, 1920: 16, emphasis added)

Hence, the most famous sentence:

Thus matter represents the refusal to think away spatial and temporal characteristics and to arrive at the bare concept of an individual entity. It is this refusal which has caused the muddle of importing the mere procedure of thought into the fact of nature. The entity, bared of all characteristics except those of space and time, has acquired a physical status as the ultimate texture of nature; so that the course of nature is conceived as being merely the fortunes of matter in its adventure through space. (Whitehead, 1920: 20)

Space and time are important "procedures of thought" for the mode of existence of acquiring knowledge along the pathways going, for instance, from sand pits to museums, but they are not to be confused with the ways "individual entities" manage to remain in existence. What Whitehead has achieved single-handedly is to overcome the impasse in which the theory of knowledge has engaged certainty production, by allowing both of them to go *their own separate ways*. End of the muddle of matter.[24] Both have to be respected, cherished, and nurtured: the ecological conditions necessary for organisms to reproduce "next to next" along continuous paths, and the ecological conditions for reference to be produced "next to next" along continuous paths. It would be a "fraud," Whitehead argues, to mix them up.

My argument is that this dragging in of the mind as making additions of its own to the thing posited for knowledge by sense-awareness is merely a way of shirking the problem of natural philosophy. That problem is to discuss the relations inter se of things known, abstracted from the bare fact that they are known . . . Natural philosophy should never ask, what is in the mind and what is in nature. (Whitehead, 1920: 30)

Here is the philosophical crossroad: one path is indicated in German: *An sich*, the other in Latin: *inter se*. The cosmological consequences of Whitehead's reparative surgery are enormous.[25] What I want to take from Whitehead is simply the possibility of giving ontological weight to what is usually defined as objective knowledge. From the very success of our development of scientific enterprises, epistemology has wrongly concluded that they were two termini—forgetting to fill in the pathways continuously—and it added that, of those two termini *only one*—the object—had some ontological import, while the other one, the subject anchor, had the mysterious ability to produce knowledge about the first as if knowledge itself had no ontological weight. Hence, the odd use of the word "representation" or "idea." Rocks and mugs and cats and mats have an ontology, but what is known about them does not. Because of this clumsy framing of the question, science students, intimidated by epistemology, have taken their own discovery of the pathways they were describing as being "merely" about human-made, mundane, word-like, discourse, without realizing that they had in effect unearthed a new, valid, sturdy, and completely mature mode of existence. They behave as if they had simply complicated or enriched the "word" side of the *same bridge* that had obsessed first empiricism, while the "world" side had remained intact or even had recessed even further from any grasp and into the Kantian *An sich*.

My claim is that, without Whitehead's reparative surgery, historians of science could never take seriously in their own discoveries that they had redirected attention to a type of vector affecting *both* the words and the worlds *inter se*. This is why, to counteract this trend, I wish to use the same expression "mode of existence" for both vectors: those for subsistence and those for reference. Provided, that is, we do not grant to "what" is known the confusing two sets of traits: *moving forward like an organism to subsist and moving forward like a reference to generate objective knowledge*. In other words, science students so far never dared to transform the chains of reference into a mode *of being*. And yet it is all quite simple: knowledge is *added* to the world; it does not suck things into representations or, alternatively, disappear in the object it knows. It is added to the landscape.

How Much Ontological Weight Has the Book of Nature?

We might now be in position to give some interesting meaning to the proposition I made at the beginning that history of science should mean the history of what is known as well as of the knowledge itself. This is the proposition that I staged in the, after all, not so provocative statement, "Newton *happens* to gravity," or "Pasteur *happens* to the microbes and the paleontologists to the bones of horses." We can summarize what we learned in this chapter by considering the same process—knowledge acquisition—viewed from two different frames of reference. The first one (see figure 4.2), which after James I have called "the somersault scheme," is characterized by (1) a vertical connection (2) established between two points—object and subject; (3) one of them moves through successive versions while the other does not; (4) the connection between the two is not marked and can be interrupted at any moment. In the second frame (see figure 4.3), which I have called "continuous," we have (1) vectors in undetermined numbers (2) flowing into the same direction of time (3) with many crossing points such that (4) the intermediary steps are continuously linked and constantly traceable.

My contention is that no realistic interpretation of knowledge production can be provided by the first frame: the only conclusion will be either that we forget entirely about the successive versions of the subject side—history of science should be rated X—or else that we abandon all hope to know "for sure," and we wallow in various schools of idealism and subjectivism. If this last view is correct, then the curators of the Gallery were wrong or disingenuous to put in parallel the lines of horses through evolution and the successively revised versions by paleontologists of this evolution. They should be kicked out of the museum as dangerous relativists, revisionists, and social constructivists. They are mere pawns in the Bushists' war on science, they are crypto-Derridians embedded into collections of fossilized bones to pervert good, positivist American schoolchildren.

However, a realistic version of knowledge production may be provided by the second frame of reference because no attempt is made there to confuse the movements of horses in evolution and the circulation of bones into paleontological pathways, and yet there are enough shunts, enough points of articulation, to generate many provisional termini for knowledge to be certified—that is, rectified, equipped with instruments, corrected by colleagues, guaranteed by institutions, and "directed" as Fleck said. More importantly, the pathways that connect the intersection points have continuous, recognizable, documentable material shapes. Is it plausible that, after thirty years of science studies, a sturdy correspondence theory of truth might finally be within grasp? (Remember I am patterning the metaphor after the metropolitan lines, not the teleportation version.)

What authorizes me to say that the second frame is better? This is the crux of the matter. In the first frame, all the attention is concentrated on two loci: the object intact out there, and the subject that has shifting versions "in there." In the second frame, the two anchors have disappeared: there is no longer one subject and there is no longer one object. Instead there are threads woven by the crisscrossing pathways. How could I take this second version as being more realistic? It is like saying that a Picasso portrait is more realistic than a Holbein or an Ingres. Well, but it might be, that is the whole point. Because what is now made fully visible in the second frame—and that is the ground for my claim—are the knowledge acquisition pathways that are generating, as so many byproducts, successive temporally marked versions of the objects and the subjects—now in the plural. It might be, I agree, a great loss not to be able to hold fast any more to the two termini of the object and subject. But consider, I beg you, the gain: the long and costly paths necessary to produce objective knowledge are now fully highlighted. The choice is now clear, and the question is for the reader to decide what is to be favored most. Do you prefer to highlight object and subject with the immense danger of opening a mysterious gap in between, the famous "out there" with the risk that skeptics will soon swarm in that gap much like crocodiles in a swamp ready to swallow you whole? Or do you prefer to deemphasize the questionable presence of object and subject and to underline the practical pathways necessary to nurture the production of objective knowledge? Now this is, in my eyes at least, what relativism should always provide: a clear choice between what you gain and what you lose depending on which frame of reference you decide to cling to.[26] Now you may choose.

The reason for my own choice is that it offers a fresh solution to the difficulty I mentioned earlier: the quasi-impossibility, after years of epistemology and then of science studies, to focus satisfactorily both on the mundane human, discourse-based aspects of science and the nonhuman, unfabricated, object-based aspects of the same activity. The reason for this impossibility was the choice of an inappropriate frame of reference—it's like in movies where, a full century after Lumière, in shooting a dialog between two characters, the cameraman still cannot focus simultaneously on the foreground and the background, even though our eyes, outside the movie theater, do it at once with no effort at all. But if you accept for one minute to see the fabric of science through the second frame, the two elements snap into focus at once: it becomes perfectly true to say that science is *not* manmade, even though it takes a lot of work to carry a bone from a sand pit to a Museum, a lot of colleagues to rectify what you say about it, a lot of time to make sense of your data, and a well-endowed institution to keep scientific truth valid. The bones have been made to behave in a completely different mode of existence that is just as foreign to the ways ideas behave in our mind as to the ways horses galloped on the great plains.[27]

An additional benefit of the second frame is that it squares nicely with the usual requirements of the philosophy of mathematics: mathematical constructs have to be nonhuman and yet constructed, just like the pathways I am highlighting which are badly handled if you try to hold them between objects out there and ideas in there—and the situation is even worse if you try a little bit of both. All mathematicians are alternatively Platonists and constructivists, and rightly so. And yet they have to work every day, to consummate, as the saying goes, enough loads of coffee cups to figure out theorems and to construct a world that in itself has the mysterious quality of being applicable to the real world. Those requirements are contradictory only if the first frame is applied, but not in the second, since being able to establish connections on paper between objects is precisely the service rendered, from the time of the Babylonians to today, to the knowledge acquisition pathways. Is not allowing the transportation through deformation without deformation—that is the invention of *constants*—what mathematics is all about?[28] And is this not exactly what is required to "lay down," so to speak, the networks necessary to make the solar system, the bones, the microbes and all the phenomena movable, transportable, codable in a way that makes objective knowledge possible? Objects are not made to exist "out there" before one of those pathways has been continuously, "next to next" as James said, filled in by mathematical grids. But it is entirely true to say that once they are uploaded into those pathways, stars, planets, bones, and microbes *become* objective and generate objectivity in the minds of those occupied to welcome, to lay out, or to install them.[29] Objective knowledge is not first in the minds of scientists who, then, turn to the world and marvel at how their ideas "fit" with the entities out there: objective knowledge is what circulates and then *grants* the entity seized by the networks another mode of existence and *grants* the minds seized by them a level of objectivity no human ever dreamed of before the seventeenth century—or rather they *dreamed* of it in earlier times but not before the collective, instrumented, and material pathways of scientific organizations were fully in place.[30]

Such is the great fallacy of those who imagine that objective science is the daughter of "human curiosity since the dawn of time" and that there is a direct epistemological line from Lucy looking over the savannah with upright posture to the Hubble telescope.[31] No, the laying down of long-range networks allowing for the shunting of many entities into objectivity-making trajectories is a contingent history, a new feature in world history, that did not need to be invented and that could be still *disinvented* if enough Bushists have their way and are able to destroy the practical conditions allowing those pathways to be continuously maintained. Is this not a way to respect the historicity of science and the objectivity of its results in a more productive fashion than what was possible in the first framework with its endless series of perilous artifacts? Especially important to me: is this not a better way to respect the ways to nurture the fragile ecological matrix necessary to add to the world the mode of existence of objective knowledge? What I have never understood about epistemologists is how, with their teleportation scheme, they would convince the people to invest in the devising, upkeep, and enlargement of the very humble means necessary to know something with objectivity. In spite of my reputation as a "social constructivist," I have always considered myself as one of those who tried to offer another realistic version of science against the absurd requirements of epistemology that could only have one consequence: skepticism. Here, as everywhere, relativity offers, in the end, a sturdier grasp than absolutism.

The operation I have offered in this chapter as a more plausible solution to an old problem is simply to reload with ontological weight the knowledge pathways instead of considering them, as we so often do even in science studies, as another and better version of "the mind facing the object."

I am actually saying nothing out of the ordinary; this is exactly what was designated, with great philosophical accuracy, by the very metaphor that Galileo had revisited: the book of nature is written in mathematical terms.[32] This mixed metaphor renewing the Bible points at exactly the same problem as the one I have proposed: yes, it is a book—and now Gingerich (2004) has shown how realistically this book pathway metaphor can be taken[33]—and, yes, it is the book in which a few of nature's movements forward can be welcomed, transported, calculated, made to behave in new ways. But the metaphor breaks down very rapidly if we don't consider under which ontological condition nature can be made to be written about in mathematical format. The Book of Nature metaphor provides the exact interpretation for this amazing event of the seventeenth century, known as "the scientific revolution": some features of the passage of nature became shunted and loaded into pathways, so that they provide them with a new mode of existence: they became objective. This is why any history of the trajectory of stars in time has to include, as one of their intersections, Copernicus and Galileo.

But this is also why, according to this view, their new post-seventeenth century existence as objects does not *allow anyone to withdraw* from the world other modes of existence that might have different pathways, different requirements for their own continuation into existence. The tissue of experience, what James called the *pluriverse*, is woven with more than one thread; this is why it is granted to us with such a dappled, glittering aspect—an aspect that has been enhanced somehow by the "revised version" offered by the curators of the Natural History Museum gallery. After all, I was simply trying here to understand the healthy meaning of the labels in the gallery of evolution I started with.

Notes

I warmly thank Beckett W. Sterner for editing my English. Isabelle Stengers, Michael Lynch, and Gerard de Vries proposed many more emendations than those I was able to carry through. Thanks to Adrian Johns and Joan Fujimura for allowing me to test this argument on their friends and colleagues.

1. I leave aside in this piece that almost all scientific disciplines, in recent times, have shifted from a Parminedian to an Heraclitean version: every science is now narrated in a way that takes time into account, from the Big Bang to the history of Earth geology or Earth climate. In this sense, the narrative mode made familiar to us by historians has triumphed, and physicists would tell us about the "historical emergence of particles" in the same mode. But this does not mean that those new Heraclitean versions of science will cross the path of the history of science more often than when the Earth, the Sky, and the Matter were supposed to be immutable. In other words, it is just as difficult for historians of cosmology to link their time narratives with those of physicists than it would have been in the time of Laplace when the cosmos had not yet any intrinsic history. This is what makes this public display so telling—and what explains the continuing success of the late Stephen Jay Gould, who was one of the rare writers to link the two histories so artfully.

2. See the fierce attacks on science autonomy and public discussions as they are related in, for instance, Mooney's book (2005). Against reactionaries the temptation is always to fall back on the "good old days," but like all temptations, it should be resisted. As will become clear in this chapter, there are many other ways to fight perverted skepticism than ardent positivism.

3. This is why I don't consider here the evolutionary epistemology that tries to replace the notion of "fit" between a representation and the world by a neo-Darwinian model in which organisms would be blindly "fitting" their environment. Naturalizing (or biologizing) epistemology does not modify the question: it is the very notion of "fit" and "fitness" that I want to "revisit" here. "Fit" is very much a remnant in biology of Kant's philosophy of science, where "adaptation" has replaced the "construction" of the world to be known by our intellectual categories. In both cases, humans would be blind to the things in themselves. This is too implausible and unrealistic a philosophy.

4. Needless to say that this "revision" does not lead to "revisionism" and even less to "negationism." It has always seemed to me, on the contrary, that a sturdy culture of fact-making was the only way to resist the perverse inversion of positivism that is so extensive in negationism and in other types of conspiracy theories (Marcus, 1999). It is only those who recognize the fragility of fact-making who may confide safely in their solidity.

5. In case of doubt, the word "relationism" can be substituted for the loaded term "relativism," which has two opposite meanings depending on whether it is Pope Benedict XVI or Gilles Deleuze who uses it.

6. "Continuous" is a confusing term here that should be understood, in James, as contrasting only with "*salto mortale*," with the big gap between "word" and "world." So "continuous" is not used here to deny that, once you look at the "tissue of experience," you will recognize a series of small gaps, discontinuities that are due to the complete heterogeneity of their constituents. For instance, in James's own example, there is a gap between the anticipation of the dog and the warm furry sensation once the dog is there. Those tiny discontinuities have been shown in many science studies through the work of Hutchins, Latour, Lynch, and Netz. But no matter if you talk about "intellectual technologies" (Hutchins, 1995), "chains of custody" (Lynch & McNally, 2005), "chains of reference" (Latour, 1999), or "diagrams" (Netz, 2003), the succession of varied media are like pearls—discontinuous, yes, but along the same thread. The knowledge trajectory is thus *continuous* in the first meaning of the word (against the language/world distinction) but is of course *discontinuous* when the set of micro gaps—the pearls on the thread—in the making is considered. Having done much work to show the micro-discontinuity necessary for the circulation of immutable mobiles (Latour, 1990), I use in this paper the adjective "continuous" only in the first sense.

7. I call first empiricism the effort from Locke to James (excluded) to define knowledge around the invention of matters of fact, and second empiricism the efforts from James to science studies (as I see them) to develop what I call "matters of concern" (see Latour, 2004a).

8. It is true that clever novelists such as Baker (1988), artful historians such as Petroski (1990), and daring philosophers such as James himself, may put mundane artifacts to good use by unfolding what

is not revealed in the two termini of "user" and "tool." But this is precisely the problem: they have to be clever, artful, and daring—all qualities that are much too infrequent in the rest of us. It is much easier to take new unknown objects so as to render fully visible the trajectory and its retroactive process of "certification." (In many ways, the problem with James was his considerable lightness of touch: people misunderstood this lightness for superficiality.)

9. But with a new meaning that is revealed in the last section. Once again, networks are made of many small discontinuities between different media. They are continuous only in the sense that they don't attempt to jump over the abyss between words and world. The confusion between the two words "continuous" and "discontinuous" will be removed in the next section (see note 6).

10. I guess most science students will say that they "of course" take positively the social aspects they unfold in their writing, but this is because they have simplified the philosophical task enormously and left aside the ontological question. By "positive," I mean here factors that are conducive to the genesis of the durable fact of the matter itself. It is noticeable that those who have made exception to this rule, such as Pickering (1995) for instance, have been greatly influenced by pragmatism.

11. Not once, for example, does Ian Hacking (1999) even contemplate in a book that claims to bring science studies to its senses that it would be possible not to think along the gap-bridging scheme: if social then not real, if real then not social, or maybe you want "a little bit of both"? No, we want *none of it*, this is what Fleck would answer, the whole position of the problem is unrealistic.

12. According to Ilana Lowy, there is actually a possible direct connection between pragmatism and Fleck through the teaching in Warsaw of the Polish pragmatist philosopher Wladyslaw Bieganski (1857–1917)—see her foreword to the French edition (Fleck, [1934]2005).

13. You can find some more instances recently of the historicization of the objects of science, not only of our representations, in the book edited by Daston (2000) on the biography of scientific objects.

14. "In time" should mean here "in process" because there are many philosophies that obliterate even time. On this obliteration, see Stengers's work, especially her *Whitehead* (2002) and my review of it (2005).

15. It is also the limit of anti-Whiggishness in history of science. Although it is a healthy position to start an inquiry, it becomes quickly counterproductive when we have to act as if there was no asymmetry between going forward and going backward. The second scheme is clearly asymmetrical as far as the arrow of time is concerned.

16. Notice that Descartes tried to ascertain the absolute certainty of his *ego cogito* exactly at the time of the invention of collective science—another proof that philosophers are pretty bad informants for what happens in their time.

17. It is strange that such a daring philosopher as James caved in to the enemy, so to speak, and denigrated his own position by accepting to say that it was "good enough to pass muster" for practical purposes only. In that sense, pragmatism is certainly the wrong label for what I am trying to present here.

18. This was the basis of Gabriel Tarde's (republication, 1999) alternative syllogistic. Tarde, like James, like Dewey, like Bergson, was very much part of this vast movement to renew philosophy, science, and society and to absorb the shock of Darwinism, which has been very much lost during the twentieth century and that we spend so much effort in trying to retrieve.

19. I have shown elsewhere (Latour, 2004b) that an absolute, unmediated, and timeless indisputable form of knowledge could seem, in some situations, to offer a solution to an entirely unrelated problem: that of producing agreement among rival parties in the noisy, smelly and crowded *agora*, an agreement that normal procedures, proper to political debates, could not generate. This is what I have called *political epistemology*. In this interpretation, epistemology would never have aimed at fostering science

ecology but rather at introducing into politics a source of certainty that could play the role of the court of appeal in case of debates that could not be closed to the satisfaction of the parties. The funny thing is that even though it was a terrible description of science's own way of achieving certainty, it was used nonetheless—and still is—as a template, an ideal, to shame the sordid ways in which politics could provide agreement.

20. I developed this point at some length in the middle chapters of *Pandora's Hope* (1999) but without having fully grasped the notion of mode of existence I put forward here.

21. Again, we should resist the temptation here to follow evolutionary epistemology and to unify prematurely all of the components by saying that "of course" they are all "parts of nature." As I have shown elsewhere, what is wrong in naturalization is not its sturdy materialism but its premature unification (Latour, 2004b). The point has been made even more forcefully and with much greater empirical precision by Philippe Descola's major book (2005).

22. The expression "mode of existence" is from Etienne Souriau (1943), and see my commentary on this book (Latour, 2007). The question of their number and definition is the object of my present work. Mode of existence is a banal expression clearly linked to the exploration of alternative ontologies.

23. I have tried even to document this movement and the many intermediary steps through a photo essay (see Latour, 1999: chapter 2). Everything that maximizes the two opposite qualities of "immutability" and "mobility" (see Latour, 1990, and the entire book by Lynch and Woolgar [1990]).

24. On the interpretation of this book, see Stengers (2002).

25. A sizeable body of philosophers informed by science studies have taken the challenge of Whitehead, chief among them Stengers, but see also Didier Debaise (2006).

26. It should be clear from the examples that the first model is actually a *consequence* of the second when the knowledge uncertainty has stabilized to the point where it seems common sense to say that there is "a dog" here and the word "dog" there.

27. It could be interesting to see how much more reasonable is this solution than that of the "anthropic principle," which implies too much predestination for my taste. But what is nice in the anthropic principle is at least to have taken into consideration knowledge and known as events that happen to all.

28. This question has made a decisive move with the publication of Netz's book (2003), which does for Greek geometry what Shapin and Schaffer (1985) have done for the scientific revolution (even though, with some *coquetterie*, Netz claims not to want to be the Shapin of mathematical diagrams!). What he has done is to provide the first systematic *materialist* reading of formalism—but where "matter" no longer has any of the drawbacks criticized by Whitehead.

29. The reason I prefer the notion of immutable mobiles is that it includes all the practices to maintain, through the invention of constants, the contradictory features of mobility and immutability of which those achieved by geometry and mathematics are only the most obvious ones, but there are many others: labeling, collecting, keeping up, listing, digitalizing, and so on (on this wide extension of knowledge pathways, see, for instance, Bowker, 2006).

30. No one has documented this granting of objectivity to passing minds better than Ed Hutchins (1995) when he shows how the U.S. Navy might generate provisional competences to sailors with a high turnover (1995). Objectivity is what you *gain* when you subscribe to one of the highly equipped knowledge acquisition networks. Outside of them, there is no more sense in saying that you are "objective."

31. On this, see the unwittingly hilarious movie, "The Odyssey of the Species," for which Yves Coppens was the scientific advisor: Lucy walks upright because she sees the bright future of science above the grass!

32. We now have a full historical interpretation of this highly complex metaphor from Elizabeth Eisenstein's classic (1979) to Mario Biagioli (2006) through Adrian Johns (2000).

33. Gingerich never moves out of the material connections established by the successive prints of the initial drafts written by Copernicus, from the University of Frauenburg, until some aspects of the book have been sunk into the common cosmos of astronomers through the many publications, annotations, textbooks, and popular cultures. Thus, Gingerich at last gives a realistic rendering of what it means for stars and planets to become calculations on paper without losing for one second their objective weight. Or rather, it is because they are at last calculated upon that they become objective, but only as long as the knowledge acquisition pathways are kept up. Copernicus happens to the cosmos because of this new event of being calculated upon. Naturally, as soon as we revert to the discontinuous frame of reference, stars and planets become fixed, they recede to out there, and have no history.

References

Baker, Nicholson (1988) *The Mezzanine* (New York: Weidenfeld and Nicholson).

Biagioli, Mario (2006) Galileo's Instruments of Credit: Telescopes, Images, Secrecy (Chicago: The University of Chicago Press).

Bowker, Geoffrey C. (2006) *Memory Practices in the Sciences* (Cambridge, MA: MIT Press).

Coetzee, J. M. (2005) *Slow Man* (London: Vintage Books).

Daston, Lorraine (2000) *Biographies of Scientific Objects* (Chicago: University of Chicago Press).

Debaise, Didier (2006) "Un empirisme spéculatif." Lecture de Procès et Réalité (Paris: Vrin).

Descola, Philippe (2005) *Par delà nature et culture* (Paris: Gallimard).

Dewey, John (1958) *Experience and Nature* (New York: Dover Publications).

Eisenstein, Elizabeth (1979) *The Printing Press as an Agent of Change* (Cambridge: Cambridge University Press).

Fleck, Ludwig ([1934]2005) *Genèse et développement d'un fait scientifique*, trans. Nathalie Jas, preface by Ilana Löwy, postface by Bruno Latour (Paris: Belles Lettres).

Fleck, Ludwig ([1935]1981) *Genesis and Development of a Scientific Fact* (Chicago: University of Chicago Press).

Gingerich, Owen (2004) *The Book Nobody Read: Chasing the Revolution of Nicolaus Copernicus* (New York: Penguin).

Hacking, Ian (1999) *The Social Construction of What?* (Cambridge, MA: Harvard University Press).

Hutchins, Edwin (1995) *Cognition in the Wild* (Cambridge, MA: MIT Press).

James, William ([1907]1996) *Essays in Radical Empiricism* (Lincoln and London: University of Nebraska Press).

Johns, Adrian (2000) *The Nature of the Book: Print and Knowledge in the Making* (Chicago: University of Chicago Press).

Latour, Bruno (1990) "Drawing Things Together," in M. Lynch & S. Woolgar (eds), *Representation in Scientific Practice* (Cambridge, MA: MIT Press): 19–68.

Latour, Bruno (1999) *Pandora's Hope: Essays on the Reality of Science Studies* (Cambridge, MA: Harvard University Press).

Latour, Bruno (2004a) "Why Has Critique Run Out of Steam? From Matters of Fact to Matters of Concern," Symposium on the "The Future of Criticism," *Critical Inquiry* 30: 225–48.

Latour, Bruno (2004b) *Politics of Nature: How to Bring the Sciences into Democracy*, trans. Catherine Porter (Cambridge, MA: Harvard University Press).

Latour, Bruno (2005) "What is Given in Experience? A Review of Isabelle Stengers' *Penser avec Whitehead*," *Boundary 2* 32(1): 222–37.

Latour, Bruno (2007) "Pluralité des manières d'être." *Agenda de la pensée contemporaine, Printemps* 7: 171–94.

Lynch, Michael & Ruth McNally (2005) "Chains of Custody: Visualization, Representation, and Accountability in the Processing of DNA Evidence," *Communication & Cognition* 38 (3/4): 297–318.

Lynch, Michael & Steve Woolgar (eds) (1990) *Representation in Scientific Practice* (Cambridge, MA: MIT Press).

Marcus, George E. (1999) *Paranoia within Reason: A Casebook on Conspiracy as Explanation* (Chicago: University of Chicago Press).

Mooney, Chris (2005) *The Republican War on Science* (New York: Basic Books).

Netz, Reviel (2003) *The Shaping of Deduction in Greek Mathematics: A Study in Cognitive History* (Cambridge: Cambridge University Press).

Petroski, Henry (1990) *The Pencil: A History* (London: Faber and Faber).

Pickering, Andrew (1995) *The Mangle of Practice: Time, Agency and Science* (Chicago: University of Chicago Press).

Shapin, Steven & Simon Schaffer (1985) *Leviathan and the Air Pump* (Princeton, NJ: Princeton University Press).

Souriau, Etienne (1943) *Les différents modes d'existence* (Paris: Presses Universitaires de France).

Stengers, Isabelle (2002) *Penser avec Whitehead: Une libre et sauvage création de concepts* (Paris: Gallimard).

Tarde, Gabriel (1999) *La logique sociale* (Paris: Les Empêcheurs de penser en rond).

Whitehead, Alfred North (1920) *The Concept of Nature* (Cambridge: Cambridge University Press).

5 The Social Worlds Framework: A Theory/Methods Package

Adele E. Clarke and Susan Leigh Star

In this chapter, we present the social worlds framework, a form of analysis used in a wide array of STS studies. The social worlds framework focuses on meaning-making amongst groups of actors—collectivities of various sorts—and on collective action—people "doing things together" (Becker, 1986) and working with shared objects, which in science and technology often include highly specialized tools and technologies (Clarke & Fujimura, 1992; Star & Ruhleder, 1996). Social worlds are defined as "universes of discourse," shared discursive spaces that are profoundly relational (Strauss, 1978). Over time, social worlds typically segment into multiple worlds, intersect with other worlds with which they share substantive/topical interests and commitments, and merge. If and when the number of social worlds becomes large and crisscrossed with conflicts, different sorts of careers, viewpoints, funding sources, and so on, the whole is analyzed as an *arena*. An arena, then, is composed of multiple worlds organized ecologically around issues of mutual concern and commitment to action.

This framework thus assumes multiple collective actors—social worlds—in all kinds of negotiations and conflicts, committed to usually on-going participation in broad substantive arenas. The framework is relentlessly ecological, seeking to understand the nature of relations and action across the arrays of people *and things* in the arena, representations (narrative, visual, historical, rhetorical), processes of work (including cooperation without consensus, career paths, and routines/anomalies), and many sorts of interwoven discourses. The social worlds framework is particularly attentive to situatedness and contingency, history and fluidity, and commitment and change.

We begin with a brief account of the development of the concept of social worlds in the American sociological tradition of symbolic interactionism. We then demonstrate how the social worlds framework is a "theory/method package," drawing upon an understanding of perception itself as theory-driven. Next we turn to some of the key concepts generated through using the social worlds framework. Especially in studies of scientific work practices, these include boundary objects, segments, doability, work objects, bandwagons, implicated actors/actants, and cooperation without consensus. The social worlds framework has also been especially useful in studies of controversy and of disciplinary emergence, and we review this work. Recently, the social worlds framework has become the conceptual infrastructure of situational analysis (Clarke, 2005), a new extension of the grounded theory method, with which the social worlds framework has long been associated. In conclusion, we offer a brief overview of the more methodological aspects of the theory/methods package.

SOCIAL WORLDS IN THE SYMBOLIC INTERACTIONIST TRADITION

The social worlds framework has its historical roots in Chicago School of Sociology, originally based at the University of Chicago. Confusingly, the "Chicago School" left Chicago in the late 1950s. In diaspora, however, we its descendents still refer to it as the Chicago School of Sociology (*not* to be confused with the Chicago School of Economics). Initially, the Chicago School practiced an empirical, urban sociology, studying different neighborhoods and worksites of the city. The insights of pragmatist philosophers George Herbert Mead and John Dewey were folded into these small regional studies by drawing attention to meaning-making, gestures, and identities. Groups, within which individuals were situated, were regarded as "social wholes" (Thomas, 1914), making meaning together and acting on the basis of those meanings. The meanings of phenomena thus lie in their embeddedness in relationships—in universes of discourse (Mead, [1938]1972: 518) which Strauss (1978) later called social worlds.

Early sociological ecologies of these "social wholes" focused on various kinds of communities (e.g., ethnic enclaves, elite neighborhoods, impoverished slums), distinctive locales (e.g., taxi dancehalls, the stockyards), and signal events of varying temporal duration (e.g., strikes). The sociological task was to "to make the group the focal center and to build up from its discoveries in concrete situations, a knowledge of the whole . . ." (Eubank in Meltzer et al., 1975: 42). One could begin from a place or a problem. Baszanger and Dodier (1997: 16, emphasis added) have asserted:

Compared with the anthropological tradition, the originality of the first works in the Chicago tradition was that they did not necessarily integrate the data collected around a collective whole in terms of a common culture, but in terms of territory or geographic space. The problem with which these sociologists were concerned was based on human ecology: interactions of human groups with the natural environment and interactions of human groups in a given geographic milieu . . . The *main point here was to make an inventory of a space* by studying the different communities and activities of which it is composed, that is, which encounter and confront each other in that space.

These "inventories of space" often took the form of maps (see especially Zorbaugh, 1929). The communities, organizations, and kinds of sites and collectivities represented on such maps were to be explicitly viewed *in relation to the sitings or situations of one another and within their larger contexts, featuring relationality.* "The power of the ecological model underlying the traditional Chicago approach lies in the ability to focus now on the niche and now on the ecosystem which defined it" (Dingwall, 1999: 217; see also Star, 1995a). This analytic power is retained today.

In the generation following Mead, just after World War II, several analysts combined some of the traditional focus on (1) meanings/discourse as related to ethnicity and neighborhood, and (2) the search for identity in the forms of work, practice, and memory. This synthesis resulted in a sociology that was both material *and* symbolic, interactive, processual, and structural. James Carey (2002: 202) claims that Anselm Strauss, one of the practitioners from this period, had invented "a sociology of structuration before Anthony Giddens invented the word," grasping structure as emergent in ways that later informed his social worlds theory (see also Reynolds & Herman-Kinney, 2003).

During the 1950s and 1960s, researchers in this tradition continued studies of "social wholes" in new ways, shifting to studies of work, occupations, and professions, and moving from local to national and international groups. Geographic boundaries were no longer regarded as necessarily salient, and attention shifted to *shared discourses* as both making and marking boundaries. Perhaps most significantly, researchers increasingly attended to the relationships of groups to other "social wholes," *the interactions of collective actors and discourses*. In today's methodological vernacular, many such studies would be termed "multi-sited."

At this time, several Chicago School sociologists initiated the development of explicit social worlds theory—the high-modern version of studies of "social wholes" mentioned above. Social worlds (e.g., a recreation group, an occupation, a theoretical tradition) generate shared perspectives that then form the basis for collective action while individual and collective identities are constituted through commitments to and participation in social worlds and arenas (Shibutani, 1955; Strauss, 1959). Commitment is *both* predisposition to act and a part of identity construction (Becker, 1960, 1967). Strauss (1978, 1982, 1993) and Becker (1982) defined social worlds as groups with shared commitments to certain activities, sharing resources of many kinds to achieve their goals and building shared ideologies about how to go about their business. Social worlds are *universes of discourse* (Mead, [1938]1972: 518), principal affiliative mechanisms through which people organize social life.

Until the 1980s, most symbolic interactionist research focused on social worlds centered around social problems, art, medicine, occupations, and professions (e.g., Becker, 1963, 1982; Bucher & Strauss, 1961; Bucher, 1962; Bucher & Stelling, 1977; Wiener, 1981). Since the early 1980s, as more interactionists became involved in STS, the social worlds framework has been increasingly used in interactionist research in STS. Initial work on the material bases of social worlds in life sciences (Clarke, 1987; Clarke & Fujimura, 1992) led to many fruitful avenues of inquiry about the nature of tools, nonhuman components of social worlds (Latour, 1987; Suchman, 1987), and the interaction between humans and nonhumans. This in turn encouraged the exploration of infrastructure as a deeply rooted aspect of social worlds analysis. Contemporarily, infrastructures (virtual, offline, textual, and technical) are imbricated with the unique nature of each social world and, especially as scale becomes important, with arenas (Star & Ruhleder, 1996; Neumann & Star 1996; Star, 1999). Infrastructures can be understood, in a sense, as frozen discourses that form avenues between social worlds and into arenas and larger structures.

Social worlds studies have encompassed examinations of the doing of science; the organization of scientific work; and the making, distribution, and use of technology as forms of work. We review many of these below but first address an important epistemological issue.

THE SOCIAL WORLDS/ARENAS FRAMEWORK AS A THEORY/METHODS PACKAGE

In social science, from William James at the turn of the twentieth century to the present, considerable and convincing work has been done that asserts the theory-driven and socially based nature of perception. Particularly in STS, we no longer struggle with the image of some sort of *tabula rasa* as the beginning moment of research, to be gradually filled in as we encounter the "real" world. Rather, we understand that we begin with some combination of previous scholarship, funding opportunities, materials, mentorship, theoretical traditions and their assumptions, as well as a kind of deep inertia at the level of research infrastructure (Bowker, 1994; Star & Ruhleder, 1996). Such traditions and assumptions serve as root metaphors applicable to the situation of inquiry—from social worlds to actor-network theory (e.g., Law & Hassard, 1999) to ethnomethodology (e.g., Lynch, 1985). Blumer ([1969]1993: 24–25, emphases added) discussed such metaphors as follows:

The Possession and Use of a Prior Picture or Scheme of the Empirical World Under Study . . . [T]his is an *unavoidable* prerequisite for any study of the empirical world. One can *see* the empirical world only through some scheme or image of it. The entire act of scientific study is oriented and shaped by the underlying picture of the empirical world that is used. This picture sets the selection and formulation of problems, the determination of what are data, the means to be used in getting data, the kinds of relations sought between data, and the forms in which propositions are cast.

The social worlds framework is one such "prior picture or scheme."

The social worlds framework relies strongly upon George Herbert Mead's ([1927]1964, [1934]1962) key concepts of perspective and commitment—that all actors, including social worlds as collective actors, have their own perspectives, sites of work and commitments to action vis-à-vis the substantive situation/arena. As social worlds intersect or grow to become arenas, their joint courses of commitment and (inter)action are articulated through discourses. Discourses here, then, mean these assemblages of language, motive, and meaning, moving toward mutually understood *modus vivendi*—ways of (inter)acting. Perspectives, as defined by Mead to include commitments that stem from work and material contingencies, are *discourses in collective, material action*. This concept of "discourse" and its particular history are distinct from concepts of discourse analysis stemming from European phenomenology and critical theory (e.g., Jaworski & Coupland, 1999; Weiss & Wodak, 2003; Lynch & Woolgar, 1990).

The particular power of the social worlds framework is that precisely because social worlds are "universes of discourse" the framework explicitly goes beyond "the usual sociological suspects"—conventional, highly bounded framings of collective actors such as organizations, institutions, and even social movements. These "suspects" are displaced in the social worlds framework by more open, fluidly bounded, yet discourse-based forms of collective action. Analysis must take into account more problematically bounded and contingent discursive as well as organizational arrangements (Clarke, 1991). Thus, the broader *situation* is opened up for emergent and ongoing analysis (Clarke, 2005).

Researchers using this approach in STS have worked from the assumption that the social worlds framework constitutes a *theory/methods package* itself rooted in grounded theory/symbolic interactionism. Such packages[1] include a set of epistemological and ontological assumptions, along with concrete practices through which social scientists go about their work, including relating to/with one another and with the various nonhuman entities involved in the situation. This concept of theory-methods package focuses on the integral—and ultimately nonfungible—aspects of ontology and epistemology. The concept of theory/methods package assumes that ontology and epistemology are both co-constitutive (make each other up) and manifest in actual practices.

Star (1989a) demonstrated the materiality and consequentiality of such theory/methods packages in brain research. Fujimura (1987, 1988, 1992, 1996) pushed on the modes through which theory/methods packages can travel—by being widely accepted as part of a "bandwagon" effect. Such packages often travel well in science because they perform well in situations at hand, such as creating "doable problems" for research. Bowker and Star (1999) elucidate how, through classification and standardization processes, computer and information science can dramatically facilitate such travel.

For most of us using the social worlds framework, the methods "end" of the theory/methods package has been grounded theory, an approach to analyzing largely qualitative ethnographic (observational and interview) materials. Developed by Strauss and Glaser (Glaser & Strauss, 1967; Glaser, 1978; Strauss, 1987; Strauss & Corbin, 1990), it is an abductive approach in which the analyst tacks back and forth between the empirical materials and conceptual means of expressing them. Today grounded theory is one of the major approaches used in qualitative analysis globally (Clarke, 2006a,b).

Over the past twenty years, a more Straussian version of grounded theory that is more constructivist, interactionist, and reflexive has been generated (e.g., Strauss, 1987; Charmaz, 2006). Strauss was also generating his social worlds framework at the same time. Many of us in STS routinely drew upon both of these (see also Clarke & Star, 1998), and they have recently been synthesized by Clarke (2005).

The very idea of theory/methods packages assumes that "Method, then, is not the servant of theory: method actually grounds theory" (Jenks, 1995: 12). This means, of course, that theory/methods packages are both objects of interactionist science studies research and that the social worlds framework itself is a theory/methods package.

SENSITIZING CONCEPTS IN THE SOCIAL WORLDS TOOLBOX

Over the years, a toolbox of useful concepts with which to think about the relational ecologies of social worlds, arenas, and their discourses has been generated. In our framework we treat them as what Herbert Blumer ([1969]1993: 147–48) called "sensitizing concepts" (emphases added):

[T]he concepts of our discipline are fundamentally sensitizing instruments. Hence, I call them "sensitizing concepts" and put them in contrast with definitive concepts . . . A *definitive concept* refers precisely to what is *common* to a class of objects, and by the aid of a clear definition in terms of attributes or fixed bench marks . . . A *sensitizing concept* lacks such specification . . . Instead, it gives the user a general sense of reference and guidance in approaching empirical instances. Whereas definitive concepts provide prescriptions of what to see, *sensitizing concepts merely suggest directions along which to look.*

Sensitizing concepts are thus tools for doing further analysis using this theory-methods toolkit. They are not intended as ends in themselves, but as means of analytical entrée and provisional theorizing.[2] The following are the key concepts developed to date in social worlds theory:

The Social Worlds/Arenas Framework Conceptual Toolbox for Science Studies[3]

Universes of discourse	Entrepreneurs
Situations	Mavericks
Identities	Segments/subworlds/reform movements
Commitments	Shared ideologies
Bandwagons	Primary activities
Intersections	Segmentations
Particular sites	Technology(ies)
Implicated actors and actants	Boundary objects
Work objects	Boundary infrastructures
Conventions	

We next elucidate each of the concepts listed above, illustrating them with examples from STS research. In his seminal article on social worlds and arenas, Strauss argued (1978: 122) that each social world has at least one primary activity, particular sites, and a technology (inherited or innovative means of carrying out the social world's activities) and that once under way, more formal organizations typically evolve to further one aspect or another of the world's activities.[4] People typically participate in a number of social worlds simultaneously, and such participation usually remains highly fluid. *Entrepreneurs*, deeply committed and active individuals (Becker, 1963), cluster around the core of the world and mobilize those around them.

Activities within all social worlds and arenas include establishing and maintaining perceptible *boundaries* between worlds and gaining social *legitimation* for the world itself. Indeed, the very history of the social world is commonly constructed or reconstructed in discursive processes (Strauss, 1982). Of course, individual actors compose social worlds, but in arenas they commonly act as *representatives* of their social worlds, performing their collective identities (Klapp, 1972) as well as generating their careers (Wiener, 1991). For example, in the fetal surgery operating and recovery rooms, the surgeons, neonatologists, and obstetricians have often distinct and sometimes competing agendas of concern as they may see themselves as having different *work objects*—primary patients (Casper, 1994, 1998a,b)—at the same time that they are negotiating career trajectories involving others in those very rooms.

There can also be *implicated actors* in a social world and/or arena, actors silenced or only discursively present—constructed by others for their own purposes (Clarke & Montini, 1993; Clarke, 2005: 46–48). There are at least two kinds of implicated actors. First are those who are physically present but are generally silenced/ignored/made invisible by those in power in the social world or arena (Christensen & Casper, 2000; Star & Strauss, 1999). Second are those implicated actors *not* physically present in a given social world but solely discursively constructed and discursively present; they are conceived, represented, and perhaps targeted by the work of arena participants. Much of postcolonial literature focuses precisely on this matter. Neither kind of implicated actor is actively involved in the actual negotiations of self-representation in the social world or arena, nor are their thoughts or opinions or identities explored or sought out by other actors through any openly empirical mode of inquiry (such as asking them questions).

Within information technology, computer developers have been notorious for their stereotyping and disregard of the needs of computer users, classic implicated actors. Many even called these people "lusers" (Bishop et al., 2000). Currently, this trend is somewhat offset by the use of ethnographers and other social scientists in usability laboratories of major corporations for their consumer products. However, at the more custom level of technically state-of-the-art devices, more elitist practices still prevail These include "just throw it over the wall" (and let users deal with it as best they can) and the assumption in computer modeling that "I am the World" (and no one else needs to be taken into account) (Forsythe, 2001; for studies of users and their roles, see also Oudshoorn & Pinch, chapter 22 in this volume).

There can, of course, also be *implicated actants*—implicated nonhuman actors in situations of concern.[5] Like implicated humans, implicated actants can be physically *and/or* discursively present in the situation of inquiry. That is, human actors (individually and/or collectively as social worlds) routinely discursively construct nonhuman actants from those human actors' own perspectives. The analytical question here is who is discursively constructing what, how, and why?

Every complex social world characteristically has *segments*, subdivisions or subworlds, shifting as patterns of commitment alter, reorganize, and realign. Bucher (Bucher & Strauss 1961; Bucher 1962, 1988) named such fluidity and change *within* social worlds by extending social movements analysis to frame these as reform movements of various kinds undertaken by segments or subworlds within professions, disciplines, or other work organizations. Bucher called these "professions in process." Drawing on Bucher in her study of cardiovascular epidemiology, Shim (2002, 2005) found two major segments: mainstream and social epidemiologists. The latter constitute a reform segment or movement, today informing research approaches in new areas of study including health disparities and population health.

The concept of *staged intersections*—one-shot or short-term events where multiple social worlds in a specific arena come together—is Garrety's (1998) particular conceptual contribution to social worlds theory. The key feature of staged intersections is that despite the fact that this may be a one-time-only meeting for representatives of those worlds, the events can be highly consequential for the future of all the social worlds involved, for that arena—and beyond.

Fujimura (1988, 1996), in her study of the molecularization of biology, called successful versions of such reform processes "bandwagons" when they occur on a larger scale, mobilizing the commitments of many laboratories and related organizations. This mobilization placed the package of oncogene theory (on the molecular genetic origins of cancer) and recombinant DNA and other molecular biotechnological methods at the heart of that social world. This *theory/methods package* was highly transportable, marketed as a means of constructing highly doable problems in multiple research centers; well aligned with funding, organizational, material, and other constraints upon research; and a means for attacking long-standing problems in many biological disciplines. Perhaps counterintuitively, Fujimura found no grand marshal orchestrating the bandwagon but rather a cascading series of decentralized choices, changes, exchanges, and commitments, vividly demonstrating how widely distributed a social world can be (see also Star, 1997; Strübing, 1998). The difference between a bandwagon and an arena is that a bandwagon is more narrowly focused, in a "fad-like" way, on a single package. An arena is larger, encompassing debates about packages and worlds involved over a wide range of interests, boundary objects (and potentially boundary infrastructures), and temporalities (see also Wiener, 2000).

Fujimura (1987) also introduced the useful concept of *doable problems* in scientific research. Doable problems require successful alignment across several scales of work organization. These include (1) the experiment as a set of tasks; (2) the laboratory as a bundle of experiments and other administrative and professional tasks; and (3) the wider scientific social world as the work of laboratories, colleagues, sponsors, regulators and other players all focused on the same family of problems. Doability is achieved by *articulating alignment* to meet the demands and constraints imposed at all three scales simultaneously: a problem must provide doable experiments, which are feasible within the parameters of immediate constraints and opportunities in a given laboratory, and be viewed as worthwhile and supportable work within the larger scientific world.

In many modern arenas, *reform movements* have centered around processes of homogenization, standardization, and formal classifications—things that would organize and articulate the work of the social worlds in that arena in parallel ways (Star, 1989a, 1995c). Bowker and Star (1999) analyzed how the application of computer and information science programs in nursing has standardized that work, displacing some areas of discretion with strict assessments of accountability. Clarke and Casper (1996; Casper & Clarke, 1998) studied Pap smear classification systems as attempts to impose standardization in a notoriously ambiguous clinical domain across the heterogeneous worlds involved in that arena. Timmermans, Berg, and Bowker (Berg, 1997;

Timmermans, 1999; Timmermans & Berg, 1997, 2003; Berg & Timmermans, 2000; and Berg & Bowker, 1997) discussed the application of computer-based techniques to medical practices and how these produce "universalities" in medical work. And Timmermans (2006) followed on with a study of coroners' classifications of suspicious deaths. Lampland and Star (under review) have focused on comparative standardizations across people, techniques, laws, and concepts in their edited volume *Formalizing Practices*. And Karnik (1998) explored consequences of classification in the media.

Understanding boundaries has long been important to science studies (Gieryn, 1995) and has become increasingly important in the social sciences (Lamont & Molnar, 2002). In social worlds theory, Star and Griesemer (1989) developed the concept of *boundary objects* for things that exist at junctures where varied social worlds meet in an arena of mutual concern. Boundary objects can be treaties among countries, software programs for users in different settings, even concepts themselves. Here the basic social process of translation allows boundary objects to be (re)constructed to meet the specific needs or demands placed on it by the different worlds involved (Star, 1989b). Boundary objects are often very important to many or most of the social worlds involved and hence can be sites of intense controversy and competition for the power to define them. The distinctive translations used *within* different worlds for their own purposes also enable boundary objects to facilitate cooperation without consensus.

For example, in Star and Griesemer's (1989) study of a regional zoology museum founded at the turn of the twentieth century, the museum's specimens were boundary objects. There were collections of multiple specimens of each species and subspecies which, for the zoologists to find them useful, had to be very carefully tagged as to date and where collected and carefully preserved and taxidermied. Aerial temperature, humidity, rainfall, and precise habitat information on the geographic origins of specimens all were important. The mammal and bird specimens were usually killed, gathered, and sent to the museum by amateur collectors and "mercenaries" (paid collectors) of varied backgrounds. Also involved were university administrators, a powerful patron who was herself an amateur collector, curators, research scientists, clerical staff, members of scientific clubs, and taxidermists. *All* had particular concerns about the specimens that needed to be addressed and mutually articulated for the museum's collections to "work" well for *all* involved.

Thus, the study of boundary objects can be an important pathway into complicated situations, allowing the analyst to study the different participants through their distinctive relations with and discourses about the specific boundary object in question. This can help frame the broader situation of inquiry as well. The concept of boundary objects has also been extended. For example, Henderson (1999) included visual representations as "conscription devices," weaving this understanding into the analysis of work, power, and the visual practices of engineers. Her work lends a powerful, visual-based sensibility to the boundary objects idea.

Bowker and Star (1999: 313–14) recently raised the conceptual ante with the concept of "boundary infrastructures" as larger infrastructures of classification deeply institutionalized, "sunk into the built environment, . . . objects that cross larger levels of scale than boundary objects." These are often digitalized information systems that link large-scale organizations with multiple purposes and/or constituencies. "Boundary infrastructures by and large do the work that is required to keep things moving along . . . [T]hey deal with regimes and networks of boundary objects (and not of unitary, well-defined objects)." Boundary infrastructures by no means imply universal consensus, at whatever level they may be analyzed. For any individual, perspective, or locale, they can as well produce a misfit with infrastructure, called torque by Bowker and Star (1999). (The metaphor is like the twisting of steel just a bit out of alignment.) However, the torque may not be visible or perceptible to most, often until a social movement makes it so, as with disabled people and the accessibility of structures.

Most recently, Bowker's (2005) *Memory Practices in the Sciences* examined the history of information infrastructures from paper to silicon. Using geology, cybernetics, and biodiversity as case studies, Bowker analyzed the work that their information infrastructures have done in mediating the traffic between natural and social worlds. Some facets of that trafficking are memorialized—preserved in the infrastructure as memory device—while others are reconfigured or erased. Bowker vividly shows how scientific infrastructures are projecting our modes of organization onto nature at increasingly broader scales via emergent, globally used boundary infrastructures.

In sum, then, the conceptual toolbox of social worlds theory permits analyses of a full array of collective human social entities and their actions, discourses, and related nonhuman elements in the situation of concern. The key analytical power of social worlds/arenas theory, so rooted in Chicago social ecologies, is that one can take advantage of the elasticity of the concepts to analyze at multiple levels of complexity. The utility of social worlds theory for STS was recognized not only by interactionists but also by others. For example, Becker's (1982) *Art Worlds* was taught in STS courses in the 1980s.[6]

Over the years, the social worlds/arenas framework has been compared with actor-network theory (e.g., Law & Hassard, 1999; Neyland, 2006), known as ANT, a major analytical frame in STS. While we lack space here for an extended comparison, we do want to state that we view these two approaches as kindred in many ways (especially compared with earlier approaches to the study of science) and yet also as offering quite different affordances and accomplishing different analytical ends. The social worlds framework allows for the drag of history; the cumulative consequences of commitment and action over time are deeply etched. For example, Karin Garrety (1997, 1998) compared ANT and social worlds approaches in an examination of the cholesterol, dietary fat, and heart disease controversy which has extended over four decades. She found the social worlds framework allowed analysis of changes within and across worlds over time while the scientific "facts" also remained unstable and contested.

In contrast, ANT is excellent at grasping emergent connections that may or may not gel into social worlds in arenas. Networks (that are not worlds) of many kinds may also endure and fully deserve analysis. ANT is most robust at this. Because of its common lead-scientist focus on the perspective of the most powerful, ANT has been characterized as a more "executive" vision. ANT concepts of "interessement" and "obligatory points of passage" are often framed as one-way streets. In contrast, social worlds theory is insistently pluralist, seeking to analyze all the perspectives in the situation. But these approaches can also vary in the hands of different researchers and when the approaches are used for particular purposes.

In comparing Latourian ANT with social worlds/arenas, it has been said that the centralized nature of power in ANT is more French, whereas for social worlds, the pluralism of perspective is vividly inflected American. Bowker and Latour (1987) made a somewhat similar argument in their paper comparing French and Anglo-American science and technology studies. They argued that the rationality/power axis so natural to French technocracy (and explored in Foucault inter alia) is precisely that which must be proven in Anglo-American work, since, in the Anglo American context, it is usually "assumed" that there is no relationship between the two.[7]

SOCIAL WORLDS STUDIES OF CONTROVERSIES AND DISCIPLINES

Key sociological differences emerge when researchers focus on studying the work activities, organization, and discourses of social worlds in science, technology, and/or (bio)medicine rather than studying individuals. Placing work—action—in the foreground facilitates the analysis of social worlds *qua* worlds and the elucidation of the key human and nonhuman elements. For Strauss (1978), Becker (1982), and some others working with the social worlds framework (e.g., Star, Fujimura, Baszanger, Clarke, Garrety, Casper, Shim, Shostak, and others), the social worlds and arenas themselves became the units of analysis in two main genres of studies of collective discourse and action—scientific controversies and disciplines (including boundary objects and infrastructures). Here we often see the phenomenon identified by interactionists as cooperation without consensus writ large.

Clarke and Montini (1993) provide an accessible example of controversy studies by focusing on the multiple social worlds involved in the controversy surrounding use of the abortifacient RU486, also known as "the French abortion pill" in the U.S. The paper analytically places RU486 in the center and then moves through the specific perspectives on it of each of the major social worlds involved in the broader abortion/reproduction arena: reproductive and other scientists, birth control/population control organizations, pharmaceutical companies, medical groups, anti-abortion groups, pro-choice groups, women's health movement groups, politicians, Congress, the FDA, and last but not least, women users/consumers of RU486. Here Clarke initiated the concept of *implicated actors* discussed earlier. As is often the case, these were the users/consumers. Clarke and Montini showed that social worlds themselves are not at all monolithic but commonly contain extensive differences of perspective that may be more or less contentious. Moreover, contra Mol, they demonstrated how, given these different perspectives, RU486 is vividly different things to different social worlds in the arena.[8]

In their study of the controversy about whether hormone disruption is caused by exposure to synthetic chemicals in the environment, Christensen and Casper (2000) used a social worlds/arenas analytic to map the discourse in key documents (see also

Albrechtsen & Jacob, 1998). They focused on two sets of implicated actors particularly vulnerable to exposures but excluded from the possibility of scientific claims-making: farm workers and fetuses. Their focus allowed them to analyze hierarchies of knowledges and to develop policy implications of the possible future inclusion of heretofore silenced but clearly implicated actors.

A number of social worlds/arenas studies take up disciplinary and specialty emergence and competition. One of the earliest was Star's (1989a) examination of the work of late nineteenth century British neurophysiologists that investigated the contest in brain research between "localizationists" who sought to map specific regions and functions and "diffusionists" who argued for an interactive, flexible, and resilient brain model. The scientists who supported localizationist theories of brain function built a successful research program (read as social world here) through several strategies: by gaining control of relevant journals, hospital practices, teaching posts, and other means of knowledge production and distribution; by screening out those who held opposing points of view from print and employment; by linking a successful clinical program with both basic research and a theoretical model; and by uniting against common enemies with powerful scientists from other fields. Star examined the production of robust scientific knowledge through concrete practices and collective rhetorical strategies.

Clarke (1998) studied the emergence and coalescence of the American reproductive sciences across the twentieth century as an intersectional discipline dwelling in three professional domains: biology, medicine, and agriculture. She situates the emergence of this scientific social world within the larger sociocultural reproductive arena that included other key worlds including birth control, population control, and eugenics movements and strong philanthropic sponsors. Reproductive scientists coped strategically with the illegitimacy of this sexuality-laden and therefore suspect research in their negotiations with various audiences. Clarke (2000) further detailed how it was only *maverick* reproductive scientists who actually worked on contraceptive development and did so only outside university settings, largely in private research institutes supported by major philanthropists and/or pharmaceutical companies. The exclusion of women as patients and users/consumers from participation at design stages has constituted millions of women as implicated rather than agentic actors in the contraceptive arena for almost a century. This has contributed to the ongoing spread of sexually transmitted diseases, including AIDS. Such problematics of agency and choice are commonly linked to gender and race in STS.

Sara Shostak (2003, 2005) applied social worlds/arenas theory in an historical sociological analysis of the disciplinary emergence of environmental genetics as an intersectional project. She explored the changing relationships of the social worlds and segments of pharmacogenetics, molecular epidemiology, genetic epidemiology, ecogenetics, and toxicology from 1950 to 2000. Her analysis centers on the reconfiguration of these worlds and their relationships to each other in scientific and public health/policy arenas that are increasingly shaped by desires of environmental health risk assessment and "regulatory science" practitioners for reliable scientific information about gene-environment interaction. Shostak further explored the construction and consequences of new technologies (e.g., molecular biomarkers and toxicogenomics) within these worlds and their appropriation and transformation by other social worlds, including activist movements, especially in local struggles about the health effects of environmental exposures.[9]

A number of studies of disciplines and specialties using the social worlds framework emphasized a key interactionist assumption that cooperation can proceed *without* consensus, that individuals and collectivities can "set their differences aside," however temporarily and contingently, in the interests of individual or shared goals. For example, Baszanger (1998) examined the emergence of organized pain medicine as produced through the intersection of segments of multiple specialties in an international arena. Demonstrating the capacity for ongoing *disunity* within a functioning specialty (cooperation without consensus), Baszanger offered ethnographic case studies of two paradigmatically different pain clinics in France: one emphasizing analgesia and the other focusing on patient self-management through self-surveillance and particular self-disciplining practices. The segmental scientific history and theory of pain medicine were thus inscribed in clinical practices.[10]

The emergence of fetal surgery, a more rarefied specialty, was studied by Casper (1998a,b). In these still largely experimental practices, clinicians partially remove a fetus from a woman's uterus, operate for a variety of structural problems, and if it survives, replace it for continued gestation. Fetal surgery has been controversial since its inception in 1960s New Zealand and Puerto Rico using sheep and chimpanzees as animal models. Like Baszanger and Star, Casper provided detailed histories of both the laboratory science and clinical practices. Like Clarke, she found that links to other social worlds, specifically anti-abortion movements, were characteristic of and important to key actors in the social world of this emergent specialty. Building on Mead's concept of social objects, Casper (see also 1994, 1998b) developed the concept of *work objects* to describe and analyze the tangible and symbolic objects around and with which social actors work. She analyzed the relations (sometimes cooperative, sometimes vituperative, rarely if ever based on consensus) among the different practitioners involved in fetal surgery who struggle over who is the patient—mother or fetus—and who should have jurisdiction over which patient in the surgical situation and beyond.

In the field of information technology, changes in engineering design and manufacturing teams and the consequences these may have for prototypes as boundary objects were the focus of a project by Subrahmanian and colleagues (2003). Changes in the teams, they found, disrupted the *modus vivendi* that the various groups had established for cooperation (without consensus) and (re)opened debates about boundary objects per se. Gal and colleagues (2004) followed on with a fascinating study of AEC, the architecture, engineering and construction industry, focusing on how changing information technologies, which are themselves the boundary objects operating between the architectural, engineering and construction worlds, produce changes not only in the relationships among these worlds but also in the identities of those worlds.

That is, boundary objects are used not only as translation devices but also as resources for the formation and expression of professional identities. Using the example of the introduction of three-dimensional modeling technologies into building design by architect Frank Gehry, the technology that afforded the possibility of using materials in innovative ways for which he is now famous, Gal and colleagues argued that changes in one world may cascade to other worlds through shared boundary objects (see also Star, 1993, 1995b; Carlile, 2002; Walenstein, 2003). Cooperation without consensus was very much the order of the day.

Another recent social worlds study found both cooperation and consensus problematic. Tuunainen (2005) examined "disciplinary worlds colliding" in Finland when a university agronomy department focused on plant production research was pressured by the government to incorporate new modes of doing science (including molecular biology, plant physiology, horticulture, and agroecology) and to establish relations with industry. Tuunainen found the disunity of plant production research readily observable as the scientists did not create "new hybrid worlds of different disciplines" (2005: 224) but instead retained their commitments both to their disciplines of origin and to their historical organizational niches in the university.

In her study of the making of meteorology, Sundberg (2005) focuses on intersections where modeling practice meets experimentation. New and necessary components of simulation models became boundary objects shaping relations between the disciplinary segments of experimentalists and modelers. In the same vein, Halfon's (2006) analysis of the regime change from "population control" to "women's empowerment" enacted as the Cairo consensus foregrounds the scientization of *both* population policy and social movement worlds through the institutionalization of shared technical language and practices. Making and talking about demographic surveys—using the science as shared work object—offered "neutral" sites in and through which the requisite serious negotiations could and did flourish. He reveals the too often invisible work of making change in a complex world.

Last, Strübing (1998) has written on cooperation without consensus in a study of computer scientists and symbolic interactionist sociologists collaborating over a period of years, an intersection that has never been fully stabilized. A segment of the computing world focused on Distributed Artificial Intelligence (DAI) was interested in modeling and supporting spatially and temporally distributed work and decision practices, often in applied settings. The "distributed" in DAI means modeling problem-solving across space and time, conducted by many entities that in some senses had to cooperate. For example, a typical problem would be how to get computers at several locations, with different kinds of data, to return the answer to a problem, using each of their local data sets. This problem both reflected and bridged to interactionist concerns with translation issues, complex intersections, and the division of labor in large scientific projects. Strübing concluded that the sustained collaboration involved not just "the migration of metaphors" but also the mutual creation and maintenance of organizational structures for shared work—what Star (1991a) might call "invisible infrastructures."

The concepts of boundary objects, boundary infrastructures, and conscription devices are now canonically useful, central to understanding the intersections of social worlds in social worlds/arenas theory in STS and beyond. Discipline-focused studies utilizing these concepts have examined library science (Albrechtsen & Jacob, 1998), genetics, geography, and artificial intelligence. Fujimura and Fortun (1996; Fujimura, 1999, 2000) have studied the construction of DNA sequence databases in molecular biology as internationally utilized boundary infrastructures. Such databases pose fascinating challenges because they must be both constructed across multiple social worlds and serve the needs of multiple worlds.

In geography, Harvey and Chrisman (1998) examined boundary objects in the social construction of geographical information system (GIS) technology. GIS, a major innovation, requires complex relationships between technology and people because it is used not only as a tool but also as a means of connecting different social groups in the construction of new localized social arrangements. Harvey and Chrisman view boundary objects as much like geographic boundaries, separating different social groups yet at the same time delineating important points of reference between them, and stabilizing relationships through the negotiation of flexible and dynamic coherences. Such negotiations are fundamental to the construction of GIS technology, as Harvey and Chrisman illustrate in a study of the use of GIS data standards in the definition of wetlands.

In public health, Frost and colleagues (2002) used the boundary objects framework in a study of a public-private partnership project. The project brought together Big Pharma (Merck) and an international health organization (the Task Force for Child Survival and Development) to organize the donation by Merck of a drug for the treatment of river blindness endemic in 35 countries. Frost and colleagues asked how such divergent organizations could cooperate. They argued that the different meanings of key boundary objects held by the participating groups allowed them both to collaborate without having to come to consensus and to maintain their sharply different organizational missions. The main benefit was that the project itself as boundary object provided legitimacy to *all* participants *and* to the partnership per se. The Mectizan Donation Program has become a model for similar partnerships.

In sum, social worlds theory and especially the concept of boundary objects have traveled widely and been taken up since the 1980s by researchers from an array of disciplines that contribute to STS.

A NEW SOCIAL WORLDS THEORY/METHODS PACKAGE: SITUATIONAL ANALYSIS

[M]ethodology embraces the entire scientific quest and not merely some selected portion or aspect of that quest. (Blumer, [1969]1993: 24)

As noted earlier, the methods end of the social worlds theory/methods package has heretofore largely been held down by Straussian versions of the grounded theory method of data analysis (Charmaz, 2006; Clarke, 2006a; Star, 1998), including feminist versions (Clarke, 2006b). Toward the end of his career, Strauss worked assiduously on framing and articulating ways to do grounded theory analysis that included *specifying structural conditions*—literally making them visible in the analysis—along with the analysis of forms of action that traditionally centers grounded theory. To this end, Strauss (Strauss & Corbin, 1990:163) produced what he called the conditional matrix to more fully capture the specific conditions under which the action occurs. Clarke (2003; 2005) developed a sustained critique of this matrix. To accomplish similar goals she instead took Strauss's social worlds framework and used it as theoretical infrastructure for a new extension of grounded theory. Fusing it with C. Wright Mills's (1940), Donna Haraway's (1991), and others' conceptions of situated action, and with analytic concepts of discourse from Foucault and visual cultural studies, she forged an approach called "situational analysis."

In situational analysis, *the conditions of the situation are in the situation*. There is no such thing as "context." The conditional elements of the situation need to be specified in the analysis of the situation itself as *they are constitutive of it*, not merely surrounding it or framing it or contributing to it. They *are* it. Ultimately, what structures and conditions any situation is an empirical question—or set of analytic questions. Situational analysis then involves the researcher in the making of three kinds of maps to respond to those empirical questions analytically:

1. *Situational maps* that lay out the major human, nonhuman, discursive and other elements in the research situation of inquiry and provoke analysis of relations among them

2. *Social worlds/arenas maps* that lay out the collective actors, key nonhuman elements, and the arena(s) of commitment and discourse within which they are engaged in ongoing negotiations—mesolevel interpretations of the situation

3. *Positional maps* that lay out the major positions taken, and *not* taken, in the data vis-à-vis particular axes of difference, concern, and controversy around issues in discourses in the situation of inquiry.

All three kinds of maps are intended as analytic exercises, fresh ways into social science data. They are especially well suited to designing and conducting contemporary science and technology studies ranging from solely interview-based research to multisited ethnographic projects. Doing situational maps can be especially useful for ongoing reflexive research design and implementation across the life of the project. They allow researchers to track all of the elements in the situation and to analyze their relationality. All the maps can, of course, be done for different historical moments, allowing comparisons.

Through mapping the data, the analyst constructs the situation of inquiry empirically. The *situation* per se *becomes the ultimate unit of analysis*, and understanding its elements and their relations is the primary goal. By extending grounded theory to the study of discourses, situational analysis takes it around the postmodern turn. Historical, visual, and narrative discourses may each and all be included in research designs and in the three kinds of analytic maps. Drawing deeply on Foucault, situational analysis understands discourses as elements in the situation of inquiry. Discursive and ethnographic/interview data can be analyzed together or comparatively. The positional maps elucidate positions taken in discourses and innovatively allow researchers to specify positions *not* taken, allowing discursive silences to speak (Clarke, in prep.).

These innovations may be central to some of the next generation of interactionist STS studies. For example, Jennifer Fosket (forthcoming) used these mapping strategies to analyze the situatedness of knowledge production in a large-scale, multi-sited clinical trial of chemoprevention drugs. The trial *qua* arena involved multiple and quite heterogeneous social worlds: pharmaceutical companies, social movements, scientific specialties, and the FDA. The trial needed to manage not only millions of human and nonhuman objects but also credibility and legitimacy across diverse settings and in the face of conflicting demands. Mapping the arena allowed Fosket to specify the nature of relations among worlds and relations with key elements in the situation, such as tissue samples. Situational analysis is thus one example of building on the tradition of social worlds/arenas as a theory/methods package with grounded theory to produce a novel mode of analysis.

CONCLUSIONS

Since the 1980s, the social worlds framework has become mainstream in STS (Clarke & Star, 2003). Of particular note for us is the link to earlier interactionist studies of work that began from the premise that science is "just another kind of work," not special and different, and that it is about not only ideas but also materialities (see Mukerji, 1989). The social worlds framework thus seeks to examine *all* the human and nonhuman actors and elements contained in a situation from the perspectives of each. It seeks to analyze the various kinds of work involved in creating and utilizing sciences, technologies and medicines, elucidating multiple levels of group meaning-making and material involvements, commitments, and practices.

In sum, the social worlds framework as a theory/methods package enhances analytic capacities to conduct incisive studies of differences of perspective, of highly complex situations of action and position, and of the heterogeneous discourses increasingly characteristic of contemporary technosciences. The concepts of boundary objects and boundary infrastructures offer analytic entrée into sites of intersection of social worlds and to the negotiations and other work occurring there. The concepts of implicated actors and actants can be particularly useful in the explicit analysis of power. Such analyses are both complicated and enhanced by the fact that there are generally *multiple* discursive constructions of both the human and nonhuman actors circulating in any given situation. Situational analysis offers methodological means of grasping such multiplicities. The social worlds framework as a theory/methods package can thus be useful in pragmatic empirical science, technology, and medicine projects.

Notes

We are most grateful to Olga Amsterdamska, Mike Lynch, Ed Hackett, Judy Wajcman, and the ambitious anonymous reviewers for their patience and exceptionally thoughtful and helpful comments. We would also like to thank Geof Bowker, Sampsa Hyysalo, and Allan Regenstreif for generous comments and support.

1. We use the term package to indicate and emphasize the advantages of using the elements of the social worlds framework together with symbolic interactionist-inflected grounded theory. They "fit" one another in terms of both ontology and epistemology. See Star (1989a; 1991a,b; 1999) and Clarke (1991, 2005:2–5, 2006a). We do *not* mean that one can opt for two items from column A and two from column B to tailor a package, nor do we mean that one element automatically "comes with" the other as a prefabricated package. Using a "package" takes all the work involved in learning the practices and how to articulate them across time and circumstance.

2. Contra Glaser and Strauss (Glaser & Strauss, 1967; Glaser, 1978; Strauss, 1995), we do not advocate the generation of formal theory. See also Clarke (2005: 28–29).

3. On universes of discourse, see, for example, Mead (1917), Shibutani (1955); and Strauss (1978). On situations, see Clarke (2005). On identities and shared ideologies, see, for example, Strauss (1959, 1993; Bucher & Stelling, 1977). On commitments, entrepreneurs and mavericks, see Becker (1960, 1963, 1982, 1986). On primary activities, sites, and technology (ies), see Strauss (1978) and Strauss et al. (1985). On subworlds/segments and reform movements, see Bucher (1962; Bucher & Strauss, 1961) and Clarke and Montini (1993). On bandwagons and doability, see Fujimura (1987, 1988, 1992, 1996). On intersections and segmentations, see Strauss (1984). On implicated actors and actants, see Clarke and Montini (1993), Clarke (2005), Christensen and Casper (2000), and Star and Strauss (1999). On boundary objects and infrastructures, see Star and Griesemer (1989) and Bowker and Star (1999). On work objects, see Casper (1994, 1998b). On conventions, see Becker (1982) and Star (1991b). On social worlds theory more generally, see Clarke (2006c).

4. Boundaries of social worlds may cross-cut or be more or less contiguous with those of formal organizations, distinguishing social worlds/arenas theory from most organizations theory (Strauss 1982, 1993; Clarke 1991, 2005).

5. The term actant is used thanks to Latour (1987). Keating and Cambrosio (2003) have critiqued the "social worlds" perspective for minimizing the significance of the nonhuman—tools, techniques, and research materials. This is rather bizarre, since we were among the earliest in STS to write on these topics. See Clarke (1987), Star (1989a), and Clarke and Fujimura (1992), and for a broader review, Clarke and Star (2003).

6. Warwick Anderson taught Becker's book in an STS course at Harvard (personal communication, 2005).

7. Special thanks to Geof Bowker (personal communication, 7/03). See also Star (1991a,b, 1995c), Fujimura (1991), Clarke and Montini (1993), and Clarke (2005: 60–63).

8. Mol (Mol & Messman, 1996; Mol, 2002) has erroneously insisted that the interactionist concept of perspective "means" that the "same" thing is merely "viewed" differently across perspectives. On the contrary, we assert that many different "things" are actually perceived according to perspective. Moreover, actions are taken based on those perceptions of things as different. We suspect that Mol has not adequately grasped the interactionist assumption that there can be "cooperation without consensus" illustrated several times in this section, nor that perspective, from an interactionist stance, is not a cognitive-ideal concept.

9. Ganchoff (2004) examines social worlds and the growing arena of stem cell research and politics.

10. Baszanger's study goes beyond most others in the social worlds/arenas tradition by also studying patients' perceptions of and perspectives on pain medicine. Pain itself has simultaneously become a stand-alone disease label *and* an arena at the international level.

References

Albrechtsen, H. & E. K. Jacob (1998) "The Dynamics of Classification Systems as Boundary Objects for Cooperation in the Electronic Library," *Library Trends*, 47 (2): 293–312.

Baszanger, Isabelle (1998) *Inventing Pain Medicine: From the Laboratory to the Clinic* (New Brunswick, NJ: Rutgers University Press).

Baszanger, Isabelle & Nicolas Dodier (1997) "Ethnography: Relating the Part to the Whole," in David Silverman (ed) *Qualitative Research: Theory, Method, and Practice* (London: Sage): 8–23.

Becker, Howard S. (1960) "Notes on the Concept of Commitment," *American Journal of Sociology* 66 (July): 32–40.

Becker, Howard S. (1963) *Outsiders: Studies in the Sociology of Deviance* (New York: Free Press).

Becker, Howard S. ([1967]1970) "Whose Side Are We On?" reprinted in *Sociological Work: Method and Substance* (Chicago: Aldine): 123–34.

Becker, Howard S. (1982) *Art Worlds* (Berkeley: University of California Press).

Becker, Howard S. (1986) *Doing Things Together* (Evanston, IL: Northwestern University Press).

Berg, Marc (1997) *Rationalizing Medical Work: Decision-Support Techniques and Medical Practices* (Cambridge, MA: MIT Press).

Berg, Marc & Geof Bowker (1997) "The Multiple Bodies of the Medical Record: Toward a Sociology of an Artifact," *Sociological Quarterly* 38: 513–37.

Berg, Marc & Stefan Timmermans (2000) "Orders and Their Others: On the Constitution of Universalities in Medical Work," *Configurations* 8 (1): 31–61.

Bishop, Ann, Laura Neumann, Susan Leigh Star, Cecelia Merkel, Emily Ignacio, & Robert Sandusky (2000) "Digital Libraries: Situating Use in Changing Information Infrastructure," *Journal of the American Society for Information Science* 51 (4): 394–413.

Blumer, Herbert ([1969]1993) *Symbolic Interactionism: Perspective and Method* (Englewood Cliffs, NJ: Prentice-Hall, 1969; Berkeley: University of California Press, 1993).

Bowker, Geoffrey C. (1994) "Information Mythology and Infrastructure," in L. Bud-Frierman (ed), *Information Acumen: The Understanding and Use of Knowledge in Modern Business* (London: Routledge): 231–47.

Bowker, Geoffrey C. (2005) *Memory Practices in the Sciences* (Cambridge, MA: MIT Press).

Bowker, Geoffrey C. & Bruno Latour (1987) "A Booming Discipline Short of Discipline: (Social) Studies of Science in France," *Social Studies of Science* 17: 715–48.

Bowker, Geoffrey C. & Susan Leigh Star (1999) *Sorting Things Out: Classification and Its Consequences* (Cambridge, MA: The MIT Press).

Bucher, Rue (1962) "Pathology: A Study of Social Movements Within a Profession," *Social Problems* 10: 40–51.

Bucher, Rue (1988) "On the Natural History of Health Care Occupations," *Work and Occupations* 15 (2): 131–47.

Bucher, Rue & Joan Stelling (1977) *Becoming Professional* [preface by Eliot Freidson] (Beverly Hills, CA: Sage).

Bucher, Rue & Anselm L. Strauss (1961) "Professions in Process," *American Journal of Sociology* 66: 325–34.

Carey, James W. (2002) "Cultural Studies and Symbolic Interactionism: Notes in Critique and Tribute to Norman Denzin," *Studies in Symbolic Interaction* 25: 199–209.

Carlile, Paul (2002) "A Pragmatic View of Knowledge and Boundaries: Boundary Objects in New Product Development," *Organization Science* 13: 442–55.

Casper, Monica J. (1994) "Reframing and Grounding Nonhuman Agency: What Makes a Fetus an Agent?" *American Behavioral Scientist* 37 (6): 839–56.

Casper, Monica J. (1998a) *The Making of the Unborn Patient: A Social Anatomy of Fetal Surgery* (New Brunswick, NJ: Rutgers University Press).

Casper, Monica J. (1998b) "Negotiations, Work Objects, and the Unborn Patient: The Interactional Scaffolding of Fetal Surgery," *Symbolic Interaction* 21 (4): 379–400.

Casper, Monica J. & Adele E. Clarke (1998) "Making the Pap Smear into the 'Right Tool' for the Job: Cervical Cancer Screening in the USA, circa 1940–95," *Social Studies of Science* 28 (2): 255–90.

Charmaz, Kathy (2006) *Constructing Grounded Theory: A Practical Guide Through Qualitative Analysis* (London: Sage).

Christensen, Vivian & Monica J. Casper (2000) "Hormone Mimics and Disrupted Bodies: A Social Worlds Analysis of a Scientific Controversy," *Sociological Perspectives* 43 (4): S93–S120.

Clarke, Adele E. (1987) "Research Materials and Reproductive Science in the United States, 1910–1940," in Gerald L. Geison (ed), *Physiology in the American Context, 1850–1940* (Bethesda, MD: American Physiological Society): 323–50. Reprinted (1995) with new epilogue in S. Leigh Star (ed), *Ecologies of Knowledge: New Directions in Sociology of Science and Technology* (Albany: State University of New York Press): 183–219.

Clarke, Adele E. (1991) "Social Worlds/Arenas Theory as Organization Theory," in David Maines (ed) *Social Organization and Social Process: Essays in Honor of Anselm Strauss* (Hawthorne, NY: Aldine de Gruyter): 119–58.

Clarke, Adele E. (1998) *Disciplining Reproduction: Modernity, American Life Sciences, and "the Problems of Sex"* (Berkeley: University of California Press).

Clarke, Adele E. (2000) "Maverick Reproductive Scientists and the Production of Contraceptives, 1915–2000+," in Anne Saetnan, Nelly Oudshoorn, & Marta Kirejczyk (eds), *Bodies of Technology: Women's Involvement with Reproductive Medicine* (Columbus: Ohio State University Press): 37–89.

Clarke, Adele E. (2003) "Situational Analyses: Grounded Theory Mapping After the Postmodern Turn," *Symbolic Interaction* 26 (4): 553–76.

Clarke, Adele E. (2005) *Situational Analysis: Grounded Theory After the Postmodern Turn* (Thousand Oaks, CA: Sage).

Clarke, Adele E. (2006a) "Grounded Theory: Critiques, Debates, and Situational Analysis" in William Outhwaite & Stephen P. Turner (eds), *Handbook of Social Science Methodology* (Thousand Oaks, CA: Sage), in press.

Clarke, Adele E. (2006b) "Feminisms, Grounded Theory and Situational Analysis," in Sharlene Hesse-Biber (ed) *The Handbook of Feminist Research: Theory and Praxis* (Thousand Oaks, CA: Sage), in press.

Clarke, Adele E. (2006c) "Social Worlds," in *The Blackwell Encyclopedia of Sociology* (Malden MA: Blackwell), 4547–49.

Clarke, Adele E. (in prep.) "Helping Silences Speak: The Use of Positional Maps in Situational Analysis."

Clarke, Adele E. & Monica J. Casper (1996) "From Simple Technology to Complex Arena: Classification of Pap Smears, 1917–90," *Medical Anthropology Quarterly* 10 (4): 601–23.

Clarke, Adele E. & Joan Fujimura (1992) "Introduction: What Tools? Which Jobs? Why Right?" in A. E. Clarke & J. Fujimura (eds), *The Right Tools for the Job: At Work in Twentieth Century Life Sciences* (Princeton, NJ: Princeton University Press): 3–44. French translation: *La Materialite des Sciences: Savoir-faire et Instruments dans les Sciences de la Vie* (Paris: Synthelabo Groupe, 1996).

Clarke, Adele E. & Theresa Montini (1993) "The Many Faces of RU486: Tales of Situated Knowledges and Technological Contestations," *Science, Technology & Human Values* 18 (1): 42–78.

Clarke, Adele E. & Susan Leigh Star (1998) "On Coming Home and Intellectual Generosity" (Introduction to Special Issue: New Work in the Tradition of Anselm L. Strauss), *Symbolic Interaction* 21 (4): 341–49.

Clarke, Adele E. & Susan Leigh Star (2003) "Science, Technology, and Medicine Studies," in Larry Reynolds & Nancy Herman-Kinney (eds), *Handbook of Symbolic Interactionism* (Walnut Creek, CA: Alta Mira Press): 539–74.

Dingwall, Robert (1999) "On the Nonnegotiable in Sociological Life," in Barry Glassner & R. Hertz (eds), *Qualitative Sociology as Everyday Life* (Thousand Oaks, CA: Sage): 215–25.

Forsythe, Diana E. (2001) *Studying Those Who Study Us: An Anthropologist in the World of Artificial Intelligence* (Stanford, CA: Stanford University Press).

Fosket, Jennifer Ruth (forthcoming) "Situating Knowledge Production: The Social Worlds and Arenas of a Clinical Trial." *Qualitative Inquiry*.

Frost, Laura, Michael R. Reich & Tomoko Fujisaki (2002) "A Partnership for Ivermectin: Social Worlds and Boundary Objects," in Michael R. Reich (ed), *Public-Private Partnerships for Public Health*, Harvard Series on Population and International Health (Cambridge, MA: Harvard University Press): 87–113.

Fujimura, Joan H. (1987) "Constructing 'Do-able' Problems in Cancer Research: Articulating Alignment," *Social Studies of Science* 17: 257–93.

Fujimura, Joan H. (1988) "The Molecular Biological Bandwagon in Cancer Research: Where Social Worlds Meet," *Social Problems* 35: 261–83. Reprinted in Anselm Strauss & Juliet Corbin (eds) (1997), *Grounded Theory in Practice* (Thousand Oaks, CA: Sage): 95–130.

Fujimura, Joan H. (1991) "On Methods, Ontologies and Representation in the Sociology of Science: Where Do We Stand?" in David Maines (ed), *Social Organization and Social Process: Essays in Honor of Anselm Strauss* (Hawthorne, NY: Aldine de Gruyter): 207–48.

Fujimura, Joan H. (1992) "Crafting Science: Standardized Packages, Boundary Objects, and 'Translation'," in Andrew Pickering (ed) *Science as Practice and Culture* (Chicago: University of Chicago Press): 168–211.

Fujimura, Joan H. (1996) *Crafting Science: A Socio-History of the Quest for the Genetics of Cancer* (Cambridge, MA: Harvard University Press).

Fujimura, Joan H. (1999) "The Practices and Politics of Producing Meaning in the Human Genome Project," *Sociology of Science Yearbook* 21 (1): 49–87.

Fujimura, Joan H. (2000) "Transnational Genomics in Japan: Transgressing the Boundary Between the 'Modern/West' and the 'Pre-Modern/East'," in Roddey Reid & Sharon Traweek (eds), *Cultural Studies of Science, Technology, and Medicine* (New York and London: Routledge): 71–92.

Fujimura, Joan H. & Michael A. Fortun (1996) "Constructing Knowledge Across Social Worlds: The Case of DNA Sequence Databases in Molecular Biology," in Laura Nader (ed), *Naked Science: Anthropological Inquiry into Boundaries, Power, and Knowledge* (New York: Routledge): 160–73.

Gal, U., Y. Yoo, & R. J. Boland (2004) "The Dynamics of Boundary Objects, Social Infrastructures and Social Identities," Sprouts: Working Papers on Information Environments, Systems and Organizations 4 (4): 193–206, Article 11. Available at: http://weatherhead.cwru.edu/sprouts/2004/040411.pdf.

Ganchoff, Chris (2004) "Regenerating Movements: Embryonic Stem Cells and the Politics of Potentiality," *Sociology of Health and Illness* 26 (6): 757–74.

Garrety, Karin (1997) "Social Worlds, Actor-Networks and Controversy: The Case of Cholesterol, Dietary Fat and Heart Disease," *Social Studies of Science* 27 (5): 727–73.

Garrety, Karin (1998) "Science, Policy, and Controversy in the Cholesterol Arena," *Symbolic Interaction* 21 (4): 401–24.

Gieryn, Thomas (1995) "Boundaries of Science," in Sheila Jasanoff, G. Markle, J. Petersen, & T. Pinch (eds), *Handbook of Science and Technology Studies* (Thousand Oaks, CA: Sage): 393–443.

Glaser, Barney G. (1978) *Theoretical Sensitivity: Advances in the Methodology of Grounded Theory* (Mill Valley, CA: Sociology Press).

Glaser, Barney G. & Anselm L. Strauss (1967) *The Discovery of Grounded Theory: Strategies for Qualitative Research* (Chicago: Aldine; London: Weidenfeld and Nicolson).

Halfon, Saul (2006) *The Cairo Consensus: Demographic Surveys, Women's Empowerment, and Regime Change in Population Policy* (Lanham, MD: Lexington Books).

Haraway, Donna (1991) "Situated Knowledges: The Science Question in Feminism and the Privilege of Partial Perspective," in D. Haraway (ed), *Simians, Cyborgs, and Women: The Reinvention of Nature* (New York: Routledge): 183–202.

Harvey, Francis & Nick R. Chrisman (1998) "Boundary Objects and the Social Construction of GIS Technology," *Environment and Planning A* 30 (9): 1683–94.

Henderson, Kathryn (1999) *On Line and on Paper: Visual Representations, Visual Culture, and Computer Graphics in Design Engineering* (Cambridge, MA: MIT Press).

Jaworski, A. & N. Coupland (eds) (1999) *The Discourse Reader* (Routledge: London).

Jenks, Chris (1995) "The Centrality of the Eye in Western Culture: An Introduction," in C. Jenks (ed), *Visual Culture* (London and New York: Routledge): 1–25.

Karnik, Niranjan (1998) "Rwanda and the Media: Imagery, War and Refuge," *Review of African Political Economy* 25 (78): 611–23.

Keating, Peter & Alberto Cambrosio (2003) *Biomedical Platforms: Realigning the Normal and the Pathological in Late-Twentieth-Century Medicine* (Cambridge, MA: MIT Press).

Klapp, Orrin (1972) *Heroes, Villains and Fools: Reflections of the American Character* (San Diego, CA: Aegis).

Lamont, Michele & Virag Molnar (2002) "The Study of Boundaries in the Social Sciences," *Annual Review of Sociology* 28: 167–95.

Lampland, Martha & Susan Leigh Star (eds) (forthcoming) *Standards and Their Stories* (Ithaca, NY: Cornell University Press).

Latour, Bruno (1987) *Science in Action* (Cambridge, MA: Harvard University Press).

Law, John & John Hassard (eds) (1999) *Actor Network Theory and After* (Malden, MA: Blackwell).

Lynch, Michael (1985) *Art and Artifact in Laboratory Science: A Study of Shop Work and Shop Talk in a Research Laboratory* (London: Routledge & Kegan Paul).

Lynch, Michael & Steve Woolgar (eds) (1990) *Representation in Scientific Practice* (Cambridge, MA: MIT Press).

Mead, George Herbert (1917) "Scientific Method and the Individual Thinker," in John Dewey (ed), *Creative Intelligence: Essays in the Pragmatic Attitude* (New York: Henry Holt).

Mead, George Herbert ([1927]1964) "The Objective Reality of Perspectives," in A.J. Reck (ed), *Selected Writings of George Herbert Mead* (Chicago: University of Chicago Press): 306–19.

Mead, George Herbert ([1934]1962) in Charles W. Morris (ed), *Mind, Self and Society* (Chicago: University of Chicago Press).

Mead, George Herbert ([1938]1972) *The Philosophy of the Act* (Chicago: University of Chicago Press).

Meltzer, Bernard N., John W. Petras, & Larry T. Reynolds (1975) *Symbolic Interactionism: Genesis, Varieties and Criticism* (Boston: Routledge & Kegan Paul).

Mills, C. Wright (1940) "Situated Actions and Vocabularies of Motive," *American Sociological Review* 6: 904–13.

Mol, Annemarie (2002) *The Body Multiple: Ontology in Medical Practice* (Durham, NC: Duke University Press).

Mol, Annemarie & Jessica Messman (1996) "Neonatal Food and the Politics of Theory: Some Questions of Method," *Social Studies of Science* 26: 419–44.

Mukerji, Chandra (1989) *A Fragile Power: Scientists and the State* (Princeton, NJ: Princeton University Press).

Neumann, Laura & Susan Leigh Star (1996) "Making Infrastructure: The Dream of a Common Language," in J. Blomberg, F. Kensing, & E. Dykstra-Erickson (eds) *Proceedings of the Fourth Biennial Participatory Design Conference (PDC'96)* (Palo Alto, CA: Computer Professionals for Social Responsibility): 231–40.

Neyland, Daniel (2006) "Dismissed Content and Discontent: An Analysis of the Strategic Aspects of Actor-Network Theory," *Science, Technology & Human Values* 31 (1): 29–51.

Reynolds, Larry & Nancy Herman-Kinney (eds) (2003) *Handbook of Symbolic Interactionism* (Walnut Creek, CA: Alta Mira Press).

Shibutani, Tamotsu (1955) "Reference Groups as Perspectives," *American Journal of Sociology* 60: 562–69.

Shim, Janet K. (2002) "Understanding the Routinised Inclusion of Race, Socioeconomic Status and Sex in Epidemiology: The Utility of Concepts from Technoscience Studies," *Sociology of Health and Illness* 24: 129–50.

Shim, Janet K. (2005) "Constructing 'Race' Across the Science-Lay Divide: Racial Formation in the Epidemiology and Experience of Cardiovascular Disease," *Social Studies of Science* 35 (3): 405–36.

Shostak, Sara (2003) "Locating Gene-Environment Interaction: At the Intersections of Genetics and Public Health," *Social Science and Medicine* 56: 2327–42.

Shostak, Sara (2005) "The Emergence of Toxicogenomics: A Case Study of Molecularization," *Social Studies of Science* 35 (3): 367–404.

Star, Susan Leigh (1989a) *Regions of the Mind: Brain Research and the Quest for Scientific Certainty* (Stanford, CA: Stanford University Press).

Star, Susan Leigh (1989b) "The Structure of Ill-Structured Solutions: Boundary Objects and Distributed Heterogeneous Problem Solving," in L. Gasser & M. Huhns (eds) *Distributed Artificial Intelligence 2* (San Mateo, CA: Morgan Kauffmann): 37–54.

Star, Susan Leigh (1991a) "The Sociology of the Invisible: The Primacy of Work in the Writings of Anselm Strauss," in David R. Maines (ed), *Social Organization and Social Process: Essays in Honor of Anselm Strauss* (Hawthorne, NY: Aldine de Gruyter): 265–83.

Star, S. Leigh (1991b) "Power, Technologies and the Phenomenology of Conventions: On Being Allergic to Onions," in John Law (ed), *A Sociology of Monsters: Essays on Power, Technology and Domination*, [Sociological Review Monograph No. 38] (New York: Routledge): 26–56.

Star, Susan Leigh (1993) "Cooperation Without Consensus in Scientific Problem Solving: Dynamics of Closure in Open Systems," in Steve Easterbrook (ed) *CSCW: Cooperation or Conflict?* (London: Springer-Verlag): 93–105.

Star, Susan Leigh (ed) (1995a) *Ecologies of Knowledge: Work and Politics in Science and Technology* (Albany: State University of New York Press).

Star, Susan Leigh (ed) (1995b) *The Cultures of Computing* [Sociological Review Monograph] (Oxford: Basil Blackwell).

Star, Susan Leigh (1995c) "The Politics of Formal Representations: Wizards, Gurus, and Organizational Complexity," in S. L. Leigh (ed), *Ecologies of Knowledge: Work and Politics in Science and Technology* (Albany: State University of New York Press): 88–118.

Star, Susan Leigh (1997) "Working Together: Symbolic Interactionism, Activity Theory, and Information Systems," in Yrjö Engeström & David Middleton (eds), *Communication and Cognition at Work* (Cambridge: Cambridge University Press): 296–318.

Star, Susan Leigh (1998) "Grounded Classification: Grounded Theory and Faceted Classifications," *Library Trends* 47: 218–32.

Star, Susan Leigh (1999) "The Ethnography of Infrastructure," *American Behavioral Scientist* 43: 377–91.

Star, Susan Leigh & James R. Griesemer (1989) "Institutional Ecology, 'Translations' and Boundary Objects: Amateurs and Professionals in Berkeley's Museum of Vertebrate Zoology, 1907–39," *Social Studies of Science* 19: 387–420. Reprinted in Mario Biagioli (ed) (1999), *The Science Studies Reader* (New York: Routledge): 505–24.

Star, Susan Leigh & Karen Ruhleder (1996) "Steps Toward an Ecology of Infrastructure: Design and Access for Large Information Spaces," *Information Systems Research* 7: 111–34.

Star, Susan Leigh & Anselm Strauss (1999) "Layers of Silence, Arenas of Voice: The Ecology of Visible and Invisible Work," *Computer-Supported Cooperative Work: Journal of Collaborative Computing* 8: 9–30.

Strauss, Anselm L. (1959) *Mirrors and Masks: The Search for Identity* (Glencoe, IL: Free Press).

Strauss, Anselm L. (1978) "A Social World Perspective," in Norman Denzin (ed.), *Studies in Symbolic Interaction* 1: 119–28 (Greenwich, CT: JAI Press).

Strauss, Anselm L. (1982) "Social Worlds and Legitimation Processes," in Norman Denzin (ed.), *Studies in Symbolic Interaction* 4: 171–90 (Greenwich, CT: JAI Press).

Strauss, Anselm L. (1984) "Social Worlds and Their Segmentation Processes," in Norman Denzin (ed), *Studies in Symbolic Interaction* 5: 123–39 (Greenwich, CT: JAI Press).

Strauss, Anselm L. (1987) Qualitative Analysis for Social Scientists (Cambridge: Cambridge University Press).

Strauss, Anselm L. (1993) *Continual Permutations of Action* (New York: Aldine de Gruyter).

Strauss, Anselm L. (1995) "Notes on the Nature and Development of General Theories," *Qualitative Inquiry* 1 (1): 7–18.

Strauss, Anselm L. & Juliet Corbin (1990) *The Basics of Qualitative Research: Grounded Theory Procedures and Techniques* (Thousand Oaks, CA: Sage).

Strauss, Anselm, Shizuko Fagerhaugh, Barbara Suczek, & Carolyn Wiener (1985) *Social Organization of Medical Work* (Chicago: University of Chicago Press); (1997) New edition with new introduction by Anselm L. Strauss (New Brunswick, NJ: Transaction Publishers).

Strübing, Joerg (1998) "Bridging the Gap: On the Collaboration Between Symbolic Interactionism and Distributed Artificial Intelligence in the Field of Multi-Agent Systems Research," *Symbolic Interaction* 21 (4): 441–64.

Strübing, Jöerg (2007) *Anselm Strauss* (Konstanz, Germany: UVK Verlagsgesellschaft mbH).

Subrahmanian E., I. Monarch, S. Konda, H. Granger, R. Milliken, & A. Westerberg (2003) "Boundary Objects and Prototypes at the Interfaces of Engineering Design," *Computer Supported Cooperative Work (CSCW)*, 12 (2): 185–203.

Suchman, Lucy (1987) *Plans and Situated Actions: The Problem of Human-Machine Communication* (New York: Cambridge University Press).

Sundberg, Makaela (2005) *Making Meteorology: Social Relations and Scientific Practice (*Stockholm: Stockholm University Studies in Sociology, New Series 25).

Thomas, William Isaac (1914) "The Polish-Prussian Situation: An Experiment in Assimilation," *American Journal of Sociology* 19: 624–39.

Timmermans, Stefan (1999) *Sudden Death and the Myth of CPR* (Philadelphia: Temple University Press).

Timmermans, Stefan (2006) *Postmortem: How Medical Examiners Explain Suspicious Deaths* (Chicago: University of Chicago Press).

Timmermans, Stefan & Marc Berg (1997) "Standardization in Action: Achieving Local Universality Through Medical Protocols," *Social Studies of Science* 27 (2): 273–305.

Timmermans, Stefan & Marc Berg (2003) *The Gold Standard: The Challenge of Evidence-Based Medicine and Standardization in Health Care* (Philadelphia: Temple University Press).

Tuunainen, Juha (2005) "When Disciplinary Worlds Collide: The Organizational Ecology of Disciplines in a University Department," *Symbolic Interaction* 28 (2): 205–28.

Walenstein, Andrew (2003) "Finding Boundary Objects in SE and HCI: An Approach Through Engineering-oriented Design Theories," International Federation of Information Processing (IFIP) Workshop, Bridging Gaps Between SE and HCI, May 3–4, Portland, Oregon.

Weiss, Gilbert & Ruth Wodak (eds) (2003) *Critical Discourse Analysis: Theory and Interdisciplinarity* (London: Palgrave).

Wiener, Carolyn (1981) *The Politics of Alcoholism: Building an Arena Around a Social Problem (*New Brunswick, NJ: Transaction Books).

Wiener, Carolyn (1991) "Arenas and Careers: The Complex Interweaving of Personal and Organizational Destiny," in David Maines (ed) *Social Organization and Social Process: Essays in Honor of Anselm Strauss (*Hawthorne, NY: Aldine de Gruyter): 175–88.

Wiener, Carolyn (2000) *The Elusive Quest: Accountability in Hospitals (*Hawthorne, NY: Aldine de Gruyter).

Zorbaugh, Harvey (1929) *The Gold Coast and the Slum: A Sociological Study of Chicago's Near North Side* (Chicago: University of Chicago Press).

6 Feminist STS and the Sciences of the Artificial

Lucy Suchman

The past twenty years have seen an expanding engagement at the intersection of feminist scholarship and science and technology studies (STS). This corpus of research is now sufficiently rich that it invites close and more circumscribed reviews of its various areas of concentration and associated literatures. In that spirit, the aim of this chapter is to offer an integrative reflection on engagements of feminist STS with recent developments in a particular domain of science and technology, which I designate here as the sciences of the artificial.[1] Building on previous discussions relating the perspectives of feminist research to technology more broadly, the focus of this chapter is on developments at the shifting boundary of nature and artifice as it figures in relations between humans and computational machines. Central projects are those collected under the rubric of the cognitive sciences and their associated technologies, including Artificial Intelligence (AI), robotics, and software agents as well as other forms of embedded computing.[2] Central concerns are changing conceptions of the sociomaterial grounds of agency and lived experience, of bodies and persons, of resemblance and difference, and of relations across the human/machine boundary.

In framing my discussion with reference to feminist STS my aim is not to delineate the latter into a discrete subdiscipline somehow apart from science and technology studies more broadly. Not only are the interconnections—historical and conceptual—far too thick and generative to support a separation, but such territorial claims would be antithetical to the spirit of the scholarship that I have selected to review. The point of distinguishing feminist-inspired STS from the wider field of research, and the "sciences of the artificial" from technosciences more broadly, is rather to draw the boundaries of this particular chapter in a way that calls out certain focal interests and concerns. I include here work done under a range of disciplinary and methodological affiliations, most centrally feminist theory, but also the sociology of science, cultural anthropology, ethnomethodology, and information studies and design. The connecting thread for the writings that I discuss is an interest in questioning antecedents and contemporary figurings of human/technology relations through close historical, textual, and ethnographically based inquiry. The research considered here is distinguished from technology studies more broadly by a critical engagement with (1) technosciences founded on the trope of "information"; (2) artifacts that are "digital" or computationally based, (3) a lineage involving automata or the creation of machines in (a certain) image of the human and human capacities, and (4) analysis informed by, or on my reading resonant with, feminist theorizing.

I take it that a virtue of STS is its aspiration to work across disciplines in constructing detailed and critical understandings of the sociality of science and technology, both historically and as contemporary projects. Feminist scholarship, similarly, is organized around core interests and problems rather than disciplinary canons, and comprises an open-ended and heterodox body of work.[3] The aspects of feminist STS that I trace out in this chapter define a relationship to technoscience that combines critical examination of relevant discourses with a respecification of material practices. The aim is to clear the ground in order to plant the seeds for other ways of configuring technology futures.

FEMINIST STS

Certain problematics, while not exclusive to feminist research, act as guiding questions for contemporary feminist scholars engaging with technoscience. Primary among these is the ongoing project of unsettling binary oppositions, through philosophical critique and through historical reconstruction of the practices through which particular divisions emerged as foundational to modern technoscientific definitions of the real. The latter include divisions of subject and object, human and nonhuman, nature and culture, and relatedly, same and other, us and them. Feminist scholars most directly have illuminated the politics of ordering within such divisions, particularly with respect to identifications of sex and gender. A starting observation is that in these pairings the first term typically acts as the privileged referent against which the second is defined and judged.

In constituting the real, questions of resemblance and difference and their associated politics are key. The question of difference outside of overly dichotomous and politically conservative oppositions is one that has been deeply and productively engaged, particularly within feminist and postcolonial scholarship.[4] Feminist STS joins with other recent scholarship in interrogating the conceptual and empirical grounds of the collapsing but still potent boundary between those most foundational categories of science and technology, that is, nature and culture.[5] At least since Donna Haraway's famous intervention ([1985]1991), feminist scholars embrace as well the increasingly evident inseparability of subjects and objects, "natural" bodies and "artificial" augmentations. The study of those connections includes a concern with the labors through which particular assemblages of persons and things come into being, as well as the ways in which humans or nonhumans, cut off from the specific sites and occasions that enliven them, become fetishized. In the latter process, social relations and labors are obscured, and artifacts are mystified.

Feminist research shares with poststructuralist approaches, moreover, the premise that the durable and compulsory character of categorizations and associated politics of difference are reproduced through ongoing reiterations, generated from within everyday social action and interaction.[6] Correspondingly, the consequences of those re-enactments are intelligible only as the lived experiences of specifically situated, embodied persons. Taken as enacted rather than given, the status of resemblance and difference shifts from a foundational premise to an ongoing question—one to be answered always in the moment—of "Which differences matter, here?" (Ahmed, 1998: 4). As I discuss further below, this question takes some novel turns in the case of the politics of difference between nature and artifice, human and machine.

SCIENCES OF THE ARTIFICIAL

These concerns at the intersection of feminist scholarship and STS have immediate relevance for initiatives underway in what computer scientist, psychologist, economist, and management theorist Herbert Simon famously named (1969) "the sciences of the artificial." More specifically, the perspectives sketched above stand in challenging contrast to Simon's conception of relations of nature and artifice, along several dimensions. First, Simon's phrase was assembled within a frame that set the "artificial" in counterdistinction to the "natural" and then sought to define sciences of the former modeled on what he took to be the foundational knowledge-making practices of the latter. The work considered here, in contrast, is occupied with exploring the premise that the boundary that Simon's initiative was concerned to overcome—that between nature and culture—is itself a result of historically specific practices of materially based, imaginative artifice. Second, while Simon defined the "artificial" as made up of systems formed in adaptive relations between "inner" and "outer" environments, however defined, feminist STS joins with other modes of poststructuralist theorizing to question the implied separation, and functional reintegration, of interiors and exteriors that Simon's framework implies. Rather, the focus is on practices through which the boundary of entity and environment, affect and sociality, personal and political emerges on particular occasions, and what it effects. Moreover, while Simon's project takes "information" as foundational, it is the history and contemporary workings of that potent trope that forms the focus for the research considered here. And finally, while Simon's articulation of the sciences of the artificial took as its central subject/object the universal figure of "man," the work of feminist STS is to undo that figure and the arrangements that it serves to keep in place.

In this context the rise of information sciences and technologies is a moment that, under the banner of transformative change, simultaneously intensifies and brings into relief long-standing social arrangements and cultural assumptions. The stage is set by critical social histories like Paul Edwards's *The Closed World* (1996), Alison Adam's *Artificial Knowing: Gender and the Thinking Machine* (1998), N. Katherine Hayles's *How We Became Posthuman* (1999), and Sarah Kember's *Cyberfeminism and Artificial Life* (2003), which examine the emergence of information theory and the cognitive sciences during the latter half of the last century. These writers consider how the body and experience have been displaced by informationalism, computational reductionism, and functionalism in the sciences of the artificial (see also Bowker, 1993; Helmreich,

1998; Forsythe, 2001; Star, 1989a). Artifice here becomes complicated, as simulacra are understood less as copies of some idealized original than as evidence for the increasingly staged character of naturalized authenticity (Halberstam & Livingston, 1995: 5). The trope of informatics provides a broad and extensible connective tissue as well between the production of code as software, and the productive codes of bioengineering (Fujimura & Fortun, 1996; Franklin, 2000; Fujimura, 2005).

In the remainder of this chapter I consider a rich body of STS scholarship engaged in critical debate with initiatives under the banner of the sciences of the artificial. I turn first to the primary site of natural/cultural experimentation; namely, the project of engineering the *humanlike machine*, in the form of artificially intelligent or expert systems, robotics, and computationally based "software agents." For STS scholars the interest of this grand project, in its various forms, is less as a "science of the human" than as a powerful disclosing agent for specific cultural assumptions regarding the nature of the human and the foundations of humanness as a distinctive species property. I turn next to developments in the area of *human-machine mixings*, rendered iconic as the figure of the cyborg, and materialized most obviously in the case of various bodily augmentations. I then expand the frame from the figure of the augmented body to more extended arrangements of persons and things, which I discuss under the heading of *sociomaterial assemblages*. I close with a reflection on the preconditions and possibilities for generative critical exchange between feminist STS and these contemporary technoscience initiatives.

MIMESIS: HUMANLIKE MACHINES

The most comprehensive consideration to date of relations between feminist theory and the project of the intelligent machine is unquestionably Alison Adam's (1998) *Artificial Knowing: Gender and the Thinking Machine*. Adam, a historian of science working for the past twenty years within practical and academic computing, provides a close and extensive analysis of the gendered epistemological foundations of AI. Her argument is that AI builds its projects on deeply conservative foundations, drawn from long-standing Western philosophical assumptions regarding the nature of human intelligence. She examines the implications of this heritage by identifying assumptions evident in AI writings and artifacts, and more revealingly, alternatives notable for their absence. The alternatives are those developed, within feminist scholarship and more broadly, that emphasize the specificity of the knowing, materially embodied and socially embedded subject. The absence of that subject from AI discourses and imaginaries, she observes, contributes among other things to the invisibility of a host of requisite labors, of practical and corporeal care, essential to the progress of science. Not coincidentally, this lacuna effects an erasure, from associated accounts of technoscientific knowledge production, of work historically performed by women.[7]

Adam's analysis is enriched throughout by her careful readings of AI texts and projects, and two examples in particular serve as points of reference for her critique. The first, named "State, Operator, and Result" or Soar, was initiated by AI founding father

Allen Newell in the late 1980s. The aim of the project was to implement ideas put forward by Newell and his collaborator Herbert Simon in their 1972 book *Human Problem Solving*. Adam observes that the empirical basis for that text, proposed by Newell and Simon as a generalized "information processing psychology," comprised experiments involving unspecified subjects. While the particularities of the subjects are treated as irrelevant for Newell and Simon's theory, the former appear, on Adam's closer examination of the text, to have been all male and mostly students at Carnegie Mellon University. The tasks they were asked to complete comprised a standard set of symbolic logic, chess, and cryptarithmetic problems:

> All this leads to the strong possibility that the theory of human problem solving developed in the book, and which has strongly influenced not just the development of Soar but of symbolic AI in general, is based on the behaviour of a few, technically educated, young, male, probably middle-class, probably white, college students working on a set of rather unnatural tasks in a US university in the late 1960s and early 1970s. (Adam, 1998: 94)

The burden of proof for the irrelevance of these particulars, Adam points out, falls to those who would claim the generality of the theory. Nonetheless, despite the absence of such evidence, the results reported in the book were treated by the cognitive science research community as a successful demonstration of the proposition that all intelligent behavior is a form of problem solving, or goal-directed search through a "problem space." Soar became a basis for what Newell named in his 1990 book *Unified Theories of Cognition*, though the project's aims were subsequently qualified by Newell's students, who developed the system into a programming language and associated "cognitive architectural framework" for a range of AI applications (Adam, 1998: 95).

Adam takes as her second example the project "Cyc," the grand ten-year initiative of Douglas Lenat and colleagues funded by American industry during the 1980s and 1990s through the Microelectronics and Computer Technology Corporation (MCC) consortium. Where Newell aspired to identify a general model of cognitive processes independent of any particular domain, Lenat's aim was to design and build an encyclopedic database of propositional knowledge that could serve as a foundation for expert systems. Intended to remedy the evident "brittleness" or narrowness of the expert systems then under development, the premise of the Cyc project was that the tremendous flexibility of human cognition was due to the availability, in the brain, of an enormous repository of relevant knowledge. Neither generalized cognitive processes nor specialized knowledge bases, Lenat argued, could finesse the absence of such consensual, or "common sense," knowledge. Taking objects as both self-standing and foundational, Lenat and his colleagues characterized their project as one of "ontological engineering," the problem being to decide what kinds of objects there are in the world that need to be represented (Lenat & Guha, 1989:23). Not surprisingly the resulting menagerie of objects was both culturally specific and irremediably *ad hoc*, with new objects being introduced seemingly *ad infinitum* as the need arose.

Adam observes that the Cyc project foundered on its assumption of the generalized knower who, like the problem-solver figured in Soar, belies the contingent practices of knowledge making. The common-sense knowledge base, intended to represent "what everyone knows," implicitly modeled relevant knowledge on the canonical texts of the dictionary and encyclopedia. And charged with the task of knowing independently of any practical purposes at hand, the project's end point receded indefinitely into a future horizon well beyond the already generous ten years originally assigned it. More fundamentally, both the Soar and Cyc projects exemplify the assumption, endemic to AI projects, that the very particular domains of knowing familiar to AI practitioners comprise an adequate basis for imagining and implementing "the human." It is precisely this projection of a normative self, unaware of its own specificity, that feminist scholarship has been at pains to contest.

Along with its close reading of AI texts and projects, *Artificial Knowing* includes a commentary on specifically anthropological and sociological engagements with AI practice, focusing on my early critique (Suchman, 1987; see also 2007), and those of Diana Forsythe (1993a,b; see also 2001), Harry Collins (1990), and Stefan Helmreich (1998).[8] My own work, beginning in the 1980s, has been concerned with the question of what understandings of the human, and more particularly of human action, are realized in initiatives in the fields of artificial intelligence and robotics.[9] Immersed in studies of symbolic interactionism and ethnomethodology, I came to the question with an orientation to the primacy of communication, or interaction, to the emergence of those particular capacities that have come to define the human. This emphasis on sociality stood in strong contrast to my colleagues' fixation on the individual cognizer as the origin point for rational action. A growing engagement with anthropology and with STS expanded the grounds for my critique and underscored the value of close empirical investigations into the mundane ordering of sociomaterial practices. Initiatives in the participatory or cooperative design of information systems opened up a further space for proactive experiments, during the 1990s, in the development of an ethnographically informed and politically engaged design practice (Blomberg et al., 1996; Suchman, 2002a,b). Most recently, my frame of reference has been further expanded through the generative theorizing and innovative research practices of feminist scholarship. Within this feminist frame, the universal human cognizer is progressively displaced by attention to the specificities of knowing subjects, multiply and differentially positioned, and variously engaged in reiterative and transformative activities of collective world-making.

Diana Forsythe's studies, based on time spent in the Knowledge Systems Laboratory at Stanford University in the late 1980s and early 1990s, focus on questions of "knowledge acquisition" within the context of "knowledge engineering" and the design of so-called expert systems (Forsythe, 1993a,b; 2001). Considered a persistent and intractable "bottleneck" in the process of expert system building, knowledge acquisition references a series of primarily interview-based practices aimed at "extraction" of the knowledge presumed to be stored inside the head of an expert. As the metaphors suggest, the project of the intelligent machine from the point of view of the AI practitioners studied by Forsythe is imagined in terms of process engineering, the design and management of a flow of epistemological content. The raw material of knowledge is extracted from the head of the expert (a procedure resonant with the more recent trope of "data mining"), then processed by the knowledge engineer into the refined product that is in turn transferred into the machine. The problem with this process from the point of view of AI practitioners in the 1980s and early 1990s was one of efficiency, the solution a technological one, including attempts at automation of the knowledge acquisition process itself. Forsythe's critique is framed in terms of assumptions regarding knowledge implicit in the knowledge engineering approach, including the starting premise that knowledge exists in a stable and alienable form that is in essence cognitive, available to "retrieval" and report, and applicable directly to practice. In contrast she directs attention to the forms of knowing in practice that escape expert reports and, consequently, the process of knowledge acquisition. Most importantly, Forsythe points toward the still largely unexamined issue of the politics of knowledge implied in expert systems projects. This includes most obviously the laboring bodies—of scientists as well as of the many other practitioners essential to scientific knowledge making—that remain invisible in the knowledge engineers' imaginary and associated artifacts. And it includes, somewhat less obviously, the more specific selections and translations built in to the knowledge engineering project from its inception and throughout its course.

Machinelike Actions and Others

Within the STS research community it is Collins's (1990, 1995) debate with AI that is perhaps best known. Insistently refusing to take up questions of gender, power, and the like, Collins nonetheless develops a critique of AI's premises regarding the acquisition of knowledge, drawn from the Sociology of Scientific Knowledge, that has significant resonance with feminist epistemologies.[10] Building on his groundbreaking studies of the replication of laboratory science (1985), Collins demonstrates the necessity of embodied practice—formulated in his case in terms of "tacit knowledge"—to the acquisition of scientific and technical expertise. His later work develops these ideas in relation to the question of knowledge within AI and expert systems projects, with attendant distinctions of propositional and procedural, knowing that and knowing how.[11]

As Collins points out, what he designates "machine-like actions" are as likely to be delegated to humans as to be inscribed in so-called intelligent machines. This observation invites attention to the question of just which humans historically have been the subjects/objects of this form of "mechanization." Pointing to the historical relation between automation and labor, Chasin (1995) explores identifications across women, servants, and machines in contemporary robotics.[12] Her project is to trace the relations between changes in forms of machinic (re)production (mechanical to electrical to electronic), types of labor (industrial to service), and conceptions of human-machine difference. Figured as servants, she points out, technologies reinscribe the difference between "us" and those who serve us, while eliding the difference between the latter and machines: "The servant troubles the distinction between we-human-subjects-inventors with a lot to do (on the one hand) and them-object-things that make it easier for us (on the other)" (1995: 73).

Domestic service, doubly invisible because (1) it is reproductive and (2) it takes place in the household, is frequently provided by people—and of those predominately women—who are displaced and desperate for employment. The latter are, moreover, positioned as "others" to the dominant (typically white and affluent, at least in North America and Europe) populace. Given the undesirability of service work, the conclusion might be that the growth of the middle class will depend on the replacement of human service providers by "smart" machines. Or this is the premise, at least, promoted by those who are invested in the latter's development (see Brooks, 2002). The reality, however, is more likely to involve the continued labors of human service providers. Chasin's analysis of robotics in the context of service work makes clear that, given the nonexistence of a universal "human" identity, the performance of human-ness inevitably entails marks of class, gender, ethnicity, and the like. As well as denying the "smart" machine's specific social locations, moreover, the rhetorics of its presentation as the always obliging, "labor-saving device" erases any evidence of the labor involved in its operation "from bank personnel to software programmers to the third-world workers who so often make the chips" (Chasin, 1995: 75). Yet as Ruth Schwartz Cowan (1983) and others have demonstrated with respect to domestic appliances, rather than a process of simple replacement, the delegation of new capacities to machines simultaneously generates new forms of human labor as its precondition.

Situated Robotics and "New" AI

Feminist theorists have extensively documented the subordination, if not erasure, of the body within the Western philosophical canon. In *How We Became Posthuman* (1999), Katherine Hayles traces out the inheritance of this legacy in the processes through which information "lost its body" in the emerging sciences of the artificial over the last century (1999: 2).[13] Recent developments in AI and robotics appear to reverse this trend, however, taking to heart arguments to the effect that "embodiment," rather than being coincidental, is a fundamental condition for cognition.[14] The most widely cited exception to the rule of disembodied intelligence in AI is the initiative named "situated robotics," launched by Rodney Brooks in the 1980s.[15] In her generally critical review of work in AI and robotics, Alison Adam writes that developments under the heading of "situated robotics," in particular, "demonstrate a clear recognition of the way in which embodiment informs our knowledge" (1998: 149). Sarah Kember (2003) similarly sees the project of situated robotics as providing a radical alternative to the life-as-software simulationism school of Artificial Life.[16] Central to this project, she argues, is a move from the liberal humanist ideal of a self-contained, autonomous agent to an investment in "autopoesis." The latter, as formulated most famously by Maturana and Varela (1980), shifts attention from boundaries of organism and environment as given, to the interactions that define an organism through its relations with its environment. This, according to Kember, comprises recognition of life as always embodied and situated and represents "a potent resource for debating the increasingly symbiotic relation between humans and machines" (2003: 6). But what, exactly, does it mean to be embodied and situated in this context?

The first thing to note is that discoveries of the body in artificial intelligence and robotics inevitably locate its importance *vis-à-vis* the successful operations of mind, or at least of some form of instrumental cognition. The latter in this respect remains primary, however much mind may be formed in and through the workings of embodied action. The second consistent move is the positing of a "world" that preexists independent of the body. Just as mind remains primary to body, the world remains prior to and separate from perception and action, however much the latter may affect and be affected by it. And both body and world remain a naturalized foundation for the workings of mind. As Adam points out, the question as framed by Brooks is whether cognition, and the knowledge that it presupposes, can be modeled separately from perception and motor control (1998: 137). Brooks's answer is "no," but given the constraints of current engineering practice, Adam observes, the figure that results from his ensuing work remains "a bodied individual in a physical environment, rather than a socially situated individual" (1998: 136).

It is important to note as well that the materialization of even a bodied individual in a physical environment has proven more problematic than anticipated. In particular, it seems extraordinarily difficult to construct robotic embodiments, even of the so-called "emergent" kind, that do not rely on the associated construction of a "world" that anticipates relevant stimuli and constrains appropriate response. Just as reliance on propositional knowledge leads to a seemingly infinite regress for more traditional, symbolic AI, attempts to create artificial agents that are "embodied and embedded" seem to lead to an infinite regress of stipulations about the conditions of possibility for perception and action, bodies and environments. The inadequacies of physicalism as a model for bodies or worlds are reflected in Brooks's recent resort to some kind of yet to be determined "new stuff" as the missing ingredient for human-like machines (2002: chapter 8.)

The project of situated robotics has more recently been extended to encompass what researchers identify as "emotion" and "sociability."[17] These developments represent in part a response to earlier critiques regarding the disembodied and disembedded nature of intendedly intelligent artifacts but are cast as well in terms of AI's discovery of these as further necessary components of effective rationality. The most famous materializations of machine affect and sociability were the celebrity robots developed during the 1990s in MIT's AI Lab, Cog and Kismet. Cog, a humanoid robot "torso" incorporating a sophisticated machine vision system linked to skillfully engineered electromechanical arms and hands, is represented as a step along the road to an embodied intelligence capable of engaging in human-like interaction with both objects and human interlocutors. Cog's sister robot, Kismet, is a robot head with cartoon-like, highly suggestive three-dimensional facial features, mobilized in response to stimuli through a system of vision and audio sensors, and accompanied by inflective sound. Both robots were engineered in large measure through the labors of a former doctoral student of Brooks, Cynthia Breazeal. Both Cog and Kismet are represented through an extensive corpus of media renderings—stories, photographs, and in Kismet's case, QuickTime videos available on the MIT website. Pictured from the "waist" up, Cog appears as freestanding if not mobile, and Kismet's Web site offers a series of recorded "interactions" between Kismet, Breazeal, and selected other humans. Like other conventional documentary productions, these representations are framed and narrated in ways that instruct the viewer in what to see. Sitting between the documentary film and the genre of the system demonstration, or "demo," the videos create a record that can be reliably repeated and reviewed in what becomes a form of eternal ethnographic present. These reenactments thereby imply that the capacities they record have an ongoing existence, that they are themselves robust and repeatable, and that like any other living creatures Cog and Kismet's agencies are not only ongoing but also continuing to develop and unfold.[18]

Robotics presents the technoscientist with the challenges of obdurate materialities of bodies in space, and Kember maintains the possibility that these challenges will effect equally profound shifts in the onto-epistemological premises not only of the artificial but also of the human sciences.[19] But despite efforts by sympathetic critics such as Adam and Kember to draw attention to the relevance of feminist theory for AI and robotics, the environments of design return researchers from the rhetorics of embodiment to the familiar practices of computer science and engineering. Brooks embraces an idea of situated action as part of his campaign against representationalism in AI, but Sengers (in press) observes that while references to the situated nature of cognition and action have become "business as usual" within AI research, researchers have for the most part failed to see the argument's consequences for their own relations to their research objects. I return to the implications of this for the possibilities of what Agre (1997) has named a "critical technical practice" below but here simply note the associated persistence of an unreconstructed form of realism in roboticists' constitution of the "situation."

SYNTHESIS: HUMAN/MACHINE MIXINGS

Haraway's subversive refiguring of the cyborg ([1985]1991, 1997) gave impetus to the appearance in the1990s of so-called "cyborg anthropology" and "cyberfeminism."[20] Both see the human/machine boundary so clearly drawn in humanist ontologies as increasingly elusive. Cyborg studies now encompass a range of sociomaterial mixings, many centered on the engineering of information technologies in increasingly intimate relation with the body (Balsamo, 1996; Kirkup et al., 2000; Wolmark, 1999). A starting premise of these studies, following Haraway (1991: 195) is that bodies are always already intimately engaged with a range of augmenting artifacts. Increasingly the project for science and technology scholars is go beyond a simple acknowledgement of natural/artificial embodiment to articulate the specific and multiple configurations of bodily prostheses and their consequences. In this context, Jain (1999) provides a restorative antidote to any simplistic embrace of the prosthetic, in considering the multiple ways in which prostheses are wounding at the same time that they are enabling. In contrast to the easy promise of bodily augmentation, the fit of bodies and artifacts is often less seamless and more painful than the trope would suggest. The point is not, however, to demonize the prosthetic where formerly it was valorized but rather to recognize the misalignments that inevitably exist within human/machine syntheses and the labors and endurances required to accommodate them (see also Viseu, 2005).

One aim of feminist research on the intersections of bodies and technologies is to explore possibilities for figuring the body as other than either a medicalized or aestheticized object (Halberstam & Livingston, 1995: 1). A first step toward such refiguring is through critical interrogation of the ways in which new imaging and body-altering technologies have been enrolled in amplifying the medical gaze and in imagining the body as gendered, and raced, in familiar ways. Feminist research on biomedical imaging technologies, for example, focuses on the rhetorical and material practices through which figures of the universal body are renewed in the context of recent "visual human" projects, uncritically translating very specific, actual bodies as "everyman/woman" (Cartwright, 1997; Prentice, 2005; Waldby, 2000). More popular appropriations of digital imaging technologies appear in the synthesis of newly gendered and racialized mixings, most notably the use of "morphing" software in the constitution of science fiction depictions of future life forms. This same technology has been put to more pedagogical purposes in the case of the hybridized "Sim Eve," incisively analyzed by Hammonds (1997) and Haraway (1997).[21] Across these cases we find technologies deployed in the reiteration of a "normal" person/body—even, in the cases that Hammonds and Haraway discuss, an idealized mixing—against which others are read as approximations, deviations, and the like. Attention to the normative and idealized invites as well consideration of the ways in which new technologies of the artificial might be put to more subversive uses. Kin to Haraway's cyborg, the "monstrous" has become a generative figure for writing against the grain of a deeper entrenchment of normative forms (Hales, 1995; Law, 1991; Lykke & Braidotti, 1996).[22] This figure links in turn to long-standing feminist concerns with (orderings of) difference.

With respect to information technologies more widely, feminist scholars have pointed out the need for a genealogy that traces and locates now widely accepted metaphors (e.g., that of "surfing" or the electronic "frontier") within their very particular cultural and historical origins.[23] The point of doing this is not simply as a matter of historical accuracy but also because the repetition of these metaphors and their associated imaginaries have social and material effects, not least in the form of systematic inclusions and exclusions built in to the narratives that they invoke. The configurations of inclusion/exclusion involved apply with equal force and material effect to those involved in technology production. As Sara Diamond concisely states, it is still the case within the so-called high tech and new media industries that "what kind of work you perform depends, in great part, on how you are configured biologically and positioned socially" (1997: 84).

A guiding interest of feminist investigations of the "virtual" is the continued place of lived experience and associated materialities in what have been too easily characterized as "disembodied" spaces. Recent research moves away from debates over whether participants in such spaces "leave the body behind," toward the sometimes strange, sometimes familiar forms that computer-mediated embodiments take. Feminist research orients, for example, to the multiplicity, and specificity, of computer-mediated sociality. Through her various studies, Nina Wakeford promotes a conception of "cyberspace" as "not a coherent global and unitary entity but a series of performances" (1997: 53). Communications technologies commonly represented as offering "narrower bandwidths" than face-to-face co-presence, Sandy Stone (1999) observes, in their use can actually afford new spaces for expanding identity play.[24] More generally, these investigations suggest a conceptualization of encounters at the interface that opens out from the boundaries of the machine narrowly construed, to the ambient environments and transformative subject/object relations that comprise the lived experience of technological practice.

SOCIOMATERIAL ASSEMBLAGES

In the closing chapter of *Cyberfeminism and Artificial Life*, Kember asks, "So how should feminists contest the material and metaphoric grounds of human and machine identities, human and machine relations?" (2003: 176). In the remainder of this chapter I offer some at least preliminary responses to that question, based in recent efforts to reconfigure agencies at the human-machine interface, both materially and metaphorically, in ways informed by feminist theorizing. The figure of the assemblage helps to keep *associations* between humans and nonhumans as our basic unit of analysis.[25] The body of work that is now available to elaborate our understanding of sociomaterial relations as assemblages is too extensive to be comprehensively reviewed, but a few indicative examples can serve as illustration.

The surgery, with its growing entanglement of virtual mediations and material embodiments, has afforded a perspicuous research site. Minimally invasive, or "keyhole," surgery, for example, as it has developed over the past few decades, has involved a series of shifts in the gaze of the surgeon and attendant practitioners from the interior of the patient's body—formerly achieved through a correspondingly large incision—to views mediated first through microscopy and now through digital cameras and large screen monitors. Aanestad (2003) focuses on the labors of nurses, traditionally a feminized occupation, responsible for setting up the complex sociotechnical environment required for the conduct of "keyhole" surgery. Her analysis follows the course of shifting interdependencies in the surgical assemblage, as changes to existing arrangements necessitate further changes in what she names the *in situ* work of "design in configuration" (2003: 2). At the same time Prentice (2005) finds that, rather than being alienated from the patient's body through these extended mediations, surgeons accustomed to performing minimally invasive surgery experience themselves as proprioceptively shifted more directly and proximally into the operative site, with the manipulative instruments serving as fully incorporated extensions of their own acting body. As Prentice observes of these boundary transformations: "When the patient's body is distributed by technology, the surgeon's body reunites it through the circuit of his or her own body" (2005: 8; see also Goodwin, in press; Lenoir & Wei, 2002).

Myers (2005) explores the transformation of body boundaries that occurs as molecular biologists incorporate knowledge of protein structures through their engagement with physical and virtual models. Interactive molecular graphics technologies, she argues, afford crystallographers the experience of handling and manipulating otherwise intangible protein structures. The process of learning those structures involves not simply mentation but a reconfiguration of the scientist's body, as "protein modelers can be understood to 'dilate' and extend their bodies into the prosthetic technologies offered by computer graphics, and 'interiorize' the products of their body-work as embodied models of molecular structure" (in press). The result, she proposes, is a kind of "animate assemblage" of continually shifting and progressively deepening competency, enabled through the prosthetic conjoining of persons and things.

A more violent form of human-machine assemblage is evident in Schull's (2005) account of the interconnected circuitry of the gaming industry, digital gambling machine developers, machines, and gamblers in Las Vegas, Nevada. Her ethnography explores "the intimate connection between extreme states of subjective absorption in play and design elements that manipulate space and time to accelerate the extraction of money from players" (2005: 66). Values of productivity and efficiency on the part of actors in the gaming industry align with players' own desires to enter into a simultaneously intensified and extended state of congress with the machine, enabled through the progressive trimming of "dead time" from the cycles of play. As in molecular modeling, physical and digital materials are joined together to effect the resulting agencies, in this case in the form of input devices and machine feedback that minimize the motion required of players, ergonomically designed chairs that maintain the circulation of blood and the body's corresponding comfort despite the lack of movement, and computationally enabled operating systems that expand and more tightly manage the gaming possibilities. The aim of developers and players alike is that the latter should achieve "a dissociated subjective state that gamblers call the 'zone,' in which conventional spatial, bodily, monetary, and temporal parameters are suspended," as the boundary of player and machine dissolves into a new and compelling union. The point, the compulsive gambler explains, is not to win but to keep playing.

The crucial move in each of these studies is a shift from a treatment of subjects and objects as singular and separately constituted to a focus on the kinds of connections and capacities for action that particular arrangements of persons and things afford. The idea of subject/object configurations as an effect of specific practices of boundary-making and remaking is elaborated by feminist physicist Karen Barad, who proposes that stabilized entities are constructed out of specific apparatuses of sociomaterial "intra-action" (2003). While the construct of *interaction* presupposes two entities, given in advance, that come together and engage in some kind of exchange, *intra-action* underscores the sense in which subjects and objects emerge through their encounters with each other. More specifically, Barad locates technoscientific practices as critical sites for the emergence of new subjects and objects. Taking physics as a case in point, her project is to work through long-standing divisions between the virtual and the real, while simultaneously coming to grips with the ways in which material-ities, as she puts it, "kick back" in response to our intra-actions with them (1998: 112). Through her readings of Niels Bohr, Barad insists that "object" and "agencies of observation" form a nondualistic whole: it is that relational entity that comprises the objective "phenomenon" (1996: 170). Different "apparatuses of observation" enable different, always temporary, subject/object cuts that in turn enable measurement or other forms of objectification, distinction, manipulation and the like *within* the phe-nomenon. The relation is "ontologically primitive" (2003: 815), in other words, or prior to its components; the latter come about only through the "cut" effected through a particular apparatus of observation. Acknowledging the work of boundary making, as a necessary but at least potentially reconfigurable aspect of reality construction, sug-gests a form of accountability based not in control but in ongoing engagement.

SITES OF ENGAGEMENT

Among the various contemporary approaches to the study of science and technology within the social sciences, feminist research practices are marked by the joining of rigorous critique with a commitment to transformative intervention. However compelling the critique, intervention presupposes forms of engagement, both exten-sive and intensive, that involve their own, often contradictory positionings. In par-ticular, the disciplines and projects that currently dominate professional sites of technology production are narrowly circumscribed, and the expected form of engage-ment is that of service to established agendas. Reflecting upon this dilemma in an essay titled "Ethics and Politics of Studying Up" (1999[2001]), Forsythe poses the ques-tion of how we should practice an anthropology within, and of, powerful institutions that is at once critical and respectful. Respectful critique, she argues, is particularly problematic when ours are dissenting voices, in settings where anthropological affili-ations grant us marginality as much as privilege. In response to this essay, I have sug-gested that recent reconceptualizations of ethnographic practice, from distanced description to an engagement in multiple, partial, unfolding, and differentially pow-erful narratives can help recast the anthropologist's dilemma (Suchman, 1999a). This recasting involves a view of critique not as ridicule but as a questioning of basic assumptions, and of practice not as disinterested but as deeply implicated. At the same time, I would maintain that respectful critique requires the associated incorporation of critical reflection as an indigenous aspect of the professional practices in question (see Agre, 1997).

In *Cyberfeminism and Artificial Life* (2003), Kember examines the relations between two broad arenas of scholarship and technology building at the intersection of feminism and the sciences of the artificial, which she identifies as cyberfeminism and ALife respectively.[26] Kember is concerned that those whom she identifies as cyberfeminists have maintained a distanced, outsider's relation to developments in ALife. Insofar as the view has remained that of the outsider, she argues, it has remained an exclusively critical one. Rather than exemplifying a generative reworking of the boundaries of nature and culture, ALife appears to the feminist critic to reinscribe the most conservative versions of biological thinking (2003: viii). In contrast, seen from within, Kember proposes that just as feminism is internally heterogeneous and contested, so are discourses of ALife. The conditions for dialogue are provided by these endogenous debates, in her view, as long as the outcome imagined is not resolution but risk—a risk that she urges cyberfeminists to take. This raises the question of whether, or to what extent, a critical exchange must—at least if it is to be an exchange—involve a reciprocity of risk. If so, is it really, or at least exclusively, feminism that has failed to take risks across these disciplinary boundaries?[27]

Haraway proposes that it is a concern with the possibilities for "materialized refiguration" that animates the interests of feminist researchers in science and technology (1997: 23). Figuration recognizes the intimate connections of available cultural imaginaries with the possibilities materialized in technologies. The contemporary technoscience projects considered here involve particular ways of figuring together, or *configuring*, humans and machines. It follows that one form of intervention is through a critical consideration of how humans and machines are figured in those practices and how they might be figured and configured differently. The most common forms of engagement are interdisciplinary initiatives aimed at reconfiguring relations of design and use (Balsamo, in press; Greenbaum & Kyng, 1991; Ooudshorn & Pinch, 2003; Lyman & Wakeford, 1999; Star, 1995b; Suchman, [1994]1999b, 2002a,b).[28] While these developments bring researchers onto politically charged and variously compromised terrain, they open as well new spaces for theoretical and political action.

My aim in this chapter has been to draw out a sense of the critical exchange emerging in feminist-inspired STS encounters with new digital technologies and the plethora of configurations that they have materialized. This exchange involves a spectrum of engagements, from questions regarding received assumptions to dialogic interventions and more directly experimental alternatives. Theoretically, this body of research explores the rewriting of old boundaries of human and nonhuman. Politically and practically, it has implications for how we conceptualize and configure practices of information technology design and use and the relations between them. I take an identifying commitment of feminist research to be a deeper appreciation of the specific relationalities of the sociomaterial world, combined with forms of constructive engagement aimed at more just distributions of symbolic and economic reward. The moves that Haraway encourages, toward recognition of the material consequences of the figural and the figural grounds of the material, and toward a different kind of positioning for the researcher/observer, mark the spirit of feminist STS. This effort engages with the broader aim of understanding science as culture,[29] as a way of shifting the frame of analysis—our own as well as that of our research subjects—from the discovery of universal laws to the ongoing elaboration and potential transformation of culturally and historically specific practices to which we are all implicated, rather than innocently modest, witnesses.

Notes

My thanks to the editors of the *Handbook* and its reviewers, in particular Toni Robertson for her close and critical reading of early versions of this chapter.

1. Adopted from Simon, 1969. I return to a consideration of Simon's use of this phrase below. For useful overviews of feminist STS more broadly, see Creager et al., 2001; Harding, 1998; Keller, 1995, 1999; Mayberry et al., 2001; McNeil, 1987; McNeil and Franklin, 1991. For introductions and anthologies on gender and technology, see Balka & Smith, 2000; Grint & Gill, 1995; Terry & Calvert, 1997; Wajcman, 1991, 1995, 2004; and for indicative case studies, see Balsamo, 1996; Cockburn, 1988, 1991; Cockburn & Ormrod, 1993; Cowan, 1983; Martin, 1991.

2. Related areas of contemporary scholarship that are not encompassed in this chapter include artificial life, computer-mediated communication, cultural and media studies (particularly close and critical readings of science fiction and related popular cultural genre), and feminist critiques of reproductive and biotechnologies. My decision to focus the chapter more narrowly is a (regrettably) pragmatic one and a sign not of the unimportance of these areas but, on the contrary, of the impossibility of doing them justice in the space available. At the same time, I do attempt to cite some indicative points of interchange and to emphasize the interrelatedness of concerns. For critical discussions of projects in artificial life informed by feminist theory, see Adam 1998: chapter 5; Helmreich, 1998; Kember, 2003. For feminist writings in the area of computer-mediated communication and new media, see Cherny, 1996; Robertson, 2002; Star, 1995a; and on reproductive and biotechnologies, see Casper, 1998; Clarke, 1998; Davis-Floyd & Dumit, 1998; Franklin & McKinnon, 2001; Franklin & Ragone, 1998; Fujimura, 2005; Hayden, 2003; M'Charek, 2005; Strathern, 1992; Thompson, 2005.

3. I embrace here the suggestion of Ahmed et al. that "if feminism is to be/become a transformative politics, then it might need to refuse to (re)present itself as programmatic" (2000: 12).

4. For some exemplary texts, see Ahmed, 1998; Ahmed et al., 2000; Berg & Mol, 1998; Braidotti, 1994, 2002; Castañeda, 2002; Gupta & Ferguson, 1992; Law 1991; Mol, 2002; Strathern, 1999; Verran 2001.

5. See, for example, Franklin, 2003; Franklin et al., 2000; Haraway, 1991, 1997. Haraway's early writings employed the conjunctive "/" to join nature and culture together, but she subsequently erased this residual trace of dualism; see Haraway, 2000: 105.

6. For the definitive articulation of a performative approach to normativity and transgression, see Butler, 1993. See also the call of Ashmore et al. for "a rejection of a resolution of the question of relations between human and nonhuman, particularly with respect to agency, through recourse to 'essentialist ontological arguments'" (1994: 1). On the centrality of categorization practices in scientific practice and everyday action, see also Lynch, 1993; Bowker & Star, 1999.

7. This invisibility turns on the erasure of bodies, as either knowing subjects or the objects of women's labor. Historically, as Adam points out:

> Women's lives and experiences are to do with bodies, the bearing and raising of children, the looking after of bodies, the young, old and sick, as well as men's bodies in their own, and others' homes, and in the workplace. (1998: 134)

I return to the question of embodiment below, but note here Adam's point that it is this practical care of the body, in sum, that enables the "transcendence" of the mind.

8. For another early engagement, see Star 1989b.

9. Unfortunately, Adam reiterates a prevalent misreading of my argument in *Plans and Situated Actions* (1987), stating that I propose that people do not make plans but rather act in ways that are situated and contingent (Adam, 1998: 56–57). See my attempts to remedy this misunderstanding in favor of a view of planning as itself a (specific) form of situated activity (Suchman, 1993, 2007) as well as the intervention represented in Suchman and Trigg (1993) regarding AI's own situated and contingent practices. More egregiously, Adam attributes to me the position that "members of a culture have agreed, known-in-common, social conventions or behavioral norms and that these shape agreement on the appropriate relation between actions and situations" (Adam, 1998: 65). Read in context (Suchman, 1987: 63), this is instead a characterization that I offer of the position *against which* ethnomethodology is framed, proposing that rather than pregiven and stable in the way that it is assumed to be in structural functionalist sociology, "shared knowledge" is a contingent achievement of practical action and interaction. Note that this latter view has profound implications for the premises of the Cyc project as well.

10. Once again, this is not to suggest that Collins himself is engaged in feminist scholarship, but simply that his work provides some invaluable resources for others so engaged. Adam observes that, across his own writings, Collins, like the AI practitioners he critiques, presumes a universal reader-like-himself, positing things that "everyone knows" without locating that knowing subject more specifically (Adam, 1998: 65). This is consistent, she points out, with the tradition of the unmarked subject prevalent in Western moral philosophy; an implied knower who, as feminist epistemologists have argued, is only actually interchangeable with others within the confines of a quite particular and narrow membership group. Feminist epistemology, in contrast, is concerned with the specificity of the knowing subject, the '*S*' in propositional logic's '*S* knows that *p*.' "Yet," Adam observes, "asking 'Who is *S*?' is not considered a proper concern for traditional epistemologists" (1998: 77).

11. See also Dreyfus, [1979]1992.

12. The dream of machines as the new servant class comprises a translation from the robot visions of the industrial age to that of the service economy. This vision is clearly presented in innumerable invocations of the future of human-computer interactions, perhaps most notably by Brooks, 2002. For further critical discussions, see Berg, 1999; Crutzen, 2005; Gonzalez, (1995)1999; Markussen, 1995; Turkle, 1995: 45; Suchman, 2003, 2007: chapter 12.

13. See also Balsamo, 1996; Adam, 1998; Gatens, 1996; Grosz, 1994; Helmreich, 1998; Kember, 2003. For useful anthologies of writings on feminist theories of the body, see Price & Shildrick, 1999; Schiebinger, 2000.

14. For anthropological writings that made some contribution to this shift, see Suchman, 1987; Lave, 1988. For accounts from within the cognitive sciences, see also Hutchins, 1995; Agre, 1997; for overviews, see Clark, 1997, 2001, 2003; Dourish, 2001.

15. For formulations of Brooks's position written for a general reader, see Brooks, 1999, 2002. For a more extended consideration of the tropes of embodiment, sociality, and emotion in situated robotics, including an account of how a concern with the "situated" might have made its way into the MIT AI lab, see Suchman, 2007: chapter 14.

16. Kember takes as her primary exemplar roboticist Steve Grand. For a critique of Grand's latest project in situated robotics, named "Lucy the Robot Orangutan" (Grand 2003), read through the lens of Haraway's history of primatology and the "almost human," see Castañeda & Suchman, in press.

17. See, for example, Breazeal, 2002; Cassell et al., 2000; Picard, 1997. See also Castañeda, 2001; Wilson, 2002.

18. For an examination of the mystifications involved with these modes of representation, see Suchman, 2007: chapter 14.

19. Her argument here is resonant with that of Castañeda (2001).

20. See, for example, Downey & Dumit, 1997; Fischer, 1999; Hawthorne & Klein, 1999; Kember, 2003.

21. On figurings of race in online venues, see Nakamura, 2002.

22. Along with its generative connotations, the "monster," like the "cyborg," can become too easy and broad, even romanticized, a trope. Both are in need of careful analysis and specification with respect to their historical origins, their contemporary manifestations, and the range of lived experiences that they imply.

23. See, for example, Miller, 1995. Miller's focus is on the "frontier" metaphor as it invokes the need for "protection" of women and children. Who, she asks, are the absent others from whom the danger comes? The further implication, of course, is of an expansion of ownership over territories constructed as "empty" in ways that erase those "others" who have long inhabited them, albeit in different (and for those invested in the frontier), unrecognizable ways. For widely cited discussions of women's presence online, particularly in the ongoing productions, constructions, and engagements of the World Wide Web and the Internet more broadly, see Wakeford, 1997; Spender, 1996.

24. This is, of course, Sherry Turkle's position as well. See Turkle, 1995.

25. The trope of the "assemblage" has been developed within science studies to reference a bringing together of things, both material and semiotic, into configurations that are more and less durable but always contingent on their ongoing enactment as a unity. See Law, 2004: 41–42.

26. Both of these terms are defined broadly by Kember: cyberfeminism affords a general label for feminist research and scholarship concerned with information and communications technologies, artificial life or any research in artificial intelligence or robotics that, in rejecting the tenets of "good old-fashioned AI" (GOFAI), comprises what roboticist Rodney Brooks terms the "nouvelle AI" (2002: viii). This is in contrast to more circumscribed uses of the term cyberfeminism on the one hand, to reference in particular the enthusiastic hopes for networked, digital technologies; or of ALife, on the other hand, to identify the particular lines of computationalism involving the simulation of biological systems in software.

27. The risk, moreover, may not only be that of a challenge to one's deeply held beliefs. An even more dangerous possibility may be that of appropriation of one's position in the service of another, which is further entrenched, rather than reworked, in the process.

28. I have suggested ([1994]1999b, 2002a,b) that responsible design might be understood as a form of "located accountability," that would stand in contrast to existing practices of "design from nowhere." Adam (1998) unfortunately translates the latter phrase into the problem that "no one is willing to hold ultimate responsibility for the design of the system, as it is difficult to identify the designer as one single clearly identifiable individual" (1998: 79). My argument is that, insofar as no one designer does have ultimate responsibility for the design of a system or control over its effects, accountable design cannot depend on any simple idea of individual responsibility. Rather, located accountability with respect to design must mean a continuing awareness of, and engagement in, the dilemmas and debates that technological systems inevitably generate.

29. See Pickering, 1992; Franklin, 1995; Helmreich, 1998; Reid & Traweek, 2000.

References

Aanestad, Margun (2003) "The Camera as an Actor: Design-in-Use of Telemedicine Infrastructure in Surgery," *Computer-Supported Cooperative Work (CSCW)* 12: 1–20.

Adam, Alison (1998) *Artificial Knowing: Gender and the Thinking Machine* (New York: Routledge).

Agre, Philip (1997) *Computation and Human Experience* (New York: Cambridge University Press).

Ahmed, Sara (1998) *Differences That Matter: Feminist Theory and Postmodernism* (Cambridge: Cambridge University Press).

Ahmed, Sara, Jane Kilby, Celia Lury, Maureen McNeil, & Beverly Skeggs (2000) *Transformations: Thinking Through Feminism* (London and New York: Routledge).

Ashmore, Malcolm, Robin Wooffitt, & Stella Harding (1994) "Humans and Others, Agents and Things," *American Behavioral Scientist* 37(6): 733–40.

Balka, Ellen & Richard Smith (eds) (2000) *Women, Work and Computerization: Charting a Course to the Future* (Boston: Kluwer Academic).

Balsamo, Anne (1996) *Technologies of the Gendered Body: Reading Cyborg Women* (Durham, NC: Duke University Press).

Balsamo, Anne (in press) *Designing Culture: A Work of the Technological Imagination* (Durham, NC: Duke University Press).

Barad, Karen (1996) "Meeting the Universe Halfway: Realism and Social Constructivism Without Contradiction," in L. H. Nelson & J. Nelson (eds), *Feminism, Science, and the Philosophy of Science* (Norwell, MA: Kluwer): 161–94.

Barad, Karen (1998) "Getting Real: Technoscientific Practices and the Materialization of Reality," *differences: A Journal of Feminist Cultural Studies* 10: 88–128.

Barad, Karen (2003) "Posthumanist Performativity: Toward an Understanding of How Matter Comes to Matter," *Signs: Journal of Women in Culture and Society* 28: 801–31.

Berg, Anne Jorunn (1999) "A Gendered Socio-technical Construction: The Smart House," in D. Mackenzie & J. Wajcman (eds), *The Social Shaping of Technology* (Buckingham and Philadelphia: Open University Press): 301–13.

Berg, Marc & Annemarie Mol (eds) (1998) *Differences in Medicine: Unraveling Practices, Techniques, and Bodies* (Durham, NC: Duke University Press).

Blomberg, Jeanette, Lucy Suchman, & Randall Trigg (1996) "Reflections on a Work-Oriented Design Project," *Human-Computer Interaction* 11: 237–65.

Bowker, Geoffrey (1993) "How to Be Universal: Some Cybernetic Strategies, 1943–1970," *Social Studies of Science* 23: 107–27.

Bowker, Geoffrey & Susan Leigh Star (1999) *Sorting Things out: Classification and Its Consequences* (Cambridge, MA: MIT Press).

Braidotti, Rosi (1994) *Nomadic Subjects: Embodiment and Sexual Differences in Contemporary Feminist Theory* (New York: Columbia University Press).

Braidotti, Rosi (2002) *Metamorphoses: Towards a Materialist Theory of Becoming* (Cambridge: Blackwell).

Breazeal, Cynthia (2002) *Designing Sociable Robots* (Cambridge, MA: MIT Press).

Brooks, Rodney (1999) *Cambrian Intelligence: The Early History of the New AI* (Cambridge, MA: MIT Press).

Brooks, Rodney (2002) *Flesh and Machines: How Robots Will Change Us* (New York: Pantheon Books).

Butler, Judith (1993) *Bodies That Matter: On the Discursive Limits of "Sex"* (New York: Routledge).

Cartwright, Lisa (1997) "The Visible Man: The Male Criminal Subject as Biomedical Norm," in J. Terry & M. Calvert (eds), *Processed Lives: Gender and Technology in Everyday Life* (London and New York: Routledge): 123–37.

Casper, Monica (1998) *The Making of the Unborn Patient: A Social Anatomy of Fetal Surgery* (New Brunswick, NJ: Rutgers University Press).

Cassell, Justine, Joseph Sullivan, Scott Prevost, & Elizabeth Churchill (2000) *Embodied Conversational Agents* (Cambridge, MA: MIT Press).

Castañeda, Claudia (2001) "Robotic Skin: The Future of Touch?" in S. Ahmed & J. Stacey (eds), *Thinking Through the Skin* (London: Routledge): 223–36.

Castañeda, Claudia (2002) *Figurations: Child, Bodies, Worlds* (Durham, NC: Duke University Press).

Castañeda, Claudia & Lucy Suchman (in press) "Robot Visions," in Sharon Ghamari-Tabrizi (ed), *Thinking with Haraway.*

Chasin, Alexandra (1995) "Class and Its Close Relations: Identities Among Women, Servants, and Machines," in J. Halberstram & I. Livingston (eds), *Posthuman Bodies* (Bloomington: Indiana University Press): 73–96.

Cherny, Lynn & Elizabeth Reba Weise (eds) (1996) *Wired Women: Gender and New Realities in Cyberspace* (Seattle, WA: Seal Press).

Clark, Andy (1997) *Being There: Putting Brain, Body, and World Together Again* (Cambridge, MA: MIT Press).

Clark, Andy (2001) *Mindware: An Introduction to the Philosophy of Cognitive Science* (New York: Oxford University Press).

Clark, Andy (2003) *Natural-born Cyborgs: Minds, Technologies, and the Future of Human Intelligence* (Oxford and New York: Oxford University Press).

Clarke, Adele (1998) *Disciplining Reproduction: Modernity, American Life Sciences, and "the Problems of Sex"* (Berkeley: University of California Press).

Cockburn, Cynthia (1988) *Machinery of Dominance: Women, Men, and Technical Know-how* (Boston: Northeastern University Press).

Cockburn, Cynthia (1991) *Brothers: Male Dominance and Technological Change* (London: Pluto Press).

Cockburn, Cynthia & Susan Ormrod (1993) *Gender and Technology in the Making* (London: Sage).

Collins, H. M. (1985) *Changing Order: Replication and Induction in Scientific Practice* (London and Beverly Hills, CA: Sage).

Collins, H. M. (1990) *Artificial Experts: Social Knowledge and Intelligent Machines* (Cambridge, MA: MIT Press).

Collins, H. M. (1995) "Science Studies and Machine Intelligence," in S. Jasanoff, G. Markle, J. Petersen, & T. Pinch (eds), *Handbook of Science and Technology Studies* (London: Sage): 286–301.

Cowan, Ruth Schwartz (1983) *More Work for Mother: The Ironies of Household Technology from the Open Hearth to the Microwave* (New York: Basic Books).

Creager, Angela, Elizabeth Lunbeck, & Londa Schiebinger (eds) (2001) *Feminism in Twentieth Century Science, Technology, and Medicine* (Chicago: University of Chicago Press).

Crutzen, Cecile (2005) "Intelligent Ambience Between Heaven and Hell: A Salvation?" *Information, Communication Ethics and Society (ICES)* 3(4): 219–232.

Davis-Floyd, Robbie & Joe Dumit (eds) (1998) *Cyborg Babies: From Techno-Sex to Techno-Tots* (New York and London: Routledge).

Diamond, Sara (1997) "Taylor's Way: Women, Cultures and Technology" in J. Terry & M. Calvert (eds), *Processed Lives: Gender and Technology in Everyday Life* (London and New York: Routledge): 81–92.

Dourish, Paul (2001) *Where the Action Is: The Foundations of Embodied Interaction* (Cambridge, MA: MIT Press).

Downey, Gary & Joe Dumit (eds) (1997) *Cyborgs and Citadels: Anthropological Interventions in Emerging Sciences and Technologies* (Santa Fe, NM: School of American Research).

Dreyfus, Hubert ([1979]1992) *What Computers Still Can't Do: A Critique of Artificial Reason* (Cambridge, MA: MIT Press).

Edwards, Paul (1996) *The Closed World: Computers and the Politics of Discourse in Cold War America* (Cambridge, MA: MIT Press).

Fischer, Michael (1999) "Worlding Cyberspace: Toward a Critical Ethnography in Time, Space, and Theory," in G. Marcus (ed), *Critical Anthropology Now* (Santa Fe, NM: School of American Research): 245–304.

Forsythe, Diana (1993a) "Engineering Knowledge: The Construction of Knowledge in Artificial Intelligence," *Social Studies of Science* 23: 445–77.

Forsythe, Diana (1993b) "The Construction of Work in Artificial Intelligence," *Science, Technology & Human Values* 18: 460–79.

Forsythe, Diana (2001) *Studying Those Who Study Us: An Anthropologist in the World of Artificial Intelligence* (Stanford, CA.: Stanford University Press).

Franklin, Sarah (1995) "Science as Culture, Cultures of Science," *Annual Reviews of Anthropology* 24: 163–84.

Franklin, Sarah (2000) "Life Itself: Global Nature and the Genetic Imaginary," in S. Franklin, C. Lury, & J. Stacey, *Global Nature, Global Culture* (London: Sage): 188–227.

Franklin, Sarah (2003) "Re-Thinking Nature-Culture: Anthropology and the New Genetics," *Anthropological Theory* 3: 65–85.

Franklin, Sarah & Helen Ragoné (1998) *Reproducing Reproduction: Kinship, Power, and Technological Innovation* (Philadelphia: University of Pennsylvania Press).

Franklin, Sarah, Celia Lury, & Jackie Stacey (2000) *Global Nature, Global Culture* (London: Sage).

Franklin, Sarah, & Susan McKinnon (eds) (2001) *Relative Values: Reconfiguring Kinship Studies* (Durham, NC: Duke University Press).

Fujimura, Joan (2005) "Postgenomic Futures: Translations Across the Machine-Nature Border in Systems Biology," *New Genetics & Society* 24: 195–225.

Fujimura, Joan & Michael Fortun (1996) "Constructing Knowledge Across Social Worlds: The Case of DNA Sequence Databases in Molecular Biology," in L. Nader (ed), *Naked Science: Anthropological Inquiries into Boundaries, Power, and Knowledge* (London: Routledge): 160–73.

Gatens, Moira (1996) *Imaginary Bodies: Ethics, Power and Corporeality* (London: Routledge).

Gonzalez, Jennifer ([1995]1999) "Envisioning Cyborg Bodies: Notes from Current Research," in G. Kirkup, L. Janes, & K. Woodward (eds), *The Gendered Cyborg* (New York and London: Routledge): 58–73.

Goodwin, Dawn (in press) *Agency, Participation, and Legitimation: Acting in Anaesthesia.* (New York: Cambridge University Press).

Grand, Steve (2004) *Growing Up With Lucy: How to Build an Android in Twenty Easy Steps* (London: Weidenfeld & Nicolson).

Greenbaum, Joan & Morten Kyng (eds) (1991) *Design at Work: Cooperative Design of Computer Systems* (Hillsdale, NJ: Erlbaum Associates).

Grint, Keith & Rosalind Gill (eds) (1995) *The Gender-Technology Relation: Contemporary Theory and Research* (London: Taylor & Francis).

Grosz, Elizabeth (1994) *Volatile Bodies: Toward a Corporeal Feminism* (Bloomington: Indiana University Press).

Gupta, Akhil & James Ferguson (1992) "Beyond 'Culture': Space, Identity, and the Politics of Difference," *Cultural Anthropology* 7: 6–23.

Halberstam, Judith & Ira Livingston (eds) (1995) *Posthuman Bodies* (Bloomington and Indianapolis: Indiana University Press).

Hales, Mike (1995) "Information Systems Strategy: A Cultural Borderland, Some Monstrous Behaviour," in S. L. Star (ed), *Cultures of Computing* (Oxford: Blackwell): 103–17.

Hammonds, Evelynn (1997) "New Technologies of Race," in J. Terry & M. Calvert (eds), *Processed Lives: Gender and Technology in Everyday Life* (London and New York: Routledge): 107–21.

Haraway, Donna ([1985]1991) "Manifesto for Cyborgs: Science, Technology and Socialist Feminism in the 1980s," *Socialist Review* 80: 65–108. Reprinted (1991) in *Simians, Cyborgs, and Women: The Reinvention of Nature* (New York: Routledge).

Haraway, Donna (1991) *Simians, Cyborgs, and Women: The Reinvention of Nature* (New York: Routledge).

Haraway, Donna (1997) *Modest Witness@Second_Millenium.FemaleMan_Meets_OncoMouse: Feminism and Technoscience* (New York: Routledge).

Haraway, Donna (2000) *How Like a Leaf: An Interview with Thyrza Nichols Goodeve* (New York: Routledge).

Harding, Sandra (1998) *Is Science Multicultural? Postcolonialisms, Feminisms, and Epistemologies* (Bloomington: Indiana University Press).

Hawthorne, Susan & Renate Klein (1999) *Cyberfeminism, Connectivity, Critique and Creativity* (North Melbourne: Spinifex Press).

Hayden, Cori (2003) *When Nature Goes Public: The Making and Unmaking of Bioprospecting in Mexico* (Princeton, NJ: Princeton University Press).

Hayles, N. Katherine (1999) *How We Became Posthuman: Virtual Bodies in Cybernetics, Literature, and Informatics* (Chicago: University of Chicago Press).

Helmreich, Stefan (1998) *Silicon Second Nature: Culturing Artificial Life in a Digital World* (Berkeley: University of California Press).

Hutchins, Edwin (1995) *Cognition in the Wild* (Cambridge, MA: MIT Press).

Jain, Sarah (1999) "The Prosthetic Imagination: Enabling and Disabling the Prosthesis Trope" *Science, Technology & Human Values* 24: 31–53.

Keller, Evelyn Fox (1995) "The Origin, History, and Politics of the Subject Called 'Gender and Science': A First Person Account," in S. Jasanoff, G. Markle, J. Petersen, & T. Pinch (eds), *Handbook of Science and Technology Studies* (London: Sage): 80–94.

Keller, Evelyn Fox (1999) "The Gender/Science System: Or Is Sex to Gender as Nature Is to Science?" in M. Biagioli (ed), *The Science Studies Reader* (New York and London: Routledge): 234–42.

Kember, Sarah (2003) *Cyberfeminism and Artificial Life* (London and New York: Routledge).

Kirkup, Gill, Linda Janes, Kath Woodward, & Fiona Hovenden (eds) (2000) *The Gendered Cyborg: A Reader* (London: Routledge).

Lave, Jean (1988) *Cognition in Practice: Mind, Mathematics, and Culture in Everyday Life* (Cambridge and New York: Cambridge University Press).

Law, John (1991) *A Sociology of Monsters: Essays on Power, Technology, and Domination* (London and New York: Routledge).

Law, John (2004) *After Method: Mess in Social Science Research* (London and New York: Routledge).

Lenat, Douglas B. & R. V. Guha (1989) *Building Large Knowledge-based Systems: Representation and Inference in the Cyc Project* (Reading, MA: Addison-Wesley).

Lenoir, Tim & Sha Xin Wei (2002) "Authorship and Surgery: The Shifting Ontology of the Virtual Surgeon," in B. Clarke & L. D. Henderson (eds), *From Energy to Information: Representation in Science and Technology, Art, and Literature* (Stanford, CA: Stanford University Press): 283–308.

Lykke, Nina & Rosi Braidotti (eds) (1996) *Between Monsters, Goddesses and Cyborgs: Feminist Confrontations with Science, Medicine and Cyberspace* (London: Zed Books).

Lyman, Peter & Nina Wakeford (eds) (1999) "Analyzing Virtual Societies: New Directions in Methodology," *American Behavioral Scientist* 43: 3.

Lynch, Michael (1993) *Scientific Practice and Ordinary Action: Ethnomethodology and Social Studies of Science* (Cambridge and New York: Cambridge University Press).

Markussen, Randi (1995) "Constructing Easiness: Historical Perspectives on Work, Computerization, and Women," in S. L. Star (ed), *Cultures of Computing* (Oxford: Blackwell): 158–80.

Martin, Michelle (1991) *'Hello, Central?' Gender, Technology and Culture in the Formation of Telephone Systems* (Kingston, Ont.: Queen's University Press).

Maturana, Humberto & Francisco Varela (1980) *Autopoiesis and Cognition: The Realization of the Living* (Dordrecht, Netherlands, and Boston: D. Reidel).

Mayberry, Maralee, Banu Subramaniam, & Lisa Weasel (eds) (2001) *Feminist Science Studies: A New Generation* (New York and London: Routledge).

M'charek, Amade (2005) *The Human Genome Diversity Project: An Ethnography of Scientific Practice.* (Cambridge and New York: Cambridge University Press).

McNeil, Maureen (ed) (1987) *Gender and Expertise* (London: Free Association Books).

McNeil, Maureen & Sarah Franklin (1991) "Science and Technology: Questions for Cultural Studies and Feminism," In S. Franklin, C. Lury, & J. Stacey (eds), *Off-Centre: Feminism and Cultural Studies* (London: Harper Collins Academic): 129–46.

Miller, Laura (1995) "Women and Children First: Gender and the Settling of the Electronic Frontier," in J. Brook & I. A. Boal (eds), *Resisting the Virtual Life: The Culture and Politics of Information* (San Francisco: City Lights): 49–57.

Mol, Annemarie (2002) *The Body Multiple: Ontology in Medical Practice* (Durham, NC: Duke University Press).

Myers, Natasha (2005) "Molecular Embodiments and the Body-Work of Modeling in Protein Crystallography," *Social Studies of Science* 35(6): 837–66.

Nakamura, Lisa (2002) *Cybertypes: Race, Ethnicity and Identity on the Internet* (New York and London: Routledge).

Newell, Allen (1990) *Unified Theories of Cognition* (Cambridge, MA: Harvard University Press).

Newell, Allen & Herbert Simon (1972) *Human Problem Solving* (Englewood Cliffs, NJ: Prentice Hall).

Picard, Rosalind (1997) *Affective Computing* (Cambridge, MA: MIT Press).

Pickering, Andrew (1992) *Science as Practice and Culture* (Chicago and London: University of Chicago Press).

Prentice, Rachel (2005) "The Anatomy of a Surgical Simulation: The Mutual Articulation of Bodies in and Through the Machine," *Social Studies of Science* 35: 837–66.

Price, Janet & Margrit Shildrick (eds) (1999) *Feminist Theory and the Body: A Reader* (Edinburgh: Edinburgh University Press).

Reid, Roddey & Sharon Traweek (2000) *Doing Science + Culture: How Cultural and Interdisciplinary Studies Are Changing the Way We Look at Science and Medicine* (New York: Routledge).

Robertson, Toni (2002) "The Public Availability of Actions and Artifacts," *Computer-Supported Cooperative Work (CSCW)* 11: 299–316.

Schiebinger, Londa (ed) (2000) *Feminism and the Body* (Oxford: Oxford University Press).

Schull, Natasha (2005) "Digital Gambling: The Coincidence of Desire and Design," *Annals of the American Academy of Political and Social Science* 597: 65–81.

Sengers, Phoebe (in press) "The Autonomous Agency of STS: Boundary Crossings Between STS and Artificial Intelligence," *Social Studies of Science*.

Simon, Herbert (1969) *The Sciences of the Artificial* (Cambridge, MA: MIT Press).

Spender, Dale (1996) *Nattering on the Net: Women, Power and Cyberspace* (Toronto: Garamond Press).

Star, Susan Leigh (1989a) "Layered Space, Formal Representations and Long Distance Control: The Politics of Information," *Fundamenta Scientiae* 10: 125–55.

Star, Susan Leigh (1989b) "The Structure of Ill-structured Solutions: Boundary Objects and Heterogeneous Distributed Problem-solving," in L. Gasser & M. Huhns (eds), *Readings in Distributed Artificial Intelligence 2* (Menlo Park, CA: Morgan Kaufmann): 37–54.

Star, Susan Leigh (ed) (1995a) *The Cultures of Computing* (Oxford and Cambridge, MA: Blackwell).

Star, Susan Leigh (ed) (1995b) *Ecologies of Knowledge: Work and Politics in Science and Technology* (Albany: State University of New York Press).

Stone, Allucquere Rosanne (1999) "Will the Real Body Please Stand Up? Boundary Stories About Virtual Cultures," in J. Wolmark (ed), *Cybersexualities* (Edinburgh: Edinburgh University Press): 69–98.

Strathern, Marilyn (1992) *Reproducing the Future: Essays on Anthropology, Kinship and the New Reproductive Technologies* (Manchester: Manchester University).

Strathern, Marilyn (1999) *Property, Substance, and Effect: Anthropological Essays on Persons and Things* (London and New Brunswick, NJ: Athlone Press).

Suchman, Lucy (1987) *Plans and Situated Actions: The Problem of Human-Machine Communication* (New York: Cambridge University Press).

Suchman, Lucy (1993) "Response to Vera and Simon's 'Situated Action: A Symbolic Interpretation'," *Cognitive Science* 17: 71–75.

Suchman, Lucy (1999a) "Critical Practices," *Anthropology of Work Review* 20: 12–14.

Suchman, Lucy ([1994]1999b) "Working Relations of Technology Production and Use," in D. Mackenzie & J. Wajcman (eds), *The Social Shaping of Technology,* 2nd ed (Buckingham, U.K., and Philadelphia: Open University Press): 258–68.

Suchman, Lucy (2002a) "Located Accountabilities in Technology Production," *Scandinavian Journal of Information Systems* 14: 91–105.

Suchman, Lucy (2002b) "Practice-based Design of Information Systems: Notes from the Hyperdeveloped World," *Information Society* 18: 139–44.

Suchman, Lucy (2003) "Figuring Service in Discourses of ICT: The Case of Software Agents," in Eleanor Wynn, Edgar Whitley, Michael Myers, & Janice DeGross (eds), *Global and Organizational Discourses About Information Technology* (Boston: Kluwer Academic, copyright International Federation for Information Processing): 15–32.

Suchman, Lucy (2007) *Human-Machine Reconfigurations: Plans and Situated Actions*, 2nd ed. (New York: Cambridge University Press).

Suchman, Lucy & Randall Trigg (1993) "Artificial Intelligence as Craftwork," in S. Chaiklin & J. Lave (eds), *Understanding Practice: Perspectives on Activity and Context* (New York: Cambridge University Press): 144–78.

Terry, Jennifer & Melodie Calvert (eds) (1997) *Processed Lives: Gender and Technology in Everyday Life* (London and New York: Routledge).

Thompson, Charis (2005) *Making Parents: The Ontological Choreography of Reproductive Technologies* (Cambridge, MA: MIT Press).

Turkle, Sherry (1995) *Life on the Screen: Identity in the Age of the Internet* (New York and Toronto: Simon & Schuster).

Verran, Helen (2001) *Science and an African Logic* (Chicago: University of Chicago Press).

Viseu, Ana (2005) *Augmented Bodies: The Visions and Realities of Wearable Computers*, unpublished Ph.D. dissertation, University of Toronto.

Wajcman, Judy (1991) *Feminism Confronts Technology* (University Park: Pennsylvania State University Press).

Wajcman, Judy (1995) "Feminist Theories of Technology," in S. Jasanoff, G. Markle, J. Petersen, & T. Pinch (eds), *Handbook of Science and Technology Studies* (Thousand Oaks, CA, and London: Sage): 189–204.

Wajcman, Judy (2004) *Technofeminism* (Cambridge: Polity).

Wakeford, Nina (1997) "Networking Women and Grrrls with Information/Communication Technologies: Surfing Tales of the World Wide Web," in J. Terry & M. Calvert (eds), *Processed Lives: Gender and Technology in Everyday Life* (London and New York: Routledge): 51–66.

Waldby, Cathy (2000) *The Visible Human Project: Informatic Bodies and Posthuman Medicine* (London and New York: Routledge).

Wilson, Elizabeth (2002) "Imaginable Computers: Affects and Intelligence in Alan Turing," in D. Tofts, A. Jonson, & A. Cavallaro (eds), *Prefiguring Cyberculture: An Intellectual History* (Cambridge, MA: MIT Press): 38–51.

Wolmark, Jennifer (ed) (1999) *Cybersexualities: A Reader on Feminist Theory, Cyborgs and Cyberspace* (Edinburgh: Edinburgh University Press).

7 Technological Determinism Is Dead; Long Live Technological Determinism

Sally Wyatt

The story of Robert Moses and the bridges between New York and Long Island made a great impression on me as on many generations of STS students. Langdon Winner argues that Moses, city planner, deliberately allowed overpasses built during the 1920s and '30s that were too low to permit buses to go beneath them, thus excluding poor, black, and working-class people from the beaches of Long Island.[1] I first read Winner's "Do Artifacts Have Politics?" in the mid-1980s, in the first edition of *The Social Shaping of Technology* (MacKenzie and Wajcman, [1985]1999), where it is the opening chapter following the editors' introduction.[2] As I have described elsewhere (Wyatt, 2001), I am the daughter of a nuclear engineer, so I grew up knowing that technologies are political. Making nuclear power work and justifying his efforts to do so to both his family and a wider public were the stuff of my father's daily life for many years. Despite the continued political differences between me and my father, we shared an appreciation of the existence and implications of technical choices; reading Langdon Winner provided me with a way of thinking about the politics of artifacts more systematically, and perhaps enabled my father and me to discuss these politics more dispassionately. What I learned from my father was that technology indeed matters and that technical choices have consequences, though perhaps I would not have expressed it in quite such terms when I was six years old and he took me to Niagara Falls, not only to admire the water but also to look at the turbines down river. My father and I did not distinguish between the natural and the technological sublime (Nye, 1996).

In this chapter, I wish to address both the ways in which technology itself and the idea of technological determinism continue to fascinate, even if those of us in the STS community sometimes deny this fascination. In the next section, I discuss technological determinism and then turn to the "principle of symmetry" (see also table 7.1) in order to make two points, one about success and failure and the other about treating actors' and analysts' concepts symmetrically, as a way of allowing technological determinism back into our analyses. I then return to technological determinism, arguing that one way of taking it more seriously is to disentangle the different types and what work they do. I identify four different types: justificatory, descriptive, methodological, and normative (see also table 7.2).

Table 7.1.
Extending the Principle of Symmetry

Bloor (1973, 1976) on Science	Pinch & Bijker (1984) on Technology	Callon (1986) on Socio-technology	Wyatt (1998) on Method in STS
Impartial to a statement being true or false	Impartial to a machine being a success or failure	Impartial to an actor being human or nonhuman	Impartial to an actor being identified by other actors or the analyst
Symmetrical with respect to explaining truth and falsity	Symmetrical with respect to explaining success and failure	Symmetrical with respect to explaining the social world and the technical world	Symmetrical with respect to using concepts from analysts and actors
What we take to be nature is the result and not the cause of a statement becoming a true fact	*"Working" is the result and not the cause of a machine becoming a successful artifact*	Distinction between the "technical" and the "social" is the result and not the cause of the stabilization of socio-technical ensembles	*"Success" is the result and not the cause of a machine becoming a working artifact*

Source: First three columns adapted from Bijker (1995: 275).

Table 7.2
Four Types of Technological Determinism

Justificatory
• EU Information Society Forum (2000)
Descriptive
• Technology developed independently of social forces (Misa, 1988)
• Technology causes social change (Misa, 1988; Smith & Marx, 1994)
• Technology developed independently of social forces and causing social change (MacKenzie and Wajcman, [1985] 1999)
• Limited autonomy of science and technology in determining economic developments (Freeman, 1987)
Methodological
• "Look to the technologies available to societies, organizations, and so on" (Heilbroner, 1994a or b)
• "Momentum" (Hughes, 1983, 1994)
• "Society is determined by technology in use" (Edgerton, 1999)
Normative
• Decoupling of technology from political accountability (Bimber, 1994)
• Triumph of technological rationality (Winner, 1977, 1986)

TECHNOLOGICAL DETERMINISM IS DEAD, OR IS IT?

Technological determinism persists in the actions taken and justifications given by many actors; it persists in analysts' use of it to make sense of the introduction of technology in a variety of social settings; it persists in manifold theoretical and abstract accounts of the relationship between the technical and the social; it persists in the responses of policy makers and politicians to challenges about the need for or appropriateness of new technologies; and it persists in the reactions we all experience when confronted with new machines and new ways of doing things. (Examples of each of these can be found in the Back to Technological Determinism section.)

Hannah Arendt (1958: 144) wrote, "[t]ools and instruments are so intensely worldly objects that we can classify whole civilizations using them as criteria." Not only can we, but frequently we do; thus, we speak of the "stone," "iron," "steam," and "computer" ages. We also characterize nations by reference to technologies in which they have played a prominent developmental role and/or which are highly symbolic of their culture: Holland and windmills, the United States and cars, Japan and microelectronics. Robert Heilbroner (1994b) and David Edgerton (1999) argue that it is the availability of different machines that defines what it is like to live in a particular place and time. Lewis Mumford (1961) suggests that the tendency to associate whole millennia or entire nations with a single material artifact has arisen because the first academic disciplines to treat technological change seriously were anthropology and archaeology, which often focus on nonliterate societies for which material artifacts are the sole record.

[T]he stone or pottery artifact came to be treated as self-existent, almost self-explanatory objects ... These tools, utensils, and weapons even created strange technological homunculi, called 'Beaker Men', 'Double Axe Men', or 'Glazed Pottery Men' ... The fact that such durable artifacts could be arranged in an orderly progressive series often made it seem that technological change had no other source than the tendency to manipulate the materials, improve the processes, refine the shapes, make the product efficient. Here the absence of documents and the paucity of specimens resulted in a grotesque overemphasis of the material object, as a link in a self-propelling, self-sustaining technological advance, which required no further illumination from the culture as a whole even when the historic record finally became available. (Mumford, 1961: 231)

Those of us concerned with more contemporary societies have no similarly convenient excuse for such reductionist thinking. Yet the linguistic habit persists of naming whole historical epochs and societies by their dominant technological artifacts. This habit can be witnessed frequently in museums, schoolbooks, and newspapers and on television and radio.[3] Even a few years into the twenty-first century, it is still difficult to predict for which of its many new technologies the twentieth century will be remembered by future generations, yet the habit of thought and language of associating places and time periods with their technologies endures, even if causality is not always explicit. This way of thinking about the relationship between technology and society has been "common sense" for so long that it has hardly needed a label. But its critics have termed it "technological determinism," which has two parts. The first part is that technological developments take place outside society, independently of social, economic, and political forces. New or improved products or ways of making things arise from the activities of inventors, engineers, and designers following an internal, technical logic that has nothing to do with social relationships. The more crucial second part is that technological change causes or determines social change. Misa (1988) suggests that what I have presented here as two parts of a single whole are actually two different versions of technological determinism. Defining it as two different versions enables the scourges of technological determinism to cast their condemnatory net more widely by defining people like Winner and Ellul as technological determinists because they point to the inexorable logic of capitalist rationality. This is to confuse their materialism and realism with determinism. If they are to be accused of any sort of determinism, economic determinism is the more appropriate charge. I follow MacKenzie and Wajcman ([1985]1999) in defining technological determinism as having two parts, both of which are necessary,[4] and I will return to this distinction later. Over the past 25 years, STS has focused primarily on demonstrating how limited the first part of technological determinism is, usually by doing empirically rich historical or ethnographic studies demonstrating how deeply social the processes of technological development are.[5]

Technological determinism is imbued with the notion that technological progress equals social progress. This was the view of Lenin (1920) when he claimed that "Communism is Soviet power plus the electrification of the whole country" and it remains the view of politicians of all political persuasions. For example, George W. Bush, a politician very different from Lenin, is committed to missile defense, and as he stated in his 2006 State of the Union address, he sees technology as the solution to the looming energy crisis in the United States.[6] There is also a strand of very pessimistic technological determinism, associated with the work of Ellul (1980), Marcuse (1964), and the Frankfurt School generally. Historically, technological determinism means that each generation produces a few inventors whose inventions appear to be both the determinants and stepping stones of human development. Unsuccessful inventions are condemned by their failure to the dust heap of history. Successful ones soon prove their value and are more or less rapidly integrated into society, which they proceed to transform. In this way, a technological breakthrough can be claimed to have important social consequences.

The simplicity of this model is, in large part, the reason for its endurance. It is also the model that makes most sense of many people's experience. For most of us, most of the time, the technologies we use every day are of mysterious origin and design. We have no idea whence they came and possibly even less idea how they actually work. We simply adapt ourselves to their requirements and hope that they continue to function in the predictable and expected ways promised by those who sold them to us. It is because technological determinism conforms with a huge majority of people's experiences that it remains the "common sense" explanation.

One of the problems with technological determinism is that it leaves no space for human choice or intervention and, moreover, absolves us from responsibility for the technologies we make and use. If technologies are developed outside of social interests, then workers, citizens, and others have very few options about the use and effects of these technologies. This serves the interests of those responsible for developing new technologies, regardless of whether they are consumer products or power stations. If technology does indeed follow an inexorable path, then technological determinism does allow all of us to deny responsibility for the technological choices we individually and collectively make and to ridicule those people who do challenge the pace and direction of technological change.

This chapter demonstrates that we cannot ignore technological determinism in the hope that it will disappear and that the world will embrace the indeterminancy and complexity of other types of accounts of the technology-society relationship. I argue that we in the STS community cannot simply despair of the endurance of technological determinism and carry on with our more subtle analyses. We must take technological determinism more seriously, disentangle the different types, clarify the purposes for which it is used by social actors in specific circumstances. Moreover, I argue that in order to do this we have to recognize the technological determinists within ourselves.

A BRIEF AND SYMMETRICAL DETOUR

Before returning to the discussion of technological determinism, I want to digress slightly and discuss the principle of symmetry (table 7.1) in order to demonstrate two points. The first is the more conventional application of the principle of symmetry related to working and success versus nonworking and failure. The second relates to the symmetrical treatment of actors' and analysts' concepts.

First, the principle of symmetry was initially articulated by David Bloor (1973, 1976) in relation to the sociology of science. He argues that knowledge claims that are accepted as true and those that are regarded as false are both amenable to sociological explanation, an explanation that must be given in the same terms. Nature itself must not be used to justify one claim and not another: what we take to be nature is the result of something being accepted as true, not the cause. In the case of technology, the principle of symmetry suggests that successful and failed machines or artifacts need to be explained in the same, social terms. However, unequivocally successful systems do not provide such a rigorous test for Pinch and Bijker's (1984) claim that working is the result and not the cause of a machine becoming a successful artifact. For successful systems, such a claim is tautological. However, there are other, more ambiguous systems,[7] in terms of success and failure, working and nonworking, which are a better illustration of how right Pinch and Bijker are, especially if an iterative loop is added to the statement. In previous work about ICT-network systems in the U.S. and U.K. central government administrations (Wyatt, 1998, 2000), I demonstrated how such systems worked, were not successful, and no longer work. Playing the postmodern trick[8] of reversing the wording of the claim so that it becomes, "success is the result and not the cause of a machine becoming a working artifact" illustrates the significance of Callon's contribution to table 7.1, namely, his exhortation to treat the sociotechnical divide as a consequence of the stabilization of sociotechnical ensembles and not as a prior cause. One of the difficulties with the Pinch and Bijker claim about working being the result rather than the cause of a machine becoming a successful artifact is that they presume the existence of that divide in their association of success with the social world and of working with the technical world, thus presuming a binary divide between the social and the technical, whereas much of STS is concerned with demonstrating how interwoven the social and technical are with one another. Moreover, one cannot privilege the social as they do by placing "success" prior to "working." It has to be possible to reverse the claim as I have done here in order to make visible the mutual constitution of the social and the technical, but that means that successive extensions of the principle of symmetry have led us back to a position of classical realism. This should not come as a surprise. The claims of "success" and "working" have to be interchangeable to enable us to treat the social and the technical symmetrically. Rather than seeing the bottom items in the columns in table 7.1 attributed to Pinch and Bijker and Wyatt as alternatives, they need to be understood as two sides of the same coin. Neither is adequate on its own.

The second point is that actors' and analysts' concepts need to be treated symmetrically, the middle claim by Wyatt in table 7.1. Others (Bijsterveld, 1991; Martin and Scott, 1992; Russell, 1986; Winner, 1993) have pointed to the limits of "following the actors" (Latour, 1987), in particular that by doing so analysts may miss important social groups that are invisible to the actors but nonetheless important. Users are often overlooked by developers (Oudshoorn and Pinch, chapter 22 in this volume). Often it is possible to define clearly who the users are or will be. But with information networks, for example, there can be at least two sets of users. The first group is those people who are conventionally considered to be the users; employees who use the system to access information in order to perform their job tasks. In many cases, there is also a second group of users: clients or customers whom the more direct users ultimately serve with the help of the system and who have different interests.

To understand the role of users, it is important to distinguish between "real" users in the "real" world and the images of those users and their relationships held by designers, engineers, and other sorts of system builders. It is also important to be aware of "implicated users" (Clarke, 1998), those who are served by the system but who do not have any physical contact with it. Again, distinctions need to be made about their actual social relations and the images held of them. Sometimes both sorts of users are ignored during systems development, in other words, serious attempts are not always made to configure the users (cf. Woolgar, 1991), raising both methodological and normative issues.

There are problems with following the actors. Identifying all the relevant social groups as mentioned above and defining scale[9] and success can become messy or impossible if analysts are over-reliant on actors' accounts. As analysts, we have to rely on ourselves and on the research done by others to help us define our concepts and identify relevant groups. Let us continue to take seriously the principle of symmetry. If, as analysts, we allow our own categories and interpretations into the constructions of our stories, we also need to allow actors' concepts and theories to inform our accounts. Actors and analysts all have access to both the abstract and the material.

Anthony Giddens (1984) has a particular view of the double hermeneutic in social science[10]: Not only do social scientists need to find ways of understanding the world of social actors, they also need to understand the ways in which their theories of the social world are interpreted by those social actors. In other words, the ideas, concepts, theories of both social actors and social scientists need to be given space. "Follow the actors" can be rescued by recourse to the higher principle of symmetry. Actors' and analysts' identification of other actors and their interests should be treated symmetrically. But I certainly do not wish to grant the analyst the status of an omniscient, superior being. In the next and final section, I will return to the persistence of technological determinism and argue that its continued use by actors necessitates that as analysts we take it more seriously than we have done in recent years. Following Giddens (1984) means that actors' theoretical ideas need to be treated symmetrically with our own, even if they are antithetical to our deeply held views.

BACK TO TECHNOLOGICAL DETERMINISM

Within the humanities and social sciences[11] we frequently ignore the equivalent of a thundering herd of elephants when we dismiss the role of technological determinism in shaping the views and actions of actors.[12] Michael L. Smith eloquently expresses a similar view,

We scholars of technology and culture lament the stubborn tenacity of technological determinism, but we rarely try to identify the needs it identifies and attempts to address. On the face of it, our brief against this variety of superstition resembles the academy's response to creationism: How can something so demonstrably wrong-headed continue to sway adherents? (1994: 38–39)

Smith is correct to point to the importance of understanding the needs and interests served by a continued adherence to technological determinism, and I will return to that below. He is wrong, however, to dismiss technological determinism (and creationism) as wrong-headed superstition or as a form of false consciousness. Recall Bloor's (1973, 1976) original formulation of the principle of symmetry, namely, that both true and false beliefs stand in need of explanation. We need to remain impartial in our attempt to explain the persistence of technological determinism in order to understand why it continues to be regarded as true by so many people. In the previous section, I argued that the categories deployed by both actors and analysts need to be pursued in order to justify paying attention to users who might never be noticed if analysts naïvely follow the actors. Now it is time to follow the actors in their continued commitment to technological determinism.

One of the most misleading and dangerous aspects of technological determinism is its equation of technological change with progress.[13] From the many histories and contemporary case studies of technological change we know how messy and ambiguous the processes of developing technologies can be. But this is not always the perspective of actors. Some actors, some of the time, present projects as simple and straightforward. It is necessary for them to do so in order to make things happen and to justify their actions. Sometimes sociotechnical ensembles work; sometimes they do not. Including stories of systems that do not work or were not used or were not successful provides further armory in the arsenal to be used against technological determinism because such stories challenge the equation of technology with progress, though not, of course, if we have an evolutionary perspective on progress. But we should not be under any illusions that technological determinism will disappear, and we should recognize that it has a useful function for system builders.

In this section, I return to an exploration of the endurance of technological determinism—endurance in the accounts of some analysts, in the actions of system builders, as well as the justifications proffered by policy makers and other social groups. Despite all the detailed empirical work in STS about both historical and contemporary examples of the contingency of technological change and despite the nuanced and sophisticated theoretical alternatives that have been proposed, technological determinism persists. One of the dangers of simply ignoring it in the hope it will disappear is that we do not pay sufficient attention to its subtlety and variety. Sometimes it is a table upon which to thump our realist credentials; occasionally it can be a rapier to pierce the pretensions of pompous pedants. In whatever way it is used, my argument here is that we need to take it more seriously.

One of the few sustained engagements with technological determinism to be published is the collection, *Does Technology Drive History? The Dilemma of Technological Determinism*, edited by Merritt Roe Smith and Leo Marx (1994). All the contributors are professors of history at U.S. universities, and the concerns they express are largely those of historians of technology, in their relationship with other historians, and of Americans, with their historic paradigmatic equation of technology with progress and their collective but partial loss of faith with that equation. The contributors provide a valuable mapping of the terrain of meanings associated with the concept of technological determinism.

In their introduction, Smith and Marx (1994: ix–xv) suggest that technological determinism can take several forms, along a spectrum between hard and soft poles.

At the "hard" end of the spectrum, agency (the power to effect change) is imputed to technology itself, or to some of its intrinsic attributes; thus the advance of technology leads to a situation of inescapable necessity . . . To optimists, such a future is the outcome of many free choices and the realization of the dream of progress; to pessimists, it is a product of necessity's iron hand, and it points to a totalitarian nightmare. (Smith & Marx, 1994: xii)

At the pole of "soft" determinism, technology is located, "in a far more various and complex social, economic, political, and cultural matrix." (Smith & Marx, 1994: xiii) In my view, this soft determinism is vague and is not really determinism at all, as it returns us to the stuff of history, albeit a history in which technology is taken seriously.

Robert Heilbroner's famous article, "Do Machines Make History?," originally published in *Technology and Culture* in 1967, is reproduced in the collection, together with his own recent reflections on the question. He is the most avowedly technologically determinist of the contributors, in both an ontological and methodological sense. He suggests that a good place to start in the study of an unfamiliar society is to examine the availability of different machines, since this will define what it is like to live in a particular place and time (1994a: 69–70). He proposes this as a heuristic for investigation, not as a normative prescription. "[T]echnological determinism does not imply that human behaviour must be deprived of its core of consciousness and responsibility" (1994a: 74). David Edgerton makes a similar point when he argues that technological determinism must be seen as the "the thesis that society is determined by technology in use" (1999: 120), which, as he points out, allows inclusion of societies with technology but not necessarily with high rates of technological change.

Bruce Bimber picks up the theme of normative prescription. He distinguishes between three interpretations of technological determinism, what he terms "normative," "nomological," and "unintended consequences" accounts.[14] The first he associates with the work of Winner (1977), Ellul (1980), and Habermas (1971), among others, who suggest that technology can be considered autonomous and determining when the norms by which it is developed have become removed from political and ethical debates. For all the authors Bimber mentions, the decoupling of technology from political accountability is a matter of great concern. Nomological technological determinism is Bimber's very hard version: "in light of the past (and current) state of technological development and the laws of nature, there is only one possible future course of social change" (1994: 83). To make this even harder, Bimber imposes a very narrow definition of technology: artifacts only. No knowledge of production or use can be incorporated because that would allow social factors to enter this otherwise asocial world. His final category arises from the observation that social actors are unable to anticipate all the effects of technological change. However, since this is true for many other activities and does not arise from some intrinsic property of technology, Bimber dismisses this as a form of technological determinism. Bimber is concerned to rescue Karl Marx from the accusation of technological determinism. This he does by setting up these three accounts, suggesting that the nomological is the only true technological determinism and that Marx does not meet the strict criteria.[15]

Thomas Hughes returns to the spirit of the distinctions made by Smith and Marx between hard and soft determinism, albeit in different terms and with the explicit objective of establishing "technological momentum" as "a concept that can be located somewhere between the poles of technical determinism and social constructivism." For Hughes, "[a] technological system can be both a cause and an effect; it can shape or be shaped by society. As they grow larger and more complex, systems tend to be more shaping of society and less shaped by it" (Hughes, 1994: 112). On a methodological level, he suggests that social constructivist accounts are useful for understanding the emergence and development of technological systems, but momentum is more useful for understanding their subsequent growth and the acquisition of at least the appearance of autonomy.

This discussion leads me to distinguish between four types of technological determinism, which I term justificatory, descriptive, methodological, and normative (table 7.2). Justificatory technological determinism is deployed largely by actors. It is all around us. It is the type of technological determinism used by employers to justify downsizing and reorganization. It is the technological determinism we are all susceptible to when we consider how people's lives have changed in the past 200 years. It is the technological determinism (and frustration) we feel when confronted with an automated call response system. It can be found in policy documents, including the EU Information Society Forum report, which claims, "[t]he tremendous achievement of the ICT sector in the last few years, and particularly of the Internet, have practically cancelled the concept of time and distance . . . The emerging digital economy is radically changing the way we live, work and communicate, and there is no doubt about the benefits that will lead us to a better quality of life" (2000: 3). It is similar to what Paul Edwards has called the "ideology of technological determinism" (1995: 268) when he reflects on "managers' frequent belief that productivity gains and social transformation will be automatic results of computerization." (1995: 268)

Second is the descriptive technological determinism identified by MacKenzie and Wajcman ([1985]1999), Misa (1988), and Smith and Marx (1994: ix–xv). These authors eschew technological determinism as modes of explanation for themselves but certainly recognize it when they see it in others. Having recognized it, they rarely attempt to understand the reasons for it and instead focus on developing richer, more situated explanations of sociotechnical change. They simply reject technological determinism because of its inadequate explanatory power. Christopher Freeman (1987) is more assertive in his defense of this type of technological determinism, arguing that in some cases at least, technological determinism is quite a good description of the historical record.

Third is the methodological technological determinism of Heilbroner, Edgerton, and Hughes. Heilbroner reminds us to start our analyses of societies, and of smaller scale social organizations, by examining the technologies available to them. Hughes' methodological technological determinism is more analytical. But, like Heilbroner, he too is attempting to develop a tool for helping us understand the place of technology in history. In STS, that is what we are all doing—attempting to understand the role of technology in history and in contemporary social life; actor-network theory, social constructivism, history of technology, and innovation theory all take technology seriously. All of these approaches are regarded as deviant by their parent disciplines because they include technologies in their analyses of the social world. My provocation here is that our guilty secret in STS is that really we are all technological determinists. If we were not, we would have no object of analysis; our *raison d'être* would disappear. Winner hints at this obliquely at the end of the preface of *Autonomous Technology* (1977) when he writes, "there are institutions [machines] one must oppose and struggle to modify even though one also has considerable affection for them" (1977: x).

Finally, there is the normative technological determinism identified by Bimber, by Misa in his second version, and implicit in Hughes' concept of momentum. This is the autonomous technology of Langdon Winner, technology that has grown so big and so complex that it is no longer amenable to social control. It is this version of technological determinism that has resulted in the intra-STS skirmishes, in which Winner (1993) accuses constructivists of abandoning the need to render technology and technological change more accountable, and it is with this accusation in mind that I conclude.

CONCLUSION

Does Technology Drive History? ends with a moving plea from John Staudenmaier to continue to take the history of technology seriously, to treat artifacts, "as crystallized moments of past human vision ... each one buffeted by the swirl of passion, contention, celebration, grief and violence that makes up the human condition" (1994: 273). Scholars concerned with understanding the relationship between technology and society share that commitment.

In STS we study people and things, and we study images of people and things. We also need to study explanations of people and things. Just as we treat technology seriously, we must treat technological determinism seriously. It is no longer sufficient to dismiss it for its conceptual crudeness, nor is it enough to dismiss it as false consciousness on the part of actors or as a bleak, Nietzschean outlook for the future of humanity. Technological determinism is still here and unlikely to disappear. It remains in the justifications of actors who are keen to promote a particular direction of change, it remains as a heuristic for organizing accounts of technological change, and it remains as part of a broader public discourse which seeks to render technology opaque and beyond political intervention and control.

What I have done here is to delineate different types of technological determinism, not because I believe it to be an adequate framework for understanding the relationship between the social and technical worlds but because lots of other actors do, and therefore we need to understand its different manifestations and functions. Within STS, we have always treated technology seriously; we have always been concerned with the risks and dangers of autonomous technology. We are not innocent in the ways of methodological and normative technological determinism. But we can no longer afford to be so obtuse in ignoring the justificatory technological determinism of so many actors. Only by taking that type of technological determinism seriously will we be able to deepen our understanding of the dynamics of sociotechnical systems and the rhetorical devices of some decision makers.

The challenges for STS remain: to understand how machines make history in concert with current generations of people; to conceptualize the dialectical relationship between the social shaping of technology and the technical shaping of society; and to treat symmetrically the categories of analysts and those of actors even if the latter includes technological determinism, anathema to so much contemporary scholarship in the humanities and social sciences. These dialectics are unresolvable one way or another, but that is as it should be. What is important is to continue to wrestle with them. We need to take seriously the efforts to stabilize and extend the messy and heterogeneous collections of individuals, groups, artifacts, rules, and knowledges that make up our sociotechnical world. We need to continue to grapple with understanding why sometimes such efforts succeed and sometimes they do not. Only then will people have the tools to participate in creating a more democratic sociotechnical order.

Notes

I am very grateful to the editors and three anonymous reviewers for their thoughtful and provocative comments, which helped improve this chapter considerably. I am also grateful to the editors for their patience in waiting for this chapter. My father died at the end of 2005, when I should have been preparing the final version. As the reader will learn from the first paragraph, my father had an enormous influence on my own views about technology. The most difficult revisions I had to make were to the verb tenses in that paragraph. This chapter is dedicated to the memory of my father, Alan Wyatt.

1. See discussion by Joerges (1999) and Woolgar and Cooper (1999) regarding the mythic status of the Moses/Winner story. For my purposes here, it is precisely the mythic quality of the story that counts.

2. The foundational status of this piece is confirmed by its inclusion in the second edition (MacKenzie & Wajcman, [1985]1999), still in the number one spot.

3. For example, the 2005 BBC Reith lectures were given by Lord Broers, Chairman of the British House of Lords Science and Technology Committee and President of the Royal Academy of Engineers. The title of his first lecture was "Technology will determine the future of the human race." The title for the series of five lectures was "The Triumph of Technology" (see www.bbc.co.uk/radio4/reith2005).

4. Feenberg also identifies two premises on which technological determinism is based, what he calls "unilinear progress" and "determination by the base" (1999: 77). This is much the same as the two parts of MacKenzie and Wajcman, since unilinear progress refers to the internal logic of technological development and determination by the base refers to the ways in which social institutions are required to adapt to the technological "base."

5. Two early examples are Latour and Woolgar (1986) and Traweek (1988). However, the pages of *Social Studies of Science* and of *Science, Technology & Human Values* are filled with such case studies.

6. For the full text of Bush's speech, see http://www.whitehouse.gov/stateoftheunion/2006/index.html.

7. For example, it is often very difficult to evaluate clearly the success and failure of many information technology-based systems in terms of their success and failure, working and nonworking.

8. See Derrida (1976): The signified is always already in the position of the signifier, often paraphrased as X is always already Y.

9. Defining scale is not only an analytical problem facing the researcher (Joerges, 1988), it is also a practical one for the actors. The resolution of this problem is necessary for the researcher to circumscribe the object of study, but it is also a problem experienced by the actors.

10. Within philosophy, the "double hermeneutic" is used more generally to refer to the problem that social scientists have in dealing with the interpretations of social life produced by social actors themselves as well as the interpretations of social life produced by analysts.

11. I exempt historians from this criticism, especially in light of the publication of Smith and Marx (1994) and, more recently, of Oldenziel (1999), in which she carefully traces the shifting meaning of technology and the rise of technocracy in the United States.

12. Equally peculiar is the way in which technology itself is ignored. As Brey (2003) points out in his comprehensive review of literature pertaining to technology and modernity, much of the modernity literature makes, at best, only passing reference to technology. Brey argues that this is not because modernity authors do not recognize the importance of technology but rather because they see it as the means by which regulative frameworks such as capitalism, the nation state, or the family are governed and not as an institution itself. Another reason may be that social science and humanities scholars may not have the tools or the confidence to analyze technology as such, and at most are only able to critique discourses around technology.

13. See Leo Marx (1994) for a detailed historical account of the emergence of technology and its relationship to ideas of progress.

14. These are not dissimilar to Radder's (1992) distinctions between methodological, epistemological, and ontological relativism.

15. I agree with Bimber that Karl Marx was not a technological determinist, but this point has already been more than adequately made by MacKenzie (1984) in his detailed review of this literature.

References

Arendt, Hannah (1958) *The Human Condition* (Chicago: University of Chicago Press).

Bijker, Wiebe (1995) *Of Bicycles, Bakelites and Bulbs, Toward a Theory of Socio-Technical Change* (Cambridge, MA: MIT Press).

Bijsterveld, Karin (1991) "The Nature of Aging: Some Problems of 'An Insider's Perspective' Illustrated by Dutch Debates about Aging (1945–1982)," presented at the Society for Social Studies of Science conference, Cambridge, MA, November.

Bimber, Bruce (1994) "Three Faces of Technological Determinism," in M. R. Smith & L. Marx (eds), *Does Technology Drive History? The Dilemma of Technological Determinism* (Cambridge, MA: MIT Press): 79–100.

Bloor, David (1973) "Wittgenstein and Mannheim on the Sociology of Mathematics," *Studies in History and Philosophy of Science* 4: 173–91.

Bloor, David (1976) *Knowledge and Social Imagery* (London: Routledge & Kegan Paul).

Brey, Philip (2003) "Theorizing Modernity and Technology," in T. Misa, P. Brey, & A. Feenberg (eds), *Modernity and Technology* (Cambridge, MA: MIT Press): 33–72.

Callon, Michel (1986) "Some Elements of a Sociology of Translation: Domestication of the Scallops and the Fishermen of St. Brieuc Bay," in J. Law (ed), *Power, Action and Belief: A New Sociology of Knowledge?* (London: Routledge & Kegan Paul): 196–233.

Clarke, Adele (1998) *Disciplining Reproduction: Modernity, American Life and 'The Problem of Sex'* (Berkeley: University of California Press).

Derrida, Jacques (1976) *Of Grammatology*, trans. G Spivak (Baltimore, MD: Johns Hopkins University Press).

Edgerton, David (1999) "From Innovation to Use: Ten Eclectic Theses on the Historiography of Technology," *History and Technology* 16: 111–36.

Edwards, Paul (1995) "From 'Impact' to Social Process: Computers in Society and Culture," in S. Jasanoff, G. Markle, J. Petersen, & T. Pinch (eds), *Handbook of Science and Technology Studies* (Thousand Oaks: Sage).

Ellul, Jacques (1980) *The Technological System* (New York: Continuum).

Feenberg, Andrew (1999) *Questioning Technology* (London & New York: Routledge).

Freeman, Christopher (1987) "The Case for Technological Determinism" in R. Finnegan, G. Salaman, & K. Thompson (eds), *Information Technology: Social Issues: A Reader* (Sevenoaks, U.K.: Hodder & Stoughton in association with the Open University): 5–18.

Giddens, Anthony (1984) *The Constitution of Society: Outline of the Theory of Structuration* (Cambridge, U.K.: Polity Press).

Habermas, Jürgen (1971) *Toward a Rational Society: Student Protest, Science, and Politics*, trans. J. Shapiro (London: Heinemann).

Heilbroner, Robert (1994a) "Do Machines Make History?" in M. R. Smith & L. Marx (eds), *Does Technology Drive History? The Dilemma of Technological Determinism* (Cambridge, MA: MIT Press): 53–66.

Heilbroner, Robert (1994b) "Technological Determinism Revisited," in M. R. Smith & L. Marx (eds), *Does Technology Drive History? The Dilemma of Technological Determinism* (Cambridge, MA: MIT Press): 67–78.

Hughes, Thomas P. (1983) *Networks of Power: Electrification in Western Society, 1880–1930* (Baltimore, MD: Johns Hopkins University Press).

Hughes, Thomas P. (1994) "Technological Momentum," in M. R. Smith & L. Marx (eds), *Does Technology Drive History? The Dilemma of Technological Determinism* (Cambridge, MA: MIT Press): 101–14.

Information Society Forum (2000) *A European Way for the Information Society* (Luxembourg: Office for Official Publications of the European Communities).

Joerges, Bernward (1988) "Large Technical Systems: Concepts and Issues," in R. Mayntz and T. P. Hughes (eds), *The Development of Large Technical Systems* (Frankfurt: Campus Verlag): 9–36.

Joerges, Bernward (1999) "Do Politics Have Artefacts?" *Social Studies of Science* 29(3): 411–31.

Latour, Bruno (1987) *Science in Action, How to Follow Scientists and Engineers through Society* (Milton Keynes: Open University Press).

Latour, Bruno & Steve Woolgar (1986) *Laboratory Life: The Construction of Scientific Facts*, 2nd ed. (Princeton, NJ: Princeton University Press).

Lenin, Vladimir Ilyich (1921) "Notes on Electrification," February 1921, reprinted (1977) in *Collected Works*, vol. 42 (Moscow: Progress Publishers): 280–81.

MacKenzie, Donald (1984) "Marx and the Machine," *Technology and Culture* 25(3): 473–502.

MacKenzie, Donald & Judy Wajcman (eds) ([1985]1999) *The Social Shaping of Technology: How the Refrigerator Got Its Hum* (Milton Keynes, U.K.: Open University Press); 2nd ed. (1999) (Philadelphia: Open University Press).

Marcuse, Herbert (1964) *One-Dimensional Man: Studies in the Ideology of Advanced Industrial Society* (Boston: Beacon Press).

Martin, Brian & Pam Scott (1992) "Automatic Vehicle Identification: A Test of Theories of Technology," *Science, Technology & Human Values* 17(4): 485–505.

Marx, Leo (1994) "The Idea of 'Technology' and Postmodern Pessimism," in M. R. Smith & L. Marx (eds), *Does Technology Drive History? The Dilemma of Technological Determinism* (Cambridge, MA: MIT Press): 237–58.

Misa, Thomas (1988) "How Machines Make History and How Historians (and Others) Help Them To Do So," *Science, Technology & Human Values* 13(3 and 4): 308–31.

Mumford, Lewis (1961) "History: Neglected Clue to Technological Change," *Technology & Culture* 2(3): 230–36.

Oldenziel, Ruth (1999) *Making Technology Masculine. Men, Women and Modern Machines in America, 1870–1945* (Amsterdam: University of Amsterdam Press).

Nye, David (1996) *American Technological Sublime* (Cambridge, MA: MIT Press).

Pinch, Trevor & Wiebe Bijker (1984) "The Social Construction of Facts and Artefacts: Or How the Sociology of Science and the Sociology of Technology Might Benefit Each Other," *Social Studies of Science* 14: 399–441.

Radder, Hans (1992) "Normative Reflections on Constructivist Approaches to Science and Technology," *Social Studies of Science* 22: 141–73.

Russell, Stewart (1986) "The Social Construction of Artefacts: A Response to Pinch and Bijker," *Social Studies of Science* 16: 331–46.

Smith, Merritt Roe & Leo Marx (eds) (1994) *Does Technology Drive History? The Dilemma of Technological Determinism* (Cambridge, MA: MIT Press).

Smith, Michael L. (1994) "Recourse of Empire: Landscapes of Progress in Technological America," in M. R. Smith & L. Marx (eds), *Does Technology Drive History? The Dilemma of Technological Determinism* (Cambridge, MA: MIT Press): 37–52.

Staudenmeier, John (1994) "Rationality Versus Contingency in the History of Technology," in M. R. Smith & L. Marx (eds), *Does Technology Drive History? The Dilemma of Technological Determinism* (Cambridge, MA: MIT Press): 259–74.

Traweek, Sharon (1988) *Beamtimes and Lifetimes, The World of High Energy Physicists* (Cambridge, MA: Harvard University Press).

Winner, Langdon (1977) *Autonomous Technology, Technics-out-of-control as a Theme in Political Thought* (Cambridge, MA: MIT Press).

Winner, Langdon (1980) "Do Artifacts Have Politics?" *Daedalus* 109: 121–36 (reprinted in MacKenzie & Wajcman, 1985 and 1999).

Winner, Langdon (1986) *The Whale and the Reactor, A Search for Limits in an Age of High Technology* (Chicago: University of Chicago Press).

Winner, Langdon (1993) "Upon Opening the Black Box and Finding It Empty: Social Constructivism and the Philosophy of Technology," *Science, Technology & Human Values* 18(3): 362–78.

Woolgar, Steve (1991) "Configuring the User: The Case of Usability Trials," in J. Law (ed), *A Sociology of Monsters: Essays on Power, Technology and Domination* (London: Routledge): 57–99.

Woolgar, Steve & Geoff Cooper (1999) "Do Artefacts Have Ambivalence? Moses' Bridges, Winner's Bridges and Other Urban Legends in S&TS," *Social Studies of Science* 29(3): 433–49.

Wyatt, Sally (1998) *Technology's Arrow: Developing Information Networks for Public Administration in Britain and the United States*. Doctoral dissertation. University of Maastricht.

Wyatt, Sally (2000) "ICT Innovation in Central Government: Learning from the Past," *International Journal of Innovation Management* 4(4): 391–416.

Wyatt, Sally (2001) "Growing up in the Belly of the Beast," in F. Henwood, H. Kennedy, & N. Miller (eds), *Cyborg Lives? Women's Technobiographies* (York: Raw Nerve): 77–90.

8 Pramoedya's Chickens: Postcolonial Studies of Technoscience

Warwick Anderson and Vincanne Adams

During the last twenty years, scholars in science and technology studies have tried to develop a vocabulary to describe the articulation of knowledge and practice across cultures, or social worlds, and between different places. We can now recognize standardized packages, boundary objects, immutable mobiles (and more recently mutable ones), even wordless creoles and pidgins. Scientists might accumulate and invest symbolic or cultural capital, or they participate in gift exchange, sometimes systematized as a moral economy. Alternatively, they translate and enroll, making their laboratories and field sites obligatory passage points. For a while, they were also heterogeneous engineers. All these various, valiant efforts to connect particular sites of knowledge production and practice—from symbolic interactionism to actor-network theory—raise new questions as they answer old ones. Helpful as they are, such analytical frameworks have yet to capture the full range of material trans-actions, translations, and transformations that occur in making and mobilizing technoscience. As science and technology travel, they continue to elude our best efforts to track them. We still do not know enough about what causes them to depart, what happens when they arrive—and even less about the journey, rough or smooth. In what follows, we provide an analytic road map to approach this phenomenon of travel in an age of globalized science, noting the need for more connections to the histories and political relations that enable such travel. Our effort is intentionally focused on culling from materials that point to a compelling approach and is not meant as an exhaustive review of the literature or critique of wrong turns and impasses.[1]

In the past, accounting for the travels of technoscience was not really a problem. Scientific knowledge appeared to diffuse beyond the laboratory when it justifiably corresponded with nature, the result of either method or norms. Techniques and machines traveled simply because they were useful and efficient—they drove themselves, it seemed. But scholars of scientific knowledge have repeatedly demonstrated the situated character of such claims to universal validity—the many ways in which science is constitutively social and local, and always particular and political.[2] Others have figured technological innovation as a stabilization, or black-boxing, of specific socio-technical relations (Latour, 1987).

How can we then account socially and politically for the dispersion of this knowledge, for its distribution, its abundance or absence? How are locally contingent scientific and technical practices mobilized and extended? What do they do at different sites? How are they standardized or transformed? In other words, how do we recognize the complex spatialities that link a multitude of local case studies? Why is it, inquired Bruno Latour (1988: 227), that Newtonian laws of physics work as well in Gabon as in England? As Steven Shapin (1998: 6–7) observes, "we need to understand not only how knowledge is made in specific places but also how transactions occur between places."[3] Having placed the view from nowhere, how do we get from there to somewhere else? "The more local and specific knowledge becomes," writes James A. Secord (2004: 660), "the harder it is to see how it travels."[4] "If facts depend so much on . . . local features, how do they work elsewhere?" asks Simon Schaffer (1992: 23). The challenge, in a sense, is to explain the extension of formalisms in nonformalist terms and to identify the technologies and institutions that have worked in the past to delete locality. Moreover, as Steven J. Harris (1998: 271) notes, we should recognize that "how science travels has as much to do with the problem of travel in the making of science as it does with the problem of making science travel."

If we substitute "medicine," or "technology," or even "modernity," for science in Harris's formulation, parallels with recent anthropological studies become clear. Many anthropologists turned their attention to the traffic in culture some time ago, tracing the circulation and exchange of commodities, practices, and ideas across shifting social boundaries. Medical anthropologists have examined the interaction of biomedical and other health beliefs at widely scattered sites, charting the development of new medical knowledge and practice.[5] When anthropologists of technology survey their field, they find most case studies are located outside Western Europe and North America. Bryan Pfaffenberger (1992: 505; 1988), for example, has described the cross-cultural study of things as a "technological drama," where the construction of a sociotechnical system, through the local regularization, adjustment, and reconstitution of new or introduced technology, becomes a polity building process. Anthropological studies of development regimes similarly recognize the dynamics of "modernity" in the making as it is accomplished by a wide variety of technoscientific efforts. Even "failed" development may become a technoscientific accomplishment of a sort (Ferguson, 1994). Recently, many historians of colonial medicine and technology have followed anthropologists in mapping out a distinct, complex set of engagements with the "modern" (Vaughan, 1991; Arnold, 1993; Hunt, 1999; Anderson, 2003). Yet laboratory science, defetishized at its "origins," still moves around the globe as a fetish, with its social relations conveniently erased. It seems to arrive with capitalism, "like a ship," then magically arrive elsewhere, just as powerful, packaged, and intact.[6] We remain attached to the "Marie Celeste" model of scientific travel.

Anthropologists have identified alternative modernities, new modernities, and indigenous modernities, in any number of locations (Appadurai, 1991; Strathern, 1999; Sahlins, 1999). Modernity—of which technoscience is surely a part—has emerged as complexly hybrid and widely dispersed, in ways that science studies have only begun to encompass. This is far more than the multiplication of contexts or the creation of networks or arenas, more than standardization or enrollment, more than dominance or submission. Postcolonial investigations of proliferating modernities, or "development," might offer some guidance for scholars in science and technology studies, yet they are largely ignored.[7] In claiming that "we" have never been Modern, Latour reminded us of the need to unpack the social at the same time as we unpack its figurative double "science" (Latour, 1993). But in defusing such analytic distinctions, he may have missed the real action: those of us outside Paris have never had so many ways of being modern, so many different ways of being scientific![8]

The term postcolonial was thrown into the last paragraph like an incendiary device, so perhaps we should either try to justify its use or deactivate it. For Edward W. Said (1983: 241), "traveling theory"—a nicer term—was not so much a theory as a "critical consciousness," a sort of spatial sense, an awareness of differences between situations and an appreciation of resistances. What happens when something moves into a new environment? This apparently simple question gives rise to the postcolonial critique that informs the anthropology of modernity. Said discerned a point of origin, "passage through the pressure of various contexts" (1983: 227), a set of conditions through which the transplanted idea or practice must pass, and transformation through new uses. He criticized those who regard this transfer merely as "slavish copying" or "creative misreading" (1983: 236). His orientation is postcolonial primarily in the sense that it relentlessly insists on revealing the geographical predicates of Western cultural forms, the theoretical mapping and charting of territories that are otherwise hidden. As the imperialist model—which includes for Said the contemporary United States—is disassembled, "its incorporative, universalizing, and totalizing codes [are] rendered ineffective and inapplicable." That is, he wanted us to remove Western cultural forms from their "autonomous enclosures" and look at the cultural archive "contrapuntally," not "univocally," to recognize heteroglossia in colonialism (Said, 1994: 29, 28, 29). As Stuart Hall (1996) has argued, the "post" in postcolonial is in effect "under erasure," thus making available the colonial as a contemporary analytic. He believes that postcolonial studies enable a "decentered, diasporic, or 'global' rewriting of earlier nation-centered imperial grand narratives" (1996: 247)—in other words, a "re-phrasing of Modernity within the framework of globalization" (1996: 250); that is, for us, the project becomes a rephrasing of technoscience within the framework of globalization, allowing ample use of the analytical dynamics suited to a colonial era at a time when the political framework of such engagements lives on under other guises, other names.

It is Said's notion of the postcolonial as critical consciousness, as forced (and often resisted) recognition of complex and realistic spatialities—figured in identity politics, discursive formations, material practices, representations and technological possibilities—that most interests us here. We are arguing for the value of postcolonial perspectives—views from elsewhere—in science and technology studies, and not for any transcendent postcolonial theory.[9] This requires a multiplication of the sites of technoscience, revealing and acknowledging hidden geographical notations and power relations, and further study of the mechanisms and forms of travel between sites. It means we need to be sensitive to dislocation, transformation, and resistance; to the proliferation of partially purified and hybrid forms and identities; to the contestation and renegotiation of boundaries; and to recognizing that practices of science are always multi-sited (Marcus, 1998). Western categories, Gyan Prakash (1994: 3) reminds us, were disrupted in colonial travel. "The writ of rationality and order was always overwritten by its denial in the colonies, the pieties of progress always violated irreverently in practice, the assertion of the universality of Western ideals always qualified drastically." This suggests that postcolonial analysis will be as useful in Western Europe and the United States as in Madagascar and the Philippines. That is, it offers a flexible and contingent framework for understanding contact zones of all sorts, for tracking the unequal and messy translations and transactions that take place between different cultures and social positions, including between different laboratories and disciplines even within Western Europe and North America.

We may be able to view such interactions through other lenses, but a postcolonial perspective possesses the advantages of historical and geographical complexity and of political realism. We sometimes hear from scholars in science studies that postcolonial perspectives are not relevant to their work because they are not examining science in temporally postcolonial locations—we hear this even (perhaps especially) from those working on science in the United States!—but this seems to us to echo the assertions of an earlier generation of scholars who argued that because they did not study *failed* science, sociology was irrelevant to their work. Both of these significant resistances should, as they say, be worked through.

In the rest of the essay we first consider the expanding geographies of technoscience, what happens when formalisms extend to different sites, and then look at how technoscience travels, how "situated" knowledge becomes mobile. In conclusion, we attempt to reactivate the postcolonial in the hope that it will carve out some new landscapes, and allow fresh channels and flows, through science and technology studies.

GEOGRAPHIES OF TECHNOSCIENCE

To say that the place of science has become ever more important in all spheres of social life is a truism to scholars of society, to the extent that debates about what formally constitutes "science" are now focused as much on geography as on problems of epistemology. Euro-American laboratories are no longer the most important locations for the study of science.[10] Historians have long noted the importance of the field site in the production of technoscience, and the ability to conceptualize it in relation to the "laboratory" is not entirely new (Kuklick & Kohler, 1996; Kohler, 2002; Mitman, 1992). At the same time, few efforts have been made to explain the linkages between other locations of technoscientific pursuits and the laboratories that still fuel their interventionary arsenal. Classrooms, hospital and outpatient clinics, playing fields (Owens, 1985), companies (Harris, 1998), foundations (Cueto, 1997), bilateral development agencies and nongovernment organizations (Escobar, 1995; Pigg, 1997;

Gupta, 1998; Shrum, 2000; Shepherd, 2004), for example, make suitable sites for the study of technoscience when questions of "geography" are placed above those of epistemology. For this, a good deal can be gleaned from the historians of colonialism and its critics, the medical anthropologists who early on offered critiques of development, or from the historians and sociologists of education who took account of science's need to teach its terms before it could produce its truths (Pauly, 2000; Prescott, 2002).

As David N. Livingstone (2005: 100; 2004) has argued, the space of science is not just local: it is also "distributed and relational." Studies of scientific sites beyond Western Europe and North America have begun springing up during the past twenty or so years, contributing to a "provincializing" of Europe (Chakrabarty, 2000), to rethinking it as one among many equivalent and related terrains for technoscience. Historians have traced the routes of scientific expeditions and described the local entanglement and reshaping of imperial biomedical and environmental sciences. Anthropologists have analyzed widely distributed negotiations and contestation over bioprospecting, genomics, and environmental and agricultural sciences. The state and NGOs have been particularly salient in these accounts, perhaps more so than in studies of Euro-American science. David Arnold (2005a: 86) observes that "body and land" have become "exemplary sites for the understanding of colonial and postcolonial technologies."[11]

We now have a number of examples of how contemporary technoscience is globally distributed and entangled.[12] Itty Abraham (2000: 67; 1998) considers the interaction, the co-constitution, of nuclear physics and the nation-state in India, helping us view Western science differently as he reveals "the international circuits it works through and occupies." In her work on high-energy physics in Japan, Sharon Traweek (1992: 105; 1988) examines a local instantiation of "colonialist discourse in science." Joan H. Fujimura (2000) describes the reconfiguration of science and culture, of West and East, in transnational genomics research. In *Space in the Tropics*, Peter Redfield (2000; also 2002) localizes the French space program in Guiana, representing it as enmeshed in a specific geography and colonial history. For Gabrielle Hecht (2002), contemporary nuclear maps are incomplete without including places like Gabon and Madagascar, where uranium mining is conjugated with nuclearity, decolonization, and the formation of "modern" subjects. Kaushik Sunder Rajan (2005) shows that commodification in pharmaceutical genomics in the United States provides a template for the commodification of human subjects in India, by way of the same scientific languages and practices. According to Kim Fortun (2001), efforts to read the Union Carbide disaster in Bhopal, India, as an isolated event only make more visible the multi-sited, "globalized" character of science policy-making.

Yet a few years ago when Lewis Pyenson (1985; 1989; 1990) surveyed physics and astronomy in the German, Dutch, and French empires, he took pains to demonstrate that the "exact" sciences had evaded any colonial contamination. These manifestations of European civilization had diffused across the oceans, allegedly impervious to contact with local circumstances, to pollution and miscegenation. Since these sciences did not become mere imperial superstructure, Pyenson assumed that absolutely no colonial entanglement and reshaping had occurred. Paolo Palladino and Michael Worboys (1993: 99; also see Pyenson, 1993), however, countered that such "Western methods and knowledge were not accepted passively, but were adapted and selectively absorbed in relation to existing traditions of natural knowledge and religion." Techno-science is a product of local encounters and transactions. Moreover, Palladino and Worboys reminded readers that "for most of humanity, the history of science and imperialism *is* the history of science" (1993: 102, original emphasis).

"To follow [the] transformation of a society by a 'science,'" advises Latour (1988: 140), "we must look not in the home country but in the colonies." The bacteriologi-cal laboratory, he suggests, worked unimpeded to reorder colonial society. "It was in the tropics that we can imagine best what a pasteurized medicine and society are" (1988, 141). For Latour, colonial relations thus might still be cast simply in terms of dominance and submission, presenting a story of the expansion of sovereign networks of European science. In *Pandora's Hope* (1999a: 24), he takes us on a field trip to the Amazon, which "will not require too much previous knowledge." The postcolonial setting does little more than provide some exciting metaphors: Latour sees himself bringing order "to the jungle of scientific practice," and writes about "going 'native'" (1999a: 47). He announces that he will "omit many aspects of the field trip that pertain to the colonial situation" (1999a: 27).[13] Instead, he gives a vivid account of how scientists develop a proto-laboratory in the depopulated jungle, regulating the transformations and translations of phenomena in order to make them appear con-stant—that is, constantly European. More colonial *amour propre* than postcolonial analysis, Latour's brilliant exposition of the movement of invariant technoscientific objects along a reversible chain of transformations suggests that scientists can mobi-lize the world, and make their own context, so long as they remain deep in conver-sation with their European colleagues. One does wonder, however, what happened to the "native" agents and intermediaries in this supposedly virgin territory.

In contrast, for Gyan Prakash (1999: 7), colonial science becomes "a sign of Indian modernity," translated and appropriated by Hindu nationalists. In order to achieve hegemony, the dominant discourse must be distorted, yet remain identifiably scien-tific. Moreover, the technics of the nation-state become inseparable from the technics of "science gone native," or "tropicalized." While for Latour—and, indeed, for Pyenson (though for different reasons and to different effect)—technoscience produces "immutable mobiles," a postcolonial scholar like Prakash discerns ambivalence and hybridity. This is not a process of contamination or mimicry, but a matter of putting science to work in a colonial context and resisting efforts to purify its genealogy. Technoscience provides yet another stage for the colonial drama.

While Prakash sees colonial technoscience contributing to the negotiation of hybrid or alternative modernities at a specific site, others have imagined a globalized multi-cultural science. Sandra Harding, for example, has sought to use cross-cultural studies of knowledge traditions to achieve further epistemological pluralism. For Harding (1998: 8), postcolonial accounts provide "resources for more accurate and compre-hensive scientific and technological thought." "We can employ the category of the postcolonial strategically," she writes, "as a kind of instrument or method of detecting phenomena that otherwise are occluded" (1998: 16).[14] Influenced by "post-Kuhnian science studies" and feminist standpoint theory, Harding has emphasized the importance of local knowledge and called for more dynamic and inclusive global histories. But her main goal is the strengthening of modern scientific objectivity, achieving better Modernity through remedying "dysfunctional universality claims" (1998: 33). In a sense, she is trying to valorize the "nonmodern" as a form of the modern too, to broaden the compass of Modernity.

Lawrence Cohen (1994: 345) has suggested that while Harding wants to "pluralize the field of discourse," most postcolonial intellectuals pine for "an insurrectionary abandonment." The danger of *multicultural* science studies, according to Cohen, is its "mapping of difference onto an underlying hegemony." In contrast, postcolonial scholars have tried to reveal the heterogeneity and messiness of technosciences, which undercut the singularity and authority of their attendant Modernity. J. P. S. Uberoi (1984; 2002), for example, argues that Goethe's theory of colors represents a non-Newtonian system of knowledge that can represent the basis of another, nondualist modernity. In a creative deployment of Thomas Kuhn's notion of paradigm, Ashis Nandy (1995; 1988) argues that a "modern" scientist might have multiple subjectivities and sensitivities, with religious or cultural affiliation offering possibilities of alternative modernities. Shiv Visvanathan (1997) deplores the vivisection and violence of some modern science and extols the gentler, more observational aspects of science as the basis for an alternative science and modernity.[15] But the emphasis of these scholars is more on political engagement than epistemological renewal.

Some of the more densely realized stories of the contact zones of mobile knowledge practices have focused on the contemporary interactions of scientists and indigenous peoples. The work of Helen Verran, David Turnbull, and their students has been especially influential; they could be said to represent a "Melbourne-Deakin school" of postcolonial science studies, shaped by local enthusiasm for ethnohistory, and building on constructivist and feminist approaches to the study of science and technology (Watson-Verran & Turnbull, 1995). With the Yolngu people of Arnhem Land, Verran has studied the interaction of local knowledge practices, one "traditional," the other "scientific," and described "the politics waged over ontic/epistemic commitments." Her goal is not just to exploit the splits and contradictions of Western rationality: she aims toward a community that "accepts that it shares imaginaries and articulates those imaginaries as part of recognizing the myriad hybrid assemblages with which we constitute our worlds" (1998: 238). In her current research project, Verran seeks to move beyond description and to find ways in which one might do good work—such as negotiating land use—within and between the messiness, contingency, and ineradicable heterogeneity of different knowledge practices (2002).[16] Turnbull (2000: 4), similarly, has studied the "interactive, contingent assemblage of space and knowledge" in diverse settings, arguing that "all knowledge traditions, including Western technoscience, can be compared as forms of local knowledge so that their differential power effects can be compared but without privileging any of them epistemologically" (1998:

6). That is, even the most generalized technoscience, like any other practice, always has a local history and a local politics, even as the actors involved claim to be "doing global."

Postcolonial perspectives might foreground the debates over not just the accrediting processes of science but also the violence that is still accomplished in its wake, often through erasure and disenfranchisement as much as in physical trauma (Nandy, 1988; Visvanathan, 1997; Apffel-Marglin, 1990). Cori Hayden (2003) identifies the subtle institutional bonds in the international pharmaceutical and ethnobotanical industries that simultaneously recruit notions of indigenous participation and deny credit, profit, or accountability to local participants. Similarly, Vincanne Adams (2002) suggests that the field on which international pharmaceutical industries negotiate their use of traditional "scientific" medicines is an uneven one and that the terms of universalist science can be used to negotiate away local claims to intellectual property. In the end, local practitioners of medicine begin to develop uncertainty about traditional knowledge and practice not because they are presented with empirical evidence that what they do is ineffective but because the terms on which biomedical efficacy rests deny the validity of their knowledge.

The effort to think through postcolonial theory in science studies requires heightened sensitivity to the ways that not only geography, race, and class but also gender hierarchies are (re)constituted through the relations of traveling sciences. Postcolonial science studies offer a glimpse of the ways that science becomes implicated in gendered social orders. Adams and Stacy Leigh Pigg (2005), for example, show that development programs that ask women to reconceptualize their biological lives by way of family planning, safe sex, and sexual identity make science speak for and sometimes against gender hierarchies that are institutionalized in families, schools, NGO regimes, agricultural and domestic labor, and even media campaigns. Sexual identities can be offered up as new forms of subjectivity that impart a "biological" and scientific notion of self and person, at the same time that traditional regimes of gender discrimination are reconstituted by way of the logics of technoscientific development programs. Even when informed by feminist perspectives, technoscientific interventions can be deeply bound up in the reproduction of social orders that implicitly, and epistemologically, reinforce inequalities, forming a neocolonial regime of technoscientific truths that often goes unchallenged.

These are just a few examples from the emerging postcolonial ecology of technoscience, evidence of the multiplication of its various niches. Such studies represent a move away from genealogical reasoning in science studies, from simple origin and reproduction stories, toward archeological reconstructions of local knowledge practices. "Global" technoscience is becoming disaggregated analytically, appearing more as a set of variant "global projects," where globalization may be requisite or opportunistic or vicarious. "Every sociotechnical system is in principle a *de novo* construct," according to Pfaffenberger (1992: 500, 511). "People are engaged in the active technological elaboration, appropriation, and modification of artifacts as the means of coming to know themselves and of coordinating labor to sustain their lives." But if we now recognize complex sites of technoscience outside Europe and North America, what do we know about travel between these places? How do we avoid default to the old stories of the expansion of Europe and instead manage to recognize the multiple vectors of technoscience?

TRAVEL BROADENS THE MIND

In 1960, W. W. Rostow described the stages of economic growth in his "non-communist manifesto," a classic of modernization theory. Rostow emphasized the importance of science and technology in achieving a "take-off" from traditional society—indeed, the stimulus "was mainly (but not wholly) technological" (1960: 8). Science, it seemed, diffused from Europe and pooled where the ground was ready to receive it. A few years later, George Basalla amplified this diffusionist hypothesis, giving details of the phases in the spread of Western science from center to periphery. According to Basalla (1967), in phase 1, expeditions in the periphery merely provided raw material for European science; during phase 2, the derivative and dependent institutions of colonial science emerged; and sometimes, an independent and national science, called phase 3, would later develop.[17] In the 1990s, Thomas Schott (1991: 440) also asked "what conditions promote the spread and establishment of scientific activity" over the globe. Following Joseph Ben-David (1971) and Basalla, he argued that scientific ideas and institutional arrangements diffused from center to periphery. Schott attributed the success of science to widely shared cognitive norms, cosmopolitanism, and the institutionalization of collaborative relationships. "The faith in invariance of nature and in truthfulness of knowledge across places, together with the cosmopolitan orientation of the participants, created a potential for adoption of the European tradition in non-Western civilizations" (Schott, 1993: 198).[18]

Simple evolutionary and functionalist models of scientific development provoked extensive criticism: first in anthropology, as "failed" modernization directed attention to the intensification of cultural forms and preexisting labor relations of exchange (Geertz, 1963; Meillassoux, 1981), and then in science studies during the 1980s and 1990s. The critical response was inspired in part by the more general challenge of dependency theories, and world systems theory, to the older diffusionist models of modernization and development (Frank, 1969; Wallerstein, 1974).[19] Roy MacLeod, for example, disapproved of the linear and homogeneous character of diffusionist arguments and noted the lack of attention to the complex political dimensions of science. He called instead for a more dynamic conception of imperial science, the recognition of a "moving metropolis" as a function of empire, rather than a stable dichotomy of center and periphery (MacLeod, 1987).[20] David Wade Chambers also rejected Basalla's diffusionism and asked for more case studies of science in non-Western settings and for interactive models of scientific development. But Chambers warned that "without a more general framework, we sink into a sea of local histories"; he wondered about the salience of the "colonial," yet doubted at the time its explanatory power (Chambers, 1987: 314; 1993).[21]

Perhaps the most influential contemporary analysis of the mobility of technoscience comes from actor-network theory (ANT). How did Portuguese ships, asked John Law in 1986, keep their shape as they voyaged from Lisbon to distant parts of the empire? That is, how are scientific facts or practices, and technological configurations, stabilized in different places? Actor-network theory initially was meant to provide an explanation for the production of these "immutable mobiles," arguing that a series of transformations and translations across a network could keep technoscience invariant in different settings (Callon, 1986; Latour, 1987). Accordingly, the more articulations developed with human and nonhuman actors, the more stable and robust the object becomes. Society, nature, and geography are outcomes of these mobilizations, translations, and enrollments. An advanced collective, in contrast to a "primitive" one, "translates, crosses over, enrolls, and mobilizes more elements which are more intimately connected, with a more finely woven social fabric"—it makes more alliances (Latour, 1999b: 195). So "facts" are "circulating entities. They are like a fluid flowing through a complex network" (Latour, 2000: 365).[22] The focus of ANT is therefore on movement, but as Latour reminds us, "even a longer network remains local at all points" (Latour, 1993: 117). Networks are, in a sense, summed up locally, so there is no need to zoom between local and global scales.

There can be little doubt that the gestalt switch required in ANT has proven stimulating and productive for science studies. But a sort of semiotic formalism often seems to supervene on the analysis of local sites: the "local" can seem quite abstract, depleted of historical and social specificity. The structural features of the network become clear, but often it is hard to discern the relations and the politics engendered through it. Law (1999: 6) has reflected that ANT "tends to ignore the hierarchies of distribution, it is excessively strategic, and it colonizes . . . the Other." Shapin (1998: 7) criticizes "the militaristic and imperialistic language that is so characteristic of Latour's work." Some have therefore called for more "complex" spatialities than smoothly colonizing "network space."[23]

Later versions of ANT have emphasized a more varied terrain, describing the adaptation and reconfiguration of objects and practices as they travel. Annemarie Mol and Law (1994: 643) claim that the social "performs several *kinds of space* in which different operations take place." They describe regions, networks, and fluid topologies—in the last, "boundaries come and go, allow leakage or disappear altogether, while relations transform themselves without fracture" (1994: 643). Interviewing Dutch tropical doctors about anemia, Mol and Law recognized a topological multiplicity, where a network of hemoglobin measurement (with immutable mobiles held stable as they travel) mixes with fluidity, which allows "invariant transformation" (1994: 658) in clinical diagnosis. "In a fluid space," they write, "it's not possible to determine identities nice and neatly, once and for all . . . They come, as it were, in varying shades and colors" (1994: 660). Later, in a study of the Zimbabwean bush pump, Marianne de Laet and Mol (2000) explain how the object changed shape and re-formed relations from one village to the next, while staying identifiably a Zimbabwean bush pump. Mol and her colleagues thus are able to recognize the displacement, transformation, and contestation that occur when technoscience travels, the proliferation of hybrid-ity and the refashioning of places and identities that contact makes happen. They are edging, it seems, toward a postcolonial spatiality, though their reluctance to acknowl-edge this may limit the depth and scope of their analyses—that is, their ability to decolonialize (and demasculinize) ANT. The essay on tropical anemia, for example, assures readers that "least of all will we transport ourselves to Africa," yet twice the authors observe that "it is different in Africa" (Mol & Law, 1994: 643, 650, 656).[24] If only ANT could embrace its postcolonial condition, we might learn more about these resisted differences—in the Netherlands as well as Africa.

Practices of science emerge as hegemonic and, while few would suggest that scien-tific ideas are free-standing notions, external to the social practices that render them visible. Few researchers take seriously the possibility of reading technoscience as a contested practice of subjectivity and not simply "knowledge," even though this perspective takes precedence in postcolonial analytics. Pigg (2001), in contrast, specifically probes this concern, asking what it means for a Nepali villager to use the notion of AIDS or HIV in a language that has no foundation for understanding such terms or the attendant "sciences" that enable them to make one kind of sense. Nikolas Rose (1999) focuses specifically on this set of subject negotiations, which in at least some contexts become the *sine qua non* of political modernity. Adriana Petryna (2002) explains how a scientific subject is produced in a modern state that has learned to articulate its governmental possibilities through the biological life of its citizens.[25] Exploring the way apparatuses of science mediate state and subject in the Ukraine, after Chernobyl, as Petryna does, suggests an ethnographic approach that is implic-itly informed by postcolonial analytics. Technoscience becomes part of identity poli-tics, a set of problems that require holding the state and subject within the same analytical frame. While such an approach may have originated in an older tradition of dialectical materialism, a postcolonial analytic recognizes novel (or hitherto obscured) deployments of modernity and power. Postcolonial identity politics are often negotiated in and through technoscientific imaginaries that create the space for new kinds of political engagements (Cohen, 2005; Nguyen, 2005).

In part helped by ethnographic evidence of the complicated politics that emerge in a blurry field of a multitude of local scientific practices, discussion of diffusion and network construction has gradually given way to talk of contact zones (Pratt, 1992). Recently, MacLeod (2000: 6) urged again the abandonment of center-periphery models and proposed instead a study of the traffic of ideas and institutions, a recognition of reciprocity, using "perspectives colored by the complexities of contact." Secord (2004: 669) has called for the history of science to move beyond ANT to "a fully historical understanding, often informed by anthropological perspectives, with divisions of center and periphery replaced by new patterns of mutual interdependence." Such post-colonial sensitivity to contact zones might build on existing analyses of the coor-dination and management of work across different social worlds. It would include the material, literary, and social technologies used to stabilize facts at multiple sites, whether "standardized packages," the infrastructures of classification and standardization, or "heterogeneous engineering" (Fujimura, 1992; Bowker & Star, 2000; Law, 1986).[26] But it would also need to account for the dispersion of techno-science, the generation of incoherence and fragmentation, which cannot be dismissed as mere "noise" at the margins of the system (Jordan & Lynch, 1992). Moreover, a rethinking of boundary objects, infrastructures, and implicated actors in relation to postcolonial hybridity might thicken the political texture of social worlds analysis and make its "topology" more nuanced (Star & Griesemer, 1989; Löwy, 1992; Clarke, 2005).

Some scholars in science studies have already begun drawing on postcolonial anthropology to explain the interactions and transformations that take place in contact zones. In a study of the practices of experimentation, instrumentation, and theory in modern physics, Peter Galison (1997: 51, 52) has recognized the need for an analysis of "trading zones" between scientific "sub-cultures." He frames the activities of physicists in terms derived from Pacific sociolinguistics and ethnohistory, as he tracks material exchanges across scientific cultures, and observes the construal of "wordless pidgins" and "wordless creoles." The languages of colonial contact thus help explain the patterning of interactions among modern Euro-American physicists. Warwick Anderson (2000) has employed anthropological studies of material exchange to explore the shaping of identities and creation of scientific value in research into kuru, a disease afflicting the Fore people in the highlands of New Guinea. He takes studies of local New Guinea economic transactions and uses them to explain the interactions of the scientists themselves and their methods of mobilizing Fore goods and body parts as technoscience. Scientists, anthropologists, patrol officers, and Fore found themselves performing dramas of alienability, reciprocity, valuation, commitment, and identity formation. Anderson claims that his account of the discovery of the first human "slow virus" challenges us to understand "global" technoscience as a series of local economic accomplishments, each of them confused and contested. He concludes (2000: 736):

We need multi-sited histories of science which study the bounding of sites of knowledge production, the creation of value within such boundaries, the relations with other local social circumstances, and the traffic of objects and careers between these sites, and in and out of them. Such histories would help us to comprehend situatedness and mobility of scientists, and to recognize the unstable economy of "scientific" transaction. If we are especially fortunate, these histories will creatively complicate conventional distinctions between center and periphery, modern and traditional, dominant and subordinate, civilized and primitive, global and local.

CONCLUSION: CATCHING PRAMOEDYA'S CHICKENS

Postcolonial studies of technoscience will benefit from recent efforts in anthropology to theorize the "interface of local and global frames of analysis." Anna L. Tsing (1994: 279, 280), for example, urges us to conjure up an "uncanny magic," and imagine "the local in the heart of the global." We need to think about world-making flows not just as interconnections or networks but also as "the re-carving of channels and the re-mapping of the possibilities of geography" (Tsing, 2002: 453). This means becoming more sensitive to the culture and politics of "scale-making" and to emergent forms of subjectivity and agency in "global projects." Tsing (2002: 463, emphasis added) finds circulation models are too often "closed to attention to struggles over the *terrain* of circulation and the privileging of certain kinds of people as players." She argues that we should study the landscapes of circulation along with the flow, gauging how people, cultures, and things are refashioned through travel. "The task of understanding planet-wide interconnections," she writes (2002: 456), "requires locating and specifying globalist projects and dreams, with their contradictory as well as charismatic logics and their messy as well as effective encounters and translations." Surely scholars in science and technology studies are better prepared than ever before to locate and specify in this way the globalist dreams of their own subject matter—that is, to situate technoscience within differing global, or at least multi-sited, imaginaries, using postcolonial perspectives.

In his disconcerting study of technological modernity in the Dutch East Indies, Rudolf Mrázek (2002: 197) tells the story of Pramoedya Ananta Toer, a former political prisoner and a forceful novelist, a "modern and apprehensive" Indonesian. Mrázek surrounds him with hard and sticky technologies: radios, weapons, ships, roads, medicines, clothes, toilets, photographs, typewriters, concrete, cigarette lighters, gramophones, electricity, wristwatches, bicycles, and then more radios. Pramoedya encounters colonial and nationalist sportsmen, dandies, jokers, and engineers. He remembers attending radio school, when "in technical drawing, my assignment was to make a sketch of a television, at the time not generally known, and I was not seriously worried about the result of this particular test" (Pramoedya quoted in Mrázek, 2002: 209). Everything is switched on, and there is a constant buzzing in the air. Indonesia has become a modern technological project. According to Mrázek, Pramoedya "recalls himself living in technology. The rhythm of typing and the principle of the typewriter appear to organize Pramoedya's recollections of the time" (2002: 210–11).

When a group of academics asked him how he obtained writing paper in prison, Pramoedya—whose hearing is poor—replied: "I have eight chickens." A mistake? An odd turn of phrase? A joke? A figure of resistance? Or a suggestion of otherwise hidden exchange? The opacity and disruptiveness of the chickens—their haunting of technoscience—produces a postcolonial vertigo in Mrázek's text, a sense of the instability and precariousness of imagined Western discourses. How little we know, how little we can assume, once Pramoedya's modernist chickens appear on the scene.

Following Pramoedya's release, after ten years on Buru island, a steady stream of visitors comes to his house in Jakarta. "There is an upgraded smoothness in the air," writes Mrázek, in typically allusive fashion. "What can we decently do," he continues, "except to join in the trickle, now the stream, be patient, ask Pramoedya when we get to him what he thinks about all this, and then just hope that he may switch off his new Japanese hearing aids for a moment and answer our question, with the wisest of his smiles: 'I have eight chickens'" (2002: 233).

No doubt Pramoedya's chickens are hard to catch, and tough, but you can be sure they are tastier than the old epistemological chicken (Collins & Yearley, 1992).

Notes

This essay builds on (and includes a few passages from) Warwick Anderson (2002) "Postcolonial technoscience," *Social Studies of Science* 32: 643–58. We are especially grateful to Adele Clarke, who commented extensively on earlier drafts of the essay. David Arnold, Michael Fischer, Joan Fujimura, Sandra Harding, Sheila Jasanoff, Michael Lynch, Amit Prasad, Chris Shepherd, and two anonymous reviewers also provided helpful advice. Warwick Anderson drafted versions of the paper while in residence at the Rockefeller Study and Conference Center, Bellagio, and the Institute for Advanced Study, Princeton. A special issue of *Science as Culture* (2005,14[2]) on "Postcolonial Technoscience" appeared as this essay was completed.

1. There is a particular risk that this is misread as a review of the anthropology of science; for such reviews, see Franklin, 1995; Martin, 1998; Fischer, 2003, 2005.

2. On "situated knowledge," see Haraway, 1988; although Haraway, of course, is arguing about standpoint more than geography.

3. As early as 1991, Adi Ophir and Shapin were wondering: "How is it, if knowledge is indeed local, that certain forms of it appear global in domain of application? . . . If it is the case that some knowledge spreads from one context to many, how is that spread achieved and what is the cause of its movement?" (1991: 16). See also Shapin, 1995; Livingstone, 2005.

4. Secord (2004: 655) claims that the "key question" in history of science is "How and why does knowledge circulate?"

5. See, for example, Hughes & Hunter, 1970; Good, 1994; Nichter & Nichter, 1996; Kaufert & O'Neill, 1990; Langford, 2002.

6. On capitalism arriving like a ship, see Ortner, 1984. One of the reasons we have chosen the term technoscience is to link studies of science with investigations of the movements of other Western technologies and medicine. For a justification of the term, see Latour, 1987: 175.

7. None appears in Biagioli, 1999. See, for example, Escobar, 1995; Apffel-Marglin & Marglin, 1996; Gupta, 1998; Moon, 1998; Thompson, 2002. For a helpful institutionalist survey, see Shrum & Shenhav, 1995.

8. Latour (1993: 122, 100) criticizes the "perverse taste for the margins" and urges anthropology to "come home from the Tropics."

9. For an extensive review of postcolonial theories (including the work of Fanon, Said, Bhabha, Spivak, and many others) as they relate to science studies, see Anderson, 2002. Instead of performing the usual listing of theories, we have tried here a more subtle rereading of some of Said's lesser known work. For those wanting reviews of postcolonial studies we recommend Young, 1990; Thomas, 1994; Williams & Chrisman, 1994; Barker et al., 1994; Moore-Gilbert, 1997; Loomba, 1998; Gandhi, 1998; Loomba et al., 2005.

10. We use "Euro-American" as shorthand for people of European descent who happen to be in Western Europe, North America, and Australasia, not in any typological "racial" sense.

11. See also Headrick, 1981; Reingold & Rothenberg, 1987; Crawford, 1990; Krige, 1990; Home & Kohlstedt, 1991; Petitjean et al., 1992; Crawford et al., 1993; Drayton, 1995; Selin, 1997; Adas, 1997; Kubicek, 1999; McClellan & Dorn, 1999. For a collection of useful reprints, see Storey, 1996. We have omitted many of the excellent nationally circumscribed histories of colonial science, the majority of

them pertaining to Australia, South Africa, India, and the Caribbean. Among the more arresting studies of colonial botany, forestry, and geology are Brockway, 1979; Stafford, 1989; Grove, 1995; Gascoigne, 1998; Drayton, 2000; Schiebinger, 2004; Arnold 2005b.

12. A debt must also be acknowledged to historians of science who have attempted to identify and legitimize science in other traditions. Work on science and medicine in China, for example, though not framed explicitly as postcolonial, offers opportunity for such reinterpretation. See, for example, Needham, 1954; Bray, 1997; Farquhar, 1996.

13. Later, Latour (1999a: 104) describes "the scientists themselves placing the discipline in a context." Compare Raffles, 2002.

14. See also Hess, 1995; Figueroa & Harding, 2003. For a more explicitly extractive model, see Goonatilake, 1998.

15. We are grateful to Amit Prasad for his contribution to this paragraph.

16. Verran emphasizes that this need not imply purification, compromise, synthesis, or conversion. See also Agrawal, 1995; Grove, 1996; Nader, 1996; Smith, 1999; Bauschpies, 2000; Hayden, 2003. For a collection of essays exploring Smith's arguments, see Mutua & Swadener, 2004.

17. See also Raina, 1999. Adas (1989) shows how colonial regimes used technological accomplishment as a gauge of civilization and saw technology transfer as part of the "civilizing mission."

18. See also Schott, 1994; Ben-David, 1971; Shils, 1991. Good recent examples of such "institutionalist" studies are Drori et al., 2003; Schofer, 2004.

19. Much of this critique retained an implicit demarcation of center and periphery, and the economism of diffusionist models: see Joseph, 1998.

20. For a critique of "technology transfer" theories, and the assumption of a passive role for receivers of technology, see MacLeod & Kumar, 1995; Raina, 1996.

21. More recently, Chambers and Gillespie (2000: 231) have recommended investigation of the "conglomerate vectors of assemblage that form the local infrastructure of technoscience."

22. See also Latour, 1999b. As Strathern (1999: 122) suggests, questions need to be asked not about the boundedness of cultures but about the "length of networks."

23. For example, Mol & Law, 1994. Lux and Cook (1998) have attempted to integrate sociological theories of the strength of weak ties with ANT in their study of international scientific exchanges during the late-seventeenth century.

24. That all the doctors they interviewed were Dutch and did not speak an African language is mentioned only in passing, and we find out nothing about the space in which they practiced. In contrast, see Verran, 2001.

25. For other accounts of biomedical "citizenship," see Anderson, 2006; Briggs with Mantini-Briggs, 2003.

26. In his essay on metrology, Schaffer (1992: 24) claims that physics laboratories in Victorian Britain were "part of the imperial communication project."

References

Abraham, Itty (1998) *The Making of the Indian Atomic Bomb: Science, Secrecy and the Postcolonial State* (London: Zed Books).

Abraham, Itty (2000) "Postcolonial Science, Big Science, and Landscape," in Roddey Reid & Sharon Traweek (eds), *Doing Science + Culture* (New York: Routledge): 49–70.

Adams, Vincanne (2002) "Randomized Controlled Crime: Postcolonial Sciences in Alternative Medicine Research," *Social Studies of Science* 32: 659–90.

Adams, Vincanne & Stacy Leigh Pigg (eds) (2005) *Sex in Development: Science, Sexuality and Morality in Global Perspective* (Durham, NC: Duke University Press).

Adas, Michael (1989) *Machines as the Measure of Men: Science, Technology, and Ideologies of Western Dominance* (Ithaca, NY: Cornell University Press).

Adas, Michael (1997) "A Field Matures: Technology, Science, and Western Colonialism," *Technology and Culture* 38: 478–87.

Agrawal, Arun (1995) "Dismantling the Divide Between Indigenous and Scientific Knowledge," *Development and Change* 26: 413–39.

Anderson, Warwick (2000) "The Possession of Kuru: Medical Science and Biocolonial Exchange," *Comparative Studies in Society and History* 42: 713–44.

Anderson, Warwick (2002) "Postcolonial Technoscience," *Social Studies of Science* 32: 643–58.

Anderson, Warwick (2003) *The Cultivation of Whiteness: Science, Health and Racial Destiny in Australia* (New York: Basic Books).

Anderson, Warwick (2006) *Colonial Pathologies: American Tropical Medicine and Race Hygiene in the Philippines* (Durham, NC: Duke University Press).

Apffel-Marglin, Frédérique (1990) "Smallpox in Two Systems of Knowledge," in Frédérique Apffel-Marglin & Stephen A. Marglin (eds), *Dominating Knowledge: Development, Culture and Resistance* (Oxford: Clarendon Press).

Apffel-Marglin, Frédérique & Stephen A. Marglin (eds) (1996) *Decolonizing Knowledge: From Development to Dialogue* (Oxford: Clarendon Press).

Appadurai, Arjun (1991) "Global Ethnoscapes: Notes and Queries for a Transnational Anthropology," in Richard G. Fox (ed), *Recapturing Anthropology: Working in the Present* (Santa Fe, NM: School of American Research Press): 191–210.

Arnold, David (1993) *Colonizing the Body: State Medicine and Epidemic Disease in Nineteenth-Century India* (Berkeley: University of California Press).

Arnold, David (2005a) "Europe, Technology, and Colonialism in the Twentieth Century," *History and Technology* 21: 85–106.

Arnold, David (2005b) *The Tropics and the Traveling Gaze: India, Landscape, and Science 1800–1856* (Delhi: Permanent Black).

Barker, F., M. Hulme, & N. Iversen (eds) (1994) *Colonial Discourse/Postcolonial Theory* (Manchester, U.K.: Manchester University Press).

Basalla, George (1967) "The Spread of Western Science," *Science* 156 (5 May): 611–22.

Bauschpies, Wenda K. (2000) "Images of Mathematics in Togo, West Africa," *Social Epistemology* 14: 43–54.

Ben-David, Joseph (1971) *The Scientist's Role in Society: A Comparative Study* (Chicago: University of Chicago Press).

Biagioli, Mario (ed) (1999) *The Science Studies Reader* (New York: Routledge).

Bowker, Geoffrey C. & Susan Leigh Star (2000) *Sorting Things Out: Classification and its Consequences* (Cambridge, MA: MIT Press).

Bray, Francesca (1997) *Technology and Gender: Fabrics of Power in Late Imperial China* (Berkeley: University of California Press).

Briggs, Charles L. with Clara Mantini-Briggs (2003) *Stories in a Time of Cholera: Racial Profiling During a Medical Nightmare* (Berkeley: University of California Press).

Brockway, Lucile H. (1979) *Science and Colonial Expansion: The Role of the British Botanical Gardens* (New York: Academic Press).

Callon, Michel (1986) "Some Elements of a Sociology of Translation: Domestication of the Scallops and the Fishermen of St Brieuc Bay," in John Law (ed), *Power, Action and Belief: A New Sociology of Knowledge?* (London: Routledge & Kegan Paul): 196–229.

Chakrabarty, Dipesh (2000) *Provinicializing Europe: Postcolonial Thought and Historical Difference* (Princeton, NJ: Princeton University Press).

Chambers, David Wade (1987) "Period and Process in Colonial and National Science," in Nathan Reingold & Marc Rothenberg (eds), *Scientific Colonialism: A Cross-Cultural Comparison* (Washington DC: Smithsonian Institution Press): 297–321.

Chambers, David Wade (1993) "Locality and Science: Myths of Centre and Periphery," in Antonio Lafuente, Alberto Elena, & Maria Luisa Ortega (eds), *Mundialización de la ciencia y cultural nacional* (Madrid: Doce Calles): 605–18.

Chambers, David Wade & Richard Gillespie (2000) "Locality in the History of Science: Colonial Science, Technoscience, and Indigenous Knowledge," in Roy MacLeod (ed), *Nature and Empire: Science and the Colonial Enterprise* (*Osiris* 2nd series, vol. 15; Chicago: University of Chicago Press, 2001): 221–40.

Clarke, Adele E. (2005) *Situational Analysis: Grounded Theory after the Postmodern Turn* (Thousand Oaks, CA: Sage).

Cohen, Lawrence (1994) "Whodunit?—Violence and the myth of fingerprints: Comment on Harding," *Configurations* 2: 343–47.

Cohen, Lawrence (2005) "The Kothi Wars: AIDS Cosmopolitanism and the Morality of Classification," in Vincanne Adams & Stacy Leigh Pigg (eds) *Sex in Development: Science, Sexuality and Morality in Global Perspective* (Durham, NC: Duke University Press): 269–303.

Collins, Harry & Steven Yearley (1992) "Epistemological Chicken," in Andrew Pickering (ed), *Science as Practice and Culture* (Chicago: University of Chicago Press): 301–26.

Crawford, Elisabeth (1990) "The Universe of International Science, 1880–1939," in Tore Frängsmyr (ed), *Solomon's House Revisited: The Organization and Institutionalization of Science* (Canton, MA: Science History Publications): 251–69.

Crawford, Elisabeth, Terry Shin, & Sverker Sörlin (eds) (1993) *Denationalizing Science: The Contexts of International Scientific Practice* (Dordrecht, The Netherlands: Kluwer).

Cueto, Marcos (1997) "Science under Adversity: Latin American Medical Research and American Private Philanthropy, 1920–1960," *Minerva* 35: 233–45.

de Laet, Marianne & Annemarie Mol (2000) "The Zimbabwean Bush Pump: Mechanics of a Fluid Technology," *Social Studies of Science* 30: 225–63.

Drayton, Richard (1995) "Science and the European Empires," *Journal of Imperial and Commonwealth History* 23: 503–10.

Drayton, Richard (2000) *Nature's Government: Science, Imperial Britain, and the Improvement of the World* (New Haven, CT: Yale University Press).

Drori, Gili S., John W. Meyer, Francisco O. Ramirez, & Evan Schofer (eds) (2003) *Science in the Modern World Polity* (Stanford, CA: Stanford University Press).

Escobar, Arturo (1995) *Encountering Development: The Making and Unmaking of the Third World* (Princeton, NJ: Princeton University Press).

Farquhar, Judith (1996) *Knowing Practice: The Clinical Encounter of Chinese Medicine* (Boulder, CO: Westview Press).

Ferguson, James (1994) *The "Anti-Politics" Machine: "Development," Depoliticization and Bureaucratic Power in Lesotho* (Minneapolis: University of Minnesota Press).

Figueroa, Robert & Sandra Harding (eds) (2003) *Science and Other Cultures: Issues in Philosophies and Science and Technologies* (New York: Routledge).

Fischer, Michael M. J. (2003) *Emergent Forms of Life and the Anthropological Voice* (Durham, NC: Duke University Press).

Fischer, Michael M. J. (2005) "Technoscientific Infrastructures and Emergent Forms of Life: A Commentary," *American Anthropologist* 107: 55–61.

Fortun, Kim (2001) *Advocacy After Bhopal: Environmentalism, Disaster, New Global Orders* (Chicago: University of Chicago Press).

Frank, André Gunder (1969) *Capitalism and Underdevelopment in Latin America* (New York: Monthly Review Press).

Franklin, Sarah (1995) "Science as Culture, Cultures of Science," *Annual Review of Anthropology* 24: 163–84.

Fujimura, Joan H. (1992) "Crafting Science: Standardized Packages, Boundary Objects, and 'Translation,'" in Andrew Pickering (ed), *Science as Practice and Culture* (Chicago: University of Chicago Press): 168–211.

Fujimura, Joan H. (2000) "Transnational Genomics: Transgressing the Boundary Between the 'Modern/West' and the 'Premodern/East,'" in Roddey Reid & Sharon Traweek (eds), *Doing Science + Culture* (New York: Routledge): 71–92.

Galison, Peter (1997) *Image and Logic: A Material Culture of Microphysics* (Chicago: University of Chicago Press).

Gandhi, Leela (1998) *Postcolonial Theory: A Critical Introduction* (St Leonards, NSW: Allen and Unwin).

Gascoigne, John (1998) *Science in the Service of Empire: Joseph Banks, the British State, and the Uses of Science in the Age of Revolution* (New York: Cambridge University Press).

Geertz, Clifford (1963) *Agricultural Involution: The Processes of Ecological Change in Indonesia* (Berkeley: University of California Press).

Good, Byron (1994) *Medicine, Rationality and Experience: An Anthropological Perspective* (Cambridge: Cambridge University Press).

Goonatilake, Susantha (1998) *Toward Global Science: Mining Civilizational Knowledge* (Bloomington: Indiana University Press).

Grove, Richard (1995) *Green Imperialism: Colonial Expansion, Tropical Island Edens and the Origins of Environmentalism, 1600–1860* (Cambridge: Cambridge University Press).

Grove, Richard (1996) "Indigenous Knowledge and the Significance of South-West India for Portuguese and Dutch Constructions of Tropical Nature," *Modern Asian Studies* 30: 121–43.

Gupta, Akhil (1998) *Postcolonial Developments: Agriculture in the Making of Modern India* (Durham, NC: Duke University Press).

Hall, Stuart (1996) "When was 'the Post-colonial'? Thinking at the Limit," in Iain Chambers & Lidia Curti (eds), *The Post-Colonial Question: Common Skies, Divided Horizons* (London: Routledge): 242–60.

Haraway, Donna (1988) "Situated Knowledge: The Science Question in Feminism as a Site of Discourse on the Privilege of a Partial Perspective," *Feminist Studies* 14: 575–99.

Harding, Sandra (1998) *Is Science Multicultural? Postcolonialisms. Feminisms, and Epistemologies* (Bloomington: Indiana University Press).

Harris, Steven J. (1998) "Long-distance Corporations, Big Sciences, and the Geography of Knowledge," *Configurations* 6: 269–304.

Hayden, Cori (2003) *When Nature Goes Public: The Making and Unmaking of Bioprospecting in Mexico* (Princeton, NJ: Princeton University Press).

Headrick, Daniel (1981) *The Tools of Empire: Technology and European Imperialism in the Nineteenth Century* (New York: Oxford University Press).

Hecht, Gabrielle (2002) "Rupture Talk in the Nuclear Age: Conjugating Colonial Power in Africa," *Social Studies of Science* 32: 691–727.

Hess, David J. (1995) *Science and Technology in a Multicultural World: The Cultural Politics of Facts and Artifacts* (New York: Columbia University Press).

Home, R. W. & Sally Gregory Kohlstedt (eds) (1991) *International Science and National Scientific Identity: Australia between Britain and America* (Dordrecht, The Netherlands: Kluwer).

Hughes, C. C. & J. M. Hunter (1970) "Disease and 'Development' in Tropical Africa," *Social Science and Medicine* 3: 443–93.

Hunt, Nancy Rose (1999) *A Colonial Lexicon of Birth Ritual, Medicalization, and Mobility in the Congo* (Durham, NC: Duke University Press).

Jordan, Kathleen & Michael Lynch (1992) "The Sociology of a Genetic Engineering Technique: Ritual and Rationality in the Performance of the 'Plasmid Prep,' " in Adele E. Clarke & Joan H. Fujimura (eds), *The Right Tools for the Job: At Work in Twentieth-Century Life Sciences* (Princeton, NJ: Princeton University Press): 77–114.

Joseph, Gilbert M. (1998) "Close Encounters: Toward a New Cultural History of U.S.–Latin American Relations," in Gilbert M. Joseph, Catherine C. LeGrand, & Ricardo Salvatore (eds), *Close Encounters of Empire: Writing the Cultural History of U.S.–Latin American Relations* (Durham, NC, and London: Duke University Press): 3–46.

Kaufert, Patricia & John O'Neill (1990) "Cooptation and Control: The Construction of Inuit Birth," *Medical Anthropology Quarterly* 4: 427–42.

Kohler, Robert (2002) *Landscapes and Labscapes: Exploring the Lab-Field Border in Biology* (Chicago: University of Chicago Press).

Krige, John (1990) "The Internationalization of Scientific Work," in Susan Cozzens, Peter Healey, Arie Rip, & John Ziman (eds), *The Research System in Transition* (Dordrecht, The Netherlands: Kluwer): 179–97.

Kubicek, Robert (1999) "British Expansion, Empire, and Technological Change," in Andrew Porter (ed), *The Oxford History of the British Empire,* vol. 3: *The Nineteenth Century* (Oxford: Oxford University Press): 247–69.

Kuklick, Henrika & Robert Kohler (eds) (1996) "Science in the Field," *Osiris* 2nd series, vol. 11 [special issue].

Langford, Jean (2002) *Fluent Bodies: Ayurvedic Remedies for Postcolonial Imbalance* (Durham, NC: Duke University Press).

Latour, Bruno (1987) *Science in Action: How to Follow Scientists and Engineers Through Society* (Milton Keynes, U.K.: Open University Press).

Latour, Bruno (1988) *The Pasteurization of France,* trans. Alan Sheridan & John Law (Cambridge, MA: Harvard University Press).

Latour, Bruno (1993) *We Have Never Been Modern,* trans. Catherine Porter (Cambridge, MA: Harvard University Press).

Latour, Bruno (1999a) *Pandora's Hope: Essays on the Reality of Science Studies* (Cambridge MA: Harvard University Press).

Latour, Bruno (1999b) "On recalling ANT," in John Law & John Hassard (eds), *Actor-Network Theory and After* (Oxford: Blackwell): 15–25.

Latour, Bruno (2000) "A Well-articulated Primatology: Reflections of a Fellow-traveler," in Shirley C. Strum & Linda M. Fedigan (eds), *Primate Encounters: Models of Science, Gender, and Society* (Chicago: University of Chicago Press): 358–81.

Law, John (1986) "On Methods of Long Distance Control: Vessels, Navigation and the Portuguese Route to India," in John Law (ed), *Power, Action and Belief: A New Sociology of Knowledge?* (London: Routledge & Kegan Paul): 234–63.

Law, John (1999) "After ANT: Complexity, Naming and Topology," in John Law & John Hassard (eds), *Actor-Network Theory and After* (Oxford: Blackwell): 1–14.

Livingstone, David N. (2004) *Putting Science in its Place: Geographies of Scientific Knowledge* (Chicago: University of Chicago Press).

Livingstone, David N. (2005) "Text, Talk, and Testimony: Geographical Reflections on Scientific Habits," *British Journal for the History of Science* 38: 93–100.

Loomba, Ania (1998) *Colonialism/Postcolonialism* (London: Routledge).

Loomba, Ania, Suvir Kaul, Matti Bunzl, Antoinette Burton, & Jed Esty (eds) (2005) *Postcolonial Studies and Beyond* (Durham, NC: Duke University Press).

Löwy, Ilana (1992) "The Strength of Loose Concepts—Boundary Concepts, Federative Experimental Strategies and Disciplinary Growth: The Case of Immunology," *History of Science* 30: 371–96.

Lux, David S., & Harold J. Cook (1998) "Closed Circles or Open Networks: Communicating at a Distance during the Scientific Revolution," *History of Science* 36: 179–211.

MacLeod, Roy (1987) "On Visiting the 'Moving Metropolis': Reflections on the Architecture of Imperial Science," in Nathan Reingold & Marc Rothenberg (eds), *Scientific Colonialism: A Cross-Cultural Comparison* (Washington DC: Smithsonian Institution Press): 217–49.

MacLeod, Roy (2000) "Introduction," in Roy MacLeod (ed), *Nature and Empire: Science and the Colonial Enterprise* (*Osiris* 2nd series, vol. 15; Chicago: University of Chicago Press, 2001): 1–13.

MacLeod, Roy, & Deepak Kumar (eds) (1995) *Technology and the Raj: Western Technology and Technical Transfers to India, 1700–1947* (New Delhi: Oxford University Press).

Marcus, George E. (1998) *Ethnography Through Thick and Thin* (Princeton, NJ: Princeton University Press).

Martin, Emily (1998) "Anthropology and the Cultural Study of Science: Citadels, Rhizomes and String Figures," *Science, Technology & Human Values* 23: 24–44

McClellan, James & Harold Dorn (1999) *Science and Technology in World History: An Introduction* (Baltimore, MD: Johns Hopkins University Press).

Meillassoux, Claude (1981) *Maidens, Meal and Money: Capitalism and the Domestic Community* (Cambridge: Cambridge University Press).

Mitman, Gregg (1992) *State of Nature: Ecology, Community, and American Social Thought, 1900–1950* (Chicago: University of Chicago Press).

Mol, Annemarie & John Law (1994) "Regions, Networks and Fluids: Anaemia and Social Topology," *Social Studies of Science* 24: 641–71.

Moon, Suzanne M. (1998) "Takeoff or Self-sufficiency? Ideologies of Development in Indonesia, 1957–1961," *Technology and Culture* 39: 187–212.

Moore-Gilbert, Bart (1997) *Postcolonial Theory: Contexts, Practices, Politics* (London: Verso).

Mrázek, Rudolf (2002) *Engineers of Happy Land: Technology and Nationalism in a Colony* (Princeton, NJ: Princeton University Press).

Mutua, Kagendo & Beth Blue Swadener (eds) (2004) *Decolonizing Research in Cross-Cultural Contexts: Critical Personal Narratives* (Albany: State University of New York Press).

Nader, Laura (ed) (1996) *Naked Science: Anthropological Inquiry into Boundaries, Power and Knowledge* (New York: Routledge).

Nandy, Ashis (1995) *Alternative Sciences: Creativity and Authenticity in Two Indian Scientists* (Delhi: Oxford University Press).

Nandy, Ashis (ed) (1988) *Science, Hegemony and Violence: A Requiem for Modernity* (New Delhi: Oxford University Press).

Needham, Joseph (1954) *Science and Civilization in China* (Cambridge: Cambridge University Press).

Nguyen, Vinh-Kim (2005) "Uses and Pleasures: Sexual Modernity, HIV/AIDS, and Confessional Technologies in a West African Metropolis," in Vincanne Adams & Stacy Leigh Pigg (eds), *Sex in Development: Science, Sexuality and Morality in Global Perspective* (Durham, NC: Duke University Press): 245–68.

Nichter, Mark & Mimi Nichter (1996) *Anthropology and International Health: Asian Case Studies* (Amsterdam: Gordon and Breach).

Ophir, Adi & Steven Shapin (1991) "The Place of Knowledge: a Methodological Survey," *Science in Context* 4: 3–21.

Ortner, Sherry B. (1984) "Theory in Anthropology Since the Sixties," *Comparative Studies in Society and History* 26: 126–66.

Owens, Larry (1985) "Pure and Sound Government: Laboratories, Playing Fields, and Gymnasia in the Nineteenth-century Search for Order," *Isis* 76: 182–94.

Palladino, Paolo & Michael Worboys (1993) "Science and Imperialism," *Isis* 84: 91–102.

Pauly, Philip J. (2000) *Biologists and the Promise of American Life: From Meriwether Lewis to Alfred Kinsey* (Princeton, NJ: Princeton University Press).

Petitjean, Patrick, Catherine Jami, & Anne Marie Moulin (eds) (1992) *Science and Empires: Historical Studies about Scientific Development and European Expansion* (Dordrecht, The Netherlands: Kluwer).

Petryna, Adriana (2002) *Life Exposed: Biological Citizens After Chernobyl* (Princeton, NJ: Princeton University Press).

Pfaffenberger, Bryan (1988) "Fetishized Objects and Humanized Nature: Towards an Anthropology of Technology," *Man* 23(2): 236–52.

Pfaffenberger, Bryan (1992) "Social Anthropology of Technology," *Annual Review of Anthropology* 21: 491–516.

Pigg, Stacy Leigh (1997) "'Found in Most Traditional Societies': Traditional Medical Practitioners Between Culture and Development," in Frederick Cooper & Randall Packard (eds), *International Development and the Social Sciences* (Berkeley: University of California Press): 259–90.

Pigg, Stacy Leigh (2001) "Languages of Sex and AIDS in Nepal: Notes on the Social Production of Commensurability," *Cultural Anthropology* 16: 481–541.

Prakash, Gyan (1994) "Introduction: After Colonialism," in Gyan Prakash (ed), *After Colonialism: Imperial Histories and Postcolonial Displacements* (Princeton, NJ: Princeton University Press): 3–17.

Prakash, Gyan (1999) *Another Reason: Science and the Imagination of Colonial India* (Princeton, NJ: Princeton University Press).

Pratt, Mary Louise (1992) *Imperial Eyes: Travel Writing and Transculturation* (New York: Routledge).

Prescott, Heather Munro (2002) "Using the Student Body: College and University Students as Research Subjects in the United States during the Twentieth Century," *Journal of the History of Medicine and Allied Sciences* 57: 3–38.

Pyenson, Lewis (1985) *Cultural Imperialism and Exact Sciences: German Expansion Overseas, 1900–1930* (New York: Peter Lang Publishing).

Pyenson, Lewis (1989) *Empire of Reason: Exact Sciences in Indonesia, 1840–1940* (Leiden, The Netherlands: Brill Academic Publishers).

Pyenson, Lewis (1990) "Habits of Mind: Geophysics at Shanghai and Algiers, 1920–1940," *Historical Studies in the Physical and Biological Sciences* 21: 161–96.

Pyenson, Lewis (1993) "Cultural Imperialism and the Exact Sciences Revisited," *Isis* 84: 103–108.

Raffles, Hugh (2002) *In Amazonia: A Natural History* (Princeton, NJ: Princeton University Press).

Raina, Dhruv (1996) "Reconfiguring the Center: the Structure of Scientific Exchanges Between Colonial India and Europe," *Minerva* 34: 161–76.

Raina, Dhruv (1999) "From West to Non-West? Basalla's Three Stage Model Revisited," *Science as Culture* 8: 497–516.

Redfield, Peter (2000) *Space in the Tropics: From Convicts to Rockets in French Guiana* (Berkeley: University of California Press).

Redfield, Peter (2002) "The Half-life of Empire in Outer Space," *Social Studies of Science* 32: 791–825.

Reingold, Nathan & Marc Rothenberg (eds) (1987) *Scientific Colonialism: A Cross-Cultural Comparison* (Washington DC: Smithsonian Institution Press).

Rose, Nikolas (1999) *Governing the Soul: The Shaping of the Private Self* (London: Free Association Books).

Rostow, W. W. (1960) *The Stages of Economic Growth: A Non-Communist Manifesto* (Cambridge: Cambridge University Press).

Sahlins, Marshall (1999) "What Is Anthropological Enlightenment? Some Lessons of the Twentieth Century," *Annual Review of Anthropology* 28: i–xxiii.

Said, Edward W. (1983) "Traveling Theory," in *The Word, the Text, and the Critic* (Cambridge, MA: Harvard University Press): 226–47.

Said, Edward W. (1994) "Secular Interpretation, the Geographical Element, and the Methodology of Imperialism," in Gyan Prakash (ed), *After Colonialism: Imperial Histories and Postcolonial Displacements* (Princeton, NJ: Princeton University Press): 21–39.

Schaffer, Simon (1992) "Late Victorian Metrology and Its Instrumentation: A Manufactory of Ohms," in Robert Bud & Susan E. Cozzens (eds), *Invisible Connections: Instruments, Institutions and Science* (Bellingham, WA: SPIE Optical Engineering Press, 1992): 23–56.

Schiebinger, Londa (2004) *Plants and Empire: Colonial Bioprospecting in the Atlantic World* (Cambridge, MA: Harvard University Press).

Schofer, Evan (2004) "Cross-national Differences in the Expansion of Science, 1870–1990," *Social Forces* 83: 215–48.

Schott, Thomas (1991) "The World Scientific Community: Globality and Globalization," *Minerva* 29: 440–62.

Schott, Thomas (1993) "World Science: Globalization of Institutions and Participation," *Science, Technology & Human Values* 18: 196–208.

Schott, Thomas (1994) "Collaboration in the Invention of Technology: Globalization, Regions, and Centers," *Social Science Research* 23: 23–56.

Secord, James A. (2004) "Knowledge in Transit," *Isis* 95: 654–72.

Selin, Helaine (ed) (1997) *Encyclopedia of the History of Science, Technology and Medicine in Non-Western Countries* (Dordrecht, The Netherlands: Kluwer).

Shapin, Steven (1995) "Here and Everywhere: The Sociology of Scientific Knowledge," *Annual Review of Sociology* 21: 289–321.

Shapin, Steven (1998) "Placing the View from Nowhere: Historical and Sociological Problems in the Location of Science," *Transactions of the Institute of British Geographers* 23(1): 5–12.

Shepherd, Chris J. (2004) "Agricultural Hybridity and the 'Pathology' of Traditional Ways: The Translation of Desire and Need in Postcolonial Development," *Journal of Latin American Anthropology* 9: 235–66.

Shils, Edward (1991) "Reflections on Tradition, Center and Periphery, and the Universal Validity of Science: The Significance of the Life of S. Ramanujan," *Minerva* 29: 393–419.

Shrum, Wesley (2000) "Science and Story in Development: The Emergence of Non-governmental Organizations in Agricultural Research," *Social Studies of Science* 30: 95–124.

Shrum, Wesley & Yehouda Shenhav (1995) "Science and Technology in Less-developed Counties," in Sheila Jasanoff, Gerald Markel, James Peterson, & Trevor Pinch (eds), *Handbook of Science and Technology Studies* (Thousand Oaks, CA: Sage): 627–51.

Smith, Linda Tuhiwari (1999) *Decolonizing Methodologies: Research and Indigenous Peoples* (Dunedin, NZ: University of Otago Press).

Stafford, Robert A. (1989) *Scientist of Empire: Sir Roderick Murchison, Scientific Exploration, and Victorian Imperialism* (New York: Cambridge University Press).

Star, Susan Leigh & James R. Griesemer (1989) "Institutional Ecology, 'Translations,' and Boundary Objects: Amateurs and Professionals in Berkeley's Museum of Vertebrate Zoology, 1907–39," *Social Studies of Science* 19: 387–420.

Storey, William K. (ed) (1996) *Scientific Aspects of European Expansion* (Aldershot: Variorum).

Strathern, Marilyn (1999) "The New Modernities," in *Property, Substance and Effect: Anthropological Essays on Persons and Things* (London and New Brunswick, NJ: Athlone Press): 117–35.

Sunder Rajan, Kaushik (2005) "Subjects of Speculation: Emergent Life Sciences and Market Logics in the United States and India," *American Anthropologist* 107: 19–30.

Thomas, Nicholas (1994) *Colonialism's Culture: Anthropology, Travel, and Government* (Princeton, NJ: Princeton University Press).

Thompson, Charis (2002) "Ranchers, Scientists, and Grass-roots Development in the United States and Kenya," *Environmental Values* 11: 303–26.

Traweek, Sharon (1988) *Beamtimes and Lifetimes: The World of High-Energy Physicists* (Cambridge, MA: Harvard University Press).

Traweek, Sharon (1992) "Big Science and Colonialist Discourse: Building High-energy Physics in Japan," in Peter Galison & Bruce Hevly (eds), *Big Science: The Growth of Large Scale Research* (Stanford, CA: Stanford University Press, 1992): 100–28.

Tsing, Anna Loewenhaupt (1994) "From the Margins," *Cultural Anthropology* 9: 279–97.

Tsing, Anna (2002) "The Global Situation," in Jonathan Xavier Inda & Renato Rosaldo (eds), *The Anthropology of Globalization: A Reader* (Oxford: Blackwell): 453–86.

Turnbull, David (2000) *Masons, Tricksters and Cartographers: Comparative Studies in the Sociology of Scientific and Indigenous Knowledge* (Amsterdam: Harwood Academic).

Uberoi, J. P. S. (1984) *The Other Mind of Europe* (Delhi: Oxford University Press).

Uberoi, J. P. S. (2002) *The European Modernity: Science, Truth and Method* (Delhi: Oxford University Press).

Vaughan, Megan (1991) *Curing Their Ills: Colonial Power and African Illness* (Stanford: Stanford University Press).

Verran, Helen (1998) "Re-imagining land ownership in Australia," *Postcolonial Studies* 1: 237–54.

Verran, Helen (2001) *Science and an African Logic* (Chicago: University of Chicago Press).

Verran, Helen (2002) "A Postcolonial Moment in Science Studies: Alternative Firing Regimes of Environmental Scientists and Aboriginal Landowners," *Social Studies of Science* 32: 729–62.

Visvanathan, Shiv (1997) *Carnival for Science: Essays on Science, Technology and Development* (New York: Oxford University Press).

Wallerstein, Immanuel (1974) *The Modern World System* (New York: Academic Press).

Watson-Verran, Helen & David Turnbull (1995) "Science and Other Indigenous Knowledge Systems," in Sheila Jasanoff, Gerald Markel, James Peterson, & Trevor Pinch (eds), *Handbook of Science and Technology Studies* (Thousand Oaks, CA: Sage): 115–39.

Williams, P. & Laura Chrisman (eds) (1994) *Colonial Discourse and Postcolonial Theory* (New York: Columbia University Press).

Young, Robert (1990) *White Mythologies: Writing History and the West* (London: Routledge).

II Practices, People, and Places

Olga Amsterdamska

More than a quarter of a century ago, science studies scholars began shifting their attention from science as a system of ideas or beliefs produced by a social institution to a conceptualization of science as a set of practices. A theoretically and disciplinarily diverse set of laboratory and controversy studies published in the late 1970s and early 1980s offered a "naturalistic" look at what scientists are doing when they prepare, devise, or conduct their experiments; collect and interpret data; discuss, formulate, or write up their work; and agree or disagree about their findings. Some adopted the new approach because of a commitment to ethnomethodology; others had a background in anthropology and its ethnographic methods or in symbolic interactionist modes of analysis of work. Still others were inspired by Kuhn's interpretation of paradigms as exemplars, concrete practical achievements which scientists treat as models in need of further elaboration rather than as articulated systems of beliefs; by Polanyi's idea of tacit knowledge; or by the Wittgensteinian or Winchian attention to forms of life.

To a casual observer, the change might have seemed primarily methodological: social scientists developed an interest in conducting ethnographic studies in the laboratories, and began observing the mundane, everyday activities of scientists. Micro-sociological approaches focusing on situated actions supplanted the macro-sociological, structural analyses. Participant observation, interviews, and discourse analysis were used for detailed case studies of scientists at work. The specificities of the locales where research was conducted, interactions among scientists, and their engagements with material environments became objects of interest to social scientists. The titles of early works in this genre—whether Latour and Woolgar's *Laboratory Life* or Karin Knorr-Cetina's *The Manufacture of Knowledge* or Michael Lynch's *Art and Artefact in Laboratory Life*—testify to this emphasis on the processes of knowledge production rather than their products. The first results of these studies seemed largely philosophically deflationary: some of the old distinctions lost their relevance (e.g., between the context of discovery and the context of justification, external and internal factors or social and cognitive activities); and nothing uniquely scientific was happening in the laboratories.

The change in science studies was far more profound than the deceptively naïve call to "follow scientists around" would suggest. The focus on practices signaled an interest in patterned activities rather than rules, in speech and discourse rather than language as a structure, in questions about the use of instruments or ideas in a particular location and situation rather than in universal knowledge, in production and intervention rather than representation, and in science as a mode of working and doing things in and to the world rather than as a system of propositions arranged into theories. Scientists were no longer unproblematically associated with their specialties and disciplines, but were seen as engaging in a variety of interactions with a heterogeneous group of actors, including anyone from patients to laboratory assistants to funding agencies. The achievements of these practice-oriented science studies are visible in virtually every chapter of this Handbook. In this section, however, the STS focus on scientific practice becomes itself an object of reflection, elaboration, and critique.

Practice-oriented approaches to the study of science were seen from the beginning as in some respects problematic. The need to breach (or prove irrelevant) some of their limitations or constraints was often acknowledged and reiterated. For example, while studies of knowledge production emphasized the local character of the research process and of the knowledge claims made by scientists, many critics averred that scientific knowledge is, if not universal, at least translocal or global and that the focus on local practice concealed that fact. How then can the conceptual and methodological toolbox of STS be adjusted and expanded to accommodate questions about the production and reproduction of translocal scientific knowledge? Can we even talk about such knowledge? What are the consequences of STS's concrete focus on the local and historically specific for our ability to distinguish science from other kinds of knowledge, or to justify drawing a distinction between good and bad science? Are there ways to overcome the implicit normative agnosticism that came with the emphasis on the practical and the local? Similarly, practice-oriented investigation of scientific knowledge tended to emphasize the manner in which scientists "do" things, and thus intervention and experimentation were studied more intensely than the production of propositional or theoretical knowledge. But if so, is there a way to look at patterns of argumentation and rhetoric in science without abandoning the practice-oriented approach? And does practice orientation make STS researchers oblivious to larger-scale social processes, to economic, institutional, or cultural constraints and the more permanent forms of the distribution of power in society?

The essays in this section of the Handbook review a wide range of studies of the various aspects of scientific practices and suggest new ways to address these concerns about the limits of the pragmatic turn in science studies.

The first three chapters in this section draw on the resources of neighboring fields—argumentation studies and rhetoric, social epistemology, and cognitive science—to suggest how some of the perceived limitations of science studies could be overcome. Underlying these possibilities for dialogue is a shared focus on scientific practice. And so, William Keith and William Rehg review studies of scientific argumentation and rhetoric. They emphasize that rhetorical analyses of science are likely today to examine various kinds of discourses in their contexts, to study argumentation as a process as well as a product, to analyze informal rather than formal structures of argumentation, and to pay attention to the exigencies of goals, modalities, and audiences. In all these respects, these studies share the concerns and approaches of STS studies of discourse, yet at the same time they offer us tools to examine the larger communicative contexts of scientific discourse and, the authors hope, to build a bridge between "normative philosophical approaches and descriptive/explanatory sociologies of knowledge, often considered non-critical or anti-prescriptive." Concern with how practice-oriented approaches to science can develop a normative orientation is also paramount in the articles of Ronald Giere and Miriam Solomon, both of whom investigate the intersection between STS and the new practice-oriented philosophical studies of science.

In philosophy, the abandonment of the grand project of the logical reconstruction of scientific knowledge and its methodology has generated increased interest in more historically and empirically rooted approaches to knowledge and a variety of appeals to the American pragmatist tradition. At the same time, philosophers have been made particularly uncomfortable by the supposed relativist and nonevaluative attitudes of constructivist approaches to science dominant in STS. The attempt to develop a coherent normative position—to distinguish good science from bad and to develop recommendations for how science should be done—is at the heart of both Giere's review of the cognitive sciences and Solomon's review of social epistemology. Both regard science as situated practice and agree that to view it as a passive representation of the world or as a logical form is to misunderstand scientific endeavors. Moreover, although Giere does not want to lose sight of the psychological aspects of cognition, both he and Solomon emphasize the collective aspects of scientific investigation and knowledge, allow for a plurality of culturally and disciplinarily variable scientific research strategies and evaluative approaches, and formulate evaluative norms that govern communal activities rather than individual cognition or abstract systems of propositions.

While Solomon and Giere look into and beyond social studies of science from the perspectives of their own fields, Park Doing reviews laboratory studies from the "inside," asking to what extent such studies met the goals set by their authors. The most fundamental claim of early laboratory studies was the assertion that the process of construction and acceptance of scientific claims cannot be separated from their content, or that the production—shown to be driven by contingency, opportunism, political expediencies, tinkering toward success, and so on—shapes the product. In his chapter, Doing argues that while laboratory life has indeed been shown to be full of contingencies of all sorts, ethnographies of the production of scientific facts have not established how these contingencies actually affect the formulation of specific claims and their acceptance or rejection. Park Doing proposes an ethnomethodological solution to this shortcoming of the existent laboratory studies. He advocates turning to actors' accounts of the closure of controversies, while pointing out that thus far, those who have tried to explain such closure have tended to look beyond the immediate contexts of practice—to invoke the authority of disciplines or instruments.

Once science ceased to be regarded as a body of propositions, it quickly became apparent that images and other forms of visualization played an important, often organizing role in much laboratory work. They mediate both social and instrumental interactions. Accordingly, interest in visual representations entered social studies of science together with the interest in practices. Studies of the production, interpretation, and use of scientific images are reviewed in the chapter by Regula Burri and Joseph Dumit. They call for extending studies of scientific imaging practices to places beyond the laboratory walls where they not only carry the authority of science but also intersect with—reinforce, challenge, or are challenged by—other kinds of knowledge. Medical practice, with its heavy reliance on visual technologie, is one of the settings in which such interactions between different kinds of knowledge and modes of representation and seeing are particularly interesting. Studies of medical imaging allow us to ask questions about the social persuasiveness and power of images and about the role of science in the constitution of identity and seeing.

As Burri and Dumit remind us in their work on images and the authors from the Virtual Knowledge Studio (VKS) reiterate in their chapter, studies of scientific practices in laboratories have devoted much attention to the uses of instrumentation, tools, and technologies of research. Some of these studies emphasize the mediating role of instruments and technologies, while others point to their unruliness and recalcitrance in the daily work of knowledge production, to the skill and tacit knowledge which goes into dealing with instruments, to the articulation work needed to get and use "the right tool for the job," and to efforts of standardization deployed to limit uncertainty or facilitate communication among scientists working in different settings. The roles of instrumentation and technologies of research are, however, particularly wide ranging and multifaceted in the case of e-science, examined here by the VKS. The amazing heterogeneity of the uses of computers and the Internet in contemporary science—with changes in both methods and media permeating so many different aspects of scientific practice—prompts the VKS authors to advocate extending the existing focus on scientific work by trying to conceptualize it as scientific *labor*, thus incorporating the economic dimensions of instrumental practices alongside studies of practices as epistemic cultures. At the same time, e-science provides us with a unique opportunity to examine the significance of location and displacement in the practice of science.

The emergence of e-science, the globalization of communication and research technologies, and the seemingly unlimited mobility of researchers, research objects, and knowledge claims are reflected in the (seamless, virtual, fluid) "network" vocabularies used both to describe scientific practices in the Internet era and to theorize about science more generally. Network imagery shifts attention away from the constitutive roles of contexts and places but facilitates discussions of processes of the de- and re-contextualization of knowledge and the merging of micro and macro levels. And yet, as Christopher Henke and Thomas Gieryn argue in their chapter, places—as geographical and sociocultural locations and as architectural settings with specific designs and equipment—continue to matter for the practice of science by, for example, enabling and organizing face-to-face interactions among practitioners, helping to define some activities as scientific while delegitimizing others (and thus securing science its cultural authority), or organizing activities as individual and collective, visible and hidden from view, public and private. The question "where does science happen?" retains its relevance for studies of scientific practice even in the age of global networks and standardized settings.

The question of "who?"—of how to conceptualize actors and their identities—is, of course, equally central. As many of these chapters make apparent, the critiques of practice-oriented studies of science often focus on the continuing difficulty of resolving action-structure dilemmas. Since a focus on practice brings to the fore the manner in which the scientists themselves actively shape their world, bringing both order and change, many authors find it necessary to try to account for the co-constitutive character of the context or environment of practice, be it material, social, economic, or cultural. The search for such structural factors dominates the attempt of Henry Etzkowitz, Stefan Fuchs, Namrata Gupta, Carol Kemelgor, and Marina Ranga to explain the continuing low levels of women's participation in science. In contrast, Cyrus Mody and David Kaiser's study of scientific education explicitly strives to combine structural and social action approaches, appealing to Foucault and Bourdieu alongside Wittgenstein and Kuhn. Mody and Kaiser see science education not merely as reproduction of values, knowledge, and credentialed personnel but as their generation. They study learning and teaching as a process leading to both transmission of ready-made book knowledge and the development of skills and of tacit knowledge. For them, students and teachers are not simply the followers of rules and norms but politically and socially savvy actors, so that education is not just the filtering of recruits into science but an active and historically changing process of the fashioning of the moral economy of science.

The directive to study practices has widened the range of places where STS scholars now look when studying the production of scientific knowledge. From a practice perspective, every diagnostic or treatment decision by a doctor, every choice of policy by a government regulatory agency, and every user's attempt to master a new technology can be seen as part of the process of knowledge production. But if so, there was, of course, no reason to keep our eyes fixed inside the walls of laboratories, universities, research institutes, and R&D departments, and, as many essays in other sections of this Handbook testify, much justified attention has in recent years come to settle on actors who are not scientists and on areas of activity where scientific knowledge, technological know-how, and research are made to intersect with other knowledges, skills, and tasks. While productive, such a broadening of focus makes theory construction more complex and contributes to the sense that the term "practice" itself has become all-inclusive and less distinct. The chapters in this section do not share a common theory, or even a common definition of practice, but a family resemblance and a set of problems that might be a good place from which to continue thinking about science.

9 Argumentation in Science: The Cross-Fertilization of Argumentation Theory and Science Studies

William Keith and William Rehg

The STS literature offers numerous studies of scientific inquiry and communication that investigate scientific argumentation, the ways in which scientists evaluate and contest claims about the world, scientific practice, and each other. Inspired by Thomas Kuhn, historians and sociologists have trained their sights on the content of scientific argument, territory traditionally reserved to philosophers trained in formal logic. Students of rhetoric have also brought their expertise to bear on science.[1]

In this chapter we document the cross-fertilization of argumentation studies and science studies and suggest new relationships between them. As we understand it, cross-fertilization occurs when argumentation theorists and science scholars collaborate on common projects, or when a scholar from one of these two areas draws on studies from the other area. The rhetoric of science thus represents an area of science studies that was constituted by cross-fertilization.

Interdisciplinary engagement between science studies and argumentation studies is fostered by "boundary concepts" (Klein, 1996)—ideas such as "text," "discourse," "logic," "rhetoric," and "controversy"—that have some purchase in both fields. For a set of such concepts we first look to the disciplines that have informed the study of argumentation: rhetoric, speech communication, philosophy and logic, composition, linguistics, and computer science.[2] We then map existing studies of scientific argumentation according to the different contexts that govern argumentation and arguments.[3] We conclude by suggesting some avenues for further interdisciplinary cross-fertilization.

ARGUMENTATION: WHAT IS IT AND WHO STUDIES IT?

"Argument" is an odd word. In English, its meaning radically changes in different environments, even with a slight change in context: "making an argument" and "having an argument" are quite different (the first requires only one person, while the second requires at least two). Inspired by O'Keefe (1977), argumentation theorists distinguish between argument as a product and argumentation as a process. Although theorists have traditionally described and evaluated argument products independently of the specific processes (discourse, reflection, etc.) that generated them, some approaches tend to resist this separation (e.g., dialogical models of conversational arguments, rhetorical approaches). In the sciences, at any rate, arguments often appear as distinct, identifiable products (e.g., conference talks, written reports, articles) that issue from processes of inquiry and discussion—even if understanding the product depends on the process.

We normally divide the content of an argument into two parts: the conclusion or point of the argument, and the material (reasons, premises) that supports that conclusion. Beyond this general characterization, however, analyses of content diverge. Theorists differ over the kind of material or reasons—the modes of representation— that may go into an argument, and they differ over the kinds of structure needed for the product to be interpretable as an argument. These two questions of *argument constitution* affect not only how we interpret (and reconstruct) actual arguments, but they also determine how we *evaluate* arguments as valid, reasonable, or good, insofar as evaluation requires us to assess the quality of the supporting reasons (their relevance, truth, etc.) and the quality of the structural relationships between reasons and conclusion (validity, inductive strength, etc.).

"Argumentation," as a process, usually refers to a human activity involving two or more people.[4] Consequently, argumentation requires taking account of communication: Whereas arguments are often taken to be describable independently of particular instantiations or communication situations, argumentation generally must be understood in terms of these. As a communicative process, argumentation can occur in different *modalities* or *venues* of communication, which in turn affects whether the argumentation is monological or dialogical. Thus dialogical argumentation is easiest to achieve in a face-to-face modality, more difficult in public venues (conference talks, televised debates). Argumentation can also be conducted textually, through e-mail, successive letters to the editor in a publication, or journal articles that respond to each other, perhaps over a period of years. We can also imagine argument as circulating— as a set of texts and utterances that circulate through society, in different forms and modalities, modifying and being modified as they go.[5]

As a social practice, argumentation can have different *purposes or goals* (Walton, 1998): It might be aimed at inquiry (Meiland, 1989)—at the testing of statements or hypotheses, or the generation of new ones (i.e., "abduction"). Arguers may also engage in advocacy, attempting to convince others that they should change their beliefs or values. Some theorists consider conflict resolution (Keith, 1995) and negotiation to involve argumentation (Walton, 1998, chapter 4). In a less savory guise, argumentation might be part of an attempt to manipulate an audience by using deceptive arguments. Finally, argumentation lies at the heart of collective deliberation, reasoned choice-making by groups. Insofar as scientific inquiry involves modes of practical reasoning and choice, both at the local and institutional level, scientific reasoning has a deliberative component (cf. Knorr Cetina, 1981; Fuller, 2000a).

Some theorists further distinguish argumentation *procedures* from the more inclusive notion of process (e.g., Wenzel, 1990; Tindale, 1999). "Process" indicates the activity of arguing as unfolding over time, as for example in an argumentative conversation, where the argumentation involves turn-taking and thus is not locatable in any single utterance. "Procedure" usually refers to a discursive structure that normatively guides a process, determining (in part) the order in which participants speak or communicate, the allowable or relevant content at each stage, role divisions, and the like (e.g., trial procedures that govern argumentation about the guilt of a defendant).

Given the breadth of the concept of argumentation, it should come as no surprise that different disciplines take somewhat different approaches to its study. We focus here on the two traditions that have generated the largest body of reflection on scientific argument: philosophy and rhetoric.[6]

Philosophy

At mid-twentieth century, the philosophical study of argument was dominated by formal-logical approaches (e.g., logical empiricism in the philosophy of science).[7] Formal-logical models take a normative approach and treat the content of arguments as detached from social contexts and influences (for a survey, see Goble, 2001). These models typically construe the content of arguments as a sequence of propositions (or statements, or sentences)[8] some of which (the premises) have inferential or justificatory relationships to others (intermediate and final conclusions). Propositionalist approaches take different views of good argument structure. Deductivists (e.g., Karl Popper) admit as valid only those arguments whose form is truth-preserving. Because the information in the conclusion does not go beyond that in the premises, the form guarantees that true premises will generate a true conclusion invulnerable to additional information. Argument evaluation then involves assessing the logical validity of the structure and the truth (or rational acceptability) of the premises.

Dualist models accept not only deduction but also inductive arguments, that is, ampliative modes of inference whose conclusions go beyond the information in the premises. Because inductive conclusions are vulnerable to new information, they are only more or less *probably* true. Logical empiricists attempted to formalize inductive support by drawing on probability theory, which allowed them to define a quantitative "degree of confirmation" as a formal relationship between evidence sentences and the hypothesis-conclusion. Assessing the strength of an induction meant calculating this quantity for a given hypothesis relative to an acceptable set of evidence statements (see Salmon, 1967; Kyburg, 1970).

Some argumentation theorists maintain that the range of interesting yet nondeductive argument structures includes not only simple induction but also analogical arguments, inference to best explanation, casuistic reasoning, narrative, and so on (Govier, 1987; Johnson, 2000; Walton, 1989, 1998). Influential proposals of alternatives to formal logic (e.g., Naess, [1947]1966; Toulmin, 1958; Perelman and Olbrechts-Tyteca, [1958]1969), along with the informal logic and critical thinking movements (see van Eemeren et al., 1996; Johnson and Blair, 2000), have led to an increased appreciation among philosophers for "informal" methods of argument evaluation, which generally assume that arguments can be described and evaluated independently of whether or not they can be syntactically formalized.[9]

Informal inferences depend on the interrelated meanings of terms and on background information that resists complete formalization. Accordingly, arguments can also include nonlinguistic modes of representation such as symbolic or mathematical notations, various forms of pictorial representation, physical models, and computer simulations, which are common in science.[10] Because such arguments involve ampliative inferences, their conclusions are more or less "probable." Unlike formal inductive logics, however, probability is not so much quantitative as *pragmatic*, in the sense associated with notions of cogency or plausibility (Toulmin, 1958: chapter 2; Walton, 1992).[11] The level of probability or cogency typically depends on satisfying standards such as relevance, sufficiency, and acceptability. To apply such criteria, we must attend to the interpretive subtleties of argument in context.[12] The normative treatment of informal arguments is also heavily invested in the definition, identification, and criticism of fallacies. Although Aristotle famously defined a fallacy as a nonargument masquerading as an argument (*Sophistical Refutations* I), contemporary theorists differ over its definition.[13]

Many informal logicians consider their approach to be a development of the dialectical tradition of argument evaluation stemming from the ancient Greeks (in particular, Aristotle) and the medieval practice of disputation. From a dialectical perspective, cogent arguments must meet a specified burden of proof and rebut relevant challenges (Rescher, 1977; Walton, 1998; Johnson, 2000; Goldman, 1994, 1999: chapter 5). Consequently, dialectical theorists often embed their accounts of the argument product in a theory of the argumentation process as a dialogue or critical discussion that should meet certain criteria (e.g., procedures that ensure severe testing of claims, social conditions that foster open, noncoercive communication).[14] Such standards project an idealized social space, protected from "external" social-political factors, in which the community of inquirers is more likely to produce (and if possible agree on) arguments that are in some sense objectively better or more reasonable.[15]

Rhetoric

Informal and formal approaches share an emphasis on the rational use of arguments: reasons provide the conclusion with a justification or rational grounding. But we can also take a *rhetorical* perspective on arguments. Although generally associated with the study of persuasion, the rhetorical tradition—which stretches from ancient Greece to modern discourse theory in the United States and Europe—addresses a vast range of issues, some descriptive, some explanatory, some prescriptive; some concerned with the speaker's "invention" (i.e., the discovery of arguments), others with the "criticism" of texts.[16] To keep our survey manageable, we focus on two subtraditions explicitly devoted to the study of rhetoric and influential in the rhetoric of science. Both are based in U.S. universities, specifically in the disciplines of Communication (or Speech Communication, formerly Speech) and English Composition.[17]

Speech Communication Formed in U.S. universities around the teaching of public speaking and debate, this tradition foregrounds oral communication and the political context of deliberation. Much of its research is framed by an appreciation for or reaction against Aristotle's somewhat idealized account of the political speech situation:

> The species of rhetoric are three in number, for such is the number [of classes] to which the hearers of speeches belong. A speech [situation] consists of three things: a speaker, and a subject on which he speaks, and someone addressed, and the objective [*telos*] of the speech relates to the last [I mean the hearer]. Now, it is necessary for the hearer to be either a spectator [*theoros*] or a judge [*krites*], and [in the latter case] a judge of either past or future happenings. A member of a democratic assembly is an example of one judging about future happenings, a juryman an example of one judging the past. (Aristotle 1991: I.1.3, 1358a-b)

So the key elements are the speaker, the topic, the speaker's purpose, and the audience. Aristotle speaks only very indirectly of context, since he presumes that the listeners have gathered in an institutional setting such as the legislature or the court for the purpose of coming to a judgment. While Aristotle recognizes that multiple elements play a role in the process of persuasion, he devotes more attention to argument (*logos*) than to the other means of persuasion, character (*ethos*) and emotion (*pathos*).[18]

In contrast to philosophers, theorists in the U.S. speech tradition are less concerned with argument per se than with argumentation, and they focus not on dialectical exchanges intended to (dis)prove theses but on group deliberations aimed at making decisions about a course of action. Consequently, communication theorists usually position argumentation as part of a process of conviction (change in belief) or persuasion (change in action). A focus on persuasion means that arguments must take account of their contexts; they must be specific and relevant in the situation. And contexts are relative: arguments that matter in one context, no matter how "generally" valid, may not matter in another context. Persuasion also highlights the importance of audience, whose members evaluate arguments in view of their own standpoints and opinions. While rhetoricians in this tradition have done a considerable amount of innovation since the 1950s, much of it focusing on a rhetorical version of symbolic interactionism, traces of the tradition are still visible in much of the rhetoric of science literature, as in Goodnight's influential 1982 piece on "spheres of argument," which attempts to blend Aristotle with Habermas, or Campbell's many attempts to reconstruct the deliberative context for the acceptance of Darwinian theory.

English/Composition In English departments, and the field of Composition, rhetoric has typically been understood in terms of the figurative and the generic aspects of written argument. Both aspects are important in teaching college students to write. Since the audience is not physically present in writing, generic considerations are invoked to supply an appropriate context. Originally, genre referred either to literary forms (essay, short story, etc.) or to what, after Alexander Bain ([1871]1996), were called the "modes" of discourse: narration, description, exposition, and argument, which represent a fusion of style and communicative function. Argument is one of these modes, and in composition, argument was often treated as a product, similar to its treatment by philosophers. Students were taught to assemble evidence, avoid fallacies, and so forth, on the assumption that their arguments would be critically read by a "general audience."

Growing out of its eighteenth century *belles lettres* heritage, composition instruction was also attentive to verbal style or the figurative aspects of writing. It distinguished between tropes, which involve nonliteral meanings of words, and schemes, which involve unusual arrangements of words. Tropes include metaphor, metonymy, and simile, while figures include repetition ("of the people, by the people, for the people"), antithesis, and *klimax*.[19] Writing teachers understand the use of figuration with respect to different rhetorical aims: as primarily aesthetic or as strategic and functional (for example, as a way of supporting or clarifying an argument).

In both subtraditions of rhetoric, scholars situate arguments within a larger social and communicative context. Rhetorical theorists thus insist on seeing the rationality of argumentation relative to the social, cultural, and political context of the participants, such that one cannot cleanly separate the "internal" dimension of reason from its "external" context. For critical evaluation, they tend to rely on field-specific or local standards, or political ideals and norms derived from the humanistic tradition of rhetoric.[20]

ARGUMENT IN SCIENCE: WHERE AND HOW

The overlapping contexts in which arguments are made confront participants with specific "exigencies": particular goals, modalities, and audiences. Arguments are found in journals and books originate in local settings—in the laboratory, at the field site, in small groups, in notebooks—where researchers engage in conversations and private reflection. Local processes of argument making, in turn, unfold within larger discursive contexts and institutional settings, including funding agencies, interested publics, and law- and policymakers.

In this section, we organize the science-studies research according to these different contexts of argumentation. Starting with studies of argument construction at the local research site, we move to studies of wider discourse communities, a context where much of the argumentation is conducted in print and where scientific controversies typically occur. Scientific argumentation is further affected by institutional and cultural aspects of science—its "ethos," funding mechanisms, disciplinary divisions, and the like. Finally, broader nonscientific publics also participate in arguments about the sciences. Naturally, many science-studies investigations focus on more than one of these sites, since they investigate argument across multiple contexts or with multiple purposes. The schema nonetheless remains useful as a means of differentiating various sites for interdisciplinary engagement.

To identify interdisciplinary possibilities, we rely on various boundary concepts that are relevant in both areas of study. Some of these concepts we already identified in our survey of argumentation studies (logic, deduction, induction, dialectic, aspects of rhetoric, etc.); others emerge as salient concerns for science scholars (e.g., controversy, evidence, consensus).

The Local Construction of Arguments at the Research Site

Recent philosophical work on the local construction of arguments has focused on normative theories of evidence that respond to flaws in logical empiricist treatments of confirmation (see Achinstein, 2001, 2005; Taper and Lele, 2004). In a departure from the Bayesian assumptions that had informed that approach, Mayo (1996) examines the "error-statistical" methods that scientists actually use to discriminate between hypotheses and eliminate likely sources of error. Staley (2004) has refined Mayo's approach and applied it to a detailed case study of the discovery of the top quark at Fermilab. Some aspects of Staley's study, for example, his analysis of the article-writing process in a large collaboration, would certainly benefit from a deeper engagement with argumentation theory—in particular, dialectic and rhetoric (see Rehg & Staley, in press).

Feminist philosophers of science have also contributed to theories of evidence, demonstrating how local argument construction depends on broader contexts of discourse. Longino (1990, 2002) shows how evidential arguments depend on metaphysical and value-laden background assumptions, including gender biases from the broader culture. According to Keller (1983), geneticist Barbara McClintock lacked recognition until late in her career because the genetics community was simply unable to understand the sort of arguments McClintock was making or the sort of evidence she provided. Keller argues that McClintock's vision of science stood outside the rapidly growing institutional laboratory structure, and this outsider status was the source both of her creativity and of the difficulty the biology community had in understanding her contributions.

Philosophical models of evidence address both the product and process of local argument making, and their attention to substantive, contextual detail goes far beyond logical empiricism. Many philosophers now recognize that rhetoric is a necessary component of scientific argument (McMullin, 1991; Toulmin, 1995; Kitcher 1991, 1995). Nonetheless, normative theories of evidence could still benefit from a closer attention to rhetorical studies of argument construction, such as that of Blakeslee (2001). In her study of article writing in physics as a face-to-face process of audience construction, Blakeslee examined how a physics research team revised their article (intended for biologists) according to the understanding of their audience, which they acquired through local interactions with biologists.

Sociologists, anthropologists, and historians of science have also made impressive contributions to the understanding of local argumentative practices in science, although clear examples of cross-fertilization with argumentation studies remain limited.[21] Latour and Woolgar's ethnography of laboratory work ([1979]1986) approaches the laboratory as a "system of literary inscription." The authors analyze how scientists construct facts from data by working to transform qualified statements (e.g., "Smith observed evidence for x") into unqualified factual ones ("x exists"). They go on to explain scientists' behavior in terms of the quest for credibility rather than adherence to norms of method.

Some of the most detailed and rigorously descriptive studies of argumentation at the research site we owe to ethnomethodologists, whose close description of scientists' shop talk serves to reveal the local, situated rationalities of everyday scientific practice (Lynch, 1993). For example, in his study of a neurosciences lab, Lynch (1985) catalogues the ways neuroscientists reach consensus on data interpretation. Livingston (1986, 1987, 1999) applies ethnomethodology to "cultures of proving," including mathematics.[22] By tracking mathematicians through their construction of various proofs (geometrical, Gödel's proof, etc.), he hopes to show how the proof text, or "proof account," provides a set of cues, a "gestalt or reasoning," whose sense of universal, objective compulsion depends on the embodied, social practices of mathematicians. Such intensely focused studies are complemented by analyses that link laboratory interaction with the broader ethos of the science community. In her study of high-energy physics, Traweek (1988), for example, notices that effective argument in this community requires an aggressive style of communication.

Other sociologists attempt to explain how micro- and macro-sociological conditions (individual needs and goal orientations, professional and other social interests, class, etc.) affect local argument construction. MacKenzie, for example, links Karl Pearson's understanding of statistical argument with his promotion of social eugenics and, at a further remove, with class interests (MacKenzie and Barnes, 1979; MacKenzie, 1978). One of the best examples of actual cross-fertilization is Bloor's (1983, chapter 6; cf. [1976]1991, 1984) Wittgensteinian explanation of choices between competing types of logic. Because "deductive intuitions" alone underdetermine this choice, further "interests and needs," i.e., aims of the various practices in which the logical language game is embedded, codetermine the choice.

Since his collaboration with Woolgar, Latour has developed the rhetorical aspects of fact construction more fully in the context of actor-network theory (though he draws more explicitly on semiotics than rhetorical studies).[23] Latour (1987) systematically explains how scientific arguments are built through networks of texts, things, machines, inscriptions, calculations, and citations. He compares the elements of networks with rhetorical resources for turning opinions into facts: a "fact" is a claim that no one any longer has the resources to challenge with an effective counterargument. Scientists achieve this persuasive effect partly by enlisting powerful allies in their cause—as, for example, the hygiene movement in France aided Pasteur's success as a scientist (Latour 1988). Latour thus links lab-level argumentation with institutional and technological dimensions of science.

Among historical treatments of laboratory work, Galison's magisterial studies of high-energy physics, or HEP (1987, 1994), stand out for linking local argumentation with both laboratory technology and broader institutional trends. At the lab level, he shows how argumentation depends not only on theoretical commitments but also on the "material culture" of the laboratory—in particular its specific instrumental commitments. For physicists in the "image" tradition, evidential arguments depend on the analysis of visible tracks recorded in devices such as bubble chambers; physicists in the "logic" tradition employ statistical arguments based on the output of counting devices. As HEP became "big science" requiring massive material outlays and large collaborations, argumentation in the lab acquired the institutional complexity of science as a whole, forcing collaborators to develop skills at interdisciplinary communication.

The above survey indicates that a rich potential for interdisciplinary work exists for the study of local argumentation. Some of the more pressing questions here concern the implications of the various contingencies and concrete particularities of laboratory culture for the normativity of evidential arguments. In the final part of this chapter we suggest some possible interdisciplinary approaches to this issue.

Writing and Controversy: Science as Discourse Community and Field

Much of the actual cross-fertilization between argumentation theory and science studies has occurred in the study of argumentation across a given discipline or field of research, where the sciences have been treated as discourse communities. The focus here has been on argumentation in print and controversy studies. First, since the record of scientific argumentation is mostly a written one, the text is a natural place to begin analyses of arguments. Second, as qualitative sociologists have long claimed, the underlying values and assumptions of a field are most visible during moments of crisis or breaks in the normal routine (Garfinkel, 1967). In the same way, controversies in science have been attractive to argumentation researchers, since they not only display scientific argument but also in some cases reflect on it as well. To the extent that science, in its presentation as "normal science," seems transparent and unavailable to rhetorical or argument analysis, controversies provide a site of entry.

Argumentation in Print   Many of the disciplines that took the "rhetorical turn" are text-oriented (see Klein, 1996: 66–70), and so it should come as no surprise that much, perhaps most, of the work in the rhetoric of science has focused on scientific texts. Specific aims, perspectives, and foci differ. Some theorists show how scientific argument is continuous with other kinds of argument, whereas others show how it is distinctive. Many studies focus on single texts, but some authors (e.g., Myers, Campbell) touch on the process of intertextual argumentation, attempting to account for argument across a number of texts and sometimes authors. Much of the rhetorical analysis is primarily descriptive or analytic, but some studies venture explanatory or prescriptive claims.

Such a diverse range of scholarship resists neat organization. Here, we approach this body of work as attempts to account for the textual aspects of argument in relation to the discursive context and the various rhetorical conditions it imposes on persuasiveness or acceptability. Our survey aims to convey a sense of the density of the rhetorical dimensions of scientific texts: once considered as marginal, suspicious, and possibly irrelevant ornamentation in scientific argument, the rhetoric of science has emerged as epistemically central and all-pervasive, open to seemingly endless variation.[24]

Working in the genre tradition, Bazerman (1988) shows how conventions of writing help determine what can and cannot be argued, what kinds of evidence can be used, and how conclusions can be drawn. His influential analysis of the American Psychological Association *Manual of Style* shows that the changes in the guide from the 1920s through the 1980s reflect the changes in the discipline's self-understanding, as well as changes in methodology that follow from the discipline's struggles to become more empirical over time. The development of the familiar five-part structure of the research article (introduction, literature review, method, results, and discussion) made it virtually impossible for introspectionist or philosophical arguments to make their way into psychology journals.

Fahnestock (1999) provides a good example of the figurative approach. She claims that some scientific arguments are best understood by analyzing the stylistic elements—the figures and tropes—that express them. For instance, she considers the traditional figure *gradatio* (*klimax* in Greek), in which a repetition is combined with a change in degree or scale. A traditional example is "I came, I saw, I conquered," which not only uses repetition but also nests the early assertions within the expanding later assertions. Fahnestock shows that scientists use this figure to structure an argument in which an effect increases through a series of changes in experimental conditions leading to a causal conclusion.

At least two textual studies are noteworthy for their sustained historical sweep: Gross et al. (2002) track changes in the scientific article—analyzed in terms of Aristotle's distinction between style, presentation (i.e., arrangement), and argument—as it appeared in three languages (English, German, and French) from the seventeenth through the twentieth century. They attempt to explain these literary developments by drawing on evolutionary models of conceptual change in science (e.g., Hull, 1988). Atkinson (1999) combines resources from sociology of science, rhetoric, and quantitative linguistics to document shifts in generic aspects of *The Philosophical Transactions of the Royal Society* from 1675 to 1975. By tracking changes in the frequency of linguistic patterns ("registers") indicative of genre, Atkinson demonstrates the gradual emergence of various textual features (e.g., non-narrativity, abstractness) of contemporary science writing. Historians of science have also taken an interest in this kind of rhetorical analysis. Drawing heavily on the figurative and generic traditions, the contributions in Dear (1991) analyze the textual dynamics that conditioned argument and communication in a number of disciplines from the seventeenth through nineteenth centuries, including zoology, physiology, mathematics, and chemistry.

These studies show how generic and figurative elements are associated with specific discourse communities and affect the substance of scientific argument. Another broad area of research has focused on how more specific demands of audience and occasion, connected with a specific topic or controversy, shape scientific texts. Prelli (1989a), for example, approaches argument construction through the classical rhetoric of invention, a perspective that catalogues the available resources for developing arguments: the stases (potential points of disagreement) and available "lines of argument" (or commonplaces) that are reasonable for a given content, audience, and situation. In this manner, Prelli lays out an informal "topical" logic of argumentation that reveals possible grounds, based in practical reasoning, for situated audience judgment. He offers an extensive system of stases and commonplaces, and documents how scientific texts systematically respond to their argumentative burdens. For example, biologists must defend their sampling techniques, methods of analysis, judgments of the significance of the outcome, and the like. Prelli offers a perspective for comparing disparate texts as arguments and for explaining their persuasive success or failure.

Numerous case studies examine the ways that texts reflect specific audiences interested in specific issues or "occasions." Gross ([1990]1996), for example, shows how scientific texts from Newton's *Principia* to Watson's *Double Helix* adapt their arguments to the goals of the scientist, against a background of audience knowledge and assumptions. According to Gross, Newton deliberately cast his argument in a geometrical idiom to meet the expectations of his readers. Selzer (1993) presents a range of analyses from scholars from Communication and English, who comment on Stephen J. Gould and Richard Lewontin's "Spandrels of San Marcos and the Panglossian Paradigm: A Critique of the Adaptationist Programme," which is a critique of the excesses of the adaptationist program in evolutionary theory. By situating Gould and Lewontin's arguments in evolutionary debates, the essays bring out how they contingently and strategically represent the history of biology, the literature in the field, and their opponents' views. Miller (1992; cf. Fuller, 1995) argues that differences in the reception of scientific articles can be explained in terms of the Sophistic idea of *kairos*: successful articles, such as Watson and Crick's 1953 *Nature* report on the structure of DNA, position themselves at the opportune moment and place in the dynamic field of research problem-solving.

Myers (1990) provides an exceptionally detailed and broad analysis of a range of scientific texts (grant proposals, journal articles under development, popular science essays) as they are shaped by the demands of specific professional and lay audiences. Drawing both on his background in linguistics and on constructivist sociology of science, Myers wants to show how texts, *qua* texts, argue not only for their conclusions but also for their scientific status. By carefully describing how arguments emerge from the textual negotiations among authors, editors, and referees, Myers illuminates both the writing process and its products (see also Berkenkotter and Huckin, 1995; Blakeslee, 2001).

John Angus Campbell's (1990, 1995, 1997) historically informed, close textual studies of the argument strategies of Charles Darwin, show, among other things, that the structure of Darwin's notebooks is generative of argument found later in *The Origin of Species*. Campbell also demonstrates that Darwin strategically used the ambiguities in certain arguments to bridge the cultural gap between the older theological paradigm and the newer scientific one and that Darwin and his allies were clever self-promoters who employed public relations techniques to win a favorable public hearing. Campbell (1986) examines Darwin's "cultural grammar," the background assumptions unfavorable to Darwin's case, which Darwin nonetheless used to his advantage in both intellectual and popular discourse.

Finally, some authors go beyond explanation of textual arguments by setting them into a larger normative structure and critically reflecting on the possibilities for scientific argument. McCloskey—an economist deeply interested in rhetoric—shows that both historical and contemporary economic arguments are conditioned by rhetorical form and audience considerations ([1986]1998). Mathematical appearances notwithstanding, much economic argument relies on metaphors and narrative structures. McCloskey (1990) goes on to criticize the practice of economics as stunted and hypocritical: if economics were less constrained to argue exclusively in a mathematical idiom, it could contribute more effectively to understanding and resolving social-political problems.[25]

Although it is not completely systematic (Gaonkar, 1997), the analysis of argumentation in scientific texts has shown that even apparently "dry" or transparent texts have interesting argumentative features that can be usefully explicated. The textual features of scientific texts are evidently *functional*—they respond to and help create discursive situations (e.g., "proof," "evidence") and effects (e.g., acceptance to a journal, a replication or refutation) within scientific communities and the cultures that house them. Interestingly, in some cases rhetorical features of textual arguments reflect disciplinary constituencies, while in other cases they seem to constitute them, helping determine what it means to be scientific or a scientist within a given setting.

Controversy and Theory Change   Many studies described above focus on controversial texts, but argumentation during controversies and in times of theory change has also been a subject of another, distinct body of work which emerged in the wake of the rationality debates following the publication of Kuhn's *Structure of Scientific Revolutions*.[26]

Responding to Kuhn, philosophers proposed dialectical models of theoretical development.[27] Pera (1994; cf. 2000), for example, attempts to apply the dialectical tradition—which he considers the "logic" of rhetorical discourse—to the study of scientific argumentation. He articulates a kind of informal logic for science, a set of substantive and procedural rules for conducting and resolving debate (though his rules remain rather abstract from a rhetorical perspective). Kitcher (1993, 2000) analyzes controversies from an implicitly dialectical perspective. He takes "eliminative induction" as the basic argumentative strategy: in controversies, scientists try to force their opponents into positions that they cannot maintain without falling into internal inconsistencies or suffering "explanatory losses" (severe cutbacks in the scope of their claims).

As philosophers struggled to rescue science from contingency, social historians and sociologists of scientific knowledge emphasized its effects. The study of controversy provided a rich field for this project. For Collins (1983, 1985), the microanalysis of controversies brings out the contingencies that afflict inductive inference. Shapin and Schaffer's (1985) analysis of the Boyle-Hobbes debate situated the controversy in a broader macro-sociological context. Issues of social organization of the scientific community and of the polity hid behind the protagonists' opposing views about method (experimental vs. geometric), paths to consensus (publicly repeatable experiments vs. compelling deductions), and the definition of knowledge (probabilistic vs. certain). Shapin and Schaffer also show how the protagonists' different views were reflected in their different rhetorical strategies ("literary technologies").

Among sociologists, Kim's (1994) study of the Mendelian-biometrician debate in evolutionary theory goes further than most in providing an "internal" argumentation–theoretic explanation of closure. Kim analyzes the argumentative process in terms of three groups: the elite protagonists, the paradigm articulators (e.g., disciples open to theoretical conversion), and the "critical mass" of breeders and physicians who assessed the practical usability of the competing models for their own work.

These social-historical studies generally follow Kuhn in his attempt to grasp the dynamics of theory change from the perspective of the historical participants themselves, without the benefit of hindsight (Kuhn, [1962]1996; cf. Hoyningen-Huene, 1993; Golinski, 1998). In contrast to his structural macro-history of theory change, however, historians of science have tended to take a micro-historical approach to scientific controversies and describe argumentation in rich empirical detail (e.g., Rudwick, 1985; Galison, 1987).

The Institutional Structuring of Scientific Argumentation
Scientific argumentation takes place within specific institutional and disciplinary structures: in virtue of specific modes of funding, within specific organizations (university, government laboratory, corporations), via specific avenues of communication (refereed journals, conferences, etc.), involving specific modes of recognition, gate-keeping, and the like. How do these structures affect scientific argumentation? Interdisciplinary answers to this question can draw on boundary concepts that include ideas of ethos, consensus, rational dialogue, and disciplinary boundaries.

Many of the studies that address the institutional dimension of scientific argumentation are reactions to Merton's (1973) classic pre-Kuhnian sociology of the institutional "ethos" of science. Merton attributed progress in modern science to certain institutionalized "norms" or ideals that govern the behavior of scientists and make science into a rational collective endeavor: universalism (adherence to impersonal standards of evaluation), organized skepticism, disinterested pursuit of knowledge, and "communism" (a commitment to share results with the community). A key issue that arises concerns the relation between this ethos and consensus formation. Departing from the Mertonian approach to this issue, Gilbert and Mulkay (1984) regard consensus not as an objective social fact but as a context-dependent discursive construction. Scientists invoke consensus, criticize opponents, and explain disagreement by using "social accounting" methods that exploit the interpretive flexibility of meanings, membership, and beliefs.

Prelli (1989b) has explicitly linked Merton's idea of institutional ethos with Aristotle's rhetorical concept of ethos (argument from character). Prelli argues that scientists invoke "norms" such as Merton's not as general rules but as situated rhetorical *topoi*, argumentative resources for establishing (or undermining) the credibility of those whose research they want to support (or attack). Moreover, such *topoi* include the *opposites* of Mertonian ideals: Prelli's case study (the debate about whether the gorilla Koko had learned sign language) shows how reversing traditional ideals can serve to support controversial claims as "revolutionary."

Hull (1988) tests Merton's high-minded ethos against a kind of social naturalism. Digging into debates in evolutionary biology and taxonomy, Hull shows how institutional mechanisms, such as credit, lead self-interested scientists to cooperate in the production of knowledge. Solomon (2001; see also Solomon in chapter 10 in this volume) takes naturalism in a social-psychological direction by analyzing controversy in terms of the various "decision vectors"—formerly considered "biasing factors"—that actually motivate scientists to accept or reject a theory. Solomon's social epistemology[28] belongs to a growing body of critical work—pursued by philosophers, sociologists, and historians—on cultural and gender-based biases in scientific argumentation: in the interpretation of evidence, assumptions guiding theory construction, and so on (e.g., Harding, 1999; Wylie, 2002). Much of this work clearly bears on the institutional level, whose history and structures have systematically worked to discourage women in science (e.g., Potter, 2001; for an overview, see Scheibinger, 1999). Some critical proposals appeal to process norms. Longino (1990: chapter 4; 2002: 128–35), for example, develops a normative model of argumentative process that invokes idealized standards (similar to Habermas's) for the conduct and institutional organization of critical scientific discussion.

Institutional and cultural influences on arguments have also been demonstrated by Paul Edwards (1996). Edwards shows how the development of computer science and artificial intelligence research was heavily conditioned by a preference of the U.S. military (and its arm, the RAND corporation) for mathematical models based on finite sets of axioms (i.e., the "closed-world" assumption that everything relevant to the problem at hand is contained in the model of the problem). Edwards shows that this style of argument influenced both the understanding of science and development of technology in the Cold War.

Finally, a number of theorists have brought argumentation theory to bear on issues connected with disciplinary boundaries. Ceccarelli (2001) focuses on the management of disciplinary differences by examining three famous works in biology from the standpoint of audience effects. She shows how arguments and presentation styles used in two of these works (Dobzhansky's *Genetics and the Origins of Species* and Schrödinger's *What Is Life?*) are designed to extend their audiences beyond disciplinary boundaries. These books became classics precisely because their arguments "spoke" the language of more than one discipline, strategically suppressing disagreement between a descriptionist biological tradition and an analytic tradition in physics and chemistry. Taylor (1996) also applies rhetorical analysis to the question of boundaries, showing how disciplinary boundaries are created and maintained through strategies of argument. In his view, scientific argumentation belongs to an "ecosystem" of people, publications, and institutions that certify or reject arguments.

Disciplinary boundaries have also been a problem for STS because a strong sense of disciplinary incommensurability leads to a reluctance to engage "other" fields, thus thwarting interdisciplinary argument and communication (Fuller and Collier, 2004; see also Fuller 1988, 1993). Rejecting the underlying internalist assumption that arguments must be relative to fields, Fuller and Collier propose dialectical and rhetorical strategies for promoting responsible interdisciplinary dialogue. Their model also has implications for the relationship between science and politics, the fourth setting we treat here.

Public Discourse and Policy Argumentation

Scientific argumentation occurs not only in experimental, discursive, and institutional contexts within the science community but also at the interfaces of science and society. These interfaces have long been the concern of critical social theorists, such as Habermas (1971), whose attempt to situate policy argumentation within a democratic context anticipated the "argumentative turn" in policy studies (Fischer & Forester, 1993; cf. Majone, 1989; Schön & Rein, 1994; Williams & Matheny, 1995; De Leon 1997; Forester 1999). The literature is as diverse as the interfaces themselves (courtroom, bureaucracy, legislature, hospital, media venue, etc.). Here we focus on studies dealing with the prospects for democratic public involvement.

Fuller approaches issues of science and democracy by way of a critical social theory informed by constructivist sociology of science. He takes the social conditioning of science seriously but, unlike many sociologists, maintains a deep commitment to a normative critique that bridges the gap between scientific argumentation and public deliberation (Fuller, 1988, 1993; cf. Remedios, 2003). Recalling the Enlightenment ideal of science as both a path to more scientific governance and a model for democratizing society, Fuller (2000a) asks what science would look like if we held it accountable for its democratic character (or lack thereof), as we do other institutions. He thus opposes the elitist stance that served the interest of post-WWII research universities by assuring them of government funding free of public oversight (and found its philosophical justification in Kuhn [Fuller, 2000b]). If we restrict the participation in arguments about science policy and funding to experts *in* the field, then neither the public nor other scientists would be able to influence the goals of scientific research or the allotment of research money. Explicating the "liberal" versus "republican" modes of evaluating scientific argument, Fuller finds that neither is in harmony with what he calls the "mafia" tendencies of current funding processes.

Willard (1996) takes up the issue of science and democracy via Lippmann's (1925) question: Doesn't an expertise-driven society make democracy useless and counterproductive? Willard believes this problem emerges out of a mistaken liberal conception of community, where experts appear, endlessly, to be outside the democratic community. He thus proposes an "epistemics" model of scientific argument in society, which would shift the focus of debates from questions of "Who's included?" and "Who's watching the government?" to the characteristics of scientific and policy argument that make it accessible and/or controversial across multiple audiences. Willard's account takes seriously both the political content of scientific argument and the scientific relevance of political points of view.

Such proposals must face the challenge of meaningful public participation in science-intensive policy argumentation. Whereas Willard points out the importance of translation across venues, Brown (1998) invokes the advantages of narrative: scientific arguments are typically nestled within narratives and must be understood in relation to them. Brown tries to offer an account of scientific argument that would make it more accessible to democratic institutions.

Of course, work on science-and-democracy hardly exhausts the work in this area. We close with two examples of issue-focused studies.[29] Condit (1999) examines the development of genetic theory in the twentieth century in relation to its public reception. She shows that public arguments about what genetics means for society and human self-understanding interact with those in the scholarly literature. The essays in Campbell and Meyer (2003) grapple with the many sides of the creationism debate. Starting with the question of what should be taught to students in school, they delve into arguments not only about education but also about the kinds and quality of arguments for evolution, intelligent design, and creation theory—should we teach the debates or just the "right" answers?

EXPANDING INTERDISCIPLINARY RECIPROCITY: WHERE DO WE GO FROM HERE?

Our overview reveals a rich body of work concerned with scientific argumentation. We have also identified some notable examples of science-studies scholars drawing on and employing categories from the argumentation-studies literature. We believe, however, that there are important problem areas where a closer and more direct collaboration between argumentation theorists and STS scholars would prove particularly fruitful.

For example, given the increasing need of lay publics to make critical assessments of expert advice, as well as the growing interest among STS researchers in policy debate and expertise, collaborations between STS scholars and argumentation theorists might be especially interesting in this area. As we saw in the "Where and How" section, above, a number of scholars have pursued some version of what we might call "critical science studies" (CSS) (e.g., Fuller, 1988; Longino, 1990; cf. Hess, 1997: chapter 5). A collaboration between scholars interested in CSS and argumentation theorists might allow for a better integration of philosophical, rhetorical, and sociological perspectives. Here, the main challenge lies in overcoming the deep differences between normative philosophical approaches and descriptive/explanatory sociologies of knowledge, often considered noncritical or antiprescriptive. We close by suggesting three paths for circumventing such differences. The paths present increasingly strong versions of interdisciplinarity, but in each case the rhetorical perspective helps bridge the divide.

The first path allows each side to cooperate while retaining its initial stance on argumentation, by agreeing to set aside divisive philosophical commitments for the sake of a particular case. Consider, for example, the deep differences that separate critical theorists, such as Habermas, from the Strong Program in the Sociology of Knowledge. Whereas the latter takes a skeptical view of the justificatory "force" of arguments in explaining consensus formation, the former seems to believe in the intrinsic "force of the better argument."[30] In fact, neither side denies that scientists *believe* arguments can be compelling; thus, both sides can proceed on that phenomenological assumption. In effect, they would then be making a claim about the rhetorical effects of argument and then asking how consensus formation (or the lack of it) should be explained in the given case by the available arguments and other social conditions. If sociological analysis reveals that the outcome depends in its substance on social conditions, then a further critical question becomes pertinent: does knowledge of this dependence undermine our confidence in the reasonableness of the outcome? In some cases it might, in others it might not; the answer, again, depends on the rhetorical-dialectical situation, specifically, the aims of science in context (Rehg, 1999).

The second path, involving ethnomethodologists and critical theorists, challenges both sides to engage, and perhaps modify, their methodological commitments. Unlike critical social theorists and philosophers, radical ethnomethodologists strive simply to notice and perspicuously describe—but not theorize, evaluate, or criticize—the situated "methods" and local rationalities that practitioners themselves employ in their interactions (Lynch, 1993; cf. Lynch, 1997). As it turns out, however, these studies show that scientists *use* norms of method to hold one another accountable for their practices (e.g., Gilbert and Mulkay, 1984). This suggests that ethnomethodologically informed critical theorists need not abandon critique so long as they adopt the attitude of participants and contextualize argumentative norms. At a minimum, they may view idealized norms in pragmatic and rhetorical, rather than legislative, terms; seeing norms as rhetorical moves, the intelligibility of which depends on substantive features of the local context of inquiry, opens up new possibilities for critique (Rehg, 2001; cf. Prelli, 1989b). Conversely, this approach suggests the idea of a critical ethnomethodology (cf. Lynch, 1999).

A final path to the normative appreciation of scientific arguments places scientific argument more firmly into a multilayered rhetorical context that sets it in dialog with its civic and political contexts. (For an articulation of the possibilities for that dialogue, see Cherwitz, 2004, 2005a,b.) The critical theorist, that is, creates a description of argument that makes argumentation in the lab and the journals continuous with argumentation in the legislature and the public sphere. This is already a reality in politically divisive fields, such as marine ecology and forestry, and it is rapidly emerging in certain biomedical areas. Relative to a context of democratic governance and principles of social justice (Fuller, 2000a), it would be possible (in a highly nuanced way) to create a critical dialog between scientific practices and public/social values, neither determining the other. For example, the movement toward increasing attention to medical research on women, driven by a perception of unfairness and scientific inadequacy (i.e., results from clinical trials on men only cannot be easily generalized to women), shows that scientific practice can be fruitfully criticized.

These examples suggest that sociology of scientific knowledge scholars who aim primarily at descriptive and explanatory analyses of argumentation can nonetheless engage interdisciplinarily with a critical project committed to normative standards of reasonable argument. If argumentation theory can foster such surprising alliances, then greater cross-fertilization between science studies and argumentation theory is a promising prospect.

Notes

The authors thank Olga Amsterdamska and the four anonymous reviewers for their feedback on an earlier draft of this essay.

1. For anthologies on the rhetoric of science, see Simons (1989, 1990); Pera and Shea (1991); Krips et al. (1995); Gross & Keith (1997); Harris (1997); Battalio (1998).

2. The interdisciplinary character of argumentation studies is evident in conferences (e.g., International Society for the Study of Argumentation, Ontario Society for the Study of Argumentation), journals (*Argumentation, Informal Logic*), and graduate programs (e.g., University of Amsterdam). For overviews of argumentation theory, see Cox & Willard (1982), van Eemeren et al. (1996).

3. Our treatment of science studies focuses mainly on studies of mathematics and the natural sciences in various contexts, including policymaking; a thorough treatment of other areas of STS is beyond the scope of this chapter.

4. An exception is found in areas of AI that attempt to model argumentation among "intelligent agents" (see McBurney & Parsons, 2002), which, while not human, are still plural.

5. Warner (2002); this idea is also central to Latour (1987).

6. To be sure, philosophical theories of discourse have significantly influenced rhetorical analysis. We focus here mainly on the two disciplinary traditions in the United States that have produced the largest body of literature explicitly devoted to rhetorical analysis. Thus, we do not directly take up all the continental traditions in discourse theory and linguistic analysis, though some of the rhetoric of science we describe below draws on such work; for a useful overview of these traditions, see Sills & Jensen (1992). We also pass over other areas that contribute to argumentation theory, such as law, whose scholars have studied aspects of legal argumentation. Continental scholars explicitly identified with rhetoric have mostly worked in classical rhetoric and so contribute only indirectly to argumentation in science.

7. In the United States, pragmatists have produced important studies of logic and scientific argumentation (e.g., Peirce, 1931–33; Hanson, 1958), but by the 1950s their influence on philosophy departments was giving way to analytical philosophers.

8. Philosophers have traditionally understood propositions to represent the content of sentences or statements, independent of their superficial form (e.g., German or English); some philosophers, however, consider sentences or utterances, not propositions, the basic "truth-bearer" in arguments; see Kirkham (1992: chapter 2) for further details.

9. We refer here to the second of the various senses of "formal" that Barth and Krabbe (1982) distinguished: formal$_1$ (Platonic Forms), formal$_2$ (rules of syntax for using logical constants in a deductive system), formal$_3$ (rules of dialogical procedure).

10. There is a growing interest in nontextual representation and argument, both in argumentation studies (Birdsell & Groake, 1996; Hauser, 1999) and in science studies (Lynch & Woolgar, 1990; Galison, 1994; Perini, 2005; Ommen 2005).

11. Rescher (1976) attempts to formalize plausibility arguments; theorists interested in computational modeling of argument systems have attempted to formalize types of defeasible reasoning (cf. Prakken, 1997; Gilbert, 2002). Keith (2005) reconstructs Toulmin's model as nonmonotonic reasoning and amplifies the various senses of "probably" at issue.

12. For typical criteria, see Johnson & Blair (1977), Johnson (2000), Govier (2005); for contextualist approaches to relevance, see Hitchcock (1992), Tindale (1999), Walton (2004).

13. Some examples: Tindale (1999) views fallacies as bad product, procedure, or process; van Eemeren et al. (2002) as violations of the ten rules of dialogue; Walton (1996) as illicit shifts in the type of dialogue, i.e., as an argument that blocks the inherent goal of the given dialogue-type.

14. Lakatos (1976) is an example of a procedurally focused dialectical approach in the philosophy of mathematics. Criticizing the formalist approach, he reconstructs historical developments in geometry as a fictional conversation in which students argue about the "real" definition of a particular polyhedron and self-consciously challenge each other about their modes of argument and counterargument. By showing that this process conforms to a Popperian account, proof followed by refutation and further proof, Lakatos lays bare the dialectical structure of mathematical reasoning. Although Popper was a deductivist in his approach to argument analysis, his methodology of conjecture and refutation is dialectical (Lakatos 1976: 143 note 2).

15. Dialogical and deliberative democratic models often take this approach; Habermas's work is especially well known among argumentation theorists (e.g., Habermas, 1984, 1996: cf. Rehg, 2003), but see also Alexy (1990), van Eemeren et al. (1993), Bohman & Rehg (1997); on the difficulties in applying these standards, see Elster (1998), Blaug (1999).

16. The modes of rhetorical analysis are too varied to list here. For historical overviews, see Kennedy (1980), Bizzell & Herzberg (2001); on contemporary rhetoric, see Lucaites et al. (1999), Jasinski (2001); on rhetorical criticism, see Burgchardt (2000); for a treatment of the tradition of invention, see Heidelbaugh (2001). Farrell (1993) and Leff (2002) argue for a normative understanding of rhetoric.

17. This is not intended to be an exhaustive or international survey of all that has been done under the heading of "rhetoric" in the last hundred years but instead a handy interdisciplinary guide to those traditions that have influenced the study of argument in science. Nor do we suggest that rhetorical studies fully encompasses, or is encompassed by, these two disciplines.

18. As means of persuasion, ethos and pathos may count as arguments for Aristotle in a broader sense, which he distinguishes from the style and arrangement of speeches; in any case, interpretations of Aristotle remain controversial. Note also that within the European tradition, an Isocratean/Ciceronian humanism, rather than Aristotle per se, dominated university pedagogy until the nineteenth century; see Kimball (1995).

19. For example, "You can take the boy out of the country, but you can't take the country out of the boy" is an antithesis, while the word "country" is a metonym for rural culture.

20. In the critical thinking movement, domain-specific standards have been advocated by some; for discussion of the relevant debates, see Siegel (1988) and McPeck (1990). Toulmin et al. (1984) draw evaluative standards from disciplinary fields (law, science, ethics, etc.); Willard (1989, 1996) further sit-

uates incommensurable disciplinary problem-solving rationalities in contexts of democratic critique. Drawing on Foucault, McKerrow (1989) conceives a "critical rhetoric" aimed at discursive performance rather than truth. Fisher (1984, 1987) has proposed narrative as a tool for evaluating the quality of argument.

21. A major exception to this characterization is historical work on the rhetoric of experiment, e.g., Dear's (1995) rhetorically informed study of the different uses of the term *experimentum* in the seventeenth century; see also Dear (1991).

22. For other work on mathematical proof and cultures of proving, see the literature cited in Heintz (2003).

23. Meanwhile, Woolgar has gone on to elaborate the skeptical implications of social constructionism as a critique of the rhetoric of objective representation in science—a critique that applies reflexively to SSK itself (see Woolgar, 1983; 1988a,b; Ashmore, 1989). Other scholars have further developed Latour and Woolgar's ([1979]1986) literary-critical approach to laboratory work by drawing on Derridean ideas (see Lenoir, 1998).

24. Indeed, some critics charge the rhetoric of science with being vapid and uninformative (Fuller, 1995; Gaonkar, 1997). A more just criticism might be that there have typically not been well-defined research programs; many studies of scientific rhetoric, even if high quality, are motivated by "here is another interesting text."

25. Two close philosophical engagements with texts deserve mention in this section: Suppe (1998) and Hardcastle (1999) test different normative models of scientific argument through line-by-line analyses of actual articles.

26. Edited collections on the study of controversy include Engelhardt & Caplan (1987), Brante et al. (1993), Machamer et al. (2000).

27. Works such as Laudan (1977) and Shapere (1986) are at least tacitly dialectical in approach; Brown (1977) and Ackerman (1985) present their models as dialectical; for a formal approach to Kuhn's account of theory change, see Stegmüller (1976).

28. The term is due to Fuller (1988), and proponents of social epistemology include Nelson (1990); Hull (1988); Kitcher (1993, 2001); Longino (1990, 2002); Goldman (1999); Harding (1999); Kusch (2002); for a range of views, see Schmitt (1994).

29. Argumentation–theoretic case studies of the science-society interface are numerous; for some well-known examples, see Harris (1997: chapters 7–9); Waddell's study of public hearings regarding research at Harvard is well known (Waddell 1989, 1990) as is Farrell and Goodnight's (1981) study of the Three-Mile Island episode. Fabj & Sobnosky (1995) draw on Goodnight's (1982) model of argument spheres to analyze the much-studied case of AIDS treatment activism; for a detailed analysis of argumentation at NAS and NIH, see Hilgartner (2000).

30. See Barnes & Bloor (1982) and Bloor (1984) for views that suggest this reading of the Strong Program; Habermas (1984) represents the opposite view (cf. also McCarthy, 1988; Bohman, 1991). For our purposes, what matters is not so much the accuracy of these interpretations, but whether scholars who interpret their counterparts this way might still manage to collaborate.

References

Achinstein, Peter (2001) *The Book of Evidence* (Cambridge: Oxford University Press).

Achinstein, Peter (ed) (2005) *Scientific Evidence: Philosophical Theories and Applications* (Baltimore, MD: Johns Hopkins University Press).

Ackerman, Robert John (1985) *Data, Instruments, and Theory: A Dialectical Approach to Understanding Science* (Princeton, NJ: Princeton University Press).

Alexy, Robert (1990) "A Theory of Practical Discourse," in S. Benhabib & F. Dallmayr (eds), *The Communicative Ethics Controversy* (Cambridge, MA: MIT Press): 151–90.

Aristotle (1991) *On Rhetoric,* trans. G. A. Kennedy (New York and Oxford: Oxford University Press).

Ashmore, Malcolm (1989) *The Reflexive Thesis: Wrighting Sociology of Scientific Knowledge* (Chicago: University of Chicago Press).

Atkinson, Dwight (1999) *Scientific Discourse in Sociohistorical Context: The Philosophical Transactions of the Royal Society of London, 1675–1975* (Mahwah, NJ: Lawrence Erlbaum).

Bain, Alexander ([1871]1996) *English Composition and Rhetoric* (Delmar, NY: Scholars' Facsimiles and Reprints).

Barnes, Barry & David Bloor (1982) "Relativism, Rationalism and the Sociology of Knowledge," in M. Hollis & S. Lukes (eds), *Rationality and Relativism* (Cambridge, MA: MIT Press): 21–47.

Barth, E. M. & E. C. W. Krabbe (1982) *From Axiom to Dialogue: A Philosophical Study of Logics and Argumentation* (Berlin and New York: de Gruyter).

Battalio, John T. (ed) (1998) *Essays in the Study of Scientific Discourse: Methods, Practice, and Pedagogy* (Stamford, CT: Ablex).

Bazerman, Charles (1988) *Shaping Written Knowledge: The Genre and Activity of the Experimental Article in Science* (Madison: University of Wisconsin Press).

Berkenkotter, Carol & Thomas N. Huckin (1995) *Genre Knowledge in Disciplinary Communication: Cognition/Culture/Power* (Hillsdale, NJ: Lawrence Erlbaum).

Birdsell, David J. & Leo Groark (eds) (1996) Special Issues on Visual Argument, *Argumentation and Advocacy* 33: 1–39, 53–80.

Bizzell, Patricia & Bruce Herzberg (eds) (2001) *The Rhetorical Tradition* (Boston: Bedford/St. Martins).

Blakeslee, Anne M. (2001) *Interacting with Audiences: Social Influences on the Production of Scientific Writing* (Mahwah, NJ: Lawrence Erlbaum).

Blaug, Ricardo (1999) *Democracy, Real and Ideal* (Albany: State University of New York Press).

Bloor, David ([1976]1991) *Knowledge and Social Imagery* (Chicago: University of Chicago Press).

Bloor, David (1983) *Wittgenstein: A Social Theory of Knowledge* (New York: Columbia University Press).

Bloor, David (1984) "The Sociology of Reasons: Or Why 'Epistemic Factors' Are Really 'Social Factors'," in J. R. Brown (ed), *Scientific Rationality* (Dordrecht, The Netherlands, and Boston: Reidel): 295–324.

Bohman, James (1991) *New Philosophy of Social Science* (Cambridge, MA: MIT Press).

Bohman, James & William Rehg (eds) (1997) *Deliberative Democracy: Essays on Reason and Politics* (Cambridge, MA: MIT Press).

Brante, Thomas, Steve Fuller, & William Lynch (eds.) (1993) *Controversial Science* (Albany: State University of New York Press).

Brown, Harold I. (1977) *Perception, Theory, and Commitment* (Chicago: Precedent).

Brown, Richard Harvey (1998) *Toward a Democratic Science: Scientific Narration and Civic Communication* (New Haven, CT: Yale University Press).

Burgchardt, Carl R. (2000) *Readings in Rhetorical Criticism* (State College, PA: Strata).

Campbell, John Angus (1986) "Scientific Revolution and the Grammar of Culture: The Case of Darwin's *Origin*," *Quarterly Journal of Speech* 72(4): 351–67.

Campbell, John Angus (1990) "Scientific Discovery and Rhetorical Invention: The Path to Darwin's *Origin*," in H. W. Simons (ed), *The Rhetorical Turn: Invention and Persuasion in the Conduct of Inquiry* (Chicago: University of Chicago Press): 58–90.

Campbell, John Angus (1995) "Topics, Tropes, and Tradition: Darwin's Reinvention and Subversion of the Argument to Design," in H. Krips, J. E. McGuire, & T. Melia (eds), *Science, Reason, and Rhetoric* (Pittsburgh: University of Pittsburgh Press): 211–35.

Campbell, John Angus (1997) "Charles Darwin: Rhetorician of Science," in R. A. Harris, *Landmark Essays on Rhetoric of Science: Case Studies* (Mahwah, NJ: Hermagoras): 3–18.

Campbell, John & Stephen C. Meyer (eds) (2003) *Darwinism, Design and Public Education* (East Lansing: Michigan State University Press).

Ceccarelli, Leah (2001) *Shaping Science with Rhetoric: The Cases of Dobzhansky, Schroedinger and Wilson* (Chicago: University of Chicago Press).

Cherwitz, Richard (2004) "A Call for Academic and Civic Engagement," *ALCALDE*, January/February, 2004.

Cherwitz, Richard (2005a) "Citizen Scholars: Research Universities Must Strive for Academic Engagement," *The Scientist* 19 (January 17): 10.

Cherwitz, Richard (2005b) "Intellectual Entrepreneurship: The New Social Compact," *Inside Higher Ed* (March 9, 2005). Available at: http://www.insidehighered.com/views/2005/03/09/cherwitz1.

Collins, H. M. (1983) "An Empirical Relativist Programme in the Sociology of Scientific Knowledge," in K. D. Knorr Cetina & M. Mulkay (eds) *Science Observed* (London and Beverly Hills, CA: Sage): 85–140.

Collins, H. M. (1985) *Changing Order: Replication and Induction in Scientific Practice* (London and Beverly Hills, CA: Sage).

Condit, Celeste Michelle (1999) *The Meaning of the Gene: Public Debates about Human Heredity* (Madison: University of Wisconsin Press).

Cox, J. Robert & Charles Arthur Willard (eds) (1982) *Advances in Argumentation Theory and Research* (Carbondale: Southern Illinois University Press).

Dear, Peter (ed) (1991) *The Literary Structure of Scientific Argument* (Philadelphia: University of Pennsylvania Press).

Dear, Peter (1995) *Discipline and Experience: The Mathematical Way in the Scientific Revolution* (Chicago: University of Chicago Press).

De Leon, Peter (1997) *Democracy and the Policy Sciences* (Albany: State University of New York Press).

Edwards, Paul N. (1996) *The Closed World: Computers and the Politics of Discourse in Cold War America* (Cambridge, MA: MIT Press).

Eemeren, Frans H. van & Rob Groodendorst (1992) *Argumentation, Communication, and Fallacies* (Hillsdale, NJ: Lawrence Erlbaum).

Eemeren, Frans H. van, Rob Grootendorst, Sally Jackson, & Scott Jacobs (1993) *Reconstructing Argumentative Discourse* (Tuscaloosa: University of Alabama Press).

Eemeren, Frans H. van, Rob Grootendorst, Francisca Snoek Henkemans, J. Anthony Blair, Ralph H. Johnson, Erik C. W. Krabbe, Christian Plantin, Douglas N. Walton, Charles A. Willard, John Woods, & David Zarefsky (1996) *Fundamentals of Argumentation Theory* (Mahwah, NJ: Lawrence Erlbaum).

Eemeren, Frans H. van, Rob Grootendorst, & A. Francisca Snoek Henkemans (2002) *Argumentation: Analysis, Evaluation, Presentation* (Mahwah, NJ: Lawrence Erlbaum).

Elster, Jon (1998) *Deliberative Democracy* (Cambridge: Cambridge University Press).

Engelhardt, H. Tristam & Arthur L. Caplan (eds) (1987) *Scientific Controversies: Case Studies in the Resolution and Closure of Disputes in Science and Technology* (Cambridge: Cambridge University Press).

Fabj, Valeria & Matthew J. Sobnosky (1995) "AIDS Activism and the Rejuvenation of the Public Sphere," *Argumentation and Advocacy* 31: 163–84.

Fahnestock, Jeanne (1999) *Rhetorical Figures in Science* (New York: Oxford University Press).

Farrell, Thomas B. (1993) *The Norms of Rhetorical Culture* (New Haven, CT: Yale University Press).

Farrell, Thomas B. & G. Thomas Goodnight (1981) "Accidental Rhetoric: The Root Metaphors of Three Mile Island," *Communication Monographs* 48: 271–300.

Fischer, Frank & John Forester (eds) (1993) *The Argumentative Turn in Policy Analysis and Planning* (Durham, NC: Duke University Press).

Fisher, Walter R. (1984) "Narration as Human Communication Paradigm: The Case of Public Moral Argument," *Communication Monographs* 51: 1–22.

Fisher, Walter R. (1987) *Human Communication as Narration: Toward a Philosophy of Reason, Value, and Action* (Columbia: University of South Carolina Press).

Forester, John (1999) *The Deliberative Practitioner* (Cambridge, MA: MIT Press).

Fuller, Steve (1988) *Social Epistemology* (Bloomington: Indiana University Press).

Fuller, Steve (1993) *Philosophy, Rhetoric, and the End of Knowledge: The Coming of Science and Technology Studies* (Madison: University of Wisconsin Press).

Fuller, Steve (1995) "The Strong Program in the Rhetoric of Science," in H. Krips, J. E. McGuire, & Trevor Melia (eds) *Science, Reason, and Rhetoric* (Pittsburgh: University of Pittsburgh Press): 95–117.

Fuller, Steve (2000a) *The Governance of Science* (Buckingham: Open Press).

Fuller, Steve (2000b) *Thomas Kuhn: A Philosophical History for Our Times* (Chicago: University of Chicago Press).

Fuller, Steve & James H. Collier (2004) *Philosophy, Rhetoric, and the End of Knowledge:* A *New Beginning for Science and Technology Studies*, 2nd ed. (Mahwah, NJ: Lawrence Erlbaum).

Galison, Peter (1987) *How Experiments End* (Chicago: University of Chicago Press).

Galison, Peter (1994) *Image and Logic* (Chicago: University of Chicago Press).

Gaonkar, Dilip Parameshwar (1997) "The Idea of Rhetoric in the Rhetoric of Science," in A. G. Gross & W. Keith (eds) *Rhetorical Hermeneutics* (Albany: State University of New York Press): 25–85.

Garfinkel, Harold (1967) "Studies of the Routine Grounds of Everyday Activities," in *Studies in Ethnomethodology* (Englewood Cliffs, NJ: Prentice-Hall): 35–75. Reprinted in Garfinkel, Harold (1984) *Studies in Ethnomethodology* (Cambridge: Polity).

Gilbert, Michael A. (ed) (2002) Special Issue on Informal Logic, Argumentation Theory, and Artificial Intelligence, *Informal Logic* 22(3).

Gilbert, G. Nigel & Michael Mulkay (1984) *Opening Pandora's Box* (Cambridge: Cambridge University Press).

Goble, Lou (ed) (2001) *The Blackwell Guide to Philosophical Logic* (Oxford: Blackwell).

Goldman, Alvin I. (1994) "Argumentation and Social Epistemology," *Journal of Philosophy* 91: 27–49.

Goldman, Alvin I. (1999) *Knowledge in a Social World* (Oxford: Oxford University Press).

Golinski, Jan (1998) *Making Natural Knowledge: Constructivism and the History of Science* (Cambridge: Cambridge University Press).

Goodnight, G. Thomas (1982) "The Personal, Technical, and Public Spheres of Argument: A Speculative Inquiry into the Art of Public Deliberation," *Journal of the American Forensic Association* 18: 214–27.

Govier, Trudy (1987) *Problems in Argument Analysis and Evaluation* (Dordrecht: Foris).

Govier, Trudy (2005) *A Practical Study of Argument*, 6th ed. (Belmont, CA: Thomson-Wadsworth).

Gross, Alan G. ([1990]1996) *The Rhetoric of Science* (Cambridge, MA: Harvard University Press).

Gross, Alan G. & William Keith (eds) (1997) *Rhetorical Hermeneutics* (Albany, NY: SUNY Press).

Gross, Alan G., Joseph E. Harmon, & Michael Reidy (2002) *Communicating Science: The Scientific Article from the 17th Century to the Present* (New York: Oxford University Press).

Habermas, Jürgen (1971) *Toward a Rational Society*, trans. J. J. Shapiro (Boston: Beacon).

Habermas, Jürgen (1984) *Theory of Communicative Action*, vol. 1, trans. T. McCarthy (Boston: Beacon).

Habermas, Jürgen (1996) *Between Facts and Norms,* trans. W. Rehg (Cambridge, MA: MIT Press).

Hanson, Norwood Russell (1958) *Patterns of Discovery: An Inquiry into the Conceptual Foundations of Science* (Cambridge: Cambridge University Press).

Hardcastle, Valerie Gray (1999) "Scientific Papers Have Various Structures," *Philosophy of Science* 66: 415–39.

Harding, Sandra (1999) *Is Science Multicultural? Postcolonialisms, Feminisms, and Epistemologies* (Bloomington: Indiana University Press).

Harris, Randy Allen (1997) *Landmark Essays on Rhetoric of Science: Case Studies* (Mahwah, NJ: Hermagoras).

Hauser, Gerared A. (ed.) (1999) Special Issues on Body Argument, *Argumentation and Advocacy* 36: 1–49, 51–100.

Heidelbaugh, Nola J. (2001) *Judgment, Rhetoric, and the Problem of Incommensurability: Recalling Practical Wisdom* (Columbia: University of South Carolina Press).

Heintz, Bettina (2003) "When Is a Proof a Proof?" *Social Studies of Science* 33(6): 929–43.

Hess, David J. (1997) *Science Studies* (New York: New York University Press).

Hilgartner, Stephen (2000) *Science on Stage: Expert Advice as Public Drama* (Stanford, CA: Stanford University Press).

Hitchcock, David (1992) "Relevance," *Argumentation* 6: 251–70.

Hoyningen-Huene, Paul (1993) *Reconstructing Scientific Revolutions: Thomas S. Kuhn's Philosophy of Science,* trans. A. T. Levine (Chicago: University of Chicago Press).

Hull, David L. (1988) *Science as a Process* (Chicago: University of Chicago Press).

Jasinski, James (2001) *Sourcebook on Rhetoric: Key Concepts in Contemporary Rhetorical Studies* (Thousand Oaks, CA: Sage).

Johnson, Ralph H. (2000) *Manifest Rationality* (Mahwah, NJ: Lawrence Erlbaum).

Johnson, Ralph H., & J. Anthony Blair (1977) *Logical Self-Defense* (New York: McGraw-Hill).

Johnson, Ralph H. & J. Anthony Blair (2000) "Informal Logic: An Overview," *Informal Logic* 20(2): 93–107.

Keith, William (1995) "Argument Practices," *Argumentation* 9: 163–79.

Keith, William (2005) "The Toulmin Model and Non-monotonic Reasoning," in *The Uses of Argument: Proceedings of a Conference at McMaster University* (Hamilton, Ontario: OSSA): 243–51.

Keller, Evelyn Fox (1983) *A Feeling for the Organism: The Life and Work of Barbara McClintock* (New York: Freeman).

Kennedy, George A (1980) *Classical Rhetoric and Its Christian and Secular Tradition from Ancient to Modern Times* (Chapel Hill: University of North Carolina Press).

Kim, Kyung-Man (1994) *Explaining Scientific Consensus: The Case of Mendelian Genetics* (New York: Guilford).

Kimball, Bruce A. (1995) *Orators and Philosophers: A History of the Idea of Liberal Education* (New York: College Entrance Examination Board; College Board Publications).

Kirkham, Richard L. (1992) *Theories of Truth: A Critical Introduction* (Cambridge, MA: MIT Press).

Kitcher, Philip (1991) "Persuasion," in M. Pera & W. R. Shea (eds), *Persuading Science* (Canton, MA: Science History Publications): 3–27.

Kitcher, Philip (1993) *The Advancement of Science* (New York: Oxford University Press).

Kitcher, Philip (1995) "The Cognitive Functions of Scientific Rhetoric," in H. Krips, J. E. McGuire, & T. Melia (eds) *Science, Reason, and Rhetoric* (Pittsburgh: University of Pittsburgh Press): 47–66.

Kitcher, Philip (2000) "Patterns of Scientific Controversies," in P. Machamer, M. Pera, & A. Baltas (eds), *Scientific Controversies: Philosophical and Historical Pespectives* (New York: Oxford University Press): 21–39.

Kitcher, Philip (2001) *Science, Truth, and Democracy* (Oxford: Oxford University Press).

Klein, Julie Thompson (1996) *Crossing Boundaries: Knowledge, Disciplinarities, and Interdisciplinarities* (Charlottesville: University Press of Virginia).

Kleinman, Daniel Lee (Ed.) (2000) *Science, Technology, and Democracy.* (Albany: State University of New York Press).

Knorr Cetina, Karin D. (1981) *The Manufacture of Knowledge* (Oxford: Pergamon).

Krips, Henry, J. E. McGuire, & Trevor Melia (eds) (1995) *Science, Reason, and Rhetoric* (Pittsburgh: University of Pittsburgh Press).

Kuhn, Thomas S. ([1962]1996) *The Structure of Scientific Revolutions,* 3rd ed. (Chicago: University of Chicago Press).

Kusch, Martin (2002) *Knowledge by Agreement* (Oxford: Oxford University Press).

Kyburg, Henry E. Jr. (1970) *Probability and Inductive Logic* (New York: Macmillan).

Lakatos, Imre (1976) *Proofs and Refutations: The Logic of Mathematical Discovery* (Cambridge: Cambridge University Press).

Latour, Bruno (1987) *Science in Action* (Cambridge, MA: Harvard University Press).

Latour, Bruno (1988) *The Pasteurization of France,* trans. A. Sheridan & J. Law (Cambridge, MA: Harvard University Press).

Latour, Bruno & Steve Woolgar ([1979]1986) *Laboratory Life: The Construction of Scientific Facts* (Princeton, NJ: Princeton University Press).

Laudan, Larry (1977) *Progress and Its Problems* (Berkeley: University of California Press).

Leff, Michael (2002) "The Relation between Dialectic and Rhetoric in a Classical and a Modern Perspective," in F. H. van Eemeren and P. Houtlosser (eds), *Dialectic and Rhetoric* (Dordrecht, The Netherlands: Kluwer): 53–63.

Lenoir, Timothy (ed) (1998) *Inscribing Science: Scientific Texts and the Materiality of Communication* (Stanford, CA: Stanford University Press).

Lippmann, Walter (1925) *The Phantom Public* (New York: Macmillan).

Livingston, Eric (1986) *The Ethnomethodological Foundations of Mathematics* (Boston and London: Routledge).

Livingston, Eric (1987) *Making Sense of Ethnomethodology* (London: Routledge).

Livingston, Eric (1999) "Cultures of Proving," *Social Studies of Science* 29: 867–88.

Longino, Helen E. (1990) *Knowledge as a Social Process* (Princeton, NJ: Princeton University Press).

Longino, Helen E. (2002) *The Fate of Knowledge* (Princeton, NJ: Princeton University Press).

Lucaites, John Louis, Celeste Michelle Condit, & Sally Caudill (eds) (1999) *Contemporary Rhetorical Theory* (New York: Guilford).

Lynch, Michael (1985) *Art and Artifact in Laboratory Science* (London: Routledge).

Lynch, Michael (1993) *Scientific Practice and Ordinary Action* (Cambridge: Cambridge University Press).

Lynch, Michael (1997) "Ethnomethodology without Indifference," *Human Studies* 20: 371–76.

Lynch, Michael (1999) "Silence in Context: Ethnomethodology and Social Theory," *Human Studies* 22: 211–33.

Lynch, Michael & Steve Woolgar (eds) (1990) *Representation in Scientific Practice* (Cambridge, MA: MIT Press).

Machamer, Peter, Marcello Pera, & Aristides Baltas (eds) (2000) *Scientific Controversies: Philosophical and Historical Perspectives* (New York: Oxford University Press).

MacKenzie, Donald (1978) "Statistical Theory and Social Interests: A Case Study," *Social Studies of Science* 8: 35–83.

MacKenzie, Donald & Barry Barnes (1979) "Scientific Judgment: The Biometry-Mendelism Controversy," in B. Barnes & S. Shapin (eds), *Natural Order* (Beverly Hills: Sage): 191–210.

Majone, Giandomenico (1989) *Evidence, Argument, and Persuasion in the Policy Process* (New Haven, CT: Yale University Press).

Mayo, Deborah G. (1996) *Error and the Growth of Knowledge* (Chicago: University of Chicago Press).

McBurney, Peter & Simon Parsons (2002) "Dialogue Games in Multi-Agent Systems," *Informal Logic* 22(3): 257–74.

McCarthy, Thomas (1988) "Scientific Rationality and the 'Strong Program' in the Sociology of Knowledge," in E. McMullin (ed), *Construction and Constraint* (Notre Dame: University of Notre Dame Press): 75–95.

McCloskey, Deirdre N. ([1986]1998) *The Rhetoric of Economics* (Madison: University of Wisconsin Press). Originally published under the name Donald N. McCloskey.

McCloskey, Deirdre N. (1990) *If You're So Smart: The Narrative of Economic Expertise* (Chicago: University of Chicago Press).

McKerrow, Raymie E. (1989) "Critical Rhetoric: Theory and Praxis," *Communication Monographs* 56: 91–111.

McPeck, John E. (1990) *Teaching Critical Thinking* (New York: Routledge).

McMullin, Ernan (1991) "Rhetoric and Theory Choice in Science," in M. Pera & W. R. Shea (eds), *Persuading Science* (Canton, MA: Science History Publications): 55–76.

Meiland, Jack W. (1989) "Argument as Inquiry and Argument as Persuasion," *Argumentation* 3: 185–96.

Merton, Robert K. (1973) *The Sociology of Science* (Chicago: University of Chicago Press).

Miller, Carolyn R. (1992) "*Kairos* in the Rhetoric of Science," in S. P. White, N. Nakadate, & R. D. Cherry (eds), *A Rhetoric of Doing* (Carbondale: Southern Illinois University Press): 310–27.

Myers, Greg (1990) *Writing Biology: Texts in the Social Construction of Scientific Knowledge* (Madison, WI: University of Wisconsin Press).

Naess, Arne ([1947]1966) *Communication and Argument,* trans. A. Hannay (Oslo: Universitetsforlaget; London: Allen & Unwin).

Nelson, Lynn Hankinson (1990) *Who Knows: From Quine to a Feminist Empiricism* (Philadelphia: Temple University Press).

O'Keefe, Daniel J. (1977) "Two Concepts of Argument," *Journal of the American Forensic Association* 13: 121–28.

Ommen, Brett (2005) "The Rhetorical Interface: Material Addressivity and Scientific Images," presented at the National Communication Association meeting, Chicago, November 10–14.

Peirce, Charles Sanders (1931–1933) *Collected Papers,* vols. I–III, C. Hartshorne & P. Weiss (eds) (Cambridge, MA: Harvard University Press).

Pera, Marcello (1994) *The Discourses of Science,* trans. C. Botsford (Chicago: University of Chicago Press).

Pera, Marcello (2000) "Rhetoric and Scientific Controversies," P. Machamer, M. Pera, & A. Baltas (eds), *Scientific Controversies: Philosophical and Historical Pespectives* (New York: Oxford University Press): 50–66.

Pera, Marcello & William R. Shea (eds) (1991) *Persuading Science* (Canton, MA: Science History Publications).

Perelman, Chaim & Lucie Olbrechts-Tyteca ([1958]1969) *The New Rhetoric: A Treatise on Argumentation,* trans. J. Wilkinson & P. Weaver (Notre Dame, IN: University of Notre Dame Press).

Perini, Laura (2005) "The Truth in Pictures," *Philosophy of Science* 72: 262–85.

Potter, Elizabeth (2001) *Gender and Boyle's Law of Gases* (Bloomington: Indiana University Press).

Prakken, Henry (1997) *Logical Tools for Modelling Legal Argument* (Dordrecht, The Netherlands: Kluwer).

Prelli, Lawrence J. (1989a) *A Rhetoric of Science: Inventing Scientific Discourse* (Columbia, SC: University of South Carolina Press).

Prelli, Lawrence J. (1989b) "The Rhetorical Constructions of Scientific Ethos," in H. W. Simons, *Rhetoric in the Human Sciences* (London: Sage): 48–68.

Rehg, William (1999) "Critical Science Studies as Argumentation Theory: Who's Afraid of SSK?", *Philosophy of the Social Sciences* 30(1): 33–48.

Rehg, William (2001) "Adjusting the Pragmatic Turn: Ethnomethodology and Critical Argumentation Theory," in W. Rehg & J. Bohman (eds), *Pluralism and the Pragmatic Turn* (Cambridge, MA: MIT Press): 115–43.

Rehg, William (ed) (2003) Special Issue on Habermas and Argumentation, *Informal Logic* 23: 115–99.

Rehg, William & Kent Staley (in press) "The CDF Collaboration and Argumentation Theory: The Role of Process in Objective Knowledge," *Perspectives on Science*.

Remedios, Francis (2003) *Legitimizing Scientific Knowledge: An Introduction to Steve Fuller's Social Epistemology* (Lanham, MD: Lexington Books).

Rescher, Nicolas (1976) *Plausible Reasoning* (Amsterdam: Van Gorcum).

Rescher, Nicolas (1977) *Dialectics: A Controversy Oriented Approach to the Theory of Knowledge* (Albany: State University of New York Press).

Rudwick, Martin J. S. (1985) *The Great Devonian Controversy* (Chicago: University of Chicago Press).

Salmon, Wesley C. (1967) *The Foundations of Scientific Inference* (Pittsburgh: University of Pittsburgh Press).

Schiebinger, Londa (1999) *Has Feminism Changed Science?* (Cambridge, MA: Harvard University Press).

Schmitt, Frederick F. (ed) (1994) *Socializing Epistemology: The Social Dimensions of Knowledge* (Lanham, MD: Rowman & Littlefield).

Schön, Donald A. & Martin Rein (1994) *Frame Reflection: Toward the Resolution of Intractable Policy Controversies* (New York: Basic).

Selzer, Jack (ed) (1993) *Understanding Scientific Prose* (Madison: University of Wisconsin Press).

Shapere, Dudley (1986) *Reason and the Search for Knowledge* (Boston and Dordrecht, The Netherlands: Reidel).

Shapin, Steven & Simon Schaffer (1985) *Leviathan and the Air Pump: Hobbes, Boyle, and the Experimental Life* (Princeton, NJ: Princeton University Press).

Siegel, Harvey (1988) *Educating Reason* (New York: Routledge).

Sills, Chip & George H. Jensen (eds) (1992) *The Philosophy of Discourse: The Rhetorical Turn in Twentieth-Century Thought*, 2 vols. (Portsmouth, NH: Boynton/Cook-Heinmann).

Simons, Herbert W. (ed) (1989) *Rhetoric in the Human Sciences* (London: Sage).

Simons, Herbert W. (ed) (1990) *The Rhetorical Turn: Invention and Persuasion in the Conduct of Inquiry* (Chicago: University of Chicago Press).

Solomon, Miriam (2001) *Social Empiricism* (Cambridge, MA: MIT Press).

Staley, Kent W. (2004) *The Evidence for the Top Quark* (Cambridge: Cambridge University Press).

Stegmüller, Wolfgang (1976) *The Structure and Dynamics of Theories* (Berlin: Springer-Verlag).

Suppe, Frederick (1998) "The Structure of a Scientific Paper," *Philosophy of Science* 65: 381–405.

Taper, Mark L. & Subhash R. Lele (eds) (2004) *The Nature of Scientific Evidence* (Chicago: University of Chicago Press).

Taylor, Charles Alan (1996) *Defining Science* (Madison: University of Wisconsin Press).

Tindale, Christopher W. (1999) *Acts of Arguing* (Albany: State University of New York Press).

Toulmin, Stephen (1958) *The Uses of Argument* (Cambridge: Cambridge University Press).

Toulmin, Stephen (1995) "Science and the Many Faces of Rhetoric," in in H. Krips, J. E. McGuire, & T. Melia (eds), *Science, Reason, and Rhetoric* (Pittsburgh: University of Pittsburgh Press): 3–11.

Toulmin, Stephen, Richard Rieke & Allan Janik (1984) *An Introduction to Reasoning*, 2nd ed. (New York: Macmillan).

Traweek, Sharon (1988) *Beamtimes and Lifetimes: The World of High Energy Physicists* (Cambridge, MA: Harvard University Press).

Waddell, Craig (1989) "Reasonableness versus Rationality in the Construction and Justification of Science Policy Decisions: The Case of the Cambridge Experimentation Review Board," *Science, Technology & Human Values* 14 (Winter): 7–25.

Waddell, Craig (1990) "The Role of *Pathos* in the Decision-Making Process: A Case in the Rhetoric of Science Policy," *Quarterly Journal of Speech* 76: 381–400. Reprinted in Harris (1997).

Walton, Douglas N. (1989) *Informal Logic* (Cambridge: Cambridge University Press).

Walton, Douglas N. (1992) *Plausible Argument in Everyday Conversation* (Albany: State University of New York Press).

Walton, Douglas N. (1996) *A Pragmatic Theory of Fallacy* (Tuscaloosa: University of Alabama Press).

Walton, Douglas N. (1998) *The New Dialectic* (Toronto: University of Toronto Press).

Walton, Douglas N. (2004) *Relevance in Argumentation* (Mahwah, NJ: Lawrence Erlbaum).

Warner, Michael (2002) *Publics and Counterpublics* (New York: Zone Books).

Wenzel, Joseph A. (1990) "Three Perspectives on Argument," in R. Trapp and J. Schuetz (eds), *Perspectives on Argumentation: Essays in Honor of Wayne Brockriede* (Prospect Heights, IL: Waveland): 9–26.

Willard, Charles Arthur (1989) *A Theory of Argumentation* (Tuscaloosa: University of Alabama Press).

Willard, Charles Arthur (1996.) *Liberalism and the Problem of Knowledge* (Chicago: University of Chicago Press).

Williams, Bruce A. & Albert R. Matheny (eds) (1995) *Democracy, Dialogue, and Environmental Disputes* (New Haven, CT: Yale University Press).

Woolgar, Steve (1983) "Irony in the Social Study of Science," in K. D. Knorr Cetina & M. Mulkay (eds) *Science Observed* (London and Beverly Hills, CA: Sage): 239–66.

Woolgar, Steve (1988a) *Science: The Very Idea* (Chichester: Horwood; London: Tavistock).

Woolgar, Steve (1988b) (ed) *Knowledge and Reflexivity* (London: Sage).

Wylie, Alison (2002) *Thinking from Things: Essays in the Philosophy of Archaeology* (Berkeley: University of California Press).

10 STS and Social Epistemology of Science

Miriam Solomon

Social epistemology of science has been a growing, multidisciplinary field since the early 1980s. From Bruno Latour's "actor-networks" (1987) to Jean Lave and Etienne Wenger's model of learning through "peripheral participation" (1991) to Alvin Goldman's "social epistemics" (1999) to Donna Haraway's "socially situated knowledge" (1991) and Helen Longino's socially constituted "objectivity" (1990), the focus has been on socially distributed skills, knowledge, and evaluation.

In this chapter, I focus on recent ideas from the fields of epistemology and philosophy of science, mostly from the Anglo-American philosophical tradition, and present them in a manner in which, it is hoped, they are useful to those working in other areas of science studies. STS—with its abundance of historical cases documenting social aspects of scientific change—provided the most important motivation for Anglo-American philosophical work on social epistemology of science. It is time for Anglo-American Philosophy[1] to pay back STS in similar currency, with ideas that can guide new inquiry.

Continental Philosophical traditions have played an important role in the development of STS ideas. I discuss these only briefly, in part because they are not my area of expertise, but also because they are already well known in STS. From Ludwig Wittgenstein's influence on the Edinburgh Strong Program, to Jacques Derrida's, Michel Foucault's, Jurgen Habermas's, Martin Heidegger's, Edmund Husserl's, Hans-Jörg Rheinberger's, Michel Serres's, and Isabelle Stenger's influences on Bruno Latour and Karin Knorr Cetina, these influences are pervasive and ongoing. I wish there were an essay in this Handbook dedicated to this topic! The most useful resource I have found is John Zammito's *A Nice Derangement of Epistemes* (2004), but even this excellent scholarly work lacks comprehensiveness about the Continental Philosophical tradition.

I will also not say much about the ideas of Peter Galison, Bruno Latour, Jean Lave, Donna Haraway and others who work in social epistemology but whose disciplinary identifications are not in Philosophy. This is not because their work is unphilosophical (in fact, I think their work is highly philosophical and, especially, influenced by the Continental Philosophical tradition), but because their work is probably already familiar to readers of this essay.

Anglo-American Philosophers who work in the area of social epistemology of science often identify themselves as *naturalized* epistemologists. This recent tradition, inspired by W. V. Quine's "Epistemology Naturalized" (1969), takes philosophy to be continuous with other sciences of knowledge, and to be similarly accountable to observational and experimental data. Naturalistic epistemologists reject the dominant early and mid–twentieth century approaches in Anglo-American epistemology: the approaches of conceptual analysis and *a priori* investigations. Quine thought that epistemology should be "a chapter of psychology" (1969: 82), and the first wave of naturalistic epistemologists worked primarily in the areas of cognitive psychology and cognitive neuroscience. Quine's arguments for naturalism do not, however, entail individualism: they do not imply that the only relevant sciences of knowledge are those investigating individual minds. By the late 1980s, some naturalistic epistemologists realized (often with the help of case studies in early STS such as those from the Edinburgh Strong Program) that knowledge phenomena can rarely be described and understood without invoking social facts and social mechanisms. As naturalistic epistemologists, they reject individualism.

This inclusion of "the social" or "social factors" tends to be on the "thin" side (as judged by those familiar with the Continental Philosophical tradition). "The social" is understood as individuals (sometimes diverse individuals) in plural. Usually there is not much discussion of historical or cultural context, and the focus is on more universal social mechanisms. I give detailed examples of these views shortly.

In the Continental Philosophical tradition, on the other hand, cultural and social contexts have always been acknowledged. In fact, rather than argue that "the social" needs to be taken into account in addition to "the individual," the background position is that *everything* needs to be understood in social, historical, and cultural context. I am inclined to call this "socio-cultural epistemology," to distinguish it from Anglo-American "social epistemology" (but note that this is my own suggested terminology).

Not everyone in the Continental Philosophical tradition lives in non-Anglo Europe, of course. (Nor do all Anglo-American Philosophers work in English speaking countries.) Notable examples of U.S. Philosophers of science strongly influenced by the Continental tradition are Arnold Davidson (2002), Ian Hacking (e.g., 2000, 2004), Helen Longino (1990, 2001), Joseph Rouse (1987, 1996), and Alison Wylie (2002).

This paper presents several ideas, mostly from Anglo-American social epistemology, that are not yet well known in STS and may be useful for STS. For the most part, the ideas are about *normative strategies*, that is, ways of evaluating knowledge-making processes and recommendations for their improvement. For naturalistic epistemologists, normative recommendations must be realistic recommendations for improvement. As is said in the area of ethics, "ought implies can." There is no point, for example, in requiring that scientists check all their beliefs for possible mutual inconsistency, since that is humanly impossible. There is no point in requiring that humans always avoid base rate errors, errors of attribution, salience biases, or confirmation bias, since we have ample evidence from the work of Daniel Kahneman, Amos Tversky, and others that we frequently reason with these errors and biases.[2] Any naturalistic normative recommendations need to be constrained by descriptive understanding of the epistemic processes that we, in fact, use.

In some disciplines, most notably history and ethnographic studies, normative recommendations are typically eschewed. In the history of these fields, normative judgments have been fundamentally flawed by ideological bias, and it has become prudent to avoid normative judgments altogether. I would argue (and other philosophers would agree) that, eventually, normative questions should be asked. This is for at least two reasons. First, we have normative goals, such as scientific or technological success, and normative investigations can help identify better ways of achieving those goals. Second, normative recommendations are recommendations for *intervention* in scientific practices by, for example, changing the grant structure or the relationships between research groups or the regulations for corporate funding. Successful intervention (as Hacking [1983], especially, has noted) demands far more of a model than do mere observational or explanatory successes. Successful intervention is therefore an excellent test for the plausibility of our theories about scientific change. Philosophical social epistemologies that pay explicit attention to normative questions are a vital resource for STS's theorizing about science.

Some normative recommendations from recent social epistemology of science seem, frankly, obvious. Casual reflection on scientific practice seems to yield recommendations such as "criticism is a resource for improvement of theories," "division of cognitive labor is an efficient way to proceed," and "consultation with acknowledged experts improves research." Indeed, these recommendations were anticipated by Isaac Newton, John Stuart Mill, and Karl Popper. But not all are so obvious, and some of the apparently obvious ones turn out to be flawed recommendations. So I take the time to describe the whole range of normative tools and recommendations as well as to indicate the kind of evidence that supports them.

Normative tools are ways of thinking about how to evaluate scientific decision making and scientific change. Traditionally, this meant evaluating individual scientists for their rationality, or their steps toward getting true theories. Social epistemologists of science suggest different ways to evaluate what scientists do, typically looking at what the scientific community as a whole does instead of, or in addition to, what individual scientists do. These tools of evaluation may range in generality from the more local to the more universal (in time, domain, discipline, etc.).

Normative recommendations are specific recommendations for or against particular social epistemic practices in particular contexts. The kinds of practices to be considered include trust and testimony from others, reliance on authority, peer review, corporate funding, governmental funding, and the transmission of knowledge through publication, conferences, and other means. Again, the level of generality of these recommendations can vary. Of course, some normative tools need to be in place before normative recommendations can be made with confidence.

NORMATIVE TOOLS

Distribution of Cognitive Labor

The most important normative tools offered by social epistemologists of science are those that make possible evaluation at the social, rather than the individual, level. The simplest, and most frequently given, example is the strategy of evaluating the reasoning of individual scientists by investigating how it contributes to an overall beneficial division of cognitive labor. For instance, Frank Sulloway (1996) argues that personality has an influence on theory choice and that, during times of scientific revolution, some scientists will pick the more radical theory, and others the less radical theory, because of their different personalities (which he thinks are ultimately caused by the scientists' birth order situations). Sulloway acknowledges that while individual scientists may be biased, the overall effect of this division of labor is beneficial, because each competing theory will be developed. There is a reasonable balance of later born (more radical) versus first-born (more conservative) scientists, so there will be a reasonable division of labor. Eventually, Sulloway thinks, there will be sufficient evidence that one theory will triumph. But it is not possible to guess this in advance, so the division of labor is an efficient way to proceed.

Many philosophers have made the same point about division of labor while exploring different factors affecting scientists' reasoning and choices. David Hull (1988) was perhaps the first in recent years, with a claim that credit-seeking motives lead scientists to distribute their effort. The idea is that a scientist sees more promise for his or her own career in pursuing a plausible theory that does not yet have champions. If that theory is successful, the scientist will—as an early champion—take all or most of the credit rather than share it with others. Alvin Goldman's influential 1992 article (with the economist Moshe Shaked) formalizes this insight and argues that making decisions on the basis of credit seeking is almost as good as making them on the basis of pure truth seeking, if that were possible. Philip Kitcher (1990, 1993) reaches similar conclusions, presenting epistemically "sullied" individuals as trading the goal of truth for that of credit. Both Goldman and Kitcher claim that individuals will sometimes pursue theories that they evaluate as less plausible, because the likely payoff in terms of credit is higher. And both conclude that this benefits the scientific community, because the result is a good (if not ideal) division of cognitive labor.

Some philosophers have looked at other causes of the distribution of cognitive labor. Ronald Giere (1988: 277) claims that "accidents of training and experience" lead scientists to make different judgments about the plausibility of competing theories. I (Solomon [1992]) build on part of that claim, by looking more closely at Giere's example of geology after Wegener's *Origin of Continents and Oceans* (1915) and finding that so-called "cognitive biases" such as salience, availability, and representativeness are important causes of the distribution of cognitive effort.[3] Kitcher (1993: 374) suggests that cognitive diversity can be brought about by variation in judgment (e.g., in the assignment of original probabilities) as well as by the incentive of credit seeking. Thagard (1993) has argued that the same result is achieved by delays in dissemination of information, which lead to different scientists making decisions on the basis of different data sets.

Of course, not each of these accounts of the causes of division of labor is entirely correct and complete. Some combination of them probably gives the complex true story. Yet all these ideas about the division of cognitive labor have in common that they assess the decisions of individual scientists in terms of their contribution to what the scientific community does as a whole during times of dissent. From the perspective of the scientific community, it does not matter if an individual scientist reasons in a "biased" manner; what matters is how that "biased" reasoning contributes to the aggregated project.[4] These ideas also have in common a certain assumed optimism; they focus on cases in which the distribution of cognitive effort produced by different individual decisions is a good distribution. Like Adam Smith's "invisible hand," in which myriad actions of self-interest result in a good general distribution of wealth, they suggest that there is an "invisible hand of reason" such that the various causes of individual decisions result in a good overall distribution of cognitive effort. I have argued (Solomon, 2001) that, in fact, there is no "invisible hand of reason" and that examination of historical cases shows that sometimes cognitive effort is distributed well and sometimes not so well or badly.[5] Examples where cognitive effort is distributed less well, or poorly, include the early–twentieth century debate over nuclear versus cytoplasmic genetics and mid–twentieth century work on cancer viruses. In my view, cytoplasmic (nonchromosomal) genetics and viral models of cancer received less attention than they deserved. This means that normative recommendations can be made about ways to improve the distribution of cognitive effort.

Cognitive labor can be divided not only for discovery and development of new ideas but also for storage of facts, theories, and techniques that are widely accepted. Just as books contain quantities of information that no individual could retain, information is also stored in communities in ways that are accessible to most or all members of that community but could not be duplicated in each head. One important way in which this is achieved is when experts on different subjects, or with different experiences or techniques, increase the knowledge within a community. Knowledge and expertise are socially distributed. Edwin Hutchins's account of navigation (1995), in which skills and knowledge are distributed across the officers and enlisted men on board a naval vessel, is an example. David Turnbull (2000) also gives a number of examples of distributed knowledge and research, from masonry to cartography to the creation of a malaria vaccine.

A final way in which cognitive labor can be distributed is for the epistemic work that is required when scientific communities move from dissent to consensus. In traditional philosophies of science, consensus is presented as the outcome of the identical decision of each member of a scientific community. A good consensus is the result of each scientist choosing the best theory, through the same process, and a bad consensus is the result of each scientist choosing the wrong theory through a common inappropriate process. But of course, this is just the simplest model of group consensus formation, and it is one that presumes the same starting point as well as the same endpoint and the same processes of change. (Nevertheless, this simplest model is the one that most philosophers have assumed.) Other accounts of consensus formation, in which cognitive labor is distributed, include those of Hussein Sarkar (1983) and Rachel Laudan and Larry Laudan (1989), who find that different scientists may select the same theory for different good reasons. My own account in *Social Empiricism* (Solomon, 2001) finds that, just as with dissent, individual scientists may make biased and idiosyncratic decisions, yet the overall decision should be evaluated from a social perspective. I assess cases of consensus by looking at the distribution of biases (which I call "decision vectors") that operate in particular cases.

The case of consensus on continental drift—or, more precisely, on its successor, plate tectonics—illustrates the distributed nature of consensus and the need, again, to adopt a social perspective in order to make normative judgments about the appropriateness of consensus. Consensus on plate tectonics emerged during the 1960s. Historians of the period have made remarks such as "most specialists were only convinced by observations related to their specialties" (Menard, 1986: 238). The order of consensus formation was paleomagnetists, oceanographers, seismologists, stratigraphers, and then continental geologists concerned with paleontology and orogeny. Belief change occurred after observations confirming mobilism were produced in each specialty and after old data were reinterpreted to be consistent with mobilism. In addition to this general pattern, there were patterns of belief change and belief perseverance due to prior beliefs, personal observations, peer pressure, the influence of those in authority, salient presentation of data, and so forth.[6]

All these patterns can be explained in terms of decision vectors (my preferred term for what others call "biasing factors"[7]). Decision vectors are so called because they influence the outcome (direction) of a decision. Their influence may or may not be conducive to scientific success, hence the epistemic neutrality of the term. Moreover, the decision vectors leading to consensus are similar in kind, variety, and magnitude to those causing dissent in other historical cases. The result here is consensus, rather than continued dissent, because enough of the decision vectors are pulling in the same direction.

It is important to note that although there *could* have been a crucial experiment/crucial set of observations, there was, in fact, not. The paleomagnetic data, especially the data showing symmetric magnetic striations on either side of deep ocean ridges found by Vine and Matthews in 1965, could have been viewed in this way. But it was viewed as crucial data only by Vine, Matthews, and a few well-placed oceanographers. Paleomagnetists tended to be persuaded earlier, and seismologists, stratigraphers, and continental geologists somewhat later. No single experiment was "crucial" for more than a group of geologists at a time. The consensus was gradual, and distributed across subfields of geology, countries, and informal networks of communication. Consensus was complete when plate tectonics had all the empirical successes (those that permanentism, contractionism, and drift had, and more besides)—but not because any individual or individuals saw that it had all the successes. Moreover, empirical success was not sufficient to bring about consensus. Decision vectors (sometimes tied to empirical success, sometimes not) were crucial.

Consensus is not always normatively appropriate. Sometimes scientists agree on theories that do not deserve universal support. Consensus in the 1950s on the "central dogma" that DNA controls cellular processes via messenger RNA and protein synthesis left out important work on cytoplasmic inheritance, genetic transposition, genetic regulation, and inherited supramolecular structures (work of Ephrussi, Sonneborn, Sager, McClintock, and others). Sometimes scientists agree on theories that do not deserve any support at all. For example, the extracranial-intercranial bypass operation, developed in 1967 by Gazi Yasargil, became the standard treatment for stroke patients before there was any robust data for it. The operation became "the darling of the neurosurgical community" (Vertosick, 1988: 108) because it required special training and was very lucrative. The rationale was good enough: since the blood vessels feeding the scalp rarely develop atherosclerosis, it was reasoned that the suturing them to arteries that feed the brain would increase blood flow to the brain and thus help prevent strokes. But note that this is "plausible reasoning" or theoretical support, not empirical evidence. In fact, good supportive data were never produced, and the surgery fell out of favor after 1985. But that was almost 18 years of consensus on a theory with no significant empirical success.

Consensus is not the *telos* of scientific research. It is not a marker for truth or even for "happily ever after." In *Social Empiricism*, I also look at the dissolution of consensus from a social perspective, and use the same normative framework, to make normative judgments. Cases discussed include the dissolution of consensus that cold fusion in a test tube is impossible and the dissolution of consensus that ulcers are caused by excess acid.

My normative framework, "social empiricism," requires essentially the same conditions for appropriate dissent, formation of consensus, and dissolution of consensus. The normative conditions are that theories which are taken seriously should each have some associated empirical successes and that decision vectors should be fairly distributed over the theories under consideration.[8] "Social empiricism" does not privilege any "end point," but most theorists of science privilege consensus (sometimes also called "resolution" or "closure").

The normative perspective taken by all those mentioned in this subsection who work on scientific decision making and the distribution of cognitive labor—assessment at the aggregate level—is applied more widely by some philosophers of science. Effective distribution of cognitive labor is, after all, not the only normative goal or normative procedure for scientists. So in the next two subsections I present other kinds of evaluation of group and aggregated scientific work.

Epistemic Goals

Science is a complex, historically embedded activity, and its goals are not transparent. Philosophers (and others in science studies) do not agree about the goals of science. First, they disagree about the goals they discern in the activities of scientists. For example, Kitcher (1993) claims that scientists aim for approximate significant truths; Giere (1999) claims that scientists aim to produce representations that resemble the world in relevant respects; Bas van Fraassen (1980) argues that scientists merely try to make good predictions. Some philosophers (e.g., Fuller, 1993; Longino, 1990, 2001) argue that the goals of science are neither universal nor immutable.

Secondly, philosophers also disagree about what they think *ought* to be the goals of science (for naturalistic philosophers, of course, these should be close enough to the actual goals of science to be realistically achievable). Here, there are a number of competing normative views, which often include (what are traditionally regarded as) nonepistemic goals such as human flourishing, helping oppressed people, or Earth conservation. In this essay, I'm not going to take a stand on the issue.[9] Instead, I will present the different views about goals of science in terms of the different normative tools they offer.

Steve Fuller (1993) writes of a range of different epistemic goals espoused by scientific communities and argues that those goals should themselves be debated, in a democratic process, by those interested in science policy.[10] Kitcher (2001) has recently changed his views and now also argues for a democratic process of deliberation about scientific goals.

Sandra Harding (e.g., in Alcoff and Potter, 1993) advocates feminist science. Feminist science is science done from a politically engaged and critical standpoint. Of particular merit is science done "from marginalized lives," that is, by those who are politically marginalized with respect to mainstream scientific work, who also become engaged in a transformative project, politically and epistemically challenging the foundations of a field. Harding uses the normative term "strong objectivity" for such science, and she expects that it will provide the research goals and that those goals will include political goals such as improving the situation of marginalized communities. Harding's concept of "strong objectivity" is similar to Donna Haraway's "situated knowledges:" both argue that, as Haraway puts it, "vision is better from below" (Haraway, 1991: 190). They both draw on Marxist epistemic ideas and adapt them to broader political concerns. Haraway adds that objectivity comprises ". . . the joining of partial views and halting voices into a collective subject position" (Haraway, 1991: 196). Harding is more insistent that "strong objectivity" is an achievement of political reflection on one's social standpoint and does not come from social standpoint alone. Both suggest that objectivity is socially, rather than individually, achieved.

Lynn Hankinson Nelson's *Who Knows? From Quine to a Feminist Empiricism* (1990) argues that epistemic goals are achieved socially. Moreover, she claims, epistemic goals cannot be distinguished from nonepistemic values. Theories, observations, and values are interwoven, socially embodied, and all, in principle, revisable. Her epistemic framework is that of W. V. Quine's web of belief, with the community instead of the individual as the locus of the web, and with "values" expanded to include traditionally nonepistemic goals.

The most radical position on epistemic goals is one that claims that our social epistemic practices *construct* truths rather than discover them and, furthermore, negotiate the goals of inquiry rather than set them in some nonarbitrary manner. Work in the Strong Program in sociology of science during the 1970s and 1980s—notably by Barnes and Bloor, Latour and Woolgar, Shapin and Schaffer, Collins and Pinch—was frequently guided by such social constructivism. Most contemporary social epistemologists in the Anglo-American philosophical tradition were motivated by their disagreement with the social constructivist tradition, and they argue for the less radical positions. Recent work in STS (e.g., by Knorr Cetina, 1999; Sheila Jasanoff, 2004; Latour, 2005; and Andrew Pickering, 1995), while maintaining that scientific goals are negotiated, gives the material as well as the social world a voice in the negotiations. These constructivist views have thus become less radical (in the Strong Program sense of radical) and closer to work in Philosophy (both Continental and Anglo-American). They also challenge the static view of the social assumed by their predecessors. Instead, they regard scientific knowledge and scientific societies as *co-produced* by their material engagements.

Normative Procedures

There are two ways to explore normative epistemic ideas. One is to state epistemic goals (truth, empirical success, or whatever), and then to evaluate decisions and methods according to their effectiveness at achieving these epistemic goals. The other way is to evaluate the procedures themselves for their "reasonableness" or "rationality" without considering the products of the procedures. So far, I have looked at social epistemic positions that belong in the first category, that of "instrumental rationality." Longino's normative views belong, mostly, to the second category. She claims to be spelling out the meaning of "objectivity" by specifying objective procedures for all scientific communities.

Longino currently refers to her social epistemology of science as "critical contextual empiricism" (Longino, 2001). There are four socially applicable norms, which she claims are obtained by reflection on the meaning of "objectivity." These norms are

1. equality of intellectual authority (or "tempered equality," which respects differences in expertise)
2. some shared values, especially the valuing of empirical success
3. public forums for criticism (e.g., conferences, replies to papers in journals)
4. responsiveness to criticism

To the degree that a scientific community satisfies these norms and also produces theories (or models) that conform to the world, it satisfies conditions for scientific knowledge. Note that the norms are satisfied by communities, not individuals, although some individuals have to satisfy some conditions (e.g., they have to be responsive to particular criticisms).

The result of satisfying these four norms, according to Longino, is, typically, pluralism. Empirical success is the one universally shared value in scientific research, but it comes in many forms and is not sufficient to arbitrate scientific disputes. Thus, theories are underdetermined by the available evidence and more than one theory of a single domain can be empirically successful. Different theories can be empirically successful in different ways. And, Longino claims, the best criticism usually comes from scientists who are working on different theories. Hence, pluralism is the typical and the preferable state of scientific research.

Longino's account has a certain intuitive plausibility. The requirements of greater intellectual democracy would seem to give all scientists a fairer hearing, thereby increasing the chance that good ideas will be considered. It also has moral appeal. Everyone gets a chance to speak and every genuine criticism gets a response.

NORMATIVE RECOMMENDATIONS

Normative recommendations are specific recommendations for or against particular social practices in particular contexts. Sometimes, they come from application of normative tools. For example, recommendations for specific distributions of cognitive labor may come from taking a social normative stance, as discussed in the previous section. In this section, I will focus on normative recommendations that have been made after more local assessments, which may or may not be guided by general normative tools.

Epistemic practices may be individual or social. Examples of recommendations for individual epistemic practice are "define all terms clearly," "make sure you have evidence for your claims," "be bold in hypothesizing," and "avoid logical fallacies." Examples of recommendations for social epistemic practices are "trust the observations of trained co-investigators," "make sure your work is read and criticized by others," "respect the advice of authoritative persons," and "present work at conferences as well as in publications." These examples are of social epistemic recommendations to individuals; other examples make recommendations to communities, at the level of science policy. Examples here include recommendations such as "corporate funding of science leads to less reliable results than government funding," "double-blind review of scientific manuscripts is the best practice," and "serious science education should begin in middle school, if we wish to produce professional scientists."

Philosophers do not have unique expertise on matters of recommendation for epistemic practice. Recently, they have only written about some of these questions, primarily the social epistemic recommendations to individuals. Here is a brief survey.

Trust about Testimony
Several epistemologists (rather than philosophers of science) have written about trust. There is lack of consensus about the conditions under which trust of testimony from other people is appropriate. C.A.J. Coady (1992) and Martin Kusch (2002), for example, think that trust of other scientists is the default position: trust is appropriate unless there is a specific reason to distrust. Others, such as Elizabeth Fricker (2005), claim that trust has to be earned, so that an investigator needs specific justification for trusting another scientist, such as evidence that a speaker is reliable about reporting evidence and that the speaker is honest.

Goldman (2002) has an unusual, affective theory about trust. He claims that human beings are more likely to trust the beliefs of those with whom they have a prior bond of affection (this is the bonding approach to belief acceptability, or BABA). Emotion constrains the intellect. Goldman claims that this practice is reliable, when understood through evolutionary psychology: those with affective bonds will have less incentive to deceive one another than those without. BABA is, in particular, the foundation of parent-child educational interactions. Goldman does not explore the reliability or pervasiveness of BABA in the scientific community.

It would be instructive to compare this philosophical literature with STS literature on trust, such as Steven Shapin's (1994) *A Social History of Truth*. Shapin takes a descriptive historical approach, of course, rather than a normative approach, but since naturalistic epistemologists aim for realistic recommendations, they could explore whether the conditions for trust that Shapin discerns, which have to do with gender and social class (being a "gentleman" in seventeenth century England), bear any relation to the norms that epistemologists recommend. Lorraine Daston's work (e.g., 1992) on the historical development of objectivity in the nineteenth century is another rich resource for looking at changing practices of trust and testimony.

Authority

Several epistemologists (e.g., Goldman, 1999; Kitcher, 1993) have written about authority. For the most part, they have been concerned to distinguish "earned" from "unearned" authority, and to recommend that scientists rely only on the former. "Earned" authority may be measured by either the track record of the scientist or the quality of the scientist's arguments (as perceived by the evaluating scientist or layperson). Hilary Kornblith (1994) claims that societal judgments about authoritativeness can be reliable if they are based on reliable proxies such as place in a meritocratic employment system; in such a case a person's authoritativeness in the scientific community would be well calibrated to the degree that they earned their authoritative position. Kornblith differs from Goldman and Kitcher in that he does not require that each individual evaluate the authority they depend on; it is sufficient that ascribed authority within the scientific community is, in fact, correlated with past reliable work.

Longino is more skeptical about the widespread deference to authority. Her concern is that unearned authority—political advantage gained in various ways not correlated with knowledge or past performance—will play too much of a role and that marginalized voices will not be heard, let alone cultivated. In her earlier work (Longino, 1990) she calls for "equality of intellectual authority." Later on (Longino, 2001), she modifies her position to call for "tempered equality of intellectual authority," which allows for respect of differences in intellectual ability and knowledge but still demands the same standards for democratic discussion.

Criticism

The vast majority of epistemologists—starting with John Stuart Mill, C. S. Pierce, and Karl Popper—as well as scientists, recommend that scientific work be evaluated and criticized by scientists who are experts in the field. The idea is that peer scientists will help correct any errors, give criticisms that will stimulate the best arguments in response, and in the case of publication, will certify the quality of the resulting manuscript. The assumption is that critical discourse—discussion at the coffee machine, questions at conferences, reader's responses to manuscripts, and so forth—improves scientific work. The assumption is also that we can tell which discourse is "critical" and which is not (e.g., by being "just rhetorical"). Longino (1990, 2001) is a strong but typical example of an epistemologist who requires not only that scientists subject their work to criticism but also that they respond to criticism. The commitment to "criticism" and "rational deliberation" has deep philosophical roots, from Plato to Kant to Habermas and Rawls.[11]

I believe that I am unusual in my skepticism about the practice of criticism: I note research in social psychology (e.g., Kahneman et al., 1982; Nisbett & Ross, 1980) showing that criticism is often ignored, misunderstood, or stonewalled owing to mechanisms such as pride, belief perseverance, and confirmation bias[12], and particular cases in the history of science and contemporary science in which researchers are unresponsive, dismissive, or worse, to criticism (Priestley and Einstein are well-known cases; there are many others). In my view, we need to do more empirical work on the process of criticism to see what the benefits in fact are and what is reasonable to expect of researchers in terms of their response to criticism. Such work could begin, for example, with discourse analysis of transcripts of conference discussions. It is important for social epistemology to make further observations and to examine not only the effect of criticism on the individual being criticized but also the effect of witnessing criticism on the entire scientific community.[13]

Collaboration

Paul Thagard (1997) has explored the pervasiveness and the increasing rate of collaboration in physical, biological, and human sciences. He describes different kinds of collaboration, the most common of which are teacher/apprentice (typically, a professor and a graduate student), peer-similar (trained researchers in the same field), and peer-different (trained researchers in different fields engaged in an interdisciplinary project). Teacher-apprentice collaborations are intended to increase the productivity of senior researchers, at lower cost than hiring peer collaborators but with greater risk of error due to inexperience of the apprentices. They also train the next generation of senior researchers. Peer-similar collaboration can increase the power of inquiry, at least theoretically; computer simulation shows that agents who communicate with each other can solve problems more quickly than agents who do not. Peer-similar collaboration also provides a check on results and reasoning. A well-known example of peer-similar collaboration is Watson and Crick's discovery of the structure of DNA.[14] Peer-different collaboration makes interdisciplinary research possible, although without the check on results and reasoning that peer-similar collaboration provides. An example of peer-different collaboration is the work of Barry Marshall (a gastroenterologist) and Robin Warren (a pathologist) in discovery of bacteria as a primary cause of peptic ulcers.

Another epistemic benefit of collaboration, often mentioned by scientists, is that it increases motivation for many researchers who find less motivation in doing individual work. Increased motivation usually translates into increased productivity.

On the basis of this exploration of the epistemic benefits and risks of collaboration, Thagard makes several normative recommendations. For example, he argues that early career scientists in some sciences should not be encouraged to give up fruitful collaborations with their dissertation advisors in order to establish themselves as independent researchers. Thagard thinks that the epistemic losses here outweigh the benefits.

Competition

Hull (1988) was one of the first philosophers to write about competition in science.[15] He makes much of an analogy between evolutionary change and scientific change, thinking of scientific theories as competing in the same kind of way that species compete for survival. The competitive energy comes from the desire of scientists to receive credit for successful work. (Credit, in recent scientific culture, is in limited supply.) Kitcher (1990, 1993) and Goldman and Shaked (1992) both build on this, arguing that the desire for credit can lead to competitiveness between scientists that leads them to effectively distribute cognitive labor over the available scientific theories. Competitiveness can also lead scientists to work harder and be more productive than they otherwise would be.[16]

Competitiveness among scientists varies across cultures, and it changes over time. For example, in the mid-twentieth century British scientists had gentlemanly agreements to stay away from each other's research problems, whereas in the United States competing with other laboratories working on the same problem was acceptable. (This difference shows in the history of discovery of the structure of DNA, where Francis Crick was uncomfortable competing with other British scientists and James Watson saw nothing objectionable in doing so.) Recently, competition in science has become more tied to commercial interests than to credit, and secrecy has also increased. Most scientists deplore this situation, and some have organized against it. This was the case, for example, with the joint effort of university scientists (with DOE and NIH funding) to sequence DNA before Craig Venter at Celera Genomics could establish a monopoly with his proprietary data.

It is time for philosophers of science to grapple with these more complex issues of "how much competitiveness?" and "what are appropriate conditions of competition?" making recommendations so that excessive secrecy, with its associated decrease in assignment of credit, reduction in the effectiveness of cooperative and complementary work, and cutback in peer reviewed publication, will be minimized.

Dissent and Consensus

Sometimes scientists agree, and sometimes they disagree. Many philosophers (as well as scientists) think of agreement as the goal of research, and of the achievement of consensus as, ideally, an epistemic achievement. The psychotherapeutically popular term of accomplishment, "closure," is also widely used for consensus. A significant minority of philosophers, however, starting with Mill (1859) and continuing with Paul Feyerabend (1975), Longino (1990, 2001), and myself (Solomon, 2001) see the pressure to form consensus as an epistemic liability, since it can cut off promising avenues of research and/or suppress criticism. For political or practical reasons it is sometimes useful, of course, to reach consensus, at least temporarily. For example, when physicians wish to argue for insurance coverage of a procedure or a medication, they are more powerful if they can present a united front. Or if nations wish to take steps to repair environmental damage, it is best if they work with an environmental model that their scientists can agree on. When such external political and practical needs are absent, however, dissent is not a problem and, in my view, does not need to be fixed by an effort to reach consensus. Longino (1990, 2001) embraces pluralism universally. I demur and think that consensus is *occasionally* epistemically appropriate—although less often than it happens in practice (Solomon, 2001).

Diversity

It is commonplace to say that diversity is good for science. Different writers focus on different types of diversity. Sulloway (1996) writes about differences in personality, especially the difference between more radical and more conservative personalities, claiming that these differences lead to beneficial division of labor in science. Frederick Grinnell (1992) finds that diversity in experience and professional level of attainment (graduate student vs. postdoctoral fellow vs. senior researcher) leads to sufficient diversity in thought. Laudan and Laudan (1979) as well as Thomas Kuhn (1977) have speculated that different scientists apply scientific method differently, for example, by assigning different weights to valued traits such as simplicity, consistency, scope, fecundity, explanatory power, and predictive power. Kitcher (1993: 69–71) claims that there is a degree of cognitive variation among rational individuals that is "healthy" for the growth of science. Longino (1990, 2001) has argued that diversity in deeply held values is vital for scientific progress. Other feminist critics—e.g., Sandra Harding (1991), Donna Haraway (1991), and Evelyn Fox Keller (especially in her earlier work, such as 1985) have also argued for the importance of racial, class, and gender diversity in the population of scientists.

What is missing in these discussions, in my view, is a detailed exploration of which types of diversity are good for which particular sciences in which situations. All agree that diversity is, in general, good for science, because it can increase the number and variety of creative ideas and distribute cognitive labor on those ideas. No one seems to worry about "too much" diversity.

Future Topics

Philosophers are just beginning to work in applied social epistemology of science. Future topics—on which they could collaborate with science policy experts—include exploring and perhaps challenging publication practices such as digital publishing, the hierarchical organization of the scientific community, private versus public funding, university versus industry research, and processes of peer review. Experimental science is only three centuries old and has not experimented much with its social institutions. With tools such as the ones described in this essay, philosophers with imagination might see beyond the entrenched traditional practices of Western scientists and make recommendations that will benefit future research.

Notes

An early version of this paper received extensive comments from four anonymous referees and Olga Amsterdamska. I am most grateful for this feedback, which expanded my appreciation of the field. I wish I could cover in detail all the philosophical traditions that are relevant to the social understanding of science; I try to say enough to stimulate further work.

1. I use the capital "P" to identify the official discipline "Philosophy" and have in mind both Anglo-American and Continental Philosophy departments. When I say that work is philosophical (with a small "p"), I mean that it deals with foundational and/or normative issues, whether or not it is done by a member of a Philosophy department. The term "normative," as used by philosophers, covers evaluative judgments and prescriptive recommendations.

2. Gerd Gigerenzer (2002) and some others dispute this research. I do not find these contrary studies convincing, but this is not the place to work through this controversy.

3. Naomi Oreskes (1999) has a different account, attributing differences in judgment to different methodological traditions.

4. Note that I put "bias" in quotation marks; this is to mark the fact that "bias" is an epistemically negatively charged term. Because of the reflections above, such "biases" may not in fact have negative consequences for science.

5. "Well" and "badly" of course need further specification. In Solomon 2001, I have made the specification in terms of the comparative "empirical successes," so far, of the theories under consideration. Many other types of specification are possible (e.g., in terms of predictive power, or explanatory unification, or problem-solving capacity).

6. Again, Oreskes (1999) tells a different story, more coherent with Laudan and Laudan's (1979) and Sarkar's (1983) social epistemologies. This is not the place to decide between these competing accounts of the geological revolution.

7. Well, not quite. The precise definition of decision vectors in *Social Empiricism*, "Anything that influences the outcome of a decision," is wider than the scope of "biasing factors." This difference is not important here.

8. This is spelled out specifically, in terms of concepts developed in *Social Empiricism*.

9. I discuss the matter at length, and offer my own view, in Solomon 2001.

10. This is only one theme in Fuller's work, which is extensive and interdisciplinary. It should also be noted that Fuller founded the only interdisciplinary journal in this field, *Social Epistemology* (1986 to the present).

11. See Solomon (2006) for a fuller discussion.

12. As in note 2, above, the critical work of Gigerenzer (2002) should be mentioned.

13. For this point I am indebted to Olga Amsterdamska.

14. Watson and Crick were of course not peer-identical; they had different training in the same discipline and, while Watson was younger than Crick, Crick had not yet completed his PhD. For the purposes of Thagard's classification, they count as peer-similar, since they had similar professional status and could fully understand one another's work and expertise.

15. Michael Polyani (1962) anticipated more recent philosophical work on competition, the economics of science, and the self-organization of scientists. The sociologist Robert Merton (1968) argued that competition has been a feature of science since at least the seventeenth century that functions to motivate scientists to work faster and publish earlier.

16. There may be considerable individual variation in the response to a competitive environment, as well as variation depending on the strength and the nature of the competitive forces. It would be worth exploring what science loses (personnel, effort) as well as what it gains in various competitive and noncompetitive environments.

References

Coady, C.A.J. (1992) *Testimony: A Philosophical Study* (Oxford: Oxford University Press).

Daston, Lorraine (1992) "Objectivity and the Escape from Perspective," *Social Studies of Science* 22(4): 597–618.

Davidson, Arnold (2002) *The Emergence of Sexuality: Historical Epistemology and the Formation of Concepts* (Cambridge, MA: Harvard University Press).

Feyerabend, Paul (1975) *Against Method* (London: Verso).

Fricker, Elizabeth (2005) "Telling and Trusting: Reductionism and Anti-Reductionism in the Epistemology of Testimony," *Mind* 104: 393–411.

Fuller, Steve (1993) *Philosophy of Science and Its Discontents*, 2nd ed. (New York: The Guilford Press).

Giere, Ronald (1988) *Explaining Science: A Cognitive Approach* (Chicago: University of Chicago Press).

Giere, Ronald (1999) *Science Without Laws* (Chicago: University of Chicago Press).

Gigerenzer, Gerd (2002) *Adaptive Thinking: Rationality in the Real World* (New York: Oxford University Press).

Goldman, Alvin (1999) *Knowledge in a Social World* (Oxford and New York: Oxford University Press).

Goldman, Alvin (2002) *Pathways to Knowledge: Public and Private* (Oxford and New York: Oxford University Press).

Goldman, Alvin with Moshe Shaked (1992) "An Economic Model of Scientific Activity and Truth Acquisition," in Alvin Goldman (ed), *Liaisons: Philosophy Meets the Cognitive and Social Sciences* (Cambridge, MA: MIT Press).

Grinnell, Frederick (1992) *The Scientific Attitude* (New York: The Guilford Press).

Hacking, Ian (1983) *Representing and Intervening: Introductory Topics in the Philosophy of Natural Science* (Cambridge: Cambridge University Press).

Hacking, Ian (2000) *The Social Construction of What?* (Cambridge, MA: Harvard University Press).

Hacking, Ian (2004) *Historical Ontology* (Cambridge, MA: Harvard University Press).

Haraway, Donna (1991) *Simians, Cyborgs and Women: The Reinvention of Nature* (New York: Routledge Press).

Harding, Sandra (1991) *Whose Science? Whose Knowledge? Thinking from Women's Lives* (Ithaca, NY: Cornell University Press).

Harding, Sandra (1993) "Rethinking Standpoint Epistemology: What Is Strong Objectivity?" in Linda Alcoff & Elizabeth Potter (eds), *Feminist Epistemologies*. (London: Routledge): 49–82.

Hull, David (1988) *Science as a Process: An Evolutionary Account of the Social and Conceptual Development of Science* (Chicago: University of Chicago Press).

Hutchins, Edwin (1995) *Cognition in the Wild* (Cambridge, MA: MIT Press).

Jasanoff, Sheila (ed) (2004) *States of Knowledge: The Co-Production of Science and the Social Order* (London: Routledge).

Kahneman, Daniel, Paul Slovic, & Amos Tversky (eds) (1982) *Judgments Under Uncertainty: Heuristics and Biases* (Cambridge: Cambridge University Press).

Keller, Evelyn Fox (1985) *Reflections on Gender and Science* (New Haven and London: Yale University Press.)

Kitcher, Philip (1990) "The Division of Cognitive Labor," *Journal of Philosophy* 87(1): 5–22.

Kitcher, Philip (1993) *The Advancement of Science* (Oxford and New York: Oxford University Press).

Kitcher, Philip (2001) *Science, Truth and Democracy* (Oxford and New York: Oxford University Press).

Knorr Cetina, Karin (1999) *Epistemic Cultures: How the Sciences Make Knowledge* (Cambridge, MA: Harvard University Press).

Kornblith, Hilary (1994) "A Conservative Approach to Social Epistemology," in F. F. Schmitt (ed), *Socializing Epistemology: The Social Dimensions of Knowledge* (Lanham, MD: Rowman and Littlefield).

Kuhn, Thomas (1977) "Objectivity Value Judgment and Theory Choice," in Thomas Kuhn, *The Essential Tension* (Chicago: University of Chicago Press): 320–39.

Kusch, Martin (2002) *Knowledge by Agreement: The Programme of Communitarian Epistemology* (Oxford: Oxford University Press).

Latour, Bruno (1987) *Science in Action* (Cambridge, MA: Harvard University Press).

Latour, Bruno (2005) *Reassembling the Social: An Introduction to Actor-Network-Theory* (New York and Oxford: Oxford University Press).

Laudan, Rachel & Larry Laudan (1989) "Dominance and the Disunity of Method: Solving the Problems of Innovation and Consensus," *Philosophy of Science* 56(2): 221–37.

Lave, Jean & Etienne Wenger (1991) *Situated Learning: Legitimate Peripheral Participation* (Cambridge and New York: Cambridge University Press).

Longino, Helen (1990) *Science as Social Knowledge: Values and Objectivity in Scientific Inquiry* (Princeton, NJ: Princeton University Press).

Longino, Helen (2001) *The Fate of Knowledge* (Princeton, NJ and Oxford: Princeton University Press).

Menard, H. W. (1986) *The Ocean of Truth: A Personal History of Global Tectonics* (Princeton, NJ: Princeton University Press).

Merton, Robert K. (1968) "Behavior Patterns of Scientists," in R. K. Merton (1973), *The Sociology of Science: Theoretical and Empirical Investigations* (Chicago: University of Chicago Press): 325–42.

Mill, John Stuart (1859) *On Liberty*, reprinted in Mary Warnock (ed) (1962), *John Stuart Mill: Utilitarianism* (Glasgow, U.K.: William Collins Sons).

Nelson, Lynn Hankinson (1990) *Who Knows? From Quine to a Feminist Empiricism* (Philadelphia: Temple University Press).

Nisbett, Richard & Lee Ross (1980) *Human Inference: Strategies and Shortcomings of Social Judgment* (Englewood Cliffs, NJ: Prentice-Hall).

Oreskes, Naomi (1999) *The Rejection of Continental Drift: Theory and Method in American Earth Science* (New York and Oxford: Oxford University Press).

Pickering, Andrew (1995) *The Mangle of Practice: Time, Agency and Science* (Chicago: University of Chicago Press).

Polanyi, Michael (1962) "The Republic of Science: Its Political and Economic Theory," in Marjorie Greene (ed) (1969) *Knowing and Being: Essays by Michael Polanyi* (Chicago: University of Chicago Press): 49–72.

Quine, W. V. (1969) "Epistemology Naturalized," in *Ontological Relativity and Other Essays* (New York: Columbia University Press).

Rouse, Joseph (1987) *Knowledge and Power: Toward a Political Philosophy of Science* (Ithaca, NY and London: Cornell University Press).

Rouse, Joseph (1996) *Engaging Science: How to Understand Its Practices Philosophically* (Ithaca, NY: Cornell University Press).

Sarkar, Hussein (1983) *A Theory of Method* (Berkeley and Los Angeles: University of California Press).

Shapin, Steven (1994) *A Social History of Truth: Civility and Science in Seventeenth-Century England* (Chicago: University of Chicago Press).

Solomon, Miriam (1992) "Scientific Rationality and Human Reasoning," *Philosophy of Science* 59(3): 439–55.

Solomon, Miriam (2001) *Social Empiricism* (Cambridge, MA: MIT Press).

Solomon, Miriam (2006) "Groupthink versus *The Wisdom of Crowds*: The Social Epistemology of Deliberation and Dissent," *The Southern Journal of Philosophy* 44 (special issue on Social Epistemology based on the Spindel Conference, September 2005): 28–42.

Sulloway, Frank (1996) *Born to Rebel: Birth Order, Family Dynamics and Creative Lives* (New York: Pantheon Books).

Thagard, Paul (1993) "Societies of Minds: Science as Distributed Computing," *Studies in the History and Philosophy of Science* 24(1): 49–67.

Thagard, Paul (1997) "Collaborative Knowledge." *Nous* 31(2): 242–61.

Turnbull, David (2000) *Masons, Tricksters and Cartographers: Comparative Studies in the Sociology of Scientific and Indigenous Knowledge* (Amsterdam: Harwood).

Van Fraassen, Bas (1980) *The Scientific Image* (New York: Oxford University Press).

Vertosick, Frank Jr. (1988) "First, Do No Harm," *Discover* July: 106–11.

Wegener, Alfred ([1915]1966) *The Origin of Continents and Oceans,* 4th ed. Trans. J. Biram. (London: Dover).

Wylie, Alison (2002) *Thinking from Things: Essays in the Philosophy of Archeology* (Berkeley: University of California Press).

Zammito, John (2004) *A Nice Derangement of Epistemes: Post-Positivism in the Study of Science from Quine to Latour* (Chicago and London: University of Chicago Press).

11 Cognitive Studies of Science and Technology

Ronald N. Giere

The cognitive study of science and technology developed originally as a multidisciplinary mixture involving the history and philosophy of science and technology and the cognitive sciences, particularly cognitive psychology and artificial intelligence. There were some early European influences due to the work of Jean Piaget (1929), and later Howard Gruber (1981) and Arthur Miller (1986), but most of the early work was American. Beginning in the mid-1960s, inspired in part by a desire to challenge then standard philosophical claims that the scientific discovery process is fundamentally arational, Herbert Simon (1966, 1973) suggested applying the techniques of artificial intelligence to study the process of scientific discovery. This work, carried on with many collaborators in Pittsburgh, culminated in the book *Scientific Discovery: Computational Explorations of the Creative Processes* (Langley et al., 1987). It was soon followed by *Computational Models of Scientific Discovery and Theory Formation* edited by Pat Langley and Jeff Shrager (1990), which involved mostly AI researchers from around the United States. A decade later, Simon, together with a Pittsburgh psychologist, David Klahr, surveyed a quarter century of work in a review entitled "Studies of Scientific Discovery: Complementary Approaches and Divergent Findings" (Klahr & Simon, 1999). The "complementary approaches" of the title are (1) historical accounts of scientific discoveries, (2) psychological experiments with nonscientists working on tasks related to scientific discoveries, (3) direct observations of scientific laboratories, and (4) computational modeling of scientific discovery processes (see also Klahr, 2000, 2005).

In the late 1970s, a group of psychologists at Bowling Green State University in Ohio, led by Ryan Tweney, began a program of systematically studying scientific thinking, including simulated experiments in which students were to try to discover the "laws" governing an "artificial universe." Their book, *On Scientific Thinking* (Tweney et al., 1981), discussed such phenomena as "confirmation bias," the tendency of subjects to pursue hypotheses that agreed with some data even in the face of clearly negative data. This program of research, unfortunately, died owing to lack of funding. Tweney himself began a study of Michael Faraday, which continues to this day (Tweney, 1985; Tweney et al., 2005). Michael Gorman, a psychologist at the University of Virginia, also conducted experiments on scientific reasoning in simulated situations as summarized in his book *Simulating Science: Heuristics, Mental Models and Technoscientific Thinking* (1992).

Around the same time, the psychologist Dedre Gentner and her associates began developing theories of analogical reasoning and mental modeling, which they applied to historical cases such as the discovery of electricity and the distinction between heat and temperature (Gentner 1983; Gentner & Stevens, 1983). A developmental psychologist, Susan Carey (1985), began applying Thomas Kuhn's account of revolutionary change in science to the cognitive development of children in explicit opposition to Piaget's (1929) theory of developmental stages. Michelene T. H. Chi (1992), another Pittsburgh psychologist, also investigated the phenomenon of conceptual change in the context of differences in problem-solving strategies among novices and experts.

Kuhn's *Structure of Scientific Revolutions* (1962) had a tremendous impact on all of science studies and especially on the philosophy of science. In response to Kuhn's work, many philosophers of science began studying conceptual change. Kuhn himself invoked Gestalt psychology as part of his explanation of how scientific conceptual change takes place. In the 1980s, several philosophers of science began applying more recent notions from the cognitive sciences to understand conceptual change in science. In *Faraday to Einstein: Constructing Meaning in Scientific Theories* (1984), Nancy Nersessian applied notions of mental models and analogical reasoning to the development of field theories in physics in the nineteenth and early twentieth centuries. Lindley Darden's *Theory Change in Science: Strategies from Mendelian Genetics* (1991) applied techniques from artificial intelligence in an attempt to program theoretical and experimental strategies employed during the development of Mendelian genetics. Paul Thagard (1988, 1991, 2000) advocated a full-blown "computational philosophy of science" and went on to develop an account of conceptual change based on a notion of "explanatory coherence," which he implemented in a computer program. The work of Carey, Chi, Darden, Nersessian, Thagard, Tweney, and others was represented in a volume of Minnesota Studies in the Philosophy of Science, *Cognitive Models of Science* (Giere, 1992b).

There are several things to note from this brief and necessarily incomplete survey of mostly early work in what came to be recognized as the cognitive study of science and technology. First, it is mostly about science. The cognitive study of *technology* developed only later. Second, the work is quite heterogeneous, involving people in different fields appealing to different aspects of the cognitive sciences and focusing on different topics and different historical periods or figures. The cognitive study of science and technology is thus decidedly *multidisciplinary*. Third, most of this work was done with little or no recognition of contemporaneous developments in the social study of science. That changed at the end of the 1980s, as exemplified by my own book *Explaining Science: A Cognitive Approach* (1988) and an attempted refutation of "The Sociology of Scientific Knowledge" by a computer scientist, Peter Slezak (1989), in the pages of *Social Studies of Science*.

Slezak's attack was pretty much dispatched by the commentary that followed his paper. My book was not so much an attack on the social study of science as an attempt to develop an alternative (but in many ways complementary) approach, "The Cognitive Construction of Scientific Knowledge" (Giere, 1992a). This approach shares with the social study science the attempt to be fully "naturalistic," thus embracing three of the four ideals of the original Strong Program in the sociology of scientific knowledge: causality, symmetry, and reflexivity, omitting only impartiality. This is in opposition to logical empiricism and the successor constructive empiricism of Bas van Fraassen (1980) as well as the more historically oriented approaches of Lakatos (1970) and Laudan (1977). It also recognized the diversity in the then extant social study of science including Merton's structural-functional analysis, the Strong Program, the constructivism of Latour and Woolgar and of Knorr Cetina, and discourse analysis.

In what follows I survey more recent developments in the cognitive study of science and technology, focusing on a range of topics that have been studied rather than on chronology or the disciplinary bases of the contributors. This should provide a useful, though hardly complete, introduction to current work.[1]

DISTRIBUTED COGNITION

I begin with distributed cognition, which, although a relatively new topic in the cognitive sciences, offers considerable potential for constructive interaction between the communities of cognitive studies and social studies of science and technology.

The still dominant paradigm in cognitive science is characterized by the phrase "cognition is computation." This has been understood in the fairly strict sense that there is a symbol system used to construct representations, and these representations are transformed by operations governed by explicit rules. The archetypical computational system is, of course, the digital computer, which has provided a model for the human mind/brain. The assumption has been that computation, and thus cognition, is localized in a machine or an individual human body.

Even just the titles of many of the works cited above provide ample evidence that, for the first twenty-five or so years, the cognitive study of science operated mainly within the computational paradigm. This is particularly apparent in the literature on scientific discovery. On its face, this paradigm leaves little room for considerations of social interaction. When pressed by critics to account for social interaction, proponents have responded that a computational system is capable in receiving inputs containing information about the social situation and operating on these inputs to produce appropriate responses (Vera & Simon, 1993). Thus, all the cognition remains internal to the individual cognitive system.

The social study of science, of course, operates with a much more general notion of cognition according to which simply doing science is engaging in a cognitive activity. And since doing science is regarded as a social activity, cognition is automatically social. But many in the social study of science are guilty of the reverse reduction, assuming that scientific cognition takes place only on the social level, so that whatever might be going in individual heads is irrelevant to understanding cognition in science.

In recent years, a few people *within* the cognitive sciences have reached the conclusion that there is an irreducible external and social component to cognition. To facilitate understanding between these two research communities, let us say that a process is scientifically cognitive if it produces a clearly scientific cognitive *output*, namely, scientific knowledge. Remember that SSK stands for the Sociology of Scientific *Knowledge*. I now briefly describe two contributions from *within* the cognitive sciences that argue for the conclusion that processes which produce scientific knowledge take place at least in part outside a human body and, eventually, in a social community (for further elaboration, see Giere, 2006: chapter 5).

The PDP Research Group

One source of the concept of distributed cognition lies in disciplines usually regarded as being within the core of cognitive science: computer science, neuroscience, and psychology. During the early 1980s, the PDP Research Group, based mainly in San Diego, explored the capabilities of networks of simple processors thought to be functionally at least somewhat similar to neural structures in the human brain (McClelland & Rumelhart, 1986). It was discovered that what such networks do best is recognize and complete *patterns* in input provided by the environment. By generalization, it has been argued that much of human cognition is also a matter of recognizing patterns through the activation of prototypes embodied in groups of neurons whose activities are influenced by prior sensory experience.[2]

But if so, how do humans do the kind of *linear* symbol processing apparently required for such fundamental cognitive activities as using language and doing mathematics? McClelland and Rumelhart suggested that humans do the kind of cognitive processing required for these linear activities by creating and manipulating *external* representations. These latter tasks *can* be done well by a complex pattern matcher. Consider the following now canonical example. Try to multiply two three-digit numbers, say, 456×789, in your head. Few people can perform even this very simple arithmetical task. The old fashioned way to perform this task involves creating an *external representation* by writing the two three-digit numbers one above the other and then multiplying two digits at a time, beginning with 9×6, and writing the products in a specified order. The symbols are manipulated, literally, by hand. The process involves eye-hand motor co-ordination and is not simply going on in the head of the person doing the multiplying. The person's contribution is (1) constructing the external representation, (2) doing the correct manipulations in the right order, and (3) supplying the products for any two integers, which can be done easily from memory. The story is similar if the person uses an electronic calculator or a computer.

Now, what is the cognitive system that performs this task? McClelland and Rumelhart answered that it is not merely the mind/brain of the person doing the multiplication but the *whole system* consisting of the person *plus* the external physical representation. The cognitive process is distributed between a person and an external representation.

Hutchins's *Cognition in the Wild*

A second prominent source for the concept of distributed cognition is Ed Hutchins's study of navigation in *Cognition in the Wild* (1995). This is an ethnographic study of traditional "pilotage," that is, navigation near land as when coming into port. Hutchins demonstrates that individual humans may be merely components in a complex cognitive system. For example, there are sailors on each side of the ship who telescopically record angular locations of landmarks relative to the ship's gyrocompass. These readings are then passed on, for example, by the ship's telephone, to the pilothouse where they are combined by the navigator on a specially designed chart to plot the location of the ship. In this system, one person could not possibly perform all the required tasks in the allotted time interval. Only the navigator and perhaps his assistant know the outcome until it is communicated to others in the pilothouse.

In Hutchins's detailed analysis, the social structure aboard the ship and even the culture of the U.S. Navy play a central role in the operation of this cognitive system. For example, the smooth operation of the system requires that the navigator hold a higher rank than those making the sightings because he must be in a position to give orders to the others. The social system relating the human components is as much a part of the whole cognitive system as the physical arrangement of the required instruments. In general, the social system determines *how* the cognition is distributed.

We might treat Hutchins's case as merely an example of *collective cognition* (Resnick et al., 1991). A collective, an organized group, performs the cognitive task, determining the location of the ship. Hutchins's conception of distributed cognition, however, goes beyond collective cognition. He includes not only persons but also instruments and other artifacts as parts of the cognitive system. Thus, among the components of the cognitive system determining the ship's position is the navigational chart on which bearings are drawn with a ruler-like device called a "hoey." The ship's position is determined by the intersection of two lines drawn using bearings from two sightings on opposite sides of the ship. So parts of the cognitive process take place not in anyone's head but in an instrument or on a chart. The cognitive process is distributed among humans and material artifacts.

Some (Clark, 1997; Tomasello, 1999, 2003) have extended this line of thinking to include human language itself. According to this view, language is a cultural artifact that individuals employ to communicate with others. The idea that speaking and writing are a matter of rendering external what is first internal is regarded as an illusion fostered by the fact that people can learn to internalize what is communal and thus silently "talk to themselves." But the ability to communicate with others precedes and enables the ability to communicate with oneself. Thus, even the apparently solitary scientific genius constructing a new theory is operating in what is in fact a social context. The final conclusion is that cognition is both physically embodied and culturally embedded.

Experimentation

Experimentation has been one of the core concerns of the social study of science (Shapin & Schaffer, 1985; Gooding et al., 1989). From the perspective of the cognitive study of science, and extending Hutchins's analysis of navigational systems, experimentation may be thought of as the operation of distributed cognitive systems.

In *Epistemic Cultures*, Karin Knorr Cetina (1999) examines work in both high-energy physics and molecular biology. She describes the operation of the accelerator at CERN as involving "a kind of distributed cognition," by which she seems to mean "collective cognition," emphasizing the fact that the operation of the accelerator involves hundreds of people whose participation changes over time. Her descriptions of the actions of various classes of actors implicitly illustrate the operation of a particular type of social system in determining just how the cognitive task is distributed among the scientists and their instruments. By contrast, she claims that experiments in molecular biology typically involve only a single investigator working with a variety of instruments. Accordingly, she does not talk about collective cognition in molecular biology. In a cognitive analysis, however, even a single person with an instrument constitutes a distributed cognitive system. Like one of Hutchins's sailors recording coordinates for a landmark using a telescope, such a system has cognitive capabilities not possessed by a person alone.

In an essay reprinted as "Circulating Reference: Sampling the Soil in the Amazon Forest," Bruno Latour (1999) describes how pedologists (soil scientists) use a pedocomparator, a shallow shelf with small boxes arranged in orderly columns and rows. The pedologists fill the small boxes with their soil samples according to a particular protocol. Rows correspond to sample sites and columns correspond to depths. Each box is labeled with the coordinates of the grid system. When the pedocomparator is full, it allows the scientists to read a pattern directly off its arrangement of soil samples. As the soil changes from clay (which forests prefer) to sand (which savannahs prefer), it changes color. When the samples are arranged in the pedocomparator, the pattern of coloration reveals the changes in soil composition. This simple fact provides a striking illustration of the power of the distributed cognitive systems approach to capture scientific practice. The scientists and the samples of dirt arranged in a pedocomparator form a distributed cognitive system in which the interaction between the scientists and an environment structured in a particular way allows the scientists to solve a problem simply by recognizing a pattern.

Both of these examples show that, by invoking the notion of distributed cognition, one can give an account of scientific activities that is simultaneously cognitive and social. There need be no conflict between these two modes of analysis (Giere & Moffatt, 2003).

The Problem of Agency

Introducing the notion of distributed cognition into science studies is, unfortunately, not without some problems. Here I consider only one major problem, again using examples provided by Knorr Cetina and Latour.

One of the most provocative ideas in Knorr Cetina's *Epistemic Cultures* is "the erasure of the individual as an epistemic subject" in high-energy physics (1999: 166–71). She argues that one cannot identify any individual person, or even a small group of individuals, producing the resulting knowledge. The only available epistemic agent, she suggests, is the extended experiment itself. Indeed, she attributes to the experiment itself a kind of "self-knowledge" generated by the continual testing of components and procedures, and by the continual informal sharing of information by participants. In the end, she invokes the Durkheimian notion of "collective consciousness" (1999: 178–79). Here Knorr Cetina seems to be assuming that, if knowledge is being produced, there must be an epistemic subject, the thing that knows what comes to be known. Moreover, knowing requires a subject with a mind, where minds are typically conscious. But, being unable to find traditional epistemic subjects, one or more individual persons within the organization of experiments in high-energy physics, she feels herself forced to find another epistemic subject, settling eventually on the experiment itself as the epistemic subject.

Whereas Knorr Cetina is tempted to invoke a kind of super epistemic agent, it may not be an exaggeration to say that, for Latour, there is no such thing as a cognitive agent at all. There are only "actants" connected in more or less tightly bound networks, transforming material representations, and engaged in agonistic competition with other networks. Actants include both humans and nonhumans in relationships that Latour insists are "symmetric." Thus, in *Pandora's Hope* (1999: 90), Latour writes that when the physicist Joliot was trying to produce the first nuclear chain reaction, he was seeking favors from, among others, neutrons and Norwegians. There are at least two ways of understanding this symmetry from a more traditional standpoint. It might mean that we ascribe to nonhumans properties normally ascribed to human agents, such as the ability to grant favors or to authorize others to speak in their name. Alternatively, it might mean reducing human agents to the status of nonhumans, so Norwegians would be no more and no less cognitive agents than neutrons. Latour, of course, would reject both of these understandings, since he rejects the categories in terms of which they are stated.

Around 1985 Latour still agreed with the proposal in the Postscript to the second edition of *Laboratory Life* (Latour & Woolgar, 1986c: 280) that there be "a ten-year moratorium on cognitive explanations of science" with the promise "that if anything remains to be explained at the end of this period, we too will turn to the mind!" In a less frequently quoted passage, the proposal continues: "If our French epistemologist colleagues are sufficiently confident in the paramount importance of cognitive phenomena for understanding science, they will accept the challenge." This suggests that the kind of cognitive explanations being rejected are those to be found in the works of Gaston Bachelard or in more general appeals to a scientific *mentalité*. Latour seeks simpler, more verifiable, explanations. As he says (1986b: 1), "No 'new man' suddenly emerged sometime in the sixteenth century . . . The idea that a more rational mind . . . emerged from darkness and chaos is too complicated a hypothesis." Hutchins would agree. Appeals to cognitive capacities now studied in the cognitive sciences are meant to explain how humans with normal human cognitive capacities manage to do modern science. One way, it is suggested, is by constructing distributed cognitive systems that can be operated by humans possessing only the limited cognitive capacities they in fact possess. Moreover, Latour himself now seems to agree with this assessment. In a 1986 review of Hutchins's *Cognition in the Wild* (1995), he explicitly lifts his earlier moratorium claiming that "cognitive explanations . . . have been . . . made thoroughly compatible with the social explanations of science, technology and formalism devised by my colleagues and myself" (Latour, 1986a: 62). How this latter statement is to be reconciled with his theory of actants is not clear.

Here I would agree with Andy Pickering (1995: 9–20), who is otherwise quite sympathetic to Latour's enterprise, that we should retain the ordinary asymmetrical conception of human agents, rejecting both Knorr Cetina's super-agents and Latour's actants. Thus, even in a distributed cognitive system, we need not assign such attributes as intention or knowledge to a cognitive system as a whole but only to the human components of the system. In addition to placating common sense, this resolution has the additional virtue that it respects the commitment of historians of science to a narrative form that features scientists as human actors.

Laboratories as Evolving Distributed Cognitive Systems

Applying the notion of distributed cognition, Nancy Nersessian and associates (2003) have recently been investigating reasoning and representational practices employed in problem-solving in biomedical engineering laboratories. They argue that these laboratories are best construed as evolving distributed cognitive systems. The laboratory, they claim, is not simply a physical space but a problem space, the components of which change over time. Cognition is distributed among people and artifacts, and the relationships among the technological artifacts and the researchers in the system evolve. To investigate this evolving cognitive system, they employ both ethnography and historical analysis, using in-depth observation of the lab as well as research into the histories of the experimental devices used in it. They argue that one cannot divorce research from learning in the context of the laboratory, where learning involves building relationships with artifacts. So here we have a prime example of the merger of social, cognitive, and historical analyses built around the notion of distributed cognition—and in a technological context.

MODELS AND VISUAL REPRESENTATIONS

Although mental models have been discussed in the cognitive sciences for a generation, there is still no canonical view of what constitutes a mental model or how mental models function in reasoning. The majority view among cognitive scientists assimilates mental models to standard computational models with propositional representations manipulated according to linguistic rules. Here the special feature of mental models is that they involve organized sets of propositions. Work in the cognitive study of science generally follows the minority view that the mental models used in reasoning about physical systems are *iconic*. An exemplar of an iconic mental model is a person's mental image of a familiar room, where "mental image" is understood as highly schematic and not as a detailed "picture in the mind." Many experiments indicate that people can determine features of such a room, such as the number and placement of windows, by mentally examining their mental images of that room.

While not denying that mental models play a role in the activity of doing science, I would emphasize the role of *external* models, including three-dimensional physical models (de Chadarevian & Hopwood, 2004), visual models such as sketches, diagrams, graphs, photographs, and computer graphics, but also including abstract models such as a simple harmonic oscillator, an ideal gas, or economic exchanges with perfect information. External models have the added advantage that they can be considered as components in distributed cognitive systems (Giere, 2006: chapter 5).

Combining research in cognitive psychology showing that ordinary concepts exhibit a graded rather than sharply dichotomous structure, together with a model-based understanding of scientific theories developed in the philosophy of science, I (1994, 1999) suggested that scientific theories can be seen as exhibiting a cognitive as well as a logical structure. Thus, the many models generated within any general theoretical framework may be displayed as exhibiting a "horizontal" graded structure, multiple hierarchies of "vertical" structures, with many detailed models radiating outward from individual generic models.

Using examples from the 1960s revolution in geology, I argued that scientists sometimes base their judgments of the fit of models to the world directly on visual representations, particularly those produced by instrumentation (Giere, 1996, 1999). There need be no inference in the form of propositional reasoning. Similarly, David Gooding (1990) found widespread use of visual representations in science. In his detailed study of Faraday's discovery of electromagnetic induction, he argued that the many diagrams in Faraday's notebooks are part of the process by which Faraday constructed interpretations of his experimental results. Most recently, Gooding (2005) surveyed work on visual representation in science and provided a new theoretical framework, abbreviated as the PSP schema, for studying the use of such representations. In its standard form, the schema begins with a two-dimensional image depicting a Pattern. The Pattern is "dimensionally enhanced" to create a representation of a three-dimensional Structure, then further enhanced to produce a representation of a four-dimensional Process. In general, there can also be "dimensional reductions" from Process down to Structure and down again to a Pattern. Gooding illustrates use of the scheme with examples from paleobiology, hepatology, geophysics, and electromagnetism (see also Gooding, 2004).

JUDGMENT AND REASONING

There is a large literature devoted to the experimental study of reasoning by individuals, typically undergraduate subjects but sometimes scientists or other technically trained people (Tweney et al., 1981; Gorman, 1992). Here I consider first two lines of research that indicate that reasoning by individuals is strongly influenced by context and only weakly constrained by normative principles. I then describe a recent large comparative study of reasoning strategies employed by individuals in research groups in molecular biology and immunology in the United States, Canada, and Italy.

Biases in Individual Reasoning

The Selection Task One of the most discussed problems in studies of individual reasoning is the so-called selection task devised by Peter Wason in the 1960s. In a recent version (Evans, 2002), the subject is presented with four cards turned one side up and told that one side shows either the letter A or some other letter while the other side shows either the number 3 or some other number. The four cards presented have the following sides facing up: A, D, 3, 7. The subject is instructed to select those cards, and only those cards, necessary to determine the truth or falsity of the general proposition ("law") covering just these four cards: If any of these cards has an A on one side, then it has a 3 on the other side.

The correct answer is to select the card with the A on front and the card with the 7 on front. If the card with an A on front does not have a 3 on the back, the law is false. Likewise, if the card with a 7 on front has an A on the back, the law is false. The cards with a D or a 3 showing provide no decisive information, since whatever is on the back is compatible with the law in question. On average, over many experiments, only about ten percent of subjects give the right answer. Most subjects correctly choose to turn over the card with an A on front, but then either stop there or choose also to turn over the uninformative card with the 3 on front.

Many have drawn the conclusion that natural reasoning does not follow the idea long advocated by Karl Popper (1959) that science proceeds by attempted falsification of general propositions. If one were trying to falsify the stated law, one would insist on turning over the card with the 7 facing up to determine whether or not it has an A on the back. Others have drawn the more general conclusion that, in ordinary circumstances, people exhibit a "confirmation bias," that is, they look for evidence that agrees with a proposed hypothesis rather than evidence that might falsify it. This leads them to focus on the cards with either an A or a 3 showing, since these symbols figure in the proposed law.

A striking result of this line of research is that the results are dramatically different if, rather than being presented in abstract form, the proposed "law" has significant content. For example, suppose the "law" in question concerns the legal age for drinking alcoholic beverages, such as: If a person is drinking beer, that person must be over 18 years of age. Now the cards represent drinkers at a bar (or pub) and have their age on one side and their drink, either a soft drink or beer, on the other. Suppose the four cards presented with one side up are: beer, soda, 20, and 16. In this case, on the average, about 75 percent of subjects say correctly that one must turn over both the cards saying beer and age 16. This is correct because only these cards represent possible violators of the law.

This contrast is important because it indicates that socially shared conventions (or, in other examples, causal knowledge) are more important for reasoning than logical form. Indeed, Evans (2002: 194) goes so far as to claim that "The fundamental computational bias in machine cognition is the *inability* to contextualize information."

Probability and Representativeness   A battery of experiments (Kahneman et al., 1982) demonstrate that even people with some training in probability and statistical inference make probability judgments inconsistent with the normative theory of probability. In a particularly striking experiment, replicated many times, subjects are presented with a general description of a person and then asked to rank probability judgments about that person. Thus, for example, a hypothetical young woman is described as bright, outspoken, and very concerned with issues of discrimination and social justice. Subjects are then asked to rank the probability of various statements about this person, for example, that she is a bank teller or that she is a feminist and a bank teller. Surprisingly, subjects on the average rank the probability of the conjunction, feminist and bank teller, significantly higher than the simple attribution of being a bank teller. This in spite of the law of probability according to which the conjunction of two contingent statements must be lower than that of either conjunct since the individual probabilities must be multiplied.

The accepted explanation for this and related effects is that, rather than following the laws of probability, people base probability judgments on a general perception of how representative a particular example is of a general category. Thus, additional detail may increase perceived representativeness even though it necessarily decreases probability. On a contrary note, Gigerenzer (2000) argues that representativeness is generally a useful strategy. It is only in relatively contrived or unusual circumstances where it breaks down. Solomon (2001 and chapter 10 in this volume) discusses the possibility that biases in reasoning by individuals are compatible with an instrumentally rational understanding of collective scientific judgment.

Comparative Laboratory Studies of Reasoning

For over a decade, Kevin Dunbar (2002) and various collaborators have been examining scientific reasoning as it takes place, *in vivo*, in weekly lab meetings in major molecular biology and immunology labs in the United States, Canada, and Italy. In addition to tape recording meetings and coding conversations for types of reasoning used by scientists, Dunbar and colleagues have conducted interviews and examined lab notes, grant proposals, and the like. Among the major classes of cognitive activity they distinguish are causal reasoning, analogy, and distributed reasoning.

Causal Reasoning   Dunbar and colleagues found that more than 80 percent of the statements made at lab meetings concern mechanisms that might lead from a particular cause to a particular effect. But causal reasoning, they claim, is not a unitary cognitive process. Rather, it involves iterations of a variety of processes, including the use of inductive generalization, deductive reasoning, categorization, and analogy. The initiation of a sequence of causal reasoning is often a response to a report of unexpected results, which constitute 30 to 70 percent of the findings presented at any particular meeting. The first response is to categorize the result as due to some particular type of methodological error, the presumption being that, if the experiment were done correctly, one would get the expected result. Only if the unexpected result continues to show up in improved experiments do the scientists resort to proposing analogies leading to revised models of the phenomena under investigation.

Analogy Dunbar et al. found that analogies are a common feature of reasoning in laboratory meetings. In one series of observations of sixteen meetings in four laboratories, they identified 99 analogies. But not all analogies are of the same type. When the task is to explain an unexpected result, both the source and target of the analogies are typically drawn from the same or a very similar area of research so that the difference between the analogized and the actual situation is relatively superficial. Nevertheless, these relatively mundane analogies are described as "workhorses of the scientific mind" (Dunbar, 2002: 159).

When the task switches to devising new models, the differences between the analogized and actual situation are more substantial, referring to structural or relational features of the source and target. Although they found that only about 25 percent of all analogies used were of this more structural variety, over 80 percent of these were used in model construction. Interestingly, analogies of either type rarely find their way into published papers. They mainly serve as a kind of cognitive scaffolding that is discarded once their job is done.

Distributed Reasoning A third type of thinking discussed by Dunbar and associates is collective and is most common in what they call the Representational Change Cycle. This typically occurs when an unexpected result won't go away with minor modifications in the experiment and new or revised models of the system under investigation are required. In these situations they find that many different people contribute parts of the eventual solution through complex interactions subject to both cognitive and social constraints. Here causal reasoning and analogies play a major cognitive role.

Culture and Scientific Cognition Richard Nisbett (2003) has recently argued that there are deep differences in the ways Westerners and Asians interact cognitively not only with other people but also with the world. Dunbar argues that one can also see cultural differences in the way scientists reason in the laboratory. He compared the reasoning in lab meetings in American and Italian immunology labs that were of similar size, worked on similar materials, and used similar methods. Members of the labs published in the same international journals and attended the same international meetings. Many of the Italians were trained in American labs. Nevertheless, Dunbar found significant differences in their cognitive styles.

Scientists working in American labs used analogies more often than those working in the Italian laboratories. Induction or inductive generalization was also used in the

American labs more often than in the Italian labs, where the predominant mode of reasoning was deductive. In American labs, deductive reasoning was used only to make predictions about the results of potential experiments. There is some evidence that these differences in cognitive strategies among scientists in the laboratory reflect similar differences in the cultures at large.

Thus, it seems that no single cognitive process characterizes modern science and research in a given field can be done using different mixes of cognitive processes. Which mix predominates in a given laboratory may depend as much on the surrounding culture as on the subject matter under investigation.

CONCEPTUAL CHANGE

As noted at the beginning of this essay, following the publication of Thomas Kuhn's *Structure of Scientific Revolutions* (1962), conceptual change became a major topic of concern among historians, philosophers, and psychologists of science. When the cognitive revolution came along a decade later, tools being developed in the cognitive sciences came to be applied to improve our understanding of conceptual change in science. I will discuss just one ongoing program of this sort, Nancy Nersessian's Model-Based Reasoning.

Following the general strategy in cognitive studies of science, Nersessian's goal is to explain the process of conceptual change in science in terms of general cognitive mechanisms and strategies used in other areas of life. Her overall framework is provided by a tradition emphasizing the role of mental models in reasoning. Within this framework she focuses on three processes: analogy, visual representation, and simulation or "thought experimenting," which together provide sufficient means for effecting conceptual change (Nersessian, 2002a).

The Mental Modeling Framework

Extending standard notions of mental models, Nersessian claims that some models in the sciences are generic. They abstract from many features of real systems for which models are sought. An example would be Newton's generic model for gravitation near a large body in which the main constraint is that the force on another body varies as the inverse square of its distance from the larger body. This abstraction allows one eventually to think of the motion of a cannon ball and that of the Moon as instances of the same generic model.

Analogical Modeling   A considerable body of cognitive science literature focuses on metaphor and analogy (Lakoff, 1987; Gentner et al., 2001). The relationship between the source domain and the target domain is regarded as productive when it preserves fundamental structural relationships, including causal relationships. Nersessian suggests that the source domain contributes to the model building process by providing additional constraints on the construction of generic models of the target domain. The use of analogy in everyday reasoning seems to differ from its use in science, where finding a fruitful source domain may be a major part of the problem when constructing new generic models. It helps to know what a good analogy should be like, but there seems to remain a good bit of historical contingency in finding one.

Visual Modeling   The importance of diagrams and pictures in the process of doing science has long been a focus of attention in the social study of science (Lynch and Woolgar, 1990). For Nersessian, these are visual models, and she emphasizes the relationship between visual models and mental models. Visual models facilitate the process of developing analogies and constructing new generic models. Nersessian also recognizes the importance of visual models as external representations and appreciates the idea that they function as elements in a distributed cognitive system that includes other researchers. Indeed, she notes that visual models, like Latour's immutable mobiles, provide a major means for transporting models from one person to another and even across disciplines. This latter point seems now accepted wisdom in STS.

Simulative Modeling   We tend to think of models, especially visual models, as being relatively static, but this is a mistake. Many models, like models in mechanics, are intrinsically dynamic. Others can be made dynamic by being imagined in an experimental setting. Until recently, thought experiments were the best-known example of simulative modeling. Now computer simulations are commonplace. However, the cognitive function is the same. Imagining or calculating the temporal behavior of a model of a dynamic system can reveal important constraints built into the model and suggest how the constraints might be modified to model different behavior. Thought experiments can also reveal features of analogies. A famous case is Galileo's analogy based on the thought experiment of dropping a weight from the mast of a moving ship. Realizing that the weight will fall to the base of the mast provides a way of understanding why an object dropped near the surface of a spinning earth nevertheless falls straight down.

Nersessian brings all these elements together in what she calls a "cognitive-historical analysis" of Maxwell's development of electrodynamics following Faraday's and Thompson's work on interactions between electricity and magnetism (Nersessian, 2002b). This analysis shows how visual representations of simulative physical models were used in the derivation of mathematical representations (see also Gooding & Addis, 1999).

COGNITIVE STUDIES OF TECHNOLOGY

In history, philosophy, and sociology, the study of technology has lagged behind the study of science. The history of technology is now well established, but both the philosophy and sociology of technology have only recently moved into the mainstream, and in both cases there have been attempts to apply to the study of technology approaches first established in the study of science. This is apparent in the work presented in *The Nature of Technological Knowledge: Are Models of Scientific Change Relevant?* (R. Laudan, 1984) and *The Social Construction of Technological Systems: New Directions in the Sociology and History of Technology* (Bijker et al., 1987). The closest thing to a comparable volume in the cognitive study of technology, *Scientific and Technological Thinking* (Gorman et al., 2005b), has appeared only very recently, and even here, only five of fourteen chapters focus exclusively on technology rather than science. An obvious supplement would be the earlier collaboration between Gorman and the historian of technology, Bernard Carlson, and others, on the invention of the telephone (Gorman & Carlson, 1990; Gorman et al., 1993).

Gary Bradshaw's "What's So Hard About Rocket Science? Secrets the Rocket Boys Knew" (2005) can be read as a sequel to his paper in the Minnesota Studies volume on the Wright brothers' successful design of an airplane (Bradshaw, 1992). Bradshaw, who was initially a member of the Simon group working on scientific discovery, begins with Simon's notion of a "search-space." Invention is then understood as a search through a "design space" of possible designs. Success in invention turns out to be a matter of devising heuristics for efficient search of the design space. In the case of the teenaged "rocket boys" working on a prize-winning science project following Sputnik, launch-testing every combination of attempted solutions to a dozen different design features would have required roughly two million tests. Yet the boys achieved success after only twenty-five launches. Bradshaw explains both how they did it and how and why their strategy differed from that of the Wright brothers, thus revealing that there is no universal solution to the design problem as he conceives it. Contextual factors matter.

Michael Gorman's (2005a) programmatic contribution, "Levels of Expertise and Trading Zones: Combining Cognitive and Social Approaches to Technology Studies," sketches a framework for a multidisciplinary study of science and technology. He begins with Collins and Evans's (2002) proposal that STS focus on the study of experience and expertise (SEE), which, he suggests, connects with cognitive studies of problem solving by novices and experts. Collins and Evans distinguished three levels of shared experience when practitioners from several disciplines, or experts and lay people, are involved in a technological project: (1) they have no shared experience, (2) there is interaction among participants, and (3) participants contribute to developments in each other's disciplines. Gorman invokes the idea of "trading zones" to characterize these relationships, distinguishing three types of relationships within a trading zone: (1) control by one elite, (2) rough parity among participants, and (3) the sharing of mental models. Finally, he characterizes the nature of communication among participants as being (1) orders given by an elite, or (2) the development of a creole language, or (3) the development of shared meanings. He clearly thinks it desirable to achieve state 3, with participants sharing meanings and mental models and contributing to each other's disciplines. Whether intended reflexively or not, this would be a good state for multidisciplinary studies in STS itself, particularly ones involving both cognitive and social approaches.

CONCLUSION

Looking to the future, my hope is that when the time comes for the next edition of a *Handbook of Science and Technology Studies*, cognitive and social approaches will be sufficiently integrated that a separate article on cognitive studies of science and technology will not be required.

Notes

I would like to thank Olga Amsterdamska, Nancy Nersessian, and three anonymous reviewers of an earlier draft of this article for many helpful suggestions.

1. For other recent introductions, see Carruthers et al. (2002); Gorman et al. (2005); Nersessian (2005); and Solomon (chapter 10 in this volume).

2. For a philosophical introduction to this understanding of cognition, see Churchland (1989, 1996).

References

Bijker, W., T. Pinch, & T. Hughes (eds) (1987) *The Social Construction of Technological Systems: New Directions in the Sociology and History of Technology* (Cambridge, MA: MIT Press).

Bradshaw, Gary (2005) "What's So Hard about Rocket Science? Secrets the Rocket Boys Knew," in Michael E. Gorman, Ryan Tweney, David Gooding, & Alexandra Kincannon (eds), *Scientific and Technological Thinking* (Mahwah, NJ: Lawrence Erlbaum): 259–76.

Bradshaw, Gary (1992) "The Airplane and the Logic of Invention," in R. N. Giere (ed), *Cognitive Models of Science*, Minnesota Studies in the Philosophy of Science, vol. XV (Minneapolis: University of Minnesota Press): 239–50.

Carey, Susan (1985) *Conceptual Change in Childhood* (Cambridge, MA: MIT Press).

Carruthers, Peter, Stephen Stitch, & Michael Siegal (eds) (2002) *The Cognitive Basis of Science* (Cambridge: Cambridge University Press).

Chi, Michelene T. H. (1992) "Conceptual Change Within and Across Ontological Categories: Examples from Learning and Discovery in Science," in R. N. Giere (ed), *Cognitive Models of Science*, Minnesota Studies in the Philosophy of Science, vol. XV (Minneapolis: University of Minnesota Press): 129–86.

Churchland, Paul M. (1989) *A Neurocomputational Perspective: The Nature of Mind and the Structure of Science* (Cambridge, MA: MIT Press).

Churchland, Paul M. (1996) *The Engine of Reason, The Seat of the Soul: A Philosophical Journey into the Brain* (Cambridge, MA: MIT Press).

Clark, Andy (1997) *Being There: Putting Brain, Body, and World Together Again* (Cambridge, MA: MIT Press).

Collins, H. M. & R. Evans (2002) "The Third Wave of Science Studies," *Social Studies of Science* 32: 235–96.

Darden, Lindley (1991) *Theory Change in Science: Strategies from Mendelian Genetics* (New York: Oxford University Press).

de Chadarevian, Soraya & Nick Hopwood (2004) *Models: The Third Dimension of Science* (Stanford, CA: Stanford University Press).

Dunbar, Kevin (2002) "Understanding the Role of Cognition in Science: The Science as Category Framework," in Peter Carruthers, Stephen Stitch, & Michael Siegal (eds), *The Cognitive Basis of Science* (Cambridge: Cambridge University Press): 154–70.

Evans, Jonathan (2002) "The Influence of Prior Belief on Scientific Thinking," in Peter Carruthers, Stephen Stitch, & Michael Siegal (eds), *The Cognitive Basis of Science* (Cambridge: Cambridge University Press): 193–210.

Gentner, Dedre (1983) "Structure Mapping: A Theoretical Framework for Analogy," *Cognitive Science* 7: 155–70.

Gentner, Dedre & Albert L. Stevens (eds) (1983) *Mental Models* (Hillsdale, NJ: Lawrence Erlbaum).

Gentner, Dedre, Keith Holyoak, & B. Kokinov (eds) (2001) *The Analogical Mind: Perspectives from Cognitive Science* (Cambridge, MA: MIT Press).

Giere, Ronald N. (1988) *Explaining Science: A Cognitive Approach* (Chicago: University of Chicago Press).

Giere, Ronald N. (1992a) "The Cognitive Construction of Scientific Knowledge," *Social Studies of Science* 22: 95–107.

Giere, Ronald N. (ed) (1992b) *Cognitive Models of Science*, Minnesota Studies in the Philosophy of Science, vol. XV (Minneapolis: University of Minnesota Press).

Giere, Ronald N. (1994) "The Cognitive Structure of Scientific Theories," *Philosophy of Science* 61: 276–96.

Giere, Ronald N. (1996) "Visual Models and Scientific Judgment," in Brian S. Baigrie (ed), *Picturing Knowledge: Historical and Philosophical Problems Concerning the Use of Art in Science* (Toronto: University of Toronto Press): 269–302.

Giere, Ronald N. (1999) *Science Without Laws* (Chicago: University of Chicago Press).

Giere, Ronald N. (2006) *Scientific Perspectivism* (Chicago: University of Chicago Press).

Giere, Ronald N. & Barton Moffatt (2003) "Distributed Cognition: Where the Cognitive and the Social Merge," *Social Studies of Science* 33: 301–10.

Gigerenzer, G. (2000) *Adaptive Thinking* (New York: Oxford University Press).

Gooding, David (1990) *Experiment and the Making of Meaning* (Dordrecht, Netherlands: Kluwer).

Gooding, David (2004) "Cognition, Construction and Culture: Visual Theories in the Sciences," *Journal of Cognition and Culture* 4: 551–97.

Gooding, David (2005) "Seeing the Forest for the Trees: Visualization, Cognition and Scientific Inference," in Michael E. Gorman, Ryan Tweney, David Gooding, & Alexandra Kincannon (eds), *Scientific and Technological Thinking* (Mahwah, NJ: Lawrence Erlbaum): 173–218.

Gooding, David, T. Pinch, & S. Schaffer (1989) *The Uses of Experiment: Studies in the Natural Sciences* (Cambridge: Cambridge University Press).

Gooding, David & T. Addis (1999) "A Simulation of Model-Based Reasoning about Disparate Phenomena," in L. Magnani, N. J. Nersessian, & P. Thagard (eds), *Model-Based Reasoning in Scientific Discovery* (New York: Kluwer): 103–23.

Gorman, Michael E. (1992) *Simulating Science: Heuristics, Mental Models and Technoscientific Thinking* (Bloomington, IN: Indiana University Press).

Gorman, Michael E. (2005a) "Levels of Expertise and Trading Zones: Combining Cognitive and Social Approaches to Technology Studies," in Michael E. Gorman, Ryan Tweney, David Gooding, & Alexandra Kincannon (eds), *Scientific and Technological Thinking* (Mahwah, NJ: Lawrence Erlbaum): 287–302.

Gorman, Michael E., Ryan Tweney, David Gooding, & Alexandra Kincannon (eds) (2005b) *Scientific and Technological Thinking* (Mahwah, NJ: Lawrence Erlbaum).

Gorman, Michael E. & W. Bernard Carlson (1990) "Interpreting Invention as a Cognitive Process: The Case of Alexander Graham Bell, Thomas Edison, and the Telephone," *Science, Technology & Human Values* 15(2): 131–64.

Gorman, Michael E., M. M. Mehalik, W. B. Carlson, & M. Oblon (1993) "Alexander Graham Bell, Elisha Gray and the Speaking Telegraph: A Cognitive Comparison," *History of Technology* 15: 1–56.

Gruber, Howard E. (1981) *Darwin on Man: A Psychological Study of Scientific Creativity* (Chicago: University of Chicago Press).

Hutchins, Edwin (1995) *Cognition in the Wild* (Cambridge, MA: MIT Press).

Kahneman, D., P. Slovic, & A. Tversky (eds) (1982) *Judgment Under Uncertainty: Heuristics and Biases* (Cambridge: Cambridge University Press).

Klahr, David (2000) *Exploring Science: The Cognition and Development of Discovery Processes* (Cambridge, MA: MIT Press).

Klahr, David (2005) "A Framework for Cognitive Studies of Science and Technology," in Michael E. Gorman, Ryan Tweney, David Gooding, & Alexandra Kincannon (eds), *Scientific and Technological Thinking* (Mahwah, NJ: Lawrence Erlbaum): 81–96.

Klahr, David & Herbert A. Simon (1999) "Studies of Scientific Discovery: Complementary Approaches and Convergent Findings," *Psychological Bulletin* 125(5): 524–43.

Knorr Cetina, Karin (1999) *Epistemic Cultures: How the Sciences Make Knowledge* (Cambridge, MA: Harvard University Press).

Kuhn, Thomas S. (1962) *The Structure of Scientific Revolutions* (Chicago: University of Chicago Press).

Lakatos, I. (1970) "Falsification and the Methodology of Scientific Research Programmes," in I. Lakatos and A. Musgrave (eds), *Criticism and the Growth of Knowledge* (Cambridge: Cambridge University Press): 91–195.

Lakoff, George (1987) *Women, Fire, and Dangerous Things: What Categories Reveal About the Mind* (Chicago: University of Chicago Press).

Langley, Pat, Herbert A. Simon, Gary L. Bradshaw, & Jan M. Zytkow (1987) *Scientific Discovery: Computational Explorations of the Creative Processes* (Cambridge, MA: MIT Press).

Latour, Bruno (1986a) "Review of Ed Hutchins' *Cognition in the Wild*," *Mind, Culture and Activity* 3(1): 54–63.

Latour, Bruno (1986b) "Visualization and Cognition: Thinking with Eyes and Hands," *Knowledge and Society: Studies in the Sociology of Culture, Past and Present* 6:1–40.

Latour, Bruno & Steve Woolgar (1986c) *Laboratory Life*, 2nd ed. (Princeton, NJ: Princeton University Press).

Latour, Bruno (1999) *Pandora's Hope: Essays on the Reality of Science Studies* (Cambridge, MA: Harvard University Press).

Laudan, L. (1977) *Progress and Its Problems* (Berkeley: University of California Press).

Laudan, R. (1984) *The Nature of Technological Knowledge: Are Models of Scientific Change Relevant?* (Dordrecht, Netherlands: Reidel).

Lynch, Michael & Steve Woolgar (eds) (1990) *Representation in Scientific Practice* (Cambridge, MA: MIT Press).

McClelland, J. L., D. E. Rumelhart, & the PDP Research Group (eds) (1986) *Parallel Distributed Processing: Explorations in the Microstructure of Cognition*, 2 vols. (Cambridge, MA: MIT Press).

Miller, Arthur (1986*) Imagery in Scientific Thinking* (Cambridge, MA: MIT Press).

Nersessian, Nancy J. (1984) *Faraday to Einstein: Constructing Meaning in Scientific Theories* (Dordrecht, Netherlands: Nijhoff).

Nersessian, Nancy J. (2002a) "The Cognitive Basis of Model-Based Reasoning in Science," in Peter Carruthers, Stephen Stitch, & Michael Siegal (eds), *The Cognitive Basis of Science* (Cambridge: Cambridge University Press): 133–53.

Nersessian, Nancy J. (2002b) "Maxwell and the Method of Physical Analogy: Model-Based Reasoning, Generic Abstraction, and Conceptual Change," in David Malamet (ed), *Reading Natural Philosophy* (LaSalle, IL: Open Court).

Nersessian, Nancy J. (2005) "Interpreting Scientific and Engineering Practices: Integrating the Cognitive, Social, and Cultural Dimensions," in Michael E. Gorman, Ryan Tweney, David Gooding, & Alexandra Kincannon (eds), *Scientific and Technological Thinking* (Mahwah, NJ: Lawrence Erlbaum): 17–56.

Nersessian, Nancy J., Elke Kurz-Milcke, Wendy C. Newstetter, & Jim Davies (2003) "Research Laboratories as Evolving Distributed Cognitive Systems," in R. Alterman & D. Kirsch (eds), *Proceedings of the Cognitive Science Society* 25 (Hillsdale, NJ: Lawrence Erlbaum): 857–62.

Nisbett, Richard E. (2003) *The Geography of Thought: How Asians and Westerners Think Differently—and Why* (New York: The Free Press).

Piaget, Jean (1929) *The Child's Conception of the World* (London: Routledge & Kegan Paul).

Pickering, Andy (1995) *The Mangle of Practice: Time, Agency, and Science* (Chicago: University of Chicago Press).

Popper, Karl R. (1959) *The Logic of Scientific Discovery* (London: Hutchinson).

Resnick, Lauren B., John M. Levine, & Stephanie D. Teasley (eds) (1991) *Perspectives on Socially Shared Cognition* (Washington, DC: American Psychological Association).

Shapin, Steven & Simon Schaffer (1985) *Leviathan and the Air-Pump: Hobbes, Boyle, and the Experimental Life* (Princeton, NJ: Princeton University Press).

Shrager, Jeff & Pat Langley (eds) (1990) *Computational Models of Scientific Discovery and Theory Formation* (San Mateo, CA: Morgan Kaufmann).

Simon, Herbert A. (1966) "Scientific Discovery and the Psychology of Problem Solving," in Robert Colodny (ed), *Mind and Cosmos* (Pittsburgh: University of Pittsburgh Press).

Simon, Herbert A. (1973) "Does Scientific Discovery Have a Logic?" *Philosophy of Science* 49: 471–80.

Slezak, Peter (1989) "Scientific Discovery by Computer as Empirical Refutation of the Strong Programme," *Social Studies of Science* 19: 563–600.

Solomon, Miriam (2001) *Social Empiricism* (Cambridge, MA: MIT Press).

Thagard, Paul (1988) *Computational Philosophy of Science* (Cambridge: MIT Press).

Thagard, Paul (1991) *Conceptual Revolutions* (Princeton, NJ: Princeton University Press).

Thagard, Paul (2000) *How Scientists Explain Disease* (Princeton, NJ: Princeton University Press).

Tomasello, Michael (1999) *The Cultural Origins of Human Cognition* (Cambridge, MA: Harvard University Press).

Tomasello, Michael (2003) *Constructing a Language: A Usage-Based Theory of Language Acquisition* (Cambridge, MA: Harvard University Press).

Tweney, Ryan D. (1985) "Faraday's Discovery of Induction: A Cognitive Approach," in David Gooding, Frank A. J. L. James (eds), *Faraday Rediscovered* (New York: Stockton): 189–210.

Tweney, Ryan D., Michael E. Doherty, & Clifford R. Mynatt (eds) (1981) *On Scientific Thinking* (New York: Columbia University Press).

Tweney, Ryan D., Ryan P. Mears, & Christiane Spitzmuller (2005) "Replicating the Practices of Discovery: Michael Faraday and the Interaction of Gold and Light," in Michael E. Gorman, Ryan Tweney, David Gooding, & Alexandra Kincannon (eds), *Scientific and Technological Thinking* (Mahwah, NJ: Lawrence Erlbaum): 137–58.

van Fraassen, B. C. (1980) *The Scientific Image* (Oxford: Oxford University Press).

Vera, A. & H. A. Simon (1993) "Situated Cognition: A Symbolic Interpretation," *Cognitive Science* 17: 4–48.

12 Give Me a Laboratory and I Will Raise a Discipline: The Past, Present, and Future Politics of Laboratory Studies in STS

Park Doing

When Bruno Latour (1983) used the line, "give me a laboratory and I will raise the world," he was referring to the power of that entity the laboratory as it was used by Louis Pasteur to change thinking about disease and health. Latour, however, might well have been referring to the power of such entities as they have been put to use in his own world, not nineteenth-century France, but that revolutionary province in that hotly contested academic region between the republics of sociology, philosophy, history, and anthropology known as Science and Technology Studies (STS). For the key to independence for this new territory was the same as it was for Pasteur in France: the laboratory. Before the rise of this aspiring republic, laboratories had been demarcated, through a series of conquests like the one accomplished by Pasteur that Latour describes, as special places from which pure knowledge emanated. During these conquests, philosophers had asserted confidently and social scientists and historians had harmonized dutifully that the twin gendarmes of falsifiability and adherence to proper experimental controls protect knowledge made in the laboratory from the sullying dirt of the social and political world. Knowledge from the lab was apolitically, asocially, transtemporally, translocally true. But what if an advance unit of Special Forces from sociology and anthropology (enlisting some turncoats from philosophy) could manage to get inside the laboratory walls and show that there too was a political world of negotiated or coerced pacts to get along in the accepted ways, to see what should be seen? A sociology and anthropology of that hardest of hard places—the lab—and by implication of its hardest of hard productions—scientific knowledge—would leave the demarcationist philosophers with no place to hide—no epistemic quarter, as it were, in which they could incontestably make their claims for the unassailable nature of scientific knowledge and their dominion over its study.

In the late 1970s, then, inspired by and looking to powerfully cash out the programmatic claims of such fields of thought as the Strong Programme, ethnomethodology, social constructivist philosophy, phenomenology, and literary theory, ethnographic researchers began somewhat independently and simultaneously to breach a physical and epistemological barrier that had until that time proven to be impenetrable to such engagements: the laboratory.[1] The primary mission of these laboratory ethnographers, as Karin Knorr Cetina asserts in an earlier review of laboratory studies, was to explicate how local laboratory practice was implicated in the "'made' and accomplished character of technical effects" (Knorr Cetina, 1995: 141). Laboratory ethnographers could, "through direct observation and discourse analysis at the root of where knowledge is produced" (Knorr Cetina, 1995: 140), thus disclose "the process of knowledge production as 'constructive' rather than descriptive" (Knorr Cetina, 1995: 141). Such constructions, Michael Lynch points out, should thus be considered "as matters to be observed and described in the present, and not as the exclusive property of historians and philosophers of science" (Lynch, 1985: xiv).

News of their early successes spread rapidly. In San Diego, the laboratory of the eminent Jonas Salk was engaged, and the sociopolitical world was seen to be invisibly permeating the work of fact production (Latour & Woolgar, 1979). Up north (but still in California), close scrutiny of lab bench conversations and "shop talk" showed how the real-time work of science, indeed what was "seen" in a given situation, was guided by intricately choreographed social coercions and assertions (Knorr Cetina, 1981; Lynch, 1985). In Britain, and deep under the badlands of the American Midwest, scientists looking for gravity waves and solar neutrinos were also observed to rely on social enculturation to generate facts (Collins, 1985; Pinch, 1986). All these new and dangerous-to-the-old-guard studies were lauded for their epistemic derring-do as well as their attention to the details of laboratory activity. The care and love that they had for their subjects was evident and compelling. Together, they formed a corpus of new intellectual work with provocative and profound implications for both the project of intellectual inquiry and also the essence of political citizenship.

Catching the wave of excitement growing around these projects, energetic scholars then built upon the implications of this work in innovative ways, greatly helping to build up the field of Science and Technology Studies over the next three decades. Referring to the early lab studies as foundational pillars of a new discipline, these scholars analyzed episodes of science and technical expertise in a variety of societal forums outside labs while referring to the studies inside labs as a justification for their own approaches to analyzing knowledge production. Why should analysts take at face value the unmitigated truth claims made by AIDS researchers, government and industry scientists, epidemiologists, and others, when the hardest of the hard—pure laboratory science—had already been deconstructed?[2] These new writers questioned previous notions of citizenship, identity, and expertise in society, and in doing so provoked and promoted new kinds of interventions that have the potential to reconfigure current modes of access, voice, and control in society. This process has led so successfully to a built-up field that the worth of the foundational laboratory studies is taken as self-evident and their work is seen to have been accomplished. These days, few sessions at professional meetings, only a handful of journal articles, and even fewer new books are dedicated to the project of ethnographically exploring fact making in the laboratory. After all, why repeat a job that has already been done? Indeed, the job was apparently done so well that there are not even that many laboratory studies in total, despite their subsequent importance to the field. In spite of this unfolding of history, however, questions must be asked of laboratory studies in STS. Did the early lab studies really accomplish what they were purported to have accomplished? Did they, as Knorr Cetina said, show the "'made' and accomplished character of technical effects"? And, importantly, are what present studies there are now doing all that they can do?

A close look at laboratory studies in this regard leads to a sobering and halting conclusion: they have not, in fact, implicated the contingencies of local laboratory practice in the production of any *specific* enduring technical fact. If we look past the compelling, precise, and at times dazzling theorizing to the actual facts in question in the studies, we see that the fact that laboratory facts have been ethnographically demonstrated to be deconstructable has itself been black-boxed and put to use by the field of STS. Such facts are the "dark matter" of STS (the boxes are black)—ethnographically demonstrated-to-be-deconstructed facts must exist to explain the STS universe, yet they are undetectable on inspection. This chapter opens the black box of the deconstructed laboratory fact and searches for the dark matter of the STS universe in order to guide a discussion of laboratory studies in STS and call for a reengagement between ethnographic work in laboratories and the now established field of STS.

THE SHOP FLOOR OF FACTS

In *Art and Artifact in Laboratory Science: A Study of Shop Work and Shop Talk in a Research Laboratory* (1985), Michael Lynch asserts that his work is a revolutionary project of antidemarcation (in opposition to demarcationist philosophers such as Popper [1963], Merton [1973], and Reichenbach [1951], as well as public portrayals of science), telling us that the "science that exists in practice is not at all like the science we read about in textbooks," that "successful experimentation would be impossible without . . . decisions to proceed in ways not defined *a priori* by canons of proper experimental procedure," and that "a principled demarcation between science and common sense no longer seems tenable" (Lynch, 1985: xiv). Lynch then sets to work to bring out how the fluidity of judgments of sameness and difference, conversational accounts, practical limitations, and negotiations—the processes of scientific practice—play into the acceptance and rejection of reality on the laboratory floor, with the caveat, as he explains, that the study of science in practice "should be exclusively preoccupied with the production of social order, *in situ*, not with defining, selecting among, and establishing orders of relevance for the antecedent variables that impinge upon 'actors' in a given setting" (Lynch, 1985: xv). In other words, the analyst is not privileged with regard to method—knowledge comes from practice, wherever it is found (Ashmore, 1989).

Lynch's subsequent descriptions of laboratory life are quite compelling. In real time, researchers struggle to negotiate what is "understood" in the moment such that a subsequent action is justified. The descriptions of the myriad of microsocial assertions and resistances put to work in order to work is rich, and that such negotiations are part and parcel of moment-to-moment practice is apparent. But what is the relation between the working world of the laboratory floor and the status of any particular enduring fact that the laboratory is seen to have produced? Given his introductory explanations of his project, we might expect such an enduring fact to be subjected to Lynch's analysis and method in his book—yet none are thus engaged. Lynch's conclusion in this regard is direct, actually, and somewhat startling given his initial framing of the project. According to Lynch, any claims about the relation between the endurance of the factual products of the laboratory and the practice at the lab are not actually part of his project. At the end of his ethnography, he disclaims specifically that "whether agreements in shop talk achieve an extended relevance by being presupposed in the further talk and conduct of members or whether they are treated as episodic concessions to the particular scene which later have no such relevance, cannot be definitively addressed in this study" (Lynch, 1985: 256). To be clear, he then further asserts that "the possibility that a study of science might attain to an essentializing grasp of the inquiry studied is no more than a conjecture in the present study" (Lynch, 1985: 293). Lynch's study, then, is not a direct challenge to the "principled demarcation" of science. In *Art and Artifact*, we are invited to consider the possibility that the detailed and compelling dynamics of day-to-day laboratory work presented might have implications for demarcating science from other forms of life, but by Lynch's own explicit acknowledgement we are not presented with an account of how this is so for a particular fact claim: how any particular episodic agreement is, as a matter of practice, achieved as a fact with "extended relevance."

If Lynch, after outlining a method for implicating local practices and agreements in the enduring products of science, did not technically connect his ethnography to a particular enduring fact, let us look at other authors of the early laboratory studies to see if they directly accomplished the job.

INDEXICAL MANUFACTURING

In her book *The Manufacture of Knowledge: An Essay on the Constructivist and Contextual Nature of Science* (1981), Karin Knorr Cetina also takes up the challenge to ethnographically demonstrate the local construction of an epistemically demarcated fact. When explaining her project, Knorr Cetina tells us that:

> In recent years, the notion of situation and the idea of context dependency has gained its greatest prominence in some microsociological approaches, where it stands for what ethnomethodologists have called the "indexicality" of social action . . . Within ethnomethodology, indexicality refers to the location of utterances in a context of time, space, and eventually, of tacit rules. In contrast to a correspondence theory of meaning, meanings are held to be "situationally determined," dependent only on the concrete context in which they appear in the sense that "they unfold only within an unending sequence of practical actions" through the participants' interactional activities. (Knorr Cetina, 1981: 33)

The shop floor of the lab, again, is the place to find this situational world of practical action, and Knorr Cetina does indeed find it. Like Lynch, she provides compelling ingredients for a sociopolitical analysis of the technical. She astutely observes the subtle way in which power is "played" out between scientists for access and control of resources and authorship and credit (Knorr Cetina, 1981: 44–47) and convincingly argues that a series of "translations" from one context to another is the mill from which new "ideas" are generated and pursued in the course of laboratory research (Knorr Cetina, 1981: 52–62). She further asserts how larger "trans-scientific" fields are ever-present in the day-to-day activities and decisions of laboratory researchers (Knorr Cetina, 1981: 81–91). Moreover, she goes further than Lynch in pursuit of a political account of a technical fact as she follows a particular technical fact through to its culminating fixation in a scientific publication. Knorr Cetina points out that the active, situated work on the part of researchers as they negotiate the contingent, messy, lifeworld of the laboratory that she brought out with her study cannot be found in the final official published account of the episode, which reads like a high school textbook account of the scientific method (hypothesis, experiment, results, etc.). The question, again, is how, precisely, does the fact that this work took place and was subsequently erased relate to the status of the particular technical fact claimed by the scientists in their publication on that subject. Precisely how is the technical claim presented by the practitioners that "laboratory experiments showed that $FeCl_3$ compared favourably with HCl/heat treatment at pH 2–4 with respect to the amount of coagulable protein recovered from the protein water" (Knorr Cetina, 1981: 122) implicated as "situationally determined"? On this question, Knorr Cetina is also silent.

The problem is that demarcationist philosophers would agree that the context of discovery leading up to a technical claim is a mess, filled with contingent practice, intrigue, uncertainty, and judgments, just as Knorr Cetina has described. But that, in and of itself, according to them, does not mean that a claim that is finally put forth from that process is not testable and falsifiable and thereby a demarcatable technical matter. Knorr Cetina's study does not confront the demarcationists head on but instead sidesteps their distinction between contexts of discovery and proof. All scientific papers erase contingency, but not all of them "produce" facts. It's not the erasing in and of itself that coerces the acceptance of a fact claim. Knorr Cetina does not address why *this* erasing worked in *this* situation while other erasings do or have not, and that is the crux of the matter for a study that seeks to assert that knowledge production is "constructive" rather than "descriptive."

Where Knorr Cetina leaves off, however, Bruno Latour and Steve Woolgar press on in spectacular fashion. Again we must ask, though, if they really achieved what they (and subsequent others) said they did.

CONTINGENT INSCRIPTIONS

In their study of Jonas Salk's laboratory at the University of California, San Diego, *Laboratory Life: The Social Construction of Scientific Facts* (1979) (of course, later retitled to remove the "Social") Bruno Latour and Steve Woolgar (1986) explicitly set out to show how the hardest of facts could be deconstructed. Self-aware revolutionaries, Latour and Woolgar state again that the objective of their anthropological study is to take back the laboratory from the demarcationists, to show that "a close inspection of

laboratory life provides a useful means of tackling problems usually taken up by epistemologists" (Latour & Woolgar, 1979: 183). Their approach relies on the important
ethnomethodological tenet that practitioners use methods tautologically and the
analyst has no privilege in this regard. They explain to us that their project is to show
how "the realities of scientific practice become transformed into statements about how
science has been done" (Latour & Woolgar, 1979: 29); they also sound the cautionary
note of Lynch, noting that "our explanation of scientific activity should not depend
in any significant way on the uncritical use of the very concepts and terminology
which feature as part of (scientific) activity" (Latour & Woolgar, 1979: 27). Latour and
Woolgar are keenly aware, of course, that the distinction between the technical and
the social is a resource put to use by the participants they are studying, and they seek
to elucidate the process by which such ethnomethods succeed in producing facts at
the lab.

To make their point demonstrably, Latour and Woolgar focus on no small fact but
rather one that resulted in Nobel prize awards and historical prestige for a legendary
laboratory: the discovery at Salk Institute that thyrotropin-releasing factor (or
hormone) (TRF or TRH) is, in fact, the compound (in somewhat shorthand) Pyro-Glu-
His-Pro-NH$_2$. As Latour and Woolgar pursue their analysis of the discovery of the
nature of TRF(H), they never lose sight, or let us lose sight, of their antidemarcationist mission, stating and restating it many times, and the field of STS has ever since
referred to these statements of their accomplishment as foundational pillars of the discipline. But again we must ask our question: exactly where are the points at which
Latour and Woolgar's account of the "discovery" of TRF(H) as Pyro-Glu-His-Pro-NH$_2$
implicates contingent local practice in the enduring, accepted fact? Where, precisely,
does their account depart from a demarcationist line? In this regard, there are two critical points in the TRF(H) as Pyro-Glu-His-Pro-NH$_2$ story that bear close scrutiny. First
is the point at which, in the research described by Latour and Woolgar, the acceptable criteria for what counted as a statement of fact regarding TRF(H) changed among
the practitioners. Where previously isolating the compound in question was seen as
undoable, and therefore irrelevant for making statements of fact about TRF(H), owing
to the fact that literally millions of hypothalami would have to be processed, there
later came a point where the field decided that such a big science-type project was the
only way to obtain acceptable evidence of the actual structure of TRF(H). Old claims
about TRF(H) were now "unacceptable because somebody else entered the field, redefined the subspecialty in terms of a new set of rules, had decided to obtain the structure at all costs, and had been prepared to devote the energy of 'a steam roller' to its
solution" (Latour & Woolgar, 1979: 120). The success of this intervention, importantly,
"completely reshaped the professional practice of the subfield" (Latour & Woolgar,
1979: 119).

This would seem an episode ripe for antidemarcationist explanation. The criteria
for fact judging changed owing to local, contingent, and historical actions! Now the
move would be to explore why and how this happened and was sustained—why
it worked. Here, however, the authors become very quiet. As to why the researcher
who pushed the change through would go to such lengths, we are left with only a

cryptic reference to his dogged immigrant mentality. As for why his pursuit succeeded as valid, proper science, becoming the new touchstone of claims about TRF(H), rather than being seen as golem-like excess and unnecessary waste, we get this explanation:

The decision to drastically change the rules of the subfield appears to have involved the kind of asceticism associated with strategies of not spending a penny before earning a million. There was this kind of asceticism in the decision to resist simplifying the research question, to accumulate a new technology, to start bioassays from scratch, and firmly to reject any previous claims. In the main, the constraints on what was acceptable were determined by the imperatives of the research goals, that is, to obtain the structure *at any cost*. Previously, it had been possible to embark on physiological research with a semi-purified fraction because the research objective was to obtain the physiological effect. When attempting to determine the structure, however, researchers needed absolutely to rely on their bioassays. The new constraints on work were thus defined by the new research goal and by the means through which structures could be determined. (Latour & Woolgar, 1979: 124)

Here asceticism is the forceful entity doing the work, akin, actually, to a kind of Mertonian norm that the authors eschew.

Another point at which the local is crucially implicated in the subsequently "produced" fact comes at the end of the account of the emergence of TRF(H), when Latour and Woolgar describe the key episode in the making of the fact as fact—the point at which TRF(H) becomes Pyro-Glu-His-Pro-NH$_2$. The authors point to contestations over decisions about the sameness or difference of various curves obtained with a device called a chromatograph. Since the nature of TRF(H) rested on judgments of sameness and difference for the curves made with this device (as any good STSer now knows), such judgments can always be challenged. Consequently, the structure of TRF(H) appeared to be in epistemological limbo. How was this episode closed off, so that its product could endure as a scientific fact? It is at this point that Latour and Woolgar describe how an unquestionable device from physics, the mass spectrometer, carried the day. They tell us that the scientists "considered that only mass spectrometry could provide a fully satisfying answer to the problem of evaluating the differences between natural and synthetic (a compound made to be like) TRF(H). Once a spectrometer had been provided, no one would argue anymore" (Latour & Woolgar, 1979: 124). Here, then, is the critical juncture for the antidemarcationist epistemologist to go to work, at this nexus of the inscription to end all inscriptions—the mass spectrometry graph. But alas, after we have followed the journey of TRF(H) all this way, we are informed by the authors that "it is not our purpose here to study the social history of mass spectrometry." Further, we are given the very demarcationist line that "the strength of the mass spectrometer is given by the physics it embodies" (Latour & Woolgar, 1979: 146). Well, if mass spectrometry did in fact decide the day and usher in an "ontological change" for TRF(H) to become Pyro-Glu-His-Pro-NH$_2$, such that now it exists as a matter of fact rather than a contestable assertion, it should have been Latour and Woolgar's *main* purpose to analyze the technique as a "social historical" phenomenon. They are silent at precisely the point when they should be most vocal and assertive. The statement that the new definition of TRF(H) will "remain unambiguous as long as the analytical chemistry and the physics of mass spectrometry remain unaltered" (Latour & Woolgar, 1979: 148) has no analytical bite.[3]

Now, after their account of the emergence of TRF(H), Latour and Woolgar do go on to bring out many interesting and compelling ways that the reality of science is negotiated in real time, on the shop floor, in everyday work. *This* world is rife with political passions, contestations of power, ever-changing definitions of logic and proof. Referencing Harold Garfinkel, they give many compelling examples of how the day-to-day practice of science "comprises local, tacit negotiations, constantly changing evaluations, and unconscious institutional gestures," rather than standard scientific terms such as hypothesis, proof, and deduction, which are used only tautologically (Latour & Woolgar, 1979: 152). The only problem is that these discussions are *next* to the analysis of the emergence of TRF(H) as Pyro-Glu-His-Pro-NH$_2$ (described in the previous chapter of Latour & Woolgar's book), not *in* it. There is no clear route from the contingent world of the shop floor to the enduring fact of TRF(H) Pyro-Glu-His-Pro-NH$_2$ other than via the inference that, in principle, a thorough-going deconstruction along those lines could be undertaken. Again, that deconstruction has not been done for us.

The issue is the relation between contingent, local practice and the status of enduring translocal, transtemporal technical facts. And the point that Lynch is particularly cautious in this regard is worth considering carefully. In a world where method is used tautologically, at money-time what establishes that a particular fact endures? Indeed, the only time the endurance of a particular fact is specifically addressed in the three early lab studies (Lynch begs off the question, and Knorr Cetina does not address it in a specific way for the fact in question) is when Latour and Woolgar meekly gesture to such entities as "immigrant mentality" and the asceticism of making a million before spending a penny to explain how the accepted criteria for the basis of a fact claim changed, and then settle on the atomic mass spectrometer to account for how the TRF(H) controversy was eventually decided. But all these explanatory elements (immigrant mentality, asceticism, the law- embodied instrument of the mass spectrometer) go against Lynch's caveat and Latour and Woolgar's own methodological caution; they are elements taken from *outside the immediate life-world of laboratory practice*. They are forceful narrative entities, or "antecedent variables," brought in by the analyst to explain the endurance of the particular product of laboratory practice under question. In the end, the authors become decidedly unpreoccupied with the establishment of order *in situ* and instead bring in these antecedent variables to carry the day at money-time in the closing off of the contingency of a technical claim. By way of foreshadowing, let's keep in mind that the status of these entities as "social" or "nonmodern social/technical" is not salient—the important point is that they are antecedent, *ex situ* elements brought in to carry forth the narrative of deconstruction.

FALSIFIABILITY IS FALSE

There is a section in Knorr Cetina's account in which she shows how the scientists she studied themselves, in their own paper, account for their step-by-step method of discovery. She points out that there is ambiguity among the scientists as to exactly what information is necessary to include in a description of a step-by-step method, such that other scientists will be able to replicate the experiment. By showing that there is uncertainty and disagreement between the scientists (that one of the two collaborators on the paper is not sure how exactly to explain it to the *other collaborator*), Knorr Cetina implies that there is a problem in principle with the concept of an explainable, step-by-step method as the underpinning of facticity in science (Knorr Cetina, 1981: 128). Here she gives the kind of argument that Harry Collins, in his book *Changing Order: Replication and Induction in Scientific Practice* (1985), puts forward as a fundamental epistemological challenge to the demarcationists: that, in principle, there are no rules for following the rules and, therefore, there is a fundamental regress in experimental replication. (This idea is right in line with ethnomethodology—it is another way of saying that there is no way out of the situatedness of practice).

Animated by the principle of the experimenter's regress, Collins looks to a specific scientific controversy in order to empirically bring out how this dilemma is dealt with in the actual practice of doing science. When reading Collins's account of gravity wave experimenters, we find ourselves in a similar situation as with Latour and Woolgar—at the crucial juncture where controversy ends and a fact is born, we are left to wonder just how practice coerced the acceptance of this particular fact claim. One of the investigators in Collins's study had been making a claim for the detection of "high flux" gravity waves. This claim went against the prevailing theory of gravity waves and also against the results from other detectors. When an electrostatic calibrator was brought in to simulate gravity wave input, it was found that the investigator's detector was 20 times *less* sensitive than the others, and the claims for high flux gravity waves were dismissed. Collins points out that according to the experimenter's regress, the investigator could claim that the electrostatic calibrator did *not* simulate gravity waves and that the fact that high fluxes were detected with only this particular kind of detector, even though it was less sensitive to the calibrator, gave important information *about the nature of gravity waves*. Well, this is just what the investigator did, only it didn't wash. The investigator's claims in this regard were seen as "pathological and uninteresting." As Collins explains,

the act of electrostatic calibration ensured that it was henceforth implausible to treat gravitational forces in an exotic way. They were to be understood as belonging to the class of phenomena which behaved in broadly the same way as the well-understood electrostatic forces. After calibration, freedom of interpretation was limited to pulse profile rather than the quality or nature of the signals. (Collins, 1985: 105)

Collins assures us that all of this is not determined by nature. It was *the investigator* who had the agency, who "accepted constraints on his freedom" by "bowing to the pressure" to calibrate electrostatically, and thus "setting" certain assumptions beyond question. Collins asserts that the investigator would have been better served to refuse this electrostatic calibration that was so constraining. But what of this pressure on the investigator to calibrate? What gave it such force that the investigator *did* capitulate?

Where did it come from? Who controlled it? Why did it work? Here Collins is silent. There is no exploration into the means by which the dispute about the fact was closed off so that the fact endured. Again, the account reads like a conventional treatment of science—calibration settled the dispute. We are simply told by Collins that in principle the episode could have gone otherwise and been accepted as scientific.

Collins draws upon unexplored antecedent forces that compelled his investigator to comply with the electrostatic calibration to explain how high flux gravity waves were discounted. It is important to press the point here that he is just like Latour and Woolgar with regard to the project of implicating local scientific practice in the products of that practice. They both privilege something outside of the life-world of laboratory practice to explain the endurance of a particular technical fact. While each may say that the problem with the other is that they unduly privilege (respectively) the natural or the social in their explanation, the important point to understand is that *both* Collins and Latour and Woolgar (with their respective followers) have for many years gone against the admonition asserted by Lynch not to be preoccupied with "defining, selecting among, and establishing orders of relevance for the antecedent variables that impinge upon 'actors' in a given setting" (Lynch, 1985: xv). Whether it is social construction that is claimed to be demonstrated or Latour and Woolgar's (1986) later, nonmodern "construction" without the social that the theory supposedly proved, does not matter. Both camps break with the plane of practice in which method is used tautologically and bring in an element or elements from the outside to account for the endurance of the facts under question, and then argue over which is the better way to do so. These subsequent arguments have to this day not furthered the project of implicating local practice in the ontological status of any particular scientific fact.

CONFRONTING SOCIOLOGY

In his book *Confronting Nature: The Sociology of Solar Neutrino Detection* (1986), Trevor Pinch describes *in situ* the first experimental attempts to detect entities known as solar neutrinos. There are disagreements among the practitioners about what is going on, but again at a certain point, different interpretations are closed off and competing explanations are eliminated. Again, the linchpin of closure is calibration, but this time Pinch goes further than Collins, asserting that the linchpin of calibration is credibility. He then endeavors to explore this "credibility" by examining just how his experimenter was able to negotiate the relationships necessary to ward off critics of his detector. Pinch explains how the experimenter in question, Davis, would give the details of his experiment to a group of nuclear astrophysicists who were the benchmark group by which any assertion regarding solar neutrinos would be accepted, firsthand. This enabled the astrophysicists to "put their criticisms directly to him [Davis]" rather than through the medium of publication. Pinch notes that by the time a criticism did appear in print, "the battle had largely been won by Davis" (Pinch, 1986: 173). Pinch points out also that Davis was willing to go through the "ritual" of testing all sorts of "implausible" hypotheses brought forth from the astrophysicists. By taking on all comers, Davis performed "an important ritual function in satisfying the nuclear astrophysicists, and thereby boosting the credibility of his experiment." Popperian openness is *used* tautologically (Pinch, 1986: 174). Also, Davis stayed importantly within the boundary of his "acknowledged expertise," and he could do so through his informal relationship with the astrophysicists, to credible effect. As Davis himself put it, "this all started out as a kinda joint thing . . . and if you start that way you tend to leave these little boundaries in between. So I stayed away from forcing any strong opinions about solar models and they've never made much comment about the experiment" (Pinch, 1986: 173). Of course, this is performance (Pinch would say that "they" *did* make comments, and just not in print), but it is performance to effect—the effect of closure.

Here Pinch is not drawing on an outside element in the same way that Collins does to bear the epistemological burden in the account. The nuclear astrophysics group was the powerful touchstone for what counted as a proper experiment, and Pinch investigated the practical matter of the negotiation of relations of authority, such as work with the "little boundaries," which reflexively reinforced the "credibility" used to close off the contingency of a technical fact. At this point, though, we have a similar situation as with Knorr Cetina. Why did *this arrangement* with regard to little boundaries work in *this situation* as a means of demarcating a fact? Informal dialog and deft professional boundary managing, as well as performative rituals of testability, are part and parcel of practice. Why did such activities this time produce an enduring fact? As it was with the others, this question is not addressed in Pinch's study.

THE PRESENT–FUTURE OF LABORATORY STUDIES

For a lab study to give an account of a technical fact as "constructive" rather than "descriptive" in a way that is not insultingly scientistic or ironic, it must explain the endurance of a particular fact from within the discourse and practice of the practitioners—that is, in a way that does not privilege the analyst's method. In this regard, the early lab studies have been almost silent in deed, if not word. The project of wrestling with accounting for enduring legacies of practice was left off almost just as soon as laboratory studies began, despite the continuing professions of the field.[4] As the field "grew up," we should have been pressing the iconic laboratory studies (and we should be pressing lab studies now) on the points where their accounts of fact emergence might successfully have departed from the demarcationist program. Instead we have a cleavage in the field with subsequent and important anthropologies of laboratories bringing out important modalities of scientific research, but not pursuing particular episodes of fact making. The gulf between these anthropologies and the antidemarcationist lab studies has been noted by David Hess (1997) in his review of laboratory studies. Sharon Traweek's study of the Stanford Linear Accelerator (SLAC), *Beamtimes and Lifetimes: The World of High Energy Physicists* (1988), and Hugh Guster-son's *Nuclear Rites: A Weapons Laboratory at the End of the Cold War* (1996) are prominent examples in this regard. Both deliver insightful observations and reflections on the play of power, identity, and laboratory organization, especially the ways in which practitioners view and operate in these modes, but they do not address the production of a specific scientific fact.[5] Other studies of the organization of laboratory research also fall into this vein. John Law's study of a British synchrotron x-ray laboratory, *Organizing Modernity: Social Order and Social Theory* (1994), is likewise a compelling exploration into the work done by certain kinds of reflexive (to the practitioners) identities (like "cowboy" and "bureaucrat") in the operation of a scientific laboratory, which could play into fact determination, but the production of technical facts is of no interest in the study. In a similar manner, other political analyses of the organization of scientific practice, such as Knorr Cetina's (1999) "epistemic cultures" and Peter Galison's (1995) "trading zones," are also disengaged from accounting for the making of any particular scientific fact. It is possible that because the early antidemarcationist studies left off of their stated project at the outset, and because the field left off of them presuming the job had been accomplished, this has contributed to the lack of engagement between what can now be seen as two separate strands in laboratory studies.[6]

There have been some recent studies in which researchers have gone into laboratories, but the project of implicating a particular fact as situationally determined has not been advanced. Several researchers have spent time in laboratories in recent years and pushed on compelling aspects of laboratory life that could, in principle, be linked to particular fact production but are not. Sims (2005) explores how the framing modality of "safety" is at play in scientists' judgments and interpretations of instruments and equipment at Los Alamos. Roth (2005) ethnographically explores "classification activities" in practice. I explore the ways that "experience" is invoked and performed in claims over understandings of instruments and equipment at a synchrotron radiation laboratory between scientists and technicians (Doing, 2004). Mody (2001) interrogates the concept of purity as it plays into materials science researchers' conceptions of their practices, and Merz and Knorr Cetina (1997) have pursued the "practice" of theoretical physicists as they work. All these studies explore compelling sites and modalities of contingency in laboratory practice, but do not attempt to tie their analyses to a specific, enduring scientific fact claim. Other researchers have explicitly gone after particular fact claims, yet not advanced beyond the early works with regard to implicating the contingency of practice in an enduring fact claim. Kennefick (2000) has sought to explain why an account of star implosion in astronomy was not accepted, and Cole (1996) has worked to account for the dismissal of Thomas Gold's assertion that petroleum is not in fact derived from fossilized plants. These studies, like those of Collins and Pinch, push the notion that contingency is present, in principle, and give accounts of the participants' wranglings. But again, the studies leave off at wrestling with why the wrangling dynamics of *these* particular episodes did or did not result in enduring facts where in other situations such moves failed (or succeeded).

Facts have not been accounted for in laboratory studies. So many aspects of laboratory life have been ethnographically engaged: professional hierarchy, organizational identity, informal identity, gendered identity, national identity, modalities of "safety" and "purity," of risk and threat, the complex microplay of benchtop negotiations, relations to industry and commerce, ritualistic performances, and the erasing of reported contingency. Yet, none have been tied to the production of a particular, specific, and enduring fact. For lab studies to fulfill their promise for the field of STS, and to reengage the current works in progress, we have to admit what laboratory studies have *not* done, and we have to hold them, and the field, accountable. In Latour and Woolgar's account, the criteria necessary for a fact to be seen as ontologically prior *did* change (whereas earlier, a semipurified fraction would suffice for a determination based on effect; the hypothalami-intensive isolation was subsequently seen as not impossible but instead required). If the criteria changed, that means that practice was implicated in the status of the fact. What is needed is a more compelling exploration into this change than the invocation of immigrant mentality or asceticism. Why did a machine and labor-intensive methodology come to be seen as the proper way to do the experiment and justify fact claims? With Knorr Cetina, why do the erasings of contingency work in some cases to produce enduring facts and in other cases not (indeed, why wouldn't emphasizing such contingency work precisely to produce a fact in some circumstances)? With Pinch, why do similar negotiations over boundaries of expertise and rituals of intergroup interaction sometimes result in agreements over the nature of the world and sometimes not? If the touchstone for a criterion or technique of fact justification changes, how is it that that change is coerced as being valid scientifically rather than a corruption of the empirical project? As it stands, these kinds of questions have not been pressed upon the early antidemarcationist lab studies, nor have they been pursued in subsequent ethnographies with regard to any particular fact. What is needed now is for laboratory studies to press forcefully in this direction.

Latour and Woolgar (1979: 257) said that the difference between their work and the work of the subjects of study was that the latter had a laboratory. But of course, Latour and Woolgar *did* have a laboratory, and they put it to good use. Moreover the STS *field* has put *those* laboratories to good use for the past three decades. Referring to a corpus of pioneering studies that politicized that hardest of hard places and by implication the hardest of hard products—technical facts—a diverse group of scholars pressed on to consider matters of fact production in policy settings, public forums, technological controversies, medicine, and a host of other modes, using the successes of the early laboratory studies as a justification for a new approach to considerations of science and technology in society. However, there is an accounting error in STS. The undetectable dark matter of ethnographically demonstrated deconstructed laboratory facts has been invoked to balance and justify the STS universe, yet a close look at the account of any particular technical fact in laboratory studies makes us aware that actually only a few steps have been taken in implicating the contingent, performative world of local practice in the endurance of any particular fact claim. The first thing any new lab study should do is go directly for what laboratory studies have missed— a particular fact—and wrestle with how its endurance obtains within the *"in situ"* world of practice. Let's make detectable the dark matter in STS lab studies and get the books straight. I do not know just what such accounts will look like, but I do know that they should not begin with the ironic line, "Laboratory studies have shown . . .". In a recent article wrestling with the politically oppressive uptake of deconstructivist claims from STS, Bruno Latour asked, "is it enough to say that we did not really mean what we meant?" (Latour, 2004). Well, perhaps we should say, at least for now, that we did not really do what we said.

Notes

1. Some examples of the programmatic strands that early laboratory studies researchers who were explicitly interested in addressing the demarcation of scientific facts were familiar with are the Strong Programme of the Edinburgh school (Bloor, 1976; Barnes, 1974) and also with the ethnomethodological project (Garfinkel, 1967). These strands were themselves engaged with related mid-century ideas regarding reference to reality from social constructivist sociology (Berger & Luckmann, 1966), phenomenology (Schutz, 1972), linguistic philosophy (Winch, 1958; Lauer, 1958), and literary theory (Lyotard, 1954).

2. Writings on how scientific and technological expertise should interact with communities (Wynne, 1989; Epstein, 1996; Collins & Evans, 2002), government and policy makers (Jasanoff, 1990; Hilgartner, 2000; Guston, 2000), and political activism (Woodhouse, Hess, Breyman, & Martin, 2002; Moore & Frickel, 2006) and indeed notions of citizenship (Haraway, 1991) that point to laboratory studies as a foundational part of a project that considers scientific and technological knowledge as political, of which these are just some examples, are part and parcel of STS as a field.

3. In a review of the second edition of Latour and Woolgar (1986), Harry Collins (1988) criticizes the authors for what he calls a reification of the instrumentation of the spectrometer. This critique is a subset of the critique asserted in this chapter, as explained in the text.

4. It should be noted that Collins and Pinch's (1982) account of paranormal experimentation is readily included as an antidemarcationist lab study. In principle it is the same project, simply inverted, and therefore subject to the same critique asserted in this chapter.

5. It is important to note some historical studies of contemporary laboratory practice that also explicitly pursue the project of implicating that practice in the ontological status of scientific facts. Pickering (1984) notes that he can only address the antidemarcation project specifically with respect to one episode in his study—the assertion of the existence of neutral currents. He notes that the criteria of acceptance of this claim changed over time, and so is like Latour and Woolgar in this respect. In accounting for the change, Pickering asserts his concept of the interplay and registration between the theoretical and experimental communities. As with Collins, however, Pickering asserts that the choices were opportunistic for each community rather than ordained by evidence, yet those opportunities could just as easily be read from Pickering's account as opportunities based on evidence. He only asserts that in principle they were not. Galison (1987) brings out the agency of decision making on the part of contemporary particle physicists. Galison does not trace changes in criteria for this decision making, but calls for studies that might do so. Fox Keller (1983) notes that Barbara McClintock, who Fox Keller describes as employing a different kind of scientific method, was first ostracized and then recognized by the scientific community. But this recognition, according to the community, was not based on the acceptance of a new method but on the agreed-upon testable validity of McClintock's fact claims. The assertion that it was a vindication of a new method in science is Fox Keller's. As with Knorr Cetina, this interpretation does not confront demarcationist philosophers directly.

6. The trend in anthropology toward multisite ethnography has contributed to discouraging the kind of extended, on-site investigation of a particular work site practiced by the early lab studies (Marcus, 1995).

References

Ashmore, Malcolm (1989) *The Reflexive Thesis: Wrighting the Sociology of Scientific Knowledge* (Chicago: University of Chicago Press).

Barnes, Barry (1974) *Scientific Knowledge and Sociological Theory* (London: Routledge & Kegan Paul).

Berger, Peter & Thomas Luckmann (1966) *The Social Construction of Reality: A Treatise in the Sociology of Knowledge* (Garden City, NY: Doubleday).

Bloor, David (1976) *Knowledge and Social Imagery* (London: Routledge & Kegan Paul).

Cole, Simon (1996) "Which Came First: The Fossil or the Fuel?" *Social Studies of Science* 26(4): 733–66.

Collins, H. M. (1975) "The Seven Sexes: A Study in the Sociology of a Phenomenon, or the Replication of Experiments in Physics," *Sociology* 9: 205–24.

Collins, H. M. (1985) *Changing Order: Replication and Induction in Scientific Practice* (London: Sage).

Collins, H. M. (1988) "Review of B. Latour and S. Woolgar, *Laboratory Life: The Construction of Scientific Facts*" *Isis* 79(1): 148–49.

Collins, H. M. & Trevor Pinch (1982) *Frames of Meaning: The Social Construction of Extraordinary Science* (London: Routledge & Kegan Paul).

Collins, H. M. & Robert Evans (2002) "The Third Wave of Science Studies: Studies of Expertise and Experience," *Social Studies of Science* 32(2): 235–96.

Doing, Park (2004) "Lab Hands and the Scarlet 'O': Epistemic Politics and (Scientific) Labor," *Social Studies of Science* 34(3): 299–323.

Epstein, Steven (1996) *Impure Science: AIDS Activism and the Politics of Knowledge* (Berkeley: University of California Press).

Fox Keller, Evelyn (1983) *A Feeling for the Organism: The Life and Work of Barbara McClintock* (San Francisco: W. H. Freeman).

Galison, Peter (1987) *How Experiments End* (Chicago: University of Chicago Press).

Galison, Peter (1995) *Image and Logic: A Material Culture of Microphysics* (Chicago: University of Chicago Press).

Garfinkel, Harold (1967) *Studies in Ethnomethodology* (Englewood Cliffs, NJ: Prentice- Hall).

Gusterson, Hugh (1996) *Nuclear Rites: A Weapons Laboratory at the End of the Cold War* (Berkeley: University of California Press).

Guston, David (2000) *Between Politics and Science: Assuring the Integrity and Productivity of Science* (Cambridge: Cambridge University Press).

Haraway, Donna (1991) "A Cyborg Manifesto: Science, Technology, and Socialist-Feminism in the Late Twentieth Century," *Simians, Cyborgs and Women: The Reinvention of Nature* (New York: Routledge): 149–81.

Hess, David (1997) "If You're Thinking of Living in STS: A Guide for the Perplexed," in Gary Lee Downey & Joseph Dumit (eds), *Cyborgs and Citadels: Anthropological Interventions in Emerging Sciences and Technologies* (Santa Fe, NM: School of American Research Press).

Hilgartner, Stephen (2000) *Science on Stage: Expert Advice as Public Drama* (Stanford, CA: Stanford University Press).

Jasanoff, Sheila (1990) *The Fifth Branch* (Cambridge, MA: Harvard University Press).

Kennefick, Daniel (2000) "Star Crushing: Theoretical Practice and the Theoreticians Regress," *Social Studies of Science* 30(1): 5–40.

Knorr Cetina, Karin (1981) *The Manufacture of Knowledge: An Essay on the Constructivist and Contextual Nature of Science* (Oxford: Pergamon Press).

Knorr Cetina, Karin (1995) "Laboratory Studies: The Cultural Approach to the Study of Science," in Sheila Jasanoff, Gerald Merkle, James Petersen, & Trevor Pinch, *Handbook of Science and Technology Studies* (Thousand Oaks, CA: Sage): 140–66.

Knorr Cetina, Karin (1999) *Epistemic Cultures: How the Sciences Make Knowledge* (Cambridge, MA: Harvard University Press).

Latour, Bruno (1983) "Give Me a Laboratory and I Will Raise the World," in K. Knorr Cetina & M. Mulkay (eds), *Science Observed: Perspectives on the Social Study of Science* (London: Sage): 141–70.

Latour, Bruno (2004) "Why Has Critique Run out of Steam? From Matters of Fact to Matters of Concern," *Critical Inquiry* 30(2): 225–48.

Latour, Bruno & Steve Woolgar (1979) *Laboratory Life: The Social Construction of Scientific Facts* (London: Sage).

Latour, Bruno & Steve Woolgar (1986) *Laboratory Life: The Construction of Scientific Facts* (Princeton, NJ: Princeton University Press).

Lauer, Quentin (1958) *Triumph of Subjectivity: An Introduction to Transcendental Phenomenology* (New York: Fordham University Press).

Law, John (1994) *Organizing Modernity: Social Order and Social Theory* (Cambridge: Blackwell).

Lynch, Michael (1985) *Art and Artifact in Laboratory Science: A Study of Shop Work and Shop Talk in a Research Laboratory* (London: Routledge & Kegan Paul).

Lyotard, Jean-François (1954) *La Phénoménologie* (Paris: Presses Universitaires de France).

Marcus, George (1995) "Ethnography in/of the World System: The Emergence of Multi-Sited Ethnography," *Annual Review of Anthropology* 24: 95–111.

Merton, Robert K. (1973) "The Normative Structure of Science," in *The Sociology of Science* (Chicago: University of Chicago Press): 267–78.

Merz, Martina & Karin Knorr Cetina (1997) "Deconstruction in a Thinking Science: Theoretical Physicists at Work," *Social Studies of Science* 27(1): 73–111.

Mody, Cyrus (2001) "A Little Dirt Never Hurt Anyone: Knowledge-Making and Contamination in Materials Science," *Social Studies of Science* 31(1): 7–36.

Moore, Kelly & Scott Frickel (eds) (2006) *The New Political Sociology of Science: Institutions, Networks, and Power* (Madison: University of Wisconsin Press).

Pickering, Andrew, (1984) *Constructing Quarks: A Sociological History of Particle Physics* (Chicago: University of Chicago Press).

Pinch, Trevor (1986) *Confronting Nature: The Sociology of Solar Neutrino Detection* (Dordrecht, the Netherlands: D. Reidel).

Popper, Karl (1963) *Conjectures and Refutations: The Growth of Scientific Knowledge* (London: Routledge & Kegan Paul).

Reichenbach, Hans (1951) *The Rise of Scientific Philosophy* (Berkeley: University of California Press).

Roth, Wolf Michael (2005) "Making Classifications (at) Work: Ordering Practices in Science," *Social Studies of Science* 35(4): 581–621.

Schutz, Alfred (1972) *The Phenomenology of the Social World* (London: Heinemann).

Sims, Ben (2005) "Safe Science: Material and Social Order in Laboratory Work," *Social Studies of Science* 35(3): 333–66.

Traweek, Sharon (1988) *Beamtimes and Lifetimes: The World of High Energy Physicists* (Cambridge, MA: Harvard University Press).

Winch, Peter (1958) *The Idea of a Social Science and Its Relation to Philosophy* (London: Routledge & Kegan Paul).

Woodhouse, Edward, David Hess, Steve Breyman, & Brian Martin (2002) "Science Studies and Activism: Possibilities and Problems for Reconstructivist Agendas," *Social Studies of Science* 32(2): 297–319.

Wynne, Brian (1989) "Sheepfarming after Chernobyl: A Case Study in Communicating Scientific Information," *Environment* 31(2): 33–39.

13 Social Studies of Scientific Imaging and Visualization

Regula Valérie Burri and Joseph Dumit

Images are inextricable from the daily practices of science, knowledge representation, and dissemination. Diagrams, maps, graphs, tables, drawings, illustrations, photographs, simulations, computer visualizations, and body scans are used in everyday scientific work and publications. Furthermore, scientific images are increasingly traveling outside the laboratories and entering news magazines, courtrooms, and media. Today, we live in a visual culture (e.g., Stafford, 1996), which also values numbers (Porter, 1995; Rose, 1999) and science (Hubbard, 1988; Nelkin & Tancredi, 1989). Scientific images rely on these cultural preferences to create persuasive representations. The ubiquity of scientific images has raised the interest of STS scholars in studying visual representations and in exploring the visual knowledges they engender.

Visual representations in science have been studied from a variety of different theoretical and disciplinary perspectives. Philosophers of science have raised ontological questions about the nature and properties of visual representations in science and have theorized about the intersection of hermeneutics and science (among others, Griesemer & Wimsatt, 1989; Ruse & Taylor, 1991; Griesemer, 1991, 1992; Ihde, 1999). Historians of science have pointed to the importance of scientific depictions of nature for the emergence of a new concept of objectivity in the nineteenth century (Daston & Galison, 1992, 2007). They have drawn attention to visualization instruments and visual representations used in experimental systems from the Early Modern period to today (among many others see Cambrosio et al., 1993; Galison, 1997; Rheinberger, 1998; Kaiser, 2000; Métraux, 2000; Breidbach, 2002; Francoeur, 2002; Lefèvre et al., 2003; Hopwood, 2005; Lane, 2005). Other works have reconstructed the histories of (medical) visualization technologies and their introduction in the field of medicine (e.g., Yoxen, 1987; Pasveer, 1989, 1993; Blume, 1992; Lenoir & Lécuyer, 1995; Holtzmann Kevles, 1997; Warwick, 2005; Joyce, 2006). Laboratory studies have examined the use of images in the manufacture of scientific knowledge from sociological and anthropological perspectives (Latour & Woolgar, 1979; Knorr Cetina, 1981; Latour, 1986, 1987, [1986]1990; Lynch, 1985a, b, 1990, 1998; Lynch & Edgerton, 1988; Lynch & Woolgar, 1990; Knorr Cetina & Amann, 1990; Amann & Knorr Cetina, [1988]1990; Traweek, 1997; Henderson, 1999; Prasad, 2005a).

Work on visual images at the intersection of STS and other disciplines is also thriving. Scholars working in art studies proclaimed a "pictorial turn" in culture (Mitchell, 1994) and reflected on the relation of artistic and scientific images and on the regimes of representation in which they are displayed (e.g., Stafford, 1994, 1996; Jones & Galison, 1998; Elkins, forthcoming). In *Picturing Science, Producing Art*, Caroline Jones and Peter Galison (1998) staged an encounter between art theorists' analyses of the modes of interpretation and STS's ideas about social construction of scientific knowledge and technologies of production. Style and genre as understood by art historians created contexts in which laboratory practices and cultural practices could be seen to share specific aesthetic forms. Finally, cultural studies explored the intersections of scientific imagery with popular narratives and culture (e.g., Holtzmann Kevles, 1997; Lammer, 2002; van Dijk, 2005; Locke, 2005) and reflected about images of the body from a feminist perspective (e.g., Duden, 1993; Cartwright, 1995; Casper, 1998; Treichler et al., 1998; Marchessault & Sawchuk, 2000). Some of this work draws on semiotic, linguistic, psychoanalytical, and philosophical traditions of thinking about the visual and the existence of a visual language (e.g., Goodman, 1968; Arnheim, 1969; Metz, 1974, 1982; Rudwick, 1976; Barthes, 1977; Mitchell, 1980, 1987; Myers, 1990; Elkins, 1998; Davidson, [1996]1999) and about specific "techniques of the observer" (Crary, 1990; Elkins, 1994).

The body of work concerned with scientific visualizations is thus extremely diverse, and any attempt to synthesize the various strands would necessarily be reductive and selective. Because it is also a very lively area of concern, its boundaries are difficult to demarcate. Accordingly, instead of an exhaustive overview of the work done so far, this chapter outlines approaches to the social studies of scientific imaging and visualization (SIV) and raises some further questions and directions concerning the future study of visual representations in science.

IMAGING PRACTICES AND PERFORMANCE OF IMAGES

SIV asks questions such as what is the specificity of the visual as a form of (scientific) knowledge? If the visual is a special form of knowledge, understanding, and expression, how and why is it different from other forms of knowledge? In contrast to most philosophical, art historical, or linguistic studies on visual representations in science, SIV answers these questions by focusing on the social dimensions and implications of scientific images and visual knowledge rather than inquiring into their nature,[1] as has been exemplarily demonstrated by Gordon Fyfe and John Law's *Picturing Power: Visual Depiction and Social Relations* (1988). SIV follows the practice turn in social theory (Schatzki et al., 2001) by its interest in the epistemic practices of the production, interpretation, and use of scientific images.

This manner of exploring the role of visual representations in scientific activities when examining the manufacture of scientific knowledge has been one of the trademarks of laboratory studies. In his ethnomethodological studies of scientific work, for example, Michael Lynch analyzed the constitution of images and showed how specimens are modified in the laboratory and turned into visual displays for purposes of investigation (Lynch, 1985a,b, 1990, 1998; Lynch & Edgerton, 1988). Karin Knorr Cetina explored how visual representations interact with scientists' versatile discourses in everyday practice and how they work in experiments (Knorr Cetina, 1981; Knorr Cetina & Amann, 1990; Amann & Knorr Cetina, [1988]1990) while Bruno Latour has argued that images are deployed by researchers to find allies within the scientific community and create networks that stabilize their research findings (Latour & Woolgar, 1979; Latour, 1986, [1986]1990, 1987).

SIV shares these concerns with laboratory studies but extends the focus beyond scientific laboratories and communities. It asks: What happens when images travel outside academic environments and diffuse into other contexts? SIV explores the trajectories of scientific images from their production and reading through their diffusion, deployment, and adoption in different social worlds to their incorporation into the lives and identities of individuals, groups, and institutions. Following the "social life of images," SIV includes the study of both imaging practice and the performance of scientific imagery with particular attention to its visual power and persuasiveness.

Scientific images and visualizations are exceptionally persuasive because they partake in the objective authority of science and technology, and they rely on what is regarded as immediate form of visual apprehension and engagement. As Donna Haraway observed, "There are no unmediated photographs . . . only highly specific visual possibilities, each with a wonderfully detailed, active, partial way of organizing worlds." (Haraway, 1997: 177). Haraway's feminist approach treats scientific images as objectivizing gazes that appear universal and neutral while selectively privileging certain points of view and overlooking others. In the daily news, for instance, images of the earth as seen from space—originally products of the space program—are often used to invoke concern for the environment by appealing to the idea of one earth as a precious place shared by all (Haraway, 1991; Jasanoff, 2004). This earth image, even though it is highly processed, suggests the *realism* of a photograph, an unmediated (as in unaltered, immediate, direct, or true) relationship between the viewer and the object. In semiotic terms, an image of earth as seen from space, without clouds, is *hyper-real*: it is stylized, reduced in layers, and produced to correspond not with what would be seen by an astronaut but with an idealized *concept* of Nature. As such it is more compelling than a "real" picture would be.

There is a *desire to see* the truth in the visualizations of phenomena such as the whole earth, the brain in action, DNA diagrams, or global warming. The history of images in science and art, however, has shown that seeing and recognition are historically and culturally shaped (e.g., Alpers, 1983; Daston & Galison, 1992; Hacking, 1999). Foucault's ([1963]1973) analyses of medicine, madness, and prison systems demonstrated the value of historicizing *what can be seen*, through close attention to the sciences and technologies, bureaucracies, and classification systems (cf. Rajchman, 1991; Davidson, [1996]1999; Hacking, 1999; Rose, 1999).

The practice turn in STS taught us to attend to the *work* of science and technology in terms of both the processes of production and the resulting products. The challenge of SIV is now to incorporate together the work of science and technology in producing particular visual objects with the historicity of what any scientific or lay audience is given to see.

Representation in Scientific Practice, edited by Michael Lynch and Steve Woolgar (1990), has provided a touchstone for further work in this area. This collection of articles draws on laboratory studies and uses ethnographic and ethnomethodological methods to study what scientists do with words, pencils, paper, computers, technologies, and colors when dealing with images. Furthermore, the volume expanded the use of semiotic and rhetorical tools of linguistic analysis to the social study of visual objects and scientists' representational practices when using terms such as "inscription" or "evidence" (Latour & Woolgar, 1979; Rheinberger, 1997). The volume shows that visual representations cannot be understood in isolation from the pragmatic situations in which they are used. Since scientists "compose and place representations within texts, data sets, files, and conversations . . . and . . . use them in the course of a myriad of activities" (Lynch & Woolgar, 1990: viii), it is not enough to simply describe the things images depict or the meanings they reflect. We have to focus also on the textual arrangements and discursive practices within which these representations are embedded.

Lynch and Woolgar's collection is a starting point for studying the cultural embeddedness of the practices of the making and handling of visual representations and of the shaping, distributing, applying, and embodying of scientific visual knowledge. If seeing is so often believing, SIV must demonstrate how the making and using of images come together with seeing and believing in practices of scientific truth-making and perceptual convention.

In the remainder of this essay, we organize our discussion around three artificially separated topics: the production, engagement, and deployment of visualizations. When studying *production*, STS scholars examine how and by whom images are constructed by analyzing the practices, methods, technology, actors, and networks involved in the making of an image. The analysis of *engagement* focuses on the instrumental role of images in the production of scientific knowledge. Research on *deployment*, finally, refers to the use of scientific visualizations in different social milieus. It studies how images diffuse into nonacademic environments and analyzes the intersections of different forms of (visual) knowledge. Deployment also includes scientific images becoming parts of the body image of individuals and "objectively" grounding the everyday givenness of the social world. In other words, examining production means studying images as artifacts; examining engagement means analyzing the role of images as instruments in science; and examining deployment stands for focusing on how images are used outside the laboratories and how they intersect with different forms of knowledge about ourselves and our world. This analytic grid draws special attention to the interpretative openness of scientific images and to their persuasive power.

PRODUCTION

Like all artifacts, scientific images and visualizations are constructed by combinations of machines and people using concepts, instruments, standards, and styles of practice. STS offers methodological tools for telling the history of *how* a particular image—a picture in a scientific journal or a computer screen showing the visual result of a program—was created. The retrospective approach demonstrates that images result from a lengthy series of technological opportunities and constraints, negotiations, and decisions.

To illustrate image production, we offer the example of magnetic resonance images. MRI images depend on a series of decisions about the setup of the MRI machine and the data to be generated. Scientists and technicians make decisions about parameters such as the number and thickness of the cross-sectional slices, the angle they are to be taken from, and the scale or resolution of the image data. Decisions also have to be made when it comes to post-processing the images on the screen: perspectives can be rotated, contrast modified, and colors chosen for scientific publications. These specific decisions do not depend on technical and professional standards alone but also on cultural and aesthetic conventions or individual preferences (cf. Burri, 2001, and forthcoming). An MRI scan thus is not a "neutral" product but the result of a series of specific, culturally shaped sociotechnical negotiations, which imply—like any technological fabrication—processes of formalization and transformation (Lynch, 1985a, 1985b, 1990).

Examining visual displays used in scientific publications, Lynch (1990) described their pictorial space as a graphic, coordinate space, entailing a previous stage when the scientific object—whether a mouse, a cell, a brain, or an electron—had been rendered spatial and measurable, "mathematized." As this requirement is formalized, "mathematization" becomes a *necessary assumption* of the science and image that depends on it. We can thus attend to the costs and the powers of this assumption. In this case, the parts of the cell that could not be measured became unimportant to the work being done in the experiment. When mathematization is built into computerized instruments and software, the unmeasured can be completely and invisibly erased.

Who is involved in image production stages is as important a question to pursue as *how* the images are produced. While some images are produced by a single person from start to finish (even if relying on software and instruments produced by others), other images are the result of a series of hand-offs among individuals, and still others are coordinated team efforts. Schaffer and Galison have each attended closely to this labor dimension of image production. Schaffer's (1998) article "How Astronomers Mark Time" delineates the variety of different labor arrangements possible within the same community of scientists at the same time. In one case, a number of individuals were organized so that they could virtually replicate the operation of an automatic pattern-recognition machine. Above them, the scientist as expert and manager consolidated their work as the author and true interpreter of the meaning of the images they selected. Galison (1997) has shown similar processes in physics, including many instances where gender and class divisions mirrored discriminatory practices found elsewhere in industry at the time. Today this form of labor organization continues: the authors have witnessed many instances of undergraduates employed in brain imaging laboratories to perform recognition tasks that either cannot currently be automated or are cheaper to accomplish by employing undergraduates than by using specialized hardware and software.

It is important to understand who knows what, who is allowed to know, and who can actually say what he or she knows. Early x-ray technicians, for instance, professionalized in alliance with radiologists through an agreement that the technicians would learn anatomy so as to be able accurately to position x-ray machines, but they would not learn pathology in order to preserve the exclusive diagnostic ability of the radiologist-doctor (Larkin, 1978). CT scanning posed a problem to this division of epistemic labor because adjusting the CT scanner to produce diagnostically useful images *required knowing* what pathological objects such as tumors *looked like.* CT technicians therefore had to learn pathology. Barley (1984) documented instances when technicians had developed apparent visual diagnostic expertise through deep familiarity with the instrument but were not allowed (legally and conventionally) to express it and instead had to guide some less familiar radiologists indirectly to the proper conclusions. Barley noted that this type of interpretive hierarchy reversal was a contingent local phenomenon as it occurred only in one of the two hospitals he studied. Noticing local variation in who *can* read images and who is *allowed* to read them is a hallmark of STS insight (e.g., Mol & Law, 1994; Jasanoff, 1998).

Visual expertise also creates its own form of literacy and specialization. Scientific and medical illustrators have often been valued members of experimental teams, but with computer visualization, a great number of new specialties have arisen. Simulation modelers, programmers, interface designers, graphic designers, as well as computer-based electron and fluorescent microscope makers have all established distinct subdisciplines of visual science and technology necessary to most cutting-edge labs, with their own journals and professionalized career tracks. They also can move among many different disciplines—e.g., biology, chemistry, physics, engineering, and mathematics—creating not just trans-disciplinary visual and digital standards but new trading zones in visual and interactive instruments, algorithms, and concepts.

ENGAGEMENT

If studying production examines how and by whom an image is *made,* studying engagement means examining how images are *used* in the course of scientific work and are made instrumental in the production of scientific knowledge. How are images talked about? What roles do they play in this talk? What concepts do they represent, and what forms of creativity do they engender? Engagement analysis treats each visual form as an actor in its own right, actively involved in the doing of science.

In disciplines using computer visualization, hundreds of images are often produced in the course of a single experiment. Some of these images are treated as uninterpreted raw data, other images are manipulated visually in order to make data meaningful, and still other images are interpretive summaries of known meanings. In a biology experiment, for instance, a digital confocal microscope may collect data on protein changes in a moving cell in full color plus three fluorescent channels (each keyed to a different gene) over a ten-minute period. The resulting data file contains what the scientists call seven dimensions of data (three physical dimensions, time, and the three different gene activities, all located spatially). The total size of this one collection of data is over three terabytes (over 3000 gigabytes). Although many choices have been made as to what is being collected, this visual data collection is treated by the researchers as *raw data* requiring extensive processing, analysis, and interpretation as well as massive reduction in size to become meaningful at all.

A variety of data-mining techniques, qualitative visual selections, and quantitative algorithms are then applied to generate a series of different screen images. These visualizations—alternatively called models, hypotheses, maps, and simulations—are provisional and interactive. The researchers constantly tweak them, altering parameters, changing color scales, substituting different algorithms or statistical analyses. These visualizations are part of *making the data meaningful*. They are interstitial, facilitated modes of seeing and intervening. One way of analyzing this process is to investigate how images contribute to the generation of an "objectified" knowledge by reducing the uncertainty of the observations and narrowing down the interpretative flexibility of research findings (Latour, 1986, [1986]1990, 1987; Amann & Knorr Cetina, [1988]1990; Lynch & Woolgar, 1990; and Beaulieu, 2001). In some cases, the same computer interface that enables data reconfiguration also runs the experimental instruments themselves, shaping future data collection.

Finally, when a research team reaches a provisional conclusion, the same software can be used to construct a clear summary of this conclusion as a meaningful visualization of the *data as knowledge*. In this construction, the image is tweaked toward a coherence of reception, with aesthetic and scientific conventions in mind (Lynch & Edgerton, 1988). The work of Edward Tufte (1997) is important to consider here from an STS perspective as he has spent a career looking at how one can most effectively convey a complex known meaning to an audience via a scientific graph or visualization. He demonstrates the hard work and many forms of visual literacy required to create shared meanings.

Once an image becomes part of a body of knowledge, it can be used to diffuse and stabilize the knowledge and theoretical concepts it represents. As Latour and others have shown, visualizations are instruments to support and transport arguments used to convince other scientists (see also Keith & Rehg, chapter 9 in this volume). In other words, images are advantageous in "rhetorical or polemical" situations, and they help researchers find allies within the scientific community (cf. Latour, 1986, [1986]1990; Traweek, 1997; Henderson, 1999).

A consequence of any scientific image or visualization is that the representational practice involves a new conceptual space. Whether these are the two dimensions of branching trees on a piece of paper (Griesemer & Wimsatt, 1989) or the complex simulated "world" of artificial life (Helmreich, 1998), the material basis of the representation invokes its own rules, which in turn bear upon the scientific object in creative and challenging ways. In the *Visible Human Project*, for instance, a data set was created by finely slicing a frozen cadaver, photographing the cross-sections, and digitizing the images, generating a new body space with depth, volume, and colors that could be "flown through" (Waldby, 2000). Questions about the generalizability of this body space based on one individual led to a *Visible Woman Project*, and a *Visible Korean Project*, illustrating the intractable yet generative problems of universalistic projects (Cartwright, 1998). The *Visible Human Project* in turn serves as the basis for virtual simulations expected to train future generations of surgeons, creating additional questions about how to prepare them for the human variability they will face with "real" patients (Prentice, 2005). To use visualization, then, is to be disciplined by its spatial and epistemic standards (see Cussins, 1998).

Returning to the experimental engagement with scientific images, we can note their wild fecundity with respect to creativity and invention. Just as for models, paper tools, thought experiments, and diagrams, a key source of legitimacy for visualizations in science lies in their usefulness (Morgan & Morrison, 1999). Data volume alone often serves as the justification for visualizations as indispensable first steps in generating hypotheses. Learning to see with the help of diagrams and models has been documented throughout science and medicine (Dumit, 2004; Myers, forthcoming; Saunders, forthcoming). Cambrosio, Jacobi, and Keating's essay on "Ehrlich's 'Beautiful Pictures'" (1993) shows how a series of hand-drawn animations of antibodies were crucial in making such objects "visible" in the microscope. These hand-drawn images were themselves *epistemic creations*, essential tools in thinking, theorizing, and creating (see also Hopwood, 1999; Francoeur, 2000).

Attending to visualizations as *interactive* also requires attention to the researchers' bodily engagement with computers and other instruments. In direct contrast with a simplistic analysis that sees interaction on a computer screen as a form of disembodiment and of a virtual separation of person and object, Myers (forthcoming) found that protein crystallographers had to develop an intense form of embodied relationship with the complex three-dimensional (3D) objects on the screen in front them whose structure they were trying to "solve." In previous decades, 3D models of the same proteins were painstakingly built out of wire, wood, glass or plastic. But these had the physical disadvantage of being too heavy and unwieldy and were difficult to modify (Francoeur, 2002; de Chadarevian & Hopwood, 2004). Visual interactive expertise still required mentoring, Myers showed, but new forms of tacit knowledge included having good software hands and a 3D sense that crystallographers often expressed through contorting their bodies and minds in front of the screen.

DEPLOYMENT

Exploring deployment requires us to look at the trajectories of images leaving their production site, entering different social milieus, and interacting with different forms of knowledge. On the one hand, scientific images' persuasiveness depends on their being regarded as the simultaneous voice of technoscientific authority and as expressions of nature. On the other hand, the semiotic openness of images leaves many openings for contesting their meaning and calling into question their objective authority.

Outside the laboratories, scientific images intersect with a range of other items and images at a given time (Jordanova, 2004). These images and items, from science, art, mass culture, and digital media, converse with each other and with previous eras in conjuring meanings and generating significance for viewers. In *The Visible Woman: Imaging Technologies, Gender, and Science* (1998), editors Paula Treichler, Lisa Cartwright, and Constance Penley integrated feminist and cultural studies with STS by foregrounding medicine. In attempts to understand how diagnostic and public discourses interact, these essays examine the continuum of digital medical images, public health posters and films, advertisements, and photographs in order to disclose social inequalities, personal and political identities, and disciplinary and economic formations.

Images traverse scientific and nonscientific domains supporting prevailing metaphors and stories (Martin, 1987). Photographs, ultrasounds, videos, and other visualizations reinforce some narratives and disempower others. Following Emily Martin, we can understand how the so-called abstract graphs and images operate within codes that tell "very concrete stor[ies] rooted in our particular form of social hierarchy and control. Usually we do not hear the story, we hear the 'facts', and this is part of what makes science so powerful" (Martin, 1987: 197).

The lives of cells, for instance, are known to us through early attempts of graphical representation (cf. Ratcliff, 1999) and the visual narratives of film. Early cell microcinematography and motion pictures were developed in tandem, exchanging concepts, equipment, and styles (Landecker, 2005). Framing, long exposures, time-lapse photography, slow motion, and close-ups familiarize or defamiliarize our perception and understanding of events. Scientists manipulate these techniques to understand the processes in the first place and to persuade others of their reality. Similar visual tactics have been used to portray technical and scientific issues in the public sphere (Treichler, 1991; Hartouni, 1997; Sturken & Cartwright, 2001).

The journal *Nature,* for instance, has an ongoing discussion of the temptations of image manipulation (Pearson, 2005; Greene, 2005; Peterson, 2005; see also Dumit, 2004). This journal now requires researchers to explain exactly what photoshop filters and processes have been applied to published images. Many scientists complain about the unfairness of having to compete for grants or public support against "cool-looking projects" (Turkle et al., 2005). Given a top-tier visualization software program, for instance, even nonsignificant pilot data can look solid and complete (Dumit, 2000). Good simulations and visualizations are expensive, however, and given the constant increase in computer power and software complexity, the output of older computer programs looks dated even when the science behind it is cutting edge.

The antagonism between interpretative openness and persuasive authority can be well observed in scientific debates, court trials, and intersections of scientific visualizations of the body with experiential body knowledge. Visualizations of global temperature change have been attacked on the basis of data sources, selection of data, the algorithms used, how the analysis is organized, forms of presentation, exaggeration of conclusions, and captions. Oreskes (2003), Bowker (2005), Lahsen (2005), Edwards (forthcoming), among others, offer excellent STS analyses of how global climate change and biodiversity data are painstakingly produced through series of social and technical compromises and adjustments. Within the political terrain of questions concerning the reality and causes of climate change, however, the large amount of work necessary to produce images can be used against the authority of those images to show that an enemy's images are not proof but "only" constructed to look like it. Thus, STS work may be read both as rigorous analyses of large-scale capital and expert labor intensive visualizations and as strategy manuals for attacking complex data claims on the grounds of not being "pure science" enough.

The courts are another key site where visual authority is regularly and formally challenged. Exploring a famous contemporary criminal trial in the United States, Sheila Jasanoff (1998) analyzed the trial as an arena in which visual authority was created and defended. She showed how the judge's comments and rulings established whose vision would be considered as visual expertise, and in what circumstances lay vision could take precedence over expert sight. Jennifer Mnookin (1998) and Tal Golan (1998, 2004) have traced how photographs and then x-rays entered courtrooms in the United States, first under a cloud of suspicion, requiring their producers to be present to testify to their veracity. In these cases, the story told by the photographer was the evidence to be considered. Because x-rays imaged objects that were invisible to the naked eye, the x-ray image required an expert to interpret it while the jury looked at it. This category of demonstrative evidence (evidence that is shown, or demonstrated, to the jury) again highlights the interpretive tension in every image: representations are never completely self-explanatory, and they are polyvocal, or open to many meanings. Photographs, x-rays, and other medical images and computer visualizations of all kinds require captions and expert interpretations. In tension with this requirement, the relative power of images to convey stereotypical, expected, or conventional meanings is quite strong. We all think we know what a fractured bone on an x-ray should look like. This creates a visual and haptic rhetorical space in which images can convey meanings despite expert protest, and courts have to constantly manage this persuasive power of images.

The deployment of scientific images and persuasion are most striking perhaps when the images are of our own bodies and lives. Our bodies as objects of knowledge and perception are *educated* bodies, shaped by descriptions, drawings, and visualizations (Duden, 1991). We learn about our bodies during childhood and throughout our lives. This form of meta-learning Emily Martin calls "practicums," ways of learning that change how we think and act with regard to the "ideal and fit person" (Martin, 1994: 15). Using examples of cells endowed with personhood, sperm and egg stories told with diagrams, and micro-photos that are narrated, framed, and cut up, Martin emphasizes the disjunction between what is necessarily in the data and what can be done with the data.

Biomedical seeing is not only persuasive but also *entangling*. Scientific images of humans are images of us, they point deictically at us (Duden, 1993), telling us truths about ourselves. Images of disease—of viruses, bad genes, and abnormal brain scans— create and reinforce basic categories of personhood, of normal and abnormal. The everyday identification with these scientific images can be called "objective self-fashioning" (Dumit, 2004). Medical images of ourselves are deeply personal; they partake in diagnosing our illnesses and foretelling our fate. At the same time, medical images are fascinating and exciting, providing tales to tell. Biomedical technologies also materialize new types of bodies with visualizable interiors (Stacey, 1997). Such visualizations seem to imply that seeing equals curing (van Dijck, 2005). Public discourse about seeing-eye machines promises utopian futures, but rarely acknowledges how seldom these machines actually change clinical outcomes. Medical narratives are quite routinized in this respect (Joyce, 2005).

Ultrasound images, for example, have entered into the experience and trajectory of pregnancy. Feminist anthropologists and sociologists found that ultrasound imaging simultaneously provokes both hope and anxiety. The images "fast-forward" pregnancy, conveying a special reality of the fetus often before the pregnancy is physically felt by the mother (Rapp, 1998). But these insightful sociotechnical observations need to be carefully located historically and culturally. As Mitchell and Georges (1998) documented through comparative ethnographies, ultrasound functions quite differently in Greece and North America. Greek doctors do not show the images to mothers, for instance, while in the United States, images of ultrasound are often demanded, carried in wallets, posted on refrigerators. These images are also used in advertising and in antiabortion public relations campaigns. The latter images are particularly decontextualized, framing the autonomy of the fetus and therefore its personhood (Hartouni, 1997) and creating the image of the fetus as a patient and the mother therefore as womb and incubator (Casper, 1998). These brief examples show the importance of treating technoscience in each case as an ethnoscience (Morgan, 2000).

We still lack STS studies of the processes by which people are visually persuaded and of the deployment of scientific visual knowledge in other social milieus and with other forms of knowledge (though see Kress & van Leeuwen, 1996; Elkins, 1998; Sturken & Cartwright, 2001). However, there are more questions to be explored. The final section of this chapter outlines a research agenda for the social studies of scientific imaging and visualization in the future.

SOCIAL STUDIES OF SCIENTIFIC IMAGING AND VISUALIZATION: OUTLINING A RESEARCH AGENDA

One of the key problems in formulating and demarcating the field of SIV is locating and defining the specificity of the visual. How is the persuasiveness of visual images in science to be separated from the persuasiveness of textual arguments, numbers, models, and the like, especially when software allows the ready transformation and juxtaposition of these different forms? We thus need to study the status of images as "epistemic things" (Rheinberger, 1997) in the knowledge generation process: How do images serve as "boundary objects" (Star & Griesemer, 1989) and transgress disciplinary boundaries? How are symbolic meanings assigned to visual representations? How do images influence the way in which researchers think and look at things? How do images provoke changes in routine practices?

Jordanova's (2004) insight that models are *incomplete concepts* may be used as a springboard for analysis. Jordanova is referring to the creative openness of models: they are constraining and suggestive at the same time, "implying the existence of something else, by virtue of which the model makes sense. As a result there are interpretive gaps for viewers to fill in, the 'beholder's share' in Gombrich's words" (Jordanova, 2004: 446). If models are incomplete concepts, then perhaps visualizations can be thought of as *incomplete models*. Interactive visualizations are practically and immediately manipulable. There is thus a "programmer's share" in addition to a beholder's share, leaving them remarkably difficult to black-box. Researchers familiar with the field of visualization can not only recognize the programs used but also single out the great number of assumptions and algorithms deployed in their making. Visualization disputes in the climate sciences are exemplary in this regard. As Naomi Oreskes (2003) and Paul Edwards (forthcoming) have each shown, critics can quite easily and ably create *countervisualizations*, calling into question the validity of the models implied by the original ones. Another response to the openness of visualizations to manipulation is a movement among scientists to "open-source" their data, making the raw data available online for other experimental groups to download and analyze themselves.[2]

Thus, images and imaging technologies have an impact on the social organization, the institutional and disciplinary arrangements, the work culture (cf. Henderson, 1999), and the interactions between members of research communities. We need case studies of these disciplinary transformations and comparative investigations that might allow us to identify the specificity of the visual. A preliminary series of workshops conducted at MIT showed that scientists in a variety of disciplines found these topics important and worth pursuing (Turkle et al., 2005). We also need international comparisons that would complement local and historical studies (e.g., Pasveer, 1989; Anderson, 2003; Cohn, 2004; Acland, forthcoming). Ethnographic research on visual practices suggests there is very little about seeing, drawing, framing, imaging, and imagining that can be assumed to be the same across cultures (e.g., Eglash, 1999; Riles, 2001; Strathern, 2002; Prasad, 2005b).

The labor- and capital-intensive nature of imaging and visualization also requires more attention. Bourdieu's work on science and on art markets might be brought together to examine the symbolic capital of scientific images (Burri, forthcoming). Research along these lines is beginning in the fields of bioinformatics, geographic information systems (GIS), computer-generated imaging, and nanotechnology (e.g., Schienke, 2003; Fortun & Fortun, 2005). Some of this work has identified "hype," which depends in part on visual persuasiveness, as a crucial part of contemporary scientific authority (e.g., Milburn, 2002; Kelty & Landecker, 2004; Sunder Rajan, 2006). As science and technology become inextricably market- and marketing-oriented, we need more STS studies of advertising and public relations (e.g., Hartouni, 1997; Fortun, 2001; Hogle, 2001; Fischman, 2004; Greenslit, 2005).

Finally, the growing conversations and hybridization between STS and art, as a site where *counterimages* are being produced, need to be explored from the point of view of SIV. We think of counterimages as civic responses to the postulate that in some cases, the best exploration of and response to the rhetorical power of images may be visual. The series of STS-art exhibitions curated by Bruno Latour, Peter Weibel, Peter Galison, and others—including "Iconoclash" and "Making Things Public" at ZKM Karlsruhe (Latour & Weibel, 2002, 2005) and "Laboratorium" in Antwerp (Obrist & Vanderlinden, 2001)—are exemplary in this regard. Other examples are the projects of STS scholars working at the intersection of science and art.[3] Critical appraisals of all these works are an important task for SIV and STS more generally.

Notes

1. We distinguish SIV as one of the many approaches to the study of scientific images, which all together we understand as constituting the virtual field of Visual STS.

2. OME, 2005 (Open Microscopy Environment). Available at: http://www.openmicroscopy.org/.

3. For example, the works of Natalie Jeremijenko (available at: http://visarts.ucsd.edu/node/view/491/31); Chris Csikszentmihályi (available at: http://web.media.mit.edu/~csik/); Phoebe Sengers (available at: http://cemcom.infosci.cornell.edu/); Chris Kelty (available at: http://www.kelty.org/). See also the project "Information Technology and Creativity" of the U.S. National Academies (available at: http://www7.nationalacademies.org/cstb/project_creativity.html) and various initiatives at institutional levels, e.g., the Arts & Genomics Centre at the University of Amsterdam (available at: http://www.artsgenomics.org), the Artist residency project at the BIOS Centre (available at: http://www.lse.ac.uk/collections/BIOS/); the sciart project funded by the Wellcome Trust and others (available at: http://www.sciart.org); and collaborations between Caltech and Pasadena's Art Center College of Design.

References

Acland, Charles R. (forthcoming) "The Swift View: Tachistoscopes and the Residual Modern," in C. R. Acland (ed), *Residual Media: Residual Technologies and Culture* (Minneapolis: University of Minnesota Press).

Alpers, Svetlana (1983) *The Art of Describing: Dutch Art in the Seventeenth Century* (Chicago: University of Chicago Press).

Amann, Klaus & Karin Knorr Cetina ([1988]1990) "The Fixation of (Visual) Evidence," in M. Lynch & S. Woolgar (eds), *Representation in Scientific Practice* (Cambridge, MA: MIT Press): 85–121.

Anderson, Katharine (2003) "Looking at the Sky: The Visual Context of Victorian Meteorology," *British Journal for the History of Science* 36(130): 301–32.

Arnheim, Rudolf (1969) *Visual Thinking* (Berkeley: University of California Press).

Barley, Stephen R. (1984) *The Professional, The Semi-Professional, and the Machine: The Social Ramifications of Computer Based Imaging in Radiology,* Ph.D. diss., Massachusetts Institute of Technology.

Barthes, Roland (1977) *Image-Music-Text* (London: Fontana Press).

Beaulieu, Anne (2001) "Voxels in the Brain: Neuroscience, Informatics and Changing Notions of Objectivity," *Social Studies of Science* 31(5): 635–80.

Blume, Stuart S. (1992) *Insight and Industry: On the Dynamics of Technological Change in Medicine* (Cambridge, MA: MIT Press).

Bowker, Geoffrey C. (2005) *Memory Practices in the Sciences* (Cambridge, MA: MIT Press).

Breidbach, Olaf (2002) "Representation of the Microcosm: The Claim for Objectivity in 19th Century Scientific Microphotography," *Journal of the History of Biology* 35(2): 221–50.

Burri, Regula Valérie (2001) "Doing Images: Zur soziotechnischen Fabrikation visueller Erkenntnis in der Medizin," in B. Heintz & J. Huber (eds), *Mit dem Auge denken: Strategien der Sichtbarmachung in wissenschaftlichen und virtuellen Welten* (Wien, New York, and Zürich: Springer, Edition Voldemeer): 277–303.

Burri, Regula Valérie (forthcoming) "Doing Distinctions: Boundary Work and Symbolic Capital in Radiology," *Social Studies of Science.*

Burri, Regula Valérie (forthcoming) *Doing Images: Zur Praxis medizinischer Bilder.*

Cambrosio, Alberto, Daniel Jacobi, & Peter Keating (1993) "Ehrlich's 'Beautiful Pictures' and the Controversial Beginnings of Immunological Imagery," *Isis* 84(4): 662–99.

Cartwright, Lisa (1995) *Screening the Body: Tracing Medicine's Visual Culture* (Minneapolis: University of Minnesota Press).

Cartwright, Lisa (1998) "A Cultural Anatomy of the Visible Human Project," in P. A. Treichler, L. Cartwright, & C. Penley (eds), *The Visible Woman: Imaging Technologies, Gender and Science* (New York: New York University Press).

Casper, Monica J. (1998) *The Making of the Unborn Patient: A Social Anatomy of Fetal Surgery* (New Brunswick, NJ: Rutgers University Press).

Cohn, Simon (2004) "Increasing Resolution, Intensifying Ambiguity: An Ethnographic Account of Seeing Life in Brain Scans," *Economy and Society* 33(1): 52–76.

Crary, Jonathan (1990) *Techniques of the Observer: On Vision and Modernity in the Nineteenth Century* (Cambridge, MA: MIT Press).

Cussins, Charis M. (1998) "Ontological Choreography: Agency for Women Patients in an Infertility Clinic," in M. Berg & A. Mol (eds), *Differences in Medicine: Unraveling Practices, Techniques, and Bodies* (Durham, NC: Duke University Press): 166–201.

Daston, Lorraine & Peter Galison (1992) "The Image of Objectivity," in *Representations* 40: 81–128.

Daston, Lorraine & Peter Galison (2007): *Images of Objectivity* (New York: Zone Books).

Davidson, Arnold ([1996]1999) "Styles of Reasoning, Conceptual History, and the Emergence of Psychiatry," in M. Biagioli (ed), *Science Studies Reader* (New York: Routledge): 124–36.

de Chadarevian, Soraya & Nick Hopwood (2004) *Models: The Third Dimension of Science* (Writing Science) (Stanford, CA: Stanford University Press).

Duden, Barbara (1991) *The Woman Beneath the Skin: A Doctor's Patients in Eighteenth-century Germany* (Cambridge, MA: Harvard University Press).

Duden, Barbara (1993) *Disembodying Women: Perspectives on Pregnancy and the Unborn* (Cambridge, MA: Harvard University Press).

Dumit, Joseph (2000) "When Explanations Rest: 'Good-enough' Brain Science and the New Sociomedical Disorders," in M. Lock, A. Young, & A. Cambrosio (eds), *Living and Working with the New Biomedical Technologies: Intersections of Inquiry* (Cambridge: Cambridge University Press): 209–32.

Dumit, Joseph (2004) *Picturing Personhood: Brain Scans and Biomedical Identity* (Princeton, NJ: Princeton University Press).

Edwards, Paul (forthcoming) *The World in a Machine: Computer Models, Data Networks, and Global Atmospheric Politics* (Cambridge, MA: MIT Press).

Eglash, Ron (1999) *African Fractals: Modern Computing and Indigenous Design* (New Brunswick, NJ: Rutgers University Press).

Elkins, James (1994) *The Poetics of Perspective* (Ithaca, NY: Cornell University Press).

Elkins, James (1998) *On Pictures and the Words That Fail Them* (Cambridge: Cambridge University Press).

Elkins, James (forthcoming) *Six Stories from the End of Representation* (Stanford, CA: Stanford University Press).

Fishman, Jennifer R. (2004) "Manufacturing Desire: The Commodification of Female Sexual Dysfunction," *Social Studies of Science* 34(2): 187–218.

Fortun, Kim (2001) *Advocacy After Bhopal: Environmentalism, Disaster, New Global Orders* (Chicago: University of Chicago Press).

Fortun, Kim & Mike Fortun (2005) "Scientific Imaginaries and Ethical Plateaus in Contemporary U.S. Toxicology," *American Anthropologist* 107(1): 43–54.

Foucault, Michel ([1963]1973) *The Birth of the Clinic: An Archaeology of Medical Perception* (New York: Pantheon Books).

Francoeur, Eric (2000) "Beyond Dematerialization and Inscription: Does the Materiality of Molecular Models Really Matter?" *HYLE—International Journal for Philosophy of Chemistry* 6(1): 63–84.

Francoeur, Eric (2002) "Cyrus Levinthal, the Kluge and the Origins of Interactive Molecular Graphics," *Endeavour* 26(4): 127–31.

Fyfe, Gordon & John Law (eds) (1988) "Picturing Power: Visual Depiction and Social Relations," *Sociological Review Monograph* 35 (London and New York: Routledge).

Galison, Peter (1997) *Image and Logic: A Material Culture of Microphysics* (Chicago: University of Chicago Press).

Golan, Tal (1998) "The Authority of Shadows: The Legal Embrace of the X-Ray," *Historical Reflections* 24(3): 437–58.

Golan, Tal (2004) "The Emergence of the Silent Witness: The Legal and Medical Reception of X-rays in the USA," *Social Studies of Science* 34(4): 469–99.

Goodman, Nelson (1968) *Languages of Art: An Approach to a Theory of Symbols* (Indianapolis, IN: Bobbs-Merrill).

Greene, Mott T. (2005) "Seeing Clearly Is Not Necessarily Believing," *Nature* 435 (12 May): 143.

Greenslit, Nathan (2005) "Depression and Consumption: Psychopharmaceuticals, Branding, and New Identity Politics," *Culture, Medicine, and Psychiatry* 29(4).

Griesemer, James R. (1991) "Must Scientific Diagrams Be Eliminable? The Case of Path Analysis," *Biology and Philosophy* 6: 155–80.

Griesemer, James R. (1992) "The Role of Instruments in the Generative Analysis of Science," in A. Clarke & J. Fujimura (eds), *The Right Tools for the Job: At Work in Twentieth Century Life Sciences* (Princeton, NJ: Princeton University Press): 47–76.

Griesemer, James R. & William C. Wimsatt (1989) "Picturing Weismannism: A Case Study of Conceptual Evolution," in M. Ruse (ed), *What the Philosophy of Biology Is: Essays Dedicated to David Hull* (Dordrecht, Netherlands: Kluwer Academic Publishers): 75–137.

Hacking, Ian (1999) *The Social Construction of What?* (Cambridge, MA: Harvard University Press).

Haraway, Donna J. (1991) *Simians, Cyborgs, and Women: The Reinvention of Nature* (London: Free Association Books).

Haraway, Donna J. (1997) *Modest-Witness@SecondMillennium.FemaleMan-Meets-OncoMouse: Feminism and Technoscience* (New York: Routledge).

Hartouni, Valerie (1997) *Cultural Conceptions: On Reproductive Technologies and the Remaking of Life* (Minneapolis: University of Minnesota Press).

Henderson, Kathryn (1999) *On Line and on Paper: Visual Representations, Visual Culture, and Computer Graphics in Design Engineering* (Cambridge, MA: MIT Press).

Helmreich, Stefan (1998) *Silicon Second Nature: Culturing Artificial Life in a Digital World* (Berkeley: University of California Press).

Hogle, Linda F. (2001) "Chemoprevention for Healthy Women: Harbinger of Things to Come?" *Health* 5(3): 311–33.

Holtzmann Kevles, Bettyann (1997) *Naked to the Bone: Medical Imaging in the Twentieth Century* (New Brunswick, NJ: Rutgers University Press).

Hopwood, Nick (1999) "Giving Body to Embryos: Modeling, Mechanism, and the Microtome in Late Nineteenth-Century Anatomy," *Isis* 90: 462–96.

Hopwood, Nick (2005) "Visual Standards and Disciplinary Change: Normal Plates, Tables and Stages in Embryology," *History of Science* 43(141): 239–303.

Hubbard, Ruth (1988) "Science, Facts, and Feminism," *Hypatia* 3(1): 5–17.

Ihde, Don (1999) *Expanding Hermeneutics: Visualism in Science* (Chicago: Northwestern University Press).

Jasanoff, Sheila (1998) "The Eye of Everyman: Witnessing DNA in the Simpson Trial," *Social Studies of Science* 28(5–6): 713–40.

Jasanoff, Sheila (2004) "Heaven and Earth: The Politics of Environmental Images," in M. L. Martello & S. Jasanoff (eds) *Earthly Politics: Local and Global in Environmental Governance* (Cambridge, MA: MIT Press).

Jones, Caroline A., & Peter Galison (eds) (1998) *Picturing Science, Producing Art* (New York: Routledge).

Jordanova, Ludmilla (2004) "Material Models as Visual Culture," in S. D. Chadarevian & N. Hopwood (eds), *Models: The Third Dimension of Science* (Stanford, CA: Stanford University Press): 443–51.

Joyce, Kelly (2005) "Appealing Images: Magnetic Resonance Imaging and the Production of Authoritative Knowledge," *Social Studies of Science* 35(3): 437–62.

Joyce, Kelly (2006) "From Numbers to Pictures: The Development of Magnetic Resonance Imaging and the Visual Turn in Medicine," *Science as Culture* 15(1): 1–22.

Kaiser, David (2000) "Stick-figure Realism: Conventions, Reification, and the Persistence of Feynman Diagrams, 1948–1964," *Representations* 70: 49–86.

Kelty, Chris & Hannah Landecker (2004) "A Theory of Animation: Cells, Film and L-Systems," *Grey Room* 17 (Fall 2004): 30–63.

Knorr Cetina, Karin (1981) *The Manufacture of Knowledge: An Essay on the Constructivist and Contextual Nature of Science* (Oxford: Pergamon Press).

Knorr Cetina, Karin & Klaus Amann (1990) "Image Dissection in Natural Scientific Inquiry," *Science, Technology & Human Values* 15(3): 259–83.

Kress, Gunther R. & Theo Van Leeuwen (1996) *Reading Images: The Grammar of Visual Design* (London and New York: Routledge).

Lahsen, Myanna (2005) "Technocracy, Democracy and U.S. Climate Science Politics: The Need for Demarcations," *Science, Technology & Human Values* 30(1): 137–69.

Lammer, Christina (2002) "Horizontal Cuts & Vertical Penetration: The 'Flesh and Blood' of Image Fabrication in the Operating Theatres of Interventional Radiology," *Cultural Studies* 16(6): 833–47.

Landecker, Hannah (2005) "Cellular Features: Microcinematography and Early Film Theory," *Critical Inquiry* 31(4): 903–97.

Lane, K. Maria D. (2005) "Geographers of Mars: Cartographic Inscription and Exploration Narrative in Late Victorian Representations of the Red Planet," *Isis* 96(4): 477–506.

Larkin, Gerald V. (1978) "Medical Dominance and Control: Radiographers in the Division of Labour," *Sociological Review* 26(4): 843–58.

Latour, Bruno (1986) "Visualization and Cognition: Thinking with Eyes and Hands," *Knowledge and Society: Studies in the Sociology of Culture Past and Present* 6: 1–40.

Latour, Bruno ([1986]1990) "Drawing Things Together," in M. Lynch & S. Woolgar (eds), *Representation in Scientific Practice* (Cambridge, MA: MIT Press): 19–68.

Latour, Bruno (1987) *Science in Action. How to Follow Scientists and Engineers Through Society* (Cambridge, MA: Harvard University Press).

Latour, Bruno & Peter Weibel (eds) (2002) *ICONOCLASH: Beyond the Image Wars in Science, Religion and Art* (Cambridge, MA: MIT Press).

Latour, Bruno & Peter Weibel (eds) (2005) *Making Things Public: Atmospheres of Democracy* (Cambridge, MA: MIT Press).

Latour, Bruno & Steve Woolgar (1979) *Laboratory Life: The Construction of Scientific Facts* (Princeton, NJ: Princeton University Press).

Lefèvre, Wolfgang, Jürgen Renn, & Urs Schoepflin (eds) (2003) *The Power of Images in Early Modern Science* (Basel, Boston, and Berlin: Birkhäuser).

Lenoir, Timothy & Christophe Lécuyer (1995) "Instrument Makers and Discipline Builders: The Case of Nuclear Magnetic Resonance," *Perspectives on Science* 3(3): 276–345.

Locke, Simon (2005) "Fantastically Reasonable: Ambivalence in the Representation of Science and Technology in Super-Hero Comics," *Public Understanding of Science* 14(1): 25–46.

Lynch, Michael (1985a) *Art and Artifact in Laboratory Science: A Study of Shop Work and Shop Talk in a Research Laboratory* (London and New York: Routledge & Kegan Paul).

Lynch, Michael (1985b) "Discipline and the Material Form of Images: An Analysis of Scientific Visibility," *Social Studies of Science* 15(1): 37–66.

Lynch, Michael (1990) "The Externalized Retina: Selection and Mathematization in the Visual Documentation of Objects in the Life Sciences," in M. Lynch & S. Woolgar (eds), *Representation in Scientific Practice* (Cambridge, MA: MIT Press): 153–86.

Lynch, Michael (1998) "The Production of Scientific Images: Vision and Re-vision in the History, Philosophy, and Sociology of Science," *Communication & Cognition* 31: 213–28.

Lynch, Michael & Samuel Y. Edgerton Jr. (1988) "Aesthetics and Digital Image Processing: Representational Craft in Contemporary Astronomy," in G. Fyfe & J. Law (eds), *Picturing Power: Visual Depiction and Social Relations* (London and New York: Routledge): 184–220.

Lynch, Michael & Steve Woolgar (eds) (1990) *Representation in Scientific Practice* (Cambridge, MA: MIT Press).

Marchessault, Janine & Kim Sawchuk (eds) (2000) *Wild Science: Reading Feminism, Medicine, and the Media* (London and New York: Routledge).

Martin, Emily (1987) *The Woman in the Body: A Cultural Analysis of Reproduction* (Boston: Beacon Press).

Martin, Emily (1994) *Flexible Bodies: Tracking Immunity in American Culture from the Days of Polio to the Age of AIDS* (Boston: Beacon Press).

Métraux, Alexandre (ed) (2000) "Managing Small-Scale Entities in the Life Sciences," *Science in Context* 13(1).

Metz, Christian (1974) *Film Language: A Semiotics of the Cinema* (Oxford: Oxford University Press).

Metz, Christian (1982) *The Imaginary Signifier: Psychoanalysis and the Cinema* (Bloomington: Indiana University Press).

Milburn, Colin Nazhone (2002) "Nanotechnology in the Age of Posthuman Engineering: Science Fiction as Science," *Configurations* 10(2): 261–95.

Mitchell, W. J. T. (ed) (1980) *The Language of Images* (Chicago: University of Chicago Press).

Mitchell, W. J. T. (1987) *Iconology: Image, Text, Ideology* (Chicago: University of Chicago Press).

Mitchell, W. J. T. (1994) *Picture Theory: Essays on Verbal and Visual Representation.* (Chicago: University of Chicago Press).

Mitchell, Lisa M. & Eugenia Georges (1998) "Baby's First Picture: The Cyborg Fetus of Ultrasound Imaging," in R. Davis-Floyd & J. Dumit (eds), *Cyborg Babies: From Techno-sex to Techno-tots* (New York: Routledge): 105–24.

Mnookin, Jennifer (1998) "The Image of Truth: Photographic Evidence and the Power of Analogy," *Yale Journal of Law and the Humanities* 10: 1–74.

Mol, Annemarie & John Law (1994) "Regions, Networks and Fluids—Anemia and Social Topology," *Social Studies of Science* 24: 641–71.

Morgan, Lynn M. (2000) "Magic and a Little Bit of Science: Technoscience, Ethnoscience, and the Social Construction of the Fetus," in A. R. Saetnan, N. Oudshoorn, & M. Kirejczyk (eds), *Bodies of Technology: Women's Involvement with Reproductive Medicine* (Columbus: Ohio State University Press): 355–67.

Morgan, Mary S. & Margaret Morrison (eds) (1999) *Models as Mediators: Perspectives on Natural and Social Science* (Ideas in Context) (Cambridge: Cambridge University Press).

Myers, Greg (1990) *Writing Biology: Texts in the Social Construction of Scientific Knowledge* (Science and Literature Series) (Madison: University of Wisconsin Press).

Myers, Natasha (forthcoming) "Molecular Embodiments and the Body-work of Modeling in Protein Crystallography," *Social Studies of Science.*

Nelkin, Dorothy & Laurence Tancredi (1989) *Dangerous Diagnostics: The Social Power of Biological Information* (New York: Basic Books).

Obrist, Hans-Ulrich & Barbara Vanderlinden (eds) (2001) *Laboratorium* (Antwerp, Belgium: DuMont).

Oreskes, Naomi (2003) "The Role of Quantitative Models in Science," in C. D. Canham, J. J. Cole, & W. K. Lauenroth (eds), *The Role of Models in Ecosystem Science* (Princeton, NJ: Princeton University Press): 13–31.

Pasveer, Bernike (1989) "Knowledge of Shadows: The Introduction of X-ray Images in Medicine," *Sociology of Health and Illness* 11(4): 360–81.

Pasveer, Bernike (1993) "Depiction in Medicine as a Two-Way Affair: X-Ray Pictures and Pulmonary Tuberculosis in the Early Twentieth Century," in I. Löwy (ed), *Medical Change: Historical and Sociological Studies of Medical Innovation* (Paris: Colloques INSERM 220): 85–104.

Pearson, Helen (2005) "Image Manipulation CSI: Cell Biology," *Nature* 434 (21 April): 952–53.

Peterson, Daniel A. (2005) "Images: Keep a Distinction Between Beauty and Truth," *Nature* 435 (16 June): 881.

Porter, Theodore (1995) *Trust in Numbers: The Pursuit of Objectivity in Science and Public Life* (Princeton, NJ: Princeton University Press).

Prasad, Amit (2005a) "Making Images/Making Bodies: Visibilizing and Disciplining Through Magnetic Resonance Imaging (MRI)," *Science, Technology & Human Values* 30(2): 291–316.

Prasad, Amit (2005b) "Scientific Culture in the 'Other' Theater of 'Modern Science': An Analysis of the Culture of Magnetic Resonance Imaging Research in India," *Social Studies of Science* 35(3): 463–89.

Prentice, Rachel (2005) "The Anatomy of a Surgical Simulation," *Social Studies of Science* 35(6): 837–66.

Rajchman, John (1991) "Foucault's Art of Seeing," in *Philosophical Events: Essays of the '80s* (New York: Columbia University Press): 68–102.

Rapp, Rayna (1998) "Real-Time Fetus: The Role of the Sonogram in the Age of Monitored Reproduction," in G. L. Downey & J. Dumit (eds), *Cyborgs and Citadels: Anthropological Interventions in Emerging Sciences and Technologies* (Santa Fe, NM: School of American Research Press): 31–48.

Ratcliff, M. J. (1999) "Temporality, Sequential Iconography and Linearity in Figures: The Impact of the Discovery of Division in Infusoria," *History and Philosophy of the Life Sciences*, 21(3): 255–92.

Rheinberger, Hans-Jörg (1997) *Toward a History of Epistemic Things: Synthesizing Proteins in the Test Tube* (Stanford, CA: Stanford University Press).

Rheinberger, Hans-Jörg (1998) "Experimental Systems, Graphematic Spaces," in T. Lenoir (ed), *Inscribing Science: Scientific Texts and the Materiality of Communication* (Stanford, CA: Stanford University Press): 298–303.

Riles, Annelise (2001) *The Network Inside Out* (Ann Arbor: University of Michigan Press).

Rose, Nikolas S. (1999) *Powers of Freedom: Reframing Political Thought* (Cambridge: Cambridge University Press).

Rudwick, Martin J. S. (1976) "The Emergence of a Visual Language for Geological Science, 1760–1840," *History of Science*, 14: 149–95.

Ruse, Michael & Peter Taylor (eds) (1991) "Pictorial Representation in Biology," *Biology and Philosophy* 6(2): 125–294.

Saunders, Barry F. (forthcoming) *CT Suite: The Work of Diagnosis in the Age of Noninvasive Cutting* (Durham, NC: Duke University Press).

Schaffer, Simon (1998) "How Astronomers Mark Time," *Science in Context* 2(1): 115–45.

Schatzki, Theodore R., Karin Knorr Cetina, & Eike von Savigny (eds) (2001) *The Practice Turn in Contemporary Theory* (London and New York: Routledge).

Schienke, Erich W. (2003) "Who's Mapping the Mappers? Ethnographic Research in the Production of Digital Cartography," in M. Hård, A. Lösch, & D. Verdicchio (eds), *Transforming Spaces: The Topological Turn in Technology Studies* (Darmstadt, Germany: Online Conference Proceedings).

Stacey, Jackie (1997) *Teratologies: A Cultural Study of Cancer* (London and New York: Routledge).

Stafford, Barbara (1994) *Artful Science: Enlightenment, Entertainment, and the Eclipse of Visual Education* (Cambridge, MA: MIT Press).

Stafford, Barbara Maria (1996) *Good Looking: Essays on the Virtue of Images* (Cambridge, MA: MIT Press).

Star, Susan Leigh & James R. Griesemer (1989) "Institutional Ecology, 'Translations' and Boundary Objects: Amateurs and Professionals in Berkeley's Museum of Vertebrate Zoology, 1907–1939," *Social Studies of Science* 19(3): 387–420.

Strathern, Marilyn (2002) "On Space and Depth," in J. Law & A. Mol (eds), *Complexities: Social Studies of Knowledge Practices* (Durham, NC: Duke University Press).

Sturken, Marita & Lisa Cartwright (2001) *Practices of Looking: An Introduction to Visual Culture* (Oxford: Oxford University Press).

Sunder Rajan, Kaushik (2006) *Biocapital: The Constitution of Postgenomic Life* (Durham, NC: Duke University Press).

Traweek, Sharon (1997) "Iconic Devices: Toward an Ethnography of Physics Images," in G. L. Downey & J. Dumit (eds), *Cyborgs and Citadels: Anthropological Interventions in Emerging Sciences and Technologies* (Santa Fe, NM: School of American Research Press): 103–15.

Treichler, Paula A. (1991) "How to Have Theory in an Epidemic: The Evolution of AIDS Treatment Activism," in C. Penley & A. Ross (eds), *Technoculture* (Minneapolis: University of Minnesota Press): 57–106.

Treichler, Paula A., Lisa Cartwright, & Constance Penley (eds) (1998) *The Visible Woman: Imaging Technologies, Gender, and Science* (New York: New York University Press).

Tufte, Edward R. (1997) *Visual Explanations: Images and Quantities, Evidence and Narrative* (Cheshire, CT: Graphics Press).

Turkle, Sherry, Joseph Dumit, David Mindell, Hugh Gusterson, Susan Silbey, Yanni Loukissas, & Natasha Myers (2005) *Information Technologies and Professional Identity: A Comparative Study of the Effects of Virtuality*. Report to the National Science Foundation, Grant No. 0220347.

van Dijck, José (2005) *The Transparent Body: A Cultural Analysis of Medical Imaging* (Seattle: University of Washington Press).

Waldby, Catherine (2000) *The Visible Human Project: Informatic Bodies and Posthuman Medicine* (New York: Routledge).

Warwick, Andrew (2005) "X-rays as Evidence in German Orthopedic Surgery, 1895–1900," *Isis* 96: 1–24.

Yoxen, Edward (1987) "Seeing with Sound: A Study of the Development of Medical Images," in W. Bijker, T. Hughes, & T. Pinch (eds), *The Social Construction of Technological Systems: New Directions in the Sociology and History of Technology* (Cambridge, MA: MIT Press): 281–303.

14   Messy Shapes of Knowledge—STS Explores Informatization, New Media, and Academic Work

The Virtual Knowledge Studio[1]

The Internet is firmly entrenched in the daily lives of millions of people, faithfully reproducing existing patterns of power inequalities across the globe, including critical differences in access to the Internet itself (Castells, 2001; Wellman & Haythornthwaite, 2002; Woolgar, 2002b). Understanding the way people integrate the Internet into their routines and how they thereby construct what "the Internet" actually means and what it is about has attracted scholars from the very beginning of the Internet (Dutton, 1996; Ito, 1996; Walsh, 1996; Lazinger et al., 1997; Porter, 1997; OECD, 1998; Molyneux & Williams, 1999). This scholarship has, unsurprisingly, itself reproduced all conceivable positions with respect to the technology, its construction, and its impact on the world—from enthusiastic stories about the new life on the net (Rheingold, 1994; Turkle, 1995; Hauben & Hauben, 1997) to dark broodings about the undermining of civil society and civility due to an unfettered spread of either exhibitionism or privacy invasion, with empirically minded sociologists, anthropologists, and psychologists in between (Webster, 1995). These positions do not provide equal resources to Internet users for self-understanding. Techno-optimism is dominant. Technology critique is mobilized much more sparingly and often only in specific contexts (such as the fear of global Internet terrorism). Of course, users are regularly disappointed and turned off by their experience in surfing the Web (Henwood et al., 2002; Wyatt et al., 2002). This does not affect, however, the public discourse about the Internet so much—nor the sales of the hardware needed for Internet access—partly because there is always the next promise (Lewis, 2000), partly because these disappointments are backgrounded in the news media (Vasterman & Aerden, 1995). In the meantime, the chase for "the new, new thing" (Lewis, 2000) does have material effects that influence the technological, political, and economic ordering of life.

One of these effects is the iterative generation of various forms of "impact talk," a genre strongly related but not identical to particular forms of technological determinism (Bijker et al., 1989; Smith & Marx, 1994; Wyatt, 1998; Wyatt et al., 2000; see also Wyatt, chapter 7 in this volume). First-time users, as well as more advanced users, of a particular Internet-based tool can ground their experiences in a coherent story about "the impact of the Internet." Examples of this way of making sense of the Internet are very common. Grandparents are excited that they can e-mail their grandchildren overseas; librarians, having gotten used to the digital library technologies, are talking about the revolution that Google Scholar promises to deliver (finally?) in accessing scholarly information; the music industry is forced to reinvent itself in response to large-scale peer-to-peer file sharing that was impossible before the advent of the Internet; parents face new challenges in educating their young children in how to use and not to use chat channels; teachers experience the effects of cut-and-paste writing of their pupils and respond with monitoring software. In short, "impact talk" is not only a variety of philosophical perspectives—some more deterministic than others—it is also a practical discursive processing of embodied experience.[2]

"Impact talk" about the Internet is also paramount in academia with respect to the future of knowledge creation itself. Board members of traditional academies of science do not hesitate to speak about the revolutionary impact of the Internet on the conduct of science. The open-access movement has embraced the Internet as the ideal technology to undermine excessive monopolies of publishers on the provision of scientific information and to realize the old dream of the World Brain (for a recent review, see Drott, 2006). General science journals like *Nature* and *Science* regularly report on specific changes that have been made possible by the combination of research instrumentation with Internet-based communication tools in the physical and life sciences (Blumstein, 2000; Schilling, 2000; Sugden, 2002; Cech, 2003; Anon., 2004a,b, 2005a; Shoichet, 2004; Wheeler et al., 2004; Cohen, 2005; Giles, 2005; Marris, 2005; Merali & Giles, 2005; Santos et al., 2005; Walsh et al., 2005). New ways to construct the objects of research out of the combination of new and old data are key motifs here. Data sharing also promises to enable the tackling of old questions in new ways in fields as diverse as epidemiology, systems biology, world history, archeology, cognitive sciences, and language evolution (NRC, 1997, 1999; Arzberger et al., 2004; Colwell, 2002; Koslow, 2002; Marshall, 2002; Van House, 2002; Esanu & Uhlir, 2003; Kaiser, 2003; Wouters & Schröder, 2000). At the same time, data sharing stands for a different regime of managing data sets and data production and requires the mobilization of more specialized data expertise than most researchers can deliver (Wouters & Schröder, 2003). It also raises new issues in copyright law, privacy protection, quality control of scientific data, and the private-public partnerships. Every month, new Web sites are born that use new ways to communicate research results, data, and insights (*Science* and *Nature* keep track of them on a weekly basis). In fact, there is no aspect of research or research organization that does not seem to be somehow affected in the short-to-medium run by the combination of digital research instruments and the Internet (Walsh, 1996; OECD, 1998; Walsh, 1999). To sum up, these developments add up to the plausibility of the claim of a qualitatively new state of science in which informatization of research practices is central: e-science or cyberscience (Wouters, 2000; Beaulieu, 2001; Hey & Trefethen, 2002; Hine, 2002; Berman, 2003; Nentwich, 2003).

To a veteran STS scholar, these reports may not be very convincing, however. After all, the way empirical materials and facts are combined to produce a plausible story or vision of the future is never innocent and always deconstructible. And students of science and technology often profess to *hate* technological determinism and other forms of reification. Often, "impact talk" feels too close to these perspectives to be comfortably included in STS work. For the historically minded scholar, a common response has been to show historical precedents of the new, new thing. For example, the telegraph is a nice counterexample to the claimed novelty of the Internet (Woolgar, 2002b). This approach amounts to deconstructing the novelty of the claimed revolution, usually combined with a critical analysis of the purported effects. Another reaction to "impact talk" is to turn the claims about the novelty of the Internet into empirical questions. This has been done from the very beginning of the discussion about the Internet (for an overview of early work, see Dutton, 1996). This line of thinking has led to a large body of empirical work that has convincingly shown that some aspects of life have not been changed by the emergence and use of the Internet while other aspects have indeed been transformed but in a different way than claimed or expected (Wellman & Haythornthwaite, 2002). This perspective has also been productive in the discussion about the impact of information and communication technologies (ICT) on academic work (see for a recent example Gunnarsdottír, 2005). Thanks to this body of empirical work, we know quite a lot about the way the emergence of the Internet and the use of ICT has affected some aspects—but not others—of the process of knowledge production and circulation.

With some exceptions (to which we will turn later), most scholarly work about the Internet and/or informatization of academic research can be sorted in these three types of literature: impact talk, deconstruction of impact talk, and detailed empirical description. All three have produced interesting work that may be relevant to a large variety of audiences. All three have also displayed distinct intellectual problems or "deep troubles" (Collins, 2001).

Perhaps the most impressive "impact talk" analysis of informatization in science has been published by Nentwich (2003). His book is based on a large set of interviews with practicing researchers, combined with a review of the literature about the future of academic work in cyberspace. The analysis has basically reproduced actor's impact talk, placed within the context of trying to assess the potential of the Internet for science. To his credit, the author does not claim that the potential and the actual are identical but leaves this question open to follow-up empirical research. To support this research, the book proposes a three-step impact model. In the first step, ICT has an impact on the scholarly communication system. "I explain how ICT is actually shaping the move away from traditional science and research while, at the same time, developing further not least influenced by the development it has originally initiated" (Nentwich, 2003: 63–64). In the second step, these "ICT-induced changes," which together have led to "cyberscience" as qualitatively distinct from "traditional science," influence academia at large. This leads to changes in actors, structures, processes, and products. The third step consists of "indirect consequences" in the substance of research via three routes: methodology, work modes, and representation (Nentwich, 2003: 64). This model leads to a qualitative "trend extrapolation" (Nentwich, 2003: 480), in which the substitution, i.e., the more or less complete replacement of old ways of doing things by new cyber-tools, is expected across the board. To sum up, cyberscience is already a fact, and a deep one. "I hold that the increasing use of ICT in academia impacts on the very substance of what science and research produces" (Nentwich, 2003: 486). In the end, we have a linear storyline of increasing "cyberscience-ness."[3]

The problem with deconstruction of impact talk, on the other hand, can perhaps be characterized as "unworldliness." If both the woman in the street and eminent scientists agree that the Internet has made a huge difference for them, it may come across as snobbery and elitism to insist on the opposite on purely philosophical grounds.[4] Perhaps more importantly, this approach tends to draw attention away from the role of technology in research practices to such an extent that mediation technologies have become invisible in many STS analyses, even though mediation has been central to STS theorizing. This invisibility tends to be reproduced in STS textbooks (Jasanoff et al., 1995; Fuller, 1997; Sismondo, 2004).

However, there is a growing body of work that pays specific attention to the role of mediating technologies in scientific practice. For example, Cummings and Kiesler have found that the use of communication technology was not particularly useful in the coordination of multi-organizational research projects, although e-mail was used a lot and Web sites were common[5] (Cummings & Kiesler, 2005).[6] They conclude that new types of tools are needed. Shrum (2005) has developed a series of projects about science in "distant lands" and has recently called for more research on this. Bohlin (2005) has produced an analysis of scientific communication through the lens of the sociology of technology, emphasizing interpretative flexibility and variation across disciplines (see also Lazinger et al., 1997). Differences between disciplines is also a key theme of the work of the late Rob Kling (Kling & Lamb, 1996; Kling & McKim, 2000; Kling et al., 2003; Kling et al., 2002). Information infrastructures and databases have been studied with a focus on their construction in, and implications for, scientific practice (Fujimura & Fortun, 1996; Bowker & Star, 1999; Star, 1999; Van Horn et al., 2001; Bowker, 2005). The role of mediation technologies has also been the topic of historical research (e.g., Shapin & Schaffer, 1985).

Still, much work on the use of these technologies in scientific communities has been performed outside of STS, especially in the fields of social network analysis (Wasserman & Faust, 1994; Haythornthwaite, 1998; Matzat, 2004; Wellman, 2001), Internet studies (Jones, 1995; Howard, 2002), information science[7] (Barabási, 2001, 2002; Börner et al., 2005; Huberman, 2001; Thelwall, 2002a,b, 2003, 2005; Scharnhorst 2003), computer-supported collaborative work (Galegher & Kraut, 1992; Sharples, 1993), social informatics (Suchman, 1987; Hakken, 1999, 2003; Kling, 1999), and communication sciences (Jankowski, 2002; Jankowski et al., 2004; Lievrouw & Livingstone, 2002).

These studies, independently of how informative they are, often have a different theoretical ambition from STS. Although they teach us a lot about the past and present of, say, scientific publishing or the use of networked on-line microscopy, we do not necessarily gain much insight into what they mean for our understanding of knowledge making and the politics of research. This is a reason that we abstain in this chapter from an overview of what we presently know about the way new media and the Internet are being mobilized in scientific and scholarly practices. We instead try to make links between different bodies of work about the interaction between new media and academic work and the role of information and the Internet therein, to circumscribe and address particular theoretical and methodological concerns. This, we hope, will also enable us to discuss scholarly work that has moved beyond the impact/deconstruct/describe triad mentioned earlier.

THEORY AND THE INTERNET

Theoretically, the Internet and the Web seem the almost perfect materialization of key concepts in the constructivist tradition of STS, such as "seamless network," "translation," and "hybrid socio-technical networks." STS scholars may therefore be encouraged to keep using these concepts, since their empirical references have only become more visible and hence plausible. This is indeed promising but at the same time perhaps too easy. In fact, the near-perfect fit between the Internet and a variety of analytical concepts[8] should encourage us to question these concepts, the more so since they are also mobilized in the building of new sociotechnical infrastructures (Beaulieu & Park 2003). What are the implications of the use of STS concepts in the design of new infrastructures and practices for the politics of STS and indeed for the role of STS itself?

These are questions about the intersection of method and theory in STS. Charis Thompson's (2005) characterization of STS is relevant here. According to Thompson, an interest in the deep interdependence of nature and society is the most unifying element of the field. This is accompanied by a common methodological orientation.

Despite a good deal of reflexivity on the nature of data and its interpretation, there is a valorization of empirical data collection either by ethnography and participant observation, or by original contemporary or historical archival and document research . . . Synthetic, a priori, and purely interpretative methods, for example, are all viewed suspiciously if they are not bolstered by empirical work. Versions of empiricism and positivism, thought of as not requiring any interpretation (as advocated in some natural and social science methodologies), are viewed as equally suspect . . . STS thus opts for empirical methodologies that are nonetheless assumed to be interpretative. Because of the tight knit between methodological and theoretical concerns, STS often reads like empirical philosophy or an empirical case study that has been trotted out to make a theoretical point. (Thompson, 2005: 32)

This view has important implications for the study of mediation technologies in scholarly and scientific practice. If both actor and analyst perform their research in and through new digital media, this may affect not only actor practice (the topic of impact talk) but also, though perhaps less visibly, shape the theory-method intersection for the analyst. According to Timothy Lenoir, this amounts to a new epistemic regime:

Media inscribe our situation. We are becoming immersed in a growing repertoire of computer-based media for creating, distributing, and interacting with digitized versions of the world, media that constitute the instrumentarium of a new epistemic regime. In numerous areas of our daily activities, we are witnessing a drive toward the fusion of digital and physical reality; not the replacement by a model without origin or reality as Baudrillard predicted, but a new playing field of ubiquitous computing in which wearable computers, independent computational agent-artifacts, and material objects are all part of the landscape. (Lenoir, 2002: 28)

This comment has methodological consequences because new media are not transparent. "Media not only participate in creating objects of desire, they are desiring machines that shape us . . . Media inscribe our situation: it is difficult to see how we can teleport ourselves to some morally neutral ground" (Lenoir, 2002: 46). The same is true for the analyst. Since we cannot be teleported to unmediated ground, reflexive analysis of the implications of new media for our production of knowledge about the new media in knowledge creation is perhaps the only way to pay due attention to the new shapes of mediation in scholarly practices.

This sensitivity to mediation informs our exploration of two questions. First, what can we learn from emerging e-science practices for the study of science and scholarly practice and for theory and method in STS? Second, how can these insights be included in a critical interrogation of e-science? We therefore try to address the possibility of speaking about e-science and informatization while systematically breaking down the reification that is the continuing product of this speaking.

In the next, second, section we discuss the notion of scientific labor as a useful point of departure for the analysis of scientific practice. We distinguish two dimensions: labor as the source of user-value and labor as the source of exchange value. We specify these dimensions with respect to work in the context of e-science and the Internet. This brings us to the discussion, in the third section, of epistemic culture.

We think that this concept provides a productive framework to study emerging knowledge practices. We discuss to what extent e-research leads to a redefinition of our understanding of epistemic cultures. We also explore the implications for the notion of epistemic culture itself. The fourth section shifts from labor to the institutionalization of scientific labor in disciplinary formations and the role of information infrastructures therein. This brings us to disciplinary differences in the uptake of communication and information technologies. While disciplinary differences may seem rather commonsensical and old-fashioned, it is important to revisit them because of the universalizing aspects of e-science as an ideology. Disciplines, infrastructures, and institutions are also the basis for the reproduction of scientific labor and the generation of the identity of researchers and scholars. The last section will draw conclusions for a critical interrogation of informatization and e-science.

RESEARCH AS LABOR

In an industry first, Chemical Abstracts Service (CAS) demonstrated the delivery of chemical information, including structure, via live interaction using Blackberry and other handheld devices at the CAS European conference in Vienna this week. More than 20 handheld devices were used simultaneously by conference participants to retrieve hundreds of literature references as well as molecular structure and related data for specific substances in real time. CAS will be making this new mobile route to scientific databases, called CAS Mobile, available through its STN and SciFinder services in the near future.

This is certainly worthy of attention, but I hope to heaven that it doesn't mean we have to now make handheld device client software available to our patrons and support it as well! On the other hand, perhaps it means that it will all be Web compatible! (e-mail exchange, STS-L list, 20 January 2006)

This e-mail exchange contains in a nutshell some key problems of e-science with respect to the organization of scientific and technical labor. Before sketching the potential of taking labor as analytical point of departure, we need to become more intimate with the way e-science is being defined. In other words: what and where is it? This will lay the ground for a discussion of two different but perhaps complementary approaches to science as labor.

The term e-science is most popular in the United Kingdom, continental Europe, Australia, and some other parts of Asia. In the United States, other parts of Asia, and other parts of the Americas, the concept of cyberinfrastructure for research is more common. The difference between these terms is interesting. One stresses the practice of research, the other the infrastructural condition for that practice, but both concepts are understood to refer to a shared view of computationally intensive research as a qualitatively novel way of doing research. Since 2003–2004, the concept of e-research as the more generic capture of ICT-driven change of the organization and nature of research has become popular with most actors involved, sometimes in combination with the concept of cyberinfrastructure (see, for example, Goldenberg-Hart, 2004).

The UK E-Science Programme defines its topic as follows:

What is meant by e-Science? In the future, e-Science will refer to the large scale science that will increasingly be carried out through distributed global collaborations enabled by the Internet. Typically, a feature of such collaborative scientific enterprises is that they will require access to very large data collections, very large scale computing resources and high performance visualisation back to the individual user scientists. The Grid is an architecture proposed to bring all these issues together and make a reality of such a vision for e-Science. (U.K. Research Councils, 2001)

This concept of e-science stresses computational research, processing of huge data sets, video-conferencing, and collaborative research relying on digital communication channels. In terms of disciplines, physicists, computer scientists, life scientists, and some computational social sciences are dominant in e-science (Wouters & Beaulieu, 2006). Yet, the prospect of e-research has spread out across the entire scholarly community, including the interpretative social sciences and humanities. On October 15, 2004, over 100 leaders from higher education, libraries, and information technology gathered in Washington, D.C., at a forum entitled "E-Research and Supporting Cyberinfrastructure: A Forum to Consider the Implications for Research Libraries & Research Institutions," which was cosponsored by the Coalition for Networked Information (CNI) and by the Association of Research Libraries (ARL). The forum was addressed by

Atkins, author of a crucial 2003 NSF Blue Ribbon Report (Atkins et al., 2003). According to the report,

a new age has dawned in scientific and engineering research, pushed by continuing progress in computing, information and communication technology . . . The capacity of this new technology has crossed thresholds that now make possible a comprehensive "cyberinfrastructure" on which to build new types of scientific and engineering knowledge environments and organizations and to pursue research in new ways and with increased efficacy.

At the December 2004 ARL Forum, the chairperson of the Coalition for Networked Information, Clifford Lynch, stated that massive changes in scholarly practice are occurring across all disciplines. He argued that new practices, products, and modes of documenting and communicating research will have far-reaching implications for all organizations involved in managing the scholarly record and supporting the ongoing enterprise of scholarship, and that libraries in particular play a central role because they manage the record across time and across disciplines. These changes in scholarly practice will create profound changes throughout the entire system of scholarly communication, and a failure to put into place effective new support structures in response to these changes would pose tremendous risk to the enterprise of research and scholarship. "This is what is at stake when we consider how to lead our institutions in addressing these new needs," Lynch said at the Forum (Goldenberg-Hart, 2004).

Note how similar this future vision is to the prospected trajectory of cyberscience in Nentwich (2003). The implications of e-science as a prospective technology have been analyzed by Vann and Bowker (2006), building on Brown's analysis of the role of promise in science and technology policy (Brown & Michael, 2003). In the same book (Hine, 2006), an interrogation of future visions contrasting e-science as computational research with women's studies use of ICT has been developed by Wouters and Beaulieu (2006). We will come back to these issues in the last section, in which we try to sketch the outline of a critique of e-science. What concerns us here is the impact of emerging e-research investments, practices, and infrastructures on the organization of labor. The fact that librarians, researchers, information technology experts, and new hybrids such as bioinformaticians, neuroinformaticians, and visualization experts are all heavily involved, points to an important, not yet well studied phenomenon: a redefinition of what type of labor is needed to sustain these scientific and scholarly practices (but see Doing, 2004).

It is common for practitioners in science studies to refer to science as labor and to use the terms "work" and "labor" interchangeably to highlight the concrete actions undertaken by subjects in a process of knowledge creation—indeed, the process of knowledge creation itself. While science as labor in this specific sense is certainly an important sociological and historical category, it may be insufficient to the task of theorizing the mediated character of science. Instead, the distinction between two different standpoints may prove fruitful. In its attempt to emphasize the historical and social contingency of scientific knowledge, and of nature itself, much of the work on science in STS over the past three decades has been concerned to look at science as a "practice" (cf. Pickering, 1992; Star, 1992; Clark & Fujimura, 1992). STS research on practice, also often referred to as "the nitty gritty of scientific work," has highlighted the centrality of tools to knowledge production process. Here, the hope was to build an ecology of the contents of scientific knowledge as well as of the conditions of its production (Clark & Fujimura, 1992). As a concrete, situated activity, scientific work was taken to be highly mediated by the objects that enabled specific orientations to the scientists' tasks at hand. This emphasis on science as labor also had important implications for the methodologies and approaches to research in STS. Ethnographic approaches and detailed case studies of work done in particular locations became core tools for STS. Even issues of circulation of knowledge were posed in terms of adaptation to local conditions of work.

Although looking at scientific labor as practice in this sense is important and crucial to the development of methodologies that could grasp its mediated character, it is only one of its faces. In other words, the category of the "nitty gritty" of scientific labor as epistemic practice is only one dimension of scientific knowledge production as commodity-determined labor. Indeed, exclusive focus on the "nitty gritty" of scientific labor may reflect a "technicist" model of labor, which diverts researchers' attention from questions about the economic valorization of laboring action (Vann, 2004).[9] A second perspective is therefore important that enables scientific activity to be construed as a social process through which the market values of specific knowledge-producing efforts are achieved. Here, the labor of scientists emerges as an object of market exchange that is constructed through highly mediated cultural and institutional processes of qualification and inscription (cf. Callon et al., 2002; Vann, 2004). In other words, scientific labor is not only a process of constructing knowledge of objects; it is itself the result of objectification of is own practices through processes of market exchange. From the standpoint of this latter dimension of labor, which value-form theoretic Marxists refer to as its "exchange-value," the study of scientific practice would entail study of how scientific practice—as a valorized object of institutional knowledges—is itself produced (Vann & Bowker, 2006).

STS scholars are increasingly turning their attention to the ways in which information infrastructures enable the visibility and invisibility of "concrete" work in institutional settings to emerge (Star, 1999; Bowker & Star, 1999). Although much of this work does focus on the "informal, nitty gritty of largely invisible articulation work," it also brings our attention to the processes and politics of institutional coding, the ways in which the identities of institutional subjects are themselves constructed through modes of infrastructural recognition. This strand of STS literature is distinguishable from that strand which emphasizes science as epistemic practice because it actually has important affinities with another tradition of science studies from which studies of science as epistemic cultures distanced themselves: studies of scientific accounting and of scientific disciplines as labor market sectors.

Given these analytical transversals, the contemporary proliferation of digitally mediated scientific labor presents even more difficult analytical challenges than those which aim to explicate how digitization mediates "the nitty gritty of scientific practice." Indeed, when we intersect the analytical points made above with the capacities of digital techniques for mediating knowledge production, new challenges emerge for the study of the production and accountability of the identities of scientific laborers. In contexts in which technological interfaces and social and digital networks are increasingly broader and longer, many dialectical relationships, such as those of visibility and invisibility and of expert and nonexpert work, may be altered. Taking our hunches from those aspects of Marx's work that dealt with the commodity-determined labor as a duality of use-value and exchange-value, we can begin to orient ourselves to forms of implicit analytical reductionism in STS and to pursue new methodological vocabularies that bring us closer to the textures of contemporary digitally mediated scientific practice—both as site of knowledge production and as site of knowledge about production. This chapter therefore entails the methodological idea that the study of science as a process mediated by new information technologies requires a synthesis of conceptual resources that zoom in on labor in each of its dimensions because qualitatively different forms of mediation occur in each.

EPISTEMIC CULTURES IN E-SCIENCE

The notion of epistemic culture focuses on the machinery of knowledge production (Knorr Cetina, 1999). It emphasizes configurations of persons and objects rather than institutions or concepts. This approach has the advantage of making research practices and epistemic objects, rather than digital technology, central to the analysis. Research instrumentation is nevertheless important, since epistemic cultures and ensembles of research technologies define one another (Hackett, 2005: 822 note 4). The long tradition of studying the role of research instruments in STS is therefore a fruitful perspective to incorporate in the study of emerging e-science practices (Edge & Mulkay, 1976; Fleck, 1979; Fujimura, 1987; Clark & Fujimura, 1992; Fujimura, 1992; Fujimura & Fortun, 1996; Rheinberger, 1997; Benschop, 1998; Joerges & Shinn, 2001; Hackett et al., 2004; Price, 1984; Traweek, 1988; Zeldenrust, 1988).

An "epistemic cultures approach" enables an understanding of the multiple ways new technologies can be taken up, without foreclosing possibilities based on existing technological promises (Lenoir, 1997, 2002; Cronin, 2003). This approach therefore provides ways of addressing knowledge creation as varying configurations of practices around epistemic objects. Of particular value in Knorr Cetina's (1999) characterization is the inclusion of material, symbolic, and subjective worlds in her exploration of practice. Knorr Cetina carries out her task of redefining practice by the notion of knowledge objects as epistemic—as productive generators of activity, continually opening up and revealing new knowledge. This concept draws on Rheinberger's epistemic things, which Knorr Cetina defines as "any scientific objects of investigation that are at the centre of a research process and in the process of being materially defined." These objects are productive because of their defining characteristic, their ". . . changing, unfolding character . . . lack of object-ivity and completeness of being" (Knorr Cetina, 1999: 181).

As a framework to "work up" ethnographic analysis, three main categories are central in analyzing epistemic cultures. These correspond to concern with empirical, technological, and social aspects of an epistemic culture. The first typifies the way the empirical is brought in as an object of study, the second focuses on the performance of relations to the object through instruments, and the social dimension brings to the fore the emergence of relations between units in a field of knowledge (i.e., laboratories, individual researchers). In digitization, there is a reconfiguration of the object that results from repeated de-contextualization and re-contextualization steps, which feed on each other. Both decontextualization and recontextualization require work. To maintain globalized knowledge networks is particularly labor intensive—one reason for the renewed debate about the most appropriate funding model for scientific and scholarly work. Digitization affects all three dimensions of epistemic cultures through this process of de- and re-contextualization.

Epistemic cultures have been studied through "laboratory studies" involving observation, interviews, as well as the analysis of technical instruments and documents. This methodology can be adapted to the study of e-science. For example, the distributed aspects of e-science resemble the set-up of high-energy physics experiments studied by Knorr Cetina in which e-mail seemed to play an important coordinating role and computing and networked access to data are clearly present (see also Wouters & Reddy, 2003). The analyst can shape her approach based on these features of the epistemic practice, so that an "intermittently absent observer can always link up with the e-mail network, as physicists do" (Knorr Cetina, 1999: 23). The challenge of studying a center with outposts active at different times can also be met by a "scaling up" of ethnography, from one to many participant-observers. Other adaptations of traditional laboratory studies are likely to be needed, to deal not only with scale but also with mediation of laboratory work (Beaulieu, 2005; Hine, 2002), scientific communication (Fry, 2004; Hellsten, 2002), and interaction via and about infrastructure and knowledge circulation.

While it is worthwhile to use epistemic cultures to question the notion of e-science (Wouters, 2004, 2005; Wouters & Beaulieu, 2006), the framework of epistemic cultures itself may also be interrogated, in relation to e-science. It may be necessary to interrogate the concepts of epistemic cultures and see how the contexts of knowledge are shaped across cultures.

One way of investigating this issue is to consider the distinction that is foundational to Knorr Cetina's framework. Epistemic objects are not the same as the experimental systems that embed them into broader fields of scientific culture.[10] Furthermore, the relations between experiments and laboratories can be articulated in a variety of ways (via correspondence to the world, intervention in the world, or processing of signs) (Knorr Cetina, 1992). E-science may require a revisiting of these relations, since it troubles the distinction between technological objects and epistemic objects. Rheinberger states that the first are stable and black-boxed, whereas the second produce unexpected results when interrogated (Rheinberger, 1997). An object can also move (historically) from the first to the second category, so that what was once a discovery becomes a routine technique that enables other objects to be interrogated. In e-science, infrastructures for experiments can be databases, "grid" applications, or models for example. Bowker and Star have noted that electronic infrastructures may offer more flexibility with regard to the way data have to be aggregated in order to be managed (Bowker & Star, 1999). The following questions arise: Does it become more difficult to distinguish experimental objects from technological objects when they are in a similar medium, when the technologies and their content are continuous? How are such distinctions maintained?

Given the importance of this distinction between experiment and lab and its implications for methodological approaches, it is worth reflecting on the ways the case of e-science might or might not be covered by Knorr Cetina's typology of relations. In the archetypal laboratory of experimental science, the laboratory signals independence from the field, and the reconfiguration of the object as to what it is (or what can stand in for it), when it occurs, and where. When there is more "epistemic infrastructure" for experiments (modeling, data repositories, search capabilities; see Van Lente & Rip, 1998), the relation to the experiment may be different. For example, if the laboratory can be thought of as the removal from the field, it seems that the Internet can serve as both field and laboratory because of the mediating aspects of interactions on the Net (i.e., the Internet as place where people act, talk, etc., in a "natural" field but where these interactions also leave traces, so that they can be manipulated/measured/modeled as though they were taking place in a lab). This possible shift, between epistemic object and "setting" for experimentation can be seen as a complex version of the tension between an epistemic object and a technical object. Whereas Rheinberger argues that objects may move (historically) from one category to the other, Knorr Cetina argues for an unsettled status of technical objects as well (Knorr Cetina & Bruegger, 2000). Neither of these perspectives, however, fully accounts for the complex ways in which objects may shift, from technical to epistemic object and back again. Such shifts seem to characterize e-science work, where the distinctions between tool and object are not clear-cut. The ways in which these shifts occur, and the points of stabilization, are therefore important elements in understanding e-science.

The ongoing nature of shifts between technical and epistemic objects was particularly visible in a recently studied e-science project: a distributed database of genetic and epidemiological data (Ratto & Beaulieu, in press 2007). As the project team worked through the ethical, scientific, and technical issues involved in setting up this data infrastructure, new questions emerged. These questions, on such matters as bioinformatics techniques, methods of preservation of "wet" materials, and the best way to make disparate kinds of data comparable (e.g., comparing life style data to genotyped information) blended scientific and technical issues. Seen in this light, the work is less the application of technical tools to epistemic objects and more what Norris has called "tinkery business" (Norris, 1993) and Fujimura has characterized as the searching out of "doable problems" (Fujimura, 1987). Importantly, as the project progressed, the stabilization of various aspects of the database as either "scientific" or "technical" could be read as part of the disciplinary organization of the project (e.g., that the database infrastructure is a technical task for the bioinformaticians, whereas the composition of individual database records is the job of the genetic epidemiologists). This analysis points to the complex linkages between technical and epistemic objects and, as noted above, indicates that the choices about how to distinguish between them have important ramifications for the valuing of certain kinds of scientific labor. In addition, the stabilization of certain kinds of problems as "technical" was also related to the delegation of particular questions of trust or authority to various parts of the overall infrastructure, in some cases preempting other systems of due diligence. The tensions (and importance) of these choices were made obvious when technical systems such as password managers and public key encryption techniques were evaluated as possible additions to scientific oversight committees and informed consent procedures. While these technologies were not seen as replacing such systems, the possibility of such delegation is indicative of the transformation of a scientific (and social) problem into a technical issue that can have vast ramifications for the way science is carried out as well as ethically guided (Ratto & Beaulieu, in press 2007).

The social dimension of epistemic cultures is also useful for understanding possible shifts in the practice of science. The relations between units in a field have to do with the social structures in scientific work. For example, these relations can be exchanges of expert services between individuals, competition between laboratories or collaboration in experiments. One way in which relations between units have been mediated consists in the use of texts (articles and books). The status of texts as key units of scientific knowledge may be challenged in e-science, for example, when contributions to knowledge become constructions of or additions to databases or infrastructures (Bowker, 2000; Van House, 2002; Cronin, 2003). The economy of credit—for example, attribution of authorships of publications or their citation (Scharnhorst & Thelwall, 2005)—is likely to be different in e-science than in traditional sciences because social relations in e-science are also affected by the change from face-to-face to mediated communication/textuality (Hayles, 2002).

It is here that the issues involved in studying epistemic cultures bring us to questions about the institutionalization of scientific work. As noted above, the institutional configuration of e-science involves a great deal of mediation, and thus we need to pay attention to the effects of such mediation on the epistemics of representation, proof, the issue of trust, and the kinds of work that e-sciences involve, enable, distribute, emphasize, and conceal. Tracing the dimensions of scientific work requires linking the cognitive and associated practices of scientists "on the ground" to the particular institutional settings that both create and are created by them.

SCIENCE WITHOUT DISCIPLINE?

The previous section focuses on scientific practice through the lens of epistemic cultures. However, some of the issues it raises relate to the problem of disciplining scientific labor in academic institutions: "the structural and organizational conditions of scientific production" (Fuchs, 1992). The concept of the scholarly discipline is useful for exploring these issues in e-research, such as new modes of dissemination, the allocation of intellectual property, concentration of resources to leverage new forms of science, mutual dependencies, task uncertainty, and variation in work practices. Thus, the discipline as an analytic unit also introduces a comparative lens for understanding practice.

The concept of the scholarly discipline is useful for this exploration. Disciplines provide for the stabilization of heterogeneous and localized practices through an economy of practices (Lenoir, 1997). As Becher (2000) points out, the concept of academic discipline is not straightforward. Its study is a long-standing, although intermittently performed, tradition in STS (Becher, 1989; Heilbron, 2004; Lemaine et al., 1976; Weingart et al., 1983; Weingart, 2000; Whitley, 2000). The exploration of the performance of e-science labor as a professional activity may be enhanced by revisiting the notion of discipline and interdisciplinarity. This is particularly relevant, since e-science not only claims to be interdisciplinary but also promises to transform the core idea of disciplinarity. While the idea that e-science exists "beyond disciplines" might characterize the discourse about e-science better than its practice, transdisciplinarity is a strong element of e-research. In a way, it is the defining rationale of the cyberinfrastructure that must make e-science possible. The philosophy behind the Grid is crucial here (figure 14.1).

Presently, the user-scientist often has problems navigating the confusing mix of tools, data, interfaces, and protocols that make up current on-line computational and archival resources. Resolving this issue is one of the primary claims of e-science. By providing a single interface (typically Web-based) that "automatically" translates diverse databases and other informational resources through the use of a common middle layer of software often called "middleware," many e-science projects hope to standardize this diversity without losing track of the productive disciplinary and technical differences that create it. Thus, in these projects the idea of "middleware" is extended from being a purely technical solution to being seen as one possible answer to the problems to multidisciplinarity. Indeed, we can argue that an important selling point for e-science will be the extent to which complex problems can be tackled by research teams that are much more hybrid than currently possible:

In order to achieve the best results research depends increasingly on worldwide collaboration between scientists and use of distributed resources. The complexity of the science often requires multidisciplinary teams. Such collaborations of distributed scientists, brought together in a common solution-oriented goal, have been named 'Virtual Organizations (VO)'. Thus, the Grid has also been defined as an enabler for Virtual Organizations: Given the barriers to collaborative working without such technology, it is immediately clear that the quality of the e-Science environment available to such distributed teams is vital to their ability to collaborate and perform good science (Berg et al., 2003: 7).

However, this begs the question of how e-science visions and the technical possibility of sharing resources will relate to diverse research practices: common problem formulation, shared socio-cognitive styles, a shared scientific and technical language, and

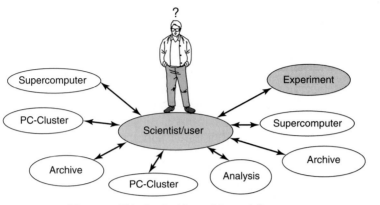

Often monolithic, "vertical," proprietary solutions

Through open, standard interfaces: flexible, adaptable,
interchangeable, multiple vendor solutions

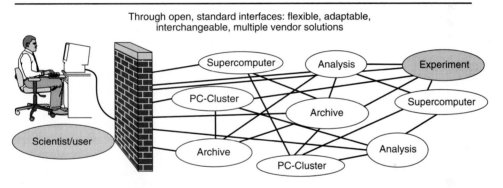

Figure 14.1
The future of e-research, from Berg, A., C. Jones, A. Osseyran, & P. Wielinga (2003) *e-Science Park Amsterdam* (Amsterdam: Science Park Amsterdam). Available at: http://www.wtcw.nl/nl/projecten/eScience.pdf.

enough commonality in reputational control mechanisms to allow researchers seriously to invest in collaboration and the creation and testing of new infrastructures. The rich STS tradition of the study of the institutionalization of research, mechanisms of reputational control, and the emergence and stabilization of disciplinary structures is relevant here. This body of work may be helpful to e-researchers, either in the natural or in the social sciences, who might otherwise run the risk of being end-users of an infrastructure, and its related diffusion model, imbued with a set of assumptions relating to what constitutes valid practice in one particular scientific tradition, namely, high-energy physics, that does not necessarily translate across epistemic cultures. To STS researchers, this perspective may clarify the potential of older elements of its theoretical and methodological tradition and protect against a progressivist model of STS methodology in which only the latest "turn" is worthy of attention.

Questions about reputation, communication, and new modes of authorship and authority are paramount in such explorations of disciplinarity and of e-science as labor. According to Whitley (2000), two main factors influence the reputational control over scientific labor: the degree of horizontal concentration of employment units, resources, audience, and reputations; and the vertical integration of goals, problems, and techniques. Vertical integration relates to the community-wide practices in epistemic cultures, whereas horizontal concentration relates to resource and accountability infrastructures. Reputational control determines how identities of scientific labor force are produced and can be highly influential in legitimating or marginalizing scientific labor.

A concrete example is the role of informal technical communication in technologically driven intellectual work. Knorr Cetina's notion of technical gossip provides one way of understanding the construction of researcher identities through reputational networks around the object of research. She describes technical gossip as an evaluative and personal discourse that interweaves "report, commentary and assessment regarding technical objects and regarding the relevant behavior of persons" (Knorr Cetina, 1999: 205). Reproducing a "personalized ontology" that transcends organizational boundaries such as experiments, experimental groups, and institutions, technical gossip is an informal infrastructure for development of interpersonal recognition. This is obviously not limited to intellectual communities that are organized around experimental or laboratory apparatus. Reflecting on the generation of new insights by geographers, Passmore (1998) stresses the centrality of gossip to the development of new ideologies. He observes that in geography, rumor communicated through the structure of personal networks is influential in building reputations and determining the position an individual holds in relation to the research front of a field. On the Internet, technical gossip may become embodied in specialized forms of communication such as Weblogs, Wikipedia, and collaboratories by which it may become more visible and perhaps acquire new roles (Hine, 2002; Mortensen & Walker, 2002, Thelwall & Wouters, 2005).

New modes of communication and quality control that challenge the traditional scientific communication system, such as digital preprint archives[11], open-access journals, and open peer review have emerged within a handful of disciplines whose work organization fits their perceived benefits. For example, Knorr Cetina's (1999) notion of technical gossip and of confidence pathways may explain saturation in the uptake of computer-mediated communication technologies for informal communication within the high-energy physics community, which has not been mirrored in other physical science disciplines, such as chemistry (Walsh & Bayma, 1996). Technical gossip and confidence pathways in experimental high-energy physics, albeit often formalized in collaboration-based quality control mechanisms (Traweek, 1988), create a high degree of interpersonal recognition that has partly displaced traditional forms of peer review. In intellectual fields where mutual dependence between scientists is not a necessary factor in making valid contributions to science and interpersonal recognition is a function of the formal publication system, rather than informal mechanisms such as "technical gossip" and "confidence pathways," open-access models of communication can be perceived as a threat to intellectual property (Fry, 2006b). Furthermore, data repositories, which are increasingly becoming linked to the submission of articles in visions of e-science knowledge management, threaten the notion of both authorship and ownership. This would explain why Kling et al. (2002) observed resistance to the digital preprint model of publishing by workers in medicine (see also Bohlin, 2004).

More generally, the role of e-journals and digital forms of scholarly communication and publication varies strongly by field (Kling & McKim, 2000; Fry, 2006a). This holds for the use of ICT in research as such. E-science is not oblivious to these disciplinary variations. On the contrary, because e-science represents a particularly capital-intensive mode of research, it is even more susceptible to the disciplinary and local regimes of control over resources and reputations.

In short, the role of scientific and scholarly disciplines and subdisciplines will become more rather than less important in the development of informational infrastructures. This makes a revisiting of older STS work on disciplines perhaps pertinent although we will have to devote more attention to the disunity of science (Lenoir, 1997).

INTERROGATING INFORMATIZATION AND E-SCIENCE

E-science is clearly not a unified phenomenon. On the one hand, it is a specific, local process emerging from cooperative work in computer science, physics, and hardware production. It makes possible new computational forms of research on huge and diverse data sets. This goal requires massive standardization of software tools, information infrastructures and databases, and embedded digital research instruments. It attempts to formalize information structures and operations in knowledge fields. E-science is highly capital intensive. It is supported by a discourse coalition of widely different actors. Its prospects are promoted in a future-oriented discourse and logic even in the face of, and perhaps precisely because of, its repeated shortcomings (Lewis, 2000; Vann & Bowker, 2006). It can be analyzed as a creative, yet colonizing movement occupying new territories in the sciences, social sciences, and humanities. On the other hand, e-science is embedded in a context of very diverse initiatives taken by individual scholars, librarians, artists, and amateurs. The boundary between e-science and this diffuse array of informatization action is fuzzy. This brings us back to the beginning of this chapter: in what ways are scholars actually incorporating the Internet and ICT in their practices?

One way to answer this question is to list the forms that e-research and cyber-stuff take. Our list includes Web-accessible databases; digital libraries; virtual research instruments; virtual reality objects; simulations; multi-sited games; Web surveys; videoconferencing; CUSeeMe cameras; speech technology; search engines; crawlers; network analysis tools; Web pages; annotated Web pages; Web site analysis; MUDs and MOOs; Wikipedia; Weblogs; Weblog analysis software; maps and map overlays; Google

Earth; Geographical Information Systems; semantic Web structures; portals; e-mail lists; multimedia publications; traditional publications as pdf files; history repositories; on-line digitized collections; clinical trial databases; Crackberry databases; specialized databases of annotated and standardized raw research data; ontologies; robotic agents; text parsing; artificial life forms; monitoring systems logging every action in cyberspace; spyware; podcasts; and standards, standards, standards.

Another way is to consider the way different fields have taken up the Internet either in their methodology or in their theoretical or topical research agendas. The most obvious move, of course, is to look at sciences that take Internet-related phenomena as research objects. This has been done on a truly huge scale in the social sciences, information and computer science, and the humanities. According to the editors of *Academia and the Internet:*

> The Internet and its impact on society has been a matter, quite appropriately, of focus by scholars across disciplines. We are not here assessing *whether* the Internet has had impact; it is a starting assumption of this book *that* it has had a substantial impact and has already affected people, societies and institutions. (Nissenbaum & Price, 2004: xi)

This expectation of impact has stimulated an increasing body of work with a huge variety of conceptualizations of the Internet, theoretical frameworks, and analytical questions.[12] This work includes, among others, studies of on-line social networks, virtual communities, identity formation in virtuality, the construction and analysis of the digital divide, trust and civic involvement (Barry, 2001), surveys of Internet use and time spent on-line, e-commerce and on-line auctions, the political economy of information (Mosco & Wasko, 1988; Shapiro & Varian, 1999; Lyman & Varian, 2000; David, 2004), distance learning, gender relationships, science in developing countries (Shrum, 2005), data practices, the World Wide Web and cultural theory (Herman & Swiss, 2000), on-line eroticism, and flex working. There is actually no good way to group this expanding work together except by the somewhat superficial observation that somehow something called or related to the Internet is involved in the research topic.

A more interesting question is whether the new topics of investigation resulting from the inclusion of Internet in everyday life have also influenced the conceptual or theoretical structure or apparatus of research fields studying them. This is far less clear. Obviously, the Internet has been taken up in the sociological theories of the information society (Webster, 1995; Castells, 1996; Slevin, 2000; Castells, 2001; Poster, 2001). In economics, there is a debate about the economics of information with a new emphasis on public goods and the commons (Mosco & Wasko, 1988; David, 2004). In cultural theory, scholars have taken the Web as the embodiment of postmodernism. Overall, the literature points to an incorporation of Internet issues within existing disciplinary structures instead of a proliferation of new fields (with Internet research as a possible exception although it is not yet clear whether this can be called a field in the traditional sense). "Throughout, however, accommodation and change occurred within traditional disciplines, and research concerning the Internet and its impact on society was established, to a greater or lesser extent, within existing debates, existing structures, and existing thematic approaches" (Nissenbaum & Price, 2004: x). Nissenbaum and Price go so far as to claim that the increased attention to the Internet has led to "a retreat back into disciplinary folds" compared with the interdisciplinary wave of work in the 1990s. This may also point to the resilience of existing disciplinary paradigms.

Researchers do report, however, an uptake of the Internet as platform for new social science and humanities methodologies. Early adopters were quick to establish the potential of the Internet for qualitative data collection. This may even be one of the defining elements of the field of Internet research: a shared fascination with the methodological potential of the Internet and/or the World Wide Web. It may explain how it is that the Association of Internet Research has come to bring together postmodernists focusing on intertextuality with social network analysts aiming to explain causal relationships in human networks, a rather unusual combination (Consalvo et al., 2004). Two examples may make this clear. Writing on hypertextuality, George Landow has proposed that digital text as a technology represents the embodiment of postmodernist theory (Landow, 1992). This perspective has been influential in cultural theories of the Internet and the Web, ranging from literary theories to explorations of aesthetics (Hjort, 2004). Landow sees hypertext as "the natural fulfillment" of postmodern literary theory. The literary theorist Marie-Laure Ryan takes a subtly different position. She does not see the convergence as a form of media determinism but as a reminder that "available technologies affect the use as well as the theorizing of already available technologies" (Ryan, cited in Hjort, 2004: 211). In other words, hypertext might be put to very different uses in a world in which the literary elite would value "plot, character and coherence." The key question is how scholarly and scientific methodologies are being influenced by mediating technologies. The second example embodies this in a different way. The Pew Internet and American Life project has surveyed the use of the Internet in the United States since March 1, 2000. It monitors the use, penetration, and appreciation of media in society. Methodologically, the project is innovative in its scale. By using the Internet as channel for the data collection, the project succeeds in collecting large representative samples with a diversity that enables a meaningful breakdown by gender, age, and Internet experience (Jankowski et al., 2004). Because it uses a daily sample design, the survey allows respondents to register fresh experiences and is therefore seen as more accurate than conventional surveys. Comparable use of the Internet as both a source of new data and a new source of old types of data has been reported in virtually all fields of the social sciences, information science, and the newly emerging field of Webometrics (Almind & Ingwersen, 1997; Aguillo, 1998; Ingwersen, 1998; Thelwall, 2000, 2004, 2005; Björneborn & Ingwersen, 2001; Scharnhorst, 2003; Scharnhorst et al., 2006).

To sum up, we can take it as empirically established that mediating technologies are influencing scholarly and scientific methodologies (see also Reips & Bosnjak, 2001). For example, Kwa has found that despite its uncertainties, modeling techniques have caught on in climate research because by use of these techniques climate can be visualized so effectively (Kwa, 2005). This claim is no "creeping technological determinism" (Lenoir, 2002), because the influence of technology on methods is driven by interaction embedded in practices. It would be too simple, however, to claim that new media need new methodologies, which would also amount to a form of method-oriented technological determinism. Most modifications in methodology, however interesting or promising, are based on already existing research designs and methods (Jankowski et al., 2004).

This brings us back to e-science in the more restricted sense. Methodological innovation is the central promise that e-science seems to hold for social scientists and humanists. A new body of work, sometimes labeled as e-social science and e-humanities, is currently being created. In 2005, the first International Conference on E-Social Science took place in Manchester, U.K., and it promises to become a yearly happening and showcase. In the United States and elsewhere, coalitions have formed to create new interactions between humanists and the digital,[13] trying to combine critical deconstruction with constructive development of new ways of performing scholarship (Ang & Cassity, 2004). Although the critical element is lively and well (e.g., Woolgar, 2002a), this body of work is dominated by tool development and infrastructure building (Proceedings, 2005; Anon., 2005b).

How can STS interact with this social science and humanities agenda, given its strength in empirically grounded theoretical work? We have tried to sketch some key questions that may inform a developing research agenda for STS.

Perhaps we should first point to the value of disrespecting the boundaries around e-science. By insisting on contextualized analysis and not accepting a narrow definition of e-science, STS may help infuse the debates about e-social science and e-humanities with discourses and experiences that would otherwise not become part of those debates. We think that it is also pertinent for STS inquiry itself that its agenda in this area is not restricted by dominant views on e-science and the future of research. In other words, let us keep things messy.

Second, we have tried to sketch some emerging analytical lines of work that may be of value here. We have shown how the concept of epistemic culture helps us to ask crucial questions about the networked practices that are increasingly bundled together under the notion of e-science. By paying attention to the role of epistemic objects and experimental settings, the core business of collectively producing inscriptions in scientific and scholarly research is highlighted. The notion of epistemic culture is a powerful one because it may bridge the analysis of day-to-day practice with the study of the processes of institutionalization that are based on and constrained by these practices. We also drew on the notion of disciplines as conservative institutions that carry a tension between the need to produce novel results on the one hand and the stability required to monopolize knowledges as markets. Digitally mediated knowledge practices seem to invite us to see both epistemic culture and discipline as two intertwined analytical perspectives, a stance that is not very common in STS—if only because the notion of epistemic culture has been developed on the basis of a critique of the notion of the discipline (Knorr Cetina, 1981).

We have also focused on the analysis of scientific labor, which includes *inter alia* the labor of technicians and support staff. On the one hand, we wish to zoom in on the net-*work*: the work that produces the networks and maintains them. The production as well as circulation of inscriptions is key here. In this analysis we find the epistemic culture a productive analytical device. On the other hand, renewed attention to the value production by scientific labor seems pertinent. Analyzing science as a value-producing and circulating process is especially productive because it enables the analysis of the creation and sustenance of markets for scientific results, expertise, and, not least, scientific labor itself.

This approach also relates to the analysis of inscription. Inscriptions do not move by themselves, nor are they self-producing. They are the product of labor. But it is in inscriptions that labor manifests itself, both in its capacity of producing use-value and in its capacity of producing exchange value. Increasingly, these traces are embedded in a digital medium that is itself composed of the same type of inscriptions, produced by similar labor at an earlier point in time. Therefore, institutionalization is itself the product of labor, and digital institutions are nothing but recurring patterns of circulation of inscription. Seen in this light, e-science may invite us to take a step beyond the received STS analysis of science as inscription activity (for a more extended discussion of this point, see Wouters, 2006). In Latour and Woolgar's (1979) analysis of the laboratory, scientists were obsessed by the frantic production of inscriptions. Research instruments were created to enable the large-scale, routinized production of these traces. Yet, the scientists themselves and their institutions were still separate from these inscriptions, although they derived their meaning and identity from them.

Informatization can be interpreted as the reflexive reinscription of research in and on itself. All actors are literally embodied in bundles of inscriptions that perform highly circumscribed operations on each other (Lenoir, 2002). We would like to stress that this is different from the notion of dematerialization. Neither human nor animal nor machine bodies disappear as performing work. On the contrary, the reflexive self-inscription has huge implications for what it means to be implicated in knowledge production. This attention to the semiotics of information, labor, and its material forms in digital practices and tools is, we think, an interesting emerging line of work in the history and sociology of science (Lenoir, 1997, 2002; Rheinberger, 1997; Kay, 2000; Thurtle & Mitchell, 2002; Beaulieu, 2003; Mitchell & Thurtle, 2004) that may inform ethnographic and historical case study work in STS.

Interaction between media and methodology is complex. It is not simply that new media need new methods. However, the new mediation technologies do influence methods and ways of working, including our own methods and work. We have tried to exemplify this by discussing how our analysis speaks back to the notion of epistemic culture. We have discussed how a crucial distinction in that analytical framework seems undermined in digital media: the distinction between epistemic object and experimental system. Perhaps more importantly, we suggest that this mode of analysis may help us understand better the interplay between scholarly identity, research infrastructures, and practice in and through the organization of labor. This may be a productive basis for critically interrogating informatization and e-research as both promising practice and problematic ideology.[14]

Notes

1. The Virtual Knowledge Studio is a research center of the Royal Netherlands Academy of Arts and Sciences, based in Amsterdam. This chapter was written by Paul Wouters, Katie Vann, Andrea Scharnhorst, Matt Ratto, Iina Hellsten, Jenny Fry, and Anne Beaulieu. E-mail: paul.wouters@vks.knaw.nl. Since November 2005, Fry has been at the Oxford Internet Institute, Oxford, U.K.

2. For a critical discussion of impact talk and a plea to focus rather on implications, see Woolgar (2002a).

3. An additional problem with generic impact talk is noted by Hakken (2003: 187): "Indeed, specialization is so extensive as to make very difficult any meaningful discourse on general knowledge creation."

4. John Zammito has characterized this as a problem for the value of empirical research in STS by quoting Willard Quine in his critical analysis of STS contructivism: "To disavow the very core of common sense, to require evidence for that which both the physicist and the man in the street accept as platitudinous, is no laudable perfectionism; it is a pompous confusion" (Zammito, 2004: 275).

5. The experience that there are at least as many technology-related impediments to knowledge creation as stimuli has also been noted by Hakken (2003: 203).

6. The role of ICT in scientific communication has been a focus in both STS and information studies. See, for example, Voorbij (1999), Cronin and Atkins (2000), Kling and McKim (2000), Borgman and Furner (2002), Fry (2004), Bohlin, (2004), Heimeriks (2005) and the Annual Review of Information Science and Technology (ARIST) series.

7. In information science, we can differentiate between analyses about the "impact" of ICT on traditional scholarly practices (collaboration, publishing behavior [Lawrence 2001, Wouters & de Vries 2004]), the emergence of new scholarly practices (e-mail, chat, on-line peer review), and new ways of studying scholarly practices (both using Web data [hyperlinks] as well as digitized bibliometric data [Chen & Lobo 2006]).

8. In a similar way, postmodern literary researchers were encouraged by the invention of hypertext (Landow, 1992). Social network analysts tend to see the Internet as a new source of social network data (Park, 2003) and sociologists claim that the Internet is both embodiment and proof of the thesis of the network society (Castells, 2001).

9. Vann (2004) has identified a similar "technicism" and analytical reduction on the part of some contemporary theorists of "immaterial" and/or "emotional" labor, and discusses its affinities with a particular strand of Marxian theory.

10. The interaction between the broader fields and locally configured action is not worked out fully by Knorr Cetina, but others have suggested how this might be investigated (Lynch, 1990; Beaulieu, 2005).

11. For an example, see the "endorsement policy" of the physics preprint archive (http://arxiv.org/new/) (17 January 2004):

> ArXiv was developed to be, and remains, a means for specific communities of scientists to exchange information. Moderators and the arXiv administrative team have worked behind the scenes to ensure that content is appropriate to the user communities. The growth in number of submissions to arXiv necessitates an automated endorsement system. Current members of arXiv scientific communities will have the opportunity to endorse new submitters. This process will ensure that arXiv content is

relevant to current research while controlling costs so we can continue to offer free and open Web access to all.

12. Recent reviews and exemplary cases of this work can be found in Wellman and Haythornthwaite (2002), Nissenbaum and Price (2004), Miller and Slater (2000), Abbate (1999), Slevin (2000), Bakardjieva and Smith (2001), Barnett et al. (2001), Castells (2001), DiMaggio et al. (2001), Poster (2001), Bar-Ilan and Peritz (2002), Henwood et al. (2002), Van Zoonen (2002), Barjak (2003), and Chadwick and May (2003). Only a relatively small part of this work has happened to come together in the series of conferences by the Association of Internet Researchers (AoIR) (Jones, 1995, 1999).

13. In the United States, the Humanities, Arts, Science, and Technology Advanced Collaboratory (HASTAC, pronounced "haystack") aims to promote the creative use of technology in the humanities and arts (http://www.hastac.org/). In Europe, several "computing and humanities" research centers have developed in the past decades, although with varying degrees of success. See for recent overviews of this approach in the literature Breure et al. (2004) and the Proceedings of the XVI Conference of the Association for History and Computing, 2005 (Anon., 2005b).

14. For example, the consequences of e-research for time management and speed control in research may be an interesting area of normatively oriented STS research. See Pels (2003) for a plea for "unhastening science."

References

Abbate, J. (1999) *Inventing the Internet* (Cambridge, MA: MIT Press).

Aguillo, I. F. (1998) "STM Information on the Web and the Development of New Internet R&D Databases and Indicators," in *Proceedings Online Information Meeting 98, London, Learned Information* 1998: 239–43.

Almind, T. & Ingwersen, P. (1997) "Informetric Analyses on the World Wide Web: Methodological Approaches to 'Webometrics,'" *Journal of Documentation* 53: 404–26.

Ang, I. & E. Cassity (2004) *Attraction of Strangers: Partnerships in Humanities Research* (Sydney: Australian Academy of the Humanities).

Anon. (2004a) "Making Data Dreams Come True," *Nature* 428: 239.

Anon. (2004b) "Virtual Observatory Finds Black Holes in Previous Data," *Nature* 429: 494–95.

Anon. (2005a) "Let Data Speak to Data," *Nature* 438: 531.

Anon. (2005b) *Humanities, Computers & Cultural Heritage*, Proceedings of the XVI Conference of the Association for History and Computing, Amsterdam.

Arzberger, P., P. Schroeder, A. Beaulieu, G. Bowker, K. Casey, L. Laaksonen, D. Moorman, P. Uhlir, & P. Wouters (2004) "Science and Government: An International Framework to Promote Access to Data," *Science* 303: 1777–78.

Atkins, D., K. Droegemeier, S. I. Feldman, H. Garcia-Molina, M. L. Klein, D. G. Messerschmitt, P. Messina, J. P. Ostriker, & M. H. Wright (2003) "Revolutionizing Science and Engineering Through Cyberinfrastructure: Report of the National Science Foundation Blue-Ribbon Advisory Panel on Cyberinfrastructure" (Washington, DC, National Science Foundation). Available at: http://dlist.sir.arizona.edu/897/.

Bakardjieva, M. & R. Smith (2001) "The Internet in Everyday Life—Computer Networking from the Standpoint of the Domestic User," *New Media & Society* 3(1): 67–83.

Barabási, A.-L. (2001) "The Physics of the Web," *Physics World,* July, Available at: http://physicsweb.org/articles/world/14/7/9/1.

Barabási, A.-L. (2002) *Linked: The New Science of Networks* (Cambridge, MA: Perseus Publishing).

Bar-Ilan, J. & B. C. Peritz (2002) "Informetric Theories and Methods for Exploring the Internet: An Analytical Survey of Recent Research Literature," *Library Trends* 50(3): 371–92.

Barjak, F. (2003) *The Internet in Public Science* (Brussels: SIBIS).

Barnett, G. A., B. Chon, H. Park, & D. Rosen (2001) "An Examination of International Internet Flows: An Autopoietic Model," presented at the annual conference of International Communication Association, May 2001, Washington, DC.

Barry, A. (2001) *Political Machines: Governing a Technological Society* (London: Athlone Press).

Beaulieu, A. (2001) "Voxels in the Brain: Neuroscience, Informatics and Changing Notions of Objectivity," *Social Studies of Science* 31(5): 635–80.

Beaulieu, A. (2003) "Review of Semiotic Flesh: Information and the Human Body," by Philip Thurtle & Robert Mitchell. Resource Center for Cyberculture Studies. Available at: http://rccs.usfca.edu/bookinfo.asp?ReviewID=271&BookID=217

Beaulieu, A. (2004) "From Brainbank to Database: The Informational Turn in the Study of the Brain," *Studies in History and Philosophy of Biological and Biomedical Sciences* 35(2): 367–90.

Beaulieu, A. (2005) "Sociable Hyperlinks: An Ethnographic Approach to Connectivity" in C. Hine (ed), *Virtual Methods: Issues in Social Research on the Internet* (Oxford: Berg): 183–98.

Beaulieu, A. & H. Park (eds) (2003) "Internet Networks: The Form and the Feel," *Journal of Computer Mediated Communication* 8(4), special issue. Available at: http://jcmc.indiana.edu/vol8/issue4/.

Becher, T. (1989) *Academic Tribes and Territories: Intellectual Inquiry and the Culture of Disciplines* (Buckingham, U.K.: SHRE & Open University Press).

Benschop, R. (1998) "What Is a Tachistoscope? Historical Explorations of an Instrument," *Science in Context* 11(1): 23–50.

Berg, A., C. Jones, A. Osseyran, & P. Wielinga (2003), *e-Science Park Amsterdam* (Amsterdam: Science Park Amsterdam). Available at: http://www.wtcw.nl/nl/projecten/eScience.pdf.

Berman F., G. Fox, & T. Hey (eds) (2003) *Grid Computing: Making the Global Infrastructure a Reality* (Chichester, West Sussex, U.K.: Wiley): 9–50.

Bijker, W. E., T. P. Hughes, & T. Pinch (eds) (1989) *The Social Construction of Technological Systems* (Cambridge, MA: MIT Press).

Björneborn, L., & P. Ingwersen (2001) "Perspectives of Webometrics," *Scientometrics,* 50(1): 65–82.

Blumstein, A. (2000) "Violence: A New Frontier for Scientific Research," *Science* 289: 545.

Bohlin, I. (2004) "Communication Regimes in Competition: The Current Transition in Scholarly Communication Seen Through the Lens of the Sociology of Technology," *Social Studies of Science* 34(3): 365–91.

Borgman, C. & J. Furner (2002) "Scholarly Communication and Bibliometrics," *Annual Review of Information Science and Technology* 36: 3–72.

Börner, K., L. Dall'Asta, W. Ke, & A. Vespignani (2005) "Studying the Emerging Global Brain: Analyzing and Visualizing the Impact of Co-Authorship Teams," *Complexity* 10(4): 57–67.

Bowker, G. (2000) "Biodiversity Datadiversity," *Social Studies of Science,* 30(5), 643–84.

Bowker, G. (2005) *Memory Practices in the Sciences* (Cambridge, MA: MIT Press).

Bowker, G. & S. L. Star (1999) *Sorting Things Out: Classification and its Consequences* (Cambridge, MA: MIT Press).

Breure, L., O. Boonstra, & P. Doorn (2004) "Past, Present and Future of Historical Information Science," *Historical Social Research/Historische Sozialforschung* 29(2): 4–132.

Brown, N. & M. Michael (2003) "A Sociology of Expectations: Retrospecting Prospects and Prospecting Retrospects," *Technology Analysis & Strategic Management* 15(1): 3–18.

Callon M., C. Méadel, & V. Rabeharisoa (2002) "The Economy of Qualities," *Economy and Society* 31(2): 194–217.

Castells, M. (1996) *The Rise of the Network Society* (Cambridge, MA: Blackwell).

Castells, M. (2001) *The Internet Galaxy: Reflections on the Internet, Business, and Society* (Oxford: Oxford University Press).

Cech, T. (2003) "Rebalancing Teaching and Research," *Science* 299: 165.

Chadwick, A. & C. May (2003) "Interaction between States and Citizens in the Age of the Internet: e-Government in the United States, Britain, and the European Union," *Governance* 16(2): 271–300.

Chen, C. & N. Lobo (2006) "Analyzing and Visualizing the Dynamics of Scientific Frontiers and Knowledge Diffusion," in C. Ghaoui (ed), *Encyclopedia of Human-Computer Interaction* (Hershey, PA: Idea Group Reference): 24–30.

Clark, A. & J. Fujimura, (1992) *The Right Tools for the Right Job: At Work in 20th-Century Life Sciences* (Princeton, NJ: Princeton University Press).

Cohen, J. (2005) "New Virtual Center Aims to Speed AIDS Vaccine Progress," *Science* 309: 541.

Collins, R. (2001) *The Sociology of Philosophies: A Global Theory of Intellectual Change* (Cambridge, MA: Belknap Press of Harvard University Press).

Colwell, R. (2002) "A Global Thirst for Safe Water: The Case of Cholera," 2002 Abe Wolman Distinguished Lecture, National Academies of Science, 25 January 2002, Washington, DC.

Consalvo, M., N. Baym, J. Hunsinger, K. B. Jensen, J. Logie, M. Murero, et al. (2004) *Internet Research Annual: Selected Papers from the Association of Internet Researchers Conferences 2000–2002 (Digital Formations, 19)* (New York: Peter Lang Publishing Group).

Cronin, B. (2003) "Scholarly Communication and Epistemic Cultures, Keynote Address, Scholarly Tribes and Tribulations: How Tradition and Technology are Driving Disciplinary Change" (Washington, DC: ARL).

Cronin, B. & H. B. Atkins (2000) *The Web of Knowledge: A Festschrift in Honor of Eugene Garfield* (Medford, NJ: Information Today).

Cummings, J. & S. Kiesler (2005) "Collaborative Research Across Disciplinary and Organizational Boundaries," *Social Studies of Science* 35(5): 703–22.

David, P. (2004) "Economists and the Net: Problems of a Policy for a Telecommunications Anomaly," in M. E. Price & H. Nissenbaum (eds), *The Academy & the Internet* (New York: Peter Lang Publishing Group): 142–68.

De Boer, H., J. Huisman, A. Klemperer, B. van der Meulen, G. Neave, H. Theisens, et al. (2002) *Academia in the 21st Century: An Analysis of Trends and Perspectives in Higher Education and Research* (Den Haag, Netherlands: Adviesraad voor het Wetenschaps-en Technologiebeleid).

DiMaggio, P., E. Hargittai, W. R. Neuman, & J. P. Robinson (2001) "Social Implications of the Internet," *Annual Review of Sociology* 27: 307–36.

Doing, P. (2004) " 'Lab Hands' and the 'Scarlet O': Epistemic Politics and (Scientific Labor)," *Social Studies of Science* 34(3): 299–324.

Drott, C. M. (2006) "Open Access," *Annual Review of Information Science and Technology* 40: 79–109.

Dutton, W. H. (1996) *Information and Communication Technologies: Visions and Realities* (Oxford: Oxford University Press).

Edge, D. O. & M. J. Mulkay (1976) *Astronomy Transformed: The Emergence of Radio Astronomy in Britain* (New York: Wiley).

Esanu, J. & P. Uhlir (2003) *The Role of Scientific and Technical Data and Information in the Public Domain* (Washington, DC: National Academies Press).

Fleck, L. (1979) *Genesis and Development of a Scientific Fact* (Chicago: University of Chicago Press).

Fry, J. (2004) "The Cultural Shaping of ICTs within Academic Fields: Corpus-based Linguistics as a Case Study," *Literary and Linguistic Computing* 19(3): 303–19.

Fry, J. (2006a) "Coordination and Control across Scientific Fields: Implications for a Differentiated e-science," in C. Hine (ed), *New Infrastructures for Knowledge Production: Understanding e-science* (Hershey, PA: Idea Group).

Fry, J. (2006b) "Scholarly Research and Information Practices: A Domain Analytical Approach," *Information Processing & Management* 42: 299–316.

Fuchs, S. (1992) *The Professional Quest for Truth: A Social Theory of Science and Knowledge* (Albany: State University of New York Press).

Fujimura, J. (1987) "The Construction of Doable Problems in Cancer Research," *Social Studies of Science* 17(2): 257–93.

Fujimura, J. (1992) "Crafting Science: Standardized Packages, Boundary Objects, and Translation," in A. Pickering, *Science as Practice and Culture* (Chicago and London: University of Chicago Press): 168–211.

Fujimura, J. & M. Fortun (1996) "Constructing Knowledge Across Social Worlds: The Case of DNA Sequence Databases in Molecular Biology," in L. Nader, *Naked Science: Anthropological Inquiry into Boundaries, Power and Knowledge* (New York: Routledge): 160–73.

Fuller, S. (1997) *Science* (Minneapolis: University of Minnesota Press).

Galegher, J. & R. Kraut (1992) *Computer-mediated Communication and Collaborative Writing: Media Influence and Adaptation to Communication Constraints,* Computer Supported Cooperative Work (Toronto: ACM Press).

Giles, J. (2005) "Online Access Offers Fresh Scope for Bud Identification," *Nature* 433: 673.

Goldenberg-Hart, D. (2004) "Libraries and Changing Research Practices: A Report of the ARL/CNI Forum on E-Research and Cyberinfrastructure," Associations of Research Libraries (ARL) Bimonthly Report No. 237.

Gunnarsdottír, K. (2005) "Scientific Journal Publications: On the Role of Electronic Preprint Exchange in the Distribution of Scientific Literature," *Social Studies of Science* 35(4): 549–80.

Hackett, E. (2005), "Essential Tensions. Identity, Control and Risk in Research," *Social Studies of Science* 35(5): 787–826.

Hackett, E., D. Conz, J. Parker, J. Bashford, & S. DeLay (2004) "Tokamaks and Turbulence: Research Ensembles, Policy and Technoscientific Work," *Research Policy* 33(5): 747–67.

Hakken, D. (1999) *Cyborgs@Cyberspace? An Ethnographer Looks to the Future* (New York, Routledge).

Hakken, D. (2003) *The Knowledge Landscapes of Cyberspace* (New York, Routledge).

Hauben, R. & M. Hauben (1997) *Netizens: On the History and Impact of Usenet and the Internet* (New York: Wiley).

Hayles, N. K. (2002) "Material Metaphors, Technotexts, and Media-specific Analysis" in *Writing Machines* (Cambridge, MA: MIT Press): 18–33.

Haythornthwaite, C. & B. Wellman (1998) "Work, Friendship, and Media Use for Information Exchange in a Networked Organization," *Journal of the American Society for Information Science* 49(12): 1101–14.

Heilbron, J. (2004) "A Regime of Disciplines: Towards a Historical Sociology of Disciplinary Knowledge," in C. Camic & H. Joas (eds), *The Dialogical Turn: New Roles for Sociology in the Postdisciplinary Age* (Lanham, MD: Rowman & Littlefield), 23–42.

Heimeriks, G. (2005) "Knowledge Production and Communication in the Information Society: Mapping Communications in Heterogenous Research Networks," PhD diss., University of Amsterdam.

Hellsten, I. (2002) *The Politics of Metaphor: Biotechnology and Biodiversity in the Media* (Tampere, Finland: Acta Universitatis Tamperensis; 876, Tampere University Press).

Henwood, F., S. Wyatt, A. Hart, & J. Smith (2002) "Turned on or Turned off? Accessing Health Information on the Internet," *Scandinavian Journal of Information Systems* 14(2): 79–90.

Herman, A. & T. Swiss (eds) (2000) *The World Wide Web and Contemporary Cultural Theory* (New York and London: Routledge).

Hey, T. & A. E. Trefethen (2002) "The UK e-Science Core Programme and the Grid," *Future Generation Computer Systems* 18(8): 1017–31.

Hine, C. (2002) "Cyberscience and Social Boundaries: The Implications of Laboratory Talk on the Internet," *Sociological Research Online* 7(2), U79–U99.

Hine, C. (ed) (2006) *New Infrastructures for Knowledge Production: Understanding e-science* (Hershey, PA: Idea Group).

Hjort, M. (2004) "Aesthetic Approaches to the Internet and New Media," in M. E. Price & H. Nissenbaum (eds), *The Academy & the Internet* (New York: Peter Lang Publishing Group): 229–61.

Howard, P. (2002) "Network Ethnography and the Hypermedia Organization: New Organizations, New Media, New Methods," *New Media & Society* 4(4): 551–75.

Huberman, B. A. (2001) *The Laws of the Web: Patterns in the Ecology of Information* (Cambridge, MA: MIT Press).

Ingwersen, P. (1998) "The Calculation of Web Impact Factors," *Journal of Documentation* 54(2): 236–43.

Ito, M. (1996) "Theory, Method, and Design in Anthropologies of the Internet," *Social Science Computer Review* 14(1): 24–26.

Jankowski, N., S. Jones, K. Foot, P. Howard, R. Mansell, S. Schneider, & R. Silverstone (2004) "The Internet and Communication Studies," in M. E. Price & H. Nissenbaum (eds), *The Academy and the Internet* (New York: Peter Lang Publishing Group): 197–228.

Jankowski, N. W. & O. Prehn (eds) (2002) *Community Media in the Information Age: Perspectives and Prospects* (Cresskill, NJ: Hampton Press).

Jasanoff, S., G. Merkle, J. Petersen, & T. Pinch (eds) (1995) *Handbook of Science and Technology Studies* (Thousand Oaks, CA: Sage).

Joerges, B. & T. Shinn (2001) *Instrumentation Between Science, State, and Industry* (Dordrecht, Netherlands: Kluwer Academic Publishers).

Jones, S. G. (1995) *Cybersociety* (London: Sage).

Jones, S. (1999) *Doing Internet Research: Critical Issues and Methods for Examining the Net* (Thousand Oaks, CA: Sage).

Kaiser, J. (2003) "NIH Sets Data Sharing Rules," *Science* 299: 1643.

Kay, L. E. (2000) *Who Wrote the Book of Life? A History of the Genetic Code* (Stanford, CA: Stanford University Press)

Kling, R. (1999) "What is Social Informatics and Why Does It Matter?" *D-lib Magazine* 5(1). Available at: http://www.dlib.org/dlib/january99/kling/01kling.html.

Kling, R. & R. Lamb (1996) "Analyzing Visions of Electronic Publishing and Digital Libraries," in G. B. Newb & R. M. Peek (eds), *Scholarly Publishing: The Electronic Frontier* (Cambridge, MA: MIT Press): 17–54.

Kling, R. & G. McKim (2000) "Not Just a Matter of Time: Field Differences and the Shaping of Electronic Media in Supporting Scientific Communication," *Journal of the American Society for Information Science* 51(14): 1306–20.

Kling, R., G. McKim, & A. King (2003) "A Bit More to IT: Scholarly Communication Forums as Sociotechnical Interaction Networks," *Journal of the American Society for Information Science and Technology* 54(1): 47–67.

Kling, R., L. Spector, & G. McKim (2002) "Locally Controlled Scholarly Publishing via the Internet: The Guild Model," *Journal of Electronic Publishing* 8(1). Available at: http://www.press.umich.edu/jep/08-01/kling.html.

Knorr Cetina, K. (1981) *The Manufacture of Knowledge: An Essay on the Constructivist and Contextual Nature of Science* (Oxford: Pergamon).

Knorr Cetina, K. (1992) "The Couch, the Cathedral, and the Laboratory: On the Relationship between Experiment and Laboratory Science," in A. Pickering (ed), *Science as Practice and Culture* (Chicago and London: University of Chicago Press): 113–38.

Knorr Cetina, K. (1999) *Epistemic Cultures: How the Sciences Make Knowledge* (Cambridge, MA: Harvard University Press).

Knorr Cetina, K. & U. Bruegger (2000) "The Market as an Object of Attachment: Exploring Postsocial Relations in Financial Markets," *Canadian Journal of Sociology* 25(2): 141–68.

Koslow, S. H. (2002) "Sharing Primary Data: A Threat or Asset to Discovery?" *Nature Reviews Neuroscience* 3(4): 311–13.

Kwa, C. (2005) "Local Ecologies and Global Science: Discourses and Strategies of the International Geosphere-Biosphere Programme," *Social Studies of Science* 35(6): 923–50.

Landow, G. P. (1992) *Hypertext 2.0: The Convergence of Contemporary Critical Theory and Technology* (Baltimore and London: Johns Hopkins University Press).

Latour, B. & S. Woolgar (1979) *Laboratory Life: The Social Construction of Scientific Facts* (Beverly Hills, CA: Sage).

Latour, B. & S. Woolgar (1986) *Laboratory Life: The Construction of Scientific Facts*, 2nd ed. (Princeton, NJ: Princeton University Press).

Lawrence, S. (2001) "Online or Invisible?" *Nature* 411: 521.

Lazinger, S., J. Bar-Ilan, & B. Peritz (1997) "Internet Use by Faculty Members in Various Disciplines: A Comparative Case Study," *Journal of the American Society for Information Science* 48(6): 508–18.

Lemaine, G., R. MacLeod, M. Mulkay, & P. Weingart (1976) *Perspectives on the Emergence of Scientific Disciplines* (The Hague: Mouton-Aldine).

Lenoir, T. (1997) *Instituting Science: The Cultural Production of Scientific Disciplines* (Stanford, CA: Stanford University Press).

Lenoir, T. (2002) "The Virtual Surgeon," in P. Turtle & P. Howard (eds), *Semiotic Flesh: Information and the Human Body* (Seattle: University of Washington Press): 28–51.

Lewis, M. (2000) *The New New Thing: A Silicon Valley Story* (New York, Norton).

Lievrouw, L. & S. Livingstone (2002) *The Handbook of New Media: Social Shaping and Consequences of ICTs* (London: Sage).

Lyman, P. & H. R. Varian (2000) *How Much Information?* Available at: http://www2.sims.berkeley.edu/research/projects/how-much-info-2003/.

Marris, E. (2005) "Free Genome Databases Finally Defeat Celera," *Nature* 435: 6.

Marshall, E. (2002) "DATA SHARING: Clear-Cut Publication Rules Prove Elusive," *Science* 295: 1625.

Matzat, U. (2004) "Academic Communication and Internet Discussion Groups: Transfer of Information or Creation of Social Contacts?" *Social Networks* 26(3): 221–55.

Merali, Z. & J. Giles (2005) "Databases in Peril," *Nature* 435: 1010–11.

Miller, D. & D. Slater (2000) *The Internet: An Ethnographic Approach* (Oxford and New York: Berg).

Mitchell, R. & P. Thurtle (eds) (2004) *Data Made Flesh. Embodying Information* (New York: Routledge)

Molyneux, R. E. & R. V. Williams (1999) "Measuring the Internet," *Annual Review of Information Science and Technology* 34: 287–339.

Mortensen, T. & J. Walker (2002) "Blogging Thoughts: Personal Publication as an Online Research Tool," in A. Morrison (ed), *Researching ICTs in Context* (Oslo, Norway: Intermedia, University of Oslo): 249–79.

Mosco, V. & J. Wasko (eds) (1988) *The Political Economy of Information* (Madison: University of Wisconsin Press).

National Research Council (NRC) (1997) *Bits of Power: Issues in Global Access to Scientific Data* (Washington, DC: National Academies Press).

National Research Council (NRC) (1999) *A Question of Balance: Private Rights and the Public Interest in Scientific and Technical Databases.* (Washington, DC: National Academies Press).

Nentwich, M. (2003) *Cyberscience: Research in the Age of the Internet* (Vienna: Austrian Academy of Sciences Press).

Nissenbaum, H. & M. Price (eds) (2004) *Academia and the Internet* (New York: Peter Lang Publishing Group).

Norris, K. S. (1993) *Dolphin Days: The Life and Times of the Spinner Dolphin* (New York: Avon Books).

OECD (1998) *The Global Research Village* (Paris: OECD).

Park, H. W. (2003) "What Is Hyperlink Network Analysis? New Method for the Study of Social Structure on the Web," *Connections* 25(1): 49–61.

Passmore, A. (1998) "Geogossip," *Environment and Planning A* 30(8): 1332–36.

Pels, D. (2003) *Unhastening Science: Autonomy and Reflexivity in the Social Theory of Knowledge* (New York: Routledge).

Pickering, A. (1992) *Science as Practice and Culture* (Chicago: University of Chicago Press).

Porter, D. (1997) *Internet Culture* (London: Routledge).

Poster, M. (2001) *What Is the Matter with the Internet?* (Minneapolis: University of Minnesota Press).

Price, D. J. de Solla (1984) "The Science/Technology Relationship: The Craft of Experimental Science, and Policy for the Improvement of High Technology Innovation," *Research Policy* 13(1): 3–20.

Proceedings of the First International Conference on e-Social Science, Manchester, 22–24 June 2005. Available at: http://www.ncess.ac.uk/events/conference/2005/papers/.

Ratto, M. & A. Beaulieu (in press 2007) "Banking on the Human Genome Project," *Canadian Review of Sociology and Anthropology/Revue Canadienne de Sociologie* 44(2).

Reips, U.-D. & M. Bosnjak (eds) (2001) *Dimensions of Internet Science* (Langerich, Germany: Pabst Science Publishers).

Rheinberger, H.-J. (1997) *Toward a History of Epistemic Things: Synthesizing Proteins in the Test Tube* (Stanford, CA, Stanford University Press).

Rheingold, H. (1994) *The Virtual Community: Homesteading on the Electronic Frontier* (New York: HarperPerennial, or http://www.rheingold.com/vc/book/).

Santos, C., J. Blake, & D. J. States (2005) "Supplementary Data Need to Be Kept in Public Repositories," *Nature* 438: 738.

Scharnhorst, A. (2003) "Complex Networks and the Web: Insights from Nonlinear Physics," *Journal of Computer-Mediated Communication* 8(4). Available at: http://jcmc.indiana.edu/vol8/issue4/scharnhorst.html.

Scharnhorst, A. & M. Thelwall (2005), "Citation and Hyperlink Networks," *Current Science* 89(9): 1518–23.

Scharnhorst, A., P. Van den Besselaar, & P. Wouters (eds) (2006) "What Does the Web Represent? From Virtual Ethnography to Web Indicators," *Cybermetrics* 10(1). Available at: http://www.cindoc.csic.es/cybermetrics/articles/v10i1p0.html.

Schilling, G. (2000) "The Virtual Observatory Moves Closer to Reality," *Science* 289: 238–39

Shapin, S. & S. Schaffer (1985) *Leviathan and the Air Pump: Hobbes, Boyle and the Experimental Life* (Princeton, NJ: Princeton University Press).

Shapiro, C. & H. R. Varian (1999) *Information Rules: A Strategic Guide to the Network Economy* (Boston: Harvard Business School Press).

Sharples, M. (ed) (1993) *Computer Supported Collaborative Writing* (London: Springer-Verlag).

Shoichet, B. (2004) "Virtual Screening of Chemical Libraries," *Nature* 432: 862–65.

Shrum, W. (2005) "Reagency of the Internet, or How I Became a Guest for Science," *Social Studies of Science* 35(5): 723–54.

Sismondo, S. (2004) *An Introduction to Science and Technology Studies* (Malden and Oxford: Blackwell Publishing).

Slevin, J. (2000) *The Internet and Society* (Cambridge: Polity Press).

Smith, M. R. & L. Marx (eds) (1994) *Does Technology Drive History? The Dilemma of Technological Determinism* (Cambridge, MA: MIT Press).

Star, S. L. (1992) "Craft vs. Commodity, Mess vs. Transcendence: How the Right Tool Became the Wrong One in the Case of Taxidermy and Natural History," in A. Clark & J. Fujimura (1992) *The Right Tools for the Right Job: At Work in 20th-Century Life Sciences* (Princeton, NJ: Princeton University Press): 257–86.

Star, S. L. (1999) "The Ethnography of Infrastructure," *American Behavioral Scientist* 43(3): 377–91.

Suchman, L. A. (1987) *Plans and Situated Actions: The Problem of Human-Machine Communication* (Cambridge: Cambridge University Press).

Sugden, A. (2002) "Computer Dating," *Science* 295: 17.

Thelwall, M. (2000) "Web Impact Factors and Search Engine Coverage" *Journal of Documentation* 56(2): 185–89.

Thelwall, M. (2002a) "A Comparison of Sources of Links for Academic Web Impact Factor Calculations," *Journal of Documentation* 58: 60–72.

Thelwall, M. (2002b) "The Top 100 Linked Pages on UK University Web Sites: High Inlink Counts Are Not Usually Directly Associated with Quality Scholarly Content," *Journal of Information Science* 28(6): 485–93.

Thelwall, M. (2003) "Can Google's PageRank Be Used to Find the Most Important Academic Web Pages?" *Journal of Documentation* 59: 205–17.

Thelwall, M. (2004) "Scientific Web Intelligence: Finding Relationships in University Webs," *Communications of the ACM* 48(7): 93–96.

Thelwall, M. (2005) *Link Analysis: An Information Science Approach* (San Diego: Academic Press).

Thelwall, M. & P. Wouters (2005) "What's the Deal with the Web/Blogs/the Next Big Technology: A Key Role for Information Science in e-Social Science Research?" *Lecture Notes in Computer Science* 3507: 187–200.

Thompson, C. (2005) *Making Parents: The Ontological Choreography of Reproductive Technologies* (Cambridge, MA: MIT Press).

Thurtle, P. & R. Mitchell (eds) (2002) *Semiotic Flesh. Information and the Human Body* (Seattle: University of Washington Press)

Traweek, S. (1988) *Lifetimes and Beamtimes: The World of High Energy Physics* (Cambridge, MA: Harvard University Press).

Turkle, S. (1995) *Life on the Screen: Identity in the Age of the Internet* (New York: Simon & Schuster).

U.K. Research Councils (2001) *About the UK e-Science Programme.*

Van Horn, J. D., J. S. Grethe, P. Kostelec, J. B. Woodward, & J. A. Aslam (2001) "The Functional Magnetic Resonance Imaging Data Center (fMRIDC): The Challenges and Rewards of Large-scale Databasing of Neuroimaging Studies," *Philosophical Transactions of the Royal Society of London B: Biological Sciences* 356(1412): 1323–39.

Van House, N. A. (2002) "Digital Libraries and Practices of Trust: Networked Biodiversity Information," *Social Epistemology* 16(1): 99–114.

Van Lente, H., & A. Rip (1998) "Expectations in Technological Developments: An Example of Prospective Structures to Be Filled in by Agency," in C. Disco & B. E. van der Meulen (eds), *Getting New Technologies Together* (Berlin and New York: Walter de Gruyter): 203–29.

Vann, K. (2004) "On the Valorisation of Informatic Labour," *Ephemera: Theory and Politics in Organization* 4(3): 242–66.

Vann, K. & G. C. Bowker (2006) "Interest in Production: on the Configuration of Technology-bearing Labors for Epistemic IT," in C. Hine (ed), *New Infrastructures for Knowledge Production: Understanding e-science* (London: Information Science Publishing): 71–97.

Van Zoonen, L. (2002) "Gendering the Internet: Claims, Controversies and Cultures," *European Journal of Communication* 17(1): 5–23.

Vasterman, P. & O. Aerden (1995) *De context van het nieuws* (Groningen, Netherlands: Wolters Noordhoff).

Voorbij, H. (1999) "Searching Scientific Information on the Internet: A Dutch Academic User Survey," *Journal of the American Society for Information Science* 50(7): 598–615.

Walsh, J. P. (1999) "Computer Networks and The Virtual College," *OECD STI Review* 24: 49–77.

Walsh, J. & T. Bayma, (1996) "Computer Networks and Scientific Work," *Social Studies of Science* 26: 661–703.

Walsh, J., C. Cho, & W. Cohen (2005) "The View from the Bench: Patents, Material Transfers and Bio-medical Science," *Science* 309: 2002–3.

Wasserman, S. & K. Faust (1994) *Social Network Analysis: Methods and Applications,* vol. 8 (Cambridge: Cambridge University Press).

Webster, F. (1995) *Theories of the Information Society* (New York: Routledge).

Weingart, P. (2000) "Interdisciplinarity: The Paradoxical Discourse," in P. Weingart & N. Stehr (eds), *Practising Interdisciplinarity* (Toronto: University of Toronto Press): 25–45.

Weingart, P., L. R. Graham, & W. Lepenies (eds) (1983) *Functions and Uses of Disciplinary Histories: Sociology of the Sciences Yearbook*, vol. VIII (Dordrecht, Netherlands: Reidel).

Wellman, B. (2001) "Computer Networks as Social Networks," *Science* 293: 2031–34.

Wellman, B. & C. Haythornthwaite (2002) *The Internet in Everyday Life* (Malden, MA: Blackwell).

Wheeler, Q. D., P. H. Raven, & E. O. Wilson (2004) "Taxonomy: Impediment or Expedient?" *Science* 303: 285.

Whitley, R. (2000) *The Intellectual and Social Organization of the Sciences* (Oxford: Oxford University Press).

Woolgar, S. (2002a) "Five Rules of Virtuality," in S. Woolgar (ed.) *Virtual Society? Technology, Cyberbole, Reality* (Oxford: Oxford University Press): 1–22.

Woolgar, S. (2002b) *Virtual Society? Technology, Cyberbole, Reality* (Oxford: Oxford University Press).

Wouters, P. (2000) "Cyberscience: The Informational Turn in Science," lecture at the Free University, Amsterdam, 13 March.

Wouters, P. (2004) *The Virtual Knowledge Studio for the Humanities and Social Sciences @ the Royal Netherlands Academy of Arts and Sciences* (Amsterdam: Royal Netherlands Academy of Arts and Sciences).

Wouters, P. (2005) *The Virtual Knowledge Studio for the Humanities and Social Sciences*, Proceedings of the First International Conference on e-Social Science, Manchester, June 22–24.

Wouters, P. (2006) "What Is the Matter with e-science? Thinking Aloud about Informatisation in Knowledge Creation," *Pantaneto Forum*, July 2006. Available at: http://www.pantaneto.co.uk/issue23/wouters.htm.

Wouters, P. & A. Beaulieu (2006) "Imagining e-science Beyond Computation," in C. Hine (ed), *New Infrastructures for Knowledge Production: Understanding e-science* (London: Information Science Publishing): 48–70.

Wouters, P. & R. de Vries (2004) "Formally Citing the Web," *Journal of the American Society for Information Science and Technology* 55(14): 1250–60.

Wouters, P. & C. Reddy (2003) "Big Science Data Policies," in P. Wouters & P. Schröder (eds) (2003), *Promise and Practice in Data Sharing* (Amsterdam: NIWI-KNAW): 13–40.

Wouters, P. & P. Schröder (eds) (2000) *Access to Publicly Financed Research: The Global Research Village III* (Amsterdam: NIWI-KNAW).

Wouters, P. & P. Schröder (eds) (2003) *Promise and Practice in Data Sharing* (Amsterdam: NIWI-KNAW).

Wyatt, S. (1998) "Technology's Arrow: Developing Information Networks for Public Administration in Britain and the United States," PhD diss., University of Maastricht.

Wyatt, S., G. Thomas, & T. Terranova (2002) "They Came, They Surfed, They Went Back to the Beach: Conceptualizing Use and Non-Use of the Internet," in S. Woolgar, *Virtual Society? Technology, Cyberbole, Reality* (Oxford: Oxford University Press): 23–40.

Wyatt, S., F. Henwood, N. Miller, & P. Senker (2000) *Technology and In/equality: Questioning the Information Society* (London: Routledge).

Zammito, J. H. (2004) *A Nice Derangement of Epistemes: Post-positivism in the Study of Science from Quine to Latour* (Chicago: University of Chicago Press).

Zeldenrust, S. (1988) *Ambiguity, Choice, and Control in Research* (Amsterdam: Amsterdam University Press).

15 Sites of Scientific Practice: The Enduring Importance of Place

Christopher R. Henke and Thomas F. Gieryn

Science in the twenty-first century is seemingly a world of perpetual motion. Scientists, specimens, instruments, and inscriptions race around the world on jets and through digitized communications, largely unfettered by the drag of distance or physical location. In an era when the globalization of science has never been more apparent, it seems almost anachronistic for us to suggest that "place" continues to matter a great deal for the practices and accomplishments of science. Our task in this chapter is to show that globalized science is at the same time emplaced science: research happens at identifiable geographic locations amid special architectural and material circumstances, in places that acquire distinctive cultural meanings. We seek to go beyond a mere listing of the various (and sometimes surprising) places where science happens, in an attempt to theorize *how* the material and geographic situations of research are sociologically consequential for institutionalized activities that appear, at a glance, to depend so little on them. In fact, the global standardization of research facilities shows how both the brick-and-mortar of material infrastructure as well as the symbolic understandings that privilege some places as authoritative sites for knowledge-construction actually *enable* the mobility of science all around. Place, ironically, achieves the appearance of placelessness.

Whether or not place matters for science—and how—has long been debated in STS.[1] These discussions have moved through four waves, and we suggest the need for a fifth. In the first wave, positivist and rationalist philosophers of science found little cause to examine the specific places where science occurs (Reichenbach, 1938; Popper, 1959; Hempel, 1966). However situated the actual practices of scientists might be, what mattered most from this perspective was the abstract, universal, and placeless character of scientific truth at the end of the day. The laws of gravity worked the same everywhere; even if scientists in different locations disagreed for a time about the content of those laws, persuasive evidence and compelling theory would eventually rub out geographical differences in belief. In wave one, science epitomized a "view from nowhere" (Nagel, 1989), disciplined into a single eye by method, instrumentation, techniques, and logic.

The second wave began with a recognition that this supposed "God trick" (Haraway, 1991) was a philosophical conceit rather than an adequate empirical account of how scientists construct legitimate knowledge. Beginning in the 1970s, STS ethnographers moved into the laboratory, discovering context-specific contingencies that shaped how scientists differently interpreted data, used machines, conducted experiments, and judged validity (Collins, 1974; Latour & Woolgar, 1979; Knorr Cetina, 1981; Lynch, 1985). The supposed placeless and transcendent character of scientific claims was no longer seen as a philosophical necessity but as a discursive accomplishment. Wave two discourse analysts showed how scientists routinely excise circumstantial "modalities" of specific places from their texts, leaving the appearance that the facts came straight from Nature (Latour & Woolgar, 1979; Gilbert & Mulkay, 1984). Although laboratory ethnographers established the irreducibly local character of scientific knowledge-making, conceptual interest in the laboratory as a place was minimal. The lab became an analytical resource—a means for deconstructing the "view from nowhere"—rather than a topic of interest in its own right.

By the 1990s, a third wave of research was well under way, in which STS scholars produced case studies of historically-changing sites of science, revealing the different geographic and material preconditions of making legitimate knowledge. By comparing the various settings where science happened, it was possible to discern how distinctive epistemic regimes were constituted in and through the situated, material conditions of inquiry. For example, the ancient agora in Athens was a place where privileged males could decide truth and virtue through public argumentation (Sennett, 1994)—in stark contrast to cloistered monasteries (Noble, 1992) and the secluded Renaissance studio (Thornton, 1997; Ophir, 1991), where solitude and contemplation were seen as necessary for scholarly pursuits. In the early modern period, the growing epistemic significance attached to "witnessing" collections of specimens accompanied the rise of museums, which were initially located in wealthy households and then in more accessible stand-alone buildings (Findlen, 1994). Similarly, the later importance of witnessing experimental apparatuses moved from the "gentleman's house" (Shapin & Schaffer, 1985; Shapin, 1988) to specialized laboratories in the nineteenth century (Gooday, 1991; Schaffer, 1998). By analyzing the shifting links between the place deemed appropriate for science and the creation of legitimate knowledge, studies from wave three provide rich materials for answering a signal question in STS: what must the construction of legitimate natural knowledge be like such that these kinds of places—located at this spot, built to these designs—fit the bill?

At about the same time, actor-network theory (ANT) offered conceptual perspectives that—in an emerging wave four—could suggest a diminished role for place in STS. To be sure, ANT directs attention to the nonhuman materialities at "centers of calculation" such as Pasteur's Parisian laboratory and the public arenas where he demonstrated his anthrax vaccine (Latour, 1983, 1988). And yet, it is the *transit* of Pasteur (and his research materials) from farms to labs to sites of public display that carries the most explanatory weight in Latour's explanation of the pasteurization of France. This insight has led some to give greater attention to "immutable mobiles" (and, more recently, "mutable mobiles") than to the seemingly static and emplanted centers of calculation. Emphasis is placed on the mobility or "flows" of heterogeneous actants through networks and, in particular, on the fluidity or malleability of substances as they move about—thereby diminishing the apparent significance of the specific geographical places where the actants pass through or end up. For Callon and Law, "circulation has become more important than fixed positions" (2004: 9), and this idea finds further support in social and cultural theory more generally, as in Manuel Castells' "network society" (2000) or David Harvey's (1990) arguments for the compression of space (and time) in postmodernity. As Frederic Jameson puts it, "the truth of experience no longer coincides with the place in which it takes place" (1988: 349).

We have no quarrel with recent STS attention to mobilities and fluidities, but these properties of technoscientific actants do not warrant abandoning the investigation of materially-situated and symbolically-encrusted "nodes," the places that serve as endpoints for the links comprising heterogeneous networks in the ANT approach. There is still a great deal to be learned about laboratories, field-sites, and museums as places of science—however unmoving they might now seem to be—and we argue that the initiative to fold places into non-geographic networks actually overlooks important features useful for explaining *how* science travels. Our fifth wave seeks to be more theoretical than wave three, as it tries to identify precisely how place has consequence for scientific knowledge and practices, and why a focus on geographic location and situated materialities can enlarge our understanding of science in society. We discuss (1) why science clumps geographically in discrete spots, (2) how the material architecture of laboratories resolves certain tensions inherent in the juxtaposition of the ordinary practice of science and its imagery or public understanding, and (3) how the emplacement of science creates opportunities for resistance to its cultural authority.

LOCATING SCIENCE

The stuff of science circulates swiftly and globally, but not unendingly. For all its obvious mobilities and fluidities (Mol & Law, 1994; Callon & Law, 2004), science alights at universities, laboratories, field stations, libraries, and other centers of calculation (Latour, 1987). And when scientific practices stay put for a while, an interesting geographical pattern emerges: science is not randomly or evenly distributed all over the skin of the earth. Rather, the activities and wherewithal of scientists are clustered together in discrete locations recognizable as centers where most science happens. It is provocative to say that the whole world must become a laboratory in order for it to be known scientifically (Latour, 1999: 43), but it is also sloppy. The map looks more like an archipelago, islands of science vastly different from the surrounding sea.[2] "Natural knowledge is constructed in specifically designed and enclosed space" (Golinski, 1998: 98).

Why does science disperse geographically into clumps? In this respect, science is much like any large-scale productive activity, such as making cars or making money: having certain people, machines, archives, and raw materials reliably close at hand is simply a more effective way to do business. Economists have described "agglomeration efficiencies" (Marshall, 1890)—gains in productivity that result from gathering together at a common geographic location the diverse constituent elements of an activity. At first glance, however, capitalism today does not evince agglomeration: corporate moguls jet everywhere, representing clients and investors from all over the world; transactions involving millions of dollars or Euros are made in the flick of a keystroke by currency traders "in fields of interaction that stretch across all time zones" (Knorr Cetina & Bruegger, 2002: 909); core assumptions about the economic theories underlying markets are understood more or less in the same way here and everywhere; factories, offices, and outsourced jobs flow from country to country, seeking greater profitability. What could be more "global" or "mobile?" And yet Saskia Sassen (2001: 5) finds that this globalization of economic activities generates "global cities" (New York, London, and Tokyo), specific places where corporate headquarters huddle together around the geographically centralized financial and specialized service functions on which they depend—lawyers, accountants, programmers, telecommunications experts, and public relations specialists. The "extremely dense and intense information loop" afforded by "being in a city" "still cannot be replicated fully in electronic space" (2001: xx). It is premature, Sassen suggests, to conclude that innovations in information, communication, and transportation technologies have the capacity "to neutralize distance and place" (2001: xxii)—and that is as much the case for science as for corporate capitalism.

Science clusters at discrete places because geographic proximity is vital for the production of scientific knowledge and for the authorization of that knowledge as credible (Livingstone, 2003: 27). "Place" enables copresence among people, instruments, specimens, and inscriptions (Bennett, 1998: 29). Particle accelerators, colliders, and detectors in high-energy physics illustrate the necessity—but also the difficulties—in gathering up scientific instruments at a common location (Galison, 1997; Knorr Cetina, 1999). Pieces of a detector may be built at scattered sites, just as the scientists involved with an experiment may corporeally reside at CERN, SLAC, or Fermilab only intermittently and for short durations. To cast experimental high-energy physics, therefore, as transient science misses the significance of the destination toward which the machines (and their tenders) eventually move. New particles could not be found without the precise temporary commingling of accelerators, detectors, and computers on site (no matter how much analysis of the data subsequently happens at universities often far away from the accelerator). Still, success at melding sophisticated machines is rarely automatic and typically hard-won for social and technical reasons: what happens at the destination laboratory in high-energy physics is described as "breaking components out of other ontologies and of configuring, with them, a new structural form" (Knorr Cetina, 1999: 214).

The "magnet" attracting science to a discrete place may also be a collection of specimens unrivaled in the world. Linnaeus's botanical taxonomy appears, curiously, as an eighteenth-century achievement of an already globalized science. Linnaeus himself traveled from Uppsala to Lapland (for collecting), and more consequentially to Holland, where an immense number of plant species had been gathered from around the world at botanical gardens in and around Leiden. For some historians, this movement of plants and scientists is key: Linnaeus's achievement "does not depend solely on the cascades of inscriptions produced, gathered, and reproduced within any one particular 'center of calculation'" because "the very possibility of that taxonomy presupposed the formation of a worldwide system of plant circulation mediating a plurality of sites of knowledge production, both peripheral and central, in which 'stable' and 'variable' features could fall apart" (Müller-Wille, 2003: 484). So much analytic attention is given to this "vast network of translation and exchange" that the locus of arrival becomes a trivial after-effect. Without a doubt, historical studies of collecting and transporting specimens have enriched our understanding of field sciences by expanding the cast of characters involved in science and by showing the mutability of research materials as they move from periphery to center (Drayton, 2000; Schiebinger, 2004; Schiebinger & Swan, 2005; Star & Griesemer, 1989). Still, Linnaeus did not need to travel to China or the Americas—just Leiden, because that is where the plants converged. He was as dependent, for example, on George Clifford's careful gardeners and passion for collecting as he was on the traders and sailors who procured the plants and got them safely to Holland, and there is little merit in diminishing the consequentiality of the former just to raise curiosity about the latter. Leiden mattered (Stearn, 1962) because Linnaeus's taxonomic efforts depended on the affordances of the Dutch gardens: "spaces in which things are juxtaposed," making them "already virtually analyzed" (Foucault, 1970: 131). With the concentration of so many botanical species at Leiden, and with their classificatory plantings, Linnaeus's gaze was impossible to achieve almost anywhere else in the world.

On other occasions, the accumulation of people at a place serves as its own magnet—attracting still more scientists to that spot. Even in sciences without much need for unique massive instruments or an incomparable collection of specimens, geographical clustering occurs. Folk wisdom depicts mathematicians as an especially peripatetic bunch of scientists—always scurrying from university to university to share ideas up-close and personally, a pattern of work and "flow" that reaches back to the late nineteenth century. Between 1900 and 1933, Göttingen was the place to be for cutting-edge mathematics. Felix Klein and David Hilbert were there, and "what made Göttingen probably the most eminent center of mathematics in the word—until 1933—was the unrivaled inspiring atmosphere among the numerous young mathematicians who flocked to Göttingen from everywhere" (Schappacher, 1991: 16). The place was a "cauldron of activity" with a "highly competitive atmosphere" where "even budding geniuses, like Norbert Wiener and Max Born, could be scarred by the daunting experience of facing the hypercritical audiences that gathered at the weekly meetings of the Göttingen Mathematical Society" (Rowe, 2004: 97). For early twentieth-century mathematicians, if you could make it in Göttingen, you could make it anywhere. The city assembled the most formidable audience that fresh mathematical ideas might ever face—and those that survived carried a widely respected geographic seal of approval (Warwick, 2003). Thus, some *places* ratify scientific claims.

The clumping of mathematics in centers like Göttingen is explained in part by the "thick" interactions enabled uniquely by face-to-face proximity. Boden and Molotch

(1994) suggest that the rich contextual information accompanying talk and gesture in close-up encounters is important for judging the reliability and authenticity of what others are saying (or implying). This, in turn, is vital for the development of trust on which scientific practices significantly depend (Shapin, 1994: xxvi, 21). Indeed, that sense of trust seems especially difficult to achieve among collaborators in the absence of face-to-face interaction (Handy, 1995; Olson & Olson, 2000: 27; Finholt, 2002; Cummings & Kiesler, 2005; Duque et al., 2005). In his analysis of physicists who study gravitational waves, Harry Collins (2004: 450–51) writes:

> As the Internet expands, more and more people are saying that it is time to put an end to these expensive little holidays for scientists in pleasant places. But conferences are vital. The chat in the bars and corridors is what matters. Little groups talk animatedly about their current work and potential collaborations. Face-to-face communication is extraordinarily efficient—so much can be transmitted with the proper eye contact, body movement, hand contact, and so forth. This is where tokens of trust are exchanged, the trust that holds the whole scientific community together.

Copresence at a place is also vital for the transfer of tacit knowledge: "experiments are matters of the transfer of skills among the members of a community," so that "the knowledge and skill . . . [are] embodied in their practices and discourse and [can]not be . . . 'read off' from what could be found in print, but [are] located in the uniqueness and extent of their experience" (Collins, 2004: 388, 608). Collins's "enculturation model" fits the Göttingen mathematicians: David Rowe contends that developments at Göttingen began to institutionalize an "oral culture" among mathematicians, in which "to keep abreast of it one must attend conferences or workshops or, better yet, be associated with a leading research center where the latest developments from near and far are constantly being discussed." Echoing Collins, Rowe suggests that it is "probably impossible to understand" print versions of the latest proof "without the aid of an 'interpreter' who already knows the thrust of the argument through an oral source" (Rowe, 1986: 444; Merz, 1998).

But what if Klein, Hilbert, and the Göttingen Mathematical Society had had access to video teleconferencing, which would seem to capture much of the contextual thickness of copresence? Göttingen might then have become just a node on a network of hook-ups, with no geographical location of any special significance (being there would matter less). Or maybe not: the coagulation of mathematicians at Göttingen also afforded a high probability of chance encounters with other experts, unexpected meetings that sometimes yield creative solutions or, at least, previously unimagined problems (Allen, 1977; Boden & Molotch, 1994: 274). Unplanned meetings sometimes take place in "trading zones," which Peter Galison (1997) has described (in his history of high-energy physics) as physical sites where theorists, experimentalists, and engineers run into each other—and, via emergent "contact languages" or "pidgins," collaboratively exchange ideas and information whose meaning may be different from one subculture to the next. Although Fermilab created a joint experimental-theoretical seminar every Friday, "More frequent are informal meetings 'in offices on the third floor of the Central Laboratory and at the Cafeteria, Lounge and airports'" (Galison, 1997: 829). At MIT's Radiation Lab, "engineers and physicists worked within sight of one another," and its "success was directly related to the creation of such common domains in which action could proceed . . ." (1997: 830). By contrast, video teleconferencing is an arranged and scheduled interaction: you need to plan in advance who is expected to phone in, and when. But in theoretical physics, Merz suggests that "interaction should not be forced, it should just happen . . . casual, non-final, provisory, informal" (1998: 318). Further research is needed to decide whether chance discoveries in science are as likely to emerge from video teleconferencing as from physical copresence in what Merton and Barber identify as "'serendipitous microenvironments' . . . where diverse scientific talents were brought together to engage in intensive sociocognitive interaction" (2004: 294).

MATERIALIZING SCIENCE

The point of Anne Secord's celebrated paper, "Science in the Pub: Artisan Botanists in Early Nineteenth-Century Lancashire" (1994), is to show that science *cannot* happen in a pub. Secord avoids contradiction by consistently using adjectives to modify botany or science: those who gathered at the pubs to talk about plants were "artisan" or "working-class" practitioners, and their societies were "local." It is surely the case, as Secord says, that these working men and women bought botanical treatises, tried to grow the best gooseberry, learned some Linnaean nomenclature, inspected plants on pub benches, and provided useful specimens to gentlemen who practiced "'scientific botany'" (1994: 276). Moreover, they saw themselves as doing botany and as contributing to botanical knowledge (and not just as collectors of specimens). Still, their "science" requires adjectives or scare-quotes. Secord is appropriately constructivist in seeking the contested meanings of such distinctions as professional versus popular science in the emerging practices of historically-situated people—she refrains from imposing timeless boundaries by analytic fiat (1994: 294; Gieryn, 1999). Whatever those working class Lancashire botanists thought they were up to, the evaluation of their activities by those who then (and later) had greater power to solidify the boundaries of science put them on the outside—not just because of their social class or lack of Latin and other refinements, but because of the places where they gathered: pubs.

Legitimate knowledge requires legitimizing places. The rising cultural authority of science through the nineteenth century (and beyond) depended in part on geographic and architectural distinctions between those places deemed appropriate for science and those that were not. The pub—along with other quotidian places where almost-science or pseudoscience occurred—was epistemically delegitimated, as Secord (1994: 297) suggests:

[S]cientific practice became increasingly associated with specific sites from which "the people" were excluded. By defining the laboratory and the experimental station as the sites of legitimation of botany and zoology from the mid-nineteenth century (and thereby increasing their status), the place of science became strictly defined and popular science was marginalized.

By *materializing* scientific investigation in buildings distinctively different from other kinds of places, assumptions get made far and wide about the credibility of the results—and how that credibility may depend on real or imagined circumstances of production. Science elevates its cultural authority as the purveyor of legitimate natural knowledge by making its places of provenance into something *unlike* everywhere else. Putting science in the pub was an "exercise in denigration" (Ophir & Shapin, 1991: 4), and sometimes just having liquor nearby was sufficiently degrading. In 1852, Thomas Thomson reported that sharing a building for his new and excellent chemical laboratory with a "whiskey shop which occupies the ground floor does not accord with what one would expect from the University of Glasgow" (Fenby, 1989: 32).

But what kind of architecture now secures epistemic authority? A hint is found in Secord's story: a handloom weaver named John Martin gave a moss specimen to William Wilson, gent., who passed it along to his friend William Jackson Hooker, then professor of botany at Glasgow (1820–1841) and later the first director of the Botanical Garden at Kew. Hooker was pleased and asked Wilson to investigate the possibility of Martin's coming to work in Hooker's herbarium. Wilson had initially seen Martin as "addicted to neatness" (Secord, 1994: 288), but a visit to his working-class cottage convinced him otherwise (1994: 290):

"I did not find that neatness which I expected," he reported to Hooker, and he was puzzled that there were few outward signs of *"order & arrangement"* when Martin's mind seemed to be "very well regulated" and he was "an original & patient thinker" (emphasis in original). Martin's plant specimens were "rather carelessly mixed in the leaves of a copy of Withering & in other Books, which are not so clean as I expected."

Pubs are also disorderly, not especially clean, and just as indicative of a material disposition unsuited for real science as Martin's messy cottage.

Order and *arrangement* have become markers of sites where genuine science occurs. The design of laboratories—through the material arrangement of its spaces and physical fixtures—achieves types of *control* not commonly found in other places. Foucault could easily have been thinking about the scientific laboratory when he wrote, about "heterotopias" in general: "their role is to create a space that is other, another real space, as perfect, as meticulous, as well arranged as ours is messy, ill-constructed and jumbled" (1986: 27). But Foucault may not have spent much time in actual labs: most give the appearance of being packed to the rafters with stuff, strewn about in disarray, giving off the impression that everywhere somebody is in the middle of something. Orderliness and cleanliness describe the laboratory as it exists in widely-shared cultural imageries—assumptions about what such places must be like in order to unlock the secrets of Nature. Emphatically, sites of science are both quotidian work places (not always meticulous) and authorizing spaces (purifying and logical), and, we suggest, the coexistence of these disparate states depends on architectural manipulations and stabilizations of three apparent antinomies: public and private, visible and invisible, standardized and differentiated. Laboratory sites simultaneously materialize both ends of these three polarities, in intricate ways that are consequential both for the productive efficiency of scientific knowledge and for the cultural authority of science as an institution and profession. How so?

Science is, at once, public *and* private (Gieryn, 1998). On one level, scientific work is an oscillation between intense communal interaction and solitude. Both the public and private aspects of inquiry have, at different times, been connected to the credibility and authenticity of resulting knowledge (Shapin, 1991). In Greek and Medieval thought, solitude was a means to prevent the corruption of thought by minimizing interference from others and enabling unmediated contact with the source of genuine wisdom—reclusive monks found God in the hermitage, Montaigne later found truth in the loneliness of his tower library (Ophir, 1991), Thoreau retreated to the "wilderness" of Walden Pond (Gieryn, 2002), and Darwin withdrew to Down House (Golinski, 1998: 83; Browne, 2003). Seclusion has its epistemic risks: delusion perhaps, parochialism, or secrecy (none contribute much to the pursuit of legitimate natural knowledge). So, starting from the early modern period, science also parades its public character: claims must be shared (Merton, 1973), experiments must be witnessed (Shapin, 1988), collaboration is increasingly required, and conferences become necessities. The scientific life these days is marked by intermittent solitude (for reflection, for creative bursts unfettered by the doubts of others) amid sustained collective efforts; the public side of science speeds the production of knowledge via efficient divisions of labor and, at the same time, secures credibility through the authentication of claims by informed audiences. This all gets *built-in*: the Salk Institute of Biological Studies in La Jolla, California, designed by modernist hero Louis Kahn in the early 1960s, has two dramatically different kinds of spaces. The architect Moshe Safdie (1999: 486) worked on the project:

> Kahn was obsessed with how he might create a space that would enhance the creative activity of scientists. He was impressed with the fact that scientific activity today requires solitude and collaboration. This led him to develop the basic scheme for the Salk: places for solitude reaching forward into a long courtyard from the places for collective work, the great, flexible laboratories.

Architecture manages the jointly public and private character of science work: "space . . . articulates exactly this double need for the individual and the collective aspects of research" (Hillier & Penn, 1991: 47).

Science is also "public" in its active engagement with constituencies outside the profession. Laboratories could not exist without financial support by corporate and government investors, creating an implicit *quid pro quo*: space for science yields knowledge and technologies vital for making profits, legitimating policy, and improving civil society. And yet the ability of science to deliver the goods is assumed to depend on its autonomy from direct interference by these constituencies—a different sense of "private." This ideology also gets materialized in the architecture of science buildings. The Cornell Biotechnology Building in Ithaca, New York, constructed in the mid-1980s, was designed to provide a welcoming space for diverse constituencies and beneficiaries while at the same time building-in a sequestration of research activities

(Gieryn, 1998). The place was built for a number of "publics": Cornell students, the taxpayers of New York State, and corporations with interests in biotechnology. During the design process, these publics were defined as a risk and a threat to the safe and autonomous pursuit of knowledge, even as they were acknowledged to be the *raison d'être* for the $34 million project. Architecture provided a solution to this social problem: a "beachhead" for the public is created in the atrium lobby, conference and seminar rooms, small cafés, and some administrative offices, giving constituencies a symbolic and material place in the building. In a "bubble diagram" drawn up early in the design process, a thick black horizontal bar separates PUBLIC from PRIVATE. Above the bar, the entering public is routed to conference rooms or administrative offices; below the bar is a list of research groups (drosophila, eukaryotes, prokaryotes, etc.) and support facilities (plant growth rooms, animal rooms, etc.). The bar on paper gets materialized as an inconspicuous door off the inviting lobby (with straw mats by Alexander Calder)—without way-finding signs—and leading to a hallway whose utilitarian finishes, unfamiliar machineries, and strange odors suggest a "backstage" (Goffman, 1959) where the uninvited are made to feel out of place. Jon Agar finds the same pattern with Britain's radio telescope at Jodrell Bank: "the spectacle needed spectators, but the public needed to be held back," and "a key tool in achieving this distancing was this discourse of interference: the identification of unwanted visitors as disturbing" (1998: 273).

Sites of science also manage juxtapositions of the visible and invisible. Laboratories create enhanced environments where it becomes possible to see things not visible elsewhere (Knorr Cetina, 1999). Accelerators and detectors enable high-energy physicists to see quarks (Pickering, 1984; Galison, 1997), arrays of centrifuges and PCR machines enable molecular biologists to see precise segments of DNA (Rabinow, 1996), a vat of dry-cleaning fluid in a mile-deep cave enables physicists to see massless solar neutrinos (Pinch, 1986), and astrophysicists on earth manipulate a space telescope to see stars as never before (Smith, 1989). "The laboratory is the locus of mechanisms and processes which can be taken to account for the success of science," accomplished by its "detachment of the objects from a natural environment and their installation in a new phenomenal field defined by social agents" (Knorr Cetina, 1992: 166, 117).

However, even as laboratories render natural objects visible, they make the observing practices of scientists invisible—or, at least, incomprehensible—to all but the few knowing experts. Visitors to the Cornell Biotechnology Building are steered away from research spaces by an environment coded as "public not welcome here." And yet, the "success of science" as a *privileged and authoritative* eye on nature depends on the transparency of the process of scientists' seeing—in principle, scientific practice is assumed to be open for all to view (secrecy pollutes credibility). Golinski (1998: 84) puts it this way:

[T]he laboratory is a place where valuable instruments and materials are sequestered, where skilled personnel seek to work undisturbed, and where intrusion by outsiders is unwelcome . . . On the other hand, what is produced there is declaredly "public knowledge"; it is supposed to be valid universally and available to all.

For this reason, the Stanford Linear Accelerator hosts tour groups to make its activities visible to anybody. It is not apparent what those visitors actually see: "most visitors on these tours arrived wanting to be awed rather than informed . . . [and] often behaved as though they had been granted a special dispensation to see the inner sanctum of science and its most learned priests" (Traweek, 1988: 23). The James H. Clark Center, designed in 2003 for the Bio-X initiative at Stanford by noted British architect Norman Foster, opens up working laboratory spaces to full view through floor-to-ceiling external glass walls on three stories. The stunning new building has attracted tours and random visitors who confront signs, pasted all over the glass: "Experiments in progress—no public tours" and "Please do not ask to open the door!!!" The Clark Center suggests that what Ophir and Shapin found in the seventeenth-century house of experiment gets materialized still: the "site is at one and the same time a mechanism of social exclusion and a means of epistemically constituting conditions of visibility" (1991: 14).

Finally, the materialization of science in buildings plays both sides of yet another fence: standardization and differentiation. The Lewis Thomas Laboratory at Princeton University, completed in 1986, is very much like every other university molecular biology building of the same vintage and, at the same time, is architecturally unique. Almost nothing in the list of functional spaces (research labs, offices, support facilities, seminar rooms) or in the arrangements of benches, desks, sinks, and fume hoods within a research lab or in the infrastructural guts of the place (wiring, piping, conduits) makes the Lewis Thomas Lab stand out from its peers. It is as if the biotech building itself had been cloned, at universities all over (Gieryn, 2002). Neo-institutional theory from sociology predicts that *bureaucratic* structures in research organizations will become increasingly isomorphic (DiMaggio & Powell, 1991; Meyer & Rowan, 1991), but the same social processes may also cause a homogenization of the *physical spaces* that house such activities. Safety codes and requirements of the Americans with Disabilities Act coerce architects to conform to an approved legal standard. Professional trade associations such as Tradelines, Inc., bring architects and university facilities managers together at international conferences where design innovations are given either a "thumbs up" or a "thumbs down," creating a normative context in which few designers decide to go against the grain. Peripatetic scientists remember desirable features from a lab they visited recently and implore architects to design just the same thing for their proposed new building—a kind of mimesis. Moreover, a measure of institutional legitimacy is secured when a lab looks much like all the other successful labs elsewhere (indeed, the very presence of a laboratory legitimates some fields as genuinely scientific—like psychology [O'Donnell, 1985: 7] or physics [Aronovitch, 1989]—in their early days).

Importantly, these social processes responsible for the standardization of laboratory design are analytically distinct from their epistemic consequences. "The wide distribution of scientific knowledge flows from the success of certain cultures in creating and spreading standardized contexts for making and applying that knowledge" (Shapin, 1995: 7). With the rubbing out of idiosyncratic design elements, scientific laboratories become generic "placeless places" (Kohler, 2002), enabling scientists to presume that the "ambient" conditions in a laboratory here are equivalent to those anywhere else. This homogenization of space is vital to the flow of scientists, scientific instruments, specimens, and inscriptions from site to site: geographical location may change, but the mobile unit finds itself "at home" on arrival, in a set of circumstances not dramatically different from those where it started out. Ironically, the very "circulation" of scientific claims and objects is dependent on the materialization of equivalent standardized places where science settles down. For us, this signals the continuing *importance* of place for science, rather than its evisceration. Moreover, research on "situated activity" (Suchman, 1987, 1996, 2000; Lave, 1988)—paying attention "to the ways the body and local environment are literally built into the processing loops that result in intelligent action" (Clark, 1997: xii; Hutchins, 1995; Goodwin, 1994, 1995)—invites the possibility that standardized work spaces in laboratories could foster a routinization of bodily activities even as scientists migrate from one university to another. In this respect, STS interest in the importance of "embodied" or "tacit" knowledge is really only half the equation; practices get routinized in part by taking place in standardized spaces.

Still, "placeless" places are not necessarily "faceless" ones. Laboratories also materialize identities for *different* social categories, groups, or organizations, and so their designs seek to differentiate "us from them." The facade of one side of the Lewis Thomas Building shows a beige and white checkerboard—a signature feature of postmodern marquee architects Robert Venturi and Denise Scott-Brown, who were hired to provide Princeton with a building that would signal the University's commitment to molecular biology and elevate its national reputation in this field. A building "just like any other" would hardly have succeeded in luring top biologists to Princeton. MIT hired celebrity architect Frank Gehry to design its recently-completed Stata Center (2004). Gehry's "controlled chaos,"[3] a wonderful jumble of boxes tilted and askew, clad in brick and titanium, would seem to have little bearing on the very orderly artificial intelligence, logic, and computer science going on inside. But MIT now has "a Gehry," and when it comes to competition for scientific talent and institutional prestige, the difference is everything.[4] In the past, laboratories assumed different symbolic skins to announce other kinds of cultural significance. Nineteenth-century science buildings at British universities draped themselves in Gothic referents as a visible sign of the respectability of experimental research, semiotically aligning the activities inside with monkish purity and devotion while distinguishing them from the pursuit of lucre expected in factories (Forgan, 1998). These days, when the line between pure research and applied research for profit is difficult to locate, corporate labs (Knowles & Leslie, 2001) and university labs may be almost indistinguishable in their external appearances (or may even be co-located).

Even the insides of science buildings differentiate social groups and assert identities—more through location and restricted access than through ornament or infrastructure. At SLAC, the top floor is for theorists and directors while the basement is for instrument shops (Traweek, 1988); at the Lewis Thomas Laboratory, mouse people had space demarcated from that occupied by scientists using yeast or worms (Levine, 1999). Galison writes that "in the floor plans we are seeing far more than pragmatically situated air ducts; we are witnessing a physicalized architecture of knowledge" (1997: 785). From the nineteenth century to now, there has been a shift in laboratory architecture, from an emphasis on the unity of science to an emphasis on disciplinary differences. Earlier assertions calling for needed "juxtapositions" of all the sciences "[were] in the main supplanted by the vocabulary of separation and of specialization, which meant the creation of separate, purpose-built architectural spaces, with all the functional differentiation in plan, construction, and equipment which attended increasing specialization in scientific research and education" (Forgan, 1998: 213).

Other sites of science reproduce fundamental societal distinctions, such as gender. Eleanor Annie Lamson's contributions to the geophysical understanding of the Earth's density and structure were diminished as a result of *where* she did her research (Oreskes, 1996). Lamson, associate astronomer at the U.S. Naval Observatory, stayed on land to process data on marine gravity, data that had been collected—in part—during expeditions using submarines (mobile laboratories). "But only men went to sea. Only the men's work could be cast as a heroic voyage to 'conquer the earth's secrets.' Therefore, only men appeared in the public eye" (Oreskes, 1996: 100). This spatialized sexism has a long history: distinguished callers at Aldrovandi's sixteenth-century Italian museum were asked to record their presence at this privileged site for witnessing nature, but "he did not ask [women] to sign the visitors' book" (Findlen, 1999: 30), recalling an even more ancient pattern of female exclusion from monastic intellectual life (Noble, 1992). Findlen (1999: 50) believes that these gendered differentiations of space for knowledge-making had lasting consequence for the presence of women in science:

[These configurations] established important preconditions for the public understanding of scientific space, as museums and laboratories emerged from the homes of aristocrats and gentlemen to enjoy a new autonomy. Such institutions, even when divested of their former location, continued to incorporate a host of assumptions about the appropriateness of women in sites of knowledge.

In turn, materiality also served as a boundary marker for cultural change when, at the Radium Institute of Vienna (1910s), "women working on radioactivity succeeded in acquiring 'a laboratory bench of their own,' indicating a shift in political importance of the role of women in science" (Rentetzi, 2005: 305).

CONTESTING SCIENCE

Spaces for science are a powerful blend of material infrastructure and cultural iconography that lend credibility to knowledge claims. And yet the situatedness of science in discrete geographical locations creates at the same time a certain vulnerability to challenge and contestation. Latour (1983) famously wrote: "Give me a laboratory and I will raise the world." But you can also throw a rock at a laboratory, break into it, and burn it down. Much as Foucault (1980) argues that the exercise of power always goes hand-in-hand with resistance, the very materiality of scientific sites makes them good targets, a kind of "contested terrain" (Edwards, 1979) where actors with divergent interests have something to dig in to and hold on to, in both the literal and figurative senses. The capacity of physical sites to authorize knowledge claims is never automatic or permanent; credibility emerges instead from a "negotiated order" (Maines, 1982; Fine, 1984), where scientific spaces become the loci for resistance and the negotiation of consent.

Scientists themselves assert that they have a unique and privileged way of seeing the places of knowledge-making, a view that is generally uncontested by nonscientists:

The "doubling" of space in the places of knowledge means that two people looking at the same spot on the ground . . . might construe two different objects. And this "double vision" would flow from the fact that the one person is an officially competent and authorized inhabitant of the space while the other is a visitor or a support worker. Nor do modern sensibilities regard this phenomenon as anything out of the ordinary (Ophir & Shapin, 1991: 14).

On their own turf, the scientist's vision is hegemonic, trumping other ways of seeing. But scientists sometimes find themselves on other terrain, where their understanding of place is less privileged, and where nonscientists seek to establish their own authority over its representation. Recent STS studies of "field sciences" have found examples of this kind, where the boundaries between scientific and other ways of knowing places—as well as the boundary between laboratory and field itself—are blurred and contested (Bowker, 1994; Kuklick & Kohler, 1996; Henke, 2000; Kohler, 2002). The potential for conflict revolves around the materiality of places, and especially the place-bound interests that actors may have in particular sites—often quite different from scientists' interests in the same place.

Farmers, for example, have particular ways of growing crops that represent a kind of investment, a commitment to the interface of place and practice that structures their modes of production and colors their perceptions of new agricultural techniques. Henke (2000) has studied University of California "farm advisors"—scientists employed by the University but stationed in specific farm communities, charged with improving the production practices of local farmers. Farm advisors frequently use an experimental technique called a "field trial" to demonstrate to the local farm community the advantages of a new agricultural method or technology. These demonstration trials are often conducted on a farmer's own land because farmers simply do not accept "immutable mobiles"; they are more likely to trust results that take into account the local contingencies of their own place (climate, soil types, cultural practices, etc.). The overall objective of the field trial, then, is to adjust an experimental mode of knowing place to one that accords with farmers' ways of seeing their land. In effect, the field trial represents farm advisors' attempts to negotiate a compromise that will incorporate both the standardization of experimental practice and farmers' prejudice for place-bound data.

These kinds of negotiations trouble an easy attribution of epistemic authority in the field. When science seeks to shape places in the field, as in examples of applied science, other actors may be empowered through their own place-based knowledge. One way to explore these divergent "ways of knowing" place is through the study of environmental hazards. A canonical example in STS is Brian Wynne's (1989) study of negotiations between British government experts and sheep farmers jointly dealing with the effects of radioactive fallout from the Chernobyl power plant explosion. In the aftermath of the disaster, experts dispatched to the affected sheep farming area in Cumbria "assumed that scientific knowledge could be applied without adjusting to local circumstances," which greatly damaged their credibility with the sheep farmers (Wynne, 1989: 34). The story is similar to other conflicts over environmental risk, where the place-bound, experiential knowledge of local actors—variously described as "lay persons," "citizens," or "activists"—challenges the reductionist and supposedly universal techniques deployed by experts for assessing hazards (Martin, 1991; Tesh, 2000). Rejecting models of risk perception that posit a divide between fundamentally rational and irrational modes of perceiving risk,[5] many of these studies focus on the knowledge of place that comes from a long-term, bodily residence in a specific site. These "bodies in protest" (Kroll-Smith & Floyd, 1997; Beck, 1992) argue for the credibility of a more informal knowledge, one grounded in experience and place.

At the same time, many of these studies also show that communities responding to environmental hazards work to ally themselves with experts or to gain their own formal expertise in the methods of environmental risk assessment (Macnaghten & Urry, 1998; Fischer, 2000; Allen, 2003). The work of these "expert-activists" (Allen, 2003) makes an interesting comparison to Henke's case of applied agricultural science. On one hand, the University of California farm advisors tried to balance the formal and universalizing methods of science with an acquired knowledge of the specific geographical places where farmers grow their crops. On the other hand, communities that challenge expert assessments of local environmental risks sometimes choose to augment their own experiential and embodied understanding of place with more technical and institutionally-credentialed methods of measuring hazards. In each case, there is the potential for a fully "double vision" of place, drawing on both the scientific and the experiential—indeed, for this reason, some STS scholars have begun to deconstruct the very divide between "expert" and "lay" understandings (Tesh, 2000; Frickel, 2004; Henke, 2006).

Interestingly, scientists engaged specifically in field studies have historically faced their own problems of credibility, brought into high relief when their research was contrasted (often unfavorably) to laboratory experiments in which the relevant variables may be far easier to control. Laboratories and field sites have their distinctive epistemic virtues as places where legitimate natural knowledge gets made, leading to contestations between the rival "truth-spots" in disciplines as varied as biology (Kohler, 2002) and urban studies (Gieryn, 2006): labs maximize precision and control, but the field seems less of a contrivance and closer to the way Nature (or Society) really is. However, just as the distinctions between expert and lay understandings have been obscured almost beyond recognition, so too is the figure of "laboratory vs. field-sites" something more than a simple opposition of cultural practices and epistemic legitimations. Gieryn (2006) finds that members of the Chicago School of urban studies (1900–1930) constructed the city as *both* a laboratory *and* a field site. They oscillate (in their texts) between making Chicago into a specimen sliced and diced for statistical analysis and making Chicago into a found place best understood ethnographically through patient and absorbing long walks. Kohler suggests that by the 1950s, a variety of borderland or hybrid sciences had emerged in biology that drew variously on the epistemic virtues of both lab and field: "Traffic between laboratory and field no longer necessarily involved passage across a cultural frontier, or even physical movement from field to laboratory or vice versa" (2002: 293). Contestations over those places most suitable for making scientific knowledge need not persist forever.

A more graphic kind of emplaced contestation over the cultural authority of science will probably be more difficult to resolve. Whether science is located in the laboratory or in the field, the materiality and geographic specificity of places where research is conducted gives protesters a concrete target to attack, as in the case of break-ins, vandalism, and outright destruction of experimental places. However dramatic such assaults on science might be, they have received little systematic attention in STS. One well-known example happened on August 24, 1970, when activists opposed to U.S. involvement in the Vietnam War set off a bomb at Sterling Hall, on the University of Wisconsin campus in Madison. The building housed the Army Math Research Center, and the attack was designed to disrupt research allegedly focused on the development of new weapons. The explosion killed graduate student Robert Fassnacht and, prior to the Oklahoma City bombing in 1995, represented the largest bomb blast set off as a form of domestic protest in the United States (Bates, 1992; Durhams & Maller, 2000). More recently, animal rights groups have destroyed laboratory equipment as they liberated mice and monkeys from what is, for them, inhumane experimental handling (Lutherer & Simon, 1992). These attacks on the places of science show how impossible it is to sequester research from political turmoil, although cage rooms at animal labs are routinely shrouded in security and surveillance systems worthy of a bank vault.

Field sites for testing transgenic crops have also been frequent sites of protest and vandalism, at least since the technology was field-tested and became commercially available in the 1990s. Environmental activists concerned about possible hazards in the transfer of genes from transgenic to non-transgenic organisms—or potential disruption of ecosystems more broadly—have opposed the release of genetically engineered organisms "into the wild." At field-trial sites across the world, protesters have destroyed transgenic crops in an attempt to prevent the spread of genetic materials beyond the borders of the site (Cooper, 2000; Anon., 2002). These sometimes violent interventions center political attention on the boundary between a supposedly controlled space inside the laboratory and the unpredictability of placing research in the field. A Greenpeace press release condemning New Zealand's decision to allow field trials of transgenic crops asserts, "The only safe place for genetic research is a properly contained laboratory" (Greenpeace, 2001).[6] There is much irony in that assumption: laboratories and experimental field sites are designed and built to bring wild nature under control, to render specimens docile and compliant with the instruments and theoretical ambitions of the scientist. And yet, by putting science in a place—by giving an available and material home to the process of knowledge-making—sites are created that cannot render docile and compliant those human specimens who have cause to challenge the means, aims, and authority of science.

CONCLUSION

Martin Rudwick's map of 1840 London begins to suggest how the places of science have changed. With labels pointing to major scientific institutions and to residences of scientists important for the Great Devonian Controversy, the map plainly indicates the "small scale of scientific London" (Rudwick, 1985: 35). Everybody that one needed to talk to, every book or specimen to consult, every association meeting to attend, was (almost literally) just around the corner. How different things are today: relevant experts and major research centers are now scattered throughout the world, scientists collaborate with those on another continent (but not always face to face), and they analyze data from places they have never been themselves—sometimes gathered by remote sensing devices and then digitized and stored in a computer whose location does not really seem to matter. Do these changes signal that place itself has become less vital for an understanding of science in society? We think not.

Paradoxically, these historical changes in the siting of scientific inquiry could make the production of new knowledge easier and faster but make it more difficult to trust the received results. So much confidence in the credibility of scientific claims stems from widely shared assumptions about *where* processes of discovery and justification take place, and about the people, instruments, specimens, inscriptions and infrastructure assembled right *there*. As distal observations are increasingly mechanized and as data are increasingly standardized and made instantly (and anonymously) available to scientists everywhere, legitimate concerns about the "chain of custody" arise: exactly where did these data come from, who was present at their initial construction or later manipulation, and who ultimately is accountable for their validity? As scientists disperse themselves globally and replicate laboratories hither and yon, questions of credibility will grow (not cease): was the experiment done in architectural circumstances that enabled the collective witnessing and scrutiny that is (for some) the touchstone of scientific objectivity? Ironically, place was once thought to pollute the credibility of science—merely local knowledge was parochial and idiosyncratic and thus untrustworthy. Now that the production of scientific knowledge has gone global with a vengeance (the view today is from Everywhere), place will reassert its significance for science as ratifier of authenticity and trust.

Notes

1. For reviews of this literature, see Ophir & Shapin (1991) and Livingstone (2003). Two recent special issues of journals in the history of science have focused on geographical topics; cf. Dierig et al. (2003) and Naylor (2005). Our depiction of this research as comprising a series of "waves" borrows from Law and Mol (2001). Gieryn (2000) reviews the interdisciplinary literature on "place."

2. Andrew Barry has usefully distinguished "sites of calculation" from more encompassing but discontinuous "zones of circulation," where artifacts, technologies, and practices are "comparable and connectable" (2001: 203). But before STS researchers rush to collapse sites into zones (or worse, networks), we suggest the need for a better understanding of why, how, and when those discrete sites are consequential for science.

3. Gehry is quoted in Joyce (2004: xiii).

4. Raiding other universities for scientific talent is hardly new: "Thomson's Glasgow personified and incorporated the solution to these puzzles, and several dons decided the obvious course would be to hire him for Cambridge. Thomson turned down the offer. Space and resources were what counted: 'the great advantages I have here with the new College, the apparatus and the assistance provided, the convenience of Glasgow for getting mechanical work done, give me means of action which I could not have in any other place'" (Schaffer, 1998: 157).

5. As in, for example, Douglas and Wildavsky (1983) and Margolis (1996).

6. Even laboratory-based research on transgenic crops has been targeted by activists. The best-known example is probably the fire set at the office of Michigan State University researcher Catherine Ives in 1999 by members of the Earth Liberation Front (Earth Liberation Front, n.d.; Cooper, 2000).

References

Agar, Jon (1998) "Screening Science: Spatial Organization and Valuation at Jodrell Bank," in C. Smith & J. Agar (eds), *Making Space for Science: Territorial Themes in the Shaping of Knowledge* (London: Macmillan): 265–80.

Allen, Barbara L. (2003) *Uneasy Alchemy: Citizens and Experts in Louisiana's Chemical Corridor Disputes* (Cambridge, MA: MIT Press).

Allen, Thomas J. (1977) *Managing the Flow of Technology: Technology Transfer and the Dissemination of Technological Information within the R&D Organization* (Cambridge, MA: MIT Press).

Anon. (2002) "GM Crop Protestors Released," BBC News, August 19, 2002. Available at: http://news.bbc.co.uk/2/hi/england/2203220.stm.

Aronovitch, Lawrence (1989) "The Spirit of Investigation: Physics at Harvard University, 1870–1910," in F.A.J.L. James (ed), *The Development of the Laboratory: Essays on the Place of Experiment in Industrial Civilization* (New York: American Institute of Physics): 83–103.

Barry, Andrew (2001) *Political Machines: Governing a Technological Society* (New York: Athlone Press).

Bates, Tom (1992) *Rads: The 1970 Bombing of the Army Math Research Center at the University of Wisconsin and Its Aftermath* (New York: Harper Collins).

Beck, Ulrich (1992) *Risk Society: Towards a New Modernity* (London: Sage).

Bennett, Jim (1998) "Projection and the Ubiquitous Virtue of Geometry in the Renaissance," in C. Smith & J. Agar (eds), *Making Space for Science: Territorial Themes in the Shaping of Knowledge* (London: Macmillan): 27–38.

Boden, Deirdre & Harvey L. Molotch (1994) "The Compulsion of Proximity," in R. Friedland & D. Boden (eds), *NowHere: Space, Time and Modernity* (Berkeley: University of California Press): 257–86.

Bowker, Geoffrey C. (1994) *Science on the Run: Information Management and Industrial Geophysics at Schlumberger, 1920–1940* (Cambridge, MA: MIT Press).

Browne, E. Janet (2003) *Charles Darwin: The Power of Place* (Princeton, NJ: Princeton University Press).

Callon, Michel & John Law (2004) "Guest Editorial," *Environment and Planning D: Society and Space* 22: 3–11.

Castells, Manuel (2000) *The Rise of the Network Society*, 2nd ed. (Malden, MA: Blackwell).

Clark, Andy (1997) *Being There: Putting Brain, Body, and World Together Again* (Cambridge, MA: MIT Press).

Collins, H. M. (1974) "The TEA Set: Tacit Knowledge and Scientific Networks," *Science Studies* 4: 165–86.

Collins, Harry (2004) *Gravity's Shadow: The Search for Gravitational Waves* (Chicago: University of Chicago Press).

Cooper, Michael (2000) "Wave of 'Eco-terrorism' Appears to Hit Experimental Cornfield" (*The New York Times*, July 21, 2000).

Cummings, Jonathon N. & Sara Kiesler (2005) "Collaborative Research Across Disciplinary and Organizational Boundaries," *Social Studies of Science* 35: 703–22.

Dierig, Sven, Jens Lachmund, & J. Andrew Mendelsohn (2003) "Introduction: Toward an Urban History of Science," *Osiris* 18: 1–20.

DiMaggio, Paul J. & Walter W. Powell (1991) "The Iron Cage Revisited: Institutional Isomorphism and Collective Rationality in Organizational Fields," in P. J. DiMaggio & W. W. Powell (eds), *The New Institutionalism in Organizational Analysis* (Chicago: University of Chicago Press): 63–82.

Douglas, Mary & Aaron Wildavsky (1983) *Risk and Culture: An Essay on the Selection of Technical and Environmental Dangers* (Berkeley: University of California Press).

Drayton, Richard (2000) *Nature's Government: Science, Imperial Britain, and the "Improvement" of the World* (New Haven, CT: Yale University Press)

Duque, Ricardo B., Marcus Ynalvez, R. Sooryamoorthy, Paul Mbatia, Dan-Bright S. Dzorgbo, & Wesley Shrum (2005) "Collaboration Paradox: Scientific Productivity, the Internet, and Problems of Research in Developing Areas," *Social Studies of Science* 35: 755–85.

Durhams, Sharif & Peter Maller (2000) "30 Years Ago, Bomb Shattered UW Campus," *Milwaukee Journal Sentinel*, August 20, 2000.

Earth Liberation Front, North American Press Office (n.d.) "Frequently Asked Questions about the Earth Liberation Front." Available at: http://www.animalliberationfront.com/ALFront/ELF/elf_faq.pdf.

Edwards, Richard (1979) *Contested Terrain: The Transformation of the Workplace in the Twentieth Century* (New York: Basic Books).

Fenby, David (1989) "The Lectureship in Chemistry and the Chemical Laboratory, University of Glasgow, 1747–1818," in F.A.J.L. James (ed), *The Development of the Laboratory: Essays on the Place of Experiment in Industrial Civilization* (New York: American Institute of Physics): 22–36.

Findlen, Paula (1994) *Possessing Nature: Museums, Collecting, and Scientific Culture in Early Modern Italy* (Berkeley: University of California Press).

Findlen, Paula (1999) "Masculine Prerogatives: Gender, Space, and Knowledge in the Early Modern Museum," in P. Galison & E. Thompson (eds), *The Architecture of Science* (Cambridge, MA: MIT Press): 29–58.

Fine, Gary Alan (1984) "Negotiated Orders and Organizational Cultures," *Annual Review of Sociology* 10: 239–62.

Finholt, Thomas A. (2002) "Collaboratories," *Annual Review of Information Science and Technology* 36: 73–107.

Fischer, Frank (2000) *Citizens, Experts, and the Environment: The Politics of Local Knowledge* (Durham, NC: Duke University Press).

Forgan, Sophie (1998) "'But Indifferently Lodged.' Perception and Place in Building for Science in Victorian London," in C. Smith & J. Agar (eds), *Making Space for Science: Territorial Themes in the Shaping of Knowledge* (London: Macmillan): 149–80.

Foucault, Michel (1970) *The Order of Things: An Archaeology of the Human Sciences* (New York: Pantheon Books).

Foucault, Michel (1980) *Power/Knowledge: Selected Interviews and Other Writings, 1972–1977* (New York: Pantheon Books).

Foucault, Michel (1986) "Of Other Spaces," *Diacritics* 16: 22–27.

Frickel, Scott (2004) "Scientist Activism in Environmental Justice Conflicts: An Argument for Synergy," *Society and Natural Resources* 17: 369–76.

Galison, Peter (1997) *Image and Logic: A Material Culture of Microphysics* (Chicago: University of Chicago Press).

Gieryn, Thomas F. (1998) "Biotechnology's Private Parts (and Some Public Ones)," in A. Thackray (ed), *Private Science: Biotechnology and the Rise of the Molecular Sciences* (Philadelphia: University of Pennsylvania Press): 219–53.

Gieryn, Thomas F. (1999) *Cultural Boundaries of Science: Credibility on the Line* (Chicago: University of Chicago Press).

Gieryn, Thomas F. (2000) "A Space for Place in Sociology," *Annual Review of Sociology* 26: 463–96.

Gieryn, Thomas F. (2002) "Three Truth-Spots," *Journal of the History of the Behavioral Sciences* 38(2): 113–32.

Gieryn, Thomas F. (2006) "City as Truth-Spot: Laboratories and Field-Sites in Urban Studies," *Social Studies of Science* 36: 5–38.

Gilbert, G. Nigel, & Michael Mulkay (1984) "Experiments Are the Key: Participants' Histories and Historians' Histories of Science," *Isis* 75: 105–25.

Goffman, Erving (1959) *The Presentation of the Self in Everyday Life* (New York: Doubleday).

Golinski, Jan (1998) *Making Natural Knowledge: Constructivism and the History of Science* (Cambridge: Cambridge University Press).

Gooday, Graeme (1991) "'Nature' in the Laboratory: Domestication and Discipline with the Microscope in Victorian Life Science," *British Journal for the History of Science* 24: 307–41.

Goodwin, Charles (1994) "Professional Vision," *American Anthropologist* 96(3): 606–33.

Goodwin, Charles (1995) "Seeing in Depth," *Social Studies of Science* 25: 237–74.

Greenpeace (2001) "GE Free NZ Now up to the People," Press release, October 30, 2001. Available at: http://www.poptel.org.uk/panap/latest/peower.htm.

Handy, Charles (1995) "Trust and the Virtual Organization," *Harvard Business Review* (May/June): 40–50.

Haraway, Donna J. (1991) *Simians, Cyborgs, and Women: The Reinvention of Nature* (New York: Routledge).

Harvey, David (1990) *The Condition of Postmodernity: An Enquiry into the Origins of Cultural Change* (Cambridge, MA: Blackwell).

Hempel, Carl G. (1966) *Philosophy of Natural Science* (Englewood Cliffs, NJ: Prentice-Hall).

Henke, Christopher R. (2000) "Making a Place for Science: The Field Trial," *Social Studies of Science* 30: 483–512.

Henke, Christopher R. (2006) "Changing Ecologies: Science and Environmental Politics in Agriculture," in S. Frickel & K. Moore (eds), *The New Political Sociology of Science: Institutions, Networks, and Power* (Madison: University of Wisconsin Press): 215–43.

Hillier, Bill & Alan Penn (1991) "Visible Colleges: Structure and Randomness in the Place of Discovery," *Science in Context* 4: 23–49.

Hutchins, Edwin (1995) *Cognition in the Wild* (Cambridge, MA: MIT Press).

Jameson, Fredric (1988) "Cognitive Mapping," in C. Nelson & L. Grossberg (eds), *Marxism and the Interpretation of Culture* (Champaign-Urbana: University of Illinois Press): 347–57.

Joyce, Nancy E. (2004) *Building Stata: The Design and Construction of Frank O. Gehry's Stata Center at MIT* (Cambridge, MA: MIT Press).

Knorr Cetina, Karin (1981) *The Manufacture of Knowledge: An Essay on the Constructivist and Contextual Nature of Science* (Oxford: Pergamon Press).

Knorr Cetina, Karin (1992) "The Couch, the Cathedral, and the Laboratory: On the Relationship between Experiment and Laboratory in Science," in A. Pickering (ed), *Science as Practice and Culture* (Chicago: University of Chicago Press): 113–38.

Knorr Cetina, Karin (1999) *Epistemic Cultures: How the Sciences Make Knowledge* (Cambridge, MA: Harvard University Press).

Knorr Cetina, Karin & Urs Bruegger (2002) "Global Microstructures: The Virtual Societies of Financial Markets," *American Journal of Sociology* 107: 905–50.

Knowles, Scott G. & Stuart W. Leslie (2001) "'Industrial Versailles': Eero Saarinen's Corporate Campuses for GM, IBM, and AT&T," *Isis* 92: 1–33.

Kohler, Robert E. (2002) *Landscapes and Labscapes: Exploring the Lab-Field Border in Biology* (Chicago: University of Chicago Press).

Kroll-Smith, Steve & H. Hugh Floyd (1997) *Bodies in Protest: Environmental Illness and the Struggle over Medical Knowledge* (New York: New York University Press).

Kuklick, Henrika & Robert E. Kohler (eds) (1996) *Science in the Field*, *Osiris* 11.

Latour, Bruno (1983) "Give Me a Laboratory and I Will Raise the World," in K. Knorr Cetina & M. Mulkay (eds), *Science Observed: Perspectives on the Social Study of Science* (London: Sage): 141–70.

Latour, Bruno (1987) *Science in Action: How to Follow Scientists and Engineers Through Society* (Cambridge, MA: Harvard University Press).

Latour, Bruno (1988) *The Pasteurization of France* (Cambridge, MA: Harvard University Press).

Latour, Bruno (1999) *Pandora's Hope: Essays on the Reality of Science Studies* (Cambridge, MA: Harvard University Press).

Latour, Bruno & Steve Woolgar (1979) *Laboratory Life: The Construction of Scientific Facts* (Princeton, NJ: Princeton University Press).

Lave, Jean (1988) *Cognition in Practice: Mind, Mathematics and Culture in Everyday Life* (New York: Cambridge University Press).

Law, John & Annemarie Mol (2001) "Situating Technoscience: An Inquiry into Spatialities," *Environment and Planning D: Society and Space* 19: 609–21.

Levine, Arnold J. (1999) "Life in the Lewis Thomas Laboratory," in P. Galison & E. Thompson (eds), *The Architecture of Science* (Cambridge, MA: MIT Press): 413–22.

Livingstone, David N. (2003) *Putting Science in Its Place: Geographies of Scientific Knowledge* (Chicago: University of Chicago Press).

Lutherer, Lorenz Otto & Margaret Sheffield Simon (1992) *Targeted: The Anatomy of an Animal Rights Attack* (Norman: University of Oklahoma Press).

Lynch, Michael (1985) *Art and Artifact in Laboratory Science: A Study of Shop Work and Shop Talk in a Research Laboratory* (Boston, MA: Routledge).

Macnaghten, Phil & John Urry (1998) *Contested Natures* (Thousand Oaks, CA: Sage).

Maines, David (1982) "In Search of Mesostructure: Studies in the Negotiated Order," *Urban Life* 11: 267–79.

Margolis, Howard (1996) *Dealing with Risk: Why the Public and the Experts Disagree on Environmental Issues* (Chicago: University of Chicago Press).

Marshall, Alfred (1890) *Principles of Economics* (London: Macmillan).

Martin, Brian (1991) *Scientific Knowledge in Controversy: The Social Dynamics of the Fluoridation Debate* (Albany: State University of New York Press).

Merton, Robert K. (1973) *The Sociology of Science: Theoretical and Empirical Investigations* (Chicago: University of Chicago Press).

Merton, Robert K. & Elinor Barber (2004) *The Travels and Adventures of Serendipity: A Study in Sociological Semantics and the Sociology of Science* (Princeton, NJ: Princeton University Press).

Merz, Martina (1998) "Nobody Can Force You When You Are Across the Ocean: Face-to-Face and E-mail Exchanges Between Theoretical Physicists," in C. Smith & J. Agar (eds), *Making Space for Science: Territorial Themes in the Shaping of Knowledge* (London: Macmillan): 313–29.

Meyer, John W. & Brian Rowan (1991) "Institutional Organizations: Formal Structure as Myth and Ceremony," in P. J. DiMaggio & W. W. Powell (eds), *The New Institutionalism in Organizational Analysis* (Chicago: University of Chicago Press): 41–62.

Mol, Annemarie & John Law (1994) "Regions, Networks and Fluids: Anaemia and Social Topology," *Social Studies of Science* 24: 641–72.

Müller-Wille, Staffan (2003) "Joining Lapland and the Topinambes in Flourishing Holland: Center and Periphery in Linnaean Botany," *Science in Context* 16(4): 461–88.

Nagel, Thomas (1989) *The View from Nowhere* (New York: Oxford University Press).

Naylor, Simon (2005) "Introduction: Historical Geographies of Science—Places, Contexts, Cartographies," *British Journal for the History of Science* 38: 1–12.

Noble, David F. (1992) *A World Without Women: The Christian Clerical Culture of Western Science* (New York: Knopf).

O'Donnell, John M. (1985) *The Origins of Behaviorism: American Psychology, 1870–1920* (New York: New York University Press).

Olson, Gary M. and Judith S. Olson (2000) "Distance Matters," *Human-Computer Interaction* 15(2&3): 139–78.

Ophir, Adi (1991) "A Place of Knowledge Recreated: The Library of Michel de Montaigne," *Science in Context* 4(1): 163–89.

Ophir, Adi & Steven Shapin (1991) "The Place of Knowledge: A Methodological Survey," *Science in Context* 4: 3–21.

Oreskes, Naomi (1996) "Objectivity or Heroism? On the Invisibility of Women in Science," *Osiris* 11: 87–113.

Pickering, Andrew (1984) *Constructing Quarks: A Sociological History of Particle Physics* (Chicago: University of Chicago Press).

Pinch, Trevor (1986) *Confronting Nature: The Sociology of Solar-Neutrino Detection* (Dordrecht, Netherlands: D. Reidel).

Popper, Karl R. (1959) *The Logic of Scientific Discovery* (London: Unwin Hyman).

Rabinow, Paul (1996) *Making PCR: A Story of Biotechnology* (Chicago: University of Chicago Press).

Reichenbach, Hans (1938) *Experience and Prediction: An Analysis of the Foundations and the Structure of Knowledge* (Chicago: University of Chicago Press).

Rentetzi, Maria (2005) "Designing (for) a New Scientific Discipline: The Location and Architecture of the Institut für Radiumforschung in Early Twentieth-Century Vienna," *British Journal for the History of Science* 38: 275–306.

Rowe, David E. (1986) "'Jewish Mathematics' at Göttingen in the Era of Felix Klein," *Isis* 77: 422–49.

Rowe, David E. (2004) "Making Mathematics in an Oral Culture: Göttingen in the Era of Klein and Hilbert," *Science in Context* 17: 85–129.

Rudwick, Martin J. S. (1985) *The Great Devonian Controversy: The Shaping of Scientific Knowledge among Gentlemanly Specialists* (Chicago: University of Chicago Press).

Safdie, Moshe (1999) "The Architecture of Science: From D'Arcy Thompson to the SSC," in P. Galison & E. Thompson (eds), *The Architecture of Science* (Cambridge, MA: MIT Press): 475–96.

Sassen, Saskia (2001) *The Global City: New York, London, Tokyo* (Princeton, NJ: Princeton University Press).

Schaffer, Simon (1998) "Physics Laboratories and the Victorian Country House," in C. Smith & J. Agar (eds), *Making Space for Science: Territorial Themes in the Shaping of Knowledge* (London: Macmillan): 149–80.

Schappacher, Norbert (1991) "Edmund Landau's Göttingen: From the Life and Death of a Great Mathematical Center," *The Mathematical Intelligencer* 13: 12–18.

Schiebinger, Londa (2004) *Plants and Empire: Colonial Bioprospecting in the Atlantic World* (Cambridge, MA: Harvard University Press).

Schiebinger, Londa & Claudia Swan (eds) (2005) *Colonial Botany: Science, Commerce and Politics in the Early Modern World* (Philadelphia: University of Pennsylvania Press).

Secord, Anne (1994) "Science in the Pub: Artisan Botanists in Early Nineteenth-Century Lancashire," *History of Science* 32: 269–315.

Sennett, Richard (1994) *Flesh and Stone: The Body and the City in Western Civilization* (New York: W. W. Norton).

Shapin, Steven (1988) "The House of Experiment in Seventeenth-Century England," *Isis* 79: 373–404.

Shapin, Steven (1991) "'The Mind in Its Own Place': Science and Solitude in Seventeenth-Century England," *Science in Context* 4: 191–218.

Shapin, Steven (1994) *A Social History of Truth: Civility and Science in Seventeenth-Century England* (Chicago: University of Chicago Press).

Shapin, Steven (1995) "Here and Everywhere: Sociology of Scientific Knowledge," *Annual Review of Sociology* 21: 289–321.

Shapin, Steven & Simon Schaffer (1985) *Leviathan and the Air-Pump: Hobbes, Boyle, and the Experimental Life* (Princeton, NJ: Princeton University Press).

Smith, Robert W. (1989) *The Space Telescope: A Study of NASA, Science, Technology, and Politics* (New York: Cambridge University Press).

Star, Susan Leigh & J. R. Griesemer (1989) "Institutional Ecology, 'Translations,' and Boundary Objects: Amateurs and Professionals in Berkeley's Museum of Vertebrate Zoology, 1907–1939," *Social Studies of Science* 19: 387–420.

Stearn, William T. (1962) "The Influence of Leyden on Botany in the Seventeenth and Eighteenth Centuries," *British Journal for the History of Science* 1: 137–59.

Suchman, Lucy (1987) *Plans and Situated Actions: The Problem of Human-Machine Communication* (New York: Cambridge University Press).

Suchman, Lucy (1996) "Constituting Shared Workspaces," in Y. Engeström & D. Middleton (eds), *Cognition and Communication at Work* (New York: Cambridge University Press): 35–60.

Suchman, Lucy (2000) "Embodied Practices of Engineering Work," *Mind, Culture, and Activity* 7(1&2): 4–18.

Tesh, Sylvia Noble (2000) *Uncertain Hazards: Environmental Activists and Scientific Proof* (Ithaca, NY: Cornell University Press).

Thornton, Dora (1997) *The Scholar in His Study: Ownership and Experience in Renaissance Italy* (New Haven, CT: Yale University Press).

Traweek, Sharon (1988) *Beamtimes and Lifetimes: The World of High Energy Physics* (Cambridge, MA: Harvard University Press).

Warwick, Andrew (2003) *Masters of Theory: Cambridge and the Rise of Mathematical Physics* (Chicago: University of Chicago Press).

Wynne, Brian (1989) "Sheepfarming after Chernobyl: A Case Study in Communicating Scientific Information," *Environment* 31: 10–15, 33–39.

16 Scientific Training and the Creation of Scientific Knowledge

Cyrus C. M. Mody and David Kaiser

Remember the classic science studies parlor game "Awkward Student" (Collins, 1992)? One player pretends to be a teacher, the other a pupil. The teacher provides some basic instruction, then the student makes things awkward by stubbornly finding "correct" but non-common-sensical ways to follow that instruction. The teacher then adds more rules to the basic instruction to try to make it impossible for the student to provide awkward answers. As a thought experiment, Awkward Student demonstrates the interpretive flexibility inherent in experimental practice. No description of an experimental setup can ever be complete enough that it will be safe from "awkward" misreadings by replicators. There is always room for disagreement about whether one experimenter has awkwardly or faithfully replicated the technique of another.

The Awkward Student was a heuristic cornerstone of studies of scientific controversies in that it illustrated that both sides of a controversy could reasonably believe they were correctly following directions; hence, no asocial criterion could adjudicate between them. The other students in the thought experiment rely on social determinants (the authority of the teacher, the Awkward Student's status as friend, geek, clown, etc.) to decide whom to believe; similarly, scientists in a controversy must use social cues such as trust, class, nationality, gender, and age to help them decide who has done an experiment correctly. Almost all the early controversy studies, though, focused not on student-scientists, but on disputes between well-established peers, researchers in the prime of their careers (Collins, 1975, 1998; Pickering, 1984; Pinch, 1986; MacKenzie, 1990; Shapin & Schaffer, 1985). This made sense at the time in that controversies between peers were the "hard case" in which both sides could equally command authority, respect, and resources (hence, peer controversies were a more reliable test of the ideas of the new sociology of science).[1]

We suggest, though, that the pedagogical setting of the Awkward Student should be taken seriously. The student is "awkward" because he or she defies both common sense and the norms for the behavior of respectful pupils. Yet the student is also creating a kind of knowledge by pointing out alternative interpretations. Thus, in the classroom, like the laboratory, knowledge is simultaneously taught and created. Indeed, this interpretation of the Awkward Student is perhaps more faithful to its Wittgensteinian roots than to the uses to which its creator (Harry Collins) and other exponents of controversy studies have put it.[2] Even a quick review of Wittgenstein's life and work (Wittgenstein, 1953; Monk, 1990; Cavell, 1990) reveals the centrality of practices of pedagogy, training, schooling, and upbringing in his ethical and philosophical thought. Wittgenstein saw education as a site where meanings and values are generated, not just conferred.

Nor is Wittgenstein alone among the forebears of STS for having articulated now-overlooked insights about pedagogy. Most prominently, Thomas Kuhn and Michel Foucault both accorded pedagogy an important place in their analyses of science. Science for Kuhn and Foucault was not merely a positive, cumulative body of facts, but a thicket of practices, tools, and relationships that must be learned in order to be lived.[3] Foucault, of course, focused on the architecture and bureaucracy of pedagogy, on the physicality and ubiquity of regimes of surveillance that co-produce knowledge along with subjects who know and are known (Foucault, 1977, 1994). Kuhn (1962), meanwhile, drew attention to the tools and time scales of training, to the ways textbooks, problem sets, and the succession of student cohorts generate "normal science."

Building on the work of Wittgenstein, Kuhn, and Foucault, we interpret pedagogy broadly in this essay, not merely as formalized classroom teaching techniques—although these are certainly important—but rather as the entire constellation of training exercises through which novices become working scientists and engineers. This pedagogical dimension has been an important but underemphasized ingredient in many STS narratives. The classic stories of STS often unfold in modern research universities or other settings where teaching and training are overt, even primary, institutional motivations (e.g., Collins, 1974; Galison, 1987; Woolgar, 1990; Lynch, 1985b). Yet knowledge-*making* is the primary focus of these studies; their protagonists' roles as teachers and/or students are subordinated to (or invisible beside) their roles as researchers.

At the same time a more Mertonian-institutionalist strand of history and sociology of science has analyzed the mechanics and evolution of scientific training (Rossiter, 1982, 1986; Kohler, 1987; Owens, 1985). These latter studies have artfully shown how pedagogical institutions can mirror and drive wider cultural change, and how training regimes structure and organize the colleges (invisible and otherwise) of science. Knowledge, though, is usually taken as an unproblematic product in these Mertonian stories. These authors acknowledge that new understandings of the world emerge from pedagogical settings, but how institutional structure and pedagogical imperative shapes the content of scientific knowledge is left unexplored.

We will make a stronger claim that synthesizes these two literatures. It is no accident that modern scientific knowledge is tied to teaching and training. If science and technology studies returns to Wittgenstein, Kuhn, and Foucault and makes pedagogy a central analytic category, this coincidence disappears and fruitful connections emerge. By bringing training in focus, we can see that, even in ostensibly nonpedagogical settings, teaching and research activities are mutually reliant. The exigencies of one activity strongly inform the practice and content of the other. What scientists know of the world is a product of culturally driven decisions about whom to teach, what knowledge to validate by passing on, how to use pedagogy in the pursuit of social interests, and how to organize education. The tools of science are closely bound to the tools of pedagogy. Since nature is equivocal about its representation, the question of which instrument, image, or equation to use is often answered by asking "which tool most facilitates pedagogy? Which representation is most easily passed on, or most adequately manufactures a new generation that adheres to the vision and values of current practitioners?"

We follow on a growing literature that makes explicit these connections between research and pedagogy (Olesko, 1991; Leslie, 1993; Kohler, 1994; Dennis, 1994; Warwick, 2003; Kaiser, 2005a,b). Historians and sociologists of education have also offered important insights into the connections between training, practice, and social values in science (Geiger, 1986, 1993; Solomon, 1985; Hofstadter, 1963; Clark, 1993, 1995). Here, as elsewhere, science studies faces the paradox—science and technology are cultural activities and thus share features with other human endeavors, yet they also have (or have been accorded) a distinctive domain of practice and knowledge that presents analytical peculiarities. This essay charts a course between these alternatives, showing where ideas about pedagogy can be imported into science studies and where we must forge our own vocabulary. For brevity and coherence, we focus on the modern period, although pedagogical issues certainly were not absent from earlier periods.[4] Likewise, we concentrate on science and technology disciplines rather than on medicine although some path-breaking work on medical education should still inspire new work on the topic (Starr, 1982; Ludmerer, 1985; Bosk, 1979; Rosenberg, 1979).

REPRODUCTION

Two principal questions lurk behind all decisions regarding scientific training: why and how? Why should a society expend so much capital and effort to train new generations of a technical workforce, and how should their training proceed? Neither question has an automatic answer rising above the vagaries of time and place. In this section, we take up some prominent responses to the "why" question, as gleaned from recent studies. We turn to the "how" in the sections that follow.

Since at least the middle of the nineteenth century, nearly all practicing scientists and engineers have gone through some kind of formal training; the past century and a half has seen the decline of the "gentlemanly amateur" of science. Naturally the forms of training have varied across time and place, as well as across the evolving disciplinary map (Kaiser, 2005b). Yet the necessity of some form of training has emerged as the one constant across these many distinct settings. As Sharon Traweek has emphasized, scientists and engineers must always work to reproduce new generations of practitioners, replenishing the scientific workforce (Traweek, 1988; 2005).

This reproductive work takes place in all kinds of institutions, including some that are not overtly "educational." Throughout the twentieth century, for example, universities have partnered with many types of off-campus spaces to train new recruits: from exchange programs with industrial laboratories (Lowen, 1997; Slaughter et al.,

2002) to the citadels of "big science" at national laboratories (Galison, 1987, 1997; Traweek, 1988; Galison & Hevly, 1992; Westwick, 2003) to top-secret weapons laboratories (Gusterson, 1996, 2005; McNamara, 2001). In all these kinds of places, scientists and engineers work to train new members of their fields, mixing formal course work with more hands-on means of apprenticeship.

The reasons for undertaking this training are embedded within larger sociopolitical discussions. Reproduction of scientists and engineers is always a response to reproduction for: for national sovereignty or security, for economic well-being, for technological spin-offs, and so on. At the height of British imperial rule, for example, consensus emerged that Britain needed large cadres of "disciplined minds" who could staff the expanding civil service positions throughout the empire. This seemed to call for a certain kind of reproduction—one based on intense mathematical training and grueling written examinations (Warwick, 2003). Early in the nineteenth century, policymakers throughout the German states used similar arguments to encourage technical training, to build up a stock of efficient administrators (Turner, 1987). By the closing decades of the century, however, the rationale had shifted: education and industry leaders in the newly unified Germany decided that the country needed large groups of technically trained people to help manage the country's late-blooming industrialization. This called for a new type of pupil to undergo a new type of training, weakening the hold of the classically oriented *Gymnasien* and encouraging the rapid growth of *Realschulen* and *Technische Hochschulen*, with their emphases on precision measurement and the sophisticated management of error (Pyenson, 1977, 1979; Stichweh, 1984; Cahan, 1985; Fox & Guagnini, 1993; Olesko, 1991, 2005; Shinn, 2003). During the Cold War, politicians and educators in the United States, Western Europe, and the Soviet Union decided that "standing armies" of physical scientists were required; the ideological battle between East and West would be fought in the classroom, a race to create the largest "manpower" reserves in nuclear physics and allied disciplines (Ailes & Rushing, 1982; Mukerji, 1989; Krige, 2000; Kaiser, 2002; Rudolph, 2002). In all these ways, scientific training has often assumed center stage in larger debates over political economy, domestic policy, and international relations.

At the level of institutions, decisions over which equipment to build and which lines of research to support are also interwoven with decisions about which type of training to foster. Should the new recruits learn individual initiative, focused around small-scale apparatus, or teamwork sensibilities, using factory-sized equipment (Heilbron, 1992; Kaiser, 2004; Traweek, 2005)? Should new instruments and practitioners be gauged by how well they fit into a long-established academic field, or by how well they cross boundaries, merging ideas and techniques across a wide range of specialties (Mody, 2005)?

Pedagogical institutions also serve as powerful filters. They can either encourage or interrupt the flow of certain types of students—such as women and minorities—into the professional pipeline. Although some seminal work has been done on women's (often fraught) participation in modern science and technology (Keller, 1977; Rossiter, 1980, 1982, 1995; Murray, 2000; Oldenziel, 2000; Etzkowitz et al., chapter 17 in this volume) and on that of minorities and non-Westerners (Manning, 1983; Williams, 2001; Slaton, 2004; Ito, 2004; Sur, 1999; Anderson & Adams, chapter 8 in this volume), much remains to be done. Beyond narrowly demographic studies, interesting recent work has examined various meritocratic impulses and their relationship with pedagogical infrastructure, such as the movement toward standardized testing in the United States during the middle of the twentieth century (Lemann, 1999). Economists and sociologists have likewise turned to this topic with zeal of late, demonstrating in detail the persistent gaps in enrollment, retention, and advancement of women and minorities in the technical workforce (Levin & Stephan, 1998; Stephan & Levin, 2005; Hargens & Long, 2002; Preston, 1994; Pearson & Fechter, 1994; Rosser, 2004).

As Pierre Bourdieu and other historians and sociologists of education have long emphasized, therefore, generational reproduction is always based on a series of active choices and political-cultural decisions; training is never a neutral or passive activity (Bourdieu & Passeron, 1977; Bourdieu, 1988; Spring, 1989; Kliebard, 1999). Scientists and engineers mold their disciplines by pedagogically fashioning their disciples.

MORAL ECONOMIES

Historians and sociologists of education often talk of a "hidden curriculum," a series of values or norms—about proper behavior, civic duty, patriotism, and the like—that are embedded within schools' more explicit pedagogical operations (Arum & Beattie, 2000). So too is scientific and technical training shot through with decisions about values. Training is the central arena within which various communities craft and then reinforce their "moral economies"—often tacit conventions that regulate how members of their discipline should interact and behave, allocating resources, research programs, and credit (Shapin, 1991; Kohler, 1994; Daston, 1995). Young recruits learn these rules for behavior as part of their formative training; they learn what it means to be a scientist or engineer as they learn how to wield the tools of their trade. Just as responses to the "why" question—why undertake the labor-intensive task of replenishing a technical workforce?—these "how" questions show revealing variation across time and place. At stake are older generations' aspirations and expectations for new recruits' behavior, as well as up-and-coming trainees' own evolving self-image, including what they deem appropriate in their new roles (Daston & Sibum, 2003).

One thing that technical training imparts is a set of expectations or guidelines for acceptable behavior. For example, the students in Franz Neumann's nineteenth-century physics seminar in Königsberg, whom Kathryn Olesko (1991, 1995) analyzed, internalized a specific lesson about proper comportment. Neumann's students cultivated an "ethos of exactitude," learning to value rigorous error analysis above theoretical speculation. Calculating least-square deviations for discrete data points, rather than relying on graphical interpolation (which, they feared, mixed data of different degrees of quality), was more than a mathematical exercise—it became a badge of integrity. The Victorian undergraduates at Cambridge University whom Andrew Warwick (2003) has studied internalized different lessons about the scientist's proper role: success was bred from strict discipline. Unwavering mental concentration could only be achieved, they came to believe, by maintaining rigid schedules, interspersing competitive athletics (such as rowing) with coaching sessions with their mathematics tutors and several hours each day of solitary study. The reliance on individual mathematical virtuosity that this regime fostered meshed poorly with other types of training in the late nineteenth century, such as group-based on-site engineering apprenticeship, leading to bitter conflicts over what type of training—and hence what type of person—would best command the new terrain of electrical engineering (Gooday, 2004, 2005).

Often the pedagogically reinforced moral economies are deeply gendered. The post-docs in high-energy physics whom Sharon Traweek followed during the 1970s and 1980s, for example, internalized the lesson that they needed to brashly display their independence—it was no longer sufficient to complete their assigned tasks competently, as might have been expected of them as graduate students. They learned not to ask questions in front of certain people and to roundly disparage certain types of remarks from their peers (Traweek, 1988).[5]

Even the resources for research can be imbued with symbolic meaning. For example, dozens of leading American physicists worried that the rapid influx of federal funding after World War II was spoiling the values of the new generation. They cast a suspicious eye on the hordes of graduate students flooding their departments, complaining that the new recruits treated physics like a 9-to-5 job, a mere career rather than a calling. Many of the new students, meanwhile, daydreamed of parlaying their scientific training into a comfortable middle-class lifestyle. As their tools of training shifted to ever-larger group projects, their self-identity tended more and more to the practical teamworker rather than the individual *Kulturträger* (Kaiser, 2004; Hermanowicz, 1998).

Of course, the proper behavior of "practical teamworkers" has not been constant across time, place, or field. As Robert Kohler (1994) has shown, the young "drosophilists" who flourished during the early decades of the twentieth century forged a distinctive pattern of behavior, centered around the exchange of fruit fly stocks. These exchanges should always be reciprocal, never for cash; always swapped with "full disclosure" on both sides about research plans and know-how; and while research problems could be "owned," tools and materials could not be. To be a functioning member of the fruit fly genetics community meant adopting these customs and shaping one's behavior and practices accordingly. Many concerns have been voiced more recently, meanwhile, about the purported threats to long-standing scientific values by corporate interests on university campuses. Should graduate students, postdocs, and faculty learn to chase the bottom line (in the form of proprietary information controlled by industrial sponsors) or labor for the free exchange of scientific and technical information? The heat and light such questions can elicit reveals a contemporary moral economy in transition (Hackett, 1990; Mody, 2006).

Scientific and technical training thus forges communities of practitioners who share broadly similar values, norms, and self-understandings. Students must learn what it means to be a scientist or engineer—not (or not only) in the abstract, but as enacted through daily interactions within specific settings. Throughout their training they internalize these lessons, acculturating to their discipline's moral economy.

PRACTICES AND SKILLS

Studies such as Kohler's shed light on more than one aspect of the "how" question: not only how a distinct moral economy is fostered within a scientific community but also how members of that community craft research practices and pass them along to new recruits. As Kohler (1994) demonstrates, *Drosophila melanogaster* was never a research tool outside of a specific community of drosophilists and a specific set of social, political, and economic ties that these researchers forged and shared. It took a lot of work to domesticate the nascent community of fruit fly investigators to share their stocks of mutant fly varieties, communicate their findings, and regulate intellectual property claims. All the while, these same drosophilists had to work hard to domesticate a particular variation of the fly into a useful and interpretable tool. Just as moral economies are substantiated through pedagogy, so too are the tools and techniques that make up everyday scientific life.

Such a focus on scientific practices and embodied skills represents a return to a previously forgotten Kuhnian legacy. As Joseph Rouse (1987) has analyzed so clearly, two distinct visions of science reside within Thomas Kuhn's *Structure of Scientific Revolutions*. The dominant interpretation, which so exercised historians and philosophers on the book's publication, centers around conceptual worldviews, incommensurable paradigms, and the theory-ladenness of observations. Yet also lurking within Kuhn's seminal study is a focus on practices and skills—on the need to master exemplars, for example, and on the incorporation of distinctive methods within a reigning paradigm. During the past two decades, scholars in science and technology studies have capitalized on this second Kuhnian motif, developing sophisticated means of analyzing the percolation of local practices in daily scientific work (Lynch, 1985a, 1993; Collins, 1992; Shapin & Schaffer, 1985; Galison, 1987, 1997; Pickering, 1995; Fujimura, 1996; Creager, 2002; Warwick, 2003; Mody, 2005; Kaiser, 2005a). This burgeoning literature on scientific practice can be pushed further still by incorporating Kuhn's famous focus on scientific training. "Practices," after all, must be practiced.

Sometimes scientific practices are inculcated via explicit means, such as the circulation of texts. Education scholars, for example, have scrutinized how formal curricula get forged and promulgated. Major initiatives, such as the quintessential Cold War "Physical Sciences Study Commission" (PSSC) in the United States, have produced spates of new teaching materials, ranging from textbooks to exercise workbooks, films, and lecture demonstrations. The challenge always remains, of course, how to align the specific goals of their authors with those of the teachers and students who encounter these texts in the classroom (Rudolph, 2002, 2005; Donahue, 1993).

Historians of chemistry have also been at the forefront of studying elements of explicit instruction such as textbooks. Contrary to the dour view of scientific textbooks (propounded by Kuhn, among many others), these books are often much more creative than usually thought. Scientific textbooks are rarely stale repositories of finished work, or mere logical reconstructions of reigning theories. Rather, for more than two centuries textbooks have provided authors, publishers, teachers, and students a forum for intellectual and pedagogical improvisation. Several prominent chemists, such as Antoine Lavoisier, Dmitri Mendeleev, and Linus Pauling used their textbooks to formulate—not just disseminate—their new visions of chemical knowledge and practice. Scores of other textbook authors, most of whose names have not survived with the same prominence to the present day, likewise experimented with their chemical textbooks, figuring out novel ways of treating such complicated topics as atomism, classification, and valence, along with their preferred protocols for investigating them (Hannaway, 1975; Lundgren & Bensaude-Vincent, 2000; Gordin, 2005; Garcia-Belmar et al., 2005; Park, 2005). Scientists in other physical sciences have likewise fashioned their textbooks as instruments in on-going intellectual debates, deftly assembling collections of tools and techniques for ready cultivation (Olesko, 1993; Kaiser, 1998, 2005a; Warwick, 2003; Hall, 2005).

Drawing on a long line of research, leading from Michael Polanyi (1962, 1966) through Harry Collins (1974, 1992) and beyond, several STS scholars have also interrogated nontextual means by which scientists have sought to transfer research practices and skills. Scientists and engineers have fashioned several distinct methods for trying to instill in their students the "tacit knowledge" needed to become competent practitioners. Early in nineteenth-century Germany, for example, physicists like Franz Neumann and Friedrich Kohlrausch taught a new type of seminar, coordinating the seminar's curriculum with more formal lectures and creating new sets of hands-on teaching exercises that the students could work on together (Olesko, 1991, 2005). Cambridge University, meanwhile, underwent a major pedagogical realignment around the same time, shifting away from a culture of Latin oral disputations and catechetical lectures on authoritative texts to paper-based examinations. Central to these changes became the Mathematical Tripos, a grueling nine-day written examination that capped students' undergraduate studies. The text-based Tripos set in motion several further changes in the instructional milieu. Students hired private coaches who worked with ten or so fee-paying students at a time, training them to tackle progressively difficult problems. Texts mattered to the new Tripos regime but only within an elaborate framework for instilling the local coaches' tacit knowledge (Warwick, 2003).

Meanwhile new traditions of laboratory-based instruction took root throughout the United States and Great Britain, emphasizing hands-on techniques rather than rote book-learning as the key to pedagogical success (Hannaway, 1976; Owens, 1985; Kohler, 1990; Gooday, 1990; Hentschel, 2002; Rudolph, 2003). Even further removed from text-based training was the apprenticeship model adopted by Victorian engineers, which often set itself in explicit contrast with the Cambridge Tripos tradition

(Gooday, 2004, 2005). Research schools have flourished throughout the nineteenth and twentieth centuries across Europe and North America, fostering the inculcation of in-house research techniques (Servos, 1990; Geison & Holmes, 1993).

In the twentieth century, scientists and engineers in several disciplines have turned more and more to postdoctoral training. Although today the "postdoc" stage often functions primarily as a "holding pattern" for young researchers—stuck waiting for a more permanent position, carrying the largest burden of day-to-day tasks in the laboratory, often without receiving full credit for their labors (Davis, 2005)—it has not always been that way. Postdoctoral training was originally developed with several goals in mind: it was meant to allow young scientists and engineers to develop the storehouse of tacit knowledge and practical skills that they would need to launch their careers, supplementing the formal course work that had filled an increasing proportion of their graduate training. Postdoctoral training, in other words, was designed to cultivate non-text-based practices and skills. Moreover, postdoctoral appointments often drive the circulation of these tacit skills. They last only a few years, and students usually conduct their postdoctoral research neither at the institutions at which they earned their doctorates nor at the institutions in which they will establish their careers (Traweek, 1988; Assmus, 1993; Delamont & Atkinson, 2001). Hence, postdocs have been custom-designed to cultivate tacit knowledge and spread it across separate communities of practitioners, leading to "postdoc cascades" driving the transfer of skills (Kaiser, 2005a; Mody, 2005).

The result: only after extensive practice, drawing on a combination of text-based and tacit routines, do research skills become second nature for new technical trainees. Only after intense pedagogical inculcation do new recruits develop the "disciplined seeing" or "hands" of accomplished practitioners (Goodwin, 1994, 1997; Doing, 2004; Mody, 2005).

DISCIPLINE(S), POWER, AND INSTITUTIONS

Training thus generates scientific knowledge by creating the tacit skills that are an inalienable part of scientific understanding and by acclimating researchers to the tools, questions, exemplars, and outlook that constitute scientific disciplines. Controlling the levers of education can, therefore, be a powerful tool in promoting one paradigm over another, one technical culture over another. Promoters of worldviews are, at heart, seeking to realize their picture of the ideal cultural agent, the pliant subject who knows and lives in a world consistent with a particular paradigm. Training and pedagogy are imbued with the politics of competing images of the ideal practitioner; survivors of this competition determine which inherited traditions will be seen as appropriate for pedagogical propagation. Foucault (1970; 1977), for one, made this point clear in his study of "discipline" and disciplines. Pedagogy is a variety of social control; not only is knowledge power, but so is the ability to decide which social actors are suitable recipients of what institutionally enshrined knowledge. Moreover, education is not merely the transmission of knowledge; it is a license to bend its subjects to an authority's view of the world, to make them move, talk, and eventually think like "normal" citizens.[6]

Arguments about training methods are, therefore, integral to the formation of scientific disciplines and the maintenance of boundaries between them (Gieryn, 1983, 1999). Examination of the institutions of technical training, then, offers dramatic examples of how disciplinary jockeying over "jurisdictions" (Abbott, 1988) or terrains of work and expertise intersects with institutional jockeying for power and resources. For instance, as the engineering disciplines were professionalizing in the United States (and elsewhere) from the 1880s to the 1920s, many of the questions at stake in professionalization (e.g., does the engineer serve a client or the public? Who should count as an engineer? Are engineers technical experts or executives?) were played out among the faculties of schools like the Massachusetts Institute of Technology (Layton, 1971; Noble, 1977; Servos, 1980; Carlson, 1988).

Importantly, these fights had both a disciplinary and local aspect. As Christophe Lécuyer (1995) has shown, MIT's future was contested in these years by factions of faculty and local elites who organized around different notions of the relationship between science, engineering, and political reform. Some faculty members saw engineering education as a populist alternative to Harvard and other bastions of "classical studies" and constructed MIT as a "school of industrial science." Others viewed engineering as (mere) "applied science" and attempted to implement that definition by turning MIT into a research school, training its students in basic science and sending them out to apply that knowledge as engineering. Somewhat later, Dugald Jackson and his allies promoted a vision of engineering as a branch of management and pushed for an MIT that would train engineers to serve the giant research-oriented firms of emerging corporate America. The MIT that survived these disputes was forged by a later generation of professors who had trained under (and hence were loyal to) the early populist faction, yet whose careers depended on the "applied science" and "engineering as management" factions—that is, training and career arc came full circle as determinants of MIT's organization.

Thus, institutional and disciplinary politics can, through standards and curricula, make real the different visions of what a discipline is and how it relates to its competitors—technical communities compete in institutions of learning for the recruits who will make credible the disciplines' claims to work jurisdictions. Yet, as Foucault pointed out, pedagogy is not just standards and curricula; it is a process that unfolds within specific places and architectures (not limited to universities, of course) via the maintenance of specific relationships of power. Education brings its subjects within the reach of power; students are not merely taught but also watched, graded, measured, tested, punished, and otherwise surveilled and disciplined on their way to becoming full-fledged members of society and practitioners of their field. These observations have slowly filtered into STS, but often lacking Foucault's emphasis on pedagogy. Much of the past decade's interest in the architecture of science (Galison & Thompson, 1999; Gieryn, 1998; Lynch, 1991; Thompson, 2002; Hannaway, 1986; Henke & Gieryn, chapter 15 in this volume), for instance, derives from Foucault's focus on built environment, yet few STS scholars have analyzed the experimental workplace as a pedagogical site that simultaneously fosters knowledge creation and the training/disciplining of knowing subjects.[7]

A focus on pedagogy would similarly illustrate Foucault's point that knowledge and power are often most closely linked when they are most asymmetrically distributed, in particular when knowledge becomes a tool for advancing state power. For instance, in the heat of the Cold War, nuclear weapons designers at America's national laboratories gradually instituted a rich system of training in which young designers were apprenticed to their elders, with novices slowly demonstrating to their overseers that they had learned the necessary tacit skills through participation in the ritual cycle of nuclear tests (Gusterson, 1996; McNamara, 2001). Out of this system emerged fully mature designers who were enculturated to the national strategy of deterrence (a strategy many outside the labs found unfathomable) and who saw weapons science as a tool for world peace and security.

Of course, when nuclear weapons testing ended in 1992 and the pool of master designers to whom novices could be apprenticed shrank, the mesh of pedagogy and power began to unravel. Today, the U.S. nuclear establishment is obsessed with the problem of "knowledge loss" and has moved away from the informal apprenticeship model and toward classroom instruction, archiving, oral histories, and even ethnography to formalize the tacit knowledge thought to reside with older designers. Yet one result is that designers who grew up in the testing era now feel like "dinosaurs" and mourn the loss of a once vibrant training culture (Gusterson, 2005).[8] Systems of pedagogy, then, can offer a rare window on the microdomains of international politics, the emotional attachments of scientists and engineers to their epistemic cultures, and the often overlooked phenomenon of knowledge deterioration.

In other cases, pedagogical regimes put in place to reinforce national objectives and asymmetries of power can be redeployed by those whom training is meant to discipline. For instance, in colonial India, Western science was promoted by British administrators and seized on by local elites as a way to enculturate Indians to British values and practices and to preserve those elites' position in Indian society (Prakash, 1992; Raina & Habib, 2004; Chakrabarti, 2004). Institutions for training Indians in Western science—museums, schools, agricultural extension—spread rapidly through India (Tomlinson, 1998). Through these institutions, many Indians accepted Western knowledge as a yardstick of progress and took on the cultural models of a progressive, educated colonial subject through which that knowledge was exported.

Yet Indians also came to use the infrastructure of scientific training as a means for pushing the colonial state to take greater responsibility for its subjects and for building networks that subtly subverted the imperial relationship. For instance, as Ian Petrie (2004) has shown, the state's response to a series of late nineteenth century famines in India was to try to restructure rural life using the latest in Western science. The agents of these changes were, in many cases, to be young Indians sent abroad for study. After 1905, though, these men increasingly traveled to land-grant colleges in the United States, both to learn about crops (sugarcane, cotton, rice) that they would be less likely to study in Britain and to absorb Progressive models of education and culture that many Indian intellectuals found "purer and healthier" than British analogues. That is, the infrastructure of technical education allowed Indian intellectuals to construct an idea of the United States as a more wholesome alternative to the Raj—an alternative that both put pressure for reform on the British administration and forged pathways of international cooperation that widened after independence.

IMPURE PEDAGOGIES

Power relations, of course, rarely run in only one direction. As the example of colonial India shows, pedagogy's use in maintaining discipline and order is never wholly successful. This is perhaps even more the case in the hybrid, impure world of modern research. Laboratories are (and have long been) diverse sites containing participants from a variety of disciplines, at different stages in their careers, and positioned in different parts of the lab hierarchy. STS scholars have recently become fascinated by such "trading zones" (Galison, 1997) as a synecdoche for larger changes in the disciplines and the creation of a global knowledge economy.[9]

It is an obvious, though underemphasized, point that pedagogy is a continuous and pervasive aspect of such trading zones. With representatives of so many different disciplines and so many "novices" and "experts" in one place, modern research organizations are rife with pedagogy; their habitués must teach each other their skills and knowledge in order to forge even the most temporary working language. In such a situation, power relations are continually reconstructed through pedagogy. As Sally Jacoby and Patrick Gonzales (1991) have pointed out, modern research is so complex that no one can understand the entirety of even small projects; thus, as often as not, "novices" (graduate and undergraduate students) can be seen teaching the "experts" (their advisors, who may be more senior but may have lost touch with lab work and not understand the particularities of research).[10]

Moreover, numerous studies, particularly of the development of scientific instruments (Rasmussen, 1997; Bromberg, 1991; Mody, 2004) have shown that the informal training of awkward newcomers at a diverse research site can foster revolutionary insights. As has been well documented, the circulation of postdocs and other itinerant researchers allows research labs both to adopt (and reinterpret) innovations developed in the postdocs' home institution, as well as to re-export postdocs to market their adopted home's practices, knowledge, technologies, and worldview (Mody, 2005)—the so-called postdoc cascade (Kaiser, 2005a). The continual exchange of graduate students and postdocs among clusters of academic, commercial, and government researchers helps knit together instrumental communities (Slaughter et al., 2002), and the back-and-forth of students across national borders co-constructs knowledge and foreign policy (Gordin, 2005; Ito, 2005; Martin-Rovet, 1995; Martin-Rovet & Carlson, 1995). Indeed, since September 11, 2001, the global trade in postdocs and graduate students has sparked major policy debates, as visa restrictions in the United States have encouraged foreign students to seek training in other countries, and as the booming economies of China and India have allowed those nations to reimport postdocs to staff their fast-growing educational infrastructure (Anon., 2005).

We argue, then, that the heterogeneity of modern research drives discovery best when it is coupled to pervasive pedagogy. As the communities of practice literature has illustrated, organizations innovate when they contain people who need to be taught (Wenger, 2000). "Trading zones" have been successful sites of research at least in part because they always contain people who are "awkward"—newcomers who elicit instruction and who provide an insider/outsider perspective. Teaching, training, and learning push researchers to reconsider their practices and introduce mutations that advance discovery. When practices are relearned and replicated, they are (as the Awkward Student shows) never replicated in "the same" way. In much of STS, this is taken to be a problem in need of explanation, a site of disagreement and controversy. Often this is the case—teaching and learning are rarely free of disagreement—but as often as not replication offers a new way to do things, a chance to unlearn old habits while teaching new ones and to generate (not just pass on) knowledge through the interaction of novices and experts.

PAYOFFS

Methodological

For ethnographers of science and technology, attention to the pedagogical dimensions of technical work can have several methodological benefits. Most ethnographers of lab and field find that aspects of the social position of the ethnographer are fruitfully shared with the social position of students and trainees.[11] Apart from the sociologist or anthropologist, students and trainees are usually the newest entrants to a laboratory setting. Often, they have the same awkward questions as the ethnographer and many of the same difficulties in adjusting to local practices, even the same insider/outsider's critical perspective on local mores.

Thus, there is room for building significant rapport with students and trainees through these commonalities of position.[12] Although the notion of rapport has come under criticism among anthropologists in the past two decades, the building of solidarities can still be a fruitful tool in coming to understand local technical cultures. Students and trainees will often have concerns or interpretations or *sub rosa* practices that they are unwilling to exhibit to their supervisors. Since, in many labs, students and trainees perform most of the day-to-day experimental or observational work, most ethnographers will find it worthwhile to participate in the distinct subculture of student life in and around the laboratory—through, for example, intramural softball teams and departmental picnics and holiday parties (Collins, 2004; Kaiser, 2004).

In finding similarities between ethnography and pedagogy, sociologists and anthropologists not only ratify their relationships with members of the laboratory, they can also legitimate their presence and their method. As the communities of practice literature has noted, newcomers are often inducted into technical practices through "legitimate peripheral participation" (Lave & Wenger, 1991)—i.e., a kind of participant-observational status at the margins of the community's activities. Ethnographers should recognize this kind of pedagogy—so-called sitting with Nelly—as a central tool of their own practice.[13] That is, the lab and the field already contain local methods for generating and passing on something like ethnographic knowledge; ethnographers should locate these practices and incorporate them into their studies. Often, when actors' methods resonate with analysts', something interesting is at stake. In this case, the similarity of pedagogy and ethnography can be used to pry open the inevitability and universality of scientific knowledge claims. This removal of ships from bottles (Collins, 1992) has traditionally been accomplished through controversy studies—i.e., through analysis of turbulent times in which actors' disagreements belie the harmony of scientific knowledge. Yet the same turbulence can be seen more routinely and less disruptively in the continual education of newcomers in the practices of technical work.[14]

Institutional

Historians as well as sociologists and anthropologists stand to gain by elevating pedagogy to a central analytic category. In particular, a close scrutiny of pedagogy offers a means of merging insights from the quantitative Mertonian tradition in sociology of science with more recent work in a constructivist vein. Institutions and infrastructure—features that are obsessively quantified in the tradition of "scientometrics"—matter deeply to the modern sciences. Trends that often extend beyond an isolated laboratory or two can easily be missed if the focus remains exclusively on the hyper-local. Yet these institutional trends themselves are rarely the whole story—budget lines and enrollment patterns never interpret themselves; structural changes always underdetermine scientists' reactions to them. Hence the need to interrogate what gets deemed "appropriate" for pedagogical propagation in a given setting—and who gets to decide? How do the exigencies of training—with all its dependence on political economy and institutional momentum—help condition what will be deemed "teachable" and most fitting for new recruits to practice and master (Kaiser, 2002, 2004, 2006)?

Moreover, training—as a practice to which large, important institutions (universities) are dedicated, and which all institutions must do in part—is an analytic category STS should share with historians and sociologists of organizations. In particular, the New Institutionalism in sociology—with its wide-ranging exploration of "institutional isomorphism"—resonates strongly with science studies (DiMaggio & Powell, 1983; DiMaggio, 1991). Clearly, the question of how and why technical knowledge spreads and becomes standardized can mutually cast light on why and how different organizations come to resemble each other. For instance, as Annalisa Salonius (forthcoming) has shown, the norm for biomedical labs in the 1960s in much of North America was "small science"—lab groups of three or four people. In the 1980s, environmental pressures on research institutions (more competitive grants, increase in total funding, the rise of "big biology" typified by the Human Genome Project) caused a new norm for much larger labs (20 or more people) to spread. This institutional isomorphism was associated with a certain kind of "knowledge isomorphism"—biomedicine moved toward questions that could be answered by larger groups and questions that alleviated new funding and personnel pressures. Yet, perhaps more importantly, the pedagogy of larger labs triggered a more complex kind of knowledge dispersion—extended and/or multiple postdoctoral stints, once rare in the field, became more common, and young researchers spent much more of their careers moving from one institution to another, bringing with them (and often demanding) the values and practices they had learned elsewhere. Biomedicine became an epistemic community founded on mobility of people, practices, and knowledge.

Science Education and Policy

We conclude by noting that, while the study of pedagogy is only recently (re)gaining ground in science and technology studies, the insights of STS have been percolating into science education circles for some time. Primary and secondary science educators find themselves in the middle of practical conundrums about the nature of science that most STS scholars experience only second- or third-hand. For decades, the prevailing model of precollege science education was a more or less positivist (or perhaps Popperian) one. Students learned (in many cases still learn) an abstract, all-purpose "scientific method" involving the advancement and testing of hypotheses and the unproblematic transmission and replicability of experimental methods; the stories of a few exemplary scientific heroes, usually with little attention to the paradigms, practices, and wider social contexts associated with those heroes; and scientific content cleaned up and dehistoricized. In the past two decades, though, science educators have begun to use science studies to challenge this model and replace it with a more ambiguous, less triumphalist view of science.

The classroom is, after all, a messy place, and some science education scholars such as Bill Carlsen and Gregory Kelly (Kelly et al., 1993; Crawford et al., 2000), Reed Stevens (Stevens & Hall, 1997, 1998; Stevens, 2000), and Wolff-Michael Roth (Roth & McGinn, 1998) have used STS to validate and enrich that messiness in ways that may make science more transparent, more publicly accountable, and less polarizing. Students will, after all, be awkward, whether intentionally or otherwise; laboratory exercises will be irreproducible, no matter how canned the procedures; and, as the debates over creationism and "intelligent design" continue to show (Numbers, 1992; Toumey, 1991; Larson, 2004), students' locally constructed knowledge of the world will be at odds with ostensibly universal scientific knowledge handed down by technical elites. By offering a picture of science as a human, temporally and culturally situated endeavor, science and technology studies can make awkwardness in the classroom a more positive experience and can prepare students better to judge the civic contributions of science once they graduate. A properly designed science curriculum could use STS to help a wider spectrum of society appropriate science for itself and make a re-envisioned science and engineering more attractive to women and minorities (Cunningham & Helms, 1998)—or, at the very least, prepare science students more adequately for the highly social (even political) world of technical work.

STS in the classroom is not, of course, an unproblematic match. As we have tried to demonstrate, education both reflects and drives cultural values; thus schools and universities have been hotly contested battlegrounds of various culture wars, including the so-called "science wars" of the 1990s. Science educators and education scholars have furiously debated the worth of "positivist" versus "postmodern" models of science (Allchin, 2004; Turner & Sullenger, 1999). Some worry that a curriculum borrowing from STS will be unteachable or even dangerous. These debates are healthy; indeed, we encourage STS scholars to reach out and engage with the pedagogical literature more closely. After twenty years of trying, STS may find the best place for "applied science and technology studies" is in education. STS's self-image as an interdisciplinary field has so far overlooked the potential ties between STS units and education departments; we encourage this to change. Finally, looking beyond primary and secondary schools, it is already apparent that science and technology studies can influence debates about higher education. University administrators and national grant officers are starting to read the STS literature, and STS scholars are starting to contribute to long-standing arguments about the role of the university, the corporatization of pedagogy, and the commercialization of knowledge (Croissant & Smith-Doerr, chapter 27 in this volume; Mirowski & Sent, chapter 26 in this volume). As we have tried to show, the view from science and technology studies on the pedagogy of science and engineering is now sophisticated and complex and potentially of importance to educators and students alike.

Notes

1. Remember that some early controversy studies also explicitly focused on physics and mathematics as the "hard cases" that would prove the feasibility of social analysis of scientific practice.

2. Collins's articulation of the Awkward Student follows immediately on, and derives from, his discussion of Wittgenstein's views on rule-following.

3. For a recent discussion, see Warwick and Kaiser (2005).

4. See, for example, Gingerich and Westman (1988); Dear (1995), and Alder (1997).

5. See also Keller (1977, 1983, 1985).

6. We take care to note, though, that this disciplining license is neither consistently used nor successful. Many teachers offer up idiosyncratic views of the world, and many students reject what they are taught and remain "awkward." Such moments of pedagogical subversion or resistance deserve study in science, as in other realms of social practice.

7. For an exception, see Ritter's (2001) study of early American science lecture halls.

8. Though as McNamara points out, this culture may not be dying so much as reorienting to new tools and new ways of making connections between different knowledge domains. See also Clifford (1988).

9. The trading zone is a "place" where different kinds of practitioners meet and collaborate, construct local interlanguages for mediating those collaborations, and exchange artifacts, techniques, ideas, shortcuts, personnel, and other cultural materiel, and knowledge.

10. See also the other work of the team of Ochs, Jacoby, and Gonzalez: Ochs & Jacoby (1997), Ochs et al. (1994), and Ochs et al. (1996).

11. Doing (2004) nicely illustrates this by using stories from the author's own training as a synchrotron operator to politicize the notion of tacit skills. See also Latour and Woolgar's (1986) use of the perspective of an awkward technician-in-training.

12. Traweek (1988) contains probably the most self-aware explication of student-ethnographer rapport.

13. As Lave and Wenger explain, "sitting with Nelly" is shorthand for a kind of training common in cottage industries and the early industrial revolution. Newcomers to an organization received little or no formal instruction; instead, they sat next to an experienced practitioner ("Nelly"—most pieceworkers were women), observing and asking questions until they could go and replicate the work themselves. The similarity to many kinds of ethnography should be obvious.

14. Goodwin (1994, 1996, 1997) offers excellent examples by showing how the student and ethnographer are similarly unable to see and live in the same world as the adept until they have been through an extended, embodied process of perceptual realignment.

References

Abbott, Andrew (1988) *System of Professions: An Essay on the Division of Expert Labor* (Chicago: University of Chicago Press).

Ailes, Catherine P. & Francis W. Rushing (eds) (1982) *The Science Race: Training and Utilization of Scientists and Engineers, US and USSR* (New York: Crane, Russak).

Alder, Ken (1997) *Engineering the Revolution: Arms and Enlightenment in France, 1763–1815* (Princeton, NJ: Princeton University Press).

Allchin, Douglas (2004) "Should the Sociology of Science Be Rated X?" *Science Education* 88: 934–46.

Anon. (2005) "The Changing Nature of US Physics," *Nature Materials* 4: 185.

Arum, Richard & Irene Beattie (eds) (2000) *The Structure of Schooling: Readings in the Sociology of Education* (Mountain View, CA: Mayfield).

Assmus, Alexi (1993) "The Creation of Postdoctoral Fellowships and the Siting of American Scientific Research," *Minerva* 31: 151–83.

Bosk, Charles L. (1979) *Forgive and Remember: Managing Medical Failure* (Chicago: University of Chicago Press).

Bourdieu, Pierre (1988) *Homo Academicus*, trans. Peter Collier (Stanford, CA: Stanford University Press).

Bourdieu, Pierre & Jean-Claude Passeron (1977) *Reproduction in Education, Society, and Culture*, trans. Richard Nice (London: Sage).

Bromberg, Joan Lisa (1991) *The Laser in America, 1950–1970* (Cambridge, MA: MIT Press).

Cahan, David (1985) "The Institutional Revolution in German Physics, 1865–1914," *Historical Studies in the Physical Sciences* 15: 1–66.

Carlson, Bernard (1988) "Academic Entrepreneurship and Engineering Education: Dugald C. Jackson and the MIT-GE Cooperative Engineering Course, 1907–1932," *Technology and Culture* 29: 536–67.

Cavell, Stanley (1990) *Conditions Handsome and Unhandsome: The Constitution of Emersonian Perfectionism* (Chicago: University of Chicago Press).

Chakrabarti, Pratik (2004) *Western Science in Modern India: Metropolitan Methods, Colonial Practices* (Delhi: Permanent Black).

Clark, Burton R. (ed) (1993) *The Research Foundations of Graduate Education: Germany, Britain, France, United States, Japan* (Berkeley: University of California Press).

Clark, Burton R. (1995) *Places of Inquiry: Research and Advanced Education in Modern Universities* (Berkeley: University of California Press).

Clifford, James (1988) "Identity in Mashpee," in *The Predicament of Culture: Twentieth-Century Ethnography, Literature, and Art* (Cambridge, MA: Harvard University Press): 277–346.

Collins, H. M. (1974) "TEA Set—Tacit Knowledge and Scientific Networks," *Science Studies* 4(2): 165–85.

Collins, H. M. (1975) "The Seven Sexes: A Study in the Sociology of a Phenomenon, or the Replication of Experiments in Physics," *Sociology* 9 (2): 205–24.

Collins, H. M. (1992) *Changing Order: Replication and Induction in Scientific Practice* (Chicago: University of Chicago Press).

Collins, H. M. (1998) "The Meaning of Data: Open and Closed Evidential Cultures in the Search for Gravitational Waves," *American Journal of Sociology* 104(2): 293–338.

Collins, Harry (2004) *Gravity's Shadow: The Search for Gravitational Waves* (Chicago: University of Chicago Press).

Crawford, Teresa, Gregory J. Kelly, & Candice Brown (2000) "Ways of Knowing Beyond Facts and Laws of Science: An Ethnographic Investigation of Student Engagement in Scientific Practices," *Journal of Research in Science Teaching* 37(3): 237–58.

Creager, Angela (2002) *The Life of a Virus: Tobacco Mosaic Virus as an Experimental Model, 1930–1965* (Chicago: University of Chicago Press).

Cunningham, C. M. & J. V. Helms (1998) "Sociology of Science as a Means to a More Authentic, Inclusive Science Education," *Journal of Research in Science Teaching* 35(5): 483–99.

Daston, Lorraine (1995) "The Moral Economy of Science," *Osiris* 10: 2–24.

Daston, Lorraine & Otto Sibum (eds) (2003) *Scientific Personae and their Histories*, published as *Science in Context* 16: 1–269.

Davis, Geoffrey (2005) "Doctors without Orders," *American Scientist* 93(3): supplement.

Dear, Peter (1995) *Discipline and Experience: The Mathematical Way in the Scientific Revolution* (Chicago: University of Chicago Press).

Delamont, Sara & Paul Atkinson (2001) "Doctoring Uncertainty: Mastering Craft Knowledge," *Social Studies of Science* 31: 87–107.

Dennis, Michael A. (1994) "Our First Line of Defense—Two University Laboratories in the Postwar American State," *Isis* 85(3): 427–55.

DiMaggio, Paul J. (1991) "Constructing an Organizational Field as a Professional Project: U.S. Art Museums, 1920–1940," in Walter W. Powell & Paul J. DiMaggio (eds), *The New Institutionalism in Organizational Analysis* (Chicago: University of Chicago Press): 267–92.

DiMaggio, Paul J. & Walter W. Powell (1983) "The Iron Cage Revisited: Institutional Isomorphism and Collective Rationality in Organizational Fields," *American Sociological Review* 48: 147–160.

Doing, Park (2004) "Lab Hands and the Scarlet 'O': Epistemic Politics and Scientific Labor," *Social Studies of Science* 34(3): 299–323.

Donahue, David (1993) "Serving Students, Science, or Society? The Secondary School Physics Curriculum in the United States, 1930–65," *History of Education Quarterly* 33: 321–52.

Foucault, Michel (1970) *The Order of Things: An Archaeology of the Human Sciences* (New York: Vintage Books).

Foucault, Michel (1977) *Discipline and Punish: The Birth of the Prison*, trans. Alan Sheridan (New York: Vintage Books).

Foucault, Michel (1994) *The Birth of the Clinic: An Archaeology of Medical Perception*, trans. A. M. Sheridan-Smith (New York: Vintage Books).

Fox, Robert & Anna Guagnini (eds) (1993) *Education, Technology, and Industrial Performance in Europe, 1850–1939* (Cambridge: Cambridge University Press).

Fujimura, Joan (1996) *Crafting Science: A Sociohistory of the Quest for the Genetics of Cancer* (Cambridge, MA: Harvard University Press).

Galison, Peter (1987) *How Experiments End* (Chicago: University of Chicago Press).

Galison, Peter (1997) *Image and Logic: A Material Culture of Microphysics* (Chicago: University of Chicago Press).

Galison, Peter & Bruce Hevly (eds) (1992) *Big Science: The Growth of Large-Scale Research* (Stanford, CA: Stanford University Press).

Galison, Peter & Emily Thompson (eds) (1999) *The Architecture of Science* (Cambridge, MA: MIT Press).

Garcia-Belmar, Antonio, Jose Ramon Bertomeu, & Bernadette Bensaude-Vincent (2005) "The Power of Didactic Writings: French Chemistry Textbooks of the Nineteenth Century," in D. Kaiser (ed.), *Pedagogy and the Practice of Science: Historical and Contemporary Perspectives* (Cambridge, MA: MIT Press): 219–51.

Geiger, Roger (1986) To Advance Knowledge: The Growth of American Research Universities, 1900–1940 (New York: Oxford University Press).

Geiger, Roger (1993) *Research and Relevant Knowledge: American Research Universities Since World War II* (New York: Oxford University Press).

Geison, Gerald & Frederic L. Holmes (eds) (1993) *Research Schools: Historical Reappraisals*, published as *Osiris* 8: 1–238.

Gieryn, Thomas F. (1983) "Boundary-Work and the Demarcation of Science from Non-Science: Strains and Interests in Professional Ideologies of Scientists," *American Sociological Review* 48(December): 781–95.

Gieryn, Thomas F. (1998) "Biotechnology's Private Parts (and Some Public Ones)," in C. Smith & J. Agar (eds), *Making Space for Science: Territorial Themes in the Shaping of Knowledge* (London: Macmillan): 281–312.

Gieryn, Thomas F. (1999) *Cultural Boundaries of Science: Credibility on the Line* (Chicago: University of Chicago Press).

Gingerich, Owen & Robert Westman (1988) "The Wittich Connection: Conflict and Priority in Late Sixteenth-Century Cosmology," *Transactions of the American Philosophical Society* 78: 1–148.

Gooday, G. (1990) "Precision-Measurement and the Genesis of Physics Teaching Laboratories in Victorian Britain," *British Journal for the History of Science* 23(76): 25–51.

Gooday, Graeme (2004) *The Morals of Measurement: Accuracy, Irony, and Trust in Late Victorian Electrical Practice* (Cambridge: Cambridge University Press).

Gooday, Graeme (2005) "Fear, Shunning, and Valuelessness: Controversy over the Use of 'Cambridge' Mathematics in Late Victorian Electro-technology," in D. Kaiser (ed) *Pedagogy and the Practice of Science: Historical and Contemporary Perspectives* (Cambridge, MA: MIT Press): 111–49.

Goodwin, Charles (1994) "Professional Vision," *American Anthropologist* 96(3): 606–33.

Goodwin, Charles (1996) "Practices of Color Classification," *Ninchi Kagaku (Cognitive Studies: Bulletin of the Japanese Cognitive Science Society)* 3(2): 62–82.

Goodwin, Charles (1997) "The Blackness of Black: Color Categories as Situated Practice," in L. Resnick, R. Saljo, C. Pontecorvo, & B. Burge (eds), *Discourse, Tools, and Reasoning: Essays on Situated Cognition* (Berlin: Springer-Verlag): 111–42.

Gordin, Michael (2005) "Beilstein Unbound: The Pedagogical Unraveling of a Man and His *Handbuch*," in D. Kaiser (ed), *Pedagogy and the Practice of Science: Historical and Contemporary Perspectives* (Cambridge, MA: MIT Press): 11–39.

Gusterson, Hugh (1996) *Nuclear Rites: A Weapons Laboratory at the End of the Cold War* (Berkeley: University of California Press).

Gusterson, Hugh (2005) "A Pedagogy of Diminishing Returns: Scientific Involution Across Three Generations of Nuclear Weapons Science," in D. Kaiser (ed), *Pedagogy and the Practice of Science: Historical and Contemporary Perspectives* (Cambridge, MA: MIT Press): 75–107.

Hackett, Edward J. (1990) "Science as a Vocation in the 1990s: The Changing Organizational Culture of Academic Science," *Journal of Higher Education* 61: 241–79.

Hall, Karl (2005) "'Think Less about Foundations': A Short Course on Landau and Lifshitz's *Course of Theoretical Physics*," in D. Kaiser (ed), *Pedagogy and the Practice of Science: Historical and Contemporary Perspectives* (Cambridge, MA: MIT Press): 253–86.

Hannaway, Owen (1975) *The Chemists and the Word: The Didactic Origins of Chemistry* (Baltimore, MD: Johns Hopkins University Press).

Hannaway, Owen (1976) "The German Model of Chemical Education in America: Ira Remsen at Johns Hopkins (1876–1913)," *Ambix* 23: 145–64.

Hannaway, Owen (1986) "Laboratory Design and the Aim of Science: Andreas Libavius versus Tycho Brahe," *Isis* 77: 584–610.

Hargens, Lowell L. & J. Scott Long (2002) "Demographic Inertia and Women's Representation among Faculty in Higher Education," *Journal of Higher Education* 73: 494–517.

Heilbron, John (1992) "Creativity and Big Science," *Physics Today* 45(Nov): 42–47.

Hentschel, Klaus (2002) *Mapping the Spectrum: Techniques of Visual Representation in Research and Teaching* (Oxford: Oxford University Press).

Hermanowicz, Joseph (1998) The Stars Are Not Enough: Scientists—Their Passions and Professions (Chicago: University of Chicago Press).

Hofstadter, Richard (1963) *Anti-intellectualism in American Life* (New York: Knopf).

Ito, Kenji (2004) "Gender and Physics in Early 20th Century Japan: Yuasa Toshiko's Case," *Historia Scientiarum* 14(Nov): 118–36.

Ito, Kenji (2005) "The Geist in the Institute: The Production of Quantum Physicists in 1930s Japan," in D. Kaiser (ed), *Pedagogy and the Practice of Science: Historical and Contemporary Perspectives* (Cambridge, MA: MIT Press): 151–83.

Jacoby, Sally & Patrick Gonzales (1991) "The Constitution of Expert-Novice in Scientific Discourse," *Issues in Applied Linguistics* 2(2): 149–81.

Kaiser, David (1998) "A Psi Is Just a Psi? Pedagogy, Practice, and the Reconstitution of General Relativity, 1942–1975," *Studies in History and Philosophy of Modern Physics 29B* (3): 321–38.

Kaiser, David (2002) "Cold War Requisitions, Scientific Manpower, and the Production of American Physicists after World War II," *Historical Studies in the Physical and Biological Sciences* 33: 131–59.

Kaiser, David (2004) "The Postwar Suburbanization of American Physics," *American Quarterly* 56: 851–88.

Kaiser, David (2005a) *Drawing Theories Apart: The Dispersion of Feynman Diagrams in Postwar Physics* (Chicago: University of Chicago Press).

Kaiser, David (ed) (2005b) *Pedagogy and the Practice of Science: Historical and Contemporary Perspectives* (Cambridge, MA: MIT Press).

Kaiser, David (2006) "Whose Mass Is It Anyway? Particle Cosmology and the Objects of Theory," *Social Studies of Science* 36: 533–64.

Keller, Evelyn Fox (1977) "The Anomaly of a Woman in Physics," in S. Ruddick & P. Daniels (eds), *Working It Out: 23 Women Writers, Artists, Scientists, and Scholars Talk about Their Lives and Work* (New York: Pantheon): 78–91.

Keller, Evelyn Fox (1983) *A Feeling for the Organism: The Life and Work of Barbara McClintock* (San Francisco, CA: W. H. Freeman).

Keller, Evelyn Fox (1985) *Reflections on Gender and Science* (New Haven, CT: Yale University Press).

Kelly, G. J., W. S. Carlsen, & C. M. Cunningham (1993) "Science-Education in Sociocultural Context—Perspectives from the Sociology of Science," *Science Education* 77(2): 207–20.

Kliebard, Herbert (1999) *Schooled to Work: Vocationalism and the American Curriculum, 1876–946* (New York: Teachers College Press).

Kohler, Robert E. (1987) "Science, Foundations, and American Universities in the 1920s," *Osiris* 3: 135–64.

Kohler, Robert (1990) "The Ph.D. Machine: Building on the Collegiate Base," *Isis* 81: 638–62.

Kohler, Robert (1994) *Lords of the Fly:* Drosophila *Genetics and the Experimental Life* (Chicago: University of Chicago Press).

Krige, John (2000) "NATO and the Strengthening of Western Science in the Post-Sputnik Era," *Minerva* 38: 81–108.

Kuhn, Thomas S. (1962) *The Structure of Scientific Revolutions* (Chicago: University of Chicago Press).

Larson, Edward J. (2004) *Evolution: The Remarkable History of a Scientific Theory* (New York: Modern Library).

Latour, Bruno & Steve Woolgar (1986) *Laboratory Life: The Construction of Scientific Facts* (Princeton, NJ: Princeton University Press).

Lave, Jean & Etienne Wenger (1991) *Situated Learning: Legitimate Peripheral Participation* (Cambridge: Cambridge University Press).

Layton, Edwin T., Jr. (1971) *The Revolt of the Engineers: Social Responsibility and the American Engineering Profession* (Cleveland, OH: Press of Case Western Reserve University).

Lecuyer, Christophe (1995) "MIT, Progressive Reform, and 'Industrial Service,' 1890–1920," *Historical Studies in the Physical and Biological Sciences* 26: 35–88.

Lemann, Nicholas (1999) *The Big Test: The Secret History of the American Meritocracy* (New York: Farrar, Straus, Giroux).

Leslie, Stuart W. (1993) *The Cold War and American Science: The Military-Industrial-Academic Complex at MIT and Stanford* (New York: Columbia University Press).

Levin, Sharon G. & Paula E. Stephan (1998) "Gender Differences in the Rewards to Publishing in Academe: Science in the 1970s," *Sex Roles* 38(11/12): 1049–64.

Lowen, Rebecca S. (1997) *Creating the Cold War University: The Transformation of Stanford* (Berkeley: University of California Press).

Ludmerer, Kenneth (1985) *Learning to Heal: The Development of American Medical Education* (New York: Basic Books).

Lundgren, Anders & Bernadette Bensaude-Vincent (eds) (2000) *Communicating Chemistry: Textbooks and Their Audiences, 1789–1939* (Canton, MA: Science History Publications).

Lynch, Michael (1985a) *Art and Artifact in Laboratory Science: A Study of Shop Work and Shop Talk in a Research Laboratory* (London: Routledge & Kegan Paul).

Lynch, Michael (1985b) "Discipline and the Material Form of Images: An Analysis of Scientific Visibility," *Social Studies of Science* 15: 37–66.

Lynch, Michael (1991) "Laboratory Space and the Technological Complex: An Investigation of Topical Contextures," *Science in Context* 4(1): 51–78.

Lynch, Michael (1993) *Scientific Practice and Ordinary Action: Ethnomethodology and Social Studies of Science* (Cambridge: Cambridge University Press).

MacKenzie, Donald (1990) *Inventing Accuracy: A Historical Sociology of Nuclear Missile Guidance* (Cambridge, MA: MIT Press).

Manning, Kenneth (1983) *Black Apollo of Science: The Life of Ernest Everett Just* (New York: Oxford University Press).

Martin-Rovet, D. (1995) "The International Exchange of Scholars—the Training of Young Scientists Through Research Abroad—1: Young French Scientists in the United States," *Minerva* 33(1): 75–98.

Martin-Rovet, Dominique & Timothy Carlson (1995) "The International Exchange of Scholars—the Training of Young Scientists Through Research Abroad—2: American Scientists in France," *Minerva* 33(2): 171–91.

McNamara, Laura (2001) "'Ways of Knowing' About Weapons: The Cold War's End at the Los Alamos National Laboratory," Ph.D. diss., Department of Anthropology, University of New Mexico, Albuquerque.

Mody, Cyrus C. M. (2004) "How Probe Microscopists Became Nanotechnologists," in D. Baird, A. Nordmann, & J. Schummer (eds), *Discovering the Nanoscale* (Amsterdam: IOS Press): 119–33.

Mody, Cyrus C. M. (2005) "Instruments in Training: The Growth of American Probe Microscopy in the 1980s," in D. Kaiser (ed), *Pedagogy and the Practice of Science: Historical and Contemporary Perspectives* (Cambridge, MA: MIT Press): 185–216.

Mody, Cyrus C. M. (2006) "Universities, Corporations, and Instrumental Communities: Commercializing Probe Microscopy, 1981–1996," *Technology and Culture* 47: 56–80.

Monk, Ray (1990) *Ludwig Wittgenstein: The Duty of Genius* (New York: Penguin).

Mukerji, Chandra (1989) *A Fragile Power: Scientists and the State* (Princeton, NJ: Princeton University Press).

Murray, Margaret A. M. (2000) *Women Becoming Mathematicians: Creating a Professional Identity in Post–World War II America* (Cambridge, MA: MIT Press).

Noble, David (1977) *America by Design: Science, Technology, and the Rise of Corporate Capitalism* (Oxford: Oxford University Press).

Numbers, Ronald L. (1992) *The Creationists: The Evolution of Scientific Creationism* (New York: Knopf).

Ochs, Elinor, Patrick Gonzales, & Sally Jacoby (1996) " 'When I Come Down I'm in the Domain State': Grammar and Graphic Representation in the Interpretive Activity of Physicists," in E. Ochs, Emanuel A. Schegloff, & Sandra A. Thompson (eds), *Interaction and Grammar: Studies in Interactional Sociolinguistics 13* (Cambridge: Cambridge University Press): 328–69.

Ochs, Elinor & Sally Jacoby (1997) "Down to the Wire: The Cultural Clock of Physicists and the Discourse of Consensus," *Language in Society* 26: 479–505.

Ochs, Elinor, Sally Jacoby, & Patrick Gonzales (1994) "Interpretive Journeys: How Physicists Talk and Travel through Graphic Space," *Configurations* 2(1): 151–71.

Oldenziel, Ruth (2000) "Multiple-entry Visas: Gender and Engineering in the U.S., 1870–1945," in A. Canel, R. Oldenziel, & K. Zachmann (eds), *Crossing Boundaries, Building Bridges: Comparing the History of Women Engineers, 1870s–1990s* (Amsterdam: Harwood): 11–49.

Olesko, Kathryn (1991) *Physics as a Calling: Discipline and Practice in the Konigsberg Seminar for Physics* (Ithaca, NY: Cornell University Press).

Olesko, Kathryn (1993) "Tacit Knowledge and School Formation," *Osiris* 8: 16–29.

Olesko, Kathryn (1995) "The Meaning of Precision: The Exact Sensibility in Early 19th-century Germany," in M. N. Wise (ed), *The Values of Precision* (Princeton, NJ: Princeton University Press): 103–34.

Olesko, Kathryn (2005) "The Foundations of a Canon: Kohlrausch's Practical Physics," in D. Kaiser (ed), *Pedagogy and the Practice of Science: Historical and Contemporary Perspectives* (Cambridge, MA: MIT Press): 323–56.

Owens, Larry (1985) "Pure and Sound Government: Laboratories, Playing Fields, and Gymnasia in the Nineteenth-Century Search for Order," *Isis* 76: 182–94.

Park, Buhm Soon (2005) "In the 'Context of Pedagogy': Teaching Strategy and Theory Change in Quantum Chemistry," in D. Kaiser (ed), *Pedagogy and the Practice of Science: Historical and Contemporary Perspectives* (Cambridge, MA: MIT Press): 287–319.

Pearson, Willie, Jr. & Alan Fechter (eds) (1994) *Who Will Do Science? Educating the Next Generation* (Baltimore, MD: Johns Hopkins University Press).

Petrie, Ian (2004) " 'Practical Agriculture' for the Colonies: Cornell and South Asian Uplift, c. 1905–45," presentation to Agricultural History Society Symposium, Cornell University, Ithaca, NY.

Pickering, Andrew (1984) *Constructing Quarks: A Sociological History of Particle Physics* (Chicago: University of Chicago Press).

Pickering, Andrew (1995) *The Mangle of Practice: Time, Agency, and Science* (Chicago: University of Chicago Press).

Pinch, Trevor (1986) *Confronting Nature: The Sociology of Solar-neutrino Detection* (Dordrecht, Netherlands: Reidel).

Polanyi, Michael (1962) *Personal Knowledge: Towards a Post-Critical Philosophy* (New York: Harper Torchbooks).

Polanyi, Michael (1966) *The Tacit Dimension* (Garden City, NY: Doubleday).

Prakash, Gyan (1992) "Science 'Gone Native' in Colonial India," *Representations* 40(Fall): 153–78.

Preston, Anne E. (1994) "Why Have All the Women Gone? A Study of Exit of Women from the Science and Engineering Professions," *American Economic Review* 84(5): 1446–62.

Pyenson, Lewis (1977) "Educating Physicists in Germany circa 1900," *Social Studies of Science* 7: 329–66.

Pyenson, Lewis (1979) "Mathematics, Education, and the Goettingen Approach to Physical Reality, 1890–1914," *Europa* 2: 91–127.

Raina, Dhruv & S. Irfan Habib (2004) *Domesticating Modern Science: A Social History of Science and Culture in Colonial India* (New Delhi: Tulika Books).

Rasmussen, Nicolas (1997) *Picture Control: The Electron Microscope and the Transformation of Biology in America, 1940–1960* (Stanford, CA: Stanford University Press).

Ritter, Christopher (2001) "Re-presenting Science: Visual and Didactic Practice in Nineteenth-century Chemistry," Ph.D. diss., Department of History of Science, University of California, Berkeley.

Rosenberg, Charles E. (1979) "Toward an Ecology of Knowledge: On Discipline, Context, and History," in A. Oleson & J. Voss (eds), *The Organization of Knowledge in Modern America, 1860–1920* (Baltimore, MD: Johns Hopkins University Press): 440–55.

Rosser, S. V. (2004) *The Science Glass Ceiling: Academic Women Scientists and the Struggle to Succeed* (New York: Routledge).

Rossiter, Margaret (1980) " 'Women's Work' in Science, 1880–1920," *Isis* 71(258): 381–98.

Rossiter, Margaret W. (1982) "Doctorates for American Women, 1868–1907," *History of Education Quarterly* 22(2): 159–83.

Rossiter, Margaret W. (1986) "Graduate Work in the Agricultural Sciences, 1900–1970," *Agricultural History* 60(2): 37–57.

Rossiter, Margaret W. (1995) *Women Scientists in America: Before Affirmative Action, 1940–1972* (Baltimore, MD: Johns Hopkins University Press).

Roth, Wolff-Michael & M. K. McGinn (1998) "Knowing, Researching, and Reporting Science Education: Lessons from Science and Technology Studies," *Journal of Research in Science Teaching* 35(2): 213–35.

Rouse, Joseph (1987) *Knowledge and Power: Toward a Political Philosophy of Science* (Ithaca, NY: Cornell University Press).

Rudolph, John (2002) *Scientists in the Classroom: The Cold War Reconstruction of American Science Education* (New York: Palgrave).

Rudolph, John (2003) "Portraying Epistemology: School Science in Historical Context," *Science Education* 87: 64–79.

Rudolph, John (2005) "Turning Science to Account: Chicago and the General Science Movement in Secondary Education, 1905–1920," *Isis* 96: 353–89.

Salonius, Annalisa (forthcoming) "Social Organization of Work in Biomedical Research Labs: Socio-historical Dynamics and the Influence of Research Funding."

Servos, John (1980) "The Industrial Relations of Science: Chemical Engineering at MIT, 1900–1939," *Isis* 71: 531–49.

Servos, John (1990) *Physical Chemistry from Ostwald to Pauling: The Making of a Science in America* (Princeton, NJ: Princeton University Press).

Shapin, Steven (1991) " 'A Scholar and a Gentleman': The Problematic Identity of the Scientific Practitioner in Early Modern England," *History of Science* 29: 279–327.

Shapin, Steven & Simon Schaffer (1985) *Leviathan and the Air-Pump: Hobbes, Boyle, and the Experimental Life* (Princeton, NJ: Princeton University Press).

Shinn, Terry (2003) "The Industry, Research, and Education Nexus," in M. J. Nye (ed), *The Cambridge History of Science*, vol. V: *The Modern Physical and Mathematical Sciences* (Cambridge: Cambridge University Press).

Slaton, Amy (2004) "Urban Renewal and the Whiteness of Engineering: Historical Origins of an Occupational Color Line," Drexel University Dean's Seminar Series, Philadelphia, November 10.

Slaughter, Sheila, Teresa Campbell, Margaret Holleman, & Edward Morgan (2002) "The 'Traffic' in Graduate Students: Graduate Students as Tokens of Exchange Between Academe and Industry," *Science, Technology & Human Values* 27: 282–312.

Solomon, Barbara Miller (1985) *In the Company of Educated Women: A History of Women and Higher Education in America* (New Haven, CT: Yale University Press).

Spring, Joel (1989) *The Sorting Machine Revisited: National Educational Policy Since 1945* (New York: Longman).

Starr, Paul (1982) *The Social Transformation of American Medicine* (New York: Basic Books).

Stephan, Paula E. & Sharon G. Levin (2005) "Leaving Careers in IT: Gender Differences in Retention," *Journal of Technology Transfer* 30(4): 383–96.

Stevens, Reed (2000) "Who Counts What as Math: Emergent and Assigned Mathematics Problems in a Project-based Classroom," in J. Boaler (ed), *Multiple Perspectives on Mathematics Teaching and Learning* (New York: Elsevier).

Stevens, Reed & Rogers Hall (1997) "Seeing Tornado: How Video Traces Mediate Visitor Understandings of (Natural?) Spectacles in a Science Museum," *Science Education* 18(6): 735–48.

Stevens, Reed & Rodgers Hall (1998) "Disciplined Perception: Learning to See in Technoscience," in M. Lampert & M. L. Blunk (eds), *Talking Mathematics in School: Studies of Teaching and Learning* (New York: Cambridge University Press): 107–49.

Stichweh, Rudolf (1984) *Zur Entstehung des modernen Systems wissenschaftlicher Disziplinen: Physik in Deutschland, 1740–1890* (Frankfurt am Main: Suhrkamp).

Sur, Abha (1999) "Aesthetics, Authority, and Control in an Indian Laboratory: The Raman-Born Controversy on Lattice Dynamics," *Isis* 90: 25–49.

Thompson, Emily (2002) *The Soundscape of Modernity: Architectural Acoustics and the Culture of Listening in America, 1900–1933* (Cambridge, MA: MIT Press).

Tomlinson, B. R. (1998) "Technical Education in Colonial India, 1880–1914: Searching for a 'Suitable Boy,'" in S. Bhattacharya (ed), *The Contested Terrain: Perspectives on Education in India* (New Delhi: Orient Longman): 322–41.

Toumey, Christopher P. (1991) "Modern Creationism and Scientific Authority," *Social Studies of Science* 21: 681–99.

Traweek, Sharon (1988) *Beamtimes and Lifetimes: The World of High Energy Physicists* (Cambridge, MA: Harvard University Press).

Traweek, Sharon (2005) "Generating High-Energy Physics in Japan: Moral Imperatives of a Future Pluperfect," in D. Kaiser (ed), *Pedagogy and the Practice of Science: Historical and Contemporary Perspectives* (Cambridge, MA: MIT Press): 357–92.

Turner, R. Steven (1987) "The Great Transition and the Social Patterns of German Science," *Minerva* 25: 56–76.

Turner, Steven & Karen Sullenger (1999) "Kuhn in the Classroom, Lakatos in the Lab: Science Educators Confront the Nature-of-Science Debate," *Science, Technology & Human Values* 24: 5–30.

Warwick, Andrew (2003) *Masters of Theory: Cambridge and the Rise of Mathematical Physics* (Chicago: University of Chicago Press).

Warwick, Andrew & David Kaiser (2005) "Kuhn, Foucault, and the Power of Pedagogy," in D. Kaiser (ed), *Pedagogy and the Practice of Science: Historical and Contemporary Perspectives* (Cambridge, MA: MIT Press): 393–409.

Wenger, Etienne (2000) "Communities of Practice and Social Learning Systems," *Organization* 7(2): 225–46.

Westwick, Peter (2003) *The National Labs: Science in an American System, 1947–1974* (Cambridge, MA: Harvard University Press).

Williams, C. G. (ed) (2001) *Technology and the Dream: Reflections on the Black Experience at MIT, 1941–1999* (Cambridge, MA: MIT Press).

Wittgenstein, Ludwig (1953) *Philosophical Investigations*, trans. G. E. M. Anscombe (New York: MacMillan).

Woolgar, Steve (1990) "Time and Documents in Researcher Interaction: Some Ways of Making out What Is Happening in Experimental Science," in M. Lynch & S. Woolgar (eds), *Representation in Scientific Practice* (Cambridge, MA: MIT Press): 123–52.

17 The Coming Gender Revolution in Science

Henry Etzkowitz, Stefan Fuchs, Namrata Gupta, Carol Kemelgor, and Marina Ranga

When one considers woman's possibilities and her future . . . it is especially interesting to make a close study of their situation . . .[1]

SEXUAL SEPARATION OF SCIENTIFIC LABOR

Why is science, the quintessentially rational profession, pervaded by seemingly irrational gendered social arrangements (Glaser, 1964; Dix, 1987; Osborne, 1994; McIlwee & Robinson, 1992; Valian, 1999; Tri-national Conference, 2003; Commission on Professionals in Science and Technology, 2004; Rosser, 2004)? Paradoxically, an uneven co-evolution of science, gender, and society displaces universalistic norms of science with discriminatory social practices and invisibilizes these harms. (Merton, [1942]1973; Bielby, 1991; Ferree et al., 1999; Fox, 2001). By the late nineteenth century, a few women broke through gender barriers and entered the laboratory as "honorary men" but had to accept subordinate status. Like Lise Meitner, they were relegated to a basement lab, literally or figuratively (Sime, 1996). Marie Curie was putative junior partner to her husband, a fiction maintained after his death despite the award of successive Nobel prizes (Goldsmith, 2005). Nobelist Marie Goeppert Meyer was a research associate in her husband's university lab, reprising an earlier household gendered structure of science, until the shortage of male scientists during World War II allowed her to emerge as a researcher in her own right. Nevertheless, she did not receive an appropriate academic appointment to match her achievements until just before being awarded the highest scientific honor.

Despite the fact that women have entered academic science in ever larger numbers in recent years, they also leave traditional fields, in larger numbers than men, at each "critical transition" (Etzkowitz et al., 1995; National Science Foundation, 1996). Although lost to academia, women reappear in science-related occupations in the media, law, research management, and technology transfer that have opened up as a result of the increasing economic and social relevance of science. A "coming gender revolution in science" also transcends the traditional "sexual separation of labor" in science. Thus, the seemingly ineluctable negative relationship between female gender and scientific status is subject to change under conditions where there is (1) pressure from female scientists organizing to receive due recognition and reward as part of a broader feminist movement, (2) an ever tighter connection between human capital and economic development that militates against wasting human resources, and (3) the transformation of scientific work from hierarchical organizational to flat network structures in growing fields like biotechnology. Despite signs of change, inequality persists, making difficult the determination whether the proverbial glass is half full or half empty.

In societies where science is high status, women are excessively located in low status positions. Conversely, in societies where science is low status, women may be disproportionately found in high-level positions (Etzkowitz et al., 2000). Under "normal" conditions, women's opportunities for scientific achievement and reward are limited except when science itself is held in low regard. As the status of a scientific field rises or falls, the position of women changes concomitantly. When *Drosophila* genetics was a marginal emerging field in the early twentieth century, women were prominent in the "flyroom." As the field became established, women's presence diminished (Kohler, 1994). A similar phenomenon has been noted in computer programming. Indeed, when a subfield such as computer theory is central to one academic department and marginal to another, the participation of women is respectively suppressed and enhanced. This chapter provides a comparative global analysis of the condition of women in science and the potential for change, drawing on available statistical data and studies of women in science under contrasting economic, social, and academic systems.

THE EVOLUTION OF WOMEN IN SCIENCE

An entanglement of sex and gender in science can be identified in different political regimes and social structures, with rare anomalies. Men dominate the culture, organization, theories, and methods of science (see, e.g., Harding, 1991). More generally, from a radical feminist perspective, sex differences function as the basic and hierarchic principles on which all modern capitalist societies are built. However, in socialist societies, where an increase in numbers of female scientists occurred somewhat earlier than in capitalist regimes, a common gendered division of labor in science can be identified corresponding to traditional patriarchal formats. Science is thus neither an exception nor a special case regarding the general pattern of social relations despite an ideology of universalism. Indeed, the assumption of universalistic norms in science, as a taken for granted reality rather than as a goal, has blinded many scientists from facing persistent gender inequities in their profession. Indeed, in a self-study that occasioned much publicity, senior female scientists at MIT were astonished to discover that, despite equality in formal rank, their material condition differed significantly from that of male colleagues. Marked disparity in numbers, increasing with ascendance in rank, also characterized MIT although, in recent years, female participation at the undergraduate level has equalized rapidly.

A feudal social organization persists in academic science despite various industrial, social, and political transformations in the larger society. Patriarchal systems of asexual reproduction by doctoral fathers and grandfathers with social relationships of vassalage characterize doctoral and postdoctoral training in U.S.-oriented academic systems, whereas this mode extends up until the professorship in Germanic academia. A gendered separation of labor, with caste-like characteristics, carries intimations of "pollution" should boundaries between men and women in science be breached (de Beauvoir, 1952; Etzkowitz, 1971; Rosenberg, 1982). Women were allowed to participate in research but not to share equally in recognition when science was conducted in the home (Schiebinger, 1989; Abir-Am, 1991). They were "invisible scientists," helpmates to fathers, brothers, and husbands in the early stages of the scientific revolution in the eighteenth century.

Allowed in But Not to Fully Partake of Science

Expanded participation does not guarantee advancement for women in science given the caste-like social structure of science. As science was professionalized and industrialized, moving from home to laboratory, women became the personal support structure for male scientists in the home and then in the lab, a condition that persists to this day in attenuated form. Cultural lag is even more pronounced in developing and former colonized countries where modern science was often implanted as an isolated enclave, without engendering an indigenous intellectual and social revolution. A lack of social capital among women in science is accentuated by traditional gender roles that impede scientific networking and interaction. In a survey of Ghana, Kenya, and Kerala (Campion & Shrum, 2004), it was found that women have difficulty in pursuing research careers owing to restricted professional networks for women. Men, on the other hand, had greater external contacts through education and professional travel. Constrained by family ties and security concerns, women are less likely to travel or receive education abroad. In India, a social segregation norm makes interaction with men difficult, further reducing women's social capital (Gupta & Sharma, 2002).

Although participation rates vary significantly, women are almost invariably less present especially in the higher levels of scientific careers. UNESCO data show large disparities in women's participation in R&D in the mid-nineties, ranging from 15.7% in Austria and 26.4% in Spain to 39.6% in Russia, 41.4% in Bulgaria, and 44.4% in Romania (UNESCO, 1999). Further significant evidence in this respect is provided by the EU Report "She Figures 2003" (European Commission, 2003): in Europe, women remain in the minority in public research—34% in 2001, a slight increase from 32% in 1999—but their annual growth is 8% compared with 3.1% for men. Gender distribution in scientific fields across Europe appears to be characterized by strong common patterns: women scientists and engineers are in the minority across EU-15 (except for Finland) as Ph.D. graduates, researchers, senior university staff, and members of scientific boards, and they remain under-represented as researchers, particularly in the business enterprise sector, where they account for only 15%. Women are less present in engineering and natural sciences but constitute the majority in health and medical sciences, humanities, and social sciences, in both higher education and research.

Significant gender differences have also been highlighted at the decision-making level as well as in research funding, where significant differences between the success rates of women and men have been reported in the U.K., Germany, Sweden, Austria, and Hungary.

So-called countries in transition, former socialist regimes in Eastern Europe, recruited large numbers of persons, including women, to scientific professions. Nevertheless, a similar picture of gender stratification can be found in the Associated Countries of the European Union, with the exceptions of Bulgaria and Romania, where women are least represented in the higher education sector. In previous socialist societies where large numbers of women were recruited into science, traditional gender relations trumped social ideals and females were seldom allowed to hold leadership positions in science (Etzkowitz & Muller, 2000). However, especially in its decline, the system informally accommodated some of women's needs. As men left the lab in mid-afternoon for a second paid employment in Bulgaria, women also left for a second unpaid employment at home (Simeonova, 1998).

Expanded presence did not, by itself, bring about social equality for women in science, a condition that persists in the postsocialist era (Glover, 2005). A recent EU report on women scientists in the countries of Central and Eastern Europe and the Baltic States (European Commission, 2004a) concludes that women account for 38% of the scientific workforce in these countries (also called the Enwise countries). Nevertheless, the relatively larger numbers of women in science are shadowed by other findings, such as the fact that a large proportion of female scientists is employed in areas with the lowest R&D expenditure, that inadequate resources and poor infrastructure impede the progress of a whole generation of promising scientists, and that men are three times more likely to reach senior academic positions than women. The changing condition of women in science over time is uneven, and different stages in the movement toward equality can be identified in various contemporary societies and even in the same workplace.

CROSS-NATIONAL REPRESENTATION OF WOMEN IN ACADEMIC SCIENCE

The progress of women in science takes place within a broader framework of expansion of higher education and training that occurs with the growth of a knowledge economy. There have been considerable increases in women's participation and attainment in education throughout the industrialized world (Shavit & Blossfeld, 1993; Windolf, 1997). Despite this overall shift toward more equality, significant differences in the distribution of men and women across positions and fields of study continue to persist (Jacobs, 1996; Bradley & Ramirez, 1996). There is considerable variation in women's share among the professorate throughout the industrialized world. However, even in Turkey, the country with the highest proportion of female professors, the share of women academics at the highest academic positions is still below 25%. Moreover, marked differences exist between countries regarding female academics in the pipeline. In countries like Germany, the pattern suggests less openness of the academic system to women across all positions, whereas in countries like Portugal or Sweden there is a growing proportion of females in the lower positions.[2]

Women in science fare better in countries where women are more likely to work full-time as in the United States, France, Spain, and the Scandinavian countries. Whether this pattern also mirrors other influences needs further research. For example, the higher proportion of females among professors may be associated with the diffusion and enactment of more gender egalitarian beliefs in Finland or the United States. But larger shares of women in academia and science may also be due to the influence of class or social origin on educational choices, as in Turkey where high-status males were preoccupied with political leadership during the transition from the Ottoman Empire in the early twentieth century, leaving an opening for their female social peers in academia. The effect of historical ruptures was observable during the colonial war that gripped Portugal during the 1970s where the involvement of cohorts of men abroad opened unprecedented opportunities in education to women at home. Finally, cross-national variations in the proportion of women in science may also stem from variations in the "worth" of the academic and scientific enterprise (European Commission, 2000).

Although country percentages vary dramatically among disciplines, demonstrating the potential eluctability and flux of these figures, women are overall less represented in fields where physical objects, whether natural or artificial, rather than people and symbolic and social relations are the focus of attention. Table 17.1 shows the percentage of women among full professors and comparable staff (grade A) by scientific field in 2001.[3]

Overall, the proportion of female full professors is lowest in technology and engineering and highest in the social sciences and the humanities. Nevertheless, notable differences exist between and within countries. In Portugal, for example, women have relatively high shares across all disciplines with the exception of engineering and technology, excluding the natural sciences, where women account for almost a quarter of all full professors. In comparison, women are represented poorly in the highest academic disciplines in countries such as Austria, Denmark, and Germany. Other countries show a pronounced concentration of women professors in particular sciences, for example, in the medical sciences in the United Kingdom, Israel, and Finland. Some of the variance is traditionally associated with high or low status of a field, but the relationship between women's increase and timing of the status change is not always clear, as in the case of the recent increase in the participation of women in veterinary science in Sweden.

INCREASING PARTICIPATION/CONTINUED SEGREGATION

The relation between gender and scientific interests and the focus of scientific disciplines, especially when gendered topics are the focus of analysis, also needs to be unraveled. It was traditionally assumed that variation in women's participation in scientific fields was related to sexual traits. More recently, the cultural overlay on

Table 17.1
Percentage of women among full professors and comparable staff

Country	Natural Sciences	Engineering and Technology	Medical Sciences	Agricultural Sciences	Social Sciences	Humanities
Belgium	4.2	1.0	3.4	5.1	12.3	10.5
Denmark	4.2	2.8	9.8	9.8	9.7	13.3
Germany	4.6	3.2	4.0	8.0	6.8	13.7
France	15.7	6.4	8.9	n.a.	23.8	n.a.
Italy	15.0	5.2	9.5	10.2	16.8	22.9
Netherlands	3.2	2.7	5.2	7.1	7.0	14.2
Austria	3.1	1.7	7.6	9.3	6.4	11.1
Portugal	22.4	3.1	30.2	17.6	21.8	n.a.
Finland	8.3	5.2	21.3	12.8	24.7	33.2
Sweden	10.4	5.2	12.9	16.3	15.8	25.4
United Kingdom	7.7	2.3	14.5	7.9	17.8	17.9
Iceland	7.0	5.6	9.7	n.a.	9.4	6.1
Israel	6.6	4.8	16.4	0	13.6	18.9
Norway	6.9	2.8	14.2	8.9	15.3	24.3
Poland	16.1	6.8	26.2	20.0	19.2	21.0
Slovakia	10.4	2.4	9.4	4.6	10.9	12.2
Slovenia	6.0	2.8	18.3	14.0	11.5	15.8

n.a., not available.
Source: European Commission 2003a, p. 65, Table 3.2.

physical characteristics has moved to the forefront as an explanation for divergence and the production of gender inequity in science. "Territorial sex segregation" and "ghettoization," creating a separate, gendered labor market in science, developed from (1) the rise in the supply of qualified women, (2) employers' strong resistance to these women entering traditional scientific employment such as university teaching or government employment, and (3) new opportunities in scientific work but low status and behind-the-scenes, arising from the need for large staffs of assistants in research centers (Rossiter, 1982, 1995).

Not surprisingly, a strong emphasis on traditional gender relations reinforces the level of sex segregation in various systems of higher education. A comparison of 29 countries found remarkably little change in the sex segregation of fields of study between 1960 and 1990 (Bradley, 2000). The varying patterns of segregation are explained, in part, by the impact of cultural factors on the country level with the status of different types of higher education institutions. For example, there is more sex segregation in Japan, where nonuniversity institutions that are dominated by females have grown disproportionately. In Germany, female "access" is achieved through women's concentration in vocational colleges or stereotypically female fields of study (Charles & Bradley, 2002).

Dramatic differences in the condition of women in science can be identified in the United States, even in the same university. Some women advance to full professorial rank, albeit at a slower rate and in lesser proportion than their male colleagues. However, other female scientists constitute an invisible underclass of researchers. Not willing to sacrifice family to the seemingly ineluctable pressures of the front-loading of scientific careers, based on assumptions of disproportionate early achievement that is not supported by empirical evidence (Cole, 1979), they have opted to pursue two thirds–time research careers "off the books" as research associates. They seek and get their own grant support, which is officially signed off by colleagues with professorial positions. In contrast to a previous generation of female research associates who worked as assistants to men, these women in science run their own research programs but have little or no opportunity for academic advance. Nevertheless, working within the constraints of an academic system in which the tenure clock is still in tension with the biological clock, despite ameliorative measures such as time extensions, a larger number of productive female researchers exist who could quickly fill higher level positions, should they open up, without having to wait for generational change.

Movements for social and political equality have a mutually reinforcing relationship with movements for gender and racial equality that eventually influences science and higher education.

In more gender egalitarian countries like Sweden or Norway, there is a more equal distribution of degrees awarded at the university or tertiary level. Even there, however, the extent of segregation across fields of study at the tertiary level is very pronounced. Hence, egalitarian norms may diminish horizontal sex segregation in education to a lesser extent than vertical sex segregation—probably because vertical sex segregation is harder to cloak or justify than differences between men and women across fields of study (Charles & Bradley, 2002: 593).

Nevertheless, there is strong cultural lag in the impact of these movements on increasing the participation of women in science. The persistence of sex segregation across fields of study is highlighted in research on women in science. Analyzing UNESCO data for 76 countries from 1972 to 1992, Ramirez and Wotipka (2001) show that women's gains in less prestigious disciplines are positively associated with the likelihood of entry into more prestigious fields of study such as science and engineering ("incorporation as empowerment;" 2001: 243). However, the authors also concede that there are vast cross-national differences in the openness of science and engineering as a field of study and that many forms of inequalities in science and education persist despite the (global) diffusion of egalitarian norms and beliefs.

REFRACTIONS OF INEQUALITY IN SCIENTIFIC LITERATURE

The unequal gendered social structure of science is reinforced by the archival literature of science, a phenomenon that has received increased attention since the 1970s. A common conclusion of several studies of gender differences in scientific productivity, covering diverse fields and periods, was that on average, women tend to publish less than men (Zuckerman & Cole, 1975; Fox, 1983; Cole & Zuckerman, 1984; Hornig, 1987; Long, 1987, Kaplan et al., 1996; Valian, 1999; Schiebinger, 1999; Prpic, 2002), sometimes with considerable differences across sectors. Several possible explanations for this phenomenon, also called the "productivity puzzle" (Cole & Zuckerman, 1984) have been proposed, ranging from differences in personal characteristics, such as ability, motivation or dedication, to educational backgrounds and family obligations, but none of them has proven entirely accurate. More recent insights into the "productivity puzzle" point to the need to broaden the examination focus to the wider context of the social and economic organization of scientific work.

Gender differences in scientific output are hardly surprising if we take into account women's under-representation in science. Gender differences in scientific productivity are closely related to the broader differences in national social, economic, and cultural settings, especially in terms of education and R&D organization and structure of labor force. For example, the focus on the early years of the scientific career in many countries for the operation of gate-keeping mechanisms such as tenure fails to take into account the finding that the productivity peak for women tends to occur later in the career life cycle than for men. In addition to the national socioeconomic and cultural factors discussed above, other factors influencing gendered productivity include the following:

Academic Rank

Several studies report a direct relationship between productivity and academic rank. For instance, Prpic (2002) found that female scientists' publication productivity in Croatia is positively influenced by their higher position in the social organization of science. Similarly, Palomba (2004) found that the productivity of Italian researchers at CNR is generally deeply influenced by academic rank and gender differences are more marked at the top of the career ladder. Bordons et al. (2003) investigated productivity in natural resources and chemistry by gender and professional category in Spain and found that women work at lower professional ranks than men, although within the same professional category no significant differences by gender have been identified. The productivity tended to increase as the professional category improved in the two areas, but no significant differences in productivity were found between genders within each category. Distribution of females by professional categories and number of years at the institution showed a more positive picture in chemistry than in natural resources owing to a process of "feminization" begun in that area at the lowest professional categories, with female progression to the upper ranks expected to follow in the near future.

Career Stage

The evidence with regard to the influence of career stage on gendered productivity seems to be rather inconclusive. Some authors report little difference between the productivity rates of men and women at the start of their scientific careers, mostly among recent doctoral graduates, and increasing differences at later stages (Simon et al. 1967; Cole & Cole, 1973; Zuckerman & Cole, 1975). Martin and Irvine (1982) found publication performance of women Ph.D.'s in radio astronomy to be similar to that of their male peers, suggesting that the possible subsequent lack of success in women's scientific careers could not be attributed to poor performance during the early career stage of their doctoral research. On the other hand, authors like Long (1992) identified increasing gender differences in the number of publications and citations during the first decade of the career, which was reversed at later career stages—dynamics that could not be explained by collaboration patterns that appeared to be nearly identical for males and females.

Family Responsibilities

Zuckerman and Cole (1975, 1987) were among the first to provide evidence against the long-held opinion that women scientists have lower comparative productivity because of the often-conflicting career advancement and family obligations. They showed that marriage and parenthood do not affect women's publication rates; since the productivity of married as well as unmarried women declines, this cannot be attributed entirely to family responsibilities. Later studies such as Sax et al. (2002) confirmed this view, showing that factors affecting faculty research productivity are nearly identical for men and for women, and family-related variables (e.g., having dependent children) have little or no effect on research productivity. Other findings (e.g., Palomba, 2004) relate productivity to a family effect manifested in the publication peaks, which were found to appear at different stages in men's and women's careers—earlier for men (35–39 years) and later for women (45–49 years).

Scientific Field

Gender gaps in output vary greatly from field to field, and gender differentials are lower in some scientific fields, such as medicine, biology, and the sciences, and wider in other areas, such as the humanities (Palomba, 2004). Leta and Lewison's (2003) analysis of publication productivity of Brazilian researchers showed that women published most in immunology, moderately in oceanography, and least in astronomy. Nevertheless, women were less likely than men to receive fellowships to supplement their salaries, suggesting that some sexual discrimination may still be occurring in the Brazilian peer-review process.

Next to publication numbers, another frequent indicator of gendered productivity is citations. Literature evidence in this respect appears again to be rather inconclusive; some studies (e.g., Cole & Cole, 1973) find that women's papers are cited less than men's while others report the reverse tendency (Long, 1992; Sonnert & Holton, 1996; Schiebinger, 1999). Teghtsoonian (1974) finds no significant evidence that women's publications are less cited.

In terms of citation impact, a study of the 1000 most cited scientists from 1965 to 1978 (Garfield, 1981) shows that, although the average number of papers and the average number of citations per woman were lower than those per man, the women's average impact (citations divided by papers) was substantially higher. In contrast, Leta and Lewison (2003) found that men and women published similar numbers of papers, which were of similar potential impact.

One of the major problems raised by commonly used indicators of scientific productivity, such as the numbers of publications and citations, is their limited capacity to capture specific aspects of gender differences pertaining to scientific productivity, or their capacity to reflect gender biases in the wider context of the scientific environment. One example in this respect is Feller's (2004) distinction between two areas of gender bias in science: (1) bias in the system of evaluating research performance and excellence usually referred to as "equity" and (2) bias in the validity and reliability of the metrics that assess performance or excellence in different contexts. These two conceptualizations of bias can generate a matrix of four possible combinations: (a) unbiased system, unbiased metrics; (b) unbiased metrics, biased system; (c) biased metrics, unbiased system; and (d) biased metrics, biased system, where most of the literature on women in science is concentrated on (b) (e.g., Wennerås & Wold, 1997; Valian, 1999) and (d) (e.g., Schiebinger, 1999). These limitations of bibliometrics point to the need to develop an expanded set of metrics that mark the difference between performance and excellence, or between quantity and quality, and to ensure that these productivity indicators are gender neutral. Literature, however, is a lagging indicator of other changes in the social organization of science.

REFLECTIONS OF INEQUALITY IN SCIENTIFIC ORGANIZATION

The position of women in science is shaped by the role of science in society, whether as fundamental productive force or merely a cultural attribute (High/Low Science) and the gender structure of society, whether women are accepted as equals or exist in a subordinate status (High/Low Women). In a fourfold table (figure 17.1), the first cell—High Science/HighWomen—does not fully exist in any society. Nevertheless, pockets can be identified; for example, in biotechnology firms in the United States (Smith-Doerr, 2004) High Science/Low Women is the situation of female scientists in most western societies where science is an important part of societal infrastructure, with women occupying a subordinate status. A series of studies in the stratification of science, showing contradiction between Mertonian norms and the position of women in various scientific institutions and organizations, exemplify this cell (Cole & Cole, 1973; Cole, 1979; Fox, 2001; Fox, 2005; Fox & Stephan, 2001; Long & Fox, 1995. High-Women/Low Science is exemplified by the situation of women in science in many developing countries. Science is a peripheral to the economy, but female scientists typically are from upper class backgrounds and occupy a superior status. In Low Science/Low Women countries, science is underdeveloped and women's status in science is also depressed. Science becomes a central part of the development agenda as economic growth becomes more knowledge-based. As scientific professions increase in number and economic centrality, changes in gender relations lag because the struggle for positions is dominated by men.

The position of science and academia in society affects the rise of women in science in apparently contradictory ways, always linked to common conditions of gender inequality. Women have made greatest gains in participation under conditions of both

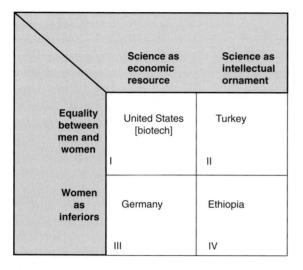

Figure 17.1
Attitudes toward women in science.

system expansion and status decline. Expanding systems of higher education, industrialization, and modernization opened up scientific education and to some extent science careers to women in Portugal and Turkey. A declining academic economy in Mexico has led to the feminization of the university as men leave for more lucrative fields. The low status of science has improved women's participation as in Turkey. Thus, even these advances reflect continuing inequalities. In Mexico, women eschew scientific networking because of family obligations (Etzkowitz & Kemelgor, 2001). The condition of women in science in most countries falls within cells 2 and 3. Countries in cell 4 are attempting to upgrade by establishing new universities (Duri, 2004). Cell 1 is a contested environment but with great potential for growth given success in the struggle of women scientists to attain equality and the need for societies to fully develop all their human capital to remain internationally competitive. Nevertheless, resistance to change arises both from internal and external sources within science and from the larger society that have cumulative and escalating effects.

UNIVERSAL ROLE OVERLOAD

Persisting gender inequality has similar effects on women in science. Germany, the United States, and India have different socioeconomic systems and span three continents. Yet, women in science face a common "triple burden" across the continents (Gupta, 2001). The problems of working in a hostile work environment result in career-related stress—the first burden. The second burden is the usual predicament of domestic responsibilities, which fall disproportionately on women. This dual burden forces women to work harder than men to prove themselves. In all countries, female scientists also carry a third burden of grappling with a deficit of social capital and the relative exclusion from strong networks. The interaction among these burdens induces "surplus anxiety" among women that is well above the normal stressors of obtaining funds, results, and recognition common to all scientists.

Family issues, predominantly seen as women's responsibility, negatively affect women's scientific and academic career opportunities. Thus, in the United States, women's personal obligations are taken into account and ignored for men when they are being hired. In Germany, women are seen as risky employees who may at least temporarily drop out (Fuchs et al., 2001; von Stebut, 2003). In India, appointment and promotion committees bring up family issues and question women's commitment to the job (Gupta, 2001).[4] The traditional extended family, still commonplace in developing countries, provides significant support for women scientists, particularly in Brazil and Mexico (Etzkowitz & Kemelgor, 2001). However, while extended family is helpful in providing greater freedom for women to work without anxiety about domestic duties, it also perpetuates the traditional stereotypes about women reflected by additional duties related to the joint family (Gupta, 2001).

Traditional gender role expectations and a rigid structure in the workplace that makes a combination of family and career difficult for women constitute barriers to women in science. Thus, in Brazil, female scientists have been held back by stereotyped images, by gendered familial obligations, and by the sexism of "old boy networks" that still control senior positions (Plonski & Saidel, 2001). In countries such as Spain, an expanding science and technology system helps in raising women's share of research positions, but they continue to be excluded from "social power." In the United Kingdom also, there is covert resistance to women in science, expressed as extremely lower levels of women in high academic and science policy positions.

Economic growth and development do not necessarily guarantee a change in the traditional social structure. In Japan, for instance, the society developing with the growth of industry between 1955 and 1975 encouraged women to be housewives. In the 1970s, growth of the service sector created a demand for a more flexible and creative workforce, but women were relegated to unstable and peripheral jobs (Kuwahara, 2001). Even economic growth combined with a strong ideology of equality has its limits. Finland exemplifies the experience of women in highly industrialized countries with strong social support systems. Here, women scientists are constrained by an inflexible scientific research system where the expected period of high research productivity coincides with the childbearing and child-raising years.

HOPE FOR CHANGE?

The connection between science and economic development is increasing, broadening participation in higher education and eventual gender equality. In the age of globalization, exchange of ideas and personnel between developed and developing countries has become important, and the transnational traffic of ideas, people, and technologies is becoming more inclusive of women. The educated urban middle class from industrializing countries, such as India, looks to the more industrialized countries for greater opportunities in terms of professional growth and monetary success. While women lag far behind men in going abroad for higher studies, their number is increasing at an accelerating rate. In 1991–92, the proportion of women students going abroad was 13.72%, which increased to 16.1% in 1998–99 (Ministry of Human Resource Development, Government of India). In absolute numbers, the number of male students increased from 5579 to 5806 in the same period (a 4% increase) and of women from 887 to 1112, a 25% increase. This indicates that educated women (and their families that allow them) are increasingly willing to break the traditional stronghold of "patrifocal" ideology and venture abroad for higher satisfaction of talents and ambitions.[5]

The relationship between enhancement of the role of science and technology in economic development and growth of female opportunities in science is paradoxically shaped by persisting gender inequalities. Since the last decade, in India, there has been a substantial increase in proportion of women in pure sciences compared with engineering. Globalization and liberalization since the 1990s in India have reduced the demand for pure sciences, since they are less lucrative and lack job potential. This has led to a trend of feminization of pure sciences, which earlier were regarded as masculine subjects (Chanana, 2001).[6] Nevertheless, the concentration of women in low status fields may have unexpected effects as the status of scientific fields shifts, for example, the physical and biological sciences in recent decades. If women can hold their position against historical trends to exclude females as previously low ranked fields rise, they may ride the winds of scientific change.

Exemplar of Change

Some have argued that the advancement of women in the professions is enhanced by strengthening procedural safeguards, relying on the apparently neutral structure of bureaucracy to promote women's rise (Reskin, 1977). Others hold that when patriarchy is embedded in hierarchy, as in science, such a strategy may fail or even prove counterproductive by providing a "veil" for discrimination (Witz, 1992). For example, behind apparently neutral academic appointment procedures where women are invited for interviews to meet formal criteria, the "old boy" network may still determine the final result, with little external scrutiny possible owing to academic freedom concerns.

Recent research suggests the efficacy of lateral, rather than hierarchical structures, for promoting the advancement of women in science and technology. Smith-Doerr's intriguing study of the biotechnology start-up and growth firm found that it offers women a flexible workplace where their contributions are acknowledged and rewarded. Moreover, biotechnology firms, with their flat organizational structures and emphasis on teamwork and cooperation, provide a better environment for women to advance. Interdisciplinary work is more open to women, and their networking skills are rewarded. She further argues that contrary to expectations that bureaucratic structures offer protection from discrimination, flexible structures serve women better than, "... a set of rules that function only as formal window dressing (Smith-Doerr, 2004: xiv). In addition, within the context of the lateral firm, young female Ph.D.'s were "... about eight times as likely to lead research in bio-tech firms ... than in university research groups or large pharmaceutical firms" included in the study (Smith-Doerr, 2004: 115).

This finding, if supported by other indicators, may augur a coming gender revolution in science. When a new field emerges at the periphery of science, women are typically well represented, as during the early days of genetics research, but were pushed out as the status of the field rose (Kohler, 1994). However, in the early twenty-first century women's beachhead into biotech is holding. Not only has their presence persisted, but women have moved up to high positions in the industry. The collegial, less hierarchical, teams characteristic of the biotech industry are similar to the "relational" research group that some women in academia have attempted to establish as an alternative model (Etzkowitz et al., 1994). The promotion of women to high positions of academic leadership in high-status academic institutions, like Chicago, Princeton, and MIT, represents another positive trend with significant potential. Nevertheless, a woman who had achieved a provost's position reflected that she had not utilized it as much as she might have to institutionalize change in gender relations in academia.

The external environment for academic science in relations with government and industry is another factor that can promote or retard change. Government funding agencies, such as the National Institutes of Health in the United States, that have made achieving results in diversity a factor in distributing funds, has raised the awareness of the need for change from "lip service" to action programs in academic departments threatened with the loss of grants. On the other hand, flexible network structures in biotech firms reduce discrimination only up to a point. The glass ceiling reappears in the firm-formation process, with women having less access than men to the venture capital needed to found firms. Various "springboard" programs to improve access of women to venture capital have had limited effect to date, although the problem has been recognized and addressed.

To achieve equality for women in science, counterproductive rules and norms with unintended negative effects on women must be revised. For example, in the United States an informal requirement that individuals must move at each early career stage—for example, from Ph.D., to postdoc, to initial position—depresses women's chances for advancement when male partners are given first preference. In Scandinavia, where continuity in position is expected, women who move may have their career chances depressed. It is not the particular rule or norm but its inflexibility that has additional negative consequences for women, especially under conditions of persisting gender inequality.

A "neutral bureaucratic" strategy may work to increase the numbers of women in science, but it is grossly inadequate to addressing the more intractable issue of promoting the rise of women in science. A more radical strategy of breaking through glass ceilings by removing the strata themselves rather than squeezing a few women past barriers is required (Wajcman, 1998). Biotechnology firms, a hybrid format between traditional academic and industrial science may point the way to achieving equality. We suggest that future research focus on such "pockets of emerging change." Suggested strategic research sites include female founders of high-tech start-ups, academic women principal investigators and their research groups; university technology transfer offices, European Union (and similar) research networks, and R&D funding agencies.

BREAKING THE DOUBLE PARADOX

A human capital paradox of lesser return from investment in women in science is nested within the so-called "European paradox" of relatively small return on R&D spend into the economy.[7]

The transformation of the role of science in society from a contributor to industrial society to the base of the knowledge economy transforms gender issues from a matter of equity to one of competitive advantage or loss (Ramirez, 2001: 367). This change has prompted political institutions to wake up to the potential of women scientists. Thus, the European Union's European Research Area contains two main aims relating to women scientists. The first can be seen as explicitly related to the bottom line of productivity, while the second, sometimes referred to as the "democratic principle" (European Commission, 2003d), is concerned with the moral arguments for equal opportunities (Glover, 2005).

Women are also viewed pragmatically as a major untapped pool that could bring about the intended growth in the knowledge economy. "Women are an under-exploited resource in research for the European Union and have a huge potential for the future of research in Europe" (European Commission, 2004b: 47). Commissioner for Research Philippe Busquin specifically linked the employment of women scientists to the 3% of GDP target and the related 2010 objective of a further 700,000 researchers, referring to retention and advancement as well as recruitment (and thus implicitly acknowledging the "democratic principle" of equal opportunities): "we will not reach the 3 percent objective if we fail to recruit, retain and promote the women who constitute an important share of Europe's pool of trained scientists" (European Commission, 2003a: 5). These "fairness" arguments are reinforced through the requirement that applicants for EC Framework funds take gender into account in terms of both project content and staffing (although the sanctions for not doing so are unclear).

Against this background, new (and old) inequalities are not only detected more rapidly, they are also increasingly perceived as unjust as well as providing a largely untapped pool that will contribute toward the bottom line of productivity. Furthermore, they are seen as a crucial component in the bid to increase public trust in science and scientists (European Commission, 2002); the Commission's view is that a more culturally diverse scientific workforce could increase public confidence in science and, perhaps, taxpayers' willingness to invest in the knowledge economy.

As the economic and social uses of science increasingly become the source of a knowledge-based economy, the issue of women in science takes a new, perhaps more promising, direction. There seems to be less resistance to women in patent law firms, university technology transfer offices, science media outlets, biotechnology firms—and other new hybrid venues of science—than in the traditional core in academia. Moreover, what is peripheral and what is core to the role of science in society is in flux. Despite persisting rigidity and resistance in old hierarchical organizations, the creation of lateral structures and bridging mechanisms with flat organizational designs may augur a more positive and central role for women in science.

As science has become a more organized endeavor, whether in the research groups of "small science" or the mega collaborations of "big science," organizational and networking skills have come to be as important to scientific success as theoretical insight and experimental skills. James Watson's path to the DNA discovery in Cambridge pubs and colleagues' data sets may be seen as an early augur of this trend (Watson, 1968). More recently, the ability to coordinate scientific networks across national and disciplinary boundaries, and the egos that compete for reward and recognition, have placed a premium on activities that were heretofore seen as peripheral to the scientific enterprise.

Some territorially distinct areas are being revalued, with significant implications for women in science and technology (Wajcman, 1991). As certain heretofore ancillary tasks relating to the economic and social uses of science become more important, so do the holders of those positions. It is noteworthy that women, whether they have actively sought positions in the new uses of science or been sidelined into them, have attained leadership roles in such venues as European Union research networks and U.S. technology transfer offices. Will women retain their prominence in emerging fields, such as technology transfer, or will past patterns hold of women being pushed out as the status of a field rises?

CONCLUSION: GENDER REVOLUTION IN SCIENCE?

The irrational gendered arrangements in the seemingly rational profession of science are a product of the correlation between the status of women in society and the status of science in society. Though this correlation is complex and varies across space and time, the discrimination against women has been most pronounced, almost everywhere, in the traditional stronghold of science, that is, in academia. This persistence across the span of a century is evidenced in the Albion Small survey in 1905 and the 2005 statement of former President Lawrence Summers of Harvard University.

A broad review of the issue of women in science was conducted a century ago, in 1905. Albion Small, the founder of the first sociology department in the United States, conducted a survey of three groups: members of the American Association for the Advancement of Science (AAAS), professors at women's colleges, and female graduate students (Nerad & Czerny, 1999). The AAAS sample reflected the common belief that men would more likely devote themselves to genuine scholarly work than women.

Prof. G. Stanley Hall, a leading psychologist, contributed his analysis that women are by nature different from men, incompetent in fields that require abstract thinking, and proposed that they be directed to scientific fields that do not emphasize such skills. The female graduate students reported that they enjoyed little intellectual contact with their instructors but were aware that their male peers often met informally with professors. Nerad and Czerny observed that "Many of the women's responses to Prof. Small's survey can still be heard echoing through the halls of modern campuses." (Nerad & Czerny, 1999: 3)

In January 2005, Lawrence Summers, President of Harvard University, addressed a National Bureau of Economic Research Conference on Diversity in Science. He suggested that "the primary barrier to women, as in other high powered jobs, is that employers demand single-minded dedication to work. He also offered a so-called, "fat tails hypothesis" of differences between men and women: that more women have average scientific ability while larger numbers of men are at the high and low ends of a scientific ability scale. His third hypothesis, which he characterized as the least significant of the three, was that "women are discriminated against or socialized as children not to go into science." Summers' first hypothesis reprises Small's summary of the attitudes of AAAS members in 1905; his second, which also included the corollary that women may have lesser innate mathematical abilities than men, replicates Hall's analysis. Finally, his third hypothesis is congruent with the experience of female graduate students in 1905 and more recently as well. The firestorm of response to Summers' remarks called forth new initiatives to improve the condition of women in science, including from his own university (Henessey et al., 2005; Etzkowitz & Gupta, 2006).

Although the situation of women in science has been the subject of intense debate in academic and political venues, there is still a notable lack of systematic, comparative, empirical research on the situation of women in science. Three reasons may account for this paucity. First, data on the representation of women across fields of study and academic positions are gathered on a regular basis, for example, by the OECD or UNESCO, but they are hardly comparable given the large differences in how systems of higher education are organized, the size of the academic and/or scientific labor market, the openness of these systems, and the rewards they provide to women at the country level (see, e.g., Jacobs, 1996 and Charles & Bradley, 2002).

Second, the focus of cross-national studies to date has been more on the academic than on the scientific labor market because data on enrollments and the representation of women across positions and fields are more accessible than data on the situation of male and female scientists outside the university sector (Fuchs et al., 2001). Finally, most data used in comparative cross-national research are at the aggregate level and cross-sectional in scope. Systematic analysis of careers in science, however, would ideally rely on longitudinal biographic information on cohorts of scientists to assess the influence of changes in labor market conditions or other institutional regulations (Mayer, 2002).

Moreover, most research on women in science, with a few notable exceptions, focuses on the traditional core rather than the newly emerging and increasingly significant peripheries. Moreover, much as software was once viewed as a "peripheral" to computer hardware, a similar restructuring of scientific roles may be at hand. In the past, women's rise in science occurred when men were not available, for example, in wartime or when discriminatory priorities based on class and ethnicity were stronger than gender concerns. However, when men again became available, women tended to disappear from the bench. Women are still less often found at the upper reaches of academic science, even as they reappear in emerging science-related professional scenes that appear to offer an enhanced environment for women.

As the role of science in society changes, the role of women in science may also be affected as individuals with training in scientific and technological disciplines are hired into law firms, technology transfer offices, newspapers, and other media.[8] Shake-up of traditional rigid organizational structures such as academic departments by new interdisciplinary fields opens the way for new people in new posts. New positions are created, such as Director of the Media X program at Stanford University, with faculty status, held by a Ph.D. in psychology who previously worked as a partner in a venture capital firm. Her job is to identify new interdisciplinary research themes, recruit companies to membership in the program, and manage a grant program targeted at faculty members.

Territorial integration is the hopeful sign in these new scientific arenas, with women often in a position of responsibility. Traditional female socialization emphasized relationship building and networking skills that have become increasingly important, both within traditional research fields increasingly dependent on long-distance collaboration and in the new venues of science that are typically networked organizations. Thus, socialization that worked against an intense focus on solitary bench work, the hallmark of traditional science, works for success in the emerging roles of science and the reformed old ones.

As developed as well as developing countries realize the potential of science to fuel growth, women scientists can no longer be ignored. Although persistent, the negative correlation between science and female gender is a historical not a biological phenomenon and is subject to revision, as is science itself. Science is changing from an ancillary activity of the industrial revolution, systematizing its production processes and providing deeper understanding of practices arrived at through trial and error, to become the fundamental source of industrial advance in the late twentieth and early twenty-first centuries (Misa, 2004; Viale & Etzkowitz, 2005).

The transformation of science from a peripheral to core societal activity calls into question the cultural lag of unequal gender, class, and ethnic relations in science, not only on principles of equity and fairness but on grounds of competitive and comparative advantage (Pearson, 1985; Tang, 1996, 1997). Leaders of political and scientific establishments now call for all brain power, including female and minority, to be mobilized in order to be competitive in the global knowledge economy. The advancement of science is increasingly dependent on women's advancement in science.

Notes

1. Simone de Beauvoir, on the "fairly large number of privileged women who find in their professions a means of economic and social autonomy" (1952: 681).

2. In the case of Belgium, there are notable differences within country, i.e., a higher share of women in science in the French-speaking than in the Flemish-speaking part of the country. Also note that the data are not differentiated by age, discipline, or type of institution.

3. Please note that the European Commission underlines that due to "differences in coverage & definitions" the data are "not yet comparable between countries" (2003a: 65).

4. "Patrifocality," coined by Mukhopadhyay and Seymour (1994), refers to a set of social institutions and associated beliefs that give precedence to men over women. It refers to a family system in an agrarian, hierarchical society in which rank depends on ritual purity that requires, among other things, control of women's sexuality.

5. Sex-Wise Number of Students Going Abroad (1991–92 to 1998–99), Indian Students/Trainees Going Abroad 1998–99, Ministry of Human Resource Development & Past Issue, Government of India.

6. About 32% of enrollment in physics in India is of women, which is quite high in the global context (Godbole et al., 2002).

7. The ERA, first mooted at the Lisbon Summit of 2000 and elaborated by the European Commission, reflects a concern that the gap between European funding of R&D and that of the United States and Japan has been widening (European Commission, 2003b: 4). The Commission attributes this to low investment by the private sector, which in Europe provides only 56% of the total financing of research versus more than two thirds in the United States and Japan (European Commission, 2003c). The EU as a whole spent only 1.94% of GDP on R&D in 2000, compared with 2.80% in the United States and 2.98% in Japan. Moreover, this "investment gap" has widened rapidly since the mid-1990s. In terms of purchasing power, the EU-U.S. divide increased markedly, from 43 billion Euros in 1994 to 83 billion Euros in 2000; and although the EU produces a larger number of graduates and Ph.D.'s in science and technology than does the United States and Japan, it employs fewer researchers: 5.4 per 1000 labor force versus 8.7 in the United States and 9.7 in Japan (European Commission, 2003c). This implies a poor return on the costs of education. There is also specific concern about a slowdown in growth: the growth rates in the EU-15 of both overall investment and overall performance in the knowledge-based economy were markedly lower in 2000–2001 than during the second half of the 1990s (European Commission, 2003c).

8. For example, women make up a majority of the staff, including senior positions and director, of the Stanford University Office of Technology Licensing and are strongly represented in the profession in general. See also http://www.autm.com.

References

Abir-Am, P. (1991) "Science Policy for Women in Science: From Historical Case Studies to an Agenda for Women in Science," *History of Science Meetings*, Madison, WI, Nov. 2.

Allmendinger, J., H. Brückner, S. Fuchs, & J. von Stebut (1999) "Eine Liga für sich? Berufliche Werdegaenge von Wissenschaftlerinnen der Max-Planck-Gesellschaft," in A. Neusel and A. Wetterer (eds), *Vielfaeltige Verschiedenheiten-Geschlechterverhaeltnisse in Studium, Hochschule und Beruf* (Frankfurt a.M./New York: Campus, S.): 193–220.

Althauser, R. & A. Kalleberg (1981) "Firms, Occupations and the Structure of Labor Markets," in I. Berg (ed), *Sociological Perspectives on Labor Markets* (New York: Academic Press).

Athena Project (2004) "ASSET 2003: The Athena Survey of Science, Engineering and Technology in Higher Education" (London: Athena Project).

Baltimore Charter for Women in Astronomy (1993) www.stsci.edu/stsci/meetings/WiA/BaltoCharter.html

Bielby, W. T. (1991) "Sex Differences in Careers: Is Science a Special Case?" in H. Zuckerman, J. R. Cole, & J. T. Bruer (eds), *The Outer Circle* (New York: Norton): 171–87.

Bordons, M., F. Morillo, M. T. Fernández, & I. Gómez (2003) "One Step Further in the Production of Bibliometric Indicators at the Micro Level: Differences by Gender and Professional Category of Scientists," *Scientometrics* 57(2): 159–73.

Bradley, Karen (2000) "The Incorporation of Women into Higher Education: Paradoxical Outcomes?" *Sociology of Education* 73: 1–18.

Bradley, Karen & Francisco O. Ramirez (1996) "World Polity and Gender Parity: Women's Share of Higher Education," *Research in Sociology of Education and Socialization* 11: 63–91.

Campion, P. & W. Shrum (2004) "Gender and Science in Development: Women Scientists in Ghana, Kenya and India," *Science, Technology & Human Values* 29: 459–85.

Carabelli, A., D. Parisi, and A. Rosselli (1999) *"Che genere" di Economista?* (Bologna: Il Mulino).

Chanana, K. (2001) "Hinduism and Female Sexuality: Social Control and Education of Girls in India," *Sociological Bulletin* 50(1): 37–63.

Charles, M. & K. Bradley (2002) "Equal but Separate? A Cross-National Study of Sex Segregation in Higher Education," *American Sociological Review* 67: 573–99.

Chu Clewell, B. & P. B. Campbell (2002) "Taking Stock: Where We've Been, Where We Are, Where We Are Going," *Journal of Women and Minorities in Science and Engineering* 3(4): 225–84.

Cole, J. (1979) *Fair Science* (New York: The Free Press).

Cole, J. & S. Cole (1973) *Social Stratification in Science* (Chicago: University of Chicago Press).

Cole, J. R. & H. Zuckerman (1984) "The Productivity Puzzle," in M. L. Maehr & M. Steinkamp (eds), *Advances in Motivation and Achievement* (Greenwich, CT: JAJ Press): 217–58.

Cole, S. (1979) "Age and Scientific Performance," *American Journal of Sociology* 84(4): 958–77.

Commission on the Advancement of Women and Minorities in Science, Engineering and Technology Development (2000) *Land of Plenty: Diversity as America's Competitive Edge in Science, Engineering and Technology.*

Commission on Professionals in Science and Technology (2004) *Professional Women and Minorities: A Total Data Source Compendium* (CPST Publications).

Davis, A. E. (1969) "Women as a Minority Group in Higher Academics," *The American Sociologist* 4: 95–98.

Dean, C. (2005) "For Some Girls, the Problem with Math Is That They're Good at It," *New York Times*, F3, February 1.

de Beauvoir, S. (1952) *The Second Sex* (New York: Knopf).

de Wet, C. B., G. M. Ashley, & D. P. Kegel (2002) "Biological Clocks and Tenure Timetables: Restructuring the Academic Timeline," Supplement to Nov. 2002 *GSA Today*, 1–7. http://www.geosociety.org/pubs/gsatoday/0211clocks/0211clocks.htm

Dix, L. (ed) (1987) "Women: Their Underrepresentation and Career Differentials in Science and Engineering," proceedings of a workshop (Washington, DC: National Academy Press).

Drori, G. S., J. W. Myer, F. O. Ramiriez, & E. Schofer (2003) *Science in the Modern World Polity: Institutionalization and Globalization* (Stanford, CA: (Stanford University Press).

Dupree, A. K. (2002) "Review of the Status of Women at STScI" (Washington), Aura. Arailalde at: www.aura-astronomy.org/nv/womensreport.pdf.

Duri, M. (2004) Ethiopian Ambassador to the United Nations, interview with Henry Etzkowitz, March 2004, New York.

Etzkowitz, H. (1971) "The Male Sister: Sexual Separation of Labor in Society," *Journal of Marriage and the Family* 33(3): 431–34.

Etzkowitz, H. & N. Gupta (2006) "Women in Science: A Fair Shake?" *Minerva* 44: 185–99.

Etzkowitz, H. & C. Kemelgor (2001) "Gender Inequality in Science: A Universal Condition?" *Minerva* 39: 153–74.

Etzkowitz, H., C. Kemelgor, & J. Alonzo (1995) "The Rites and Wrongs of Passage: Critical Transitions for Female PhD Students in the Sciences" (Arlington, VA: Report to the NSF).

Etzkowitz, H., C. Kemelgor, M. Neuschatz, B. Uzzi, & J. Alonzo (1994) "The Paradox of Critical Mass for Women in Science," *Science* 266: 51–54.

Etzkowitz, H, C. Kemelgor, & B. Uzzi (2000) *Athena Unbound: The Advancement of Women in Science and Technology* (Cambridge: Cambridge University Press).

Etzkowitz, H. & K. Muller (2000) "S&T Human Resources: The Comparative Advantage of the Postsocialist Countries," *Science and Public Policy,* August.

European Commission (2000) *Science Policies in the European Union: Promoting Excellence Through Mainstreaming Gender Equality.* European Technology Assessment Network (ETAN), "Women and Science" report (Brussels: European Commission).

European Commission (2002) *Science and Society Action Plan* (Brussels: European Commission).

European Commission (2003) *Waste of Talents. Turning Private Struggles into a Public Issues: Women and Science in the ENWISE Countries.* Co-authored by M Blagojević, M. Bundule, A. Burkhardt, M. Calloni, E. Ergma, J. Glover, D. Gróo, H. Havelková, D. Mladenič, E. Oleksy, N. Sretenova, M. F. Tripsa, D. Velichová & A. Zvinkliene (Brussels: European Commission).

European Commission (2003a) *"She Figures": Women and Science Statistics and Indicators* (Brussels: European Commission).

European Commission (2003b) *Investing in Research: An Action Plan for Europe* (Brussels: European Commission).

European Commission (2003c) *Third European Report on Science & Technology (S&T) Indicators* (Brussels: DG Research).

European Commission (2003d) *Women in Industrial Research: A Wake Up Call for European Industry* (Brussels: European Commission). Luxembourg Office for Official Publications of the European Communities.

European Commission (2004a) *Wasted Talents: The Situation of Women Scientists in Eastern European Countries*, ENWISE press conference, Brussels, January 30.

European Commission (2004b) *Key Figures 2003–2004: Towards a European Research Area* (Brussels: DG Research).

Feller, I. (2004) "Measurement of Scientific Performance and Gender Bias" in *Gender and Excellence in the Making* (Brussels: DG Research).

Ferree, M. M, B. B. Hess, & J. Lorber (eds) (1999) *Revisioning Gender: New Directions in the Social Sciences* (Thousand Oaks, CA: Sage).

Florida, R. (2005) *The Flight of the Creative Class: The New Global Competition for Talent* (New York: Harper Business).

Fox, M. F. (1983) "Publication Productivity Among Scientists: A Critical Review," *History of Science* 17: 102–34.

Fox, M. F. (1991) "Gender, Environmental Milieu, and Productivity in Science," in H. Zuckerman, J. Cole, & J. Bruer (eds) *The Outer Circle: Women in the Scientific Community* (New York: Norton): 188–204.

Fox, M. F. (2001) "Women, Science, and Academia: Graduate Education and Careers," *Gender & Society* 15: 654–66.

Fox, M. F. (2005) "Gender, Family Characteristics, and Publication Productivity Among Scientists," *Social Studies of Science* 35: 131–50.

Fox, M. F. & P. E. Stephan (2001) "Careers of Young Scientists: Preferences, Prospects, and Realities by Gender and Field," *Social Studies of Science* 31: 109–22.

Fuchs, S, J. von Stebut, & J. Allemendinger (2001) "Gender, Science and Scientific Organizations in Germany" *Minerva* 39(2): 175–201.

Garfield, E. (1981) "The 1000 Contemporary Scientists Most Cited 1965 to 1978," *Essays of an Information Scientist* 5: 269–78.

Glaser, B. (1964) *Organizational Scientists: Their Professional Careers* (Indianapolis. IN: Bobbs-Merrill).

Glover, J. (2000) *Women and Scientific Employment* (Basingstoke, U.K.: Macmillan).

Glover, J. (2005) "Highly Qualified Women in the 'New Europe': Territorial Sex Segregation," *European Journal of Industrial Relations* 11: 231–45.

Godbole R., N. Gupta, & and S. Rao (2002) "Women in Physics, Meeting Reports," *Current Science* 83: 359–61.

Goldsmith, B. (2005) *Obsessive Genius: The Inner World of Marie Curie* (New York: Norton).

Greenfield, S. (2002) "SET Fair: A Report on Women in Science, Engineering and Technology" (London: Department for Trade and Industry).

Gupta, B. M., Kumar, S. & Aggarwal, B. S. (1999) "A Comparison of Productivity of Male and Female Scientists," *Scientometrics* 45: 269–89.

Gupta, N. (2001) *Women Academic Scientists: A Study of Social and Work Environment of Women Academic Scientists at Institutes of Higher Learning in Science and Technology in India*, Ph.D. diss., Indian Institute of Technology, Kanpur.

Gupta, N. & A. K. Sharma (2002) "Women Academic Scientists in India" *Social Studies of Science* 32(5–6): 901–15.

Hacker, S. (1981) "The Culture of Engineering: Women, Workplace and Machine," *Women's Studies International Quarterly* 4:341–53.

Hanson, S. L. (1996) *Lost Talent: Women in the Sciences* (Philadelphia: Temple University Press).

Hanson, S. L., M. Schaub, & D. P. Baker (1996) "Gender Stratification in the Science Pipeline: A Comparative Analysis of Seven Countries," *Gender and Society* 10: 271–90.

Harding, S. (1986) *The Science Question in Feminism* (Ithaca, NY: Cornell University Press).

Harding, S. (1991) *Whose Science? Whose Knowledge?: Thinking from Women's Lives* (Ithaca, NY: Cornell University Press).

Hennessy, J., S. Hockfield, & S. Tilghman (2005) "Look to Future of Women in Science and Engineering," *Stanford Report* 37, February 18: 8.

Hicks, E. (1991) "Women at the Top in Science and Technology Fields. Profile of Women Academics at Dutch Universities," in Veronica Stolte-Heiskanen et al. (eds), *Women in Science: Token Women of Gender Equality* (Oxford: Berg Publishers): 173–92.

Hirschauer, S. (2003) "Wozu, Gender Studies? Geschlechterdifferenzierungsforschung zwischen politischem Populismus und naturwissenschaftlicher Konkurrenz," *Soziale Welt* 54: 461–82.

Hornig, L. S. (1987) "Women Graduate Students," in L. S. Dix (ed), *Women: Their Underrepresentation and Career Differentials in Science and Engineering* (National Academy Press): 103–22.

Ibarra, R. (1999) "Multicontextuality: A New Perspective on Minority Underrepresentation in SEM Academic Fields," *Making Strides* 1(3): 1–9.

Ibarra, R. (2000) *Beyond Affirmative Action: Reframing the Context of Higher Education* (Madison: University of Wisconsin Press).

Jacobs, Jerry A. (1996) "Gender Inequality and Higher Education," *Annual Review of Sociology* 22: 153–85.

Kaplan, S. H., L. M. Sullivan, K. A. Dukes, C. F. Philips, R. P. Kelch, & J. G. Schaler (1996) "Sex Differences in Academic Advancement," *New England Journal of Medicine* 335: 1282–89.

Kohler, R. (1994) *Lords of the Fly: Drosophila Genetics and the Experimental Life* (Chicago: University of Chicago Press).

Kohlstedt, S. (2005) "Nature, Not Books: Scientists and the Origins of the Nature-Study Movement in the 1890's," *ISIS* 96: 324–52.

Kulis, S. S. & K. Miller-Loessi (1992) "Organizational Dynamics and Gender Equity. The Case of Sociology Departments in the Pacific Region," *Work and Occupations* 19: 157–83.

Kuwahara, M. (2001) "Japanese Women in Science and Technology," *Minerva* 39(2): 203–16.

Leta, J. & G. Lewison (2003) "The Contribution of Women in Brazilian Science: A Case Study in Astronomy, Immunology and Oceanography," *Scientometrics* 57(3): 339–53.

Long, J. S. (1987) "Problems and Prospects for Research on Sex Differences in the Scientific Career," in L. S. Dix (ed), *Women: Their Underrepresentation and Career Differentials in Science and Engineering* (National Academy Press): 157–69.

Long, J. S. (1992) "Measures of Sex Differences in Scientific Productivity," *Social Forces* 71: 159–78.

Long, J. S. & M. F. Fox (1995) "Scientific Carreers: Universalism and Particularism," *Annual Review of Sociology* 21: 45–71.

MacLachlann, A. *Graduate Education: The Experience of Women and Minorities at the University of California, Berkeley, 1980–1989* (Washington, DC: National Association of Graduate and Professional Students).

Martin, B. R. & J. Irvine (1982) "Women in Science—The Astronomical Brain Drain," *Women's Studies International Forum* 5(1): 41–68.

Matthies, H., E. Kuhlmann, M. Oppen, & D. Simon (2001) *Karrieren und Barrieren im Wissenschaftsbetrieb: Geschlechterdifferente Teilhabechancen in Ausseruniversitaeren Forschungseinrichtungen* (Berlin: Edition Sigma).

Mayer, K. U. (2002) "Wissenschaft als Beruf oder Karriere?" in Wolfgang Glatzer, Roland Habich, & Karl Ulrich Mayer (eds), *Sozialer Wandel und Gesellschaftliche Dauerbeobachtung* (Opladen: Leske & Budrich): 421–38.

McIlwee, J. & J. Robinson (1992) *Women in Engineering: Gender, Power and Workplace Culture* (Albany: State University of New York Press).

Merton, R. K. ([1942]1973) "The Normative Structure of Science," in *The Sociology of Science. Theoretical and Empirical Investigations* (Chicago, London: The University of Chicago Press).

Merton, R. K. (1973) *The Sociology of Science. Theoretical and Empirical Investigations* (Chicago, London: The University of Chicago Press).

Misa, T. (2004) *Leonardo to the Internet: Technology and Culture from the Renaissance to the Present* (Baltimore, MD: Johns Hopkins University Press).

Mukhopadhyay, C. C. & S. Seymour (eds) (1994) *Women, Education and Family Structure in India* (Boulder, CO: Westview Press).

Muller, C. (2003) "The Under-representation of Women in Engineering and Related Sciences: Pursuing Two Complementary Paths to Parity." A Position Paper for the National Academies Government University Industry Research Roundtable Pan-Organizational Summit on the U.S. Science and Engineering Workforce (Washington DC: National Academies Press).

National Research Council (2001) *From Scarcity to Visibility: Gender Differences in the Careers of Doctoral Scientists and Engineers* (Washington, DC: National Academies Press).

National Science Foundation, Science Statistics (1996) *Women, Minorities and Persons with Disabilities in Science and Engineering*. Available at: www.nsf.gov.

National Science Foundation, Science Statistics (2004) *Women, Minorities and Persons with Disabilities in Science and Engineering*.

Nerad, M. & J. Cerny (1999) "Widening the Circle: Another Look at Women," *Graduate Students Communicator* 32(6): 2–7.

Nerad, M. & J. Czerny (1999b) "Post-doctoral Career Patterns, Employment Advancement and Problems," *Science* 285: 1533–35.

Nobel, D. F. (1992) *A World Without Women: The Christian Clerical Culture of Western Science* (Oxford: Oxford University Press).

Osborne, M. (1991) "Status and Prospects of Women in Science in Europe," *Science* 263: 1389–91.

Palomba, R. (2004) "Does Gender Matter in Scientific Leadership?" in *Gender and Excellence in the Making* (European Commission: DG Research): 121–25.

Pearson, W. (1985) *Black Scientists, White Society, and Colorless Science: A Study of Universalism in American Science* (Millwood, NY: Associated Faculty Press).

Plonski, G. A. & R. G. Saidel (2001) "Gender, Science and Technology in Brazil," *Minerva* 39(2): 175–201.

Postrel, V. (2005) "Economic Scene. Some Economists Say the President of Harvard Talks Just Like One of Them," *New York Times* 240205 C2.

Prpic, K. (2002) "Gender and Productivity Differentials in Science," *Scientometrics* 55(1): 27–58.

Ramirez, F. O. (2001) "Frauenrechte, Weltgesellschaft und die gesellschaftliche Integration von Frauen," in Bettina Heintz (ed), *Geschlechtersoziologie, Sonderheft 41 der Koelner Zeitschrift für Soziologie und Sozialpsychologie* (Opladen: Westdeutscher Verlag): 356–97.

Ramirez, F. O. & C. M. Wotipka (2001) "Slowly But Surely? The Global Expansion of Women's Participation in Science and Engineering Fields of Study, 1972–1992," *Sociology of Education* 74: 231–51.

Reskin, B. (1977) "Scientific Productivity and the Reward Structure of Science," *American Sociological Review* 42: 491–504.

Riska, E & K. Wegar (eds) (1993) *Gender, Work and Medicine* (London: Sage).

Rosenberg, Rosalind (1982) *Beyond Separate Spheres: Intellectual Roots of Modern Feminism* (New Haven, CT: Yale University Press).

Rosser, S. (2004) *The Science Glass Ceiling: Academic Women Scientists and the Struggle to Succeed* (New York: Routledge).

Rossiter, M. (1982) *Women Scientists in America: Struggles and Strategies to 1940* (Baltimore, MD: Johns Hopkins University Press).

Rossiter, M. (1995) *Women Scientists in America: Before Affirmative Action 1940–1972* (Baltimore, MD: Johns Hopkins University Press).

Sax, L. J., L. S. Hagedoorn, M. Arredondo, & F. A. Dicrisili (2002) "Faculty Research Productivity: Exploring the Role of Gender and Family-related Factors," *Research in Higher Education* 43: 423–46.

Schiebinger, L. (1989) *The Mind Has No Sex? Women in the Origins of Modern Science* (Cambridge, MA: Harvard University Press).

Schiebinger, L. (1999) *Has Feminism Changed Science?* (Cambridge, MA: Harvard University Press).

Science and Technology for Women in India, <http://dst.gov.in/>. Accessed on June 27, 2005.

Shavit, Y. & H. Blossfeld (1993) *Persistent Inequality: Changing Educational Attainment in Thirteen Countries* (Boulder, CO: Westview Press).

Sime, R. (1996) *Lise Meitner* (Berkeley: University of California Press).

Simeonova, K. (1998) Bulgarian Academy of Sciences, interview with Henry Etzkowitz.

Simon, R. J., S. M. Clark, & K. Galway (1967) "The Woman PhD: A Recent Profile," *Social Problems* 15: 221–36.

Smith-Doerr, L. (2004) *Women's Work: Gender Equality vs. Hierarchy in the Life Sciences* (Boulder, CO: Lynne Rienner).

Sonnert, G. & G. Holton (1995a) *Who Succeeds in Science? The Gender Dimension* (New Brunswick, NJ: Rutgers University Press).

Sonnert, G. & G. Holton (1995b) *Gender Differences in Science Careers: The Project Access Study* (New Brunswick, NJ: Rutgers University Press).

Sonnert, G. & G. Holton (1996) "Career Patterns of Women and Men in the Sciences," *American Scientist* 84: 67–71.

Summers, L. H. (2005) "Remarks at NBER Conference on Diversifying the Science & Engineering Workforce," Harvard University, Cambridge, MA, January 14.

Tabak, F. (1993) "Women Scientists in Brazil: Overcoming national, social and professional obstacles." Paper presented at the Third World Organzation of Women Scientists, Cairo, January.

Tang, J. (1996) "To Be or Not To Be Your Own Boss: A Comparison of White, Black, and Asian Scientists and Engineers," in Helena Z. Lopata & Anne E. Figert (eds), *Current Research on Occupations and Professions*, vol. 9 (Greenwich, CT: JAI Press): 129–65.

Tang, J. (1997) "Evidence For and Against the 'Double Penalty' Thesis in the Science and Engineering Fields," *Population Research and Policy Review* 16(4): 337–62.

Teghtsoonian, M. (1974) "Distribution by Sex of Authors and Editors of Psychological Journals: Are There Enough Women Editors?" *American Psychologist* 29: 262–69.

Tri-national Conference: UK/France/Canada (2003) "The Role of the National Academies in Removing Gender Bias" (London: Royal Society), July.

UNESCO (1999) "Science and Technology" in *UNESCO Statistical Yearbook* (Paris: UNESCO, Bernan Press).

U.S. Department of Commerce, Office of Technology Policy (1999) *The Digital Workforce: Building Infotech Skills at the Speed of Innovation* (Washington, DC: OTP, June 1999).

Valian, V. (1999) *Why So Slow? The Advancement of Women* (Cambridge, MA: MIT Press).

Viale, R. & H. Etzkowitz (2005) "Polyvalent Knowledge: The 'DNA' of the Triple Helix," theme paper presented at Triple Helix V Conference. Available at: http://www.triplehelix5.com/programme.htm.

von Stebut, J. (2003) *Eine Frage der Zeit? Zur Integration von Frauen in die Wissenschaft. Eine empirische Untersuchung der Max-Planck-Gesellschaft* (Opladen: Leske & Budrich).

Wajcman, J. (1991) *Feminism Confronts Technology* (College Park: Pennsylvania State University Press).

Wajcman, J. (1998) *Managing Like a Man: Women and Men in Corporate Management* (College Park: Pennsylvania State University Press).

Watson, J. (1968) *The Double Helix* (New York: Norton).

Wax, R. (1999) "Information Technology Fields Lacks Balance of Gender," *Los Angeles Times*, January 19, 1999.

Wennerås, C. & A. Wold (1997) "Nepotism and Sexism in Peer-Review," *Nature* 22: 341–43

Whalley, P. (1986) *The Social Production of Technical Work* (Basingstoke: Macmillan).

Wilde, C. (1997) "Women Cut through IT's Glass Ceiling," *Information Week*, January 20, 1997.

Wilson, F. (1996) "Research Note: Organizational Theory: Blind and Deaf to Gender?" *Organization Studies* 5: 825–42.

Wilson, R. (1999) "An MIT Professor's Suspicion of Bias Leads to a New Movement for 'Academic Women'," *Chronicle of Higher Education* (46) (3 December): A16–18.

Windolf, P. (1997) *Expansion and Structural Change: Higher Education in Germany, the United States, and Japan, 1870–1990* (Boulder, CO: Westview Press).

Witz, A. (1992) *Professions and Patriarchy* (London: Routledge).

Xie, Y. & K. A. Shauman (2003) *Women in Science: Career Processes and Outcomes* (Cambridge: Harvard University Press).

Zuckerman, H. (1987) "Persistence and Change in the Careers of Men and Women Scientists and Engineers," in Dix, L. S. (ed), *Women: Their Underrepresentation and Career Differentials in Science and Engineering* (Washington, DC: National Academy Press): 127–56.

Zuckerman, H. & J. R. Cole (1975) "Women in American Science," *Minerva* 13: 82–102.

Zuckerman, H. & J. R. Cole (1987) "Marriage, Motherhood and Research Performance in Science," *Scientific American* 256: 119–25.

III   Politics and Publics

Edward J. Hackett

There was a time when science and technology occupied a realm of genius and wizardry, a world apart that "the public" viewed with awe and admiration. In that earlier time, decisions having to do with science or technology were the prerogative of experts who would make them in the public interest but without the public's involvement. That time has passed, or perhaps never really happened, and STS research of recent years has changed our understanding of the engagement of science and technology with politics and publics. Today, decisions involving science and technology are understood to be inherently political: various publics are involved in different ways with science and technology, and the responsible conduct of a career in science demands consideration of matters of ethics and values that had previously been held to one side. Chapters in this section explore the changing dimensions and dynamics of the relationship among science, technology, and medicine and their politics and publics.

Steven Shapin begins this section by asking what people might mean when they claim that "science made the modern world." This simple question launches an inquiry into the foundations of scientific authority that asks how pervasive and deeply engrained in the public mind are scientific knowledge and patterns of thought. Reviewing a range of empirical studies of the general public, Shapin finds uneven commitment to the canonical scientific method and outlook (that is, a critical, empirical, demystifying approach to inquiry), and little evidence that substantive scientific knowledge is widely understood. Scientists themselves, in fact, demur from claims to ultimate truth or morality. At best, it seems, the public authority of science rests upon a general notion of the independence and integrity of science, and these qualities are now jeopardized by increasingly close connections of science with the production of wealth and projection of power. We're left to wonder if the modern world is the unmaking (or unmasking?) of science.

Massimiano Bucchi and Federico Neresini take an inclusive view of public engagement with science, a phenomenon that for them includes public involvement in setting research agendas, making decisions, shaping policy, and co-producing scientific knowledge. Bucchi and Neresini contend that the "deficit model" of public understanding of science is undermined by the many different ways publics engage science and technology, which in turn demands that we devise new ways of characterizing these relationships. To this end they analyze public involvement along two principal axes, one defined by *intensity* of interaction, or how deeply the public can shape the content of science and the organization of scientific work, the other by *spontaneity*, or the degree to which the public is invited to participate, with end points anchored by engagements initiated by scientists near one pole and protest movements near the other. We now understand that interactions between experts and publics are fluid and dynamic; and that this invites systematic, empirical study of the circumstances, qualities, and consequences of such interactions.

David Hess, Steve Breyman, Nancy Campbell, and Brian Martin explain how social movements are powerful democratizing forces that shape science and technology and are themselves shaped by their cultural, historical, and social contexts. Drawing illustrations from social movements concerned with health, environment, peace, and information-media, Hess and his colleagues delineate reciprocal influences that are simultaneously cooperative and conflictual. Among the challenging avenues for future research they sketch, perhaps the boldest calls for scholarship that transcends academic requirements and promotes the interests of democracy.

Steven Epstein's chapter takes a complementary approach, selecting health social movements as important in their own right and also "good to think with" about a spectrum of questions concerning knowledge, technology, social organization, power, and the like. We learn that the formation and continuity of health social movements are influenced by the density of social interaction, circumstances of the group and its social context, and the communication technologies available to members. For example, stigmatizing illnesses may reduce social interaction within a group, while the use of the Internet has reduced face-to-face interaction. The embodied and experiential knowledge of patients and their families complements and challenges credentialed expert knowledge, and the two combine in ways that have powerful but unpredictable consequences. Taken together, health social movements have transformed the understanding and management of diseases, shaped research and technologies, and influenced policies and markets. Comparisons across cases and cultures, studies of the life course and dynamics of movements and their diffusion across national borders, and attention to social inequalities, health disparities, and the ambiguities of membership are all promising topics for inquiry.

New ideas and understandings about the design and use of technologies pose analytic challenges for Nelly Oudshoorn and Trevor Pinch. To meet those challenges, they develop a conceptual vocabulary that both expresses the hybrid identities of technology's producers and users and represents their entangled and ambiguous roles in the process of making technologies. Borrowing from the conceptual lexicons of five academic literatures (innovation studies, the sociology of technology, feminist studies, semiotics, and media and cultural studies), Oudshoorn and Pinch build a framework for thinking through the reciprocal influences that are endemic to contemporary STS analysis: technological designs are completed in their use, yet uses are built into design; technologies resemble scripts or laws that guide social behavior, yet are also shaped by their creators' values and actions; technologies are gendered at conception and in use at the "consumption junction," yet are endowed with sufficient interpretative flexibility to be different things to different people; consumption is an act of production, resistance a dimension of use. From this conceptual language for representing entangled identities and reciprocal interactions emerges a stimulating collection of problems for empirical research.

Ethical precepts are the strongest expressions of public values and interests, and analysis that stops short of engaging ethical concerns may be considered timid or incomplete. Deborah Johnson and Jameson Wetmore urge scholars to engage with ethical issues in their analyses, and illustrate how to do so by reconsidering research about sociotechnical systems and the relationship between technology and society. Johnson and Wetmore reject determinisms grounded in nature, science, or the autonomy of technology, placing agency and responsibility squarely, but not solely, with engineers: engineers' work is embedded within sociotechnical systems of production and consumption, and those systems, simultaneously and somewhat paradoxically, both limit and extend the ethical responsibilities of engineers. Engineers are not the sole actors in the sociotechnical system, so their latitude is somewhat constrained by other elements of the system. Yet within a sociotechnical network, engineers' responsibilities are also enlarged because their work must now take account of others—engineers, scientists, users, policy makers, and, of course, ethicists, who are all active members of the system.

Alan Irwin employs the term "governance" to describe interactions among science, technology, and politics, replacing the term "science policy" and its relatives with a concept deeper in meaning and richer in research implications. Viewed in this light, the study of scientific and technological governance becomes central to STS research, replacing linear thinking about science and policy with a conceptualization that embraces hybrid identities, fluid interactions, and reciprocal influences. He develops these ideas in the form of complementary principles, illustrated by material drawn from case studies. For example, concerns about democratization now lead to questions about democracy; governance is not mechanical and sure-handed but instead is characterized by uncertainty and doubt; and expertise and power are understood to form or constitute one another in their various arenas of interaction. Consequently, we come to question the received view of sciences and markets as "neutral, fixed, and objective entities," a systematic skepticism that brings STS scholarship into productive engagement with studies of power, inequality, globalization, technological innovation, and development.

Experts and expertise are counterparts to the lay public and its generalized knowledge, but recent STS scholarship challenges facile constructions of these categories, and instead examines the social behavior that creates the categories and sustains boundaries between them. Robert Evans and Harry Collins review STS critiques of expertise and augment them by discussing alternative models of decision making that value the generalist knowledge of amateurs and the lay public (heuristics, low-informational rationality). They point out that all such arguments for lay input to expert decisions hinge on deciding who is an expert and in discerning differences between types of expertise: expertise as an attribute, acquired through socialization and interaction, is essential to such decisions and discernments. The authors systematize their perspective in a "Periodic Table of expertises," that organizes sources and characteristic modes of expertise in orderly rows and columns that associate them with implications for practical action and research.

18 Science and the Modern World

Steven Shapin

Science Made the Modern World, and it's science that shapes modern culture. That's a sentiment that gained currency in the latter part of the nineteenth century and the early twentieth century—a sentiment that seemed almost too obvious to articulate then and whose obviousness has, if anything, become even more pronounced over time. Science continues to Make the Modern World. Whatever names we want to give to the leading edges of change—globalization, the networked society, the knowledge economy—it's science that's understood to be their motive force. It's science that drives the economy and, more pervasively, it's science that shapes our culture. We think in scientific terms. To think any other way is to think inadequately, illegitimately, nonsensically. In 1959, C. P. Snow's *Two Cultures and The Scientific Revolution* complained about the low standing of science in official culture, but he was presiding not at a funeral but at a christening. In just that very broad sense, the "science wars" have long been over and science is the winner.

In the 1870s, Andrew Dickson White, then president of Cornell, wrote about the great warfare between science and what he called "dogmatic theology" that was being inexorably won by science.[1] In 1918, Max Weber announced the "disenchantment of the world," conceding only that "certain big children" still harbored reservations about the triumph of amoral science (Weber, [1919]1991: 142). Some years earlier, writing from the University of Chicago, Thorstein Veblen described the essential mark of modern civilization as its "matter of fact" character, its "hard headed apprehension of facts." "This characteristic of western civilization comes to a head in modern science," and it's the possession of science that guarantees the triumph of the West over "barbarism." The scientist rules: "On any large question which is to be disposed of for good and all the final appeal is by common consent taken to the scientist. The solution offered by the scientist is decisive," unless it is superseded by new science. "Modern common sense holds that the scientist's answer is the only ultimately true one." It is matter-of-fact science that "gives tone" to modern culture (Veblen, 1906: 585–88). This is not an injunction about how modern people *ought* to think and speak but Veblen's description of how we *do* think and speak.

In 1925, Alfred North Whitehead's *Science and the Modern World* introduced the historical episode that "made modernity," which had not *yet* been baptized as "the

Scientific Revolution": it was "the most intimate change in outlook which the human race had yet encountered . . . Since a babe was born in a manger, it may be doubted whether so great a thing has happened with so little stir." What started as the possession of an embattled few had reconstituted our collective view of the world and the way to know it; the "growth of science has practically recoloured our mentality so that modes of thought which in former times were exceptional, are now broadly spread through the educated world." Science "has altered the metaphysical presuppositions and the imaginative contents of our minds . . ." Born in Europe in the sixteenth and seventeenth centuries, its home is now "the whole world." Science, that is to say, travels with unique efficiency: it is "transferable from country to country, and from race to race, wherever there is a rational society" (Whitehead, [1925]1946: 2).

The founder of the academic discipline called the history of science—Harvard's George Sarton—announced in 1936 that science was humankind's *only* "truly cumulative and progressive" activity, so if you wanted to understand progress towards modernity, the history of science was the only place to look (Sarton, 1936: 5). The great thing about scientific progress was—as was later said and often repeated—that "the average college freshman knows more physics than Galileo knew . . . and more too than Newton" (Gillispie, 1960: 9). Science, Sarton (1948: 55) wrote, "is the most precious patrimony of mankind. It is immortal. It is inalienable." When, toward the middle of the just-past century, the Scientific Revolution was given its proper name, it was, at the same time, pointed to as the moment modernity came to be. Listen to Herbert Butterfield in 1949, an English political historian, making his one foray into the history of science:

[The Scientific Revolution] outshines everything [in history] since the rise of Christianity and reduces the Renaissance and Reformation to the rank of mere episodes, mere internal displacements, within the system of medieval Christendom. Since it changes the character of men's habitual mental operations even in the conduct of the non-material sciences, while transforming the whole diagram of the physical universe and the very texture of human life itself, it looms . . . large as the real origin of the modern world and of the modern mentality . . . . (Butterfield, 1949: vii–viii)

Butterfield's formulation was soon echoed and endorsed, as in this example from the Oxford historian of science A. C. Crombie:

The effects of the new science on life and thought have . . . been so great and special that the Scientific Revolution has been compared in the history of civilisation to the rise of ancient Greek philosophy in the 6th and 5th centuries B.C. and to the spread of Christianity throughout the Roman Empire . . . . (Crombie, [1952]1959: vol. 1, p. 7)

And by 1960 it had become a commonplace—Princeton historian Charles Gillispie (1960: 8) concurring that modern science, originating in the seventeenth century, was "the most . . . influential creation of the western mind." As late as 1986, Richard Westfall—then the dean of America's historians of science—put science right at the heart of the modern order: "For good and for ill, science stands at the center of every dimen-

sion of modern life. It has shaped most of the categories in terms of which we think
. . ." (Westfall, 1986).

Evidence of that contemporary influence and authority is all around us and is unde-
niable. In the academy, and most especially in the modern research university, it is
the natural sciences that have pride of place and the humanities and social sciences
that look on with envy and, sometimes, resentment. In academic culture generally,
the authority of the natural sciences is made manifest in the long-established desire
of many forms of inquiry to take their place among the "sciences": social science,
management science, domestic science, nutrition science, sexual science. Just because
the designation "science" is such a prize, more practices now represent themselves as
scientific than ever before. The homage is paid from the weak to the strong: students
in sociology, anthropology, and psychology commonly experience total immersion in
"methods" courses, and while chemists learn how to use mass spectrometers and
Bunsen burners, they are rarely exposed to courses in "scientific method." The
strongest present-day redoubts of belief in the existence, coherence, and power of the
scientific method are found in the departments of human, not of natural, science.

Moreover, though it may be vulgar to mention such things, one index of the authority of science in academic culture is the distribution of cash, a distribution that
seems—crudely but effectively—to reflect public sensibilities about which forms of
inquiry have real value and which do not. The National Science Foundation and the
National Institutes of Health distribute vastly more money to natural scientific
research than the National Endowment for the Humanities does to its constituents.
Statistics firmly establish pay differentials between academic natural scientists and
engineers and their colleagues in sociology and history departments, and the "summer
salary" instituted by the National Science Foundation early in its career was one
explicit means of ensuring this result in a Cold War era when the "scarcity" of physicists and chemists, but not of, say, art historians, was a matter of political concern.
These days it is more likely the "opportunity cost" argument that justifies this
outcome, even if it means that not just scientists and engineers but also academic
lawyers, physicians, economists, and business school professors now command higher
salaries.[2] Many scientists and engineers are now the apples of their administrators'
eyes because their work brings in government and corporate funding, with the attendant overheads on which research universities now rely to pay their bills. Finally, the
ability of university administrators to advertise to their political masters how their
activities help "grow the local economy," spinning off entrepreneurial companies,
transferring technology, and creating high-paid, high-tax jobs, all support the increasing influence of science and engineering in the contemporary research university. In
the 1960s, social and cultural theorists—following Habermas—began to worry about
what they called a "technocracy," in which decisions properly belonging in the public
sphere, to be taken by democratically elected and democratically accountable politicians, were co-opted by a cadre of scientific and technical experts—as the saying is,
"on top" rather than "on tap." Even though that worry seems to have been allayed
by more recent concern with political interference in scientific judgments, a recent

New Yorker magazine piece complaining about the Bush Administration's attack on the autonomy of science blandly asserted the primacy of science as the leading force of modern historical change: "Science largely dictated the political realities of the twentieth century" (Specter, 2006: 61).

Sixty years after Hiroshima, and over a century after General Electric founded the first industrial research laboratory, it is almost too obvious to be pointed out that it is the natural sciences that are now so closely integrated into the structures of power and wealth, and not their poorer intellectual cousins. It is science that has the capacity to deliver the goods wanted by the military and by industry, and not sociology or history, though some obvious qualifications need to be made—not all the natural sciences do this—and there was a period, early in the post–World War II world, when there were visions of how the human sciences might make major contributions to problems of conflict, deviance, strategic war-gaming, the rational conduct of military operations and weapons development, and the global extension of benign American power. Few observers disagree when it is said that science has changed much about the way we live now and are likely to live in the future: how we communicate, how long we are likely to live and how well, whether any of the crucial global problems we now confront—from global warming to our ability to feed ourselves—are likely to be solved—indeed, what it will mean to be human.

Some time about the middle of the just-past century, sociologists noted an exponential increase in the size of the scientific enterprise. By any measure, almost everything to do with science was burgeoning: in the early 1960s, it was said that 90 percent of all the scientists who had ever lived were then alive and that a similar proportion of all the scientific literature ever published had been published in the past decade. Expenditures on scientific research were going up and up, and, if these trends continued—which in the nature of things they could not—every man, woman, child, and dog in the United States would be a scientist and every dollar of the Gross Domestic Product would be spent on the support of science (Price, [1963]1968: 19). By these and many other measures, it makes excellent sense to observe that science *is* constitutive of the Modern World. And so it's hard to say that claims that Science Made the Modern World or that Science is constitutive of Modern Culture are either nonsense or that they need massive qualification. Nevertheless, unless we take a much closer look at such claims, we will almost certainly fail to give any worthwhile account of the Way We Live Now.

Do we live in a scientific world? Assuming that we could agree on what such a statement might mean, there is quite a lot of evidence that we do not now and never have. In 2003, a Harris poll revealed that 90 percent of American adults believe in God, a belief that, of course, is not now, and never was, in any necessary conflict with whatever might be meant by a scientific mentality. But 82 percent believe in a physical Heaven—a belief that is—perhaps predictably, just because Heaven is so much more pleasant than The Other Place—13 percent more popular than a belief in Hell; 84 percent believe in the survival of an immaterial soul after death, and 51 percent in the reality of ghosts. The triumph of science over religion trumpeted in the late nineteenth century crucially centered on the question of whether or not supernatural spiritual agencies could intervene in the course of nature, that is to say, whether such things as miracles existed. By that criterion, 84 percent of American adults are unmarked by the triumph of science over religion that supposedly happened over a century ago. These responses are not quite the same thing as the "public ignorance of science" (or "public misunderstanding of science") so frequently bemoaned by leaders of the scientific community. For that, you'll want statistics on public beliefs about things like species change or the Copernican system. Such figures are available: 57 percent of Americans say they believe in psychic phenomena, such as ESP and telepathy, that cannot be explained by "normal means."[3] Americans are often said to be more credulous than Europeans, but comparative statistics point to a more patchy state of affairs. Forty percent of Americans said astrology is "very" or "sort of" scientific, while 53 percent of Europeans that it was "rather scientific." Americans did somewhat better than Europeans in grasping that the Earth revolves around the Sun and not the other way: 24 percent of Americans got that wrong compared with 32 percent of Europeans, and only 48 percent of Americans believed that antibiotics killed viruses compared with 59 percent of Europeans. Unsurprisingly, the "Darwin question" is flunked by more Americans than Europeans: 69 percent of Europeans, but only 52 percent of Americans, agreed that "Human beings developed from earlier species of animals" (National Science Foundation, 2001; European Commission, 2001). A still more recent transnational survey published in *Science* shows that, when asked the same question, Americans yielded the second-lowest rate of acceptance (now 40 percent) of all 34 countries polled—above only Turkey (Miller et al., 2006). If you believe the Gallup pollsters, then in 2005 the percentage of Americans who agreed with the more specific and loaded statement that "Man has developed over millions of years from less advanced forms of life [and] no God participated in this process" was 12 percent, encouragingly up from 9 percent in 1999.[4]

Whitehead's *Science and the Modern World* was based on the Lowell Lectures given at Harvard by a newly minted professor of philosophy, and perhaps that context is relevant to his assertion that scientific modes of thought "are now broadly spread through the educated world." Perhaps we can conclude that there is now, just as there always has been, a big gulf between "the educated world" and the unwashed and unlettered. But Whitehead was quite aware that the Galilean-Newtonian "revolution" was the possession of only a very small number of people and that their beliefs bore slight relationship to those of the peasantry in Sussex, much less in Serbia or Siam. Although a number of twentieth-century scholars loosely referred (and refer) to science-induced tectonic and decisive shifts in "our" ways of thinking, or to those of "the West," Whitehead, addressing his Harvard audience, confined himself to "the educated world." So it must, then, be relevant that the 84 percent of contemporary Americans who profess belief in miracles does indeed drop when the responses of only those with *postgraduate degrees* are considered, that is to say, not just who are college educated but have master's or doctoral degrees. The percentage of *these* elites who say they believe in miracles is *only* 72 percent and the percentage of college graduates who agree with the

Gallup poll's version of Darwinian evolution is 16.5 percent. The possibility remains that we can still make some distinction of the general sort that Whitehead intended: suppose that "science" is what's believed at Harvard and Haverford that's not believed at, say, Oral Roberts. Maybe that's right, but that's not *quite* what Whitehead said.

Perhaps, then, we should find some statistics about what *scientists* believe. A survey conducted in 1916 found that 40 percent of randomly selected American scientists professed belief in a personal God. This was a surprise to the author of the report, and he expressed his confidence that the figure would surely drop as education spread (Leuba, 1916). But it has not. In a survey published in *Nature* in 1997, an identical 40 percent of American scientists counted themselves as believers in God, with only 45 percent willing to say they did not believe (Radford, 2003; Larson & Witham, 1997). Those wanting to get the figure of scientists believing in a personal God or human immortality under 10 percent will have to accept a 1998 survey confined to members of the National Academy of Sciences, while the mathematicians among this elite were the most likely to believe, at about 15 percent (Larson & Witham, 1998). Scientists, of course, are leading the charge in the recent American defense of Darwinism in the classroom, but according to the Gallup poll, only a bare majority of *them*—55 percent—actually assent to the poll's version of Darwinian evolution.[5]

There is no reason to fetishize a Harris, Gallup, or any other systematic attitude survey. We do not know with any great specificity what people might *mean* when they say they believe in miracles (or, indeed, astrology), and the inadequacy of any simple-minded juxtaposition of "scientific" versus "fundamentalist" beliefs is indicated by the soaring popularity of stem cell research, even among evangelical Christians who are widely supposed to be against tampering with God-given human life. Religiosity seems to bear on embryo destruction in abortion in a way it does not in stem cell research.[6] And, if it were thought that religiosity translates into a "don't mess with God's Nature" attitude, then Americans again are much more favorably disposed toward genetically modified foods than are Western Europeans or Japanese.[7] The legal scholar Ronald Dworkin has recently pointed out—without evidence, but plausibly enough—that not a lot should be inferred about overall attitudes to scientific expertise from evangelicals' doubts about Darwinism:

Almost all religious conservatives accept that the methods of empirical science are in general well designed for the discovery of truth . . . . They would not countenance requiring or permitting teachers to teach, even as an alternate theory, what science has established as unquestionably and beyond challenge false: that the sun orbits the earth or that radioactivity is harmless, for example.[8] (Dworkin, 2006: 24)

But it still seems safe to say that the great majority of the people professing belief in things like miracles have been presented with multiple articulations of what it might mean to "think scientifically" and thinking miracles happen is understood not to be part of the scientific game.[9] Quite a lot of the people saying they believe in miracles, like quite a lot of the people saying that human beings were specially created by a divine agency, must be well aware that they are, in so saying, poking one in the eye of scientific authority. And so one thing we cannot sensibly mean when we say that we live in a Scientific Age or that Science Made the Modern World is that scientific beliefs have got much grip on the modern mind writ large. That just isn't true. Maybe, if we mean anything legitimate at all by saying such things, we mean that the *Idea of Science* is widely held in respect. That seems plausible enough. Consider the litany of complaints from high scientific places about "public ignorance of science"— complaints that often are inspired by such statistics as those just cited. These complaints can actually help establish the esteem in which science is held in our culture. It's been some time since I heard anyone gain a public platform for complaining about "public ignorance of sociological theory" or "public ignorance of the novels of Mrs. Gaskell." Nor do official worries about the proliferation of pseudo-science or junk science necessarily bear on the authority of science. Consider present-day concerns over "Intelligent Design" and "Creation Science," but note that these represent themselves as forms of science, not as nonscience or as antiscience. Advocates of Intelligent Design want it taught in *science* classrooms. From a pertinent perspective, the problem today is not antiscience but a contest for the proper winner of the designation "science." That's a sign that the label "science" is a prize very much worth having. A writer in *The New York Times* (Holt, 2005), referring to the apparent upsurge in evangelical Christianity, recently announced that "Americans on the whole do not seem to care greatly for science," but such conclusions are not well grounded. American faith in the power of science—or, more accurately, of science and technology—has been, and continues to be, enormous. In the late 1950s, surveys showed that a remarkable 83 percent of the U.S. public reckoned that the world was "better off" because of science and only a negligible 2 percent thought it was "worse off" (Withey, 1959).[10] Amid anxieties about "increasing public skepticism toward science," various surveys conducted in the 1970s—phrasing their questions somewhat differently—purported to find a decline in approval (to between 71 and 75 percent, with a negative assessment rising to between 5 and 7 percent)—though few other modern American institutions could hope to come close to that level of public favor (Pion & Lipsey, 1981: 304, table 1).[11] In the most recent survey, Americans expressed a "great deal" of confidence (42 percent) in the scientific community and significantly less in the banking system (29 percent), the presidency (22 percent), and, tellingly, organized religion (24 percent).[12] The Pew Research Center's Global Attitudes Project discovered that 19 percent of Americans surveyed recently accounted "Science/Technology" to be the "greatest achievement" of the U.S. government during the course of the twentieth century—more than twice as many as those who pointed to civil rights and more than three times as many as those giving the prize to the social security system. In the public mind, science and technology are endowed with colossal power: about 80 percent of Americans think that within the next fifty years science will ("probably/ definitely") deliver cures for cancer and AIDS and will "improve [the] environment," compared with just 44 percent who believe that Jesus Christ will reappear on Earth during that period (Kohut & Stokes, 2006: 60, 86).

Suppose we concede that scientific *beliefs*—or at least beliefs of the sort approved of at Harvard—are not very widely distributed in modern culture. This means that the authority of science—the sense that we live in a scientific age—has to reside in something other than the widespread *understanding* of particular scientific facts or theories, no matter how important, foundational, or elementary they may be. This would be quite a concession in itself, and we should reflect a lot more on what it means. But can't we nevertheless say that the authority and influence of science reside in something other than shared beliefs, something that nevertheless "belongs to" science? Consider, again, the notion of the Idea of Science. I've given some reasons to think that the Idea of Science confers authority, even if a range of specific scientific beliefs do not possess authority. What might be meant by the Idea of Science? There are difficulties in saying much about such an Idea. If we want to talk about the Idea of Science apart from specific beliefs, then we probably are pointing at some notion of scientific method. Scientists—and, more importantly, philosophers of science—have been identifying, celebrating, and propagating the scientific method for a long time—arguably at least as far back as the time of Descartes, Newton, and Boyle. It's that universal, rational, and effective method which has been said to account for the power of science and to mark it out from other modes of inquiry lacking such a method. As the recent *New Yorker* piece announced, "The scientific method has come to shape our notion of progress and of modern life" (Specter, 2006: 61).

The problem is that there is not now, and never has been, a consensus about what such a method is.[13] The first two entries for "scientific method" that Google gave me opted for observation before the formulation of an explanatory hypothesis, followed by experimental tests of the hypothesis, though that account excludes all those sciences which are not experimental, for example, geology, meteorology, and many forms of evolutionary biology.[14] The current Wikipedia entry makes reference to the views of Thomas Kuhn, who, like Karl Popper, Imre Lakatos, and Paul Feyerabend, famously doubted whether theory-free observation ever occurred. *Science* magazine has usefully addressed the question by annotating a number of scientific papers to show the scientific method at work.[15] A "pragmatical scheme" of that seven-step method is provided, starting with "define the question," going through "analyze the data," and concluding with "publish results," but it's hard to look at this list without concluding that—"perform experiment" apart—its directions can be found in any kind of systematic inquiry pretending to rigor, and not just in science.[16] Other entries early in the Google list give deductive, rather than inductive, inference pride of place and omit references to experiment.[17] Some make reference to "proof" or "confirmation" of a hypothesis; others point out—following Popper—that one can never prove but only disprove the validity of a hypothesis. Few bother to cite T. H. Huxley's ([1854]1900: 45) view that science is "nothing but trained and organised common sense" or that of the Nobel Prize–winning immunologist Peter Medawar (1967: 132) that "The scientific method does not exist."

In fact, if the authority of science—the way in which it is supposed to mark modernity—resides in some idea of scientific method, that would be as much as saying not just that academic *philosophy of science* rules the roost, but that some specific *version* of philosophy of science was the most authoritative form of modern culture. Somehow that doesn't seem right. The authority of philosophers in our culture doesn't come close to the authority of scientists. Much the same sort of argument, I think, applies to any Idea of Science that flows from identifying shared *conceptual* content. The Unity of Science movement of the early and middle part of the twentieth century arose out of a worry that, while science *must*, of course, be conceptually unified, no one had yet definitively shown what the basis of that unity was. That situation has not changed, and although scientists these days seem not to be much worried about "unity," leading-edge philosophers of science are now increasingly writing books and papers taking the *"disunity"* of science as their subject (e.g., Dupré, 1993; Rosenberg, 1994; Cartwright, 1999; Galison & Stump, 1996).

I doubt that searching for some stable and plausible Idea of Science is going to get us very far in trying to describe the authority of science in the modern world, or in showing that science *does* have such authority. But, if I'm right, we're beginning to see the shape of a real problem: science, we say, marks modernity—it enjoys unique authority—but that authority does not seem to consist either in lay possession of any specific set of scientific beliefs—no matter how elementary or fundamental—nor in any stable sense of the method scientists supposedly used to guarantee the power of their knowledge. Should we just agree that science has very little to do with Modern Culture—bizarre as that might sound—*or* that the authority of science resides in something besides knowledge of its beliefs or methods?

It seems that if we want to talk about the authority of science in the Modern World, we can't sensibly talk about our culture's knowledge of scientific beliefs or our grasp of some notion of Method. What seems to be essential is not knowing *science* but knowing where to look for it, knowing who are the relevant authorities, knowing that we can and should assent to what they say, *that* we can and should trust them in their proper domains. Pragmatically, there's a lot to recommend this state of affairs: it's unfortunate that the ideas of both Darwinian evolution and the heliocentric system have not taken better root in our culture, but, in general, *no one* can know very much of science, and so knowing who the relevant experts are is sufficient in the great majority of cases. This applies to scientists as well as the laity: even plant physiologists are likely to have a deficient knowledge of astrophysics, and a cardiologist is going to go to a neurologist if she has persistent headaches. Expertise isn't considered to be fungible: it comes in various special flavors. And so knowing where to look for the relevant experts has to involve some notion of relevant expertise, of relevant authority.

When we say that our task is recognizing the experts in their proper domains, what *are* those domains? Putting the question that way identifies a sense in which scientific authority is now not greater but clearly much *less* than it once was. Consider what philosophers—following G. E. Moore in the first years of the twentieth century—call "the Naturalistic Fallacy." That fallacy is believing something that is impossible, moving logically from an "is-statement"—a description of how things are in the world—to an "ought-statement"—a prescription of how things should be. Put another way, science is one thing, morality another; and you should not think of deducing what's good from what is. But the Naturalistic Fallacy is not just about a philosopher's boundary; during the course of the twentieth century, very many scientists publicly insisted that they possessed no special moral standing and that questions of what ought to be done—for example, about the consequences of their own work—were not their preserve. As Edward Teller (1950) put it, it was the scientist's job to discover the laws of nature, not to pronounce on whether the laws permitting nuclear fusion ought to be mobilized for the construction of a hydrogen bomb. You would think that Oppenheimer would have disagreed with such a sentiment, but on this point he was at one with Teller (see, for example, Oppenheimer, 1965: 272).

Scientists—it was widely insisted by modern scientists themselves—possessed no particular moral authority. It was once assumed they did; now it was not. If moral authority is what you want, you should go to some other sort of person, and that's why the late Stephen Jay Gould (1997) referred to science and religion as "non-overlapping magisteria." That division of labor between natural experts and ethical experts is now institutionalized, accepted almost as a matter of course. Yet it leads to a pervasive awkwardness in contemporary culture. Just as so many social and political decisions increasingly come to draw on massive amounts of specialized expertise—even to understand what they're *about*—so it is accepted that those who know most should accept radical restrictions on having consequential opinions about *what ought to be done*. Here, the up-curve of the reach of science in our social and political life meets the down-curve of scientists' acknowledged moral authority. Who are they, such that we can trust them—not just to know more about their specialized bits of the world but to do the right thing?

"The scientist is not a priest." That's another way of identifying the limited authority of the modern scientist, and the nonpriestly status of the scientist was much insisted on throughout the twentieth century by scientists themselves. At the same time, and perhaps responding to what was seen as the *increasing* cultural authority of science during the course of the century, the scientific community was accused of becoming "the new priesthood" and scientists as "the new brahmins" (e.g., Lapp, 1965; Klaw, 1968). An essay in the *Bulletin of the Atomic Scientists* about immediate postwar Congressional engagements with science noted that, after Hiroshima,

[S]cientists became charismatic figures of a new era, if not a new world, in which science was the new religion and scientists the new prophets . . . . Scientists appeared to [politicians] as superior beings who had gone far ahead of the rest of the human race in knowledge and power . . . . Congressmen perceived scientists as being in touch with a supernatural world of mysterious and awesome forces whose terrible power they alone could control. Their exclusive knowledge set scientists apart and made them tower far above other men. (quoted in Hall, [1956]1962: 270–72)

It's a tension that remains unresolved: science is our most powerful form of knowledge; it's scientists—or at least those pretending to be scientists—that are turned to when we want an account of how matters stand in the natural world. But, however esoteric their knowledge is, it is not scientists who decide what ought to be done. For those decisions—and there are an increasing number of them that are potentially world-changing—it's politics as usual.

Knowing where to look for the relevant experts also involves some notion of what it is they know. In the early modern period, a common cultural distinction was made between mathematics and natural philosophy. Philosophy was understood as the search for Truth, for the realities behind appearances, for the real causal structure of the world. Mathematics, by contrast, was taken as the quest for regular patterns of natural relationships, such that you could use the resulting knowledge to predict and control, without necessarily taking a bet on what the world was really like. Copernicus was acting as a mathematician when he stipulated that the heliocentric system was to be regarded as a predictive tool, and Galileo was blurring the boundaries between mathematics and philosophy when he defied the Vatican in asserting the physical reality of Copernicanism.[18] I mention this old chestnut, just because it may have significance for our current problem of identifying who the relevant experts are.

At least from the early twentieth century, very many scientists—physicists, of course, but not just physicists—publicly asserted that they were not, so to speak, in the Truth Business.[19] Their task, it was insisted, was not metaphysics; it was not discovering ultimate realities. It was, rather, finding out what "works": what picture of nature was maximally coherent, with existing theories and evidence, and what picture of nature would allow scientists most powerfully to predict and control. Pragmatism was one version of such a sensibility, but so were those positions called operationalism, conventionalism, and phenomenalism. In 1899, the Johns Hopkins physicist Henry Rowland (1899: 13), making no allusions to pragmatism or to any other formal philosophy of science, explicitly contrasted the scientific with the "vulgar" or "ordinary crude" mind: the scientist alone properly appreciated that "There is no such thing as absolute truth and absolute falsehood."[20] By the 1920s, Albert Einstein ([1929]1954) was reminding the general reader that "It is difficult even to attach a precise meaning to the term 'scientific truth,'" its semantics varying radically according to context of use.[21] And C. P. Snow (1961: 257) surely spoke for most scientists when he bumptiously stipulated that "By *truth*, I don't intend anything complicated . . . I am using the word as a scientist uses it. We all know that the philosophical examination of the concept of empirical truth gets us into some curious complexities, but most scientists really don't care." The scientist was properly to be understood not on the model of the philosopher but on the model of the engineer and technician. Our culture used to insist on massive differences between science and technology and between the role of the scientist and that of the engineer. It's a distinction that now makes less and less sense: we're all engineers now, and the authority of science is increasingly based not on what scientists know but on what they can help make happen. It's a distinction that increasingly resonates in the public culture: an NSF survey in 1976 revealed that government funding of science was overwhelmingly popular but that only 9 percent of the respondents wanted any of their tax dollars used to support basic research (Pion & Lipsey, 1981: 308 [table IV] and 309).

What difference does it make to the public authority of science if scientific knowledge is just what works and if the scientist is understood as an aid to the technologist? First, at one time it was believed that a world saturated with technology would not only be a modernized world but a secularized world. That turned out to be spectacularly untrue. The mere presence of advanced technology in a society seems to have little or nothing to do with how people think and what they value: some of the world's Web wizards are *jihadis*, and there seems to be no conflict between computer skill and religious fundamentalism. We should be clear about another thing: engineers seem to include as many morally admirable people as any other group of professionals; some are more admirable than some scientists I know. But it's the institutions we're talking about here, and what virtues and authority are associated with the institutions. The technologist supplies what society wants; the scientist used to give society what it didn't know it wanted. That's a simplification, but, I think, a useful one: corporations, governments, and the military enlist experts in the natural world overwhelmingly on the condition that they can assist them in achieving useful goals—wealth, health, and power. During the course of the twentieth century, the enterprise called science was effectively enfolded in the institutions dedicated to the production of wealth and the projection of power. That's where we started, and that's one way of describing the success of science in modernity. But one of the conditions of that success is, at the same time, a problem for the authority of science in the modern world.

Modern scientists are not priests. Their expertises are not fungible—either one form of technical expertise into another or technical expertise into moral authority. What the modern scientist *may* have left as a basis of authority is a kind of independence and a resulting notion of integrity. Yet the enfolding of science into the institutions of wealth-making and power-projecting makes that independence harder to recognize and acknowledge. And when scientific knowledge becomes patentable property, then the independence of science from civic institutions becomes finally invisible. We've gone some way in these directions—but not yet all the way, so it's not a bad moment to reflect on where we've come from and where we might be going.

I started by recalling how easy it once was to talk about science as an independent cause of modernity, as modernity's characteristic form of culture and as its distinct master authority. It's not so easy now. And one reason it's not so easy is that our ability to recognize relevant experts, and to recognize their independent authority, has become more and more problematic. The success of science has created its successor problem. That problem—the problem of the independent authority of science in our modern world—may be a problem *for* science, but, more importantly, it's a problem *in* our modern order of things. The place of science in the modern world is just the problem of *describing* the way we live now: what to believe, whom to trust, what to do.

Notes

1. White (1876), and then developed as White (1896). White was following in the tradition of John William Draper, whose *History of the Conflict Between Religion and Science* (1874) similarly announced the inevitable triumph of science over religion.

2. See, for example, Hollinger (2000).

3. Available at: http://www.cbsnews.com/stories/2002/04/29/opinion/polls/main507515.shtml.

4. Available at: http://www.pollingreport.com/science.htm *and* http://www.unl.edu/rhames/courses/current/creation/evol-poll.htm.

5. Available at: http://www.religioustolerance.org/ev_publi.htm. (These figures are from a poll conducted in November 1991.)

6. According to Kohut & Stokes (2006: 61), "In 2004, by a 52 percent to 34 percent margin, Americans said it was more important to conduct such research, which might result in new cures for human diseases, than to avoid destroying the potential life of embryos. Two years earlier, only a plurality of Americans supported stem-cell research (43 percent in favor to 38 percent against)."

7. Available at: http://pewglobal.org/commentary/display.php?AnalysisID=66.

8. We can set aside without comment, as an instance of a lawyer's scientific naivete, the fact that much radioactivity is indeed "harmless."

9. See, e.g., Turner (1974).

10. Etzioni and Nunn (1974) argue convincingly that the public mind makes little, if any, distinction between science and technology.

11. The National Opinion Research Center (NORC) has compiled time-series data on public confidence in various institutions. The data show a decline in confidence in science from the early 1960s to the late 1970s, but this follows a drop in confidence for *all* major public institutions, and the decline for science was notably *less* than it was for others (Pion & Lipsey, 1981: 307).

12. Figures quoted in Holt (2005: 25), from an NORC survey conducted between August 2004 and January 2005.

13. See, for example, Shapin (2001).

14. Available at: http://teacher.pas.rochester.edu/phy_labs/AppendixE/AppendixE.html *and* http://physics.ucr.edu/~wudka/Physics7/Notes_www/node5.html.

15. For example, available at: http://www.sciencemag.org/feature/data/scope/keystone1/.

16. "1. Define the question; 2. Gather information and resources; 3. Form hypothesis; 4. Perform experiment and collect data; 5. Analyze data; 6. Interpret data and draw conclusions that serve as a starting point for new hypotheses; 7. Publish results." Available at: http://en.wikipedia.org/wiki/Scientific_method.

17. Available at: http://www2.selu.edu/Academics/Education/EDF600/Mod3/sld001.htm.

18. See, for example, Dear (1995); Westman (1980).

19. The material in this and the next several paragraphs is included in Shapin (forthcoming: chapters 2–3).

20. For a pertinent Hopkins context to Rowland's remarks, see Feldman and Desrochers (2004: 117–18).

21. For Einstein's early operationalism, influenced by Mach, see Holton (1972).

References

Butterfield, Herbert (1949) *The Origins of Modern Science, 1300–1800* (London: G. Bell).

Cartwright, Nancy (1999) *The Dappled World: A Study of the Boundaries of Science* (Cambridge: Cambridge University Press).

Crombie, A. C. ([1952]1959) *Medieval and Early Modern Science*, 2 vols., revised 2nd ed. (Garden City, NY: Doubleday Anchor Books).

Dear, Peter (1995) *Discipline and Experience: The Mathematical Way in the Scientific Revolution* (Chicago: University of Chicago Press).

Draper, John William (1874) *History of the Conflict Between Religion and Science* (New York: D. Appleton).

Dupré, John (1993) *The Disorder of Things: Metaphysical Foundations of the Disunity of Science* (Cambridge, MA: Harvard University Press).

Dworkin, Ronald (2006) "Three Questions for America," *New York Review of Books* 53(14): 24–30.

Einstein, Albert ([1929]1954) "Scientific Truth" in *Ideas and Opinions* (New York: Crown Publishers): 261–62.

Etzioni, Amitai & Clyde Nunn (1974) "The Public Appreciation of Science in Contemporary America," *Daedalus* 103(3): 191–205.

European Commission (2001) "Europeans, Science and Technology" (Eurobarometer 55.2, December). Available at: http://europa.eu.int/comm/public_opinion/archives/eb/ebs_154_en.pdf.

Feldman, Maryann & Pierre Desrochers (2004) "Truth for Its Own Sake: Academic Culture and Technology Transfer at Johns Hopkins University," *Minerva* 42: 105–26.

Galison, Peter & David J. Stump (eds) (1996) *The Disunity of Science: Boundaries, Contexts, and Power* (Stanford, CA: Stanford University Press).

Gillispie, Charles Coulston (1960) *The Edge of Objectivity: An Essay in the History of Scientific Ideas* (Princeton, NJ: Princeton University Press).

Gould, Stephen Jay (1997) "Nonoverlapping Magisteria—Evolution versus Creationism," *Natural History* 106(2): 16–25.

Hall, Henry S. ([1956]1962) "Scientists and Politicians," in Bernard Barber and Walter Hirsch (eds), *The Sociology of Science* (New York: Free Press; orig. publ. in *Bulletin of the Atomic Scientists* [February 1956]: 269–87).

Hollinger, David A. (2000) "Money and Academic Freedom a Half-Century after McCarthyism: Universities and the Force Fields of Capital," in Peggie J. Hollingsworth (ed), *Unfettered Expression: Freedom in American Intellectual Life* (Ann Arbor: University of Michigan Press): 161–84.

Holt, Jim (2005) "Madness About a Method: How Did Science Become So Contentious and Politicized?" *New York Times Magazine* (December 11): 25, 28.

Holton, Gerald (1972) "Mach, Einstein, and the Search for Reality," in G. Holton (ed), *The Twentieth-Century Sciences: Studies in the Biography of Ideas* (New York: Norton): 344–81.

Huxley, Thomas Henry ([1854]1900) "On the Educational Value of the Natural History Sciences," in *Collected Essays, vol. III, Science and Education: Essays* (New York: D. Appleton): 38–65.

Klaw, Spencer (1968) *The New Brahmins: Scientific Life in America* (New York: William Morrow).

Kohut, Andrew & Bruce Stokes (2006) *America Against the World: How We Are Different and Why We Are Disliked* (New York: Times Books/Henry Holt).

Lapp, Ralph E. (1965) *The New Priesthood: The Scientific Elite and the Uses of Power* (New York: Harper & Row).

Larson, Edward J. & Larry Witham (1997) "Scientists Are Still Keeping the Faith," *Nature* 386: 435–36.

Larson, Edward J. & Larry Witham (1998) "Leading Scientists Still Reject God," *Nature* 394: 313.

Leuba, J. H. (1916) *The Belief in God and Immortality: A Psychological, Anthropological and Statistical Survey* (Boston: Sherman, French).

Medawar, Peter B. (1967) *The Art of the Soluble* (London: Methuen).

Miller, Jon D., Eugenie C. Scott, & Shinji Okamoto (2006) "Public Acceptance of Evolution," *Science* 313(5788): 765–66.

National Science Foundation (2001) *Survey of Public Attitudes toward and Understanding of Science and Technology, 2001*, Division of Science Resources Statistics. Available at: http://www.nsf.gov/statistics/seind04/c7/fig07-06.htm.

Oppenheimer, J. Robert (1965) "Communication and Comprehension of Scientific Knowledge," in Melvin Calvin et al. (eds), *The Scientific Endeavor: Centennial Celebration of the National Academy of Sciences* (New York: Rockefeller University Press): 271–79.

Pion, Georgine M. & Mark W. Lipsey (1981) "Public Attitudes toward Science and Technology: What Have the Surveys Told Us?" *Public Opinion Quarterly* 45: 303–16.

Price, Derek J. deSolla ([1963]1968) *Little Science, Big Science* (New York: Columbia University Press, 1968).

Radford, Tim (2003) "'Science Cannot Provide All the Answers': Why Do So Many Scientists Believe in God?" *Guardian* (London, September 4). Available at: http://www.guardian.co.uk/life/feature/story/0,13026,1034872,00.html.

Rosenberg, Alexander (1994) *Instrumental Biology, or the Disunity of Science* (Chicago: University of Chicago Press).

Rowland, Henry A. (1899) "The Highest Aim of the Physicist," Presidential Address Delivered at the Second Meeting of the American Physical Society, October 28, 1899, *Bulletin of the American Physical Society* 1: 4–16; *Science* 10(258): 825–33.

Sarton, George (1936) *The Study of the History of Science* (Cambridge, MA: Harvard University Press).

Sarton, George (1948) "The History of Science," in *The Life of Science: Essays in the History of Civilization* (New York: Henry Schuman): 29–58.

Shapin, Steven (2001) "How to Be Antiscientific," in Jay A. Labinger & Harry Collins (eds), *The One Culture? A Conversation about Science* (Chicago: University of Chicago Press): 99–115.

Shapin, Steven (forthcoming) *Science as a Vocation: Scientific Authority and Personal Virtue in Late Modernity*.

Snow, C. P. (1959) *The Two Cultures and the Scientific Revolution* (Cambridge: Cambridge University Press).

Snow, C. P. (1961) "The Moral Un-neutrality of Science," *Science* 133(3448): 256–59.

Specter, Michael (2006) "Political Science: The Bush Administration's War on the Laboratory," *New Yorker* (13 March): 58–69.

Teller, Edward (1950) "Back to the Laboratories," *Bulletin of the Atomic Scientists* 6(3): 71–72.

Turner, Frank M. (1974) "Rainfall, Plagues, and the Prince of Wales: A Chapter in the Conflict of Science and Religion," *Journal of British Studies* 13: 46–65.

Veblen, Thorstein (1906) "The Place of Science in Modern Civilization," *American Journal of Sociology* 11: 585–609.

Weber, Max ([1919]1991) "Science as a Vocation," in H. H. Gerth & C. Wright Mills (trans. and eds), *From Max Weber: Essays in Sociology* (orig. publ. from a 1918 speech) (London: Routledge): 129–56.

Westfall, Richard S. (1986) "The Scientific Revolution," *History of Science Society Newsletter* 15(3). Available at: http://www.clas.ufl.edu/users/rhatch/pages/03-Sci-Rev/SCI-REV-Home/05-RSW-Sci-Rev.htm.

Westman, Robert S. (1980) "The Astronomer's Role in the Sixteenth Century: A Preliminary Study," *History of Science* 18: 105–47.

White, Andrew Dickson (1876) *The Warfare of Science* (New York: D. Appleton).

White, Andrew Dickson (1896) *A History of the Warfare of Science with Theology in Christendom*, 2 vols (New York: D. Appleton).

Whitehead, A. N. ([1925]1946) *Science and the Modern World: Lowell Lectures* (London: Scientific Book Club).

Withey, Stephen B. (1959) "Public Opinion about Science and Scientists," *Public Opinion Quarterly* 23: 382–88.

19 Science and Public Participation

Massimiano Bucchi and Federico Neresini

Whatever happened to the heroes?[1]

A group of activists protest against GMOs outside a biotechnology research institute. The citizens of a region vote in a referendum on a new waste disposal facility. A patients' association compiles a large database of the symptoms and clinical evolution of a rare genetic disease. A group of citizens is invited to discuss the issue of embryo stem cell research and produce a final document to be submitted to policy makers.

What do these examples have in common? Are they all in their own way expressions of a profound change in the terms and conditions under which scientific knowledge is produced, discussed, and legitimated?

Public participation in science is an emerging phenomenon with uncertain boundaries, and the difficulties of defining it are compounded by the fact that it has simultaneously become a key focus of social mobilization, policy initiatives, and scholarly analysis. Moreover, a plurality of points of view and motives of interest for public participation can be identified within each of these areas.

However, for our purposes here, public participation may be broadly defined as the diversified set of situations and activities, more or less spontaneous, organized and structured, whereby nonexperts become involved, and provide their own input to, agenda setting, decision-making, policy forming, and knowledge production processes regarding science (Callon et al., 2001; Rowe & Frewer, 2005).

This chapter seeks to (1) provide an overview of the emergence of the phenomenon and theme of public participation in science, (2) define a general interpretative framework with which to map its various manifestations, and (3) outline the possible driving forces behind it as well as its potential impact in terms of changes in the production of scientific knowledge. Specific types of public participation are dealt with in the following chapters.

THE "DEFICIT MODEL OF PUBLIC UNDERSTANDING OF SCIENCE" AND ITS DISCONTENTS

Although antecedents can be traced back to long-standing debates on participatory democracy that have touched science and technology issues since the 1970s (see, e.g., Dickson, 1984), the theme of public participation with regard to science has come with new force to attention in conjunction with the crisis of the so-called deficit model of public understanding of science (Wynne, 1991, 1995). This model has emphasized the public's inability to understand and appreciate the achievements of science—owing to prejudicial public hostility as well as to misrepresentation by the mass media—and adopted a linear, pedagogical and paternalistic view of communication to argue that the quantity and quality of the public communication of science should be improved. To recover this deficit, public and private bodies—especially since the mid-1980s—have launched schemes aimed at promoting public interest in and aware-ness of science. These initiatives have ranged from "open days," which have become a routine feature at most laboratories and research institutions, to science festivals or training courses in science journalism.[2]

Despite their variety, these activities, as well as the studies conceived within the framework of the deficit model, share certain assumptions and features, namely,

1. the assumption that *public understanding of science* largely coincides with *scientific literacy*, i.e., with the ability to understand science "correctly" as it is communicated by the experts, which is measured by appropriate questions on scientific methods and contents;

2. the assumption that this understanding, once achieved, guarantees favorable atti-tudes toward science and technological innovation;

3. the tendency to problematize the relationship between science and the public only as regards the latter term of the relationship, i.e., the public.

Especially since the early 1990s, however, these assumptions have been strongly crit-icized on several grounds. For example, it has been pointed out that the equation between public understanding and the ability to answer questions about science has long restricted the discussion to the somewhat tautological observation that members of the public do not reason in the same way as professional scientists. This has prompted the question about whether surveys of scientific literacy are actually mea-suring "the degree of the public's social conformity to a stereotype held by scientists of a 'scientifically literate public'" (Layton et al., 1986: 38).

Also disputed is the linkage among exposure to science in the media, level of knowl-edge, and a favorable attitude toward research and its applications. As regards biotech-nologies, for example, recent research has shown a substantial degree of skepticism and suspicion even among the sections of the population most exposed to scientific communication and best informed about biotechnological topics (Bucchi & Neresini, 2002). In general, therefore, it does not seem that the opposition of certain sectors of the general public to particular technical-scientific innovations is due solely to the presence of an information deficit. Rather, the phenomenon requires more systematic and detailed analysis.

More in general, the disjunction between expert and lay knowledge cannot be reduced to a mere information gap between experts and the general public as envisaged by the deficit model. Lay knowledge is not an impoverished or quantitatively inferior version of expert knowledge; it is qualitatively different. Factual information is only one ingredient of lay knowledge, in which it interweaves with other elements (value judgments, trust in the scientific institutions, the person's perception of his or her ability to put scientific knowledge to practical use) to form a corpus no less sophisticated than specialist expertise.[3]

Critics of the deficit model have also pointed out that these are complex matters difficult to grasp with large-scale surveys. This criticism has prompted the use of ethnographic methods and discourse analysis tools to produce a series of in-depth studies of specific cases of (mis)understanding of scientific questions by nonspecialists. The use of ethnographic rather than quantitative survey methods, a definition of the relationship between science and the public that is not abstract but locally situated, and a conception of both expert and lay knowledge as socially and culturally contingent are some of the key features of the approach known as "critical/interpretative public understanding of science" (Wynne, 1995; Michael, 2002).

One of the studies based on this approach has detected, for instance, a difference between a conception of "science in general"—used by nonspecialists as a distancing device whereby science is defined as "other" from oneself—and a conception of "science in particular" used in practical settings. From this perspective, the notion itself of public "ignorance" is difficult to define. A group of electricians working at the Sellafield nuclear reprocessing plant in the United Kingdom gave the researchers various reasons for their lack of interest—contrary to expected—in acquiring scientific information about the risks of irradiation. First, the electricians believed that interesting themselves in the scientific aspects of irradiation would have caught them in a chain of pointless argument and discussion. Secondly, they feared that being confronted by uncertainties and probabilistic estimates of risks would cause them alarm, or even panic, and would therefore be dangerous. Thirdly, the electricians said that there were other workers at the plant who possessed the information; any active effort on their part to acquire it would have undermined the trust and authority relations established in the workplace (Michael, 1992). In other cases, scientific information may be ignored by the public as irrelevant to their needs, or simply because they distrust the source, believing it to represent interests other than their own. Thus, "technical ignorance becomes a function of social *intelligence,* indeed of an *understanding* of science in the sense of its institutional dimensions" (Wynne, 1995: 380).

A classic example of the gap between expert and lay knowledge is provided by Brian Wynne's study of the "radioactive sheep" crisis that erupted in certain areas of Britain in 1986, following the Chernobyl nuclear plant accident in Russia. British government experts long minimized the risk that sheep flocks in Cumberland had been contaminated by irradiation. However, their assessments proved to be wrong and had to be drastically revised, with the result that the slaughter and sale of sheep was banned in the area for two years. The farmers for their part had been worried from the outset, because they had direct knowledge based on everyday experience (which the scientific experts dispatched to the area by the government obviously did not possess) of the terrain, of water run-off, and of how the ground could have absorbed the radioactivity and transferred it to plant roots. This clash between the abstract and formalized estimates of the experts and the perception of risk by the farmers caused a loss of confidence by the latter in the government experts and their conviction that official assessments were vitiated by the government's desire to "hush up" the affair (Wynne, 1989).

According to some scholars, experts themselves may reinforce the representation of the public as "ignorant." During a study on communication between doctors and patients in a large Canadian hospital, a questionnaire was administered to patients to assess their level of medical knowledge. At the same time, the doctors were asked to estimate the same knowledge for each patient. The three main findings were decidedly surprising. While the patients proved to be reasonably well informed (providing an average of 75.8 percent of correct answers to the questions asked of them), less than half the doctors were able to estimate their patients' knowledge accurately. Thirdly, this estimate was in any case not utilized by the doctors to adjust their communication style to the information level that they attributed to the patients. In other words, the fact that a doctor realized that a patient found it difficult to understand medical questions or terms did not induce him or her to modify his or her explanatory manner to any significant extent. The patients' lack of knowledge—the authors of the study somewhat drastically conclude—appeared in many cases to be a self-fulfilling prophecy, for it was the doctors who, by considering the patients to be ignorant and making no attempt to make themselves understood, rendered them effectively ignorant (Segall & Roberts, 1980).

HYBRID FORUMS AND THE CO-PRODUCTION OF SCIENTIFIC KNOWLEDGE

More recently, various studies have reported the advent of a new form of interaction between nonexperts and scientific knowledge. Levels of communication and social actors external to the research sphere may, in certain circumstances, play a significant role in the definition and accreditation of scientific knowledge (Irwin & Wynne, 1996, Bucchi, 1998). These forms have been interpreted as representing a major change not only with regard to the deficit model but also with regard to its critical or interpretative version. According to Callon, for instance, the critical/interpretative version of public understanding of science shifts the priority from "the education of a scientifically illiterate public" to the need and right of the public to participate in the discussion, on the assumption that "lay people have knowledge and competencies which enhance and complete those of scientists and specialists"; however, both models are seen as sharing "a common obsession: that of demarcation. [The first model], in a forceful way, and [the second model], in a gentler, more pragmatic way, deny lay people any competence for participating in the production of the only knowledge of any value: that which warrants the term 'scientific.'" (Callon, 1999: 89). On this basis, Callon invokes the need for another, more substantial shift to a model of knowledge co-production in which the role of nonexperts and their local knowledge can be conceived as neither an obstacle to be overcome by virtue of appropriate education initiatives (as in the deficit model) nor as an additional element that simply enriches professionals' expertise (as in the critical-interpretative model) but rather as essential for the production of knowledge itself. Expert and lay knowledge are not produced independently in separate contexts to later encounter each other; rather, they result from common processes carried forward in "hybrid forums" in which both specialists and nonspecialists can actively interact (Callon et al., 2001).

Medical Research and the Active Role of Patient Organizations

One area in which this co-production has been particularly visible is the area of medical research, where patient organizations have become increasingly active in shaping the agenda of research in fields of their concern.

Particularly well known and carefully studied is the case of AIDS research, where methods to test the effectiveness of drugs, and the term itself chosen to denote the disease (which was changed from the initial Gay-related immunodeficiency disease [GRID] under pressure from American homosexual associations) were negotiated with activists and patients' associations (Grmek, 1989; Epstein, 1996). In the mid-1980s, AIDS patients participating in clinical trials of the AZT drug (then considered a likely cure for the disease) developed a marked ability to contribute to and influence the experimental procedure—for example, by learning to recognize placebos and refusing to take them—thereby accelerating approval of the drug by the U.S. Food and Drug Administration (FDA). Human trials of another drug, aerosol pentamidine, used to treat an AIDS-related disease, *Pneumocystis carinii* pneumonia, were conducted by groups of activists after scientists had refused to do so. In 1989 the use of aerosol pentamidine was approved by the FDA, which for the first time in its history authorized marketing of a drug solely on the basis of data collected by means of community-based experimentation (Epstein, 1995).

Another configuration of knowledge co-production in the biomedical field is offered by the case of the French Muscular Dystrophy Association (AFM). AFM was founded in 1958, at a time when muscular dystrophies were "orphan diseases," rare genetic pathologies largely neglected by specialists who considered putting effort into their study and care unrewarding and too much at risk of failure. By actively promoting and performing the collection of clinical data, conducting surveys among patients, and establishing a genetic bank, AFM managed to create the body of knowledge lacking on these diseases, establishing muscular dystrophies as a fully legitimate object of scientific inquiry as well as of public concern. AFM advocacy has had profound consequences on research in this field, redefining the professional trajectories of specialists themselves—so that researchers working for AFM are at the same time geneticists and pediatricians and thus combine research with daily therapeutic experience—or launching new research lines or structures, for example, Genethon, which was founded by AFM in 1990 to identify the genes responsible for diseases like MD, an inquiry that neither the public nor the private research sector had been able or willing to pursue (Callon & Rabeharisoa, 1999; see also Bourret, 2005).

Public Mobilization on Technoscience Issues

Public participation is also manifest in the increasingly frequent cases of public mobilization on technoscience issues. Especially since the second half of the 1990s, in more or less organized forms, citizens have demanded closer involvement in decisions concerning the development of research and technical-scientific innovation. Most active in this field have been the "new social movements" and NGOs. These two phenomena are in fact closely interconnected: NGOs have provided crucial organizational support for new social movements and have been their main channel of recruitment (Della Porta et al., 1999; Diani, 1995), while the latter have given the former visibility and enabled them to intervene more effectively in decision-making (Della Porta & Tarrow, 2004). Nor should we overlook the local forms of protest usually centered on protecting health and the environment and opposed to the siting of installations that they deem dangerous in their area (waste disposal facilities, electricity power lines and booster stations, power plants). Although it is not yet certain whether these forms of mobilization pertain to the new social movements, they express a clear public demand for involvement in issues with high technical-scientific content.

Relations between new social movements and science have always been characterized by a marked ambivalence. According to social movement theorists (Touraine, 1978, 1985; Melucci at al., 1989; Melucci, 1996; Castells, 1997), the distinctive features of such movements are the ways in which they construct an individual and collective identity, define the adversary, and structure a vision of the world put forward as an alternative to the dominant one. It is evident that science and technology are bound up with each of these three features.

On the one hand, science and technology are often an integral part of the "enemy" against which the new social movements mobilize. They are viewed as instruments of the dominant power and as responsible for the perverse effects of globalization, especially so now that the connection between scientific research and economic interests is increasingly apparent. This is a connection that STS has repeatedly analyzed, and from various standpoints (Etzkowitz, 1990; Funtowicz & Ravetz, 1993; Ziman, 2000). Once again the case of biotechnologies is paradigmatic: the science sustained by—and therefore subservient to—the multinationals is seen as threatening the future of the environment (destruction of biodiversity), jeopardizing human health (harmful emissions), and increasing the third world's dependence on the industrialized countries (it erodes the social bases of small-scale farming in the developing countries [Shiva, 1993]). For these reasons, it is an enemy to be fought against.

Yet science and technology are also resources for the identity, organization, and action of the new movements themselves. In fact, criticisms of the present model of development and the dominant economic paradigm base themselves on data concerning the depletion of environmental and social resources furnished by scientific analyses and forecasts (Moore, 1995; Yearley 1995: 458, 461). Moreover, not only do the new movements rely heavily on the latest communication technologies (e.g., Internet and mobile phones) to organize their activities (Castells, 1997: 117, 142) but they also exploit more traditional media to gain access to the public arena and to exert political pressure (Castells, 1997: 86–89, 116, 129–30, 141).

This "ambiguous, deep connection with science and technology" (Castells, 1997: 123; Yearley, 1992) enables the new social movements, especially the environmentalist ones, to play a significant part in the production of scientific knowledge itself. As Yearley has pointed out, this participation may take place at various levels. First, a number of NGOs have set up laboratories and research facilities in order to have their own scientists produce independent scientific research (Yearley, 1995: 462). Also to be mentioned are the "science shops" set up by universities or networks of NGOs so that research projects can be commissioned from those universities, or from other research bodies, on the basis of recommendations by civil society. Launched by Dutch universities during the 1970s, the science shop system has been introduced in various other European countries, as well as in Eastern Europe, the United States, Canada, Israel, Malaysia, and South Korea, where the initial intent to "reorient science toward social needs" has been declined in various ways. In the case of the more mature Dutch experience, "science shops started out as a counterculture phenomenon in the 1970s, but by the end of 1980s, most had become regular elements of university organization" (Wachelder 2003: 253–54). In certain respects, science shops can be viewed as a "bottom up" interpretation of what is termed community-based research, especially when they are intended to promote public participation not only in specific research projects but also more generally in research policies (Sclove, 1998). But the new social movements and NGOs participate directly in the production of scientific knowledge also by orienting research work in accordance with their beliefs (regarding experiments on animals, for example). They do so by seeking to disseminate certain theories (certainly the best-known of which is the Gaia theory) and to condition decisions on research policy (Yearley, 1995: 469–77). Finally, the new social movements have on occasion put themselves forward as the champions of a "true science," urging what they regard as a scientific community in league with political and economic power to regain its neutrality and independence—"the science of life versus life under science" (Castells, 1997: 127)—or promoting alternative paradigms, such as those proposed by Capra (1975), Morin (1977), or Prigogine and Stengers (1979).

Making Science in the Court
A third important interaction between expert and lay knowledge lies in the area of law. Over the past decade, a technicist view of the application of scientific knowledge in legal practice has been replaced by one where the law not only utilizes the tools of scientific research but actively participates in it, for example, by defining what can be patented as a scientific discovery, who can be considered a scientific expert, or even what counts as "scientific proof" (Mackenzie, 1993; Jasanoff, 1995). Scholars who have

analyzed this process describe the settings where laws are devised, and especially where they find interpretation by the courts, as ones of co-production between science and law in a context characterized by an "erosion of faith in legal processes and institutions" and by "the public's often expressed distrust of technical experts and their undemocratic authority" (Jasanoff, 1995: 4).

Although science and law have numerous important features in common, they also differ considerably, especially in the methods used to verify facts. However, these differences do not prevent them from collaborating in designing frameworks and definitions that shape society and transform it, especially in relation to scientific and technological developments that impact on well-established social relations. Thus, courtrooms are used for experiments to solve the problem of arriving at socially endorsed decisions without delegating to experts—whether judicial or scientific—the task of unequivocally establishing what is to be taken as fact, on what bases, and by what procedures. For example, the Daubert decision by the U.S. Supreme Court, while acknowledging the importance of peer review by the expert community, reaffirmed the central role of judges in evaluating a certain piece of information as "scientific evidence" in a trial (Solomon & Hackett, 1996). Legislation is currently being enacted that enables citizens to object to the standard neurological criteria used to establish death—cerebral in the New Jersey Health Statute—in case it conflicts with individual religious beliefs, thus making room for a sort of "pluralist" definition of it (New Jersey Statutes, 1991; Tallacchini, 2002). The concept of democracy here acquires a meaning quite different from the traditional one: no longer a majoritarian decision-making process driven by hegemonic expert knowledge but rather the scrutiny of a range of options using a procedure that gives a decision greater transparency without hampering its effectiveness. Hence, the law as practiced in the U.S. courts has contributed significantly to the construction of a civic culture of science, both by revealing how the opinions of experts differ and by evincing "their underlying normative and social commitments in ways that permit intelligent evaluation by lay persons" (Jasanoff, 1995: 215).

Users and the Shaping of Technology

As described in detail in chapter 22 of this Handbook, technology is yet another terrain where the involvement of different actors in knowledge creation is of increasing importance. Numerous studies have shown the various ways in which users are involved not only in the implementation of technologies but also in their design, and in developing the knowledge that makes them possible. In fact, artifacts can be reinterpreted, adapted, and in certain cases actually reinvented by users; their needs and point of views can be incorporated in the design process itself (Pinch & Oudshoorn, 2003; Kent, 2003; Eglash et al., 2004). Especially in cases in which user-innovation communities come into being—for example, the free and open source software movement—participation by users in innovation processes goes well beyond the adaptation of initial projects to the needs of the final users, giving rise to the co-production of new knowledge embodied in technological artifacts (von Hippel, 2005). Also of increasing importance are organizations set up both to protect consumers and to influence innovation processes so that they more closely reflect their needs. Technology deserves attention also insofar as it frequently represents a background for the other forms of interaction described here—think, for instance, to the importance that recent media technologies have acquired in collective mobilization processes, or to the role of the Internet in facilitating the gathering and circulation of information among patients and families affected by rare genetic diseases.

FORMAL INITIATIVES PROMOTING PUBLIC PARTICIPATION IN SCIENCE

Additionally as a result of these and other cases, the attention of scholars and practitioners has concentrated on a variety of schemes to promote public involvement in issues concerning science and technology. Particularly since the mid-1990s, local, national, and international public institutions as well as NGOs in many countries have devoted significant effort to creating opportunities for citizen participation with regard to potentially controversial science and technology issues such as genetically modified (GM) food, genetic testing, transport technology, and ozone depletion. Political institutions have also started to consider "citizen participation" a necessary policy provision in the field of research and innovation, with special regard to highly sensitive fields like biotechnologies, the siting of radioactive waste disposal facilities, or more in general sustainable development.[4] Indeed, we can trace, for instance, the history of the policy framing of such relationships by analyzing the linguistic shifts in documents and funding schemes of national and international institutions: from "public awareness of science" to "citizen involvement," from "communication" to "dialogue," from "science and society" to "science in society."[5] In some countries, Switzerland for example, specific agencies have been established to undertake "participatory technology assessment" of upcoming innovations on behalf of parliaments or governments (Joss & Bellucci, 2002).

The promotion of public participation in the area of science and technology is often justified by the sponsoring institutions in terms of enhanced citizenship and democratic participation. This rationale is sometimes expressed in the more sophisticated argument that advances in research and innovation are challenging the standard forms and procedures of democracy, requiring new forums and opportunities in which complex technoscience issues can be addressed without sacrificing the needs of contemporary democracy. However, not infrequently present is a more or less implicit expectation by those same sponsors that opportunities for participation will prevent heated public controversies on sensitive issues related to science and technology and restore otherwise declining public trust in science. Indeed, a number of initiatives in the area of science and public participation, like the wide-ranging "GM nation" debate conducted in the United Kingdom in 2003, have been launched after significant public mobilization on a particular issue (Jasanoff, 2004a). This expectation—in this or its even more cynical version where participation is seen simply as providing stronger public legitimation for decisions already taken (Callon et al., 2001)—may in some cases be expressed so explicitly that participants and commentators begin to suspect that the institutions promoting public participation initiatives see them, to some extent, as the "prosecution of the deficit model by other means," a more subtle way of disciplining—in a Foucaultian sense—citizenship to make it more suitable to comply with technoscience advancements (Foucault, 1975).[6]

Sponsored initiatives for public participation in science have taken a variety of forms in terms of

1. the nature and number of participants, how they are selected, the time frame, and the geographic scale;
2. the method by which the public input is gathered;
3. the extent to which this input may be binding for policy decisions;
4. the type of issue at stake (Rowe & Frewer, 2000).

Participants may, for instance, be stakeholder representatives, as in the case of the "negotiated rule making" exercise, or ordinary citizens selected—according to certain criteria—to represent the public, as in the "consensus conference" (a model first experimented with in Denmark in the late 1980s) (Joss, 1999; Joss & Bellucci, 2002). The number of participants may be fairly small—as in the case of citizens' juries—or quite large (public opinion surveys). Events may last a few minutes (public opinion surveys) or several months (public hearings, negotiated rule-making); geographical scale may range from the very local to national and (more rarely) transnational. The methods used to obtain public input may be multiple choice questions, moderated or free discussion by participants, the questioning of expert witnesses, or presentations by agency representatives, expert witnesses, or stakeholders. Participant input may be strictly binding (as in the case of referenda) or simply offered to policy makers as additional support for their decisions (consensus conferences, citizens' juries) or even no more than opinions that may well not be included in the final recommendations (public hearings). The themes addressed may be very general topics (in 1996 a consensus conference was organized in Denmark on "The Consumption and Environment of the Future") or single issues (genetic testing, GM food, cloning) (Joss, 1999; Joss & Bellucci, 2002). The questions asked of participants may concern the implementing of specific decisions at the local level (e.g., choosing the most appropriate site for a new waste disposal facility) or the devising of broad, long-term scenarios (e.g., the future of transportation). Table 19.1 sets out some of the most widespread forms of public participation elicited by a sponsor.

Evaluation of these initiatives has not yet provided clear indications, for several reasons. One is the lack of unambiguous definitions of key concepts like "effectiveness," which can be articulated on several dimensions, as well as from the different perspectives of the actors directly involved, of those who are in some way affected, or even of those excluded from the initiative. The criteria employed to assess participatory initiatives have referred both to their public "acceptability" (e.g., representativeness, independence, early involvement of the public, influence on policy,

Table 19.1

Some of the Most Widespread Forms of Public Participation in Science Elicited by a Sponsor

Participation Method	Nature of Participants	Time Scale/Duration	Characteristics/Mechanism
Referenda	A significant proportion of national or local population	Vote cast at single point in time	Vote is usually choice of one of two options. All participants have equal influence. Final outcome is binding.
Public hearings or inquiries	Interested citizens, limited in number	May last many weeks or months or even years	Presentations by agencies regarding plans in open forum. Public may voice opinions but have no direct impact on recommendation.
Public opinion surveys	Large sample of population	Single event, lasting a few minutes	Input gathered through a questionnaire administered face-to-face, by telephone, via post, or e-mail.
Negotiated rule-making	Small number of representatives of stakeholders groups	Uncertain; usually lasting days to months	Working committee of stakeholder representatives (and from sponsor). Consensus required on specific question.
Consensus conference	Generally, 10–16 members of the public, selected as representative	Preparatory demonstrations and lectures to inform panelists about topic, then 3-day conference	Lay panel with independent facilitator questions expert witnesses chosen by stakeholder panel. Meetings open to wider public. Conclusions on key questions made via report or press conference.
Citizen's jury/panel	Generally, 12–20 members of the public selected as representative	Generally involve meetings over a few days	Lay panel with independent facilitator questions expert witnesses chosen by stakeholder panel. Meetings not generally open. Conclusions on key questions made via report or press conference.

Source: Rowe and Frewer 2000, pp. 8–9.

transparency) and to "process" considerations relative to their design and implementation (e.g., the availability to participants of the resources necessary for their task, clear definition of the task, structured decision-making, cost-effectiveness).[7]

SCIENCE AND PUBLIC PARTICIPATION: A PROPOSED INTERPRETATIVE FRAMEWORK

The proliferation and variety of participatory mechanisms and the problem of finding common definitions has been seen as reflecting the *statu nascenti* instability of the field and as being at least partially responsible for the difficulty of deciding "what works best and when," that is, of assessing the effectiveness of each specific technique (Rowe & Frewer, 2004, 2005). Nevertheless, since participatory initiatives first made their appearance, attempts have been made to categorize them on the basis of such dimensions as objectives, type of participants, and the extent to which the procedure is structured. In a recent study, Rowe & Frewer (2005) draw up a typology of participatory mechanisms with a view to evaluating their effectiveness. The authors consider a general aim of "public engagement" to be "maximizing the relevant information flow (knowledge and/or opinions) from the maximum number of relevant sources and transferring this efficiently to the appropriate receivers" (Rowe & Frewer, 2005: 263). Depending on where the emphasis in the process is placed, three broad categories of public engagement can be thus identified:

• *public communication*, "maximizing the relevant information flow from the sponsor . . . to the maximum number of relevant population";
• *public consultation*, "maximizing the relevant information flow from the maximum number of the relevant population and . . . transferring it to the sponsor";
• truly *public participation*, "maximizing the relevant information from the maximum number of all relevant sources and transferring it . . . to the other parties" (Rowe & Frewer, 2005: 254–55).

Differences between specific participatory procedures can then be related to a series of variables associated with the above objectives (e.g., maximization of relevant participants, maximization of relevant information from participants). This typology has several advantages: most notably, it highlights similarities and differences between mechanisms and thereby paves the way for conceptual clarification and thorough impact evaluation. For instance, consensus conferences, citizens' juries, and action planning workshops can be treated as a homogeneous cluster of participatory forms, since they all involve a controlled selection of participants, facilitated elicitation, an open response mode, and unstructured group output (Rowe & Frewer, 2005: 281). Nonetheless, the typology may not fully respond to our purposes here, for a series of reasons.

First, it anchors public engagement to a notion of information flow—described as a rather mechanical process of "transfer"—which seems largely to reprise the limits of the deficit model and traditional communication paradigms, the main difference being that it envisages the possibility of two-way transfer (i.e., not only from the sponsor/experts to the participants but also from participants to the sponsor/experts).[8] However, hybrid forums often involve not only the exchange of information among the actors concerned but also the negotiation and production of new identities (Callon, 1999).

Second, defining relevance as a key concept for the typology is only unproblematic if a specific point of view is adopted. Who defines which information is relevant? Who defines which population is relevant? Is it the sponsor promoting the specific participatory initiative? The potential participants? In the case of muscular dystrophy patient associations, the relevant groups did not exist until thorough interaction between them and the experts became possible; just like the disease, they became visible and relevant only through this interactive process (Callon, 1999).

This brings to the fore a third, and probably more substantial, shortcoming of the typology: the fact that it is limited only to mechanisms actively promoted by a sponsor.

In that we have instead adopted a broader definition of participation, we propose an interpretative framework able to account also for "spontaneous" participatory forms, i.e., those not deliberately elicited by a sponsor in all the varieties that were briefly outlined above: public mobilization and protests, patient associations shaping the research and care agenda, community-based research.

This framework is partly based on the one used by Callon and colleagues (2001) to classify hybrid forums and adopts one of its key dimensions: the *intensity* of cooperation among different actors in knowledge production processes (Callon et al., 2001: 175). While intensity should, of course, be understood as a continuum, some key gradations can be identified, what in Callon et al. correspond to "access points" where nonexperts can intervene. One such point is the moment when laboratory results are "translated" to real-life situations, which is a crucial stage in the stabilization of scientific knowledge (Callon et al., 2001: 89ff.). At that point, contradictions and conflicts may emerge between specialist and lay knowledge, with nonexperts questioning the extent to which laboratory data can be applied to their own specific situation. This was, for example, the case of people living close to the Sellafield nuclear reprocessing site, who used data collected by themselves to contradict the reassuring statistics of experts on the number of leukemia cases in their area and eventually obtained an official enquiry (Wynne, 1996), and the case of the Cumbria sheep farmers whose concrete experience of the peculiarity of Cumbrian soil gainsaid predictions based on expert models that the contamination would soon disappear (Wynne, 1989).

A second and more substantial degree of participation corresponds to the access point offered by what Callon and colleagues (2001) call "the definition of the research collective," for instance, when members of AIDS patient associations managed to gain involvement in the design of experiments and drug trial tests, thereby broadening the research collective to include nonresearchers.

The public may even participate in the initial recognition of research problems, for example, by making a particular event or series of events leave the limbo of happenstance and enter the realm of problems warranting expert interest and attention. The public may also accumulate the initial stock of knowledge required to make professional research possible and worthwhile. For instance, in the 1980s it was the action by Woburn, Massachusetts, residents in gathering by themselves epidemiological data and information on a suspiciously high number of childhood leukemia cases in their area that eventually persuaded MIT to initiate a research program that uncovered genetic mutations caused by trichloroethylene (Brown & Mikkelsen, 1990). Similarly, the mobilization of patient associations like the French AFM has been crucial in prompting fruitful research on genetic diseases.

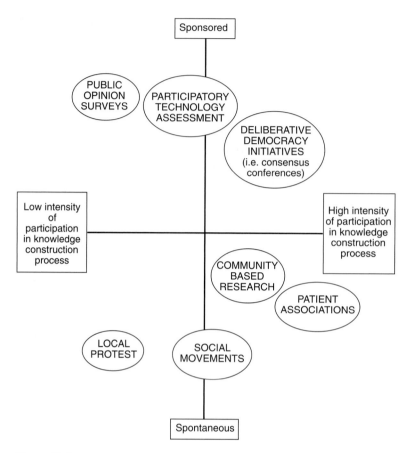

Figure 19.1
A map of public participation in science and technology.

The second axis of our diagram plots the extent to which public participation is elicited by a sponsor: what could be defined, with a certain amount of simplification, as the degree of *spontaneity* of public participation. Here again, the variable should be viewed as a continuum, with the participatory initiatives described by Rowe and Frewer at the upper end of the axis and protest movements and research activities of patient and resident organizations at the lower end. Figure 19.1 gives a graphical representation of the space defined by these two dimensions, together with some illustrative examples.

A wide variety of forms and cases of public participation can be mapped in this space. The upper left quadrant comprises forms typically elicited by a sponsor and characterized by low-intensity participation by nonexperts in knowledge production, e.g., a public opinion survey. The lower left quadrant contains spontaneous mobilizations that do not significantly impact the dynamics of research, e.g., residents' protests against the decision to locate a radioactive waste site in their area. The lower right quadrant includes "spontaneous" forms of knowledge co-production, such as those exemplified by the Woburn residents or by the AFM. Finally, a participatory initiative like a consensus conference on a science issue organized by a sponsoring institution can be placed in the upper right quadrant (high degree of elicitation, high degree of intensity).

Over time, public participation with regard to a certain issue may move along one or both dimensions: for instance, when a public protest induces an institutional sponsor to organize a consensus conference or a citizen panel, or when patient families initially get together to lobby research institutions or drug companies and in the long run decide to establish their own research facilities.

The "open-endedness" of public participation is also emphasized in this interpretative framework. By open-endedness is meant that the output of public participation is rarely entirely predictable on the basis of its structural features or on the basis of the sponsor's objectives; a public protest, for instance, may lead to renegotiation of a consensual decision, just as a participatory initiative originally designed to produce a consensus document may bring to light and radicalize conflicting positions, both among actual participants and—especially when the conflicts are reported by the media—in the broader public arena. Some degree of apprehension for this open-endedness may be regarded as a key factor accounting for the sometimes resurgent temptation, on the part of research bodies and other institutions, to tame unruly public participation through formal initiatives.

The interpretative framework outlined above could be integrated by at least two sets of considerations.

First, the use of inevitably broad labels such as "nonexperts" or "lay public" should not lead us to flatten the intrinsic variety of citizens' involvement and their significantly differentiated capability and interest to shape knowledge production processes. Indeed, some of the most intense examples of participation actually involve highly motivated, highly informed groups—"quasi-experts" among nonexperts, so to speak—that leave large parts of the public potentially disenfranchised. Sponsored and institutionalized forms of participation are by definition selective, and even those aimed at the widest possible involvement—such as the voting referendum—entail a substantial degree of self-selection. In other words, the question of "who participates" remains open for future research not less than the question of "which forms of participation."

It might also be tempting to overemphasize the most intense forms of participation as well as to interpret the different analytical models of interactions among experts and the public as a chronological sequence of stages in which the emerging participatory form obscures the previous ones—e.g., with the critical version plainly obliterating the deficit one or the co-production version plainly substituting for the critical one. Such tendency might be related, among other factors, to how the theme of public participation has been framed at the same time as a policy issue and as a scholarly issue—something to encourage as well as something to study—in a way that may well go back to the long-standing debate on commitment versus neutrality within STS (Cozzens & Woodhouse, 1995; Ashmore & Richards, 1996; Woodhouse et al., 2002; Lengwiler, 2004).

However, it is clear that not only may different expert-public interaction patterns gradually shift one into another but that even at a given time, a certain technoscientific theme may witness different configurations of such interaction, involving different groups of experts and nonexperts. Accordingly, our proposed interpretative framework seeks to account for the simultaneous coexistence of different patterns of participation that may coalesce depending on specific conditions and on the issues at stake—from the "zero degree" of participation entailed by the deficit model to the most substantial forms of cooperation. In this light, rather than "which model of participation accounts best" for expert-public interactions, one of the key questions becomes "under what conditions do different forms of public participation emerge?"

CONCLUDING REMARKS: THE END OF EXPERTS?

Whence derives public participation in science, and where is it heading? In other words, is it possible to identify the factors responsible for its increasing salience—either in its more spontaneous, grass-roots version or as institutionally driven—and is it possible to determine its impact on the dynamics of knowledge production?

An argument frequently adduced to justify the relevance and promotion of forms of public participation in science is the growing frequency with which contemporary democracies must take decisions on issues imbued with science and technology elements: BSE ("mad cow disease"), radioactive waste, GM food, stem cells, nanotechnology. Contemporary democracies are considered ill equipped to deal with such issues, and their current institutional arrangements are seen as unable to withstand a powerful injection of technoscience. "Innovation in natural knowledge and in its technological applications demands a corresponding capacity for social innovation" (Jasanoff, 2004a: 91). However, complex science-related decisions were not uncommon in the past: take, for instance, the decision by the U.S. government to build the first nuclear weapon during World War II or the decision to allow the introduction of pesticides in agriculture—decisions potentially not less controversial than that of introducing GM organisms. Why was there no call for more public participation on those occasions? The answer may be that a number of conditions have changed in the meantime, so that by now largely implausible is the situation that Snow termed "closed politics" (Snow, 1961: 56), where policy makers consult their own trusted scientific experts away from public scrutiny.

One set of conditions may be the increasingly pervasive role of the mass media in questioning not only policy decisions in this area but more specifically the connection between expertise and policy making. They thus substantially shape—in accordance with their own rationales and production routines rather than with those of the scientific community—the selection and legitimation of scientific experts in the public arena (Peters, 2002). Seen in this light, the role of the media is part of the general process eroding the perceived neutrality and *super partes* status of scientific expertise also visible in the action of environmental movements and judicial forums over the past few decades (Jasanoff, 1995; Yearley, 1995; Lynch & Jasanoff, 1998; Jasanoff, 2003). The growing public perception of scientific expertise as interest-laden and unable to provide the scientific community with a consensual voice is damaging the credibility of traditional decision-making arrangements that involve only experts and policy makers (Bucchi & Neresini, 2004b).[9] This perception is accentuated by the mobilization of researchers in the public arena—when they protest against budget cuts or against state regulation of certain research fields, or simply advocate greater public concern with science—increasingly paralleled by the presence of citizens within research laboratories.[10] This suggests an ironical and somewhat paradoxical generalization of the above-mentioned open-endedness—the citizen pressures for more participation that have contributed to undermining the deficit approach may have been stimulated, among other things, by scientists' advocacy of that selfsame approach. Additionally, over the past decades, a series of issues and events—from nuclear accidents to BSE—have contributed to make not just public opinion but also sectors of the expert community particularly sensitive to the implications of technoscience.[11] Pressures for public participation in science can also be viewed as part of more general criticism of the capacity of traditional democracies to represent and include citizens' points of view when addressing global challenges, with crucial decisions being more and more taken at levels not directly subject to citizens' influence. This is the "democratic deficit" that is frequently a matter of concern with regard to, for instance, European or international institutions.[12] Science is obviously not extraneous to such challenges, for it highlights and fosters new processes of inclusion and exclusion that shape the meaning itself of citizenship (Jasanoff, 2004a, 2005).

Callon et al. (2001) account for both specific and general processes by interpreting the proliferation of "hybrid forums" as questioning the double delegation on which democracy used to be founded: to wit, the delegation of knowledge about the world to professional scientists and the delegation of knowledge about the sociopolitical collective to professional politicians. The fragility of this double delegation, originally intended to confine uncertainty within specific institutions where it could be appropriately dealt with (laboratories on the one hand, parliaments on the other), is revealed by the endemic uncertainty and innumerable controversies that arise in connection with science and technology. Whence derives the need to move from a democracy of delegation to a "technical democracy, or more exactly to make our democracies capable of absorbing the debates and controversies provoked by the rapid advances of science and technology" (Callon et al., 2001: 23–24).

In similar vein, Latour has called for the institution of a "Parliament of things"— where both nature and social collectives can be thoroughly explored—and for the design of new "rules of the method" to guide the sociotechnical experiments—like global warming, BSE, GM food—now blurring the boundaries between experts and nonexperts as well as the boundaries of laboratories, which expand to encompass houses, farms, and hospitals (Latour, 2004).

It seems difficult to predict that the development of public participation in science will result in the outright disappearance of professional experts and their replacement by widespread socially diluted knowledge. One reason for this, besides those already mentioned, is that the model of knowledge co-production, undoubtedly common-place in certain areas of biomedical and environmental research, does not seem equally applicable in other fields of scientific enquiry such as theoretical physics (Callon, 1999). Moreover, a significant number of the initiatives envisaged and the positions taken up by policy makers and representatives of the scientific community with regard to public participation reproduce an idea of a public that must be suitably "involved" to forestall uncontrolled mobilization.[13]

Public participation warrants attention not only because it may be a solution to a decisional impasse on technoscience issues or to a crisis of representativeness, but also because it exposes the inevitably political nature of current dilemmas. Both the technocratic option ("leave it to the experts") and the ethical option ("leave it to the conscience of individual users or producers") have for long been used to confine—respectively, in the domain of expertise and ethics—tensions that arise in areas typically pertaining to politics. Participatory experiences highlight, among other things, a growing endeavor to bring back into mainstream democratic politics those transformations driven by science and the economy that modernity sought to exclude from it (Beck, 1992). Defining which politics and which democracy, however, is far from straightforward.

On this view, public participation, with its variety of expressions, is not merely a *response*—albeit undoubtedly more sophisticated than that set forth by the deficit model—either to the need for democracy to keep pace with the headlong advance of science or to the need for science to adjust to public pressures and demands.[14] If the "anaesthetization" of politics by the massive injection of technoscientific expertise has been not sufficient to deal with crucial dilemmas, this is not a reason to expect that those same dilemmas will be solved simply by injecting democratic arrangements into science, especially if democracy is defined with its most simplistic meaning of "major-ity voting." Democracy, like science, cannot be taken as given; just as the latter is transformed by the entry of citizens into research laboratories, so the former is trans-formed when, for instance, scientists protest in public or propose a "compromise" on the public funding of stem cell research (Shaywitz & Mellon, 2004).). From a broader historical perspective, this role of science in the development of democracy has been significantly documented as regards not only the mere supplying of technical com-petences for policy deliberation but also the shaping of democracy's argumentative styles.[15]

Moreover, emerging forms of knowledge co-production challenge the conceptual-ization of expert and lay knowledge as discrete sites (in a unilinear or dialogic communicative relationship, respectively, according to the deficit model and the critical approach) as well as a model of interaction centered on the individual (as a cognizing individual in the deficit model, and as a socialized individual in the critical approach). On the one hand, those conceptualizations impede understanding of the variety of lay knowledge and the conflicts internal to it; on the other, they make such knowledge impervious to general cultural dynamics. As a matter of fact, the citizen's identity itself increasingly includes substantial elements coming from science, mediated, for example, by practices like consumption (Michael, 1998, 2002).

Straightforward confrontations between experts and the public are being replaced by unstable, hetereogeneous "ethno-epistemic assemblages" of experts, citizens, patients, stakeholders, and human and nonnonhuman actors (Irwin & Michael, 2003). Yet "participation" should not be reified as a circumscribed, static event—nor, in the perspective of certain institutions sponsoring participatory activities, as a prerogative that can be switched on and off at will. Rather, it should be viewed as a process that fluidly assumes different contingent configurations. A certain notion of the relationship between professional experts and the public—for example, as segregated categories in the deficit model or as inextricably intertwined as in the co-production model—is in itself a result of, and not a precondition for, the struggles, negotiations, and alliances taking place in those configurations. Nonexperts' interactions with technoscience are embedded in a wider "civic epistemology," which in turn includes participatory styles as one of its central elements (Jasanoff, 2005).

In broader terms, public participation is today one of the key dynamics at the core of the co-evolutionary, co-production processes (Nowotny et al., 2001; Jasanoff, 2004b, 2005) redefining the meanings of science and the public, knowledge and citizenship, expertise and democracy.

Notes

1. The Stranglers, "No More Heroes," UA records, 1977.

2. For an overview, see, for instance, OECD (1997), European Commission (2002).

3. See also Cooter and Pumfrey (1994).

4. See, for instance, the European Directive 2001/18/EC on the release of GMOs into the environment (European Commission, 2001b) or the UN document "Agenda 21" (available at: http://www.un.org/esa/sustdev/documents/agenda21/index.htm).

5. Possible examples include the European "Science and Society" action plan (2001a) and the U.K. House of Lords Science and Technology Committee *Science and Society Report* (2000).

6. See also Levidow and Marris (2001), Irwin (2004).

7. For an overview, see Rowe and Frewer (2000, 2004); for analyses of specific cases see also Guston (1999), Mayer and Guerts (1998), Marris and Joly (1999), Irwin (2001).

8. For a critique of the "transfer" communication paradigm with regard to science communication, see, for instance, Bucchi (2004).

9. Conflicts of interests and their public exposure have become a growing concern for the scientific communities. See, for instance, van Kolfschooten (2002).

10. See for example the case study of recent researchers' mobilization in Bucchi & Neresini (2004a).

11. With regard to the expert community, recall, for instance, the forming of groups like the Union of Concerned Scientists or the debate that in 1974 led molecular biologists to propose a moratorium on recombinant DNA experiments (cf. Berg et al., 1974).

12. For an overview of this theme with particular regard to EU institutions, see, for example, Burns and Andersen (1996). A more specific discussion concerning science and technology governance is in Levidow and Marris (2001).

13. A striking recent example is offered by the concluding report of the European Group of Life Sciences (2004): "One lesson to emerge after a decade of controversies (GM food, stem cells, reproductive technologies . . .) is that research, development and innovation can hardly prosper in the face of social opposition to science."

14. It has been argued, in fact, that enlarging the number of actors involved in knowledge production and negotiation reduces the efficiency of decisional processes, making them slower, more muddled, and ultimately fragile (Stehr, 2001).

15. This is, for example, the thesis of Ezrahi (1990), who considers the emergence itself of democracy as indebted to the "attestive" style of the experimental method, as opposed to the "celebratory" style of absolutist regimes.

References

Andersen, Svein S. & Tom Burns (1996) "The European Union and the Erosion of Parliamentary Democracy: A Study of Post-Parliamentary Governance" in S. Andersen & K. A. Eliassen (eds), *European Union— How Democratic Is It?* (London: Sage): 227–51.

Ashmore, Malcolm & Evelleen Richards (1996) (eds) "The Politics of SSK: Neutrality, Commitment and Beyond," special issue of *Social Studies of Science* 26(2).

Beck, Ulrich (1992) *Risk Society: Towards a New Modernity* (London: Sage).

Berg, Paul, D. Baltimore, H. W. Boyer, S. N. Cohen, R. W. Davis, D. S. Hogness, D. Nathans, R. Roblin, J. D. Watson, S. Weissman, & N. D. Zinder (1974) "Potential Biohazards of Recombinant DNA Molecules" (letter), *Science* 185(148): 303.

Bourret, Pascale (2005) "BRCA Patients and Clinical Collectives: New Configurations of Action in Cancer Genetics Practices," *Social Studies of Science* 35(1): 41–68.

Brown, Phil & Edwin Mikkelsen (1990) *No Safe Place: Toxic Waste, Leukemia, and Community Action* (Berkeley: University of California Press).

Bucchi, Massimiano (1998) *Science and the Media: Alternative Routes in Scientific Communication* (London: Routledge).

Bucchi, Massimiano (2004) "Can Genetics Help Us Rethink Communication? Public Communication of Science as a 'Double Helix,'" *New Genetics and Society* 23(3): 269–83.

Bucchi, Massimiano & Federico Neresini (2002) "Biotech Remains Unloved by the More Informed," *Nature* 416: 261.

Bucchi, Massimiano & Federico Neresini (2004a) "Science Against Politics or the Politicization of Science? Research Agencies and the Debate over Research," in S. Fabbrini & V. Della Sala (eds) *Italian Politics: Italy between Europeanization and Domestic Politics* (Oxford: Berghan): 133–49.

Bucchi, Massimiano & Federico Neresini (2004b) "Why Are People Hostile to Biotechnologies?" *Science* 304: 1749.

Callon, Michel (1999) "The Role of Lay People in the Production and Dissemination of Scientific Knowledge," *Science, Technology & Society* 4(1): 81–94.

Callon, Michel & Vololona Rabeharisoa (1999) *Le Pouvoir des Malades, l'AFM et la Recherche* (Paris: P.E.M.).

Callon, Michel, Pierre Lascoumes, & Yannick Barthe (2001) *Agir dans un Monde incertain: Essai sur la démocratie Technique* (Paris: Editions de Seuil).

Capra, Fritjiof (1975) *The Tao of Physics* (Boulder: Shambhala Publications).

Castells, Manuel (1997) *The Power of Identity* (Malden, MA: Blackwell).

Cooter, Roger & Stephen Pumfrey (1994) "Separate Spheres and Public Places: Reflections on the History of Science Popularization and Science in Popular Culture," *History of Science* 32(3): 237–67.

Cozzens, Susan E. & Edward J. Woodhouse (1995) "Science, Government, and the Politics of Knowledge," in S. Jasanoff, G. E. Markle, J. C. Petersen, & T. Pinch (eds), *Science Technology and Society Handbook* (Thousand Oaks, CA: Sage): 533–53.

Della Porta, Donatella & Sidney Tarrow (eds) (2004) *Transnational Protest and Global Activism* (New York: Roman & Littlefield).

Della Porta, Donatella, Hanspeter Kriesi, & Dieter Rucht (eds) (1999) *Social Movements in a Globalizing World* (New York: St. Martin's Press).

Diani, Mario (1995) *Green Networks: A Structural Analysis of the Italian Environmental Movement* (Edinburgh: Edinburgh University Press).

Dickson, David (1984) *The New Politics of Science* (New York: Pantheon Books).

Eglash, Ron, Jennifer Croissant, Giovanna Di Chiro, & Rayvon Fouché (eds) (2004) *Appropriating Technology: Vernacular Science and Social Power* (Minneapolis: University of Minnesota Press).

Epstein, Steven (1995) "The Construction of Lay Expertise: AIDS Activism and the Forging of Credibility in the Reform of Clinical Trials," *Science, Technology & Human Values* 20(4): 408–37.

Epstein, Steven (1996) *Impure Science: AIDS, Activism and the Politics of Knowledge* (Berkeley: University of California Press).

Etzkowitz, Henry (1990) "The Second Academic Revolution: The Role of the Research University in Economic Development," in S. E. Cozzens, P. Healey, A. Rip, & J. Ziman (eds), *The Research System in Transition* (Boston: Kluwer Academic): 109–24.

European Commission (2001a) European "Science and Society" Action Plan. Available at: http://europa.eu.int/comm/research/science-society/pdf/ss_ap_en.pdf.

European Commission (2001b), "European Directive 2001/18/EC" on the release of GMOs into the environment. Available at: http://ec.europa.eu/environment/biotechnology/pdf/dir2001_18.pdf.

European Commission (2002) *Benchmarking the Promotion of RTD Culture and Public Understanding of Science* (Luxembourg: Office for Official Publications of the European Communities).

European Group of Life Sciences (2004)) *Conclusions of the European Group of Life Sciences (EGLS) at the Termination of Its Mandate* (2000–2004). Available at: http://europa.eu.int/comm/research/life-sciences/egls/pdf/conclusions_egls.pdf.

Ezrahi, Yaron (1990) *The Descent of Icarus: Science and the Transformation of Contemporary Democracy* (Cambridge, MA: Harvard University Press).

Foucault, Michel (1975) *Surveiller et Punir: Naissance de la Prison* (Paris: Gallimard).

Funtowicz, Silvio & Jerome Ravetz (1993) "Science for the Post-normal Age," *Futures* 25(7): 739–55.

Grmek, Mirko D. (1989) *Histoire du SIDA* (Paris: Payot).

Guston, David H. (1999) "Evaluating the First U.S. Consensus Conference: The Impact of Citizens' Panel on Tele-communications and the Future of Democracy," *Science, Technology & Human Values* 24(4): 451–82.

Irwin, Alan (2001) "Constructing the Scientific Citizen: Science and Democracy in the Biosciences," *Public Understanding of Science* 10(1): 1–18.

Irwin, Alan (2004) "Expertise and Experience in the Governance of Science: What Is Participation For?" in G. Edmond (ed), *Expertise in Law and Regulation* (Aldershot, Hampshire, U.K.: Ashgate): 32–50.

Irwin, Alan & Mike Michael (2003) *Science, Social Theory and Public Knowledge* (Maidenhead, U.K.: Open University Press/McGraw-Hill).

Irwin, Alan & Brian Wynne (eds) (1996) *Misunderstanding Science? The Public Reconstruction of Science and Technology* (Cambridge: Cambridge University Press).

Jasanoff, Sheila (1995) *Science at the Bar: Law, Science, and Technology in America* (Cambridge, MA: Harvard University Press).

Jasanoff, Sheila (2003) "(No) Accounting for Expertise," *Science and Public Policy* 30(3): 157–62.

Jasanoff, Sheila (2004a) "Science and Citizenship: A New Synergy," *Science and Public Policy* 31(2): 90–94.

Jasanoff, Sheila (2004b) *States of Knowledge: The Knowledge: The Co-production of Science and Social Order* (London: Routledge).

Jasanoff, Sheila (2005) *Designs on Nature. Science and Democracy in Europe and the United States* (Princeton, NJ: Princeton University Press).

Joss, Simon (ed) (1999) *Public Participation in Science and Technology*, special issue of *Science and Public Policy* XXVI: 290–374.

Joss, Simon & Sergio Bellucci (eds) (2002) *Participatory Technology Assessment: European Perspectives* (London: Centre for the Study of Democracy).

Kent, Julie (2003) "Lay Experts and the Politics of Breast Implants," *Public Understanding of Science* 12(4): 403–22.

Latour, Bruno (2004) "Von 'Tatsachen' zu 'Sachverhalten': Wie sollen die neuen kollektiven Experimente protokollier werden?" in Schmidgen Henning, Geiger Peter, & Dierig Sven (eds) *Kultur im Experiment* (Berlin: Kultuverlag Kadmos): 17–36.

Layton, D., A. Davey, & E. Jenkins (1986) "Science for Specific Purposes," *Studies in Science Education* 13: 17–40.

Lengwiler, Martin (2004) "Shifting Boundaries between Science and Politics: New Insights into the Participatory Question in Science Studies," *Technoscience* 20(3): 2–5.

Levidow, Les & Claire Marris (2001) "Science and Governance in Europe: Lessons from the Case of Agricultural Biotechnology," *Science and Public Policy* 28(5): 345–60.

Lynch, Michael & Sheila Jasanoff (1998) (eds) "Contested Identities: Science, Law and Forensic Practice," special issue of *Social Studies of Science* 28(5–6).

MacKenzie, Donald (1993) "Negotiating Arithmetic, Constructing Proof: The Sociology of Mathematics and Information Technology," *Social Studies of Science* 23(1): 37–65.

Marris, Claire & Joly Pierre-Benoit (1999) "Between Consensus and Citizens: Public Participation in Technology Assessment in France," *Science Studies* 12(2): 3–32.

Mayer, Igor & Jac Guerts (1998) "Consensus Conference as Participatory Policy Analysis: A Methodological Contribution to the Social Management of Technology," in P. Wheale, R. Von Schomberg, & P. Glasner (eds), *The Social Management of Genetic Engineering* (Aldershot, Hampshire, U.K.: Ashgate): 279–302.

Melucci, Alberto (1996) *Challenging Codes* (Cambridge and New York: Cambridge University Press).

Melucci, Alberto, John Keane, & Paul Mier (eds) (1989) *Nomads of the Present: Social Movements and Individual Needs in Contemporary Society* (London: Century Hutchinson).

Michael, Mike (1992) "Lay Discourses of Science: Science-in-General, Science-in-Particular, and Self," *Science, Technology & Human Values* 17(3): 313–33.

Michael, Mike (1998) "Between Citizen and Consumer: Multiplying the Meanings of Public Understanding of Science," *Public Understanding of Science* 7(4): 313–28.

Michael, Mike (2002) "Comprehension, Apprehension, Prehension: Heterogeneity and the Public Understanding of Science," *Science, Technology & Human Values* 27(3): 357–78.

Moore, Kelly (1995) "Organizing Integrity: American Science and the Creation of Public Interest Organizations, 1955–1975," *American Journal of Sociology* 101: 1592–627.

Morin, Edgar (1977) *La Méthode: La Nature de la Nature* (Paris: Editions de Seuil).

New Jersey Statutes Annotated (1991), Declaration of Death, 26:6A-5.

Nowotny, Helga, Scott Peter & Michael Gibbons (2001) *Re-Thinking Science: Knowledge and the Public in an Age of Uncertainty* (Cambridge: Polity Press).

OECD (1997) *Promoting Public Understanding of Science and Technology* (Paris: OECD). Available at: http://www.oecd.org/dataoecd/9/28/2754562.pdf.

Peters, Hans Peter (2002) "Scientists as 'Public Experts,'" *Tijdschrift voor Wetenschap, Technologie en Samenleving* 10(2): 39–42.

Pinch, Trevor & Nelly Oudshoorn (2003) (eds) *When Users Matter: The Co-Construction of Users and Technologies* (Cambridge: MIT Press).

Prigogine, Ilya & Isabelle Stengers (1979) *La nouvelle Alliance: Métamorphose de la Science* (Paris: Gallimard).

Rowe, Gene & Lynn J. Frewer (2000) "Public Participation Methods: A Framework for Evaluation," *Science, Technology & Human Values* 25(1): 3–29.

Rowe, Gene & Lynn J. Frewer (2004) "Evaluating Public Participation Exercises: A Research Agenda," *Science, Technology & Human Values* 29(4): 512–56.

Rowe, Gene & Lynn J. Frewer (2005) "A Typology of Public Mechanisms," *Science, Technology & Human Values* 30(2): 251–90.

Sclove, Richard E. (1998) "Better Approaches to Science Policy," *Science* 279(5355): 1283.

Segall, A. & L. W. Roberts (1980) "A Comparative Analysis of Physician Estimates and Levels of Medical Knowledge Among Patients," *Sociology of Health and Illness* 2(3): 317–34.

Shaywitz, D. & D. Mellon (2004) "How to Resolve America's Stem Cell Dilemma," *The Financial Times* (London) October 22.

Shiva, Vandan A (1993) Mono Cultures of the Mind. *Perspectives on Biodiversity and Biotechnology* (London: Zed Books).

Snow, Charles P. (1961) *Science and Government* (Cambridge, MA: Harvard University Press).

Solomon, Shana M. & Edward J. Hackett (1996) "Setting Boundaries between Science and Law: Lessons from Daubert v. Merrel Dow Pharmaceuticals, Inc." *Science, Technology & Human Values* 21(2): 131–56.

Stehr, Nico (2001) *The Fragility of Modern Societies: Knowledge and Risk in the Information Age* (London: Sage).

Tallacchini, Mariachiara (2002) "The Epistemic State: The Legal Regulation of Science," in C. M. Mazzoni, (ed), *Ethics and Law in Biological Research* (Dordrecht, Netherlands: Kluwer): 79–96.

Touraine, Alain (1978) *La Voix et le Regard* (Paris: Editions de Seuil).

Touraine, Alain (1985) "An Introduction to the Study of Social Movements," *Social Research* 52: 749–88.

U.K. House of Lords (2000), *Science and Society: Report of the Select Committee on Science and Technology*. Available at: http://www.parliament.the-stationery-office.co.uk/pa/ld199900/ldselect/ldsctech/38/3801 .htm.

van Kolfschooten, F. (2002) "Can You Believe What You Read?" *Nature* 416: 360–63.

Von Hippel, Eric (2005) *Democratizing Innovation* (Cambridge, MA: MIT Press).

Wachelder, Joseph (2003) "Democratizing Science: Various Routes and Visions of Dutch Science Shops," *Science, Technology & Human Values* 28(2): 244–73.

Wynne, Brian (1989) "Sheep Farming after Chernobyl: A Case Study in Communicating Scientific Information," *Environment Magazine* 31(2): 10–15, 33–40.

Wynne, Brian (1991) "Knowledges in Context," *Science, Technology & Human Values* 16(1): 111–21.

Wynne, Brian (1995) "Public Understanding of Science," in S. Jasanoff, G. E. Markle, J. C. Petersen, & T. Pinch (eds), *Handbook of Science Technology Studies* (London: Sage): 361–89.

Wynne, Brian (1996) "May the Sheep Safely Graze? A Reflexive View of the Expert-Lay Knowledge Divide," in S. Lash, B. Szerzynski, & B. Wynne (eds), *Risk, Environment and Modernity* (London: Sage): 44–83.

Woodhouse E., D. Hess, S. Breyman, & B. Martin (2002) "Science Studies and Activism: Possibilities and Problems for Reconstructivist Agendas," *Social Studies of Science* 32(2): 297–320.

Yearley, Steven (1992) "Green Ambivalence about Science: Legal-rational Authority and the Scientific Legitimation of a Social Movement," *British Journal of Sociology* 43: 511–32.

Yearley, Steven (1995) "The Environmental Challenge to Science Studies," in S. Jasanoff, G. E. Markle, J. C. Petersen, & T. Pinch (eds), *Handbook of Science Technology Studies* (London: Sage): 357–79.

Ziman, John M. (2000) *Real Science* (Cambridge: Cambridge University Press).

20 Science, Technology, and Social Movements

David Hess, Steve Breyman, Nancy Campbell, and Brian Martin

As the STS field has paid increasing attention to the problem of how to make our research relevant to the pressing ethical and policy issues of the day, researchers have examined how democratic participation in science and technology can be enhanced (e.g., Fischer, 2000; Sclove, 1995; Wynne, 1996). Social movements are one of the main pathways toward increased democratic participation, and consequently their study has come to occupy increasing attention among STS researchers. Social movements enhance public participation in scientific and technical decision-making, encourage inclusion of popular perspectives even in specialized fields, and contribute to changes in the policy-making process that favor greater participation from nongovernmental organizations and citizens generally.

As researchers informed by STS embark on studies of social movements, they draw on a well-developed body of empirical studies and theory on social movements. Although some currents of general social movement studies, in particular feminist research, exhibit a sophisticated understanding of the social shaping/social construction hypothesis that is continuous with the STS field, in general the central focus of the existing literature on social movements has not been issues of expertise, knowledge, and technology design. As a result, STS perspectives extend the social movements literature by bringing a sophisticated understanding of how the knowledge-making process works in science and how the politics of expertise and technology design play out in various political arenas.

An additional contribution that STS can make to social movement studies, and vice-versa, returns to the history of a current in the STS field that developed out of reform movements within science that sought to link scholarship to partisan and activist goals (Martin, 1993; Woodhouse et al., 2002). The "reconstructivist" current can provide a helpful corrective to both the social movement and STS literatures, which activists tend not to read or use, by posing the question of how research that follows a social justice–oriented agenda is different from research based on a scholar-directed agenda. Just as social movements shape and are shaped by their environment, so social movement researchers shape and are shaped by theirs. The key question in movements for social justice is who does the shaping?

BACKGROUND ON SOCIAL MOVEMENT THEORY

Social movements can be distinguished from several other types of collective, intentional efforts to promote or resist social change. Although all definitions need to acknowledge the fuzziness of categories, the key features are broad scope (unlike networks of activists or single campaigns), extra-institutional strategies such as protest (unlike advocacy groups), a goal of fundamental social change (unlike interest groups), and a challenge to elites or established organizations (unlike elite-based reforms and campaigns). Some social movements embody a challenge from socially or economically disenfranchised groups, but other social movements include diverse coalitions of people who share specific causes (such as breast cancer patients or open-source programmers). Likewise, social movements may seek benefits beyond the immediate interests of their membership; examples include peace movements, human rights movements, and in many cases environmental movements.

Contemporary social movement theory departs from one of three major traditions—resource mobilization theory, frame analysis, and political process/political opportunity theory—but we also address a continental, historical sociological tradition. Resource mobilization theory is the oldest of the three frameworks, and it was influential in the 1970s and 1980s in the anglophone countries (McCarthy & Zald, 1977). Resource mobilization theory focuses on strategy, agency, and organizations, and it examines problems such as building mass membership, competition among social movement organizations, and growth trajectories. From this perspective, science and technology are viewed as one of many potential resources that a movement can access. Frame analysis focuses on questions of meaning, the ways in which movement leaders must define issues to attract adherents, and the processes of frame diffusion (Benford & Snow, 2000). From the frame analysis perspective, science and technology enter into the ways in which issues are defined and made credible to potential supporters. Political process/opportunity theory draws attention to the structural conditions that make it possible for social movements to mobilize, and frequently the studies adopt a comparative perspective (Kitschelt, 1986; McAdam, 1982; Tilly, 1978). From the perspective of political opportunity structures, science and technology can shape the conditions of possibility, including risks and hazards, that create spaces for mobilization and enhance or diminish the success of a movement. In the 1990s some leading social movement researchers developed a fourth, synthetic framework that brought together resource mobilization, framing, and political opportunity structures (McAdam et al., 2001). The "contentious politics" framework shifted the focus of attention toward various processes and mechanisms that occur within and across movements and other forms of political contention.

Theorists working in a continental tradition of social movement theory, which was especially prominent in the 1980s and early 1990s, drew attention to the change of goals, targets, and repertoires of post-1960s movements (Habermas, 1987; Melucci, 1980; Touraine, 1992). Research on "new social movements" involved a debate over the extent to which the environmental, women's, gay-lesbian-bi, and related movements were in some way fundamentally different from older, class-based movements. For example, some claimed that new social movements emphasized lifestyle and identity change over state-oriented political protest. Although many researchers, particularly in the anglophone tradition, were skeptical of the value of such contrasts (e.g., Pichardo, 1997), new social movement theory appealed to those seeking a language for emergent forms of collective action different from those of the modern labor movement, and some theorists developed integrated approaches (e.g., Klandermans & Tarrow 1988; Taylor & Whittier, 1992). New social movement theory also developed another distinction—between state-oriented targets and non-state targets—that has been influential. Of particular relevance to STS researchers are movements that have non-state targets such as science, medicine, and industries (Moore, 1999).

New social movement theory provides a historical perspective on social movements by suggesting that they have changed in part as a reaction to the colonization of the life-world or because the central societal conflict has shifted away from class struggle to issues of democracy (Habermas, 1987; Tourraine, 1992). However, other approaches in historical sociology, such as the risk society thesis of Beck (1999) or constructivist variants/critiques of it (Wynne, 1996), may be more suitable for understanding why science and technology issues have become increasingly salient in the social movements from the mid-twentieth century to the present. As the perception of increased risks and hazards associated with industrial technology has increased, to some degree new social movements have also become "risk movements" (Halfmann, 1999). Another explanation of the changing repertoires and targets of social movements is that neoliberal policies are undercutting the social understanding by which citizens support the state in return for services and protections such as security and health. Certainly, under neoliberal governance the state has relinquished much of the regulatory potential that would help control and reduce the risks associated with emergent technologies. Social movements, along with some courageous scientists, have stepped into the void. When employing such historical explanations, scholars also need to attend to the contingency and variability of social movement mobilizations. For example, the episodic trajectory of peace movements suggests the difficulty of developing overarching explanations for the increased salience of science and technology to social movements.

MAPPINGS OF SCIENCE, TECHNOLOGY, AND SOCIAL MOVEMENTS

The triangle of science, technology, and social movements can be mapped according to the locus of change. One locus of change involves reform movements or countermovements within scientific fields (Nowotny & Rose, 1979). Science is rarely characterized by a Kuhnian paradigm (Fuller, 2000); instead, researchers tend to be organized in networks that compete with each other for control of resources such as funding, major academic departments, and professional associations and journals. Much of the history of science documents those struggles and the displacement of one network by another, and the sociology of science has also studied such processes through research on specialty group formation (Mullins, 1972) and the dynamics of actor-networks in science (Latour, 1988). Emergent research suggests that networks and research fields are sometimes connected to broader social movements, such as environmentally oriented reform movements within the natural sciences or feminist reform movements within primatology (Frickel, 2004a; Frickel & Moore, 2005; Haraway, 1989). Research now underway is exploring the dynamics of "scientific and intellectual reform movements" and how social movement theory can be relevant to understanding them (Frickel & Gross, 2005).

Another locus of change involves the adoption and reconfiguration of technology by social movements. Ruling elites have long used information management strategies to maintain their positions, including their monopolization of the means of communication and their suppression of challengers. In turn, social movements also develop media and communication strategies to circumvent control, and in some cases specific social movements or grassroots campaigns develop around media and information reform. Social movement organizations such as Greenpeace have specialized in media-oriented events (Dale, 1996; Mattelart, 1980; Raboy, 1984; Scalmer, 2002), and access to new information technologies, especially the Internet, has also facilitated social movement organization. Social movements' use of the Internet is one of the few areas where the much vaunted but rarely realized "democratic promise" of the Internet is at least partially borne out. Web sites and listservs never sleep; they are available twenty-four hours a day to anyone who has the equipment and infrastructure to access or post to them (Breyman, 2003). For example, the Internet has allowed the global women's movement to become a truly transnational movement, not through an inherent politics of the technology but because the Internet can be used in both instrumental and expressive ways (Moghadam, 2005; Stienstra, 2000). Although information technologies are the most widely used new technology that social movements have adopted and modified, environmental organizations have also adopted new biotechnologies to document problems such as environmental contamination. In some cases the new biotechnologies have divided movements because they create opportunities for activists in the form of new tools for documenting risks and exposures, but they also individualize and medicalize scientific research, thereby making it more difficult for activists to make claims of environmental causation (Shostak, 2004).

A third locus of change involves scientists who enter the political arena, often in collaboration with social movements, to oppose policies supported by elites and advocate alternatives. Political action by scientists has occurred throughout the twentieth century, but in the late 1960s and early 1970s various social-responsibility-in-science and radical science groups emerged, such as the British Society for Social Responsibility in Science and Science for the People in the U.S. (Beckwith, 1986; Biggins, 1978; Moore, 1996, 2006; Moore & Hala, 2002). The radical science movement's critiques covered both political and epistemological dimensions of science, drew inspiration from some revolutionary societies, and proposed an alternative: people's science (Arditti et al., 1980; Moore, 2006; Science for the People, 1974). The movement affected STS scholarship, although the STS debt to the movement is seldom acknowledged (Martin, 1993). The tradition of social responsibility in science continues today, embodied in at least four major organizational forms (Frickel, 2004b): (1) boundary organizations (Guston, 2001), which are located in universities or government agencies and mediate scientific, political, and industrial worlds; (2) public interest science organizations (Moore, 1996, 2006), which are located outside the government and overtly aligned with social movements; (3) professional scientific associations, which defend scientists' autonomy, including that of dissident scientists, and sometimes take political positions (Moore, 1996); and (4) grassroots support organizations, which are social movement organizations, rather than organizations of scientists, that draw on scientific expertise to develop critiques of and promote alternatives to existing government and industry policies.

Scientists who work with social movements find that their relations can become tense and involve complex negotiated settlements. Some scientists seek to maintain the role of the disinterested researcher who shuns visibility and attempts to produce peer-reviewed knowledge on a controversial issue. However, even scientists who adopt such neutral strategies can rapidly find themselves at the center of unwanted and highly public controversies for which they are ill prepared (Allen, 2003, 2004). The existence of a social movement has also tended to increase the surveillance and levels of suppression of scientists whose work can aid the movement (Martin, 1999). At the same time, social movement activists sometimes view their alliances with scientists with ambivalence partly because of the independence of the scientists and the unpredictability of the research generated by scientists (Yearley, 1992). In some cases scientists may help social movements by developing research programs and technologies that have some correlation with the ends of social movements, but they may do so on the basis of a *quid pro quo* or an offer to develop a research program or technology that may not be exactly what the social movement wanted (Clarke, 1998, 2000). In short, scientists' concern with autonomy is frequently a source of tension between them and social movements.

The remainder of this essay focuses on how social movements have influenced the development of modern science and technology through epistemic and technological change (Eyerman & Jamison, 1991; Jamison, 2001a,b). Nineteenth- and twentieth-century social movements tended to flourish during periods of economic decline, yet they often contained seeds of innovation that were developed in subsequent expansionary periods (Jamison, 2006). For example, well before the most recent wave of influential social movement activity, in the early nineteenth century a branch of the labor movement, the Luddites, developed a politics of technology that challenged the imposition of capitalist control over the labor process during Britain's industrialization (Thompson, 1963; see also MacKenzie & Wajcman, 1999: part II). In contrast with the popular use of the term Luddite today to describe machine-breaking activities, the original Luddites had a comprehensive and sophisticated program (Binfield, 2004; Fox, 2002; Sale, 1995), and the Luddite tradition has influenced some contemporary STS researchers (Noble, 1993; cf. Hedman, 1989).

Table 20.1

Oppositional and Alternative Social Movements

Social Movement	Oppose Existing Technologies	Develop Alternative Science and Technology
Health	Antismoking, antivaccine	Health-care access, embodied health movements
Environmental	Antinuclear, anti-GM food, environmental justice	Organic food, recycling and remanufacturing, green chemistry
Peace/weapons	Disarmament	Nonviolent defense
Information/media	Media reform	Alternative media, open source

Social movements today continue to be challengers, producers, and sometimes advocates of science and technology. Social movements challenge research priorities, professional practices, research methods, technology development, market developments, risk assessments, and public policy by renegotiating what counts as science for the purposes of governance. They do so through various roles, including those of entrepreneurial brokers, movement intellectuals, and custodians of local knowledge (Jamison, 2001a,b). Social movement organizations develop alliances with scientists or scientific organizations, hire scientists and occasionally contract for research, and draw on their own lay and local knowledge of issues that involve science and technology (Epstein, 1996; Moore, 2005). The movements may emerge to oppose specific research agendas or technology trajectories, and they may also develop in support of alternatives (table 20.1). We focus on health, environmental, peace, and information/media movements, partly because our collective knowledge is specialized in those four areas and partly because they have mounted the clearest epistemic challenges to the direction of science and society.

HEALTH SOCIAL MOVEMENTS

Prior to the last decades of the twentieth century, when huge disease-based patient advocacy movements emerged around AIDS and breast cancer, the primary popular mobilizations in the health arena were based on increasing access to health care (e.g., health insurance and government programs) and public health works (e.g., sanitation systems). In the late twentieth century, social movements responsive to the movements for civil rights and women's rights developed wings specifically directed towards increasing access to health care, changing the quality of health care, and reforming the caring professions. For example, U.S. women mobilized to gain greater access to reproductive technologies and control over reproduction (Clarke, 1998; Wajcman, 1991). Health reform was a cornerstone of early civil rights organizing in the United States during segregation, and a "medical" civil rights movement emerged in the 1950s to push for racial integration of the medical professions as well as community health initiatives (Smith, 1995). The women's health movement, which developed in close conjunction with the movement for sexual self-determination and the reproductive rights movement, established a clinical infrastructure that increased women's access to woman-friendly health care (Morgen, 2002).

There are many possible categorizations of health social movements (see chapter 21 in this volume); we focus here on a category that Brown & Zavestoski (2004: 685–86) have called embodied health movements, which address "disease, disability, or illness experience by challenging science on etiology, diagnosis, treatment, and prevention." Primary examples of embodied health social movements are those based on disease, such as the breast cancer movement, and those based on therapies, such as the complementary and alternative medicine (CAM) movement or the antivaccination movement. Embodied health social movements problematize the biological body, challenge existing scientific and medical knowledge, and involve collaborations between activists and scientists and health professionals (Brown et al., 2004a).

An intense focus on the biosocial body emerged in the context of the second wave women's movement, which linked self-identity, health, sexuality, and reproductive status (Boston Women's Health Book Collective, 1971). That focus, which was unique to health-related and sexual rights social movements, provided a model as well as an organizing base for HIV/AIDS, breast cancer, and other mobilizations around specific diseases. The AIDS, breast cancer, CAM, and feminist health movements developed extensive epistemic challenges to health research in arenas such as clinical trials methods, alternative therapies, and the modernization of research funding to include patient advocates (Epstein, 1996; Hess, 2004a; Treichler, 1996; Klawiter, 2002). Research on embodied health social movements has some parallels with environmental and other technology-oriented movements, so some of the findings can be generalized to other social movements where science and technology issues are salient.

Embodied health social movements often face and challenge a "dominant epidemiological paradigm" based on a biomedical model widely believed to represent consensus knowledge about a disease, its etiology, and its treatment (Zavetoski et al., 2001; see also Clarke & Olesen, 1999; Kroll-Smith & Floyd, 1997). Some movements have challenged diagnostic criteria as well as disease categories such as homosexuality (Fausto-Sterling, 2000; Terry, 1999) or schizophrenia (Crossley, 1998, 2006), and others have challenged the safety of standard preventative or therapeutic measures such as vaccines (Blume, 2006). The challenges are particularly acute in cases of presumptive diseases—such as postpartum depression (Taylor, 1996), Gulf War–related diseases (Zavestoski et al., 2001), and multiple chemical sensitivity (Dumit, 2006)— where there is no expert consensus regarding the existence of the disease, in contrast with diseases for which the existence is undisputed, such as breast cancer. In the case of breast cancer activism, the goal has centered on the less epistemically challenging issues of increasing research spending on treatment, diversifying treatment choices, developing greater access to treatment choices (Casamayou, 2001; Lerner, 2001) and, to a lesser extent, promoting prevention through nutrition and reduced exposure to carcinogenic chemicals (Epstein et al., 1998). Such activism has yielded significant changes in the "regimes of practice" that breast cancer patients experience in the clinical setting (Klawiter, 2004). The medicalization of breast cancer prevention has embroiled the movement in scientific and regulatory controversies over the value of the use of drugs such as tamoxifen in "at risk" healthy women. Analysis of social movement action on this issue has necessitated a broadened theoretical framework that includes the pharmaceutical industry, regulatory policy, design controversies over clinical trials, clinical standards differences, and the doctor-patient relationship (Fosket, 2004; Klawiter, 2002; Wooddell, 2004).

The various movements for complementary and alternative medicine usually involve scientific controversies over the etiology and treatment of recognized diseases, but they provoke intense political confrontations with the medical profession, regulators, and medical research community (Johnston, 2004). The movement for CAM cancer therapies in the United States exhibits two general features shared with other pro- or alternative "technology- and product-oriented movements," such as movements for sustainable agriculture, renewable energy, and open source software: (1) opposition to a specific technology or product combined with support for an alternative, and (2) a mix of grassroots social movement and advocacy organizations with professional and/or industrial reform movements that involve scientists and/or entrepreneurs (Hess, 2005, 2007). Professional reform movements generally do not use extra-institutional strategies, but they are often sympathetic with social movements that do, even if they operate at some distance from them (Frickel, 2004a; Hoffman, 1989; Woodhouse & Breyman, 2005). The organizational mixture of the CAM movement is one factor behind the medical mainstream's range of organizational responses, which include avoidance, compromise, acquiescence, manipulation, and defiance (Goldner, 2004).

Over time, many health social movements, like other social movements, undergo diversification and transformation. Sometimes countermovements develop, or movements emerge on both sides of a long-standing controversy, as in the case of pro- and anti-fluoridation networks (Martin, 1991; McNeil, 1957). Often movements divide into accommodationist and radical wings; the former organizations tend toward professionalized advocacy rather than grassroots activism. The pharmaceutical industry has provided significant funding for U.S. breast cancer organizations, leading to the possibility of organizational capture, while at the same time the growth of private breast cancer research foundations has created opportunities for, and potential conflicts among, lay funders and scientist researchers (Gibbon, 2003).

Another effect of the diversification and transformation of health social movements is that in some cases, such as the AIDS movement, social movement leaders undergo an "expertification" process (Epstein, 1996). The crossing of lay-expert divisions has continued to attract attention in the study of health social movements. In the U.S. breast cancer movement, the diversification of organizations across class and ethnic divisions was accompanied by organizational conflict between long-standing staff, who acquired various forms of expertise, and newcomers, who possessed new and different knowledges (Hoffman, 2004). In interactions with scientists, health social movement organizations play a role of discriminating between science and nonscience that is similar to the state-funded boundary organizations described by Guston (2001), but the organizations push the boundaries of science in new directions and challenge identities and interests on both sides of the lay-expert divide (Brown et al., 2004a; Ganchoff, 2004). Those interactions emphasize the mutual learning that occurs among patients, researchers, and clinicians in "reflexive organization" (Rabeharisoa & Callon, 2004). Some activists make the transition from the "narrow-band" competence of lay expertise, which is largely "interactional" expertise in Collins's terms (2002), by assembling networks of researchers to produce biomedical knowledge or by obtaining more education so that they become professional researchers (Hess, 2004a). Institutionally and historically, in the United States a process of "medical modernization"—which recognizes the legitimacy of participation from patient representatives in funding decisions—has tended to replace the previous strategy of suppression of dissident scientist/activist coalitions that coincided with a paternalistic, transmission model of biomedical knowledge (Hess, 2004a).

In addition to diversifying lay-expert divisions through hybridization, health social movements have also undergone fragmentation in social composition that has typically accompanied growth and alliances across social categories. The original AIDS movement in the United States was largely middle-class, male, and white, but over time it struggled with new issues as the social address opened up to African Americans and women (Epstein, 1996). Likewise, antismoking campaigns have struggled with the politics of extension to ethnic communities in California and with the politics of national cultural differences as the campaigns extended outward from the English-speaking countries (Reid, 2005). In some cases, antitobacco and other antidrug movements have also become linked to other social justice issues such as structural inequality and gender equity (Campbell, 2000; Nathanson, 1999; Oaks, 2001). The heterogeneity of participants in the U.S. disability rights and reproductive rights movements led to the formation of "divided interests" in the reproductive technologies arena (Rapp, 1999). Although health social movements can fracture around gender, racial-ethnic categories, sexualities, categories of age and ability, and class-based identities, recognition of differences and health disparities can also stimulate greater attention to "culturally competent" health care provision; gender, age, and ability equity; and the inclusion of formerly stigmatized identities such as sex workers and persons with alcoholism, drug addiction, and AIDS (Campbell, 2000; Stoller, 1998). Social movements have exerted pressure for mechanisms to ensure greater accountability among "markets" composed of users, consumers, and patients and the government agencies, health care providers, scientific researchers, and technological designers that supply these markets (Clarke, 1998; Oudshoorn & Pinch, 2003). Finally, movements to promote or limit the use of specific reproductive technologies arise to address the diversity of power-laden cultural contexts in which health-care decisions are made (Briggs, 2003; Sen & Snow, 1994).

ENVIRONMENTAL MOVEMENTS

Many scholars now recognize that *the* environmental movement is, like other social movement categories, a diverse sociological entity. Historical studies generally delineate a major transition during the 1960s from a focus on preservation and conservation to industrial pollution, and in the United States and some other countries during the 1980s there was a second shift to a focus on environmental justice (Dowie, 1995; Gottlieb, 1993; Kline, 1997). Organizations tend to focus on one of the three types of environmental action, but many have mixed goals that reflect the influence of all three waves. In many countries, striking divisions have emerged between the government-oriented, insider, advocacy organizations and the proliferation of struggles at the grassroots level around environmental justice. There is also tremendous diversity across world regions and even within wealthy Western regions. For example, in Europe there has been a relatively stronger policy articulation of environmental concerns than in the United States, where green or left-wing parties have been much more marginalized in electoral politics.

Of the various opposition movements within the broader environmental movement that target mainstream science and technology, the worldwide movements against nuclear power and genetically modified (GM) food provide two examples of how movements challenge scientific knowledge and emergent technologies, particularly around issues of risk and safety. Activists have proceeded, independently of STS critiques of technological determinism, on the assumption that nuclear power is not inevitable (Smith & Marx, 1994; Winner, 1977). Activists and STS scholars alike developed a critique of the politics of design around nuclear power: it is expensive, potentially dangerous, dependent on experts, and thus antagonistic to democratic society (Patterson, 1977; Perin, 2004; Winner, 1986; Woodhouse & Morone 1988). Likewise, campaigns against GM foods have challenged industrial, scientific, and government assurances of safety (Bauer & Gaskell, 2002; Purdue, 2000). Although activists have sometimes been drawn into a debate with experts over the risks of GM food in Europe, India, and other world regions, they also utilize frames beyond the science of risk and safety. For example, they frequently frame the debate and protest events around concerns with globalization and U.S. food hegemony (Harper, 2004; Heller, 2001; Shiva, 2000). In addition to the comparative effectiveness of frames of food politics, differences in industrial structure help account for the different degrees of success of movements against GM food (Schurman, 2004).

Environmental movements not only challenge the epistemic assurances of governments and scientists but also encourage the development of alternatives. In the 1970s, proponents of appropriate technology—sometimes also called alternative technology or intermediate technology—argued that technologies embodied elite political values, and they developed and promoted technologies appropriate for communities (Kleiman, forthcoming). In poorer countries, appropriate technology ideally required low capital; used local resources; was labor intensive and small scale; could be controlled by villagers; and could be controlled, produced, and modified by villagers in ways that brought people together and were environmentally sound (Darrow & Saxenian, 1986). There have been many debates about the politics of appropriate technology (Boyle et al., 1976; Illich, 1973; Kleiman, forthcoming; Lovins, 1977; Riedijk, 1986; Willoughby, 1990); the key point is that the movement drew attention to the politics of technology design (Winner, 1986). The legacy of the appropriate technology movement today is, in developing countries, one of low-tech, locally controlled development projects, and, in wealthy countries, advocacy around renewable energy and sustainable agriculture.

Renewable energy and sustainable agriculture gradually grew from social movements into industries with associated scientific research programs. For example, wind energy in Denmark was once a social movement, but over time it was mainstreamed (Jamison et al., 1990). As the control of design shifted from lay users to professionals oriented toward industrial production on wind farms, the scale of the technology increased (Jørgensen and Karnøe, 1995). The transformations of technology design involve a process of "complementarization" or redesign that adapted alternative, movement-based technologies to fit into existing portfolios of industrial production technologies and industry products (Hess, 2005). Likewise, the organic food movement developed an alternative form of scientific knowledge that challenged dominant research programs and combined lay-expert knowledges (Hassanein, 1999). Over time, organic food production underwent industrialization, and a portion of the movement became mainstreamed, but the grassroots side of the movement regrouped around the antiglobalization politics of local, sustainable agriculture (Guthman, 2004; Hess, 2004b). The organic food movement also played a significant role in the mobilization against GM food, another indication of the fluidity of movements that oppose some forms of technology and support alternatives for other forms (Reed, 2002). Similar changes occurred with the recycling movement, which in some places began as a grassroots movement and was subsequently incorporated into the waste industry (Pellow, 2002; Scheinberg, 2003; Weinberg et al., 2000).

More generally, the environmental movement underwent a change from activism to brokerage, and protest politics shifted toward the development of green business networks (Jamison, 2001b). By the 1990s, a new polarization had also emerged between the ecological modernization frame of green business and the environmental justice orientation of grassroots activists (Hård & Jamison, 2005; Mol, 2000; Pellow 2002; Pellow & Park 2002). As environmentalism underwent professionalization and industrialization, "object conflicts" developed over definitions of what the technology/product should be. The conflicts took place in three arenas: research agendas, consumer decisions and loyalties, and standards set by regulatory agencies or industrial groups (Hess, 2004b). Clashes over regulatory standards can also involve a movement's environmental values versus the health and safety values of state agencies (Henderson, 2006). The processes of institutionalizing environmental social movement goals has also led to a "systematic discounting" of efforts by activists and advocates to build corporate responsibility goals into legislation and corporate policies, as occurred in the case of the failure to respond completely to the calls for reform in the wake of the Bhopal disaster (Fortun, 2001).

In addition to problems that occur with industrialization, activists also encounter problems in their efforts to work with scientists and other social movements. As activists and environmental professionals work together, many have become convinced of the need for heterogeneity in environmental problem-solving models (Di Chiro, 2003, 2004). By recognizing the different bases of lay and scientific knowledges, activists and scientists may develop deliberative processes that allow for synergy between lay and expert knowledges (Breyman, 1993; Brown & Mikkelsen, 1990; Carson & Martin, 2002; Fischer, 2000). In building cross-movement bridges, issues of expertise and design have been salient in the relations between environmental justice and sustainability groups (Agyeman et al., 2003), civil rights and urban transportation design reformers (Bullard et al., 2004), and labor and environmental coalitions (Burgmann & Burgmann, 1998; Gould et al., 2004; Grossman & Daneker, 1979; Mundey, 1981; Roddewig, 1978; Obach, 2002; Rose, 2000). Likewise, the environmental breast cancer movement (a wing of the larger breast cancer movement that focuses on environmental factors such as endocrine-disrupting chemicals) has allied with the environmental justice movement (Ley, forthcoming). The two movements may each be in a "steering" or "guiding" role with respect to the broader breast cancer and environmental movements of which they are a part (Brown et al., 2004b). Likewise, food-based politics provide a point of connection between health and environmental movements (Cohen, 2005; Hess, 2002).

PEACE, INFORMATION, AND OTHER MOVEMENTS

Although the epistemic politics of health and environmental movements have dominated the intellectual landscape for STS-related scholarship, other movements have engaged in epistemic challenges to science and technology. For example, with the increasing role of technology in warfare, peace movements have grappled with issues of expertise, technology design, and antiwar tactics. There has been some study of the social shaping of military technologies, for example, the machine gun (Ellis, 1975), airplanes (Schatzberg, 1994), missile guidance systems (MacKenzie & Spinardi, 1995), and computing (Edwards, 1996). Particular types of weapons, especially those that are deemed inhumane, have long generated special disgust and consequent attempts to abolish or regulate them. Examples include antipersonnel weapons such as dumdum bullets and land mines, biological and chemical weapons, nuclear weapons, and "nonlethal" weapons (Gusterson, 1996; Prokosch, 1995; Rappert, 2003).

Of opposition efforts, antinuclear weapons movements have been most prominent. Some scientists raised concerns about nuclear weapons from the very beginning, with the *Bulletin of the Atomic Scientists* serving as an ongoing platform for debate and critique. Popular opposition expanded in the late 1950s with concerns about radioactive fallout. Official reassurances were challenged by a few dissident scientists, of whom Linus Pauling (1958) was most prominent during that period. The movement faded in the early 1960s, especially following the 1963 Partial Test Ban Treaty. Beginning about 1979, a second phase of the global antinuclear weapons movement blossomed, with an associated expansion of social analysis. In the 1980s, a number of scientists and writers painted doomsday scenarios, including "nuclear winter," and concluded that the survival of the human species could not be guaranteed (Ehrlich et al., 1985; Schell, 1982). This is a prominent example of science deployed in service of a social movement; the scientists presumed that they had a special mandate to intervene in policy debates because they had access to scientific knowledge (Eden, 2004; Martin, 1988). The debate about nuclear winter vanished from scientific and public sight after the end of the Cold War, despite the persistence of nuclear arsenals, showing the way that international affairs, as well as social movements, can affect research agendas and the saliency of policy issues (Breyman, 1997).

The nonviolence movement is in part a component of the peace movement, but it has also influenced other social movements such as the environmental, antiracist, and feminist movements. Nonviolent action—such as noncooperation, strikes, boycotts, fasts, and setting up alternative institutions—challenges oppressive systems and offers an alternative to violence. In relation to peace issues, the nonviolence movement has focused on social and psychological dimensions of resistance to oppression and aggression. Nonviolence provides an alternative agenda for research and development, for example in the design of communication systems and technological systems (such as energy, industry, and agriculture) for survival in the case of attack (Martin, 1997, 2001). A nonviolence agenda points both to different technologies—for example, network communication forms such as telephone, fax, and e-mail rather than centralized media such as television and radio, usually the first targets in military coups—and to different, more participatory research methods. Social movements have not adopted this approach explicitly, but in many cases they are proceeding along parallel lines. For example, the appropriate technology movement sets criteria for technology that mesh perfectly with technological specifications for nonviolent resistance.

Information and media reform movements also target issues of technology design. At the most basic level, literacy campaigns have been a constant of some social movement agendas, and basic literacy education continues to be a site for contesting class domination among the poor (Freire, 1972). Where compulsory schooling is the norm, adult education is a more common source for information-oriented resistance (Lovett et al., 1983). However, literacy campaigns can be a double-edged sword. For example, the emerging STS-inspired research on the digital divide documents how computer illiteracy has been overestimated, and in fact many persons with limited income utilize information-technology skills in their work. For them, the role of computer-based surveillance is a more salient issue (Eubanks, 2006; Monahan, 2005, 2006).

Another strand of information and social movements focuses on media reform. Opposition movements stretch back to the commercialization of U.S. radio in the 1920s and to the development of public broadcasting in subsequent decades, the history for which varies significantly across countries (McChesney, 1993). In the 1990s, a new wave of media reform took off when an international coalition of church,

education, and media-related NGOs joined together to protest the increasing concentration and commercialization of the mass media (Free Press, 2003; Goodale, 1996).
Alternative media also has a long history, but the 1960s social movements spurred the
creation of alternative radio and print media, often oriented to local markets and self-
identified as part of a "community media movement" (Downing, 1984; Pierce, 2003).
The organizations have been subject to problems of burn-out or bureaucratization, as
well as a drift toward commercialization (Castells, 2001; McChesney, 2001). Social
movement activists have often made hopeful statements about the Internet, and some
have suggested that its design, which is interactive and decentralized in contrast with
broadcast and print media, is inherently liberatory. However, the Internet is also
subject to political control (Privacy International and the GreenNet Educational Trust,
2003), and the relationship between the Internet and democracy remains a topic of
empirical study (Fortier, 2001; Kalathil & Boas, 2003). Current debates on the democratic potential of the Internet were preceded by a prior generation of debates on
"computerization movements." Some touted the advantages of widespread computer
use in the workplace, home, and schools, a view that challenges the prevalent idea
that the introduction of computing was entirely driven by technical or market considerations. In contrast, other groups, including representatives of traditional social
movements, developed a dystopian view of the computer, such as environmentalists
who saw them as sources of alienation (Hakken & Andrews, 1993; Kling & Iacono,
1988; Mander, 1984).

In the 1980s and 1990s, some leaders of the 1960s counterculture helped rethink
the computer from a symbol of the "system" to a symbol of liberation, and those ideas
spread through a network of people around the *Whole Earth Catalog* (Turner, 2005) as
well as in experiments around community informatics (Cohill & Kavenaugh, 2000;
Gurstein, 2000). The community informatics projects faced problems of organizational
viability that were similar to the volunteer media experiments of community radio
and alternative newspapers. To survive, some of the community informatics projects
institutionalized as nonprofit organizations and sought the support of local governments or foundations (Castells, 2001; Schuler, 2000). A second wave of voluntary,
Internet-based alternative organizations emerged during the Seattle demonstrations
against the World Trade Organization in 1999, when the Independent Media Centers
(IMC), or Indymedia, movement was launched (Morris, 2004; Pierce, 2003). The computer programmers who designed Indymedia software were motivated by a desire to
construct a system that allowed open access for publishing while restricting the potential for central control, and its ethic of open access had some similarities to two other
Internet-based reform movements: open source and open content.

Unlike the Indymedia movement and community informatics, the "free/libre open
source software" (FLOSS) movement is more oriented toward the politics of software
design, and its alternative code can be used by governments, corporations, and
activists alike. The reform movement mobilizes volunteer programmers partly by offering a system of credit and recognition that has similarities to the scientific reward
system (Kelty, 2001). One of the key transitions (and divisions) in the FLOSS move-

ment's history was its reframing from the term "free software," associated with activist Richard Stallman, to "open source" and a more business-oriented perspective (Bretthauer, 2002). The best-known product is the operating system Linux, which has become competitive with commercial programs (Moody, 2002; Weber, 2004). As commercialization has progressed, object conflicts have developed around maintaining the original GNU license structure versus more commercially oriented license structures developed by proprietary software firms such as Microsoft (Hess, 2005). In the open-content movement (the provision of free information in the public domain, including scientific journal articles), conflicts have developed between copyright holders, especially media and publishing companies, on the one side and scientists, librarians, hackers, and consumers on the other (Poynder, 2004). Hackers also challenge emergent digital global property rights regimes through the development of code that allows users to swap files or break encryption codes. They view their work as civil disobedience, given that corporations and governments have prosecuted their activism as criminal violations of intellectual property laws (Postigo, 2005).

CONCLUSIONS

Social movement organizations that emerge from grassroots grievances frequently challenge consensus scientific knowledge, official assessments of safety and risk, and the technology trajectories developed by elites in industry and the state. They seek alliances with scientists and already established interest groups as well as with entrepreneurs and the business sector. Yet, relations among social movements, scientific research networks, and business organizations are frequently beset by conflict as much as cooperation. At a technical level, the success of alternative technologies and products comes at the cost of a complementarization process in which the more politically charged design elements and social organizational innovations drop out. At the discursive level, social movements must often pitch critical alternatives in a language that reflects the dominant "governing mentalities" that prevail in a particular policy arena in order to be heard as credible (Campbell, 2000). As a result, some social movements that seek changes in science and technology issues often find their goals incorporated at a technical level but at a cost of severing the technical goals from the broader political and justice goals. In summary, social movements, scientists, and entrepreneurs are uneasy allies and partners, and alliances sometimes shift into conflict and hostility—or they simply drift in different directions—even as they generate new research programs, technologies, and material culture.

Regarding topics for further exploration of the uneasy partnerships involved in social movements, science, and technology, several questions emerge from our review, among them the following: Is it true that issues of science and technology have become more salient in social movements, and, if so, what are the historical explanations? How does the science, technology, and movement interface vary across not only time but also space? To what extent do comparative differences become less important as movements become more globalized? How do science and technology issues work in conservative and antidemocratic movements (which were not the focus of this essay)?

Before charting an agenda for the study of STS and social movements, we suggest that it would be valuable to step back and return to the broader issue of science, technology, and democracy that was raised at the beginning of this essay. If mapping social movements is to be more than an academic enterprise, if that work is meant to contribute to the success of democratic social movements, then the first question might be how can the study of science, technology, and social movements be configured in a way that is of value to activists? Does the goal require a shift in methods, such as moving toward participant-action research? Those questions return to one of the originary strands of STS, when portions of the interdisciplinary field were closely connected to scientific and technological reform movements.

Note

We thank Phil Brown and Andrew Jamison for comments on a draft of the essay, Steve Epstein and an anonymous reviewer for detailed comments on the draft, and Andy Holtzman and Anna Salleh for helpful discussions.

References

Agyeman, Julian, Robert Bullard. & Robert Evans (eds) (2003) *Just Sustainabilities: Development in an Unequal World* (Cambridge, MA: MIT Press/Earthscan).

Allen, Barbara (2003) *Uneasy Alchemy: Citizens and Experts in Louisiana's Chemical Corridor Disputes* (Cambridge, MA: MIT Press).

Allen, Barbara (2004) "Shifting Boundary Work: Issues and Tensions in Environmental Health Science in the Case of Grand Bois, Louisiana," *Science as Culture* 13(4): 429–48.

Arditti, Rita, Pat Brennan, & Steve Cavrak (eds) (1980) *Science and Liberation* (Boston: South End Press).

Bauer, Martin & George Gaskell (eds) (2002) *Biotechnology: The Making of a Global Controversy* (Cambridge, U.K.: Cambridge University Press).

Beck, Ulrich (1999) *World Risk Society* (Cambridge, U.K.: Polity Press).

Beckwith, Jon (1986) "The Radical Science Movement in the United States," *Monthly Review* 38(3): 118–28.

Benford, Robert & David Snow (2000) "Framing Processes and Social Movements: An Overview and Assessment," *Annual Review of Sociology* 26: 611–39.

Binfield, Kevin (ed) (2004) *Writings of the Luddites* (Baltimore, MD: Johns Hopkins University Press).

Biggins, David (1978) "Social Responsibility in Science," *Social Alternatives* 1(3): 54–60.

Blume, Stuart (2006) "'Anti-Vaccination Movements' and their Interpretation," *Social Science and Medicine* 62(3): 628–42.

Boston Women's Health Book Collective (1971) *Our Bodies, Our Selves: A Course by and for Women* (Boston: New England Free Press).

Boyle, Godfrey, Peter Harper, & the editors of *Undercurrents* (eds) (1976) *Radical Technology* (London: Wildwood House).

Bretthauer, David (2002) "Open Source Software: A History," *Information Technology and Libraries* 21(1): 3–10.

Breyman, Steve (1993) "Knowledge as Power: Ecology Movements and Global Environmental Problems," in R. Lipschutz & K. Conca (eds), *The State and Social Power in Global Environmental Politics* (New York: Columbia University Press): 124–57.

Breyman, Steve (1997) *Movement Genesis: Social Movement Theory and the West German Peace Movement* (Boulder, CO: Westview Press).

Breyman, Steve (2003) "Moyens de Communications, Mobilisation Rapide et Actions Préventives Contre la Guerre: Analyse de la Mobilisation Pacifiste Actuelle en Occident contre la Guerre en Irak," *EcoRev* 12(April): 37–44.

Briggs, Laura (2003) *Reproducing Empire: Race, Sex, Science, and U.S. Imperialism in Puerto Rico* (Berkeley: University of California Press).

Brown, Phil & Edwin Mikkelsen (1990) *No Safe Place: Toxic Waste, Leukemia, and Community Action* (Berkeley: University of California Press).

Brown, Phil & Stephen Zavestoski (eds) (2004) "Social Movements and Health: An Introduction," *Sociology of Health and Illness* 26(6): 679–94.

Brown, Phil, Stephen Zavestoski, Sabrina McCormick, Brian Mayer, Rachel Morello-Frosch, & Rebecca Gasior (eds) (2004a) "Embodied Health Movements: Uncharted Territory in Social Movement Research," *Sociology of Health and Illness* 26(6): 1–31.

Brown, Phil, Rachel Morello-Frosch, & Rebecca Altman (2004b) "Emerging Collaborations Between Breast Cancer Advocacy and Environmental Justice," presentation at the meeting on Science, Technology, and the Environment, Rensselaer Polytechnic Institute, Troy, NY.

Bullard, Robert, Glenn Johnson & Angel Torres (2004) *Highway Robbery: Transportation Racism and New Routes to Equity* (Cambridge, MA: South End Press).

Burgmann, Meredith & Verity Burgmann (1998) *Green Bans, Red Union: Environmental Activism and the New South Wales Builders Labourers' Federation* (Sydney: University of New South Wales Press).

Campbell, Nancy (2000) *Using Women: Gender, Drug Policy, and Social Justice* (New York: Routledge).

Carson, Lyn & Brian Martin (2002) "Random Selection of Citizens for Technological Decision Making," *Science and Public Policy* 29(2): 105–13.

Casamayou, Maureen (2001) *The Politics of Breast Cancer* (Washington, DC: Georgetown University Press).

Castells, Manuel (2001) *The Internet Galaxy: Reflections on the Internet, Business, and Society* (Oxford: Oxford University Press).

Clarke, Adele (1998) *Disciplining Reproduction: Modernity, American Life Sciences, and the Problem of Sex* (Berkeley: University of California Press).

Clarke, Adele (2000) "Maverick Reproductive Scientists and the Production of Contraceptives, 1915–2000+," in A. Saetnan, N. Oudshoorn, & M. Kirejczyk (eds), *Bodies of Technology* (Columbus: Ohio State University Press): 37–89.

Clarke, Adele & Virginia Olesen (1999) *Revisioning Women, Health, and Healing* (New York: Routledge).

Cohen, Maurie (2005) "Sustainable Consumption, American Style: Nutrition Education, Active Living, and Financial Literacy," *International Journal of Sustainable Development and World Ecology* 12(3): 407–18.

Cohill, Michael & Andrea Kavanaugh (eds) (2000) *Community Networks: Lessons Learned from Blacksburg, Virginia* (Cambridge, MA: Artech House).

Collins, Harry (2002) "The Third Wave of Science Studies: Studies of Expertise and Experience," *Social Studies of Science* 32(2): 235–96.

Crossley, Nicholas (1998) "R. D. Laing and the British Anti-Psychiatry Movement: A Socio-Historical Analysis," *Social Science and Medicine* 47(7): 877–89.

Crossley, Nicholas (2006) "The Field of Psychiatric Contention in the U.K.: 1960–2000," *Social Science and Medicine* 62(3): 552–63.

Dale, Stephen (1996) *McLuhan's Children: The Greenpeace Message and the Media* (Toronto: Between the Lines).

Darrow, Ken & Mike Saxenian (eds) (1986) *Appropriate Technology Sourcebook* (Stanford, CA: Volunteers in Asia).

Di Chiro, Giovanna (2003) "Steps to an Ecology of Justice: Women's Environmental Networks Across the Santa Cruz River Watershed," in V. Scharff (ed), *Seeing Nature Through Gender* (Lawrence: University of Kansas Press): 282–319.

Di Chiro, Giovanna (2004) "'Living is for Everyone': Border Crossings for Environment, Community, and Health," *Osiris: Journal of the History of Science Society* 19: 112–29.

Dowie, Mark (1995) *Losing Ground: American Environmentalism at the Close of the Twentieth Century* (Cambridge, MA: MIT Press).

Downing, John (1984) *Radical Media: The Political Experience of Alternative Communication* (Boston: South End Press).

Dumit, Joseph (2006) "Illnesses You Have to Fight to Get: Facts as Forces in Uncertain, Emergent Illnesses," *Social Science and Medicine* 62(3): 577–90.

Eden, Lynn (2004) *Whole World on Fire: Organizations, Knowledge, and Nuclear Weapons Devastation* (Ithaca, NY: Cornell University Press).

Edwards, Paul (1996) *The Closed World: Computers and the Politics of Discourse in Cold War America* (Cambridge, MA: MIT Press).

Ehrlich, Paul, Carl Sagan, Donald Kennedy, & Walter Roberts (1985) *The Nuclear Winter: The Cold and the Dark* (London: Sidgwick and Jackson).

Ellis, John (1975) *The Social History of the Machine Gun* (London: Croom Helm).

Epstein, Samuel, David Steinman & Suzanne Levert (1998) *The Breast Cancer Prevention Program* (New York: Macmillan).

Epstein, Steven (1996) *Impure Science: AIDS, Activism, and the Politics of Knowledge* (Berkeley: University of California Press).

Eubanks, Virginia (2006) "Technologies of Citizenship: Surveillance and Political Learning in the Welfare System," in T. Monahan (ed), *Surveillance and Security: Technological Politics and Power in Everyday Life* (New York: Routledge): 89–107.

Eyerman, Ron & Andrew Jamison (1991). *Social Movements: a Cognitive Approach* (University Park: Pennsylvania State University Press).

Fausto-Sterling, Anne (2000) *Sexing the Body: Gender Politics and the Construction of Sexuality* (New York: Basic Books).

Fischer, Frank (2000) *Citizens, Experts, and the Environment* (Durham, NC: Duke University Press).

Fortier, François (2001) *Virtuality Check: Power Relations and Alternative Strategies in the Information Society* (London: Verso).

Fortun, Kim (2001) *Advocacy after Bhopal: Environmentalism, Disaster, New Global Orders* (Chicago: University of Chicago Press).

Fosket, Jennifer (2004) "Constructing 'High-Risk' Women: the Development and Standardization of a Breast Cancer Risk Assessment Tool," *Science, Technology & Human Values* 29(3): 291–313.

Fox, Nicols (2002) *Against the Machine: the Hidden Luddite Tradition in Literature, Art, and Individual Lives* (Washington, DC: Island Press/Shearwater Books).

Free Press (2003) "Free Press: About Us." Available at: http://www.freepress.net/content/about.

Freire, Paulo (1972) *Pedagogy of the Oppressed* (Harmondsworth, U.K.: Penguin).

Frickel, Scott (2004a) *Chemical Consequences: Environmental Mutagens, Scientist Activism, and the Rise of Genetic Toxicology* (New Brunswick, NJ: Rutgers University Press).

Frickel, Scott (2004b) "Just Science? Organizing Scientist Activism in the U.S. Environmental Justice Movement," *Science as Culture* 13(4): 449–70.

Frickel, Scott & Neil Gross (2005) "A General Theory of Scientific/Intellectual Movements," *American Sociological Review* 70: 204–32.

Frickel, Scott & Kelly Moore (eds) (2005) *The New Political Sociology of Science* (Madison: University of Wisconsin Press).

Fuller, Steve (2000) *Thomas Kuhn: A Philosophical History for Our Times* (Chicago: University of Chicago Press).

Ganchoff, Chris (2004) "Regenerating Movements: Embryonic Stem Cells and the Politics of Potentiality," *Sociology of Health and Illness* 26(6): 757–74.

Gibbon, Sahra (2003) "Fundraising, Citizenship, and Redemptory Knowledge: The Social Context of Lay Expertise in a Breast Cancer Charity," presentation at the annual meeting of the Society for Social Studies of Science, Atlanta.

Goldner, Melinda (2004) "The Dynamic Interplay Between Western Medicine and the Complementary and Alternative Medicine Movement: How Activists Perceive a Range of Responses from Physicians and Hospitals," *Sociology of Health and Illness* 26(6): 710–39.

Goodale, Gloria (1996) "Superseding the Hollywood 'Ministry of Culture,'" *Christian Science Monitor*, November 18.

Gottlieb, Robert (1993) *Forcing the Spring: The Transformation of the Environmental Movement* (Washington, DC: Island Press).

Gould, Kenneth, Tammy Lewis, & J. Timmons Roberts (2004) Blue-Green Coalitions: Constraints and Possibilities in the Post 9–11 Political Environment," *Journal of World-Systems Research* 10(1): 91–116.

Grossman, Richard, and Gail Daneker (1979) *Energy, Jobs and the Economy* (Boston: Alyson).

Gurstein, Michael (ed) (2000) *Community Informatics: Enabling Communities with Information and Communications Technologies* (Hershey, PA: Idea Publishing Group).

Gusterson, Hugh (1996) *Nuclear Rites: A Weapons Laboratory at the End of the Cold War* (Berkeley: University of California Press).

Guston, David (2001) "Boundary Organizations in Environmental Policy and Science: An Introduction," *Science, Technology & Human Values* 26(4): 399–408.

Guthman, Julie (2004). *Agrarian Dreams: The Paradox of Organic Farming in California* (Berkeley: University of California Press).

Habermas, Jürgen (1987) *A Theory of Communicative Action,* vol. 2 (Boston: Beacon).

Hakken, David & Barbara Andrews (1993) *Computing Myths, Class Realities: an Ethnography of Technology and Working People in Sheffield, England* (Boulder, CO: Westview).

Halfmann, Jost (1999) "Community and Life-Chances: Risk Movements in the United States and Germany," *Environmental Values* 8: 177–97.

Haraway, Donna (1989) *Primate Visions: Gender, Race, and Nature in the World of Modern Science* (New York: Routledge).

Hård, Mikael & Andrew Jamison (2005) *Hubris and Hybrids: A Cultural History of Technology and Science* (New York: Routledge).

Harper, Krista (2004) "The Genius of a Nation Versus the Gene-Tech of a Nation: Science, Identity, and Genetically Modified Food in Hungary," *Science as Culture* 13(4): 471–92.

Hassanein, Neva (1999) *Changing the Way America Farms: Knowledge and Community in the Sustainable Agriculture Movement* (Lincoln: University of Nebraska Press).

Hedman, Carl (1989) "Luddites, Hippies and Robots: Automation and the Possibility of Resistance," *Prometheus* 7(2): 273–91.

Heller, Chaia (2001) "From Risk to Globalization: Discursive Shifts in the French Debate about GMOs," *Medical Anthropology Quarterly* 15(1): 25–28.

Henderson, Kathryn (2006) "Ethics, Culture, and Structure in the Negotiation of Straw Bale Building Codes." *Science, Technology & Human Values* 31(3): 261–88.

Hess, David (2002) "The Raw and the Organic: Politics of Therapeutic Cancer Diets in the U.S.," *Annals of the Academy of Political and Social Science* 583(September): 76–97.

Hess, David (2004a) "Medical Modernisation, Scientific Research Fields, and the Epistemic Politics of Health Social Movements," *Sociology of Health and Illness* 26(6): 695–709.

Hess, David (2004b) "Organic Food and Agriculture in the U.S.: Object Conflicts in a Health-Environmental Social Movement," *Science as Culture* 13(4): 493–514.

Hess, David (2005) "Technology- and Product-Oriented Movements: Approximating Social Movement Studies and STS," *Science, Technology & Human Values* 30(4): 515–35.

Hess, David (2007) *Alternative Pathways in Science and Technology: Activism, Innovation, and the Environment in an Era of Globalization* (Cambridge, MA: MIT Press).

Hoffman, Karen (2004) "Hierarchy within Social Movements: 'Science' and Power in Pollution Prevention Activism," presentation at the conference on Science, Technology, and the Environment, Rensselaer Polytechnic Institute.

Hoffman, Lily (1989) *The Politics of Knowledge: Activist Movements in Medicine and Planning* (Albany: State University of New York Press).

Illich, Ivan (1973) *Tools for Conviviality* (London: Calder & Boyars).

Jamison, Andrew, Ron Eyerman, Jacqueline Cramer, & Jeppe Laessøe (1990) *The Making of the New Environmental Consciousness* (Edinburgh: Edinburgh University Press).

Jamison, Andrew (2001a) *The Making of Green Knowledge* (Cambridge, U.K.: Cambridge University Press).

Jamison, Andrew (2001b) "Science and Social Movements," in N. Smelser & P. Bates (eds.) *International Encyclopedia of the Social and Behavioral Sciences* (Amsterdam, New York: Elsevier) 20: 13625–28.

Jamison, Andrew (2006) "Social Movements and Science: Cultural Appropriations of Cognitive Practice," *Science as Culture* 15(1): 45–59.

Johnston, Robert (2004) *The Politics of Healing: Histories of Alternative Medicine in Twentieth-Century North America* (New York: Routledge).

Jørgensen, Ulrik & Peter Karnøe (1995) "The Danish Wind-Turbine Story: Technical Solutions to Political Visions?" in A. Rip, T. Misa, & J. Schot (eds), *Managing Technology in Society* (London: Pinter): 57–82.

Kalathil, Shanthi & Taylor Boas (2003) *Open Networks, Closed Regimes: The Impact of the Internet on Authoritarian Rule* (Washington, DC: Carnegie Endowment for International Peace).

Kelty, Christopher (2001) "Free Software/Free Science," *First Monday* 6(12). Available at: http://www.firstmonday.org/issues/issue6_12/kelty/.

Kitschelt, Herbert (1986) "Political Opportunity Structure and Political Protest: Anti-Nuclear Movements in Four Democracies," *British Journal of Political Science* 16: 57–85.

Klandermans, Bert and Sidney Tarrow (1988) "Mobilization into Social Movements: Synthesizing European and American Approaches," *International Social Movement Research* 1: 1–38.

Klawiter, Maren (2002) "Risk, Prevention, and the Breast Cancer Continuum: The NCI, the FDA, Health Activism, and the Pharmaceutical Industry," *History and Technology* 18(4): 309–53.

Klawiter, Maren (2004) "Breast Cancer in Two Regimes: The Impact of Social Movements on Illness Experience," *Sociology of Health and Illness* 26(6): 845–74.

Kleiman, Jordan (forthcoming) "The Gods Must Be Crazy," *Technology and Culture*.

Kline, Benjamin (1997) *First Along the River: A Brief History of the U.S. Environmental Movement* (San Francisco: Acada Books).

Kling, Rob & Suzanne Iacono (1988) "The Mobilization of Support for Computerization: The Role of Computerization Movements," *Social Problems* 35(3): 226–43.

Kroll-Smith, Steve & H. Hugh Floyd (1997) *Bodies in Protest: Environmental Illness and the Struggle over Medical Knowledge* (New York: New York University Press).

Latour, Bruno (1988) *The Pasteurization of France* (Cambridge, MA: Harvard University Press).

Lerner, Barron (2001) *The Breast Cancer Wars: Hope, Fear, and the Pursuit of a Cure in Twentieth-Century America* (Oxford: Oxford University Press).

Ley, Barbara (forthcoming) *Assembling Breast Cancer: Activism, Science, and the Environment.*

Lovett, Tom, Chris Clarke, & Avila Kilmurray (1983) *Adult Education and Community Action: Adult Education and Popular Social Movements* (London: Croom Helm).

Lovins, Amory (1977) *Soft Energy Paths: Toward a Durable Peace* (New York: Ballinger).

MacKenzie, Donald & Graham Spinardi (1995) "Tacit Knowledge, Weapons Design, and the Uninvention of Nuclear Weapons," *American Journal of Sociology* 101(1): 44–99.

MacKenzie, Donald & Judy Wajcman (1999) *The Social Shaping of Technology* (Philadelphia and Buckingham: Open University Press).

Mander, Jerry (1984) "Six Grave Doubts about Computers," *Whole Earth Review* (January): 10–20.

Martin, Brian (1988) "Nuclear Winter: Science and Politics," *Science and Public Policy* 15(5): 321–34.

Martin, Brian (1991) *Scientific Knowledge in Controversy: The Social Dynamics of the Fluoridation Debate* (Albany: State University of New York Press).

Martin, Brian (1993) "The Critique of Science Becomes Academic," *Science, Technology & Human Values* 18(2): 247–59.

Martin, Brian (1997) "Science, Technology and Nonviolent Action: The Case for a Utopian Dimension in the Social Analysis of Science and Technology," *Social Studies of Science* 27: 439–63.

Martin, Brian (1999) "Suppression of Dissent in Science," *Research in Social Problems and Public Policy* 7: 105–35.

Martin, Brian (2001) *Technology for Nonviolent Struggle* (London: War Resisters' International).

Mattelart, Armand (1980) *Mass Media, Ideologies and the Revolutionary Movement* (Atlantic Highlands, NJ: Humanities Press).

McAdam, Doug (1982) *Political Process and the Development of Black Insurgency* (Chicago: University of Chicago Press).

McAdam, Doug, Sidney Tarrow, & Charles Tilly (2001) *Dynamics of Contention* (Cambridge, U.K.: Cambridge University Press).

McCarthy, John & Meyer Zald (1977) "Resource Mobilization and Social Movements: A Partial Theory," *American Journal of Sociology* 82: 1212–41.

McChesney, Robert (1993) *Telecommunications, Mass Media, and Democracy* (New York: Oxford University Press).

McChesney, Robert (2001) "Pacifica: A Way Out," *The Nation,* February 12. Available at: http://www.thenation.com/doc/20010212/mcchesney.

McNeil, Donald R. (1957) *The Fight for Fluoridation* (New York: Oxford University Press).

Melucci, Alberto (1980) "The New Social Movements: A Theoretical Approach," *Social Science Information* 19:199–226.

Moghadam, Valerie (2005) *Globalizing Women: Transnational Feminist Networks* (Baltimore, MD: Johns Hopkins University Press).

Mol, Arthur (2000) "The Environmental Movement in an Era of Ecological Modernization," *Geoforum* 31(1): 45–56.

Monahan, Torin (2005) *Globalization, Technological Change, and Public Education* (New York: Routledge).

Monahan, Torin (ed) (2006) *Surveillance and Security: Technological Politics in Everyday Life* (New York: Routledge).

Moody, Glyn (2002) *Rebel Code: Linux and the Open Source Revolution* (New York: Perseus Books).

Moore, Kelly (1996) "Organizing Integrity: American Science and the Creation of Public Interest Science Organizations, 1955–1975," *American Journal of Sociology* 101: 1592–627.

Moore, Kelly (1999) "Political Protest and Institutional Change: The Anti-Vietnam War Movement and American Science," in M. Giugni, D. McAdam, & C. Tilly (eds), *How Social Movements Matter* (Minneapolis: University of Minnesota Press): 97–115.

Moore, Kelly (2005) "Powered by the People: Scientific Authority in Participatory Science," in S. Frickel & K. Moore (eds) *The New Political Sociology of Science: Institutions, Networks, and Power* (Madison: University of Wisconsin Press): 403–41.

Moore, Kelly (2006) *Doing Good While Doing Science: Social Movements and Institutional Change in American Science, 1945–1975* (Princeton, NJ: Princeton University Press).

Moore, Kelly & Nicole Hala (2002) "Organizing Identity: The Creation of Science for the People," *Research in the Sociology of Organizations* 19: 309–35.

Morgen, Sandra (2002) *Into Our Own Hands: The Women's Health Movement in the United States, 1969–1990* (New Brunswick, NJ: Rutgers University Press).

Morris, Douglas (2004) "Globalization and Media Democracy: The Case of Indymedia," in D. Schuler & P. Day (eds), *Shaping the Network Society* (Cambridge, MA: MIT Press): 325–52.

Mullins, Nicholas (1972) "The Development of a Scientific Specialty: The Phage Group and the Origins of Molecular Biology," *Minerva* 10: 52–82.

Mundey, Jack (1981) *Green Bans and Beyond* (Sydney: Angus and Robertson).

Nathanson, Constance (1999) "Social Movements as Catalysts for Policy Change: The Case of Smoking and Guns," *Journal of Health Politics, Policy, and Law* 24(3): 421–88.

Noble, David (1993) *Progress without People: In Defense of Luddism* (Chicago: Charles H. Kerr).

Nowotny, Helga & Hilary Rose (eds) (1979). *Counter-Movements in the Sciences* (Dordrecht, Netherlands: D. Reidel).

Oaks, Laury (2001) *Smoking and Pregnancy: The Politics of Fetal Protection* (New Brunswick, NJ: Rutgers University Press).

Obach, Brian (2002) "Labor-Environment Relations: An Analysis of the Relations Between Labor Unionists and Environmentalists," *Social Science Quarterly* 83(1): 82–100.

Oudshoorn, Nelly & Trevor Pinch (2003) *How Users Matter: The Co-construction of Users and Technology* (Cambridge, MA: MIT Press).

Patterson, Walter (1977) *The Fissile Society* (London: Earth Resources Research).

Pauling, Linus (1958) *No More War!* (New York: Dodd Mead).

Pellow, David (2002) Garbage Wars: The Struggle for Environmental Justice in Chicago (Cambridge, MA: MIT Press).

Pellow, David & Lisa Sun-Hee Park (2002) *Silicon Valley of Dreams: Environmental Injustice, Immigrant Workers, and the High-Tech Global Economy* (New York: New York University Press).

Perin, Constance (2004) *Shouldering Risks: The Culture of Control in the Nuclear Power Industry* (Princeton, NJ: Princeton University Press).

Pichardo, Nelson (1997) "New Social Movements: A Critical Review," *Annual Review of Sociology* 23: 411–30.

Pierce, Steve (2003) *The Community Teleport: Participatory Media as a Path to Participatory Democracy*, Ph.D. diss., Rensselaer Polytechnic Institute.

Postigo, Hector (2005) "The Subversion of Digital Copyright through Technological Means," presentation at Law and Society Conference, Las Vegas.

Poynder, Richard (2004) "Ten Years After," *Information Today* 21(9). Available at: http://www.infotoday.com/it/oct04/poynder.shtml.

Privacy International and the GreenNet Educational Trust (2003) *Silenced: Censorship and Control of the Internet,* Privacy International and the GreenNet Educational Trust. Available at: http://www.privacyinternational.org/survey/censorship/.

Prokosch, Eric (1995) *The Technology of Killing: A Military and Political History of Antipersonnel Weapons* (London: Zed Books).

Purdue, Derrick (2000) *Anti-Genetix: The Emergence of the Anti-GM Movement* (Aldershot, U.K.: Ashgate).

Rabeharisoa, Vololona & Michel Callon (2004) "Patients and Scientists in French Muscular Dystrophy Research," in Sheila Jasanoff (ed), *States of Knowledge: The Co-production of Science and Social Order* (New York: Routledge): 234–53.

Raboy, Marc (1984) *Movements and Messages: Media and Radical Politics in Quebec* (Toronto: Between the Lines).

Rapp, Rayna (1999) *Testing Women, Testing the Fetus: The Social Impact of Amniocentesis in America* (New York: Routledge).

Rappert, Brian (2003) Non-lethal Weapons as Legitimizing Forces?: Technology, Politics and the Management of Conflict (London: Frank Cass).

Reed, Matthew (2002) "Rebels from the Drown Down: The Organic Movement's Revolt against Agricultural Biotechnology," *Science as Culture* 11(4): 481–505.

Reid, Roddey (2005) *Globalizing Tobacco Control: Anti-Smoking Campaigns in California, France, and Japan.* (Bloomington: Indiana University Press).

Riedijk, Willem (1986) *Technology for Liberation: Appropriate Technology for New Employment* (Delft, Holland: Delft University Press).

Roddewig, Richard (1978) *Green Bans: The Birth of Australian Environmental Politics* (Montclair, NJ: Allanheld, Osmun).

Rose, Fred (2000) *Coalitions Across the Class Divide: Lessons from the Labor, Peace, and Environmental Movements* (Ithaca, NY: Cornell University Press).

Sale, Kirkpatrick (1995) *Rebels Against the Future: The Luddites and their War on the Industrial Revolution—Lessons for the Computer Age* (Reading, MA: Addison-Wesley).

Scalmer, Sean (2002) *Dissent Events: Protest, the Media, and the Political Gimmick in Australia* (Sydney: University of New South Wales Press).

Schatzberg, Eric (1994) "Ideology and Technical Choice: The Decline of the Wooden Airplane in the United States, 1920–1945," *Technology and Culture* 35(1): 34–69.

Scheinberg, Anne (2003) "The Proof of the Pudding: Urban Recycling in North America as a Process of Ecological Modernisation," *Environmental Politics* 12(4): 49–75.

Schell, Jonathan (1982) *The Fate of the Earth* (New York: Knopf).

Schuler, Doug (2000) "New Communities and New Community Networks," in M. Gurstein (ed), *Community Informatics: Enabling Communities with Information and Communications Technologies* (London: Idea Publishing Group): 174–89.

Schurman, Rachel (2004) "Fighting 'Frankenfoods': Industry Opportunity Structures and the Efficacy of the Anti-Biotech Movement in Western Europe," *Social Problems* 51(2): 243–68.

Science for the People (1974) *China: Science Walks on Two Legs* (New York: Avon).

Sclove, Richard (1995) *Democracy and Technology* (New York: Guilford).

Sen, Gita & Rachel Snow (1994) *Power and Decision: the Social Control of Reproduction* (Cambridge, MA: Harvard University Press).

Shiva, Vandana (2000) *Stolen Harvest: The Hijacking of the Global Food Supply* (Cambridge, MA: South End Press).

Shostak, Sara (2004) "Environmental Justice and Genomics: Acting on the Futures of Environmental Health," *Science as Culture* 13(4): 539–62.

Smith, Merritt Roe & Leo Marx (ed) (1994) *Does Technology Drive History? The Dilemma of Technological Determinism* (Cambridge, MA: MIT Press).

Smith, Susan (1995) *Sick and Tired of Being Sick and Tired: Black Women's Health Activism in America, 1890–1950* (Philadelphia: University of Pennsylvania Press).

Stienstra, Deborah (2000) "Making Global Connections Among Women, 1970–1999," in Robin Cohen and Shirin Rai (eds), *Global Social Movements* (London: Athlone): 62–82.

Stoller, Nancy (1998) *Lessons from the Damned: Queers, Whores, and Junkies Respond to AIDS* (New York: Routledge).

Taylor, Verta (1996) *Rock-a-by Baby: Feminism, Self-Help, and Postpartum Depression* (New York: Routledge).

Taylor, Verta & Nancy Whittier (1992) "Collective Identity in Social Movement Communities," in A. Morris and C. Mueller, *Frontiers of Social Movement Theory* (New Haven, CT: Yale University Press): 104–29.

Terry, Jennifer (1999) *An American Obsession: Science, Medicine, and Homosexuality in Modern Society* (Chicago: University of Chicago Press).

Thompson, E. P. (1963) *The Making of the English Working Class* (New York: Vintage).

Tilly, Charles (1978) *From Mobilization to Revolution* (Reading, MA: Addison-Wesley).

Touraine, Alaine (1992) "Beyond Social Movements?" *Theory, Culture, and Society* 9(1): 125–45.

Treichler, Paula (1996) *How to Have a Theory in an Epidemic: Cultural Chronicles of AIDS* (Durham, NC: Duke University Press).

Turner, Fred (2005) "Where the Counterculture Met the New Economy: The WELL and the Origins of Virtual Community," *Technology and Culture* 46(3): 485–512.

Wajcman, Judy (1991) *Feminism Confronts Technology* (University Park: Pennsylvania State University Press).

Weber, Steven (2004) *The Success of Open Source* (Cambridge, MA: Harvard University Press).

Weinberg, Adam, David Pellow, & Allan Schnaiberg (2000) *Urban Recycling and the Search for Sustainable Community Development* (Princeton, NJ: Princeton University Press).

Willoughby, Kelvin (1990) *Technology Choice: A Critique of the Appropriate Technology Movement* (Boulder, CO: Westview Press).

Winner, Langdon (1977) *Autonomous Technology: Technics-out-of-control as a Theme in Political Thought* (Cambridge, MA: MIT Press).

Winner, Langdon (1986) *The Whale and the Reactor: A Search for Limits in an Age of High Technology* (Chicago: University of Chicago Press).

Wooddell, Margaret (2004) "Codes, Identities, and Pathologies in the Construction of Tamoxifen as a Chemoprophylactic for Breast Cancer Risk Reduction in Healthy Women at High Risk," Ph.D. diss., Rensselaer Polytechnic Institute.

Woodhouse, Edward & Steve Breyman (2005) "Green Chemistry as Social Movement?" *Science, Technology & Human Values* 30(2): 199–222.

Woodhouse, Edward & Joseph Morone (1988) *The Demise of Nuclear Energy? Lessons for the Democratic Control of Technology* (New Haven, CT: Yale University Press).

Woodhouse, Edward, David Hess, Steve Breyman, & Brian Martin (2002) "Science Studies and Activism: Possibilities and Problems for Reconstructivist Agendas," *Social Studies of Science* 32: 297–319.

Wynne, Brian (1996) "May the Sheep Safely Graze?" in S. Lash, B. Szerszynski, & B. Wynne (eds), *Risk, Environment, and Modernity* (Thousand Oaks, CA: Sage): 44–83.

Yearley, Steven (1992) "Green Ambivalence about Science," *British Journal of Sociology*, 43(4): 511–32.

Zavestoski, Stephen, Phil Brown, Meadow Linder, Sabrina McCormick, & Brian Mayer (2001) "Science, Policy, Activism, and War: Defining the Health of Gulf War Veterans," *Science, Technology & Human Values* 27(2): 171–205.

21   Patient Groups and Health Movements

Steven Epstein

In 1990, when I began work on a Ph.D. dissertation on AIDS research and AIDS activism, I was able to find little written from an STS or allied perspective on the politics and projects of patients groups, or on health movements more generally.[1] An article by Rainald von Gizycki about cooperation between the German Retinitis Pigmentosa Society and medical researchers, published a few years earlier in a sociology of science yearbook, had seen fit to comment on the novelty of studying any such interaction "from the point of view of the nonscientist rather than the scientist." Distinguishing his work from all the other contributions to the volume, he proposed to "look at the conditions prevailing inside the nonscientific group which have made it possible to exert influence on scientists, rather than the other way around" (von Gizycki, 1987: 75).

Fifteen years later, such an intention would hardly raise eyebrows, as explorations of these topics have mushroomed. Four different journals have published special issues devoted to the epistemic and practical projects of patient groups and health movements (Bonnet et al., 1998; Brown & Zavestoski, 2004; Hess, 2004a; Landzelius & Dumit, 2006), and an edited volume also takes up this theme (Packard et al., 2004b). Several of these collective endeavors originated out of conferences that brought together scholars studying these topics in many different countries. From abortion to vaccines, from preterm babies to Alzheimer's, from intersexuality to alternative medicine, analysts now have studied an extensive and extraordinarily diverse range of cases that span the human life cycle and shed light on nearly every conceivable aspect of the politics of health, illness, and biomedical research.

Patient groups and associated health advocacy organizations pose crucial questions for scholars in the field of STS. How do "disease constituencies" arise, how do they forge "illness identities" as a collective accomplishment, and how do they use those collective identities as the basis for political mobilization? How do new developments in the biomedical sciences serve to "carve out" new groupings of individuals, in ways that provide unanticipated bases for identity formation or social affiliation? How do the actions of patients or their lay representatives change the way that medicine is practiced, health care services are distributed, biomedical research is conducted, and medical technologies are developed? What is the character of the experiential knowledge of illness possessed or cultivated by patient groups or health movements? What sorts of challenges do these lay actors pose to the authority of credentialed experts, and what kinds of alliances with professionals do they construct? What sorts of "politics of the body" do such groups put into practice, and how are bodies transformed as a result? When does health activism of this kind result in the extension of medicalized frames of understanding, and when does it contest such medicalization? How do patient groups intervene in the web of relationships that connect biomedical institutions both with the market and with the state? What are the effects of these groups on the vast social inequalities that characterize the field of health and health care? What conceptions of medical science do patient groups promote and contest, and what visions do they articulate of what it means to be healthy?

I offer different vantage points for viewing this burgeoning body of literature and its answers to the questions listed above. First, I suggest that the surge of interest in this topic within STS cannot be taken for granted, and I consider why it might be that studies of patient groups and health movements have proliferated within STS in recent years. Second, I look more closely and critically at the definitional question: Just what is the object of study here, and what are its boundaries? Third, I briefly describe the different research methods that have been used by STS scholars to study patient groups and health movements. Fourth, in place of a formal typology, I propose a number of different criteria by which we might usefully compare and contrast different patient groups and health movements. Fifth, I consider three key research questions that have emerged in relation to the emergence and functioning of these groups. Sixth, I examine the effects or consequences of patient groups and health movements. Finally, I suggest some potentially useful directions for future scholarship.

WHY THIS, WHY NOW?

The recent efflorescence of intellectual activity in relation to patient associations and health advocacy poses an interesting question in its own right, a question in the sociology of knowledge: Why the burst of scholarly attention to this topic at this particular time? No doubt it reflects, in part, the growing prominence of the phenomenon itself. On the one hand, it is worth emphasizing that group formation and activism of this kind is by no means new. In the United States, for example, voluntary national health associations such as the National Tuberculosis Association and the predecessor of the American Cancer Society were inventions of the early years of the twentieth century (Talley, 2004: 40) and self-help groups in the "12-step" mold, such as Alcoholics Anonymous, followed a few decades later (Rapp, 2000: 193) while other sorts of group-specific health activism with enduring legacies, such as women's health movements, date back to the nineteenth century (Weisman, 1998). On the other hand, many commentators have noted the sheer quantitative increase in such organizing in recent years as well as its enhanced social visibility (Katz, 1993: 1; Kelleher, 1994: 105; Rapp et al., 2001: 393; Rabeharisoa, 2003: 2127; Allsop et al., 2004: 738, 741).[2] Qualitative changes may also be heightening the salience of this social form. Rapp, Heath, and Taussig have observed the tendency toward "marriages, mergers, and traffic among these organizations" in recent years (Rapp et al., 2001: 392) while Allsop, Jones, and Baggott, in reference to the United Kingdom, have pointed to the diffusion of "shared values and norms across condition areas" as well as the emergence of a common discourse across groups (Allsop et al., 2004: 745). Although the bulk of studies to date have tracked these developments in the United States, the United Kingdom, and France, existing analyses suggest a trend that is, if not global, then well represented at least in many Western countries and in Japan (Matoba, 2002). Indeed, the shortage of analyses of patient groups and health movements in other parts of the globe almost certainly reflects a research gap to be remedied rather than an absence of the phenomenon on the ground. In fact, transnational alliances increasingly are connecting health advocates in the global "South" with their counterparts in the "North" (Whyte et al., 2002: 146–60; Bell, 2003; Barbot, 2006: 549–50; Hardon, 2006).

It is understandable, then, that this proliferation of patient groups and health movements has attracted scholarly attention, especially because of a widespread sense of their consequence—indeed, a perception that such groups often have been successful in their goals. As a number of commentators have suggested, this upsurge of health- and disease-based organizing reflects the prevalence in recent decades of more skeptical attitudes toward doctors, scientists, and other experts, trends that also have manifested themselves in new conceptions of patients' rights and renewed concerns with bioethical debates (Brown & Zavestoski, 2004: 682). Many scholars also have associated recent patient groups and health movements with the more general expansion of rights-based movements and of so-called new social movements since the 1960s (Shakespeare, 1993; Kelleher, 1994: 113; Epstein, 1995: 412–13; 1996: 20–21; Kaufert, 1998: 303; Layne, 2003: 38–39; Silverman, 2004: 361, 370; Blume, 2006: 630–31; McInerney, 2006: 654–55).[3]

At the same time, the fact that scholarly literature on patient groups and health movements has flourished in the past fifteen years also says something about the field of STS and its own trajectory of development. During that time scholarly work in STS has moved decisively "beyond the lab" to analyze—in all their messiness, variability, and volatility—the broader dimensions of public engagement with science and technology. Patient groups and health movements have proved to be more than incidental objects of attention for analysts seeking to understand how and why it is that in a technoscientific world, "without public participation, things simply fall apart" (Elam & Bertilsson, 2003: 243). Rescuing us from the vague and hopelessly undifferentiated notions of "the public" or "the public sphere" that all too often are invoked in discussions of the "public understanding of science," patient groups and health movements are—at least by comparison—specific, concrete, and locatable entities, well available for study. Moreover, the passion and moral fervor that often animates them makes them especially interesting as exemplars of the new kinds of subjectivities that STS has encompassed within its scholarly embrace.

A range of recent work of broad significance to the overall field of STS has found it useful to focus attention on the specific phenomena of patient groups and health movements in order to make more general points. Michel Callon and Vololona Rabeharisoa have treated *associations de malades* as exemplary manifestations of "concerned groups"—nonscientists conceived of as "(potentially) genuine researchers, capable of working cooperatively with professional scientists," whose dramatic growth in recent years has sparked new varieties of "research in the wild" (Callon & Rabeharisoa, 2003: 195; see also Callon, 2003). Similarly, Bruno Latour has pointed to the engagement of patient groups with biomedical research as emblematic of the "collective experiments" by which science policy is now generated (Latour, 1998). Patient groups and health movements also figure as prominent examples in the line of STS research that Sheila Jasanoff has termed "co-production" studies (Jasanoff, 2004); in the formulation of new notions of "scientific citizenship" (Elam & Bertilsson, 2003; Irwin & Michael, 2003); in the renewed emphasis on the institutional, structural, and political dimensions of science and the social order that Scott Frickel and Kelly Moore have promoted as "the new political sociology of science" (Frickel & Moore, 2005); and in the proclamation by Harry Collins and Robert Evans (2002) of a "third wave" of science studies that reconceives the nature and boundaries of expertise. From the rubric of technology studies, analysts have recognized patient groups and health movements as an important subtype of the "relevant social groups" described by Bijker, Hughes, and Pinch in their canonical work on how the trajectories of technological development are shaped (Pinch & Bijker, 1993: 30–34; Blume, 1997: 46) while the recent scholarly emphasis on "users" of technologies has provided an additional entry point for assessing the contributions of advocacy groups (Oudshoorn & Pinch, 2003). In all these ways, from all these diverse perspectives within STS, patient groups and health movements in recent years have proven remarkably "good to think with."

However, none of this is sufficient to account for the particular emphasis on questions of biomedicine, health, and illness. Patient groups and health movements have come to be of growing interest to STS scholars not just in response to broad debates about public engagement with science but also because of their centrality to the processes by which bodies, diseases, and life itself are being remade by the biomedical revolutions of recent years. On the one hand, the rise of interest in health activism reflects the more general movement of biomedical topics from the relative periphery to the very center of attention within STS over the past fifteen years. On the other hand, during those same years, as Adele Clarke and coauthors have described, medicine itself has been transformed "from the inside out" (Clarke et al., 2003). Through innovations in molecular biology, genomics, bioinformatics, and new medical technologies; through the intensification of clinical research practices; through vast increases in public and private funding for biomedical research; through the ascendance of evidence-based medicine and the growing prominence of techniques of standardization and rationalization in medicine; through the development of neoliberal approaches to health that promote new modes of governing bodies and populations; through the rapid expansion of a global pharmaceutical industry constantly searching for new markets and engaging in new ways with consumers; through the resurgence of dreams of human enhancement or perfectibility by means of biotechnologies;

and through the dominance in the United States of managed care as a system attempt-ing to rationalize and ration health care delivery, the world of medicine has to some significant degree been refashioned in ways that impinge (variably) upon the every-day experiences and practices of people around the globe (Berg & Mol, 1998; Lock et al., 2000; Rose, 2001; Franklin & Lock, 2003; Keating & Cambrosio, 2003; Rothman & Rothman, 2003; Timmermans & Berg, 2003; Conrad, 2005; Lakoff, 2005).

The increasing "disunity" of medicine (Berg & Mol, 1998; Barbot & Dodier, 2002) makes it hard work to comprehend all these shifts. Yet an analysis of patient groups and health movements is crucial for understanding the consequences of these mani-fold biomedical transformations, especially including the resistances that have arisen in response to them. We live in a world characterized by what Nikolas Rose has called "vital politics," in which "selfhood has become intrinsically somatic," and in which "biopolitics now addresses human existence at the molecular level" (Rose, 2001: 16, 18). Categories of personhood are being reconstructed by new medical technologies (Dumit, 1997), and new practices of research, care delivery, and risk profiling cut across (or remake) populations in widely divergent ways—sometimes shoring up, sometimes reconfiguring profound disparities in health care and health outcomes according to social class, race, ethnicity, gender, sexuality, region, and nation. Yet as Clarke and coauthors rightly insist, new biomedical developments cannot be understood only in top-down fashion: we must simultaneously be on the lookout for "new forms of agency, empowerment, confusion, resistance, responsibility, docility, subjugation, cit-izenship, subjectivity, and morality" that emerge from dispersed social locations in response to such changes (Clarke et al., 2003: 184).

Recent biomedical developments have thrust into view new outcroppings of agency and resistance in at least two ways that are well reflected in the new scholarship on patient groups and health movements. First, drawing on Paul Rabinow's descriptions of "biosociality" (Rabinow, 1996: 91–111), several analysts have expanded the concept of citizenship to describe the practices that link bodies, individuals, groups, and nations together—or that separate the biosocially privileged from the excluded—at the biological or genetic level (Petryna, 2002; Briggs & Mantini-Briggs, 2003; Heath et al., 2004; Rose & Novas, 2005; Epstein, 2007). Patient groups and health movements have been pivotal actors in the making and the unmaking of these new varieties of so-called biological, biomedical, biopolitical, or genetic citizenship.

Second, the diverse politics of feminism and women's health not only crisscross the new biomedical landscape but also are implicated within the rise of patient groups and health movements to an astonishing degree. In reviewing the literature in prepa-ration for writing this chapter, I quickly recognized what a hefty proportion of the recent research is devoted to understanding groups that concern themselves with women's bodies and women's health—particularly including breast cancer activism, which is now the most extensively researched of any health movement from an STS perspective, but also abortion, reproductive and contraceptive technologies, preg-nancy loss, postpartum depression, and menopause, among others.[4] Of course, this scholarly emphasis is indicative not only of the social centrality of these issues but also of the distinctive impact of feminist theory and politics on several generations of STS researchers.

   Thus, the recent scholarly interest in patient groups and health movements reflects both the growing salience of the analytical object and the larger transformations of the biosciences and the political environment at the same time as it tracks broader substantive shifts in emphases and concerns within STS. Although a genealogy of studies of patient groups would locate much of the earliest scholarship well outside the field of STS (Stewart, 1990; Borkman, 1991; Chesler, 1991; Katz, 1993; Kelleher, 1994; Lavoie et al, 1994), at present STS is perhaps the largest contributor. Yet this claim is in some respects misleading, because—just as it has become harder in general to place boundaries around the field of STS in recent years—the STS scholarship on patient groups and social movements reflects creative fusions and cross-fertilizations, especially with medical anthropology, the sociology of health and illness, and the sociological study of social movements.[5] Indeed, the new STS work on these topics may be one of the chief pathways by which the field of STS is having an influence on these other fields—as evidenced by the publication of special issues in mainstream journals of medical sociology (*Sociology of Health & Illness* and *Social Science & Medicine*). Familiar sociological and anthropological concepts such as the illness experience, the doctor-patient relationship, collective identity, and mobilization are being reworked through conversation with STS approaches. As Kelly Moore has observed, studies of activist challenges to medicine are proving an important exception to the tendency in social movement scholarship to presume that movements are worth studying only when they take on the state (Moore, 1999). More generally and more ambitiously, the study of patient groups and health movements provides STS scholars with an appropriate vehicle for explaining to scholars in other fields the broader relevance of technoscientific developments for the understanding of important theoretical concepts, such as collective identity, solidarity, personhood, and embodiment.

WHAT IS THE OBJECT?

So far I have been using the phrase "patient group" as if its meaning and referents were clear and unequivocal. Yet the body of scholarship I am reviewing here has tended to burst the bounds of the category in several ways. First, quite a few advocacy groups that have been studied under this general rubric are organized not by patients per se but by various sorts of "proxies" for patients. These may be parents, relatives, or partners in cases in which the actual patient is too young or too physically or mentally incapacitated to advance his or her own interests (Beard, 2004: 798); they may sometimes be activists who may or may not have the disease or condition in question and whose interests may not precisely coincide with the larger group of patients or users of medical technologies (Epstein, 1995, 1996: 252–53; Van Kammen, 2003); or they may be advocates speaking on behalf of broad constituencies (such as "women's health") whose interests transcend any specific disease (Epstein, 2003a). The point is not to exclude cases of these kinds as legitimate instances of the phenomenon under study but rather to call attention to the very practices of representation by which spokespersons come to stand in for a group—a task for which STS is well suited. In other cases, patienthood itself may be a murky status. In her analysis of infant loss support groups, Linda Layne has observed that it is "surprisingly difficult" to say whether there is a patient present: "By the time a loss has occurred, the embry-onic/fetal/neonatal patient is no more and the woman is no longer pregnant, and therefore no longer an active obstetrical patient" (Layne, 2006: 603). Here again, rather than quibble over who really qualifies as a patient, we would be advised to embrace elastic classifications and ask what we can learn from the juxtaposition of examples.

In practice, many analysts of patient groups simply have not found it possible or meaningful to discuss the phenomenon apart from consideration of broader cate-gories, such as "health social movements" (Brown & Zavestoski, 2004; Hess, 2004b), "consumer movements" in health (Bastian, 1998; Allsop et al., 2004; Rosengarten, 2004); the practice of organizing around "pain and loss experiences" (Allsop et al., 2004: 738); and the political projects advanced by "communities of suffering" (Packard et al., 2004a). Furthermore, the study of patient groups shares blurry boundaries with still other sorts of phenomena: science advocacy movements pressing for research on specific biomedical topics, such as stem cells (Ganchoff 2004); movements advocat-ing democratic participation in priority-setting for public funding of medical research (Dresser, 2001); ecological and environmental justice movements that have significant health implications (Pellow & Park, 2002; Allen, 2003; Brown, Mayer et al., 2003; Allen, 2004; Hess, 2004a,c; Shostak, 2004); movements for new therapeutic directions, such as efforts to advance complementary and alternative medicine (Goldner, 2001, 2004; Hess, 2004b); and movements that work with private-sector firms to develop alternative health products (Hess, 2005). In other cases, such as the French muscular dystrophy association (AFM) studied by Rabeharisoa and Callon, affinities to social movements may be less relevant than similarities to large, formal organizations: the AFM employs more than 500 workers and has an annual budget of nearly 80 million Euros (Rabeharisoa, 2003: 2130).

On the ground, the actors participating in these collectives are defining themselves in an expanding variety of ways—indeed, there is probably a complex interaction between the spread of analytical categories, on the one hand, and of the self-descriptions mobilized by the groups and movements, on the other. Rather than attempt any exclusionary boundary work, I prefer to follow both the analysts and the actors as they increasingly think outside the box of "patient groups," in the narrow sense of the term, so as to draw connections as well as contrasts across a diverse range of cases.[6] In the rest of this chapter, my use of the term "patient groups and health movements" is meant as shorthand to invoke this broader array.

Three methodological implications follow from this expansion of analytical focus. First, even while we are catholic in acknowledging diverse ways of framing the con-ceptual object, it seems important to consider how different terminological choices affect the mapping of the intellectual terrain. According to Hilda Bastian in her analy-sis of the rise of consumer advocacy in health care, "people can argue for hours over whether we are 'consumers,' 'users,' 'patients,' 'clients,' or any other term from a list of favorites," suggesting that the nomenclature is on some level arbitrary and that we can rescue words such as "consumer" from any negative associations they might bear (Bastian, 1998: 3–4). Others, however, have insisted that words inevitably bring meanings along with them. In his study of tobacco control, Roddey Reid has critically analyzed the consequences of viewing the targets of health promotion campaigns as "consumers," arguing that the displacement of more substantive notions of citizenship by a market model of social relations is consistent with the rise of neoliberal approaches to managing the health of populations (Reid, 2004). Similarly, those who consider themselves to belong to a "health movement" may be more likely to link their concerns to questions of power, participation, and democracy than those who see themselves as part of a "patient group."

Second, it becomes important to take the hybrid and boundary-crossing character of patient groups and health movements as an explicit object of study—much as researchers increasingly have been doing. Many such groups are hybrid insofar as they blur the divisions not only between "expert" and "lay," but also between "civil society," "the state," and "the market," and—of course—between "science" and "politics" (Epstein, 2001). Observation of patient groups and health movements reveals, on the one hand, cases of patients and activists behaving like scientists or doctors (Epstein, 1996; Anglin, 1997; Myhre, 2002) and, on the other hand, cases of scientists or doctors behaving like activists (McCally, 2002; Frickel, 2004) or like patients (Mykytyn, 2006). In still other cases, such as the abortion rights movement in the United States, the movement itself encompasses both lay and professional actors (Joffe et al, 2004). McCormick, Brown, and Zavestoski have proposed the term "boundary movements" to describe such blurrings (McCormick et al., 2003), thereby usefully connecting the study of these characteristics of patient groups and health movements to related STS concepts concerned with boundaries (Gieryn, 1983; Star & Griesemer, 1989; Guston, 2000). Somewhat similarly, I have proposed that Mark Wolfson's concept of "interpenetration" (developed in his study of the antitobacco movement) is helpful in describing those cases where it is systematically unclear "where the movement ends and the state begins" (Wolfson, 2001: 145; see also Epstein, 2005; Klawiter, unpublished).[7] Though not using the term, analysts have revealed state/movement interpenetration to be a defining characteristic of a number of national health advocacy groups in the United States in their formative relationship with specific branches of the National Institutes of Health (Fox, 1989; Talley, 2004: 58). A recent example is the intimate relationship between the Genetic Alliance (a super-group of genetic support groups) and the NIH's Office of Rare Diseases, established in 1993 (Rayna Rapp, personal communication).

The third methodological implication of the move beyond any narrow consideration of patient groups is the importance of locating multiple patient groups and health movements in relational terms, both diachronically and synchronically. Sometimes this has been done by examining the "diffusion" or "spillover" effects of one movement upon another (Meyer & Whittier, 1994; McAdam, 1995)—analyzing how emergent groups adopt and adapt the frames, strategies, or action repertoires of previous movements or organizations (Epstein, 1996: 12, 347–48; Karkazis, 2002: chapter 9; Brown et al., 2004: 65, 68). In other cases, STS scholars have examined the simultaneous impact of different patient groups or health movements on the same issue, as in Stefan Timmermans and Valerie Leiter's analysis of how FDA hearings on the revival of thalidomide as a legitimate treatment brought into competition the perspectives of the Thalidomide Victims Association of Canada, HIV/AIDS activists, women's health advocates, and representatives of people with leprosy (Timmermans & Leiter, 2000; for a different example, see Clarke & Montini, 1993). Most ambitiously, a number of scholars have sought to locate patient groups and health movements within "fields" of activity (in some cases borrowing on Bourdieu's general theorization of fields of practice (Bourdieu, 1985) and in other cases adopting Raka Ray's more specific concept of "fields of protest" (Ray, 1999). For example, Nick Crossley has analyzed the relatively autonomous "field of psychiatric contention" in the United Kingdom, within which "organizations variously compete, cooperate, agree, disagree, debate and take up positions relative to one another" (Crossley, 2006: 562); Maren Klawiter has described the different "cultures of action" present within the field of breast cancer activism in the San Francisco Bay Area (Klawiter, 1999, 2000) as well as transformations in the institutional field of mammography screening (Klawiter, unpublished) and syntheses across the fields of cancer activism and environmental activism (Klawiter, 2003); and Chris Ganchoff has located embryonic stem cell movements and countermovements within a larger "field of biotechnology," understood as "an imagined space within which various politicized collective illness identities exist" (Ganchoff, 2004: 760). In all these ways, scholars have been moving beyond the analysis of the patient group in isolation to examine the institutional and cultural webs in which they are multiply entangled.

METHODS

STS researchers studying patient groups and health movements have been employing an increasingly diverse mix of data sources and specific research techniques (Brown & Zavestoski, 2004: 690). These include single- and multi-sited ethnographic methods, content analysis, questionnaires, focus groups, and textual analysis. A few researchers have taken a biographical approach that emphasizes the stories of key individuals (Lerner, 2001; Klawiter, 2004), and a few have employed computerized tools of network analysis (Rabeharisoa, 2006).

Not surprisingly, given the growing significance of the Internet for the formation and maintenance of patient groups, there has been a parallel interest in obtaining and analyzing various forms of on-line data. While mostly this research has consisted of on-line ethnographic observation of listserves and newsgroups (Goldstein, 2004) or content analysis of Web sites (Novas, unpublished), others have taken less typical approaches. In her study of a breast cancer organization that emphasized web-based communication, Patricia Radin also analyzed server logs and interviewed the

Webmistress by phone, email, and in person (Radin, 2006). Scholars also have begun to experiment with the use of computerized information tools to manage the massive quantities of data available online in sources such as newsgroup archives (Dumit, 2006: 581).

For a number of scholar/activists, the study of patient groups and health movements has raised important methodological (and ethical and political) questions about conducting forms of research when the analyst is also an actor in the movement or organization. For example, in her study of the transnational controversies surrounding new contraceptive technologies, Anita Hardon "wore both hats, participating in debates [as a women's health advocate] and taking meticulous notes on the actions and reactions as a researcher." In her written text, Hardon noted each occasion on which she participated as an advocate (Hardon, 2006: 619 note 8). Layne also has considered such issues in her recent work on infant loss support groups (Layne 2003, 2006). Layne's analysis was based on extensive field research, but it additionally reflected her personal experience of having miscarried as well as her emerging role as an activist in the movement (Layne, 2006).

TYPOLOGIES, TYPOLOGIES

In an attempt to impose order on the mix of collectivities encompassed under the rubric of patient groups and health movements, researchers have suggested a number of helpful (if potentially competing) typologies. One set of distinctions that has been widely cited is that drawn by Phil Brown and coauthors between *health access movements* concerned with the equitable provision of health care services; *constituency-based health movements* that focus on the health agendas of large, socially visible groups, such as those defined by gender, race, ethnicity, or sexuality; and *embodied health movements* that "address disease, disability or illness experience by challenging science on etiology, diagnosis, treatment and prevention" (Brown et al., 2004: 52–53; see also Zavestoski et al., 2004c). This classification is useful especially for shining a spotlight on the third category—embodied health movements—which in practice has been the primary concern of STS scholars.[8]

The question is how well these categories serve to distinguish specific movements. Brown and coauthors note that their terms are only ideal types, and they acknowledge that some health movements may have characteristics of more than one category (Brown et al., 2004: 53). Still, it is striking how many real-world examples cross the typological lines. For example, Layne has described how the infant loss support movement, although in many ways emblematic of what Brown and coauthors would call an embodied health movement, has also been concerned with extending the movement to "underserved communities"—thereby taking on tasks associated with health access movements and constituency-based health movements (Layne, 2006: 605). Similarly, Alondra Nelson's excavation of the little-known history of the Black Panther party's involvement in health activism has depicted a movement whose forms of engagement with health issues cut across the tripartite division described by Brown and coauthors (Nelson, 2003). The risk, then, is that an overly rigid adherence to this typology could conceivably lead to oversimplified and inaccurate understandings of internally diverse social movements or a truncation of their complex histories—for example, imagining that "embodied" AIDS activists were not also concerned with issues of access to health care (Hoffman, 2003).

By contrast with Brown and coauthors, the typology proposed by Rabeharisoa and Callon is concerned less with the structural characteristics of groups and more with their orientations toward biomedical partnerships. Rabeharisoa and Callon divided patient associations into three kinds (all of which, however, might be seen as subtypes of embodied health movements). *Auxiliary associations* support the biomedical research process, but either leave it up to credentialed experts to decide which topics to research or else "[set] about acquiring the necessary knowledge to be able to enter into discussion with them" about research priorities. *Partner associations* play a more substantial role in the organization of medical research in ways that often include, but go beyond, directly raising funds for research. Unlike the other types, *opposing associations* simply want nothing to do with medical specialists (Rabeharisoa & Callon, 2002: 60–63). This typology is also helpful, although the somewhat elastic definition of auxiliary associations to encompass groups that acquire their own medical expertise leaves it a bit unclear exactly how they can be distinguished in practice from partner associations.

With both of these typologies, there is also the risk that taking disease-specific groups as the unit of analysis can falsely incline us to imagine an internal homogeneity and to disregard crucial, cross-cutting divisions by other categories of identity, such as race, class, gender, and sexuality (Cohen, 1999). It may sometimes be quite problematic to assume "that disease is the great leveler," as Lisa Cartwright has warned, since "the experiences and cultures of illnesses . . . are always lived through identity positions and arenas of public and professional discourse that exceed the frameworks and cultures of disease" (Cartwright, 2000: 121–22).

More generally, it may be wise to be skeptical of the idea that any single, unidimensional typology adequately can capture the variation of patient groups and health movements: each well-posed research question about patient groups will generate a unique classificatory scheme that chops up the universe of cases in a distinctive way. The point, then, is to consider what some of those important questions might be, and by that route to examine the key dimensions along which patient groups and health movements may vary. Several examples are worth considering.

Relationship to Medicalization[9]
Patient groups and health movements can be categorized according to their orientation toward the extension of medical frames of understanding. One fascinating family of cases that has been well represented in the recent literature concerns conditions such as chronic fatigue (Barrett, 2004; Dumit, 2006), fibromyalgia (Barker, 2002; Barrett, 2004), multiple chemical sensitivity (Kroll-Smith & Floyd, 1997; Dumit, 2006), Gulf War Syndrome (Zavestoski et al., 2002; Brown, Zavestoski et al., 2003; Zavestoski

et al., 2004b), post partum depression (Taylor, 1996), sick building syndrome (Murphy 2004b, 2006), and repetitive strain injuries (Bammer & Martin 1992; Arksey, 1994, 1998). These "illnesses you have to fight to get," as Joe Dumit aptly terms them (they have also been characterized as "contested emergent illnesses" [Packard et al., 2004a: 26] and as "medically unexplained physical symptoms" [MUPS; Zavestoski et al., 2004b]) have in common the demand on the part of sufferers that their mysterious conditions be publicly acknowledged as being legitimately of a medical nature. As Kristin Barker has suggested, sufferers of such conditions find themselves in a state of "epistemological purgatory"—an anxiety-provoking experience "in which they question their own sanity precisely because of their certainty about the realness of their experience in the face of public doubt" (Barker, 2002: 281). Often accused of having problems that are really just "in their head," these putative patients "cling to the biological" as a tactic of legitimation, insisting on the "realness" of their illnesses in biological terms (Taylor, 1996; Dumit, 2006: 17–18).

At the same time, a different and equally intriguing cluster of patient groups and health movements repudiate medicalization or seek to demedicalize their conditions. Studies of deaf activists who oppose the use of cochlear implants as an assault on deaf culture (Blume, 1997, 1999); of lesbian and gay liberationists who reject the definition of homosexuality as a mental disorder (Bayer, 1981); of intersex activists critical of pediatric specialists who insist on surgically resolving cases of "ambiguous" genitalia among newborns (Karkazis, 2002: chapter 9); of African-Americans with sickle-cell anemia who resist the racialization of the disease and the consequent pathologization of their racial identities (Fullwiley, 1998); of the sector of the mental patients' self-help movement that has embraced a position of "anti-psychiatry" (Morrison, 2004; Crossley, 2006); and of fat acceptance activists who challenge the discourse of an obesity "epidemic" and question epidemiological claims about the unhealthy effects of being overweight (Saguy & Riley, 2005) demonstrate the range of cultural resources that the unwillingly medicalized may bring to bear, in the hope of casting off the yoke of medical definitions and interventions and the normalizing judgments that underpin them. Studies of disability activism likewise have been keen to demonstrate the formation of new collective identities that partially or wholly reject the normalizing judgments of biomedicine about how human beings are supposed to look or behave (Shakespeare, 1993, 1999; Dowse, 2001; Rapp & Ginsburg, 2001).

Other demedicalizers, including various descendants of the feminist women's health movement, may be less concerned with resisting medical diagnoses or treatments than with asserting the capacity of women to exercise control over their own bodies (Copelton, 2004; Murphy, 2004a). Still other groups seek neither to claim an illness identity nor to reject one but rather to question or repudiate specific medical practices, such as vaccination (Blume, 2006) or vivisection (Elston, 1994). Thus, while medicalization and demedicalization as distinctive goals are useful terms for considering patient groups and health movements, in practice each term may encompass quite disparate examples. Moreover, many groups seeking to demedicalize their conditions nonetheless may invoke biomedical data and frameworks as part of their political argumentation—a tension that Heath, Rapp, and Taussig rightly label productive (Heath et al., 2004: 158). Conversely, groups that do not dispute the overall relevance of a biomedical framework for understanding their issue of concern—and that may accept that medical science "is the only (or most powerful) game in town" (Thompson, 2005: 238)—may still challenge particular medical projects or tendencies (Epstein, 1997a; Thompson, 2005; Hardon, 2006). Yet another tricky case is that of advocates for complementary and alternative medicine, who may accept medical definitions while rejecting conventional medical therapies (Goldner, 2001, 2004; Hess, 2004b). "Medicalization" and "demedicalization" capture something of what these various groups are up to, but the terms should be used with caution.

Constitution of the Group

A different way of categorizing patient groups and health movements looks critically at the group formation process: What is the pathway by which "groupness" comes into being? In some cases, groups emerge out of previously existing communities—such as military veterans in the case of people with Gulf War Syndrome (Zavestoski et al., 2002; Brown, Zavestoski et al., 2003; Zavestoski et al., 2004b) or gay communities in the case of many early AIDS activist groups (Epstein, 1995: 414–15, 1996: 10–14)—and their capacities to mobilize and their forms of engagement may be shaped significantly by those previous associations. In other cases, individuals with no previous connection to one another are inducted into group membership via biological, biomedical, or biotechnological processes that construct a new biosocial grouping—for example, associations formed by family members of people with genetic disorders (Rapp et al., 2001) or by the surviving kin of organ transplant donors (Sharp, 2001). Still other cases show the influence of corporations and markets in the constitution of groupness. For example, Carlos Novas has described how pharmaceutical company Web sites may deliberately "emulate the 'look and feel'" of sites produced by patient organizations, in an attempt to "create a sense of 'community' between affected persons and the company" (Novas, unpublished), while Kane Race (2001) and Marsha Rosengarten (2004) have analyzed the ways in which makers of antiretroviral drugs seek to shape the personal and collective identities of people with HIV/AIDS through "lifestyle" advertising. Finally, nation-states may sometimes play an important role in molding the identities of groups organized around illnesses (Larvie, 1999). These examples suggest the importance of studying the "looping effects" (Hacking, 1995: 34) by which external attributions about a group are taken up by the group and become constitutive of its members' identities. In addition, these examples suggest that collective illness identities are rarely stable over long periods of time. Not only do identities often evolve as groups embark on different biomedical and political projects, but the group's definition may itself be at stake in health controversies. For example, in her analysis of lesbian health advocacy, Sarah Wilcox has argued that the debate over lesbian health priorities in the United States in the 1980s and 1990s coincided with an equally heated debate over the boundaries of the category of "lesbian" (Wilcox, unpublished).

Social Organization

As already suggested by my discussion of what "the object" is, patient groups and health movements vary considerably in terms of the size of membership and finances; the geographic scope (local, national, or transnational); and the degree of formal organization, bureaucratization, and professionalization. In addition, some groups coexist or compete with different groups addressing the same condition, whereas others have the playing field to themselves. Still others, such as the various genetic disease support groups, may form organizational coalitions across genetic conditions (Heath et al., 2004). These basic differences have (at least loose) implications for how groups are governed, how leadership is constituted within them, how practices of participation and representation function within them, and how (or whether) new alliances are established across them.

Independence

To what degree does the group maintain an autonomous standing, and to what degree is it dependent upon, or fused with, other organizations, such as private firms, state agencies, professional associations, funding agencies, or nongovernmental organizations? For example, Orla O'Donovan has described a continuum of relations between patient groups and the pharmaceutical industry that includes such diverse orientations as "corporatist," "cautious cooperation," and "confrontational." As O'Donovan's research in Ireland suggests, these varying degrees of cooperation with, or autonomy from, industry have implications for patient group practices and sensibilities, though by no means in any automatic fashion (O'Donovan, 2007).

Militancy and Oppositionality

Yet another way of classifying patient groups and health movements focuses on the degree to which they mount militant challenges or seek to oppose the status quo. While many support groups adopt deliberately nonconfrontational styles and comply with the advice of medical professionals, others practice "organized noncompliance" (Emke, 1992) and cast their opposition in ways that Debbie Indyk and David Rier termed "self-help with a vengeance" (Indyk & Rier, 1993: 6). Often, militancy may be a consequence of urgency, as when a group confronts a fatal disease and perceives itself to be engaged in a life-and-death struggle. The oppositional character of a patient group or health movement also may conceivably depend on the degree to which laypersons rather than professionals dominate within the movement (von Gizycki, 1987: 85; Joffe et al., 2004), the extent to which the group resists being organized in a professionalized and bureaucratic fashion (Staggenborg, 1999), whether the group distances itself from the frameworks of Western allopathic medicine (Goldner, 2001, 2004; Hess, 2004b, 2005), and whether it articulates a clear alternative vision rather than simply rejecting the status quo (Hess 2004c). In addition, organizations with a genuine grassroots base may adopt a more oppositional repertoire than those with elite sponsorship—sometimes called "grass-tops" or "astroturf" advocacy (Dimock, 2003; O'Donovan, 2007). In the end, the choice made between agonistic and consensual approaches may greatly affect the kind of scientific citizenship that activists help forge (Elam & Bertilsson, 2003; see also Landzelius, 2006a).

An important question concerns the circumstances in which militancy is perceived as efficacious. For example, the AIDS activist group ACT UP became known for its radical politics and confrontational style, even if much of its success in transforming medical science stemmed from the melding of militant "outsider" and cooperative "insider" tactics (Epstein, 1996). Consequently, a number of groups, such as chronic fatigue activists, have styled themselves after ACT UP's militancy, though not always with the same success (Barrett, 2004). Other groups explicitly have sought to distance themselves from the aggressive image of ACT UP on the assumption that less "in-your-face" tactics would be *more* effective. These include mainstream breast cancer advocacy groups in the United States, which stressed their "ordinariness" and "moral worthiness" *vis-à-vis* "the public stereotype of the AIDS patient, gay, male, and radical" (Kaufert, 1998: 102; see also Myhre, 2002); parents of premature infants, who adopted metaphors of "generativity and affinity" in place of ACT UP's militaristic imagery (Landzelius, 2006a: 678); and advocacy groups for assisted reproductive technologies, which adopted a style of "motherly activism" that appeals both to the left ("reproductive choice") and to the right ("family-building") (Thompson, 2005: 238–39). The point is not that either being militant or being unthreatening is a universally efficacious tactic. Rather, different actors will perceive different strategic advantages accruing to these orientations, depending on the constraints that they face, as well as the specifics of the disease or condition in question, the stage in the movement's development, its perceived relationship to other visible movements within the "field," and the particular historical moment.

Goals

As a final example, it might be possible to construct a typology of patient groups and health movements based on the various sorts of goals that they pursue. The diverse goals of such groups include finding (or rejecting) medical cures; improving the quality of life of ill people; cultivating practical advice for the management of illness; raising funds for research; changing scientific and medical practices, priorities, or orientations; rejecting technoscientific approaches; opposing stigmatization and exclusion; and changing more diffuse cultural codes related to the meanings associated with health, illness, the body, and expertise. Needless to say, many if not most groups adopt more than one of these goals.

KEY RESEARCH QUESTIONS

Three sets of questions about patient groups and health movements stand out for the amount of attention they have received from STS scholars. These are (1) What kinds of social and technoscientific developments are implicated in the rise of patient groups and health movements? (2) How do particular aspects of the disease or condition affect the rise and trajectory of patient groups and health movements? (3) What conceptions of knowledge do these groups employ, how do they put their knowledge to use, and what kinds of expertise do they develop? I address these in turn.

Social and Technoscientific Developments

In understanding the rise of individual patient groups and health movements, several researchers have stressed the importance of studying "opportunity structures"—the political or cultural factors (more or less) external to the group itself that present it with opportunities or threats (Nathanson, 1999: 423; Goldner, 2001). These might include economic transformations and the rise of new technologies in the workplace (Bammer & Martin, 1992; Pellow & Park, 2002); the spread of political ideologies, such as neoliberalism (Crossley, 2006); changes in social and medical norms (Saguy & Riley, 2005); or changes in gender relations and gendered meanings (Montini, 1996; Klawiter, 1999, 2000; Zavestoski et al., 2004a; Gibbon 2007). Not infrequently the emergence of patient advocacy is linked to specific historical advances in biomedicine—for example, Patrick Fox pointed to a "shift [in the 1970s] in the biomedical conceptualization of Alzheimer's disease that allowed the inclusion of greater numbers of potential victims" as a crucial precursor for the development of national advocacy in relation to the disease in the United States (Fox, 1989: 59) while Landzelius has identified as a necessary precondition of the "parents of preemies" movement the relatively recent ability of neonatologists to push back the limits of the viability of fetuses to earlier and earlier gestational ages (Landzelius, 2006a: 672). It is worth noting that developments within academia *outside of* medical fields—for example, work done in the social sciences and humanities—have rarely been considered for their potential to prefigure or shape group formation or to provide health movements with critical tools. An interesting exception is Karkazis's analysis of how the founder of the Intersex Society of North America made use of the critiques of sex and gender categories that had been published by scholars such as Anne Fausto-Sterling, Suzanne Kessler, and Alice Dreger in order to contest the use of sexual surgeries in infancy to "treat" intersexuality (Karkazis, 2002: chapter 9).

One implication of attending to opportunity structures is that location matters. Several scholars have invoked Alexis de Tocqueville's well-known claims in the nineteenth century about the American propensity to form voluntary associations as part of an argument for why patient groups are so widespread in the United States (Talley, 2004: 41; Layne, 2006: 606). By contrast, Allsop, Jones, and Baggott have argued that the more centralized character of British political institutions, including the National Health Service but also a "centrally-regulated charity sector," have "encouraged the use of conventional channels" on the part of health consumerism in the United Kingdom, such as cooperative work with professional associations and close attention to the mainstream political process (Allsop et al., 2004: 751–52).

However, in a globally wired world, location doesn't *always* matter—at least not always to the same degree—and the birth and development of the Internet is another crucial background condition that explains much about how and why patient groups and health movements have taken particular forms in recent years (Gillett, 2003;

Loriol, 2003; Novas, 2003; Goldstein, 2004; Radin, 2006). Landzelius has gone so far as to call the "parents of preemies" movement a "direct descendant" of the Internet because of its historical dependence on the latter's "capacity to engender (virtual) community and to geographically untether information/disinformation" (Landzelius, 2006a: 669). Dumit has emphasized how the asynchronous character of Internet communication comes to the advantage of people suffering from conditions that make face-to-face, real-time communication more problematic (Dumit, 2006: 588). Indeed, the advent of Web-based communication may even lead to a decline in face-to-face group formation and the disappearance of "nonvirtual" groups, as Layne has documented in her study of the infant loss support movement over time (Layne, 2003, 2006: 602). Web-based interactions can have other powerful effects as well. Patricia Radin has described how specific features of breast cancer advocacy Web sites "gradually transform casual visits—'thin trust'—into the kind of 'thick trust' that generates social capital" (Radin, 2006: 597), and Diane Goldstein has analyzed how Internet-based support groups generate "their own separate and distinct medical culture" (Goldstein, 2004: 127). At the same time it is important to recall that not just access to the Internet but also the meaning that it acquires for users can vary considerably: Heath, Rapp, and Taussig, while observing how the Internet has transformed an older identity politics around health, also have warned of "the potential for a widening of the 'digital divide' in which expansion of technoscientific literacy among many increases the exclusion and isolation of those without access in both rich and poor countries" (Heath et al., 2004: 156).

Aspects of the Disease or Condition
In addition to the impact of background conditions, many factors specific to the group, or specific to the illness or social problem, can be quite important in determining the likelihood that a patient group takes shape, mobilizes, and attracts resources and public attention. The trajectory of a patient group or health movement can be shaped by whether those affected by the condition are numerous in the population or isolated and scattered, able-bodied or infirm, young or old, and socially privileged or disadvantaged (Epstein, 1995: 414; 1996: 10; Stockdale, 1999; Rabeharisoa, 2003; Allsop et al., 2004; Layne, 2006). Moreover, not every disease is equally likely to promote patient organizing. In some cases, an outbreak of illness can spark a "biographical disruption" that motivates affected individuals to become active; in other cases—Allsop, Jones, and Baggott point to the example of circulatory disease—illness "does not appear to arouse feeling of anger and resentment, or pose a threat to identity" in such a way as to promote group formation (Allsop et al., 2004: 741, 744; see also Shim, 2005: 429). At the same time, we have little understanding of why a given illness will motivate some people, but not others, to join groups or movements. Clearly, not every sufferer is equally likely even to claim an illness identity, let alone want to be enrolled in a condition-appropriate movement, yet few scholars have attempted to compare those who join patient groups with those who do not (for a partial counterexample, see Rapp, 2000: 202–4).[10]

Case studies have presented poignant depictions of the particular dilemmas confronting those who seek to organize around certain conditions. For example, Chloe Silverman has described how the stigma of autism often spills over onto the parents of autistic children to the detriment of their organizational efforts: they may be perceived as incompetent parents whose advocacy on behalf of their children therefore cannot fully be credited as reasonable (Silverman, 2004). And Renee Beard has analyzed the peculiar plight of people with Alzheimer's disease who, even when mentally competent and functional, are presumed incapable of advocating for themselves (Beard, 2004). Others have tried to generalize across cases to suggest broader patterns. In the introduction to their edited collection, Randall Packard and coauthors concluded that the rapidity of the social response to an emerging illness may depend on a range of factors, including the epidemiological significance of the condition, the availability of an unequivocal diagnostic test, the social class of the sufferers, the degree of activism, and the extent of media coverage (Packard et al., 2004b: 22–23). In their article on embodied health movements, Brown and coauthors extracted a series of predictions: that sufferers of not-yet-medicalized conditions like chronic fatigue will face an uphill battle compared with those with medically accepted diseases; that those with links to previous social movements will have an easier time mobilizing than those without such links; that members of socially disadvantaged groups, such as women and minorities, may be more inclined toward activism while being less likely to have access to the requisite resources; and that, everything else being equal, the absolute numbers of people touched by an illness will affect the chances of successful mobilization (Brown et al., 2004: 73–74).

Much work also suggests that patient groups can solidify their claims to authority when they succeed in constituting themselves as an obligatory passage point (Latour, 1987: 132) through their control over access to a resource desired by researchers, whether that be the bodies of patients who might enroll in clinical trials (Epstein, 1995: 420), blood and tissue samples (Taussig et al., 2003: 63), information about family genealogies (Nukaga, 2002: 59), or funding to conduct research (Rabeharisoa, 2003; Kushner, 2004). However, Emily Kolker is right to point out that scholars have tended to emphasize the potency of the structural resources available to patient groups and health movements while underplaying the significance of cultural resources, such as the development of distinctive, "culturally resonant frames to persuade audiences" (Kolker, 2004: 821; see also Epstein, 1997b).

Conceptions and Uses of Knowledge and Development of Expertise

Patient groups and health movements have been especially fertile sites for studying the manufacture and deployment of various sorts of informal knowledge and for the development of alternative bases of expertise. Drawing on concepts such as "local knowledge" (Geertz, 1983), "subjugated knowledges" (Foucault, 1980: 80–85), "situated knowledges" (Haraway, 1991: 183–201), and "ways of knowing" (Pickstone, 2000)—as well as on classic STS studies of knowledgeable lay groups (Wynne, 1992)—scholars (too many to list by name) have explored in considerable detail the capacities of organized collectives of lay actors to assess medical knowledge claims and engage with the practices of biomedical knowledge production.[11] Some of this work has emphasized how formal medical knowledge is often parasitic on patient experience. For example, Emma Whelan's analysis of attempts to develop standardized pain measurement tools for endometriosis has revealed that such tools can render comparable only "*accounts* of pain, not the pains themselves"; and the ineliminable character of patients' experiences has promoted "epistemic cooperation" between endometriosis support groups and researchers (Whelan, 2003: 464, 477).

More generally, much work has examined how being the sufferer of an illness—or a member of an affected community (Escoffier, 1999)—can serve as epistemic grounding for developing distinctive, embodied knowledge claims. Similarly, a number of scholars (many of them building on Brown's concept of "popular epidemiology" (Brown & Mikkelson, 1990; Brown, 1992), have described the deployment of local knowledge by community groups concerned about environmental health risks (Di Chiro, 1992; Clapp, 2002; Allen, 2003, 2004; Frickel, 2004; Spears, 2004). Such studies raise important questions about the character and utility of knowledge that grows out of the lived experience of sufferers of health risks. On the one hand, the literature amply demonstrates the practical benefit of incorporating the experiential knowledge of the patient, not only within the doctor-patient relationship but also within the researcher-subject relationship. On the other hand, most work to date has been insufficiently critical of the tendency to valorize or romanticize lived experience as a basis for reliable knowledge, or to treat experience as a sort of bedrock resistant to critical interpretation (Scott, 1991). As Michelle Murphy has observed in a study of occupational health, " 'experience' is a category of knowledge that is just as historical as other forms of knowledge . . . It is only through particular methods rooted historically in time and space that experience becomes a kind of evidence imbued with certain truth-telling qualities" (Murphy, 2004b: 202).

In addition to, or instead of, mobilizing experiential knowledge, patient groups and health movements have laid claim to the formal knowledge more typically monopolized by credentialed experts, sometimes through systematic practices of self-education or community-based education (Epstein, 1995; Dickersin & Schnaper, 1996; Anglin, 1997; Dickersin et al., 2001; Myhre, 2002). As opposed to groups that are dismissive of formal knowledge, those that learn the biomedical science relevant to their condition adopt (according to Paula Treichler, in an early analysis of AIDS treatment activism) "not . . . a resistance to orthodox science but . . . strategic conceptions of 'scientific truth' that leave room for action in the face of contradictions" (Treichler ,1991: 79; see also Treichler, 1999). The term "lay expert" (Arksey, 1994; Epstein, 1995) has been widely used to characterize the liminal or boundary-crossing qualities of those who succeed in establishing this sort of claim to formal knowledge. However, some have objected that "lay expert" is effectively a contradiction in terms (Prior, 2003), and that once patients have "crossed over," they should simply be classified as experts, though perhaps of a different sort. For example, Harry Collins and Robert Evans have suggested that patient groups may often acquire "interactional expertise" ("enough expertise to interact interestingly with participants") and may sometimes acquire "contributory expertise" ("enough expertise to contribute to the science of the field being analysed") (Collins & Evans, 2002: 254).

In practice many patient groups and health movements have combined experiential knowledge with varying degrees of mastery of formal knowledge, often producing interestingly hybrid or "translocal" (Heath, 1997: 81–82) ways of knowing or varieties of expertise. As Rabeharisoa noted in the case of the AFM, sustained interaction between the patient group and specialists meant "that 'experiential' knowledge and scientific knowledge on the disease ended up forming an indivisible whole, jointly influencing clinical profiles and trajectories of life with the disease" (Rabeharisoa, 2003: 2133). Recent work has been particularly helpful in focusing attention on the specific tools and technologies employed by patient groups and health movements in their epistemic work. For example, Yoshio Nukaga has described how genetic support groups "collect family narratives, geneaological inscriptions and family trees . . . which are first translated by genetic counselors and researchers into various forms of medical pedigrees for clinical and laboratory work, and then circulated as published pedigrees among lay and medical practitioners" (Nukaga, 2002: 59). In her analysis of the feminist women's health movement, Murphy has shifted attention away from their ideologies and toward their practical technologies, such as the plastic speculum and the menstrual extraction kit (Murphy, 2004a: 347; see also Wajcman, 2004: 123–24); while in her analysis of sick building syndrome, she has analyzed how office workers "rematerialized" the office through efficiency analyses, surveys, and other techniques (Murphy, 2004b: 196). Callon and Rabeharisoa have reconstructed the "primitive accumulation of knowledge" by AFM members who have used "proto-instruments" that include "cameras, camcorders for taking films and photos, accounts written by patients or their parents in the form of books for the general public, requested testimonies, spontaneous letters, and lectures given by patients or their relatives." As these latter scholars note, such tools permit the production of knowledge that is "formal, transportable, cumulative, and debatable"—characteristics associated with the products of more traditional biomedical research. By this pathway, laboratory research and research conducted "in the wild" are brought together in the form of new cross-fertilizations (Callon & Rabeharisoa, 2003: 197–98).

However, these successes on the part of activists in creating and employing hybrid and translocal expertise may be accompanied by a "scientization" of the social movement that can have unanticipated consequences for its trajectory. The case of AIDS treatment activism suggested that the emergence of a specialist group of activist-experts accentuated various existing divisions within the movement through the creation of a new cleavage—that between the new "lay experts" and the "lay lay" activists (Elbaz, 1992: 488) who are "left behind" in the knowledge-acquisition process (Epstein, 1995; 1996: 284–94). To the extent that facility with scientific and technical knowledge or tools becomes a *de facto* criterion for leadership within a movement, then scientization may reshape the movement, potentially reducing its participatory potential. In addition, scientization may raise the barriers to entry, making it harder to recruit new members and replenish leadership positions—an especially critical issue for health movements, sadly, as leaders not infrequently are at personal risk of illness or death (Epstein, 1996: 327, 350–53).

EFFECTS AND CONSEQUENCES

No review of patient groups and health movements would be complete without discussion of the results that they bring about, and STS researchers have considered this issue in some detail.[12] To be sure, it is important not to exaggerate the effects of patient advocacy, which may well be limited in many cases (Stockdale, 1999: 594). Still, scholars have identified a range of ways in which these groups contribute to social and biomedical change.

The Conceptualization of the Disease

Howard Kushner has described how the U.S. Tourette Syndrome Association played an influential role in promoting the conception of Tourette syndrome as an organic disease—by contrast with France, where, in the absence of a strong group of patients and their family members, Tourette syndrome is understood within a psychodynamic framework (Kushner, 2004). In another example, Stella Capek showed how a grassroots self-help group called the Endometriosis Association helped reorient conceptions of etiology away from purely endogenous causal factors and toward "a more holistic view that explores connections between the human body and a chemically toxic environment" (Capek, 2000: 345, 351–52).

Patients' Management of Their Illnesses

Although it has become common to speak of the "educated patient," only a few studies systematically have investigated how the activities of patient groups change the ways in which patients engage with their physicians, their medications (Akrich & Meadel, 2002), or their bodies. The work of Janine Barbot and Nicolas Dodier is exemplary in delineating how different HIV/AIDS groups in France have been associated with different "pragmatics of information gathering" (Barbot & Dodier, 2002) and strategies of illness management (Barbot, 2006) on the part of patients. In a recent article, Barbot constructed a typology of four varieties of educated patients—the patient as illness manager, the empowered patient, the science-wise patient, and the experimenter—and correlated each type with a different French HIV/AIDS support or advocacy group (Barbot, 2006).

Attitudes and Practices of Health Professionals

In some cases, health movements have inspired a greater sensitivity on the part of physicians and researchers, for example, in their judgments about people who are overweight (Saguy & Riley, 2005; Boero, forthcoming). In other cases, patient groups and health movements have brought about concrete changes in physician practice—though as Karkazis has noted in her analysis of intersex activism, physicians may sometimes be unwilling to concede that their embracing of new policies had anything to do with outside pressure (Karkazis, 2002: chapter 9). Finally, in cases where the social movement is itself built on an "uneasy alliance" between lay activists and medical professionals, as in the abortion rights movement, activism can result in important changes at the level of professional associations and medical education (Joffe et al., 2004: 784).

The Research Process

Examples of the impact of patient groups and health movements on biomedical research have been suggested throughout this chapter. Patient groups have raised funds for research and have doled it out to support the lines of research they deem most important, gained a "seat at the table" to make decisions about research directions, promoted ethical treatment of participants within clinical trials, attempted to police perceived ethical abuses such as conflicts of interest in research, challenged the techniques for conducting and interpreting clinical trials, helped create disease and treatment registries, organized conferences, coauthored publications, and pioneered new models of participatory research that join the efforts of lay citizens with those of experts. Other effects are less tangible but no less significant. Callon and Rabeharisoa have hinted at the new "entanglements" between patients and researchers by quoting the words of a young girl with spinal muscular atrophy speaking to a biologist: "I'm with you in your laboratory since you're working on my genes" (Callon & Rabeharisoa, 2003: 201). As David Hess has suggested, there are a range of alternative pathways along which such entanglements may proceed, including conversion experiences by researchers, biographical transformations of activists who become lay researchers, or the creation of "network assemblages" in which activists "help weave together networks of patients, funding sources, clinicians and potential researchers" (Hess, 2004b: 703–4).

Technological Trajectories

A growing body of literature has shown how patient groups and health movements, acting either as users of technologies or as their representatives, can intervene in the path of technological development. Scholars have examined these dynamics especially in relation to contraceptive technologies and abortifacients, showing how women's health advocates and organizations have altered technological scripts while asserting the priorities of bodily integrity and social justice (Clarke & Montini, 1993; Clarke, 1998, 2000; Dugdale, 2000; Bell, 2003; Van Kammen, 2003; Hardon, 2006). A different sort of example of engagement with technology was provided by Lisa Jean Moore, who analyzed how sex workers were configured by latex technologies but also configured their clients into new, "safe sex" users of these technologies (Moore, 1997).

State Policies

Johnson and Hufbauer's work, several decades ago, on how bereaved parents convinced the U.S. Congress to fund research by passing the Sudden Infant Death Syndrome Act of 1974 (Johnson & Hufbauer, 1982) is just one example of how patient groups have sought to influence public research funding priorities. But patient groups and health movements also have brought about other sorts of formal changes in state policies. Constance Nathanson has shown the significant effect of the tobacco control movement in the United States on legislation and regulatory policy (and has contrasted it with the limited impact of the gun control movement) (Nathanson, 1999); Allsop, Jones, and Baggott have described how the health consumer movement has pushed the British government to develop new procedures for cases where patients claim harm by health professionals (Allsop et al., 2004: 752); Saguy and Riley have shown how the fat acceptance movement prompted the U.S. Food and Drug Administration to postpone its approval of a kind of weight-loss surgery (Saguy & Riley, 2005: 911); and I have described how a diverse coalition of health advocates in the United States successfully pressed for new federal policies on the inclusion of women, racial and ethnic minorities, children, and the elderly as research subjects, as well as for the creation of federal offices of women's health and minority health (Epstein, 2004, 2007; see also Auerbach & Figert, 1995; Weisman, 1998).

Corporations and Markets

Probably the most frequent corporate target of patient group activity has been pharmaceutical companies. Activists concerned about issues such as drug pricing and research ethics have been able to wrest concessions from drug companies on occasion (Epstein, 1996), and recent global debates about access to medications such as antiretroviral drugs have suggested the efficacy of transnational linkages of patient groups and health movements in affecting the marketing practices of drug companies as well as their ability to enforce their patents (Whyte et al., 2002: 146–60). However, these are not the only ways in which patient groups have affected market relations. Sometimes, as in the patenting of the PXE gene described by Heath, Rapp, and Taussig, patient groups have successfully claimed intellectual property rights for themselves (Heath et al., 2004: 163–64). In addition, Hess has examined the productive ties between civil society organizations and companies promoting alternative health products under the banner of "nutritional therapeutics" (Hess, 2005). Such work may be suggestive of broader patterns by which patient groups affect the organization of industrial fields, for example, through their alliances with start-ups.

Cultural Effects

Some of the most profound and enduring effects of patient groups and health movements may sometimes be among the most diffuse and hardest to pinpoint. Such groups may have an important cultural impact simply by exposing prevailing norms and power relationships and making them available for public critique (Gamson, 1989; Löwy, 2000: 74). For example, as suggested by the disability movement and the intersex movement, health activists may seek to establish the legitimacy of different sorts of bodies or bodily experiences (Shakespeare, 1993, 1999; Dowse, 2001; Rapp & Ginsburg, 2001; Karkazis, 2002). Or, patient groups and health movements may enact public performances of bodies and diseases in ways that challenge conventional cultural codes about appropriate gender roles or sexualities (Klawiter, 1999). They also may reinterpret the historical record, for example, by attributing disease prevalence in certain groups to historical legacies of social oppression (Nelson, 2003: chapter 4). Several scholars have emphasized the "memorialization" work of advocacy groups—for example, Sahra Gibbon has described how breast cancer advocates perform acts of memorialization that connect the witnessing of loss to a new conception of research as redemption (Gibbon, 2007) while Lesley Sharp has shown how groups representing the surviving relatives of organ transplant donors have used cultural forms such as donor quilts and Web cemeteries to challenge transplant professionals' tendencies to "obliterate donors' identities" (Sharp, 2001: 125). In these various ways, patient groups and health movements, like social movements generally, are involved in reconstructing the "cultural schemas" that define the rules of the game by which key social institutions operate (Polletta, 2004).

Incorporation and Cooptation

While documenting in considerable detail the transformative effects of patient groups and health movements, the scholarly literature mostly has been careful to avoid an uncritically celebratory tone. In fact, a hallmark of recent work has been the attempt to make sense of the multivalent politics of incorporation, whereby the insights and legacies of patient advocacy are channeled back into institutionalized biomedical practice, and of cooptation, in which the radical potential of an activist critique is blunted or contained. Biomedical institutions are highly flexible and resilient (Löwy, 2000: 73)—one might say omnivorous—and the peculiar thing about the phenomenon of incorporation is that it may be hard to judge in principle whether it should be counted as victory or defeat: Does it mark the successful transformation of biomedicine by outside forces, or the taming of a radical challenge, or even both at once (Goldner, 2004: 727)?[13] Similarly, when activists come to moderate their critiques or adopt more conventional biomedical understanding, it is often hard to say whether they have allowed themselves to be co-opted or have made a well-advised shift in tactics.

Scholars have pointed to instances of outright manipulation of patients in order to co-opt them, for example, through the creation by pharmaceutical companies of "front groups" masquerading as patient advocacy groups that are intended to build demand for a company's products or garner support for drug approval (Zavestoski et al., 2004c: 274). However, this extreme case is one end of a continuum of relations to pharmaceutical companies, described by O'Donovan, that also includes many other instances in which patient groups receive pharmaceutical industry financing. O'Donovan rightly has cautioned against any automatic assumptions of a creeping "corporate colonisation," calling for detailed study of whether corporations indeed have increased their influence over patient groups' "cultures of action" (O'Donovan, 2007).

Scholars also have identified cases where activist intentions were co-opted in the process of partial implementation of their concerns—for example, as Natalie Boero has analyzed, surgeons' adaptations of the arguments of the fat acceptance movement in order to promote weight loss surgery, or, as Theresa Montini has described, the passage of breast cancer informed consent laws in the United States in ways that "actually advanced and protected the professional autonomy of physicians at the expense of patient rights" (Montini, 1991: vii; see also Montini, 1996). A more ambiguous case is the *quid pro quo* worked out between twentieth-century birth control advocates and reproductive scientists, as analyzed by Clarke: reproductive scientists agreed to devote their energies to developing birth control technologies but only on the condition that they would emphasize basic research on "modern," technologically advanced forms of contraception, to the exclusion of scientific attention to simpler chemical and mechanical means of preventing pregnancy (Clarke, 1998: 163, 200). Another tricky case—for which blunt and accusatory terms such as "cooptation" appear unhelpful—is that of AIDS treatment activists, many of whom began to soften their critiques of clinical research and regulatory practices as they learned about the complexities involved: "The more we learned, in some ways the less we were able to ask for," was how one activist expressed it (Epstein, 1996: 328). By one measure, these activists became more conservative as they became inculcated within biomedical frameworks; by another measure, they changed tactics appropriately in response to an evolving political environment and as the research trajectory, and their own understanding, advanced (Epstein, 1996: 325–28, 342–44, 1997a). At a minimum, their example suggests the benefits of studying expert knowledge in broadly Foucaultian terms—not as an inert tool to be acquired, but rather as something that reshapes the subjectivities of those who become subject to it (Foucault, 1980).

At the more benign end of the incorporation spectrum, Joffe has remarked on the legacy of key feminist principles within present-day medical practice: "Many of the ideas about abortion and other reproductive health services that were promoted by women's health activists of the 1970s—ranging from the simple (warming the gynecological instruments) to the more complex (seeing the patient as a fully participating partner)—have now been incorporated into practice at many facilities—even those that do not think of themselves as 'feminist'" (Joffe, 1999: 32). A related example is the mainstream medical incorporation of breastfeeding, a practice that health activists once had to defend (Ward, 2000). Although these cases might seem closer to what could simply be called victory, it is worth reflecting on the deletions of authorship and historical process that typically accompany even beneficent incorporations (Arksey, 1994: 464). Who remembers, decades later, that what has become the ordinary standard of care was once a radical innovation promoted by activists? Yet this act of historical forgetting may indeed have consequences: it limits the capacity of subsequent generations of activists to benefit from examples of past struggles and be inspired to imagine how current conditions might be otherwise.

Is institutionalization possible without some measure of "capture and control" (Hess, 2004b: 705; see also Hess, 2005)? Landzelius goes so far as to conclude her story of "parents of preemies" in two different ways, first suggesting the practical benefits of the movement's cooperative approach, then "pivoting" to highlight "the ways in which it embeds normative ideologies about maternity and likewise is comfortably embedded within and cocooned by them" (Landzelius, 2006a: 679–80). Another useful way forward in analysis has been suggested by Melinda Goldner, who took up the question of institutionalization by combining social movement perspectives with institutionalist approaches within sociology. In her analysis of the "dynamic inter-play" between the complementary and alternative medicine movement and Western medicine, Goldner rejected any simple conclusion about incorporation by showing how distinctive outcomes on the ground mapped onto a typology of diverse institu-tional responses to external challenge (Goldner, 2004).

DIRECTIONS FOR FUTURE RESEARCH

Studies of patient groups and health movements have reflected and propelled a cre-ative synthesis of STS perspectives, medical anthropology and sociology, social move-ment scholarship, and other fields. Having risen to prominence within STS as a means to reconsider problems of expertise and resituate the locus of scientific work, the topic of patient groups and health movements has proved a fruitful path to consider such diverse issues as embodiment, vital politics, biomedicalization, and scientific citizen-ship. I have emphasized how the study of patient groups and health movements suc-cessfully has built on concepts derived from a variety of intellectual sources. But it is also important to say that this body of work now has something to offer back to schol-ars in other domains. For example, insofar as the constitution of groups and collec-tive identities is a central issue in social theory, it would be valuable (though beyond the scope of this chapter) to think through the implications of studies of patient groups and health movements for general theoretical work on that topic: How does the intermingling of humans and nonhumans affect the pathways by which "group-ness" and identity take shape and evolve? How do the politics of expertise complicate the politics of alliance and division? Having absorbed so much from so many other fields, it will be important for practitioners in this research domain to reformulate their conclusions in ways that allow them to be returned to, and illuminate, other domains of theorization and empirical research.[14] At the same time, as scholars such as Stuart Blume (Blume & Catshoek, 2001/2002), Anita Hardon, and Phil Brown have been attempting, the conclusions from academic study of patient groups and health movements can and should be brought back to health activists themselves in sys-tematic ways, in order to fashion new alliances between scholars and activists.

I began this chapter by raising a series of questions about patient groups and health movements—among others, how they form and organize, what kinds of expertise they develop and deploy, how they affect the practice of medicine and biomedical research, and how they reshape the nexus of relations linking biomedical institutions to the market and the state. This review has suggested that scholars already have shed con-siderable light on these questions. I conclude with some brief suggestions of useful avenues of future research, particularly with the goal of addressing existing gaps.

Case Comparisons

Most work on patient groups and health movements has taken the form of detailed case studies. Yet many of the questions that I have raised in this review—about efficacy; about the virtues of typologies—could best be answered by close comparative analysis. Models of comparative work—both between different health conditions in the same country and between the same health condition in different countries—have been suggested by Barrett with the cases of fibromyalgia and chronic fatigue syndrome (Barrett, 2004), Dimock with breast cancer and prostate cancer (Dimock, 2003), Kushner on Tourette Syndrome in the United States and France (Kushner, 2004), Nathanson on antismoking versus gun control (Nathanson, 1999), Parthasarathy on breast cancer in the United States and Britain (Parthasarathy, 2003), and Brown and coauthors on Gulf War Syndrome, asthma, and environmental causes of breast cancer (Brown et al., 2002; see also Zavestoski et al., 2002).

Globalization and Transnationalism

It seems problematic that most analyses of patient groups and health movements to date have confined themselves within national borders—and all too often within the United States or Western European countries. Only a few studies have sought to study other parts of the world, to analyze the diffusion of activist frames from one country to another (Kirp, 1999), to consider the development of explicitly transnational health advocacy (Whelan, 2003; Barnes, 2005; Landzelius, 2006a; Radin, 2006), or to locate patient groups and health movements in a global geopolitical context in relation to the North-South divide (Whyte et al., 2002: 146–60; Bell, 2003; Barbot, 2006: 549–50; Hardon, 2006).

Movement/Countermovement Dynamics

Social movement scholars know that movements often provoke countermovements, and the complex engagement between the two often shapes movement trajectories in significant ways (Meyer & Staggenborg, 1996). These dynamics are worthy of further study. Examples in the existing literature on patient groups and health movements include the pro-choice and pro-life movements (Joffe, 1999; Joffe et al., 2004), the gun control movement and its well-organized opponents (Nathanson, 1999), the LGBT health movement and the Christian right (Epstein, 2003b), and the movements for and against stem cell research (Ganchoff, 2004).

Periodization

So far only a few scholars systematically have tracked patient groups and health movements through distinct phases of their evolution. Useful models include the work of Barbot (2002) and of Layne (2006), who describe different generations or phases of advocacy within a movement. A more ambitious concept is Klawiter's understanding of the relation between health movements and successive "disease regimes" (Klawiter, 2004).

Insiders and Outsiders

Goldner has called for more detailed study, especially from activists' perspectives, of precisely how members of patient groups and health movements negotiate being simultaneously an insider and an outsider *vis-à-vis* biomedical, state, and market institutions: "How does gaining institutional access blur the boundary between movements and mainstream organisations, and how does this ultimately impact upon the movement?" (Goldner, 2004: 730; see also Epstein, 1996; Moore, 1999; Hess, 2004a: 424, 2005).

Inequalities and Health Disparities

Scholars have not made as much as they might of the implications of the activities of patient groups and health movements for the reproduction or overturning of deeply rooted inequalities—by gender, race, ethnicity, sexuality, nationality, and religion, among other markers—with regard to exposure to health risk, access to health care, or social rewards more generally. (Here the work of Brown and collaborators, and of Rapp, Heath, and Taussig, stand as exceptions, as do the various studies of environmental health and environmental justice.) While the co-production of gender and technoscience is a relatively frequent theme in the literature (Gibbon, 2007), there has been much less consideration of other dimensions of difference and inequality, such as race (Nelson, 2003; Epstein, 2004; Reardon, 2004; Shostak, 2004; Epstein, 2005; Klawiter, unpublished). Nor has there been much analysis of how the absence of universal access to health care in countries such as the United States affects the agendas of patient groups and health movements. Finally, the stark social and health inequalities at the global level between "North" and "South" undoubtedly have a profound influence on the shapes, goals, and successes of patient groups and health movements emerging in different parts of the world. Consideration of these issues in future analyses would help flesh out the depiction of biomedical citizenship that has been emerging in the literature by linking it to the diverse struggles over rights and inclusion in the domain of health.

Notes

I am grateful to the editors of this volume, particularly Judy Wajcman, who had editorial responsibility for this chapter. I received thought-provoking and sometimes challenging, but much appreciated, advice from the following colleagues, who generously devoted time to reading an initial draft of this chapter: Stuart Blume, Phil Brown, Michel Callon, Adele Clarke (special thanks for the fine-tooth-comb treatment), Joe Dumit, David Hess, Katrina Karkazis, Kyra Landzelius, Alondra Nelson, Volo Rabeharisoa, Rayna Rapp, Abby Saguy, and Stefan Timmermans. I only wish it had proved possible to incorporate all their suggestions in the space available.

1. Existing sources that I encountered were Bayer, 1981, 1985; Petersen and Markle, 1981, 1989; von Gizycki, 1987; Hoffman, 1989; Brown and Mikkelson, 1990. Sources that I *failed* to learn about until later were Johnson and Hufbauer, 1982; Fox, 1989.

2. Group formation of this general sort has been visible enough in recent years to be represented in popular culture in diverse ways: for example, lay contributions to research on rare diseases were cele-

brated in the 1992 film *Lorenzo's Oil* while self-help groups were satirized as an escapist addiction in the 1999 film *Fight Club*.

3. "New social movements" is a problematic term if meant to describe a wholly unprecedented and distinct social form (Pichardo, 1997), but it is a useful concept if invoked to refer to certain tendencies and preoccupations that arguably have been more visible among movements in recent decades, including a reflexive concern with identity construction, a focus on the politics of the body, and a commitment to cultural transformation.

4. As Abby Saguy has suggested to me (personal communication), women's bodies and women's health may also be heavily implicated in the work of movements that are less overtly gendered, such as the fat acceptance activists whom she has studied.

5. Because chapter 20 in this volume focuses specifically on social movements, I have refrained from defining the term social movement, reviewing key schools of social movement scholarship, and providing references for key concepts in the social movement literature.

6. My approach is consistent with, and influenced by, that taken by Kyra Landzelius (2006b) in her introduction to the special issue in *Social Science & Medicine* that she organized with Joe Dumit. Landzelius introduces the concept of the "patient organization movement" while rendering problematic each of the three constituent terms. I am also grateful to Volo Rabehariso for her reflections on these conceptual and definitional issues.

7. More generally, on the interpenetration of social movements and state institutions, see Skrentny, 2002, especially p. 5; Goldstone, 2003: 1–24, especially p. 2.

8. An overlapping typology has been suggested by Judith Allsop and coauthors, who distinguished between "population-based" groups, "condition-based" groups, and "formal alliance organizations" (Allsop et al., 2004: 739). A somewhat more complex breakdown has been offered by Hilda Bastian (1998: 11), who identified six broad "strands" of consumer activism in the domain of health.

9. Medicalization refers to the process of taking a phenomenon not previously considered a medical issue and defining it in medical terms, adopting a medical framework to understand it, or licensing the medical profession to treat it (Conrad & Schneider, 1980; Conrad, 2005).

10. Within the broader field of social movement scholarship, Doug McAdam has emphasized the importance of comparing participants to nonparticipants (McAdam 1988).

11. Another large body of literature, particularly in medical anthropology and medical sociology, has examined how *individual* lay actors, such as patients—or sometimes the public at large—assesses and apprehends medical knowledge claims. A review of that literature would be the topic of another chapter.

12. The theme of the effects of social movements also has received renewed interest in the broader social movement literature. See, for example, Giugni et al., 1999.

13. I am grateful to Andrew Feenberg for past discussion of these issues.

14. I am grateful to Michel Callon for his suggestions regarding these issues.

References

Akrich, Madeleine, & Cecile Meadel (2002) "Prendre ses medicaments/prendre la parole: les usages des medicaments par les patients dans les listes de discussion electroniques," *Sciences Sociales et Santé* 20 (1): 89–115.

Allen, Barbara L. (2003) *Uneasy Alchemy: Citizens and Experts in Louisiana's Chemical Corridor Disputes* (Cambridge, MA: MIT Press).

Allen, Barbara L. (2004) "Shifting Boundary Work: Issues and Tensions in Environmental Health Science in the Case of Grand Bois, Louisiana," *Science as Culture* 13(4): 429–48.

Allsop, Judith, Kathryn Jones, & Rob Baggott (2004) "Health Consumer Groups in the UK: A New Social Movement?," *Sociology of Health & Illness* 26(6): 737–56.

Anglin, Mary K. (1997) "Working from the Inside Out: Implications of Breast Cancer Activism for Biomedical Policies and Practices," *Social Science & Medicine* 44 (9): 1403–15.

Arksey, Hilary (1994) "Expert and Lay Participation in the Construction of Medical Knowledge," *Sociology of Health & Illness* 16 (4): 448–68.

Arksey, Hilary (1998) *RSI and the Experts: The Construction of Medical Knowledge* (London: UCL Press).

Auerbach, Judith D. & Anne E. Figert (1995) "Women's Health Research: Public Policy and Sociology," *Journal of Health and Social Behavior* 35 (extra issue): 115–31.

Bammer, Gabriele & Brian Martin (1992) "Repetition Strain Injury in Australia: Medical Knowledge, Social Movement, and De Facto Partisanship," *Social Problems* 39(3): 219–37.

Barbot, Janine (2006) "How to Build an 'Active' Patient? The Work of AIDS Associations in France," *Social Science & Medicine* 62(3): 538–51.

Barbot, Janine & Nicolas Dodier (2002) "Multiplicity in Scientific Medicine: The Experience of HIV-Positive Patients," *Science, Technology & Human Values* 27(3): 404–40.

Barker, Kristin (2002) "Self-Help Literature and the Making of an Illness Identity: The Case of Fibromyalgia Syndrome (FMS)," *Social Problems* 49(3): 279–300.

Barnes, Nielan (2005) *Transnational Networks and Community-Based Organizations: The Dynamics of AIDS Activism in Tijuana and Mexico City,* Ph.D. diss., University of California, San Diego.

Barrett, Deborah (2004) "Illness Movements and the Medical Classification of Pain and Fatigue," in R. M. Packard, J. Brown, R. L. Berkelman, & H. Frumkin (eds), *Emerging Illnesses and Society: Negotiating the Public Health* (Baltimore, MD: Johns Hopkins University Press): 139–70.

Bastian, Hilda (1998) "Speaking up for Ourselves: The Evolution of Consumer Advocacy in Health Care," *International Journal of Technology Assessment in Health Care* 14(1): 3–23.

Bayer, Ronald (1981) *Homosexuality and American Psychiatry: The Politics of Diagnosis* (New York: Basic Books).

Bayer, Ronald (1985) "AIDS and the Gay Movement: Between the Specter and the Promise of Medicine," *Social Research* 52(3): 581–606.

Beard, Renee L. (2004) "Advocating Voice: Organisational, Historical and Social Milieux of the Alzheimer's Disease Movement," *Sociology of Health & Illness* 26(6): 797–819.

Bell, Susan E. (2003) "Sexual Synthetics: Women, Science, and Microbicides," in M. J. Casper (ed) *Synthetic Planet: Chemical Politics and the Hazards of Modern Life* (New York: Routledge): 197–211.

Berg, Marc & Annemarie Mol (eds) (1998) *Differences in Medicine: Unraveling Practices, Techniques, and Bodies* (Durham, NC: Duke University Press).

Blume, Stuart S. (1997) "The Rhetoric and Counter-Rhetoric of a 'Bionic' Technology," *Science, Technology & Human Values* 22 (1): 31–56.

Blume, Stuart S. (1999) "Histories of Cochlear Implantation," *Social Science & Medicine* 49: 1257–68.

Blume, Stuart (2006) "Anti-Vaccination Movements and Their Interpretations," *Social Science & Medicine* 62(3): 628–42.

Blume, Stuart & Geerke Catshoek (2001/2002) "Articulating the Patient Perspective: Strategic Options for Research" (Amsterdam: PatiëntenPraktijk).

Boero, Natalie (forthcoming) "Bypassing Blame: Bariatric Surgery and the Case of Biomedical Failure" in A. E. Clarke, J. Fosket, L. Mamo, J. Shim, and J. Fishman (eds), *Biomedicalization: Technoscience, Health and Illness in the U.S.* (Durham, NC: Duke University Press).

Bonnet, Doris, Michel Callon, Gerard De Pouvourville, & Vololona Rabeharisoa (1998) "Avant-Propos," *Sciences Sociales et Santé* 16(3): 5–15.

Borkman, Thomasina J. (1991) "Introduction to the Special Issue," *American Journal of Community Psychology* 19(5): 643–50.

Bourdieu, Pierre (1985) "The Genesis of the Concepts of Habitus and of Field," *Sociocriticism* 2(2): 11–24.

Briggs, Charles L. & Clara Mantini-Briggs (2003) *Stories in Times of Cholera: Racial Profiling During a Medical Nightmare* (Berkeley: University of California Press).

Brown, Phil (1992) "Popular Epidemiology and Toxic Waste Contamination: Lay and Professional Ways of Knowing," *Journal of Health and Social Behavior* 33: 267–81.

Brown, Phil & Edwin J. Mikkelson (1990) *No Safe Place: Toxic Waste, Leukemia, and Community Action* (Berkeley: University of California Press).

Brown, Phil & Stephen Zavestoski (2004) "Social Movements in Health: An Introduction," *Sociology of Health & Illness* 26(6): 679–94.

Brown, Phil, Stephen Zavestoski, Meadow Linder, Sabrina McCormick, & Brian Mayer (2003) "Chemicals and Casualties: The Search for Causes of Gulf War Illnesses," in M. J. Casper (ed) *Synthetic Planet: Chemical Politics and the Hazards of Modern Life* (New York: Routledge): 213–36.

Brown, Phil, Stephen Zavestoski, Brian Mayer, Sabrina McCormick, & Pamela S. Webster (2002) "Policy Issues in Environmental Health Disputes," *Annals of the American Academy of Political and Social Science* 584: 175–202.

Brown, Phil, Brian Mayer, Stephen Zavestoski, Theo Luebke, Joshua Mandelbaum, & Sabrina McCormick (2003) "The Health Politics of Asthma: Environmental Justice and Collective Illness Experience in the United States," *Social Science & Medicine* 57(3): 453–64.

Brown, Phil, Stephen Zavestoski, Sabrina McCormick, Brian Mayer, Rachel Morello-Frosch, & Rebecca Gasior Altman (2004) "Embodied Health Movements: New Approaches to Social Movements in Health," *Sociology of Health & Illness* 26(1): 50–80.

Callon, Michel (2003) "The Increasing Involvement of Concerned Groups in R&D Policies: What Lessons for Public Powers?" in A. Geuna, A. J. Salter, & W. E. Steinmueller (eds), *Science and Innovation: Rethinking the Rationales for Funding and Governance* (Cheltenham, UK: Edward Elgar): 30–68.

Callon, Michel & Vololona Rabeharisoa (2003) "Research 'in the Wild' and the Shaping of New Social Identities," *Technology in Society* 25(2): 193–204.

Capek, Stella M. (2000) "Reframing Endometriosis: From 'Career Woman's Disease' to Environment/Body Connections," in S. Kroll-Smith, P. Brown, & V. J. Gunter (eds), *Illness and the Environment: A Reader in Contested Medicine* (New York: New York University Press): 345–63.

Cartwright, Lisa (2000) "Community and the Public Body in Breast Cancer Media Activism," in J. Marchessault & K. Sawchuk (eds), *Wild Science: Reading Feminism, Medicine and the Media* (London: Routledge): 120–38.

Chesler, Mark A (1991) "Mobilizing Consumer Activism in Health Care: The Role of Self-Help Groups," *Research in Social Movements, Conflict and Change* 13: 275–305.

Clapp, Richard W. (2002) "Popular Epidemiology in Three Contaminated Communities," *Annals of the American Academy of Political and Social Science* 584: 35–46.

Clarke, Adele (1998) *Disciplining Reproduction: Modernity, American Life Sciences, and "the Problems of Sex"* (Berkeley: University of California Press).

Clarke, Adele E. (2000) "Maverick Reproductive Scientists and the Production of Contraceptives, 1915–2000+," in A. R. Saetnan, N. Oudshoorn, & M. Kirejczyk (eds), *Bodies of Technology: Women's Involvement with Reproductive Medicine* (Columbus: Ohio University Press): 37–89.

Clarke, Adele & Theresa Montini (1993) "The Many Faces of RU486: Tales of Situated Knowledges and Technological Contestations," *Science, Technology & Human Values* 18: 42–78.

Clarke, Adele E., Janet K. Shim, Laura Mamo, Jennifer Ruth Fosket, & Jennifer R. Fishman (2003) "Biomedicalization: Technoscientific Transformations of Health, Illness, and U.S. Biomedicine," *American Sociological Review* 68: 161–94.

Cohen, Cathy J. (1999) *The Boundaries of Blackness: AIDS and the Breakdown of Black Politics* (Chicago: University of Chicago Press).

Collins, H. M. & Robert Evans (2002) "The Third Wave of Science Studies: Studies of Expertise and Experience," *Social Studies of Science* 32(2): 235–96.

Conrad, Peter (2005) "The Shifting Engines of Medicalization," *Journal of Health and Social Behavior* 46(1): 3–14.

Conrad, Peter & Joseph W. Schneider (1980) *Deviance and Medicalization: From Badness to Sickness* (St. Louis: C.V. Mosby).

Copelton, Denise A. (2004) "Menstrual Extraction, Abortion & the Political Context of Feminist Self-Help," *Advances in Gender Research* 8: 129–64.

Crossley, Nick (2006) "The Field of Psychiatric Contention in the UK, 1960–2000," *Social Science & Medicine* 62(3): 552–63.

Di Chiro, Giovanna (1992) "Defining Environmental Justice: Women's Voices and Grassroots Politics," *Socialist Review* (October–December): 93–130.

Dickersin, Kay & Lauren Schnaper (1996) "Reinventing Medical Research," in K. L. Moss (ed) *Man-Made Medicine: Women's Health, Public Policy, and Reform* (Durham, NC: Duke University Press): 57–76.

Dickersin, Kay, Lundy Braun, Margaret Mead, Robert Millikan, Ana M. Wu, Jennifer Pietenpol, Susan Troyan, Benjamin Anderson, & Frances Visco (2001) "Development and Implementation of a Science Training Course for Breast Cancer Activists: Project Lead (Leadership, Education and Advocacy Development)," *Health Expectations* 4(4): 213–20.

Dimock, Susan Halebsky (2003) *Demanding Disease Dollars: How Activism and Institutions Shaped Medical Research Funding for Breast and Prostate Cancer*, Ph.D. diss., University of California, San Diego.

Dowse, Leanne (2001) "Contesting Practices, Challenging Codes: Self Advocacy, Disability Politics and the Social Model," *Disability & Society* 16(1): 123–41.

Dresser, Rebecca (2001) *When Science Offers Salvation: Patient Advocacy and Research Ethics* (Oxford: Oxford University Press).

Dugdale, Anni (2000) "Intrauterine Contraceptive Devices, Situated Knowledges, and the Making of Women's Bodies," *Australian Feminist Studies* 15(32): 165–76.

Dumit, Joseph (1997) "A Digital Image of the Category of the Person: PET Scanning and Objective Self-Fashioning," in G. L. Downey & J. Dumit (eds), *Cyborgs & Citadels* (Santa Fe, NM: School of American Research Press): 83–102.

Dumit, Joseph (2006) "Illnesses You Have to Fight to Get: Facts as Forces in Uncertain, Emergent Illnesses," *Social Science & Medicine* 62(3): 577–90.

Elam, Mark & Margareta Bertilsson (2003) "Consuming, Engaging and Confronting Science: The Emerging Dimensions of Scientific Citizenship," *European Journal of Social Theory* 6(2): 233–51.

Elbaz, Gilbert (1992) *The Sociology of AIDS Activism, the Case of Act up/New York, 1987–1992*, Ph.D. diss., City University of New York.

Elston, Mary Ann (1994) "The Anti-Vivisectionist Movement and the Science of Medicine," in J. Gabe, D. Kelleher, & G. Williams (eds), *Challenging Medicine* (London: Routledge): 160–80.

Emke, Ivan (1992) "Medical Authority and Its Discontents: A Case of Organized Non-Compliance," *Critical Sociology* 19(3): 57–80.

Epstein, Steven (1995) "The Construction of Lay Expertise: AIDS Activism and the Forging of Credibility in the Reform of Clinical Trials," *Science, Technology & Human Values* 20(4): 408–37.

Epstein, Steven (1996) *Impure Science: AIDS, Activism, and the Politics of Knowledge* (Berkeley: University of California Press).

Epstein, Steven (1997a) "Activism, Drug Regulation, and the Politics of Therapeutic Evaluation in the AIDS Era: A Case Study of ddC and the 'Surrogate Markers' Debate," *Social Studies of Science* 27(5): 691–726.

Epstein, Steven (1997b) "AIDS Activism and the Retreat from the Genocide Frame," *Social Identities* 3(3): 415–38.

Epstein, Steven (2001) "Biomedical Activism: Beyond the Binaries," presentation at the annual meeting of the Society for Social Studies of Science, Cambridge, MA, November 1–4.

Epstein, Steven (2003a) "Inclusion, Diversity, and Biomedical Knowledge Making: The Multiple Politics of Representation," in N. Oudshoorn & T. Pinch (eds), *How Users Matter: The Co-Construction of Users and Technology* (Cambridge, MA: MIT Press): 173–90.

Epstein, Steven (2003b) "Sexualizing Governance and Medicalizing Identities: The Emergence of 'State-Centered' LGBT Health Politics in the United States," *Sexualities* 6(2): 131–71.

Epstein, Steven (2004) "Bodily Differences and Collective Identities: The Politics of Gender and Race in Biomedical Research in the United States," *Body and Society* 10(2–3): 183–203.

Epstein, Steven (2005) "Institutionalizing the New Politics of Difference in U.S. Biomedical Research: Thinking across the Science/State/Society Divides," in S. Frickel & K. Moore (eds), *The New Political Sociology of Science: Institutions, Networks and Power* (Madison: University of Wisconsin Press): 327–50.

Epstein, Steven (2007) *Inclusion: The Politics of Difference in Medical Research* (Chicago: University of Chicago Press).

Escoffier, Jeffrey (1999) "The Invention of Safer Sex: Vernacular Knowledge, Gay Politics & HIV Prevention," *Berkeley Journal of Sociology* 43: 1–30.

Foucault, Michel (1980) *Power/Knowledge* (New York: Pantheon).

Fox, Patrick (1989) "From Senility to Alzheimer's Disease: The Rise of the Alzheimer's Disease Movement," *Milbank Quarterly* 67(1): 58–102.

Franklin, Sarah & Margaret Lock (eds) (2003) *Remaking Life & Death: Toward an Anthropology of the Biosciences*. (Santa Fe, NM: School of American Research Press).

Frickel, Scott (2004) "Just Science? Organizing Scientist Activism in the U.S. Environmental Justice Movement," *Science as Culture* 13(4): 449–69.

Frickel, Scott & Kelly Moore (eds) (2005) *The New Political Sociology of Science: Institutions, Networks and Power* (Madison: University of Wisconsin Press).

Fullwiley, Duana (1998) "Race, biologie et maladie: la difficile organisation des patients atteints de drépanocytose aux Etats-Unis," *Sciences Sociales et Santé* 16(3): 129–58.

Gamson, Joshua (1989) "Silence, Death, and the Invisible Enemy: AIDS Activism and Social Movement 'Newness,'" *Social Problems* 36(4): 351–65.

Ganchoff, Chris (2004) "Regenerating Movements: Embryonic Stem Cells and the Politics of Potentiality," *Sociology of Health & Illness* 26(6): 757–74.

Geertz, Clifford (1983) *Local Knowledge: Further Essays in Interpretive Anthropology* (New York: Basic Books).

Gibbon, Sahra (2007) *Breast Cancer Genes and the Gendering of Knowledge* (London: Palgrave Macmillan).

Gieryn, Thomas F. (1983) "Boundary Work and the Demarcation of Science from Non-Science: Strains and Interests in Professional Ideologies of Scientists," *American Sociological Review* 48: 781–95.

Gillett, James (2003) "Media Activism and Internet Use by People with HIV/AIDS," *Sociology of Health & Illness* 25 (6): 608–24.

Giugni, Marco, Doug McAdam, & Charles Tilly (eds) (1999) *How Social Movements Matter* (Minneapolis: University of Minnesota Press).

Goldner, Melinda (2001) "Expanding Political Opportunities and Changing Collective Identities in the Complementary and Alternative Medicine Movement," *Research in Social Movements, Conflicts and Change* 23: 69–102.

Goldner, Melinda (2004) "The Dynamic Interplay Between Western Medicine and the Complementary and Alternative Medicine Movement: How Activists Perceive a Range of Responses from Physicians and Hospitals," *Sociology of Health & Illness* 26(6): 710–36.

Goldstein, Diane E. (2004) "Communities of Suffering and the Internet," in R. M. Packard, J. Brown, R. L. Berkelman, & H. Frumkin (eds), *Emerging Illnesses and Society: Negotiating the Public Health* (Baltimore, MD: Johns Hopkins University Press): 121–38.

Goldstone, Jack A. (2003) "Introduction: Bridging Institutionalized and Noninstitutionalized Politics," in J. A. Goldstone (ed) *States, Parties, and Social Movements* (Cambridge, England: Cambridge University Press): 1–24.

Guston, David H. (2000) *Between Science and Politics: Assuring the Integrity and Productivity of Research* (New York: Cambridge University Press).

Hacking, Ian (1995) *Rewriting the Soul: Multiple Personality and the Science of Memory* (Princeton, NJ: Princeton University Press).

Haraway, Donna J. (1991) *Simians, Cyborgs, and Women: The Reinvention of Nature* (New York: Routledge).

Hardon, Anita (2006) "Contesting Contraceptive Innovation—Reinventing the Script," *Social Science & Medicine* 62(3): 614–27.

Heath, Deborah (1997) "Bodies, Antibodies, and Modest Interventions," in G. L. Downey & J. Dumit (eds), *Cyborgs & Citadels* (Santa Fe, NM: School of American Research Press): 67–82.

Heath, Deborah, Rayna Rapp, & Karen-Sue Taussig (2004) "Genetic Citizenship," in D. Nugent & J. Vincent (eds), *A Companion to the Anthropology of Politics* (London: Blackwell): 152–67.

Hess, David J. (2004a) "Health, the Environment and Social Movements" (guest editorial), *Science as Culture* 13(4): 421–27.

Hess, David J. (2004b) "Medical Modernisation, Scientific Research Fields and the Epistemic Politics of Health Social Movements," *Sociology of Health & Illness* 26(6): 695–709.

Hess, David J. (2004c) "Organic Food and Agriculture in the U.S.: Object Conflicts in a Health-Environmental Social Movement," *Science as Culture* 13(4): 493–513.

Hess, David J. (2005) "Technology- and Product-Oriented Movements: Approximating Social Movement Studies and Science and Technology Studies," *Science, Technology & Human Values* 30(4): 515–35.

Hoffman, Beatrix (2003) "Health Care Reform and Social Movements in the United States," *American Journal of Public Health* 93(1): 75–85.

Hoffman, Lily M. (1989) *The Politics of Knowledge: Activist Movements in Medicine and Planning* (Albany: State University of New York Press).

Indyk, Debbie & David Rier (1993) "Grassroots AIDS Knowledge: Implications for the Boundaries of Science and Collective Action," *Knowledge: Creation, Diffusion, Utilization* 15(1): 3–43.

Irwin, Alan & Mike Michael (2003) *Science, Social Theory, and Public Knowledge* (Maidenhead, PA: Open University Press).

Jasanoff, Sheila (ed) (2004) *States of Knowledge: The Co-Production of Science and Social Order* (New York: Routledge).

Joffe, Carole (1999) "Abortion and the Women's Health Movement: Then and Now (Commentary)," *Journal of the American Medical Women's Association* 54(1): 31–33.

Joffe, C. E., T. A. Weitz, & C. L. Stacey (2004) "Uneasy Allies: Pro-Choice Physicians, Feminist Health Activists and the Struggle for Abortion Rights," *Sociology of Health & Illness* 26(6): 775–96.

Johnson, Michael P. & Karl Hufbauer (1982) "Sudden Infant Death Syndrome as a Medical Research Problem since 1945," *Social Problems* 30(1): 65–81.

Karkazis, Katrina Alicia (2002) *Beyond Treatment: Mapping the Connections Among Gender, Genitals, and Sexuality in Recent Controversies over Intersexuality*, Ph.D. diss., Columbia University.

Katz, Alfred H. (1993) *Self-Help in America: A Social Movement Perspective* (New York: Twayne).

Kaufert, Patricia A. (1998) "Women, Resistance and the Breast Cancer Movement," in M. Lock & P. A. Kaufert (eds), *Pragmatic Women and Body Politics* (Cambridge: Cambridge University Press): 287–309.

Keating, Peter & Alberto Cambrosio (2003) *Biomedical Platforms: Realigning the Normal and the Pathological in Late-Twentieth-Century Medicine* (Cambridge, MA: MIT Press).

Kelleher, David (1994) "Self-Help Groups and Their Relationship to Medicine," in J. Gabe, D. Kelleher, & G. Williams (eds), *Challenging Medicine* (London: Routledge): 104–17.

Kirp, David L. (1999) "The Politics of Blood: Hemophilia Activism in the AIDS Crisis," in E. A. Feldman & R. Bayer (eds), *Blood Feuds: AIDS, Blood, and the Politics of Medical Disaster* (New York: Oxford University Press): 293–321.

Klawiter, Maren (1999) "Racing for the Cure, Walking Women, and Toxic Touring: Mapping Cultures of Action within the Bay Area Terrain of Breast Cancer," *Social Problems* 46(1): 104–26.

Klawiter, Maren Elise (2000) *Reshaping the Contours of Breast Cancer: From Private Stigma to Public Actions*, Ph.D. diss., University of California, Berkeley.

Klawiter, Maren (2003) "Chemicals, Cancer, and Prevention: The Synergy of Synthetic Social Movements," in M. J. Casper (ed) *Synthetic Planet: Chemical Politics and the Hazards of Modern Life* (New York: Routledge): 155–75.

Klawiter, Maren (2004) "Breast Cancer in Two Regimes: The Impact of Social Movements on Illness Experience," *Sociology of Health & Illness* 26(6): 845–74.

Klawiter, Maren (unpublished) "Transforming the Field of Mammographic Screening: Community Mobilization, Cultural Diversification, and Domain Expansion in an 'Interpenetrated' Social Movement."

Kolker, Emily S. (2004) "Framing as a Cultural Resource in Health Social Movements: Funding Activism and the Breast Cancer Movement in the US 1990–1993," *Sociology of Health & Illness* 26(6): 820–44.

Kroll-Smith, Steve & H. Hugh Floyd (1997) *Bodies in Protest: Environmental Illness and the Struggle over Medical Knowledge* (New York: New York University Press).

Kushner, Howard I. (2004) "Competing Medical Cultures, Patient Support Groups, and the Construction of Tourette's Syndrome," in R. M. Packard, J. Brown, R. L. Berkelman, & H. Frumkin (eds), *Emerging Illnesses and Society: Negotiating the Public Health* (Baltimore, MD: Johns Hopkins University Press): 71–101.

Lakoff, Andrew (2005) *Pharmaceutical Reason: Knowledge and Value in Global Psychiatry* (New York: Cambridge University Press).

Landzelius, Kyra (2006a) "The Incubation of a Social Movement: Preterm Babies, Parent Activists, and Neonatal Productions in the US Context," *Social Science & Medicine* 62(3): 668–82.

Landzelius, Kyra (2006b) "Introduction: Patient Organization Movements and New Metamorphoses in Patienthood," *Social Science & Medicine* 62(3): 529–37.

Landzelius, Kyra & Joe Dumit (2006) "Patient Organization Movements" (special issue), *Social Science & Medicine* 62(3): 529–792.

Larvie, Sean Patrick (1999) "Queerness and the Specter of Brazilian National Ruin," *GLQ* 5(4): 527–58.

Latour, Bruno (1987) *Science in Action: How to Follow Scientists and Engineers through Society* (Cambridge, MA: Harvard University Press).

Latour, Bruno (1998) "From the World of Science to the World of Research?," *Science* 280(5361): 208–9.

Lavoie, Francine, Thomasina Borkman, & Benjamin Gidron (eds) (1994) *Self-Help and Mutual Aid Groups: International and Multicultural Perspectives* (New York: Haworth).

Layne, Linda L. (2003) *Motherhood Lost: A Feminist Account of Pregnancy Loss in America* (New York: Routledge).

Layne, Linda L. (2006) "Pregnancy and Infant Loss Support: A New, Feminist, American, Patient Movement?," *Social Science & Medicine* 62 (3): 602–13.

Lerner, Barron H. (2001) "No Shrinking Violet: Rose Kushner and the Rise of American Breast Cancer Activism," *Western Journal of Medicine* 174 (5): 362–65.

Lock, Margaret, Allan Young, & Alberto Cambrosio (eds) (2000) *Living and Working with the New Medical Technologies: Intersections of Inquiry* (Cambridge: Cambridge University Press).

Loriol, Marc (2003) "Faire exister une maladie controversée: les associations de malades du syndrome de fatigue chronique et Internet," *Sciences Sociales et Santé* 21(4): 5–33.

Löwy, Ilana (2000) "Trustworthy Knowledge and Desperate Patients: Clinical Tests for New Drugs from Cancer to AIDS," in M. Lock, A. Young, & A. Cambrosio (eds), *Living and Working with the New Medical Technologies: Intersections of Inquiry* (Cambridge: Cambridge University Press): 49–81.

Matoba, Tomoko (2002) "A Sociological Study of Patients' Groups in Contemporary Japan," presentation at the annual meeting of the International Sociological Association, Brisbane, Australia.

McAdam, Doug (1988) *Freedom Summer* (New York: Oxford University Press).

McAdam, Doug (1995) "'Initiator' and 'Spin-Off' Movements: Diffusion Processes in Protest Cycles," in M. Traugott (ed), *Repertoires and Cycles of Collective Action* (Durham, NC: Duke University Press): 217–39.

McCally, Michael (2002) "Medical Activism and Environmental Health," *Annals of the American Academy of Political and Social Science* 584: 145–58.

McCormick, Sabrina, Phil Brown, & Stephen Zavestoski (2003) "The Personal Is Scientific, the Scientific Is Political: The Public Paradigm of the Environmental Breast Cancer Movement," *Sociological Forum* 18(4): 545–76.

McInerney, Fran (2006) "Heroic Frames: Discursive Constructions around the Requested Death Movement in Australia in the Late-1990s," *Social Science & Medicine* 62(3): 654–67.

Meyer, David S. & Suzanne Staggenborg (1996) "Movements, Countermovements, and the Structure of Political Opportunity," *American Journal of Sociology* 101(6): 1628–60.

Meyer, David S. & Nancy Whittier (1994) "Social Movement Spillover," *Social Problems* 41(2): 277–98.

Montini, Theresa Michalak (1991) *Women's Activism for Breast Cancer Informed Consent Laws*, Ph.D. diss., University of California, San Francisco.

Montini, Theresa (1996) "Gender and Emotion in the Advocacy for Breast Cancer Informed Consent Legislation," *Gender & Society* 10(1): 9–23.

Moore, Kelly (1999) "Political Protest and Institutional Change: The Anti-Vietnam War Movement and American Science," in M. Giugni, D. McAdam, & C. Tilly (eds), *How Social Movements Matter* (Minneapolis: University of Minnesota Press): 97–118.

Moore, Lisa Jean (1997) "'It's Like You Use Pots and Pans to Cook—It's the Tool": The Technologies of Safer Sex," *Science, Technology & Human Values* 22(4): 434–71.

Morrison, Linda Joy (2004) *Talking Back to Psychiatry: Resistant Identities in the Psychiatric Consumer/ Survivor/Ex-Patient Movement*, Ph.D. diss., University of Pittsburgh.

Murphy, Michelle (2004a) "Liberation through Control in the Body Politics of U.S. Radical Feminism," in L. Daston & F. Vidal (eds), *The Moral Authority of Nature* (Chicago: University of Chicago Press): 331–55.

Murphy, Michelle (2004b) "Occupational Health from Below: The Women's Office Workers' Movement and the Hazardous Office," in R. M. Packard, J. Brown, R. L. Berkelman, & H. Frumkin (eds), *Emerging Illnesses and Society: Negotiating the Public Health* (Baltimore, MD: Johns Hopkins University Press): 191–223.

Murphy, Michelle (2006) *Sick Building Syndrome and the Problem of Uncertainty: Environmental Politics, Technoscience, and Women Workers* (Durham, NC: Duke University Press).

Myhre, Jennifer Reid (2002) *Medical Mavens: Gender, Science, and the Consensus Politics of Breast Cancer Activism*, Ph.D. diss., University of California, Davis.

Mykytyn, Courtney Everts (2006) "Anti-Aging Medicine: A Patient/Practitioner Movement to Redefine Aging," *Social Science & Medicine* 62(3): 643–53.

Nathanson, Constance A. (1999) "Social Movements as Catalysts for Policy Change: The Case of Smoking and Guns," *Journal of Health Politics, Policy and Law* 24(3): 421–88.

Nelson, Alondra (2003) *Black Power, Biomedicine, and the Politics of Knowledge*, Ph.D. diss., New York University.

Novas, Carlos (2003) *"Governing 'Risky' Genes: Predictive Genetics, Counselling Expertise and the Care of the Self*, Ph.D. diss., Goldsmiths College, University of London.

Novas, Carlos (unpublished) "Managing Genomic Expectations: Creating Informed Consumers of Potential Innovations."

Nukaga, Yoshio (2002) "Between Tradition and Innovation in New Genetics: The Continuity of Medical Pedigrees and the Development of Combination Work in the Case of Huntington's Disease," *New Genetics and Society* 21(1): 39–64.

O'Donovan, Orla (2007) "Corporate Colonisation of Health Activism? Irish Health Advocacy Organisations' Modes of Engagement with Pharmaceutical Corporations," *International Journal of Health Services* 37(4).

Oudshoorn, Nelly & Trevor Pinch (eds) (2003) *How Users Matter: The Co-Construction of Users and Technology* (Cambridge, MA: MIT Press).

Packard, Randall M., Peter J. Brown, Ruth L. Berkelman, & Howard Frumkin (2004a) "Introduction: Emerging Illness as Social Process," in R. M. Packard, J. Brown, R. L. Berkelman, & H. Frumkin (eds), *Emerging Illnesses and Society: Negotiating the Public Health* (Baltimore, MD: Johns Hopkins University Press): 1–35.

Packard, Randall M., Peter J. Brown, Ruth L. Berkelman, & Howard Frumkin (eds) (2004b) *Emerging Illnesses and Society: Negotiating the Public Health*. (Baltimore, MD: Johns Hopkins University Press).

Parthasarathy, Shobita (2003) "Knowledge Is Power: Genetic Testing for Breast Cancer and Patient Activism in the United States and Britain," in N. Oudshoorn & T. Pinch (eds), *How Users Matter: The Co-Construction of Users and Technology* (Cambridge, MA: MIT Press): 133–50.

Pellow, David Naguib, & Lisa Sun-Hee Park (2002) *The Silicon Valley of Dreams: Environmental Injustice, Immigrant Workers, and the High-Tech Global Economy* (New York: New York University Press).

Petersen, James C. & Gerald E. Markle (1981) "Expansion of Conflict in Cancer Controversies," *Research in Social Movements, Conflict and Change* 4: 151–69.

Petersen, James C. & Gerald E. Markle (1989) "Controversies in Science and Technology," in D. E. Chubin and E. W. Chu (eds), *Science Off the Pedestal: Social Perspectives on Science and Technology* (Belmont, CA: Wadsworth): 5–18.

Petryna, Adriana (2002) *Life Exposed: Biological Citizens after Chernobyl* (Princeton, NJ: Princeton University Press).

Pichardo, Nelson (1997) "New Social Movements: A Critical Review," *Annual Review of Sociology* 23: 411–30.

Pickstone, John V. (2000) *Ways of Knowing: A New History of Science, Technology and Medicine* (Chicago: University of Chicago Press).

Pinch, Trevor J. & Weibe E. Bijker (1993) "The Social Construction of Facts and Artifacts: Or How the Sociology of Science and the Sociology of Technology Might Benefit Each Other," in W. E. Bijker, T. P. Hughes, & T. J. Pinch (eds), *The Social Construction of Technological Systems: New Directions in the Sociology and History of Technology* (Cambridge, MA: MIT Press): 17–50.

Polletta, Francesca (2004) "Culture in and Outside Institutions," in D. J. Myers & D. M. Cress (eds), *Authority in Contention: Research in Social Movements, Conflicts and Change* 25: 161–83.

Prior, Lindsay (2003) "Belief, Knowledge and Expertise: The Emergence of the Lay Expert in Medical Sociology," *Sociology of Health & Illness* 25(3): 41–57.

Rabeharisoa, Vololona (2003) "The Struggle against Neuromuscular Diseases in France and the Emergence of the 'Partnership Model' of Patient Organisation," *Social Science & Medicine* 57(11): 2127–36.

Rabeharisoa, Vololona (2006) "From Representation to Mediation: The Shaping of Collective Mobilization on Muscular Dystrophy in France," *Social Science & Medicine* 62(3): 564–76.

Rabeharisoa, Vololona & Michel Callon (2002) "The Involvement of Patients' Associations in Research," *International Social Science Journal* 54 (171): 57–63.

Rabinow, Paul (1996) *Essays on the Anthropology of Reason* (Princeton, NJ: Princeton University Press).

Race, Kane (2001) "The Undetectable Crisis: Changing Technologies of Risk," *Sexualities* 4(2): 167–89.

Radin, Patricia (2006) " 'To Me, It's My Life': Medical Communication, Trust, and Activism in Cyberspace," *Social Science & Medicine* 62(3): 591–601.

Rapp, Rayna (2000) "Extra Chromosomes and Blue Tulips: Medico-Familial Interpretations," in M. Lock, A. Young, & A. Cambrosio (eds), *Living and Working with the New Medical Technologies: Intersections of Inquiry* (Cambridge: Cambridge University Press): 184–208.

Rapp, Rayna & Faye Ginsburg (2001) "Enabling Disability: Rewriting Kinship, Reimagining Citizenship," *Public Culture* 13(3): 533–56.

Rapp, Rayna, Deborah Heath, & Karen-Sue Taussig (2001) "Genealogical Dis-Ease: Where Hereditary Abnormality, Biomedical Explanation, and Family Responsibility Meet," in S. Franklin & S. McKinnon (eds), *Relative Values: Reconfiguring Kinship Studies* (Durham, NC: Duke University Press): 384–409.

Ray, Raka (1999) *Fields of Protest: Women's Movements in India* (Minneapolis: University of Minnesota Press).

Reardon, Jennifer (2004) "Decoding Race and Human Difference in a Genomic Age," *differences: A Journal of Feminist Cultural Studies* 15(3): 38–65.

Reid, Roddey (2004) "Tensions within California Tobacco Control in the 1990s: Health Movements, State Initiatives, and Community Mobilization," *Science as Culture* 13(4): 515–37.

Rose, Nikolas (2001) "The Politics of Life Itself," *Theory, Culture & Society* 18(6): 1–30.

Rose, Nikolas & Carlos Novas (2005) "Biological Citizenship," in A. Ong & S. J. Collier (eds), *Global Assemblages: Technology, Politics, and Ethics as Anthropological Problems* (Malden, MA: Blackwell): 439–63.

Rosengarten, Marsha (2004) "Consumer Activism in the Pharmacology of HIV," *Body and Society* 10(1): 91–107.

Rothman, Sheila M. & David J. Rothman (2003) *The Pursuit of Perfection: The Promise and Perils of Medical Enhancement* (New York: Pantheon Books).

Saguy, Abigail C. & Kevin W. Riley (2005) "Weighing Both Sides: Morality, Mortality, and Framing Contests over Obesity," *Journal of Health Politics, Policy and Law* 30(5): 869–923.

Scott, Joan (1991) "The Evidence of Experience," *Critical Inquiry* 17(4): 773–97.

Shakespeare, Tom (1993) "Disabled People's Self-Organisation: A New Social Movement?," *Disability, Handicap and Society* 8(3): 249–64.

Shakespeare, Tom (1999) " 'Losing the Plot'? Medical and Activist Discourses of Contemporary Genetics and Disability," *Sociology of Health & Illness* 21(5): 669–88.

Sharp, Lesley A. (2001) "Commodified Kin: Death, Mourning, and Competing Claims on the Bodies of Organ Donors in the United States," *American Anthropologist* 103(1): 112–33.

Shim, Janet K. (2005) "Constructing 'Race' across the Science-Lay Divide: Racial Formation in the Epidemiology and Experience of Cardiovascular Disease," *Social Studies of Science* 35(3): 405–36.

Shostak, Sara (2004) "Environmental Justice and Genomics: Acting on the Futures of Environmental Health," *Science as Culture* 13(4): 539–62.

Silverman, Chloe (2004) *A Disorder of Affect: Love, Tragedy, Biomedicine, and Citizenship in American Autism Research, 1943–2003*, Ph.D. diss., University of Pennsylvania.

Skrentny, John David (2002) *The Minority Rights Revolution* (Cambridge, MA: Harvard University Press).

Spears, Ellen Griffith (2004) "The Newtown Florist Club and the Quest for Environmental Justice in Gainesville, Georgia," in R. M. Packard, J. Brown, R. L. Berkelman, & H. Frumkin (eds), *Emerging Illnesses and Society: Negotiating the Public Health* (Baltimore, MD: Johns Hopkins University Press): 171–90.

Staggenborg, Suzanne (1999) "The Consequences of Professionalization and Formalization in the Pro-Choice Movement," in J. Freeman & V. Johnson (eds), *Waves of Protest: Social Movements since the Sixties* (Lanham, MD: Rowman & Littlefield): 99–134.

Star, Susan Leigh & James R. Griesemer (1989) "Institutional Ecology, 'Translations' and Boundary Objects: Amateurs and Professionals in Berkeley's Museum of Vertebrate Zoology, 1907–39," *Social Studies of Science* 19: 387–420.

Stewart, Miriam J. (1990) "Expanding Theoretical Conceptualizations of Self-Help Groups," *Social Science & Medicine* 31(9): 1057–66.

Stockdale, Alan (1999) "Waiting for the Cure: Mapping the Social Relations of Human Gene Therapy Research," *Sociology of Health & Illness* 21(5): 579–96.

Talley, Colin (2004) "The Combined Efforts of Community and Science: American Culture, Patient Activism, and the Multiple Sclerosis Movement in the United States," in R. M. Packard, J. Brown, R. L. Berkelman, & H. Frumkin (eds), *Emerging Illnesses and Society: Negotiating the Public Health* (Baltimore, MD: Johns Hopkins University Press): 39–70.

Taussig, Karen-Sue, Rayna Rapp, & Deborah Heath (2003) "Flexible Eugenics: Technologies of the Self in the Age of Genetics," in A. H. Goodman, D. Heath, & M. S. Lindee (eds), *Genetic Nature/Culture: Anthropology and Science Beyond the Two-Culture Divide* (Berkeley: University of California Press): 58–76.

Taylor, Verta (1996) *Rock-a-by Baby: Feminism, Self-Help, and Postpartum Depression* (New York: Routledge).

Thompson, Charis (2005) *Making Parents: The Ontological Choreography of Reproductive Technologies* (Cambridge, MA: MIT Press).

Timmermans, Stefan & Marc Berg (2003) *The Gold Standard: The Challenge of Evidence-Based Medicine and Standardization in Health Care* (Philadelphia: Temple University Press).

Timmermans, Stefan & Valerie Leiter (2000) "The Redemption of Thalidomide: Standardizing the Risk of Birth Defects," *Social Studies of Science* 30(1): 41–71.

Treichler, Paula A. (1991) "How to Have Theory in an Epidemic: The Evolution of AIDS Treatment Activism," in C. Penley & A. Ross (eds), *Technoculture* (Minneapolis: University of Minnesota Press): 57–106.

Treichler, Paula A. (1999) *How to Have Theory in an Epidemic: Cultural Chronicles of AIDS* (Durham, NC: Duke University Press).

Van Kammen, Jessica (2003) "Who Represents the Users? Critical Encounters Between Women's Health Advocates and Scientists in Contraceptive R&D," in N. Oudshoorn & T. Pinch (eds), *How Users Matter: The Co-Construction of Users and Technology* (Cambridge, MA: MIT Press): 151–71.

Von Gizycki, Rainald (1987) "Cooperation Between Medical Researchers and a Self-Help Movement: The Case of the German Retinitis Pigmentosa Society," in S. Blume (ed) *The Social Direction of the Public Sciences* (Dordrecht, Netherlands: D. Reidel): 75–88.

Wajcman, Judy (2004) *Technofeminism* (Cambridge: Polity).

Ward, Jule DeJager (2000) *La Leche League: At the Crossroads of Medicine, Feminism, and Religion* (Chapel Hill: University of North Carolina Press).

Weisman, Carol S. (1998) *Women's Health Care: Activist Traditions and Institutional Change* (Baltimore, MD: John Hopkins University Press).

Whelan, Emma (2003) "Putting Pain to Paper: Endometriosis and the Documentation of Suffering," *Health* 7(4): 463–82.

Whyte, Susan Reynolds, Sjaak van der Geest, & Anita Hardon (2002) *Social Lives of Medicines* (Cambridge: Cambridge University Press).

Wilcox, Sarah (unpublished) "Framing AIDS and Breast Cancer as Lesbian Health Issues: Social Movements and the Alternative Press."

Wolfson, Mark (2001) *The Fight against Big Tobacco: The Movement, the State, and the Public's Health* (New York: Aldine de Gruyter).

Wynne, Brian (1992) "Misunderstood Misunderstandings: Social Identities and Public Uptake of Science," *Public Understanding of Science* 1: 281–304.

Zavestoski, Steve, Phil Brown, Meadow Linder, Sabrina McCormick, & Brian Mayer (2002) "Science, Policy, Activism, and War: Defining the Health of Gulf War Veterans," *Science, Technology & Human Values* 27(2): 171–205.

Zavestoski, Stephen, Sabrina McCormick, & Phil Brown (2004a) "Gender, Embodiment, and Disease: Environmental Breast Cancer Activists' Challenges to Science, the Biomedical Model, and Policy," *Science as Culture* 13(4): 563–86.

Zavestoski, Stephen, Phil Brown, Sabrina McCormick, Brian Mayer, Maryhelen D'Ottavi, & Jaime C. Lucove (2004b) "Patient Activism and the Struggle for Diagnosis: Gulf War Illnesses and Other Medically Unexplained Physical Symptoms in the U.S.," *Social Science & Medicine* 58(1): 161–75.

Zavestoski, Stephen, Rachel Morello-Frosch, Phil Brown, Brian Mayer, Sabrina McCormick & Rebecca Gasior Altman (2004c) "Embodied Health Movements and Challenges to the Dominant Epidemiological Paradigm," in D. J. Myers & D. M. Cress (eds), *Authority in Contention: Research in Social Movements, Conflicts and Change* 25: 253–78.

22   User-Technology Relationships: Some Recent Developments

Nelly Oudshoorn and Trevor Pinch

USERS MOVE CENTER STAGE

A number of different strands of scholarship have increasingly drawn attention to the importance of understanding user-technology relations. Our overview focuses in particular on the conceptual vocabulary developed within several different approaches and the similarities and differences between them. The main areas reviewed here include the following: *innovation studies*, in which the notion of "lead users" has been important; the *sociology of technology* and in particular the social construction of technology approach, which has drawn attention to the part played by users as relevant social groups and as agents of technological change; *feminist studies of technology*, where feminist historians of technology and others have focused on neglected household technologies and have developed important new concepts; *semiotic approaches*, which includes the notion of "configuring users" and of "scripts"; and *media and cultural studies* approaches, where the consumption and "domestication" of technologies have proved to be analytically useful.

Of course, no review of this sort can include all the work on users within STS, never mind allied disciplines (for interesting approaches in the philosophy of technology that draw attention to users as part of a general phenomenology of technology see Don Ihde [1990][1], but the above strands have been central within STS over the last decade. We conclude by pointing to the specific ideas that STS scholars are generating in understanding the blurring of the boundaries between production and consumption. The most dramatic manifestation is in information technology, where movements like "open source" software and distributed expertise systems such as the on-line encyclopedia Wikipedia (see chapter 37 in this volume) fully involve users as providers of content (for a similar case for on-line newspapers, see Boczkowski, 2004). Proponents of such changes see this as a fundamental reorganization of production and consumption in late capitalism, and some writers (e.g., von Hippel, 2005) see these trends as a new movement in the democratization of technology.[2]

INNOVATION STUDIES

Within innovation studies it has long been assumed that product innovations are mainly developed by product manufacturers. This piece of conventional wisdom has been turned on its head. While leading scholars like Nathan Rosenberg (1982) and Bengt-Åke Lundvall (1988) have shown growing recognition of the importance of users, it is the detailed research carried out by Eric von Hippel and his students that has been particularly influential. In one of his first studies von Hippel (1976) showed how users innovate new products in the fast-changing scientific instrument industry. Von Hippel studied four families of scientific instruments: the gas chromatographer, nuclear magnetic resonance spectrometer, ultraviolet absorption spectrophotometer, and transmission electron microscope. He examined how these instruments were first developed and how improvements came about. He found that it was often the users of these instruments who made the key innovations: "it is almost always the user, not the instrument manufacturer, who recognizes the need, solves the problem via an invention, builds a prototype and proves the prototype's value in use" (von Hippel, 1976: 227). Indeed later in-depth studies of the commercialization of the tunneling electron microscope and atomic force microscopes (Lenoir & Lecuyer, 1997; Mody, 2006) confirm this point—such instruments are often innovated by start-up companies when scientists have seen a novel application and form a business to exploit it. After three decades of work, in a variety of product and service industries, von Hippel (2005: 2) concludes that users are the "first to develop many and perhaps most new industrial consumer products." He and his students (e.g., Shah, 2005) examine examples ranging from information technologies, such as the Apache Software used by most Web servers, to extreme sports technologies such as kite surfers. The case studies are both of individual users and firms. The users who come up with the innovations, the "lead users," often go on to freely share their innovations so that other users can adopt, comment on, and improve on them. Manufacturers in turn will often commercialize these user-driven innovations.

Von Hippel was interested in his early work in how innovating firms can better do their market research to identify lead users. More recently he advocates using lead users as trialists within an iterative process so that technologies and markets are simultaneously constructed in interaction with each other. The model he develops, as is typical for the field of innovation studies, uses quantitative aggregative data. He combines this with the sort of case study methodology often found at business schools. This makes the user studies of von Hippel and his colleagues harder to integrate with the other STS approaches we discuss below, where "thick description" and more ethnographically inspired methods are the norm. There is less concern in the approaches of these economists with understanding how users conceive of their interaction with technology and with following user practices than in providing models, graphing quantifiable aspects of the process, and following trends over time (e.g., von Hippel, 2005). Interestingly, however, von Hippel advocates and brings to fruition the most interventionist approach that we will encounter in this review. His program, based at the MIT Sloan School, actively seeks to democratize innovation. Von Hippel forms and nurtures partnerships with "lead users" and offers instructional kits, training videos and joint research projects so that companies might better identify lead users and help them innovate. He regularly holds seminars with lead users, providing a forum with a unique blend of industry experience and academic concerns.

Users need not always be inherently innovative, as Hoogma and Schot (2001) caution in their study of the introduction of electric vehicles in two cities in France and Switzerland. Hoogma and Schot call for a sensitive interactive environment for the adaptation of some radical new technologies, such as electric vehicles, so that users' own preconceptions do not prevent them from taking advantage of the innovation. Scholars within innovation studies are increasingly dropping the term "users." While "lead users" often self-identify, there is clearly an issue about how users with no voice are represented. For example, Rose (2001) examines a range of different meanings attached to users (what he calls "user representations and articulations") in the innovation of vaccines in the United States. Here the end-users are often children who are not given a voice and hence are represented by other agents, such as parents, pharmaceutical companies, the Centers for Disease Control (CDC), the U.S. Food and Drug Administration (FDA), and public health clinics. This raises the interesting issue that users may represent other groups as end-users while at the same time promoting their own interests. Such work is leading to calls among innovation studies scholars for a reappraisal of the different parts played by demand, markets, and users in technological innovation (Coombs et al., 2001; Roharcher, 2005). This work thus increasingly pays attention to the interests of users and nonusers and who gets to represent users across the innovation process. Case studies of the European Information and Communication Technology sector (Williams et al., 2005) informed by the "social shaping of technology" approach (see below) have drawn attention to the importance of "social learning" in the innovation process. Such social learning is carried out by interaction between suppliers and users through the many diverse links in the innovation process. Much social learning, particularly for mass-produced consumer goods, is carried out through intermediaries (see below), which complicates the picture of flow from users to designers and vice versa. It seems that with the entry of the more sociologically inclined work there is an important opportunity for common interests to emerge among economists and sociologists working in innovation studies.

THE SCOT APPROACH: USERS AS AGENTS OF TECHNOLOGICAL CHANGE

Over the last two decades the maxim that "users matter" has become evident in a number of different areas of technology studies. The old view of users as passive consumers of technology has largely been replaced and along with it the linear model of technological innovation and diffusion. This has led to increased discussion of the social shaping of technology (MacKenzie & Wajcman, 1985; Williams et al., 2005). One of the first approaches to draw attention to users was the Social Construction of Technology (SCOT).

Pinch and Bijker (1984) in defining the SCOT approach conceived of users as one of the "relevant social groups" who played a part in the construction of a technology. Different social groups, including users groups, could construct radically different meanings of a technology, known as a technology's "interpretative flexibility." In a well-known study of the development of the bicycle, it was argued that social groups like elderly men and women gave a new meaning to the high-wheeled bicycle as the "unsafe" bicycle, and this helped pave the way for the development of the safety bicycle. The SCOT approach specifies a number of different "closure mechanisms," social processes whereby interpretative flexibility is curtailed. Eventually stabilization of a technology occurs, interpretative flexibility vanishes, and a predominant meaning and use emerges (Bijker & Pinch, 1987; Bijker, 1995a). The connection between designers and users was made more explicit with Bijker's notion of a technological frame (Bijker, 1995a). This term is rather akin to Kuhn's notion of a paradigm. Users, designers, and intermediaries can be said to share a technological frame associated with a particular technology, for example, electric lighting. The frame provides heuristics as to how users should interact with the technology such that the technology and user become part of a common "form of life."

Many classic SCOT studies were of the early stages of technologies—how new artifacts like bicycles or fluorescent lighting and materials like Bakelite moved from interpretative flexibility to stability. Relevant social groups were seen early on as the shaping agents, and only later with notions such as "sociotechnical ensembles" did SCOT fully embrace the idea of the co-construction or mutual shaping of social groups as well as technologies (Bijker, 1995b). SCOT was rightly criticized for its rather cavalier attitude toward users—it closed down the problem of users too early and did not show how users could actively modify stable technologies (Mackay & Gillespie, 1992). An attempt to remedy this deficit was offered by Kline and Pinch (1996) with their study of how a stable technology, the Model T automobile, could be appropriated and redesigned by groups such as farmers in the rural United States who turned their cars into sources of stationary power for washing machines, threshers, and the like for home and agricultural use. Kline and Pinch referred to such users as "agents of technological change."

Users have, however, remained a key element of SCOT studies. Revisiting the design stage of a technology in their recent empirical study of "contexts of use" within the early synthesizer industry, Pinch and Trocco (2002) argue that the successful synthesizer manufacturer, Robert Moog, developed new forums for learning from and interacting with his users. By understanding what users wanted and getting their feedback at all stages, Moog was able to constantly improve his synthesizer designs. For instance, he constructed a factory studio where he employed studio musicians whom he could use as guinea pigs in the development of new synthesizer modules.

The strength of SCOT is that it focuses on user practices and forums where the input of users can be studied. As with the turn to users in innovation studies, SCOT explores how the boundaries between design and use, between production and consumption, are blurred. The range of technologies studied include not only products and services but also on occasion large-scale technological systems that may have an adverse impact, for example, an environmental impact, on users (e.g., Beder, 1991). The unit of analysis in SCOT is the social group, which means that less attention is placed on individual users such as the lead users identified within innovation studies. SCOT deals with power. For instance, Bijker (1995a) offers a general semiotic theory of power, and Kline and Pinch (1996) discuss the specifics of the relative power of the Ford motor company *vis-à-vis* individual auto dealers. In general, though, SCOT, with the methodological priority it gives to social groups, has not paid as much attention to the diversity of users, the exclusion of users, and the politics of nonuse or restricted use as the feminist approaches we next consider.

FEMINIST AND PHILOSOPHICAL APPROACHES: DIVERSITY AND POWER

It is perhaps no surprise that feminist scholars have played a leading role in drawing attention to users. Their interest in users reflects a concern with the potential problematic consequences of technologies for women and the absence of women in historical accounts of technology. Since the mid-1980s, feminist historians have pointed to the neglect of women's role in the development of technology. Because women have been historically under-represented as innovators of technology and historians of technology often focused exclusively on the design and production of technologies, the history of technology came to be dominated by stories about men and their machines. Feminist historians suggested that a focus on users and use, instead of on engineers and design, would enable historians to go beyond histories of men inventing and mastering technology (Wajcman, 1991; Lehrman et al., 1997). In response to this criticism, users were gradually included in the research agenda of historians of technology. This "turn to the users" can be traced back to Ruth Schwartz Cowan's exemplary research of user-technology relations. Cowan's notion of the *consumption junction*, defined as "the place and time at which the consumer makes choices between competing technologies" (Cowan, 1987: 263) was a landmark concept. Cowan argued that a focus on consumers and the network relations in which they are embedded enables historians and sociologists of technology to improve their understanding of the unintended consequences of technologies in the hands of users. A focus on users enriches the history of technology with a better understanding of the successes and failures of technologies (Cowan, 1987: 279). STS scholars were urged to follow technologies all the way to the users (Rapp, 1998: 48). An exemplary study is Cynthia Cockburn and Suzan Ormrod's book on the microwave in the United Kingdom, including an extensive analysis of the design, production, testing (in special test kitchens), and marketing as well as the use of the new technology (Cockburn & Ormrod, 1993).[3] Thus, Cockburn and Ormrod draw an interesting distinction between "brown" and "white" goods as a way at getting at the gendering processes. Brown goods, such as VCRs, are designed, marketed, and sold to be fun and sexy for mainly male users, while white goods, such as washing machines, are designed, marketed, and sold in a more prosaic way for mainly women users. The microwave is interesting because it initially appeared as a "brown good" that enticed male consumers but later became another "white good"—part of the infrastructure of the household but no longer an object to get excited about. While different microwaves are not marketed to different female and male users, interestingly van Oost (2003) shows how in the case of shavers, the gendering goes further with different shavers being designed and marketed in very different ways for male and female users.

Gender studies, like technology studies in general, reflects a shift in the conceptualization of users from passive recipients to active participants. Whereas in the early feminist literature, women's relation to technology had been conceptualized predominantly in terms of victims of technology, the scholarship of the last two decades has emphasized women's active role in the appropriation of technology. This shift in emphasis was explicitly articulated in the first feminist collection of historical research on technology, *Dynamos and Virgins Revisited*, published in 1979, which included a section on "women as active participants in technological change" (Lehrman et al., 1997: 11).[4] Granting agency to users, particularly women, can thus be considered as a central concept in the feminist approach to understanding user-technology relations.

Another key concept in feminist studies of technology is the notion of diversity. As has been suggested by Cowan, users come in many different shapes and sizes (Cowan, 1987). Medical technologies, for example, incorporate a wide variety of users including patients, health professionals, hospital administrators, nurses, and patients' families. So, who is the user? This question is far from trivial. The very act of identifying specific individuals or groups as users may facilitate or constrain the actual role groups of users are allowed to play in shaping the development and use of technologies. Different groups involved in the design of technologies may have different views of who the user might, or should, be and these different groups can mobilize different resources to inscribe their views in the design of technical objects (Saetnan et al., 2000; Oudshoorn et al., 2004). To make things even more complicated, these different types of users don't necessarily imply homogeneous categories. Gender, age, socioeconomic, and ethnic differences can all be relevant. Because of this heterogeneity, not all users will have the same position in relation to a specific technology. For some, the room for maneuver will be great; for others, it will be slight. Feminist sociologists thus emphasize the diversity of users (see, for instance, the work of Susan Leigh Star [1991] on nonstandard users of information technologies) and encourage scholars to pay attention to differences in power relations among the multiple actors involved in the development of technology.

To capture the diversity of users[5] and the power relations encapsulating users and other actors in technological development, feminist sociologists have differentiated between *end-users, lay end-users,* and *implicated actors.* End-users are "those individuals and groups who are affected downstream by products of technological innovation" (Casper & Clarke, 1998). Lay end-users have been introduced to highlight some end-users' relative exclusion from expert discourse (Saetnan et al., 2000: 16). Implicated actor is a term introduced by Adele Clarke to refer to "those silent or not present but affected by the action" (Clarke, 1998: 267). This concept includes two categories of actors: "those not physically present but who are discursively constructed and targeted by others" and "those who are physically present but who are generally silenced/ignored/made invisible by those in power" (Clarke, 2005). All three terms reflect the long-standing feminist concern with the potential problematic consequences of technologies for women and include an explicit political agenda: the aim of feminist studies is to increase women's autonomy and their influence on technological development. A detailed understanding of how women as end-users or implicated actors matter in technological development may provide information useful in the empowerment of women or spokespersons of women, such as social movements and consumer groups.

The implicated actor concept also reflects a critical departure from actor-network approaches (see below) in technology studies. Feminists have criticized the sociology of technology, particularly actor-network theory, for the almost exclusive attention it gives to experts and producers and the preference it gives to design and innovation in understanding sociotechnical change.[6] This "executive approach" pays less attention to nonstandard positions, including women's voices (Star, 1991; Clarke & Montini, 1993: 45; Clarke, 1998: 267). Moreover, this approach implicitly assumes a specific type of power relations between users and designers in which designers are represented as powerful and users as disempowered relative to experts. Feminist sociologists suggest that the distribution of power among the multiple actors involved in sociotechnical networks should be approached as an empirical question (Lie & Sørensen, 1996: 4, 5; Clarke, 1998: 267; Oudshoorn et al., 2005). The notion of implicated actor has thus been introduced to avoid silencing invisible actors and actants and to include power relations explicitly in the analysis of user-expert relations.

Another important concept in the feminist vocabulary is the notion of *cyborg*. Donna Haraway has introduced this term to describe how by the late twentieth century we have become so thoroughly and radically merged and fused with technologies that the boundaries between the human and the technological are no longer impermeable. The cyborg implies a specific configuration of user-technology relations in which the user emerges as a hybrid of machine and organisms in fiction and as lived experience. Most importantly, Haraway has introduced the cyborg figure as a politicized entity. Cyborg analyses aim to go further than merely the deconstruction of technological discourses. In her well-known "cyborg manifesto" (1985), Haraway invites us to "question that which is taken as 'natural' and 'normal' in hierarchic social relations" (Haraway, 1985: 149). Her interest in cyborgs (and the contested subjectivities in her more recent work on animal-human hybridity around dog-human relationships, Haraway, 2003) is not to celebrate the fusion of humans and technology but to subvert and displace meanings in order to create alternative views, languages, and practices of technosciences and hybrid subjects.[7] In the last decade, the cyborg concept (popularized in science fiction as well) has resulted in an extensive body of literature, which describes the constitution and transformation of physical bodies and identities through technological practices.[8]

The feminist approach melds well with the SCOT approach in looking at processes whereby gender shapes social groups and artifacts. Its emphasis on the diversity of users and excluded or disempowered users does, however, offer new analytical tools for studying groups and individuals without a social group built around the shared meaning of an artifact. The methods used—ethnography, history, and "thick description"—also have more in common with SCOT than with the economists' innovation studies. The range of technologies studied can also be different. Feminism has always been concerned with the body and medical technologies. The turn to cyborgs and "cyborg anthropology" (Downey & Dumit, 1997) offers a new analytical vocabulary built around the body whereby excluded voices and negotiations of the boundaries between technologies and bodies can be studied. The body of the user appears within this approach as within none of the others reviewed here. Lastly, feminists wish to intervene in the politics of technology. Their goal is rather different, however, from the interventions of the innovation researchers in business schools as exemplified by von Hippel. Their desire is to change technology not so as to produce more innovations or to better identify user-driven innovations but rather to bring about the wider goals of political emancipation.

SEMIOTIC APPROACHES TO USERS: CONFIGURATION AND SCRIPT

An important new aspect for understanding user-technology relations has been introduced by scholars in STS who have extended semiotics—the study of how meanings are built—from signs to things. We focus here on two central concepts: "configuring the user" and "scripts." We start with configuring the user.

Exploring the metaphor of machine as text, Steve Woolgar has introduced the notion of the user as reader to emphasize the interpretative flexibility of technological objects and the processes that delimit this flexibility (Woolgar, 1991: 60). Although the interpretative flexibility of technologies and questions concerning the closure or stabilization of technology had already been addressed in SCOT, Woolgar focused attention on the design processes, which delimit the flexibility of machines, rather than on the negotiations between relevant social groups. He suggested that how users "read" machines is constrained because the design and the production of machines entails a process of configuring the user (Woolgar, 1991: 59). He shows this in particular in the case of a new personal computer where the sorts of interaction between the user and the computer are configured during testing with a particular user in mind. In this approach, the testing phase of a technology is portrayed as an important location to study the co-construction of technologies and users. In contrast to the approaches discussed thus far, this semiotic approach draws attention to users as represented by designers.

In recent debates, the notion of the configuration of users by designers has been extended to capture the complexities of designer-user relations more fully. Several authors have criticized Woolgar for describing configuration as a one-way process in which the power to shape technological development is merely attributed to experts in design organizations. They have suggested that the configuration processes can work both ways: "designers configure users, but designers in turn, are configured by both users and their own organizations" (Mackay et al., 2000: 752). This is increasingly the case in situations where designer-user relations are formalized by contractual arrangements (Mackay et al., 2000: 744). The capacity of designers to configure users can be further constrained by powerful groups within organizations who direct the course of design projects. In large organizations, for instance, designers usually have to follow specific organizational methods or procedures, which constrain design practices (Mackay et al., 2000: 741, 742, 744; Oudshoorn et al., 2004).

Another criticism and extension of the configuration approach is to question who is doing the configuration work. In Woolgar's studies, configuration work was restricted to the activities of actors within the company who produced the computers. Several authors have broadened this view of configuration to include other actors and to draw attention to the configuration work carried out by journalists (Oudshoorn, 2003), public sector agencies and states (Rose & Blume, 2003), policy makers, patient advocacy groups who act as spokespersons of users (van Kammen, 2000, 2003; Epstein, 2003; Parthasarathy, 2003), and other organizations and people who serve as mediators between producers and consumers, including consumer organizations (Schot & de la Bruheze, 2003), salespeople (Pinch, 2003), and clinical trials researchers (Fishman, 2004). Equally important, recent studies have shown how configuration work may also include the construction of identities for spokespersons of the technology themselves, namely, managers, firms, and engineers (Summerton, 2004: 488, 505). These studies illustrate that a thorough understanding of the role of users in technological development requires a methodology that takes into account the multiplicity and diversity of users, spokespersons of users, and locations where the co-construction of users and technologies takes place. From this perspective, technological development emerges as a culturally contested zone where users, patient advocacy groups, consumer organizations, designers, producers, salespeople, policymakers, and intermediary groups create, negotiate, and give differing, sometimes conflicting forms, meanings, and uses to technologies (Oudshoorn & Pinch, 2003). This scholarship adds a much needed richness in conceiving how the politics of users become manifest in today's technologically mediated state.

A second central notion in the semiotic approaches to user-technology relations is the concept of "script." Madeleine Akrich and Bruno Latour, in theorizing relationships between users and technology, use this term to capture how technological objects enable or constrain human relations as well as relationships between people and things. Akrich suggests that in the design phase technologists anticipate the interests, skills, motives, and behavior of future users. Subsequently, these representations of users become materialized into the design of the new product. As a result, technologies contain a script (or scenario): they attribute and delegate specific competencies, actions, and responsibilities to users and technological artifacts. Technological objects may thus create new, or transform or reinforce existing, "geographies of responsibilities" (Akrich, 1992: 207, 208). Rooted in actor network theory, Akrich and

Latour's work challenges social constructivist approaches in which only people are given the status of actors. Latour and Akrich have gone on to develop an extensive terminology to elaborate their "semiotics of machines" (Akrich & Latour, 1992).

In the last decade, feminist scholars have extended the script approach to include the gender dimensions of technological innovation. Adopting the view that technological innovation requires a renegotiation of gender relations and the articulation and performance of gender identities, Dutch and Norwegian feminists have introduced the concept of *genderscript* to capture all the work involved in the inscription and de-inscription of representations of masculinities and femininities in technological artifacts (Berg & Lie, 1993; Hubak, 1996; van Oost, 1995, 2003; Oudshoorn, 1999; Oudshoorn et al., 2002, 2004; Rommes et al., 1999; Spilkner & Sørensen, 2000). This scholarship emphasizes the importance of studying the inscription of gender into artifacts to improve our understanding of how technologies invite or inhibit specific performances of gender identities and relations. Technologies are represented as objects of identity projects, which may stabilize or destabilize hegemonic representations of gender (Oudshoorn, 2003; Saetnan et al., 2000; Crofts, 2004). Oudshoorn's 2003 book on the development of the male contraceptive pill is a good example of this approach. This book describes how the "feminization" of contraceptive technologies created a strong cultural and social alignment of contraceptive technologies with women and femininity and not with men and masculinity, which brings the development of new contraceptives for men into conflict with hegemonic masculinity. The development of new contraceptives for men thus required the destabilization of conventionalized performances of masculinity. Equally important, the genderscript approach drastically redefines the problem of exclusion of specific groups of people from technological domains and activities. Whereas policy makers and researchers have defined the problem largely in terms of deficiencies of users, genderscript analyses draw attention to the design of technologies (Oudshoorn et al., 2004; Rommes et al., 1999). These studies make visible how specific practices of configuring the user may lead to the exclusion of specific users.[9]

At first glance, the script approach seems to be similar to Woolgar's approach of configuring the user: both are concerned with understanding how designers inscribe their views of users and use in technological objects. A closer look, however, reveals important differences. Although both approaches deal with technological objects and designers, the script approach makes users more visible as active participants in technological development. Akrich in particular is aware that a focus on how technological objects constrain the ways in which people relate to things and to one another easily can be misunderstood as a technological determinist view that represents designers as active and users as passive. To avoid this misreading, she emphasizes the reciprocal relationship between objects and subjects and explicitly addresses the question of the agency of users (Akrich, 1992: 207). Akrich and Latour capture the active role of users in shaping their relationships to technical objects with the concepts of *subscription, de-inscription,* and *antiprogram.* Antiprogram refers to the users' program of action that is in conflict with the designers' program (or vice versa). Thus, the seat belt of the car is designed to restrain the user, but the user may have an antiprogram of refusing to wear the seat belt. Subscription, and its opposite, de-inscription, are used to describe the reactions of human (and nonhuman) actors to "what is prescribed and proscribed to them" and refer, respectively, to the extent to which they underwrite or reject and renegotiate the prescriptions (Akrich & Latour, 1992: 261). For example, for a while in the 1970s some cars were designed not to start unless the car seat belt was first fastened. Thus, a user fastening the seat belt is undergoing "subscription." But if a user finds a way of fooling the car into starting without the seat belt being fastened (say, by jamming a piece of metal into the seat belt attachment), the user is performing "de-inscription."

In contrast to Woolgar's work on configuring the user, script analyses thus conceptualize both designers and users as active agents in the development of technology. Compared to domestication theory (discussed in the next section), however, the script approach gives more weight to the world of designers and technological objects. The world of users, particularly the cultural and social processes that facilitate or constrain the emergence of users' antiprograms, remains largely unexplored within actor network approaches. More recently, this imbalance has been repaired to some extent by the work of scholars who have extended actor-network theory to include the study of subject-networks. These studies aim to understand the "attachment" between people and things, particularly but not exclusively between disabled people and assistive technologies, and to explore how technologies work to articulate subjectivities (Callon & Rabeharisoa, 1999; Moser, 2000; Moser & Law, 1998, 2003).[10] This scholarship conceptualizes subjects in the same way as actor-network theorists previously approached objects. Subject positions such as disability and ability are constituted as effects of actor-networks and hybrid collectives. More recently, Callon (forthcoming) in his study of patient organizations built around muscular dystrophy has gone on to consider "concerned groups" that are disenfranchised from modern consumer societies. He identifies groups that have lost all representation as "orphaned groups," who might be users who made the choice of a standard that was abandoned in favor of another that is not necessarily better or more efficient, or patients suffering from a disease in which both researchers and pharmaceutical laboratories have lost interest. He refers to "hurt groups" as groups of users that have been impacted adversely by issues of pollution and food safety, what might in more traditional economic analyses be referred to as groups impacted by externalities.

CULTURAL AND MEDIA STUDIES APPROACHES: CONSUMPTION AND DOMESTICATION

In contrast to the approaches to user-technology relations we have discussed thus far, scholars in cultural and media studies have acknowledged the importance of studying users from the very beginning. Whereas historians and sociologists of technology have chosen technology as their major topic of analysis, cultural and media studies have focused their attention primarily on users and consumers. Their central thesis is that technologies must be culturally appropriated to become fully functional. This scholarship has been inspired by Bourdieu's (1984) suggestion that consumption has become more central in the political economy of late modernity. Consequently, human relations and identities are increasingly defined in relation to consumption rather than production. In his study of differences in consumption patterns among social classes, Bourdieu defined consumption as a cultural and material activity and argued that the cultural appropriation of consumer goods depends on the "cultural capital" of people (Bourdieu, 1984).[11]

Feminist historians have also been important actors in signaling the relevance of studying consumption rather than production (McGaw, 1982). Feminists have long been aware of the conventional association and structural relations of women with consumption as a consequence of their role in the household and as objects in the commodity exchange system (de Grazia, 1996: 7). Whereas early feminist studies focused on the (negative) consequences of mass consumption for women, more recent studies address the question of whether women have been empowered by access to consumer goods. They conceptualize consumption as a site for the performance of gender and other identities.[12] The notion of consumption as a status and identity project has been further elaborated by Baudrillard (1988), who criticizes the view that the needs of consumers are dictated, manipulated, and fully controlled by the modern capitalist marketplace and by producers, as has been suggested by Adorno, Marcuse, and Horkheimer of the Frankfurt School (Adorno, 1991; Horkheimer & Adorno, ([1947]1979; Marcuse, 1964). Following Baudrillard, cultural and media studies emphasize the creative freedom of users to "make culture" in the practice of consumption as well as their dependence on "the culture industries" (Adorno, 1991), not because they control consumers but because they provide the means and the conditions of cultural creativity (Storey, 1999: xi). This scholarship portrays consumers as "cultural experts" who appropriate consumer goods to perform identities, which may transgress established social divisions (du Gay et al., 1997: 104; Chambers, 1985).

Semiotic approaches to analyzing user-technology relations have also come to the fore in cultural and media studies. One of the leading scholars in this field, Stuart Hall, has introduced the *encoding/decoding* model of media consumption (Hall, 1973). This model aims to capture both the structuring role of the media in "setting agendas and providing cultural categories and frameworks" as well as the notion of the "active viewer, who makes meaning from signs and symbols that the media provide" (Morley, 1995: 300). In the last two decades, the symbolic and communicative character of consumption has been extensively studied in cultural and media studies. Consumption fulfills a wide range of social and personal aims and serves to articulate who we are or who we would like to be, it may provide a symbolic means to create and establish friendship and to celebrate success, it may serve to produce certain lifestyles, it may provide the material for daydreams, and it may be used to articulate social difference and social distinctions (Bocock, 1993; du Gay et al., 1997; Lie & Sørensen, 1996; Mackay, 1997; Miller, 1995; Storey, 1999). Compared with technology studies, cultural and media studies thus articulate a perspective on user-technology relations, which emphasizes the role of technological objects in creating and shaping social identities, social life, and culture at large.[13]

A key concept developed in this tradition is the notion of *domestication*. Roger Silverstone has coined this term to describe how the integration of technological objects into daily life literally involves a "taming of the wild and a cultivation of the tame." Silverstone and Haddon (1996) looked at how new information technologies like computers were introduced into the home environment. A computer could be "tamed," for instance, by using it in a familiar setting (such as in the kitchen), by covering the screen with self-stick notes, or by choosing a screen-saver showing a photograph of a family member. New technologies have to be transformed from being unfamiliar, exciting, and possibly threatening things to familiar objects embedded in the culture of society and the practices and routines of everyday life (Silverstone & Hirsch, 1992; Lie & Sørensen, 1996). Domestication processes include symbolic work, where people create symbolic meanings of artifacts and adopt or transform the meanings inscribed in the technology; practical work, where users develop a pattern of usage to integrate artifacts into their daily routines; and cognitive work, which includes learning about artifacts (Lie & Sørensen, 1996: 10; Sørensen et al., 1994). In this approach, domestication is defined as a dual process in which technical objects as well as people may change. The use of technological objects may change the form and practical and symbolic functions of artifacts, and it may enable or constrain performances of identities and negotiations of status and social position (Silverstone et al., 1989; Lie & Sørensen, 1996).[14]

Domestication approaches have enriched our understanding of user-technology relations by elaborating the processes involved in consumption. In *Consuming Technologies*, Roger Silverstone and colleagues have specified four different phases of domestication: *appropriation, objectification, incorporation,* and *conversion.* Appropriation refers to the moment at which a technical object is sold and individuals or households become the owners of the product or service (Silverstone et al., 1992: 21). Objectification is a concept to describe processes of display that reveal the norms and principles of the household's sense of itself and its place in the world (Silverstone, 1992: 22). Incorporation is introduced to focus attention on the ways in which technological objects are used and incorporated into the routines of daily life. Finally, conversion describes the processes in which the use of technological objects shapes relationships between users and people outside the household (Silverstone, 1992: 25). In this process, artifacts become tools to make status claims and express a specific life style to neighbors, colleagues, family, and friends (Silverstone & Haddon, 1996: 46).

Although at first sight, the concepts of domestication and decoding or de-inscription may be considered as synonymous, there is an important difference. By specifying the processes involved in the diffusion and use of technology, domestication approaches take the dynamics of the world of users as their point of departure. Decoding and de-inscription, on the other hand, give priority to the design context in order to understand the emergence of user-technology relations. Compared with semiotic approaches, domestication approaches emphasize the complex cultural dynamics in which users appropriate technologies (Silverstone & Haddon, 1996: 52). In contrast, semiotic approaches tend to define users as isolated individuals whose relationship to technology is restricted to technical interactions with artifacts (Silverstone & Haddon, 1996: 52).

Most importantly, cultural and media studies inspire us to transcend the artificial divide between design and use. This scholarship has drastically reconceptualized the traditional distinction between production and consumption by reintroducing Karl Marx's claim that the process of production is not complete until users have defined the uses, meanings, and significance of the technology: "consumption is production" (Marx [1857–58]1980: 24). They describe design and domestication as "the two sides of the innovation coin" (Lie & Sørensen, 1996: 10).

THE BLURRING OF PRODUCTION AND CONSUMPTION

The research on user-technology relationships in the different fields we have discussed emphasizes the creative capacity of users to shape technological development in all phases of technological innovation. This view has inspired scholars to argue that the boundaries between design and use are largely artificial (Suchman, 1994, 2001; Silverstone & Haddon, 1996: 44; Lie & Sørensen, 1996: 9, 10; Williams et al., 2005). What is more, users can have multiple identities. In addition to being users, they can perform activities and identities traditionally ascribed to designers.[15] This blurring of the boundaries between design and use is something that cultural commentators have noticed. For instance, reflecting on significant changes in the economy and culture of the late 1970s, including the emergence of self-help movements, do-it-yourself trends, customized production, and new production technologies, Alvin Toffler (1980), one of the gurus of the information technology revolution, introduced the notion of the "prosumer." He coined this term to highlight that consumers are increasingly involved in services and tasks once done for them by others, which draws them more deeply into the production process (Toffler, 1980: 273). According to Toffler, this "basic shift from the passive consumer to active prosumer" changes the very nature of production: production increasingly shifts from the market sector based on production for exchange to the "prosumption sector" characterized by production for use. The rise of the prosumer thus has the potential to change the entire economic system (Toffler, 1980: 283).

Within STS, several scholars have introduced new concepts to avoid *a priori* dichotomization of design and use. James Fleck has enriched the sociological vocabulary for understanding the dynamics of technological development with the notion of "innofusion" (Fleck, 1988). He introduced this term to emphasize that processes of innovation continue during the process of diffusion.[16] In a similar vein, Eric von Hippel has introduced the concept of innovation user, or user/self-manufacturer (von Hippel, 2002: 3). Von Hippel argues that user innovation networks, which he defines as "user nodes interconnected by information transfer links which may involve face-to-face, electronic or any other form of communication," can function completely independently of manufacturers (von Hippel, 2002: 2). This user-led innovation pattern is in contrast to innovation processes led by "innovation manufacturers," who share their innovation by selling it to the marketplace. Lastly, Hugh Mackay and colleagues have introduced the term designer-users to capture the role of users in user-centered design methods that became fashionable in information technology companies in the United Kingdom in the 1990s. In design approaches such as rapid application development, users are involved from the outset of the development process as part of collaborative teams involving both designers and users, going beyond traditional divisions of labor (Mackay et al., 2000: 740).[17]

Whereas the authors discussed thus far aim to avoid dualistic conceptualizations of the relationship between design and use, others try to go beyond traditional representations of user-technology relations by bringing nonuse and resistance to the fore. Several authors have argued that a focus on use alone is insufficient to capture the complexities of user-technology relations. An adequate understanding of user-related sociotechnical change also requires a detailed analysis of nonuse and resistance. Although resistance to technology is an old topic, recent scholarship challenges common perceptions and theoretical understandings that view it as irrational or heroic. Instead of representing resistance and nonuse as irrational, heroic, or involuntary actions, these scholars argue that such reactions to technology should in some circumstances be considered as perfectly reasonable choices shaping the design and (de)stabilization of technologies. As Ron Kline suggests, resistance can be considered as a common feature of the processes underlying sociotechnical change. Acts considered as resistance by promoters, mediators, and users are crucial aspects of the creation of new technologies and social relations (Kline, 2003). In a similar vein, recent scholarship has challenged common understandings of nonuse (Laegran, 2003; Wyatt, 2003; Summerton, 2004). In modernist discourse, nonuse is portrayed as a deficiency and an involuntary act. Challenging this view, Sally Wyatt and colleagues reconceptualized the category of nonuse to include the voluntary and involuntary aspects of nonuse (Wyatt, 2003; Wyatt et al., 2003). Their preliminary taxonomy identifies four different types of nonusers: resisters (people who have never used the technology because they do not want to); rejectors (people who do not use the technology anymore because they find it boring or expensive, or because they have alternatives); the excluded (people who have never used the technology because they cannot get access for a variety of reasons); and the expelled (people who have stopped using the technology involuntarily because of cost or the loss of institutional access). These studies warn us to avoid the pitfalls of implicitly accepting the rhetoric of technological progress, including a worldview in which adoption of new technologies is the norm. This scholarship urges us to take seriously nonusers and former users as relevant social groups in shaping sociotechnical change.

CONCLUSION: NEGLECTED USERS AND USERS AS NO RESPECTERS OF BOUNDARIES

Adam Smith writing in *The Wealth of Nations* in 1776 talked about "the invention of a great number of machines which facilitate and abridge labour, and enable one man to do the work of many." He went on to note that "a great part of the machines made use of in those manufactures in which labor is most subdivided, were originally the invention of common workmen, who, being each of them employed in some very simple operation, naturally turned their thoughts toward finding out easier and readier methods of performing it" (Smith, 1776: 11–13). This reminds us not only of the long and largely hidden inventive endeavors of "common" people, but also of an important class of users that most STS studies have not yet focused sufficient attention on. These are factory workers and people who are users of machines and processes in the realm of production. Nearly all the recent STS work on users has been on technologies of consumption. Although in early STS there was a strong emphasis on studying production (Winner, 1977; Nobel, 1984), modern scholarship has shifted toward studying consumption technologies. The time is ripe to repair this imbalance. Indeed, it could be argued that the work on users gives us a new lens through which to look at production. Much has been written in the older vein of scholarship about the de-skilling debate initiated by Braverman (1975)—looking at the inventive skills of workers and how, whether, and by what means they have been harnessed to capitalist production and who has benefited might provide an interesting way of returning to some of the old debates over the labor process (e.g., Cockburn, 1983).

Other lacunae in the work surveyed here are apparent. There is a vast literature on social movements and medical sociology studies of patients' groups that has barely been touched on here. The current debates about whether hospitals, health insurance companies, or national health systems are the actual users of high technologies such as MRI reminds us that institutions (including the state, the military, and corporations) as well as individuals are important users. These institutional actors if reconceptualized as user groups might offer another avenue to understanding users and their struggles to redefine technoscientific practices. We have also skated over much of the important literature on "user-centered design" in the area of information and computer technologies and the work on computer cooperatives.

What of the future? It is clear that users come in all guises and that the notion of the user is an important probe for examining all sorts of diverse areas of technoscience. For example, in the study of model organisms initiated by Robert Kohler (1994) we find attention now being paid to users. Karen Rader (2004) in her book on the development of the mouse as model organism for genetics draws on the user literature in technology studies. She shows that the geneticist C. C. Little, who produced most of the mice used in post-war genetics, was acutely aware of his users and actively recruited new users who might make use of his standardized laboratory mouse. In addition, the examination of users in emerging areas of nanotechnology and the genome might pay dividends. Users appear everywhere across the spectrum of technoscience, and often someone who is in one context a producer of, say, new knowledge will be a user of, say, techniques and knowledge produced elsewhere. Indeed, this returns us to an old and fundamental point in the sociology of science—that the main reward in science is producing something that can be used by other scientists (Mulkay, 1976). The turn to users (and indeed intermediaries and mediation junctions [Oldenziel et al., 2005; Williams, Stewart, & Slack, 2005]) and their multiple identities is thus, as we have argued above, an opportunity to address within a single context issues and approaches that have often been pursued in multiple contexts and have spawned different bodies of literature. Users are no respecters of boundaries, and studying users forces the analyst also to cross boundaries. Throughout this review we have tried to point to links between often disparate bodies of literature, links that if pursued in future research might lead to a new synthesis and new approaches in the field of STS as a whole.

Notes

1. Other research traditions not covered here include the design literature, including human-computer interaction research and user-involvement methods, and psychological research on the adoption of new technologies. For a critical review of the user-centered design literature, see Garrety and Badham (2004). For an overview of social psychological studies, particularly the uses and gratifications theory and social cognitive approaches to understanding user-technology relations, see Ruggiero (2000) and LaRose et al. (2001). For a critical analysis of the models developed to study the acceptance of technology by users, see Ventakesh et al. (2003).

2. An earlier version of this chapter was published in Oudshoorn and Pinch (2003).

3. Examples of more recent studies of the "consumption junction" include Oldenziel (2001) and Klawiter (2004).

4. For an overview of feminist studies of technology, see Faulkner (2000), Lehrman et al. (2003), and Wajcman (1991, 2004).

5. Friedman, for example, has introduced a typology of users of computer systems that includes six different types: patrons (the initiators of the technology), clients (for whom the system is intended and designed), design inter-actors (who are involved in the design process), end-users (who operate the system), maintenance or enhancement inter-actors (those involved in the further evolution of the technology), and secondary users (individuals who are displaced, de-skilled, or otherwise affected (Friedman, 1989: 184, 185). See Mackay et al. (2000) for a discussion of taxonomies of users introduced by other scholars.

6. See, for example, Lohan (2000). Similar criticism has been articulated in STS studies (Mackay & Gillespie, 1992).

7. For a more detailed discussion of the politics of Haraway's cyborg figure, see Prins (1995) and Moser (2000).

8. See, for instance, Thompson (2005), Downey and Dumit (1997), Gray (1995), Henwood et al. (2001).

9. Script approaches are not only adopted by feminist scholars but also are used by researchers interested in rethinking user involvement in design in order to enhance sustainable technologies. For an exemplary study, see Jelsma (2003).

10. See Gomart and Hennion (1999) and Bakardjieva (2005) for studies of "subject-networks" that focus on other domains—the attachment of music amateurs and drug users—and the relations users establish with the Internet.

11. The early roots of this view can be traced back to the tradition of the anthropological study of material culture, most notably the work of Mary Douglas and Baron Isherwood (1979).

12. See Lehrman et al. (1997) for an overview of this literature. Inspired by feminist scholars, historians have extensively studied the history and culture of what is familiarly called consumer society, a concept introduced to identify the emergence of a specific type of market society, the Western capitalist system of exchange. Dutch historians of technology, for example, have written detailed accounts of the active role of intermediary organizations such as consumer groups in the emergence of the consumer society in the twentieth century, which they describe in terms of a coevolution of new products and new users (Schot & de la Bruheze, 2003). See Storey (1999: chapter 1) for a discussion and overview of the historical accounts of the birth and the development of a consumer society.

13. See Lury (1996) for a discussion of the different views of the relationship between consumption and identity.

14. See McCracken (1988) for an exemplary study of the symbolic work involved in appropriating consumer technologies. For an exemplary study of the emotional and social work involved in domesticating the Internet, see Bakardjieva (2005). Bakardjieva has suggested that "warm experts," a term she introduced to refer to people who are already familiar with the technology and are part of the user's life world, such as close friends, are important to facilitate the domestication of the Internet. Warm experts act as "an intermediary between the world of technology and the new user's personal world (Bakardjieva, 2005). Other studies of domestication include Frissen (2000), Katz and Rice (2002), Ropke (2003), Schroeder (2002), Slooten et al. (2003).

15. For exemplary studies of the multiple identities of users, see Lindsay (2003).

16. See Lieshout et al. (2001) for a detailed analysis of "innofusion" processes in the introduction of multimedia in education. See also Douthwaite (2001), who has developed an "innovation by users" model to analyze the iterative processes among users and between users and designers.

17. Although user participation has become more central in information technology and computer development, particularly in the field of human-computer interaction, the actual contribution of users to the development of IT systems is often restricted (Mackay et al., 2000: 748; Suchman, 2001).

References

Adorno, T. W. (1991) *The Culture Industry: Selected Essays on Mass Culture*, ed. J. M. Bernstein (London: Verso).

Akrich, M. (1992) "The De-scription of Technical Objects," in W. Bijker & J. Law (eds), *Shaping Technology—Building Society: Studies in Sociotechnical Change* (Cambridge, MA: MIT Press): 205–24.

Akrich, M. & B. Latour (1992) "A Summary of a Convenient Vocabulary for the Semiotics of Human and Nonhuman Assemblies," in W. Bijker & J. Law (eds), *Shaping Technology—Building Society: Studies in Sociotechnical Change* (Cambridge, MA: MIT Press): 259–64.

Bakardjieva, M. (2005) *Internet Society: The Internet in Everyday Life* (London: Sage).

Baudrillard, J. (1988) "Consumer Society," in Mark Poster (ed), *Selected Writings* (Cambridge, MA: Polity Press): 29–56.

Beder, Sharon. (1991) "Controversy and Closure: Sydney's Beaches in Crisis," *Social Studies of Science* 21(2): 223–56.

Berg, A. J. & M. Lie (1993) "Feminism and Constructivism: Do Artifacts Have Gender?" *Science, Technology & Human Values* 20: 332–51.

Bijker, W. E. (1995a) *On Bikes, Bicycles and Bakelite* (Cambridge, MA: MIT Press).

Bijker, W. E. (1995b) "Sociohistorical Technology Studies," in S. Jasanoff, G. E. Markle, J. C. Petersen, & T. J. Pinch (eds), *Handbook of Science and Technology Studies* (Thousand Oaks, CA, London, and New Delhi: Sage): 229–56.

Bijker, W. E. & Pinch, T. J. (1987) "The Social Construction of Facts and Artifacts: Or How the Sociology of Science and the Sociology of Technology Might Benefit Each Other," in W. E. Bijker, T. P. Hughes, & T. J. Pinch (eds), *The Social Construction of Technological Systems: New Directions in the Sociology and History of Technology* (Cambridge, MA: MIT Press): 17–50.

Bocock, R. (1993) *Consumption* (London: Routledge).

Boczkowski, P. (2004) *Digitizing the News: Innovation in Online Newspapers* (Cambridge, MA: MIT Press).

Bourdieu, P. (1984) *Distinction: A Social Critique of the Judgement of Taste* (London: Routledge).

Braverman, Harry, (1975) *Labor and Monopoly Capital: The Degradation of Work in the Twentieth Century* (New York: Monthly Review Press).

Callon, M. (forthcoming) "Economic Markets and the Rise of Individualism: From Prosthetic Agencies to 'Habilitated' Agencies," in T. Pinch & R. Swedberg (eds), *Living in a Material World* (Cambridge, MA: MIT Press).

Callon, M. & V. Rabeharisoa (1999) "La Leçon d'Humanité de Gino," *Reseaux* 17:189–233.

Casper, M. & A. Clarke (1998) "Making the Pap Smear into the Right Tool for the Job: Cervical Cancer Screening in the United States, c. 1940–1995," *Social Studies of Science* 28(2): 255–90.

Chambers, I. (1985) *Urban Rhythms: Pop Music and Popular Culture* (Basingstoke, U.K.: Macmillan).

Clarke, A. (1998) *Disciplining Reproduction: Modernity, American Life and 'The Problem of Sex'* (Chicago: University of Chicago Press).

Clarke, A. E. (2005) *Situational Analysis: Grounded Theory After the Postmodern Turn* (Thousand Oaks, CA: Sage).

Clarke, A. E. & T. Montini (1993) "The Many Faces of RU 486: Tales of Situated Knowledges and Technological Contestations," *Science, Technology & Human Values* 18(1): 42–78.

Cockburn, C. (1983) *Brothers: Male Dominance and Technological Change* (London: Pluto Press).

Cockburn C. & S. Ormrod (1993) *Gender and Technology in the Making* (London: Sage).

Coombs, R., K. Green, A. Richards, & V. Walsh (eds) (2001) *Technology and the Market: Demand, Users and Innovation* (Cheltenham and Northampton, U.K.: Edward Elgar).

Cowan, R. S. (1987) "The Consumption Junction: A Proposal for Research Strategies in the Sociology of Technology," in W. E. Bijker, T. P. Hughes, & T. J. Pinch (eds), *The Social Construction of Technological Systems: New Directions in the Sociology and History of Technology* (Cambridge, MA: MIT Press).

Crofts, L. (2004) "Virtual Gender: Technology, Consumption and Identity," *Gender Work and Organization* 11(1): 116–19.

De Grazia, V. (1996) "Introduction," in V. de Grazia with E. Furlough (eds) *The Sex of Things: Gender and Consumption in Historical Perspective* (Berkeley, Los Angeles, and London: University of California Press): 1–10.

Douglas, M. & B. Isherwood (1979) *The World of Goods* (London: Allen Lane).

Douthwaite, T. (2001) *Enabling Innovation: A Practical Guide to Understanding and Fostering Technological Change* (London: Zed Books).

Downey, G. L. & J. Dumit (eds) (1997) *Cyborgs and Citadels: Anthropological Interventions in Emerging Sciences and Technologies* (Santa Fe, NM: School of American Research Press).

Du Gay, P., S. Hall, L. Janes, H. Mackay, & K. Negus (1997) *Doing Cultural Studies. The Story of the Sony Walkman* (Thousand Oaks, CA, London, and New Delhi: Sage).

Epstein, S. (2003) "Inclusion, Diversity, and Biomedical Knowledge-making: The Multiple Politics of Representation," in N. Oudshoorn & T. Pinch (eds), *How Users Matter: The Co-Construction of Users and Technologies* (Cambridge, MA: MIT Press): 173–93.

Faulkner, W. (2000) "The Power and the Pleasure? A Research Agenda for 'Making Gender Stick' to Engineers," *Science, Technology & Human Values*, 25: 87–119.

Fishman, J. R. (2004) "Manufacturing Desire: The Commodification of Female Sexual Dysfunction," *Social Studies of Science* 34(2): 187–218.

Fleck, J. (1988) "Innofusion or Diffusiation? The Nature of Technological Development in Robotics," Edinburgh PICT Working Paper 7, Edinburgh.

Friedman, A. (1989) *Computer System Development: History, Organisation and Implementation* (Chichester, U.K.: John Wiley).

Frissen, V. (2000) "ICTs in the Rush Hour of Life: Acceptance, Use and Meanings of ICTs in 'Busy' Households," *Information Society* 16(1): 65–76.

Garrety, K., & R. Badham (2004) "User-centered Design and the Normative Politics of Technology," *Science, Technology & Human Values* 29: 191–212.

Gomart, E., & A. Hennion (1999) "A sociology of Attachment: Music Amateurs, Drug Users," in *Actor Network Theory and After*. J. Law & J. Hassard (eds) (Oxford: Blackwell and Sociological Review): 220–47.

Gray, C. H. (ed) (1995) *The Cyborg Handbook* (New York and London: Routledge).

Hall, S. (1973) "Encoding and Decoding in the Television Discourse," CCCS Stencilled Occasional Paper No. 7, Centre for Contemporary Cultural Studies, University of Birmingham, U.K.

Haraway, D. (1985) "Manifesto for Cyborgs: Science, Technology and Socialist Feminism in the 1980s," *Socialist Review* 80: 65–108, reprinted as "A Cyborg Manifesto: Science, Technology and Socialist Feminism in the Late Twentieth Century," in D. Haraway (1991) *Simians, Cyborgs and Women: The Reinvention of Nature* (London: Routledge): 149–83.

Haraway, D. (2003) *The Companion Species Manifesto: Dogs, People and Significant Otherness* (Chicago: University of Chicago Press).

Henwood, F., H. Kennedy, & N. Miller (eds) (2001) *Cyborg Lives? Women's Technobiographies* (York, U.K.: Raw Nerve Books).

Hoogma, R. & J. Schot (2001) "How Innovative are Users? A Critique of Learning-by-Doing and -Using," in R. Coombs, K. Green, A. Richards, & V. Walsh (eds), *Technology and the Market: Demand, Users and Innovation* (Cheltenham and Northampton, U.K.: Edward Elgar): 216–33.

Horkheimer, M. & T. W. Adorno ([1947]1979) *Dialectic of Enlightenment* (London: Verso).

Hubak, M. (1996) "The Car as a Cultural Statement: Car Advertising as Gendered Socio-technical Scripts," in M. Lie & K. H. Sørenson (eds) *Making Technology our Own? Domesticating Technology into Everyday Life* (Oslo, Oxford, and Boston: Scandinavian University Press): 171–201.

Ihde, D. (1990) *Technology and the Lifeworld: From Garden to Earth* (Bloomington and Indianapolis: Indiana University Press).

Jelsma, J. (2003) "Innovating for Sustainability: Involving Users, Politics and Technology," *Innovation* 16(2): 103–16.

Jenkins, R. V. (1975) "Technology and the Market: George Eastman and the Origins of Mass Amateur Photography," *Technology and Culture* 16: 1–19.

Katz, J. & R. E. Rice (2002) *Social Consequences of Internet Use: Access, Involvement and Interaction* (Cambridge, MA: MIT Press).

Klawiter, M. (2004) "The Biopolitics of Risk and the Configuration of Users: Clinical Trials, Pharmaceutical Technologies, and the New Consumption Junction," Presentation at the 4S-EASST conference, Paris, August 25–28.

Kline, R. R. (2003) "Resisting Consumer Technology in Rural America: The Telephone and Electrification," in N. Oudshoorn & T. Pinch (eds), *How Users Matter: The Co-Construction of Users and Technologies* (Cambridge, MA: MIT Press): 51–67.

Kline, R. & T. Pinch (1996) "Users as Agents of Technological Change: The Social Construction of the Automobile in the Rural United States," *Technology and Culture* 37: 763–95.

Kohler, R. (1994) *Lords of the Fly: Drosophila Genetics and the Experimental Life* (Chicago: University of Chicago Press).

Laegran, A. (2003) "Escape Vehicles? The Internet and the Automobile in a Local/Global Intersection," in N. Oudshoorn & T. Pinch (eds), *How Users Matter: The Co-Construction of Users and Technologies* (Cambridge, MA: MIT Press): 81–103.

LaRose, R., D. Mastro, & M. S. Eastin (2001) "Understanding Internet Usage—A Social Cognitive Approach to Uses and Gratifications," *Social Science Computer Review* 19: 395–413.

Lehrman, N. E., A. P. Mohun, & R. Oldenziel (1997) "The Shoulders We Stand on and the View from Here: Historiography and Directions for Research," *Technology and Culture* 38: 9–30.

Lehrman, N. E., R. Oldenziel, & A. P. Mohun (eds) (2003) *Gender & Technology: A Reader* (Baltimore and London: Johns Hopkins University Press).

Lenoir, T., & C. Lecuyer (1997) "Instrument Makers and Discipline Builders: The Case of Nuclear Magnetic Resonance," In T. Lenoir (ed) *Instituting Science: The Cultural Production of Scientific Disciplines* (Stanford, CA: Stanford University Press): 239–92.

Lie, M. & K. H. Sørenson (1996) *Making Technology Our Own? Domesticating Technology into Everyday Life* (Oslo, Oxford, and Boston: Scandinavian University Press).

Lieshout, M., T. Egyedi, & W. E. Bijker (eds) (2001) *Social Learning Technologies: The Introduction of Multimedia in Education* (Aldershot, U.K.: Ashgate).

Lindsay, C. (2003) "From within the Shadows: Users as Designers, Producers, Marketers, Distributors, and Technical Support," in N. Oudshoorn & T. Pinch (eds), *How Users Matter: The Co-Construction of Users and Technologies* (Cambridge, MA: MIT Press): 29–51.

Lohan, M. (2000) "Constructive Tensions in Feminist Technology Studies," *Social Studies of Science* 30: 895–916.

Lundvall, B.-A. (1988) "Innovation as an Interactive Process: From User-Producer Interaction to the National System of Innovation," in G. Dosi, C. Freemna, R. Nelson, et al. (eds) *Technical Change and Economic Theory* (London: Pinter): 349–69.

Lury, C. (1996) *Consumer Culture* (London: Polity).

Mackay, H. (1997) *Consumption and Everyday Life* (London: Sage/Open University).

Mackay, H. & G. Gillespie (1992) "Extending the Social Shaping of Technology Approach: Ideology and Appropriation," *Social Studies of Science* 22: 685–716.

Mackay, H., C. Crane, P. Beynon-Davies, & D. Tudhope (2000) "Reconfiguring the User: Using Rapid Application Development," *Social Studies of Science* 30: 737–59.

MacKenzie, Donald & Judy Wajcman (1985) *The Social Shaping of Technology* (Milton Keynes, U.K.: Open University Press).

Marcuse, H. (1964) *One Dimensional Man* (London: Routledge).

Marx, K. ([1857–58]1980) *Marx's Grundrisse,* ed. David McLellan (London: MacMillan).

McCracken, G. (1988) *Culture and Consumption: New Approaches to the Symbolic Character of Consumer Goods and Activities* (Bloomington: Indiana University Press).

McGaw, J. (1982) "Women and the History of American Technology," *Signs* 7: 798–828.

Miller, D. (ed) (1995) *Acknowledging Consumption: A Review of New Studies* (New York and London: Routledge).

Mody, C. (2006) "Universities, Corporations, and Instrumental Communities: Commercializing Probe Microscopy, 1981–1996" *Technology and Culture* 47: 56–80.

Morley, D. (1995) "Theories of Consumption in Media Studies," in *Acknowledging Consumption: A Review of New Studies* (New York and London: Routledge): 296–329.

Moser, I. (2000) "Against Normalisation: Subverting Norms of Ability and Disability," *Science as Culture* 9: 201–40.

Moser, I. & J. Law (1998) "Materiality, Textuality, Subjectivity: Notes on Desire, Complexity and Inclusion," *Concepts and Transformations: International Journal of Action Research and Organizational Renewal* 3: 207–27.

Moser, I., & J. Law (2003) "'Making Voices': New Media Technologies, Disabilities, and Articulation," in G. Liestol, A. Morrison, & T. Rasmussen (eds), *Digital Media Revisited: Theoretical and Conceptual Innovation in Digital Domains* (Cambridge, MA: MIT Press): 491–520.

Mulkay, M. (1976) "Norms and Ideology of Science," *Social Science Information* 15: 637–56.

Oldenziel, R. (2001) "Man the Maker, Woman the Consumer: The Consumption Junction Revisited," in A. N. H. Creager, E. Lunbeck, & L. Schiebinger (eds), *Feminism in Twentieth Century Science, Technology, and Medicine (Women in Culture and Society)* (Chicago: University of Chicago Press): 128–49.

Oldenziel, Ruth, A. de la Bruhez, & O. de Wit (2005) "Europe's Mediation Junction: Technology and Consumer Society in the 20th Century," *History and Technology* 21: 107–39.

Oudshoorn, N. (1999) "On Masculinities, Technologies and Pain: The Testing of Male Contraceptives in the Clinic and the Media," *Science, Technology & Human Values* 24: 265–89.

Oudshoorn, N. (2003) *The Male Pill: A Biography of a Technology in the Making* (Durham, NC, and London: Duke University Press).

Oudshoorn, N. & T. Pinch (eds) (2003) "How Users and Non-users Matter" in N. Oudshoorn & T. Pinch (eds), *How Users Matter: The Co-Construction of Users and Technologies* (Cambridge, MA: MIT Press).

Oudshoorn, N., M. Brouns, & E. van Oost (2005) "Diversity and Distributed Agency in the Design and Use of Medical Video-Communication Technologies," in H. Harbers (ed): *Inside the Politics of Technology* (Amsterdam: Amsterdam University Press): 85–109.

Oudshoorn, N., M. Lie, & A. R. Saetnan, (2002) "On Gender and Things: Reflections on an Exhibition on Gendered Artefacts," *Women's Studies International Forum* 25(4): 471–83.

Oudshoorn, N., E. Rommes & M. Stienstra, (2004) "Configuring the User as Everybody: Gender and Cultures of Design in Information and Communication Technologies," *Science, Technology & Human Values* 29: 30–64.

Parthasarathy, S. (2003) "Knowledge Is Power: Producing Genetic Testing for Breast Cancer and the Civic Individual in the United States and Britain," in N. Oudshoorn & T. Pinch (eds), *How Users Matter: The Co-Construction of Users and Technologies* (Cambridge, MA: MIT Press).

Pinch, T. J. (2003) "Giving Birth to New Users: How the Minimoog Was Sold to Rock 'n' Roll," in N. Oudshoorn & T. Pinch (eds), *How Users Matter: The Co-Construction of Users and Technologies* (Cambridge, MA: MIT Press).

Pinch, T. J. & W. E. Bijker (1984) "The Social Construction of Facts and Artifacts: Or How the Sociology of Science and the Sociology of Technology Might Benefit Each Other," *Social Studies of Science* 14: 399–431.

Pinch, T. & F. Trocco (2002) *Analog Days: The Invention and Impact of the Moog Synthesizer* (Cambridge, MA: Harvard University Press).

Prins, B. (1995) "The Ethics of Hybrid Subjects: Feminist Constructivism According to Donna Haraway," *Science, Technology & Human Values* 20: 352–67.

Rader, Karen (2004) *Making Mice: Standardizing Animals for American Biomedical Research, 1900–1955* (Princeton, NJ: Princeton University Press).

Rapp, R. (1998) "Refusing Prenatal Diagnosis: The Meanings of Bioscience in a Multicultural World," *Science, Technology & Human Values* 23: 45–70.

Rohracher, R. (ed) (2005) *User Involvement in Innovation Processes: Strategies and Limitations from a Socio-Technical Perspective* in A. Bamme, P. Baumgartner, W. Berger, & E. Kotzman (eds) (Wien: Profil Verlag Technik- und Wissenschaftsforschung/Science and Technology Studies 44).

Rommes, E., E. van Oost, & N. Oudshoorn (1999) "Gender and the Design of a Digital City," *Information Technology, Communication and Society* 2(4): 476–95.

Ropke, I. (2003) "Consumption Dynamics and Technological Change—Exemplified by the Mobile Phone and Related Technologies," *Ecological Economics* 45: 171–88.

Rose, D. A. (2001) "Reconceptualizing the User(s) of—and in—Technological Innovation: The Case of Vaccines in the United States," in R. Coombs, K. Green, A. Richards, & V. Walsh (eds) *Technology and the Market: Demand, Users and Innovation* (Cheltenham and Northampton, U.K.: Edward Elgar): 68–88.

Rose, D. A. & S. Blume (2003) "Citizens as Users of Technology: An Exploratory Study of Vaccines and Vaccination," in N. Oudshoorn & T. Pinch (eds), *How Users Matter: The Co-Construction of Users and Technologies* (Cambridge, MA: MIT Press).

Rosenberg, N. (1982) *Inside the Black Box: Technology and Economics* (Cambridge and New York: Cambridge University Press).

Ruggiero, T. E. (2000) "Uses and Gratifications Theory in the 21st Century," *Mass Communication & Society* 3(1): 3–37.

Saetnan, A., N. Oudshoorn, & M. Kirejczyk (eds) (2000) *Bodies of Technology: Women's Involvement with Reproductive Medicine* (Columbus: Ohio State University Press).

Schot, J. & A. A. de la Bruheze (2003) "The Mediated Design of Products, Consumption and Consumers in the Twentieth Century," in N. Oudshoorn & T. Pinch (eds), *How Users Matter: The Co-Construction of Users and Technologies* (Cambridge, MA: MIT Press).

Schroeder, R. (2002) "The Consumption of Technology in Everyday Life: Car, Telephone and Television in Sweden and America in Comparative-Historical Perspective," *Sociological Research Online* 7(4). Available at: www.socresonline.org.uk/7/4/schroeder.html.

Shah, Sonali (2005) "Open beyond Software" in D. Cooper, C. Diboan, & M. Stone (eds), *Open Sources 2.0: The Continuing Evolution* (Sebastopol, CA: O'Reilly Media): 339–60.

Silverstone, R. (2000) "Under Construction: New Media and Information Technologies in the Societies of Europe," paper prepared for the European Media Technology and Everyday Life Network (EMTEL 2).

Silverstone R. & L. Haddon (1996) "Design and the Domestication of Information and Communication Technologies: Technical Change and Everyday Life," in R. Silverstone, & R. Mansell (eds), *Communication by Design: The Politics of Information and Communication Technologies* (Oxford: Oxford University Press): 44–74.

Silverstone, R. & E. Hirsch (eds) (1992) *Consuming Technologies: Media and Information in Domestic Spaces* (London: Routledge).

Silverstone, R., E. Hirsch, & D. Morley (1992) "Information and Communication Technologies and the Moral Economy of the Household," in R. Silverstone & E. Hirsch (eds), *Consuming Technologies: Media and Information in Domestic Spaces* (London: Routledge): 15–32.

Silverstone, R., D. Morley, A. Dahlberg, & S. Livingstone (1989) "Families, Technologies, and Consumption: The Household and Information and Communication Technologies," CRICT discussion paper, Brunel University.

Slooten, I., E. Rommes, & N. Oudshoorn (eds) (2003) *Strategies of Inclusion. Gender and the Information Society: A Qualitative Interview Study of Female User Experiences.* Report of EU-IST program SIGIS No. ST-2000–26329 (Trondheim, Norway: NTNU).

Smith, Adam (1776) *The Wealth of Nations* (London: Methuen).

Spilkner, H., & K. H. Sørenson (2000) "A ROM of One's Own or a Home for Sharing? Designing the Inclusion of Women in Multimedia," *New Media & Society* 3: 268–85.

Star, S. L. (1991) "Power, Technology and the Phenomenology of Conventions: On Being Allergic to Onions," in J. Law (ed), *A Sociology of Monsters: Essays on Power, Technology and Domination* (London and New York: Routledge): 26–55.

Storey, J. (1999) *Cultural Consumption and Everyday Life* (London, Sydney, and Auckland: Arnold).

Suchman, L. (1994) "Working Relations of Technology Production and Use," *Computer Supported Cooperative Work (CSCW)* 2: 21–39.

Suchman, L. (2001) "Human/Machine Reconsidered," Introduction to 2nd rev. ed. of *Plans and Situated Actions: The Problem of Human-Machine Communication.* (Cambridge: Cambridge University Press).

Summerton, J. (2004) "Do Electrons Have Politics? Constructing User Identities in Swedish Electricity," *Science, Technology & Human Value*, 29: 486–511.

Thompson, Charis (2005) *Making Parents: The Ontological Choreography of Reproductive Technologies* (Cambridge, MA: MIT Press).

Toffler, A. (1980) *The Third Wave* (New York: Morrow).

Van Kammen, J. (2000) "Do Users Matter?" in A. Saetnan, N. Oudshoorn, & M. Kirejczyk (eds), *Bodies of Technology: Women's Involvement with Reproductive Medicine* (Columbus: Ohio State University Press): 90–123.

Van Kammen, J. (2003) "Who Represents the Users? Critical Encounters Between Women's Health Advocates and Scientists in Contraceptive R&D" in N. Oudshoorn & T. Pinch (eds), *How Users Matter: The Co-Construction of Users and Technologies* (Cambridge, MA: MIT Press): 151–73.

van Oost, E. (1995) "Over 'Vrouwelijke' en 'Mannelijke' Dingen," in M. Brouns, M. Verloo, & M. Grunell (eds), *Vrouwenstudies in de Jaren Negentig: Een Kennismaking Vanuit Verschillende Disciplines* (Bussum: Couthinho): 287–310.

van Oost, E. (2003) "Materialized Gender: How Shavers Configure the Users' Femininity and Masculinity," in N. Oudshoorn & T. Pinch (eds), *How Users Matter: The Co-Construction of Users and Technologies* (Cambridge, MA: MIT Press).

Ventakesh, V., M. G. Davis, G. B. Davis, & F. D. Davis (2003) "User Acceptance of Information Technology: Toward a Unified View," *MIS Quarterly* 27(3): 425–78.

Von Hippel, E. (1976) "The Dominant Role of Users in the Scientific Instrument Innovation Process," *Research Policy* 5: 212–39.

Von Hippel, E. (2002) "Horizontal Innovation Networks: Innovation by and for Users," MIT Sloan School of Management Working Paper No. 4366–02.

Von Hippel, E. (2005) *Democratizing Innovation* (Cambridge, MA: MIT Press).

Wajcman, J. (1991) *Feminism Confronts Technology* (Cambridge, MA: Polity Press).

Wajcman, J. (2004) *TechnoFeminism* (Cambridge and Malden, MA: Polity Press).

Williams, Robin, James Stewart, & Roger Slack (2005) *Social Learning in Technological Innovation: Experimenting with Information and Communication Technologies* (Cheltenham, U.K., and Northampton, MA: Edward Elgar).

Winner, Langdon (1977) *Autonomous Technology: Technics-out-of-Control as a Theme in Political Thought* (Cambridge, MA: MIT Press).

Woolgar, S. (1991) "Configuring the User: The Case of Usability Trials," in J. Law (ed), *A Sociology of Monsters: Essays on Power, Technology and Domination* (London: Routledge).

Wyatt, S. (2003) "Non-Users Also Matter: The Construction of Users and Non-Users of the Internet," in N. Oudshoorn & T. Pinch (eds), *How Users Matter: The Co-Construction of Users and Technologies* (Cambridge, MA: MIT Press).

Wyatt, S., G. Thomas, & T. Terranova (2003) "They Came, They Surfed, They Went Back to the Beach: Conceptualizing Use and Non-use of the Internet," in S. Woolgar (ed), *Virtual Society? Get Real! Technology, Cyberhole, Reality* (Oxford: Oxford University Press): 23–41.

23   STS and Ethics: Implications for Engineering Ethics

Deborah G. Johnson and Jameson M. Wetmore

With some exceptions, STS scholars seem largely to avoid taking explicit normative stances. It is not uncommon to hear STS scholars trained in the social sciences claim that their job is to illuminate the social processes by which arguments achieve legitimacy rather than to use their understanding of those processes to establish the legitimacy of their own arguments or positions. This reluctance to take an explicit normative stance has been noted and critiqued by several STS scholars. Most prominently, Bijker (1993) argued that STS began on the path of critical studies, took a break from being proscriptive in order to build a firm base of knowledge, and now needs to get back to the original path. "Seen in this perspective, the science and technology studies of the 1980s are an academic detour to collect ammunition for the struggles with political, scientific, and technological authorities" (Bijker, 1993: 116). In the same year, Winner published his "Upon Opening the Black Box and Finding It Empty" in which he critiques STS theory on several grounds including "its lack of and, indeed, disdain for anything resembling an evaluative stance or any particular moral or political principles that might help people judge the possibilities that technologies present" (Winner, 1993: 371). Despite these promptings, STS scholarship of the last decade only rarely seems to involve explicit normative analysis.

This avoidance of normative analysis has manifested itself in many ways. First, it has had the obvious consequence that many STS scholars have shied away from making recommendations for change that might improve the institutions of science and engineering. Second, it has created an atmosphere in which it can be tempting to hide the normativity that is often implicit in STS analysis. And third, it has caused many scholars to be quite leery of exploring or even being associated with the field of ethics.

While the first two consequences certainly warrant further discussion, it is the final consequence that is the spark for this chapter. Our goal is to lower some of the barriers between the fields (and scholars) of ethics and STS. Despite the incongruence that is commonly assumed, the goals of STS and ethics are compatible in a number of ways. Even if STS scholars do not wish to take explicitly normative stances, they can still make important contributions to ethical inquiry. Scholarship in the field of ethics is not exclusively directed at generating and defending prescriptive conclusions; rather, a major thrust of the field is to engage in normative dialogue and to critically and reflexively explore and evaluate alternative actions and avenues for change. Using moral concepts and theories, ethics scholarship provides perspectives on the world that are useful in envisioning potential actions, appraising the possible consequences of these actions, and evaluating alternative social arrangements. In a similar manner, STS concepts and theories provide illuminating analyses of the social processes that constitute science and technology and the social institutions and arrangements of which science and technology are a part. Many of these analyses have ethical implications that are not commonly discerned; some also point to possibilities for new institutional arrangements, decision-making processes, and forms of intervention. In this way, STS concepts and theories have the potential to contribute to ethical perspectives and point the way to positive change.

Of course, the proof is in the pudding. The aim of this chapter is to illustrate how STS concepts and theories can be used to enrich normative analysis. To do this, we will focus on the fairly young field of engineering ethics. Scholarship in the field of engineering ethics critically examines the behavior of engineers and engineering institutions; identifies activities, practices, and policies that are morally problematic (or exemplary); and alerts engineers to a wide range of situations in which they might be caught up. Some engineering ethicists go so far as to make recommendations as to what engineers should do individually or collectively when faced with moral dilemmas.

STS has developed in parallel with engineering ethics over the past few decades. While there are few formal ties between the two fields, a number of scholars contribute to both. These scholars have begun the process of fleshing out the ways in which STS insights about the nature of technology, technological development, and technical expertise can inform engineering ethics. STS concepts, theories, and insights in these areas shed new light, we will argue, on engineering practice and open up new avenues for ethical analysis of engineering. In this chapter, we identify and develop further avenues in which STS can inform scholarship in engineering ethics and transform normative analysis of engineering.[1]

THE DEVELOPMENT OF ENGINEERING ETHICS

While we cannot provide a complete history of the field of engineering ethics, a quick overview of some of the important themes and trends provides a starting point for our discussion. Engineering professional societies first proposed codes of ethics in the nineteenth century, but it seems fair to say that the field of engineering ethics in the United States largely developed during the second half of the twentieth century in response to increasing concern about the dangers of technology.[2] A sequence of events starting with use of the atomic bomb in World War II, continuing with the Three Mile Island disaster, the Ford Pinto case, and the explosion at Bhopal, generated a significant concern in the media and the public about the effects of technology on human well-being. After decades of seemingly unmitigated praise, many Americans began to wonder if technology wasn't "biting back" and making us pay (in negative consequences) for the improvements it had provided.

Corporations and governments received a fair amount of blame for these events. For instance, the U.S. government was denounced for its promotion of DDT, and Ford Motor Company and Union Carbide were targets of substantial criticism as well as lawsuits for the fatalities linked to defects in their products and facilities. But a number of social critics, engineering professional associations, and the popular media also began to question the role of engineers in these catastrophes. They scrutinized the conduct of engineers and suggested that there were a number of problems, both in the way engineers behave and in their relationships to employers and clients. In this context, it seemed clear that more careful attention needed to be given to the ethical and professional responsibilities of engineers.

In response to this need, by the early 1980s, an academic field that has come to be known as "engineering ethics" had begun to form. It was built by scholars and practitioners from many different fields including philosophy, history, law, and engineering. Despite their varied backgrounds, however, most believed that concepts and theories from philosophical ethics could be useful in understanding the circumstances of engineers and assist them in making decisions in the face of difficult situations. This approach was in part inspired by the newly developing fields of medical ethics and bioethics.[3] Scholars building the field of engineering ethics contended that ethical theory and training in ethics would allow engineers to see the ethical aspects of their circumstances and help them identify the right choice and course of action with rigor and justification rather than with "gut" feeling or intuition. Like the other emerging fields of applied ethics, they saw a dose of ethical theory as a promising antidote for the temptations and pressures of the workplace. Thus, a significant part of the field of engineering ethics was dedicated to applying philosophical concepts and theories such as Kant's categorical imperative, utilitarianism, and distributive justice to issues faced by engineers.[4]

A major concern of the field was to identify the ethical issues, problems, and dilemmas that engineers commonly face in their careers. In large part because the traditional subjects of moral theory and moral analysis are institutional arrangements and social relationships, scholars looked to the organizational context of engineering and the social relationships that constitute engineering practice. Through this lens, the importance of the business context in which engineering is practiced was most salient. Scholars in the field typically portrayed the engineer as an ethical actor who had to make complicated decisions within the institutional arrangements of a corporation. The business environment was most commonly illustrated with case studies that focused on the description and analysis of disasters such as the Ford Pinto fuel tank explosions, the crashes (and near crashes) of DC-10 passenger jets, and the Bhopal chemical leak.[5] These case studies emphasized that individual engineers had to mediate their technical knowledge with institutional pressures, the demands of their employers, their professional codes of ethics, and the expectation that they protect the public. If an engineer mismanaged these demands, the results could be disastrous.

The idea that business is the context of engineering and that business generates or is wrapped up in most engineering ethics problems permeates much of the work of the time. Indeed, much of the literature of the 1980s and early 1990s can be seen as digesting the implications of engineering being practiced in the context of business interests. This emphasis can be seen in Kline's summary of the major issues that form the core of engineering ethics texts in the United States, which largely focuses on business-related interests including conflicts of interest, whistle-blowing, trade secrets, and accepting gifts (Kline, 2001–2: 16). Numerous case studies were developed that asked engineers to consider how they would act when confronted by a dilemma wherein a business interest came into conflict with the public good or some sort of professional norm.[6] To be sure, the field of engineering ethics was not and never has been monolithic but has been largely concerned with the social circumstances of engineers and the business decisions being made in the production of technologies. Scholars in the field have appropriately focused on disasters, unsafe products, and dangers to human health and well-being.

The field of engineering ethics has succeeded in illuminating an array of situations in which engineers often find themselves and provided concepts and frameworks with which to think through these situations. The literature in the field now includes two classic textbooks devoted to the topic that are updated every few years (Martin & Schinzinger, 2004; Harris et al., 2005), a handful of additional textbooks (Pinkus et al., 1998; Mitcham & Duval, 2000; Herkert, 2000; Schinzinger & Martin, 2000), and a number of somewhat more specialized single author books including Unger (1994), Whitbeck (1998), Davis (1998), and Martin (2000). The American Society for Engineering Education (ASEE) and the Association for Practical and Professional Ethics (APPE) regularly host sessions on topics of importance in the field at their annual meetings, and a special workshop on "Emerging Technologies and Ethical Issues in Engineering" was recently sponsored by the National Academy of Engineering (NAE, 2004). For the last decade, research in the area has been published in a journal devoted specifically to the field, *Science & Engineering Ethics*.

An important factor promoting and supporting the field was a change in the accreditation requirements for undergraduate engineering programs. In 2000, the U.S. Accreditation Board for Engineering and Technology (ABET) specified that to be accredited, institutions must demonstrate eleven outcomes, one of which states that their students must attain "an understanding of professional and ethical responsibility" (ABET, 2004). This has sparked the development of new programs and courses and, in turn, the development of new materials.

As the field has matured, the scope and topics that engineering ethics scholars and teachers address has also begun to expand. The relationship between engineering professionalism and business practices is still deemed to be of vital importance. But scholars in the field are beginning to think about the ethical implications of engineering through new lenses and in new places. Topics such as the public understanding of engineering and the value-laden character of design are increasingly being incorporated into the field.

As a number of scholars have already found (see Goujon & Dubreuil, 2001; van de Poel & Verbeek, 2006a), the observations, case studies, and theories developed in STS can play an important role in expanding the scope and insights of engineering ethics. STS opens up new ways to understand the processes of engineering and the effects its products have on the world. This chapter is written with an eye to promoting and escalating the turn to STS, as well as to encouraging STS scholars to take up the task of addressing ethical issues in engineering. A more robust infusion of STS concepts and theory could inform, enlighten, and transform the field of engineering ethics in significant ways.

To illustrate the links between STS and engineering ethics and the potential for cross-fertilization, we are going to focus on two core STS ideas. Engineering ethics scholars have begun to use these ideas in a variety of ways, and our aim is to demonstrate how they provide a basis for an STS-informed analysis of the responsibilities of engineers. The first of these ideas is the STS discussion about the relationship between technology and society—a discussion that examines the ideas of technological determinism and the social shaping of technology. The second idea is that of "sociotechnical systems"—that since social and technical aspects of the world are intimately interwoven and change in concert, sociotechnical systems should be the unit of analysis in technology and engineering studies. These two ideas provide a picture of engineering practice in which engineers are not isolated and not the only actors in technological development. This picture, in turn, provides the foundation for an account of the responsibilities of engineers.

THE RELATIONSHIP BETWEEN TECHNOLOGY AND SOCIETY

Much of STS scholarship is concerned with understanding the technology-society relationship and accounting for the forms, meanings, success, and effects of technologies. At the core of this concern is a debate about technological determinism. While multiple definitions and forms of technological determinism are described and then contested by STS scholars, technological determinism seems to involve two key tenets.[7] The first is the claim that technology develops independently from society. According to this claim, technological development either follows scientific discoveries—as inventors and engineers "apply" science in some straightforward, step-by-step manner—or it follows a logic of its own, with new invention deriving directly from previous inventions. Either way, technological development is understood to be an independent activity, separate from social forces. STS scholars have countered this idea with numerous theories and case studies, arguing and demonstrating that technological development is not isolated and that its character and direction are shaped by a variety of social factors and forces (Bijker et al., 1987).

A second major tenet of technological determinism is that technology (when taken up and used) "determines" the character of a society. The STS response to this tenet is complicated; while most scholars in the field agree that "determines" is too strong a term to describe how technology affects society, some concede that technology is, nevertheless, an important, and even powerful, force in shaping society, whereas others deny even this. In either case, there seems to be agreement that the important flaw of technological determinism is its failure to recognize that society shapes technology. Except for those few who believe solely in social determinism, there seems to be consensus around the claim that there is valence (influence, shaping) in both directions. Indeed, the claim that technology and society co-produce each other—that technology shapes and is shaped by society—seems to be a canon of STS theory (Jasanoff, 2004a). As Jasanoff puts it, technology "both embeds and is embedded in social practices, identities, norms, conventions, discourses, instruments and institutions—in short, in all the building blocks of what we term the *social*" (Jasanoff, 2004b: 3).

What are the implications of co-production for engineering ethics? Perhaps the best place to begin is with the rejection of the notion that technology develops in isolation and according to its own logic for this suggests that *the work of engineers* and especially the *decisions that engineers make* do not have an independent logic of their own and are not dictated by science or nature. This insight is of key importance for engineering ethics. A view of engineering as an isolated activity in which engineers are figuring out what nature will allow, tends to push much of the work of engineers out of the purview of engineering ethics. If what engineers do is determined by nature, there is no room for ethics, value judgments, or moral responsibility; engineers do only what is possible, i.e., what nature dictates.

An STS-informed account of engineering practice, however, opens up this black box of engineering practice and contends that engineers have a good deal of latitude (power, influence, discretion) in what they create. They manipulate nature, but they can and do manipulate it in this or that way because they are pursuing and responding to various pressures, interests, and values. STS studies demonstrate that there is rarely (if ever) an objectively best design solution to a given problem. Rather, engineers choose from a range of possible solutions based on the fit of each solution with a broad set of criteria and values.[8] Engineering practice—from problem definition and the weighing of alternatives through to final design specifications—requires engineers to balance and trade-off technical feasibility, legal constraints, values such as privacy and accessibility, consumer appeal, fit with other technologies, and much more. This understanding of engineering practice suggests that nearly every decision an engineer makes is not simply a detached technical decision but has ethical and value content and implications. Acknowledging that engineers make value judgments when they make technical decisions suggests that the responsibilities of engineers are broad in scope. At a minimum, it means that engineers cannot deflect responsibility for what they do by hiding behind the shield of "the dictates of nature," at least, not to the extent that might be allowed under a determinist view.

Although STS accounts of the technology-society relationship show that engineers are doing much more than designing devices and although this points to engineers having responsibility for a broader domain, in other ways STS accounts seem to shift responsibility away from engineers. STS descriptions of technological development reveal the wide range of other individuals and groups who influence technological development before, alongside, and after engineers complete their work. This includes those with whom engineers interact at work—representatives of business such as managers, CEOs, and marketing departments—as well as lawmakers, regulators, consumer groups, judges, and others. These actors may have a direct impact on technology by banning or rejecting certain devices, setting standards for the design of particular technologies, funding specific areas of research, granting patents, and so on. In a less direct, but still powerful way, these actors may shape the perception and meaning of a technology through marketing, media, or demonstrations. For instance, STS accounts have been particularly helpful in pointing to the role of users in technological development. Scholars have shown how users can take up an artifact and find meanings and uses that never occurred to the engineers who designed it (Pinch & Bijker, 1987; Oudshoorn & Pinch, 2003) or effectively re-design technology through work-arounds (Pollock, 2005). From the perspective of engineering ethics, the role and influence of these other actors in technological development mean that engineers cannot be considered wholly responsible for technology and its effects. At least some responsibility falls to these other actors.

Engineers are important (perhaps even dominant) players in technological development, but the scope of their responsibility is limited. Their responsibilities are broader than what is suggested by the picture of isolated engineers following nature, broader than what is suggested by the view of engineers as designers of neutral devices that others can choose (or not) to use. Nevertheless, the other actors involved in technological development also have responsibilities with regard to the same technology.

Although these two insights about the responsibilities of engineers may appear contradictory in that one suggests that engineers have *more* power and the other *less*, they are not. Indeed, the complexity to which they point indicates the potential of STS to provide a new foundation for engineering ethics. The point is not to determine whether engineers have more or less responsibility but rather what kind of responsibilities are appropriate to their practice. The STS-informed account suggests that to be effective and responsible, engineers must recognize the values that influence their work, the ways their work influences values, and the other actors involved in this process. Recognizing these aspects of their work allows engineers to be better and more responsible engineers. That is, attention to the values at work in their endeavors and to the full array of other actors affecting and being affected by their endeavors can lead engineers to design more effectively and have more control over the effects of the technologies they develop. For example, when engineers recognize the effects of their work on marginalized groups, they can design to produce inequities or to avoid negative effects on particular groups.[9] Of course, recognition of the values they affect doesn't mean that engineers will automatically design for socially beneficial values. On the other hand, recognition of those being affected by their work and the values being shaped is a precursor to better design. Law (1987) uses the term "heterogeneous engineer" to capture the idea that successful engineers must master and manage many factors beyond the technical. Such a view of engineering provides a robust foundation for engineering ethics.

SOCIOTECHNICAL SYSTEMS

To further understand and conceptualize an STS-informed account of the responsibilities of engineers, let us now consider a second STS idea—"sociotechnical systems." Sociotechnical systems is the generic name we use to refer to the complex systems of social and technical components intertwined in mutually influencing relationships that STS scholars often take as their unit of analysis. The concept of sociotechnical system acknowledges that attempts to understand a device or a social practice (institution, relationship, etc.) as an independent entity are misleading. To treat either as a separate unit is to abstract it from reality. Focusing on an artifact alone can cause us to bracket and black-box (and push out of sight) all the social practices and social meanings that pragmatically make the artifact a "thing." Vice versa, focusing on a social practice tends to bracket and black-box aspects of the natural and artifactual world that shape the social arrangement or practice at issue.[10] A focus on sociotechnical systems, however, helps us see the ways in which artifacts, social practices, social relationships, systems of knowledge, institutions, and so on are bound together and interact with each other in complex ways. The concept of sociotechnical systems can be used to understand and analyze what it is that engineers help create and sustain— the products of engineering.

The notion can also be used to conceptualize and understand the work of engineers, that is, engineering can itself be understood to be a sociotechnical system. Engineers work with numerous actors (nonengineers as well as other engineers) in institutional contexts with a variety of formal and informal social practices; they use artifacts and manage relationships with all the other actors involved. The sociotechnical systems of which engineers are a part produce, maintain, and give meaning to technology. Insofar as the notion of sociotechnical systems helps us understand both the products and the work of engineers, it provides a foundation for engineering ethicists.

When it comes to understanding the products of engineering, the idea of sociotechnical systems works in parallel with STS accounts of the technology-society relationship. Rejecting technological determinism allows us to see that engineers are doing more than following the dictates of nature in isolation from interests, influences, and values and shows us that engineers are juggling natural phenomena, pressures from other actors, legal constraints, and interests and values of their own and others. A focus on sociotechnical systems points in much the same direction, showing that engineering practice is not isolated and that engineers are doing more than following the dictates of nature. Engineers are not simply building devices; they are building sociotechnical systems consisting of artifacts together with social practices, social arrangements, and relationships.

While the importance of artifacts should not be disregarded, especially when it comes to engineering, a significant portion of STS scholarship argues that a focus on material objects as the fruits of engineering is misleading. Airplanes, electric power plants, the Internet, refrigerators, and playpens *are* complexes of artifacts together with social arrangements, social practices, social relationships, meanings, and institutions. Because sociotechnical systems include social practices, relationships, and arrangements, and since ethics is generally understood to be about social interactions and arrangements, the connection between ethics and engineering comes clearly into view. In building (or contributing to the building of) sociotechnical systems, engineers are building society. Through the lens of sociotechnical systems, it is much easier to see the numerous ways in which engineering is a moral and political endeavor. Building sociotechnical systems means building arrangements of people, what people do, and the way they interact with one another. Engineers contribute to building the quality and character of lives, the distribution of benefits and burdens, what people can and can't do, the risks of everyday life, and so on.

Even if we think of engineers as attending primarily to the artifactual component of sociotechnical systems, the artifacts that engineers design function in relation to the other artifactual and human parts; that is, the artifacts that engineers design require, depend on, and influence social practices, social relationships, and social arrangements. STS scholars have noted that even inanimate objects can play a key role in shaping a sociotechnical system. They suggest that the design of artifacts functions as a form of legislation or script for humans. Latour's (1992) discussion of his experience with seat belts and speed bumps and Akrich's (1992) case studies of artifacts from the developed world being introduced in developing countries point to the ways in which artifacts (designed by engineers) influence human behavior. Thus, even if it seems that engineers have primary responsibility for the artifactual component of sociotechnical systems, their work cannot be separated from the moral and political domain.[11] STS has shown that artifacts shape and even sometimes dictate social behavior. Through the lens of sociotechnical systems ethics and engineering are not distant domains; they are seamlessly intertwined.

Shifting from the products of engineers to the processes of engineering, engineers help to create sociotechnical systems by means of sociotechnical systems. Sociotechnical systems, such as the automobile, missile defense systems, and computers, are the result of processes that involve not only engineers but numerous actors, institutions, and organizations including policy makers, lawyers, marketing professionals, corporations, regulatory bodies, and ultimately users. Within the sociotechnical systems of development, production, and design, engineers must use their technical knowledge, manipulate artifacts, and communicate and coordinate with many other actors and groups including engineers and nonengineers.[12] Using the notion of sociotechnical system, engineers are framed as critical nodes in networks of people and things that influence and are influenced by one another.

When we flesh out the implications of this focus on sociotechnical systems for ethics, the picture of engineering practice that we get is similar to that which came to light in our analysis of the co-production thesis; that is, both analyses point to an expanded and a more narrow view of the responsibilities of engineers. When we understand engineers to be building sociotechnical systems, we see that their responsibilities go far beyond that of designing neutral devices that society can choose or not choose to use, and yet at the same time, we see that engineers' responsibilities must work in conjunction with the responsibilities of other actors who are involved in the development of sociotechnical systems. While seeming to dilute the responsibility of engineers, the fact that many other actors are involved in the process (in addition to engineers) does not justify engineers in saying, for example, "I design the technology, and others have to worry about whether to use it and how it affects society." If engineers are building sociotechnical systems, their designs and decisions must take into account the social practices, social arrangements, and social relationships as well as the artifacts that are part of the sociotechnical system. The fact that many other actors are involved in the process indicates not that engineers have a lessened responsibility but rather that they have a different sort of responsibility—a responsibility to communicate and coordinate with the other actors. To be sure, communicating and working with others has in the past been recognized as an important component of engineering but not as a responsibility.[13] Acknowledging both that engineers are building sociotechnical systems and working in sociotechnical systems goes a long way toward showing that communication and coordination are not just important skills for engineers to have but are crucial responsibilities.

When the work of engineers is understood to be part of a sociotechnical system, the traditional notion of engineering expertise is somewhat disrupted. While engineering expertise traditionally has been focused on the so-called technical aspects of their work (Porter, 1995), a focus on sociotechnical systems suggests that engineering endeavors involve much more than statistics, measurements, and equations. Successful engineering requires an understanding of the extant artifactual and social world in which devices and machines will have to fit. Engineering knowledge must fit together with other forms of knowledge. Engineering expertise is not simply in "the technical" but in integrating the "technical" with many other kinds of knowledge. Engineers are experts because they have the ability to design products that take into account and mesh with a complex world of people, relationships, institutions, and artifacts. When engineers keep in mind the values and politics that are promoted (or weakened) by their creations, they are more likely to have the effects they intend.

Thus, in broad terms, viewing the products of engineering as sociotechnical systems shows that engineers have a responsibility to consider the character of the world they are building. Viewing engineering as a sociotechnical system points to engineers having responsibilities appropriate to their interactions within the system. Engineers should be understood to be nodes in a complex network in which it is crucial for each part to interact and communicate effectively with the other parts.

This brief discussion is only a beginning. We do not claim to have drawn out all the implications of using the notion of sociotechnical systems to understand the products and processes of engineering. Rather, we have tried to demonstrate the potential usefulness of the concept. Much more work needs to be done to realize the full potential of reconceptualizing the products and processes of engineering as sociotechnical systems.

CONCLUSION

The complexity of engineering practice has long been recognized by scholars in the field of engineering ethics. Indeed, the central issue of engineering ethics might, arguably, be said to be figuring out how engineers can and should responsibly manage this complexity. STS theory is helpful to engineering ethicists precisely because it provides ways to understand, conceptualize, and theorize this complexity. Framing the product and processes of engineering as sociotechnical systems makes visible the ways in which they are both combinations of technical and social components. The thrust of our analysis has been to show that as the complexities of engineering practice come into clearer focus by means of STS concepts and theories, so do the responsibilities of engineers. As the complexities of engineering practice are more sharply delineated, ideas for change to improve engineering practice also come into view.

Our chapter takes two STS ideas—that of the co-production of technology and society and that of sociotechnical systems—and shows that they have important implications for understanding the responsibilities of engineers. Although the analysis seems to point to a weaker account of the responsibilities of engineers insofar as it shows that many other actors are involved in the production of sociotechnical systems, it provides a picture of engineering practice in which engineers are seen to be doing more than designing neutral devices. Indeed, engineers are shown to be (with others) building the world in which we all live, a sociotechnical world. The analysis does more than merely deny the isolation of engineers; we use the STS co-production thesis and the notion of sociotechnical systems to develop a picture of the products and processes of engineering. Our analysis suggests the importance of engineers considering the ways in which their designs will influence values, politics, and relationships, while simultaneously recognizing, responding to, and helping shape the wide variety of actors that impact the design, use, and meaning of technology. We do not claim to have done all there is to be done with the connections between STS and ethics. More work is needed to realize the full potential of STS concepts and theories to contribute to the field of engineering ethics and to make the world a better place in which to live.

Notes

The research for this chapter was supported by the National Science Foundation Award No. 0220748.

1. We are not the first to attempt to do this. We have benefited enormously from the work of Lynch and Kline (Lynch & Kline, 2000; Kline, 2001–2), Brey (1997), Herkert (2004), van de Poel and Verbeek (2006a), and others.

2. For a historical account of the development of codes of conduct in the engineering professional associations, see Layton (1971), Pfatteicher (2003).

3. See Baum (1980).

4. For example, the anthology edited by Baum and Flores (1980) (arguably the first engineering ethics reader) includes chapters that make use of Rawls, Kant, and utilitarianism.

5. For example, the first two issues of the journal *Business & Professional Ethics*, which was created by Robert J. Baum in 1981, opened with articles on the Pinto and DC-10 cases (DeGeorge, 1981; Kipnis, 1981; French, 1982).

6. For example, the 1989 video, *Gilbane Gold*, produced by the National Institute for Engineering Ethics (NIEE, 1989) focused on this theme, posing the question whether the engineer depicted should blow the whistle on his employer. NIEE's more recent video, *Incident at Morales* (NIEE, 2003) centered on this theme but has the additional element of arising in an international context.

7. For a more thorough account of technological determinism, see Bimber (1994) and more generally Smith and Marx (1994).

8. Whitbeck (1998) makes this point saliently and notes how engineering design problems and ethical problems are alike in this respect.

9. For an analysis of design with marginalized groups in mind, see Nieusma (2004).

10. Latour (1992) cautions sociologists to avoid this mistake.

11. A number of scholars at the Technical University at Delft have been analyzing technologies as a system to demonstrate the role that engineers play in promoting (or denying) certain values (van de Poel, 2001; van Gorp & van de Poel, 2001; Devon & van de Poel, 2004; van der Burg & van Gorp, 2005).

12. Vinck's account (2003) of a young engineer's discoveries during his first real-world job illustrates this nicely. Vinck explains that in order to design what at first appears to be a reasonably simple component turns out to be enormously complex because it must be coordinated with the work being done by numerous other engineers. To be effective, the young engineer has to figure out how to get information from others, convince them to listen to his concerns, and ultimately redesign his small component several times to ensure that it will fit with other parts of the overall product.

13. See, for example, Herkert (1994), which emphasizes the importance of communicating with the public.

References

Accreditation Board for Engineering and Technology (ABET), Engineering Accreditation Commission (2004), Criteria for Accrediting Engineering Programs (Baltimore, MD: ABET).

Akrich, Madeleine (1992) "The De-Scription of Technical Objects," in W. E. Bijker & J. Law (eds), *Shaping Technology/Building Society: Studies in Sociotechnical Change* (Cambridge, MA: MIT Press): 205–24.

Baum, Robert J. (1980) *Ethics and the Engineering Curriculum: The Teaching of Ethics*, vol. 7 (Hastings-on-Hudson, NY: Hastings Center).

Baum, Robert J. & Albert Flores (eds) (1980) *Ethical Problems in Engineering* (Troy, NY: Center for the Study of the Human Dimensions of Science and Technology).

Bijker, Wiebe (1993) "Do Not Despair: There Is Life after Constructivism," *Science, Technology & Human Values* 18(1): 113–38.

Bijker, Wiebe E., Thomas P. Hughes, & Trevor Pinch (eds) (1987) *The Social Construction of Technological Systems* (Cambridge, MA: MIT Press).

Bimber, Bruce (1994) "Three Faces of Technological Determinism," in Merritt Roe Smith & Leo Marx (eds), *Does Technology Drive History? The Dilemma of Technological Determinism* (Cambridge, MA: MIT Press): 80–100.

Brey, Philip (1997) "Philosophy of Technology Meets Social Constructivism," *Techne: Journal of the Society for Philosophy and Technology* 2(3–4): 56–79.

Davis, Michael (1998) *Thinking Like an Engineer* (New York: Oxford University Press).

De George, Richard T. (1981) "Ethical Responsibilities of Engineers in Large Organizations: The Pinto Case," *Business and Professional Ethics Journal* 1: 1–14.

Devon, Richard & Ibo van de Poel (2004) "Design Ethics: The Social Ethics Paradigm," *International Journal of Engineering Education* 20(3): 461–9.

French, Peter (1982) "What Is Hamlet to McDonnell-Douglas or McDonnell-Douglas to Hamlet: DC-10," *Business and Professional Ethics Journal* 1: 1–13.

Goujon, Philippe & Bertrand Hériard Dubreuil (eds) (2001) *Technology and Ethics: A European Quest for Responsible Engineering* (Leuven, Belgium: Peeters).

Harris, Charles E., Michael S. Pritchard, & Michael J. Rabins (2005) *Engineering Ethics: Concepts and Cases*, 3rd ed (Belmont, CA: Wadsworth).

Herkert, Joseph (1994) "Ethical Risk Assessment: Valuing Public Perception," *IEEE Technology & Society Magazine* 13: 4–10.

Herkert, Joseph (ed) (2000) *Social, Ethical, and Policy Implications of Engineering: Selected Readings* (New York: IEEE Press).

Herkert, Joseph R. (2001) "Microethics, Macroethics, and Professional Engineering Societies," in *Emerging Technologies and Ethical Issues in Engineering* (Washington, DC: National Academies Press): 107–14.

Jasanoff, Sheila (ed) (2004a) *States of Knowledge: The Co-Production of Science and Social Order* (London: Routledge).

Jasanoff, Sheila (2004b) "The Idiom of Co-Production," in S. Jasanoff (ed), *States of Knowledge: The Co-Production of Science and Social Order* (London: Routledge): 1–12.

Kipnis, Kenneth (1981) "Engineers Who Kill: Professional Ethics and the Paramountcy of Public Safety," *Business and Professional Ethics Journal* 1: 77–91.

Kline, Ronald (2001–2) "Using History and Sociology to Teach Engineering Ethics," *IEEE Technology and Society Magazine* (winter): 13–20.

Latour, Bruno (1992) "Where Are the Missing Masses? The Sociology of a Few Mundane Artifacts," in W. E. Bijker & J. Law (eds), *Shaping Technology/Building Society: Studies in Sociotechnical Change* (Cambridge, MA: MIT Press): 225–58.

Law, John (1987) "Technology and Heterogeneous Engineering: The Case of Portuguese Expansion," in W. E. Bijker, T. P. Hughes, & T. Pinch (eds) (1987) *The Social Construction of Technological Systems* (Cambridge, MA: MIT Press): 111–34.

Layton, Edwin T., Jr. (1971) *The Revolt of the Engineers: Social Responsibility and the American Engineering Profession* (Cleveland: Press of Case Western Reserve University).

Lynch, William & Ronald Kline (2000) "Engineering Practice and Engineering Ethics," *Science, Technology & Human Values* 25: 195–225.

Martin, Mike W. (2000) *Meaningful Work: Rethinking Professional Ethics* (New York: Oxford University Press).

Martin, Mike W. & Roland Schinzinger (2004) *Ethics in Engineering,* 4th ed. (New York: McGraw-Hill).

Mitcham, Carl & R. Shannon Duval (2000) *Engineering Ethics* (Upper Saddle River, NJ: Prentice Hall).

National Academy of Engineering (NAE) (2004) "Emerging Technologies and Ethical Issues in Engineering," papers from a workshop, October 14–15, 2003 (Washington, DC: National Academies Press).

National Institute for Engineering Ethics (NIEE), National Society of Professional Engineers (1989) "Gilbane Gold: A Case Study in Engineering Ethics" (video), Great Projects Film Company. (Available from NIEE, Texas Tech University, Box 41023, Lubbock, TX 79409.)

National Institute for Engineering Ethics (NIEE), Murdough Center for Engineering Professionalism (2003) "Incident at Morales: An Engineering Ethics Story" (video), Great Projects Film Company. (Available from NIEE, Texas Tech University, Box 41023, Lubbock, TX 79409.)

Nieusma, Dean (2004) "Alternative Design Scholarship: Working Toward Appropriate Design," *Design Issues* 20(3): 13–24.

Oudshoorn, Nelly & Trevor Pinch (2003) *How Users Matter: The Co-Construction of Users and Technology* (Cambridge, MA: MIT Press).

Pfatteicher, Sarah K. A. (2003) "Depending on Character: ASCE Shapes Its First Code of Ethics," *Journal of Professional Issues in Engineering Education and Practice* 129(1): 21–31.

Pinch, Trevor J. & Wiebe Bijker (1987) "The Social Construction of Facts and Artifacts," in W. Bijker, T. P. Hughes, & T. Pinch (eds) *The Social Construction of Technological Systems* (Cambridge, MA: MIT Press): 17–50.

Pinkus, Rosa Lynn B., Larry J. Shuman, Norman P. Hummon, & Harvey Wolfe (1998) *Engineering Ethics: Balancing Cost, Schedule and Risk—Lessons Learned from the Space Shuttle* (New York: Cambridge University Press).

Pollock, John L. (2005) "When Is a Work Around? Conflict and Negotiation in Computer Systems Development," *Science, Technology & Human Values* 30(4): 1–19.

Porter, Theodore M. (1995) *Trust in Numbers: The Pursuit of Objectivity in Science and Public Life* (Princeton, NJ: Princeton University Press).

Schinzinger, Roland & Mike W. Martin (2000) *Introduction to Engineering Ethics* (Boston: McGraw-Hill).

Smith, Merritt Roe & Leo Marx (eds) (1994) *Does Technology Drive History? The Dilemma of Technological Determinism* (Cambridge, MA: MIT Press).

Unger, Stephen H. (1994) *Controlling Technology: Ethics and the Responsible Engineer*, 2nd ed. (New York: Wiley).

Van de Poel, Ibo (2001) "Investigating Ethical Issues in Engineering Design," *Science and Engineering Ethics* 7: 429–46.

Van de Poel, Ibo & Peter-Paul Verbeek (2006a) "Ethics and Engineering Design," (special issue), *Science, Technology, & Human* Values 31(3).

Van de Poel, Ibo & Peter-Paul Verbeek (2006b) "Ethics and Engineering Design" (editorial), *Science, Technology, & Human Values* 31(3): 223–36.

Van der Burg, Simone & Anke van Gorp (2005) "Understanding Moral Responsibility in the Design of Trailers," *Science & Engineering Ethics* 11(2): 235–56.

Van Gorp, Anke & Ibo van de Poel (2001) "Ethical Considerations in Engineering Design Processes," *IEEE Technology and Society Magazine* (fall): 15–22.

Vinck, Dominique (2003) "Socio-Technical Complexity: Redesigning a Shielding Wall," in D. Vinck (ed), *Everyday Engineering: An Ethnography of Design and Innovation* (Cambridge, MA: MIT Press): 13–27.

Whitbeck, Caroline (1998) *Ethics in Engineering Practice and Research* (New York: Cambridge University Press).

Winner, Langdon (1993) "Upon Opening the Black Box and Finding It Empty," *Science, Technology, & Human Values* 18(3): 362–78.

24 STS Perspectives on Scientific Governance

Alan Irwin

One recurrent theme within Science and Technology Studies concerns the interrelationship between science, technology, and the world of political power and institutional action. Bruno Latour introduces his discussion of the "proliferation of hybrids" by noting in the context of a newspaper report on the ozone layer that "[t]he same article mixes together chemical reactions and political reactions. A single thread links the most esoteric sciences and the most sordid politics" (Latour, 1993: 1). As Sheila Jasanoff has observed also, "Science and technology permeate the culture and politics of modernity" (Jasanoff, 2004b: 1). Numerous media stories, governmental actions, and public discussions provide further evidence of the close interconnection between science, technology, and politics in contemporary life. Whether concerning the future of nuclear energy, debates over stem cell research, or controversy over climate change, it seems that everywhere science and technology is operating in a context of political uncertainty, public debate, and societal decision-making. A key question for STS concerns the nature of this relationship between science, technology, and the operation of politics. How should this be understood? What new insights has STS to offer? How, in particular, can STS scholars explore this relationship in an empirically open and symmetrical fashion—and without reducing either science or politics to fixed and essentialist categories?

In this chapter, I argue that the study of scientific and technological governance is at the core of STS and, moreover, that STS has successfully challenged the conventional assumption that scientific governance is merely about "speaking truth to power." Instead, STS scholars have developed and refined long-running debates about science, technology, and democracy but also generated new ways of conceptualizing the relationship between, for example, science, technology, and public policy. Concepts such as boundary work (Gieryn, 1983, 1999), co-production (Jasanoff, 2004a), and sociotechnical networks (Wetmore, 2004) have been proposed, discussed, and empirically refined. Such treatments typically problematize the conventional assumption that what is "scientific" or "political" (or "factual" or "objective") can be straightforwardly identified and ring-fenced. Instead, the very demarcation of certain entities and discussion points as "scientific" or "political" represents a key element in scientific governance and an essential component in STS analysis.

STS studies therefore not only problematize the relationship between science, technology, and political decision-making. They also raise more fundamental questions about the very manner in which decisions are represented and "framed" (Roth et al., 2003) and about the often implicit sociocultural assumptions that operate within these representations and framings. For this reason, I argue in this chapter that the term scientific governance is preferable to the more conventional formulation of science and technology policy, since the latter risks taking for granted what STS research has specifically attempted to question. Rather than assuming that, for example, "science for policy" can be separated from "policy for science"—or that "science policy" exists apart from wider social, cultural, and technical processes—STS perspectives on scientific governance open up the very definition of such categories as "science" and "policy" to critical reflection and empirical exploration. In this spirit, STS accounts have frequently emphasized the fluidity of cognitive and institutional boundaries and the hybridity of the issues at hand. In that way, the relationship between science and political decision-making is not fixed but open to multiple problem definitions and framings.

It should also be noted that in recent years there has been a marked academic and institutional trend away from the language of "science and technology policy" and toward "scientific (and technological) governance." As Fuller (2000) observes, this suggests also a shift from focusing primarily on the business of government (narrowly defined) toward recognition of the wider and more informal accountability relations at work within scientific institutions. "Governance" can be taken to imply that the development and control of science and technology is not simply a matter for government or "the state" (Rose, 1999: 16–17). Instead it is necessary to include the activities of a much wider range of actors—including industry, scientific organizations, public and pressure groups, consumers, and the market. "Governance" encompasses the range of organizational mechanisms, operational assumptions, modes of thought, and consequential activities involved in governing a particular area of social action—in this case, relating to the development and control of science and technology. Viewed in this way, governance is not simply about a defined set of bureaucratic and scientific institutions but also the wider activities of governing and, indeed, self-governing (Barry, 2001; Dean, 1999). The implication is that national governments no longer have the ability to direct society toward specific goals. Instead, they must play a part within de-centered networks and shifting assemblages of power.

The key point about "scientific governance" therefore is that it cannot be squeezed into a single institutional or processual definition. This perspective also reflects a characteristic methodological preference within STS to "follow the actors" rather than make categorical judgments in advance. Instead, and as discussed in this chapter, the study of scientific governance is broadly concerned with the relationship between science, technology, and political power—with special emphasis on democratic engagement, the relationship between "scientific" and wider social concerns, and the resolution of political conflict and controversy.

One possible disadvantage of this broader notion of governance is that it risks encompassing an unwieldy and ill-defined range of actors so that it potentially becomes hard to specify what is *not* included in scientific governance: the mass media, campaigning groups, the market? This is especially the case when, from a public perspective, issues highly relevant to scientific governance (for example, the social desirability of new products) may not entirely (if at all) be a matter of science and technology in themselves but instead can be wrapped up with much wider cultural framings and interpretations (for example, general feelings of social exclusion and disenfranchisement). Equally, governance has in recent years (and especially in European policy circles) become a fashionable term (e.g., CEC, 2001) with often imprecise meaning. Furthermore, the language of "governance" inevitably carries ideological baggage (linked to other contemporary constructions such as "globalization" and political debates over the welfare role and wider responsibilities of the state) and therefore must be employed in a skeptical and cautious fashion (for a discussion of the possible dangers of STS falling prey "to the technological scenarios of power elites," see Elzinga, 2004). Nevertheless, the notion of "scientific governance" can raise far-reaching questions about the very "governmentality" of science and technology—and encourage fresh thinking about what has previously been represented as a "weak and rather fragmented" area within STS (Elzinga & Jamison, 1995: 572). For this reason it has (on balance) been preferred here.

In general support of this STS approach to scientific governance, many authors have drawn attention to the changing context to such matters in contemporary life and, especially, the manner in which the previous world of science and technology policy has been challenged and undermined by wider social changes. Merelman (2000), for example, suggests that a series of recent shifts in what he presents as "the culture of technology" have helped transform the American liberal democratic state. Merelman argues that belief in science and belief in democracy coexisted and, indeed, positively reinforced one another from Thomas Jefferson and the Founding Fathers until the 1950s. As Ezrahi (1990) has influentially argued, however, efforts to employ science and technology in support of a particular model of liberal democracy went into decline in the late twentieth century. Drawing especially on European experience, Ulrich Beck (1992) presents a world where faith in science and progress has given way to a wider social malaise. Science has become a source of uncertainty, and political institutions struggle to retain widespread credibility. Addressing related questions, Jasanoff (2005) has argued that an international shift has taken place toward the fracturing of nation-state authority. In this situation, "the 'old' politics of modernity—with its core values of rationality, objectivity, universalism, centralization, and efficiency—is confronting, and possibly yielding to, a 'new' politics of pluralism, localism, irreducible ambiguity, and aestheticism in matters of lifestyle and taste" (2005: 14). One consequence is that older, more limited notions of "making policy" have (at least in analytical terms) given way to a broader conception of "governing" the complex processes of sociotechnical change.

Characteristically, STS research into scientific governance has emphasized a number of general points:

• That knowledge cannot be separated from the contexts of its development and implementation. Instead, it should be seen as contingent, situated, contextualized, and open to different framings and perspectives (Collins & Pinch, 1998; Latour & Woolgar, 1979; Wynne, 1989).

• That policy-making must be seen as much more than the simple addition of "values" to objective expertise (e.g., Gonçalves, 2000). Instead, the interaction between politics and the natural world should be viewed as a more active (if often implicit) process of defining the boundaries between the public and the private, the nature of citizenship, and the role of the state (Jasanoff, 2004a; Parthasarathy, 2004).

• That claims to "democracy" and to "public opinion" should similarly be viewed in contextual and contingent terms. Rather than simply advocating "democracy," the question instead is "what form of democracy?" and "from whose perspective?" Equally, STS research has considered the political and cognitive constraints on "democratization" (Irwin, 1995, 2001a; Hagendijk, 2004; Rayner, 2003).

• That governance processes are often characterized by uncertainty, doubt, and indeterminacy (Beck, 1992; Nowotny, 2003; Wynne, 2002). In this situation, STS research has emphasized the importance of public trust in institutions and the need for political agencies to recognize alternative framings of policy dilemmas (e.g., Zavestoski et al., 2002).

• That the study of scientific governance is not concerned with the interaction between two separate processes ("expertise" and "power") but precisely the manner in which knowledge of the natural world and political action have become mutually embedded and co-constituted. As Mary Douglas has put it, "the view of the universe and a particular kind of society holding this view are closely interdependent. They are a single system. Neither can exist without the other" (Douglas, 1980: 289). We are not dealing here simply with the separate, bureaucratically defined domain of "science and technology policy" but rather with the very constitution of social and natural order in specific governance contexts.

In order to reflect on and develop such broad insights, in the following sections I rapidly consider a range of specific STS perspectives on scientific governance. Before doing so, however, I emphasize that these are not presented as *alternative* approaches to the study of scientific governance. Instead, they should be seen as sharing key principles and in that sense being *complementary* (even if each raises particular questions and issues). Taken together, they suggest something of the dynamism and diversity of STS perspectives on scientific governance. One great strength of STS studies in this field is that they generally present detailed empirical studies rather than operating at the level of conceptual generality. Accordingly, it will be necessary (and hopefully enlightening) to offer a series of brief case studies and empirical examples to establish each of the different perspectives.

Such case studies should not be seen merely as *illustrations* of STS perspectives but in many ways as representing the essential means through which analysis has been developed—in that sense, they *are* the analysis. I begin with what has been one of the most influential perspectives on these issues within STS: the notion of boundary work. Following this, I consider the contribution to the understanding of scientific governance made by such concepts as co-production, framing, networks and assemblages, and situated knowledges. These individual, albeit strongly overlapping, concepts will then be brought together in a discussion of one of the central overarching themes in STS: the relationship between scientific governance and scientific democracy.

BOUNDARY WORK

In her 1990 study of the U.S. Science Advisory Board (SAB), Jasanoff considers how a body under considerable political pressure has managed to maintain a reputation for independence and scientific credibility. One essential element has been the Board's ability to maintain a distinction between its own declared focus on the *scientific* aspects of risk assessment and what it has successfully represented as separate *nonscientific* matters of policy and rule making. In line with the STS perspectives summarized above, Jasanoff notes that "there is little support for the existence of such a boundary in the literature on risk assessment" (Jasanoff, 1990: 97). Indeed, a substantial social scientific literature has challenged the notion that the "science" and "values" of risk can be separated in this fashion (Gillespie et al., 1982; Irwin, 2001b; Wynne, 2002). Nevertheless, the distinction identified by Jasanoff has remained an (often implicit) feature of institutional risk assessment across the world (for one U.K. example, see RCEP, 1998).

In a separate mapping of the demarcation line between U.S. politics and science, David Guston (1999a) has considered the part played by the National Institutes of Health Office of Technology Transfer (OTT). Exploring the operation of the OTT in some depth, Guston considers how this body has mediated between politicians and scientists, bringing together the politically motivated call for patent applications and licensing agreements with the work of laboratory scientists who may seek patents and royalties but lack commercial experience and awareness. Through the OTT, commercial and research goals are fused, scientists and nonscientists are brought together, and a combined "scientific and social order" is created that does not destabilize the wider relationship between politics and science.

In the context of increasingly contested relations between science-led "progress" and public groups, it is perhaps unsurprising that bioethics bodies have been established internationally. Potentially, such bodies could be threatening to scientific institutions asked to defend their priorities and working practices on political, moral and ethical grounds. However, and as Susan E. Kelly has examined in the case of the U.S. Human Embryo Research Panel, this represents an area where the "co-constitutive nature of science, ethics, and publics is both veiled and engaged" (Kelly, 2003: 351). In particular, Kelly argues that bioethics bodies have been used as a means of appearing responsive to public concerns while protecting the autonomy of science. Specific ideas of "consensus building" have served to restrict the form and content of public challenges to scientific expertise. In that way, bioethics panels appear to serve as effective "border guards" protecting scientific institutions from wider scrutiny while presenting an apparently open posture in the face of external demands.

These three case studies illustrate well the diversity of contemporary scientific governance—even if as stated they certainly do not cover all groups active in the contested relationship between science and politics (with non-governmental organizations and also industry relatively absent; see, respectively, Jamison, 2001; Eden, 1999). More pertinently, however, all three cases can be presented as illustrations of *boundary work* (for other cases, see Agrawala et al., 2001; Kinchy & Kleinman, 2003; Jasanoff, 1987; Miller, 2001; Shackley & Wynne, 1996). As Gieryn has discussed this, the epistemic authority of science is not an "always-already-there feature of social life, like Mount Everest . . . but rather is enacted as people debate (and ultimately decide) where to locate the legitimate jurisdiction over natural facts" (Gieryn, 1999: 15). In each of the above cases, boundary work operates around the margin between what is considered to be "science" and "nonscience." Typically within STS analyses, boundary work preserves the integrity and autonomy of science in the face of external challenges. As Guston discusses the related concept of "boundary organizations," these help define the terms of engagement between science and politics—opening up possibilities for exchange, being responsive to separate audiences, but without necessarily threatening the "purity" of either domain (Guston, 2001). In Gieryn's terms, the edges of science represent endlessly contested terrain. The right to declare a certain rendition of nature as "true" represents the *outcome of* (rather than, as it is more usually presented, the *input to*) specific, contextually defined social and institutional processes (see also Grint & Woolgar, 1997).

One great merit of the boundary work approach to scientific governance is that it encourages a contextual, flexible, and empirically focused understanding of specific cases but without forcing the analysis into preordained categories of "science" and "politics." Instead, the very processes whereby certain issues, arguments, and phenomena come to be defined as either "science" or "nonscience" constitute an important and integral part of STS investigation. Potentially also, boundary work can encompass both the actual decision processes around a specific case or issue and the manner in which these cases and issues come to be defined (or "framed") as problems in the first place. One potential difficulty of the boundary work concept, however, is that (depending on the interpretation) it can be taken to imply that, although "science" and "politics" come together in such examples (or that they are permeable around "the edges"), ultimately they are analytically separable and relatively static. Such an assumption would preclude discussion of wider questions of the interpenetration of science and culture or the manner in which the nature of politics has been transformed by the scientific enterprise (and, of course, vice versa—Gibbons et al., 1994; Nowotny et al., 2001). In these circumstances, several STS scholars (including those such as Guston and Jasanoff who have worked with boundary concepts) have developed the concept of "co-production" as a means of capturing not simply the inter-relationship between two separate entities but also the manner in which natural and social orders are both generated together and mutually embedded (Latour, 1987; Jasanoff, 2004a).

CO-PRODUCTION

In their study of "knowledge and political order" in the case of the European Environmental Agency (EEA), Waterton and Wynne consider not simply the boundary between science and politics but also the manner in which certain idioms of natural science have been central to the struggle over the institutional and political order of modern Europe (Waterton & Wynne, 2004). In an empirical study of the interplay of geology and policy in U.S. debates over nuclear waste disposal at Yucca Mountain, Allison Macfarlane (2003) examines the ways in which the political context has affected the kinds of technical assumptions being made about safe disposal but also how political understandings have evolved in response to deepening scientific complexities. In cases such as these, we appear to be discussing a more fundamental and mutual embedding between science and politics than the concept of boundary work necessarily conveys. In Macfarlane's terms, "scientific knowledge cannot be separated from politics and associated policies. Rather they co-evolve in response to each other" (2003: 789). For Waterton and Wynne, the EEA case study reflects a larger battle over European political identity: between a centralized European superstate with requirements for standardization and harmonization across cultures and the more exploratory aspiration for a European state where the uncertainties and contingencies of political action are brought clearly into view and reflected upon. In each case, visions of the natural and human world are being created and disseminated—reinforcing also the point that scientific governance cannot operate apart from larger questions of state authority, political identity, and personal liberty (see Waterton & Wynne 1996).

Jasanoff has built on such cases to argue that "we gain explanatory power by thinking of natural and social order as being produced together" (Jasanoff, 2004b: 2). Such an approach has of course wide relevance across STS. However, for the study of scientific governance it suggests a number of particular points.

First, it offers a fresh perspective on political power. As Jasanoff puts this, the concept of co-production highlights "the often invisible role of knowledges, expertise, technical practices and material objects in shaping, sustaining, subverting or transforming relations of authority" (2004b: 4). One implication is that (in a manner inspired loosely by Foucault—e.g., 1998) power does not reside in particular institutions and social actors but may be co-produced within particular governance practices, sociotechnical interactions, and cognitive assumptions.

Second, it moves STS studies away from determinist ideas that, for example, social controversies around science are "really" all about politics or that complex areas of innovation can be reduced to "social" construction (as if "politics" or the "social" were unproblematically given). Instead, science and society are co-produced, each "underwriting the other's existence" (see also Irwin, 2001b, for a discussion of the "co-construction" of the social and the natural). Rather than presenting either "science" or "politics" as preeminent, the point is to avoid both social and scientific determinism.

Third, the "co-productionist idiom" suggests a number of specific pathways for analysis: pathways such as the making of identities, of institutions, of discourses, and of representations. In each of these areas, the argument is that we can witness "the constant interplay of the cognitive, the material, the social and the normative" (Jasanoff, 2004c: 38). Within all four of these categories, tacit models of human and natural agency are played out, suggesting once again that, within apparently mundane areas of scientific governance, more widely significant, but often unexplored and unchallenged, sociotechnical processes may be at work. As Lynch has put it in discussion of the membership categories of "expert" and "scientist" in courtroom testimony, "the co-production of legal authority/scientific knowledge is a relentless and rather subtle undertaking. It is not simply a matter of slamming together two global sectors of a public sphere . . . or of projecting disciplinary order into a lifeworld" (Lynch, 2004: 177). As is also suggested by the concept of boundary work, governance categories are "not boxes with stable boundaries between inside and outside" (2004: 178). Instead, these are the contextual products of "moment-to-moment, institutionally embedded, discursive interaction" (2004: 178; see also Edmond, 2002; Lynch & Cole, 2005).

FRAMING GOVERNANCE

Building on these STS concepts of boundary work and co-production, a key aspect of scientific governance concerns the framing of the issues under consideration. To return to Macfarlane's example of nuclear waste disposal, should this be constructed as a "technical" problem of finding the most suitable geological conditions or a wider question of social equity and economic dependency? Although the selection of issues for discussion and the decisions about what evidence counts as "relevant" in a given context may at first glance appear to be straightforward bureaucratic matters, STS scholars have suggested that such judgments can prove profoundly influential over the governance discussion that follows and, more particularly, serve as a powerful limitation on the possibilities for democratic and open debate. As Macfarlane suggests with specific reference to the Yucca Mountain controversy, a different policy process might well have produced different scientific evidence and different practical outcomes. More generally, Wynne (2002) has argued that the characteristic framing of contested areas of science and technology as matters of "risk" (to be judged primarily by scientific analysis) serves to exclude wider cultural and political discussions (see also Jones, 2004)—and very often to present public views as not simply irrelevant, ignorant, and uninformed but also as fundamentally irrational (Irwin & Wynne, 1996).

Several forms of "framework" analysis are employed within STS. One influential perspective originates from the work of Douglas (1982) and has been further developed by Douglas and Wildavsky (1983) and Schwarz and Thompson (1990; see also Horst, 2005). As applied, for example, by Charles Lockhart (2001) to environmental activism, three rival orientations can be identified: *individualistic* framings of nature as a resilient cornucopia, *egalitarian* interpretations of nature as being in delicate balance, and *hierarchical* perspectives on nature as requiring institutional protection. To these three categories cultural theorists often add a fourth: the *fatalists* who find themselves both cynical and sidelined, sceptical of those who claim to know better but lacking the basis (or the motivation) for counterargument. The implication of "cultural theory" as applied to scientific governance issues is that different, often incommensurable, perspectives will generally emerge—with a consequent battle over who should set the terms of the discussion. It also follows that apparently universal terms such as "democracy" will take on very different meanings according to cultural perspective (setting individuals free to make their own choices, setting a path toward an assumed collective interest, or setting rules of appropriate conduct and behavior?).

A second, and more contextually flexible, approach to framing takes its inspiration from Erving Goffman ([1974]1986). As discussed by Roth and colleagues (2003), Goffman's sociological conception of "framing" directs STS to the dynamics of "collective action frames" and the ways in which social actors "mobilize and counter-mobilize ideas and meanings" (Roth et al., 2003: 10). In a detailed study of a U.S. Food and Drug Administration (FDA) regulation on the sale of tobacco products to minors, Roth and co-workers explore the manner in which the language of science came to be employed as a "master frame." As these authors put it, science was the "grammar" within which the FDA's presentation was couched. In sharp contrast, public commentary on the proposed rule employed what Roth and colleagues present as ideological, economic, and political frames to warrant their arguments. Thus, more than two fifths of the opposing commentary was expressed in the language of freedom of choice rather than science. The key point is that these counterframes did not *overtly* contest the FDA's use of science as a master frame in a direct conflict according to the Administration's chosen terms of engagement. Instead, the master frame was *tacitly* contested by those highlighting the consequences of the proposed rule for "ideological" matters such as freedom of choice and personal privacy. One outcome from these contested framings was that the proposed rule failed to establish credibility with concerned citizens.

In making these observations about the framing of governance disputes, it is important for STS analysis that it does not neglect the manner in which frameworks of meaning can also be embedded in material practices and technological artifacts (as the large STS literature on the social construction of technology would strongly reinforce: Bijker, 1997; Bijker et al., 1997). Framing therefore does not simply take place at the end of the technological or scientific development process and at the point of implementation (when governance issues generally become public) but can be embedded in the artifacts and products themselves—hence also the current interest in "upstream" engagement with science and technology (Wilsdon & Willis, 2004, Wilsdon et al., 2005).

In a discussion of the manner in which social choices can be embedded in apparently neutral technological systems, Levidow (1998) observes that such systems serve to reify social relations: "(b)oth people and nature are disciplined according to various models of the socio-natural order, as if this choice arose from the nature of things" (1998: 213). Thus, what he presents as the strategy of "biotechnologizing agriculture" leads to a situation of socioeconomic dependency within which "problems" come to be defined in specific ways: as a matter for further research and development, as the focus for risk-benefit calculation, and as a stimulus to new "genetic fixes." Attempts at public debate inevitably conform to this "genetic-pesticide treadmill," since there is little or no space within the accepted framework for alternative presentations. Whether the problem is world hunger or environmental degradation, the solution is genetically modified (GM) agriculture (see also Levidow & Carr, 2000; Levidow, 2001). Although it may be that this analysis exaggerates the inability of oppositional groups to build counterframes, it does draw attention to the manner in which precommitments and implicit understandings can be embedded not just in specific governance discussions but also in the very objects and artifacts that are at the center of decision-making. Co-production is not simply a matter of explicitly defined science and politics but also of the manner in which the material and technological world is shaped in accordance with often implicit and undefended visions of order and progress. Equally, one important aspect of boundary work is the manner in which the pre-framing of public discussion can be restricted from external scrutiny (for an earlier discussion of this issue, see Crenson, 1971).

NETWORKS AND ASSEMBLAGES

In emphasizing the extent to which matters of scientific governance are sites for the construction (or co-construction) of different material and social orders, and also the manner in which power is not held in one location but is instead (like knowledge) an outcome of specific sociotechnical encounters, STS perspectives draw strongly on actor-network theory (Latour, 1987, 1993, 1999; Law, 2002; Michael, 2000; Mol & Law, 1994). Put in the most general terms, actor-network theory directs analytical attention to the manner in which both human and nonhuman actors are enrolled in the construction of sociotechnical systems. Rather than presenting governance issues in terms of institutional processes and human agency alone, the treatment of scientific governance in terms of sociotechnical networks allows a more fluid and "hybridized" exploration of the development, emergence, and resolution of particular problems—and without "purifying" discussion into such predetermined categories as "science" or "politics." In a related development, the notion of "ethno-epistemic assemblage" has been proposed (Irwin & Michael, 2003) as a heuristic device intended to capture the emergent coalitions and alliances that draw simultaneously on both "cognitive/epistemic" and "cultural/ethical" concerns and that ultimately embed themselves within particular forms of scientific governance. As originally proposed, the concept of ethno-epistemic assemblage (EEA) offers a way of "investigating how the blurring of science and society . . . might entail rather surprising new resources and methods" (2003: 113). Put more concretely, the term draws attention to the emergence of new hybrid entities and new admixtures of science and society. Rather than attempting to close down or reinforce the conventional boundaries between the human and the nonhuman, the political and the scientific, government and the public, the EEA concept is to be open to the transgression of boundaries and the flexible construction of actants. In that sense, EEAs "are a means of *expanding the range* of entities, actors, processes and relations that get blurred and mixed up" (2003: 114).

The major power of these sociotechnical network approaches (just like the other, related concepts considered in this chapter) resides in their interpretation and unraveling of particular cases and issues, in other words, in their *explanatory* potential. Indeed, STS perspectives on scientific governance generally serve to challenge the distinction between "theoretical work" and "empirical observation." On that basis, Wetmore's study of automobile safety in the United States can be used as a means of briefly assessing the explanatory merits of network-based approaches but also as a way of exploring the "governmentality" of science and technology in contemporary society.

Wetmore (2004) sets out to document the development of, and debates around, technologies designed to mitigate the safety problems caused by "irresponsible" motorists. What is notable in this analysis is that the innovation of safety technologies such as the air bag becomes inseparable from (re)definitions of risk and the allocation of responsibilities between actors. Ultimately, the development of a new approach to automobile safety (based on crashworthiness rather than crash avoidance—see also Irwin, 1985) required the successful enrollment of a diversity of social groups, including the insurance industry, medical professionals, the President of the United States, politicians, and the media. This also meant that when the technology of air bags was subsequently held open to question (on the grounds that it might actually be the cause of fatalities and injuries), a "strange coalition" of automakers and safety advocates fought to oppose further change. On the one hand, the prevailing definition of automotive risk constrained and shaped the possible technologies that could be used. On the other, specific technological systems can be seen as playing an essential role within the emergent "web of responsibility" (Wetmore, 2004).

Viewed from an EEA perspective, such an empirical study suggests something of the dynamic context to governance debates—a context that can get lost in the focus on specific controversies and disputes. Equally, we witness in a complex case such as this the manner in which coalitions and alliances are not necessarily fixed or preconfigured but can be open to shifting movements and new emergences. Governance from this perspective is only partly about "policy making" but also inherently uncontrollable, and certainly unplanned, sociotechnical dependencies and attributions. Instead of a single locus at which governance decisions are taken (and where power "ultimately" resides), Wetmore's case suggests the distributed and "de-centered" character of governance. Rather than operating according to a model of "control," such a case hints at the fragmentation of responsibilities and of knowledge, the collapse of clear distinctions between categories of actants (whether public or private, technological or institutional), and the cross-cutting nature of the interactions and assemblages that lie at the heart of sociotechnical change. Indeed, such an example might suggest the potential "ungovernability" of sociotechnical actors and networks (Black, 2002). Once we move beyond the modernist paradigm of science and technology as being amenable to centralized, rational control, scientific governance is revealed as a much more challenging, but also more intellectually intriguing, process.

SITUATED KNOWLEDGES

One important issue within STS treatments of scientific governance concerns the kinds of contribution "engaged citizens" are (or should be) able to make within scientific governance processes: as political actors only or as legitimate sources of knowledge and understanding? Drawing on STS-based criticisms of what has become known as the "public understanding of science" (Wynne, 1995; Irwin, 1995; Irwin & Wynne, 1996; Zavestoski et al., 2002; Roth et al., 2004), we can certainly suggest that an epistemological line is conventionally drawn (or at least attempted) by government and other institutions between areas where public engagement is seen to be legitimate (typically, issues of ethics and values) and those where it is not (generally, matters requiring specialist knowledge and expertise). The definition of "legitimate knowledge" within governance disputes also represents an important form of boundary work—though one that may be driven as much by preexisting cultural and cognitive commitments to "good science" as by deliberate political manipulation (even if such manipulation cannot be ruled out: Abraham, 1995). As has been suggested by the previous discussion, this epistemological dividing line (or ring-fencing) has been severely challenged by STS analyses (see also Funtowicz & Ravetz, 1993).

The point here is neither to romanticize alternative forms of knowledge (Irwin & Michael 2003) nor to replace science as a means of understanding the physical world but to suggest that scientific institutions can in practice be socially *and* epistemologically insulated from wider questioning and debate (Fischer, 2000). In terms of the relationship between science and democracy (to which we will shortly return), the implication is that citizens wishing to engage with emerging areas of science and technology may be frustrated by institutions that relegate their expressed concerns to a secondary "nonfactual" level. Meanwhile, outside groups can find it difficult to challenge the framing assumptions made by policy makers—for example, the assumption that "laboratory" science will directly map onto "real world" conditions of application and enactment (Wynne, 1996). Some STS scholars have attempted to reestablish a distinction between "science" and "politics"—or between experience and expertise (Collins & Evans, 2002, 2003; see also Jasanoff, 2003a; Rip, 2003; Wynne, 2003). However, the construction of a barrier between "public" and "expert" assessments (whether on the part of sociologists or policy makers) appears to represent a turning away from the often contested realities of decision-making under conditions of social and technical uncertainty (Irwin, 2004).

Meanwhile, the notion that groups of citizens may bring relevant forms of knowledge and expertise to scientific governance processes has been widely reflected in STS studies (Brown, 1987; Epstein, 1996; Kerr et al., 1998). Bloor (2000), for example, has considered the "instrumental" relationship between one group of coal miners and scientific understandings of pneumoconiosis (black lung). On the one hand, faced with an intransigent scientific and legal establishment, miners were able to "draw their own conclusions" concerning the link between coal dust and pulmonary disease. On the other, miners employed a variety of strategies to secure compensation claims—including the duping and "buying" of expert witnesses. "Knowledgeability" in such cases is not simply a matter of "knowing" that there may be a connection between certain forms of exposure and patterns of mortality/morbidity. It is also a matter of "knowing" how to "work the system"—as the miners were successfully able to demonstrate when the link between coal dust and pneumoconiosis eventually became accepted in legal and scientific terms. Furthermore, knowledge in such contexts becomes an active process of sense-making and co-production—what we can term "knowledging" (Irwin et al., 1999, Irwin 2001b). The concept of "situated knowledge" (or "citizen science"), therefore, should not be taken to suggest a passive body of expertise into which governance processes can simply tap. Instead, and as the concept of co-production would also imply, forms of knowledge and understanding are contextually generated and simultaneously embody understandings of both the natural and social worlds. In this, of course, lay understandings may not be so different from scientifically validated knowledge.

SCIENCE AND DEMOCRACY: EXPLORING THE "NEW" SCIENTIFIC GOVERNANCE

The relationship between science and democracy represents a classic theme not simply within the study of scientific governance but within STS more generally. David Edge (1995) has linked what he termed "the democratic impulse" within STS to the political upheaval of the 1960s, including the social impact of the Vietnam War and the rise of the civil rights, feminist, and environmentalist movements (see also Rose & Rose, [1969]1977). Issues of science and democracy had earlier been raised by Bernal (1939), Hogben (1938), and the Association of Scientific Workers (1947; see also Werskey, 1978). Notably, Dorothy Nelkin in a series of influential publications from the 1970s onward (e.g., 1977, 1992) placed questions of science and democracy at the core of STS studies of technical decision-making in situations characterized by public controversy.

This theme of science and democracy has remained alive throughout the subsequent development of STS and has been explicitly reflected on and considered by many of the scholars whose work has featured in this chapter. Furthermore, political initiatives aimed at facilitating democratic engagement have become more widespread since the late 1990s, especially, but not exclusively, in a European context. In this section, I look particularly at these experiments in the "new" scientific governance—drawing broadly on the concepts presented so far. As we will see, one consequence of the move from straightforward advocacy of greater democratic control over science and technology to the empirical examination of specific, practical initiatives has been to raise questions about the meaning of democracy in this context, the limits to "democratization" (Lahsen, 2005), and the relationship between "public engagement" initiatives and the wider operation of scientific governance (Guston, 1999b; Wachelder, 2003).

In broad terms, "new" approaches to governance characteristically emphasize the need to rebuild trust in regulatory institutions (DTI, 2000; Council for Science and Technology, 2005; RCEP, 1998), to operate in a more open and transparent manner (Phillips et al., 2000; CEC, 2002), to "engage" with the wider publics (Royal Society/Royal Academy of Engineering, 2004), and—as the House of Lords Select Committee put it—to create a situation in which direct dialogue becomes a "normal and integral part" of the policy process (House of Lords Select Committee on Science and Technology, 2000: 43). Some countries have a well-established history of engagement activities—notably, Denmark and The Netherlands (Hagendijk & Irwin 2006). For others—notably, the UK—this is relatively new ground and represents an interesting phase in science-governance relations. Having made that point, however, we should also note that there has been widespread scepticism concerning the actual conduct and effectiveness of initiatives to date.

To offer one significant example of the "new governance" in action, the United Kingdom's "GM Nation?" debate over the commercial growing of genetically modified crops took place during summer 2003 and involved a large series of events at regional, county, and local levels. A Web site to the debate received 2.9 million hits. Around 37,000 feedback forms were returned (Understanding Risk Team, 2004). On that basis alone, this counts as the largest exercise in public engagement in that country (GM Nation?, 2003).

Despite this impressive scale, the debate was not in general judged to be a success. In a highly critical report, a House of Commons committee concluded that "[i]t is profoundly regrettable that the open part of the process, far from being a "public debate," instead became a dialogue mainly restricted to people of a particular social and academic background. The greatest failure of the debate is that it did not engage with a wider array of people" (House of Commons, Environment, Food, and Rural Affairs Committee, 2003: 15). Lack of time and money were blamed by the cross-party group of MPs for this alleged failure. In a series of criticisms that are more widely symptomatic of initiatives aimed at engaging the publics, the GM debate has been presented as failing to dispel the suspicion that it was primarily a legitimatory exercise in support of the Government's decision to proceed with selective commercialization, lacking clarity in terms of the Government's aims and objectives for the consultation, suffering generally from a difficult relationship between the debate steering board and government, and taking place far too late in the technology-development process (Council for Science and Technology, 2005).

This example suggests that the practice of public engagement is at least as important as the underlying principles (Irwin, 2006a). There is also a suggestion in many official statements that—despite the increased attention that has been recently paid—the messages from social science have so far only partly been assimilated by policy makers (Hagendijk, 2004; Marris et al., 2001). Although institutions have often sought to distance themselves from the old deficit model of science-public relations (wherein the public was seen primarily as a target for communication efforts designed to increase their understanding of, and hence it is assumed social support for, science), there is still a tendency for the new climate of dialogue to be seen as a means of persuading the publics that further science and technological innovation is necessary and, indeed, the only rational way forward (Blair, 2002). Given this apparent reluctance to acknowledge public questioning of institutional priorities (or to acknowledge that "rationality" can be a contested territory), the possibilities for science-public dialogue can appear quite restricted.

We are therefore confronted with a situation in which talk of public engagement and more open relations with the public has become increasingly common in Europe (Hagendijk et al., 2005). Meanwhile, such initiatives as have taken place lead often to further debate and disagreement (Horst, 2003; Irwin, 2001a) and to raised awareness of their limitations. On the one hand, this has encouraged some discussion of the best form and timing of future exercises (Council for Science and Technology, 2005; Wilsdon & Willis, 2004; Wilsdon et al., 2005). On the other, it suggests the need for greater appreciation and understanding of what can and cannot be achieved by specific (often "stand-alone") initiatives, especially when such initiatives fail to challenge (or often even acknowledge) underlying institutional processes and the assumptions according to which they operate. More generally, there remains an underlying (and unresolved) question about the relationship between science, democracy, and the marketplace. The current tendency is for deliberative democracy simply to add a layer of "public debate" to existing institutional processes without acknowledging potential tensions between scientific innovation and democratic engagement or considering larger questions of what it would mean to engage democratically with science and technology (see, e.g., Hagendijk & Irwin, 2006; Jasanoff, 2003b). Equally, public engagement risks being seen by policy makers as a brief phase before "business as usual" can recommence. Certainly, there is a tendency to present "science" and "the market" as neutral, fixed, and objective entities with "the public" seen as a softer (and more pliable) domain of values, emotions, and opinions. If we now consider these questions of democratic engagement with science and technology in terms of the specific concepts presented in this chapter, a number of more particular points emerge:

Practical initiatives in science and democracy represent an important site for active *boundary work*, especially in defining what counts as a matter for "good science" and what as a legitimate topic for broader public debate. In the case of the GM Nation exercise, this meant that the "scientific" dimensions of the issue were kept at arm's length from the "economic" and "public" strands. Equally, the very construction of "the public" represented an important manifestation of boundary practice—with, for example, "stakeholders" seen as problematic entities in what was portrayed as an attempt to reach out to the larger, "more representative," publics.

Elements of *co-construction* are very apparent in such debates. In this case, wider constructions of the debate as a matter of national agricultural policy in the face of globalization processes (and specifically the global reach of North American companies) contrasted sharply with the official portrayal of the debate as primarily a matter of the social and environmental impact of one category of agricultural products. The definition of what counts as relevant science is closely bound up with these visions of the natural and social worlds: testing the environmental impact of selected crops is very different from considering the wider impacts on British agricultural policy of commercializing GM crops.

In these circumstances, the *framing* of democratic exercises becomes crucial. Is it about the science of environmental impact or the ethics of intensive agriculture, the economics of technological innovation or the future of farming, the protection of vulnerable nations or the cure of malnutrition and disease? Disagreements about what the discussion should even be about jostled alongside the broader political uncertainty over where and how the discussion should take place. Equally, control over the debate framework (in other words, what gets to be debated) constitutes a significant form of political power.

In terms of *network and assemblage analysis*, it can certainly be seen that the GM issue crosses national boundaries, involves new forms of political action, and both creates and is created by shifting social and political alliances of a relatively novel kind (supermarkets and activist groups, scientists and agrochemical companies, consumers and farming organizations). Such encounters crisscross established political institutions and throw up new alliances and contests—as when, for example, linkages with groups in the developing world were used both to justify GM agriculture and to condemn it.

Most characteristically, *situated forms of knowledge* are granted diminished importance within institutional decision-making processes. For example, the expertise and insight of farmers as to the likely administration of such crops (specifically with regard to issues of cross-contamination) were down-played within the scientific appraisal. Instead, safety and environmental tests frequently operate on the basis of implicit, often unchallenged, assumptions about the operation of procedures in the social world. Thus, there can be a large difference between the controlled conditions of farm-scale trials and the more routine operation of agricultural practice. The construction of "science" within such initiatives generally excludes the robust social testing of such underlying assumptions. In further illustration of the potential significance of situated knowledges, one participant in a 2006 Mali citizens jury argued that "GM crops are associated with the kind of farming that marginalizes the mutual help and co-operation among farmers and our social and cultural life" (IIED, 2006). Such broad social assessments are generally neglected, or entirely ignored, within technocratically based governance processes.

DISCUSSION

All of the accounts considered here suggest a change in our understanding of scientific governance. Indeed, even the move from the narrower language of "policy" to the more expansive concept of "governance" signals a rethinking of traditional boundaries and categorizations. From a world where "science and technology policy" could be thought of as a discrete domain of bureaucratic and technical practice in which distinctions could safely be maintained between "science" and "politics," STS perspectives open up a more complex territory in which the boundary between science and politics becomes blurred, agency does not simply reside in human actors, and problems spill over between categories.

As a consequence of this discussion, at least two important, and inter-connected, issues arise for STS research into scientific governance. These concern the relationships, on the one hand, with wider social scientific treatments of policy and politics and, on the other, with the critical analysis of governance and practice. How does the STS exploration of scientific governance connect with larger social scientific discussions of policy, power, and politics? Meanwhile, does the adoption of a constructivist idiom within STS mean that the subdiscipline loses its critical edge and indeed the very "democratic impulse" that stimulated its development?

In considering the first of these issues, it is noteworthy that STS explorations have generally operated at some analytical distance from "mainstream" social scientific and political discussions of such topics as globalization, socioeconomic inequality, and political economy. Of course, and as has been noted, one of the main strengths of STS scholarship is that it deals in specific and contextually defined matters rather than the broad generalizations of "grand theory." To offer one notable example, talk of "globalization" characteristically adopts a deterministic and totalizing perspective that is far from the more nuanced, fine-grained, and empirical approach of STS research into cultures of scientific governance (Held et al., 1999). STS scholars have therefore treated the concept of "globalization" with understandable caution, preferring instead to view this as a particular social and political construction employed by groups of actors according to their own cultural and institutional perspectives. Far from being a coherent or unidirectional force, globalization should instead be seen as an interaction (or negotiation) between different ethno-epistemic assemblages and between different forms of sociotechnical construction. As Yearley (1996) has argued, the study of globalization should be seen also as the study of certain "universalizing discourses."

Rather than studying globalization as an external phenomenon, therefore, STS perspectives suggest the need to examine the shifting constructions of globalization—as expressed, for example, through the "universalizing" claims of science, economics, and political institutions—in suitably open-minded and symmetrical terms (Irwin, 2006b). Importantly, and as this chapter has strongly suggested, such an intellectual approach should also apply to the notion of "scientific governance." Especially given the increasing adoption of the language of governance by national and international organizations, it is essential that STS scholars should be alert to the implications of this categorization, treat claims to the "new" scientific governance with great caution, and view the social definition of "governance" as a constructed category. With both "globalization" and "governance," however, the point is not simply to abandon such terms—or the issues and questions they suggest—but instead to engage with them in an appropriately contextual and reflexive fashion.

In terms of the relationship to wider social science, this sensitivity to contextual definitions, hybrid assemblages, and shifting discourses could be presented as a weakness, as a failure to come to grips with "real" issues of power, political economy, and inequality. The argument of this chapter is that, far from representing a turning away from such issues, STS perspectives actually allow a closer and more open examination of contested understandings, and in that way bring more rather than less "reality" into our accounts (see also Jasanoff, 1999). However, in arguing the significance of an STS perspective, it is also important for scholarship in this field to remain alert to wider academic discussions and to both contribute to and learn from other areas of social scientific inquiry.

One significant challenge ahead is therefore to enhance the intellectual and methodological engagement between STS and the larger social and political sciences. On the one hand, this will involve a greater willingness for STS to connect broadly with social scientific discussions and to ensure that STS research is not marginalized as "interesting qualitative work" (or as bringing "color" to the "black and white" representations of macro social science). On the other, this will necessarily involve an openness to larger social scientific debates and alternative theoretical perspectives. Potential areas where STS studies of scientific governance could benefit from creative engagement across the social sciences include economic and market analysis of the processes of technological innovation (including elements of political economy); the analysis of politics, power, and governmentality (a core theme across the social sciences as a whole); wider treatments of socioeconomic change, including matters of inequality and disadvantage; and global studies of the relationship between development and scientific/technological innovation. The point is not to abandon the theoretical and empirical developments outlined in this chapter but rather to engage critically with such issues—without falling back into deterministic and essentialist ways of thinking. This process of creative engagement is not without its difficulties and possible threats, but it is essential for the future vitality of STS research.

Such matters lead directly to our second concluding question: has increased methodological and interpretive sophistication blunted the critical edge of STS? It is certainly true that STS research of the sort discussed here does not claim to "have the answers" in terms of political analysis and practical intervention. Certainly, there is nothing in STS scholarship that represents a tool kit for "how to do governance better" (see also Edmond & Mercer, 2002; Lynch & Cole, 2005). Equally, the requirement for analytical symmetry precludes the easy adoption of political "sides." Instead, concepts such as co-production, boundary work, and situated knowledge are intended to encourage new ways of thinking, enhance analytical possibilities, and move discussion forward from general sloganizing about science and democracy or else the fatalistic view that science and technology are pre-determined.

STS scholarship cannot itself prescribe what is best for the development of science and technology in nations that consider themselves to be democratic. However, it does have a significant contribution to make in refining and developing our understanding of current governance processes, testing out alternative possibilities for democratic intervention, and pointing to the constraints on current exercises and initiatives. Far from blunting the critical edge of STS, it is this commitment to innovative scholarship and the challenging of institutional and epistemological boundaries that *brings* STS its critical edge. For this reason also, the material presented in this chapter should be seen as a stimulus to new perspectives and areas of inquiry and certainly not as a fixed canon or academic end point.

The study of scientific governance is also a provocation to STS at a more general level. Certainly, the argument of this chapter has been that, far from representing a mere application of established STS theories and methods, the study of scientific governance constitutes a major intellectual and practical challenge in its own right. No longer confined to the rigid structures of "science and technology policy," STS perspectives on scientific governance open up new possibilities for the mutually embedded relationship between practice and understanding.

References

Abraham, John (1995) *Science, Politics and the Pharmaceutical Industry: Controversy and Bias in Drug Regulation* (London: UCL/St. Martin's Press).

Agrawala, Shardul, Kenneth Broad, & David Guston (2001) "Integrating Climate Forecasts and Societal Decision Making: Challenges to an Emergent Boundary Organisation," *Science, Technology & Human Values* 26(4): 454–77.

Association of Scientific Workers (1947) *Science and the Nation* (Harmondsworth, U.K.: Penguin).

Barry, Andrew (2001) *Political Machines: Governing a Technological Society* (London: Athlone Press).

Beck, U. (1992) *Risk Society: Towards a New Modernity* (London: Sage).

Bernal, John D. (1939) *The Social Function of Science* (London: Routledge).

Bijker, W. E. (1997) *Of Bicycles, Bakelites, and Bulbs: Towards a Theory of Sociotechnical Change* (Cambridge, MA: MIT Press).

Bijker, W. E., T. Hughes, & T. Pinch (eds) (1997) *The Social Construction of Technological Systems: New Directions in the Sociology and History of Technology* (Cambridge, MA: MIT Press).

Black, Julia (2002) "Critical Reflections on Regulation," discussion paper, Centre for the Analysis of Risk and Regulation, London School of Economics and Political Science.

Blair, Tony (2002) "Science Matters," April 10. Available at: www.number-10.gov.uk/output/Page1715.asp.

Bloor, Michael (2000) "The South Wales Miners Federation: Miners' Lung and the Instrumental Use of Expertise, 1900–1950," *Social Studies of Science* 30(1): 125–40.

Brown, Phil (1987) "Popular Epidemiology: Community Response to Toxic Waste-Induced Disease in Woburn, Massachusetts," *Science, Technology & Human Values* 12(3/4): 78–85.

Collins, H. M. & Robert Evans (2002) "The Third Wave of Science Studies: Studies of Expertise and Experience," *Social Studies of Science* 32(2): 235–96.

Collins, H. M. & Robert Evans (2003) "King Canute Meets the Beach Boys: Responses to the Third Wave," *Social Studies of Science* 33(3): 435–52.

Collins, H. M. & Trevor Pinch (1998) *The Golem at Large* (Cambridge, MA: Cambridge University Press).

Commission of the European Communities (CEC) (2001) *European Governance: A White Paper* (Brussels: CEC).

Commission of the European Communities (CEC) (2002) *Science and Society: Action Plan* (Luxembourg: CEC).

Council for Science and Technology (2005) *Policy Through Dialogue: Informing Policies Based on Science and Technology,* March (London: Council for Science and Technology).

Crenson, M. A. (1971) *The Un-Politics of Air Pollution* (Baltimore, MD: Johns Hopkins University Press).

Dean, M. (1999) *Governmentality: Power and Rule in Modern Society* (London: Sage).

Department of Trade and Industry (DTI) (2000) *Excellence and Opportunity: A Science and Innovation Policy for the 21st Century* (London: H. M. Stationery Office).

Douglas, Mary (1980) "Environments at Risk," in J. Dowie & P. Lefrere (eds), *Risk and Chance* (Milton Keynes, U.K.: Open University Press): 278–96.

Douglas, Mary (ed) (1982) *Essays in the Sociology of Perception* (London: Routledge & Kegan Paul).

Douglas, Mary & Aaron Wildavsky (1983) *Risk and Culture* (Berkeley: University of California Press).

Eden, Sally (1999) "'We Have the Facts': How Business Claims Legitimacy in the Environmental Debate," *Environment and Planning A* 31(7): 1259–309.

Edge, David (1995) "Reinventing the Wheel," in Sheila Jasanoff, Gerald E. Markle, James C. Petersen, & Trevor Pinch (eds), *Handbook of Science and Technology Studies* (London: Sage): 3–23.

Edmond, Gary (2002) "Legal Engineering: Contested Representations of Law, Science (and Non-Science) and Society," *Social Studies of Science* 32(3): 371–412.

Edmond, Gary & David Mercer (2002) "Conjectures and Exhumations: Citations of History, Philosophy and Sociology of Science in U.S. Federal Courts," *Law and Literature* 14(2): 309–66.

Elzinga, Aant (2004) "Making Science and Technology Studies Relevant for Technology Policy: Gains and Losses," *Social Studies of Science* 34(6): 949–56.

Elzinga, Aant & Andrew Jamison (1995) "Changing Policy Agendas in Science and Technology," in Sheila Jasanoff, Gerald E. Markle, James C. Petersen, & Trevor Pinch (eds), *Handbook of Science and Technology Studies* (London: Sage): 572–97.

Epstein, Steven (1996) *Impure Science: AIDS, Activism and the Politics of Knowledge* (Berkeley: University of California Press).

Ezrahi, Yaron (1990) *The Descent of Icarus: Science and the Transformation of Contemporary Democracy* (Cambridge, MA: Harvard University Press).

Fischer, Frank (2000) *Citizens, Experts, and the Environment: The Politics of Local Knowledge* (Durham, NC: Duke University Press).

Foucault, Michel (1998) *The Will to Knowledge: The History of Sexuality,* vol. 1 (London: Penguin).

Fuller, Steve (2000) *The Governance of Science* (Buckingham, U.K.: Open University Press).

Funtowicz, S. O. & J. Ravetz (1993) "Science for the Post-Normal Age," *Futures* 25(7): 739–55.

Gibbons, M., C. Limoges, H. Nowotny, S. Schwartzman, P. Scott, & M. Trow (1994) *The New Production of Knowledge* (London: Sage).

Gieryn, Thomas F. (1983) "Boundary-Work and the Demarcation of Science from Non-Science: Strains and Interests in Professional Ideologies of Scientists," *American Sociological Review* 48: 781–95.

Gieryn, Thomas F. (1999) *Cultural Boundaries of Science: Credibility on the Line* (Chicago: University of Chicago Press)

Gillespie, B., D. Eva, & R. Johnston (1982) "Carcinogenic Risk Assessment in the U.S.A. and U.K.: The Case of Aldrin/Dieldrin," in B. Barnes & D. Edge (eds), *Science in Context: Readings in the Sociology of Science* (Milton Keynes, U.K.: Open University Press): 303–35.

GM Nation? (2003) *The Findings of the Public Debate.* Available at: www.gmpublicdebate.org.uk.

Goffman, Erving ([1974]1986) *Frame Analysis* (Boston: Northeastern University Press).

Gonçalves, Maria Eduarda (2000) "The Importance of Being European: The Science and Politics of BSE in Portugal," *Science, Technology & Human Values* 25(4): 417–48.

Grint, Keith & Steve Woolgar (1997) *The Machine at Work: Technology, Work and Organisation* (Cambridge: Polity Press).

Guston, David H. (1999a) "Stabilising the Boundary Between U.S. Politics and Science: The Role of the Office of Technology Transfer as a Boundary Organisation," *Social Studies of Science* 29(1): 87–111.

Guston, David H. (1999b) "Evaluating the First U.S. Consensus Conference: The Impact of the Citizens' Panel on Telecommunications and the Future of Democracy," *Science, Technology & Human Values* 24(4): 451–82.

Guston, David H. (2001) "Boundary Organizations in Environmental Policy and Science: An Introduction," *Science, Technology & Human Values* 26(4): 399–408.

Hagendijk, R. P. (2004) "The Public Understanding of Science and Public Participation in Regulated Worlds," *Minerva* 42(1): 41–59.

Hagendijk, R., P. Healey, M. Horst, & A. Irwin (2005) *Report on the STAGE Project: Science, Technology and Governance in Europe.* Available at: www.stage-research.net.

Hagendijk, R. & Alan Irwin (2006) "Public Deliberation and Governance: Engaging with Science and Technology in Contemporary Europe," *Minerva* 44(2): 167–84.

Held, D., A. McGrew, D. Goldblatt, & J. Perraton (1999) *Global Transformations: Politics, Economics, Culture* (Cambridge: Polity Press).

Hogben, Lancelot (1938) *Science for the Citizen: A Self-Educator Based on the Social Background of Scientific Discovery* (London: Allen & Unwin).

Horst, Maja (2003) *Controversy and Collectivity: Articulations of Social and Natural Order in Mass-Mediated Representations of Biotechnology,* Ph.D. Series 28, Copenhagen Business School, Doctoral School on Knowledge and Management, Department of Management, Politics and Philosophy.

Horst, Maja (2005) "Cloning Sensations: Mass-Mediated Articulation of Social Responses to Controversial Biotechnology," *Public Understanding of Science* 14(2): 185–200.

House of Commons, Environment, Food and Rural Affairs Committee (2003) *Conduct of the GM Public Debate*, Eighteenth Report of Session 2002–2003 (London: H. M. Stationery Office).

House of Lords, Select Committee on Science and Technology (2000) *Science and Society* (London: H. M. Stationery Office).

International Institute for Environment and Development (IIED) (2006) Press Release: "African Farmers Say GM Crops Are Not the Way Forward." Available at: www.iied.org/mediaroom/releases/290106.html.

Irwin, Alan (1985) *Risk and the Control of Technology* (Manchester, U.K.: Manchester University Press).

Irwin, Alan (1995) *Citizen Science: A Study of People, Expertise and Sustainable Development* (London: Routledge).

Irwin, Alan (2001a) "Constructing the Scientific Citizen: Science and Democracy in the Biosciences," *Public Understanding of Science* 10(1): 1–18.

Irwin, Alan (2001b) *Sociology and the Environment: A Critical Introduction to Society, Nature and Knowledge* (Cambridge: Polity Press).

Irwin, Alan (2004) "Expertise and Experience in the Governance of Science: What Is Public Participation For?" in G. Edmond (ed), *Expertise in Law and Regulation* (Aldershot, U.K.: Ashgate): 32–50.

Irwin, Alan (2006a) "The Politics of Talk: Coming to Terms with the 'New' Scientific Governance," *Social Studies of Science* 36: 299–320.

Irwin, A. (2006b) "The Global Context for Risk Governance: National Regulatory Policy in an International Framework," in B. Bennett & G. Tomossy (eds), *Globalization and Health: Challenges for Health Law and Bioethics* (Amsterdam: Springer): 71–85.

Irwin, Alan & Mike Michael (2003) *Science, Social Theory and Public Knowledge* (Maidenhead, U.K.: Open University Press).

Irwin, Alan & Brian Wynne (eds) (1996) *Misunderstanding Science? The Public Reconstruction of Science and Technology* (Cambridge: Cambridge University Press).

Irwin, Alan, Peter Simmons, & Gordon Walker (1999) "Faulty Environments and Risk Reasoning: The Local Understanding of Industrial Hazards," *Environment and Planning A* 31: 1311–26.

Jamison, Andrew (2001) *The Making of Green Knowledge* (Cambridge: Cambridge University Press).

Jasanoff, Sheila (1987) "Contested Boundaries in a Policy-Relevant Science," *Social Studies of Science* 17(2): 195–230.

Jasanoff, Sheila (1990) *The Fifth Branch: Science Advisers as Policymakers* (Cambridge, MA: Harvard University Press).

Jasanoff, Sheila (1999) "STS and Public Policy: Getting Beyond Deconstruction," *Science, Technology & Society* 4(1): 59–72.

Jasanoff, Sheila (2003a) "Breaking the Waves in Science Studies," *Social Studies of Science* 33(3): 389–400.

Jasanoff, Sheila (2003b) "Technologies of Humility: Citizen Participation in Governing Science," *Minerva* 41(3): 223–44.

Jasanoff, Sheila (ed) (2004a) *States of Knowledge: The Co-Production of Science and Social Order* (London: Routledge).

Jasanoff, Sheila (2004b) "The Idiom of Co-Production," in Sheila Jasanoff (ed), *States of Knowledge: The Co-Production of Science and Social Order* (London: Routledge): 1–12.

Jasanoff, Sheila (2004c) "Ordering Knowledge, Ordering Society," in Sheila Jasanoff (ed) *States of Knowledge: The Co-Production of Science and Social Order* (London: Routledge): 13–45.

Jasanoff, Sheila (2005) *Designs on Nature: Science and Democracy in Europe and the United States* (Princeton, NJ: Princeton University Press).

Jones, Kevin E. (2004) "BSE and the Phillips Report: A Cautionary Tale About the Update of 'Risk'," in Nico Stehr (ed), *The Governance of Knowledge* (London: Transaction): 161–86.

Kelly, Susan E. (2003) "Public Bioethics and Publics: Consensus, Boundaries and Participation in Biomedical Science Policy," *Science, Technology & Human Values* 28(3): 339–64.

Kerr, Ann, Sarah Cunningham-Burley, & Amanda Amos (1998) "The New Human Genetics: Mobilizing Lay Expertise," *Public Understanding of Science* 7(1): 41–60.

Kinchy, Abby J. & Daniel Lee Kleinman (2003) "Organizing Credibility: Discursive and Organizational Orthodoxy on the Borders of Ecology and Politics," *Social Studies of Science* 33(6): 869–96.

Lahsen, Myanna (2005) "Technology, Democracy and U.S. Climate Politics: The Need for Demarcation," *Science, Technology & Human Values* 30(1): 137–69.

Latour, Bruno (1987) *Science in Action: How to Follow Scientists and Engineers Through Society* (Cambridge, MA: Harvard University Press).

Latour, Bruno (1993) *We Have Never Been Modern* (London: Harvester Wheatsheaf).

Latour, Bruno (1999) *Pandora's Hope: Essays on the Reality of Science Studies* (Cambridge, MA: Harvard University Press).

Latour, Bruno & Steve Woolgar (1979) *Laboratory Life: The Social Construction of Scientific Facts* (Beverly Hills, CA: Sage).

Law, John (2002) *Aircraft Stories: Decentering the Object in Technoscience* (Durham, NC: Duke University Press).

Levidow, L. (1998) "Democratizing Technology: Or Technologizing Democracy? Regulating Agricultural Biotechnology in Europe," *Technology in Society* 20: 211–26.

Levidow, Les (2001) "Precautionary Uncertainty: Regulating GM Crops in Europe," *Social Studies of Science* 31(6): 842–74.

Levidow, Les & Susan Carr (2000) "Unsound Science? Transatlantic Regulatory Disputes over GM Crops," *International Journal of Biotechnology* 2(1/2/3): 257–73.

Lockhart, Charles (2001) "Controversy in Environmental Policy Decisions: Conflicting Policy Means or Rival Ends?" *Science, Technology & Human Values* 26(3): 259–77.

Lynch, Michael (2004) "Circumscribing Expertise: Membership Categories in Courtroom Testimony," in Sheila Jasanoff (ed), *States of Knowledge: The Co-Production of Science and Social Order* (London: Routledge): 161–80.

Lynch, Michael & Simon Cole (2005) "Science and Technology Studies on Trial: Dilemmas of Expertise," *Social Studies of Science* 35(2): 269–311.

Macfarlane, Allison (2003) "Underlying Yucca Mountain: The Interplay of Geology and Politics in Nuclear Waste Disposal," *Social Studies of Science* 33(5): 783–807.

Marris, Claire, Brian Wynne, Peter Simmons, & Sue Weldon (2001) *Public Perceptions of Agricultural Biotechnologies in Europe*, final report of the PABE research project funded by the European Communities, December. Available at: www.lancs.ac.uk/depts/ieppp/pabe/

Merelman, Richard M. (2000) "Technological Cultures and Liberal Democracy in the United States," *Science, Technology & Human Values* 25(2): 167–94.

Michael, Mike (2000) *Reconnecting Culture, Technology and Nature: From Society to Heterogeneity* (London: Routledge).

Miller, Clark (2001) "Hybrid Management: Boundary Organizations, Science Policy, and the Environmental Governance in the Climate Regime," *Science, Technology & Human Values* 26(4): 478–500.

Mol, Annemarie & John Law (1994) "Regions, Networks and Fluids: Anaemia and Social Topology," *Social Studies of Science* 24: 641–71.

Nelkin, Dorothy (1977) *Technological Decisions and Democracy: European Experiments in Public Participation* (London: Sage).

Nelkin, Dorothy (ed) (1992) *Controversy: Politics of Technical Decisions* (Newbury Park, CA: Sage).

Nowotny, Helga (2003) "Democratising Expertise and Socially Robust Knowledge," *Science and Public Policy* 30(3): 151–56.

Nowotny, H., P. Scott, & M. Gibbons (2001) *Re-Thinking Science: Knowledge and the Public in an Age of Uncertainty* (Cambridge: Polity Press).

Parthasarathy, Shobita (2004) "Regulating Risk: Defining Genetic Privacy in the United States and Britain," *Science, Technology & Human Values* 29(3): 332–52.

Phillips, Lord, J. Bridgeman, & M. Ferguson-Smith (2000) *The BSE Inquiry: The Report—The Inquiry into BSE and Variant CJD in the United Kingdom* (London: H. M. Stationery Office).

Rayner, Steve (2003) "Democracy in the Age of Assessment: Reflections on the Roles of Expertise and Democracy in Public-Sector Decision Making," *Science and Public Policy* 30(3): 163–70.

Rip, Arie (2003) "Constructing Expertise: In a Third Wave of Science Studies?" *Social Studies of Science* 33(3): 419–34.

Rose, Hilary & Steven Rose ([1969]1977) *Science and Society* (London: Alan Lane).

Rose, Nikolas (1999) *Powers of Freedom: Reframing Political Thought* (Cambridge: Cambridge University Press).

Roth, Andrew L., Joshua Dunsby, & Lisa A. Bero (2003) "Framing Processes in Public Commentary on U.S. Federal Tobacco Control Regulation," *Social Studies of Science* 33(1): 7–44.

Roth, Wolff-Michael, Janet Riecken, Lilian Pozzer-Ardenghi, Robin McMillan, Brenda Storr, Donna Tait, Gail Bradshaw, & Trudy Pauluth Penner (2004) "Those Who Get Hurt Aren't Always Being Heard: Scientist-Resident Interactions over Community Water," *Science, Technology & Human Values* 29(2): 153–83.

Royal Commission on Environmental Pollution (RCEP) (1998) *Setting Environmental Standards 21st Report* (London: H. M. Stationery Office).

Royal Society/Royal Academy of Engineering (2004) *Nanoscience and Nanotechnologies: Opportunities and Uncertainties,* Royal Society Policy Document 19/04 (London: Royal Society).

Schwarz, M. & M. Thompson (1990) *Divided We Stand: Redefining Politics, Technology and Social Choice* (Hemel Hempstead, U.K.: Harvester Wheatsheaf).

Shackley, Simon & Brian Wynne (1996) "Representing Uncertainty in Global Climate Change Science and Policy: Boundary-Ordering Devices and Authority," *Science, Technology & Human Values* 21(3): 275–302.

Understanding Risk Team (2004) *An Independent Evaluation of the GM Nation? Public Debate About the Possible Commercialisation of Transgenic Crops in Britain, 2003,* Understanding Risk Working Paper 04-02, February. Available at: www.risks.org.uk.

Wachelder, Joseph (2003) "Democratizing Science: Various Routes and Visions of Dutch Science Shops," *Science, Technology & Human Values* 28(2): 244–73.

Waterton, C. & B. Wynne (1996) "Building the European Union: Science and the Cultural Dimensions of Environmental Policy," *Journal of European Public Policy* 3(3): 421–40.

Waterton, Claire & Brian Wynne (2004) "Knowledge and Political Order in the European Environment Agency," in Sheila Jasanoff (ed), *States of Knowledge: The Co-Production of Science and Social Order* (London: Routledge): 87–108.

Werskey, Gary (1978) *The Visible College: A Collective Biography of British Scientists and Socialists of the 1930s* (London: Allen & Unwin).

Wetmore, Jameson M. (2004) "Redefining Risks and Responsibilities: Building Networks to Increase Automobile Safety," *Science, Technology & Human Values* 29(3): 377–405.

Wilsdon, J. & R. Willis (2004) *See-Through Science: Why Public Engagement Needs to Move Upstream* (London: Demos).

Wilsdon, J., Brian Wynne, & Jack Stilgoe (2005) *The Public Value of Science: Or How to Ensure That Science Really Matters* (London: Demos).

Wynne, Brian (1989) "Frameworks of Rationality in Risk Management: Towards the Testing of Naïve Sociology," in J. Brown (ed), *Environmental Threats: Perception, Analysis and Management* (London: Belhaven Press): 33–47.

Wynne, Brian (1995) "Public Understanding of Science," in Sheila Jasanoff, Gerald E. Markle, James C. Petersen, & Trevor Pinch (eds), *Handbook of Science and Technology Studies* (London: Sage): 361–88.

Wynne, Brian (1996) "Misunderstood Misunderstandings: Social Identities and Public Uptake of Science," in Alan Irwin & Brian Wynne (eds) (1996) *Misunderstanding Science? The Public Reconstruction of Science and Technology* (Cambridge: Cambridge University Press): 19–46.

Wynne, Brian (2002) "Risk and Environment as Legitimatory Discourses of Technology: Reflexivity Inside Out?" *Current Sociology* (May) 50(3): 459–77.

Wynne, Brian (2003) "Seasick on the Third Wave? Subverting the Hegemony of Propositionalism," *Social Studies of Science* 33(3): 401–17.

Yearley, Steven (1996) *Sociology, Environmentalism, Globalization: Reinventing the Globe* (London: Sage).

Zavestoski, Stephen, Phil Brown, Meadow Linder, Sabrina McCormick, & Brian Mayer (2002) "Science, Policy, Activism and War: Defining the Health of Gulf War Veterans," *Science, Technology & Human Values* 27(2): 171–205.

25 Expertise: From Attribute to Attribution and Back Again?

Robert Evans and Harry Collins

Expert: a person who has extensive skill or knowledge in a particular field
Layman or laywoman: a person who does not have specialized or professional knowledge of a subject[1]

Expertise projects a one-dimensional shadow in the Science and Technology Studies literature. Although the social interactions and institutions through which expert status is awarded or denied have been the subject of much scrutiny, the field has surprisingly little to say about what expertise actually is.[2] This is because STS has tended to favor a relational theory of expertise, in which expertise refers to, and is warranted by, one's position in a network of other actors rather than a substantive theory of expertise, in which the nature of expertise itself is the object of investigation.[3] Although emphasizing the ways in which expertise is attributed may be politically progressive—and it is certainly the case STS has done much to highlight the boundary work that underpins expert status[4]—the retrospective nature of the work makes it difficult for STS scholars to intervene in real time or real life.

This narrowness of perspective is becoming difficult to ignore. As the two definitions given in the epigraph make clear, having expertise is inextricably linked to the possession of knowledge about some domain, whereas to be a layperson is to lack such knowledge. The problem is that the distinction between expert and nonexpert does not map neatly onto the boundaries of social institutions. Indeed, one of the most important outcomes of STS work has been to highlight the expertise and knowledge that exist outside the mainstream scientific community. As a result, we now know that expertise is often partial, that experts frequently emphasize some aspects of a problem but overlook others, and that, even if we could find the right experts, they may not have the answers.

If it is accepted that expert knowledge remains an important input to decision-making but that experts might be found anywhere, just how are relevant experts to be identified? STS can avoid the problem and focus on the "downstream" explanation of how expert status is attributed or denied in society, but then it gives up any ambition to have special expertise about expertise. In this chapter, we explore the possibilities of "upstream" analyses of expertise as well as downstream. We look at the potential for new areas for research that could contribute more actively to the wider society as well as the existing state of the field.

NATURE OF EXPERTISE

The starting point for STS is that expertise is social and performative. Being an expert involves familiarity with the formal aspects of knowledge along with the capacity to act and respond to circumstances. An expert has the tacit, social, and cultural knowledge needed for the performance of the expertise. Expertise belongs to individuals and communities. Communities provide the meaning and the means to acquire and maintain expertises.[5] Expertise is, among other things, social fluency within a form-of-life.[6]

Relativism, Symmetry, and Incommensurable Worldviews

Because expertise is shared, transmitted, and validated by a community, judgments about what is to count as a competently performed observation or a correct inference have to be *agreed*. For example, the meanings of experiments and the conclusions drawn from them are interpretations sanctioned by the relevant scientific communities. Scientific knowledge may be directed toward the universal, but it cannot entirely escape time and social space. If this is correct, then the distinction between "scientific" knowledge and "lay" or "local" knowledge loses definition. At worst, there is no distinction between the expert and the layperson. At best, expertise must be more widely distributed than it was thought to be under a more universalistic notion of science. In either case, it is no surprise that expertise has turned out to be more contestable and contested. There may not even be agreement about what counts as the relevant domain of expertise in respect of a contested decision. In some circumstances, what comes to matter is not just the identification of expertise but also the mechanisms through which competing claims to expertise are tested and held accountable. The challenge for STS is to find its own position within these debates. Is the role of STS to describe how controversies are resolved, or is it to intervene in real time? If the latter, what is STS's warrant? Both approaches find support in the literature.

BOUNDARIES AND PARTICIPATION

The term boundary work captures the idea that achieving and maintaining scientific and technological credibility involves creating and policing boundaries (see Gieryn, 1983, 1999). A standard critique is that the traditional boundaries between experts and nonexperts remain strong in the wider society even though they have been shown to be permeable by STS. The knowledge of the unaccredited "laity" has been ignored because of its position on the wrong side of the expert boundary. An accessible example of the many studies informed by this view is Alan Irwin's (1995) analysis of the treatment of U.K. farmworkers by an expert committee tasked with examining the safety of an organophosphate pesticide.[7] The committee concluded the chemical was safe so long as it was used properly. The farmworkers, who believed that their health was being harmed by the chemical, rejected this conclusion. For them it was crucial that it was not possible to use the chemical properly because the infrastructure, facilities, and other conditions needed for "safe use" were not available in the fields where they worked. Even if the science showed the chemical was safe in the lab, once taken onto the farm this conclusion no longer held. Unfortunately for the farmworkers, the expert committee did not take their view into account and their evidence was dismissed as anecdotal or unreliable.

STS as Conservative Critique

This case provides a clear example of how STS might seek to make a difference. STS research can show how the boundary between the laboratory and the outside world, or between closed and open systems more generally, is important given the social and contextual contingencies of knowledge. Once it is recognized that the laboratory is important precisely because it allows scientists a great deal of control, it becomes clear that moving to real-life settings, such as a farm or other community, reduces this control and introduces new complexities (Latour, 1983). This is not to say that the science is no longer relevant, but that it is no longer enough. Science may be useful, but it needs to be complemented by the expertise of those with experience of the settings in which it is to be applied.

In such cases, the criticisms made by the so-called lay groups meet the scientific assessment head-on. They challenge the evidence that has been collected and suggest alternative sources of data or methods of analysis. Other examples can be found in the nuclear and other industrial protest movements, where campaigners routinely collect their own data on emissions and leaks, in order to challenge official claims and rhetoric (see, e.g., Welsh, 2000). In other cases, opposition groups might argue that important variables have been omitted and that the conclusions drawn are invalid. For example, they may emphasize some local environmental feature that has been overlooked or challenge the assumption that the infrastructure and institutions available are adequate to perform the tasks required.[8]

Although these critiques are powerful, emphasizing the knowledge held outside the scientific community is an essentially conservative critique of the over-reliance on science in decision-making. The implication is simply that, in such settings, the expertise of those with direct experience deserves more weight than it has traditionally been given.

STS as Radical Critique

A more radical interpretation of the same case studies is also available. In this view, the *expertise* of the disaffected groups remains important, but what is stressed is their status as *citizens*. These nonscientist-but-nonetheless-knowledgeable participants are seen as being both "specialists" and "ordinary" at the same time and come closer to the idea of a "lay expert."[9] The more complex characterization arises because the concerns that these groups articulate challenge both the science itself and the motives, values, and assumptions that lie beneath it—the whole worldview. Questions about values shift the focus from the scientific and the technical to the distribution of resources or the lifestyle choices implied by technological decisions, even those that can be made to appear sound in a narrow context. What is at stake is the moral as much as the natural order. A simple example would be choosing a baby's sex. If this choice were to be made possible by developments in medical genetics, many would still argue that even doing the science was morally and socially undesirable.

Most cases are less clear-cut, however. Technological risks and uncertainties are inextricably mixed with concerns about ultimate value or utility. The debate is not just (or even) about the limitations of expertise but about entire research agendas. For example, those opposed to further developments in genetic testing and screening may question their economic, political, and moral consequences by stressing the way in which they reinforce existing inequalities (e.g., allowing the affluent or powerful to enhance their children's genetic inheritance); create new forms of discrimination (e.g., a return of eugenics via the "deselection" of embryos seen as likely to have a disease or disability); and/or presume the desirability of increased industrialization, commodification, and control (e.g., by implying that it is proper to choose or design humans).[10]

The latter kinds of argument underpin the more radical STS critique and the more troubling use of the notion of "lay expertise." By drawing attention to the ways in which science, like all knowledge, is intimately bound up with particular sets of institutions and relations of power, domination, and control, STS has shown how choices are never purely technical but always, and at the same time, about the kind of society that is implicated in the preservation and use of science and technology. When experts of all kinds, citizens and scientists, make appeals to wider sociopolitical communities, they are speaking not just as experts but also as political agents. Treating scientific and citizen experts symmetrically has the effect of undermining both so that what is on offer becomes a choice between competing sociotechnical futures.

To adapt the typology put forward by Functowicz and Ravetz (1993) in their discussion of post-normal science, we could say that the conservative critique of STS sees controversies around expertise as falling into the middle category of "professional consulting," in which what is at stake is the application of science rather than its relevance. In contrast, the more radical critique sees controversies as rather closer to the idea of post-normal science, in which the "traditional domination of 'hard facts' over 'soft values' has been inverted." In such circumstances:

Only a dialogue between all sides, in which scientific expertise takes its place at the table with local and environmental concerns, can achieve creative solutions to such problems, which can then be implemented and enforced (1993: 749–51).

Although Functowicz and Ravetz clearly intend such arrangements to apply only when either the uncertainty or the stakes associated with a decision are particularly high— the quote given above follows the example of the predicted rise in sea level caused by global climate change—the STS perspective generalizes it to controversy more generally. As expertise is contested so uncertainty is increased and, as concerns about the dangerous precedent that may be set are articulated, a controversy can be made to move from the arena of professional consultancy to that of post-normal science. In this sense, the stakes and uncertainty implicated in a controversy are part of what the protagonists are trying to establish. The more radical interpretation of STS thus has the effect of questioning the extent to which this choice should be left to scientific experts or even to experts at all.

In posing these questions, STS is drawing on a combination of description and democratic prescription. The description comes from the observation of controversies, in which

alliances form between fragments of the public and factions within scientific institutions such that new science-lay hybrid assemblages can be said to emerge and act as the core antagonistic actors in a particular controversy. Thus, we should be sensitive to the possibility that it is not the "obvious" or "unitary" constituencies of public, or scientific, or government actors that are key to understanding a given case, but admixtures of these. (Irwin & Michael, 2003: 142)

The democratic prescription arises as a result of the need to find some way of managing these competing combinations of scientific, policy, and lay actors. To the extent that such groups are composed of competing interest groups they speak "to" rather than "for" the public. If their coalition-building activities are seen through the lens of political rights, representation, and civil liberties, then the range of legitimate participants increases, and in particular, the role of lay citizens becomes central. As Wynne (2003: 411) puts it:

To the extent that public meanings and the imposition of problematic versions of these by powerful scientific bodies is the issue, then the proper participants are in principle every democratic citizen and not specific sub-populations qualified by dint of specialist experience-based knowledge.

Deliberative and Participatory Processes

There is now a considerable body of evidence suggesting that these ideas are being accepted. In the UK and EU the effectiveness of these arguments can be seen in the range of policy documents that recognize the importance of soliciting opinions from stakeholders, concerned citizens, and the wider public (e.g., RCEP, 1998; House of Lords, 2000; POST, 2001; Hargreaves & Ferguson, 2001; Gerold & Liberatore, 2001; OST, 2002; Wilsden & Willis, 2004; CST, 2005). In the United States the practice is also well entrenched, with Jasanoff (2003: 397) reporting that in "regulatory decision-making, for example, all federal agencies are required by law to engage the public at least by offering notice of their proposed rules and seeking comment."[11]

The argument for increasing participation thus seems to have been won, at least in principle, and the problems now relate to the practical issues of how and when to organize such participation and what to do with the outputs of participatory events when they are completed.[12] Again, STS research has important implications for how participation might be organized and for what purpose. In making these arguments, STS proceeds from a diagnosis of the more traditional public inquiry as overly restrictive in its terms of reference, day-to-day operation, and deference to established expertise, to the advocacy of more deliberative and/or participatory processes (Wynne, 1982, 1995; Rip, 1986; Nowotny et al., 2001). STS scholars have now developed a range of alternative prescriptions for processes that might offer a new and more inclusive politics of innovation (Rip et al., 1995; Grin et al., 1997). Although the specific formats vary, most operate around the generic pattern of a consensus conference in which a panel of citizens is empowered to select and question experts in order to make recommendations about a particular topic.[13] Within STS these trends are captured in the range of literature that now addresses the importance of participation and the need to reconfigure the relationships between science and society. Again, although there is some diversity between approaches, a common theme emerges: science and technology need to be made more accountable and responsive to the wider society, and one way to do this is through the increased participation of users, stakeholders and citizens.

The outcome has been the development of new ways of thinking about and doing the management of technology and science in society. Whether this has improved the way these decisions have been made is open to debate and critical inquiry. For what it is worth, our view is that it has. Recognizing that decisions about controversial and uncertain science are also decisions about social institutions, risks, and values has made these decisions more complicated. Nevertheless, recognizing this complexity does at least encourage the scrutiny and debate needed to ensure that decisions are informed by a wider range of expert and democratic opinion.

All that said, it is important to note the way these changes enhance the status of the lay citizen. In the initial, technocratic case there was nothing but scientific expertise and its overextension. In the conservative critique, there was a more limited scientific expertise complemented by the expertise of relevant groups from the wider society. In the more radical critique, the extent of participation—in principle it is open to everyone—means that the idea of expertise can no longer help us understand what different participants bring to the process. If participation becomes a mass exercise, then the expertise required must be ubiquitous and certainly very different from that held by specialist communities such as scientists and farmworkers. Is this the right way to go, or is there still a question to be answered about the extent to which participants in a deliberative or participatory process need substantive expertise to take part? To the extent that they do, participation cannot be a mass exercise. Conversely, if expertise is not required, then mass participation becomes possible but, in becoming so, undermines the core STS idea of socialization as the preeminent method for acquiring expertise and hence participation. The link between expertise and participation remains the Achilles heel in the relationship between STS and wider decision-making.

ALTERNATIVES TO STS

Because STS has a social model of knowledge it implies that extensive socialization is needed for individuals to acquire expertise. The dilemma of participation is that, precisely because of this, lay citizens cannot be experts, and even experts are only experts in some narrow domain. This is why participatory processes, in which learning can occur, seem more consistent with STS than mass democracy. There are, however, other approaches that stress the value of generalism rather than of specialism.

Amateurism

It is possible to argue that expertise is not always a good thing. If becoming an expert means becoming socialized into a specialist community, then becoming an expert means running the risk of becoming blinkered in one's outlook. In contrast, remaining an amateur allows one to avoid the narrow perspective of the single group, discipline, or paradigm and thus to see the "bigger picture."[14] Perhaps the most institutionalized version of this defense of nonexpertise was to be found in the British Civil Service, where civil servants traditionally served as "generalists," deliberately being moved between jobs to avoid being "captured" by particular departments. Seen this way, the intentional avoidance of specialist expertise brings several benefits. For example, "generalist" civil servants can

- Resist calls for expensive alternations made by specialists on unimportant aesthetic or technical grounds.
- Free scarce specialists from nonspecialist work, thus permitting the most economical use of the time of specialists.
- Act as coordinators of work involving more than one type of specialist.
- Set specialist matters in the context of Ministerial policy and Departmental practice.
- Use their own skills to synthesize and summarize the views of specialists and other administrators.[15]

This defense of nonexpertise resonates with several aspects of the STS literature. For example, in Bijker's (1995) theory of sociotechnical change, one of the theoretical concepts introduced is that of "inclusion" within a technological frame.[16] Those with a high degree of inclusion are those who are most highly socialized into a particular paradigm, while those with a low degree of inclusion are those who lack this experience and/or commitment. Significantly, radical innovations often come from those with a lower degree of inclusion precisely because they retain the ability to see things in a new or different way—to "think outside the box" as current jargon might put it. In contrast, those with a high degree of inclusion are too committed to or bound up with the standard paradigm to interpret events in any other way.

There are also some similarities with the idea of boundary objects (Starr & Griesemer, 1989) and boundary shifters (Pinch & Trocco, 2002). Again the insight of STS has been that, because there are different perspectives, negotiation and flexibility rather than top-down authority are needed if productive working relationships are to be sustained. Boundary objects provide a mechanism for reducing the expertise needed to participate in the use of a technology and, in the paradigm case, allow communities to collaborate despite the differences in their expertise, goals, and interests.[17] In

such a context, having an administrator who is not a member of a specialist technical community may well be an advantage if the aim is to create a shared definition
of the problem (i.e., boundary object) or to have at least one person who can act as
a go-between for the different specialist groups (i.e., boundary shifter).

To return briefly to the Civil Service, when these issues were, in fact, examined by
the Fulton Committee in the late 1960s, the arguments in favor of generalists and
amateurs were not persuasive (Fulton, 1968). Instead, the committee recommended
reforms that integrated specialists and high-level administrators much more closely.
What an STS trained person asked to advise the government on a similar problem
would say today is an interesting thought-experiment. Ironically, it seems likely that
the STS purist would find themselves defending the value of the Oxbridge educated
classicist against the imposition of more technocratic specialist framings. The difficulty, if there is one, emerges when the STS person is asked to specify more accurately
the type of generalist that is required—are they to be restricted to the Oxbridge elite
or not? If not, what are the qualities the new entrants should possess? In short, just
what is the difference between an "acceptable generalist" and someone with "no
relevant knowledge or experience"?

Heuristics

Having adequate knowledge upon which to base decisions is also a key concern of the
economics literature, where markets are typically modeled on the assumption that
economic agents have access to full information and then make rational choices that
maximize their returns given a set of clear and unambiguous preferences. Although
many economists would deny that their models are meant to be taken literally, these
assumptions have provided a model of decision-making that has been generalized to
a wide range of settings.[18] What is more, because it can be shown mathematically that
decisions taken this way are optimal (in the sense that they maximize financial
returns), then observed deviations from these assumptions suggest that the way to
improve outcomes is to re-engineer social processes so that the "barriers" to economic
efficiency are removed.[19]

While changing society to match the theory is clearly one response to economic
theorizing, others (typically psychologists rather than economists) have tried to
develop approaches that can explain the observed behaviors. Perhaps the most
common approach to this problem is to try to articulate the heuristics used in
making decisions under uncertainty, with the leading contributions coming from
Daniel Kahneman and his collaborator Amos Tversky (Kahneman et al., 1982;
Kahneman & Tversky, 1996). Research in this tradition attempts to make explicit
the heuristics that people use to make judgments that, in the economic sense of rational behavior, lead to suboptimal outcomes. These rules of thumb include strategies
such as the "law of small numbers" through which data from small samples are transferred to large samples, the use of "cultural" rather than "statistical" representativeness in making judgments about individuals, and the tendency to take decisions
individually rather than over a longer term sequence. In each case, the outcome is

that individuals—both in real life and in experimental conditions—reveal a systematic tendency to make decisions in ways that contradict the fundamental principles of probability and, therefore, do not conform to the predictions of standard economic theory.

It is worth noting, however, that this literature is not without its controversies. The work of Kahneman and Tversky has been extensively critiqued by Gigerenzer (1991, 1993, 1994), who argues that many everyday heuristics work almost as well as formal mathematical models and that many of the apparently suboptimal results proposed by Kahneman and Tversky can be seen as rational if the question posed is interpreted in a different but equally legitimate way. In essence, Gigerenezer's claim is that Kahneman and Tversky overemphasize the logical structure of the problem and overlook the importance of its content. These criticisms are rejected by Kahneman and Tversky.[20] STS is not forced to take a stand on this issue, but it is clear that the emphasis on context suggests that many will be sympathetic to Gigerenezer's critique, even if they also accept that heuristics, in the sense of some rule of thumb or judgment by which to simply complex information, are likely to be essential in both mundane and specialist domains.[21]

Low Information Rationality and the Miserly Citizen

If heuristics provide a way of simplifying complex information, how are we to understand decision-making in the absence of information? This problem is particularly acute for the political science literature that deals with voting behavior, in which the situation seems very different from the standard STS case study, where the focus is often the exclusion of informed or expert citizens by established interest groups. In the case of elections and other democratic processes, the danger is that a disinterested or uninformed public will undermine the legitimacy of institutions based on mass participation. In short, if democracy is about the exercise of informed choice, then is a process still a democratic one when the choices are made on the basis of little or no information?

Although many see the outcome of this info-rich/info-poor divide as a dystopian future of increasing stratification and inequality, there are those who question this conclusion. In this more positive interpretation, the negative consequences of not having *full information* are offset by the ability of individuals to make good decisions on the basis of *simple and widely available information*. Thus, for example, in the case of electoral choices, Kuklinksi and Hurley (1994: 730) argue that, rather than requiring encyclopedic knowledge of complex issues, problems, and debates, "ordinary citizens can make good political judgements even when they lack general political acumen or information about the issues at hand by taking cues from political actors." Similarly, Lupia and McCubbins (1998: 9) argue that "by forming simple and effective strategies about what information to use and how to use it, people can make the same decisions they otherwise would if they were expert." Thus, to give a simple example, it has been found that accurate inferences about academic standards and school safety can be made by parents on the basis of simple indicators like how clean a school is, whether there is graffiti on the walls, and whether or not there are broken windows (Schneider et al., 1999). In these situations, access to specialist or technical expertise is not a barrier to good decision-making, implying that the need for expertise to be everywhere has, perhaps, been overstated.

In many ways these ideas of "low information" rationality (Popkin, 1991) resonate with the much older idea of "satisficing" put forward by Herbert Simon. Simon argued that rather than constantly seeking to maximize their returns, organizations must (and do) settle for less. Because they have limited amount of information about the future, and acquiring more is costly, organizations must act on the basis of uncertain and incomplete data. As a result, their decisions are based on what Simon called a "bounded rationality" in which organizations "satisfice" rather than "maximize" by setting targets that are acceptable if achieved but that are adjusted if they are not. In this way, although the outcome is, in some sense suboptimal, in the context of the firm it is also a rational choice in the sense that acquiring the extra information to reach the optimal decision is too costly.[22]

Finally, it is worth noting that, although low information rationality theories sound like a defense of the citizen found in the STS literature in which local and personal knowledge provides the basis of informed critique, there is a difference. The STS view is that there is some expertise being displayed—even if it is in something like "folk sociology"—whereas the low information route highlights the short cuts being taken.[23] This is particularly apparent in the approaches to political preference formation that take their lead from Mary Douglas's cultural theory, in which an individual's position in the grid-group typology provides an over-arching framework through which events are filtered and preferences formed. As a result, people who possess only "inches of facts" are able to "generate miles of preferences" because "they don't actually have to work all that hard" (Wildavsky, 1987: 8). This is not to say that these preferences are always correct, or that they cannot be changed through deliberation.[24] It does, however, reinforce the STS tendency noted above to see technical matters as political and cultural, with trust in institutions thus emerging as a key dimension. More negatively it also suggests that, by appealing to cultural values, those in power have the potential to frame debates and position themselves in ways that polarize debate rather than promote dialogue. If this is the case, then the optimism of those who think there are easy ways to make good decisions may turn out to be misplaced.

STS IN ACTION OR STS INACTION?

The previous section discussed a number of alternative approaches to expertise drawn from across the social sciences. In each case the distinction was made between having expertise and not having expertise. In some cases this was seen as having negative consequences and in others as a less serious problem, but in all the distinction so often blurred in STS, between knowing and not knowing, was central. In these final sections, we return to the field of STS and the challenge raised at the beginning of the chapter, namely, how to construct STS as a critical discipline.

By emphasizing the underdetermination of interpretation by data, STS shows how different expert positions can be consistent with the available evidence yet incommensurable with each other. The problem is what follows from this. To the extent that STS shows that each position is equally reasonable or potentially open to challenge it intervenes indirectly by making evidence of disagreement more public. A more direct form of this intervention, however, would be to try to create the circumstances in which the kind of deconstruction and dialogue that STS carries out can be incorporated more routinely in the institutions and procedures through which such controversies are played out.[25] This work may be very public or operate behind the scenes, but the aim is usually to show how the aims of the process would be better met if STS advice was acted upon. Examples of STS interventions of this kind include the following:

• Analyses of legal practices: These have ranged from analyses of the ways expert witnesses are identified, selected, and their expert credibility established or challenged in cross examination to direct participation in legal proceedings, either as an expert witness or through the provision of *amicus curiae* briefs setting out key issues or concerns.[26]

• Contributions to the Public Understanding of Science (PUS) or Public Engagement with Science and Technology (PEST): While not challenging the fundamental idea that science has a duty to communicate with the wider society, STS studies have had quite a bit to say about how this should be done. In particular, STS has been highly critical of the deficit model and has championed a more dialogical approach. The effects can be seen in the gradual shift away from dissemination as the provision of simplified research summaries to consultation and more deliberative and participatory forums.[27]

• Contributions to the regulation, planning and management of science and technology: As with its contributions to PUS and PEST, STS contributions to debates about risk assessment and management have not challenged the basic idea that there are risks associated with science and technology. Instead the aim has been to show how current practices must be reformed so as to include new classes of risk identified by STS.[28]

Although this is a coherent and intellectually defensible position, it does raise some problems when applied in practice.[29] For example, in the case of debates about the reality of climate change, the scientific status of intelligent design, or the safety of vaccines such as MMR, what role does symmetrical STS have to play? In one sense it is already involved, because those involved in the arguments are making claims about the nature of sound science and expertise. In another sense, however, it cannot be involved because it sees all parties as essentially similar. STS research may describe what is going on, making visible what has traditionally been invisible, but the conclusions that follow from this remain a matter for others to resolve. In some ways this follows from the diffidence inherent in the constructivist agenda, which makes it difficult to assert that STS knowledge about knowledge can be seen as more than one account among many, but it is not inevitable. As noted earlier, there is a range of policy initiatives drawing on STS research, and STS researchers, seeking to promote new and more inclusive ways of managing controversial technological innovation. In these initiatives, STS is clearly being put into action and, in doing so, is opening up the domain of participatory and deliberative methods as a new site for STS research and theorizing.

EXPERTISE AS REAL

In the final section of this chapter, we set out a more prescriptive or normative approach to the burgeoning area of STS research that aims to reform the ways in which decisions about science and technology get made. In these cases, STS seems to have a lot to offer, with the sociological conception of knowledge in particular providing a way of analyzing the qualities that different participants might bring to more inclusive decision-making.

The basic idea is simple—knowledge is acquired by socialization, so expertise is acquired through a prolonged period of interaction within the relevant community and is revealed through the quality of those interactions.[30] One consequence is that acquiring expertise is neither all attribution nor a flip-flop process. It is possible to think of a continuum of knowledge states, ranging from ignorance to complete expertise and of individuals moving between these states over time. It is also possible to distinguish between the ways different kinds of expertise are distributed. Thus, for example, some sorts of expertise (e.g., speaking and writing a natural language) will be so widely distributed as to be ubiquitous. Others, like milking cows or growing stem cells, will be restricted to such small groups that they are seen as esoteric expertises. Similarly, while some expertise will be about substantive domains, other kinds of expertise might operate at a meta level, providing the criteria and skills needed to make judgments about the expertise held by others. All these distinctions, and the categories they give rise to, are summarized in the table that we have referred to as "the periodic table of expertises" (figure 25.1) and explained at length elsewhere (Collins & Evans, 2002, 2007; Collins, 2004a,b; Evans, 2004). Here we concentrate on some main points.

In the row labeled specialist expertises (i.e., expertise in some substantive domain such as carpentry or chemistry), an individual's expertise can range from "beer-mat expertise," which corresponds to knowing the kinds of facts that might be put on the coasters provided in bars, to contributory expertise, which corresponds to being able to contribute fully to the work of the relevant community.[31] Within this scheme, the two most important distinctions are the distinction between primary source knowledge and interactional expertise and between interactional expertise and contributory expertise.

Ubiquitous expertises					
Dispositions				Interactive ability / Reflective ability	
Specialist expertises	*Ubiquitous tacit knowledge*			*Specialist tacit knowledge*	
	Beer-mat knowledge	Popular understanding	Primary source knowledge	Interactional expertise	Contributory expertise
				Polimorphic / *Mimeomorphic*	
Meta-expertises	*EXTERNAL*		*INTERNAL*		
	Ubiquitous discrimination	Local discrimination	Technical connoisseurship	Downward discrimination	Referred expertise
Meta-criteria	Credentials		Experience		Track record

Figure 25.1
The Periodic Table of Expertises.

• The distinction between primary source knowledge and interactional expertise marks the transition from expertises that rely on widely distributed tacit knowledge to expertises that rest on tacit knowledge specific to the group in question. Thus, someone with interactional expertise would be able to pass in conversational settings as a fully fledged member of the group, whereas someone whose knowledge consisted only of that which was made explicit in written works—e.g., primary source knowledge— would not. It should be noted, however, that because interactional expertise is acquired over time, prolonged and sustained interaction within the expert community is required before an individual can pass as a native member of the community under determined interrogation.

• The distinction between interactional and contributory expertise corresponds to the distinction between being able to talk fluently about a domain of expertise and being able to contribute to it. In other words, while someone with maximum interactional expertise would be able to talk like a native member of the community, he or she would have no proficiency in practical tasks. Contributory expertise signifies that a person has both the conceptual and practical expertise held by the group, whereas someone with interactional expertise possesses only the former.

The second row of the table describes the meta-expertises needed to make judgments about the substantive expertise of others. There is an important distinction between meta-expertises that are "internal" and those which are "external":

• Internal meta-expertise denotes those judgments that require some kind of socialization within the community. Thus, the judgments labeled technical connoisseurship, downward discrimination, and referred expertise all require the person who exercises them to have some experience that allows them to appreciate the criteria used by those they judge. Thus, for example, a connoisseur of wine or art would typically be familiar with the conventions and techniques of wine-making or painting without necessarily being a wine-maker or artist.

• External meta-expertise denotes those judgments that are possible even if the individual has no socialization within the relevant expert community. In effect, these refer to the application of more or less ubiquitous standards to specific substantive domains. The idea of local discrimination highlights the case in which some communities will have experiences that will shape their views about the trustworthiness or credibility of specific experts that are not widely shared even though the criteria invoked draw on general rather than substantive knowledge.

The usefulness of distinguishing between different kinds of experts lies in the more nuanced response it offers to the apparent trade-off between expertise and participation. If it is accepted that it is impossible for everyone to be an expert about everything, then some form of categorization is needed. Similarly, if STS is to continue to contribute to debates about participation and regulation, then separating the expert from the nonexpert will be crucial, not to exclude the latter but to explain why the nonexpert lay citizen may be more valuable than is generally thought. For example, if deliberative or participatory models are to include ordinary citizens in the oversight and regulation of science, this cannot be justified on the basis of their specialist expertise (by definition, the typical citizen must know very little about any esoteric field). Instead, lay participation is warranted via the idea of meta-expertise, particularly ubiquitous and local discrimination, which use more generic social knowledge and skills to put political and moral preferences into action (Evans and Plows, 2007).

If this is the case, then our categorization of expertise suggests three lines of research than can be pursued in addition to the traditional STS case studies documenting the resolution of technoscientific controversy.

1. The categorization of expertise itself: While the basic structure of figure 25.1 seems to fit with core STS commitments, the distinctions need to be tested more fully. We have already adapted the Turing test methodology, in which hidden participants try to convince a judge that they possess a particular expertise, to test the idea of interactional expertise and the importance of socialization in its acquisition. Initial results based on color-blindness show that individuals with interactional expertise are indistinguishable from those with contributory expertise, whereas those without interactional expertise are easy to spot.[32]

2. Case studies in participation: Deliberative and participatory methods are becoming increasingly common in the regulation, funding, and oversight of science, but what do they achieve? Given that participatory decision-making and consultation exercises are now taking place in many countries and encompassing many different topics, there is an emerging data set in participatory practice that can be used to evaluate and test the adequacy of the different approaches. For example, how do deliberative and participatory methods differ, do different processes suit different kinds or combinations

of expertise, how much participation is necessary, what are the practical implications of making such events routine, and how might they be evaluated?

3. Experiments in expertise and participation: Finally, and perhaps most ambitiously, it is possible to design experiments in participatory decision-making and consultation that will test these and other ideas of expertise directly. In some respects, the literature of constructive technology assessment, consensus conferences, and interactive technology assessment all represent attempts to use STS to rethink and reshape decision-making. In terms of figure 25.1, the experiments we would most like to see are those which examine the capacity of nonexpert citizens to evaluate complex science and the kinds of interventions that are most helpful in promoting this behavior. Experiments need not be limited to this domain, however. It should also be possible to investigate how experts judge other experts, how experts judge citizens, and how elected decision-makers evaluate and combine competing forms of evidence from different expert communities.

CONCLUSIONS

The idea of expertise is central to modern life and to contemporary STS. Understanding expertise as the product of socialization into a community demonstrates both the utility of expertise and its weakness. Experts may be the best people to decide certain matters of fact, but they are not necessarily the best people to make value judgments about the utilization of that knowledge. Conversely, lay citizens are not experts, but this is also their weakness and their strength. While they are not best placed to answer those questions that belong more properly within esoteric expert communities, precisely because they lack such membership, they are, paradoxically, the best placed to make the crucial judgments about what should be done with such knowledge. Understanding and contributing to the interplay between these expert and citizen concerns provides one STS (Science and Technology Studies) with a key role in the future development of the other STS—(Science, Technology and Society).

Notes

1. Source for both definitions: Collins English Dictionary. The Mirriam-Webster on-line dictionary provides the following definitions for the same two words:

Expert: one with the special skill or knowledge representing mastery of a particular subject

Layman/woman: a person who does not belong to a particular profession or who is not expert in some field

2. Examples of such early sociology of science include Mannheim (1936) and the essays reprinted in Merton (1973). Contemporary science studies can be seen as a reaction to, and rejection of, this viewpoint, with prominent early critiques given by Bloor (1973, 1976) and Mulkey (1979). That said, however, is should be noted that the idea of science as a special kind of knowledge has not gone away, with many of the contributions to the so-called science wars (e.g., Gross & Levitt, 1994; Koertge, 2000) essentially re-making this claim.

3. The denial of expert status is clearly illustrated in the chapter on courtroom science in Barnes and Edge (1982) and in the more recent experience of Simon Cole as he attempted to defend his own status as expert (Lynch & Cole, 2005). In a similar way, the status of expert is conferred when such attributions are seen as legitimate, with the concept of boundary work being used to highlight the constructed nature of such categorizations. See, for example, Gieryn (1983, 1999) or Eriksson (2004) for a more contemporary case study.

4. An indicative, but by no means complete, list of relevant studies would include Arksey (1998), Epstein (1996), Gieryn (1999), Irwin and Wynne (1996), Welsh (2000), Jasanoff (1990, 1995), and Wynne (1982).

5. This is particularly clear in educational settings such as universities, where the aim of degree programs is to train students in the skills and knowledge associated with a particular discipline and the assessments and marking criteria used operationalize what displaying expertise means.

6. This is the argument from Wittgenstein's philosophy that, even though we cannot articulate the rules by which we know how to carry on a sequence in the correct way or follow a rule properly, the fact that we can tell when we have made a mistake shows that there are rules involved. Socialization into a group provides the mechanism through which these rules are internalized, but the size of the group itself can vary enormously. For example, when considering natural languages, the relevant form-of-life might be all English- or Chinese-speaking people. In contrast, when considering a specialized form of expertise, then the relevant form-of-life might be the members of two or three research laboratories, the residents of a small village, or the workers in a factory. The idea of expertise as social fluency is the same in each case, however.

7. For other examples, see note 4.

8. This is a particular concern in regulatory disputes, where specific standards of accuracy or supervision have to be maintained if the risk assessment is to be valid. Examples include the attempts to prevent the spread of BSE by removing all traces of potentially infected tissue in the abattoir (something that was seen as impractical by the workers) and the difficulties created through the cull of farm animals in response to foot-and-mouth disease (the armed forces were eventually required to provide logistical expertise, and the effects of the policy on tourism and hence the local economy was overlooked). Other examples are nuclear power and GM foods. For a wide range of academic perspectives on the social science approach to risk, see Krimsky and Golding (1992), Irwin and Wynne (1996), and Yearley (2000).

9. Rather asymmetrically, however, the citizen status of scientists is not usually invoked. Clearly, scientists are citizens too, but this seems to be swamped by their role as scientist/expert. Thus, interests, ambitions, and desires of scientists (government or industry) are mapped onto those of the state/capital while nonscientist interests get mapped onto the "people."

10. All these concerns are routinely raised by civil society groups critical of developments in medical genetics.

11. A recent example is the area of nanotechnology, in which the "21st Century Nanotechnology Research and Development Act," which was signed by President Bush in December 2003, requires "public input and outreach to be integrated into the Program by the convening of regular and ongoing public discussions, through mechanisms such as citizens' panels, consensus conferences, and educational events." Available at: http://frwebgate.access.gpo.gov/cgi-bin/getdoc.cgi?dbname=108_cong_public_laws&docid=f:publ153.108.

12. The controversies over genetically modified crops have key sites for both practical efforts to "do" public participation in a wide range of countries and for STS research. For example, public consultations have been held in (at least) the United Kingdom, The Netherlands, Denmark, Austria, India, and New Zealand. A review of these events was recently published in *Science, Technology & Human Values* (see Rowe & Frewer, 2005).

13. There are many examples of these approaches, which vary in scale, duration, the importance attached to reaching a "unanimous" verdict, and the opportunities given to the citizen panel to influence the selection of the topic and the recruitment of experts. A summary of these participatory events can be found in Rowe and Frewer (2005).

14. This is, of course, the standard way of thinking about social science field work—to go native is to lose the ability to see any other point of view, whereas to retain one's academic identity is to retain the ability to put the participants' actions into a different context.

15. Abridged from Fulton, 1968: 58.

16. There are also some parallels with the idea of "weak ties," since civil servants less tied to one department or perspective might be more receptive to ideas or knowledge from outside the Departmental network.

17. The paradigm case is Starr and Griesemer (1989). Similar issues arise in the context "trading zones" developed by Galison (1997) although here expertise is partially shared as a new language or pidgin develops. For more on trading zones and collaboration, see Gorman (2002) and Ribeiro (2007).

18. These include, for example, the prisoners' dilemma and game theory as well as microeconomic studies of academic career paths, marriage, labor markets, and criminal behavior, most notably in the work of Nobel Laureate Gary Becker. See, for example, Becker (1976) or the collection of Becker's essays edited by Febrero and Schwartz (1996).

19. By far the best example from within the STS literature is Donald MacKenzie's analysis of the rise and fall of long-term capital management and the Black-Scholes equation that transformed financial markets (see MacKenzie 2006).

20. The exchange can be found in *Psychological Review*. See Kahneman and Tversky (1996) and Gigerenzer (1996).

21. Examples of the use of heuristics in specialist domains such as the invention of the airplane can be found in Bradshaw (1992) while the simulation of such heuristics is described in Kulkarni and Simon (1988).

22. Available at: http://cepa.newschool.edu/het/profiles/simon.htm. See also Simon (1979).

23. There is some overlap here with Shapin's (1995) argument about the evaluation of proxies.

24. Examples of cases where deliberation appears to move opinions away from those originally informed by grid-group positions, if only for the course of the process, are given in Lindeman (2002) and Gastil and Levine (2005).

25. Examples of the suggestions for reconfiguring the relationship between science and society can be found in Wilsdon and Willis (2004), Rip et al., (1995), Functowicz and Ravetz (1993), Hajer (1995), Beck (1992), Giddens (1990), and Nowotny et al. (2001).

26. The role of expertise and science in the legal system is analyzed in Smith and Wynne (1989) and Jasanoff (1992). Simon Cole's experiences as an expert witness are analyzed in Lynch and Cole (2005). For another example of a direct intervention, see the *amicus curiae* brief to the WTO filed by Jasanoff et al. Available at: http://csec.lancs.ac.uk/wtoamicus/index.htm [accessed 28 February, 2007].

27. See, for example, policy documents such as Gerold and Liberatore (2001), House of Lords (2000), and Parliamentary Office of Science and Technology (2001). A review of one such attempt in the U.K.—the GM Nation Debate—is available as Horlick-Jones et al. (2004).

28. See, for example, Wynne (1995), Rip et al. (1995), and Renn et al. (1993).

29. These issues are addressed in the special issue of *Science, Technology & Human Values* (Winter 2005) on demarcation socialized; see Lahsen (2005) in particular.

30. Note that there are no guarantees here—interaction is a necessary but not sufficient condition.

31. In the United Kingdom, beer mats were produced as part of the campaign against the single European currency. Each beer mat reproduced six "facts" about the Euro that were intended to put the campaign message in a clear and concise manner. Examples of the statements made on the beer mats include "Unemployment in the euro countries is double ours" and "The euro countries pay £1,900 per household more than us in tax every year."

32. In practice the methodology is quite complex. For a description of our own work on this topic, see refs to working paper and Artificial Experts. For more details of our own work on this topic, including both a discussion of the Turing Test and descriptions of our experiments based on this idea, see Collins (1990) and Collins et al. (2006). Further applications of this approach can be found in Collins (2008).

References

Arksey, Hilary (1998) *RSI and the Experts: The Construction of Medical Knowledge* (London: UCL Press).

Barnes, Barry & David Edge (eds) (1982) *Science in Context: Readings in the Sociology of Science* (Milton Keynes, U.K.: Open University Press).

Beck, Ulrich (1992) *Risk Society: Towards a New Modernity* (London: Sage).

Becker, Gary S. (1976) *The Economic Approach to Human Behavior* (Chicago: University of Chicago Press).

Bijker, Wiebe E. (1995) *Of Bicycles, Bakelite, and Bulbs: Toward a Theory of Sociotechnical Change* (Cambridge, MA: MIT Press).

Bloor, David (1973) "Wittgenstein and Mannheim of the Sociology of Mathematics," *Studies in the History and Philosophy of Science* 4: 173–79.

Bloor, David (1976) *Knowledge and Social Imagery* (London: Routledge & Kegan Paul).

Bradshaw, Gary (1992) "The Airplane and the Logic of Invention," in Ronald N. Giere (ed), *Cognitive Models of Science* (Minneapolis: University of Minnesota Press): 239–50.

Collins, H. M. (1990) *Artificial Experts: Social Knowledge and Intelligent Machines* (Cambridge, Mass., London: MIT Press).

Collins, H. M. (2004a) "Interactional Expertise as a Third Kind of Knowledge," *Phenomenology and the Cognitive Sciences* 3(2): 125–43.

Collins, H. M. (2004b) "The Trouble with Madeleine," *Phenomenology and the Cognitive Sciences* 3(2): 165–70.

Collins, H. M. (ed) (2008) "Case Studies of Expertise and Experience," special issue of *Studies in History and Philosophy of Science* 39(1).

Collins, H. M. & Robert Evans (2002) "The Third Wave of Science Studies: Studies of Expertise and Experience," *Social Studies of Sciences* 32(2): 235–96.

Collins, H. M. & Robert Evans (2007) *Rethinking Expertise* (Chicago: The University of Chicago Press).

Collins, H. M., Robert Evans, Rodrigo Ribeiro, & Martin Hall (2006) "Experiments with Interactional Expertise," *Studies in the History and Philosophy of Science Part A* 37(4): 656–74.

Council for Science and Technology (CST) (2005) *Policy Through Dialogue* (London: CST). Available at: www2.cst.govuk/cst/reports. Accessed 28 February 2007.

Epstein, Steven (1996) *Impure Science: AIDS, Activism, and the Politics of Knowledge* (Berkeley: University of California Press).

Eriksson, Lena (2004) "From Persona to Person: The Unfolding of an (Un)Scientific Controversy," Ph.D. diss., Cardiff University.

Evans, Robert (2004) "Talking About Money: Public Participation and Expert Knowledge in the Euro Referendum," *British Journal of Sociology* 55(1): 35–53.

Evans, Robert & Alexandra Plows (2007) "Listening Without Prejudice? Re-Discovering the Value of the Disinterested Citizen," *Social Studies of Science*, 37(6).

Febrero, Ramon & Pedro S. Schwartz (eds) (1996) *The Essence of Becker* (Stanford, CA: Hoover Institution Press).

Fulton, Lord (1968) *The Civil Service,* vol. 2*: Report of a Management Consultancy Group*—Evidence submitted to the Committee under the Chairmanship of Lord Fulton 1966–1968 (London: H. M. Stationery Office).

Functowicz, Silvio O. & Jerry R. Ravetz (1993) "Science for the Post-Normal Age," *Futures* 25: 739–55.

Galison, Peter (1997) *Image and Logic: A Material Culture of Microphysics* (Chicago: University of Chicago Press).

Gastil, J. & P. Levine (eds) (2005) *The Deliberative Democracy Handbook* (San Francisco: Jossey-Bass).

Gerold, R. & A. Liberatore (2001) *Report of the Working Group "Democratising Expertise and Establishing Scientific Reference Systems"* (European Commission). Available at: http://europa.eu.int/comm/governance/areas/group2/report_en.pdf.

Giddens, Anthony (1990) *The Consequences of Modernity* (Cambridge: Polity Press).

Gieryn, Thomas F. (1983) "Boundary Work and the Demarcation of Science from Non-Science: Strains and Interests in Professional Interests of Scientists," *American Sociological Review* 48: 781–95.

Gieryn, Thomas F. (1999) *Cultural Boundaries of Science: Credibility on the Line* (Chicago: University of Chicago Press).

Gigerenzer, G. (1991) "How to Make Cognitive Illusions Disappear: Beyond 'Heuristics and Biases,'" in W. Stroebe & M. Hewstone (eds), *European Review of Social Psychology,* vol. 2 (Chichester, U.K.: Wiley): 83–115.

Gigerenzer, G. (1993) "The Bounded Rationality of Probabilistic Mental Models," in K. I. Manktelow & D. E. Over (eds), *Rationality* (London: Routledge): 284–313.

Gigerenzer, G. (1994) "Why the Distinction Between Single Event Probabilities and Frequencies Is Relevant for Psychology and Vice Versa," in G. Wright & P. Ayton (eds), *Subjective Probability* (New York: Wiley): 129–62.

Gigerenzer, G. (1996) "On Narrow Norms and Vague Heuristics: A Reply to Kahneman and Tversky (1996)," *Psychological Review* 103(3): 592–96.

Gorman, Michael (2002) "Levels of Expertise and Trading Zones," *Social Studies of Science* 32(6): 933–38.

Grin, J., H. van de Graaf, & R. Hoppe (1997) *Technology Assessment Through Interaction: A Guide* (The Hague, Netherlands: Rathenau Institute).

Gross, Paul R. & Norman Levitt (1994) *Higher Superstition: The Academic Left and Its Quarrels with Science* (Baltimore, MD: Johns Hopkins University Press).

Hajer, M. A. (1995) *The Politics of Environmental Discourse: Ecological Modernisation and the Policy Process* (Oxford: Clarendon).

Hargreaves, Ian & Galit Ferguson (2001) *Who's Misunderstanding Whom? Bridging the Gulf of Understanding Between the Public, the Media and Science* (Swindon, U.K.: Economic and Social Research Council).

Horlick-Jones, Tom, John Walls, Gene Rowe, Nick Pidgeon, Wouter Poortinga, & Tim O'Riordan (2004) *A Deliberative Future? An Independent Evaluation of the GM Nation? Public Debate About the Possible Commercialisation of Transgenic Crops in Britain, 2003.* Understanding Risk Working Paper 04-02, University of East Anglia. Available at: http://www.uea.ac.uk/env/pur/gm_future_top_copy_12_feb_04.pdf.

House of Lords (2000) Science and Society: Select Committee on Science and Technology, Session 1999–2000, Third Report, HL Paper 38, London.

Irwin, Alan (1995) *Citizen Science: A Study of People, Expertise and Sustainable Development* (London: Routledge).

Irwin, Alan & Mike Michael (2003) *Science, Social Theory and Public Knowledge* (Maidenhead, U.K.: Open University Press/McGraw-Hill).

Irwin, Alan & Brian Wynne (eds) (1996) *Misunderstanding Science? The Public Reconstruction of Science and Technology* (Cambridge: Cambridge University Press).

Jasanoff, Sheila (1990) *The Fifth Branch: Science Advisors as Policymakers* (London: Harvard University Press).

Jasanoff, Sheila (1992) "What Judges Should Know About the Sociology of Science," *Jurimetrics* 32: 345–59.

Jasanoff, Sheila (1995) *Science at the Bar: Law, Science, and Technology in America* (Cambridge, MA: Harvard University Press).

Jasanoff, Sheila (2003) (2003) " 'Breaking the Waves in Science Studies; Comment on H.M. Collins and Robert Evans, 'The Third Wave of Science Studies,' " *Social Studies of Science* 33(3): 389–400.

Kahneman, D. & A. Tversky (1996) "On the Reality of Cognitive Illusions," *Psychological Review* 103(3): 582–91.

Kahneman, D., P. Slovic, & A. Tversky (1982) *Judgement Under Uncertainty: Heuristics and Biases* (Cambridge: Cambridge University Press).

Koertge, Noretta (ed) (2000) *A House Built on Sand: Exposing Postmodernist Myths About Science* (New York: Oxford University Press).

Krimsky, Sheldon & Dominic Golding (eds) (1992) *Social Theories of Risk* (Westport, CT: Praeger).

Kuklinski, J. H. & N. L. Hurley (1994) "On Hearing and Interpreting Political Messages," *Journal of Politics* 56(3): 729–51.

Kulkarni, D. & H. A. Simon (1988) "The Processes of Scientific Discovery: The Strategy of Experimentation," *Cognitive Science* 12(2): 139–75.

Lahsen, Myanna (2005) "Technocracy, Democracy, and U.S. Climate Politics: The Need for Demarcations," *Science, Technology & Human Values* 30(1): 137–69.

Latour, B. (1983) "Bring Me a Laboratory and I Will Raise the World," in Karin Knorr Cetina & Michael Mulkay (eds) (1983) *Science Observed: Perspectives on the Social Study of Science* (London: Sage): 141–70.

Lindeman, Mark (2002) "Opinion Quality and Policy Preferences in Deliberative Research: Political Decision Making," *Deliberation and Participation* 6: 195–221.

Lupia, A. & M. McCubbins (1998) *The Democratic Dilemma: Can Citizens Learn What They Need to Know?* (Cambridge: Cambridge University Press).

Lynch, Michael & Simon Cole (2005) "Science and Technology Studies on Trial: Dilemmas of Expertise," *Social Studies of Science* 35(2): 269–311.

Mackenzie, Donald (2006) *An Engine, Not a Camera: How Financial Models Shape Markets* (Cambridge, MA: MIT Press).

Mannheim, Karl (1936) *Ideology and Utopia: An Introduction to the Sociology of Knowledge,* trans. Louis Wirth & Edward Shils (New York: Harcourt, Brace & World).

Merton, Robert K. (1973) *The Sociology of Science: Theoretical and Empirical Investigations* (Chicago: University of Chicago Press).

Mulkay, M. (1979) *Science and the Sociology of Knowledge* (London: Allen & Unwin).

Nowotny, Helga, Peter Scott, & Michael Gibbons (2001) *Re-Thinking Science: Knowledge and the Public in an Age of Uncertainty* (Cambridge: Polity Press).

Office of Science and Technology (OST) (2002) *The Government's Approach to Public Dialogue on Science and Technology* (London: OST). Available at: http://www.ost.gov.uk/society/public_dialogue.htm.

Parliamentary Office of Science and Technology (POST) (2001) *Open Channels: Public Dialogue in Science and Technology,* Report No. 153, March (London: H. M. Stationery Office).

Pinch, Trevor & Frank Trocco (2002) *Analog Days: The Invention and Impact of the Moog Synthesizer* (Cambridge, MA: Harvard University Press).

Popkin, S. L. (1991) *The Reasoning Voter* (Chicago: University of Chicago Press).

Renn, O., T. Webler, H. Rakel, P. C. Dienel, & B. Johnson (1993) "Public Participation in Decision Making: A Three-Step Procedure," *Policy Sciences* 26: 189–214.

Ribeiro, Rodrigo (2007) "The Language Barrier as an Aid to Communication," *Social Studies of Science* 37(4).

Rip, Arie (1986) "Controversies as Informal Technology Assessment," *Knowledge: Creation, Diffusion, Utilization* 8(2): 349–71.

Rip, Arie, Thomas J. Misa, & Johan Schot (eds) (1995) *Managing Technology in Society: The Approach of Constructive Technology Assessment* (London: Pinter).

Rowe, Gene & Lynn J. Frewer (2005) "A Typology of Public Engagement Mechanisms," *Science, Technology & Human Values* 30(2): 251–90.

Royal Commission on Environmental Pollution (RCEP) (1998) *21st Report: Setting Environmental Standards: Cm 4053* (London: RCEP).

Schneider, Mark, Melissa Marchall, Christine Roch, & Paul Teske (1999) "Heuristics, Low Information Rationality and Choosing Public Goods: Broken Windows as Shortcuts to Information About School Performance," *Urban Affairs Review* 34(5): 729–41.

Shapin, S. (1995) "Cordelia's Love: Credibility and the Social Studies of Science," *Perspectives on Science* 3: 255–75.

Simon, H. A. (1979) "Rational Decision Making in Business Organizations," *American Economic Review* 69: 493–513.

Smith, Roger & Brian Wynne (eds) (1989) *Expert Evidence: Interpreting Science in the Law* (London: Routledge).

Star, Susan Leigh & James R. Griesemer (1989) "Institutional Ecology: 'Translations' and Boundary Objects: Amateurs and Professionals in Berkeley's Museum of Vertebrate Zoology, 1907–1939," *Social Studies of Science* 19(3): 387–420.

Welsh, Ian (2000) *Mobilising Modernity: The Nuclear Moment* (London: Routledge).

Wildavsky, A. (1987) "Choosing Preferences by Constructing Institutions: A Cultural Theory of Preference Formation," *American Political Science Review* 81(1): 3–21.

Wilsdon, James & Rebecca Willis (2004) *See-Through Science: Why Public Engagement Needs to Move Upstream* (London: Demos). Available at: http://www.demos.co.uk/catalogue/paddlingupstream/.

Wynne, Brian (1982) *Rationality and Ritual: The Windscale Inquiry and Nuclear Decisions in Britain* (Chalfont St. Giles, U.K.: British Society for the History of Science).

Wynne, B. (1995) "Technology Assessment and Reflexive Social Learning: Observations from the Field of Risk," in Arie Rip, Thomas J. Misa, & Johan Schot (eds) (1995) *Managing Technology in Society: The Approach of Constructive Technology Assessment* (London: Pinter): 19–36.

Wynne, Brian (2003) "Seasick on the Third Wave? Subverting the Hegemony of Propositionalism," *Social Studies of Science* 33(3): 401–17.

Yearley, Steven (2000) "Making Systematic Sense of Public Discontents with Expert Knowledge: Two Analytical Approaches and a Case Study," *Public Understanding of Science* 9: 105–22.

IV Institutions and Economics

Olga Amsterdamska

In his famous 1962 essay on "The Republic of Science," Michael Polanyi appealed to a model of the free market as a metaphor for relations among scientists. Like Adam Smith's entrepreneurs, scientists were best able to contribute to the efficient growth of scientific knowledge when, working as individuals and unconstrained by extrinsic demands or regulations, they competed with each other in seeking solutions to the most important scientific problems. Polanyi's model of science as a form of economic exchange was meant to be understood metaphorically and not literally. He envisioned the competition and trade in scientific findings as taking place only among scientists themselves, not between science and other social institutions such as industry or the state. In his view, scientists alone were best able to judge the importance of a scientific problem or the excellence of its solution. Using the metaphor of a free market, Polanyi defended science's (need for) autonomy.

Robert Merton's 1942 conceptualization of science as an institution governed by a distinct set of norms was based more on the ideals of a democratic state with a liberal constitution than on those of a market where agents advance collective goals by pursuing individual interests. Just as a well-functioning democracy depends on its citizens having equal rights before the law and freedom of speech, so also, according to Merton, the institution of science requires that new knowledge claims be made public, open to criticism, and subject to disinterested judgment in terms of impersonal, universalistic criteria. Freedom of expression, openness of the public realm, and universalism are shared values in both institutional spheres. Both Polanyi's and Merton's institutional accounts claimed a profound cultural or ideological affinity between science and a major modern social institution. In both cases this affinity was invoked to define the historical essence of science as an institution, and the analogy involved an implicit claim for the cultural superiority of science, which was seen not only as cognitively superior because of its method but also as a socially or culturally superior instantiation of the best—liberal and democratic—political and economic values and principles. In both cases, this social and cultural superiority of science translated into a justification for its need to maintain cognitive and social independence.

As the essays in this section of the Handbook illustrate, institutional analysis of science is still a central concern to STS. Macro-scale, structural analyses of the organization of science underpin policy studies, work on the economics of science, and studies of relations between science and other social institutions. And yet, the assumptions underlying these more recent institutional attempts are quite different from those of Merton or Polanyi. The chapters that follow examine the institution of science as historically changeable rather than as an expression of a single dominant structure or ethos assuring the proper fulfillment of its functions; they regard relations between science and other institutions in terms of evolving cultural, epistemic, or social differences, power inequalities, and potential conflicts; and rather than establishing the conditions for science's autonomy, they examine the links between the organization and location of scientific practices and the nature of science's outputs. They are also motivated by a different set of social and political concerns than those underlying the classical analyses of Merton or Polanyi.

Having abandoned the idea that the proper functioning of science depends on the distinctive and unique social organization of the scientific community, institutional analyses of science have turned to the history of relations between science, the state, corporations, and universities. But while few would quarrel with the identification of these institutions as key forces in shaping the functioning of science since the late nineteenth century, how the history of these interactions is to be written and what are the relevant aspects of today's configuration remain a matter of an ideological as well as scholarly debate. As Philip Mirowski and Esther-Mirjam Sent show in their chapter, historiography is often shaped by the authors' attitude toward contemporary changes in the political economy of science.

The nature of and consequences of the profound transformations of the organization of science that began in the 1980s and accelerated after the end of the Cold War have been discussed in the literature in terms of the transition from Mode I to Mode II of the organization of research, or as a change from the Cold War to a competitiveness regime, or in terms of the increasing commercialization and globalization of science. These changes are addressed here directly in two chapters: one by Jennifer Croissant and Laurel Smith-Doerr and the other by Mirowski and Sent. Croissant and Smith-Doerr point to the need to study the history of university–industry relations in the United States in the context of changing state involvement in funding as well as regulation of science, and they pay close attention to legislation governing state funding of universities and research and that governing intellectual property. They then show that the intended and unintended consequences of this legislation structured the intensity and form of university–industry relations.

Mirowski and Sent's history of the economics of science is more inclusive, distinguishing among three successive regimes of scientific organization in the United States in terms of the structure of corporations, government policies toward industry and toward science, the funding of science, the history of higher education institutions, changes in how research is conducted, and pivotal scientific problems and concerns. In Mirowski and Sent's view, the novelty of the most recent post–Cold War regime consists not of the emergence of commercialization or the globalization of research as such, but of the changed meanings and forms that these processes have assumed today. For example, under the current regime, commercialization and globalization involve the weakening of in-house corporate labs and the outsourcing and privatizing of research—a change that is more specific than the simple establishment of closer ties between science and commercial activities that is sometimes described as characteristic of the post–Cold War period. At the same time, Mirowski and Sent insist that these changes in the organization of science are deeply consequential for the kind of scientific knowledge that is produced. Concern with relations between the institutional or organizational settings and the character of the knowledge or artifacts created in these settings is one of the distinguishing features of the new institutional analysis in STS.

For instance, the consequences of large-scale changes in international relations or the geopolitical situation and the perceptions of threat for the development of military technologies (as well as for their reconstruction as objects of STS) play a central role in Brian Rappert, Brian Balmer, and John Stone's review of STS work in this area. Arguing that these geopolitical considerations are mediated by local bureaucratic arrangements, competition among different services, and domestic politics, Rappert, Balmer, and Stone show how constructivist approaches in the social shaping of technology tradition have helped to illuminate the development of new weapons and their production, testing, uses, and evaluation, and suggest how various technologies co-construct our understanding of risks, security threats, and political dangers.

A process of co-construction is also at the heart of Andrew Lakoff's study of the pharmaceutical industry. Lakoff locates the production of drugs at the intersection of the pharmaceutical industry, markets, professional groups, government regulatory agencies, and patient organizations, and shows how these various institutions and groups participate in simultaneously reconfiguring knowledge about medications, their effects and uses, and knowledge about disorders and diseases. For example, in the case of psychoactive medications, changes in the regulatory system, such as the introduction of a requirement that new drugs be shown to be active against specific conditions, work in tandem with moves toward new classificatory systems and diagnostic practices in psychiatry requiring new descriptions and specifications of disorders and diseases. A drug's action, its safety, and its effects are then constructed simultaneously with the disease and a pharmaceutical firm's business strategy.

Lakoff's analysis brings out the fact that understanding the production and use of science and technology requires us to follow their paths through multiple institutions, groups, and settings. In the cases he examines, however, collaboration between these institutions and groups appears to be largely harmonious and interests convergent. In her analysis of the interactions of law and science, Sheila Jasanoff reminds us that this is by no means always the case. On the one hand, interactions between science and law (like relations between other institutions, whether medical, legal, or political) are becoming ever more complex and multifaceted, while on the other hand, the two institutions are culturally and epistemologically different and their claims to authority can sometimes clash or compete. Jasanoff's chapter unites many of the features of the new forms of institutional analysis discussed here: she reviews work on the history of encounters between science and technology and law, describes how their cultural and epistemological authority is reflected and legitimated in their different fact- and order-making practices and discourses, and examines the ways in which interactions between science and law take shape in different settings and arenas. Focusing on how facts and concepts (such as evidence, proof, and reason, but also justice, identity, or legitimacy) are (co-)constituted in law and science (and through their encounters), Jasanoff insists on the normative consequences of knowledge-making practices and the need for STS better to examine these "hidden normativities."

Concern with the normative consequences of conceptual choices is also paramount in Susan Cozzens, Sonia Gatchair, Kyung-Sup Kim, Gonzalo Ordóñez, and Anupit Supnithadnaporn's review of recent work on science and development. They show that different disciplinary understandings of "development" and of its goals and methods can have profound social, political, and economic consequences. Adopting Amartya Sen's definition of development as freedom, Cozzens and her colleagues distinguish between what they call the human development project and the competitiveness project, and show how different perspectives on development conceptualize the role of science and technology. The authors examine (a) the current STS approaches that emphasize the cultural clash between Western science and local knowledges; (b) studies stemming from the new growth theory that emphasize the role of the state in promoting appropriate economic policies; and (c) work relying on innovation systems approaches that emphasizes learning in individual firms working in a global environment. Each of these approaches highlights the role of different institutions, relies on a different political or economic philosophy, and sees different roles for science and technology in the development project. Each also offers a somewhat different understanding of the *goals* and not just the *means* of development.

Jasanoff's and Cozzens and colleagues' reflections suggest how the normative concerns and implications of science and technology's institutional engagements make it no longer possible to focus only on science's institutional autonomy as it was understood at the time of Merton or Polanyi. Having found normativity embedded in the concepts and practices through which science and technology engage with other social institutions, contemporary STS has opened up a difficult new research agenda for the institutional analysis of science's engagements with politics, culture, economy, and society.

26 The Commercialization of Science and the Response of STS

Philip Mirowski and Esther-Mirjam Sent

Claims about the proper method for writing the history of science are simultaneously claims about the relations between the producers and consumers of scientific knowledge.[1]

MONEY CAN'T BUY ME TRUTH?

It is not hard nowadays to find people who harbor strong opinions about the contemporary commercialization of science, primed and willing with very little prompting to recount some anecdote about the travails or triumphs of Viridiana Jones in the Temple of Mammon. First off, there are the motley ranks of Cassandras, who, significantly enough, tend to have a soft spot for the Good Old Virtues of the Mertonian norms and bewail the prospect of expulsion from the prelapsarian Garden.[2] They lament that once there may have been an invisible college, chorused sweetly in concert in the quest for truth, but now there are only feckless individual entrepreneurs scrabbling for the next short-term contract. "Who will now defend the virtue and purity of science?" they wail. By contrast, there also stand the massed phalanx of neoclassical economists, science policy specialists, and their bureaucratic allies, who by and large tend to reverse the valences but nevertheless engage in much the same forms of discourse. For them, most scientists in the "bad old days" had been operating without sufficient guidance from their ultimate patrons, the corporate pillars of the economy; but luckily, with a bit of prodding from the government, a friendly nudge from their university's intellectual property officer, plus a few dollars more waved in their directions, scientists have been ushered into an era that appreciates the compelling logic of "technology transfer." At the risk of caricature, one might summarize their central task as the gathering of empirical data in order to argue that the expanding modern commercialization of scientific research has turned out to be "inevitable," with the corollary that little evidence exists that it has "significantly changed the allocation of university research efforts" (Nelson, 2001: 14).[3] Admittedly, many of these purveyors of glad tidings would still regard themselves as defending the preservation of an "optimal" sphere of research reserved for open public science and pure unfocused curiosity (a "separate but equal" doctrine applied to unspecified portions of the university), however much they would also avow that the economy must constitute the ultimate arbiter of scientific success in this more rational regime of organization. The history of science for them is simply divided into an Age of Confusion when "open science" had unaccountably been mistakenly conflated with the whole of science, fostering a lack of understanding of the efficient organization of systems of innovation, and our own current Age of Free Enterprise, when we see the true situation of pervasive ownership with clarity. This kind of crude "before and after" discourse has also come to dominate much of the contemporary science policy literature, which is filled with euphemisms like "technology transfer" and "democratically responsive science," which seek to reconcile the harsh authority of the almighty dollar with the delicate sensibilities of those otherwise inclined to resist the advent of the End of History. It has become fashionable of late to pillory Vannevar Bush for his invention of the notion of the pipeline "linear model" that situated "applied science" as the downstream result of "basic science"; now we are all supposed to know better.[4]

This rather superficial stage 1/stage 2 narrative, be it upbeat or downbeat, has little to do with the actual histories of the sciences. Sometimes this has become a problem in some sectors of STS as well, as we discuss below in the section "Alternative Market Models of the Conduct of Scientific Research." Part of the problem arises because STS has only very recently begun to come to grips with the phenomenon of commercialization, lagging behind the Cassandras and the science policy bureaucrats by perhaps a decade or more. The "commercialization of science" turns out to be a heterogeneous phenomenon, resisting simple definition. Consequently, many contemporary discussions of the commercialization of science have proved deeply unsatisfying, tethered as they are to totemic monolithic abstractions of Science and The Market pushing each other around in Platonic hyperspace. Indeed, some historians have long sought to remind their readers of what one collection (Gaudilliere & Lowy, 1998) calls "The Invisible Industrialist" who occupied the interstices of numerous laboratories and frequented the hallways of universities since the middle of the nineteenth century. Yet, in rejecting the false polarities of the neo-Mertonians on the one hand and the economic apologists for the modern era on the other, it would appear that the denizens of science studies have of late run a very different risk of denying that there has been any significant change whatsoever in scientific protocols; hence, important structural differences are overlooked that might be traced to alterations in the ways in which science has been paid for and accommodated within the economy over long stretches of time. One recent instance of this sort of attitude has been expressed by Steven Shapin (2003: 19):

Throughout history, all sorts of universities have "served society" in all sorts of ways, and, while market opportunities are relatively novel, they do not compromise academic freedom in a way that is qualitatively distinct from the religious and political obligations that the ivory tower universities of the past owed to the powers in their societies.

A cruder version of this orientation was captured in interview transcripts with the chair of an electrical engineering department (in Slaughter et al., 2004: 135):

You have to accept the fact that it [research] is going to be driven by the people who give you the money. [If] the state gives us money, they tell us what to do. [If] NSF gives us the money, they tell us what research they want done. [If] DoD gives us the money, [its] the government . . . Why is it any different with industry? I see no difference whatsoever.

Yet another manifestation is the attempt by the Paris school of Bruno Latour and Michel Callon to reduce the economy to just another instance of the laboratory, as a prelude to erasing all ontological differences between scientific and economic activity, while chanting, "we have never been modern!"[5] Strangely, this widespread ahistorical insistence on "the way things have always been" in science in its coexistence with the economy dates back to the supposed godfather of social studies of science, Thomas Kuhn.[6] In a little-read set of comments on a pivotal conference on the relationship of industrial R&D to science held at Minnesota in 1960, he insisted that "the two activities, science and technology, have very often been almost entirely distinct," and indeed, that "historically, science and technology have been relatively independent enterprises," going back as far as classical Greece and Imperial Rome! As a historian, Kuhn felt impelled to admit that,

Since 1860 . . . one finds that characteristic twentieth century institution, the industrial research laboratory . . . Nevertheless, I see no reason to suppose that the entanglements, which have evolved over the last hundred years, have at all done away with the differences between the scientific and technological enterprises or with their potential conflicts.[7]

The indisputable fact that scientists and their institutions have always and everywhere been compelled to "sing the prince's tune when taking the prince's coin" in one form or another does not imply that the evident modern trend toward the escalated and enhanced commercialization of science need not or will not alter the makeup of the supposedly invariant "scientific community," not to mention the nature of the "outputs" of the research process. Furthermore, the underappreciated fact that the political economy of the sciences in America has been transformed from top to bottom at least twice over the past century has yet to be correlated with the types of science that have been performed in the manner that has become the trademark of science studies —that is, fine-grained studies of the interaction of forms of organization with the stabilization of knowledge claims—or indeed, the ways we tend to think about the successful operation (or conversely, the pathologies) of the "scientific community." This sort of agenda was called for in the perceptive paper of Michael Aaron Dennis in 1987, but his entreaty has yet to be sufficiently heeded.

Close on the heels of the enunciation of the Hessen thesis in the 1930s[8] and the subsequent Cold War anti-Marxian backlash against it, most appeals to economic structures as conditioning factors in the production of science simply dropped out of postwar theoretical discourse within science studies. As Dennis has written about American historians, the manner of "solving the problem of providing for the support of the material foundations of science—salaries, labs, instruments—effectively eviscerated the possibility of anything even remotely resembling the materialist historiographies of science that had developed between the wars" (1997: 16). Something similar seems to have happened in Europe as well. The postwar political shift in the philosophy of science also played a part in repressing such questions (Mirowski, 2004a,b). Consequently, as the next great transformation of research was taking place in the 1980s, science studies was instead turning its attention to micro-scale studies of laboratory life, ignoring how the laboratory's macro-scale relationship to society was being reengineered all around, not to mention the shift in those paying for all those DNA sequencers and inscription devices.[9] The qualitative effects of the panoply of market activities on scientific research thus remain an open issue.

Curiously, expressions of concern over the potential impact of economic incentives on science have instead become the province of groups who have tended to set themselves up in opposition to STS. Predictably, they frequently wind up their exercises by concluding that commercialization has not drastically changed contemporary science. Positing the invariance of the end-state from the mode of production of knowledge has become a veritable industry among those anxious to provide reassurance that their "social epistemology" underwrites an invisible hand story in the sphere of scientific research: as they phrase it, that epistemically sullied motives (which are then abruptly conflated with "social influences") do not threaten the goals of science.[10] These attitudes have taken root in the science policy community and a segment of the philosophy of science (Mirowski, 2004b, 2005) and pervade discussion of commercial research in business schools.[11]

A different approach to the "new economics of science" explores the possibility that alternative forms of the commercialization of science actually have indelibly shaped both the practice of research and the contours of whatever it is that we encounter at the end of the process (Mirowski & Sent, 2002). A key variable turns out to be the ways in which that protean entity "the laboratory" was appropriated and reconstructed by higher education, corporations, and the government over the twentieth century, a point first made by Dennis (1987) and recently propounded by Pickering (2005). In addition, the modern phenomenon of globalization tends to undermine earlier nationalist and parochial approaches to the problem of the economics of science and the notion that there might persist "national systems of innovation" (Drahos & Braithwaite, 2002; Drori et al., 2003). These issues will be the topic of the section "Three Regimes of Twentieth Century Science Organization." Another crucial variable is the way in which the divide between "public" and "private" conceptions of knowledge has shifted in the recent past and how that has fed back on the rationales for various actors in their exercise of the governance of science (Slaughter & Rhoades, 2004). The section "Alternative Market Models of the Conduct of Scientific Research" is an overview of this problem.

Many different groups have entered the fray in asserting their expertise to frame discussions of the modern commercialization of science. Examples can be found in such far-flung enterprises as literary criticism (Newfield, 2003; Miyoshi, 2000), medical schools (Angell, 2004), library science (Scheiding, unpublished), education schools (Apple, 2005; Slaughter & Rhoades, 2004), and popular journalism (Press & Washburn,

2000; Shreeve, 2004; Dillon, 2004; Judson, 2004; Washburn, 2005). Some political theorists have attempted to adapt the "social contract" literature in politics to discussions of regime change (Guston & Kenniston, 1994; Hart, 1998). Some fields (e.g., "knowledge management" specialists in business schools, intellectual property lawyers in law schools, and political economists in science policy units) highlight certain facts about the changing status of science but neglect other equally salient facts, say, from legal history, the politics of education, the annals of military procurement, or international trade policy. Other scholars, by suggesting that advanced economies were becoming increasingly "weightless," would graduate to a third stage of capitalism consisting almost exclusively of the service sector, or indeed disengage from gross physical production processes altogether. Of course, most people recognized that much of that talk bordered on delusional, but it nevertheless managed to appear sensible (or at least fashionable) by engaging in locutions such as the "Information Society" or the "New Knowledge Economy."[12] Frequently, appeal to this supposed novel entity served as a prelude to subsumption of science under a more general theory of the "marketplace of ideas" (Foray, 2004; Feldman et al., 2002; Mirowski, forthcomingB).

One might justifiably wonder if the cacophony of voices adds up to much more than a generalized atmosphere of anxiety. If STS is to claim to stake out a distinctive approach to the phenomenon of the modern commercialization of science, then it will need to make a fateful choice between casting the "constructivist" stance as one treating the entirety of science as just another form of marketing (Woolgar, 2004) and stressing the essential historical instability of the commercial/communal binary as instantiated in actual concrete practice. In this chapter, we stand as advocates of the latter position. Hence, we outline *one version* of an STS approach to commercialization in the section "Three Regimes of Twentieth Century Science Organization" and then contrast it to some other versions in the section "Alternative Market Models of the Conduct of Scientific Research."

Once the ground has been prepared in the former section by an analytical scheme of temporal periodization (albeit one grounded primarily in the American context), we then point out the differing meanings of the commercialization of science under each individual regime. Although market considerations were never absent from the laboratory or the classroom, the modern commercialization movement can in no way be considered a "return" to anything like the interwar science promoted by Jazz Age captains of industry.[13] Modern science has turned out to be a qualitatively different phenomenon because it has been grounded in profound historical transformations in the corporation, the university, and government, with consequences for their respective initiatives to exercise control in the organization and funding of science. We offer the limited exercise of this chapter more as a preliminary exemplar than a definitive template for research into other countries in other eras; a future task of STS might be to report similar species of watersheds in other disparate culture areas.[14] Whether or not that comes to pass, the other question raised by this chapter is, will the multiplicity of social trajectories of the provisioning of science tend to converge to a single, worldwide model of commercialized, globalized science in the twenty-first century? If the response is posed in the affirmative, then should we also expect the intellectual rationales for a particular mode of commercialized science to similarly be winnowed to a few simplified narratives of "scientific success"? Supposing that turns out to be the case, then one begins to appreciate the challenge that a neoliberal "new economics of science" poses to the future of STS. If broad generalizations about the commercial character of science start to attain plausibility, then they will exist because corporations and governments and INGOs have been engaged in a concerted project of standardization spanning national and cultural and disciplinary boundaries.

THREE REGIMES OF TWENTIETH-CENTURY SCIENCE ORGANIZATION

STS scholars have been wary of reifying the concept of "science" as a transcultural transhistorical category, and for good reason. The more we learn about scientists and their livelihoods, the more we come to appreciate the sheer diversity of their activities, the vast compass of their societal locations, and the multitude of ways their findings have become stabilized and accredited as knowledge. What keeps this daunting multiplicity from defeating analysis for STS scholars is the dominance of certain identifiable institutional structures involved in organizing scientific inquiry in the modern period. Scientists have not subsisted as a purely self-organized discourse community, contrary to the rhetoric dominant during the Cold War era. Rather, they have always been enmeshed in complicated alliances with and exclusions from some of the dominants institutions of our era: primarily, the commercial corporation, the state, and the university.[15]

The story of the quotidian activities of the scientist always presumes some social scaffolding of material support, which in the modern epoch has been most frequently built up from corporate, governmental, and educational (CGE) elements. Furthermore, various individual scientific fields will be experiencing relative growth or stagnation, depending on the particular historical configurations of their own intellectual trajectories, in combination with the levels of encouragement provided by the CGE sectors. To render this set of propositions more concrete, we provide in table 26.1 a schematic outline of the three regimes of science funding and organization in the United States in the twentieth century, based on our reading of the relevant economic and social history as well as the contributions of historians of science. To keep the historical sketch from becoming unwieldy, we have restricted the table to indications of CGE developments that would have direct bearing on the constitution of the "laboratory" in scientific research; considerations of length preclude extension of the CGE analysis to, say, clinical medicine, field sciences, or purely abstract mathematical endeavors (although we believe these would be amenable to similar periodization). The purpose is to elevate to consciousness the fact that the corporation, the legal framework, and the university have not been static over time, and that their alterations can be directly related to the ways in which scientists have made their livelihoods and pursued research agendas promoted by their immediate patrons. Thus, contrary to the prognostications of social scientists, no single "market" governed the evolution of

Table 26.1

American Regimes of Science Organization in the Twentieth Century

Period, Regime	Corporation	Government Corporate Policy	Government Science Policy	Science Managers	Higher Education	Pivotal Disciplinary Science
1890 to WWII Captains of erudition regime	Evolving 1895–1904 great merger movement: Chandlerian firm of "Visible Hand." Innovation of in-house R&D labs to control competition.	Massive expansion of corporate prerogatives. Corporations become legal agents; patents a major strategic tool. Beginning of antitrust. Employers own research of employees.	Almost nonexistent. NRC formed as trade association lobby for natural sciences. General suspicion of government involvement. NRE fails. Wartime patent bounty.	Charismatic PhD directs corporate labs. Foundation officers run few elite university grant programs (on corporate principles).	Mostly elite liberal arts. Research subordinate to pedagogy. Science not a major priority. Foundations attempt reform. Labs founded.	Chemistry, electrical engineering
WWII to 1980 Cold War regime	M-form, conglomerate diversification. R&D units as semi-autonomous revenue earners (due to military contracts). Regulatory capture.	Corporate powers augmented; antitrust strengthened. Intellectual property (IP) weakened. Military contracts as industrial policy.	Huge expansion of Federal military funding and control. Military promotes basic science to defeat enemies. National labs. NSF as nonmilitary face of "pure" science.	Military primary science managers for research universities, think tanks, national labs, corporate contract research. "Peer review" a secondary institution.	Mass education at expanded research universities. Integrated teaching/research. Produce democratic citizens: academic freedom as political statement.	Physics, operations research, formal logic
1980– ? Globalized privatization regime	Breakdown of Chandlerian model. Retreat from vertical integration, diversification. Corporations outsource R&D, spin off in-house labs.	Transnational trade agreements expand corporate powers to circumvent national control. Antitrust weakened; IP vastly expanded.	Privatization of publicly funded research: Bayh-Dole act, etc. Kill Office of Technology Assessment. Science just one political resource among many.	Globalized corporate officers control universities, hybrids, contract research organizations, start-ups.	Stock up human capital for those who can pay. Only entrepreneurs are free. Severing of the teaching-research connection.	Biomedicine, genetics, computer science, economics

science in America; rather, there have been multiple formats of provisioning, embedded within larger structures.

The designations provided in the table for the various regimes are predicated on popular characterizations found in the existing historical literature. The "captains of erudition" regime is so designated in honor of Thorstein Veblen (1918), who wrote one of the earliest descriptions of the American research university as becoming subject to specific corporate organizational principles; it also bows in the direction of the dominant American school of business history based on the work of Alfred Chandler.[16] The label indicates an elitist and closed corporate model of the organization of science. The Cold War regime is a label regularly used to designate what many now portray as a fleeting interlude of military dominance over science management in the period beginning in World War II.[17] The terminology of "globalization" is not so much an appeal to a fashionable concept in contemporary social theory as it is an insistence on a set of factors indispensable for an understanding of the forces that drive the current wave of commercialization of science.

The Genealogy of the American Laboratory

Laboratories were not something that just naturally appeared in the American landscape: they had to be built, and to be able to subsist as more than ephemeral entities, they had to be integrated into some sector of the economic infrastructure. Unlike the situation in Europe, large-scale laboratory science did not originate in the university sector in America. Rather, from the outset, it was very much a commercial initiative.

The broad outlines of the rise of the industrial research laboratory are now well known.[18] Everyone concedes that its origins are to be found in continental Europe, primarily but not exclusively in Germany, and that it was initially located in large firms engaged predominantly in what has become known as the "second industrial revolution": chemicals, electrical machinery, railways, and pharmaceuticals. An earlier vintage of historiography tended to assert that the "science-based industries" simply summoned an implicit exigency to incorporate research activities within their ambit, in both Germany and the United States, but modern historians have since grown more cautious, realizing that the ingredients to explain the appropriation of what had previously been a specialized pedagogical device for industrial purposes would be found in a strange brew of state policy toward advanced education; ideologies of state-building and political rectitude; the rise of various notions of intellectual property; the conditions that gave rise to large and powerful corporations in particular political settings; and the ambition to exert control over burgeoning transnational mass markets in clothing, transport, and communications, electrical equipment, and patent medicines. Whereas most manufacturing firms had long made provisions for internal quality control, routine testing, and incremental process improvement, an innovation arose around the 1870s to expand the purview of these specialized corporate arms into patent protection, the bureaucratization of trade secrets and the generation of novel processes and products. It resembled a phase transition between the periodic use of the sciences for corporate purposes to something approaching the institution of bureaus dedicated to *doing* science for corporate purposes. The distinction was not always sharp, the results were not often that immediately striking, and the transition was not always conscious.

The rise of the industrial laboratory was the consequence of an American pincers movement: on the one hand, a push to bureaucratize and industrialize (or vertically integrate backward, as economists might say) something that heretofore had been conceived as the ineffable capacity of the individual genius, and on the other, a pull to adapt a purpose-built academic social formation to corporate imperatives that itself had only recently been stabilized in specialized educational settings for pedagogical purposes. Michael Dennis correctly points out that when later nineteenth century American figures made their pleas for "pure science," they did not refer to some notion of disembodied science carried on for its own sake, nor to an imaginary autarkic scientific community defending its prerogatives, but rather to a pedagogical ideal for a species of hands-on higher education where teaching and research were combined in a setting relatively sheltered from commercial considerations. *Pace* Bruno Latour, the issue was not whether the denizens of laboratories or their proxies "circulated" in the wider world but rather whether laboratories themselves were a robust phenomenon that could be severed from the nascent research university and successfully grafted onto the multidivisional corporation. The wrenching estrangement of the laboratory from its teaching functions constituted so dramatic a departure from its conceptual origins in the later nineteenth century that it was not hard to find any number of academics expressing scorn for the newfangled industrial laboratories and their spiritually debased inhabitants, disparaging the public confusion of untutored tinkerer-inventors with real "scientists." Yet it would be an anachronism to read these as indicative of some transcendental incompatibility of science and commerce, as Kuhn did. Rather, it makes more sense to approach them as symptoms of conflicts attendant on institutional innovations in the construction of both the public and private spheres, artifacts still in their early stages.

The Captains of Erudition Regime One of the most salient differences between the German situation and its American counterpart circa 1900 was that, by and large, the academic research laboratory did not substantially predate the rise of the industrial laboratory in the United States.[19] Higher education in the natural sciences and the social sciences was acknowledged to have been superior in the German setting at the beginning of the twentieth century; it was also recognized as having attained an unprecedented level of state-sponsored centralization. The German university had pioneered the research seminar and the research laboratory; by contrast, the pedagogical research laboratory had not yet become solidly established in American universities, which were predominantly devoted to moral uplift and liberal arts education for a narrow elite, although the forms this assumed were widely decentralized and diverse.[20] As David Noble put it, in the nineteenth century "shop culture" was deemed opposed to "school culture" (1979: 27); if anything, the universities lagged behind firms when it came to building and staffing labs. Indeed, far from being transplanted bodily from an academic to a corporate context in the United States as it had been in Germany, the American scientific laboratory was built up almost from scratch, modulo some Germanic inspiration, more or less simultaneously at both sites. For instance, as early as 1881, American Bell Telephone experimented with the location of a new physics laboratory, offering Harvard University the money to build it, as long as "professors could use university laboratories in work for private companies" (Guralnick in Reingold, 1979: 133). MIT's fabled Research Lab for Applied Chemistry, originally intended to carry out industrial research, dated from 1908. Since dedicated university laboratories were rare, the academic/commercial distinction was less than distinct. Yet the siting of industrial research on college campuses often proved less than satisfactory for its patrons, mostly owing to perceived insufficiency of corporate control (Lecuyer, 1995: 64), redoubling the formation of in-house laboratories. This made for an unusual political economy of science in early twentieth-century America, going some distance toward explaining a certain impression of "exceptionalism" in the culture of science that one encounters among many commentators (Wright, 1999) and one that contributed to the fact that American scientific research achieved an advanced level of one kind of commercialization far more quickly than did any other country by the 1930s. It also coincided with the successful elevation of a subset of the natural sciences to world-class status for the first time in the United States, thereby raising the intriguing prospect of the existence of multiple institutional paths to the fortification of a research base in the course of economic development of national systems of research.

Science in the American university system had gained a foothold comparatively late, around the beginning of the Erudition regime.[21] The highly decentralized character of the American higher education sector at first posed an obstacle to the development of a scientific curriculum, although it would later prove a boon. While later historians might point with pride to the earlier founding of Harvard's Lawrence School, the Yale Sheffield School, or the Massachusetts Institute of Technology, the impact of these and other educational institutions on actual practices of research and the shape of American science were slim to negligible prior to the 1890s. The impetus for the change in regimes originated instead mostly from within the corporate sector, initially in the creation of a new kind of in-house laboratory for commercialized science, but later in the export of corporate protocols and funding structures to some handpicked research universities, by way of the instrumentality of a few activist foundations. Hence, our brief overview necessarily begins with a fly-over of the relevant background history of the corporation.

American historians of technology have tended to lean on the work of Alfred Chandler, and in particular his book *The Visible Hand* (1977), to provide the framework within which they situate their understandings of the rise of commercialized science. This turn of events has been slightly incongruous, partly because Chandler devotes very little explicit discussion to the role of industrial laboratories in his history, but also because it is sometimes predicated on a fairly old-fashioned technological determinism (Chandler, 2005a). Set against an earlier literature that approached the corporation as a nexus of power growing dangerously out of control, Chandler portrayed the rise of the large American corporation around 1900 as a rational organizational response to technological imperatives of high-throughput capital-intensive patterns of production, found primarily in the newer science-based industries, which could only be made viable through the parallel construction and organization of mass markets on an unprecedented scale. Chandler praised the Jazz Age mega-corporation for adopting centralized bureaucratic managerial structures and vertically integrating backward into inputs and forward into sales, advertising, and market research. Although he did lightly touch on the rise of the industrial laboratory (e.g., 1977: 425–33), it is treated as just another exemplar of the line-and-division managerial structure to which Chandler sought to attribute the success of firms such as Standard Oil, General Electric, and DuPont. Hence, Chandler did not so much proffer an explanation of the rise of the industrial research laboratory as mutely point to one necessary bureaucratic prerequisite for its coming into existence. Some industries could have sought to "integrate backward" into research, except for the inconvenient fact that in most cases there were no preexistent stable structures for them to integrate backward into.

The Chandlerian narrative as manifest in science studies (Smith, 1990) should therefore be supplemented by legal and political considerations, which Chandler largely shunned. The limited liability corporation, far from being an established fixture on the American scene, had just undergone a period of substantial judicial fortification at the end of the nineteenth century owing to the infamous Santa Clara nondecision extending Fourteenth Amendment rights to corporations (Nace, 2003), the race to the bottom of states to liberalize corporate charters, and the unprecedented merger movement of 1895–1904. This sudden arrogation and consolidation of power had not gone unnoticed and had begun to provoke a countermovement beginning with the Sherman Antitrust Act of 1890 and continuing with the Clayton Act of 1914, and it provoked political movements hostile to corporate dominance of the economy in the Progressive Era. The rise of the American industrial laboratory should be situated in this context to appreciate some of its more distinctive characteristics as well as its impact on academic science.

The standard popular account portrays the *fin de siècle* industrial lab as a sort of factory of innovation, churning out gadgets that became new products or improved production processes on demand for the corporate hierarchy. This was the image promoted by the Scripps Science News Service, the very first corporate-backed "public relations of science" initiative, which began in 1921 (Tobey, 1971: chapter 3). But the more recent literature resists this tendency to frame the lab either as a straightforward invention factory or as some university-science-department-in-exile,[22] and for good reasons. The prime directive behind many of the innovations growing out of the large corporation was the drive to control markets, render unforeseen events manageable, and stifle external competition. As the government began to block direct attempts at market control such as explicit cartels, pools, and other tied arrangements through its initiatives, such as antitrust prosecutions, the locus of corporate control began to shift to indirect arenas such as intellectual property, the imposition of technical standards, and the like. One primary reason that large corporations turned their attention to bringing scientific research within their walls in this period is that "invention and innovation were effective defenses against antitrust suits" (Hart, 2001: 926) and that patents in particular but intellectual property in general were conceived as the best and most effective means of controlling competition in the early twentieth century (Noble, 1979: 89). This trend was actively promoted by certain U.S. government policy moves, such as the seizure by the Alien Property Administration of German patents in 1919 and their licensure to American firms under highly favorable terms (Mowery, 1981: 52; Steen, 2001). As both case law and legislation were slanted in the direction of integrated corporate organization instead of interfirm cartels (or other features of the German model[23]),

> legal doctrine inadvertently spurred corporate consolidation, and the consolidated corporations in turn, enhanced their investments in R&D . . . The birth of the central corporate laboratories in this period . . . [is] therefore in part the product of antitrust law" (Hart, 2001: 927).

Legal redefinitions of intellectual property and clearer stipulations as to who might assert claims over the fruits of scientific research were heavily conditioned by the shifting needs of the fortified corporation. In a move with untold consequences for the future organization of science, corporations managed to have the case law with respect to employee inventions shifted away from older labor–theoretic notions of the fruits of individual genius and toward a presumption of employers' ownership of *anything* an employee might do or invent. Prior to the 1880s, the standard default rule was that rights to inventions were vested in employees; but first, through the creation of the doctrine of "shop right" in the 1880s to 1910s, and afterward, through a series of judicial decisions that made direct reference to corporate research laboratories, the presumption of ownership was shifted decisively to the firm itself (Fish, 1998). Corporate initiatives then fed back on general cultural images: by the early 1920s, American court decisions began appealing to the apparently commonly accepted notion that invention and science was a "collective" and not an individual phenomenon.[24] As a sign of the times, Nobelist Robert Millikan began to complain in the 1920s that the German research university did not sufficiently respect the collective character of scientific research (Tobey, 1971: 219). However, the convenient notion of the "collectivity" was not to be allowed to exude too far outside the firm's boundaries (as in the writings of Thorstein Veblen) for that might bring back the dreaded world of cartels, patent pools, plunderbunds, and trusts. The legal bias against cross-firm combinations and joint ventures bore direct consequences for the existence and viability of corporate labs that might try to escape from the tentacles of corporate bureaucracy. While free-standing independent industrial labs were also founded in this period, they never caught on or expanded to the extent that in-house industrial research did; unlike some of the largest in-house labs, they never conducted any world-class science; moreover, they undertook contract work that did not mimic that of the big corporate labs but was most often subordinate and supplementary to them.[25] Thus, even though the research process was clearly becoming commercialized, it was not rendered so thoroughly fungible to the extent of being freely outsourced by its corporate sponsors. (The modular "marketplace of ideas" turns out to be a much more recent phenomenon.) Hence, the particular form assumed by contract research in America was (and continued to be) heavily conditioned by industrial policy and intellectual property conventions.

After the first generation of the captains of industry had built or consolidated their massive industrial corporations and retired, or otherwise cashed out some of their gains, they or their family members decided to devote some funds to philanthropy (or perhaps merely engage in tax avoidance) through the creation of various foundations: the Russell Sage Foundation (1907), the Carnegie Corporation (1911), and the Rockefeller Foundation (1913) are some of the better known. Assistance to higher education had become part of their agenda, but serious questions arose as to the most appropriate way to pursue this goal. At first, grants were patterned on other philanthropic practices, and when it came to academic recipients, they were pitched to essentially provide temporary individual outdoor relief to indigent or otherwise needy scholars. However, just as in the case of intellectual property, by the 1920s the focus on the isolated individual as the monad of science funding had gone out of fashion, and attention turned to the targeted application of funds to provide research endowments for continuing programs, reorient whole disciplines, and build new institutions. It was consistent with this vision that the grants were overwhelmingly channeled to private universities and structured to concentrate "excellence" in a few powerful institutions. As Robert Kohler put it most succinctly, "The large foundations were . . . carrying business methods and managerial values from the world of large corporations into academic science" (1991: 396). In everything from recasting the research grant as a contract that imposed certain standards of bureaucratic accountability, to imposing the line-and-division managerial structure on university administrations and departments, to encouraging the creation of teams of researchers, the corporate officers who staffed the large foundations tended to foster the standards and practices of the large American corporation within their targeted flagship research universities. As E. B. Croft of Bell Labs put it,

It might appear that it would tend to destroy the initiative of the individual; that it would make it difficult to properly assign the credit and give the reward to the individual worker. These are all problems of administration that have had to be worked out. First of all we must establish in the individuals a state of mind, which leads them to really believe that their best results are attained through cooperation with others (Noble, 1979: 119).

Harvard and Chicago would be coaxed and inspired to become the AT&T and Standard Oil of American higher education, surrounded by smaller and relatively insignificant rivals who had not learned the lessons of building a permanent and successful managerial hierarchy, and not inconsequentially, a strong research capacity. Colleges would face the choice of emphasizing liberal-arts pedagogy or aspiring to technical expertise in research. Consequently, the scientific research laboratory was propagated throughout the academic landscape as the necessary accessory to the mature corporate business plan.

Foundation managers allied themselves with the small but growing numbers of academics . . . who realized that [corporate] organization and management were good ways to keep ahead of the pack in the increasingly crowded and competitive world of basic research (Kohler, 1991: 400).

The fact that much of the structure of the American academic science laboratory was inspired by that of the industrial research lab did not imply that academic scientists uniformly sought to mimic their industrial brethren, however. Even as the social structure of laboratories was becoming patterned on corporate social structures, the academic scientists still lauded the university laboratory as a pedagogical ideal existing separate and apart from commercial pressures, but also from government subsidy. Yet this quest for "purity" only exacerbated the problem of who precisely would fund and manage the research carried on under that banner. The nagging tension between science beholden to special interests versus science in pursuit of the public interest proved a challenge to those who apprehended the "erudition" dynamic as a danger to democracy, such as Walter Lippmann, Thorstein Veblen, and John Dewey (Mirowski, 2004b). The foundations were increasingly targeting their funds to support specific research projects in a limited portfolio, or else professionalized arenas of higher education such as medical schools, and could not be expected to bear the burden of the health of the whole gamut of sciences, much less the careers of the next generation of scientists. The National Research Council (NRC), established in 1916 as a sort of trade association to lobby for the support of the natural sciences, actually opposed direct government subvention of researchers (Noble, 1979: 155). The NRC-backed drive to institute a National Research Fund, which would derive its endowment from corporate subscriptions, failed miserably in the period 1926–1932 (Tobey, 1971: chapter 7). Robert Millikan was denouncing federal support for the sciences at private universities as late as 1937 (Lowen, 1997: 33); it remained minuscule. Outside of a few private universities favored by the foundations, the problem of sustained privatized care and maintenance of a diversified academic research capacity was not solved by the supposedly collectivized community of researchers or by its corporate patrons. It would not be solved until World War II.

Nevertheless, American laboratories for the first time in their history were able to produce some world-class science under the erudition regime. Whether the Nobel Prizes were for work originated in the academic sector, as Theodore Richards's chemistry prize in 1914 or Robert Millikan's physics prize of 1923, or from within the burgeoning industrial sector, as that of Irwin Langmuir of GE in 1932 or C. J. Davisson of Bell Labs in 1937, there was a certain American style of research that traced a part of its lineage to the corporate inspiration of the laboratories. European commentators noted a certain empiricist temper regnant, a kind of phenomenological exploration well suited to teams of researchers, infused with an experimental and accounting mentality as contrasted with a rationalist orientation. German world dominance in both physics and chemistry were still widely acknowledged in this period. Electrical engineering, however, found its center of gravity shifting westward by the 1930s. Nevertheless, America's deficiencies with regard to theoretical imagination were a common theme of opprobrium emanating from the older and cultured precincts of Continental Europe. Chemistry, probably the most lavishly supported of the natural sciences in America in this era, itself produced no radical changes in fundamental doctrines (Mowery, 1981: 104). One might therefore conclude that the corporate orientation of American science did indeed influence the types of research performed in this era as well as some of the results produced. More to the point, when larger cultural movements felt impelled to come to terms with the world-historical significance of the advancement of science, most frequently it was European science that served as their reference point.[26]

The Cold War Regime   The fact that American science was utterly transformed in World War II, and then persisted in that novel economic format throughout the Cold War, is a widespread conviction hardly requiring defense at this late date,[27] but it does tend to get confused with another notion—that mostly this was due to the rise of "big science"—the idea that postwar science organization was driven by scale effects, in much the same way that Chandler asserted that the structure of the modern corporation has also been driven by scale effects.[28] But concentration on abstract size and its quantification, a tendency often associated with Derek de Solla Price and the scientometric movement, serves in a way conformable with Cold War trends to lend itself to technological determinism. There is no doubt that the constitution of huge teams devoted to the production of a particular weapon or device, such as the MIT Radiation Lab, the Manhattan Project, or Lawrence's cyclotron, could not help but provoke revisions in the way American culture would apprehend the nature of the "laboratory" in the postwar period. Science seemed increasingly to be organized around "gadgets," as the denizens of Los Alamos called the Bomb, and the devices were Big along almost any dimension one would care to assess: reactors, accelerators, space vehicles, von Neumann's room-sized computers, and so forth.

Yet, before we become blinded by the shiny surfaces, blinking lights, and phalanxes of bench scientists, it becomes necessary to direct our attention to some rather more pedestrian aspects of the quotidian prosecution of postwar science, namely, the myriad of ways in which the government, primarily but not exclusively in the guise of the military, transposed and inverted the previous understanding of the relationship between science and industry characteristic of the interwar period. The military, responding to a relative vacuum in science policy in the immediate wake of World War II, moved to retain access to the scientists who had done so much in helping them win the last conflict; then when other governmental agencies were eventually brought into play, the political situation dictated that military innovations and military funding would remain the dominant consideration in science organization. The American government had destabilized the presumptions that ruled prior to 1940, and in altering its stance toward both industrial and science policy, it compelled both the corporation and the university to revise the ways in which science would be carried on in their precincts. This was the era of the now derided "linear model"—the assertion that innovations in "pure science" were necessary formal prerequisites for advances in "applied science"—and that both made their way in an orderly fashion down the pipeline until "technological development" resulted in the new products that drove capitalist expansion.[29] Under the triple imperatives of classification, rationalization, and projection of ideological superiority, the military refined the "purity" of the laboratory in a different crucible. As an unintended consequence, the change in regime underwrote a conviction, almost a dogma, that science and commerce should never mix, even though this flew in the face of a previous generation's experience. Gaining a better perspective on the Cold War regime will go some distance in dispelling the fruitless standoff between the neo-Mertonians and the economic enthusiasts, and the stage 1/stage 2 mind-set with which we began this paper.

The wartime experience of the OSRD/NDRC and the immediate postwar debates over civilian versus military control of science have been superbly covered by the present generation of historians, so it need not be recapitulated here. What perhaps has been missing from these accounts is the ways in which the militarization of science had an impact on the previous regime of corporate science, as well as the ways in which the American university was forced to reorient itself in order to occupy the space cleared for it within the postwar settlement. The most obvious alteration was the intrusion of the government as the third and now largest player in the funding and management of science, but this implied something more than slinging largesse at a few favored natural sciences. It involved subscribing to a tenet that politicians were often loathe to admit, given their redoubled allegiance to the virtues of market organization: that the federal government was in the business of picking winners and losers in the realm of technological development by running a *sub rosa* industrial policy under the auspices of the military, which included promotion of a very different set of practices than had held sway before the war regarding intellectual property and antitrust. Meanwhile, the corporation was growing in power and reach, given that many of its European competitors had been hobbled by the war. Both the government and the corporations were impressed by the efficacy of science in winning the late war; it was taken as a given that it would also play a pivotal role in winning the Cold War.

The Cold War is now regarded as the golden age of the Chandlerian firm. The line-and-division mode of management had proven its mettle during the war; through the 1970s the roster of the hundred largest American corporations displayed amazing stability; since a certain equilibrium had been reached in the control of their core markets, the new watchword became "diversification." Dominant firms in mature industries sought to grow by buying up new product lines and moving into newer industries, and the M-form, or multidivisional bureaucratic managerial structure, spread throughout the corporate sector (Lamoreaux et al., 2003). As corporations became less tied to single product lines or nominally related competencies, the role of the corporate laboratory began to shift. Industrial science still assumed many of the functions it had done prior to World War II, such as routine testing and product improvement. Yet the increasingly multidivisional or conglomerate nature of the firm dictated that each division should become its own profit center and that funds would be allocated within the firm according to criteria applicable across all divisions. Here is where the military takeover of science policy came into play. Not only did military funding come to dominate academic science, but it also reorchestrated a major portion of industrial or commercial science (Graham, 1985).

Because the American military did not set out with deliberate forethought and intention to become commander-in-chief of science policy in America but rather found itself backing into the commitment fitfully and by degrees, it had to be flexible about experimenting with various methods to fund and manage the scientists whom it wished to keep on retainer and, in the process, invented many new configurations of laboratories. Many point to the Manhattan Project as the first decisive American military experiment with science organization. Although the original OSRD contracts were run through universities as the research entities, soon it was decided that the industrial-scale centrifuges and uranium enrichment research at Clinton, Tennessee, and the Hanford Site would be contracted out to private firms—in that case, DuPont. The postwar legacy institutions at Oak Ridge, Los Alamos, Argonne, and Brookhaven were set up as something else that had been resisted throughout the previous regime: government-run "national labs" funded directly by the Atomic Energy Commission (Westbrook, 2003). Other sorts of research were deemed to require something other than a university or corporate setting, and so the Air Force and the Ford Foundation concocted a university campus without students or faculty combined with a nonprofit Santa Monica beachfront resort at RAND in 1948 and thereby innovated the think tank. Finally, in the critical areas of aerospace, electronics, and missile development, it was decided that R&D had best be done on a strictly commercial basis, and there the military took the fateful step down the road of subsidizing corporate R&D in areas where it believed there was a compelling national interest in maintaining supremacy at the forefront of research.[30]

The dramatic reorientation of the in-house corporate lab from an internally oriented product development agency to an external research contractor had profound implications. First and most significantly, the ability to attract military funds reconciled the corporate lab with the M-form corporation in that the lab could (and often did) justify its divisional status by capturing its own streams of external revenue. However, for this to happen, the corporate science lab had to be brought into line with the rather different protocols of accounting, control, and intellectual property propounded by their military patrons. Recently, Glen Asner made the interesting argument that a series of accounting, tax code, and procurement regulations imposed by the military over the 1950s "provided incentives for the corporations to restructure their research programs on the basis of the linear model" (2004). For example, the Procurement Act of 1947 effectively perpetuated the wartime innovation of the cost-plus contract in the realm of military R&D. The Department of Defense did not mind funding what would be dubbed "basic research" in the aftermath of World War II, because their regulations concerning overhead would putatively allow them to control the mix of basic and applied as they saw fit, and the 1954 tax code revisions allowed accelerated write-offs of new investments in research infrastructure, which DoD sought to encourage. Here we observe that the basic/applied distinction, far from mapping preset divisions between universities and industry, was inscribed in the very contracts that propagated it, largely through a myriad of nearly invisible stipulations concerning the economic provisioning of research.[31] Far from mere boondoggles, these practices had the dual effects of allowing a greater degree of disjuncture of the research of the corporate lab from the activities of other divisions of the same corporation, while at the same time allowing the lab to be structured more along the lines of the university. (The fact that the model had historically come full circle undoubtedly rendered the transition easier.) Corporate labs were consolidated at locations remote from production facilities on campus-style settings, often justified by levels of secrecy and classification also demanded by the military. Scarce postwar research personnel were often courted with promises of university lifestyles and a fair amount of autonomy with regard to research agendas. Bell Labs, Xerox Parc, IBM Yorktown Heights,[32] RCA Sarnoff, Westinghouse Pittsburgh, Merck Rahway, and other labs became powerhouses of basic research, often enjoying substantial autonomy in setting their own research agendas. "A two-class system (military and nonmilitary) developed, with the best and brightest concentrated in the military class" (Hounshell, 1996: 49). And the investment began to pay off in a more "academic" modality: between 1956 and 1987 twelve corporate scientists won Nobel Prizes (Buderi, 2000: 110). Was it therefore so very odd that even the community of corporate scientists came to subscribe to something like the linear model, since everything seemed inclined to ratify its existence?

Although it was not the intention of the American military to render the industrial research lab transformed so that it would more closely resemble the university science facility, it was their intent to channel research in such a manner as to conduct what has been sometimes called a "stealth industrial policy."[33] Specialists in funding agencies like the ONR, AEC, and DARPA thought they could predict which industries were making use of cutting-edge science to produce the technologies of the future; under the imperative of national security, they could justify their interventions to make their own predictions come true. Their successes in the areas of quantum electronics, solid-state physics, and computers are well known, but there were also significant initiatives in pharmaceuticals, radiobiology, meteorology, and catalysis. Not only did the government back select horses in the derby; they dabbled in equine husbandry as well. Through a combination of intentionally weakened intellectual property rules and fortified antitrust practices, they sought to breed a corporation better suited to withstand the chill winds of the Cold War.

The American military had publicly pledged its troth to the magic of the market but generally was not willing to entrust mission-critical aspects of weapons development or considerations of national security to the vagaries of the free market. The postwar innovation of systems management was constructed to *plan* invention (Johnson,

2002). In particular, the Cold War regime witnessed a policy of striking mitigation of intellectual property rights in areas where the military was directly involved in science management. Starting with the Atomic Energy Act of 1946, the government asserted a policy to retain patent rights deriving from military-funded research, but only to make any such inventions that arose available to American firms on a nonexclusive royalty-free basis.[34] The policy was both chauvinistic, in the sense that national security dictated the subsidy of American firms, and also antimonopolistic, in the sense that national security would be compromised if the military were to become inordinately dependent on any single firm. Such considerations also governed the "second source rule" promulgated by the Department of Defense, which conveyed the intellectual property surrounding critical weapons systems or military technologies to a second competitor firm, so that the fortunes of no single producer would constitute a bottleneck.

Not only was the military skeptical of the virtues of strong protection of intellectual property in frontier science, but so too were the economic experts that (for a time) dominated antitrust policy in the United States. In the 1940s the Department of Justice adopted the position that one of the more deleterious effects of monopoly was the suppression of technological innovation, and it filed suits against some of the nation's most high-technology companies of the time, such as DuPont, Alcoa, IBM, and General Electric. Compulsory licensing of patents became for the first time a common element in antitrust settlements (Hart, 2001: 928). The effect of these policies, in concert with military regulations, was to induce firms to pull back to some extent from acquiring the promising technologies of would-be competitors, or to play down the aggressive pursuit of patent infringement cases against major rivals, and to pour more of their resources into their own in-house labs. The result, under the banner of national security, was an oxymoronic regime of relatively open science hedged round by classification and secrecy.

It is through this Cold War lens that we can better understand the ways in which academic scientists could come to believe in the independence and isolation of their ivory tower. The military was convinced that encouragement of a certain format of higher education was an indispensable complement to the protection of national security. In stark contrast with the prior erudition regime, postwar public policy was aimed at sustained subsidy of academic science beyond the narrow scope of few private universities, although those fortunate few also benefited immeasurably under the new regime. Indeed, it might be suggested that only during the Cold War did the totality of economic sectors embrace higher education as an exercise in American nation-building, with all that might imply: mass education, a diversified research base, a democratic ideology, open science, and the open propagation of research results. The military played a major role in fostering this system, primarily through the innovation of overhead payments on research grants but also through more fleeting initiatives such as the GI Bill and generous fellowships. The objective was to fuse teaching and research into a single symbiotic system, held together by the glue of generous funding.

It was a fateful decision early on at the OSRD to keep a high proportion of contract research tied to university settings and to reconcile university administrators to that fact with lavish subsidy. Vannevar Bush arbitrarily proposed overhead payments of 50 percent of labor costs for university research grants (although his real allegiance was demonstrated by the 100 percent rate proposed for corporations); although the magnitudes of the subsidy were the subject of some controversy during the war, universities learned to deal with the inconveniences of having to subject these payments to bureaucratic accountability and oversight (Gruber, 1995). Although some university administrators were convinced that the postwar period would return rather quickly to the erudition regime's dependence on industrial contract research, other more visionary captains were impressed by the sheer magnitude of military largesse. As Robert Hutchins of the University of Chicago admitted in a memo in June 1946, "It seems likely that within the next five years the Government will become, directly and indirectly, the principal donor of the University."[35] Those who were willing to go along with the drastic shift in patronage thereby stood a chance of stealing up on their rather more hallowed and prestige-laden competitors. MIT notoriously took advantage of the opportunity to climb the league tables (Leslie, 1993). In 1946, Stanford managed to accumulate military contracts that were twice the value of its contract research during the entirety of World War II (Lowen, 1997: 99).

It may seem that the saga of the Cold War regime could be sketched entirely without consideration of the role of the private foundations, but this would not be altogether valid. Older foundations continued programs of academic subsidy, and a few new players, like the gargantuan Ford Foundation, came on the scene (Raynor, 2000). However, a government crackdown on the use of foundations as tax shelters in 1950, combined with the fact that even the largest foundations could not begin to match the magnitude of impact of the federal government on higher education and science, meant that most foundations scaled back their ambitions concerning the management of science in this era. For instance, in 1960 the Ford Foundation was channeling more support to American universities than the NSF was, but by 1970 it had all but withdrawn from the support of academic science (Geiger, 1997: 171). Foundations became notorious for their fickle initiatives, which could disappear with each executive change; they were no longer participants in science management for the long haul.[36]

Hence, the American Cold War regime was largely structured as a concertedly nationalized system of science, but one whose ideological significance was so highly charged that it had to be presented as an autonomous and autarkic invisible college of stalwart stateless individuals who need pay no heed to where the funding and institutional support for all their pure research was coming from. "Purity" had become conflated with "freedom" and "democracy"; "science" stood as the embodiment of all three states of virtue; and American science organization was promoted as a rebuke to the Soviet machine, but equally it was thought to stand as reproof to anyone who sought to make science submit to an imperious political master.[37] It was only within the Cold War regime that "academic freedom" really seemed to possess sufficient *gravitas* to actually be used in an effective defense of academic tenure—something we can now appreciate in the era of its disappearance. The researcher had only to answer to his disciplinary peers, or in the last instance, to his individual conscience, and feel an enlightened disdain for the hurly-burly of the marketplace—at least until the DARPA grants officer came to call.

The Globalized Privatization Regime The advent of the globalized regime of privatized science was not heralded in an unmistakable way by war or depression, as were the previous regimes. A superficial perspective might seek the watershed at the fall of the Berlin Wall, since, after all, it was the dramatic event that signaled the cessation of the Cold War. However, if we triangulate corporate evolution, educational transformation, and government policy, the inauguration of the privatization regime in America would have to be located a decade or so earlier.

Economic historians, legal scholars, and science studies researchers all tell the story in somewhat different ways, but it is significant that all trace the metamorphosis back to roughly 1980.[38] The trigger seemed to be the widespread conviction that the United States had lost ground to international competitors during the oil crisis and economic slowdown of the late 1970s. Although there was substantial disagreement over the causes of the supposed sclerosis, an array of initiatives were crafted to defeat the diverse culprits sapping America's economic dominance. One major candidate for economic reform was the organizational structure of the Chandlerian corporation (Lamoreaux et al., 2003, 2004; Langlois, 2004). Various participants had become convinced that the huge managerial conglomerate had become too unwieldy to effectively compete in the world market in the 1970s, and the 1980s were the era of hostile takeovers, leveraged buyouts, and shareholder attacks on the top management of large corporations. In response, there was a significant retreat from diversification within firms, with one calculation suggesting that by 1989 firms had divested themselves of as much as 60 percent of acquisitions made outside their core business between 1970 and 1982 (Bhagat et al., 1990). There was also a retreat from previous levels of vertical integration in industries like automobiles, computers, telecommunications, and retail. Consequently, corporations began to equate agility and nimbleness with repudiation of hierarchical managerial control of process, and with it the M-form paradigm, and thus sought to reengineer the supply chain to depend to a greater extent on market coordination.[39] Networks of subcontracts began to displace ownership ties as modes of organization; venture capital began to channel investment into start-up firms. Labor-intensive heavy manufacturing was outsourced to low-wage countries. Moreover, the roster of America's largest corporations underwent severe shakedown, after having enjoyed relative stability for the previous sixty years. The lumbering giants were prodded into defensive action, which was widely interpreted as a return to market methods of coordination (Langlois, 2004).

Another important initiative was deployed in the arena of organization and control of international trade. In a far-sighted mobilization, a handful of representatives of corporations located in high-tech industries such as pharmaceuticals, semiconductors, computers, and entertainment formed the International Intellectual Property Alliance in 1984, for the purpose of linking issues of intellectual property to larger trade nego-tiations.[40] They succeeded beyond their wildest ambitions, using the Uruguay Round of negotiations over the General Agreement on Tariffs and Trade to impose U.S. stan-dards and levels of intellectual property protection on developed and developing countries alike and to enforce them with trade sanctions through the World Trade Organization. TRIPs (trade-related aspects of intellectual property rights) came into force on January 1, 1995, and has implanted the basic legal premises of the global-ization regime to all corners of world, refashioning academic and corporate activity in the interim.[41] Although focused on the seemingly narrow legal playing field of intel-lectual property, TRIPs might be regarded as one facet of an even larger concerted political movement to weaken the prerogative of national governments to exert regulatory control over the corporate entities within their boundaries, all in the name of liberalization of trade and protection of foreign investment. In any event, U.S. manufacturing capacity was shifted to lower-wage countries in search of a quick productivity boost, and manufacturing job losses accelerated from the late 1980s (Burke et al., 2004).

These major restructurings of the corporate sector coincided with a crisis in the sphere of higher education. After 1975, enrollments in U.S. higher education ceased to grow for the first time in U.S. history, while cash-strapped states began to contract their funding (Geiger, 2004: 22 ff.). The military, under pressure to reduce funding of projects not immediately relevant to its mission, had been attempting to withdraw from many of its commitments to the funding of academic science in the 1970s, so universities suffered a double deficit, with no end in sight. To maintain graduate enrollments, many departments in the sciences began to admit rising proportions of foreign students (NSB, 2004: 5–25). While this had salutary effect on the rather parochial atmospheres of American university towns, it also had the deleterious effect of revealing the essential bankruptcy of the Cold War justification of education as serving the objectives of state-building. Many of the students in technical areas were not citizens of the United States, and periodically some politician would ask what uni-versities were doing training the work force of potential competitors at American expense. But more to the point, the whole idea of an informed citizenry and skilled workforce began to lose salience as more and more production activity was shifted overseas and corporate managerial cadres became more international. The university was losing its grip on its previous social *raison d'être*, even as it remained the preferred path for individual economic advancement. It also, in an ironic twist, was revamped in a Chandlerian direction, even as many corporations were fleeing that organizational model in droves. Significant aspects of faculty governance were diminished or dis-mantled altogether (Geiger, 2004: 25) and were replaced with top-heavy managerial hierarchies that multiplied divisions, institutes, and other offices, often in the name of rationalization and cost-saving. University finances were more directly addressed by replacing tenured faculty with temporary labor and part-time teachers, reversing the Cold War tendency to unite teaching and research as mutually reinforcing activities.

Then there was the overt political attempt to bring the hobbled universities more into line with the newly reengineered corporation. It has become *de rigueur* for commentators noting the commercialization of science to bow in the direction of the Bayh-Dole Act of 1980 as a major turning point in the treatment of intellectual property in the United States, because it allowed universities and small businesses to retain title to inventions made with federal R&D funding and to negotiate exclusive licenses.[42] Actually, the historical situation with regard to intellectual property was much more complex, and yet, the end result was almost a complete reversal of practices under the Cold War regime. First off, universities had been permitted on a piecemeal basis to patent federally funded research via individual institutional patent agreements since 1968 (Mowery et al., 2004: 88). Only in 1983 was Bayh-Dole style permission extended to large corporations, their real intended beneficiaries, by a Ronald Reagan Executive Memo—the better to fly under journalistic radar (Washburn, 2005: 69). Second, Bayh-Dole was only one bill in a sequence of legislation throughout the 1980s that expanded the capacity of corporations to engage in novel forms of collaborative research while capturing and controlling their products (Slaughter & Rhoades, 2002: 86). For instance, the Stevenson-Wydler Act of 1980 opened the door to commercialization of research performed at the national laboratories. The National Cooperative Research Act (NCRA) of 1984 shielded corporations from antitrust prosecution when engaged in joint research projects. The National Technologies Transfer Act of 1989 allowed federally sponsored research facilities to spin-off previously classified research to private firms. Over the same period corporations sought and won numerous laws to strengthen both patent and copyright, and in 1982 they managed to have a special Court of Appeals in the Federal Circuit dedicated to patent cases. The scope of what has been deemed susceptible to patent in America has been progressively broadened, and challenges to the legitimacy of patents have become less successful.[43] The very notion of a public sphere of codified knowledge has been rolled back at every point along its perimeter, initially by blurring the lines between public and private property. This hyper-restrictive system of intellectual property has then been exported to the rest of the world under the aegis of the WTO and the World Intellectual Property Organization, as outlined above.

The concerted fortification of intellectual property was accompanied by the weakening of antitrust policy—an exact reversal of the Cold War regime. Absolution was not just granted in the specific case of the NCRA but more generally under the influence of the Chicago School of law and economics, monopoly was increasingly downgraded as a source of inefficiency or political danger in the viewpoint of the Justice Department (Hart, 2001; Hemphill, 2003; van Horn, unpublished). The doctrine was propounded that monopoly was not necessarily harmful to innovation (even in the case of *United States vs. Microsoft*), that size of R&D budget was not correlated with demonstrated ability to innovate, and that good products win out in the end, no matter what the industry structure. In any event, defenders could point to the increasing resort to cross-licensing and joint ventures to suggest that there was no return to the bad old days of trusts and patent pools (Caloghirou et al., 2003). Rather, a fortified and unfettered corporate sector free to contract for research when and where it saw fit was thought to be one of the best prophylactics against upstart foreign producers and looming national economic decline.

The cumulative consequence of all these convergent vectors was a fateful restructuring of the American corporation and the most important revision in the organization of science within the regime of globalized privatization: the relative demise of the in-house corporate research labs, and the spreading practice of the outsourcing of corporate research.[44] It is here, and not in any vague shift in the *Zeitgeist* or narrative of the rationalization of technology transfer, that we find the root cause of the new model of commercialization of science in the twenty-first century. Although each trend we identified above was not deliberately attuned by itself to bring about the destruction of the in-house corporate lab, each contributed to its demise. It is important to understand the ways in which the withdrawal of the military from science management, the perceived failure of the Chandlerian firm, the push to globalize the neoliberal Washington consensus, and the crisis of higher education all converged on the corporate lab.

Pundits in business schools often attribute the passing of the large corporate lab to the supposed empirical observation that big in-house research labs do not deliver the goods (Anderson, 2004), usually accompanied by reference to some neoliberal doctrine that in the long run healthy science resists being planned, but this superficial analysis ignores the fact that the labs had been weaned from their internalist parochial commercial orientations by military contracts during the Cold War (Graham, 1985). The corporate labs had been permitted to maintain their external orientation and unfettered curiosity and campus ambiance as long as they were revenue centers for the firm, but when the military withdrew from the organization and funding of basic research, the semiautonomous corporate lab became a liability. In a more forgiving environment, perhaps they might have been reoriented more concertedly to the development side of R&D and persuaded to renounce the linear model of technological change, but by the 1990s they ran up against the anti-Chandlerian movement to divest the firm of its extraneous product lines and scale back on vertical integration. In many corporations, the research division was a prime candidate for downsizing or spin-off, and that is precisely what happened throughout the 1990s. RCA Sarnoff was first sold off to SRI International, and soon thereafter spun off as Sarnoff Corporation in 1987. AT&T slashed research at Bell Labs starting in 1989, only to spin off the remnant as Lucent in 1996 (Endlich, 2004). Westinghouse Pittsburgh was first decimated and then sold off to Siemens. Research divisions disappeared altogether at firms such as U.S. Steel and Gulf Chevron. By 1995 IBM had eliminated a third of its research budget, essentially gutting its flagship Yorktown Heights facility; other units, such as its Zurich laser group, were spun off as separate firms. After the merger of Hewlett-Packard and Compaq and the spin-off of Agilent, the renowned HP Labs were slated for reorganization and downsizing. The historian Robert Buderi, who has been most concerned to document this phenomenon, admits that research directors regarded it as a "research bloodbath" in the late 1980s and 1990s (Buderi, 2000: 22), but he has sought

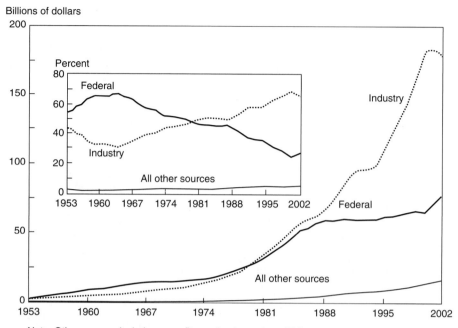

Note: Other sources include nonprofit, academic, and non-U.S. government.

Figure 26.1
U.S. R&D by source of funds: 1953–2002.

to paint the bloodletting as a proscription for both corporate and scientific health. The problem with this diagnosis is that it is too narrowly focused on the individual firm in isolation, and it ignores the larger system of the funding and organization of science. Buderi writes, "We now see less basic research going on. IBM does not chase magnetic monopoles anymore, but should it have done so in the first place?" (2002: 249) This presumes someone somewhere else will take up the chase for magnetic monopoles, and someone else will worry about where and how that will happen. But this question of who organizes which science to what ends is precisely the debate that is glaring in its absence in the globalized privatization regime.

The downsizing and expulsion of in-house corporate labs have not implied a corresponding contraction of private funding of research and development in America; quite the contrary. In a pattern that has been mimicked with a lag in other countries, in the United States, federal R&D expenditures as a proportion of the total R&D has declined continuously since the late 1960s, while the proportion of R&D expenditure originating in the industrial sector has increased from the same period, surpassing the federal proportion around 1980 (NSB, 2004: 1–11) (figure 26.1).

If corporate labs are being slashed, how could this be? The resolution of these seemingly contrary trends is that the increased volume of research is being performed outside the boundaries of the corporations funding it. Some of it is being performed in other corporations purpose-built for research under the new regime, while the rest is increasingly performed in academic and hybrid settings, like research parks and quasi-academic start-ups. It is precisely at this juncture that the other historical trends described above of the globalization of corporate trade and investment, and the crisis of the research universities, come into their own.

The breakdown of the Chandlerian model of the hierarchical integrated firm has prompted the nagging question, why integrate R&D into the firm when you can buy it externally and reduce costs by doing so? But that question presumes that dependable R&D is a distinct fungible commodity in a well-developed market, one so competitive that it can lower the costs relative to doing it in each firm. A major thesis of this chapter is that, no matter how "commercialized" science may or may not have been in the previous American science regimes, until recently it was this state of affairs that was uniformly absent. The strengthening of intellectual property, the weakening of both domestic antitrust and the ability of foreign governments to counter corporate policies, the capacity to shift research contracts to lower wage and easier regulatory environments and therefore engage in regulatory arbitrage, the availability of low-cost real-time communication technologies, and the presence of an academic sector which was willing to be restructured to surrender control of research to its corporate paymasters—all these were necessary prerequisites to seriously countenance the corporate outsourcing of research on a mass scale.

The globalization of corporate R&D is a characteristic hallmark of the new regime. Of course, multinational companies headquartered in smaller countries like the Netherlands and Switzerland have internationalized their R&D activities essentially from their inception; but the more striking trend is the international outsourcing of research across the board since the 1980s (Reddy, 2000: 52). Just as with recourse to academic capacity, global outsourcing tends to be concentrated in a few industries, such as pharmaceuticals, electrical machinery, computer software, and telecommunications equipment. Nevertheless, surveys within these industries reveal a sharp increase in research carried out beyond the home country's boundaries from the 1960s to the 1990s (Kuemmerle, 1999). A more recent survey by the Economist Intelligence Unit reveals the globalization of R&D gathering pace over the 1990s, with over half the respondents indicating that they would expand their overseas R&D investment in the next three years. When queried as to the major considerations governing their decision, the most popular responses cited strong protection of intellectual property, lower costs and the tapping of indigenous research capacities. It is the access to lower wage labor in the context of an academic infrastructure, which is disengaged from any corporate obligations to provide ongoing structural support for local educational infrastructure, that explains the shift in research funding to countries like China, India, Brazil, and the Czech Republic (European Commission, 2003–2004: 9). Outsourcing demands disengagement from earlier narratives of the ways in which capitalist firms and economic development depend on scientific advance. Another way to cut costs is to absolve the firm from nationalist appeals to help support scientific infrastructure, accompanied by improved opportunities to further reduce or avoid corporate taxation.

Approaching the commercialization of science from this angle profoundly revises the usual narrative of the privatization of modern academic science as a straightforward case of cash-strapped universities following the money, albeit with a few nagging qualms concerning the propriety of telling corporations only what they want to hear.[45] Rather, a new STS historiography might be proposed wherein many of the novel institutions of globalized privatized research were first pioneered *outside* the academic sector *per se*, especially as adjuncts to the modification and reengineering of the modern corporation, and only then foisted on universities, themselves forced to react to these benchmark citadels of the new globalization regime in their own internal restructuring of scientific research.[46] Government revisions of policies with regard to intellectual property or educational subsidy may have constituted incentives for changes in the universities, but they could not unilaterally impose the structure of the new regime. While legislation such as the Bayh-Dole Act was enabling, it should not be confused with the *cause* of the modern privatization of science, which was instead attributable to the larger shift in the nexus of science management and funding.[47]

Indeed, one of the great unspoken presuppositions of modern commentators on the commercialization of science is that either the scientist or the community at large is still capable of choosing "how much" public open science one wants to preserve, while leaving the remainder to be covered by the private sector.[48] Contrary to this presumption that one can rationally choose a menu in any combination from columns A and B, rather once the institutional structures of the globalized privatization regime have been put in place, the very character and nature of public science is irreversibly transformed. The recent rivalry between Celera and the public Human Genome Project is an illustration. In a fascinating journalistic account of the race, James Shreeve clearly finds Craig Venter a more compelling protagonist than Francis Collins, and he equally clearly subscribes to modern neoliberal doctrine that the free market produces better research more cheaply than does the hierarchical managerial model, and yet almost in spite of himself, demonstrates in numerous ways that once the commercialized Celera entered the arena, the public genome project found itself buffeted and transformed in mutations beyond its control. For instance, because it was committed to open science, the faster the Human Genome Project went, the more it ended up helping Celera beat it to the finish line (Shreeve, 2004: 198). In another instance, no matter how loudly Venter trumpeted that Celera was not costing the taxpayer a single dollar, if anything his project depended on public subsidy in ways more elaborate but more dubious than did the public genome project, in that at least the latter was subject to some forms of public accountability that Celera could effortlessly flout. Knowledge was secondary for Celera; it was fencing off the genome that trumped everything else. "The key was for Celera to be proactive, to grab as much potential intellectual property as possible and sort out later who really owns what. Celera was getting a late start" (Shreeve, 2004: 231). We might also worry that the quality of the "finished" genome

was substantially degraded by the various stratagems induced by the public/private
rivalry carried on in the glare of journalistic scrutiny. The ultimate irony was that Craig
Venter the consummate entrepreneur still conceded too much of his proprietary information to other scientists to suit his own paymasters and was unceremoniously ejected
from Celera in January 2002, after the news spotlight had moved on.

Toward a More Global Perspective   Thus far we have offered predominantly "American"
perspectives on the economics of science. This does not preclude a parallel consideration of science funding and organization in other countries, making allowances for
their own special historical circumstances. (It gives pause to think how our narrative
might have differed if America had suffered widespread indigenous devastation in
World War II, for instance.) In fact, some forms of anecdotal and narrative evidence
concerning alternative national idioms of the organization of research support our
periodization concerning the various regimes for Europe as well. For example, the
European version of the "captains of erudition" regime has been comprehensively surveyed by Fox & Guagnini (1999); in some instances (Germany) the academic lab pre-
dated the industrial lab, in others they were simultaneously constructed. A specific
illustration is NatLab, the Philips Physics Laboratory in the Netherlands, which was
established in 1914 (Boersma, 2002). Its founding director, Gilles Holst, innovated an
academic environment by organizing lectures by top scientists and stimulating congress participation and academic publications by the laboratory's own scientists. In
addition, NatLab significantly shaped the pedagogy of the technical physics degree at
the Technical University of Delft in the Netherlands. For a European version of the
Cold War regime, we could readily point to CERN, the European Organization for
Nuclear Research, which is the world's largest particle physics laboratory. It was created
at the height of the Cold War, in 1954, in an effort to rebuild European physics to its
former grandeur, to reverse the putative brain drain of the brightest and the best to
the United States, and coincidentally, consolidate postwar European integration
(Pestre & Krige, 1992). Postwar state-building and the nurture of indigenous research
capacity joined hands in ways very similar to those found in the United States. It could
be noted that (for example) investments in nuclear processing technology and aerospace (Concord) were perhaps less driven by military managers than in the United
States, although this would require further scrutiny. The end result was, however,
similar to the framework outlined herein.

  As for the globalized privatization regime, the case for parallel developments
throughout the world is even easier to make. The logic of the spread of multinational
corporations would seem to suggest the possibility of something like convergence of
diverse "national systems of innovation" to a relatively uniform "advanced" transnational model of the commercialization of science, especially since barriers between differing economic systems collapsed after the fall of the Berlin Wall. European
corporations have experienced their own twilight of the Chandlerian firm;[49] NATO
and indigenous member militaries have withdrawn from previous levels of an active
role as science manager. The outsourcing of corporate R&D has become a global

phenomenon (Narula, 2003: chapter 5). At the same time, there has been a trend toward convergence in university systems over the last two decades.

In response to a common perception that the United States had been outpacing Europe in science and technology, the European Union has been on the forefront of fostering the reengineering of institutions of research. Worried about the so-called European paradox, which is an observed discrepancy between the role of Europe in global scientific production and its place in the production of patented inventions, it has stepped up its involvement in science and technology policy (Larédo & Mustar, 2001). Since economic development has become a dominant issue of the EU agenda and because the pursuit of competitiveness is a major *raison d'être* of the Union, a shift may be observed from more traditional science and technology policy toward innovation policy (Borrás, 2003). In fact, the transfer of powers from the national to the European level and the knowledge production-appropriation-exploitation-consumption chain view adopted by the EU despite increasing public resistance may themselves be seen as artifacts of the globalized privatization regime. In the process, government, or national, laboratories such as the CNRS in France and the Max Planck Gesellschaft in Germany are perceived as losing ground to universities and encouraged to promote or renew their links with the economy and with society (Larédo & Mustar, 2001). These changes must further be situated in the context of budgetary retrenchments and privatization programs of neoliberal governments, an obsession with the knowledge-based economy, and the hasty replication of Bayh-Dole style legislation throughout the EU.

On the pedagogical front, the aim has been to move higher education in Europe toward a more "transparent" system and to create a European Higher Education Area attractive to the rest of the world.[50] Critics have argued that this Area has no coherent pedagogical or intellectual basis, is concertedly antiegalitarian, and its framework is mostly a money-saving measure geared to reduce severe overcrowding by getting students out of the classroom and into the workforce. On the research front, the avowed aim of the European Union has been to become the most competitive and dynamic knowledge-based economy in the world. These efforts originated with the Lisbon European Council of 2000, which sought to bridge the gap with the United States and Japan by coordinating member research activities and laying the foundation for a common science and technology policy across the European Union (João Rodrigues, 2002, 2003).[51] The goal is to create a European Research Area, which is referred to officially as "an internal knowledge market," considered to be an R&D equivalent of the "common market" for goods and services, and intended to establish "European added value," by 2010. In the process, the European Union is to increase its global expenditure on research to 3 percent of GDP—one and one-half times the current level—by 2010 and to encourage the share of research funded by business to rise at the same time.

The globalized privatization regime has created a radically transformed science landscape in Europe (Larédo & Mustar, 2001), with shifting CGE alliances. Calls for a renewed social contract of science, for value for money, and similar neoliberal themes originated in the United Kingdom and spread to the Continent soon thereafter. Bell-wether reforms in the United Kingdom include the Research Assessment Exercise (RAE) and the Teaching Quality Assessment (TQA) (Hargreaves Heap, 2002).[52] The former was introduced in 1986 as a mechanism of control, to enable U.K. higher education funding bodies to distribute public funds for research selectively on the basis of neolib-eral criteria, by making universities accountable for their use of public money. It is conducted roughly every four years, and since 1992 about £5 billion of research funds, which constitutes the bulk of the research component of the so-called block grant, has been allocated through the RAE. The effect of the RAE has been to concentrate funds at a few elite institutions as well as to redirect research activity in a more com-mercial direction. Since the survival of universities now depends on their RAE score, most effort at British universities goes into augmenting scores, while teaching gets neglected.

Evidence of convergence to a transnational model of the commercialization of science abounds in the EU. For instance, after two decades that witnessed an exodus of top students and scholars as well as a decrease in government support per student by 15 percent, Germany now plans to form a group of ten American-style elite uni-versities in 2006 and award almost $30 million a year for five years to increase their competitiveness and quality (Bernstein, 2004; Hochstettler, 2004). Its attempts at American-style reform further involve the establishment of alumni organizations to raise money, selectivity in admission of students, and payment of faculty salary based on performance, running counter to Germany's long-held egalitarian ideal.

It is even more significant to observe the effects of the regime of globalized priva-tization on the country that had been used to justify the Bayh-Dole reforms in the United States in the 1980s. In the period of its perceived technical superiority, Japan displayed none of the trappings of the neoliberal marketplace of ideas: universities could not own patents, IP was handled informally between firms and professors, and direct industry funding of university research was small. It was no accident that state control of science funding and education were reminiscent of the Cold War pattern: they had been imposed by the American Occupation authority. However, when the Japanese economy went into a stall in the 1990s, the very same system of science man-agement that had been praised for its postwar economic success was then used as an explanation of its stagnation. Invidious comparisons with the neoliberal regime abounded, leading to Japan adopting many globalization devices, such as a 1998 law encouraging the opening of technology licensing offices at universities, a 1999 Japanese version of the Bayh-Dole Act, and a loosening of restrictions on civil servants participating in new commercial start-ups.

Further experiments in privatization of research in Japan have ventured even beyond their supposed American inspiration (Brender, 2004; Miyake, 2004). In a sig-nificant departure, as of April 2004, national universities are being themselves priva-tized into independent administrative agencies, tantamount to being transformed into independent public corporations. University-industry relationships, which had been previously based on a dense web of informal personal ties, were being rooted out and forced to be recast as more "market-mediated" structures. This has essentially ignored the fact that patents and IP had been used by Japanese corporations in very different ways than in America; for instance, antitrust had never played the role in Japan that we described above in the U.S. regimes. Nevertheless, the neoliberal notion of an institution-free marketplace of ideas seemed to overwhelm such considerations.

The neoliberal ascendancy has since spread throughout Asia. Many Chinese universities have rushed to nurture start-up firms under pressure of the country's open-market policy, straining the national budget and causing drastic cuts in university funding. Already some Chinese universities are operating "companies" on their own. Education reform in Japan, now emulated in successful economies like China, while presented as a way to give universities more autonomy, is mostly an excuse to reduce state financial support. Changing governance structure plunges universities into an uncharted and ill-conceived era of competition, where government tries to have it both ways in cutting subsidies but retaining control by setting up evaluation committees and advisory boards. These are global symptoms of what has been called the modern corporate audit culture (Apple, 2005).

In short, there appears to be a concerted global effort to imitate the putative advantages of the United States regime in advanced science and technology, while at the same time limiting the amount of government support and reducing the accompanying bureaucracy. Universities have responded to these incentives by raising money from the private sector, conducting more privatized research, and shifting their attention away from teaching. In the process, they frequently have encountered cultural barriers due to the fact that non-American universities have performed a different societal role than those in the United States and exist in a somewhat different CGE environment. We will return to the consequences of these global developments for our analysis of the shifting alliances among the commercial corporation, the state, and the university in the conclusion. Before doing so, we now turn to the responses of the STS community to the developments outlined above.

ALTERNATIVE MARKET MODELS OF THE CONDUCT OF SCIENTIFIC RESEARCH

Ever since the field of science studies broke away from Mertonian sociology of science, its adherents have proved reticent when it comes to discussing macro-scale structural attributes of science, which of course includes the economics of science. This made it difficult for STS to coherently discuss the framework of the Cold War regime during that era, although interest has revived in that topic in more recent times. Nevertheless, back then, the task of providing an overview of the Cold War regime devolved instead to the science policy community and thus, to a great extent, to neoclassical economists. The major proponents of market models of science from the 1950s to the 1980s were a group of economists primarily associated with RAND during that era. It was they who introduced the now-pervasive habit of treating knowledge as if it were production of a "thing," on a par with any other commodity in their analytical framework. It was they who constructed the statistics of national science policy and they who proposed models of science as an input to economic growth (Godin, 2005: chapter 15). Yet these analysts did not think of themselves as market fundamentalists but rather styled themselves as left-leaning defenders of the necessity of sustaining public science through public subsidy and maintaining academic science as effectively removed from industrial R&D. They achieved this by means of the analytical construct of the "public good."[53]

The artifice of the "public good" was one (but not the only) conceptual attempt within the tradition of neoclassical economics to justify the intrusion by the government into the marketplace by insisting that there were a few anomalous "commodities" that did not possess the standard attributes expressed in orthodox economic models. In particular, these goods would be produced at "zero marginal cost," which would suggest that standard equilibrium pricing (where price = marginal cost) would lead to the underprovision of the good, or worse, to no production at all. Often the public good was saddled with further anomalous characteristics, such as "nonrival consumption" (the condition that my consumption of the good would not diminish or otherwise hamper consumption of the same good) and "nonexcludability" (the producer could not prevent you from also using the good through standard property rights), which were cited to buttress the stipulation that markets would fail in providing adequate levels of the good relative to public demand. One still encounters extended rhapsodies on the special character of that market item called "information" or "knowledge," which are little more than unwitting repetitions of the original Cold War doctrine.[54] Although there is no necessary analytical connection, the terminology of "public good" has been frequently contrasted with "private knowledge" and used to suggest that the "public commons" or our scientific birthright is being violated and encroached on by a nefarious enclosure movement (Boyle, 2000; Lessig, 2001, 2004; Nelson, 2004).

There are plenty of reasons to think that the concept of the "public good" was never a very useful or effective tool with which to understand the economics of science in any era, much less the current one. Although it was often cited to justify the lavish public subsidies of scientific research in America during the Cold War, it mainly served to distract attention from the military and chauvinistic motives for science funding, not to mention the intricate ways in which corporate organization and academic science were intermeshed and imbricated, as described in the previous section. The treatment of knowledge as a fungible thing was also the thin end of the wedge of the neoliberal attack on putative distortions in the "marketplace of ideas," a thrust that ran counter to the prevailing portrayal of science as an activity that transcended mundane political economy. After all, in economics, public good theory only maintained that it was "inconvenient" or "inefficient" to privatize some portion of knowledge production, *not* that the institutions of scientific research would be fundamentally undermined or corrupted in crucial respects by commodification. Such notions fell well beyond the pale of any version of American economics. And then one could only maintain the fiction that science produced a tangible "thing" by not looking too closely at the actual practice of research in its social context: the idea of

"scientific method" as a free-standing technology indifferently portable to any situation was the obverse of this image. But the irony we intend to highlight here is that,
just as the "public good" concept was losing its prior rationale—both because of the
transition from the Cold War to the globalized privatization regime and because of a
trend within neoclassical economics away from treating knowledge as a thing and
toward treating the agent as information processor (Mirowski, 2002)—some segments
of the science studies community began to pick it up and adapt it for their own
purposes.

Science studies scholars were not particularly quick off the mark to notice that the
funding and organization of science had been undergoing profound transformation.[55]
By the mid-90s, it had become commonplace to observe that the average scientific
career was experiencing deformations—lengthening of the period after the doctorate
but prior to first academic position, greater bureaucratic surveillance, more soft money
positions displacing tenure-track openings, increased incidence of joint authorship (or
no credit at all), a "productivist" ethos—and that this might tend to undermine the
Mertonian portrait of scientific norms (Ziman, 1994). At that juncture, at least two
groups of scholars began to write about changes in the "mode of production" of
knowledge leading to a postacademic or revolutionary kind of science. One group has
become known as proponents of "mode 1/mode 2" analysis, while the other is retailed
under the rubric of the "triple helix."

The first appearance of the mode 1/mode 2 characterization of modern science was
the multi-authored *New Production of Knowledge* (Gibbons et al., 1994). The book did
not contain a systematic empirical survey of concrete science in any particular culture
area but rather a discursive set of observations about what it felt like to pursue a
research career in the present in what was clearly assumed to be an American or European setting (mode 2), while comparing it to a past situation (mode 1), which the
authors clearly thought would be fresh in the memories of most of their readers. That
book pointed to phenomena such as a weakened university structure, the general
erosion of the power of scientific disciplines, the atrophy of peer control as internal
guidance system, the rise of interdisciplinary research teams, and the demise of the
self-sufficient laboratory. This first book did not focus attention on the commercialization of the university *per se*, but instead portrayed research in general as becoming
forced to be more responsive to external interests and concerns. A second volume by
a subset of the previous authors, *Re-thinking Science* (Nowotny et al., 2001), ventured
further in the direction of casting mode 2 as a change in the epistemological presumptions of the actors. This time, they were prompted by critics to acknowledge some
events like the demise of the Cold War and the passage of the Bayh-Dole Act, but
mode 2 was cast in cultural categories, such as the existence of a "new form of economic rationality" (2001: 37) or a postulation that "the rising tide of individualism
in society has now reached scientific communities" (2001: 103), rather than specific
concrete economic institutions or practices. A later contribution to a symposium on
their work (Nowotny et al., 2003: 186–87) led to the following condensed characterization of mode 2:

- "Mode 2 knowledge is generated within a context of application . . . [which] is different from the process of application by which 'pure' science generated in theoretical/experimental environments is applied."

- Mode 2 is marked by "transdisciplinarity, by which is meant the mobilization of a range of theoretical perspectives and practical methodologies."

- "Much greater diversity of the sites at which knowledge is produced, and the types of knowledge produced."

- "The research process can no longer be characterized as an 'objective' investigation of the natural (or social) world . . . traditional notions of 'accountability' have had to be radically revised."

- "Clear and unchallengeable criteria, by which to determine quality, may no longer be available."

In this overview, Nowotny and colleagues also assert that their scheme provided "a more nuanced account than either of the two standard [alternatives]—characterizing commercialization as a threat to scientific autonomy (and so, ultimately, to scientific quality); and as the means by which research is revitalized in both priorities and uses" (Nowotny et al., 2003: 188). Elsewhere, one of the authors maintained that mode 2 "does not represent yet another attempt to cajole universities into behaving more like businesses" (Gibbons, 2003: 107). While the later emendations did indeed complicate the earlier versions of their argument, there was no denying the fact that "knowledge" was still being treated as a thing and a product, and that the authors had maintained a mildly positive stance toward the modern developments.

The triple helix (3H) thematic has not been so extensively codified in any particular text as the "mode 1/mode 2" doctrine, but its themes were spread throughout numerous special journal issues and edited volumes, which tended to derive from conferences convened by Henry Etzkowitz and Loet Leydesdorff. The triple in 3H referred to the insistence on looking simultaneously at the three sectors of industry, government, and academia and their interactions—a precept at first resembling our own STS analysis above.[56] However, and more to the point, Etzkowitz in particular argued that universities were experiencing a "second academic revolution," the first being the incorporation of the research function alongside the teaching function, and the second being purportedly the reconciliation of economic development with those prior two functions. "The organizing principle of the Triple Helix is the expectation that the university will play a greater role in society as entrepreneur" (Etzkowitz, 2003: 300). He envisions the genesis of an "entrepreneurial university" capable of carrying out all the requisites of commercialization without in any way impugning teaching or research, and he has repeatedly pointed to MIT as the exemplar of this novel form (Etzkowitz, 2002). In 3H, sheer entrepreneurial zeal is proposed to overcome many of the scruples that have dogged the commercialization of science: "In this information-based economy, knowledge can be a public and a private good in one and the same time" (Etzkowitz et al., 2000: 327). It appears that the new regime blurs even the institutional distinctions that would have seemed central to 3H, such as corporation/university: "the university and the firm are each assuming tasks that were once largely the province of the other" (Leydesdorff & Etzkowitz, 1998: 203). This permits adherents of 3H to be coy about whether universities are urged to simply adapt to new demands, or instead, universities and corporations as structures are converging to some single new institutional entrepreneurial entity (Shinn, 2002). By contrast to our account, the metaphorical language of "dynamics" in fact absolves most proponents of 3H from delving in detail into what sets the educational sector apart from the government or the firm in the first place, either in structure or in functions.

Both mode 2 and 3H authors have acknowledged that their "paradigms" are effectively pitted in analytical competition with one another; in one or two places, they also admit that they intend their work to "pose a challenge to STS."[57] Some science studies scholars have acknowledged the challenge, and in so doing they have not been happy with what they encountered. In a series of sharply critical commentaries, they have found both mode 2 and 3H wanting as both history and contemporary science policy, and have asked themselves what therefore accounts for the widespread attention paid to these literatures.[58] While most have indicted the mode 2 literature as lacking any demonstrable empirical component, the situation within 3H is more complex.

The mode 2 authors argue in ways more reminiscent of philosophers of science, identifying an altered "epistemology" without paying too much attention to which specific agents may experience this epiphany, much less dissecting the institutions that might foster it. The 3H authors by contrast have encouraged specific research into questions of science policy and education in a wide array of countries and culture areas (often presented at their biannual conferences), but the sense of the critics is that it still does not add up to a coherent analysis. Instead, globalization is treated as a benign diffusion of an entrepreneurial spirit to universities in the periphery (Etzkowitz, 2003: 297). For instance, when Etzkowitz writes specifically about MIT, it is "less a history than a brief for the university as an engine of economic development" (Bassett, 2003: 769). Elsewhere, 3H authors briefly note that corporate R&D is increasingly outsourced, but they reveal little curiosity about the forms it takes or what causes may be behind it (Etzkowitz et al., 1998: 55). Intellectual property issues are left insufficiently explored.

What then has proved so attractive about mode 2 and 3H? In our opinion, and that of the critics, they both provide a convenient big tent for authors who seek to "legitimate a neo-corporatist vision of the world" (Shinn, 2002: 608). In the case of mode 2, this tends to be addressed to higher education bureaucrats and scholars in the humanities located in the hegemonic developed nations, whose fears need to be assuaged:

Can the universities enter into this new closer relationship with industry and still maintain their status as independent autonomous institutions dedicated to the public good? The answer must be in the affirmative (Gibbons, 2003: 115).

In the case of 3H, in our experience it tends to appeal more to scholars located in developing countries, or else scholars located in peripheral areas of the developed world. They tend to be much more directly active in local science policy and cannot take acceptance of the neoliberal dogma of free market globalization for granted to quite the degree that their counterparts in America can manage. Their research infrastructure does not enjoy the self-confident reputation of the major world universities, and therefore those individuals find they must pay closer attention to the manner in which local governmental units and multinational corporations openly impact their attempts to provide some adequate level of high-quality education and research. Nonetheless, their activities require a generic theoretical analysis that does not appear to be too closely shackled to any specific local conditions, be they legal niceties, local educational customs, or distinct nationalist aspirations for alternative development paths. The generic character of the analysis and lack of legal specificity will help facilitate publication in hegemonic (often English language) journals or other outlets that might otherwise remain closed to the foreign scholars. While they often must appear responsive to their local constituencies who regard globalization with great suspicion, in the final analysis they are put in the difficult position of expressing qualified endorsement of commercialization initiatives, which are often imposed from without; for example, WTO-mandated changes in intellectual property, government-mandated cuts in public education expenditures, and multinationals contracting for research with a limited cadre of entrepreneurial scientists in their targeted areas of interest.[59] In this manner, 3H has become just another symptom of the globalization that has made itself felt in every university, corporate lab, and government research facility throughout the world over the past two decades.

Both mode 1/mode 2 and 3H exhibit the same drawback we have identified in the introduction to this chapter: they adopt a stark before/after approach to modern developments in the funding and organization of science and then inflate them into all-purpose doctrines that ultimately provide a generic imposition of the neoliberal mindset on their local higher education infrastructure, and, if pertinent, any government-organized scientific research capacity. They serve up more palatable versions of a neoclassical economics of science, leached of all the actual technical content, which if more openly espoused might both repulse and dismay the sorts of clientele whom they are pledged to serve. In a caricature of the neoliberal economist, they end up simply presuming that any marketized science whatsoever inevitably enhances freedom, expands choice, encourages extended participation, and improves overall welfare.

This is not the place to summarize and critique the rise of neoliberal theories of the economy and the state since World War II.[60] It will suffice to suggest that neoliberalism differs from its classical predecessor through its transcendence of the classical liberal tension between the self-interested agent and the state by reducing both state and market to the identical flat ontology of the neoclassical model of the economy. "Freedom" is thus conflated with entrepreneurial activity, and state functions are "rationalized" by reducing them to market relationships. Hence, the blurring of distinctions between university and corporation, or "public versus private" science, found in both mode 2 and 3H are derivative representations of the larger neoliberal agenda. This dictated that education should no longer serve Humboldtian ideals of creating a solid citizenry and fostering cultural development but rather should be treated as just another fungible commodity (Friedman, 1962). Since poor people would never be able to afford much of it, and certainly very little "higher education," it follows that they would be relegated to role of passive consumer, while a corporate class of experts would effectively define and steer scientific research.[61] The kind of science that would sustain a research infrastructure is that which would be responsive to the needs of corporate customers (who, conveniently, had in the interim become more interested in outsourcing their R&D). Neoliberalism is very much a top-down project, while under its sway "democracy" has been redefined to encompass pro-corporate "free market" policies interspersed with highly stylized and commercialized "elections."

The new economics of science that we believe is better oriented to conform with STS research in general accepts that some form of economic underpinnings have always shaped the organization and management of scientific research, but that because there is no such thing as a generic market, there has never anywhere existed a fully constituted "marketplace of ideas."[62] Since markets are plural and do not produce identical results either over time or between various cultural areas (much as the nature of commercialized science in America differed dramatically between the three regimes identified in the previous section), it becomes all the more imperative to specify in detail the fine structure of operation of each of the major players in the course of the organization of science: universities, corporations, and governments. The disaggregation of science into its component structures (laboratories, clinics, field stations, classrooms, libraries, conferences) and the disaggregation of its managers into diverse agents (academics, corporate officers, government representatives, corporate trustees) is the first step toward constructing a sociologically aware account of the economics of science, and not, as has been the tendency of mode 2 and 3H, to lump them together into a homogeneous entrepreneurial agent. Who pays whom, and who answers what to whom, has consequences for the sorts of knowledge fostered. It behooves analysts to pay closer attention to who performs the labor in the laboratory under which diverse circumstances, to ask how findings are published or otherwise promulgated, to trace the flows of physical items between laboratories and corporations, to itemize the forms of attribution and audit that are brought to bear, and to inquire what institutions and customs constitute and sustain "the author." Where and in what manner the various components of the university become commercialized (Slaughter & Rhoades, 2004; Kirp, 2003) matter as much for the health of science as do other more obvious variables, such as the identities of the various scientific fields presumed to hold the greatest promise for conceptual advance and commercial development. The recent innovation of the commodification of "research tools" in particular bears profound implications for nearly every aspect of scientific research, most of which have not been adequately explored by neoliberal-influenced science policy analysts.[63] It does not further comprehension to simply presume that science as a whole is a production process that extrudes a thing-like entity called "knowledge"; indeed, the attempt to reify and commoditize information is itself an artifact of the modern privatization regime—a process that can never succeed in its entirety, because complete codification and control of a reified information would paralyze scientific inquiry.

This statement is not hyperbole. Take for instance, the assertion found in 3H and elsewhere that the university and the corporation are converging to a single commercial entity. While it is certainly the case that universities may be observed to behave like corporations in progressively more elaborate ways, ranging from the exploitation of trademarks to outsourcing wage-intensive functions, very few universities in the United States are willing to altogether relinquish their special nonprofit status and the range of perquisites that attach to their educational location in the national infrastructure. The very few that do, such as the University of Phoenix, have clearly opted out of maintaining a serious research capacity and therefore are little better than digital diploma mills. Most commercialization of existing universities in the 1990s was due to a relatively small cadre of natural science faculty in alliance with some entrepreneurial academic administrators who wanted to exploit commercial opportunities while still enjoying all the fruits of their nonprofit status. Their bonanza has been the Trojan horse that opened up the rest of the university to a whole range of neoliberal "reforms." If, contrary to all present evidence, universities eventually really did become corporations, one might anticipate that the myriad functions now combined on one campus would fragment. Libraries would disappear (Kirp, 2003: 114); expensive vocational schools (like teaching hospitals) would devolve as separate units; down-market research for hire would migrate to contract research organizations located in separate research parks, along with "technology transfer offices"; theaters and concert halls would go it alone; low-cost "distance education" would decamp in search of cheap foreign labor; and dormitories would be sold off as public housing.

The fundamental crux of the modern economics of science is that, contrary to boosters such as Bok (2003), Baltimore (2003), and Thursby & Thursby (2003), it may be that the current configuration of the commercialization of science is neither stable nor viable. Corporations are interested in academic science as long as it cuts costs—that is, as long as it still receives the panoply of subsidies that accrue to it in its nonprofit status. As Slaughter & Rhoades (2004: 308) put it, "Academic capitalism in the new economy involves a shift, not a reduction, in public subsidy." But the mere fact of commercializing university research puts that status and those cost advantages in jeopardy. Already, state legislators in the United States expect their flagship universities to "float on their own bottoms"; universities in the United Kingdom are expected to attract foreign students. The imposition of "revenue center management" to units within the university begins the process of restructuring there that has already occurred in the post-Chandlerian corporation (Kirp, 2003: 115–28). The more that teaching is disengaged from research and farmed out to nonfaculty and other migrant labor, the less of a political case can be made that integrated expensive specialized research facilities should be maintained. And why would anyone want to leave a large cash legacy as a bequest to a private university that was a profitable corporation? No one leaves their fortune to General Motors (even if it wouldn't mind a little assistance). The more that natural science faculty become enfolded into their corporate roles, holding down two or more day jobs, the less they will be willing to cross-subsidize their poorer "colleagues" in the social sciences and humanities. How much longer can an increasingly privatized and balkanized educational sector expect to receive any state or philanthropic subsidy on an ongoing basis?

The current beneficiaries of the commercialization of academic science may very well be destroying the goose that laid their golden egg.

CONCLUSION

Interest in the commercialization of science has increased exponentially with the dissolution of the Cold War, the decline in military funding, hostility toward government interference, public skepticism about the telos of science, questions about the accountability of scientists, and the push to develop connections between business and science. Going beyond acknowledgment that science finds itself in a new phase of organization and retrenchment, our claim is that alternative forms of funding and organization have shaped both conduct and content of science throughout its history, characterized by shifting alliances among the commercial corporation, the state, and the university.

In the modern globalized privatization regime, the neoliberal perspective would suggest that the national research systems were merely responding to a uniform market pressure to render their academic sectors more efficient, but this analysis would miss too much of the concerted activity that results in epoch-making transformations. This, indeed, has been the problem with much of the prognoses of economists and that sector of STS which has followed in their footsteps. Modern commercialized science is profoundly different from early twentieth century science, almost as much as it diverges from the previous Cold War regime. We advocate a version of STS that explores the impact of economic institutions on the full gamut of scientific practices, ranging from the definition of the legitimate "scientific author" to the effects of conflation of epistemic efficacy with pecuniary profitability. In contrast to previous microsociologies of laboratory life, we are not as concerned with how actants "circulate" throughout society as with how viable laboratories are *constructed and maintained*.

In particular, we would like to close by suggesting that it is necessary to expand our previous CGE analysis to take into account a fourth class of actors in the modern world system. Science has not only been promoted by firms, governments, and universities, but increasingly in the twentieth century it has also been organized and funded by international agencies that have propagated commercialization and standardization of research practices and institutions. Here, Elzinga (2001) makes a helpful distinction between internationalization and globalization. The former is predicated on trust and solidarity, whereas the latter is driven by for-profit motives. The former involves the multiplication and expansion of cross-border linkages of communication and

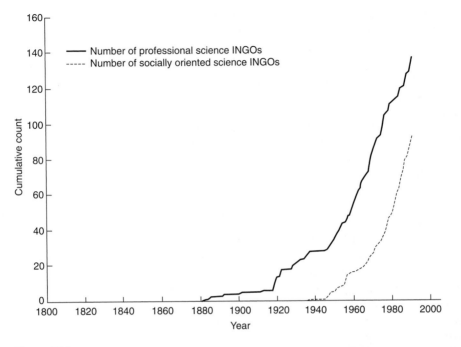

Figure 26.2
Cumulative Foundation of Science INGOs, 1870–1990.

collaboration in science and technology. The latter concerns the interconnections between large industrial corporations regarding research in the precompetitive phase as well as technological alliances. The former is fostered by scientific non- and inter-governmental organizations (scientific NGOs and scientific IGOs) as well as nongovernmental civic society organizations (civic NGOs). The latter is supported by transnational corporations. Elzinga introduces a further distinction between autoletic and heteroletic organizations. The former serve science as an end, such as scientific NGOs. The latter are created and sustained by governmental action, such as scientific IGOs.[64]

For our purposes, the various agencies may be divided into three classes: (1) the World Trade Organization, which has spread and enforced standardized rules of intellectual property and trade in services under the guise of providing a stable platform for international trade (Drahos & Braithwaite, 2002; Sell, 2003); (2) the United Nations, which through UNESCO and WIPO has promoted international science policy; and (3) a raft of international nongovernmental organizations (INGOs), which have played a crucial role in the spread of the globalized privatization regime (Drori et al., 2003).

An earlier Mertonian approach tended to treat science as subsisting beyond or outside of politics, but nothing reveals the obsolescence of this belief better than an inventory of the means by which some pivotal scientific institutions have been spread by these international organizations. Some entities, of course, are merely the international arms of national professional organizations of scientists, and as such conform to prior Mertonian images of the self-organization of science. But increasingly after World War II, these have been augmented by politically oriented INGOs that combine both scientists and laypersons into activist groups seeking to spread a model of "best-practice science" in the third world in the name of economic and political development.

The activities of these INGOs go some distance in explaining how it is possible that corporations can begin to take advantage of a globalized standardized research capacity in such a wide range of cross-cultural settings, as outlined above. The pervasive similarities of science policies in almost all of the developed world and now, increasingly, those parts of the developing world where corporate R&D is moving in the near future (Economist Intelligence Unit, 2004) are due in large part to the work of INGOs in propagating a relatively generic culture of commercialized research within local national education systems and government bureaus of science policy. The implications of this push to standardization extend far beyond the simple spread of something like a disembodied "scientific world view" or the export of the Internet to previously untapped areas. For instance, standardization of scientific institutions and the delegitimation of local knowledge has turned out to be a necessary prerequisite for the globalization of for-profit higher education (Morey, 2004) as well as the harbinger of the outsourcing of much routine scientific labor to low-wage countries. The high-wage university sector is therefore the next major economic area slated for another serious round of downsizing, cost-cutting, and outsourcing from the advanced metropolitan countries to low-wage areas, and STS scholars will appreciate that this will have unprecedented effects on the content of scientific research just over the horizon.

The analysis of the new regime of globalized privatization will depend on the theoretical orientation that is accessed in order to understand what is becoming a wide-ranging and pervasive phenomenon. Most existing work based on neoclassical economics has ignored or misunderstood many of the phenomena covered in this chapter. In its place, we advocate a new political economy of science, which joins up with recent developments in STS, to produce an independent analysis of the effects of commercialization on the practice of modern science.

Notes

Mirowski would like to thank audiences at Universidad de la Republica, Montevideo Uruguay, the Cornell Conference on Science for Sale, the University of Pennsylvania Department of the History and Sociology of Science, the New School CEPA series, and ICAS at New York University for providing important insights that helped in the writing of this paper. Sent is grateful for the feedback of participants of the workshop on Science, Democracy, and Economics organized by the Urrutia Elejalde Foundation on Economics and Philosophy (FUE) and the Spanish National Open University (UNED). She received especially helpful feedback from Tiago Santos Pereira. Mirowski and Sent are both grateful to the anonymous referees as well as to Edward Hackett for insightful suggestions.

1. Dennis (1997:1).

2. See, e.g., Brown (2000), Hollinger (2000), Miyoshi (2000), Newfield (2003), Krimsky (2003), Monbiot (2003), Washburn (2005).

3. For similar assessments: the "growing commercial engagement has not, thus far, altered the research culture of universities, so as to privilege applied orientations at the expense of basic science" (Owen-Smith & Powell, 2003: 1696); "Science today is only a short way down the path to becoming a toady of corporate power" (Greenberg, 2001: 3); do we see "universities compromising their core values [?] . . . at least at the major research universities, their revenue-enhancing activities have not seriously distorted such values" (Baltimore, 2003: 1050); "There is evidence to suggest that university licensing facilitates technology transfer with minimal effects on the research environment" (Thursby & Thursby, 2003: 1052). Richard Nelson has tended to become more skeptical of this position over time, however (see Nelson, 2004).

4. Some recent evidence begins to suggest that Bush may not have been as directly responsible for these ideas as has been previously thought, but that they might instead be traced to the economist Paul Samuelson, who helped draft *Science, The Endless Frontier* (see Samuelson 2004: 531).

5. See Callon (1998; in Mirowski and Sent, 2002; in Barry and Slater, 2003), and for critique, Mirowski and Nik-Khah (forthcoming). We might include Capshew and Rader (1992) in this tendency to see "Big Science" back to the seventeenth century although they do have the good grace to notice: "Aside from a surge of interest in the rise of industrial research, surprisingly little attention has been directed towards the economics of science. . . . The organization and production of knowledge in such contexts bears more than a passing resemblance to Big Science" (1992: 24).

6. In this respect, as in many others, we owe our awareness to the politically retrograde influence of Kuhn on the questions posed in science studies to Steve Fuller (see Fuller, 2000; Mirowski, 2004a: chapter 4).

7. The first two quotes are from NBER, 1962: 452 and the third from NBER, 1962: 454–55.

8. Named after Boris Hessen, this was an argument that the content of Newtonian mechanics owed much to the artisan traditions and economic structures of the seventeenth century. For modern research on the subsequent suppression of Hessen's work and attendant economic considerations, see McGuckin, (1984), Chilvers (2003), Mayer (2004).

9. This problem of this blind spot of science studies approaches to the laboratory has recently been noticed by Kleinman (2003: chapter 5). There is the further complication that early works such as Latour and Woolgar (1979) and Bourdieu (2004) made extensive use of economic metaphors while essentially ignoring any substantive economic structures. The *curiosum* of resort to economic metaphors while avoiding the economy is discussed in Pels (2005). Nevertheless, the range of reasons behind the neglect of economic factors by STS in the 1970s to 1990s cannot be covered in this venue. The few exceptions to this trend are covered in the next section.

10. The most prominent advocate of such a position in philosophy is Philip Kitcher (1993), but it can also be found as far back as the work of Friedrich Hayek in the 1930s, which gives some idea of its neoliberal origins. It is also showing up in the analysis of higher education: see Geiger (2004), Kirp (2003), Apple (2005). For further contemplation and critique on this issue, see Mirowski (2004a,b; forthcoming).

11. See, e.g., Cohen and Merrill (2003), Leydesdorff and Etzkowitz (2003), Tijssen (2004), Fuller (2002), and recent issues of the *Journal of Technology Transfer*. Few are aware that such neoliberal attitudes toward science date back to the 1950s, when members of the Mont Pelerin Society first conceived of the ambition to "get the state out of the knowledge industry." For discussion, consult Friedman (1962: chapter 6) and Walpen (2004).

12. For some economic examples, see Danny Quah in Vaitilingham (1999), Shapiro and Varian (1999), Powell and Snellman (2004). A Google search on the term "Knowledge Economy" in October 2005 produced an amazing 1,690,000 hits.

13. We believe it is important to counter such claims as "the weakening of university-industry research linkages during a significant proportion of the postwar [WWII] period was the real departure from the historical trend" (Mowery et al., 2004: 195 note 15), and "The so-called Mode 2 is not new; it is the original of science before its academic institutionalization in the 19th century" (Leydesdorff & Etzkowitz, 1998: 116). STS should be committed to explanation of what is especially characteristic of the new social structures of science, in comparison to other structures of longer duration.

14. Indeed, some of the best modern work in STS exploring the interface of science and political economy takes an explicitly comparative perspective (see, e.g., Daemmrich, 2004; Wright, 1994; Jasanoff, 2005; Larédo & Mustar, 2001). We do not subject the European laboratory to the same detailed regimes analysis as the American situation owing to space and research constraints. However, we do try to indicate some of the modern implications of the analysis for the European context in the last segment of the section "Three Regimes of Twentieth Century Science Organization."

15. We signal here our awareness of the literature that seeks to acknowledge their importance under the rubric of the "triple helix." We briefly evaluate its contribution in the section "Alternative Market Models of the Conduct of Scientific Research," showing how the regimes analysis differs from their narrative. Our conclusion complicates the analysis by adding a fourth actor—the international agency.

16. "The organizational approach to understanding big business developed by Alfred Chandler, Jr., has given historians a framework within which to place the research laboratory" (Smith, 1990: 121). The Chandlerian approach to business history is best represented in Chandler (1977, 2005b).

17. "The postwar R&D system, with its large well-funded universities and Federal contracts with industry, had little or no precedent in the pre-1940 era and contrasted with the structure of research systems in other postwar industrial countries. In a very real sense, the United States developed a postwar R&D system that was internationally unique" (Mowery & Rosenberg, 1998: 12).

18. See, e.g., Fox and Guagnini (1999), Shinn (2003), Hounshell (1996), Buderi (2000: chapter 2), Mowery (1981, 1990), Swann (1988), Smith (1990), Pickering (2005).

19. According to Smith (1990: 124), "broad-ranging research in German dye companies began in 1890 . . . yet these research programs do not appear to have been a model that American companies emulated." As regards the German dye industry, Pickering (2005: 389) suggests, "from the late 1870s, scientific research was itself uprooted from its academic mooring and, for the first time, set down in the middle of industry."

20. A partial exception to this generalization is the agricultural experiment station attached to state land-grant universities from the 1870s onward. Their indeterminate status as scientific research poles is discussed by Charles Rosenberg in Reingold (1979). An attempt to project the extension service onto the industrial sector in the form of a bill to establish "engineering experimental stations" at land-grant colleges in 1916 was easily defeated (Tobey, 1971: 40; Noble, 1979: 132).

21. See, e.g., Stanley Guralnick in Reingold (1979).

22. See Wise (1980), Reich (1985), Dennis (1987), Hounshell (1996).

23. The early twentieth-century American stress on suppression of interfirm governance of markets may account for the substantial differences in the German and American climate, which were treated as a puzzle by Mowery (1990: 346).

24. See, e.g., the 1921 *Wireless Specialty* case (Fish, 1998: 1176). Furthermore, the metaphor that compared exploration of the laws of nature to exploration for minerals or gas deposits, which would

dominate Kenneth Arrow's Cold War "economics of innovation" (NBER, 1962) was already present in the 1911 *National Wire* decision (Fish, 1998: 1194). The legal acceptance of the ideal of collective science in the 1920s then dovetails quite nicely with recent claims that the "theoretical" treatments of the scientific community as a distinctly social entity in philosophy and sociology find their origins in the 1930s (Jacobs, 2002; Mirowski, 2004b).

25. See Mowery (1981) and the claim by Mowery (1990: 347) that "independent research organizations do not appear to have substituted for in-house research."

26. A good example is the way that various philosophical/cultural attempts to reconcile science with democracy or science with industry tended to struggle with Einstein's theory of relativity. For an account of these struggles, see Tobey (1971: chapter 4).

27. See Mowery and Rosenberg (1998), Leslie (1993), Kleinman (1995), Lowen (1997), Morin (1993).

28. Examples of these sorts of arguments can be found in Capshew and Rader (1992), Galison and Hevly (1992), and in the otherwise insufficiently appreciated pioneer researcher in economic themes of science organization in Ravetz (1971).

29. For a description of how the linear model was precipitated out of the Cold War context, see Kline (1995), Mirowski and Sent (2002). For a brief history of its existence as a statistical category, consult Godin (2003). For the vicissitudes of its modern manifestations, see Calvert (2004). For a new perspective upon its provenance as the brainchild of the economist who also promoted the economic concept of the "public good," see Samuelson (2004: 531), Mirowski (forthcomingB).

30. See, e.g., Forman (1987), Graham (1985), and Hounshell (1996: 47–50). As Colonel Norair Lulejian said in a 1962 speech: "Can we for example plan and actually schedule inventions? I believe this can be done in most instances, provided we are willing to pay the price, and make no mistake about it, the price is high" (quoted in Johnson, 2002: 19).

31. The irony of this story is that what was rendered more "real" in the industrial sector was simultaneously further eroded in the academic sector. See, e.g., Lowen (1997: 140): "By the mid-1950s the claims that the programs of basic and applied research [at Stanford] were entirely distinct and that the applied research program was not affecting the academic program were largely rhetorical."

32. IBM stands out from these other corporations as forming its in-house research capacity rather late, creating a "pure science" department only in 1945. On this unusual history, see Akera (2002). The Yorktown Heights facility was opened only in 1960.

33. See Hart (1998: 227–29) and Teske and Johnson (1994).

34. See Westwick (2003: 51). The policy was not uniformly applied to industrial contractors, but the second-source rule often mitigated any commercial advantage that the firm might enjoy from keeping patent rights. Interestingly, the supposedly public-spirited University of California resisted the AEC rule and was defeated.

35. Quoted in Gruber (1995: 265).

36. This is not to say that foundations had no lasting effects in the Cold War regime. The Ford Foundation established the dominant model for American business schools in this era while Rockefeller University played a crucial role in the academic development of molecular biology (Kay, 1993). The relationship of science to state power in this era is astutely discussed in Ezrahi (1990).

37. These issues are discussed further in Hollinger, 1990; Mirowski, 2004b.

38. See, for economic historians, Lamoreaux et al. (2003: 405); for legal scholars, Boyle (2000), Lessig (2001, 2004), McSherry (2001); for historians of education, Geiger (2004: 3), Matkin (1990: 22), Slaughter and Rhoades (2004), Kirp (2003), Apple (2005); for politics, Krimsky (2003: 30–31), Mirowski

and van Horn (2004c). Washburn (2005: chapter 3) documents the political maneuvers leading up to the passage of the Bayh-Dole Act.

39. Interestingly, Chandler himself does not entirely agree with this assessment. See Chandler (2005a).

40. The identity of these vanguard industries is crucial for our narrative because it is widely recognized that "in most industries, university research results play little if any role in triggering new industrial R&D projects" (Mowery et al., 2004: 31). The activist firms were located in that small group of industries that did make extensive use of academic research.

41. See the discussions in Drahos and Braithwaite (2002) and Sell (2003).

42. See, e.g., Slaughter and Rhodes (2002), Geiger (2004), Miyoshi (2000), Krimsky (2003), Washburn (2005), Mowery et al. (2004). Mowery et al. (2004: 94ff) do us the service of pointing out the fallacies behind the rhetoric one finds in European contexts, to the effect that if those countries would only institute their own versions of Bayh-Dole, they would automatically reap untold benefits of escalated technology transfer.

43. For the post-1980 degradation of the U.S. patent system, see Kahin (2001) and Jaffe and Lerner (2004). For more general consideration of the fallout from extension of intellectual property, see Lessig (2001), Drahos and Braithwaite (2002), Sell (2003), Mirowski (2004a: chapter 6).

44. This sea change is documented and discussed in Anderson (2004), Buderi (2000, 2002), Economist Intelligence Unit (2004), Reddy (2000), Chesbrough (2001), Berman (2003), Markoff (2003).

45. Examples of this narrative can be found in Bok (2003), Geiger (2004), Krimsky (2003), Nelson (2001), Owen-Smith and Powell (2003), Thursby and Thursby (2003), Kirp (2003). Economists have been especially prominent in spreading this narrative.

46. One signal characteristic of each of the three regimes discussed in this paper is the insistence on the extent to which American higher education had been responding to innovations originating largely outside their purview, as well as the relative absence of first-mover advantage when it comes to the reorganization of science. Here is one place where the history of science and the history of education need to be brought into closer dialogue.

47. This is the major drawback in the premises governing Mowery et al. (2004), Krimsky (2003), Nelson (2004), and the literature that restricts itself narrowly to considerations of intellectual property such as McSherry (2001).

48. One major representative of this position in the science policy literature has been the widely cited paper of Paul David and Partha Dasgupta on their version of a "new economics of science." See the summary in Mirowski and Sent (2002) and David (2003, 2004). For an incisive critique of their distinction between "tacit" and "codified" knowledge, which they correlate with open versus commercial science, see Nightingale (2003) and Mirowski (forthcomingB).

49. Although national differences in corporate governance structures may still be a factor: see Djelic (1998), Djelic and Quack (2003), Guena et al. (2003).

50. See http://europa.eu.int/comm/education and Bernstein (2004).

51. See http://europa.eu.int/comm/research.

52. See http://www.hero.ac.uk.

53. The neoclassical approach to science, along with some of the founding documents, can be found in Mirowski and Sent (2002, especially pp. 38–43). The economists' role in creating a new political identity for the university is mentioned in Godin (2003: 67). Their more specific concerns when it came to weapons procurement are outlined in Hounshell (2000). The history of neoclassical economics in

confronting epistemological problems is outlined in Mirowski (forthcomingA). On the economists' role in the Bush report, see note 4 and Mirowski (forthcomingB).

54. See, e.g., Foray (2004: chapter 5), Guena et al. (2003), David (2003), Washburn (2005: 62), Shi (2001). The genealogy of the "economics of information" is covered in Mirowski (forthcomingA).

55. One notable exception was Dickson and Noble (1981). An early wake-up call was Slaughter and Rhoades (2002; first published in 1996).

56. The invocation of the "helix" appears to have been a mere rhetorical figure that allows the authors to introduce what they consider to be evolutionary considerations as well as a spate of terminology found more frequently at the Santa Fe Institute, such as "co-evolution," "lock-in," and other forms of complex nonlinear dynamics. More significantly, the tradition seems to avoid discussion of explicit economic concepts and modalities, as we do in the section "Three Regimes of Twentieth Century Science Organization."

57. The "pose a challenge" quote comes from Etzkowitz et al. (1998: xii). "Another purpose of Triple Helix discourse is to turn science studies from a constructivism narrowly focused on micro-processes" (Etzkowitz, 2003: 332). The critical agenda of one of the mode 2 authors can be gleaned from Guggenheim and Nowotny (2003).

58. The critiques summarized in this paragraph are taken from the following sources: Delanty (2001), Elzinga (2004), Shinn (2002), Pestre (2003), Bassett (2003), Ziedonis (2004).

59. The use of 3H terminology to backtrack from previously skeptical analyses of the commercialization of research can be observed in Campbell et al. (2004), for instance.

60. See, however, Barry et al. (1996), Rose (1999), Walpen (2004), Apple (2005), Mirowski and Plehwe (forthcoming). As mentioned in note 11, the neoliberal push to privatize knowledge production and knowledge conveyance dates back to the 1950s.

61. The segment of the STS community concerned to critique the "Public Relations of Science" movement will recognize here one of their major *bêtes noirs*. Comprehension of neoliberalism gives a different perspective on controversies over genetically modified foods, stem cell research, the safety of pharmaceuticals, and a host of other contemporary issues in science studies.

62. The types of economic theory that support this vision are discussed in Mirowski (2007).

63. See the discussion of research tools by Walsh et al. (2003, 2005) and the critique of that work in David (2003). Further important work on the commercialization of research tools is cited in Eisenberg (2001) and Streitz and Bennett (2003).

64. Elzinga (2001) further, insightfully, notes that the fact that scientific knowledge remains highly concentrated where it is first created, despite the moves toward internationalization and globalization, contrasts sharply with the notion of science as a public good.

References

Akera, Atsushi (2002) "IBM's Early Adaptation to Cold War Markets," *Business History Review* 76: 767–802.

Anderson, Howard (2004) "Why Big Companies Can't Invent," *Technology Review* May: 56–59.

Angell, Marcia (2004) *The Truth About the Drug Companies* (New York: Random House).

Apple, Michael (2005) "Education, Markets and an Audit Culture," *Critical Quarterly* 47: 11–29.

Asner, Glen (2004) "The Linear Model, the U.S. Department of Defense, and the Golden Age of Industrial Research," in Karl Grandin, Nina Wormbs, and Sven Widmalm (eds), *The Science-Industry Nexus* (Sagamore: Science History Publications).

Baltimore, David (2003) "On Over-Weighting the Bottom Line," *Science* 301: 1050–51.

Barry, Andrew & Don Slater (2003) "Technology, Politics and the Market: An Interview with Michel Callon," *Economy and Society* 31: 285–306.

Barry, Andrew, Thomas Osborne, & Nikolas Rose (eds) (1996) *Foucault and Political Reason: Liberalism, Neoliberalism and the Rationalities of Government* (London: UCL Press).

Bassett, Ross (2003) "Review of Etzkowitz: *MIT and the Rise of Entrepreneurial Science*," *Isis* 94(4): 768–69.

Berman, Dennis (2003) "At Bell Labs, Hard Times Take Toll on Pure Science," *Wall Street Journal,* May 23: A1.

Bernstein, Richard (2004) "Germany's Halls of Ivy Are Needing Miracle-Gro," *New York Times,* May 9.

Boersma, Kees (2002) *Inventing Structures for Industrial Research: A History of the Philips Natlab, 1914–1946* (Amsterdam: Askant).

Bok, Derek (2003) *Universities in the Marketplace: The Commercialization of Higher Education* (Princeton, NJ: Princeton University Press).

Borrás, Susana (2003) *The Innovation Policy of the European Union: From Government to Governance* (Cheltenham, U.K.: Edward Elgar).

Bourdieu, Pierre (2004) *Science of Science and Reflexivity* (Cambridge: Polity Press).

Boyle, James (2000) "Cruel, Mean or Lavish? Economic Analysis, Price Discrimination, and Digital Intellectual Property," *Vanderbilt Law Review* 53: 2007–39.

Brender, Alan (2004) "In Japan, Radical Reform or Same Old Subservience? National Universities Wonder How Much Freedom They Will Be Given Under Looser Government Oversight," *Chronicle of Higher Education,* March 12.

Brown, James R. (2000) "Privatizing the University," *Science* 290: 1701.

Buderi, Robert (2000) *Engines of Tomorrow* (New York: Simon & Schuster).

Buderi, Robert (2002) "The Once and Future Industrial Research," in Albert H. Teich, Stephen D. Nelson, & Stephen J. Lita (eds), *AAAS Science and Technology Policy Yearbook:* 245–51.

Burke, James, Gerald Epstein, & Choi Minsik (2004) *Rising Foreign Outsourcing and Employment Losses in U.S. Manufacturing, 1987–2002,* PERI Working Paper No. 89, University of Massachusetts.

Busch, Lawrence, Richard Allison, Craig Harris, Alan Rudy, Bradley T. Shaw, Toby Ten Eyck, Dawn Coppin, Jason Konefal, & Christopher Oliver (2004) *External Review of the Collaborative Research Agreement Between Novartis and the Regents of the University of California* (East Lansing Michigan State University Institute for Food and Agricultural Standards).

Callon, Michel (ed) (1998) *The Laws of the Markets* (Oxford: Blackwell).

Caloghirou, Yannis, Stavres Ionnides, & Nicholas Vontoras (2003) "Research Joint Ventures," *Journal of Economic Surveys* 17: 541–70.

Calvert, Jane (2004) "The Idea of Basic Research in Language and Practice," *Minerva* 42: 251–68.

Campbell, Eric, Joshua Powers, David Blumenthal, & Brian Biles (2004) "Inside the Triple Helix: Technology Transfer and the Commercialization of the Life Sciences," *Health Affairs* (January–February) 23(1): 64–76.

Capshew, James & Karen Rader (1992) "Big Science: Price to the Present," *Osiris* 2nd series 7: 3–25.

Chandler, Alfred (1977) *The Visible Hand* (Cambridge, MA: Harvard University Press).

Chandler, Alfred (2005a) "Response to the Symposium," *Enterprise and Society* 6: 134–37.

Chandler, Alfred (2005b) *Shaping the Industrial Century* (Cambridge, MA: Harvard University Press).

Chesbrough, Hank (2001) "Is the Central R&D Lab Obsolete?" *Technology Review* April.

Chilvers, C. (2003) "The Dilemmas of Seditious Men: The Crowther-Hessen Correspondence," *British Journal for the History of Science* 36(4): 417–35.

Cohen, Wesley & Stephen Merrill (eds) (2003) *Patents in the Knowledge-Based Economy* (Washington, DC: National Academies Press).

Daemmrich, Arthur (2004) *Pharmacopolitics* (Chapel Hill: University of North Carolina Press).

David, Paul (2003) "The Economic Logic of Open Science and the Balance Between Private Property Rights and the Public Domain in Scientific Information," in *The Role of Scientific and Technical Data and Information in the Public Domain* (Washington, DC: National Academies Press): 19–34.

David, Paul (2004) "Can Open Science Be Protected from the Evolving Regime of IPR Protection?" *Journal of Institutional and Theoretical Economics* 160(1): 9–34.

Delanty, Gerard (2001) *Challenging Knowledge: The University in the Knowledge Society* (Buckingham, U.K.: Open University Press).

Dennis, Michael (1987) "Accounting for Research," *Social Studies of Science* 17: 479–518.

Dennis, Michael (1997) "Historiography of Science: An American Perspective," in John Krige & Dominique Pestre (eds) (1997), *Science in the 20th Century* (Amsterdam: Harwood): 1–26.

Diamond, Arthur M. (1994) "The Economics of Science," discussion paper dated February 13 for National Science Foundation Conference, Washington, DC, January 1995. Appendix of comments by various experts dated December 14, 1994.

Dickson, David & David Noble (1981) "By Force of Reason: The Politics of Science and Technology Policy," in Tom Ferguson & Joel Rogers (eds), *The Hidden Election: Politics and Economics in the 1980 Presidential Campaign* (New York: Pantheon): 260–312.

Dillon, Sam (2004) "U.S. Slips in Status as a Hub of Higher Education," *New York Times*, December 21: A1.

Djelic, Marie-Laure (1998) *Exporting the American Model* (New York: Oxford University Press).

Djelic, Marie-Laure & Sigrid Quack (eds) (2003) *Globalization and Institutions: Rewriting the Rules of the Economic Game* (Cheltenham, U.K.: Edward Elgar).

Drahos, Peter & John Braithwaite (2002) *Information Feudalism: Who Owns the Knowledge Economy?* (New York: New Press).

Drori, Gili, John Meyer, Francisco Ramirez, & Eric Schofer (2003) *Science in the Modern World Polity: Institutionalization and Globalization* (Stanford, CA: Stanford University Press).

Economist Intelligence Unit (2004) *Scattering the Seed of Invention: The Globalization of Research and Development*. Available at: http://www.eiu.com.

Eisenberg, Rebecca (2001) "Bargaining over the Transfer of Research Tools," in R. Dreyfuss, H. First, & D. Zimmerman (eds), *Expanding the Boundaries of Intellectual Property* (Oxford: Oxford University Press): 223–49.

Elzinga, Aant (2001) "Science and Technology: Internationalization," in Neil J. Schmelser & Paul B. Baltes (eds), *International Encyclopedia of the Social and Behavioural Sciences,* vol. 20 (Elsevier: Amsterdam): 13633–38.

Elzinga, Aant (2004) "The New Production of Reductionism in Models Relating to Research Policy," in Karl Grandin, Nina Wormbs, & Sven Widmalm (eds), *The Science-Industry Nexus: History, Policy, Implications* (Sagamore Beach, MA: Science History Publications): 277–304. (Symposium paper presented at Nobel Symposium, Swedish Royal Academy of Sciences, Stockholm, November, 2002.)

Endlich, Lisa (2004) *Optical Illusions: Lucent and the Crash of Telecom* (New York: Simon & Schuster).

Etzkowitz, Henry (2002) *MIT and the Rise of Entrepreneurial Science* (London: Routledge).

Etzkowitz, Henry (2003) "Innovation in Innovation: The Triple Helix in University-Industry-Government Relations," *Social Science Information* 42(3): 293–337.

Etzkowitz, Henry & Loet Leydesdorff (2000) "The Dynamics of Innovation: A Triple Helix of University-Industry-Government Relations," *Research Policy* 29(2): 109–23.

Etzkowitz, Henry, Andrew Webster, & P. Healey (eds) (1998) *Capitalizing Knowledge: New Intersections of Industry and Academia* (Albany: State University of New York Press).

Etzkowitz, Henry, Andrew Webster, Christiane Gebhardt, & Branca Terra (2000) "The Future of the University and the University of the Future: Evolution of Ivory Tower to Entrepreneurial Paradigm," *Research Policy* 29: 313–30.

European Commission (2003–2004) *Towards a European Research Area: Science, Technology, and Innovation: Key Figures 2003–2004.* Available at: ec.europa.eu/research.

Ezrahi, Yaron (1990) *The Descent of Icarus* (Cambridge, MA: Harvard University Press).

Feldman, Maryann, Albert Link, & Donald Siegel (2002) *The Economics of Science and Technology* (Boston: Kluwer).

Fish, Catherine (1998) "Removing the Fuel of Interest from the Fires of Genius," *University of Chicago Law Review* 65: 1127–98.

Foray, Dominique (2004) *The Economics of Knowledge* (Cambridge: MIT Press).

Forman, Paul (1987) "Beyond Quantum Electronics," *Historical Studies in the Physical Sciences* 18: 149–229.

Fox, Robert & Anna Guagnini (1999) *Laboratories, Workshops and Sites* (Berkeley: University of California Press).

Friedman, Milton (1962) *Capitalism and Freedom* (Chicago: University of Chicago Press).

Fuller, Steve (2000) *Thomas Kuhn: A Philosophical History for Our Times* (Chicago: University of Chicago Press).

Fuller, Steve (2002) *Knowledge Management Foundations* (London: Butterworth).

Galison, Peter & Bruce Hevly (eds) (1992) *Big Science* (Stanford, CA: Stanford University Press).

Gaudilliere, Jean-Paul & Ilana Lowy (eds) (1998) *The Invisible Industrialist* (London: Macmillan).

Geiger, Roger (1997) "Science and the University: Patters from U.S. Experience," in John Krige & Dominique Pestre (eds) (1997), *Science in the 20th Century* (Amsterdam: Harwood): 159–74.

Geiger, Roger (2004) *Knowledge and Money* (Stanford, CA: Stanford University Press).

Gibbons, Michael (2003) "Globalization and the Future of Higher Education," in Gilles Breton & Michel Lambert (eds), *Universities and Globalization* (Quebec: UNESCO): 107–16.

Gibbons, Michael, Camille Limoges, Helga Nowotny, S. Schwartzman, Peter Scott, & Martin Trow (1994) *The New Production of Knowledge* (London: Sage).

Godin, Benoit (2003) "Measuring Science: Is There Basic Research Without Statistics?" *Social Science Information* 42(1): 57–90.

Godin, Benoit (2005) *Measurement and Statistics on Science and Technology* (London: Routledge).

Graham, Margaret (1985) "Corporate Research and Development: The Latest Transformation," *Technology in Society* 7: 179–95.

Greenberg, Daniel (2001) *Science, Money and Politics* (Chicago: University of Chicago Press).

Gruber, Carol (1995) "The Overhead System in Government-Sponsored Academic Science: Origins and Early Development," *Historical Studies in the Physical and Biological Sciences* 25(2): 241–68.

Guena, Aldo, Ammon Salter, & Edward Steinmueller (eds) (2003) *Science and Innovation* (Cheltenham, U.K.: Edward Elgar).

Guggenheim, Michael & Helga Nowotny (2003) "Joy in Repetition Makes the Future Disappear," in Bernward Joerges & Helga Nowotny (eds), *Social Studies of Science and Technology: Looking Back, Looking Ahead* (Dordrecht, Netherlands: Kluwer): 229–58.

Guston, David & Kenneth Kenniston (eds) (1994) *The Fragile Contract* (Cambridge: MIT Press).

Hargreaves Heap, Shaun (2002) "Making British Universities Accountable: In the Public Interest?" in Philip Mirowski & Esther-Mirjam Sent (eds) (2002), *Science Bought and Sold* (Chicago: University of Chicago Press): 387–411.

Hart, David (1998) *Forged Consensus: Science, Technology and Economic Policy in the U.S., 1921–53* (Princeton, NJ: Princeton University Press).

Hart, David (2001) "Antitrust and Technological Innovation in the U.S.: Ideas, Institutions, Decisions and Impacts, 1890–2000," *Research Policy* 30: 923–36.

Hemphill, Thomas (2003) "Role of Competition Policy in the U.S. Innovation System," *Science and Public Policy* 30: 285–94.

Hochstettler, Thomas John (2004) "Aspiring to Steeples of Excellence at German Universities," *Chronicle of Higher Education,* July 30.

Hollinger, David (1990) "Free Enterprise and Free Inquiry: The Emergence of Laissez-Faire Communitarianism in the Ideology of Science in the U.S.," *New Literary History* 21: 897–919.

Hollinger, David (2000) "Money and Academic Freedom a Half Century After McCarthyism: Universities Amid the Force Fields of Capital," in Peggie Hollingsworth (ed), *Unfettered Expression* (Ann Arbor: University of Michigan Press): 161–84.

Hounshell, David (1996) "The Evolution of Industrial Research in the U.S.," in Richard Rosenbloom & William Spencer (eds), *Engines of Innovation* (Cambridge, MA: Harvard Business School Press): 13–85.

Hounshell, David (2000) "The Medium Is the Message," in T. Hughes & A. Hughes (eds), *Systems, Experts and Computers* (Cambridge, MA: MIT Press): 255–310.

Jacobs, Struan (2002) "The Genesis of the 'Scientific Community,'" *Social Epistemology* 16: 157–68.

Jaffe, Adam & Josh Lerner (2004) *Innovation and Its Discontents* (Princeton, NJ: Princeton University Press).

Jasanoff, Sheila (2005) *Designs on Nature* (Princeton, NJ: Princeton University Press).

João Rodrigues, Maria (ed) (2002) *The New Knowledge Economy in Europe: A Strategy for International Competitiveness* (Cheltenham, U.K.: Edward Elgar).

João Rodrigues, Maria (2003) *European Policies for a Knowledge Economy* (Cheltenham, U.K.: Edward Elgar).

Johnson, Stephen B. (2002) *The Secret of Apollo* (Baltimore, MD: Johns Hopkins University Press).

Judson, Horace (2004) *The Great Betrayal: Fraud in Science* (New York: Harcourt).

Kahin, Brian (2001) "The Expansion of the Patent System," *First Monday* 6(1).

Kay, Lily (1993) *The Molecular Vision of Life* (New York: Oxford University Press).

Kirp, David (2003) *Shakespeare, Einstein and the Bottom Line* (Cambridge, MA: Harvard University Press).

Kitcher, Philip (1993) *The Advancement of Science* (New York: Oxford University Press).

Kleinman, Daniel (1995) *Politics on the Endless Frontier* (Durham, NC: Duke University Press).

Kleinman, Daniel (2003) *Impure Cultures: University Biology and the World of Commerce* (Madison: University of Wisconsin Press).

Kline, Ronald (1995) "Constructing Technology as Applied Science," *Isis* 86: 194–221.

Kohler, Robert (1991) *Partners in Science* (Chicago: University of Chicago Press).

Krige, John & Dominique Pestre (eds) (1997) *Science in the 20th Century* (Amsterdam: Harwood).

Krimsky, Sheldon (2003) *Science in the Private Interest* (Lanham, MD: Rowman & Littlefield).

Kuemmerle, Walter (1999) "Foreign Direct Investment in Industrial Research in the Pharmaceutical and Electronics Industries," *Research Policy* 28: 179–93.

Lamoreaux, Naomi, Daniel Raff, & Peter Temin (2003) "Beyond Markets and Hierarchies," *American Historical Review* 108: 404–33.

Lamoreaux, Naomi, Daniel Raff, & Peter Temin (2004) "Against Whig History," *Enterprise and Society* 5: 376–87.

Langlois, Richard (2004) "Chandler in the Larger Frame," *Enterprise and Society* 5: 355–75.

Larédo, Philippe & Philippe Mustar (eds) (2001) *Research and Innovation Policies in the New Global Economy: An International Comparative Analysis* (Cheltenham, U.K.: Edward Elgar).

Lecuyer, Christophe (1995) "MIT, Progressive Reform and Industrial Science," *Studies in the History of Physical and Biological Sciences* 26(1): 35–88.

Leslie, Stuart (1993) *The Cold War and American Science* (New York: Columbia University Press).

Lessig, Lawrence (2001) *The Future of Ideas* (New York: Random House).

Lessig, Lawrence (2004) *Free Culture*. Available at: http://www.free-culture.org.

Leydesdorff, Loet & Henry Etzkowitz (1998) "The Triple Helix as a Model for Innovation Studies," *Science and Public Policy* 25(3): 195–203.

Leydesdorff, Loet & Henry Etzkowitz (2003) "Can 'the Public' Be Considered a Fourth Helix in University-Industry-Government Relations?" *Science and Public Policy* 30(1): 55–61.

Lowen, Rebecca (1997) *Creating the Cold War University* (Berkeley: University of California Press).

Matkin, Gary (1990) *Technology Transfer and the University* (New York: Macmillan).

Mayer, Anna-K. (2004) "Setting up a Discipline: British History of Science and 'the End of Ideology,' 1931–48," *Studies in the History and Philosophy of Science* 35: 41–72.

McGucken, William (1984) *Scientists, Society and the State* (Columbus: Ohio State University Press).

McSherry, Corynne (2001) *Who Owns Academic Work? Battling for Control of Intellectual Property* (Cambridge, MA: Harvard University Press).

Mirowski, Philip (2002) *Machine Dreams: Economics Becomes a Cyborg Science* (New York: Cambridge University Press).

Mirowski, Philip (2004a) *The Effortless Economy of Science?* (Durham, NC: Duke University Press).

Mirowski, Philip (2004b) "The Scientific Dimensions of Social Knowledge," *Studies in the History and Philosophy of Science A* (June) 35: 283–326.

Mirowski, Philip (2004c) "Caveat Emptor: Rethinking the Commercialization of Science in America," paper presented to HSS meetings, Austin, TX. Available at: http://hssonline.org.

Mirowski, Philip (2005) "Hoedown at the OK Corral: More Reflections on the 'Social' in Current Philosophy of Science," *Studies in the History and Philosophy of Science* 36(4): 790–800.

Mirowski, Philip (2007) "Markets Come to Bits," *Journal of Economic Behavior and Organization* 63: 209–42.

Mirowski, Philip (forthcomingA) "Why There Is (as Yet) No Such Thing as an Economics of Knowledge," in Harold Kincaid (ed), *The Philosophy of Economics* (Oxford: Oxford University Press).

Mirowski, Philip (forthcomingB) *SciMart: The New Economics of Science* (Cambridge, MA: Harvard University Press).

Mirowski, Philip & Dieter Plehwe (eds) (forthcoming) *The Making of the Neoliberal Thought Collective* (Cambridge, MA: Harvard University Press).

Mirowski, Philip & Edward Nik-Khah (forthcoming) "Markets Made Flesh," in Donald MacKenzie (ed), *Performativity in Economics* (Princeton, NJ: Princeton University Press).

Mirowski, Philip & Esther-Mirjam Sent (eds) (2002) *Science Bought and Sold* (Chicago: University of Chicago Press).

Mirowski, Philip & Rob Van Horn (2005) "The Contract Research Organization and the Commercialization of Science," *Social Studies of Science* 35(4): 503–48.

Miyake, Shingo (2004) "Universities Get Taste of Business World," *Nikkei Weekly,* July 26.

Miyoshi, Masao (2000) "Ivory Tower in Escrow," *Boundary 2* 27: 7–50.

Monbiot, George (2003) "Guard Dogs of Perception: Corporate Takeover of Science," *Science and Engineering Ethics* 9: 49–57.

Morey, Ann (2004) "Globalization and the Emergence of For-Profit Higher Education," *Higher Education* 48: 131–50.

Morin, Alexander (1993) *Science Policy and Politics* (Englewood Cliffs, NJ: Prentice-Hall).

Mowery, David (1981) "The Emergence and Growth of Industrial Research in American Manufacturing, 1899–1945," Ph.D. diss., Stanford University, Stanford, CA.

Mowery, David (1990) "The Development of Industrial Research in United States Manufacturing," *American Economic Review, Papers and Proceedings* 80: 345–49.

Mowery, David & Nathan Rosenberg (1998) *Paths of Innovation* (New York: Cambridge University Press).

Mowery, David, Richard Nelson, Bhaven Sampat, & Arvids Ziedonis (2004) *Ivory Tower and Industrial Innovation* (Stanford, CA: Stanford University Press).

Nace, Ted (2003) *Gangs of America* (San Francisco: Berrett-Koehler).

Narula, Rajneesh (2003) *Globalization and Technology* (Cambridge: Polity Press).

National Bureau of Economic Research (NBER) (1962) *The Rate and Direction of Inventive Activity* (Princeton, NJ: Princeton University Press).

National Science Board (NSB) (2004) *Science and Engineering Indicators 2002* (Arlington, VA: National Science Foundation). Available at: www.nsf.gov/sbe/srs/seind02.

Nelson, Richard (2001) "Observations on the Post Bayh-Dole Rise of Patenting at American Universities," *Journal of Technology Transfer* 26: 13–19.

Nelson, Richard (2004) "The Market Economy and the Scientific Commons," *Research Policy* 33: 455–71.

Newfield, Christopher (2003) *Ivy and Industry* (Durham, NC: Duke University Press).

Nightingale, Paul (2003) "If Nelson and Winter Are Only Half Right About Tacit Knowledge, Which Half?" *Industrial and Corporate Change* 12(2): 149–83.

Noble, David (1979) *America by Design* (New York: Oxford University Press).

Nowotny, Helga, Peter Scott, & Michael Gibbons (2001) *Re-Thinking Science: Knowledge and the Public in an Age of Uncertainty* (Cambridge: Polity Press).

Nowotny, Helga, Peter Scott, & Michael Gibbons (2003) "Mode 2 Revisited," *Minerva* 41: 175–94.

Owen-Smith, J. & W. Powell (2003) "The Expanding Role of Patenting in the Life Sciences," *Research Policy* 32(9): 1695–1711.

Pels, Dick (2005) "Mixing Metaphors: Politics or Economics of Knowledge?" in Nico Stehr & Volker Meja (eds), *Society and Knowledge,* 2nd ed. (New Brunswick, NJ: Transaction Publishers): 269–98.

Pestre, Dominique (2003) "Regimes of Knowledge Production in Society: Towards a More Political and Social Reading," *Minerva* 41: 245–61.

Pestre, Dominique & John Krige (1992) "Some Thoughts on the Early History of CERN," in Peter Galison & Bruce Hevly (eds), *Big Science: The Growth of Large-Scale Research* (Stanford, CA: Stanford University Press): 78–99.

Pickering, Andrew (2005) "Decentering Sociology: Synthetic Dyes and Social Theory," *Perspectives on Science* 13: 352–405.

Powell, Woody & Kaisa Snellman (2004) "The Knowledge Economy," *Annual Review of Sociology* 30: 199–220.

Press, Eyal & Jennifer Washburn (2000) "The Kept University," *Atlantic Monthly,* March: 39–54.

Ravetz, Jerome (1971) *Scientific Knowledge and Its Social Problems* (Oxford: Oxford University Press).

Raynor, Gregory (2000) "Engineering Social Reform: The Ford Foundation and Cold War Liberalism, 1908–1959," Ph.D. diss., New York University.

Reddy, Prasada (2000) *The Globalization of Corporate R&D* (London: Routledge).

Reich, Leonard (1985) *The Making of American Industrial Research* (Cambridge: Cambridge University Press).

Reingold, Nathan (ed) (1979) *The Sciences in the American Context* (Washington, DC: Smithsonian Institution Press).

Rodriguez, Maria Joao (2002) *The New Knowledge Economy* (Cheltenham: Elgar).

Rodriguez, Maria Joao (2003) *Economic Policies for the Knowledge Economy* (Cheltenham: Elgar).

Rose, Nikolas (1999) *Powers of Freedom* (Cambridge: Cambridge University Press).

Samuelson, Paul (2004) "An Interview with Paul Samuelson," *Macroeconomic Dynamics* 8: 519–42.

Scheiding, Thomas (unpublished) "Publish and Perish," Ph.D. diss., University of Notre Dame, Notre Dame, IN.

Sell, Susan (2003) *Private Power, Public Law: The Globalization of Intellectual Property Rights* (New York: Cambridge University Press).

Shapin, Steven (2003) "Ivory Trade," *London Review of Books,* September 11: 15–19.

Shapiro, Carl & Hal Varian (1999) *Information Rules* (Cambridge, MA: Harvard Business School Press).

Shi, Yanfei (2001) *The Economics of Scientific Knowledge* (Cheltenham, U.K.: Edward Elgar).

Shinn, Terry (2002) "The Triple Helix and the New Production of Knowledge," *Social Studies of Science* 32(4): 599–614.

Shinn, Terry (2003) "Industry, Research and Education," in Mary Jo Nye (ed), *Cambridge History of Science*, vol. 5 (New York: Cambridge University Press): 133–53.

Shreeve, James (2004) *The Genome War* (New York: Knopf).

Slaughter, Sheila & Gary Rhoades (2002) "The Emergence of a Competitiveness R&D Policy Coalition and the Commercialization of Academic Science," in Philip Mirowski & Esther-Mirjam Sent (eds) (2002), *Science Bought and Sold* (Chicago: University of Chicago Press): 69–108.

Slaughter, Sheila & Gary Rhoades (2004) *Academic Capitalism and the New Economy* (Baltimore, MD: Johns Hopkins Press).

Slaughter, Sheila, Cynthia Archerd, & Teresa Campbell (2004) "Boundaries and Quandaries: How Professors Negotiate Market Relations," *Review of Higher Education* 28(1): 129–65.

Smith, John Kenly (1990) "The Scientific Tradition in American Industrial Research," *Technology and Culture* 31: 121–31.

Steen, Kathryn (2001) "Patents, Patriotism, and Skilled in the Art: U.S.A. vs. the Chemical Foundation," *Isis* 92: 91–122.

Streitz, Wendy & Alan Bennett (2003) "Material Transfer Agreements: A University Perspective," *Plant Physiology* 133: 10–13.

Swann, John (1988) *Academic Scientists and the Pharmaceutical Industry* (Baltimore, MD: Johns Hopkins University Press).

Teske, Paul & Renee Johnson (1994) "Moving Towards an American Industrial Technology Policy," *Policy Studies Journal* 22: 296–311.

Thursby, J. & M. Thursby (2003) "University Licensing and the Bayh-Dole Act," *Science* 301: 1052.

Tijssen, Robert (2004) "Is the Commercialization of Scientific Research Affecting the Production of Public Knowledge?" *Research Policy* 33: 709–33.

Tobey, Ronald (1971) *The American Ideology of National Science, 1919–30* (Pittsburgh, PA: University of Pittsburgh Press).

Vaitilingham, Romesh (ed) (1999) *The Economics of the Knowledge-Driven Economy* (London: Department of Trade and Industry).

Van Horn, Robert (unpublished) "The Rise of the Chicago School of Law and Economics," Ph.D. diss., University of Notre Dame, Notre Dame, IN.

Veblen, Thorstein (1918) *The Higher Learning in America* (New York: Heubsch).

Walpen, Bernhard (2004) *Die Offenen Feinde und Ihre Gesellschaft* (Heidelberg: VSA Verlag).

Walsh, John, Ashish Arora, & Wesley Cohen (2003) "Effects of Research Tool Patents and Licensing on Biomedical Innovation," in Wesley Cohen & Stephen Merrill (eds) (2003), *Patents in the Knowledge-Based Economy* (Washington, DC: National Academies Press): 285–340.

Walsh, John, Charlene Cho, & Wesley M. Cohen (2005) "View from the Bench: Patents and Material Transfers," *Science,* September 25, 309: 2002–3.

Washburn, Jennifer (2005) *University, Inc.* (New York: Basic Books).

Westwick, Peter (2003) *The National Labs: Science in an American System, 1947–74* (Cambridge, MA: Harvard University Press).

Wise, George (1980) "A New Role for Professional Scientists in Industry," *Technology and Culture* 21: 408–29.

Woolgar, Steve (2004) "Marketing Ideas," *Economy and Society* 33: 448–62.

Wright, Gavin (1999) "Can a Nation Learn? American Technology as a Network Phenomenon," in Naomi Lamoreaux, M. Daniel, G. Raff, & Peter Temin (eds), *Learning by Doing in Markets, Firms and Countries* (Chicago: University of Chicago Press): 295–326.

Wright, Susan (1994) *Molecular Politics* (Chicago: University of Chicago Press).

Ziedonis, Arvids A. (2004) "Review of *MIT and the Rise of Entrepreneurial Science,*" *Research Policy* 33: 177–78.

Ziman, John (1994) *Prometheus Bound* (Cambridge: Cambridge University Press).

27 Organizational Contexts of Science: Boundaries and Relationships between University and Industry

Jennifer L. Croissant and Laurel Smith-Doerr

Although there has always been commerce between the ivory tower and the for-profit lab, these two realms have seen unprecedented overlap since the 1980 U.S. Bayh-Dole act. In this chapter we explore issues in the interdisciplinary literature on university-industry research relationships (UIRRs), such as the relationships between academic and market logics and values, exchange in personnel, transfers of intellectual property, and governance issues. We look at the broader political and economic contexts and at global connections that have made academic-industry relationships commonplace worldwide. Our review also examines the effects of blurred boundaries between university and industry on scientific careers and on economic development. In addition, we trace the path of scholarship on the relations between academy and industry. For example, UIRR, once a common acronym, is now disappearing as "mode 2," "helical," and "academic" capitalism models of research organization gain prominence. And consider that prior to the 1980s, "technology transfer" meant international "development" efforts to transfer technology from "first-world" to "third-world" countries, not the transfer of state-subsidized knowledge products to private enterprise.

Exchanges between industrial firms and university researchers have rapidly expanded in both number and kind over the past twenty-five years. While U.S. federal support of research has seemingly reached a plateau,[1] industrial firms have stepped up their interest in universities. Despite the large NIH (National Institutes of Health) budget—almost $24 billion in 2004—only about 36 percent of proposals are funded, including grant renewals (Stephan, 2003). So perhaps it is not surprising that at the top 50 U.S. research universities, 28 percent of life scientists receive industry funding (Blumenthal et al., 1996). Across all sciences and engineering fields, federal research funding declined by 9.4 percent in real dollars per academic researcher between 1979 and 1991 (Cohen et al., 1998). Again, at the top 50 U.S. universities, 43 percent of faculty report receiving gifts from industry in the form of equipment, materials, support for students, or travel (Campbell et al., 1998). Early-stage technology ventures, like the companies spun off from university technology licenses, are funded mostly by industry. Between 31 and 48 percent of spinoffs and start-ups receive their funding from industry, whereas only 2 to 4 percent receive it from universities (Auerswald &

Branscomb, 2003: 75). Overall, industry funding as a percentage of the total support for university research was 2.6 percent in 1970, 3.9 percent in 1980, and 7.1 percent in 1994 (Henderson et al., 1998).[2] While still modest in terms of total revenues, this near tripling of university support is nonetheless significant. Furthermore, academic research and development performance increased slightly from 10 percent in the 1970s to 13 percent of total R&D expenditures, or approximately USD $36 billion ($33 billion in 1996 constant dollars) (National Science Board, 2004). The relative importance of different sectors in these totals, with growth shifting to biotechnology, has altered significantly. It is clear that UIRRs are important sources of revenue and present significant issues for universities and public life (Slaughter & Rhoades, 2004; Busch, 2000; Slaughter & Leslie, 1997; Bok, 2003; Bowie, 1994; Cole et al., 1994).

We begin with scholarship that attempts to explain the origins of commercialization in the university, move to research that tries to quantify the relations between university and industry, and then examine studies that locate UIRRs geographically. Next we consider the consequences of UIRRs in reviewing the debate about whether UIRRs add up to a new knowledge society or require a new kind of legal regulation. We conclude with a discussion of new directions for both expanding and synthesizing the literature on UIRRs.

CONTEXTS AND ANTECEDENTS

To write the full history of UIRRs would require attention to several categories of scholarship: the history of universities and university systems; the history of research, such as a general history of science and its antecedents in craft work; the history of various research locations, whether industrial or national laboratories or out in the field; international historical and comparative investigations; and also a history of state policies and university-state relations. Such a comprehensive project is of course beyond the scope of this chapter. Even a study focused on a single institution in the United States since the twentieth century fills a monograph, as do other reviews of the topic (e.g., Mowery et al., 2004; Etzkowitz, 2002; Leslie, 1993; Mirowski & Sent, 2002).

However, a basic review of the trajectory of UIRRs over time permits us to outline central research and policy issues. The first pivotal moment for UIRRs in the United States was the passage of the Morrill Act in 1862. This act set aside public lands to be used or sold in the western states, for the establishment of

at least one college where the leading object shall be, without excluding other scientific and classical studies, and including military tactics, to teach such branches of learning as are related to agriculture and mechanic art . . . in order to promote the liberal and practical education of the industrial classes in the several pursuits and professions of life (Morrill Act, Public Law 37–108, U.S. Statutes at Large 12 [1862]: 503).

While the Morrill Act focused on teaching and instruction, it grounded that instruction in the agricultural and productive enterprises of the time and paved the way for a series of supporting acts. International influences are also important: the Humboldt model of the university from Germany, including a classical orientation to liberal arts and an integration of research and teaching in mass university settings, came to represent one ideal (see Geiger, 1986; Lenoir, 1998). As U.S. institutions both embraced and transformed this model of the university, local connections with specific industrial activities and general connections to systems of capital shaped university instruction and research (Noble, 1977).

The Hatch Act of 1887, a descendent of the Morrill Act, established the agricultural experiment station system and led to the agricultural extension services (Hatch Act, U.S. Code 361, Statute 440 [1887]:313[24]). The various Acts and those that followed facilitated the development of the extensive public university system in the United States, which paralleled the continuing development of the private universities and technological institutes. As Noble (1977) argues, these were also facilitated by the professionalization of science and technology and the alignment of the professions with the emerging structures of corporate capitalism. It is in part because of these early alignments that we can say that recent concern about commercialization and university research is not so much new as it is more reflective of changes in degree, scope, and scale of the relations of intellectual production (Croissant & Restivo, 2001; Weiner, 1986; Rossman et al., 1934).

The second pivotal process for UIRRs was the establishment of the contemporary federal research administration (see Geiger, 1993). The span of a generation saw the founding of the National Science Foundation (NSF; 1950), the National Laboratory system (1943), the Defense Advanced Research Projects Agency (DARPA; 1958), and the Department of Energy (DOE; 1971). Although the National Institutes of Health was founded in 1887, it faced reorientation and reorganization similar to other federal research initiatives following World War II, especially with President Nixon's declaration of a "war on cancer" in 1971. This second phase represents the implementation of a specific linear model of scientific research, encapsulated by Vannevar Bush in his [1945]1990 report framing what would become the NSF. Investment in basic science, directed by the scholarly community, would provide foundational knowledge that would circulate in the public domain, spurring specific technical innovation through private industrial research and development or implementation through national or military laboratories.[3]

We treat 1980 and the passage of the 1980 Patents and Trademark Law Amendments Act, known as Bayh-Dole, as the third pivotal phase. Although the centrality of Bayh-Dole and its exact causal role in the surge in patenting and licensing activities has been debated (Mowery et al., 2004), there is nonetheless an undisputable inflection point marking a rise in patenting activities across universities in the early 1980s. Bayh-Dole allowed universities to retain property rights to inventions made with federal funding. The rationale for the change was in part to encourage industry to take the risks of developing fundamental ideas through exclusive licensure and to provide financial incentives and rewards for facilitating the development of new technologies and products.

Prior to Bayh-Dole, claims to intellectual property (IP) support by federal research dollars were negotiated on a case-by-case basis, with the default ownership of the patent resting with the funding agency. Institutional Patent Agreements (IPAs) developed as a mechanism in the late 1960s to alleviate the case-by-case assignment of IP rights. Many institutions did not have in-house IP management expertise, such as the nearly ubiquitous "technology transfer office" in contemporary major universities.[4] Instead, many contracted with Research Corporation for these services, which may have in part masked the pre-Bayh-Dole surge in patenting, since rights were not assigned to individual institutions but through the research corporation (Mowery et al., 2004). Bayh-Dole both reflected a growing interest in IP generated from rapidly increasing federal research support following World War II and also facilitated the explosive growth of university patenting emerging in the late 1980s.

In conjunction with the changes in federal policies and laws, other experiments in UIRRs were undertaken (see Geiger 2004, especially chapter 5). The NSF started the first Science and Technology Center in 1978, with the first Engineering Research Center founded in 1984 (Adams et al., 2001), and these became the center of a more general initiative in Industry University Cooperative Research Centers (IUCRCs), and a specific program with individual state (S/IUCRCs) involvement for regional economic development was founded in 1991 by the NSF. In 2005, the fifty I/UCRCs mentioned above had been sponsored by the NSF, with industrial support totaling 10 to 15 times the amount contributed to the NSF.[5] State-level projects facilitated the emergence of research parks and various forms of business incubators. As we discuss below, research parks, especially, have been linked to university locales, with the Stanford Industrial Park providing the center of gravity for what would become Silicon Valley in 1953, followed by Research Triangle Park in North Carolina in 1958.[6] Several waves of development of research parks have since followed.

Given that the three pivotal moments defining the trajectory of UIRRs in the United States are all moments of government interaction, it is not clear that there has been sufficient study of the role of the state, policy formation, and policy implementation in the UIRR literature.[7]

UIRRs also connect to other issues often ignored by researchers in science and technology studies: individual and corporate philanthropy, commercialization and finance in other university functions (such as food services, instrumentation, data systems, and printing), and the emergence of intellectual property and profit-driven concerns arising from instructional activities, such as copyrighted course materials and distance education (Slaughter & Rhoades, 2004). The field of higher education has largely ignored research and knowledge production as one of the things that happens in universities (in favor of studying student development, administration, and finance), while science and technology studies as a discipline has largely ignored the details and mechanisms in the organizational and institutional constitutive elements in knowledge production.

METRICS AND ECONOMICS: THE RIGHT JOB FOR THE TOOLS?

The quantification of university-industry research relationships is a central task of the economics of innovation, itself a subdiscipline that has grown substantially only since the 1990s. To provide a simple illustration of this trend, we searched the EconLit database (indexed by the American Economic Association) for literature that included the terms "university" (and related variants) and "innovation" in their abstracts. In 1984, zero manuscripts appeared, while the index catalogs 49 pieces in 1994 and 57 in 2004. The main interest in universities in this literature (and the closely related management of innovation and knowledge management fields) is as sources of economic growth.

Traditionally, economics views university knowledge as a "commons:" a public resource appropriable by any firm (Arrow, 1962; Nelson, 1962). Because investments in basic knowledge cannot be fully recaptured by the research unit, these should be publicly funded in universities. Thus, policies among universities and firms to increase their intellectual property holdings present a classic tragedy of the commons problem—individual organizations may benefit to the cost of economic growth in general (Nelson, 2004). Innovation economists themselves have recognized the problem in assuming that knowledge "spillovers" can be absorbed by any firm with equal ease. Firms are active learners (Nelson & Winter, 1982; Cohen & Levinthal, 1990; Feldman, 1994). Models thus increasingly try to assess how firms become active partners in creating and incorporating knowledge developed by universities or other firms (Laursen & Salter, 2004; Patrucco, forthcoming; Malo & Geuna, 2000; Knoedler, 1993). From this literature we know that the number of academic citations in industrial patents increased threefold through the mid-1990s (Narin et al., 1997). However, there has been a trend to cite older patents and to less often cite recent university patents (Mowery et al., 2004; Thursby & Thursby, 2004). Is more recent academic knowledge less useful, or have firms grown wary of academics trying to compete in commercial efforts?

While more neoclassically oriented microeconomists limit their perspective on knowledge spillovers to cases in which firms unintentionally leak information that has value to other firms (Griliches, 1990), economic geographers have adopted the term to mean the transfer of knowledge from universities to (usually high-tech) firms (Acs et al., 1992; Florida, 2002; Audretch & Stephan, 1996; Bresnahan & Gambardella, 2004). It may also be that a sector orientation rather than individual firm methodology (Malerba, 2002) may be necessary to capture the dynamics of competition, technology transfer, and innovation systems.

Throughout this literature, the way that UIRR issues can be measured or modeled with numbers shapes the debate. Consider a few prominent examples. The so-called Yale survey (Klevorick et al., 1995) of technology managers in large Fortune 1000 U.S. firms has a list of measures that constitute what the literature considers *the* ten channels by which knowledge flows from the university to industry (including patents/licensing, publications, conferences, hiring, graduate students). These U.S. surveys are how we know that patents are not considered a very important means of technology transfer (ranked eight of ten; see Cohen et al., 2002). But, for example, field observations of actual collaborations might reveal other dimensions of learning that are not on the list (let alone that other questions should be asked). European large-scale cross-national surveys that include questions on university-industry relations have catchy acronyms like PACE (Arundel & Geuna, 2004) and KNOW (Fontana et al., 2004) or otherwise indicate their promise for informing innovation (e.g., Community Innovation Surveys; see Mohnen & Hoareau, 2003).

The role of universities in measures of global knowledge production is an example of perhaps the most problematic assumptions in econometric models. An article by Jones (2002) assumes that the number of science and engineering graduates from universities in the United States, Japan, Germany, the United Kingdom, and France is a valid measure of the global knowledge base, a disembodied stock of knowledge that can be used by any for-profit firm in the world. While such assumptions permit the creation of parsimonious economic models, they entail a level of socially encoded construction that seems to require neoclassical microeconomics training (MacKenzie & Millo, 2003; Knorr Cetina & Preda, 2004; Callon, 1998; Jones & Williams, 1998).

Economists of innovation have expanded the quantitative analysis in many ways beyond strict neoclassical assumptions, for example, by paying attention to the channels of communication between university and industry. Network models have been offered as a more empirically sophisticated, nuanced approach to collaboration between university and industry (Owen-Smith & Powell, 2001; Geuna et al., 2003; Waluszewski, 2004; Giuliani & Bell, 2005; Gambardella & Malerba, 1999). Here, knowledge flow is not a one-way street; traffic goes in both directions between industry and university. This approach to innovation is also more interdisciplinary, including sociologists as well as economists.

The emphasis in this literature tends to be on the benefits to the economy. For example in a report to the U.K. government on UIRRs (Scott et al., 2001), eight benefits were identified (new information, skilled graduates, scientific networks, problem solving capacity, new instrumentation, new firms, social and regulatory knowledge, access to unique facilities) without a corresponding list of potential costs. Analyses that quantify the relationships between firms and public research organizations are still limited either to available archived data on formal collaboration or to small samples on informal ties.

While there are gaps in knowledge about UIRRs caused by disciplinary specializations, there also are gaps because there is a lack of methodological integration across studies of UIRRs. There have been numerous specific, case-based studies, either historical or contemporary (Hackett, 2001), plus impressionistic accounts based on reports in the popular press, scientific editorials and reports, and occasionally government documents on issues surrounding UIRRs. There are also ethnographic and qualitative studies of work practices and reasoning, such as Kleinman's (2003) study of how the framework of commercial agriculture and IP concerns subtly shape the direction and content of knowledge in a plant pathology laboratory, and studies of IP production practices (McCray & Croissant, 2001; Kaghan, 2001). Colyvas and Powell (2006) demonstrate that Technology Transfer Offices (TTOs) have become increasingly institutionalized in universities. The Stanford TTO had idiosyncratic responses to invention disclosures, then became increasingly standardized as the commercialization process was legitimated during the 1980s, until technology transfer became fully institutionalized in the mid-1990s.

There are also numerous surveys of technology managers in industry (Mansfield, 1991; Cohen et al., 2002) and in universities (Association of University Technology Managers' regular survey of members), of professors (Louis et al., 2002; Owen-Smith & Powell, 2001; Blumenthal et al., 1996; Campbell & Slaughter, 1999; Campbell, 1995), of administrators (Slaughter, 1993), and of students (Hackett, 1990). The study of research integrity and scientific misconduct is a growth industry, but as yet, the role of industrial funding has received scant attention in that area (but see Campbell, 1995; Holleman, 2005). However, the largest growing modes of inquiry on UIRRs have been based on quantitative analyses of patents and licensing agreements. This is both a reflection of the real power of patenting as a means of capturing rents from intellectual work and also a constraint on the scope of analysis. When one has a hammer, all the world becomes a nail. In many ways, the subdiscipline of innovation economics represents an advance beyond the strict neoclassical assumptions of the firm as a contextless black box, but it is still limited to posing questions that can be hammered at by econometric tools.

LOCATION, LOCATION, LOCATION?

Location matters in real estate *and* in UIRRs. We can also talk metaphorically about "location" in that UIRRs are demonstrating effects on social location, namely, systems of institutional stratification. What are the effects *of* location on UIRR activity, as in, how does propinquity, the principle of spatial proximity leading to fruitful intellectual and economic interactions, operate (Geiger, 2004: 205)? And what are the effects *on* location, in both geographic and social terms? For example, the prestige and value of undergraduate education echoes the market-intensified institutional rankings competition (Owen-Smith, 2001), with the "differential returns to selective schools . . . both a social fact and a self-fulfilling prophecy"; thus, one outcome of UIRRs is change in the social "location" of universities in stratified systems (Geiger, 2004:262).

Although research centers are largely located within universities, they are a particular institutional location that has not received the scholarly attention that might be warranted, as noted above, by the significant federal and industrial resources put into them. The National Science Foundation and other agencies involved in collaborative centers have expected the centers to become self-financing from industry. A survey of IUCRC members noted that those firms reporting find economic benefit to their activities (Adams et al., 2001). However, it is not clear whether IUCRCs such as the Engineering Research Centers (ERCs) can successfully separate from state and federal support (Feller, 2002), and the separation of research centers and research parks from government funds has not proved to be problem free (Newcomer, 2001; Fischer, 1999; Geiger, 2004).

Research parks and incubators have been a particular locus of UIRRs, although generally treated in a separate literature. While not all research parks or incubators are affiliated with universities or colleges, this is a prevalent arrangement. In North America, in 2004, there were more than 195 research parks in 40 states or provinces, with an average employment of 3,399 persons, 41 companies per facility, and an average capital investment of USD $186,280,327. In a 2003 survey of park leadership, 83 percent of the parks were not-for-profit entities, with 62 percent providing a business incubator component, 70 percent established with public funds, and 61 percent using public funds to attract tenants.[8] They provide, in the way that foundations do, an arms-length mechanism for universities to undertake economic development.

Nurturing spinoffs from university activities is sometimes seen as the main job of science parks (Stankiewicz, 1994). Stankiewicz argues they "attempt to create an institutional space for activities that do not quite fit into the established structures of academia and business" (1994:102). Spinoffs are large in number, but few grow large or survive long. What this survival means is ambiguous: it depends on the goals of the spinoffs. A small firm bought by a larger interest represents a "failure" of the spinoff in its own terms but also potentially represents the redistribution of technology. It is also not clear whether spinoffs should transform into manufacturing or consulting concerns on a larger scale, fundamentally changing the interest of the university in the firm. This requires defining "success" in ways that reflect the goals of innovators and institutions. Faculty spinoffs do represent the potential problem noted by Adams et al. (2001) of "input diversion," whereby faculty reduce their on-campus commitment for company-related research, masking costs and touching on one of what Johnson (2001) identifies as several potential conflicts of interest or conflicts of commitment. Link and Scott (2003) suggest that a research park has a measurable tendency to shift the academic mission of its parent institution toward more applied curricula.

Universities and firms most often engage in research collaboration on a local basis. Do local UIRRs add up to global innovation? Local political, technological, and geographical bases foster collaboration between scientists and engineers in universities and companies. Saxenian (1996) describes how Frederick Terman at Stanford University provided more than academic mentorship to students like Hewlett and Packard. Terman established Stanford Industrial Park as a training ground for the new information technology industry, where the idea of open boundaries and collaboration across organizations stimulated Silicon Valley. The Valley became a place where high-tech engineers felt more of a regional identity than loyalty to particular employing organizations. And, of course, Silicon Valley is a larger-than-life icon for politicians and lobbyists who would emulate its success through UIRRs in their own regions.[9]

What is not clear, however, is to what extent geographic clusters can be reproduced, to what extent the clusters can be successful, particularly in the absence of state support, and to what extent new communications technology may make proximity less important (Hawkes, 1997). Massey and colleagues (1992) cogently argue that the beneficial effects of research parks are largely overstated, particularly considering the continued state subsidies, such as the absence of property taxes, which undergird their success. A study in the United Kingdom by Siegel and colleagues (2003) suggests that firms associated with a research park do have slightly better research productivity than other firms, but his study is based primarily on a count measure (of patents, for example) rather than on a value measure of productivity.

In the way that "quality," meaning breadth or generality with the number of citations indicating importance, apparently is declining with the rapid influx of new universities into the IP arena (Henderson et al., 1998; Mowery & Ziedonis, 2002; Sampat et al., 2003), similarly, the rapid expansion of research parks and the emulation of models of Silicon Valley for regional development have definitely led to real estate booms in college towns but have had ambiguous effects on regional development (see Knapp, 1998). In the same way that developing a research park next to a nonprestigious institution may not provide economic benefit to the firms or the region, academic stratification also affects the perceived value of intellectual property.

The stratification of institutions by prestige is only partially independent of research and capitalization of intellectual property. In 1991, the top 20 institutions received about 70 percent of the total patents awarded (Henderson et al., 1998). Eighty percent of federal R&D resources go to 100 institutions, and the top ten receive 21 percent of the total disbursements (Owen-Smith, 2001; Schultz, 1996, Chubin & Hackett, 1990). The awarding of public research dollars imperfectly maps onto achievement in patenting (Owen-Smith, 2001). As time has passed since Bayh-Dole, the stratification of universities in either research or patenting has stabilized, so that since the 1990s there is a strong correlation between success in both "public" and "private" science, and a stable patenting elite has emerged. This points to a Matthew effect (Merton, 1988; Merton & Zuckerman, 1968) of accumulative advantage for IP activities at the institutional level, where organizations with prior experience with IP activities are more able to reap the economic benefits.

Feller and co-workers (2002) suggest that success in patenting leads to differential strategies for equity positions. Both the most elite and the least successful institutions are assuming the risks of taking equity positions in start-up companies in lieu of direct compensation from licensing, while the middle tier has not actively pursued this strategy. It appears that more successful institutions can afford to delay, or possibly never receive at all, profit-taking from IP, while the least successful institutions are behaving as if they have nothing left to lose in the race to capitalize new knowledge.

Internationally, IP considerations and economic geography also led to continued issues with institutional stratification. For example, Bakouros and colleagues (2002) suggest that the three Greek technology parks are not particularly successful in either incubating new businesses or in making connections with nearby universities. This has led to a proliferation of studies trying to determine exactly what factors make research parks work.

The United States is viewed as exceptional in its ability to develop innovations through UIRRs. The size and heterogeneity of goals within the American university system contribute to this exceptionalism. For example, the land-grant colleges founded in the nineteenth century provided a focus on practical research problems and inclusion of wider populations in higher education in comparison to the centralized, elite scholarship in Europe during that time (Rosenberg & Nelson, 1994). There was also room in the growing, diverse American system for research universities like Johns Hopkins, founded on the ideal of discovering basic scientific knowledge and academic publication (Feldman & Desroches, 2004). But if pressures to commercialize since Bayh-Dole are creating homogeneity in the American university system, the United States might find itself in a success trap. Formalizing UIRRs could kill their innovative spark. When successful, UIRRs often display elements of path dependence—early events matter disproportionately in the development of a technology—that cannot be planned in a standard fashion. Take, for example, the commercialization of FM sound modulation for synthesizer music during the 1970s to 1980s (Nelson, 2005). The Stanford musician and computer programmer who developed the technology happened to contact Yamaha while a top R&D executive was visiting California, and they met with him immediately. One of Stanford's most lucrative licenses and Yamaha's biggest selling synthesizers came out of the resulting UIRR.

FROM UNIVERSITY-INDUSTRY TIES TO NEW KNOWLEDGE PRODUCTION SOCIETY: CAN WE GET THERE FROM HERE?

Journalists and scholars in a variety of disciplines routinely describe the twenty-first century as the era of the knowledge society, and they often highlight the central role that the increased collaboration between universities and industry plays in this new global knowledge production (Stehr, 1994). Does the current state of connectedness between universities and industrial firms represent a broad societal change, and if so, what does it mean? The literature has considered implications at every level, from individual scientists' careers to globalization in knowledge commerce.

A Vocation from the Norms

The vocation of science reflects the institutional order of the university; this is a basic assumption in the sociology of science literature. Sociology of science led by Robert K. Merton (1973) in the 1940s and 1950s tried to understand social conformity (perhaps reflecting popular American currents at the time). Merton outlined an ethos consisting of four norms ideally universal among scientists for a functional system of science within the university. Communism (later communalism), one of the norms, called for scientists to publicly communicate their results. Industry would impede scientific communication by enforcing secrecy in order to secure the economic benefits of findings. Of course, Merton was idealistic considering he wrote the norms piece during World War II when physics was closely tied to the defense industry (Shorett et al., 2003).

Recent literature still finds scientists expressing Mertonian norms, even if for-profit values appear alongside them. In Cho and colleagues' (2003) survey of lab researchers, 85 percent reported that patenting reduces the amount of information shared between scientists. Owen-Smith and Powell's (2001; Powell & Owen-Smith, 1998) interviews with life scientists, however, reveal the addition of new norms around the scientist-entrepreneur role. Ziman (1994) argues that proprietary knowledge and other "postacademic" norms signal a shift from science as vocation to job. Hackett (1990) explicitly takes on the science as vocation problem and finds tensions between the classic norms and the new realities of crushing research funding burdens for some university scientists and adjunct teaching positions for others, while Louis et al. (2002), Campbell et al. (1998), and Blumenthal et al. (1996) all record concerns about secrecy and delayed publication and other normative conflicts in their surveys of life scientists. Scientists largely still focus on publications over patenting (Agrawal & Henderson, 2002), and commercialization efforts are tied to existing research rather than changing research focus (Thursby & Thursby, 2002; Colyvas et al., 2002). Still, there are some reports that academic scientists involved in commercialization efforts become more secretive (Blumenthal et al., 1997; Campbell et al., 2002; Walsh & Hong, 2003).

Can seemingly contradictory norm systems peacefully coexist? Sennett (1998) is pessimistic; he claims that flexibility and rapid change in the new economy forbids the creation of narratives that give people's careers meaning. But scientists are resilient; they readily make meaning out of newness itself in their work environment (Smith-Doerr, 2005). It is precisely the more complex, interconnected organizational structures that offer scientists greater possibilities for ignoring institutional tensions. Smith-Doerr (2005) illustrates this in the context of a biotech firm, where scientists gloss over the tensions between traditional academic norms and discourse about the innovative commercial setting in which they work. Kleinman's (2003) research among university scientists demonstrates another dimension of biotechnology that hides the tension between university and industry. He shows that firms' interests are usually formulated in indirect ways rather than as direct exchanges of money for secrecy. Late-twentieth-century biotechnology is a visible contrast to the story of mid-twentieth-century golden ageism: industry provides scientists a seemingly more open research environment than academe (Rabinow, 1996; Owen-Smith & Powell, 2001; Rayman, 2001; Smith-Doerr, 2004). The question of the existence and structure of scientists' alleged distaste for industry is less examined.

Certainly, the argument that Mertonian norms present an ideal type rather than a description of reality is a familiar refrain from Mitroff (1974) onward. But the discord in the "golden age" story, that of the constraints on scientists in earlier commercial contexts (Kornhauser, 1962; Noble, 1977; Hounshell & Smith, 1988) is missing, and the fable seems to continue to be told without much criticism. For example, the Mertonian normative structure continues to be articulated in primers for students (National Academy of Science, 1995) and undergirds major dimensions of the emerging research on research misconduct industry. However, Shapin (2004) articulates precisely such a critique by exploring an overlooked literature: postwar industrialists trying to figure out how to manage scientists. Although they wrote during the 1940s to 1960s when Merton and his students (Barber, 1952; Hagstrom, 1965) were creating the idea that scientists socialized in academia would chafe at the secrecy and lack of autonomy in industry, the managers had a very different picture of the problems of scientists in industry. Shapin argues that the industrialists' complaints had nothing to do with role conflict between academic ideals and industrial labs. In contrast, industrialists worried about how they could motivate scientists to keep up with the cutting edge in their field through publishing. Still, in Shapin's data we do not hear the voices of scientists themselves, only their managers.

Models, Modes, and Metaphors: Creators and Critics of Categorization

UIRRs are thought to have larger implications beyond effects on how individual scientists should regard their careers (or vocations). Post-1980 collaboration between universities and industries is often categorized as a break from the "linear model" of knowledge production. In the linear model, fundamental knowledge from the university is developed and applied to technology by industry. Callon (2003) prefers the phrase "Cold War institutional configuration" to describe the viewpoint from Vannevar Bush's endless frontier to Kenneth Arrow's science as public good. He argues that this configuration was a boundary-making claim, a linear narrative to try to separate university research from the market, rather than reality. Yet the new organization of science contrasted to the linear model is variously termed mode 2 (Gibbons et al., 1994), academic capitalism (Slaughter and Leslie, 1997), and the triple helix model (Etzkowitz et al., 1998; Etzkowitz & Leydesdorff, 1997). Gibbons and his five coauthors (1994) famously categorized the earlier linear model as mode 1 in contrast to mode 2. Mode 1 knowledge is bounded in academic disciplines, hierarchically organized, and separated into discovery or application. According to Gibbons et al. (1994), mode 2 knowledge focuses on problems to transcend disciplines, communicates through dense networks to innovate, and creates tension because standardization of scientific competence occurs alongside heterogeneous sources of information. Similarly, Etzkowitz and Leydesdorff use the metaphor of the triple helix to categorize new knowledge production as a break from the old linear model. The triple helix represents the progressing spiral of interconnected communication between university, industry, and government (Etzkowicz et al., 1998; Etzkowicz & Leydesdorff, 2000). The triple helix of university-industry-government means that the institutions are converging through their connections: the university is becoming more entrepreneurial, industry more knowledge based, and government more relational. The result is the formation of firms and regional advantages. Fujigaki and Leydesdorff (2000) argue that better quality control operates through communication in project development networks instead of leaving quality as a post-project issue. A series of Triple Helix meetings has encouraged scholarship applying the model to different settings, for example, to Japan and Singapore (Baber, 2001), where an increasing number of transdisciplinary fields seem to fit the model.

As the Cold War crept to its conclusion, the discourse of science policy, especially the justification for federal investment in research, shifted accordingly to a "competitiveness" agenda that reflected the convergence of business and policy interests in the face of renewed Japanese and European economies (Slaughter, 1996) and the interests of universities in securing funding and status through commercialization (Slaughter, 1993). Stokes (1997) destabilizes the "basic" and "applied" research distinction. Knowledge activities can be variously categorized in relation to their focus on searching for fundamental principles and generalizing frameworks, or situated in relation to problems derived by constituencies external to the primary intellectual community, or neither, or both. The two-by-two table of "use-inspired basic research," "pure basic," and "pure applied" research allows for consideration of what knowledge might be prioritized under what funding regimes. Unlike mode 2 knowledge production, which, in its original formulation (Gibbons et al., 1994), privileges a wholesale move to transdisciplinary and "applied" research, in Stokes's framework there remains a coherent space for research tied neither to market logics nor to models that privilege research providing theoretical generalization, that is, humanistic study and other forms of inquiry.

Although the academic capitalism model portrays a similar historical categorization of knowledge production to mode 2 and triple helix, there is a basic disagreement about whether, in the end, the change represents cost or benefit to the university and society in general. The normative position is often revealed in semantics, not surprisingly. The helix is the icon of biotechnology research, itself the poster child for the profitability of UIRRs. Academic capitalism proclaims its neo-Marxist orientation. In this model, academic capitalism (Slaughter & Leslie, 1997; Slaughter & Rhoades, 2004) is caused by economic globalization, funding cuts, information technology, and neoliberal deregulation. This university context causes a shift in academic norms toward a hypercompetitive opportunism. From the growing competition for federal dollars, the university faces a potentially Faustian bargain with industry (Bok, 2003; Cole, 1994). For example, university scientists' decisions about where to direct graduate student research are affected by the pressure to produce results to sustain industry funding (Slaughter et al., 2002).

Perhaps the categorization of knowledge production draws attention (both favorable and not) because it lends itself so easily to policy-making justifications. Collectively, models of sharp change in the production of knowledge (particularly the decreasing autonomy of the university) have provided fodder for rethinking the governance of science. For instance, a policy implication of a new mode of knowledge production would be that universities should train Ph.D.'s across different locations, not just within university labs (Rip, 2004). The arguments to regulate applications and formulate rules of ownership in science also become easier to make (Stehr, 2004).

In any event, reactions among STS scholars to models of knowledge production—particularly the normatively positive ones—have taken a sharply critical tone. The mode 2 and triple helix models are criticized for treating knowledge itself as a black box easily handed off between university and industry scientists. The lack of epistemological sensibilities, such as how knowledge is constructed, is seen as a major weakness by STS (Baber, 1998). Critics argue that the models present research taking place outside of institutions in an amorphous environment. Cyberspace and knowledge generation seem to be conflated, when in fact knowledge is "sticky" in its tacitness, social efforts, and infrastructure requirements and is not the same as information bytes. Elzinga (2004) argues that this decontextualized view of knowledge is used to sell the "universal" triple helix model to third world countries. Here is where the heart of the

STS criticism seems to lie—that knowledge is embedded in social contexts, in relationships between people, and with materials. Free-flowing information in a global knowledge society—the triple helix/mode 2 picture—misses this embeddedness (Grundmann, 2004).

The models also are said to be "presentist." Without historical context, these models tend to promote neoliberal privatization and globalization (Drori et al., 2003; Pestre, 2000). The neoliberalism encoded in the models gets translated into national policies. The critics even suggest the radical discontinuity change thesis is an agenda to earn advocate authors more funding (Godin, 1998; Shinn, 1999). Shinn (2002) goes so far as predicting that these models will be more an intellectual fad than an enduring contribution to scholarship.

Donald MacKenzie (2004) reminds STS scholars also to turn a constructivist eye on ourselves. Theoretical models in themselves do not have necessary political implications; people construct the implications. Yet theoretical models (e.g., mode 2) do have force, they can become self-fulfilling prophecies. The authors of mode 2 themselves note how their book has been used by management schools and nurses, among others, to legitimize reorganizing their workplaces (Nowotny et al., 2003).

In sum, reactions to the new organization of knowledge production are mixed. Some laud increased interaction between university and industry for stimulating innovation, the main driver of a healthy economy (e.g., Libecap, 2005; Zemsky et al., 2005; McKelvey et al., 2004; Florida, 2002; Stephan & Audretch, 1999; Etzkowicz et al., 1998; Lundvall, 1992; Hodges, 2001). Others decry the loss of academic freedom and values changes in the university (e.g., Nelson, 2004; Slaughter & Rhoades, 2004; Croissant & Restivo, 2001; Washburn, 2005). Still others criticize scholarship that defines current university-industry relations as a big shift, when history evinces similar trends (Pestre, 2000; Shinn, 2002). Despite sharp disagreements, as a whole, the large and growing literature on knowledge production demonstrates that academics are becoming more self-reflexive about the role of the university in society. Policy (i.e., for research funding) that requires scientists to be more broadminded is one example of this reflexive turn (Rip, 2004). Scientists, even if they never actually lived in an ivory tower, are now explicitly expected to be more than specialists and to reflect on the broader implications of their work.

CONCLUSIONS

In the study of technology transfer, the neophyte and the veteran researcher are easily distinguished. The neophyte is the one who is not confused (Bozeman, 2000: 627).

Our review of literature on the organization of university-industry ties has been largely disconnected: each substantive section (histories, metrics, models, locations) in many ways stands alone. This kind of divide reflects the literature itself. Any "conversation" between different streams of literature (e.g., between economics or management and the history of technology) seems to take the form of criticism, particularly over models of historical trends. To what extent do we need exacting quantitative data to know whether the mingling of industry/university/government for science and technology (e.g., triple helix model) is better policy than maintaining more heterogeneous traditional institutional divisions (MacKenzie, 2004)?

We need to be able to sort out the relationships between causes and effects, and between policies and their implementation, and continue the interrogation of framing assumptions. For example, for the proactive literature on UIRRs, whether for the siting of research parks or choosing the most productive IP policy, there needs to be more substantial discussion of the implications of UIRRs and of research on UIRRs: How to do we do it "better"—and how are we defining better? For example, do patents or research centers or research parks "work," and what is meant by "work," and for whom?

As Martin (1998) suggests, there are a lot of assumptions largely untested in framing an economic system based on securing intellectual property through patents. For example, in some industries such as microelectronics, the life of the technology is far shorter than even the time spent to secure the patent, let alone the life of the patent, so trade secrets and speed to market are the best guarantors of collecting economic rents for an innovation. Patents are sometimes used to block innovation. Patents can be engineered-around. Patents are, sometimes, merely a license to sue.[10]

The legal mechanism that is most often called upon to mediate between university and industry interests is patent law. Eisenberg's prescient 1987 article argues that the element of disclosure in patents can actually lead to freer scientific communication than in publishing. She uses the example of the 1980 debate in the *Journal of Biological Chemistry* to make her case (Eisenberg, 1987). Under debate was the policy to require that the biological materials used for any published article had to be made available for public use (such as placement in a culture collection center). Scientists did not want to do this for reasons unrelated to commercial motivation; the main concern was that replicability of their publications would be taken up by free-riders, and authors would lose the exclusive benefits of their discovery for publishing. Eisenberg argues that patenting allows scientists exclusive rights and overrides the incentive to hoard findings until all relevant publications are out. Patenting was indeed the NIH's approach to the human genome in the early 2000s in order to keep information in the public domain. The monopoly provided by patents has served as a spur to innovation, in public research organizations as well as industrial labs, in part because of its self-fulfilling character.

Legal approaches to UIRRs begin with the notion of scientific knowledge as "the commons." Basic research provides a public good on which everyone can draw to create new ideas or products. Heller and Eisenberg (1998) argue that the problem with excessive patenting in the sciences is not the classic tragedy of the commons where the resource is used up, but rather an "anticommons" problem. Privatization of upstream knowledge (by universities and firms alike) deters downstream development, such as lifesaving biomedical innovation (Nelson, 2001). In response, Lessig (2004) has developed a "creative commons" licensing scheme for the Internet, including making his book *Free Culture* available online.[11] (Lessig believes that litigation by big businesses to capture revenues for creative content available on-line is comparable to farmers' nonsensical lawsuits in the early years of air travel for chickens killed by planes flying in "their" airspace. To Lessig, the Internet is like airspace, a public commons that society benefits from using freely. Universities, presumably, would be advocates for a more open, creative commons.

Describing a recent legal development, Eisenberg (2003) explains the probable effects of the *Madey v. Duke University* ruling on patent enforcement in universities. The ruling strips academic scientists of the traditional research exemption from patent infringement that had been normative since Justice Story's 1813 opinion that experiments undertaken for pure philosophical curiosity are exempt from intellectual property disputes. Interestingly, rather than stemming from firms arguing that universities are invading commercial turf and should pay their way, *Madey v. Duke* was a battle internal to academia. Professor Madey brought intellectual property from Stanford to Duke when he was hired there. After being fired from Duke, Madey wanted to continue with grants based on the IP, which Duke disputed. The court felt that the "business" of universities is now so tied to grant writing and intellectual property that the "research exemption" no longer holds true.

However, the real problem of the legal patent system (particularly in the United States), according to Jaffe and Lerner (2004) is that the system encourages the proliferation of low-quality patents. The United States has developed a court system in which the patent holder almost always wins. This means so-called patent trolls are rewarded for buying patents that can block organizations' daily work. For example, the owner of the "call center" series of patents regularly receives million dollar settlements from credit card companies so they can avoid possible interruptions in their customer service. The implications of the broken patent system for universities are that patenting becomes increasingly costly—either to pursue infringements or even more so to avoid purposeful blockages.

Patent rights frameworks point to another important question: how does a literature seemingly concerned with the problems of privilege have implications for people other than first-world scientists and academics? For example, the current international treaty on intellectual property (Trade-Related Aspects of Intellectual Property Rights, or TRIPs) protects core nations' and large corporations' property rights but at the expense of access to basic products such as medicines for poor people in less well developed countries (Drahos & Mayne, 2002). HIV/AIDS activists are concerned with the ways in which intellectual property regimes may slow down research, raise its costs, and also inhibit either replication or critical examination of new knowledge.[12] Sustainable development, as opposed to a narrowly market-driven mentality in UIRRs, might be promoted at a local level by cooperation between universities and industry and community groups (Forrant & Pyle, 2002), but the research on UIRRs and IUCRCs suggests that small firms and nonprofit agencies may not have sufficient resources to compete for access to university knowledge and facilities. With "glocalization" being the tendency of global knowledge and systems to be modified when brought into contact with local particularities (Ritzer, 2004; Robertson, 1995), we may not have to worry too much about some sort of hegemonic application of a triple helix or any other model to third world contexts. The implementation of science, whether measured in terms of publications, students, professionals, or the presence of national ministries, has been an uneven process (Schofer, 2004; Jang, 2004; Finnemore, 1993), and it has been difficult enough to transfer models from the United States to Europe (Owen-Smith et al., 2002; Mali, 2000; Balazs & Plonski, 1994). For example, investment in scientific research for national economic development produces a significant *negative* effect on economic growth (Drori et al., 2003) in the short term, whereas investment in the scientific labor force provides a positive benefit. This points again to a larger critique of IP as the primary strategy for economic development. But unless policies for workforce development are imaginatively implemented, with university funding related to national economic well-being in global society, the commoditization of higher education may only exacerbate the divide between rich and developing nations, and within any given nation. It appears, then, that the transformation of "technology transfer" from an international development issue to an econometric one for universities has come full circle.

Along with the international implications of UIRRs, those in various social movements, such as feminism or environmentalism, or patient advocacy groups, and those scholars in the humanities and social sciences not close to the market should remain interested in UIRRs. For example, consider whether traditional academic and industrial division of labor is really diversifying. Homogeneity among scientists results from traditional hierarchy where hidden stratification lies (Smith-Doerr, 2004). Gender, race, and class issues seem to be left out of most university-industry discussions. Not only are scientists predominately white and male, but it seems those who study university-industry research may be as well. Two edited volumes published in 2004 on the implications of UIRRs are suggestive: the Grandin et al. (2004) volume includes 20 male contributors and 6 females (one from a developing nation), and the Stehr (2004) volume has 17 contributors, all men. Beyond how to include diversity among participants, where can concerns such as exploitation of indigenous resources or the "unfaculty" be addressed? It seems that in both the traditional university system and the new, more commercially savvy version, those who are "subjects" outside the system (e.g., indigenous people's groups, see Reardon, 2005) and the "temps" inside (see Hackett, 1990) are consistently giving more than they receive from science, so why should they care about the change?

For those not in the sciences, research, and by extension, instruction, in activities not close to the market may suffer from neglect or explicit cutbacks. Yet it is precisely the teaching function of the university that Fuller (2004) argues can be the source of "creative destruction of social capital" that accrues to scientists engaged in UIRRs. Intense UIRR involvement also opens up universities and science more generally to intense scrutiny, leading to questions about the public trust of science, and worries about the general legitimacy of the institution (Rampton & Stauber, 2001; Weisbrod, 1998). Commercialization is at the center of public concerns about science, such as concerns about genetically modified food or drug safety, and is also one of the factors shaping social movements' access or more general public participation in and access to knowledge production (Frickel & Moore, 2005). Discussions about UIRRs must move beyond disciplinary boundaries and even the boundaries of the institutions of knowledge production and governance into far broader circulation. Future research needs to address issues of diversity in UIRR activities and theoretical and empirical inquiry about UIRRs and notions of the public good. The role of legitimacy and isomorphism elements in changes in UIRR and organizations should also be explicitly studied. Because the destabilization of the current organization of universities comes from a variety of sources, these will require sophisticated quantitative and qualitative tools. The needs for empirical work that Agrawal (2001) identified in the economics literature on UIRRs include comparing industries in their capacities for absorbing knowledge, understanding technology transfer that happens outside the formal TTO and patents, and investigating cases in which knowledge spillovers are not geographically localized.

Notes

The second author acknowledges the support of the Robert Schuman Centre for Advanced Studies during the writing of this piece. At the center she would especially like to thank Rikard Stankiewicz and Aldo Geuna, who organized the European forum on Universities and Innovation Systems, and other colleagues in the forum who tried to teach her the details of innovation economics: Elisa Giuliani, Stephane Malo, Philippe Moguerou, Pier Paolo Patrucco, and Petri Rouvinen. We would also like to thank the reviewers and editors of this volume for their encouragement and recommendations. Any errors or omissions in this review were committed by the authors.

1. At least federal research dollars funneled through traditional scientific and technological agencies seem to have reached a plateau. Savage (1999) suggests that the twenty-fold growth in academic earmarks between 1980 and 1996 to $327,808,000 and a total of around $5.1 billion represents a new, non-peer-reviewed mechanism for generating institutional revenues for selected universities.

2. There was a slight drop-off of total funds and percentage support registered in 2002, but this did not disrupt the overall trend.

3. The consolidation of the federal research bureaucracy also corresponds with the "massification" of higher education in terms of rapidly increasing undergraduate enrollments, diversification of the student body, and emergence of social movements and new disciplinary initiative impacts on the university. See Kerr (1963) and Kernan (1997) for pieces that might serve as "bookends" to that literature.

4. The move to the development of in-house expertise for universities can be traced by the formation and growth of AUTM, the Association of University Technology Managers. Founded in 1974, AUTM registered attendance at the 1998 annual meeting of 600, and 1760 in 2004. More information is available at: http://:www.autm.net/events/meetings/annual2004.cfm. Similarly, the Technology Transfer Society was founded in 1975 with a broader constituency. Available at: http://millkern.com/washtts .doc/national.html.

5. See http://www.nsf.gov/pubs/2001/nsf01110/nsf01110.html for a report by SRI International surveying the impacts of the S/IUCRC program, and http://www.nsf.gov/eng/iucrc/directory/overview.jsp for an overview from the National Science Foundation.

6. See Denise Drescher (1998) "Research Parks in the United States: A Literature Review." Available at: http://www.planning.unc.edu/courses/261/drescher.litrev.htm.

7. But see Lenoir (1998) for an examination of the case of nineteenth-century Germany.

8. Association of University Research Parks, "Critical Role and Economic Impact of University Research Parks," Congressional Breakfast, Hyatt Regency Hotel, Washington, DC, March 4, 2004. Available at: http://www.aurp.net/about/critical_role.ppt.

9. See, for example, the Michigan biotechnology initiative. Available at: http://www.michbio.org.

10. Related are critiques of copyright for IP protection. Vaidhyanathan (2001) argues that the spread of a specific international regime of intellectual property instantiated in U.S. law stifles creativity and harms nonpropertarian cultural traditions, such as the African-American ethos of borrowing in blues traditions, and new experiments in sampling and pastiche in rap and hip-hop.

11. Available at: http://free-culture.org/freecontent/.

12. "History of Changing IP Policies," Yale AIDS Network, April 19, 2003. Available at: http://www.yale.edu/aidsnetwork /Spring%202003%20Univ%20IP%20History.ppt#1.

References

Acs, Zoltan J., David B. Audretsch, & Maryann P. Feldman (1992) "Real Effects of Academic Research: A Comment," *American Economic Review* 82: 363–67.

Adams, J. D., E. P. Chiang, & K. Starkey (2001) "Industry-University Cooperative Research Centers," Journal of Technology Transfer 26(1–2): 73–86.

Agrawal, Ajay (2001) "University-to-Industry Knowledge Transfer: Literature Review and Unanswered Questions," *International Journal of Management Reviews* 3: 285–302.

Agrawal, A. & R. Henderson (2002) "Putting Patents in Context: Exploring Knowledge Transfer from MIT," *Management Science* 48(1): 44–60.

Arrow, K. J. (1962) "Economic Welfare and the Allocation of Resources for Inventions," in R. R. Nelson (ed), *The Rate and Direction of Inventive Activity: Economic and Social Factors* (Princeton, NJ: Princeton University Press): 609–25.

Arundel, A. & Aldo Geuna (2004) "Proximity and the Use of Public Science by Innovative European Firms," *Economics of Innovation and New Technology* 13: 559–80.

Audretsch, David B. & Paula E. Stephan (1996) "Company-Scientist Locational Links: The Case of Biotechnology," *American Economic Review* 86: 641–52.

Auerswald, Philip E. & Lewis M. Branscomb (2003) "Start-Ups and Spin-Offs: Collective Entrepreneurship Between Invention and Innovation," in D. M. Hart (ed), *The Emergence of Entrepreneurship Policy: Governance, Start-Ups, and Growth in the U.S. Knowledge Economy* (Cambridge: Cambridge University Press): 61–91.

Baber, Zaheer (1998) "Science and Technology Studies After the 'Science Wars'," *Southeast Asian Journal of Social Science* 26: 113–20.

Baber, Zaheer (2001) "Globalization and Scientific Research: The Emerging Triple Helix of State-Industry-University Relations in Japan and Singapore," *Bulletin of Science, Technology and Society* 21: 401–8.

Bakouros, Yiannis L., Dimitry C. Mardas, & Nikos C. Varsakelis (2002) "Science Park: A High-Tech Fantasy? An Analysis of Greece," *Technovation* 22: 123–28.

Balazs, Katalin & Guilherme Ary Plonski (1994) "Academic-Industry Relations in Middle-Income Countries: East Europe and Ibero-America," *Science and Public Policy* 21: 109–16.

Barber, Bernard (1952) *Science and the Social Order* (New York: Collier).

Blumenthal, David, E. G. Campbell, M. S. Anderson, N. Causino, & K. S. Louis (1997) "Withholding Research Results in Academic Life Science: Evidence from a National Survey of Faculty," *Journal of the American Medical Association* 277(15): 1224–28.

Blumenthal, David, N. N. Causino, E. Campbell, & K. S. Lewis (1996) "Relationships Between Academic Institutions and Industry in the Life Sciences: An Industry Survey," *New England Journal of Medicine* 334: 368–73.

Bok, Derek Curtis (2003) *Universities in the Marketplace: The Commercialization of Higher Education* (Princeton, NJ: Princeton University Press).

Bowie, Norman E. (ed) (1994) *University-Business Partnerships: An Assessment* (Lanham, MD: Rowman & Littlefield).

Bozeman B. (2000) "Technology Transfer and Public Policy: A Review of Research and Theory," *Research Policy* 29: 627–55.

Bresnahan, T. & Alfonso Gambardella (eds) (2004) *Building High-Tech Clusters: Silicon Valley and Beyond* (Cambridge: Cambridge University Press).

Busch, Lawrence (2000) *The Eclipse of Morality: Science, State, and Market* (New York: Aldine de Gruyter).

Bush, Vannevar ([1945]1990) *Science: The Endless Frontier* (Washington, DC: National Science Foundation).

Callon, Michel (1998) *The Laws of the Markets* (London: Blackwell).

Callon, Michel (2003) "The Increasing Involvement of Concerned Groups in R&D Policies: What Lessons for Public Powers?" in Aldo Geuna, J. Ammon Salter, & W. Edward Steinmueller (eds), *Science and Innovation: Rethinking the Rationales for Funding and Governance* (Cheltenham, U.K.: Edward Elgar): 30–68.

Campbell, E. D., B. R. Clarridge, M. Gokhale, & L. Birenbaum (2002) "Data Withholding in Academic Genetics: Evidence from a National Survey," *Journal of the American Medical Association* 287: 473–80.

Campbell, E. G., K. S. Louis, & D. Blumenthal (1998) "Looking a Gift Horse in the Mouth: Corporate Gifts Supporting Life Sciences Research," *Journal of the American Medical Association* 279: 995–99.

Campbell, Teresa Isabelle Daza (1995) "Protecting the Public's Trust: A Search for Balance Among Benefits and Conflicts in University-Industry Relationships," Ph.D. diss., University of Arizona, Tucson.

Campbell, Teresa & Sheila Slaughter (1999) "Faculty and Administrator Attitudes Toward Potential Conflicts of Interest, Commitment, and Equity in University-Industry Relations," *Journal of Higher Education* (May–June) 70(3): 309–32.

Cho, Mildred K., S. Illangasekare, M. A. Weaver, D.G.B. Leonard, & J. F. Merz (2003) "Effects of Gene Patents and Licenses on the Provision of Clinical Genetic Testing Services," *Journal of Molecular Diagnosis* 5: 3–8.

Chubin, Darryl E. & Edward J. Hackett (1990) *Peerless Science: Peer Review in U.S. Science Policy* (Albany: State University of New York Press).

Cohen, Wesley & D. Levinthal (1990) "Absorptive Capacity: A New Perspective on Learning and Innovation," *Administrative Science Quarterly* 35: 128–52.

Cohen, Wesley M., Richard R. Nelson, & John P. Walsh (2002) "Links and Impacts: The Influence of Public Research on Industrial R&D," *Management Science* 48: 1–23.

Cohen, W., R. Florida, L. Randazzese, & J. Walsh (1998) "Industry and the Academy: Uneasy Partners in the Cause of Technological Advance," in R. Noll (ed), *Challenges to the Research University* (Washington, DC: Brookings Institution): 171–200.

Cole, Jonathan R. (1994) "Balancing Acts: Dilemmas of Choice Facing Research Universities," in Jonathan R. Cole, Elinor G. Barber, & S. R. Graubard (eds), *The Research University in a Time of Discontent* (Baltimore, MD: Johns Hopkins University Press): 1–36.

Cole, Jonathan R., Elinor G. Barber, & Stephen R. Graubard (eds) (1994) *The Research University in a Time of Discontent* (Baltimore, MD: Johns Hopkins University Press).

Colyvas, Jeannette A. & Walter W. Powell (2006) "Roads to Institutionalization: The Remaking of Boundaries Between Public and Private Science," *Research in Organizational Behavior* 27: 305–53.

Colyvas, Jeanette, Michael Crow, Annetine Gelijns, Roberto Mazzoleni, Richard R. Nelson, Nathan Rosenberg, & Bhaven N. Sampat (2002) "How Do University Inventions Get into Practice?" *Management Science* 48(1): 61–72.

Croissant, Jennifer & Sal P. Restivo (2001) *Degrees of Compromise: Industrial Interests and Academic Values* (Albany: State University of New York Press).

Drahos, Peter & Ruth Mayne (eds) (2002) *Global Intellectual Property Rights: Knowledge, Access, and Development* (London: Palgrave Macmillan).

Drori, Gili, John W. Meyer, Francisco O. Ramirez, & Evan Schofer (2003) *Science in the Modern World Polity: Institutionalization and Globalization* (Stanford, CA: Stanford University Press).

Eisenberg, Rebecca S. (1987) "Proprietary Rights and the Norms of Science in Biotechnology Research," *Yale Law Journal* 97: 177–231.

Eisenberg, Rebecca (2003) "Patent Swords and Shields," *Science* 299: 1018–19.

Elzinga, Aant (2004) "The New Production of Reductionism in Models Relating to Research Policy," in Karl Grandin, Nina Wormbs, & Sven Widmalm (eds), *The Science-Industry Nexus: History, Policy, Implications* (Sagamore Beach, MA: Science History Publications/USA): 277–304.

Etzkowitz, Henry (2002) *MIT and the Rise of Entrepreneurial Science* (London and New York: Routledge).

Etzkowitz, Henry & Loet Leydesdorff (eds) (1997) *Universities and the Global Knowledge Economy: A Triple Helix of University-Industry-Government Relations* (London: Cassell).

Etzkowitz, Henry & Loet Leydesdorff (2000) "The Dynamics of Innovation: From National Systems and 'Mode 2' to a Triple Helix of University-Industry-Government Relations," *Research Policy* 29: 109–23.

Etzkowitz, Henry, Andrew Webster, & Peter Healy (1998) *Capitalizing Knowledge: New Intersections of Industry and Academia* (Albany: State University of New York Press).

Feldman, M. P. (1994) *The Geography of Innovation* (Boston: Kluwer).

Feldman, M. P. & P. Desroches (2004) "Truth for Its Own Sake: Academic Culture and Technology Transfer at Johns Hopkins University," *Minerva* 42: 105–26.

Feller, Irwin (2002) "Impacts of Research Universities on Technological Innovation in Industry: Evidence from Engineering Research Centers" (with Catherine Ailes & J. David Roessner) *Research Policy* 31: 457–74.

Feller, Irwin, Maryann Feldman, Janet Bercovitz, & Richard Burton (2002) "Equity and the Technology Transfer Strategies of American Research Universities," *Management Science* 48: 105–21.

Finnemore, Martha (1993) "International Organizations as Teachers of Norms: The United Nations Educational, Scientific, and Cultural Organization and Science Policy," *International Organization* 47(4): 565–97.

Fischer, Howard (1999) "Lawmaker Wants Schools to Sell Research Parks," *Arizona Business Gazette* 119(3)(January 21): 1.

Florida, Richard (2002) *The Rise of the Creative Class* (New York: Basic Books).

Fontana, Roberto, Aldo Geuna & M. Matt (2004) "Firm Size and Openness: The Driving Forces of University-Industry Collaboration," in Y. Calaoghirous, A. Constantelou, & N. S. Vonortas (eds), *Knowledge Flows in European Industry: Mechanisms and Policy Implications* (London: Routledge).

Forrant, Robert & Jean L. Pyle (2002) "Globalization, Universities and Sustainable Human Development," *Development* 45: 102–6.

Frickel, Scott & Kelly Moore (eds) (2005) *The New Political Sociology of Science: Institutions, Networks, and Power* (Madison: University of Wisconsin Press).

Fujigaki, Yuko & Loet Leydesdorff (2000) "Quality Control and Validation Boundaries in a Triple Helix of University-Industry-Government: 'Mode 2' and the Future of University Research," *Social Science Information* 39: 635–55.

Fuller, Steve (2004) "In Search of Vehicles for Knowledge Governance: On the Need for Institutions That Creatively Destroy Social Capital," in N. Stehr (ed), *The Governance of Knowledge* (New Brunswick, NJ: Transaction): 41–78.

Gambardella, A. & F. Malerba (eds) (1999) *The Organization of Economic Innovation in Europe* (Cambridge: Cambridge University Press).

Geiger, Roger L. (1986) *To Advance Knowledge: The Growth of American Research Universities, 1900–1940* (New York: Oxford University Press).

Geiger, Roger L. (1993) *Research and Relevant Knowledge: American Universities Since World War II* (New York: Oxford University Press).

Geiger, Roger L. (2004) *Knowledge and Money: Research Universities and the Paradox of the Marketplace* (Stanford, CA: Stanford University Press).

Geuna, Aldo, Ammon J. Salter, & W. Edward Steinmueller (2003) *Science and Innovation: Rethinking the Rationales for Funding and Governance* (Cheltenham, U.K.: Edward Elgar).

Gibbons, Michael, Camille Limoges, Helga Nowotny, Simon Schwartzman, Peter Scott, & Martin Trow (1994) *The New Production of Knowledge: The Dynamics of Science and Research in Contemporary Societies* (London and Thousand Oaks, CA: Sage).

Giuliani, Elisa & M. Bell (2005) "The Micro-Determinants of Meso Learning and Innovation: Evidence from a Chilean Wine Cluster," *Research Policy* 34: 47–58.

Godin, Benoit (1998) "Writing Performative History: The New New Atlantis?" *Social Studies of Science* 28: 465–83.

Grandin, Karl, Nina Wormbs & Sven Widmalm (eds) (2004) *The Science-Industry Nexus: History, Policy, Implications* (Sagamore Beach, MA: Science History Publications/USA).

Griliches, Zvi (1990) "Patent Statistics as Economic Indicators: A Survey," *Journal of Economic Literature* 28: 1661–1707.

Grundmann, Reiner (2004) "Concluding Observations: Free Flow of Information or Embedded Expertise? Notes on the Regulation of Knowledge," in Nico Stehr (ed), *The Governance of Knowledge* (New Brunswick, NJ: Transaction): 269–86.

Hackett, Edward J. (1990) "Science as a Vocation in the 1990s: The Changing Organizational Culture of Science," *Journal of Higher Education* 61(3): 241–79.

Hackett, Edward J. (2001) "Organizational Perspectives on University-Industry Research Relations," in Jennifer Croissant & Sal P. Restivo (eds), *Degrees of Compromise: Industrial Interests and Academic Values* (Albany: State University of New York Press): 1–22.

Hagstrom, Warren O. (1965) *The Scientific Community* (Carbondale: Southern Illinois University Press).

Hawkes, Nigel (1997) "What Use Is a Science Park?" *The Times* (February 10): 15.

Heller, Michael & Rebecca S. Eisenberg (1998) "Can Patents Deter Innovation? The Anticommons in Biomedical Research," *Science* 280(5364): 698–701.

Henderson, Rebecca, Adam B. Jaffe, & Manuel Trajtenberg (1998) "Universities as a Source of Commercial Technology: A Detailed Analysis of University Patenting, 1965–1988," *Review of Economics and Statistics* 80(1): 119–27.

Hodges, D. A. (2001) "University-Industry Cooperation and the Emergence of Start-Up Companies," Public Symposium at Research Institute of Economy, Trade, and Industry, Tokyo. Available at: http://www.rieti.go.jp/en/events/01121101/doc.html. Also available at: http://andros.eecs.berkeley.edu/~hodges/UIC&ESUC.pdf.

Holleman, Margaret Ann Phillippi (2005) "Effects of Academic-Industry Relations of the Professional Socialization of Graduate Science Students," Ph.D. diss., University of Arizona, Tucson.

Hounshell, David A. & John Kenly Smith, Jr. (1988) *Science and Corporate Strategy: DuPont R&D, 1902–1980* (New York: Cambridge University Press).

Jaffe, Adam & Josh Lerner (2004) *Innovation and Its Discontents: How Our Broken Patent System Is Endangering Innovation and Progress and What to Do About It* (Princeton, NJ: Princeton University Press).

Jang, Yong Suk (2004) "The Worldwide Founding of Ministries of Science and Technology, 1950–1990," *Sociological Perspectives* 43(2): 247–70.

Johnson, Deborah G. (2001) "Conflicts of Interest and Industry-Funded Research: Chasing Norms for Professional Practice in the Academy," in Jennifer Croissant & Sal P. Restivo (eds), *Degrees of Compromise: Industrial Interests and Academic Values* (Albany: State University of New York Press): 185–98.

Jones, Charles I. (2002) "Sources of U.S. Economic Growth in a World of Ideas," *American Economic Review* 92: 220–39.

Jones, Charles I. & John C. Williams (1998) "Measuring the Social Return to R&D," *Quarterly Journal of Economics* 113(4): 1119–35.

Kaghan, William N. (2001) "Harnessing a Public Conglomerate: Professional Technology Transfer Managers and the Entrepreneurial University," in Jennifer Croissant & Sal P. Restivo (eds) (2001) *Degrees of Compromise: Industrial Interest and Academic Values* (Albany: State University of New York Press): 77–101.

Kernan, Alvin (ed) (1997) *What's Happened to the Humanities?* (Princeton, NJ: Princeton University Press).

Kerr, Clark (1963) *The Uses of the University* (Cambridge, MA: Harvard University Press).

Kleinman, Daniel Lee (2003) *Impure Cultures: University Biology and the World of Commerce* (Madison: University of Wisconsin Press).

Klevorick, A. K., R. C. Levin, R. R. Nelson & S. G. Winter (1995) "On the Sources and Significance of Interindustry Differences in Technology Opportunities," *Research Policy* 24: 185–205.

Knapp, Kevin (1998) "A Suburb Pulls the Plug on Its High-Tech Dreams," *Crain's Chicago Business* 21(22): 4–6.

Knoedler, Janet T. (1993) "Market Structure, Industrial Research, and Consumers of Innovation: Forging Backward Linkages to Research in the Turn-of-the-Century U.S. Steel Industry," *Business History Review* 67(1): 98–139.

Knorr Cetina, Karin D. & Alex Preda (2004) *The Sociology of Financial Markets* (Oxford: Oxford University Press).

Kornhauser, William (1962) *Scientists in Industry: Conflict and Accommodation* (Berkeley: University of California Press).

Laursen, K. & A. Salter (2004) "Searching Low and High: What Types of Firms Use Universities as a Source of Innovation?" *Research Policy* 33: 1201–15.

Lenoir, Timothy (1998) "Revolution from Above: The Role of the State in Creating the German Research System, 1810–1910," *American Economic Review* 88(2): 22–27.

Leslie, Stuart W. (1993) *The Cold War and American Science: The Military-Industrial-Academic Complex at MIT and Stanford* (New York: Columbia University Press).

Lessig, Lawrence (2004) *Free Culture: How Big Media Uses Technology and the Law to Lock Down Culture and Control Creativity* (New York: Penguin Press).

Libecap, Gary (ed) (2005) *University Entrepreneurship and Technology Transfer 16* (Storrs, CT: JAI/Elsevier).

Link, Albert N. & John T. Scott (2003) "U.S. Science Parks: The Diffusion of an Innovation: Effects on the Academic Missions of Universities," *International Journal of Industrial Organization* 22: 1323–56.

Louis, Karen Seashore, Lisa M. Jones, & Eric G. Campbell (2002) "Sharing in Science," *American Scientist* 90(4): 304–8.

Lundvall, Bengt-Ake (ed) (1992) *National Systems of Innovation: Towards a Theory of Innovation and Interactive Learning* (London: Pinter).

MacKenzie, Donald (2004) "Relating Science, Technology, and Industry After the Linear Model (Commentary)," in Karl Grandin, Nina Wormbs, & Sven Widmalm (eds), *The Science-Industry Nexus: History, Policy, Implications* (Sagamore Beach, MA: Science History Publications/USA): 305–12.

MacKenzie, Donald & Yuval Millo (2003) "Constructing a Market, Performing Theory: The Historical Sociology of a Financial Derivatives Exchange," *American Journal of Sociology* 109: 107–45.

Malerba, Franco (2002) "Sectoral Systems of Innovation and Production," *Research Policy* 31: 247–64.

Mali, Franc (2000) "Obstacles in Developing University, Government and Industry Links: The Case of Slovenia," *Science Studies* 13: 31–49.

Malo, Stéphane & Aldo Geuna (2000) "Science-Technology Linkages in an Emerging Research Platform: The Case of Combinatorial Chemistry and Biology," *Scientometrics* 47: 303–21.

Mansfield, E. (1991) "Academic Research and Industrial Innovation," *Research Policy* 20: 1–12.

Martin, Brian (1998) *Information Liberation: Challenging the Corruptions of Information Power* (London: Freedom Press).

Massey, D., P. Quintas, & D. Wield (1992) *High-Tech Fantasies: Science Parks in Society and Space* (London: Routledge).

McCray, W. Patrick & Jennifer L. Croissant (2001) "Entrepreneurship in Technology Transfer Offices: Making Work Visible," in Jennifer Croissant & Sal P. Restivo (eds), *Degrees of Compromise: Industrial Interest and Academic Values* (Albany: State University of New York Press): 55–76.

McKelvey, Maureen, Annika Rickne, & Jens Laage-Hellman (eds) (2004) *The Economic Dynamics of Modern Biotechnology* (Cheltenham, U.K.: Edward Elgar).

Merton, Robert K. (1973) "The Normative Structure of Science," in N. W. Storer (ed), *The Sociology of Science: Theoretical and Empirical Investigations* (Chicago: University of Chicago Press): 267–78.

Merton, Robert K. (1988) "The Matthew Effect in Science. 2. Cumulative Advantage and the Symbolism of Intellectual Property," *Isis* 79: 606–23.

Merton, Robert K. & Harriet K. Zuckerman (1968) "The Matthew Effect in Science: The Reward and Communication Systems of Science Are Considered," *Science* 199(3810)(January 5): 55–63.

Mirowski, Philip & Esther-Miriam Sent (2002) *Science Bought and Sold: Essays in the Economics of Science* (Chicago: University of Chicago Press).

Mitroff, Ian I. (1974) *The Subjective Side of Science: A Philosophical Inquiry into the Psychology of the Apollo Moon Scientists* (Amsterdam: Elsevier).

Mohnen, P. & C. Hoareau (2003) "What Type of Enterprise Forges Close Links with Universities and Government Labs? Evidence from CIS 2," *Managerial and Decision Economics* 24: 133–45.

Mowery, David C. & Arvids A. Ziedonis (2002) "Academic Patent Quality and Quantity Before and After the Bayh-Dole Act in the United States," *Research Policy* 31: 399–418.

Mowery, David C., Richard C. Nelson, Bhaven N. Sampat, & Arvids A. Ziedonis (2004) *Ivory Tower and Industrial Innovation: University-Industry Technology Transfer Before and After the Bayh-Dole Act* (Stanford, CA: Stanford Business Books).

Narin, F., K. S. Hamilton, & D. Olivastro (1997) "The Increasing Linkage Between U.S. Technology and Public Science," *Research Policy* 26(3): 317–30.

National Academy of Science (1995) *On Being a Scientist: The Responsible Conduct of Research* (Washington, DC: National Academy of Science).

National Science Board (2004) *Science and Engineering Indicators* (Washington, DC: National Science Foundation).

Nelson, A. J. (2005) "Cacophony or Harmony? Multivocal Logics and Technology Licensing by the Stanford University Department of Music," *Industrial and Corporate Change* 14: 93–118.

Nelson, Richard R. (ed) (1962) *The Rate and Direction of Inventive Activity* (Princeton, NJ: Princeton University Press).

Nelson, Richard R. (2001) "Observations on the Post Bayh-Dole Rise in Patenting at American Research Universities," *Journal of Technology Transfer* 26: 13–19.

Nelson, Richard R. (2004) "The Market Economy and the Scientific Commons," *Research Policy* 33: 455–71.

Nelson, R. & S. Winter (1982) An Evolutionary Theory of Economic Change (Cambridge, MA: Harvard University Press).

Newcomer, Jeffrey L. (2001) "Your Space or Mine? Organizational Interactions in the Development of a Two-Arm Robotic Testbed," in Jennifer Croissant & Sal P. Restivo (eds), *Degrees of Compromise: Industrial Interests and Academic Values* (Albany: State University of New York Press): 199–224.

Noble, David F. (1977) *America by Design: Science, Technology, and the Rise of Corporate Capitalism* (New York: Knopf).

Nowotny, Helga, Peter Scott & Michael Gibbons (2003) "Mode 2 Revisited: The New Production of Knowledge," *Minerva* 41: 179–94.

Owen-Smith, Jason D. (2001) "New Arenas for University Competition: Accumulative Advantage in Academic Patenting," in Jennifer Croissant & Sal P. Restivo (eds), *Degrees of Compromise: Industrial Interests and Academic Values* (Albany: State University of New York Press): 23–54.

Owen-Smith, Jason & Walter W. Powell (2001) "Careers and Contradictions: Faculty Responses to the Transformation of Knowledge and Its Uses in the Life Sciences," in Steven P. Vallas (ed), *Research in the Sociology of Work: 10: The Transformation of Work* (Greenwich, CT: JAI Press): 109–40.

Owen-Smith, J., M. Riccaboni, F. Pammolli, & W. W. Powell (2002) "A Comparison of U.S. and European University-Industry Relations in the Life Sciences," *Management Science* 48: 24–43.

Patrucco, P. P. (forthcoming) "Collective Knowledge Production, Costs and the Dynamics of Technological Systems," *Economics of Innovation and New Technology.*

Pestre, Dominique (2000) "The Production of Knowledge Between Academies and Markets: A Historical Reading of the Book *The New Production of Knowledge*," *Science, Technology and Society* 5: 169–81.

Powell, Walter W. & Jason Owen-Smith (1998) "Universities and the Market for Intellectual Property in the Life Sciences," *Journal of Policy Analysis and Management* 17(2): 253–77.

Rabinow, Paul (1996) *Making PCR: A Story of Biotechnology* (Chicago: University of Chicago Press).

Rampton, Sheldon & John Stauber (2001) *Trust Us, We're Experts: How Industry Manipulates Science and Gambles with Your Future* (New York: Jeremy P. Tarcher/Putnam).

Rayman, Paula M. (2001) *Beyond the Bottom Line: The Search for Dignity at Work* (New York: Palgrave).

Reardon, Jenny (2005) "Creating Participatory Subjects: Science, Race and Democracy in a Genomic Age," in S. Frickel & K. Moore (eds), *The New Political Sociology of Science* (Madison: University of Wisconsin Press): 351–77.

Rip, Arie (2004) "Strategic Research, Post-Modern Universities and Research Training," *Higher Education Policy* 17: 153–66.

Ritzer, George (2004) *The Globalization of Nothing* (London: Sage).

Robertson, R. (1995) "Globalization: Time-Space and Homogeneity-Heterogeneity," in M. Featherstone, S. Lash, & R. Robertson (eds), *Global Modernities* (London: Sage): 25–44.

Rosenberg, Nathan & Richard R. Nelson (1994) "American Universities and Technical Advance in Industry," *Research Policy* 23: 323–48.

Rossman, Joseph, F. G. Cottrell, A. W. Hull, & A. F. Woods (1934) *Science 79: The Protection by Patents of Scientific Discoveries* (New York: American Association for the Advancement of Science, Publication 0P-01).

Sampat, Bhaven N., David C. Mowery, & Arvids A. Ziedonis (2003) "Changes in University Patent Quality After the Bayh-Dole Act: A Re-Examination," *International Journal of Industrial Organization* 21(9): 1371–90.

Savage, James D. (1999) *Funding Science in America: Congress, Universities and the Politics of the Academic Pork Barrel* (Cambridge: Cambridge University Press).

Saxenian, AnnaLee (1996) *Regional Advantage: Culture and Competition in Silicon Valley and Route 128* (Cambridge, MA: Harvard University Press).

Schofer, Evan (2004) "Cross-National Differences in the Expansion of Science, 1970–1990," *Social Forces* 83(1): 215–48.

Schultz, J. (1996) "Interactions Between University and Industry," in F. B. Rudolph & L. W. McIntire (eds), *Biotechnology* (Washington, DC: Joseph Henry Press): 131–46.

Scott, Alister, Grove Steyn, Aldo Geuna, Stefano Brusoni & Ed Steinmueller (2001) "The Economic Returns to Basic Research and the Benefits of University-Industry Relationships: A Literature Review and Update of Findings," Report for the Office of Science and Technology, DTI, U.K. Science and Technology Policy Research, University of Sussex, Brighton. Available at: http://www.sussex.ac.uk/spru/documents.

Sennett, Richard (1998) *The Corrosion of Character: The Personal Consequences of Work in the New Capitalism* (New York: W. W. Norton).

Shapin, Steven (2004) "Who Is the Industrial Scientist? Commentary from Academic Sociology and from the Shop-Floor in the United States, ca. 1900 to ca. 1970," in Karl Grandin, Nina Wormbs, & Sven Widmalm (eds), *The Science-Industry Nexus: History, Policy, Implications* (Sagamore Beach, MA: Science History Publications/USA): 337–63.

Shinn, Terry (1999) "Change or Mutation? Reflections on the Foundations of Contemporary Science," *Social Science Information* 38: 149–76.

Shinn, Terry (2002) "The Triple Helix and New Production of Knowledge: Prepackaged Thinking on Science and Technology," *Social Studies of Science* 32: 599–614.

Shorett, Peter, Paul Rabinow, & Paul R. Billings (2003) "The Changing Norms of the Life Sciences," *Nature Biotechnology* 21: 123–25.

Siegel, Donald S., Paul Westhead, & Mike Wright (2003) "Assessing the Impact of University Science Parks on Research Productivity: Exploratory Firm-Level Evidence from the United Kingdom," *International Journal of Industrial Organization* 21(9): 1357–70.

Slaughter, Sheila (1993) "Beyond Basic Science: Research University Presidents' Narratives of Science Policy," *Science, Technology & Human Values* 18(3): 278–302.

Slaughter, Sheila (1996) "The Emergence of a Competitiveness Research and Development Policy Coalition and the Commercialization of Academic Science and Technology," *Science, Technology & Human Values* 21(3): 303–39.

Slaughter, Sheila & Larry Leslie (1997) *Academic Capitalism: Politics, Policies, and the Entrepreneurial University* (Baltimore, MD: Johns Hopkins University Press).

Slaughter, Sheila & Gary Rhoades (2004) *Academic Capitalism and the New Economy: Markets, State, and Higher Education* (Baltimore, MD: Johns Hopkins University Press).

Slaughter, Sheila, Teresa I. D. Campbell, Peggy Holleman, & Edward Morgan (2002) "The Traffic in Students: Graduate Students as Tokens of Exchange Between Industry and Academe," *Science, Technology & Human Values* 27(2): 282–313.

Smith-Doerr, Laurel (2004) *Women's Work: Gender Equality vs. Hierarchy in the Life Sciences* (Boulder, CO: Lynne Rienner).

Smith-Doerr, Laurel (2005) "Institutionalizing the Network Form: How Life Scientists Legitimate Work in the Biotechnology Industry," *Sociological Forum* 20(2): 271–99.

Stankiewicz, Rikard (1994) "University Firms: Spin-Off Companies from Universities," *Science and Public Policy* 21: 99–107.

Stehr, Nico (1994) *Knowledge Societies* (London and Thousand Oaks, CA: Sage).

Stehr, Nico (ed) (2004) *The Governance of Knowledge* (New Brunswick, NJ: Transaction).

Stephan, Paula E. (2003) "Commentary (New Actor Relationships)," in Aldo Geuna, Ammon J. Alter, & W. Edward Steinmueller (eds), *Science and Innovation: Rethinking the Rationales for Funding and Governance* (Cheltenham, U.K.: Edward Elgar): 233–36.

Stephan, Paula & David Audretsch (eds) (1999) *The Economics of Science and Innovation* (Cheltenham, U.K.: Edward Elgar).

Stokes, Donald E. (1997) *Pasteur's Quadrant: Basic Science and Technological Innovation* (Washington, DC: Brookings Institute Press).

Thursby, Jerry G. & Marie C. Thursby (2002) "Who Is Selling the Ivory Tower? Sources of Growth in University Licensing," *Management Science* 48(1): 90–105.

Thursby, J. G. & M. C. Thursby (2004) "Are Faculty Critical? Their Role in University-Industry Licensing," *Contemporary Economic Policy* 22: 162–78.

Vaidhyanathan, Siva (2001) *Copyrights and Copywrongs: The Rise of Intellectual Property and How It Threatens Creativity* (New York: New York University Press).

Walsh, John P. & Wei Hong (2003) "Secrecy Is Increasing in Step with Competition" (Letter), *Nature* 422(6934): 801–2.

Waluszewski, Alexandra (2004) "How Social Science Is Colored by Its Research Tools: Or What's Behind the Different Interpretations of a Growing 'Biotech Valley'?" in Karl Grandin, Nina Wormbs, & Sven Widmalm (eds), *Science-Industry Nexus: History, Policy, Implications* (Sagamore Beach, MA: Science History Publications/USA): 93–118.

Washburn, Jennifer (2005) *University, Inc.: The Corporate Corruption of Higher Education* (New York: Basic Books).

Weiner, Charles (1986) "Universities, Professors, and Patents: A Continuing Controversy," *Technology Review* 1: 33–43.

Weisbrod, Burton (1998) *To Profit or Not to Profit? The Commercial Transformation of the Nonprofit Sector* (New York: Cambridge University Press).

Zemsky, Robert, Gregory R. Wegner, & William F. Massy (2005) *Remaking the American University: Market-Smart and Mission-Centered* (New Brunswick, NJ: Rutgers University Press).

Ziman, John (1994) *Prometheus Bound: Science in a Steady State* (Cambridge: Cambridge University Press).

28   Science, Technology, and the Military:
Priorities, Preoccupations, and Possibilities

Brian Rappert, Brian Balmer, and John Stone

Everything has changed.

The relationships between science, technology, and the military have been an important topic of public and political debate throughout the twentieth, and into the twenty-first, centuries (Edgerton, 1990; Mendelsohn, 1997). Since at least World War II, a substantial percentage of the world's scientific and technological personnel and resources have been committed to defense-related endeavors. Nevertheless, despite the continuous importance attached to such efforts by governments, questions have always been asked about the effectiveness of, and the ends served by, military R&D. Developments in international affairs have also produced misgivings about whether the basic assumptions underpinning such expenditure remained sound. Most recently, the attacks of 9/11 have led to widespread suggestions that "everything has changed"—not least with regard to perceptions of security threats and the legitimacy accorded to the use of military force. This, in turn, has produced counterclaims to the effect that very little has in fact altered.

   This chapter reviews STS analyses of the relationships between science, technology, and the military since the publication of the first *Handbook* in 1977. It does so with particular reference to the question of how notions of change and continuity have been marshaled in attempts to understand the place and purpose of science and technology in military matters. It highlights how perceptions of what is unique and common in international affairs have pervaded analyses of the relationships between science, technology, and the military. In considering these matters, this chapter also engages with the manner in which the priorities and perspectives of STS have transformed over time.

SCIENCE, TECHNOLOGY, WAR, AND THE FORMATION OF STS

A reading of Harvey Sapolsky's (1977) chapter "Science, Technology and Military Policy" tells us much about the international context of the late 1970s, in addition to the state of the study of science and technology at the time. The chapter was written during the Cold War, when the dynamics of the competition between the United

States and the Soviet Union dominated strategic thinking. It is against this backdrop that Sapolsky identified the great importance attached to managing a basically open-ended process of military-technological innovation and the consequences that might flow from it.

Sapolsky surveyed many of the major policy issues associated with harnessing science and technology for military purposes and with the impact of military R&D on science. Much of the chapter focuses on topics such as the challenges associated with managing technological change, the organization of military R&D, and the return to civilian and military sectors from this expenditure. Advanced weaponry was understood to be increasingly integrated into systems designed for multiple purposes and for use in novel military environments. The result was a steady drive toward complexity in weapons development that brought both technological and political forms of uncertainty, thereby confounding attempts to improve the weapons acquisition process (Perry, 1970; Leitenberg, 1973). Sapolsky also dedicated significant attention to debates about the institutional arrangements associated with military R&D efforts, to the existence or otherwise of a "military-industrial complex" (e.g., Lieberson, 1971), and to those arrangements that might generate more production competition (Kurth, 1971). The importance of basic research to weapons programs was becoming increasingly questioned at the time—a development that was undercutting the status of scientists and engineers as advisors on defense-related issues (e.g., Smith, 1966; Boffey, 1975). However, one area in which scientists, individually and collectively, remained central was in international efforts to establish and police arms control agreements. As part of this contribution, the potential for setting international limits to applied military research was identified as an area of emerging discussion (Ruina, 1971).

As such, Sapolsky's contribution to the 1977 *Handbook* focused principally on the policy issues associated with military R&D, and it did so in the main from a political science perspective. Only passing indication is provided of the notion that military science and technology might itself be problematized—as exemplified in Sapolsky's (1977: 453) comment that "new weapons, it would seem, are less the product of technological forces than they are of institutional and socio-political factors."

In the 1995 edition of the STS *Handbook*, Wim Smit identified both differences and commonalities between the international contexts of the late 1970s and early 1990s, along with the corresponding priorities in the analyses of science and technology. The subtitle of his chapter—"Relations in Transition"—signaled an assessment that the associations between science, technology, and the military were located somewhere between a recent past dominated by the Cold War and a future as yet uncertain.

Five far-reaching changes from the time of Sapolsky's chapter were identified as relevant to the harnessing of science and technology for military purposes and to understanding the effects of military R&D on the character of scientific and technological developments. One was the shifting place of universities within military R&D, a process that was particularly pronounced in the United States. Between the mid-1970s and mid-1980s military funding for universities increased almost threefold, with the increase being spread unevenly across the disciplines. This development was accompanied by policy allocation questions about who received what resources and to what end, while the increasing importance placed on such research also sparked wide-ranging contention about its desirability. Moral and political debates centered on the compatibility of military pursuits with the goals of the university (Dickson, 1984; Kevles, 1978)—disputes often relying on the attributions of science as objective, neutral, and value-free that subsequently have been questioned in STS. More conventional concerns over the efficacy of military-related research encouraged discussions about whether the strings attached to military funding (e.g., publication restrictions) jeopardized the conditions (e.g., openness, skepticism) that made science productive—another topic about which STS questioned many widely voiced presumptions. At the time controversy also attended the matter of whether the priorities of academic researchers were being shaped by military funding, yet another debate that was steeped in problematic assumptions regarding the "natural" course of science and the distinctions between "basic" or "applied" research. In relation to the place of universities, Smit suggests that STS was just beginning to pose interesting new questions about whether scientists or the military determined funding agendas, whether developments in the United States were exerting a global influence on research directions, and whether military funding could influence theories in science (Forman, 1987; Gerjuoy & Baranger, 1989).

The other four important changes identified by Smit centered on the exploitation of science and technology for military ends. First, the integration of civilian and military technology became a high-profile topic with the end of the Cold War. Reflecting (on) the times, Smit (1995: 618) commented that "one thing is already clear: military budgets and forces will be substantially reduced both in the United States and in all European countries." Not surprisingly, therefore, those countries with large military expenditures were beginning to ask demanding questions about the future composition of military-industrial capabilities. The search for so-called dual-use technology became a major preoccupation for many Western governments. This was especially so in the United States, where this policy came to abate calls for the large-scale conversion of defense-manufacturing capabilities (Branscomb et al., 1992) and where it intersected with calls for "agile" and "postfordist" production practices. Organizations such as the Advanced Research Projects Agency were assigned a critical role in leveraging military funding and expertise for civilian innovation. Against this background, STS studies were seen as informing the manner in which distinctions were drawn between "civilian," "military," and "dual-use" technology as well as the compatibility of "civilian" and "military" ends (Elzinga, 1990; Gummett, 1990, 1991; Gummett & Reppy, 1988; Irvine & Martin, 1984). As part of this contribution, historical case studies indicated the important role played by military imperatives in shaping the character of civilian technology (e.g., in lasers [Seidel, 1987], transistors [Misa, 1985], and fission reactor design [Hewlett & Holl, 1989]) as well as in facilitating civilian manufacturing and production technologies (e.g., Smith, 1985; Noble, 1985).

A second area of important change identified by Smit in which STS was seen to be relevant to policy discussions was an emerging interest in steering the direction of military R&D (Greenwood, 1990; Woodhouse, 1990; Smit, 1991). A related area where traditional policy concerns were being rethought was that of arms-racing behavior. Here numerous studies had already suggested the multiple dynamics at work in the development and procurement of weapons that might provide lessons for post–Cold War attempts to rein in technologies and R&D (Buzan, 1987; Ellis, 1987). Indeed there had been a marked shift since the 1970s away from "action-reaction" models of the arms race, which treated states as simply responding to technological developments among their competitors, to "domestic structure" models, which included internal political, economic, and social factors as explanations of arms-racing behavior (Buzan & Herring, 1998).

These new understandings were themselves underpinned by the fifth major development discussed by Smit: an emergent appreciation in STS of the processes underpinning the development of technology. Key studies examining the process of weapon innovations (e.g., MacKenzie, 1990; Gummett & Reppy, 1990; Kaldor, 1982) were complementing wider efforts to comprehend the social construction, or shaping, of technology. This development was accompanied by a critique of existing linear models of innovation wherein technologies flowed unproblematically from scientific discoveries—a critique that, in turn, undermined prevailing definitions of "science" and "technology." In their place emerged an analytical approach within STS that stressed the need to treat science, technology, and society as constituting "seamless webs," or sociotechnical networks, rather than as distinct entities. Such an approach questioned the distinctions employed in past analyses, such as Sapolsky's distinction between "technological" and "political" forms of uncertainty in the weapons-acquisition process.

This brief account of the previous two *Handbook* chapters indicates the scope of issues they covered along with the priorities attached to them. As already suggested, notions of change and continuity pervaded efforts to examine the relations between science, technology, and the military. Thus, while the policy issues associated with the harnessing of science and technology to military purposes, along with the impact of military R&D on the organization of science, were major themes in both chapters, they were addressed in relation to different circumstances. Change itself could be regarded as part of the usual run of developments in the sense that military R&D was understood to be conducted in an ever-changing security environment, which, in turn, justified continuing attention to these major themes.

The two previous chapters can also be characterized according to how they portrayed the study of science and technology. Sapolsky's review concerned itself squarely with what analysts were saying about the issues associated with science, technology, and the military, rather than problematizing science and technology themselves, and their boundaries with society, economy, politics, and the military. The studies cited by Sapolosky were drawn from a variety of fields, but principally that of political science. Smit adopted similar substantive preoccupations, although he explicitly identified an emerging multidisciplinary field of STS that was bringing some distinctive perspectives to bear on traditional concerns, especially in "opening the black box" of technology.

STS has continued to develop since the early 1990s, and the rest of this chapter brings that story up to date. In doing so, it dispenses with a review of general policy issues associated with science, technology, and the military. With the continuing development of the field, it focuses instead on the question of what specifically STS has had to say on such matters. This approach demands an engagement with the question of what counts as STS work as well as what distinctive contribution "it" has made to the understanding of science, technology, and the military. Designations of the field and determinations about what counts as a rightful contribution to the literature are mutually defined. This issue is particularly salient in the case of STS because it cannot be demarcated by reference to a single or limited set of theories, methods, or topics of investigation. If science and technology are understood as distributed and heterogeneous practices, specifying neat boundaries to STS itself becomes problematic. So overall, the remainder of this chapter seeks to review the prominence, priorities, and purposes of recent STS themes associated with science, technology, and the military while posing questions about the manner in which this is done.

STS AND THE MILITARY: FURTHER CONTINUITY, MORE TRANSITIONS

In the years since the publication of Smit's chapter, events have continued to challenge previous assumptions and agendas regarding security and military matters. The decline of Cold War ideological divisions led many commentators to forecast the arrival of a new era characterized by the triumph of liberal democracy and market economics (e.g., Fukuyama, 1992). Nevertheless, the 1990s witnessed a growing preoccupation among Western governments with "rogue" states that were judged to transgress the standards of the international community and "failed" states that proved unable to endure the realignments of political and economic power consequent on the end of the Cold War. Efforts to resolve these new problems led to a variety of military responses, ranging from peacekeeping operations, via the coercive use of force, to fully fledged invasions aimed at regime change.

Against this backdrop, contributors to the field of STS have continued to explore traditional concerns associated with the harnessing of science and technology to military ends and with the impact of military R&D on scientific development. Not surprisingly, however, they have also turned their attention to topics hitherto unexamined and the "they" that makes up this corner of STS has come to incorporate new scholars from such areas as anthropology and cultural studies.

As far as the shifting role of universities within military R&D is concerned, Smit (1995) identified three traditional concerns: moral and political issues, divergence between military aims and university missions, and the influence of military research on the direction of science and technology (e.g., Wright, 1991; Edgerton, 1996; Kaiser, 2004). Subsequent analysis of these concerns has taken place in the light of two important developments. First, newly opened archival sources and a small measure of post–Cold War transparency have enabled the investigation of military research establishments as social settings for the production of knowledge (e.g., Bud & Gummett, 1999; Forman & Sanchez-Ron, 1996). Second, construing the relations between science and the military as a "seamless web" has enabled STS scholars to revisit topics that had previously been defined as "nonepistemological" and that had been confined to institutional sociology of science or science policy. Thus, the well-worn topic of ethics in military research has recently been reinvigorated by constructivist studies of scientists' moral frameworks in relation to military projects. Thorpe (2004b), for example, argues that as Oppenheimer wrestled with the morality of authoring the atomic bomb, he was equally troubled by a more general shift in the scientific profession toward narrow and blinkered specialization, thus precluding moral deliberation and reflection and moving scientists away from their former role as "universal intellectuals." Other studies of the justifications provided by universities and scientists for engaging in military research and development activities have contended that it is too simplistic to view these justifications as involving the suppression or abandonment of a Mertonian normative framework and that they should be regarded as part of the professional ideology and culture of the weapons laboratory (Balmer, 2002; Reppy, 1999; Gusterson, 1998).

Previous concerns about the compatibility of university and military research have been carried through into a wider examination of the cultures of different R&D efforts. What might be termed "weapons cultures" have been identified as constituted by relations of secrecy—relations that allow for forms of moral regulation, that facilitate the development of peculiar moral economies, and that legitimize particular avenues of research or research practices. Compartmentalization, and the strict organization of time within the Manhattan Project, for example, arguably closed off opportunities to reflect on ethical concerns (Thorpe, 2004a). Secrecy, while not confined to military institutions, is a defining element of the space in which death becomes a routine goal of the research process. As a result, it has been argued that secrecy is not simply about restricted flows of, or access to, information. Instead, scholars have suggested that we should attend to the ways in which secrecy is talked about by scientists (Dennis, 1999), how it changes scientific authorship (Gusterson, 2003), and how it becomes constitutive of social identities (Wright & Wallace, 2002). For instance, secrecy affects how military researchers see themselves as scientists. They "become weapons scientists rather than, simply, scientists," claims Gusterson (1998: 89), which in turn affects how they relate to their families and the rest of society. Closer to the heart of traditional SSK concerns, in recent analyses secrecy has been treated as co-produced with knowledge as particular practices of establishing and maintaining secrecy are constituted alongside particular types of experimental and field practices. For example, in the wake of public exposure of a fishing trawler to biological warfare agents in 1952, the practices of secrecy and the transformation of the accident into a monitoring experiment were thoroughly intertwined (Balmer, 2004). This productive dimension of secrecy also becomes apparent when breached secrecy produces differing interpretations of what exactly has become known and by whom (Masco, 2002; Kaiser, 2005; Balmer, 2006).

Turning to the influence of the military on the direction of scientific and techno-logical change, historical studies have emerged in recent years that demonstrate how the imperatives of war affected science. These include Mirowski's charting of the shift of physicists and economists into military-funded operations research during World War II, and the consequences of this migration for the disciplinary landscape of postwar economics (Mirowski, 1999, 2001). War also provided impetus and direction to biological research. Prior to World War II, plant auxins were conceptualized as growth stimulators, but the advent of war encouraged them to be seen as potential "killers" within the context of anticrop warfare efforts (Rasumussen, 2001). At a more biographical level, Galison (1998) has argued that the types of visualizable solutions and theorizing valued in war-time Los Alamos had a lasting effect on Richard Feynman's personal ways of working and theorizing, encouraging the development of Feynman diagrams.

Cold War concerns also meant that military patronage acted as an important influ-ence on the development of science and technology. As mentioned, recent studies on this topic within STS have largely shifted away from concerns over whether the mil-itary distorted the "natural" trajectory of science. Instead they have framed their analysis in terms of charting the effects of military patronage without reference to any counterfactual, pure trajectory (Cloud, 2003; Dennis, 2003; Barth, 2003). The military has been shown, in the United States at least, to have profoundly influenced the char-acter of entire universities such as MIT and Stanford (Leslie, 1993; Lowen, 1997). Mil-itary funding and goals have been shown to have affected academic disciplines besides the well-recognized areas of physics and engineering, with recent studies in this vein covering the earth sciences (Doel, 2003; Harper, 2003; Oreskes, 2003; Barth, 2003), the social sciences (Mirowski, 2001; Lowen, 1997; Solovey, 2001), and even ornithology (MacCleod, 2001). A number of these commentators have described the coexistence of secret military research alongside civil research (see also van Keuren, 2001) and have noted how secrecy has created the impression of a civil-military separation that belies the intimate connections and interface zones built up between the two sectors in the course of the Cold War (Cloud, 2001; Dennis, 1994).

The organization of military institutions in war and peace has also been shown to affect the direction of research. In addition to the studies of the military influence on technology cited earlier, Eden (2004) has focused on the organizational "framing" of scientific knowledge. Eden drew on social studies of science and organizational theory to argue that the phenomenon of nuclear bomb damage was framed by military plan-ners and scientists in a manner that focused attention on blast rather than fire damage. Incendiary damage was regarded as far less predictable, and those (such as fire-protection officers) working in the "fire damage frame" were largely marginalized. Eden's work has complemented institutional studies of the roles played by scientific advisers to the military. These studies have built on the seminal work by Gilpin (1962), by drawing on recent insights from STS on the nature of scientific advice in order to show how both expert advice and advisors are constituted in particular social and political contexts (Balmer, 2001; Thorpe, 2002).

In recent years, little attention has been given to the integration of, or conversion between, civilian and military technology in STS. The frequent use of force as a tool of foreign policy since the early 1990s means that many of the initial hopes for a large-scale conversion of military industries to civilian ones have been dashed. Nevertheless, Martin (1993, 1997, 2001) has made proposals for the replacement of military force and equipment with nonviolent forms of self-defense. Attention has also been paid to the challenges associated with disposing of some of the more dangerous products of Cold War military competition and the manner in which disposal decisions have been shaped by political as well as technical considerations. Macfarlane (2003) shows that the ostensibly scientific process of selecting a site for the long-term storage of U.S. nuclear waste has also been a political one but that the political dimension has itself been influenced by the character of the scientific knowledge brought to bear on the problem. For Macfarlane, therefore, the selection process has been attended by the "co-production" of politics and science. In his study of the U.S. chemical weapons disposal program, Futrell (2003) demonstrates the ability of the public to exert a positive influence over the formulation of policy on highly technical matters. According to Futrell, public involvement can produce decisions that not only enjoy greater political legitimacy but are also technically superior to decisions made by technical experts alone.

In his contribution to the 1995 edition of the *Handbook*, Smit identified the emerging literature on the social construction of technology as leading to wide-ranging changes in our understanding of the development of technological systems. The processes associated with the development and acquisition of weapons have long been understood in terms that might loosely be described as "constructivist." Simplistic notions of new weapons following unproblematically from developments in technology have never been a conspicuous feature of the weapons-acquisition literature. On the contrary, many Cold War studies drew attention to the salient roles played by service rivalries, and by competing bureaucratic and domestic political interests, in the weapons-acquisition process (e.g., Armacost, 1969; Halperin, 1972). As such, they have been retrospectively claimed as important instances of the "social shaping" of military technology (Mackenzie & Wajcman, 1999: 347). Clearly, however, seminal constructivist works, such as Mackenzie's (1990) analysis of missile-guidance systems—which demonstrated that the development of the technology and its subsequent technological trajectory were by no means inevitable—have informed more recent studies of the origins of new weapons (e.g., Farrell, 1997; Spinardi, 1994).

In his study of the development of British nuclear weapons at Aldermaston, Spinardi (1997) claims that weapons designers exerted considerable, but not unilateral, influence over military requirements because they were in a strong position to make technical judgments about what it was possible to create. Looking further back in time, Moy (2001) has shown how the U.S. Army Air Corps and the Marine Corps of the interwar period developed distinctive technologies (for precision bombing and amphibious operations, respectively) that reflected their political interests and cultural values. In this regard, the Army Air and Marine Corps did not so much predict the effects that emerging technology would autonomously exert on the future of military operations as develop technologies which permitted them to conduct specific operations that accorded with their bureaucratic interests.

Other studies in the constructivist tradition have concentrated on military-related technologies in order to understand the intertwining of the technical, the political, and the social (Edwards, 1996); to question the distinctions drawn between modernism and postmodernism (Law, 2002); and to map the peculiarities of innovation in military settings (Abbate, 1999). Weber (1997) has undermined the notion that military technologies are inherently masculine by highlighting the manner in which contingent processes in the design of U.S. military aircraft cockpits subsequently worked to exclude women from becoming pilots. Constructivist analyses have continued to extend research into the topic of conventional and unconventional arms races identified by Smit. Grin (1998) emphasized that military-technological innovation occurs within "networks of organizations," and explored the possibilities that this suggested for guiding the development of military technology in politically desirable directions.

Constructivists have also challenged widespread presumptions about the permanence of scientific and technological knowledge. When examining arms-control efforts in the first STS *Handbook*, Sapolsky (1977: 461) offered the disheartening remark that "knowledge once created is indestructible"; MacKenzie and Spinardi (1996) subsequently contended that the tacit (as opposed to formalized) knowledge required in the production of nuclear weapons means that they must be reinvented in new programs in nontrivial ways. The need for reinvention through the acquisition of tacit skills not only limits the proliferation of nuclear weapons but suggests that without practicing their skills, those once competent to create them might lose their ability.

Turning from the acquisition of weapons to the effects they create, we note that military technology has often been considered to provide "hard cases" for substantiating constructivist approaches. If constructivism can be relevant to the study of such topics, then surely (the argument goes) it can be relevant elsewhere. Collins and Pinch's (1998) analysis of the debates surrounding the performance of the U.S. Patriot missile system during the 1991 Gulf War revealed wide scope for negotiation in appraising the effectiveness of weapons technology. The presence of disagreements over whether, and by what criteria, Patriot was successful (disagreements that continued into the 2003 Iraq War [Postol, 2004]) clearly illustrates how efforts to measure the effectiveness of technology operate as a social activity. Yet despite such challenges to conventional thinking, Grint and Woolgar (1997), using weapons as their exemplar, have argued that many constructivists still cling to the notion that certain core capabilities of technology exist that can been known independently from acts of interpretation. By shifting the terms of this debate away from what is true, and toward how we know what is true, others have furthered the work of Grint and Woolgar by using it to question analytical dichotomies between relativist and realist approaches in STS as well as the "disposal strategies" employed in attempts to establish rules for the acceptability of force (Rappert, 2001, 2005).

Many recent contributions to the STS literature do not fit neatly into the categories created in the previous editions of the *Handbook*. These contributions embrace such disparate concerns as the construction of history in accounting for actions during the Iran-Contra hearings (Lynch & Bogen, 1996), the "disenrollment" of humans and technologies from a dominant actor-network in the nuclear submarine industry (Mort & Michael, 1998), and the social construction of Gulf War–related illness (Brown et al., 2001; Zavestoski et al., 2002). Mindell (2000) has explored the effects associated with the introduction of new technologies on military personnel's experience of war. The oft-stated requirement to minimize bloodshed in contemporary warfare has led Rappert (2003a) to examine the proliferation of so-called nonlethal weapons, with a view to subjecting the notion that capabilities inhere in weapons themselves to a constructivist unpicking. The non-Western militarization of science has also begun to attract scholarly attention within STS and related fields (Abraham, 1998; Gerovitch, 2001, 2002; Holloway, 1996).

Following a more general trend toward a focus on the body in the social sciences and humanities, the field of STS has also belatedly taken some interest in the relationship between the military and the body.[1] In her study of how the body parts of atomic-bomb victims were eventually repatriated from the United States to Japan, Lindee demonstrates how science played a crucial role in constituting a particular power/knowledge nexus through "filing systems, autopsy protocols and diplomatic negotiations surrounding a collection of sectioned and dispersed human bodies" (Lindee, 1999: 377). The body parts, she argues, became "frozen in a special state of victimization." As part of the "spoils of war," they were removed from Japan for scientific research, providing a material instantiation of U.S. victory. Their repatriation occurred only in the context of the return of political power to Japan. Nevertheless, both countries' interest in the body parts remained primarily scientific, the motivation being to "study, slice and display," rather than (say) to allow the Japanese public an opportunity to mourn. In a similar vein, Gusterson has argued that the scientific practices involved in calculating the extent and effects of the bombings at Hiroshima and Nagasaki, such as using photographs of injured victims taken from behind or in very close focus, conceal the whole, suffering bodies. Carrying his argument forward to the 1991 Gulf War, he contends that the technocratic military discourse of weapons and machines fighting other weapons and machines provided a powerful way of making the bodies of war victims disappear. This discourse contributed to the image that the war was being fought without killing and lent it "a surreal air of simulation" (Gusterson, 2004: 73).

On the theme of simulation, Lenoir and der Derian have charted the generally sporadic, but increasingly planned, links between academics, the entertainment industry, and the military (Lenoir, 2000; der Derian, 2001). They respectively adopt the terms "military-entertainment complex" and "military-industrial-media-entertainment network" to capture the military's increasing use of simulated environments and scenarios for training purposes, at a time when "military technology, which once trick-led down for civilian use, now often lags behind what is available in games, rides and movie special effects" (Lenoir, 2000: 328).[2] With the same technologies used to prepare for military missions and for gaming, the boundary between fantasy and reality, they suggest, is becoming increasingly blurred. This blurring through simulation does, however, have historical precedents. Ghamari-Tabrizi (2000, 2005) has charted the role of RAND social scientists in creating role-playing scenarios and human-machine sim-ulations (where whole command centers would be built to create an elaborate exer-cise such as a nuclear attack) for the military during the Cold War. Although the scenarios were meant to be objective and realistic, Ghamari-Tabrizi reveals the dynam-ics of the debates that occurred over what counted as objective and realistic, or more precisely, which elements of a simulation were relevant to its realism and objectivity and which were superfluous or trivial. Ultimately, she argues, with no actual emer-gencies to compare against, settling these debates became a matter of intuition and judgment over what would happen outside a simulated situation.

Participants in simulation tests could be regarded as cyborg experimental subjects. And, with the ending of the Cold War, there has been increased documentation of a wide variety of military research on humans (Moreno, 2001). From a science studies perspective, Mitchell (2003) has documented how soldiers' bodies became contested objects for military science. After experiments involving soldiers crawling across the sites of atomic explosions became public, the British Ministry of Defence claimed that no compensation would be forthcoming because the research had not been conducted on bodies. Rather, it had been conducted on the soldiers' clothing. Civilian bodies can equally be reconfigured as experimental material for military purposes (Balmer, 2003, 2004; Crease, 2003), as was evidenced by large-scale spraying of nonpathogenic bac-teria across the county of Dorset in the United Kingdom during the 1960s (Balmer, 2003). Although the military regarded the civilians as being outside the experiment, considerations of safety and secrecy with respect to the public still influenced its conduct.

These diverse contributions to the literature of STS attest to its expansion well beyond the policy concerns that dominated the previous editions of the *Handbook*. Yet defining the distinct contribution that STS has made to "the" understanding of the relationships between science, technology, and the military is difficult. Develop-ments in other fields concerned with military matters have proceeded in parallel but with few direct linkages to STS (but see Herrera, 2003). Since the early 1990s, ideational-based approaches to international relations have been challenging the tra-ditional power- or interest-based approaches to explaining political affairs. An impor-tant component of this challenge has been the identification of the roles played by social norms in constituting identities and regulating behavior. Likewise, in the study of the conduct of warfare and the operation of military forces, norm-based approaches have questioned naturalistic, rationalistic, and deterministic assumptions about the development of weaponry and also sought to consider how notions of self-identity and appraisals of technology are co-constituted. Thus, it has been argued that many developing nations procure high-tech weaponry not because of capability needs defined by strategic calculation but because of identity considerations about what it means to be a modern state (Eyre & Schuman, 1996). Studies of the taboos against the use of nuclear weapons (Tannenwald, 1999) and the development of chemical weapons (Price, 1997) have elaborated on the processes by which particular weapons became stigmatized to such an extent that their use is rarely seriously contemplated. Both Tannenwald and Price argue that acquiescence in these taboos—which were chiefly promulgated by Western industrialized states—constituted a means of defining what it meant to be "civilized." Yet with the exception of Price (1997), this body of norms work in international relations makes little explicit reference to the other analyses of technology mentioned in this chapter.

SECURITY IN THE POST-9/11 WORLD

The events of 9/11 and its aftermath have posed important questions for policy-makers, scholars, and members of the public regarding the degree of change and continuity in international security and what must be done as a result. Global terrorism and the proliferation of "weapons of mass destruction" (WMD) now dominate many policy and popular discussions, while the use of military force has become a recurring—albeit not uncontested—aspect of Western policy. As a result, Smit's (1995) prediction that "military budgets and forces will be substantially reduced both in the United States and in all European countries" has not come to pass. On the contrary, at $419.3 billion the Bush administration's defense spending request for fiscal year 2006 is 41 percent larger than it was in 2001. Of this, approximately $69.4 billion (16.6 percent) is allocated to the research, development, testing, and evaluation activities associated with the creation of new capabilities (Whitehouse, 2005).

Bijker (2003) has posed the question of whether technological societies are changing because of 9/11, and if so, how? As of yet, however, only a few writers have addressed the post-9/11 situation through the lenses of STS. Each, in different ways, suggests that the purportedly radical newness of the current security situation is contestable. Thus, the complexity of many Western technological systems has not suddenly become a risk factor, the (ir)rationality underpinning terrorist acts is not wholly alien to the West, and the U.S. responses to the threat of international terrorism and the anthrax attacks following 9/11 are by no means without historical precedents. Winner (2004) argues that by focusing on the enemy "out there," policy-makers and social scientists may ignore the fact that the vulnerability of many large technological systems is not inevitable but is instead a consequence of previous choices. A nuclear power plant, for example, is a result of prior political and technological choices—choices that make it a more likely and "unforgiving" terrorist target than a wind farm. Turning to terrorists themselves, Gusterson (2001) draws attention to the dominant discourse of "othering" that represents the terrorist threat as an irrational, premodern threat to the rational West. Gusterson claims that this discourse is incorrect;

that terrorists—in assessing the number of deaths and the degree of spectacle required to achieve their goals—have embraced the same calculative, managerial rationality as weapons designers and war planners. Taking a more historical perspective, Jenkins (2002) argues that the U.S. response to the 9/11 attacks is not without precedent. During the interwar period, he argues, a combination of politicians, scientists, and engineers, along with business and the military, joined forces to create the idea of a threat to the U.S. public posed by "outlaw" states armed with aircraft and chemical weapons. Their purpose in creating this fear was to bolster their own position as guardians of U.S. security—an approach that for Jenkins anticipates certain aspects of the current "war on terror." With respect to the anthrax attacks that followed in the wake of 9/11, King analyzes the continuities and discontinuities in the responses of the authorities. Concerns expressed during the attacks about borders, civil liberties, and surveillance, were symptomatic, he suggests, of "American concerns about global social change [being] refracted through the lens of infectious disease" (King, 2003: 435).

Whatever assessment one makes of the "newness" of the post-9/11 situation, in the future much scope will exist for revisiting issues raised in the previous section regarding the harnessing of science and technology, and the effects of military R&D on science and technology. At the time of writing (especially, albeit not solely, in the United States), security and military issues are particularly salient. In the West's search for security against terrorists and WMD, the military has become increasingly integrated and coordinated with other political and social institutions. Science has not been immune from this. With the growing attention given to bioterrorism, for instance, billions of dollars are being dedicated to R&D against possible bioagents (Wright, 2004). Yet, alongside such funding, it has also been contended that developments in the life sciences might facilitate the construction of biological weapons. This has led to a situation in which scientists are increasingly expected to find ways of regulating their behavior and controlling the future application of their work (Rappert, 2003b, 2007). Additionally, in relation to fears about terrorism, many Western countries are witnessing the merger of security and public health concerns (Guillemin, 2005) and an increasing turn to surveillance technologies along with attendant concerns over the infringement of civil liberties (Caplan & Torpey, 2001; Lyon, 2003).

Science and technology, therefore, remain central features of the changing security and military landscape, even if this landscape remains a rather peripheral concern of the STS community. Another important point is that the relatively few new studies of the relationships between science, technology, and military concerns discussed in this chapter have benefited from the post–Cold War situation with regard to secrecy. As mentioned earlier, since the 1995 edition of the STS Handbook, scholars have enjoyed access to new archival sources along with greater openness on the part of policy-makers and other actors in relation to sensitive issues. To the extent that information is becoming accessible, it is possible to revisit previous presumptions in the field, such as the degree of unanimity between government, armed forces, and contractors in the military-industrial complex (Scranton, 2004). Yet, military-related topics remain an area where access continues to be a critical methodological and political issue and where, at the time of writing, it is still not clear to what extent current modest, but improved, levels of transparency in some countries are but a transient phenomenon.[3]

The limited interrelationships between STS and other fields concerned with the military and security remains unfortunate. Additional inroads continue to be made, such as Edgerton's (2006) study of the history of military technology and Rappert's (2006) attempt to reframe efforts to prohibit weapons under international humanitarian law through an examination of the latter's classification schemas. Yet overall, the integration of STS within many traditional fields of study remains limited. For instance, "Strategic Studies" and its cognate fields potentially have much to learn from recent developments noted in this chapter. Awareness of the social content of science and technology has not been a conspicuous feature of strategic studies analyses, which have been prone to grant them autonomous or deterministic qualities. Exceptions exist, a notable example being Freedman's (1998) analysis of the changing nature of warfare associated with the exploitation of information technologies by modern armed forces—a phenomenon commonly termed the revolution in military affairs. According to Freedman, the influence of information technologies on the character of warfare has itself been contingent on the broader political and strategic contexts in which it has occurred, the implication being that there is nothing inevitable about such developments. In relation to more traditional weapons, Stone (2000, 2002) contends that national variations in the design of tanks have historically been a function of differences in the military doctrines that govern their employment in war. These differences in doctrine have also been important for shaping attitudes toward the threat posed by new antitank systems. The interpretation placed on such threats has in large part rested on the precise roles and missions that tanks have been expected to conduct. According to Stone, therefore, debates about the future of the tank should accommodate doctrinal, as well as technical, considerations. Many of the STS analyses identified in this chapter could provide excellent means with which strategists might seek to build on positions such as these, but this will be possible only if transparency and access are adequate. In the meantime, the United States in particular is moving along a path of extensive military-technical innovation, the consequences of which are uncertain.

Notes

1. History of medicine has also developed this theme in new studies of military medicine; for a review, see Bourke (2000).

2. This post–Cold War shift from military-led solutions to commercial-led solutions for military problems, as investment in military R&D and wider defense budgets were cut, has also been charted in the field of computer security (MacKenzie & Pottinger, 1997). For a fuller discussion of the relationship between civil and military R&D see Branscomb et al. (1992), and on the shift from "spin-off" to "dual-use" terminology, see Mollas-Gallart (1997) and Cowan and Foray (1995).

3. The U.S. National Archives and Records Administration (NARA), for example, announced that "in light of the terrorist events of September 11, we are re-evaluating access to some previously open archival materials and reinforcing established practices on screening materials not yet open for research." Available at: http://archives.gov/research_room/whats_new/notices/access_and_terrorism .html.

References

Abbate, Janet (1999) *Inventing the Internet* (Cambridge, MA: MIT Press).

Abraham, Itty (1998) *The Making of the Indian Atomic Bomb: Science, Secrecy and the Postcolonial State* (London: Zed Books).

Armacost, Michael (1969) *The Politics of Weapons Innovation: The Thor-Jupiter Controversy* (New York: Columbia University Press).

Balmer, Brian (2001) *Britain and Biological Warfare: Expert Advice and Science Policy: 1935–1965* (Basingstoke, U.K.: Palgrave).

Balmer, Brian (2002) "Killing 'Without the Distressing Preliminaries': Scientists' Defence of the British Biological Warfare Programme," *Minerva* 40: 57–75.

Balmer, Brian (2003) "Using the Population Body to Protect the National Body: Germ Warfare Tests in the U.K. After WWII," in J. Goodman, A. McElligott & L. Marks (eds), *Useful Bodies: Humans in the Service of Medical Science in the Twentieth Century* (Baltimore, MD: Johns Hopkins University Press): 27–52.

Balmer, Brian (2004) "How Does an Accident Become an Experiment? Secret Science and the Exposure of the Public to Biological Warfare Agents," *Science as Culture* 13(2): 197–228.

Balmer, Brian (2006) "A Secret Formula, a Rogue Patent and Public Knowledge About Nerve Gas: Secrecy as a Spatial-Epistemic Tool," *Social Studies of Science* 36: 691–722.

Barth, Kai-Henrik (2003) "The Politics of Seismology: Nuclear Testing, Arms Control and the Transformation of a Discipline," *Social Studies of Science* 33(5): 743–81.

Bijker, Wiebe E. (2003) "The Need for Public Intellectuals: A Space for STS," *Science, Technology, & Human Values* 28(4): 443–50.

Boffey, P. (1975) *The Brain Bank of America* (New York: McGraw-Hill).

Bourke, Joanna (2000) "Wartime," in R. Cooter & J. Pickstone (eds), *Medicine in the Twentieth Century* (London: Harwood): 589–600.

Branscomb, Lewis, John Alic, Harvey Brooks, Ashton Carter, & Gerald Epstein (1992) *Beyond Spinoff: Military and Commercial Technologies in a Changing World* (Boston: Harvard Business School Press).

Brown, Phil, Stephen Zavestoski, Sabrina McCormick, Meadow Linder, Joshua Mandelbaum, & Theo Luebke (2001) "A Gulf of Difference: Disputes over Gulf War–Related Illnesses," *Journal of Health and Social Behavior* 42(3): 235–57.

Bud, Robert & Philip Gummett (eds) (1999) *Cold War, Hot Science: Applied Research in Britain's Defence Laboratories, 1945–1990* (London: Harwood).

Buzan, Barry (1987) *An Introduction to Strategic Studies: Military Technology and International Relations* (Basingstoke, U.K.: Macmillan).

Buzan, Barry & Eric Herring (1998) *The Arms Dynamic in World Politics* (Boulder, CO: Lynne Rienner).

Caplan, Jane & John Torpey (eds) (2001) *Documenting Individual Identity: The Development of State Practices in the Modern World* (Princeton, NJ: Princeton University Press).

Cloud, John (2001) "Imaging the World in a Barrel: CORONA and the Clandestine Convergence of the Earth Sciences," *Social Studies of Science* 31(2): 231–51.

Cloud, John (2003) "Special Guest-Edited Issue on the Earth Sciences in the Cold War (Introduction)," *Social Studies of Science* 33(5): 629–33.

Collins, Harry & Trevor Pinch (1998) "A Clean Kill?" in *The Golem at Large: What You Should Know About Technology* (Cambridge: Cambridge University Press): 7–29.

Cowan, Robin & Dominique Foray (1995) "Quandaries in the Economics of Dual-Use Technologies and Spillovers from Military to Civilian Research and Development," *Research Policy* 24: 851–68.

Crease, R. (2003) "Fallout Issues in the Study, Treatment, and Reparations of Exposed Marshall Islanders," in R. Figueroa & S. Harding (eds), *Science and Other Cultures: Issues in Philosophies of Science and Technologies* (London: Routledge): 106–28.

Dennis, Michael Aaron (1994) "'Our First Line of Defense': Two Laboratories in the Postwar American State," *Isis* 85(3): 427–55.

Dennis, Michael Aaron (1999) "Secrecy and Science Revisited: From Politics to Historical Practice and Back," in J. Reppy (ed), *Secrecy and Knowledge Production*, Cornell University Peace Studies Program, Occasional Paper No. 23.

Dennis, Michael Aaron (2003) "Earthly Matters: On the Cold War and the Earth Sciences," *Social Studies of Science* 33(5): 809–19.

Der Derian, James (2001) *Virtuous War: Mapping the Military-Industrial-Media-Entertainment Network* (Boulder, CO: Westview Press).

Dickson, David (1984) *The New Politics of Science*, 2nd ed. (Chicago: University of Chicago Press).

Doel, Ronald E. (2003) "Constituting the Postwar Earth Sciences: The Military's Influence on the Environmental Sciences in the U.S.A. After 1945," *Social Studies of Science* 33(5): 635–66.

Eden, Lynn (2004) *Whole World on Fire: Organizations, Knowledge and Nuclear Weapons Devastation* (Ithaca, NY: Cornell University Press).

Edgerton, David (1990) "Science and War," in R. C. Olby & M.J.S. Hodge (eds), *Companion to the History of Modern Science* (London: Routledge): 934–45.

Edgerton, David (1996) "British Scientific Intellectuals and the Relations of Science and War in Twentieth-Century Britain," in P. Forman & J. M. Sanchez-Ron (eds), *National Military Establishments and the Advancement of Science: Studies in Twentieth Century History* (Dordrecht, the Netherlands: Kluwer).

Edgerton, David (2006) *Warfare State: Britain, 1920–1970* (Cambridge: Cambridge University Press).

Edwards, Paul (1996) *The Closed World: Computers and the Politics of Discourse in Cold War America* (Cambridge, MA: MIT Press).

Ellis, John ([1975]1987) *The Social History of the Machine Gun* (New York: Pantheon; London: Cresset Library).

Elzinga, Aant (1990) "Large-Scale Military Funding Induces Culture Clash," *Space Policy*, August: 187–94.

Eyre, Dana & Mark Schuman (1996) "Status, Norms and the Proliferation of Conventional Weapons," in P. Katzenstein (ed), *The Culture of National Security* (New York: Columbia University Press): 79–113.

Farrell, Theo (1997) *Weapons Without a Cause: The Politics of Weapons Acquisition in the United States* (New York: St. Martin's Press).

Forman, Paul (1987) "Behind Quantum Electronics: National Security as a Basis for Physical Research in the United States, 1940–1960," *Historical Studies in Physical and Biological Sciences* 18(1): 149–229.

Forman, Paul & Jose Sanchez-Ron (eds) (1996) *National Military Establishments and the Advancement of Science and Technology* (Dordrecht, the Netherlands: Kluwer).

Freedman, Lawrence (1998) "The Revolution in Strategic Affairs," *Adelphi Paper 318* (London: Oxford University Press for the International Institute of Strategic Studies).

Fukuyama, Francis (1992) *The End of History and the Last Man* (New York: Free Press).

Futrell, Robert (2003) "Technical Adversarialism and Participatory Collaboration in the U.S. Chemical Weapons Program," *Science, Technology, & Human Values* 28: 451–82.

Galison, Peter (1998) "Feynman's War: Modelling Weapons, Modelling Nature," *Studies in the History and Philosophy of Modern Physics* 3: 391–434.

Gerjuoy, Edwards & Elizabeth Urey Baranger (1989) "The Physical Sciences and Mathematics" [Special Issue: Universities and the Military] *Annals of the American Academy of Political and Social Science* 502(1): 58–81.

Gerovitch, Slava (2001) "'Mathematical Machines' of the Cold War: Soviet Computing, American Cybernetics and Ideological Disputes in the Early 1950s," *Social Studies of Science* 31(2): 253–87.

Gerovitch, Slava (2002) *From Newspeak to Cyberspeak: A History of Soviet Cybernetics* (Cambridge, MA: MIT Press).

Ghamari-Tabrizi, Sharon (2000) "Simulating the Unthinkable: Gaming Future War in the 1950s and 1960s," *Social Studies of Science* 30(2): 163–223.

Ghamari-Tabrizi, Sharon (2005) *The Worlds of Herman Kahn: The Intuitive Science of Thermonuclear War* (Cambridge, MA: Harvard University Press).

Gilpin, Robert (1962) *American Scientists and Nuclear Weapons Policy* (Princeton, NJ: Princeton University Press).

Greenwood, Ted (1990) "Why Military Technology Is Difficult to Restrain," *Science, Technology, & Human Values* 14(4): 412–29.

Grin, John (1998) "Bloodless War or Bloody Non-Sense?" Presentation for EASST General Conference on Cultures of Science and Technology, Lisbon, Portugal, October 1.

Grint, Keith & Steve Woolgar (1997) *The Machine at Work* (Cambridge: Polity Press).

Guillemin, Jeanne (2005) *Biological Weapons* (New York: Columbia University Press).

Gummett, Philip (1990) "Issues for STS Raised by Defence Science and Technology Policy," *Social Studies of Science* 20(3): 541–58.

Gummett, Philip (ed) (1991) *Future Relations Between Defence and Civilian Technology: A Report for the [U.K.] Parliamentary Office of Science and Technology, SPSG Review Paper No. 2* (London: Science Policy Support Group).

Gummett, Philip & Judith Reppy (eds) (1988) *The Relations Between Defence and Civil Technologies* (Dordrecht, the Netherlands: Kluwer).

Gummett, Philip & Judith Reppy (1990) "Military Industrial Networks and Technical Change in the New Strategic Environment," *Government and Opposition* 25(3): 287–303.

Gusterson, Hugh (1998) *Nuclear Rites: A Weapons Laboratory at the End of the Cold War* (Berkeley: University of California Press).

Gusterson, Hugh (2001) "The McNamara Complex," *Anthropological Quarterly* 75(1): 171–77.

Gusterson, Hugh (2003) "The Death of the Authors of Death: Prestige and Credibility Among Nuclear Weapons Scientists," in Mario Biagioli & Peter Galison (eds), *Scientific Authorship: Credit and Intellectual Property in Science* (London: Routledge): 281–308.

Gusterson, Hugh (2004) *People of the Bomb: Portraits of America's Nuclear Complex* (Minneapolis: University of Minnesota Press).

Halperin, Morton (1972) "The Decision to Deploy the ABM: Bureaucratic and Domestic Politics in the Johnson Administration," *World Politics* 25(1): 62–95.

Harper, Kristine (2003) "Research from the Boundary Layer: Civilian Leadership, Military Funding and the Development of Numerical Weather Prediction (1946–55)," *Social Studies of Science* 33(5): 667–96.

Herrera, Geoffrey (2003) "Technology and International Systems," *Millennium* 32(3): 559–93.

Hewlett, Richard & Jack Holl (1989) *Atoms for Peace and War, 1953–1961* (Berkeley: University of California Press).

Holloway, David (1996) *Stalin and the Bomb: The Soviet Union and Atomic Energy, 1939–1956* (New Haven, CT: Yale University Press).

Irvine, John & Ben Martin (1984) *Foresight in Science: Picking the Winners* (London: Pinter).

Jenkins, Dominick (2002) *The Final Frontier: America, Science and Terror* (London: Verso).

Kaiser, David (2004) "The Postwar Suburbanization of American Physics," *American Quarterly* 56: 851–88.

Kaiser, David (2005) "The Atomic Secret in Red Hands? American Suspicions of Theoretical Physicists During the Early Cold War," *Representations* 90(1): 28–60.

Kaldor, Mary (1982) *The Baroque Arsenal* (London: Abacus).

Kevles, Daniel (1978) *The Physicists: The History of a Scientific Community in Modern America* (New York: Knopf).

King, Nicholas (2003) "The Influence of Anxiety: September 11, Bioterrorism and American Public Health," *Journal of the History of Medicine* 58: 433–41.

Kurth, J. (1971) "A Widening Gyre: The Logic of American Weapons Procurement," *Public Policy* 19: 373–403.

Law, John (2002) *Aircraft Stories* (London: Duke University Press).

Leiberson, S. (1971) "An Empirical Study of Military Industrial Linkages," *American Journal of Sociology* 76(4): 562–84.

Leitenberg, M. (1973) "The Dynamics of Military Technology Today," *International Social Science Journal* 25(3): 336–57.

Lenoir, Tim (2000) "All but War Is Simulation: The Military-Entertainment Complex," *Configurations* 8: 289–335.

Leslie, Stuart W. (1993) *The Cold War and American Science: The Military-Industrial-Academic Complex at MIT and Stanford* (New York: Columbia University Press).

Lindee, Susan (1999) "The Repatriation of Atomic Bomb Victim Body Parts to Japan: Natural Objects and Diplomacy," *Osiris* 13: 376–409.

Lowen, Rebecca (1997) *Creating the Cold War University: The Transformation of Stanford* (Berkeley: University of California Press).

Lynch, Michael & David Bogen (1996) *The Spectacle of History: Speech, Text and Memory at the Iran-Contra Hearings* (Durham, NC: Duke University Press).

Lyon, David (2003) "Technology Vs. 'Terrorism': Circuits of City Surveillance Since September 11th," *International Journal of Urban and Regional Research* 27(3): 666–78.

Macfarlane, Allison (2003) "Underlying Yucca Mountain: The Interplay of Geology and Policy in Nuclear Waste Disposal," *Social Studies of Science* 33(5): 783–807.

MacKenzie, Donald (1990) *Inventing Accuracy: A Historical Sociology of Nuclear Missile Guidance* (Cambridge, MA: MIT Press).

MacKenzie, Donald & Graham Spinardi (1996) "Tacit Knowledge and the Uninvention of Nuclear Weapons," in Donald MacKenzie (ed), *Knowing Machines* (Cambridge, MA: MIT Press): 215–60.

MacKenzie, Donald & G. Pottinger (1997) "Mathematics, Technology, and Trust: Formal Verification, Computer Security and the U.S. Military," *IEEE Annals of the History of Computing* 19(3): 41–59.

MacKenzie, Donald & Judy Wajcman (1999) (eds) *The Social Shaping of Technology*, 2nd ed. (Buckingham, U.K.: Open University Press).

MacLeod, Roy (2001) "'Strictly for the Birds': Science, the Military, and the Smithsonian's Pacific Ocean Biological Survey Program," *Journal of the History of Biology* 34: 315–52.

Martin, Brian (1993) *Social Defence, Social Change* (London: Freedom Press).

Martin, Brian (1997) "Science, Technology and Nonviolent Action: The Case for a Utopian Dimension in the Social Analysis of Science and Technology," *Social Studies of Science* 27: 439–63.

Martin, Brian (2001) *Technology for Nonviolent Struggle* (London: War Resisters' International).

Masco, Joseph (2002) "Lie Detectors: On Secrets and Hypersecurity in Los Alamos," *Public Culture* 14(3): 441–67.

Mendelsohn, Everett (1997) "Science, Scientists and the Military," in John Krige & Dominique Pestre (eds), *Science in the Twentieth Century* (Reading, U.K.: Harwood): 175–202.

Mindell, D. (2000) *War, Technology and Experience Aboard the USS* Monitor (London: Johns Hopkins University Press).

Mirowski, Philip (1999) "Cyborg Agonistes: Economics Meets Operations Research in Mid-Century," *Social Studies of Science* 29(5): 685–781.

Mirowski, Philip (2001) *Machine Dreams: Economics Becomes a Cyborg Science* (Cambridge: Cambridge University Press).

Misa, Thomas (1985) "Military Needs, Commercial Realities, and the Development of the Transistor, 1948–1958," in Merritt Roe Smith (ed), *Military Enterprise and Technological Change* (Cambridge, MA: MIT Press).

Mitchell, Glen (2003) "See an Atomic Blast and Spread the Word: Indoctrination at Ground Zero," in J. Goodman, A. McElligott, & L. Marks (eds), *Useful Bodies: Humans in the Service of Twentieth Century Medicine* (Baltimore, MD: Johns Hopkins University Press): 133–64.

Molas-Gallart, Jordi (1997) "Which Way to Go? Defence Technology and the Diversity of 'Dual-Use' Technology Transfer," *Research Policy* 26: 367–85.

Moreno, Jonathan (2001) *Undue Risk: Secret State Experiments on Humans* (London: Routledge).

Mort, Maggie & Mike Michael (1998) "Human and Technological 'Redundancy': Phantom Intermediaries in a Nuclear Submarine Industry," *Social Studies of Science* 28(3): 355–400.

Moy, Timothy (2001) *War Machines: Transforming Technologies in the U.S. Military, 1920–1940* (College Station: Texas A&M University Press).

Noble, David (1985) "Command Performance: A Perspective on the Social and Economic Consequences of Military Enterprise," in M. R. Smith (1985), *Military Enterprise and Technological Change: Perspectives on the American Experience* (Cambridge, MA: MIT Press).

Oreskes, Naomi (2003) "A Context of Motivation: U.S. Navy Oceanographic Research and the Discovery of Hydrothermal Vents," *Social Studies of Science* 33(5): 697–742.

Perry, R. (1970) *A Review of System Acquisition Experience* (Santa Monica, CA: RAND).

Postol, Ted (2004) "An Informed Guess About Why Patriots Fired upon Friendly Aircraft and Saw Numerous False Missile Targets During Operation Iraqi Freedom" (Boston: MIT Security Studies Program, April 20).

Price, Richard (1997) *The Chemical Weapons Taboo* (Ithaca, NY: Cornell University Press).

Rappert, Brian (2001) "The Distribution and the Resolution of the Ambiguities of Technology: Or Why Bobby Can't Spray," *Social Studies of Science* 31(4): 557–92.

Rappert, Brian (2003a) *Non-Lethal Weapons as Legitimizing Forces? Technology, Politics and the Management of Conflict (*London: Frank Cass).

Rappert, Brian (2003b) "Coding Ethical Behaviour: The Challenges of Biological Weapons," *Science and Engineering Ethics* 9(4): 453–70.

Rappert, Brian (2005) "Prohibitions, Weapons and Controversy: Managing the Problem of Ordering," *Social Studies of Science* 35(2): 211–40.

Rappert, Brian (2006) *Controlling the Weapons of War: Politics, Persuasion and the Prohibition of Inhumanity* (London: Routledge).

Rappert, Brian (2007) *Biotechnology, Security and the Search for Limits* (London: Palgrave).

Rasmussen, Nicholas (2001) "Plant Hormones in War and Peace: Science, Industry, and Government in the Development of Herbicides in 1940s America," *Isis* 92(2): 291–316.

Reppy, Judith (ed) (1999) *Secrecy and Knowledge Production,* Cornell University Peace Studies Program, Occasional Paper No. 23.

Ruina, J. (1971) "Aborted Military Systems, " in B. T. Field, G. W. Greenwood, G. W. Rathgens, & S. Weinberg (eds), *Impact of New Technologies on the Arms Race* (Cambridge, MA: MIT Press).

Sapolsky, Harvey (1977) "Science, Technology and Military Policy," in Ina Spiegel-Rösing & Derek de Solla Price (eds), *Science, Technology, and Society* (London: Sage).

Scranton, Philip (2004) "Technology-Led Innovation: The U.S. Jet Propulsion System," Presentation to University of Exeter Business History Centre/Centre for Medical History Seminar, November 4.

Seidel, Robert (1987) "From Flow to Glow: A History of Laser Research and Development," *Historical Studies in the Physical and Biological Sciences* 18(1): 111–47.

Smit, Wim A. (1991) "Steering the Process of Military Technological Innovation," *Defense Analysis* 7(4): 401–15.

Smit, Wim (1995) "Science, Technology and the Military: Relations in Transition," in Sheila Jasanoff, Gerald E. Markle, James C. Petersen, & Trevor Pinch (eds), *Handbook of Science and Technology Studies* (London: Sage): 598–626.

Smith, B. (1966) *The RAND Corporation* (Cambridge, MA: Harvard University Press).

Smith, Merritt Roe (1985) *Military Enterprise and Technological Change: Perspectives on the American Experience* (Cambridge, MA: MIT Press).

Solovey, Mark (2001) "Project Camelot and the 1960s Epistemological Revolution: Rethinking the Politics–Patronage–Social Science Nexus," *Social Studies of Science* 31(2): 171–206.

Spinardi, Graham (1994) *From Polaris to Trident: The Development of U.S. Fleet Ballistic Missile Technology* (Cambridge: Cambridge University Press).

Spinardi, Graham (1997) "Aldermaston and British Nuclear Weapons Development: Testing the 'Zuckerman Thesis,'" *Social Studies of Science* 27(4): 547–82.

Stone, John (2000) *The Tank Debate: Armour and the Anglo-American Military Tradition* (Amsterdam: Harwood).

Stone, John (2002) "The British Army and the Tank," in T. Farrell & T. Terriff (eds), *The Sources of Military Change: Culture, Politics, Technology* (Boulder, CO: Lynne Rienner): 187–204.

Tannenwald, Nina (1999) "The Nuclear Taboo," *International Organization* 53(3): 433–68.

Thorpe, Charles (2002) "Disciplining Experts: Scientific Authority and Liberal Democracy in the Oppenheimer Case," *Social Studies of Science* 32(4): 527–64.

Thorpe, Charles (2004a) "Against Time: Scheduling, Momentum, and Moral Order at Wartime Los Alamos," *Journal of Historical Sociology* 17(1): 31–55.

Thorpe, Charles (2004b) "Violence and the Scientific Vocation," *Theory, Culture and Society* 21: 59–84.

Van Keuren, David K. (2001) "Cold War Science in Black and White: U.S. Intelligence Gathering and Its Scientific Cover at the Naval Research Laboratory, 1948–62," *Social Studies of Science* 31(2): 207–29.

Weber, Rachel (1997) "Manufacturing Gender in Commercial and Military Cockpit Design," *Science, Technology, & Human Values* 22(2): 235–53.

White House Office of Management and Budget (2005). Available at: http://www.whitehouse.gov/omb/budget/fy2006/defense.html.

Winner, Langdon (2004) "Trust and Terror: The Vulnerability of Complex Socio-Technical Systems," *Science as Culture* 13(2): 155–72.

Woodhouse, Edward J. (1990) "Is Large-Scale Military R&D Defensible Theoretically?" *Science, Technology, & Human Values* 15(4): 442–60.

Wright, Susan (1991) *Preventing a Biological Arms Race* (Cambridge, MA: MIT Press).

Wright, Susan (2004) "Taking Biodefense Too Far," *Bulletin of the Atomic Scientist* November/December: 58–66.

Wright, S. & D. Wallace (2002) "Secrecy in the Biotechnology Industry: Implications for the Biological Weapons Convention," in Susan Wright (ed), *Biological Warfare and Disarmament: New Problems/New Perspectives* (Lanham, MD: Rowman & Littlefield).

Zavestoski, Stephen, Phil Brown, Meadow Linder, Sabrina McCormick, & Brian Mayer (2002) "Science, Policy, Activism and War: Defining the Health of Gulf War Veterans," *Science, Technology, & Human Values* 27(2): 171–205.

29 The Right Patients for the Drug: Pharmaceutical Circuits and the Codification of Illness

Andrew Lakoff

The development and circulation of pharmaceuticals has received increasing attention from scholars of science, technology, and medicine. This work has analyzed the significance of pharmaceuticals from a number of vantage points: in terms of global consumer capitalism, as health interventions mediated by expertise, as objects of governmental regulation, as sources of hope, and as sites of political struggle. This chapter considers the significance of such work for questions of concern to science and technology studies and, in turn, asks what the perspective of STS might offer to analysts of the pharmaceutical industry and pharmaceutical use. A central goal for STS has been to describe how social context structures the effects of a given technology (MacKenzie & Wajcman, 1985). This work has shown that such effects do not inhere in the object itself but rather are shaped in the dynamic interaction between the goals of designers, the needs of users, and the constraints of the artifact. Following this line of inquiry, the question for pharmaceuticals is, through what social and political, as well as biochemical, means do they achieve their "effects"?

Like the operations of other techno-scientific objects, pharmaceutical effects take form in relation to the heterogeneous networks that shape them. Such effects are not embedded in the drug itself; rather, they arise at the nexus of chemical substance with governmental regulation, biomedical expertise, commercial interest, and patient experience. In this sense, the study of pharmaceutical circulation follows a number of insights from STS. Chemical substances become authorized medications as they are embedded in associations of experts, institutions, government regulations, business strategies, and patient advocates. However, studies of pharmaceutical circuits differ from many STS-based studies of heterogeneous networks in their emphasis on the distinctive significance of pharmaceuticals as biomedical innovations. These studies are often less focused on general questions such as "how are techno-scientific objects stabilized?" or "how are users configured?" than on the specific kinds of transformations that pharmaceuticals effect. They emphasize the political and ethical problems posed by current means of intervening in human bodies, and how these problems are approached in different contexts, according to diverse expert systems and under various regimes of health governance.

Social studies of pharmaceuticals range across these various "stages" of pharmaceutical elaboration—from laboratory and clinical research, to the work of advertisers to influence consumption, to physicians' practices of diagnosis and prescription, to their use and interpretation by patients. To approach the issues posed by pharmaceutical circulation, it is useful to keep in mind what is distinctive to pharmaceuticals. First, they are meant to heal: their use is part of an expert system for diagnosing and treating illness. Second, their movement into bodies is linked to their value, which is protected by intellectual property regimes at both national and transnational levels. Third, they are regulated commodities: their safety and efficacy are monitored by experts and agencies linked to the state. Thus, pharmaceuticals operate at the intersection of biomedical expertise, commercial interest, and governmental regulation. Tensions and conflicts among the values at stake in these divergent arenas generate a number of the key contemporary struggles around pharmaceutical development and circulation.

To become authorized medications, chemical substances must enter a system that is engaged in the technical administration of human life. In this sense, pharmaceuticals are *biopolitical* artifacts. That is, they are technical innovations that pose problems around how life should be understood and managed, and thus they provoke novel ethical and political quandaries (Collier & Lakoff, 2004). As Nikolas Rose (2001) argues, "biopolitics now addresses human existence at the molecular level: it is waged about molecules, amongst molecules, and when the molecules themselves are at stake." A number of key questions for social studies of pharmaceuticals surround the politics of drug circulation: Into which bodies should drugs go? Who should be able to prescribe them, and on what grounds? How can we know what they "do" to these bodies? How much should they cost, and who should pay for them?

Such questions can be broadly conceived as being concerned with problems of access to pharmaceutical circuits. I use the image of circuitry to emphasize the role of regulatory norms and technical standards in channeling the flow of medication into some bodies and not others. These circuits function both to include and to exclude patients from access to medication. Struggles over inclusion and exclusion arise around conflicting demands of health and profit, as mediated by government policy and professional norms. The central political—and technical—issues around pharmaceutical circulation are typically either questions of too *much* access (to lifestyle drugs in the North, for example) or too *little* access (to lifesaving drugs in the South). Such controversies not only include "demand side" issues such as marketing and pricing but also reach back to "supply"—to the drug development process, as in decisions over which illnesses are to be targeted as markets for new medications, which groups will be included in clinical trials or what will count as evidence of drug-related risk.

Controversies over patient exclusion from pharmaceutical circuits involve debate over whether medications, or knowledge concerning their effects, should be made more broadly available. These issues arise from the conviction among many critics that private sector drug development restricts access to necessary public goods. Private sector genomics research geared toward drug development raises the question of what limits should be placed on the patentability of genetic data (Boyle, 1997). Multilateral trade agreements that require the protection of patented medications pose the issue of the conditions under which nation-states can transgress intellectual property regimes in the name of public health emergencies or a nascent "human right" to medication (Biehl, 2004). In an era in which much research and development is conducted in the private sector, the question arises of how the development of drugs for conditions that do not have viable markets—whether AIDS patients in sub-Saharan Africa or heroin addicts in France—can be encouraged (Nguyen, 2005; Lovell, 2006). Meanwhile, controversies have arisen over access to knowledge about harmful side effects of medications generated in company-sponsored clinical trials but not made publicly available (Healy, 2006).

Like controversies over exclusion, struggles around inclusion also arise at the nexus of professional expertise, private-sector interest, and government regulation. But these controversies imply a problem of too much access—the overmedicalization of health. Such struggles often signal concern about the erosion of the authority of specialists in determining what counts as a condition that requires pharmaceutical treatment—as in the issue of direct-to-consumer advertising. They stem from the observation that private sector drug development efforts are for the most part targeted at relatively affluent consumers with chronic, but often not life-threatening conditions that promise life-long treatment, rather than at conditions that cause mortality among those who cannot afford patented medication. Relatedly, such efforts seek to medicate conditions that might better be addressed through changes in lifestyle. Thus, critics of the global pharmaceutical industry point to a gap between wealthy and poor countries, whereby patients from wealthy countries are overmedicated and those from poor countries are left in a condition of benign neglect (Petryna & Kleinman, 2006). Some analysts argue that the increasing demand, for the sake of profit, to circulate large quantities of drugs into bodies encourages the medicalization of social and psychic problems (Dumit, 2002). STS research has looked into questions such as whether advertising techniques generate illegitimate demand (Lakoff, 2004), and how the numerical threshold should be defined for illnesses with flexible diagnostic criteria, such as hypertension or depression. Other work has analyzed controversies over who should be included in trials for new drugs, and how the politics of such inclusion relates to broader and controversial forms of social classification such as race, gender, and age (Epstein, 2003; Fullwiley, 2007).

The emphasis of this chapter is on how a sociotechnical system for circulating pharmaceuticals as regulated commodities brings humans into its circuit of operations, and in so doing, reconfigures knowledge both about illness and about the effects of medication. I focus on coding techniques as critical sites for the analysis of how the pharmaceutical system functions, how it provokes ethical and political tensions, and how experts seek to resolve those tensions. Through coding, diverse domains—the market, health, and regulation—are brought into communication with one another. The chapter begins by showing how the effects of chemical substances are underdetermined by their chemical structure. It then follows three examples of coding systems that work to bring illness and medication into a shared space of calculable intervention. The first coding technique examined is the development of a new system for classifying illness so that diagnostic practice might be standardized across space and over time. The second is the use of rating scales in clinical trials both to define illness populations and measure their improvement—or lack of improvement—over the course of a trial. And the third is an effort by pharmaceutical and biotech industry strategists to refine these measuring techniques through genomics research. Overall, these means of codifying illness are designed to help pharmaceuticals find "the right patients for the drug." In turn, such coding techniques bring new kinds of humans into being.

UNDERDETERMINATION

To circulate in the regulated system of biomedicine, a drug is supposed to operate according to its model of the relation between illness and intervention. According to this model—"disease specificity"—illnesses are stable entities that exist outside of their embodiment in particular individuals and that can be explained in terms of specific causal mechanisms that are located within the sufferer's body. Disease specificity is a tool of administrative management. It makes it possible to gather populations for large-scale research, to mandate clinical practice through the institution of treatment protocols, and more generally, to rationalize health practice (Timmermans & Berg, 2003). At the intersection of individual experience and bureaucratic administration, disease specificity "helps to make experience machine readable" (Rosenberg, 2002).

The regulatory norm that guides pharmaceutical intervention in biomedicine is one of targeted effects: a given drug should work directly on a specific disease. Thus, for example, an "antidepressant" should directly treat "depression." However, psychopharmaceuticals do not clearly fit this model of targeted intervention. In the case of these drugs, both the putative effects of medication and the characteristics of its target illness population are subject to interpretation. This means that the achievement of specificity involves a process of mutual adjustment between illness and intervention. An illness comes gradually to be defined in terms of the intervention to which it "responds." The goal of linking a drug directly to diagnosis draws together a variety of projects among professionals, researchers, and administrators to craft new techniques of representation and intervention. These projects range from diagnostic standardization and the generation of clinical protocols to drug development and molecular genetics.

As psychopharmaceuticals illustrate, medications are objects whose effects are underdetermined by the biochemical characteristics of the substances themselves. The effects that a given drug produces depend, at least in part, on the milieu of expertise into which it enters. In this sense, drugs are instruments whose function is shaped by the form of rationality in which they are deployed; they are the means to various possible ends (Gomart, 2000; Schull, 2006). The history of psychopharmacology research demonstrates that the targeted effects attributed to these medications in contemporary biomedicine are not built into the medications themselves. In the 1950s, a series of major breakthroughs was made in the field, especially the development of the first generation of antidepressants and antipsychotics. In the context of overcrowded hospitals and the critique of psychiatric institutions, these drugs seemed to provide an answer to a number of problems, and their use spread rapidly. It became possible to transfer patients from asylums to community-based care and to expand the use of psychotherapy to psychotic patients (Grob, 1991).

This was a period in which social and psychodynamic models of mental illness were predominant in cosmopolitan psychiatry. The introduction of the new drugs did not immediately shift expert knowledge toward a biomedical model of targeted chemical intervention into specific disease. Rather, these substances inspired diverse readings, depending on the system of knowledge into which they entered. Initially the new medications were folded into the task of providing social and psychodynamic therapies. For social psychiatry, they were best seen as tools that were of use in developing forms of group therapy as part of the larger goal of reintegrating institutionalized patients into communities. Meanwhile, psychoanalytic work on psychosis flourished, since delusional symptoms could now be managed by medications that left patients' consciousness intact so that analysis could be practiced with them (Swain, 1994).

Pharmacology researchers sought to integrate drug effects into existing forms of expertise. The predominance of psychoanalysis in cosmopolitan psychiatry sparked an initial attempt to integrate these substances into dynamic models of the mind. In a 1957 conference in Zurich, experimenters in psychopharmacology met to compare notes on their results with the new drugs. The organizer of the conference, Nathan Kline, was a psychodynamic psychiatrist as well as a clinical drug researcher. "Are pharmacologic theories in contradiction to everything we have learned about psychodynamics?" he asked. "All the evidence is in the opposite direction. What is needed is integrating concepts that might provide possible pathways of linkage between the two sets of facts" (Kline, 1959: 18).

The diverse contributions to the conference volume illustrated these researchers' efforts to align the effects of the new drugs with psychodynamic models of human behavior. As one psychoanalyst argued: "It is time for us to treat [the patient's] personality and character structure with knowledge of the effects of drugs on the structures to be treated" (Kline, 1959: 309). The analyst argued that drugs did not have direct effects on the ego but affected the energy available to the psychic structure. He wrote about a patient who, feeling better after the administration of medication, wanted to discontinue psychotherapy: "It was explained to him that the relief was in symptoms only, and would not and could not eliminate the cause" (Kline, 1959: 312). Drugs thus operated on the surface, not on the depths, of the condition—but work on the depths, which depended on the transference relation, might be facilitated by the medication.

For these experts, the new medications assisted in the task of working on psychic structure. In his contribution to the volume, Kline wrote of the varying psychodynamic effects of these drugs: while reserpine allowed for the breakthrough of fairly deep material, chlorpromazine strengthened repressive mechanisms. However, both were useful as tools in the effort to perform psychoanalysis with psychotic patients:

"chlorpromazine and reserpine make it possible to quiet the schizophrenic sufficiently so that he can enter into psychoanalysis and tolerate the temporary threats of id inter- pretations" (Kline, 1959: 484). The effect of the drugs was to reduce the quantity of instinctive drive, or psychic energy, and so lessen the necessity of defense against unac- ceptable impulses. Thus, drug dosage could be manipulated to further the analytic process: "When the analysis loses its momentum the dosage can be reduced until suf- ficient psychic pressure once again builds up. In this way the rate of analytic progress can be regulated by the analyst" (Kline, 1959).

Kline's volume exemplifies the underdetermined character of medications' effects, from the perspective of expertise. As these early speculations indicate, the ideal of the contemporary biomedical paradigm, in which chemical interventions directly work on brain-based disorders, was only one way the understanding of these drugs could unfold. There was no direct line from the discovery of psychopharmaceuticals to the rise of the new biomedical psychiatry two decades later. Rather, the drugs provoked questions that were answered in terms of predominant forms of knowledge. However, as the regulatory and commercial system for drug circulation transformed, they came to take on effects directly targeted at biochemical abnormalities.

THE RE-CODIFICATION OF ILLNESS

The mid-century psychodynamic understanding of how mood and behavioral med- ications worked is strikingly different from the premise of contemporary biomedical psychiatry, in which medication targets a specific neurochemical deficiency in order to correct a brain-based illness. The story of how these drugs acquired targeted effects involves two linked processes: on the one hand, government regulation required that pharmaceuticals be proved to have targeted effects in order to circulate in the bio- medical system; on other hand, to demonstrate such effects, researchers had to be able to classify disorders in a standardized way. Thus, both the effects of drug treatment and ways of seeing illness had to be reconfigured in order to achieve specificity.

Changes in the drug regulatory system were critical to this process. In 1962, the U.S. Congress amended FDA legislation to require that all new medications be tested for both safety and efficacy according to randomized, controlled trials (Marks, 1997). This was a key event in shaping psychopharmaceuticals into agents with targeted effects. As Thomas Hughes (1987) argues, for a radical invention to circulate widely within a technical system, it must "embody" the economic, political, and social characteristics that will enable its survival in use. For the new drugs to be proved effective accord- ing to biomedical criteria, they had to target specific disease entities.

Under the new legislation, to be marketed to and prescribed by physicians, chemi- cal interventions had to be measurable in terms of efficacy across populations of com- parable patients. Clinical psychopharmacology researchers thus needed groups of homogeneous patients on whom to test the new substances. However, diagnostic prac- tice at the time was notoriously unreliable between different clinical observers: what one psychiatrist read in the symptoms of a patient might be understood quite differently by another. This hampered efforts to quantitatively measure the efficacy of interventions; without consistent diagnostic practice, there was no way to ensure that clinical studies were being applied to the same type of patient. In response to this need for homogeneous patient populations for research, clinical psychiatry researchers designed rating scales and questionnaires that would codify illnesses along the model of specificity—as discrete entities that corresponded to targeted therapeutic interventions. One example of a coding mechanism that helped make it possible to test medications on consistent patient populations was the Hamilton Depression Rating Scale (1960).

Once the regulation of pharmaceuticals according to the guidelines of randomized controlled trials was put in place, the development of diagnostic standards became "all but inevitable," as David Healy (1996) puts it. This process of standardization was initially important for research purposes rather than in the clinic.[1] Clinicians in the United States—most of whom were working according to individualizing psychodynamic models—could ignore such diagnostic criteria and rating scales. However, in the early seventies a widely publicized comparative study of diagnostic practices indicated that U.S. psychiatrists were significantly out of sync with international norms. Shortly thereafter, third-party payers began to demand that doctors defend their treatment strategies with consistent protocols whose effectiveness had been demonstrated according to professionally sanctioned criteria. Such pressures, as well as a desire to improve psychiatry's status within medicine, led the American Psychiatric Association (APA) to set limits to the interpretive autonomy of its members (Wilson, 1995; Young, 1996). Diagnostic practices were the focus of this effort.

The 1980 agreement by the APA on a new edition of the *Diagnostic and Statistical Manual of Mental Disorders* (DSM-III) put in place a set of standards regulating diagnosis according to the model of disease specificity. As a standards regime, DSM-III sought to produce functionally comparable results across disparate domains. Its primary goal was reliability: if the same person went to two different clinics, he or she should receive the same diagnosis at each site. Based on directly observable traits, and ostensibly atheoretical, the new diagnostic standards structured a broader system of communication. Rating scales based on questionnaires were refined to measure norms of functionality, making it possible for different observers to use the same criteria in coming to a diagnostic evaluation. DSM-III, as a coding system devised to bring heterogeneous individuals into a shared space of calculability, was as an effort to achieve universality by enforcing similar expert practice across space.

DSM-III was a standardizing but also a dynamic system: its categories were evolving rather than fixed, and its authors set up a committee-based structure within the profession for testing and revising its definitions (Kirk & Kutchins, 1992). The point was to delimit a set of rules for negotiating future standards. The enactment of this system for generating and refining standards can be understood as a process of professional normalization: "normalization produces not objects but procedures that will lead to some general consensus regarding the choice of norms and standards" (Ewald, 1990: 148). Such normative procedures do not only constrain; they also generate new objects of knowledge and forms of identity.[2] In this case, a process of mutual adjustment between drug and diagnosis, intervention and illness, has generated new definitions of pathology and thus of normality as well.

Despite many critiques, especially by experts opposed to the model of specific disease entities, DSM in its newly revised versions has continued to evolve and to attain strength. The manual has extended to new sites because of its capacity to make behavioral pathology transferable across social domains. Its standards have multiple possible uses: in gathering epidemiological data, developing treatment algorithms, and claiming insurance benefits. Disease categories thus are "objects that are able both to travel across borders and maintain some sort of constant identity" (Bowker & Star, 1999). This constant identity—of "disease specificity"—enables DSM to function as a connective tissue for biomedical psychiatry, linking populations as they are forged in multiple domains: the clinic, insurance, and scientific research.

FINDING PLACEBO RESPONDERS

The standardization of diagnosis helped drug developers define a target population and market to guide research and development efforts. However, the definitions it produced have not been sufficient to stabilize the relation between diagnosed patient and targeted intervention. The development of medications for conditions with unclear boundaries, such as depression, continues to pose problems in clarifying who is a member of an illness population. This section looks at how drug developers work to implement specificity in the case of antidepressants. It describes how substances in development "seek" the users on whom they can achieve specific effects. I focus on the placebo effect in antidepressant trials as a problem that points toward the limitations of current coding techniques.

The problem of the legitimacy of antidepressant use is an exemplary "controversy of inclusion." Critics of recent increases in the diagnosis of depression and the use of antidepressants suggest that "depression" has come to be a general term for a number of disparate forms of suffering whose only commonality is that they seem to respond to antidepressant medication. This phenomenon, they argue, explains apparent recent increases in the prevalence of depression in North America and Europe (Borch-Jacobsen, 2002; Pignarre, 2001). This critique points to the question of whether medication can legitimately delineate illness entities: does a successful response to a drug tell us what the person was suffering from? Given that medications in the class known as antidepressants have a wide range of potential effects, their capacity to alleviate symptoms in a specific case does not necessarily mean that depression is the specific illness entity being treated.

The antidepressant market grew tremendously between 1980, when SSRIs were first introduced, and 2005, reaching $20 billion per year in the United States alone.[3] Meanwhile, companies were engaged in intensive research for new molecules that would allow them to make up for the loss of patent protection on earlier drugs. A large number of experimental molecules were being tested on a shrinking pool of experimental subjects. Clinical trial results had high stakes, including for the value of major companies' stock. In this context, problems in moving new molecules from laboratory to market were the focus of increasing attention from drug developers.

In pharmaceutical development, diverse elements—chemistry, animal experiment, business strategy, statistics, illness experience—are brought together into a space of calculability in which the efficacy of medication can be demonstrated. This space is structured by the central task of moving chemical substances from the laboratory to the marketplace. To enter the market, drugs must demonstrate evidence of efficacy and nontoxicity. Clinical trials are the means of clearing this hurdle. What must be shown, given regulatory norms, is a targeted effect: the substance must quantitatively improve the condition of a population of subjects with a specific illness. Ambiguous illnesses, such as depression, raise challenges to this effort, first, in determining who should be included in the group of subjects, and second, in determining how to measure improvement of the conditions. Coding techniques, such as diagnostic protocols and symptom rating scales, bring heterogeneous patients and uncertain drug effects into this shared space of calculability.

For the process to be successful, two forms of uncertainty must be managed: a drug must have a targeted effect, and its patients must have a specific illness. The regulative principle of biomedicine—disease specificity—guides the work of integrating substance and target. In psychopharmacology this process is especially visible and problematic. Drugs that work on mood, behavior, and thought cannot be measured through physiological markers of efficacy, such as a decrease in blood pressure or PSA level. Nor can criteria for inclusion in a behavioral illness category be determined through organic markers. The coding techniques that are designed to address this problem generate illness identities, such as the "depressed person." However, the legitimacy of these relatively new "human kinds" (Hacking, 2002) remains a subject of controversy.

Biomedicine seeks to order persons into classifications based on specific disease entities that are presumed to exist outside the particular manifestation of illness in the individual. Drug regulation operates along this model, assuming that such populations should then be treatable by the same kind of intervention—in this case, antidepressants. However, it remains unclear whether patients classified as depressed all share the same disease.

The goal of the clinical trial is to test the efficacy of the drug on a population with a specific illness. The key technology for assembling populations for antidepressant clinical trials and measuring their response is the standardized rating scale—the "gold standard" is the Hamilton Depression Rating Scale. Such scales attempt to produce equivalency by turning stories into numbers—to translate subjective experience into something collectively measurable. In this sense they are similar to DSM checklists in their attempt to produce stable illness collectives defined by measurable symptoms.

For drug developers it is the drug, rather than the depressed patient, that serves as a stable reference point. Consider their use of the term "signal detection" to describe the goal of the trial. Here the drug is already presumed to have efficacy—that is, a signal to transmit—and the question is how to pick it up. If measuring devices record the signal, the patients' role is to transmit it. The task is then to find the right patients—those on whom the drug shows demonstrable effect. Drug developers are skeptical about the capacity of the standard rating scales to produce a consistent patient population for testing. From painful experience, they have learned that patients admitted under these criteria often vary tremendously in their response to drugs and to placebos. In addition, rating scales are applied inconsistently by raters at trial sites. Attempts to standardize the application of the scales, such as video training sessions and site audits, seem not to have improved trial success rates. On the one hand, researchers want to standardize raters' behavior as much as possible in order to glean consistent data—but then there is the danger of "dampening the signal" by failing to note clinical signs not measured by the rating scales. Yet if the researchers focus too closely on clinical observation, they might create a greater placebo response because of the attachment that would then form between the rater and the subject.

From the perspective of drug developers, when trials fail it is not that the drug does not work but that noise has crept into the signal detection process. The most pernicious and obstinate source of noise is the placebo effect. The placebo effect is unpredictable and seemingly unmanageable, and it costs drug companies hundreds of millions of dollars in failed trials and delayed or shelved compounds. Since the placebo response rate in depression trials is usually at least 30 percent, and that of antidepressants not much higher—often around 50 percent—it can even seem to impugn the efficacy of established and marketed drugs, used as active comparators in trials of novel compounds. Worse, it seems that the placebo response rate has been increasing in recent years. Drug developers have tried many things to reduce placebo response without at the same time reducing treatment response but have been frustrated by its intransigence.

Although advocates of alternative medicine have begun to see the placebo effect as a possible source of new forms of therapy, for drug companies it remains an impediment to proving efficacy and bringing new drugs to market (Guess et al., 2002). For drug developers, understanding the operations of the placebo effect is thus a matter of necessity: they are forced to confront it, given the exigencies of regulatory guidelines. As part of these efforts, they have located a number of possible candidates for the locus of apparent placebo effect: the patient, the healer, the measuring device, and the illness itself.

A first important distinction they make is between artifactual and real placebo response. There are at least two kinds of artifactual placebo response. One has to do with the motives of raters at the trial site. If the site is under pressure to rapidly enroll patients, raters may inflate their scores at the beginning of the trial. If those in the placebo group then show improvement, this may be due to more accurate later measurement. A second potential source of artifactual placebo response is statistical "regression to the mean"—which happens when the patient has a rapidly fluctuating course of illness and enrolls in the trial when it is at its worst. Then, if the illness improves over the course of the trial, the rater again sees what looks like, but is not really, a placebo response.

Like artifactual response, real placebo response can be attributed to the characteristics of either the trial site or the patients. One of its possible causes has to do with "investigator behavior." Here drug researchers presume a certain understanding of what the placebo response is: it is based on hope, expectation, or an attachment to the healer or to the treatment itself. Thus, if the site-based investigators contracted by the drug developers perform what is termed "covert psychotherapy" or in some other way give those who have been assigned a placebo the sense that they are being helped in any way, an unwanted placebo response can be induced. The co-inventor of one well-known rating scale has argued that such "non-specific supportive contact" can even include the filling out of forms, if these are meant to reassure patients about their participation in the trial. He advises sternly, "Patients who are overly sensitive to reassurance need to be identified and if possible excluded" (Montgomery, 1999).

This advice relates to a more general set of strategies based on the hypothesis that the source of excessive placebo response is the presence of a certain class of patients who are overly susceptible to placebo. Drug developers here shift the locus of potentiality in the trial from the drug to the patient. Instead of seeking to test the drug on an established category of patients, they seek to find the right patients for the drug. As one epidemiologist complained in an interview, "The biggest problem is getting the right patients." Who are they? "No one knows, but there are a lot of different ideas." There are some possible clues to placebo susceptibility, such as duration of illness, family history, and age of onset. But these "don't hang together from one trial to the other. Things disappear as you look at them more closely."

The most salient subpopulations to be delineated before the trial begins are drug responders and placebo responders. Subject populations are here seen as potential transmitters of drug efficacy. As one expert writes, "samples selected for trials should be able to deliver a predicted response to drug and not to placebo." Unfortunately, standardized diagnostic criteria are "not up to the task of distinguishing between clear drug responsive patients and placebo responders" (Montgomery, 1999).

It is only possible to be a certain kind of person—such as a multiple personality or hysteric—in a certain time, place, and social setting (Hacking, 1986; Young, 1996). In what context is it possible, then, to be a "placebo responder"? The figure of the placebo-responder has been a mysterious, much sought-after type in clinical pharmacology. Attempts to delineate the characteristics of placebo-responders began after the recognition of the importance of the placebo effect in the years after World War II, when the double-blind, randomized, controlled trial (RCT) was accepted as the means to police fraudulent medications. In the rationale of the RCT, the placebo effect was both an epistemological necessity and a practical obstacle to showing true drug efficacy. If one could ascertain before the trial which patients were likely to respond to a placebo and to thereby contaminate the results, one could ostensibly eliminate them from the trial beforehand and improve the chances for a successful trial.

In the 1950s, placebo researchers used personality questionnaires and Rorschach tests to characterize the typical placebo-reactor in postoperative pain trials. When asked, "What sort of people do you like best?" placebo reactors were more likely to respond, "Oh, I like everyone." They were more often active churchgoers and had less formal education than nonreactors (Lasagna et al., 1954). As for the Rorschach results, the researchers wrote: "the reactors are in general individuals whose instinctual needs are greater and whose control over the social expression of these needs is less strongly defined and developed than the non-reactors." In the 1970s, researchers found that placebo reactors scored higher on the "Social Acquiescence Scale," based on agreement with proverbs such as "obedience is the mother of success"; "seeing is believing"; and "one false friend can do more harm than 100 enemies" (McNair et al., 1979).

The line of research linking placebo response to suggestibility as a personality characteristic eventually faded, and more recently researchers have focused on somatic rather than psychological factors in looking at depression treatment and the placebo effect. They hypothesize that milder severity of illness, a more rapidly fluctuating course, and certain kinds of somatic complaints correlate with placebo response (Schatzberg & Kraemer, 2000). Recent studies have used brain imaging techniques to distinguish placebo from antidepressant response. One explanation that has been presented for the increasing placebo response rate is that less severely ill patients are being used more often, given the shortage of patients for clinical trials (Montgomery, 1999). But efforts to operationalize these criteria generate other problems, such as limiting the potential indication of the approved drug or extending the length of the trial in the search for more appropriate patients.

Given the difficulty in trying to root out the placebo effect by addressing its underlying causes, antidepressant researchers have turned to a more pragmatic approach that might be called the "mousetrap technique," after the play within a play Hamlet staged in order to goad his father's murderer into revealing his guilt. Here experimenters in effect stage the trial before it actually begins, giving all patients placebo for a week. They then eliminate from the trial those who responded to the placebo (Quitkin et al., 1998). With this approach, it does not matter why patients respond to placebo, nor does the knowledge or technique of the investigator matter. One simply needs to know which patients have responded in order to eliminate them. However, these efforts have also proved disappointing—placebo response rates during these run-in periods tend to be low, and so it has not been possible to eliminate most of the potential placebo responders. Then, in the actual trial, other subjects continue to respond to the placebo, drowning out the signal of drug efficacy and undermining the trial.

MEDICATION-RESPONSE PROFILES

How might recent developments in the life sciences address this problem of identifying ideal treatment responders? The hope is to transcend the subjectivity of rating scales as means of delineating coherent populations by finding physiological measures. One approach is to use a gene-based diagnostic test as a coding mechanism for distinguishing among otherwise heterogeneous groups of subjects. In this section I describe efforts at the intersection of business strategy and molecular biology to redefine patients according to medication response.

Beginning in the late 1990s, pharmacogenomics gained prominence in discussions of the practical implications of the completion of the genome project. The technology underpinned a projected future of personalized medicine, in which gene chips would guide physicians to the most appropriate pharmaceutical intervention, bypassing wasteful medication trials and avoiding harmful side effects. It was directed toward characterizing, at a genomic level, distinctive medication-response phenotypes. In the process, a new way of grouping people, according to their "medication-response profile," emerged.[4]

The vision of personalized medicine as a goal of genomics research crystallized in the late 1990s. We can take as an example an influential report released in 1999 by a major consulting firm. The report argued that the pharmaceutical industry was facing a period of rapid transformation: the industry's classic strategy of mass marketing blockbusters to broad segments of the population was entering into crisis owing to patent expirations, the lack of replacement products in the pipeline, and changes in the "healthcare environment," such as price controls from third party payers (Boston Consulting Group, 1999). There were new opportunities as well, coming from two new developments: first, the rise of a new, educated health care consumer who demanded tailored treatment; and second, technological innovations resulting from the Human Genome Project. The report warned that pharmaceutical players faced a growing threat from agile new biotech firms branching into pharmaceutical development. It proposed a solution that would meet these new demands and threats, shaping a post–blockbuster pharmaceutical economy: personalized medicine, using the technological platform of pharmacogenomics. The report described how the technology would be used to predict patient responses to medication: "Companies will, in essence, be able to predict which patients are likely to respond to which 'suites' of medications. With this capability, a pharmaceutical company will have the opportunity to market to specific patient subgroups."

In this vision of the future, drugs would be targeted at subpopulations of patients who were determined to be genetically responsive to these medications. Diagnostics would thus be linked directly to therapeutics through the technological platform of pharmacogenomics. One pharmaceutical strategist described his company's plan for personalized medicine: "We are focused on integrating genomics-based diagnostics and therapeutics with the ultimate vision of linking the right drug to the right patient."[5] Thus, personalized medicine promised to rationalize the management of the population's health according to a norm of pharmacological specificity. The goal was to operationalize human genetic variation by matching patients to the most appropriate pharmaceutical intervention—to directly link illness populations to market segments, calibrating health need and consumer demand. Since it proposed to break up current illness categories and reformulate subpopulations in terms of medication response, what the technology aimed for is better described as "segmented" rather than "personalized" medicine.

In the short term, drug companies were interested in a more immediate application: using the technology as part of a drug development program geared to increase productivity by bringing drugs more quickly to market. In order to meet Wall Street growth expectations, analysts estimated that the major pharmaceutical companies had to introduce three to five new chemical entities per year (Norton, 2001). But research pipelines seemed to be running dry, and new drug applications were slowing. The pharmaceutical industry claimed that it was spending five to eight hundred million dollars and eight to ten years per new drug. The difficulty of demonstrating efficacy through clinical trials was one widely cited reason for why drug development was a slow and expensive process. Clinical trials for new drugs required tremendous numbers of patients to demonstrate safety and efficacy and had a high failure rate—which was in part blamed on the heterogeneity of patient populations in these trials. Given limited patent lifetimes, companies calculated the cost of delays in market approval in the millions of dollars per day. In this context, biotech firms pitched genomics-based diagnostics to pharmaceutical companies as a technical solution to the problem of the inefficiency of the clinical trial process—as a way to locate responder populations through a simple blood test.

The potential usefulness of pharmacogenomics in drug development was a result of the centrality of the specificity model to pharmaceutical circulation. Here genomics responded to a need for better ways of stratifying populations in clinical trials. By using pharmacogenomics to forge populations for experiment, drug developers could screen patients in terms of potential drug response before the trial began. This would cut down on adverse reactions and improve the odds of running a successful trial. Rather than developing targeted drugs for individual patients, it was a matter of finding the right patients for a given drug. As one analyst envisioned: "Pharmacogenomic profiling can be used to stratify trials based on patients who are most likely to benefit from therapy" or else to exclude the "poor metabolizer type" from trials (Norton, 2001: 183, 192). In other words, if it were known beforehand which patients were going to respond to the drug, the trial would have a better chance of success.

This was especially the case in developing medications for psychiatric illness. As one leader in applying genomics to pharmaceutical development wrote: "Patient groups who have vaguely defined phenotypes that are more difficult to categorize by objective criteria, such as depression, could be studied more efficiently using medicine response profiles as selection variables" (Roses, 2000: 863). As a tool for gathering homogeneous populations for clinical drug research, the development of pharmacogenomics followed a similar logic to that which animated initial diagnostic standardization efforts in psychiatry. The regulatory demand for evidence of specific efficacy was helping drive a series of efforts to more directly couple pharmaceutical intervention and diagnostic target, as key to lock.

Even if the specificity model was not an adequate description of how psychopharmaceuticals worked in relation to mental illnesses, genomics technology sought to make the model more accurate. Pharmacogenomics served as a possible mechanism of adjustment between drug and disease entity, a way of calibrating intervention more closely to illness. Here, the adjustment between the drug's effects and the characteristics of its target population was not due to the development of more directly targeted drugs. Rather, the crucial element of the adjustment process was to occur at the diagnostic level. The drug remained stable while the target shifted in relation to it. In other words, the specificity model was being built into the technological platform—the model was in a sense being *made* more accurate, not by finding the perfect pharmacological key to fit the illness but by changing the very nature of the lock into one that, by definition, matched the key.

What to make of the prospect of locating genetic markers indicating potential responsiveness to medications? Such markers would indicate an internal relationship of potentiality to an external substance already in circulation, or one still to be invented. The idea was to turn the genome into a technology for guiding drug intervention. At the same time, pharmacogenomics would transform the need that such medication addressed. The target of the drug would no longer be an illness per se but rather an inherited capacity to respond to the drug. The technology was especially intriguing in the case of psychiatric disorders because, while it posed the possibility of delineating an organic basis for these amorphous conditions, it bypassed the question of the coherence of classical illness entities. The delineation of new subpopulations had the potential to once again transform the practice of diagnosis. In a world of gene-chip-based diagnostic tests in the clinic, the broad categories that govern psychiatric practice might be broken down in terms of medication response, so that diagnostic questions would appear no longer as—"is it bipolar disorder or schizophrenia?" but as—"is it a lithium or an olanzapine response profile?" Obviously, this would transform patient identity as well.

CONCLUDING REMARKS

Research in the social studies of pharmaceuticals indicates that one should not analyze drugs as isolated substances but rather look at the connections they require in order to circulate as authorized medication, and how these connections structure the particular transformations they effect on bodies and minds. Thus, pharmaceuticals do not work by themselves but function as elements of a broader system that both encourages and constrains their circulation. This system is technical not only in the sense that it requires expert knowledge to develop and evaluate the operations of pharmaceuticals but also because a number of administrative and regulatory techniques govern their production and circulation. The system also is political in that decisions on how to regulate the circulation of pharmaceuticals are critical to the particular effects that they achieve on individual bodies and minds. Moreover, a politics of health determines who is integrated into networks of pharmaceutical circulation, and how. As we have seen, the regulatory logic of disease specificity, in combination with imperatives of health and profit, structures the mutual adaptation of drug effects and illness definition, and in the process integrates humans into a complex sociotechnical system.

Notes

1. As the creators of the Research Diagnostic Criteria (RDC) wrote, "a major purpose of the RDC is to enable investigators to select relatively homogeneous groups of subjects who meet specified diagnostic criteria" (Spitzer et al., 1978).

2. As Pierre Macherey (1998) puts it, "the norm 'produces' the elements on which it acts as it elaborates the procedures and means of this action."

3. As one pharmaceutical company executive wrote, "Pharmacogenetics will enable individuals to be classified according to their likely response to a medicine" (Roses, 2000: 860). An industry analyst (Sadee, 1998) writes that pharmacogenomics heralds "the therapeutic management of individual patients."

4. John Maragnore, senior vice president of strategic product development, Millennium (in *BioIT World*, n.d.).

5. See Waldby (2001) for a similar argument with respect to the technical efficacy of the "central dogma" in molecular biology.

References

Aronowitz, Robert (1999) *Making Sense of Illness: Science, Society, and Disease* (Cambridge: Cambridge University Press).

Biehl, Joaõ (2004) "The Activist State: Global Pharmaceuticals, AIDS, and Citizenship in Brazil," *Social Text* 22(3): 105–32.

Bodewitz, Henk J.H.W., Henk Buurma, & Gerard H. de Vries (1987) "Regulatory Science and the Social Medicine of Trust in Medicine," in Wiebe E. Bijker, Thomas P. Hughes, & Trevor J. Pinch (eds), *The Social Construction of Technological Systems: New Directions in the Sociology and History of Technology* (Cambridge, MA: MIT Press): 243–60.

Borch-Jacobsen, Mikkel (2002) "Prozac Nation," *London Review of Books* July 9: 18–19.

Boston Consulting Group (1999) "The Pharmaceutical Industry into Its Second Century: From Serendipity to Strategy," January: 51–56.

Bowker, Geoffrey, & Susan Leigh Star (1999) *Sorting Things Out: Classification and Its Consequences* (Cambridge, MA: MIT Press).

Boyle, James (1997) *Shamans, Software and Spleens: Law and the Construction of the Information Society* (Cambridge, MA: Harvard University Press).

Collier, Stephen J. & Andrew Lakoff (2004) "On Regimes of Living," in Aihwa Ong & Stephen J. Collier (eds), *Global Assemblages: Technology, Politics and Ethics as Anthropological Problems* (Malden, MA: Blackwell).

Conrad, Peter (1985) "The Meaning of Medications: Another Look at Compliance," *Social Science and Medicine* 20(11): 29–37.

Corrigan, Oonagh P. (2002) "A Risky Business: The Detection of Adverse Drug Reactions in Clinical Trials and Post-Marketing Exercises," *Social Science and Medicine* 55: 497–507.

Dumit, Joseph (2002) "Drugs for Life," *Molecular Interventions* 2: 124–27.

Ehrenberg, Alain (1998) *La Fatigue d'être Soi: Dépression et Société* (Paris: Odile Jacob).

Epstein, Steven (2003) "Inclusion, Diversity, and Biomedical Knowledge Making: The Multiple Politics of Representation," in Nelly Oudshoorn & Trevor Pinch (eds), *How Users Matter: The Co-Construction of Users and Technology* (Cambridge, MA: MIT Press): 173–90.

Ewald, François (1990) "Norms, Discipline and the Law," *Representations* 30: 138–61.

Fishman, Jennifer (2004) "Manufacturing Desire: The Commodification of Female Sexual Dysfunction," *Social Studies of Science* 34(2): 187–218.

Fraser, Mariam (2001) "The Nature of Prozac," *History of the Human Sciences* 14(3): 56–84.

Fullwiley, Duana (2007) "The Molecularization of Race: Institutionalizing Racial Difference in Pharmacogenetics Practice," *Science as Culture* 11(1): 1–29.

Gomart, Emilie (2002) "Methadone: Six Effects in Search of a Substance," *Social Studies of Science* 32(1): 93–135.

Grob, Gerald N. (1991) *From Asylum to Community: Mental Health Policy in Modern America* (Princeton, NJ: Princeton University Press).

Guess, Harry A., Arthur Kleinman, John W. Kusek, & Linda W. Engel (eds) (2002) *The Science of the Placebo: Toward an Interdisciplinary Research Agenda* (London: BMJ Books).

Hacking, Ian (1987) "Making up People," in Thomas Heller, Morton Sosna, & David E. Wellbury (eds), *Reconstructing Individualism: Autonomy, Individuality, and the Self in Western Thought* (Stanford, CA: Stanford University Press): 222–36.

Harrington, A. (2002) " 'Seeing' the Placebo Effect: Historical Legacies and Present Opportunities," in H. A. Guess, A. Kleinman, J. Kusak, & L. Engel (eds), *The Science of the Placebo: Toward an Interdisciplinary Research Agenda* (London: BMJ Books).

Healy, David (1996) *The Antidepressant Era* (Cambridge, MA: Harvard University Press).

Healy, David (2006) "The New Medical Oikumene," in Adriana Petryna, Andrew Lakoff, & Arthur Kleinman (eds), *Global Pharmaceuticals: Ethics, Markets, Practices* (Durham, NC: Duke University Press).

Hedgecoe, Adam & Paul Martin (2003) "The Drugs Don't Work: Expectations and the Shaping of Pharmacogenetics," *Social Studies of Science* 33(3): 327–64.

Hughes, Thomas P. (1987) "The Evolution of Large Technological Systems," in Wiebe E. Bijker, Thomas P. Hughes, & Trevor J. Pinch (eds), *The Social Construction of Technological Systems: New Directions in the Sociology and History of Technology* (Cambridge, MA: MIT Press).

Kirk, Stuart A. & Herb Kutchins (1992) *The Selling of DSM: The Rhetoric of Science in Psychiatry* (New York: Aldine de Gruyter).

Kleinman, Arthur (1988) *Rethinking Psychiatry: From Cultural Category to Personal Experience* (New York: Free Press).

Kline, Nathan S. (ed) (1959) *Psychopharmacology Frontiers: Proceedings of the Psychopharmacology Symposium* (London: J. & A. Churchill).

Lakoff, Andrew (2000) "Adaptive Will: The Evolution of Attention Deficit Disorder," *Journal of the History of the Behavioral Sciences* 36(2): 149–69.

Lakoff, Andrew (2006) *Pharmaceutical Reason: Knowledge and Value in Global Psychiatry* (Cambridge: Cambridge University Press).

Lakoff, Andrew & Stephen J. Collier (2004) "Ethics and the Anthropology of Modern Reason," *Anthropological Theory* 4(4): 419–34.

Lasagna, Louis, Frederick Mosteller, John M. von Felsinger, & Henry K. Beecher (1954) "A Study of the Placebo Response," *American Journal of Medicine* 16(6): 770–79.

Lovell, Anne M. (2006) "Addiction Markets: The Case of High-Dose Buprenorphine in France," in Adriana Petryna, Andrew Lakoff, & Arthur Kleinman (eds), *Global Pharmaceuticals: Ethics, Markets, Practices* (Durham, NC: Duke University Press).

MacKenzie, Donald & Judy Wajcman (1985) *The Social Shaping of Technology* (Philadelphia: Open University Press).

Marks, Harry M. (1997) *The Progress of Experiment: Science and Therapeutic Reform in the United States, 1900–1990* (New York: Cambridge University Press).

McNair, D. M., G. Gardos, D. S. Haskell, & S. Fisher (1979) "Placebo Response, Placebo Effect, and Two Attributes," *Psychopharmacology* 63: 245–50.

Montag, Warren (ed) (1998) *In a Materialist Way: Selected Essays by Pierre Macherey* (New York: Verso).

Montgomery, S. A. (1999) "The Failure of Placebo-Controlled Studies," *European Neuropsychopharmacology* 9: 271–76.

Nguyen, Vinh-Kim (2005) "Antiretroviral Globalism, Biopolitics, and Therapeutic Citizenship," in *Global Assemblages: Technology, Politics and Ethics as Anthropological Problems* (Malden, MA: Blackwell).

Nichter, Mark & Nancy Vuckovic (1994) "Agenda for an Anthropology of Pharmaceutical Practice," *Social Science and Medicine* 39(11): 1509–25.

Norton, Ronald (2001) "Clinical Pharmacogenomics: Applications in Pharmaceutical R&D," *Drug Discovery Today* 6(4) (February): 180–85.

Palsson, Gisli & Paul Rabinow (1999) "Iceland: The Case of a National Human Genome Project," *Anthropology Today* 15(5): 14–18.

Petryna, Adriana (2006) "The Human Subjects Research Industry," in Adriana Petryna, Andrew Lakoff, & Arthur Kleinman (eds), *Global Pharmaceuticals: Ethics, Markets, Practices* (Durham, NC: Duke University Press).

Pignarre, Philippe (2001) *Comment la Depression est devenue une Epidemie?* (Paris: La Découverte).

Rose, Nikolas (2001) "The Politics of Life Itself," *Theory, Culture and Society* 18(6): 1–30.

Rosenberg, Charles (2002) "The Tyranny of Diagnosis: Specific Entities and Individual Experience," *Milbank Quarterly* 80(2): 237–60.

Roses, Allen D. (2000) "Pharmacogenomics and the Practice of Medicine," *Nature* 405(15): 857–65.

Sadee, Wolfgang (1998) "Genomics and Drugs: Finding the Optimal Drug for the Right Patient," *Pharmaceutical Research* 15(7): 959–63.

Schatzberg, Alan F. & Helena C. Kraemer (2000) "Use of Placebo Control Groups in Evaluating Efficacy of Treatment of Unipolar Major Depression," *Biological Psychiatry* 47(8): 745–47.

Schull, Natasha (2006) "Machines, Medication, Modulation: Circuits of Dependency and Self-Care in Las Vegas," *Culture, Medicine, and Psychiatry* 30(2): 223–47.

Spitzer, R. L., J. Endicott, & E. Robins (1978) "Research Diagnostic Criteria: Rationale and Reliability," *Archives of General Psychiatry* 35(6): 773–82.

Swain, Gladys (1994) *Dialogue avec l'Insensé: Essais d'Histoire de la Psychiatrie* (Paris: Gallimard).

Timmermans, Stefan & Marc Berg (2003) *The Gold Standard: The Challenge of Evidence-Based Medicine and Standardization in Health Care* (Philadelphia: Temple University Press).

Van der Geest, Sjaak, Susan Reynolds Whyte, & Anita Hardon (1996) "The Anthropology of Pharmaceuticals: A Biographic Approach," *Annual Review of Anthropology* 25: 153–78.

Waldby, Catherine (2001) "Code Unknown: Histories of the Gene," *Social Studies of Science* 31(5): 779–91.

Wilson, Mitchell (1995) "DSM-III and American Psychiatry: A History," *American Journal of Psychiatry* 150(3): 399–410.

Young, Allan (1996) *The Harmony of Illusions: Inventing Post-Traumatic Stress Disorder* (Princeton, NJ: Princeton University Press).

30 Making Order: Law and Science in Action

Sheila Jasanoff

In the subtitle of his 1987 book *Science in Action*, Bruno Latour articulated what has become a guiding methodological prescription for the field of STS: the best way to understand the scientific enterprise is to "follow scientists and engineers through society" (Latour, 1987). Simple to state, that injunction has proved not so simple in practice. The pathways that scientists, and their close kin in medicine and engineering, trace through society in modern times have grown increasingly complex. No longer even notionally restricted to laboratories or field stations,[1] scientists and scientifically trained professionals are as likely to make their appearance in corporate boardrooms, university administrations, legislative hearings, advisory committees, and courts of law. Traffic on the highway between law and science has grown particularly dense and its patterns, if any, correspondingly hard to decipher. Not only are technical experts implicated in ever more varied legal proceedings, but many of the key institutions of modernity—health care, environmental protection, insurance, education, security, financial markets, intellectual property, and criminal justice—demand an intense and ongoing collaboration between the institutions of law and those of science and technology (Jasanoff, 1995). Elucidating that interaction has become a distinct project of STS research; this chapter describes the main results of that undertaking.

Following scientists through society may lead, indeed has led, STS scholars into spaces that are emphatically not those of science; but that strategy alone cannot lay bare the interactive dynamics of two institutions that, perhaps more than any other, are responsible for making order, and guarding against disorder, in contemporary societies. At issue, after all, is not only how scientists produce facts for legal use but also how science supports ideas of causality, reason, and justice in the law, and how scientific experts supplement the work of jurists, advocates, and other actors engaged in the project of securing social stability and order. To get at that deeper level of understanding, STS scholarship has had to expand its theoretical repertoire and adopt methods that go well beyond close readings of what scientists do in or out of their own workplaces. What emerges from the growing literature on science and law is, in effect, a "*stronger* program"[2] for looking at science and technology in their wider social, cultural, and political contexts. Three sets of presumptions mark this newly contextualized study of science in relation to law, humanity's other most indispensable instrument of authority-making.

First, STS scholars have recognized a need for greater symmetry in exploring the processes and practices of science and the law. So thoroughly are these institutions enmeshed that close investigation of various dimensions of legal practice (e.g., evidentiary hearings, advisory committee meetings, patent litigation), and the actors who engage in them, is as likely to shed light on the production of scientific knowledge as studies of laboratory science-in-the-making or of scientific controversies. Put differently, the law is now an inescapable feature of the conditioning environment that produces socially embedded—or so-called Mode 2 (Gibbons et al., 1994)—science. Accounts of the development of science are incomplete without taking on board the shaping influence of legal imperatives and imaginations, and of necessity the work of legal practitioners and institutions. The law, moreover, operates with its own ideas of facticity and truth that are not identical to those of the sciences. How facts are contested and established in various legal contexts neither blindly conforms to, nor determines, similar processes in science (Jasanoff, 2005). A comprehensive "social history of truth" (Shapin, 1994) for the modern world cannot be written from starting points originating within science and technology; we need equally to follow law-work and law-workers as they pass into and through the workplaces of science and technology.

Second, the divergent cultural attributes and ambitions of law and science raise distinctive questions not only about the relationship of power and knowledge but also about the methods by which to study them. Law's language is human language, a prime achievement of culture, situated in both place and history; the social study of law and of legal cultures has tended to be similarly situated within national research traditions, permitting relatively little communication across different legal systems (consider, e.g., Leclerc, 2005; Latour, 2002; Hermitte, 1996). Science's language, as the presumed language of nature, claims a kind of universality that transcends culture, time, and place. In practice, moreover, English has gradually established itself as the *lingua franca* of science, facilitating communication among scientists, and all those who claim to act on the basis of science, wherever they are located. Science studies, to some extent, partakes of that same catholicity: academic communities in STS are sooner defined by their shared objects or periods of study (e.g., genomics, early modern science, or risk assessment) than by researchers' cultural or linguistic origins. Following legal and scientific practitioners as they interact, then, entails asymmetries of method and interpretation, on the part of analysts as well as actors, that pose significant challenges for STS.

Third, the perspectives gained on science-law interactions by simply following practitioners in either domain are necessarily limited. Scientists and lawyers move about in their professional worlds in accordance with well-established conceptions of their roles and missions; even reflexivity, a part of each institution's modes of thought, operates within circumscribed interpretive conventions. To gain analytic purchase on their epistemologies and practices, the very definition of these institutions must, to some extent, be put in play in ways that may surprise and even estrange practitioners. Just as science no longer happens exclusively within laboratories, so law too unfolds in settings far beyond the courtroom. Needed for insight into these dispersed yet mutually sustaining activities are sets of theoretical or conceptual lenses that do not precisely replicate the self-understandings of either institution. In part, the lenses used in recent STS work to look behind the public performances of law and science have derived from established disciplines, such as anthropology, history, or sociology; in part, theoretical approaches have also evolved in more organic and inductive ways from work done within STS itself.

Any literature review of the kind undertaken here necessarily performs its own boundary work, most importantly through decisions about what to include and what to exclude from coverage. On the assumption that STS, as a relative newcomer to the social sciences,[3] is still advancing through critical encounters with neighboring areas of scholarship, this chapter includes relevant works from fields such as the anthropology and sociology of law, legal history and philosophy, and law and society. This strategy not only provides a fuller, more textured account of current social science conversations concerning law, science, and technology, but it also helps contextualize the contributions that are distinctive to STS. The juxtaposition of STS writing with that emanating from legal scholarship and practice is particularly illuminating. In contrast to science, which perennially sheds its history, the law advances by openly reflecting on and continually reincorporating its own past performance; it is, in this respect, possibly the most reflexive of modern social institutions. The relationship of science and law is one area that has engaged the legal system's reflective capacities, and comparing the results of that self-analysis with analyses by STS scholars, who write from standpoints in neither science nor law, helps bring the latter's insights into sharper relief.

This chapter reviews the STS literature on law, science, and technology under four linked, yet analytically separable, headings: *engagements*, *authority*, *epistemology*, and *culture*. The section on engagements traces the relationship of science and technology with the law as a historical phenomenon and an emerging field of academic inquiry. The theme of authority addresses the varying discourses and registers in which analysts have sought to represent the authorization (and, sometimes, the destabilization) of science within the legal system, and, to a more limited extent, of the law within scientific communities. Epistemology, a central concern of STS, refers in this chapter to the law's contributions to making and unmaking scientific facts and to shaping the processes of fact-making. Finally, the heading of culture brings together a heterogeneous and still developing body of work on the varying guises in which science-law interactions play out across divergent legal and political arenas. The chapter concludes with reflections on productive future directions for STS research on science, technology, and the law.

ENGAGEMENTS: PHILOSOPHY AND HISTORIOGRAPHY

Two institutions, both intimately concerned with rules and order, cannot help but influence each other's discourses and prerogatives. Interactions between science and law have evolved over long ages and on many levels, from the constitutive and conceptual to the mundane and instrumental. For the law not only concerns itself after the fact with remedying the harmful consequences of scientific and technological change. Additionally, and perhaps more significantly, it provides an envelope of social order within which new epistemic constructs and technological objects are constantly fitted out with recognizable meanings and normative implications. No account of the engagements between science and the law can be complete, therefore, without considering the ways in which changes in our knowledge of nature, and in our ability to manipulate nature through technology, challenge and respond to some of the basic categories of legal thought.

A central node of engagement is the concept of "law" itself. Since the beginnings of the scientific revolution, the word *law* has been used to denote both regularities discerned in nature and rules by which religious or secular authorities govern human behavior. That semantic convergence has not gone unnoticed in writing about science or the law, although scholarship concerning each domain has proceeded mostly in disregard of the other. The lack of contact is especially notable given that assumptions about nature and science have long underwritten the authority of the law, just as legal ideas about codes and norms have made their way into descriptions of science (Merton, [1942]1973). A full-blown engagement between STS and legal scholarship concerning each field's presumptions and findings, as seen by the other field, promises much, though it has yet to come into being.

For legal philosophers in the "natural law" tradition, the regularities of nature provide the strongest possible warrant for legislating morality: people should be required to behave in certain ways, in this view, because it is "natural" to do so, and science can aid us in discovering what is natural. Succinctly put, "there are certain principles of human conduct, awaiting discovery by human reason, with which man-made law must conform if it is to be valid" (Hart, 1961: 182). Legal positivists contest this unproblematic derivation of moral prescriptions from the descriptive facts of nature (Waldron, 1990: 32–34). For them, the law is nothing more than what a sovereign authority decrees as the right rules of conduct; in the positivist tradition, law can be valid even if it permits behavior that is, in some sense, against nature, although this is infrequent in an era when sovereigns are routinely held to nonarbitrary, rational, and scientific modes of justification (for an arguable breakdown in such accountability, see Mooney, 2005).

Neither legal positivism nor natural law takes practice centrally into account in building theory. But, paralleling the sociological turn in STS, the law has produced its share of scholars who take their cues about the nature of their enterprise from what practitioners do rather than from what philosophers say. Justice Oliver Wendell Holmes ([1881]1963: 5) most famously captured the spirit of legal realism in his much quoted line, "The life of the law has not been logic; it has been experience." Lon Fuller's influential exploration of the morality of law resonates even more closely with themes in the sociology of scientific knowledge. Like Robert K. Merton ([1942]1973), whom he did not cite, Fuller (1969: 46–91) posited that the law contains its own inner morality that must be adhered to in a functioning legal system. Fuller's eight principles of legality can be considered in this respect akin to Merton's four familiar norms of science (communalism, universalism, disinterestedness, and organized skepticism). Fuller recognized, too, the parallels between theories of science and the law. In concluding the 1969 revision of his 1964 work, Fuller (1969: 242) pointed to philosophers of science such as Michael Polanyi and Thomas Kuhn who had oriented their field away from concept and logic "toward a study of the actual processes by which scientific discoveries are made."[4] He urged on his legal colleagues a similar "analysis of the social processes that constitute the reality of the law." By concurrently probing both sets of realities, constructivist STS research on law and science follows Fuller's injunction more completely than he himself might have imagined possible.

Out of the pages of theoretical texts, natural law thinking continues to guide judicial decision making, especially in controversies around the life sciences. In 2005, for example, the U.S. Supreme Court abolished the death penalty for defendants under 18 years of age, partly on the ground that minors are more vulnerable than adults to irresponsible behavior and less in control over their immediate surroundings.[5] Similarly, legal decisions on the use of reproductive technologies reflect underlying notions of what constitutes natural modes of kinship or naturally gendered behavior (Hartouni, 1997).[6] Natural law ideals also permeate the thinking of twenty-first-century ethical analysts who have argued for strict legal controls on embryo research, human reproductive cloning, and genetic therapies that would alter human germ lines (Fukuyama, 2002). By contrast, *Roe v. Wade*,[7] the controversial 1973 U.S. Supreme Court decision on abortion, can be seen more as a repository of positivist thinking, in that it gave greater weight to women's constitutionally guaranteed autonomy, under the rubric of privacy, than to arguments about the fundamental sanctity of fetal life. German constitutional law, which accords human dignity to the fetus from the moment when sperm and egg cells fuse, and hence to a naturalistic conception of the origins of human life, stands in sharp contrast to *Roe*'s doctrinal position.

While legal thinkers have turned to nature for firm moral warrants, early modern scientists perceived nature itself as being ruled by law. The very idea of science, according to Evelyn Fox Keller, extended the idea of domination within human societies, through law, outward to encompass the human domination of the universe. Science, in this sense, was law transposed—from social to natural order. Thus, Keller writes, the concept of laws of nature "introduces into the study of nature a metaphor indelibly marked by its political origins"; laws, whether of nature or the state, "are historically imposed from above and obeyed from below" (Keller, 1985: 130). That view of scientific law accords well with the intuitions of legal positivists concerning the authority of law, but it has been complicated by work such as Fuller's that brings society back into explanations of the law's normative power.

Focusing on the origins of experimental practice in Restoration England, Shapin and Schaffer (1986: 99–107, 326–28) describe a more complex dynamic. They note that in this period of unusual political and philosophical ferment the precise basis for scientific and legal authority, as well as the relationship between them, were simultaneously at stake. Would science, as the experimentalists associated with Robert Boyle believed, advance by adopting the common law's procedural device of witnessing, thereby creating well-defined, practice-based communities of trust? Or did the reliance on human witnesses, as Thomas Hobbes was convinced, substitute fallibility for truth, thus subverting the stability of an order ruled from on high? As we will see, that early controversy, locally resolved in Restoration England in favor of the experimentalists, continues to provide a surprisingly pertinent analytic frame for contemporary debates about fact-making in law and science.

Engagements between law and science occur not only at the level of institutional legitimation but also in the ongoing work of legal dispute resolution. Turning to specific sites of engagement, we note that the law's need for scientific facts is not new. From almost as far back as we can go in history, judges have sought to establish the facts of nature in order to secure a basis for exercising their own normative authority. A Talmudic story tells how a physician's testimony exonerated a woman of trumped-up adultery charges when he testified that the white residue on her bed sheets was egg white, not semen (Jasanoff, 1995: 42). By the late eighteenth century, the industrializing world began generating increasing numbers of controversies that could not be resolved without determining the facts about one or another natural phenomenon. Facts, or more properly claims about facts, were brought into the courtroom by specially skilled expert witnesses. The 1782 English case *Folkes v. Chadd*, in which a court formally accepted the testimony of party-employed engineers, and more generally approved the use of expert witnesses, involved a dispute over what caused the silting up of Wells Harbor in Norfolk (Golan, 2004). Throughout the nineteenth century, partisan experts claiming specialist knowledge on matters ranging from pollution to accidents to murder streamed into common law courts in England and the Anglophone world. In civil law countries, the state retained the right to call expert witnesses, but there, too, legal dispute resolution came to depend more and more on the use of technical experts (Leclerc, 2005).

The engagement between law and medicine, illustrated by the Talmudic tale above, has a particularly long and consequential history, rich enough to constitute its own burgeoning subfield in the history of medicine (Clark & Crawford, 1994). For centuries, physicians have offered their expert opinions to help resolve legal controversies involving such issues as abortion, infanticide, murder, criminal intent, mental competence, medical malpractice, and injuries from toxic substances ("toxic torts"). Through these interactions, medical authority has helped to underwrite discriminations that are essential to the implementation of the law and, indeed, to sustaining the notions of lawfulness and legality. Equally, emerging expert communities, such as forensic psychiatrists (Golan, 2004; Eigen, 1995; Smith, 1981), medical examiners (Timmermans, 2006), toxicologists, and radiologists, have consolidated their professional identity and social authority by offering their specialized knowledge as supports to legal decision-making. With the rise of the regulatory state and of biopower since the nineteenth century, medical expertise has increasingly been drawn into projects of governmentality, that is, into underwriting practices of clinical and population-based oversight that enable institutions to control people on national and even imperial scales (Stoler, 2002; Bowker & Star, 1999; Foucault, 1973, 1978, 1979, 1994).

Rounding out this section on engagements, we should note that science and technology not only assist in resolving legal disputes but also participate in producing them. Since the 1970s, the dominant conceptual frame for dealing with the harmful or destabilizing effects of technological innovation has been that of risk. Famously articulated by Ulrich Beck (1992), through his identification of the "risk society" as a radically new cultural formation, the trope of risk draws attention to the ungovernable dimensions of science and technology, and by extension to the law's role in managing the unmanageable through discourses of justice and reason (Jasanoff, 1995, 1990; Wynne, 1982). But as STS scholarship has importantly shown, the assumption of linearity, almost of determinism, built into the risk paradigm—casting science and technology as proactive and law as reactive—needs to be substantially revised and augmented in favor of more interactive accounts that pay attention to the innovative capacities of both institutions.

AUTHORITY AND COMPETITION

As prime custodians of the "is" and the "ought" of human experience, science and the law wield enormous power in society. Each plays a part in deciding how things are in the world, both cognitively and materially; each also helps shape how things and people should behave, by themselves and in combination. Competition often marks the interactions of law and science, not only in testing the limits of scientific autonomy and self-regulation (see, e.g., Kevles, 1998) but also in areas where the boundary between the legal and the scientific spheres of influence is itself at stake. Several narrative traditions have evolved for describing the relationship between law and science in these contested regions, of which five deserve particular attention: *law lag, culture clash, crisis, deference,* and *co-production.* The first four are encountered primarily in writing by members and critics of the legal profession; only the last is specifically a product of STS. Each frame organizes the law-science relationship in distinctive ways, highlighting and backgrounding different aspects. Each, in consequence, carries implications for legal reform, though not all have received equal attention from legal, scientific, or political actors—for reasons that themselves call out for scholarly analysis.

Together, the four themes of law lag, culture clash, crisis, and deference illustrate not only the centrality of the law-science relationship as a subject of reflection and analysis, especially in common-law countries, but also the fertility of contemporary legal culture in accounting for its interactions with science and technology. STS scholarship, then, stands at a double disadvantage in commenting on the relationship of law and science. Both sides offer the critical outsider a case of what anthropologists have called "studying up" (Nader, 1969). Both, to start with, are institutions of power, and a large part of that power lies, for each, in having commandeered, in the eyes of society, a privileged, almost monopolistic, position from which to explicate its own workings.[8] Scientists are seen as the most competent commentators on science, as lawyers are on law. Each institution, moreover, shores up the other's status. Neither legal nor scientific practitioners seem inclined to probe too deeply each other's claims concerning the authority of their respective epistemic and normative practices. In this sense, the two are involved in a subtle dynamic of co-production, the major narrative frame that analysts standing outside both domains have invoked to describe their complex, yet mutually sustaining, choreography.

Law Lag

The notion of the law lag can be traced back to the influential early-twentieth-century American sociologist William F. Ogburn (1957, [1922]1950), who argued that interconnected cultural institutions, such as science and law, develop at uneven paces, so that the slower are necessarily out of step with the quicker. Differential rates of innovation, accumulation, and diffusion produce, in turn, a constant need for adjustment between leading and lagging institutions. Ogburn was careful to locate the capacity for innovation in both technology and society, repudiating a simple-minded technological determinism, but he nonetheless viewed science and technology as modernity's prime movers of social change. In 1933, as chair of the President's Research Committee on Social Trends (1930–1933), Ogburn prepared for Herbert Hoover a report identifying irregular change as the chief source of social problems and advocating for better statistical data as the basis for solving the problems. That vision was informed by a positivistic model of knowledge accumulation that has been largely abandoned by post-Kuhnian social scientists. Reviewing Ogburn's contribution a half-century later, the sociologist Neil Smelser took issue with the suggestion that science inevitably leads the law. Smelser noted that even in *Brown v. Board of Education*,[9] the U.S. Supreme Court's seminal desegregation decision, the much-touted input from the social sciences did not alter legal thought but only helped substantiate a moral consensus that had been building for years (Smelser, 1986: 30–31).

Nonetheless, the perception that the law lags behind advances in science and technology dominates academic and popular writing and frequently surfaces in legal opinions. Discursive constraints provide part of the explanation. The law's rhetoric of justification is primarily backward-looking, relying on enacted rules and established judicial precedents. Judges may interpret the law as written, but they stray into dangerous territory if they are seen too openly to be making the law: "A judge disguises new ideas as old in order to enhance their social acceptability" (Goldberg, 1994: 19).

Science, by contrast, unabashedly embraces innovation. Continually erasing its own history as it moves forward, today's scientific knowledge ruthlessly casts aside yesterday's rejected theories and discarded truths. Reward structures in science consistently favor novelty. Nobel prizes are awarded for original discoveries, not for more elegant replications of others' work, and patents cannot be granted for inventions whose basic principles are already known to skilled experts. Not surprisingly, then, scientific inventiveness comes across as an inexorable, agenda-setting force to which the law responds only by reaction. Indeed, legal practitioners are among the most enthusiastic disseminators of the law lag narrative. Thus, in the landmark U.S. life patenting decision, *Diamond v. Chakrabarty*, Chief Justice Warren Berger observed that "legislative or judicial fiat as to patentability will not deter the scientific mind from probing into the unknown any more than Canute could command the tides."[10] Nothing in the opinion marked this case as a moment of exceptional legal inventiveness, when a court in effect recognized a novel type of commodity and thereby opened the door to new forms of hype and hope, investment, research, and material manipulation, with huge consequences for society.

Culture Clash

Less deterministic than the law lag narrative, the culture clash frame focuses on the discrepant aims of law and science as the chief sources of conflict between them. In Steven Goldberg's (1994) telling, the clash originates in science's commitment to progress, whereas the law's primary concern is with process. Consequences that flow from this difference include the legal system's commitment to building consensus, or at the very least to airing diverse points of view, whereas science pursues the nature of reality, come what may. Peter Schuck (1993), an analyst of administrative and tort law, triangulates the story of the culture clash by bringing politics into the picture as a third culture. All three, in Schuck's account, are characterized by their distinctive values, their incentives and techniques, and their biases and orientations. On the value axis, Schuck (while citing constructivist ideas from STS) associates science with a core commitment to truth and falsifiability, law with justice, and politics with process (see also Schuck, 1986). For Schuck, as for Goldberg, the culture clash model rests on an unproblematized notion of institutional boundaries, without taking on board the practices discussed by STS scholars through which these boundaries are erected and maintained (Hilgartner, 2000; Gieryn, 1999; Jasanoff, 1990) or the purposes served by such boundary-making.

Crisis

A third narrative invoked to frame law-science relations, particularly in the United States, is that of crisis. This can be seen as an extreme and highly reductive version of the clashing cultures narrative—one that depicts the relationship between the two institutions as pathological in its failure to produce socially desirable outcomes.

Despite Alexis de Tocqueville's early observation that, in America, political conflicts are routinely translated into legal ones, there is continual hand-wringing over the nation's propensity to settle issues through contestation rather than cooperation. America's "litigious society" (Lieberman, 1981) has been blamed for the spiraling cost of medical insurance, and more specifically for a "malpractice crisis" resulting from irresponsible lawsuits and runaway jury awards against physicians. Statistical analysis complicates that reading. A counternarrative put forward by many health policy analysts holds that, although only a small fraction of those filing claims have been negligently injured, an equally small fraction of those who have been so injured ever file claims. Further, the rise and fall of insurance premiums may have more to do with cycles in investment by insurance companies than with malpractice claims (Sage & Kersh, 2006). This school of policy analysts views with skepticism reform ideas that seek to deter malpractice lawsuits, arguing that these measures alone will not advance the important goals of preventing error and efficiently compensating deserving patients. For STS scholars, the more interesting issue is why the images of a litigation explosion and of uncontrollable jury verdicts persist with such diehard energy despite years of unsubstantiating quantitative research (Vidmar, 1995).

For some, the crisis narrative centers less on the economic costs of litigation than on the threat to science. Law, according to these critics, encourages the production of "junk science" (Huber, 1991)—science that does not meet the scientific community's minimum standards of validity, even though it passes muster with juries and judges. Proponents of this view attribute the law's uncritical reception of scientific claims to many factors: jury ignorance and confusion; mercenary and unprofessional expert witnessing; lax admissibility standards; and the lawyer's ethos of privileging victory above the truth. The rhetoric of "junk science" rests, in short, on a tacit sociology of knowledge that differs substantially, as we see in the next section, from the epistemological accounts of STS scholars (see also David Nelken's comparison of the "trial pathologies" approach with more constructivist approaches in Freeman & Reece, 1998: 14–18). As a powerful contribution to the sociology of error, however, this story line constitutes at once a challenge to STS and an object of possible study for the field.

Among the more careful contributions to the crisis genre is the analysis of science in the U.S. breast implant lawsuits by Marcia Angell (1996), former executive editor of *The New England Journal of Medicine*. Angell brought together the culture clash and crisis narratives in arguing that the law's adversarial zeal, coupled with high financial stakes, produced a settlement based on nonexistent evidence and the consequent withdrawal of a product that many women found beneficial or enabling. Particularly troubling to Angell (1996: 28–29) was the attempt to reach scientific conclusions through adversarial methods that she deemed contrary to science's reliance on co-operation and on "the slow accumulation of evidence from many sources." In her view, an unholy alliance between law and the news media captured public opinion and generated political pressure to support scientifically untenable public policy. Angell's account makes it unnecessary to ask, as the strong program's symmetry principle requires (Bloor, 1976: 7), how it is that strong countervailing beliefs arose, and were sustained, among those seeking compensation for injuries related to breast implants (for such accounts, see below and Jasanoff, 2002). Angell writes off those beliefs through a sociology of error that dismisses presumptively unscientific knowledge as needing no further querying.

Deference
What the "junk science" narrative lacked in methodological rigor, it made up for in political persuasiveness. In the 1990s, this line of criticism laid the conceptual ground

for the fourth major interpretive frame applied to authority conflicts between law and science, namely, deference—more specifically, deference by courts toward science and scientists. A U.S. Supreme Court decision signaled the shift. In the 1993 case of *Daubert v. Merrell Dow Pharmaceuticals, Inc.*,[11] the Court announced that judges should act as gatekeepers in contests over the admissibility of scientific evidence. Their task should be to make sure that only evidence meeting scientific standards of validity and reliability is admitted in court. Although judges already had the power to exclude expert testimony under the Federal Rules of Evidence,[12] *Daubert* took that rarely exercised prerogative and transformed it, in effect, into an affirmative obligation. To guide judges in making the necessary discriminations between valid and invalid science, the Court offered four nonexclusive criteria (testability, peer review, error rate, general acceptance), which are further discussed in the next section.

The deference that *Daubert* mandated in principle, however, turned out in practice to legitimate the free exercise of judicial discretion. Post-*Daubert* judges deferred to an idea of science influenced by their own culturally conditioned understandings of the scientific method, filtered through the demands of courtroom practice. Thus, to function as a legal norm, the notion of scientific reliability mandated by the Federal Rules of Evidence must be translated into tests that trial judges can easily follow. The Supreme Court offered four such explicit tests, but already beginning with *Daubert*'s rehearing,[13] federal judges showed that they were ready, at need, to make up new admissibility criteria beyond those proposed by the high court. In that particular case, the Ninth Circuit Court of Appeals proposed the additional rule that "litigation science," or science generated exclusively for the purpose of pursuing a lawsuit, should not be admitted (see below).

Daubert spurred a small industry in scientific education for judges, along with efforts by some organizations, such as the American Association for the Advancement of Science, to create lists of reliable scientists for use as court-appointed experts. As a significant by-product, the Federal Judicial Center (FJC, [1994]2000), the research and education agency created in 1967 to improve judicial administration, produced a massive *Reference Manual on Scientific Evidence*. Intended as a desktop guide for federal judges, the book contained general articles on scientific evidence and the law, as well as articles on specific types of technical evidence, such as economics, statistics, DNA typing, and engineering practice. In keeping with the theme of deference, the FJC recruited David Goodstein, physicist and vice-provost of the California Institute of Technology (Caltech), to write a chapter entitled "How Science Works." Goodstein critically reviewed the theories of three classic philosophers of science—Bacon, Popper, and Kuhn—before offering his own account of the sociology of scientific knowledge (FJC, [1994]2000: 67–82). In uncritically disseminating a scientist's-eye view of science for legal audiences, the *Manual* helped reinforce a particular understanding of science's relationship to the law. It is a small illustration of the broader proposition that the image of science the law defers to is importantly a construct of the legal process itself. The law serves in this respect as a site and an instrument of co-production.

Co-production

The framework of co-production draws attention to the simultaneous formation of social and natural order in knowledge societies. Many STS studies have shown that what one knows in science significantly depends on prior or concurrent choices about how one chooses to know it (Jasanoff, 2004, 2005; Latour, 1993; Shapin & Schaffer, 1986); the "is" and the "ought" of human experience are in this way inextricably linked, as are epistemology and metaphysics. STS scholars have made numerous contributions to legal studies by pointing out sites and processes of co-production when science interacts with the law. The crucial figure of the "expert witness," for example, is a product of science's historical engagements with the law (Golan, 2004; Mnookin, 2001; Cole, 2001). And "evidence," as the law's distinctive contribution to knowledge-making, is a hybrid product conforming to legal as well as scientific criteria of reliability. In both constructing and reinforcing dominant social understandings of expertise and evidence, legal spaces operate at one and the same time as epistemic spaces, a point we return to below.

But the co-productionist interplay of law and science does not end there. As agents of power, law and science also collaborate in sustaining wider understandings of how society works, including ideas of the human self and agency, the market, and the collective good. In a variety of decisions concerning biotechnology, for instance, U.S. courts have favored the party who appears to be the innovator or initiator of change, often also the party with more resources and greater capacity to bring innovations into economic and social circulation. Two California cases illustrate the point. In *Moore v. Regents of the University of California*,[14] the state supreme court ruled that patients possess no property rights in their cells or tissues; accordingly, physician-researchers need not share with their patients the profits from discoveries based on excised biological materials (Boyle, 1996). In *Johnson v. Calvert*,[15] the court held that a gestational surrogate who had carried another woman's genetic offspring to term could not claim to be the baby's "natural mother." The surrogate's role in sustaining fetal life was thereby reinscribed by law as that of a paid service provider (Jasanoff, 2001; Hartouni, 1997), while the genetic mother, as the party intending to procreate, retained the conventional rights of motherhood.

It is instructive to set the co-productionist account of law-science relations beside two powerful schools of thought originating in legal studies that also connect the law's normative aims ("ought") to understandings of how the world works ("is"). The first is "law and economics." Beginning in the 1960s, leading scholars in this tradition argued for reforms that would conceptualize social problems, from negligence to industrial risks (Breyer, 1993), in economic terms and seek to provide optimally efficient legal solutions. Guido Calabresi (1970), dean of Yale Law School and subsequently a federal judge, contributed to the economic analysis of tort law in ways that closely relate to institutionalist modes of thought in STS. His *Tragic Choices* (Calabresi & Bobbitt, 1978) was a classic exploration of how legal institutions permit competing measures of the value of human life to coexist in society, so that tragic contradictions are kept from public review and acknowledgement. Though related in spirit to concerns about identity, representation, and discourse in STS work on co-production (Jasanoff, 2004: 39–41), Calabresi's brilliant institutional analysis never explicitly engaged with the law's role in producing authoritative social knowledge.

In a still more provocative application of economic thinking to the law, federal judge and law professor Richard Posner (1992) argued that even human sexuality can usefully be studied as rational behavior, and that this framework might lead to more liberal, less interventionist regulatory approaches toward issues such as contraception, abortion, surrogacy, and homosexual conduct. Posner distinguished his analysis from that of "social constructionists," epitomized for him by Foucault. Unlike constructionists, economists, he maintained, are anti-utopian. Their tendency is to assign "less weight to power, exploitation, malice, ignorance, accident, and ideology as causes of human behavior and more to incentives, opportunities, constraints, and social function" (Posner, 1992: 30). For Posner, pervasive ignorance about the facts of sexual preferences, behaviors, and their consequences is the primary obstacle to rational rule-making about sex. He overlooked the co-productionist point that the law, in promoting a fact-based, rational choice approach, may create the very ontologies of rational and irrational behavior that it presumptively seeks to uphold—as, for example, when the California supreme court characterized the gestational surrogate in *Johnson v. Calvert* as a rational economic agent and service provider, not a would-be mother (see also Hacking, 1995, 1999).

More sensitive to the social and epistemic foundations of the law, and therefore closer in spirit to ideas of co-production, is the work of critical legal studies (CLS) scholars and other legal analysts who are interested in the law's ordering functions. CLS flourished as a left-oriented school of thought in American law schools in the 1980s but largely disappeared as an organized movement by the century's end (Kairys, 1998; Unger, 1986). In its heyday, however, CLS destabilized the authority of legal rule-making in much the same way that the sociology of scientific knowledge (SSK) attacked the authority of scientific fact-making. By debunking the power of legal reasoning to justify practical rule application, CLS scholars engaged in the same kind of "unmasking" that Ian Hacking (1999: 53–54) identified as one version of constructionism in STS. Behind formal legal argument, CLS analysts discovered concealed interests and ideology, much as the SSK analysts of the Edinburgh school uncovered interests on both sides in their symmetrical studies of scientific controversies.

The CLS project with regard to the law paralleled that of STS with regard to science in other important ways: in its focus on the indeterminacy of rules (cf. contingency and rule-following in STS), its emphasis on contradictions and dualities that legal doctrine cannot resolve (cf. interpretive flexibility in STS), and its awareness that the law does not simply respond to social needs but creates the very conditions from which those needs arise (cf. rejection of the correspondence theory of truth in STS). The "Critics," as they were called, conceptualized the law, just as STS scholars conceive of science, "as one of many cultural institutions that are constitutive of consciousness, that help delimit the world, make only certain thoughts sensible, thus 'legitimating' existing social relations" (Kelman, 1987: 244). Despite all this commonality, no systematic effort at intellectual bridge-building developed between STS and CLS during its most active years. In part, the failure to connect may have reflected a dearth of doctrinal synthesis on both sides. Caught up in their own unsolved theoretical dilemmas, and in confrontations with scientists and mainstream legal thinkers, respectively, neither STS nor CLS scholars found much occasion to talk across yet less familiar intellectual divides, although such moves might have led to sustained and productive conversation. In part, the strength of legal professionalization kept even radical critique from spilling out of the law's familiar discursive spaces, such as law journals. And "studying up" in STS did not, until well into the 1990s, generally include the interactions of science and the law, except, as noted, in the history of medicine (but see Smith & Wynne, 1989).

Encounters between these two critical traditions have much to offer to science studies as well as to the law. An STS-inflected analysis of Martha Minow's (1990) work on making social differences illustrates the possibilities. Throughout that work, Minow is concerned with the law's role in making demarcations that matter to how a society treats its most vulnerable members—the disabled, the mentally incompetent, the dying, women, and children. Like STS scholars of her intellectual generation, Minow recognizes that differences between the normal and the abnormal are constructed and that acts of demarcation are far from epistemologically neutral. In particular, she lists five assumptions about differences commonly made in the law that closely parallel STS observations about the essentializing of social categories: differences are intrinsic; they are defined in relation to an unacknowledged reference point; they appear standpoint neutral; their legal treatment either ignores some perspectives or presumes that all will be fairly represented by the judge; and they rest on the naturalization of existing social and economic arrangements (Minow, 1990: 50–74).

To support her constructivist analysis of legal demarcations, Minow draws broadly from poststructuralist work in the social sciences and humanities, but her reliance on science studies is limited to feminist theory (e.g., Keller's 1983 study of Barbara McClintock) and philosophy of science. Accordingly, she does not explore how categories come into being in particular cultures, are kept in place through social practices, or become embedded in material technologies. Further, in advocating a social relations approach to resolving "dilemmas of difference," she assumes a degree of fixity and invariance in social groups and identities that STS scholars have rightly called into question. At the same time, Minow's study of legal categorization is far more sensitive to possibilities for activism and social change than most canonical work in STS. Never interested in the phenomenon of demarcation for its own sake, Minow is most concerned to show how line-drawing in the law affects the allocation of rights and obligations; and she demonstrates how acknowledging the non-neutrality of the resulting demarcations might open up processes of governmentality in institutions such as schools and hospitals. That grounded attention to the normative consequences of knowledge-making practices has been missing in much STS work on the epistemic foundations of social difference (for some exceptions, see *Social Studies of Science* 1996, vol. 26).

EPISTEMOLOGY: LAW'S KNOWLEDGE

In the modern world, it is not only science's obligation but increasingly also that of the law to find out how nature works and to settle contested facts for its own purposes. By 1887, when Arthur Conan Doyle published *A Study in Scarlet*, forensic science was already well established as a distinct branch of knowledge in Britain. Dr. John Watson first encounters Sherlock Holmes in a hospital laboratory, most likely in 1881, in the process of conducting a chemical test for trace quantities of human blood. Watson admires the chemistry but innocently wonders what use such a test could have. Unrestrained by false modesty, his future flat mate and biographical subject exclaims, "Why, man, it is the most practical medico-legal discovery for years." Holmes is a natural scientist engaged in curiosity-driven research, but the questions that propel him derive from the law.

Both science and law are committed to ascertaining the facts of the matter as accurately as possible; indeed, the law's capacity to render justice depends on finding the right facts and finding them right (see, e.g., Lazer, 2004, on legal conflicts over DNA fingerprinting). The authority of both institutions depends, as Hobbes so well recognized, on appeals to transcendental truths; neither can allow itself to be seen as subjective, arbitrary, or mired in the specificities of particular cases. Yet both include among their procedural devices the systematic, if socially bounded, capacity for questioning that Boyle and his adherents cultivated within their communities of "virtual witnesses" (Shapin & Schaffer, 1986: 55–60) and that Merton (1942) three centuries later famously termed "organized skepticism." How each institution strikes a balance between the contingency of fact-making and claims to transcendence, and how the fact-finding practices of each interact with, support, or destabilize the practices of the other, have been focal points of STS inquiry—adding a distinctive and neglected dimension to work done by legal scholars and analysts on these topics.

To contextualize the insights of STS literature into the law's ways of knowing, it is helpful to begin with work on science and evidence done by other students of the law. Much of that analysis focused, especially from the early 1990s onward, on the law's capacity to distinguish reliable from unreliable science and on the impact of the Supreme Court's trilogy of evidence rulings. At stake in these writings is the very nature of the adjudicatory process, with associated struggles for authority between science and the law, and between judges and juries. Huber's (1991) blistering attack on "junk science" opened the door to a series of books arguing that the courtroom was no place for establishing scientific truths. Judges (Foster & Huber, 1997), juries (Sanders, 1998), and the culture of litigation (Faigman, 1999; Angell, 1996) were all held responsible for the inability of courts to find facts as scientists wish them to be found. As we have seen, these works contributed to the sense of crisis surrounding the law's relations with science and lent weight to the Court's call for deference to science in *Daubert*.

Not all legal scholars agreed with these bleak assessments, however, and a counter-literature of sorts also began to form. Students of jury behavior, for example, challenged the crisis proponents' claim that juries are swayed by emotion, and by the possibility of reaching into the deep pockets of corporations and hospitals, into making irresponsibly large awards in tort actions (Vidmar, 1995). Evidence scholars argued that, under the guise of deferring to science, *Daubert* and its progeny provided a powerful rationale for judges to usurp the jury's role and silently alter the burden of proof in tort cases, making it more difficult for plaintiffs to win or even pursue their cases in court (Berger, 2001). Implicit in these works, too, was a growing sense that *Daubert's* deference model not only increases judicial discretion (Solomon & Hackett, 1996; Jasanoff, 1995) but also subtly deflects courts from their normative concern with rendering justice (Jasanoff, 2005).

While legal scholars debated the crisis narrative, STS attention turned for the most part to investigating the nature of the knowledge produced in legal settings. Although science arguably plays a more pervasive role across the broad domain of public health, safety, and environmental regulation than it does in trials, the topic of "regulatory science"—science done or applied in support of governmental policy (Jasanoff, 1990: 76–80)—has attracted less attention from STS researchers (but see Daemmrich, 2004; Bal & Halffman, 1997). This work remains significant, however, for its careful demonstrations of the state's self-legitimation through boundary-drawing between science and politics and through discourses of risk that represent uncertainty as manageable by the regulatory state (Abraham & Reed, 2002). STS investigations helped focus attention on so-called boundary organizations, bodies such as expert advisory committees whose primary function is to maintain a clear demarcation between the authority of experts and political decision-makers (Guston, 2001). The political utility of this research became apparent in 2003–2004, when both proponents and opponents of the U.S. Office of Management and Budget's efforts to control the peer review of regulatory science cited Jasanoff's (1990) work on expert advisory committees.[16]

If boundary work was of paramount interest in studies of regulatory science, expertise emerged as the concept to watch in work on litigation-related science. Following early demonstrations that the law actively constructs the scientific facts that it presumes to "find" (Jasanoff, 1995), as well as the persons (expert witnesses) whom it regards as competent to represent those facts, STS researchers looked in more detail at the making of specific bodies of knowledge within the law, such as fingerprinting (Cole, 2001) and DNA profiling (Lynch & Jasanoff, 1998). Forensic sciences like these owe their existence to the law's need for unambiguous identification, especially in criminal cases, but also in areas such as paternity testing and immigration. However, establishing the facts of the matter demands more than determining a witness's expert status, and legal proceedings often serve as sites for the construction of highly *ad hoc*, technical narratives of causation. In one *outré* example, a Dutch court had to reconstruct the facts of a woman's death caused by a ballpoint pen lodged entirely within her brain case. The question before the court was whether the pen had entered her eye through a freak accident occasioned by a fall, or had been intentionally shot in with a cross-bow by a murder suspect. Lacking any precedents for choosing between these two bizarre and unlikely possibilities, the presiding judge had to determine within the four corners of the case what counted as a valid experimental demonstration of the cause of death and who was expert enough to speak authoritatively about it (Bal, 2005).

STS work on expertise vigorously takes issue with the "junk science" charge that proponents of the crisis narrative popularized so effectively. Instead of focusing on aggregate jury behavior, as sociologists of law have done, STS researchers have tended to look more closely at judicial reasoning, as offering textually grounded insights into legal epistemology. Painstakingly reviewing the Bendectin cases, one of which led to the *Daubert* ruling, Gary Edmond and David Mercer (2000) showed how the symmetrical approach of science studies undercuts the sociology of error story told by legal scholars such as Joseph Sanders. In Edmond and Mercer's reconstruction, the "favor epidemiology" rule that judges used to dispose of Bendectin cases prior to trial emerges as an artifact of judicial reasoning rather than an exogenous scientific consensus deferentially relied upon by the courts. Edmond's equally detailed analysis of briefs and judicial opinions in another U.S. Supreme Court evidence decision, *Kumho Tire Co. v. Carmichael*,[17] presents that case as a form of "judicial literary technology" that stabilized a particular social representation of expertise while crafting a new admissibility standard for nonscientific expert testimony (Edmond, 2002).

Going beyond writing about law and science, STS scholars of varied backgrounds have also participated as experts in legal proceedings, and, ironically in a field often criticized for relativism, this engagement has prompted reflexive discussion of the epistemic authority of STS. In most instances, interventions were designed to offer courts and judges a more nuanced interpretation of how science works and how it relates to legal or political decision-making (Jasanoff, 1992). Thus, in 1992, a group of historians and sociologists of science filed an *amicus curiae* ("friend of the court") brief in *Daubert*; similarly, in 2005, a group of academics submitted a brief to the World Trade Organization in a case involving the European Union's allegedly unlawful moratorium against genetically modified crops exported by U.S. producers (Winickoff et al., 2005). In possibly the most salient such intervention, philosophers of science testified against, and in one case for,[18] alternatives to the theory of evolution in cases challenging the teaching of evolution in U.S. schools (Quinn, 1984). In these instances, the philosophy and sociology of science were mobilized to establish that doctrines such as creationism and intelligent design were not scientific but rather were religiously inspired. The legitimacy of STS itself as a domain of expertise came into question when Simon Cole sought to testify, on the basis of his STS training, that fingerprinting was not a science within the terms set forth by *Daubert* (Lynch & Cole, 2005; see also Cole, 2005). Judicial skepticism toward Cole's qualifications underscored the field's still-emergent status but also the importance of specific technical skills as a basis for claiming expertise in court. In all of these cases, powerfully socialized, mainstream judicial views of the nature of science withstood the critical insights offered by STS academics.

SCIENCE, LAW, AND CULTURE

The idea of law may be universal, but the ways in which the law functions in any society are culturally specific, and that specificity can be observed in the law's inter-actions with science and technology. In turn, those workings help shape the evolu-tion of both knowledge and norms, imparting distinctive flavors to what a society wishes to know for purposes of securing social order. Forensic science, for example, develops and operates differently in a common-law, jury-based legal system from the way it functions in legal cultures descended from Roman law, in which the judge acts as the principal fact-finder (Leclerc, 2005; Bal, 2005). Similarly, regulatory science has developed differently in political systems that address uncertainty and produce consensus through disparate institutional mechanisms (Jasanoff, 2005; Winickoff et al., 2005; Porter, 1995; Brickman et al., 1985). More generally, social understandings of evidence and proof, the aims of advocacy, the nature of expertise, and indeed the status of science in relation to politics and power are all refracted through the lenses of the law. STS scholarship has illuminated some aspects of this complex dynamic, but the interplay of culture, science, and law remains an underdeveloped zone of academic analysis.

Given the field's abiding concern with epistemology, it is not surprising that the production of evidence for courtroom use has attracted particular attention in STS research. That work indicates that competing accounts of disputed facts may arise even within so-called inquisitorial systems, in which the parties do not control the pro-duction of evidence as of right, and judges are responsible for ensuring that relevant points of view are fairly heard (Leclerc, 2005; van Kampen, 1998). In French civil law, for example, a general right to contest the other side's claims underwrites discovery and disclosure rules that may not, in practice, be far different from those of common law systems (Leclerc, 2005: 312–22). In the Dutch ballpoint pen case cited above, the efforts of the suspect's father to show that death could have been accidental led to the production of tests beyond those conducted by the Dutch Forensic Institute (NFI). Counterexpertise in that case helped destabilize the absolute certainty of the prose-cution's story and exonerated the suspect; yet, the very contingency of the outcome underscores the Dutch courts' continued reliance on the neutrality of expertise and deference to the NFI as the legitimate source of forensic knowledge.

Preoccupied with the concept of expertise, and more generally with the problem of demarcation, STS scholars have not on the whole attempted to integrate their con-cerns for epistemological issues with sustained investigations of how the law's knowl-edge-making capacities relate to deeper cultural ideas (and ideals) of reason and normativity. Thick descriptions of legal controversies can sooner be found in works by legal scholars, such as Schuck's (1993) account of the Vietnam veterans' epic lawsuit against manufacturers of Agent Orange, or Jonathan Harr's (1995) compelling story of a trial lawyer's ultimately disappointing crusade on behalf of alleged water pollution victims in Woburn, Massachusetts. A notable exception is Marie-Angèle Hermitte's study of France's contaminated blood scandal, in which large numbers of hemophiliacs were infected with the AIDS virus in the 1980s. In a deeply sociological and historical account, she traces how French notions of solidarity among all citizens led to decisions that, in the name of making no invidious distinctions between social groups, inflicted fatal harm one of the nation's most medically vulnerable populations (Hermitte, 1996).

Investigations of the interactions between science and the law might be expected to add nuance to Foucault's grand narrative of governmentality by revealing culturally specific ways in which modern societies come to know the subjects who are governed. As yet, however, STS analysis of law-science interactions has tended to focus on in-depth studies of individual cases or institutions rather than on varying practices across cultures or political systems. A case in point is Latour's study of France's highest administrative court, the Conseil d'Etat, in which he applied ethnographic methods to showing how judicial actors construct legal objectivity and truth, much as his earlier studies focused on the making of truth in science (Latour, 2002). Missing in this briskly demystifying treatment of legal epistemology, however, was an analysis of what, if anything, makes the Conseil d'Etat's intuitions about facticity and legality specifically French.

Cross-national comparison has provided one means of interrogating the role of culture in shaping law-science interactions and their consequences. Although such research is in its infancy, STS work points to intriguing connections between styles of legal thought and the practices and cultures of public reason. For example, what counts as objective in the construction of argument and proof in the public sphere is importantly conditioned by legal assumptions concerning impartiality, transparency, truthfulness, and expertise. Thus, the vulnerability of decision-makers in America's particularly legalistic, and hence open and adversarial, political culture correlates with a wide-ranging preference for impersonal, mathematical modes of justification—the "view from nowhere"—in policy domains (Porter, 1995; Brickman et al., 1985). More generally, legal traditions appear both to reflect and reinforce the forms of "civic epistemology," that is, institutionalized public expectations concerning the state's knowledge-making practices, which prevail in contemporary democracies (Jasanoff, 2005).

CONCLUSION

Since the early 1990s, STS scholars have increasingly turned to science's interactions with the law as a fruitful field of study. A growing body of work attests to the productivity of these investigations, both as an extension of core concerns in STS with the construction of facts and truth and as a means of exploring the social relations of science and technology from standpoints outside the conventional spaces of scientific or technological activity. The law has emerged from these inquiries as a research site of paramount significance for STS. Not only are legal proceedings instrumental in producing and certifying new types of scientific knowledge, but the very building blocks of public reason are formed in engagements between science and the law, for example, notions of expertise, objectivity, evidence, and proof.

Intellectually rewarding as this territory has proved to be, it remains in some respects rocky. The science wars of the 1990s pointed to some of the dangers of "studying up," especially as the social sciences sought to create new, autonomous ways of describing scientific and technological activity. Socializing epistemology proved to be no easy task. STS analysts faced a two-fold challenge. They had to find meaningful ways of redescribing scientists' interactions with nature, imbuing those processes with new social meaning; and they had to break the monopoly that scientists had long enjoyed as the only actors authorized to produce trustworthy accounts of the nature of their activities. Law, too, has enjoyed a similar double monopoly—first, by controlling the language in which legal products must be written to be recognized as law, and second, by guarding the professional right to tell the rest of society how the law "really works."

The dominant narratives used in framing law-science interactions show how far STS remains from winning the two-fronted struggle of studying up with respect to law as well as science. Four of the five dominant story lines discussed in this chapter emanate primarily from lawyers and scientists rather than from STS scholars, while the fifth— co-production—remains in the domain of actor's language, understood by specialists, but with little resonance as yet for legal scholars, lawmakers, or wider society. Science studies, when all is said and done, continues to function as an agonistic field, in which analytic prowess and disciplinary insight by no means suffice to ensure that STS insights and findings will circulate to audiences outside the field.

To gain that wider hearing, STS research will have to reach beyond its parochial, field-specific, epistemological concerns and find new ways to engage with sympathetic critics of the law, both within and outside the circles of formal legal scholarship. STS analysts have been most sensitive thus far to the law's role in making scientific facts and in drawing the boundaries between legitimate and illegitimate expertise. In Latour's terms, it is the law's role in producing "indifference" that has attracted the most sustained interest; and, not surprisingly perhaps, judges, as the supreme text-writers of the law, have commanded more diligent attention than other less forceful and sometimes less articulate players, such as lawyers, juries, and litigants themselves. As we have seen, the focus on epistemology has led some STS scholars into playing active roles in the legal system, most visibly as actual or would-be expert witnesses on behalf of science, but also, less visibly, as advisers and educators to the elites of the law, in the trial bar, advisory committees, regulatory agencies, and the judiciary. But these *ad hoc* and personal encounters only skim the surface of the field's potential for constructive critique. With modernity's two most important ordering institutions as their objects of study, STS analysts of science and the law are uniquely positioned to explore and question the hidden normativities underpinning the demarcations that matter in contemporary society. These, as the CLS movement and its intellectual descendants most cogently argued, are the divides that consistently separate the weak from the strong, the rich from the poor, the disabled from the competent, and the socially marginal from the powerful and privileged.

Relentlessly concerned with the law's epistemic authority, STS students of science, technology, and the law have been on the whole less attentive to the law's magisterial role in constructing and maintaining justice, legitimacy, and constitutional order—and, of course, in holding at bay the disruptive forces of injustice, illegitimacy, and disorder. Nor has STS systematically explored the interplay of law and science with cultural notions of self-hood, kinship, exchange, or community that introduce subtle differences into the kinds of modernity that we, as modernity's inhabitants, experience in our everyday lives. To the inquiring mind, these are not omissions but openings. Through them, future STS research can be expected to push forward to new levels of insight, by bringing within its investigative reach not only law's fabrication of knowledge but also its power to establish order and justice in the world.

Notes

I would like to thank Rafael Munagorri for valuable comments on a draft of this article.

1. Leading scientists and inventors have from the beginnings of the scientific revolution been enmeshed in webs of patronage and power, so that the notion of the disinterested ivory tower scientist is something of a myth—one that justifies science's claim to self-governance (Biagioli, 1993; Latour, 1988; Shapin & Schaffer, 1986).

2. See Bloor (1976) for a statement of the "strong program" for research in science studies. The aims of the strong program were to shed light on the sociology of scientific knowledge-making and so to challenge the notion that science advances through logic and through direct correspondence between nature and scientists' observations.

3. The 2001 *International Encyclopedia of Social and Behavioral Sciences* included for the first time a set of entries under the heading of Science and Technology Studies (Smelser & Baltes, 2001).

4. Interestingly, Kuhn was not mentioned in Fuller's original 1964 text, nor was he cited for any particular ideas in the 1969 revision.

5. In keeping with the spirit of legal positivism, the Court also cited a growing national and international consensus against the death penalty for minors. *Roper v. Simmons*, 543 U.S. 541 (2005).

6. In the United States, leading decisions that incorporate natural law understandings of gender and motherhood include *In the Matter of Baby M*, 109 N.J. 396 (1988), declaring surrogacy contracts to be invalid as a matter of law and policy in New Jersey; and *Johnson v. Calvert*, 5 Cal. 4th 84 (1993), holding that a gestational surrogate had no parental rights and that the genetic mother was the natural mother under California law.

7. *Roe v. Wade*, 410 U.S. 113 (1973).

8. It is worth noting, for instance, that no other STS scholar served on the joint AAAS-ABA National Conference of Lawyers and Scientists during my six years of service on that body. Similarly, apart from the historian of science Daniel Kevles, no other STS scholar sat on the National Academy of Science Committee on Science, Technology and the Law in its first six years. The membership of both bodies consisted of professional lawyers and scientists. Even I, of course, had a law degree.

9. *Brown v. Board of Education of Topeka, Kansas*, 347 U.S. 483 (1954).

10. *Diamond v. Chakrabarty*, 447 U.S. 303 (1980), p. 317.

11. *Daubert v. Merrell Dow Pharmaceuticals, Inc.*, 509 U.S. 579 (1993).

12. Rule 702 of the Federal Rules of Evidence stipulates that expert testimony is admissible only if it "is the product of reliable principles and methods." The corollary is that testimony based on

unreliable principles and methods can be excluded. After *Daubert*, parties may challenge each other's proffered testimony, and judges are required to determine whether it meets the test of reliability.

13. *Daubert v. Merrell Dow Pharmaceuticals, Inc.*, 43 F. 3d 1311 (9th Cir. 1995).

14. *Moore v. Regents of the University of California*, 51 Cal. 3d 134 (1990).

15. *Johnson v. Calvert*, 5 Cal. 4th 84 (1993).

16. See Office of Management and Budget, Final Information Quality Bulletin for Peer Review, December 15, 2004. My own December, 2003, comments on the guidelines as originally proposed made it clear that my arguments were inconsistent with OMB's aim of centralizing control over regulatory peer review.

17. *Kumho Tire Co. v. Carmichael*, 526 U.S. 137 (1999).

18. The philosopher of science Steve Fuller testified in favor of "intelligent design" (ID) as an alternative to evolution in a widely watched federal case in Pennsylvania. In *Kitzmiller v. Dover Area School District*, 400 F. Supp. 2d 707 (M.D. Pa. 2005), Judge John E. Jones concluded that, contrary to proponents' claims, ID was rooted in religious beliefs.

References

Abraham, John & Tim Reed (2002) "Progress, Innovation and Regulatory Science," *Social Studies of Science* 32(3): 337–69.

Angell, Marcia (1996) *Science on Trial: The Clash of Medical Evidence and the Law in the Breast Implant Case* (New York: Norton).

Bal, Roland (2005) "How to Kill with a Ballpoint: Credibility in Dutch Forensic Science," *Science, Technology & Human Values* 30(1): 52–75.

Bal, Roland & W. Halffman (eds) (1997) *The Politics of Chemical Risk: Scenarios for a Regulatory Future* (Dordrecht, Netherlands: Kluwer).

Beck, Ulrich (1992) *Risk Society: Towards a New Modernity* (London: Sage).

Berger, Margaret (2001) "Upsetting the Balance Between Adverse Interests: The Impact of the Supreme Court's Trilogy on Expert Testimony in Toxic Tort Litigation," *Law and Contemporary Problems* 64: 289.

Biagioli, Mario (1993) *Galileo, Courtier: The Practice of Science in the Culture of Absolutism* (Chicago: University of Chicago Press).

Bloor, David (1976) *Knowledge and Social Imagery* (Chicago: University of Chicago Press).

Bowker, Geoff C. & Susan Leigh Star (1999) *Sorting Things Out: Classification and Its Consequences* (Cambridge, MA: MIT Press).

Boyle, James (1996) *Shamans, Software, and Spleens: Law and the Constitution of the Information Society* (Cambridge, MA: Harvard University Press).

Breyer, Stephen (1993) *Breaking the Vicious Circle: Toward Effective Risk Regulation* (Cambridge, MA: Harvard University Press).

Brickman, Ronald, Sheila Jasanoff, & Thomas Ilgen (1985) *Controlling Chemicals: The Politics of Regulation in Europe and the United States* (Ithaca, NY: Cornell University Press).

Calabresi, Guido (1970) *The Cost of Accidents* (New Haven, CT: Yale University Press).

Calabresi, Guido & Philip Bobbitt (1978) *Tragic Choices* (New York: W. W. Norton).

Clark, Michael & Catherine Crawford (1994) *Legal Medicine in History* (Cambridge: Cambridge University Press).

Cole, Simon A. (2001) *Suspect Identities: A History of Fingerprinting and Criminal Identification* (Cambridge, MA: Harvard University Press).

Cole, Simon A. (2005) "Does 'Yes' Really Mean Yes? The Attempt to Close Debate on the Admissibility of Fingerprint Testimony," *Jurimetrics* 45(4): 449–64.

Daemmrich, Arthur A. (2004) *Pharmacopolitics: Drug Regulation in the United States and Germany* (Chapel Hill: University of North Carolina Press).

Edmond, Gary (2002) "Legal Engineering: Contested Representations of Law, Science (and Non-Science) and Society," *Social Studies of Science* 32(3): 371–412.

Edmond, Gary & David Mercer (2000) "Litigation Life: Law-Science Knowledge Construction in (Bendectin) Mass Toxic Tort Litigation," *Social Studies of Science* 30(2): 265–316.

Eigen, Joel Peter (1995) *Witnessing Insanity: Madness and Mad-Doctors in the English Court* (New Haven, CT: Yale University Press).

Faigman, David (1999) *Legal Alchemy: The Use and Abuse of Science in the Law* (New York: W. H. Freeman).

Federal Judicial Center (FJC) ([1994]2000) *Reference Manual on Scientific Evidence,* 2nd ed. (Washington, DC: FJC).

Foster, Kenneth R. & Peter W. Huber (1997) *Judging Science* (Cambridge, MA: MIT Press).

Foucault, Michel (1973) *Madness and Civilization: A History of Insanity in the Age of Reason* (New York: Vintage Books).

Foucault, Michel (1978) *The History of Sexuality* (New York: Pantheon).

Foucault, Michel (1979) *Discipline and Punish* (New York: Vintage Books).

Foucault, Michel (1994) *The Birth of the Clinic: An Archaeology of Medical Perception* (New York: Vintage Books).

Freeman, Michael & Helen Reece (eds) (1998) *Science in Court* (London: Dartmouth).

Fukuyuma, Francis (2002) *Our Posthuman Future: Consequences of the Biotechnology Revolution* (New York: Farrar, Strauss & Giroux).

Fuller, Lon ([1964]1969) *The Morality of Law* (New Haven, CT: Yale University Press).

Gibbons, Michael, Camille Limoges, Helga Nowotny, Simon Schwartzman, Peter Scott, & Martin Trow (1994) *The New Production of Knowledge* (London: Sage).

Gieryn, Thomas F. (1999) *Cultural Boundaries of Science: Credibility on the Line* (Chicago: University of Chicago Press).

Golan, Tal (2004) *Laws of Men and Laws of Nature: The History of Scientific Expert Testimony in England and America* (Cambridge, MA: Harvard University Press).

Goldberg, Steven (1994) *Culture Clash: Law and Science in America* (New York: New York University Press).

Guston, David H. (ed) (2001) "Boundary Organizations in Environmental Policy and Science" (Special Issue), *Science, Technology & Human Values* 26(4).

Hacking, Ian (1992) "World-Making by Kind-Making: Child Abuse for Example," in Mary Douglas & David Hull (eds), *How Classification Works: Nelson Goodman Among the Social Sciences* (Edinburgh: Edinburgh University Press): 180–213.

Hacking, Ian (1995) *Rewriting the Soul: Multiple Personality and the Sciences of Memory* (Princeton, NJ: Princeton University Press).

Hacking, Ian (1999) *The Social Construction of What?* (Cambridge, MA: Harvard University Press).

Harr, Jonathan (1995) *A Civil Action* (New York: Random House).

Hart, H.L.A. (1961) *The Concept of Law* (Oxford: Oxford University Press).

Hartouni, Valerie (1997) *Cultural Conceptions: On Reproductive Technologies and the Remaking of Life* (Minneapolis: University of Minnesota Press).

Hermitte, Marie-Angèle (1996) *Le Sang et le Droit: Essai sur la Transfusion sanguine* (Paris: Seuil).

Hilgartner, Stephen (2000) *Science on Stage: Expert Advice as Public Drama* (Stanford, CA: Stanford University Press).

Holmes, Oliver Wendell ([1881]1963) *The Common Law* (Boston: Little, Brown).

Huber, Peter W. (1991) *Galileo's Revenge: Junk Science in the Courtroom* (New York: Basic Books).

Jasanoff, Sheila (1990) *The Fifth Branch: Science Advisers as Policymakers* (Cambridge, MA: Harvard University Press).

Jasanoff, Sheila (1992) "What Judges Should Know About the Sociology of Science," *Jurimetrics* 32(3): 345–59.

Jasanoff, Sheila (1995) *Science at the Bar: Law, Science, and Technology in America* (Cambridge, MA: Harvard University Press).

Jasanoff, Sheila (2001) "Ordering Life: Law and the Normalization of Biotechnology," *Politeia* 17(62): 34–50.

Jasanoff, Sheila (2002) "Science and the Statistical Victim: Modernizing Knowledge in Breast Implant Litigation," *Social Studies of Science* 32(1): 37–70.

Jasanoff, Sheila (2004) *States of Knowledge: The Co-Production of Science and Social Order* (London: Routledge).

Jasanoff, Sheila (2005) "Law's Knowledge: Science for Justice in Legal Settings," *American Journal of Public Health* 95(S11): S49–S58.

Kairys, David (ed) (1998) *The Politics of Law: A Progressive Critique*, 3rd ed. (New York: Basic Books).

Keller, Evelyn Fox (1983) *A Feeling for the Organism: The Life and Work of Barbara McClintock* (New York: W. H. Freeman).

Keller, Evelyn Fox (1985) *Reflections on Gender and Science* (New Haven, CT: Yale University Press).

Kelman, Mark (1987) *A Guide to Critical Legal Studies* (Cambridge, MA: Harvard University Press).

Kevles, Daniel J. (1998) *The Baltimore Case: A Trial of Politics, Science, and Character* (New York: W. W. Norton).

Latour, Bruno (1987) *Science in Action: How to Follow Scientists and Engineers Through Society* (Cambridge, MA: Harvard University Press).

Latour, Bruno (1988) *The Pasteurization of France* (Cambridge, MA: Harvard University Press).

Latour, Bruno (1993) *We Have Never Been Modern* (Cambridge, MA: Harvard University Press).

Latour, Bruno (2002) *La Fabrique du Droit: Une Ethnographie du Conseil d'Etat* (Paris: La Découverte).

Lazer, David (ed) (2004) *DNA and the Criminal Justice System: The Technology of Justice* (Cambridge, MA: MIT Press).

Leclerc, Olivier (2005) *Le Juge et l'Expert: Contribution à l'Etude des Rapports entre le Droit et la Science* (Paris: Librarie generale de Droit et de Jurisprudence).

Lieberman, Jethro K. (1981) *The Litigious Society* (New York: Basic Books).

Lynch, Michael & Simon Cole (2005) "Science and Technology Studies on Trial: Dilemmas of Expertise," *Social Studies of Science* 35(2): 269–311.

Lynch, Michael & Sheila Jasanoff (eds) (1998) "Contested Identities: Science, Law and Forensic Practice" (Special Issue), *Social Studies of Science* 28(5–6).

Merton, Robert K. ([1942]1973) "The Normative Structure of Science," in R. K. Merton (ed), *The Sociology of Science: Theoretical and Empirical Investigations* (Chicago: University of Chicago Press): 267–78.

Minow, Martha (1990) *Making All the Difference* (Ithaca, NY: Cornell University Press).

Mnookin, Jennifer (2001) "Scripting Expertise: The History of Handwriting Identification Evidence and the Judicial Construction of Expertise," *Virginia Law Review* 87: 1723–1845.

Mooney, Chris (2005) *The Republican War on Science* (New York: Basic Books).

Nader, Laura (1969) "Up the Anthropologist: Perspectives Gained from Studying Up," in Dell Hymes (ed), *Reinventing Anthropology* (New York: Pantheon): 285–311.

Ogburn, William F. ([1922]1950) *Social Change with Respect to Culture and Original Nature* (Gloucester, MA: P. Smith).

Ogburn, William F. (1957) "Cultural Lag as Theory," *Sociology and Social Research* 41: 167–74.

Porter, Theodore M. (1995) *Trust in Numbers: The Pursuit of Objectivity in Science and Public Life* (Princeton: Princeton University Press).

Posner, Richard (1992) *Sex and Reason* (Cambridge, MA: Harvard University Press).

Quinn, Philip (1984) "The Philosopher of Science as Expert Witness," in James T. Cushing, C. F. Delaney, & G. M. Gutting (eds), *Science and Reality: Recent Work in the Philosophy of Science* (Notre Dame, IN: University of Notre Dame Press): 32–53.

Sage, William H. & Rogan Kersh (eds) (2006) *Medical Malpractice and the U.S. Health Care System* (Cambridge: Cambridge University Press).

Sanders, Joseph (1998) *Bendectin on Trial: A Study of Mass Tort Litigation* (Ann Arbor: University of Michigan Press).

Schuck, Peter H. (1986) *Agent Orange on Trial: Mass Toxic Disasters in the Courts* (Cambridge, MA: Harvard University Press).

Schuck, Peter H. (1993) "Multi-Culturalism Redux: Science, Law, Politics," *Yale Law and Policy Review* 11(1): 1–46.

Shapin, Steven (1994) *A Social History of Truth: Civility and Science in 17th Century England* (Chicago: University of Chicago Press).

Shapin, Steven & Simon Schaffer (1986) *Leviathan and the Air Pump: Hobbes, Boyle, and the Experimental Life* (Princeton, NJ: Princeton University Press).

Smelser, Neil J. (1986) "The Ogburn Vision Fifty Years Later," in *Commission on Behavioral and Social Sciences and Education, Behavioral and Social Science: 50 Years of Discovery* (Washington, DC: National Academies Press): 21–35.

Smelser, Neil J. & Paul Baltes (eds) (2001) *International Encyclopedia of Social and Behavioral Sciences* (Oxford: Elsevier).

Smith, Roger (1981) *Trial by Medicine: The Insanity Defense in Victorian England* (Edinburgh: University of Edinburgh Press).

Smith, Roger & Brian Wynne (eds) (1989) *Expert Evidence: Interpreting Science in the Law* (London: Routledge).

Solomon, Shana & Edward Hackett (1996) "Setting Boundaries Between Science and Law: Lessons from *Daubert v. Merrell Dow Pharmaceuticals, Inc.,*" *Science, Technology & Human Values* 21(2): 131–56.

Stoler, Ann L. (2002) *Carnal Knowledge and Imperial Power: Race and the Intimate in Colonial Rule* (Berkeley: University of California Press).

Timmermans, Stefan (2006) *Postmortem: How Medical Examiners Explain Suspicious Deaths* (Chicago: University of Chicago Press).

Unger, Roberto (1986) *The Critical Legal Studies Movement* (Cambridge, MA: Harvard University Press).

Van Kampen, Petra T. C. (1998) *Expert Evidence Compared: Rules and Practices in the Dutch and American Criminal Justice System* (Antwerpenen Groningen, Netherlands: Intersentia Rechtswetenschappen).

Vidmar, Neil (1995) *Medical Malpractice and the American Jury: Confronting the Myths About Jury Incompetence, Deep Pockets, and Outrageous Damage Awards* (Ann Arbor: University of Michigan Press).

Waldron, Jeremy (1990) *The Law* (London: Routledge).

Winickoff, David, Sheila Jasanoff, Lawrence Busch, Robin Grove-White, & Brian Wynne (2005) "Adjudicating the GM Food Wars: Science, Risk, and Democracy in World Trade Law," *Yale Journal of International Law* 30: 81–123.

Wynne, Brian (1982) *Rationality and Ritual: The Windscale Inquiry and Nuclear Decisions in Britain* (Chalfont St. Giles, U.K.: British Society for the History of Science).

31   Knowledge and Development

Susan E. Cozzens, Sonia Gatchair, Kyung-Sup Kim, Gonzalo Ordóñez, and
Anupit Supnithadnaporn

Like democracy, development is an essentially contested concept, with too much tied up in its meaning to allow it ever to settle into one form. The word invokes process and direction, and invites the question: development toward what? For a peasant in India, development may mean steady food, the assurance of staying on the land, and fewer children dying young. For a World Bank official, the peasant's dream would appear in statistics on poverty alleviation and reduction in child mortality. To an industrialist, development may mean business survival and personal wealth; to an economist, growth in gross domestic product; and to a politician, jobs, popularity, and power.

Amartya Sen (2000) defines development as freedom. Freedom is central to the process of development, he argues, both because "achievement of development is thoroughly dependent on the free agency of people" (freedom as means) and because it provides a yardstick for measuring progress (freedom as end). Development as freedom means human beings gaining the capability to achieve their own goals in their own contexts.

The substantive freedoms include elementary capabilities like being able to avoid such deprivations as starvation, undernourishment, escapable morbidity, and premature mortality, as well as the freedoms that are associated with being literate and numerate, enjoying political participation and uncensored speech, and so on (Sen, 2000: 3).

For freedom in this sense, the fundamental difference between the global North[1] and the global South is that many more people in the South are poor. Nearly a third of the population of developing countries lives in absolute poverty on less than $1 per day (Chen & Ravallion, 2004). Life spans in the poorest nations are half those in affluent ones, and developing countries bear the main burden of such major diseases as AIDS, tuberculosis, and malaria (Task Force on HIV/AIDS, 2004). Environmental conditions contribute to poor health, for example, through lack of clean water and adequate sanitation, and poverty contributes to environmental degradation as the rural poor strain natural resources such as forests and land in an attempt to eke out a living (Vosti & Reardon, 1997). Part of the literature on science, technology, and development focuses on ways research and innovation can contribute to the solutions to these problems of everyday life, an approach we can call the *human development project*.

Another part of the development discussion focuses on providing the resources to address the human development challenge through economic growth. National mastery of new technologies, and in particular information technology, is often seen as the key. In this view, the flow of information on a global basis is the lifeblood of the new economy. The worst economic fate is not to be at the periphery of the global network, but to be irrelevant to it, in what Castells (1996) calls the "black holes of the Information Economy." The knowledge industries—in those emerging areas that hold a temporary monopoly position by being at the cutting edge—are portrayed as the main sources of wealth today and in the future. In this view, whole geographic regions (e.g., Europe versus North America) vie to win the competition in the churn and change of the contemporary industrial scene. Indeed, contemporary theories of economic growth place technological innovation right in the heart of the growth process. The strong role of technology in maintaining markets for national industries, both domestically and internationally, is thus often seen as a second main challenge in using science and technology for development, the *competitiveness project*.

Immersed in the second project, many observers find it easy to lose sight of the first; yet making lives better is the essence of development as freedom. Whether poor, comfortable, or wealthy, most citizens of the global South do not think about "science" or "technology" in the abstract, although they use or buy electricity, water, medicine, televisions, and mobile phones that are part of what STS would call sociotechnical systems. Living technology rather than analyzing it, most people in the South ask primarily how it helps them, their families, their regions, and their countries.

What do the published literatures on science and technology for development have to offer to actors in the global South who are seeking to use science or technology to achieve development as freedom? This chapter does not present a comprehensive view—the literature is too vast for that, even when we focus our attention on what has been published since the last edition of this *Handbook* (Shrum & Shenhav, 1994). But we at least try to raise research questions grounded in the concepts alive today at the intersection of science and technology studies, economic growth theory, and innovation systems research.

The first section of the chapter introduces these three perspectives. The second section applies them to interpret examples of practical development problems: education, innovation policies, and learning firms. The final section outlines some key questions for an actor-centered, knowledge–pluralistic research agenda on science and technology in the development process.

THREE PERSPECTIVES

Science and Technology Studies

Over the decade since the last edition of this *Handbook*, the social sciences have been flooded with analysis of processes of change in the world system, often under the rubric of globalization (Worthington, 1993). Globalization has many meanings, but the predominant approach defines it as the distribution of productive processes across countries on a global scale, a process that is transforming livelihoods in some developing countries while leaving others untouched. Comparisons abound between the current wave of globalization and earlier ones, including the epic migrations of the turn of the century. In this wave, it is capital, not labor, which is moving. For the first time in world history, there is mutual trade in manufactured goods between the core and the up and coming semiperiphery (Ghose, 2003).

Two changes identified as technological are often portrayed as the drivers of the current dynamic: the falling cost of transportation and the rising capability of computer-mediated communication (Ghose, 2003). Some observers attribute fundamental importance to the spread of communication networks (Castells, 1996). Sociologists have examined patterns of urbanization in this newly connected world (Sassen, 2002), and political scientists, while not abandoning the study of change in national governance patterns, have begun to analyze such emerging institutions of global governance as the World Trade Organization and the new set of rules it is negotiating in the global knowledge economy.

The STS literature includes stories that take place in the global South but does not try to add them up into an account of changing macro structures in the world economy or a coherent theory of development. Instead, the stories highlight particular actors and the forms of knowledge they bring into particular interactions, shedding light on the dynamics that create new patterns. The STS literature is not monolithic in approach: methods range from standard survey research (Campion & Shrum, 2004) to network studies (Shrum, 2000) to discourse analysis (Hecht, 2002), but the dominant approach is narrative. Yet there are some themes that appear across the various writings that may constitute an STS approach to the topic.

Most often, the actors portrayed in the STS stories belong to the global scientific community. So, for example, we find studies of women scientists (Campion & Shrum, 2004; Gupta & Sharma, 2002) and universities (Sutz, 2003) in the South. Sometimes the stories confirm conventional trajectories. For example, Velho and Pessoa (1998) describe Brazil's ambitions in international research, leading to the decision to invest in a synchrotron light source. Lomnitz and Cházaro (1999) lament the lack of understanding of the roles of computer scientists in the basic research-oriented reward system of Mexican universities. Others describe new configurations, like Shrum's account (2000) of nongovernmental agricultural research organizations.

The relationships between scholars in the North and South receive attention in the STS literature, for example, in Solovey's work (2001) on Project Camelot. Some articles reflexively consider the knowledge status of scholars from the North in their observer roles in the South, or as Shrum (2005) puts it, "reagent" roles (see also Verran, 2001). Similarly, in their "love" for the Zimbabwean bush pump, de Laet and Mol (2000) explore "new ways of 'doing' normativity."

The juxtaposition and conflict between different forms of knowledge is the most common theme in STS stories set in the global South. For example, Lei (1999) describes the exclusion of Chinese traditional medicine from emerging networks of "Western-style" doctors in China in the 1920s and 1930s. Postcolonial science carries the echoes of previous power relationships into the present (e.g., Adams, 2002; see also Dubow, 2000). Traditional and local are not always overcome by "modern" and Northern, however. In Verran's (2002) account, environmental scientists eventually come to respect aboriginal regimes of burning in the Australian bush. And farmers, engineers, and social activists jointly design the Baliraja Memorial Dam in India (Phadke, 2002).

Given these themes, the STS literature implicitly portrays globalization as a process of knowledge confrontations. "Professional" or "scientific" knowledge carries the privilege of the North into the definitions that shape life in the South. It tangles with other ways of framing and addressing issues, particularly those rooted in the knowledge of poor or indigenous people. By treating the various forms of knowledge symmetrically, the STS approach draws attention to the asymmetries in power that privilege one form of knowledge over another. STS stories include a broad set of actors, especially highlighting civil society and marginalized groups, and features their categories and knowledge. The STS literature thus highlights certain questions with regard to development projects: Whose project is it? What knowledge do the various actors bring to the interaction? Whose knowledge gets respect and deference? What are the outcomes of the project for the everyday lives of the people involved?

In the background of the STS stories are the practical problems of development as freedom, for example, AIDS (Karnik, 2001), rural energy (Gorman & Mehalik, 2002), fertility (Oudshoorn, 1997), and Chagas disease (Coutinho, 1999). The STS contribution to development is the freedom to envision both problems and solutions in local ways, without the imposition of the categories used in the sciences or technologies of the North.

New Growth Theory

Economics provides another strand of thinking about science and technology in the global South. Unlike STS, where Southern stories are not set apart, economics has a subfield for "development" and a branch of theory, growth theory, that gives a particular account of the process. In that account, nations are the central actors, with governments (usually called "the State") playing the central role. Like STS scholars, the economists themselves also play roles, since they provide analysis and advice to both national governments and the international banks, but they pay less attention to their own roles and seldom subject them to scrutiny in their work, except in autobiographical mode [e.g., Stiglitz's *Globalization and Its Discontents* (2002) on his experiences at the World Bank and Sachs's *End of Poverty* (2005)].

Growth theory traces its roots back to Adam Smith and his analysis of the role of division of labor in expanding economic activity, and to Karl Marx, who saw capitalists and production technology ("the means of production") as the driving forces of change in the economy. Classical growth theory attributes economic expansion to the accumulation of land, labor, and capital. Examining this claim in light of emerging data, scholars in the 1950s noted that the combination of these three did not explain all the variance in growth, and Solow added the hypothesis that the rest of the variation (the "residual") was due to technological change. Neoclassical growth theory, as this hypothesis came to be called, did not delve into the sources of technological change, but rather treated its result as a public good that was available to all nations and businesses alike (Solow 1956, 1957). Technology was "exogenous" in this theory. The creative capitalist had disappeared, replaced by a faceless process of technological change. Neoclassical growth theory was compatible with traditional modernization theories; countries that were "behind" could "catch up," since technology acquisition costs much less than technology development.

The most influential family of contemporary growth theories changes this picture by treating technology as endogenous, that is, as the result of deliberate economic choice, on the part of either private firms or the State (Romer, 1990). Those who invest in developing new technology earn economic rewards, because they hold a temporary monopoly over the means of doing something new and more productive. Economic growth results from the increasing returns associated with new knowledge. As a result, while returns diminish in the physical economy, they increase in the newly named knowledge economy (Cortright, 2001).

This new growth theory observes that knowledge-based economies tend toward monopolistic competition (Cortright, 2001). Because knowledge has increasing returns (continuously declining marginal costs), leading firms tend to build up insurmountable advantages and new entrants face the difficult prospect of starting out with much higher costs than their established competitors. History matters: once a technology is locked in, it is harder for competitors to replace it. Institutions matter: dynamic organizational adjustment to changing circumstances is required for continuing progress. Place matters: local institutions and cultures shape knowledge flows, and tacit knowledge is important.

All these factors suggest that once a country or region has a significant knowledge advantage, it will be difficult for another country or region to catch up. A country that has almost no technical base now might easily conclude that it will never get into the game. New growth theory's primary recommendation for such a country is to increase human capital, that is, increase its total knowledge and creativity through education. Another endogenous approach to growth, evolutionary economics (Nelson & Winter 1982), calls attention to still other opportunities inherent in the knowledge economy. The creative destruction of the market, that is, the continual appearance of new industries that supersede previous ones, opens up possibilities for countries to concentrate their resources in specific areas and leapfrog over competitors in old industries, finding places for themselves in new ones (Schumpeter, 1942). Technological change is at the root of this process, and it is thus worthwhile for countries to invest in the capabilities necessary to ride the next wave when it comes. Some of the most prominent examples of technology-based growth success have followed this path, including the Asian "tigers" such as Korea, Taiwan, and Singapore.

The actors in these accounts of science and technology in development are quite distinct from those in the STS literature on the same topic. Private firms ultimately produce the growth, but the theorists provide advice primarily to governments, urging them to create the conditions for firms to act. Knowledge is again an object of conflict, but this time it is the new knowledge embodied in innovations that forms the basis of competition, with various institutions contending over the ground rules for ownership and profits. Development is not freedom in Sen's sense for these theorists. Industrial growth can coexist with persistent poverty in the absence of redistributional mechanisms for the wealth generated. New growth theory concerns itself with the accumulation of wealth in a country or region; use of that wealth for human development is someone else's project.

Innovation Systems

A third line of recent research on science, technology, and development draws on the concepts of evolutionary economics and traces them concretely into networks of actors and the relations among them in developing countries. This is the burgeoning work on innovation systems, which has three main types: national, regional (subnational), and sectoral (product-specific). The concepts were introduced and developed by Freeman (1982), Nelson (1993), and Lundvall (1992) and have been developed by Edquist (1997) at the national level by Braczyk et al. (2003) at the regional level, and by Malerba (2004) at the sectoral level. Several recent volumes explore applications specifically in the context of developing countries (Cassiolato et al., 2003; Muchie et al., 2004; Baskaran & Muchie, 2006).

An innovation system consists of elements and their relationships (Edquist, 1997); it is a network of actors, like the ones found in actor-network theory in STS (Callon, 1999; Latour, 1987). The three usual categories of actors discussed are firms, government, and research institutions, including public sector laboratories and universities. The concept has no problem accommodating new forms of actors, for example, the nongovernmental research organizations Shrum (2000) describes, or hybrid forms such as university-associated research parks. Likewise, in principle, civil society organizations could be included, but in practice they seldom appear in the stories innovation systems researchers tell (their "case studies"). Nonetheless, firms are at the center of the networks, and a healthy innovation system is one in which firms are in the lead. Many forms of relationships appear in the stories, from competition through exchange and collaboration. The network can have multiple levels or subnetworks. For example, it can incorporate governmental actors at regional, national, and supranational levels (e.g., the European Union). Sectors could form subareas of a regional network.

As in the previous two perspectives, knowledge plays a central role in the concept of the innovation system. The life process of an innovation system is learning, which involves accessing, accumulating, and applying knowledge (single-loop learning), reacting to changes in the environment (double-loop learning), and using internally generated knowledge to transform the environment (triple-loop learning) (OECD, 2002). The value of the network in the system is that it increases learning through interaction and sharing. Everyone is supposed to learn in an innovation system: individuals, firms, other institutions, and the system itself. In principle, the sources of knowledge can be as heterogeneous as the actors involved in the network, although again in practice, organizational, business, and technical knowledge are privileged in innovation system accounts.

What is the project of an innovation system? Implicitly, the goal is growth. This focus is clear in the central role given to firms. While the centrality of learning might seem to create an affinity with the development-as-freedom approach, and although the founders claim to be analyzing societal learning processes (Johnson & Lundvall, 2003), the innovation systems literature devotes little concrete attention to whether learning extends beyond the network of firms, government agencies, and laboratories. Innovation systems could easily be elite in composition; the concept does not require otherwise. Likewise, the concept is neutral on whether the systems are oriented toward socially constructive or destructive technologies (e.g., vaccines or weapons systems). The literature on regional innovation systems has a geographically redistributive slant, exploring the ways that less wealthy regions could become wealthier. Likewise, the application of the concept to countries in the global South also supports an economic catch-up agenda (Johnson & Lundvall, 2003). But very few scholars of innovation systems have emphasized the importance of innovation for social productivity or poverty alleviation (for exceptions, see Arocena & Senkar, 2003; Arocena & Sutz, 2001, 2003; Sutz, 2003).

Summary

Each of these three literatures, then, peers into life in the South through a different lens, with some version of knowledge playing a key role in each. New growth theory focuses on the role of the State in assuring the conditions for economic growth through monopoly over new commercially important knowledge (we call this angle knowledge as growth). The innovation systems approach focuses on firms and their learning processes, asking how these can be enhanced by incentives and interactions with other institutions (knowledge as learning). The STS literature follows the science and technology institutions of the North as they encounter and engage knowledge produced in other contexts in the South, with a focus on the empowerment of civil society organizations and marginalized groups (knowledge as confrontation). None of the perspectives explicitly takes development-as-freedom as its goal nor explores concretely how the approach would contribute to meeting the basic needs of the world's population.

APPLICATIONS

The various development paradigms that have appeared in succession over the past half-century (Gore, 2000) have shared an assumption of strategy and action: some set of actors in the context of a poor country should take a specified set of steps toward "development." The paradigms direct various policy prescriptions to this assumed, but often unnamed, set of actors. But the three perspectives we have outlined above identify a multi-actor space, one in which the interests of civil society, the State, and private firms may not coincide, and surely do not necessarily add up to development as freedom. This section examines three common development tasks assigned by the current paradigm to these various actors and analyzes the prospects for the contribution of each to development as freedom.

Education as Freedom

Standing at the civil society corner of figure 31.1, we examine the process of education. The importance of education, in particular, science and technology education, as a means of augmenting productivity, increasing innovation, and solving social problems is a recurrent theme in the literature on education and on development (Lewin, 2000a,b; UNESCO, 2004; Watson et al., 2003). The economic success of Japan and more recently the East Asian tigers are cited as examples in which the emphasis on education has paid great dividends in the countries' development efforts (Mingat, 1998). Scientific and technological education in developing countries faces constraints, such as insufficient teachers, inadequate skills, lack of equipment (Sane, 1999) and inadequate access due to poverty and poor student interest in study and careers in science or engineering (UNESCO, 2004). Education policy, observers point out, needs to take other factors into account, such as macroeconomic and trade policies, institutions (legal and political systems), factor endowments, and sociocultural environment (Banerjee, 1998; Hunter & Brown, 2000).

Education is central to the concept of development as freedom. The application of the concept goes well beyond programs that address basic needs, such as training in cleaner production (Huhtala et al., 2003) and strengthening the research capability in reproductive health problems (Benagiano & Diczfalusy, 1995). The world's citizens

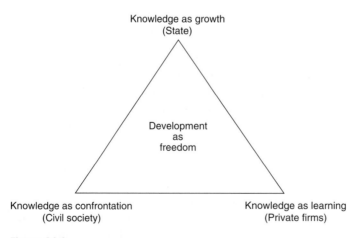

Figure 31.1
The development triangle.

need education to achieve both stable livelihoods and political voice; education is thus seen as a significant contributor to increasing democracy, social justice, and individual empowerment (Kyle, 1999; Zahur et al., 2002). Since literacy is fundamental to achieving these goals, the policy directions of international agencies have often suggested that developing countries should increase the share of their public expenditure on primary education (Curtin & Nelson, 1999).

New growth theory, in contrast, puts more emphasis on higher education and technical skills to feed the innovation process. Viewed from the angle of State action, resources allocated to education become "investments in human capital" rather than Sen's means and ends of freedom. Economists struggle with the difficulties in identifying the true social returns across primary, secondary, and tertiary levels of education (Birdsall, 1996; Heyneman, 2003; Vlaardingerbroek, 1998), and views on the efficacy and relevance of recommended programs are mixed (Curtin & Nelson, 1999; Heyneman, 2003). Nonetheless, investments in higher education appear to economists to be at least as crucial as those at the primary level, setting up difficult trade-offs for policy makers.

The "human capital" terminology suggests that government policies should not only address the direct provision of education and training but should also facilitate and encourage the private sector to play active roles, as several model economies in Asia have done. The educational strategies of the Asian tigers provide valuable insights for other developing countries, but the contextual underpinnings have to be taken into consideration (Kuruvilla et al., 2002). The role of government cannot be static, as illustrated by the changing role of the Korean government with respect to R&D and training (Hee & Soo, 1997).

Like the STS approach, the literature on science education acknowledges the confrontation between forms of knowledge shaped in the North and those shaped in the South. The literature often describes developing countries as facing "challenges" resulting from the conceptualization of science from a Western European perspective, which imposes changes in the worldview, culture, and behavior of students including cognitive learning and the use of language (Gray, 1999; Jegede, 1997; Lewin, 2000b). Some authors note that developing countries need to adapt Northern learning to make science education relevant to local culture and context (Bajracharya & Brouwer, 1997; Brown-Acquaye, 2001; Gray, 1999). Others point out that formal and informal science education both contribute to the popularization of science and to building a "scientific culture," and urge public/private partnerships in the establishment of science centers are thought to play an important role (Tan & Subramaniam, 2003). A symmetrical approach to knowledge would take each of these activities as a site for neutral epistemological research, but such studies are rare while the rush to adopt Northern approaches is common.

STS researchers following "big science" into the global South would find it being put to use in attracting young people to technical careers. The regional Centres for Space Science and Technology Education in Africa (Abiodun, 1993; Balogun, 2002) are described as contributing to human capital and the process of development.

Supporters anticipate that the regions will not only reap direct benefits of education in space science and technology but will also derive benefits from research associated with curriculum development, pedagogy, and delivery methods in space education (Andreescu et al., 1997; Hsiao et al., 1997; Kasturirangan, 1997; Lang, 2004). "Whose projects are such centers?" STS analysts would be likely to ask. The same question could be asked of institutions such as the Third World Academy of Science (TWAS) and of efforts to establish national academies of science (Guinnessy, 2003). These activities promote the exchange of information through networks of cooperation and scientific excellence while at the same time extending the power and prestige of Northern science into Southern institutions.

A broader concept of societal learning, more akin to the innovation systems approach, is implied in the literature on training strategies that encompass building skills for design and development (Alic, 1995), management and technical know-how related to the technology acquisition (Alp et al., 1997), as well as technical skills for using advanced equipment and machinery. Economists point out that threshold levels of absorptive capacity are needed to maximize the benefits of technological investments and capital flows (Borensztein et al., 1998; Eicher, 1999; Keller, 1996), which impact skills building and knowledge flows (Lall, 2002; Reddy, 1997). In studies of education in Malaysia and Korea, Snodgrass notes that while education may be seen as a necessary condition for economic growth, it is not a sufficient condition. For education to boost growth, the demand for educated or skilled labor must also increase (Snodgrass, 1998). In addition to building lower level skills for increasing productivity and efficiency, higher level skills in management, political leadership, and bureaucracy are required (Rodrigo, 2001). In this view, skills are built not only in the formal education system but also through on-the-job experience, or *learning by doing;* however, higher level skills are more difficult to acquire in this way (Rodrigo, 2001).

In summary, education is valued from all three corners of the triangle, but it does not necessarily bring freedom at the core. If education is a top-down process of infusing Northern science and its concepts into more and more people in the South, its contributions to freedom are important but limited. If education is undertaken as part of a societal learning process, however, weaving together new and old insights into a locally defined and controlled process of change, both innovation and freedom could be strengthened.

Innovation Policies

For developing countries, the creative destruction of a global, knowledge-based economy has created an unstable and uncontrollable environment (Hipkin, 2004). New growth theory stresses that technological innovation may be the only way to survive and prosper in today's world (Sikka, 1997). Toward this end, commonly recommended State actions include investing in research and development (R&D), creating the conditions for foreign direct investment, and strengthening intellectual property policies. All these steps are problematic, however, from the viewpoints of civil society, learning firms, and development as freedom.

Investment in National R&D   As with the tradeoff between primary and tertiary education, public investments in R&D are the site of conflict between the human development and competitiveness projects. The resources available are modest: at the aggregate level, R&D expenditure as a percentage of gross national product (GNP) in developing countries is still much lower than that observed in industrialized countries (Bowonder & Satish, 2003). Additionally, national R&D intensity tends to increase in line with per capita income (Mitchell, 1999). In developing countries, R&D expenditures by higher education institutions and government agencies are far higher than R&D spent by private firms. In theory, this could be an advantage for engaging civil society and developing a capacity for learning with regard to local problems. But in practice, these groups are seldom included in the discussion about research agenda, and the effort is continually undermined by the pull of research agendas from the North (Sutz, 2003).

According to the literature, the prospects are not much better for using national R&D spending to stimulate the learning process in industry. Particularly in the area of biotechnology, governments in developing countries have played major roles in pursuing R&D because the private sector is too weak to lead the way in accessing the new tools and technologies (Byerlee & Fischer, 2002). R&D activities in many developing countries address local needs that are not of broad international significance (Albuquerque, 2000). As seen in a study of domestic patent data in Brazil, there is a higher share of individual patents rather than of company patents. University R&D in Latin America has not been particularly relevant to the needs of industries (Arocena & Sutz, 2001), partly because the connection between industries and universities is usually weak.

Foreign Direct Investment (FDI)   Foreign direct investment has been seen by most developing countries as a shortcut not only to economic benefits but also to acquiring capacity for technological innovation (Sjoholm, 1999). Arze and Svensson (1997) claim that over time technology from FDI and domestic innovative capacities are interdependent. As an example, in Indonesia, spillovers from FDI can be found in certain manufacturing sectors, as reflected by the increasing productivity of locally owned firms. Moreover, the larger the technology gaps between domestic and foreign firms, the larger the spillovers. Nonetheless, the literature notes that the positive effects of FDI are not automatic. Host country characteristics and supporting policies including fiscal incentives, available skilled workers, and competitive environment are important factors in facilitating spillovers to local domestic firms (Blomstrom & Kokko, 2001; Lall, 1995).

The effects of FDI in developing countries are far from proved. The study of Uruguay by Kokko and Zejan (2001) has shown some evidence that the presence of FDI has no apparent impact on local productivity except for increasing the chances of exporting by local firms. On the contrary, instead of benefiting local firms in terms of technology transfer, FDI can create a competitive environment, resulting in pressure on local firms to increase their efficiency (Okamoto, 1999). Furthermore, clustered FDI is significantly better than dispersed FDI, particularly in terms of transferring technology (Thompson, 2002). Developing countries also need to be concerned about the issue of higher unemployment when adopting new technologies (Diwan & Walton, 1997).

Studies of firms and industries that are considering locating in developing countries have shown that several aspects of local contexts are competitive priorities for them, including labor availability, level of local competition, government laws and regulation, and market dynamism (Badri, 2000). However, local firms in some developing countries have a hard time coping with these new entrants into the domestic market. For example, there is decreasing room for locally owned companies in automobile industries in South Africa (Barnes & Kaplinsky, 2000). In other cases, emerging multinationals based in developing countries enjoy global success by fostering continual cross-border learning to help them move up the value chain (Barlett & Ghosal, 2000). In the absence of competition from foreign firms, it was difficult for local firms in India to develop their technological capabilities to penetrate the global market, with the result that most stayed inefficient (Bowonder, 1998).

The literature stresses that globalized competition has transformed production systems for both developed and developing countries (Fleury, 1999). In particular, multinational corporations have not only invested in manufacturing plants for the production of their product but also in R&D in places they consider appropriate. Studies have indicated the positive influence of such remote R&D facilities in many developing countries like Brazil, China, and Taiwan (Bowonder, 2001). On the other hand, if developing countries rely excessively on high-tech industry outsourced from foreign firms, they might discourage domestic firms from taking on more complex projects or moving up to higher levels of the product value chain, as illustrated in India by the success of software industry (D'Costa, 2002) and the failure of the hardware industry (Khan, 2001).

Intellectual Property Policies   A third example of a recommended policy that is likely to have mixed results is the new agreement on Trade-Related Intellectual Property Rights (TRIPS), negotiated through the World Trade Organization. The literature on the implications of TRIPs for developing countries has been negative or cautiously neutral at best (Correa, 1998, 2000; Hoekman et al., 2002; South Centre, 1997; UNCTAD, 1996). Some analysts maintain that stronger intellectual property rights (IPRs) will ultimately help developing countries through increased technological activities domestically and enhanced technological inflows from abroad. Critics say this claim is merely in the interests of developed countries and assert that stronger IPR protection would benefit only industrialized countries and the companies that export IPR-based technologies (Bronckers, 1994; Dealmeida, 1995).

The debate over TRIPS raises explicitly the issue of "voice" for the South—the national-level version of the STS issue of voice for marginalized groups within developing countries. For example, some observers have criticized TRIPs for excessive representation of private business interests of developed countries (Sell & Prakash, 2004), for additional bilateral pressure for heightened IPR protection beyond what is required under the TRIPs agreement (Drahos, 2001), for structural weakness in the ability of developing countries to participate in the WTO judicial process (Shaffer, 2004), and for ineffective measures for developing countries to sanction developed countries under the WTO system (Bronckers & van den Broek, 2005; Subramanian & Watal, 2000).

TRIPs creates another direct tradeoff between the human development and competitiveness projects. Drug prices are expected to rise under TRIPs because of increased requirements for patenting pharmaceuticals. This price rise could have far-reaching implications for global public health by exacerbating limited access to essential drugs to treat major diseases in poor countries, diseases that account for millions of deaths (Attaran, 2004; Perez-Casas et al., 2001; Scherer & Watal, 2002; Subramanian, 1995; Wagner & McCarthy, 2004). A fierce conflict has erupted over HIV/AIDS medications. On the one side are developed countries, backed by their multinational pharmaceutical companies, and on the other, some developing countries such as South Africa, Brazil, and Thailand (Bond, 1999; Schuklenk & Ashcroft, 2002; Sell & Prakash, 2004). The TRIPs agreement will hurt the pharmaceutical industry in some countries (including India) by prohibiting their manufacturing generic drugs as inexpensively as in the pre-TRIPs era (Watal, 2000). Many developing countries are also concerned about the lack of research on drugs for diseases prevalent in their countries (Grabowski, 2002; Kremer, 2002; Lanjouw & Cockburn, 2001; Mahoney et al., 2004).

Since many developing countries heavily rely on their agricultural sector, the TRIPs requirements regarding plant varieties and plant breeders' rights have also been controversial (Macilwain, 1998; Srinivasan, 2003, 2004). Several problematic post-TRIPs developments in biotechnology have intensified the concerns of many developing countries about theft of traditional knowledge, a problem that particularly affects indigenous communities. Local knowledge that has been held for centuries by indigenous communities and the general public may end up as part of the intellectual property of developed countries. This new phenomenon, dubbed "biopiracy," has generated its own literature with regard to pharmaceuticals (Hamilton, 2004; Timmermans, 2003), plant varieties (Macilwain, 1998; Srinivasan & Thirtle, 2003), and biodiversity (Bhat, 1996; Brechin et al., 2002; Kate & Laird, 1999; Posey & Dutfield, 1996). This topic is rife with knowledge confrontation, and it is surprising that not more STS literature has been devoted to this topic.

The TRIPS agreement has also come under criticism for its negative impact on global public goods (Maskus & Reichman, 2004). Any incursion of private knowledge into public knowledge reduces the global capacity to learn and therefore to innovate, so from the viewpoint of innovation systems, the incentives IPRs provide for innovation are at least partially outweighed by the costs in loss of available information.

Summary Overall, then, the growth prescriptions for the State corner of the development triangle hold little that is helpful in the other two corners. The stunning absence of attention to development as freedom in these prescriptions speaks volumes about the disconnect between currently fashionable national poverty reduction strategy papers and the dominant ideas about overall economic growth.

Learning Firms

In the literature on innovation processes in private industry, the landscape of development as viewed from the corner of private firms is crowded with networks and alliances with other firms, either inside or outside the country in question. The occasional publicly supported research institution appears, but the State is off in the distance as a network facilitator, and civil society appears only as "markets" or "customers" for goods and services. Employment and labor issues are invisible.

As suggested by Ernst et al. (1998) and Arnold et al. (2000), to obtain technology, firms face the choice of *creating their own technology* or *acquiring technology from outside*. To create technology of their own, firms need capacities ranging from adapting and reverse engineering to developing their own prototype technology by performing their own R&D. In acquiring technology from outside, the firm faces further choices in selecting, adopting, and implementing technologies. The adoption/adaptation process must be a knowledge confrontation but again one that the STS literature has not studied.

In fact, according to innovation theory, firms are not inclined to innovate on their own without receiving any knowledge, skills, technical support, methods, and instruments from outside. Rather, innovative firms are thought to be embedded in a complex network of relationships with customers, suppliers, research institutes, industry associations, and so on. Some scholars (Porter, 1990) refer to this interdependence as a "cluster." The new-tech agglomeration in Beijing, for example, seems to contain all the necessary elements of entrepreneurship: small firms, new firm formation, and innovativeness. Nonetheless, there are also weaknesses in that cluster, including limited direct global linkages with multinational firms and restraints on networking with state-owned institutions and firms (Wang & Wang, 1998). As this example illustrates, the literature on clusters tends to focus on learning from other firms rather than from local communities.

The literature in this area often identifies the State in the South as too weak to sustain the diffusion process (see, e.g., Conceicao & Gibson, 2001; Di Benedetto & Calantone, 2003). In a weak State, decentralized decision-making leads to duplication of efforts and hence reduces learning opportunities. The steel industry in India illustrates the effects of this fragmentation (D'Costa, 1998). Likewise, some firms in the auto industry in India failed to adopt best practices, resulting in poor performance (Diwan & Walton, 1997).

For firms in developing countries, both learning and imitating are primary capabilities affecting technological progress (Gao & Xu, 2001), as illustrated by the video/compact disk industry in China. Moreover, the learning processes of small and large firms in the same industry of the same country can be quite different. The case of color television manufacturers in China has shown that the one that focused on the local market was less successful than the one concentrating on the export market

(Xie & Wu, 2003). In addition, the study of a Chinese firm originally spun off from a government-supported research institute indicates an evolutionary pattern of path-dependency, from sales to distribution and service activities, to manufacturing product, process design, and finally R&D (Xie & White, 2004).

Studies reveal that clustering and networking help small and medium entrepreneurs (SMEs) improve their competitiveness. In short, Humphrey and Schmitz (1996) have offered the "triple C" concept—customer-oriented, collective, and cumulative. Despite suffering from competition with their counterparts in developed countries, high-tech manufacturing firms in developing countries need to identify proper technical strategies to flourish in times of national growth. These strategies include (1) using market opportunities or growth consistent with the firm's capacity and competitive advantage, (2) continually expanding the business to acquire expertise and capital enabling increasingly sophisticated processes, and (3) cooperating with technical forerunners (Wang & Pollard, 2002).

Governments in developing countries can facilitate networking in many ways. For example, China's Shanghai-Volkswagen (SVW) developed vertical networks among its suppliers because the Shanghai government encouraged it to promote outsourcing and extend supplier networks across the entire country. On the contrary, in the case of Proton in Malaysia, vertical networks did not occur among suppliers because the government limited the networking range (Yoshimatsu, 2000).

Summary The literature on firm-level innovation in developing countries is rather narrowly focused on issues of company survival in a global competitive environment. Company survival is necessary for growth, and growth is helpful in human development, but neither assures that development as freedom will be reached. There is plenty of discussion of dynamics along the State–private firm edge of the development triangle but virtually none on dynamics along the civil society–private firm edge. Given the analysis in the previous section of the weakness of the State in mediating between the human development and competitiveness agendas, dynamics on the third edge (between civil society and the State) do not provide any immediate hope for uniting knowledge and learning with development as freedom, unless a broader concept of innovation and learning is adopted. We turn now to this possibility.

RESEARCH AGENDA

This review has deliberately juxtaposed three literatures that are not usually brought together. The mostly economic literatures on knowledge as growth and knowledge as learning conventionally overlap and complement one another. But they are both mutually invisible to the literature on knowledge as confrontation, that is, the writings in the field of science and technology studies on developing countries. The development triangle (see figure 31.1) therefore may not unite the three themes but rather capture their mutual neglect. The STS literature neglects business; the literature on the developmental State neglects civil society, at least when it deals with innovation

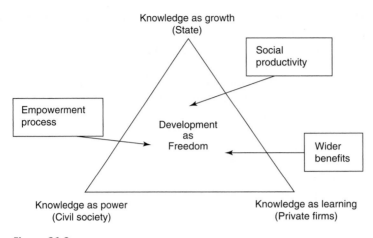

Figure 31.2
Moving toward the center of the development triangle.

policies; and the literature on learning and competence building systems by and large neglects the contributions of civil society. They all neglect development as freedom.

So are these three topics irrelevant to reaching development as freedom? Surely not. Economic growth is a necessary if not sufficient condition for improving everyday lives for the world's poor, although as Sen himself shows, it is not the accumulation of wealth that matters for health and education but rather how wealth is used. Learning is certainly a crucial process for human beings to free themselves from disease and illiteracy and to achieve open speech and participation. Part of that learning must take place in the workplace. Likewise, no matter how seemingly abstract the categories used to analyze them, knowledge confrontations have real consequences.

To increase its contributions to development as freedom, however, each of these three literatures must move its intellectual agenda closer to the center of the development triangle (figure 31.2). The literature on knowledge as growth has a close cousin in the literature on growth and inequality, and growth and human development. It needs to pay attention to these concepts and break loose from the narrow confines of the competitiveness project to embrace a broader concept of social productivity. How much more quality of life will the citizens of a country gain from a particular public investment? The answer is not captured in standard economic measurement, but it should be. The literature on learning and competence building systems needs to live up to its own ambitions to consider societal learning processes, not just those in private firms. Innovation can move in many directions. Rather than remaining silent on the direction of technological change, this literature needs to articulate the kinds of learning that would orient private industry toward businesses with wider social benefits,

Finally, the STS literature needs to engage with the real world of knowledge in development. It is not enough to follow the actors from research institutions in the North to those of the South, and in particular to follow oneself in that role. Standing on the side of marginalized communities is an excellent vantage point for analyzing knowledge confrontations that matter. STS needs to actively look for and carefully study success stories in the transfer of power through knowledge, with the goal of informing the practice of those at the bottom.

In this chapter, we have tried, like Sen, "to present, analyze and defend a particular approach to development, seen as a process of expanding substantive freedoms that people have" (Sen, 2000: 297). We have viewed science and technology as forms of knowledge and learning and explored several ways that they contribute to the process of achieving development as freedom. We hope that the next chapter on this topic in the next *Handbook* will be able to celebrate progress toward that goal.

Note

1. We define the global South as the middle- and low-income countries of Africa, Asia, and Latin America, and the global North as the high-income nations of the world. For the former group, we also sometimes use the terms developing world or developing countries. The transition countries of Eastern Europe and the former Soviet Union offer a different set of development experiences, which are not discussed here.

References

Abiodun, A. A. (1993) "Centers for Space Science and Technology Education: A United Nations Initiative," *International Journal of Remote Sensing* 14(9): 1651–58.

Adams, V. (2002) "Randomized Controlled Crime: Postcolonial Sciences in Alternative Medicine Research," *Social Studies of Science* 32(5–6): 659–90.

Albuquerque, E.D.E. (2000) "Domestic Patents and Developing Countries: Arguments for Their Study and Data from Brazil (1980–1995)," *Research Policy* 29(9): 1047–60.

Alic, J. A. (1995) "Organizational Competence: Know-How and Skills in Economic Development," *Technology in Society* 17(4): 429–36.

Alp, N., B. Alp, & Y. Omurtag (1997) "The Influence of Decision Makers for New Technology Acquisition," *Computers and Industrial Engineering* 33(1–2): 3–5.

Andreescu, D., M. I. Piso, & M. Nita (1997) "Postgraduate Training for Space Science and Technology Education," in *Problems of Space Science Education and the Role of Teachers* (Oxford: Pergamon Press): 1375–78.

Arnold, E., M. Bell, J. Bessant, & P. Brimble (2000) *Enhancing Policy and Institutional Support for Industrial Technology Development in Thailand: The Overall Policy Framework and the Development of the Industrial Innovation System* (Washington, DC: World Bank).

Arocena, R. & P. Senkar (2003) "Technology, Inequality and Underdevelopment: The Case of Latin America," *Science, Technology & Human Values* 28(1): 15–33.

Arocena, R. & J. Sutz (2001) "Changing Knowledge Production and Latin American Universities," *Research Policy* 30(8): 1221–34.

Arocena, R. & J. Sutz (2003) "Knowledge, Innovation and Learning: Systems and Policies in the North and in the South," in J. E. Cassiolato & M. Maciel (eds), *Systems of Innovation and Development: Evidence from Brazil* (Cheltenham, U.K.: Edward Elgar): 291–310.

Arze, M. C. & B. W. Svensson (1997) "Developing of International Competitiveness in Industries and Individual Firms in Developing Countries: The Case of the Chilean Forest- Based Industry and the Chilean Engineering Firm Arze, Recine and Asociados," *International Journal of Production Economics* 52(1–2): 185–202.

Attaran, A. (2004) "How Do Patents and Economic Policies Affect Access to Essential Medicines in Developing Countries?" *Health Affairs* 23(3): 155–66.

Badri, M. W. (2000) "Operations Strategy, Environmental Uncertainty and Performance," *Omega-International Journal of Management Science* 28(2): 155–73.

Bajracharya, H. & W. Brouwer (1997) "A Narrative Approach to Science Teaching in Nepal," *International Journal of Science Education* 19(4): 429–46.

Balogun, E. E. (2002) "Education and Training in Space Science and Technology in Developing Countries," *Physica Scripta* T97: 24–27.

Banerjee, D. (1998) "Science, Technology and Economic Development in India: Analysis of Divergence in Historical Perspective," *Economic and Political Weekly* 33(20): 1199–1206.

Barlett, C. A. & S. Ghosal (2000) "Going Global: Lessons from Late Movers," *Harvard Business Review* 78(2): 132–42.

Barnes, J. & R. Kaplinsky (2000) "Globalization and the Death of the Local Firms? The Automobile Components Sector in South Africa," *Regional Studies* 34(9): 797–812.

Baskaran, Angathevar & Mammo Muchie (2006) *Bridging the Digital Divide: Innovation Systems for ICT in Brazil, China, India, Thailand, and Southern Africa* (London: Adonis & Abbey).

Benagiano, G. & E. Diczfalusy (1995) "Research on Human Reproduction and the United Nations," *South African Medical Journal* 85(5): 370–73.

Bhat, M. G. (1996) "Trade-Related Intellectual Property Rights to Biological Resources: Socioeconomic Implications for Developing Countries," *Ecological Economics* 19(3): 205–17.

Birdsall, N. (1996) "Public Spending on Higher Education in Developing Countries: Too Much or Too Little?" *Economics of Education Review* 15(4): 407–19.

Blomstrom, M. & A. Kokko (2001) "Foreign Direct Investment and Spillovers of Technology," *International Journal of Technology Management* 22(5–6): 435–54.

Bond, P. (1999) "Globalization, Pharmaceutical Pricing, and South African Health Policy: Managing Confrontation with U.S. Firms and Politicians," *International Journal of Health Services* 29(4): 765–92.

Borensztein, E., J. De Gregorio, & J. W. Lee (1998) "How Does Foreign Direct Investment Affect Economic Growth?" *Journal of International Economics* 45(1): 115–35.

Bowonder, B. (1998) "Industrialization and Economic Growth of India: Interactions of Indigenous and Foreign Technology," *International Journal of Technology Management* 15(6–7): 622–45.

Bowonder, B. (2001) "Globalization of R&D: The Indian Experience and Implications for Developing Countries," *Interdisciplinary Science Reviews* 26(3): 191–203.

Bowonder, B. & N. G. Satish (2003) "Is Economic Liberalisation Stimulating Innovation in India?" *Interdisciplinary Science Reviews* 28(1): 44–53.

Braczyk, H.-J., P. Cooke, & M. Heidenreich (eds) (2003) *Regional Innovation Systems: The Role of Governances in a Globalized World* (London: Routledge).

Brechin, S. R., P. R. Wilshusen, C. L. Fortwangler, & P. C. West (2002) "Beyond the Square Wheel: Toward a More Comprehensive Understanding of Biodiversity Conservation as Social and Political Process," *Society and Natural Resources* 15(1): 41–64.

Bronckers, M. (1994) "The Impact of TRIPS: Intellectual Property Protection in Developing Countries," *Common Market Law Review* 31(6): 1245–81.

Bronckers, M. & N. van den Broek (2005) "Financial Compensation in the WTO: Improving the Remedies of WTO Dispute Settlement," *Journal of International Economic Law* 8(1): 101–26.

Brown-Acquaye, H. (2001) "Each Is Necessary and None Is Redundant: The Need for Science in Developing Countries," *Science Education* 85(1): 68–70.

Byerlee, D. & K. Fischer (2002) "Accessing Modern Science: Policy and Institutional Options for Agricultural Biotechnology in Developing Countries," *World Development* 30(6): 931–48.

Callon, M. (1999) "Actor Network Theory: The Market Test," in J. Law & J. Hassard (eds), *Actor Network Theory and After* (Oxford: Blackwell): 181–95.

Campion, P. & W. Shrum (2004) "Gender and Science in Development: Women Scientists in Ghana, Kenya and India," *Science, Technology & Human Values* 29(4): 459–85.

Cassiolato, J. E., H. Lastres, & M. Maciel (eds) (2003) *Systems of Innovation and Development: Evidence from Brazil* (Cheltenham, U.K.: Edward Elgar).

Castells, M. (1996) *The Rise of the Network Society* (Oxford: Blackwell).

Chen, S. & M. Ravallion (2004) *How Have the World's Poorest Fared Since the Early 1980s?* (Washington, DC: World Bank).

Conceicao, P. & D. V. Gibson (2001) "Knowledge for Inclusive Development: The Challenge of Globally Integrated Learning and Implications for Science and Technology Policy," *Technological Forecasting and Social Change* 66(1): 1–29.

Correa, C. M. (1998) *Implementing the TRIPS Agreement: General Context and Implications for Developing Countries* (Penang, Malaysia: Third World Network).

Correa, C. M. (2000) *Intellectual Property Rights, the WTO and Developing Countries: The TRIPS Agreement and Policy Options* (London: Zed Books).

Cortright, J. (2001) *New Growth Theory, Technology and Learning: A Practitioner's Guide* (Washington, DC: U.S. Economic Development Administration).

Coutinho, M. (1999) "Ninety Years of Chagas' Disease: A Success Story at the Periphery," *Social Studies of Science* 29(4): 519–49.

Curtin, T.R.C. & E.A.S. Nelson (1999) "Economic and Health Efficiency of Education Funding Policy," *Social Science and Medicine* 48(11): 1599–611.

D'Costa, A. P. (1998) "Coping with Technology Divergence Policies and Strategies for India's Industrial Development," *Technological Forecasting and Social Change* 58(3): 271–83.

D'Costa, A. P. (2002) "Software Outsourcing and Development Policy Implications: An Indian Perspective," *International Journal of Technology Management* 24(7–8): 705–23.

De Laet, M. & A. Mol (2000) "The Zimbabwe Bush Pump: Mechanics of a Fluid Technology," *Social Studies of Science* 30(2): 225–63.

Dealmeida, P. R. (1995) "The Political Economy of Intellectual Property Protection: Technological Protectionism and Transfer of Revenue Among Nations," *International Journal of Technology Management* 10(2–3): 214–29.

Di Benedetto, C. A. & R. J. Calantone (2003) "International Technology Transfer Model and Exploratory Study in the People's Republic of China," *International Marketing Review* 20(4): 446–62.

Diwan, I. & M. Walton (1997) "How International Exchange, Technology and Institutions Affect Workers: An Introduction," *World Bank Economic Review* 11(1): 1–15.

Drahos, P. (2001) "Bits and Bips: Bilateralism in Intellectual Property," *Journal of World Intellectual Property* 4(6): 791–808.

Dubow, S. (ed) (2000) *Science and Society in Southern Africa* (Manchester, U.K.: Manchester University Press).

Edquist, C. (ed) (1997) *Systems of Innovation: Technologies, Institutions, and Organizations* (New York: Pinter).

Eicher, T. S. (1999) "Training, Adverse Selection and Appropriate Technology: Development and Growth in a Small Open Economy," *Journal of Economic Dynamics and Control* 23(5–6): 727–46.

Ernst, D., T. Ganiastos, & L. Mytelka (eds) (1998) *Technological Capabilities and Export Success: Lessons from East Asia* (London: Routledge).

Fleury, A. (1999) "The Changing Pattern of Operations Management in Developing Countries: The Case of Brazil," *International Journal of Operations and Production Management* 19(5–6): 552–64.

Freeman, C. (1982) *The Economics of Industrial Innovation* (Cambridge, MA: MIT Press).

Gao, S. J. & G. Xu (2001) "Learning, Combinative Capabilities and Innovation in Developing Countries: The Case of Video Compact Disk (VCD) and Agricultural Vehicles in China," *International Journal of Technology Management* 22(5–6): 568–82.

Ghose, A. K. (2003) *Jobs and Incomes in a Globalizing World* (Geneva: International Labor Office).

Gore, C. (2000) "The Rise and Fall of the Washington Consensus as a Paradigm for Developing Countries," *World Development* 28(5): 789–804.

Gorman, M. E. & M. M. Mehalik (2002) "Turning Good into Gold: A Comparative Study of Two Environmental Invention Networks," *Science, Technology & Human Values* 27(4): 499–529.

Grabowski, H. (2002) "Patents, Innovation and Access to New Pharmaceuticals," *Journal of International Economic Law* 5(4): 849–60.

Gray, B. V. (1999) "Science Education in the Developing World: Issues and Considerations," *Journal of Research in Science Teaching* 36(3): 261–68.

Guinnessy, P. (2003) "Academies Seek to Promote Scientific Excellence in Developing Countries," *Physics Today* 56(10): 32–35.

Gupta, N. & A. Sharma (2002) "Women Academic Scientists in India," *Social Studies of Science* 32(5–6): 901–15.

Hamilton, A. C. (2004) "Medicinal Plants, Conservation and Livelihoods," *Biodiversity and Conservation* 13(8): 1477–517.

Hecht, G. (2002) "Rupture-Talk in the Nuclear Age: Conjugating Colonial Power in Africa," *Social Studies of Science* 32(5–6): 691–727.

Hee, C. H. & K. J. Soo (1997) "Transition of the Government Role in Research and Development in Developing Countries: R&D and Human Capital," *International Journal of Technology Management* 13(7–8): 729–43.

Heyneman, S. P. (2003) "The History and Problems in the Making of Education Policy at the World Bank 1960–2000," *International Journal of Educational Development* 23(3): 315–37.

Hipkin, I. (2004) "Determining Technology Strategy in Developing Countries," *Omega-International Journal of Management Science* 32(3): 245–60.

Hoekman, B. M., P. English, & A. Mattoo (eds) (2002) *Development, Trade, and the WTO: A Handbook* (Washington, DC: World Bank).

Hsiao, F. B., W. L. Guan, C. P. Chou, & C. T. Su (1997) "Establishing a Web Server System for Space: Major Student Education and Training in Microsatellite Technology Via Internet," in S. C. Chakravarty, J.-L. Fellows, K. Kasturirangan, & M. J. Rycroft (eds), *Problems of Space Science Education and the Role of Teachers* (Oxford: Pergamon Press): 1365–73.

Huhtala, A., J. J. Bouma, M. Bennett, & D. Savage (2003) "Human Resource Development Initiatives to Promote Sustainable Investment," *Journal of Cleaner Production* 11(6): 677–81.

Humphrey, J. & H. Schmitz (1996) "The Triple C Approach to Local Industrial Policy," *World Development* 24(12): 1859–77.

Hunter, W. & D. S. Brown (2000) "World Bank Directives, Domestic Interests, and the Politics of Human Capital Development," *Comparative Political Studies* 33(1): 113–43.

Jegede, O. J. (1997) "School Science and the Development of Scientific Culture: A Review of Contemporary Science Education in Africa," *International Journal of Science Education* 19(1): 1–20.

Johnson, B. & B.-A. Lundvall (2003) "Promoting Innovation Systems as a Response to the Globalizing Learning Economy," in J. E. Cassiolato & M. Maciel (eds), *Systems of Innovation and Development: Evidence from Brazil* (Cheltenham, U.K.: Edward Elgar): 141–84.

Kaplinsky, R. & J. Barnes (2000) "Globalization and the Death of the Local Firms? The Automobile Components Sector in South Africa," *Regional Studies* 34(9): 797–812.

Karnik, N. (2001) "Locating HIV/AIDS and India: Cautionary Notes on the Globalization of Categories," *Science, Technology & Human Values* 26(3): 322–48.

Kasturirangan, K. (1997) "Relevance and Challenges of Space Science Education in Developing Countries," in S. C. Chakravarty, J.-L. Fellows, K. Kasturirangan, & M. J. Rycroft (eds), *Problems of Space Science Education and the Role of Teachers* (Oxford: Pergamon Press): 1329–33.

Kate, K. T. & S. A. Laird (1999) *The Commercial Use of Biodiversity: Access to Genetic Resources and Benefit-Sharing* (London: Earthscan).

Keller, W. (1996) "Absorptive Capacity: On the Creation and Acquisition of Technology in Development," *Journal of Development Economics* 49(1): 199–227.

Khan, M. U. (2001) "Indicators of Techno-Management Capability Building in Indian Computer Firms," *Journal of Scientific and Industrial Research* 60(9): 717–23.

Kokko, A. & M. Zejan (2001) "Trade Regimes and Spillover Effects of FDI: Evidence from Uruguay," *Weltwirtschaftliches Archiv-Review of World Economics* 137(1): 124–49.

Kremer, M. (2002) "Pharmaceuticals and the Developing World," *Journal of Economic Perspectives* 16(4): 67–90.

Kuruvilla, S., C. L. Erickson, & A. Hwang (2002) "An Assessment of Singapore Skills Development System: Does It Constitute a Viable Model for Other Developing Countries?" 30(8): 1461–76.

Kyle, W. C. (1999) "Science Education in Developing Countries: Challenging First World Hegemony in a Global Context," *Journal of Research in Science Teaching* 36(3): 255–60.

Lall, S. (1995) "Employment and Foreign Investment: Policy Options for Developing Countries," *International Labour Review* 134(4–5): 521–40.

Lall, S. (2002) "Linking FDI and Technology Development for Capcity Building and Strategic Competitiveness," *Transnational Corporations* 11(3): 39–88.

Lang, K. R. (2004) "An Education Curriculum for Space Science in Developing Countries," *Space Policy* 20(4): 297–302.

Lanjouw, J. O. & I. M. Cockburn (2001) "New Pills for Poor People? Empirical Evidence after GATT," *World Development* 29(2): 265–89.

Latour, B. (1987) *Science in Action: How to Follow Scientists and Engineers Through Society* (Cambridge, MA: Harvard University Press).

Lei, S. H.-L. (1999) "From *Changshan* to a New Anti-Malarial Drug," *Social Studies of Science* 29(3): 323–58.

Lewin, K. M. (2000a) *Linking Science Education to Labor Markets: Issues and Strategies*. Available at: http://www1.worldbank.org/education/scied/documents/Lewin/labor.pdf.

Lewin, K. M. (2000b) *Mapping Science Education Policies in Developing Countries*. Available at: http://www1.worldbank.org/education/scied/documents/Lewin/Mapping.pdf.

Lomnitz, L. A. & L. Cházaro (1999) "Basic, Applied and Technological Research: Computer Science and Applied Mathematics at the National Autonomous University of Mexico," *Social Studies of Science* 29(1): 113–34.

Lundvall, B.-Å. (ed) (1992) *National Systems of Innovation: Towards a Theory of Innovation and Interactive Learning* (London: Pinter).

Macilwain, C. (1998) "When Rhetoric Hits Reality in Debate on Bioprospecting," *Nature* 392(6676): 535–40.

Mahoney, R. T., A. Pablos-Mendez, & S. Ramachandran (2004) "The Introduction of New Vaccines into Developing Countries: III. The Role of Intellectual Property," *Vaccine* 22(5–6): 786–92.

Malerba, F. (2004) *Sectoral Systems of Innovation: Concepts, Issues and Analyses of Six Major Sectors in Europe* (Cambridge: Cambridge University Press).

Maskus, K. E. & J. H. Reichman (2004) "The Globalization of Private Knowledge Goods and the Privatization of Global Public Goods," *Journal of International Economic Law* 7(2): 279–320.

Mingat, A. (1998) "The Strategy Used by High-Performing Asian Economies in Education: Some Lessons for Developing Countries," *World Development* 26(4): 695–715.

Mitchell, G. R. (1999) "Global Technology Policies for Economic Growth," *Technological Forecasting and Social Change* 60(3): 205–14.

Muchie, M., P. Gammeltoft, & B.-A. Lundvall (2004) *Putting Africa First: The Making of an African Innovation System* (Aalborg, Denmark: Aalborg University Press).

Nelson, R. R. (ed) (1993) *National Innovation Systems: A Comparative Analysis* (New York: Oxford University Press).

Nelson, R. R. & S. G. Winter (1982) *An Evolutionary Theory of Economic Change* (Cambridge, MA: Harvard University Press).

Okamoto, Y. (1999) "Multinationals, Production Efficiency, and Spillover Effects: The Case of the U.S. Auto Parts Industry," *Weltwirtschaftliches Archiv-Review of World Economics* 135(2): 241–46.

Organisation of Economic Co-operation and Development (OECD) (2002) *Dynamising National Innovation Systems*. Available at: www.oecd.org.

Oudshoorn, N. (1997) "From Population Control Politics to Chemicals: The WHO as an Intermediary Organization in Contraceptive Development," *Social Studies of Science* 27(1): 41–72.

Perez-Casas, C., E. Herranz, & N. Ford (2001) "Pricing of Drugs and Donations: Options for Sustainable Equity Pricing," *Tropical Medicine and International Health* 6(11): 960–64.

Phadke, R. (2002) "Assessing Water Scarcity and Watershed Development in Maharashtra, India," *Science, Technology & Human Values* 27(2): 236–61.

Porter, M. (1990) *The Competitive Advantage of Nations* (New York: Free Press).

Posey, D. A. & G. Dutfield (1996) *Beyond Intellectual Property Rights: Towards Traditional Resource Rights for Indigenous Peoples and Local Communities* (Ottawa, Canada: International Development Research Centre).

Reddy, P. (1997) "New Trends in Globalization of Corporate R&D and Implications for Innovation Capability in Host Countries: A Survey from India," *World Development* 25(11): 1821–37.

Rodrigo, G. C. (2001) *Technology, Economic Growth and Crises in East Asia* (Northampton, U.K.: Edward Elgar).

Romer, P. M. (1990) "Endogenous Technological Change," *Journal of Political Economy* 98: S71–S102.

Sachs, J. (2005) *The End of Poverty: Economic Possibilities for Our Time* (New York: Penguin Press).

Sane, K. V. (1999) "Cost-Effective Science Education in the 21st Century: The Role of Educational Technology," *Pure and Applied Chemistry* 71(6): 999–1006.

Sassen, S. (ed) (2002) *Global Networks: Linked Cities* (New York: Routledge).

Scherer, F. M. & J. Watal (2002) "Post-TRIPS Options for Access to Patented Medicines in Developing Nations," *Journal of International Economic Law* 5(4): 913–39.

Schuklenk, U. & R. E. Ashcroft (2002) "Affordable Access to Essential Medication in Developing Countries: Conflicts Between Ethical and Economic Imperatives," *Journal of Medicine and Philosophy* 27(2): 179–95.

Schumpeter, J. (1942) *Capitalism, Socialism, and Democracy* (New York: Harper and Row).

Sell, S. K. & A. Prakash (2004) "Using Ideas Strategically: The Contest Between Business and NGO Networks in Intellectual Property Rights," *International Studies Quarterly* 48(1): 143–75.

Sen, A. (2000) *Development as Freedom* (New York: Anchor Books).

Shaffer, G. (2004) "Recognizing Public Goods in WTO Dispute Settlement: Who Participates? Who Decides?" in K. Maskus & J. Reichman (eds), *International Public Goods and Transfer of Technology Under a Globalized Intellectual Property Regime* (Cambridge: Cambridge University Press): 884–908.

Shrum, W. (2000) "Science and Story in Development: The Emergence of Non-Governmental Organizations in Agricultural Research," *Social Studies of Science* 30(1): 95–124.

Shrum, W. (2005) "Reagency of the Internet: Or How I Became a Guest for Science," *Social Studies of Science* 35(5): 723–54.

Shrum, W. & Yehuda Shenhav (1994) "Science and Technology in Less-Developed Countries," in Sheila Jasanoff, Gerald Markle, James Petersen, & Trevor Pinch (eds) *Handbook of Science and Technology Studies* (Thousand Oaks, CA: Sage): 627–51.

Sikka, P. (1997) "Statistical Profile of Science and Technology in India and Brazil," *Scientometrics* 39(2): 185–95.

Sjoholm, F. (1999) "Technology Gap, Competition and Spillovers from Direct Foreign Investment: Evidence from Establishment Data," *Journal of Development Studies* 36(1): 53–57.

Snodgrass, D. R. (1998) "Education in Korea and Malaysia," in H. S. Rowen (ed), *Behind East Asian Growth: The Political and Social Foundations of Prosperity* (London: Routledge): 165–84.

Solovey, M. (2001) "Project Camelot and the 1960s Epistemological Revolution," *Social Studies of Science* 31(2): 171–206.

Solow, R. M. (1956) "A Contribution to the Theory of Economic Growth," *Quarterly Journal of Economics* 70: 65–94.

Solow, R. M. (1957) "Technical Change and the Aggregate Production Function," *Review of Economics and Statistics* 39: 312–20.

South Centre (1997) *The TRIPS Agreement: A Guide for the South* (Geneva: South Centre).

Srinivasan, C. S. (2003) "Concentration in Ownership of Plant Variety Rights: Some Implications for Developing Countries," *Food Policy* 28(5–6): 519–46.

Srinivasan, C. S. (2004) "Plant Variety Protection, Innovation, and Transferability: Some Empirical Evidence," *Review of Agricultural Economics* 26(4): 445–71.

Srinivasan, C. S. & C. Thirtle (2003) "Potential Economic Impacts of Terminator Technologies: Policy Implications for Developing Countries," *Environment and Development Economics* 8(1): 187–205.

Stiglitz, J. (2002) *Globalization and Its Discontents* (New York: W. W. Norton).

Subramanian, A. (1995) "Putting Some Numbers on the TRIPS Pharmaceutical Debate," *International Journal of Technology Management* 10(2–3): 252–68.

Subramanian, A. & J. Watal (2000) "Can TRIPS Serve as an Enforcement Device for Developing Countries in the WTO?" *Journal of International Economic Law* 3(3): 403–16.

Sutz, J. (2003) "Inequality and University Research Agendas in Latin America," *Science, Technology & Human Values* 28(1): 52–68.

Tan, L.W.H. & R. Subramaniam (2003) "Science and Technology Centers as Agents for Promoting Science Culture in Developing Countries," *International Journal of Technology Management* 25(3): 413–26.

Task Force on HIV/AIDS (2004) *Report of Task Force on HIV/AIDS, Malaria, TB, and Access to Essential Medicines* (4 parts) (New York: United Nations).

Thompson, E. R. (2002) "Clustering of Foreign Direct Investment and Enhanced Technology Transfer: Evidence from Hong Kong Garment Firms in China," *World Development* 30(4): 837–89.

Timmermans, K. (2003) "Intellectual Property Rights and Traditional Medicine: Policy Dilemmas at the Interface," *Social Science and Medicine* 57(4): 745–56.

United Nations Conference on Trade and Development (UNCTAD) (1996) *The TRIPS Agreement and Developing Countries* (New York: United Nations).

United Nations Educational, Scientific, and Cultural Organization (UNESCO) (2004) *Science and Technology Education Programme 2004–2005*. Available at: http://portal.unesco.org/education/en/file_download.php/d38be6780ed77b8cc338d12aa8139854STE+Programme+2004–2005.pdf.

Velho, L. & O. Pessoa (1998) "The Decision-Making Process in the Construction of the Synchrotron Light National Laboratory in Brazil," *Social Studies of Science* 28(2): 195–219.

Verran, H. (2001) *Science and an African Logic* (Chicago: University of Chicago Press).

Verran, H. (2002) "A Postcolonial Moment in Science Studies," *Social Studies of Science* 32(5–6): 729–62.

Vlaardingerbroek, B. (1998) "The 'Myth of Greater Access' in Agrarian LDC Education and Science Education: An Alternative Conceptual Framework," *International Journal of Educational Development* 18(1): 63–71.

Vosti, S. A. & T. Reardon (1997) *Sustainability, Growth, and Poverty Alleviation: A Policy and Agroecological Perspective: Food Policy Statement No. 25* (Washington, DC: International Food Policy Research Institute).

Wagner, J. L. & E. McCarthy (2004) "International Differences in Drug Prices," *Annual Review of Public Health* 25(1): 475–95.

Wang, J. & J. X. Wang (1998) "An Analysis of New Tech Agglomeration in Beijing: A New Industrial District in the Making," *Environmental and Planning A* 30(4): 681–701.

Wang, T. & R. Pollard (2002) "Selecting a Technical Strategy for High-Tech Enterprises in Developing Countries: A Case Study," *International Journal of Technology Management* 24(5–6): 648–55.

Watal, J. (2000) "Pharmaceutical Patent, Prices and Welfare Losses: Policy Options for India under the WTO TRIPS Agreement," *World Economy* 23(5): 733–52.

Watson, R., M. Crawford, & S. Farley (2003) *Strategic Approaches to Science and Technology in Development*. Available at: http://www-wds.worldbank.org/servlet/WDSContentServer/WDSP/IB/2003/05/23/000094946_03051404103334/Rendered/PDF/multi0page.pdf.

Worthington, R. (1993) "Introduction: Science and Technology as a Global System," *Science, Technology & Human Values* 18(2): 176–85.

Xie, W. & S. White (2004) "Sequential Learning in a Chinese Spin-Off: The Case of Lenovo Group Limited," *R&D Management* 34(4): 407–22.

Xie, W. & G. S. Wu (2003) "Difference Between Learning Processes in Small Tigers and Large Dragons: Learning Processes of Two Color TV (CTV) Firms Within China," *Research Policy* 32(8): 1463–79.

Yoshimatsu, H. (2000) "The Role of Government in Jump-Starting Industrialization in East Asia: The Case of Automobile Development in China and Malaysia," *Issues and Studies* 36(4): 166–99.

Zahur, R., A. C. Barton, & B. R. Upadhyay (2002) "Science Education for Empowerment and Social Change: A Case Study of a Teacher Educator in Urban Pakistan," *International Journal of Science Education* 24(9): 899–917.

V Emergent Technosciences

Judy Wajcman

Section V of the Handbook covers some of the new trends and interests that have emerged and are now capturing the attention of STS scholars worldwide. As more and more of contemporary life is mediated by technology, STS scholars have touched on nearly every aspect of human activity. Indeed, if the ever-increasing numbers attending STS conferences (such as the Society for Social Studies of Science [4S] and the European Association for the Study of Science and Technology [EASST]) are an indication of interest, then the field is certainly flourishing and extending its range. We cannot possibly hope to cover all the ground that our community has traversed, so we present here a selective range of topics that at least represent the main trends.

A growing interest in medical technologies—technologies for the body—is a most striking feature of the recent history of STS. Three out of seven chapters in this concluding section are in this area. And, notably, neither industrial machinery nor workplace technologies are distinct subjects for chapters. While these areas were formative for STS, newer developments in technoscience have naturally generated more interest, enthusiasm, and anxiety. This is not an entirely new area for STS: for example, feminist scholarship on medical technologies goes back to at least the 1970s, with critiques of male medical expertise and the extensive literature on new reproductive techniques. As Donna Haraway's image of the cyborg so presciently captured some years ago, biotechnologies have the potential to transform the relations between the self, the body, and machines. The ability to modify "nature," and the implications of this for rethinking our standard cultural categories of nature and culture, is a central theme in the first four chapters of this section.

The first three chapters, then, deal with the new fields of biomedical technoscience. As Adam Hedgecoe and Paul Martin remark in the first chapter, in the wake of the Human Genome Project there is a widespread belief that new genomic knowledge is transforming medicine and STS scholars have prioritized these novel technological developments. In the past 15 years, the expansion of genomic knowledge and technologies has brought many disciplines closer to STS. For example, medical anthropologists, sociologists, and bioethicists have been drawn to STS perspectives and concepts to further their own research interests. One theme they examine, which is very pertinent to all the contributions in this section of the Handbook, is that of the "dynamics of expectations," through which hope, promises, and hype help construct "the future" as a resource to shape innovation and sociotechnical change in the present. In other words, such expectations are performative in the way they help mobilize the future into the present and become embedded in postgenomic biotechnologies. An STS perspective casts a skeptical eye on revolutionary claims made for such developments and reflects on how social science itself participates in the making of sociotechnical futures. As they say, there is still a very large gap between scientific claims about dramatic changes in biomedicine and what actually happens in clinical practice. Finally, they argue that empirically grounded research that is aware of its political implications is crucial in this area, where issues of governance and regulation are hotly contested. Future research will hopefully provide a better understanding of why expectations about genomics have been realized in some areas but not others.

In Linda Hogle's overview, medical technologies are seen to include not only various devices, instruments, therapies, and procedures used for diagnostic, therapeutic, rehabilitative, preventive, or experimental purposes, but also the informational, organizational, economic, and political systems in which they exist. This broad perspective reveals the operation of power relations in determining how technologies will be used and by whom. Drawing on the now quite substantial STS studies, Hogle illustrates how medical technologies affect and are affected by identity and subjectivity, standardization processes, local meanings and use, and global exchange systems. For example, studies of imaging and information technologies for clinical decision making demonstrate how truth claims become established as facts. In this way, technologies can be thought to actively intervene in the situations in which they are used, for example, to stabilize notions of normality. A crucial issue here has been the extent to which bionics and prosthetics, for example, potentially shift received cultural notions of the body and produce new identities and subjectivities.

If one thing is clear, it is that twenty-first century medical techniques, forms of knowledge, and practices will continue to be a central concern of STS scholars.

Margaret Lock's chapter takes up similar themes of how biomedical technologies are shifting the boundaries of the "natural" body and creating news forms of biosociality and subjectivity. She approaches these issues via a detailed examination of two particular biomedical technologies: organ procurement and transplantation, and genetic testing for single gene disorders and complex diseases. In the case of organ transplants, the biologies of self and other are hybridized, and this raises difficult issues for identity, and poses unique questions about the ownership of the human body and body parts. Genetic test results potentially create molecular genealogies that destabilize kinship and ethnic ties based on a shared heredity. In both cases, the reconfiguration of conventional boundaries challenges normative expectations about embodiment, identity, and relationships of individuals to familial and other social groups. Concerns about genetic determinism, the ever-growing tendency to distinguish between people on the basis of their genetic makeup (reinforcing racial and other forms of inequality), has been a troubling issue here, although some authors stress the positive effects

of these new group identities. Ethnographic studies make it clear that these new biomedical technologies are producing "hybrid, postmodern bodies, fluid subjectivities,
and shifting human collectivities" that will provide much material for studies in the
future.

The significance of environmental topics for the STS community has grown with
startling rapidity in the last decade. Two major areas of lively debate are those concerned with humanly induced climate change and genetically modified crops. As
Steven Yearley notes, these areas are critical for STS not just as research topics in themselves but because studying them affords key insights into the status of "the natural"
in advanced modernity. Bringing the lens of STS to bear on these environmental issues,
he argues, enables us to see that the very business of "knowing nature" shapes the
knowledge that results. In turn, this process decisively influences how effective or not
such knowledge is in other public contexts. STS literature on climate change, for
example, illustrates the complexity of the relationship between knowledge and policy
advising. Risk and safety assessment, and the interpretation of scientific evidence in
an international perspective, has also been key to debates about genetically modified
crops and food. It is a particularly interesting case for STS scholars because public resistance has resulted in governments, particularly in Europe, initiating various public
consultations over the introduction of the technology. As such, it represents the paramount exercise in "technology assessment" of the last decade.

STS research on environmental topics, then, indicates how contemporary societies
have responded to the challenge of knowing nature authoritatively. Such research is
at the forefront of STS engagement with policy, social theory, and social change.

One of the most exciting developments to have come out of STS in recent years is
the development of the social studies of finance. The preeminent and increasingly
influential position occupied by financial institutions in modern societies makes this
a topic of great significance. Alex Preda's chapter recounts how since the mid-1990s
scholars have become interested in applying STS conceptualizations of the link
between scientific knowledge and practical action to the way financial information in
trading rooms is standardized and established. One of the lasting contributions of this
research is to show that "price data—regarded as unproblematic both by financial economics and by economic sociology—is constituted in a web of interactions involving
both human actors and technological artifacts." Some of this research has directly
applied the fieldwork tradition of laboratory studies to the trading room, while
another branch has been more directly influenced by actor-network theory, exploring
how economic theory itself has a performative character. The social studies of finance,
centered on researching one of the most important actors in the world economy, certainly promises to continue to be a major field of engagement for STS scholars in the
future.

Our penultimate chapter deals with the proliferation of information and communication technologies that now mediate most aspects of the way we live. The Internet and, even more so, mobile phones are in widespread use throughout the world.
According to Pablo Boczkowski and Leah Lievrouw, while these technologies have

been central to the development of communication studies, it is only in the last decade or so that media and information technologies have become a major research focus in STS. There are now many important bridges between these once-distinct disciplines. The authors argue that many scholars in communication studies have been drawn to STS concepts, such as interpretative flexibility and the social shaping or social construction of technology, in order to theorize the distinctive sociotechnical character of media and information technologies. This has been particularly apt as various problematics, such as notions of causality, the process of technological development, and the social consequences of technological change, have preoccupied scholars in both STS and communication studies. For example, while STS initially tended to focus on the processes of design and production, communication studies has been more concerned with consumption. Both fields are now interrogating the links between production and consumption, developing concepts that connect these two spheres. As the authors conclude, the distinctive domain of media and information technologies, that are at once "cultural material and material culture," will be a much richer field for the increasingly fruitful dialogue now taking place between STS and communication studies.

Nanotechnology is an emergent field concerned with phenomena at very small scales that is taking form at the interstices of several fields of science and engineering. In the Handbook's closing chapter, Daniel Barben, Erik Fisher, Cynthia Selin, and David H. Guston explore the distinctive challenges for research, assessment, and policy posed by this new and complex endeavor. The approach they describe centers around real-time analysis of the technology and "anticipatory governance" of the direction and practice of research. Amalgamating STS scholarship, stakeholder groups, policy makers, and the public with the communities of scientists and engineers will fuse the "is" and the "ought" of STS far upstream in the R&D process. The consequences of this convergence are difficult to anticipate because the research itself will reshape the object of analysis and because nanotechnology is continually changing on its own and through interaction with biotechnology, information technology, and cognitive science. Whatever the uncertainties, nanotechnology may exemplify the blended intellectual, political, and ethical challenges confronting STS over the next several years.

32 Genomics, STS, and the Making of Sociotechnical Futures

Adam M. Hedgecoe and Paul A. Martin

We can now see a future where the doctor will swab a few cells from inside your cheek, put them into a DNA-sequencing machine and a computer will spit out a complete reading of your unique genetic makeup—all 30,000 or so genes that make you who you are.[1]

In the wake of the Human Genome Project (HGP), governments throughout the developed world are promoting biotechnology in general and genomics in particular as a key part of the New Economy. For example, the 2003 U.K. genetics white paper "Our Inheritance, Our Future—Realising the Potential of Genetics in the NHS" explicitly placed increased genetic knowledge within the framework of economic development (Department of Health, 2003). At the same time as offering hope of economic prosperity, advances in genetics and genomics have provoked widespread social concerns, which have been institutionalized in new regulatory regimes and research programs. Over the past 15 years the expansion of knowledge and technologies that can be classed as "genomic" has gone hand-in-hand with the growth of literature commenting on these developments. Thus, bioethics, anthropology, medical sociology, and STS (among others) have produced extensive analyses and critiques, challenging the way these technologies have developed, the assumptions that underpin their use and the impact they may have on both individuals and societies.

The idea that new genomic knowledge is in the process of transforming biomedicine is now widely held and becoming increasingly taken for granted; see the previously mentioned genetics white paper as an example. The transformative power of genomics is a dominant view in debates around the biomedical science and plays an important role in informing and shaping social science research agendas. For example, the United Kingdom's social science funding council, the ESRC, is funding three genomics centers the justification of which is explicitly transformative: "Recent leaps in the scientific study of genes and our growing ability to manipulate the genomes of plants, animals and humans far outstrips our understanding of the social and economic consequences of genomics" (Anon., 2005; see also Harvey et al., 2002).

However, there are increasing doubts about the extent to which many of these supposedly "revolutionary" technical and social changes are actually happening in practice (Nightingale & Martin, 2004). To date, there has been relatively little progress in translating scientific advances into the clinic, and the scale of social transformation remains limited to particular niches. This lack of progress has important implications for the study of genomics and raises difficult questions for the social sciences. In this chapter, we want to take stock of how STS scholars have analyzed and understood the development and impact of genomics; what has STS taught us about these changes, and what have we learned about STS in the process? In doing this we want to draw on recent work on the dynamics of expectations to reflect on the role of social science in the making of sociotechnical futures.

SCOPE OF OUR ANALYSIS

When attempting to tackle these questions we are faced with two crucial and potentially divisive questions: what is genomics? And what is STS? Genomics seems to have mutated (pun intended) from Thomas Roderick's 1986 definition of "the scientific discipline of mapping, sequencing, and analysing genomes" (Hieter & Boguski, 1997) to anything involving industrial scale sequencing of genetic material. Our answer to this question is largely chronological: our time frame is the decade and a half since the start of the HGP in the late 1980s, although we also include a brief discussion of earlier work on biotechnology and genetics to provide a context for the analysis. Technically, our focus is on the science and technologies involved in the elaboration and investigation of the human genome, and we have deliberately excluded a number of related technologies, such as reproductive technologies, cloning, and stem cells. That said, to discuss postgenomic genetic testing without drawing on the extensive literature on "traditional" monogenic conditions, or technologies with considerable histories, such as amniocentesis, seems too strict. Thus, our focus is a pragmatic rather than an ideological one.

In terms of defining the scope of STS work, some responses to the first STS *Handbook* suggested that ours is an intellectual community that resists classification and the drawing of boundaries. While it may be straightforward to exclude some areas of the literature as clearly being "bioethics" (nonempirical statements about the application of technology) or legal scholarship, what are we to make of anthropology? A number of anthropologists have made significant contributions to the field while at the same time resisting close identification with STS. Our solution has been to work on a case-by-case basis, without drawing wider conclusions about a particular author's disciplinary affiliations, using on-line databases and manual searches to identify papers and books that might fit our criteria. In defining work as STS, we looked for a number of (hopefully uncontroversial) assumptions: an empirical focus, an interest in the content of scientific and technological developments rather than just their application, an awareness of the political implications of such work, and how it challenges conventional authority. For some authors, STS approaches are the only way to get to grips with the threat genomics poses to sociology's coherence as a discipline (Delanty, 2002), whereas for others STS barely registers as a way of exploring these issues (Willis,

1998). We doubt this chapter will make the first case, but we hope it will convincingly refute the second.

UNDERSTANDING THE ROLE OF EXPECTATIONS IN THE DEVELOPMENT OF GENOMICS

Recent work on the sociology of expectations (Brown et al., 2000) has explored the dynamic patterns by which hope, promises, and hype help construct "the future" as a resource to shape innovation and sociotechnical change in the present. Research has explored how expectations of technology are perfomative in mobilizing the future into the present through particular cultural metaphors, narrative scripts, and promissory agendas (Van Lente, 1993; Michael, 2000; Wyatt, 2000), but it has also examined the way in which expectations have become embedded and embodied in a range of sociotechnical artifacts, including postgenomic biotechnologies (Hedgecoe & Martin, 2003). This work had shown that expectations can be at their greatest and most technologically deterministic during the early stages of the development of a new technology (Akrich, 1992).

The development of biotechnology and genomics can be understood as being part of the construction of a sociotechnical regime centered on new biological knowledge of the human body. High expectations are required to mobilize the large number of actors and considerable resources needed to bring new therapies, diagnostics, clinical practices, industries, and governance regimes into being, given the long lead times and major social, cultural, organizational, political, and cultural transformations that may be required. Thus far, the construction of heterogeneous sociotechnical networks around emerging technologies such as genetic testing has been uneven, with some changes occurring rapidly and other elements of the emerging regime proving intractable. Rather than presenting a full discussion of this emerging approach to science and technology, this chapter uses concepts from the dynamics of expectations literature to reflexively explore what STS studies of genomics tells us about the possible role of social science in the creation of particular sociotechnical futures.

CONTEXTUAL VERSUS TRANSFORMATIONAL APPROACHES

In the process of conducting our review of the literature for this chapter, it became clear that the work in this area fits into two broad styles of thinking about genomics (and before that, genetics and biotechnology), which we have labeled the contextual and the transformational. Although these two approaches can be clearly distinguished, they are better thought of as "voices" or "registers" than as rigid forms of discourse. The same author may think in a transformational way in one article and contextually in another. Our aim is not to dichotomize the STS literature on genomics into two competing camps but rather to explore the way in which we have collectively described and analyzed these technological developments.

The transformational approach to genomics sees scientific and technological developments as somehow revolutionary, as encapsulating a break from what has gone before. Such changes are proposed at both a technical and a social level. Although such a stance can obviously be taken by supporters of genomics, this need not be the case. Certainly a key feature of many early critiques of biotechnology that is still present in much contemporary work is an oppositional stance based on fear of the potential power of these emerging technologies, this sense of power being shared with the advocates of the technology. The transformational position heightens the expectation that biotechnology and genomics will become widespread and pervasive technologies, summed up in the view that "It is evident that direct, personal encounters with genetic practice are becoming inescapable" and that "The meanings of citizenship, kinship, public participation, personal identity, social belonging and health are all being reworked in the light of genetic technologies" (Waldby, 2001: 781).

In contrast to this view, others are concerned that "there is a danger that this focus on transformations is resulting in less attention being paid to the static and reactionary aspects of the 'new' genetics and its wider social context, particularly the ways in which autonomy, participation and uncertainty might be undermined by new developments in genetic testing, screening and public consultation" (Kerr, 2003b: 44). This point of view questions the revolutionary nature of genomics, both in technical and social terms. It highlights the historical roots that underpin modern techniques and the darker elements, such as eugenics and racism, that still lurk in the discourse and actions of modern genomics.

From within the transformational position, the powerful effects of genomics on society need not be seen as negative. For example, Novas and Rose (2000: 507) argue it is possible that

Genetic forms of thought have become intertwined within ethical problematizations of how to conduct ones' life, formulate objectives and plan for the future in relation to genetic risk. In these life strategies, genetic forms of personhood make productive alliances and combinations with forms of selfhood that construct the subject as autonomous, prudent, responsible and self-actualizing.

Such transformations may extend beyond the realm of science and society and into social science itself with the concept of biomedicalization, a core theme of which is the increased explanatory role offered to genomics in the modern world, offering "a bridging framework for new conversations across speciality divides within sociology and more broadly across disciplinary divides within the social sciences" (Clarke et al., 2003: 184). From the contextual position, this kind of work, however nuanced, has "bought into" claims about the originality (both moral and scientific) of postgenomic research. For the contextualist, there is a need to root the HGP and its subsequent technologies in longer term historical developments; from this perspective, it is far harder to draw a line between the "old" eugenics and the "new" genetics. Of course it is perfectly possible for those writing from the transformational position to be aware of the reflexive impact of their discourse: "in the sense that any advertising is good advertising, our project here cannot help but constitute and promote biomedicalization" (Clarke et al., 2003: 184).

In addition, thinking in terms of the contextual and transformational registers introduces a welcome element of reflexivity into our consideration of expectations, since these particular ways of thinking about genomics tend to encourage specific expectations about that technology's future development. As Michael Fortun (2005: 157) notes when describing his previous research written in a contextual voice, "I read the future then pretty much in terms of reproducing the past: more genetic reductionism, more biologization of human conditions better understood as "social" or at least not simply "genetic", more stigmatisation (or worse) of the ab-normal." Fortun's dissatisfaction with his previous position (and with much STS writing on genomics as a whole) leads him to embrace the transformative nature of both genomic science and the social and ethical discussions taking place about it in the scientific community. Although this is a refreshingly reflexive contribution to the debate over the normative aspects of STS and its relationship with bioethics, to embrace the transformational to the exclusion of the contextual is to eschew an important theme in STS scholarship.

Our point is not to suggest that there is a right or wrong approach to these issues. Indeed, one of our aims is to show that the strength of STS's work on genomics lies to some extent in the discipline's ability to contain these two different ways of looking at the same technology. It may be tempting to interpret our description of the transformational approach as a normative one, as a criticism of those STSers who "buy into" the revolutionary claims of the promoters of genomics. Perhaps it is. If so, it is worth noting that as authors, the majority of our own work has been within the transformational register, focusing on the social changes brought about by new genetic technologies, as well as the way in which such novel, transformational technologies emerge. If such work is the mark of overly credulous dupes, then, with Dorothy Nelkin and Paul Rabinow as "co-transformnationalists" (to pick just two), at least we are in good company.

STS AND CONSTRUCTION OF THE GENOMIC FUTURE

To practically undertake this analysis, the STS literature on genomics will be broken down under a number of headings related to specific domains, institutions, and groups of actors, including the following:

- Production of new scientific knowledge (scientists)
- Application of new knowledge in the clinic (clinicians, researchers, and practitioners)
- Commodification and commercial exploitation of genomic knowledge (industry)
- Representation and culture of genomics (media, publics)
- Creation of new genomic identities (patients, patient groups, citizens)
- Governance of genomics (government, regulators)

It must be stressed that these domains are overlapping and interconnected but represent the social locations and processes where genomic science and technology are being created and used, including the laboratory, the firm, the clinic, the media, and the self. In each of these domains we ask the following questions of the STS literature: What sociotechnical expectations and transformations are being associated with the rise of genomics? What is seen as new and specific to genomics, and what is believed to be the extent of sociotechnical change? Our analysis cannot be comprehensive given the space available but will cover the main work in each of these areas. In doing so we wish to highlight the way in which different authors have approached ideas about the future and have in themselves (re)produced particular expectations of genomics.

PRODUCTION OF NEW SCIENTIFIC KNOWLEDGE

It should not be surprising that a major focus of STS inquiry has been the production of new scientific knowledge nor, given the focus of this chapter, that the central site of this knowledge production is the Human Genome Project ("biology's moon shot") and the technologies that surround it. In terms of the way in which we structure our analysis, the HGP is a good place to start, since the tensions between the transformational and the contextual strands of STS thinking are so clear.

Of course, the most obvious way to view the HGP is to opt out of this dichotomy and see it as an interesting example of the social organization of science, an approach adopted by Brian Balmer in his exploration of the policy and mapping strategies employed by different countries (Balmer, 1994, 1995, 1996a) and the way in which the HGP itself becomes a boundary object between different social worlds (Balmer, 1996b). Although this work tends to focus on the situation in the United Kingdom, and to a lesser extent in Australia, similar stories can be told about the organization of U.S.-based genome research (Hilgartner, 2004). Similarly, Jenny Reardon's work on the Human Genome Diversity Project takes the ethical and political controversies that surrounded this research as its raw material (rather than conclusion) (Reardon, 2001), providing a case study through which to explore broader issues surrounding genomics and racial science (Reardon, 2004), a topic that has been taken up by a number of anthropologists (McCann-Mortimer et al., 2004; Ventura Santos & Chor Maio, 2004).

At the same time others have sought to explore modern genomic initiatives by setting them within broader historical contexts (Allen, 1997), tying into approaches that emphasize the HGP's deep intellectual roots in the development of genomic metaphors and the nature of new biological knowledge. These kinds of ideas are explored in historical work such as Lilly Kay's investigation of the funding and organization of early genetic "big science" (Kay, 1993) and the ideological changes that took place in the mid-twentieth century that made sequencing the human genome an intellectual possibility (Kay, 1995, 2000). These themes chime with Evelyn Fox Keller's unpicking of the idea of a genetic code and the way its attendant information metaphor has developed and shaped scientists' approaches to human genetics (Keller, 1995; 2001) More recently, Michael Fortun has challenged both the historical myths of the HGP's origin and its classification as "big science" (Fortun, 1997).

Yet the predominant theme within the broader literature on the HGP is one of transformation, of how genomic knowledge is revolutionizing the biological sciences, and this strand of thought is well represented in STS literature. Sometimes this is explicit (Glasner, 2002), but more often it is STS's way of telling interesting stories revolving around specific technologies that tends to give the impression that fundamental changes are taking place. While this does not have to be the case (Keating & Cambrosio, 2004), studies of individual genomic technologies tend to be transformational rather than contextual. For example, research in STS has also explored the development of individual laboratory techniques crucial to the success of the HGP, such as polymerase chain reaction (PCR) (Rabinow, 1996b; Jordan & Lynch, 1998), as well as looking at the role of IT in genome mapping, automation, and the impact of the genome on practices of intellectual property (Hilgartner, 1995; Hine, 1995; Fujimura, 1999; MacKenzie, 2003; Keating et al., 1999). Complementing this focus on technologies is work that explores the way in which new social organizations, such as patient groups, have become central to the generation of genomic knowledge (Rabinow, 1999; Kaufman, 2004).

On a broader level, Joan Fujimura's work has explored how the idea of cancer genetics took root and developed in the research community, and how these intellectual changes were reified in terms of laboratory practice and equipment (Fujimura, 1987, 1988, 1996). Similarly, the impact of genomics on nonmedical science, such as forensics, has received considerable attention (Jasanoff, 1998; Lynch, 1998, 2002; Derksen, 2000; Halfon, 1998; Daemmrich, 1998).

One possible consequence of STS scholars' tendency to focus on individual technologies (derived perhaps from the importance of case study approaches) is prioritization of the transformational theme at the expense of the contextual. The kinds of expectations raised by scientific debate about the genome project and its reshaping of biomedical science and our understanding of ourselves are implicitly supported by social science research that explores novel technological developments. The long historical origins of metaphors such as the genetic code are overwritten by the emphasis on the role of new digital technologies in the progress of the HGP. Although we accept the argument that there have been high expectations in the scientific domain surrounding genome research, underpinned by the creation of large amounts of new knowledge, the scale of industrialization taking place, and the level of public funding for biological research, at the same time it is not clear to us that they necessarily need to be seen in terms of a revolution in the life sciences (and their social and ethical implications). As we shall see later on, the kinds of transformative expectations generated by discussion of the HGP rarely translate smoothly into clinical practice.

COMMODIFICATION AND COMMERCIAL EXPLOITATION OF GENOMIC KNOWLEDGE

Perhaps surprisingly, STS scholars have paid considerably less attention to the commercial exploitation of new genetic and genomic knowledge. Some of the earliest work on biotechnology from a science and society perspective placed the problem of the commodification of life at the center of analysis (Yoxen, 1983), but this concern has not been carried forward in any programmatic manner. Instead the main focus of analysis has been on changes in the knowledge production system associated with the development of the life sciences in general and the rise of genomics and "big biology" in particular.

Etzkowitz and Webster in the last edition of the *Handbook* described the increasing capitalization of knowledge across science as a whole but concluded that this change was no more than an extension of earlier patterns (Etzkowitz & Webster, 1995). However, in contrast to this view, others have argued that in the case of the life sciences the development of biotechnology and genomics is bringing about a major shift in the social relations of knowledge production, through the patenting of genes and life forms, increasing linkages with industry, restricted public access to information (Hilgartner, 1998), and greater secrecy (Wright & Wallace, 2000). This is seen as having a number of important consequences, including new work practices and relationships (Owen-Smith & Powell, 2001), the shifting of public-private boundaries, changes in the reward system (Packer & Webster, 1996), and a corrosion of the norms of "good science," leading to major conflicts of interest (Andrews & Nelkin, 2001). At one extreme, it can be argued, the publicly funded elements of the HGP can be seen as working within the bounds and interests of international pharmaceutical research (Loeppky, 2005).

A relatively small amount of STS scholarship has examined the commercial exploitation of new genomic knowledge and the development of the genomics sector within the biotechnology industry (Glasner & Rothman, 2004). Perhaps the most detailed exploration of the commercial development of genomics has come from outside the STS community, in the work of Paul Rabinow who has shown how modern biotechnological techniques, such as PCR, were developed and how firms are exploiting new genomic technologies (Rabinow, 1996b; Rabinow & Dan-Cohen, 2005). What Rabinow brings to these studies is an unwillingness to assume the worst of commercial entities and scientists, an unwillingness to adopt the sort of "majoritarian" perspective that some would say characterizes too much STS scholarship (Fortun, 2005). Sunder Rajan has analyzed the creation of value within genomics—the dynamics of knowledge flow—and argues that this represents a new form of biocapitalism (Sunder Rajan, 2003, 2005) based in particular on the speculative creation of particular genomic futures (Fortun, 2001). In contrast, more detailed case studies of specific industries, technologies, and firms have stressed the local nature of genomic knowledge production and use and the key role of existing networks and institutions. In particular, the development and integration of complex heterogeneous IT networks within the pharmaceutical industry has both enabled and constrained the commercial development of genomics (Groenewegen & Wouters, 2004). Attempts to commercially develop a diagnostic test for bipolar disorder were seriously undermined by problems in stabilizing this disease category (Lakoff, 2005). In the case of the commercialization of gene therapy, firm strategies were shaped by particular visions of their potential application and how these could be aligned with established markets and clinical practices (Martin, 1995).

As in the previous section, it appears that much STS work in this domain has been concerned with mapping the transformation of knowledge production as well as the emergence of new forms of commodities and exchange. In doing so, it shares the dominant set of expectations promoted by innovators, which are claimed to signal a clear break with the past. However, as mentioned in the introduction, there is increasing doubt about the extent to which this transformation is occurring (Nightingale & Martin, 2004). Once the focus of research shifts from the production and commodification of scientific knowledge to the development of industries, firms, and technologies, it appears that change is much more constrained. New firms and artifacts are embedded in existing sociotechnical networks, and the translation and application of genomic knowledge is contingent, locally specific, and highly dependent on the mobilization and co-construction of a wide range of other actors and resources. A similar picture emerges from studies looking at the use of genomics in the clinic.

APPLICATION OF NEW KNOWLEDGE IN THE CLINIC

One of the main arguments used to justify the HGP was the impact it would have on clinical practice. Partly this revolved around "cures" for various diseases, such as cancer, but a strong theme within biomedical debates has been the impact of genomics on the classification of disease, the main hope being a "splitting" of common disorders (like cancer or heart disease) into a number of (partly) genetically defined subtypes (Bell, 1997). STS's engagement with classification can be traced back to Edward Yoxen's early work in this area (Yoxen, 1982) and has resulted in a focus on a number of conditions. For example, cystic fibrosis (CF) has proved a lively topic of investigation, with debate over the role of genetics in CF classification (Kerr, 2000, 2005; Hedgecoe, 2003) serving as a microcosm of the debates between the transformational and contextual positions as well as the social shaping of CF screening programs (Koch & Stemerding, 1994) and gene therapy for the condition (Stockdale, 1999. STS scholars have also looked at the role of genetics in the etiology of schizophrenia (Hedgecoe, 2001; Turney & Turner, 2000), spinal muscular atrophy (Gaudillière, 1998), heart disease (Hall, 2004), polycystic kidney disease (Cox & Starzomski, 2004) and Marfan syndrome (Heath, 1998). What has clearly come out of this work is that the genetic reclassification of disease is inherently unpredictable and expectations of clarity and simplicity are bound to be disappointed. Even cystic fibrosis, ostensibly the simplest of monogenic disorders, has become a classificatory mess, expanding over time to incorporate neighboring conditions, with its boundaries retreating in some contexts and remaining extended in others.

Similar complexity can be found in the case of genetic testing for breast cancer with attention focusing on the BRCA1 and BRCA2 genes. This work has taken a traditional trope of scientific commentary, the "race" to discover a gene, and unpacked not just how that process took place (Dalpe et al., 2003) but also the broader clinical impact of that competition (cashed out in terms of patents and testing licenses) both for patients, their representatives, and clinical practice in the cancer clinic (Parthasarathy, 2003; Bourret, 2005) and for national differences over genetic privacy (Parthasarathy, 2004) and genetic testing cultures (Parthasarathy, 2005; Gibbon, 2002). If the case of the BRCA1 and 2 genes and breast cancer is to be seen as encapsulating expectations concerning genetic testing for complex diseases, then as in cystic fibrosis, STS teaches caution about simplistic assumptions regarding the delivery of such testing. A related point is made by Nelis (2000) in a comparative study of the management of uncertainty in genetic testing services in the Netherlands and the United Kingdom, where she argued that the construction of expectations and the management of the future are shaped by the structure of the local networks.

In focusing on specific technologies (rather than conditions), research has revealed just how much effort it takes to get a new form of testing or therapy into the clinic (Martin, 1999; Hedgecoe, 2003; Hedgecoe & Martin, 2003; Hedgecoe, 2004). Partly this may be because of the tendency of STS research (unlike, say, medical sociology) to focus on knowledge at the expense of practice, yet even when a clinical intervention has been available for some time, there is still considerable flexibility over how it is seen in the lab, in the clinic, and by patients (Rapp, 2000). New molecular techniques are incorporated into existing clinical practices rather than sweeping them aside in a revolution (Nukaga, 2002). The range of conditions explored in this work and the limitations faced by these technologies when they enter the clinic highlight the point that very few of the expectations that were used to justify the HGP have been realized to date, with almost all the new clinical techniques restricted to established genetic niches.

REPRESENTATION AND CULTURE OF GENOMICS

It is when debates around genomics leave the lab, clinic, or boardroom and enter the broader culture and public discourse that they become the most overtly political. In the case of public understanding of science, the expectations about genomics raised are different from at other sites. Rather than there being expectations about science and technology, in the case of PUS, the expectations concern people's reactions and behavior toward science and technology. If we view the "deficit model" of PUS as constructing typical expectations about how people will react toward genomics, then, given STS's historical role in challenging this model and the high profile human genetics has in public debate, we should not be surprised to see work in this area undermine and question simplistic beliefs about how the public will respond to genomics.

Perhaps the clearest evidence that within STS the transformational approach to genomics can be highly critical of developments in science and technology lies in Dorothy Nelkin's sustained critique of the way in which modern genetics is portrayed in the media and popular culture (Nelkin, 1994; Nelkin & Lindee, 1995, 1999). Clearly written from a position that takes developments in modern genetics as somehow different from what has gone before, Nelkin's work, and that of other scholars like Abby

Lippman (Lippman, 1994, 1998), can be criticized for lack of historical depth and methodological problems (Condit, 1999, 2004) but not for political urgency and critical drive.

Of course, it is perfectly possible to carry out a historically rooted analysis of the cultural representation of genetics and produce a critical piece of work (Turney, 1998; Smart, 2003), and overall STS researchers have tended to stay close to the discipline's qualitative roots, eschewing the survey approach often used by other social sciences to study this area (Davison et al., 1997). Of particular note is the extensive work done by Anne Kerr and colleagues who have used interviews and focus groups to explore the different ways in which geneticists (Cunningham-Burley & Kerr, 1999; Kerr et al., 1997, 1998a) and nonscientists (Kerr et al., 1998b,c) view developments in genetics and associated ethical issues. This rigorous empirical basis has provided a foundation for a subsequent critique of the way in which some social theorists have engaged with human genetics (Kerr & Cunningham-Burley, 2000) and developing concepts around the political life of human genomics (Kerr, 2003a,b,c). A core element of this and other work (Barns et al., 2000; Irwin, 2001) is to incorporate nonscientist opinion on genetics into discussions about the development of this technology, showing not only that members of the public are capable of *understanding* complex scientific concepts but also that they can contribute in a meaningful way to debates around the regulation of these new technologies.

When facing expectations about genomics, public and professional cultures tend to divide, with the concerns of professionals (both scientific and non-STS-based social scientists) being rooted in traditional models of the public and technology, with ethical expectations marginalized and simplistic solutions suggested. To some extent it might be seen as a failure that the public culture emphasized through STS for so many years has had such a low profile among practicing scientists, yet whether we take a transformational or contextual position, the increased presence of genomics in the press and public discourse seems assured. That STS shows how scientists and policy-makers who refuse to reorient their expectations in accordance with how the public reacts engender resistance and even failure (Robins, 2001) provides an opportunity for work in this area to feed directly into political discussion over how societies might respond to new technologies. Contextualists might take the opportunity to highlight the public's fears of, for example, racialized science (Duster, 2001) while, as the next section discusses, transformationalists may show how particular groups of nonexperts adapt and adopt genetic knowledge to serve their own social needs. The point is less about whether genomics is transformative of wide cultures and publics but rather that this context presents STS scholars working in this area with a unique opportunity to engage with public political debate.

CREATION OF NEW GENOMIC IDENTITIES

One of the most influential recent strands of argument in the field of the social studies of the life sciences concerns ideas of biosociality (Rabinow, 1996a) and biological citizenship. While the origins of these ideas may be see as formally lying outside the realm of STS, they have shaped much of the debate on the creation of new genomic identities.

The initial focus of work in this area arose from studies of new reproductive technologies and the development of genetic testing services for mainly rare monogenic conditions (Rapp, 1998, 2000). Research has recently started to look at more common complex disorders. Finkler, drawing on the experience of women who have a hereditary risk of breast cancer, argues that the presentation of research findings has led to a new genetic determinism, the medicalization of kinship, and changing ideas about the significance and meaning of kinship (Finkler, 2000; Finkler et al., 2003). In particular, she shows how the experience of the new genetics can transform a healthy person into a patient without symptoms and places increasing emphasis on biological rather than social determinants of health and illness. However, writing from within the contextual approach, Kerr has criticized studies of this sort for lacking empirical evidence and overemphasizing the role of genetics as a consequence of giving too much weight to the role of biological knowledge in shaping life choices (Kerr, 2004).

In contrast to seeing the new genetics as largely negative in its consequences for an individual's sense of self, Novas and Rose argue that knowledge of genetic risk does not generate fatalism but induces new relations to oneself and one's future, and a new set of obligations and biological responsibilities (Novas & Rose, 2000). This in turn is creating new individual and collective identities such as those embodied in patient groups for muscular dystrophy or Huntington's disease. These can challenge ideas of stigma and exclusion, as well as dominant medical discourses. Rabinow has called this creation of new subjectivities "biosociality," as distinct from Foucault's concept of biopower in which life and its mechanisms are calculable and this knowledge is used to discipline both bodies and populations. This perspective is explicitly transformative, distancing as it does modern genomics from traditional concerns about eugenics. This does not mean that there are not ethical issues, of course, simply that they are of a new kind (Rose, 2001).

It should be noted that other nonmedical genetic technologies, such as the development of genetic ancestry testing, are also creating new forms of collective and individual identity (Tutton, 2004; Nash, 2004). Following on from this, it is argued that the emergence of new identities based on ideas of genetic susceptibility and risk, and the embodied disciplines and representations of rights and responsibilities that are being co-constructed through new screening and public health programs, constitute a new form of biological or genetic citizenship (Rose & Novas, 2004). Through the fulfillment of the duties to know and manage genetic risk in order to protect themselves and their families, individuals are seen as constructing themselves as healthy and responsible citizens (Petersen, 2002; Polzer et al., 2002). Hovering between these two positions is work like that of Taussig, Rapp, and Heath, who, in their research on the "Little People of America" patient group, explore a range of technological interventions (such as surgery or genetic testing) using the concept of "flexible eugenics" to point out the positive and negative options for self-identity that arise from genetic technologies (Taussig et al., 2003). Similarly, Callon and Rabeharisoa note a number of ways in which people resist the imposition of such genomic identities (Callon & Rabeharisoa, 2004).

Thus, we suggest that while new genetic and genomic knowledge can be seen as helping constitute distinct new forms of identity, subjectivity, and citizenship, the extent to which these transformations are happening outside very tightly defined niches (patient groups for rare genetic diseases) or represent a clear break with the past remains unclear. As such, we feel that STS scholars ought to display caution with regard to expectations vis-à-vis genomics' impact on social identity.

GOVERNANCE OF GENOMICS

Research on the governance and regulation of genomic technologies has been fundamentally shaped by earlier work on the ethical, legal, and social issues (ELSIs) raised during the controversies surrounding the development of recombinant DNA (rDNA) and biotechnology, and the political response to these concerns. With a few notable exceptions (Nelkin & Tancredi, 1980, 1994; Duster, 1990), little of this work was from an STS perspective, most of it having a largely normative agenda that critiqued the potential hazards and social problems caused by emerging genetic technologies. There have also been important national differences between the United States and European states in terms of political and institutional responses and also in the type of scholarship that has been funded in this area. Broadly speaking, U.S. ELSI research has been dominated by bioethicists and lawyers, while in the United Kingdom social scientists have played the key role. One consequence is a relative lack of U.S. STS studies in this area.

During the 1980s and early '90s many of the institutional mechanisms and regulatory regimes designed to control early rDNA research and first-generation biotechnology products were established, and a number of STS scholars have analyzed their creation in detail (Bennett et al., 1986; Wright, 1994, 1996; Gottweis, 1995, 1998). This is important work but, strictly speaking, lies beyond the scope of this chapter. In contrast, significantly less attention has been given to more recent changes in these regimes brought about by the turn to genomics and the development of new technologies, such as genetic screening and gene therapy. In looking at the broad field of genomics and postgenomics, Gottweis has argued that "... the science of genomics is introducing a number of fundamental transformations in the practice of modern biology and medicine, in pharmaceutical industry, in society and culture" (Gottweis, 2005: 202). He goes on to suggest that there is a gap between this challenge and official policy responses, which might ultimately lead to a crisis of confidence in medical biotechnology.

The small body of work that examines the governance of genomics in more detail is mainly United Kingdom–based. Salter and Jones have studied recent changes in the overall regime governing human genetics in the United Kingdom. In particular, they have charted the creation of a complex system of statutory regulatory bodies and nonstatutory expert advisory committees. This system was reconfigured following what was constructed as a major crisis of trust following the public rejection of genetically modified food in the late 1990s and has adopted a discourse of open government, based on the language of public engagement and greater transparency, as a legitimating strategy (Jones & Salter, 2003). In a similar study of the regulation of human genetics at the EU level, Salter and Jones (2002) have shown that similar pressures have forced policy-makers to engage with a greater range of stakeholders and publics, as well as placing more emphasis on the role of expert bioethicists in mediating disputes. An important recent addition to the literature is Jasanoff's three country comparison of the governance and regulation (including informal forms such as bioethics) of biotechnology, which provides an important basis for future STS work in this area (Jasanoff, 2005).

There have also been studies of the governance of specific genomic and genetic technologies, including genetic databases (Martin, 2001; Petersen, 2005) and genetic testing (Martin & Frost, 2003) in the United Kingdom, as well as a comparative U.S./U.K. study of genetic privacy (Parthasarathy, 2004). In particular, these have shown how specific innovations are co-constructed with regulatory regimes and how they are shaped by local political, cultural, and institutional factors. Considerable attention has been paid to exploring the new forms of governance and public engagement that seem to have become associated with genetics and biotechnology in the United Kingdom over the last decade (Tutton et al., 2005; Kerr, 2004; Purdue, 1999). This research suggests that, while important changes have occurred in the way in which the public is constructed and engaged by policy-makers, established power relations continue to be reproduced. Furthermore, the narratives of choice and responsibility that are a common hallmark of policy discussions in this area are seen to frame the problems associated with new genetic technologies in ways that shift attention away from broader questions of social priorities and the goals of scientific research (Kerr, 2003c). Furthermore, Anne Kerr argues that it is premature to talk about a new form of genetic citizenship, as many questions remain unanswered about how the new rights and responsibilities of different actors are defined and exercised in practice (Kerr, 2003a).

It therefore appears that while genomics has been associated with some significant changes to institutional arrangements governing biotechnology, it has not prompted a completely new regime. Additional important drivers of change can be identified, including loss of public trust, and this has led to new policy discourses and experiments in public engagement. The difficulty in breaking down established divisions of expertise and institutional barriers casts doubt over the idea that we are seeing new forms of citizenship emerge.

CONCLUSION

Through a review of the STS literature on genomics the aim of this chapter was to answer two broad questions: What sociotechnical expectations and transformations are being associated with the rise of genomics? What is seen as new and specific to genomics, and what is believed to be the extent of sociotechnical change? As we might expect from a discipline that teaches us to question the apparently straightforward facts presented by science, work on STS forces us to challenge the assumptions that underpin even such obvious questions. To some extent the presence of a strong contextual perspective in STS scholarship, questioning claims about the transformational impact of genomics, raises doubts not just about the wording of these two questions but about the nature of this chapter itself. As STSers, our natural instinct may well be to assume novelty on the part of scientific and technological developments, in terms of both technical change and social and ethical impact. Yet the discipline's strong links to the history of science provide a conduit through which contextual assumptions can flow, challenging the automatic belief that every technological development implies a revolution. We accept that our own backgrounds mean that the contributions of sociologists are perhaps overemphasized, but we feel that the picture of STS scholarship painted in this chapter should be broadly recognizable to people working in this field.

As noted earlier, the point of this chapter is not to adjudicate between these different ways of looking at genomics. The richness of debate, variety of case studies, and rigor of research in this area stems in part from the existence of these different ways of seeing the same material. Rather, we would like to agree with Taussig, Rapp, and Heath and suggest that, with regard to the social implications of genomics, "a working knowledge of the political history of eugenics gives us reason for pessimism of the intellect, but an ethnographic perspective on the openness of these practices may give some cause for optimism of the will" (Taussig et al., 2003: 72–73). Taking a broader approach, Andrew Webster links the perceived novelty of genomics within larger social trends, namely, the more "liquid" nature of modern society, with its flexible boundaries and wide range of possible new configurations. One effect of such a context is to move away from the idea of "genomics as intrinsically and necessarily transformative . . . allow[ing] us to turn our attention to the ways in which genomics research is or could be articulated in society to close off or open up 'possibilities'" (Webster, 2005: 237).

What is clear is that much STS scholarship, of whatever kind, maintains a skeptical stance toward scientific claims about genomics, justifying this position with detailed and closely argued empirical studies. The expectations raised at the launch of the Human Genome Project have yet to be realized in any significant sense in the clinic, and it is far from certain that the impact of genomics on industry or personal identity will stretch as far as some commentators claim. That said, in certain of the domains outlined above, particularly those relating to the production of new scientific knowledge, genomics has proved to be transformational. Perhaps what we need now is an understanding of why it is that expectations about genomics are being realized in some areas and not in others.

Note

1. Tony Blair, MP, "Science Matters," speech to the Royal Society, April 10, 2002.

References

Akrich, M. (1992) "The De-Scription of Technical Objects," in W. Bijker & J. Law (eds), *Shaping Technology/Building Society: Studies in Sociotechnical Change* (Cambridge, MA: MIT Press).

Andrews, L. & D. Nelkin (2001) *Body Bazaar: The Market for Human Tissue in the Biotechnology Age* (New York: Crown).

Anon. (2005) Editorial, *ESRC Genomics Network Newsletter* 2: 3. Available at: http://www.cesagen.lancs.ac.uk/resources/newsletter/networknewsletter.htm.

Balmer, Brian (1994) "Gene Mapping and Policy Making: Australia and the Human Genome Project," *Prometheus* 12(1): 3–18.

Balmer, Brian (1995) "Transitional Science and the Human Genome Mapping Project Resource Centre," *Genetic Engineer and Biotechnologist* 15(2&3): 89–97.

Balmer, Brian (1996a) "Managing Mapping in the Human Genome Project," *Social Studies of Science* 26(3): 531–73.

Balmer, Brian (1996b) "The Political Cartography of the Human Genome Project," *Perspectives on Science* 4(3): 249–82.

Barns, I., R. Schibeci, A. Davison, & R. Shaw (2000) "'What Do You Think About Genetic Medicine?' Facilitating Sociable Public Discourse on Developments in the New Genetics," *Science, Technology & Human Values* 25(3): 283–308.

Bell, John (1997) "Genetics of Common Disease: Implications for Therapy, Screening and Redefinition of Disease," *Philosophical Transactions of the Royal Society of London B* 352: 1051–55.

Bennett, D., P. Glasner, & D. Travis (1986) *The Politics of Uncertainty: Regulating Recombinant DNA Research in Britain* (London: Routledge & Kegan Paul).

Bourret, P. (2005) "BRCA Patients and Clinical Collectives: New Configurations of Action in Cancer Genetics Practices," *Social Studies of Science* 35(1): 41–68.

Brown, N. & M. Michael (2003) "A Sociology of Expectations: Retrospecting Prospects and Prospecting Retrospects," *Technology Analysis and Strategic Management* 15: 3–18.

Brown, N., B. Rappert, & A. Webster (eds) (2000) *Contested Futures: A Sociology of Prospective Technoscience* (Aldershot, U.K.: Ashgate).

Brown, N. (2003) "Hope Against Hype: Accountability in Biopasts, Presents and Futures," *Science Studies* 16(2): 3–21.

Callon, M. & V. Rabeharisoa (2004) "Gino's Lesson on Humanity: Genetics, Mutual Entanglements and the Sociologist's Role," *Economy and Society* 33(1): 1–27.

Calvert, J. (2004) "Genomic Patenting and the Utility Requirement," *New Genetics and Society* 23(3): 301–12.

Clarke, A., L. Mamo, J. R. Fishman, J. K. Shim, & J. R. Fosket (2003) "Biomedicalization: Technoscientific Transformations of Health, Illness, and U.S. Biomedicine," *American Sociological Review* 68: 161–94.

Condit, Celeste (1999) *The Meanings of the Gene: Public Debates About Human Heredity* (Madison: University of Wisconsin Press).

Condit, Celeste (2004) "The Meaning and Effects of Discourse About Genetics: Methodological Variations in Studies of Discourse and Social Change," *Discourse and Society* 15(4): 391–407.

Cox, S. M. & R. C. Starzomski (2004) "Genes and Geneticization? The Social Construction of Autosomal Dominant Polycystic Kidney Disease," *New Genetics and Society* 23(2): 137–66.

Cunningham-Burley, Sarah & Anne Kerr (1999) "Defining the 'Social': Towards an Understanding of Scientific and Medical Discourse on the Social Aspects of the New Human Genetics," *Sociology of Health and Illness* 21: 647–68.

Daemmrich, A. (1998) "The Evidence Does Not Speak for Itself: Expert Witnesses and the Organization of DNA-Typing Companies," *Social Studies of Science* 28(5/6), (Oct.–Dec.): 741–72.

Dalpe, R., L. Bouchard, A. J. Houle, & L. Bedard (2003) "Watching the Race to Find the Breast Cancer Genes," *Science, Technology & Human Values* 28(2): 187–216.

Davison, A., I. Barns, & R. Schibeci (1997) "Problematic Publics: A Critical Review of Surveys of Public Attitudes to Biotechnology," *Science, Technology & Human Values* 22(3): 317–48.

Delanty, G. (2002) "Constructivism, Sociology and the New Genetics," *New Genetics and Society* 21(3): 279–289.

Department of Health (2003) *Our Inheritance, Our Future: Realising the Potential of Genetics in the NHS* (London: H. M. Stationery Office).

Derksen, L. (2000) "Towards a Sociology of Measurement: The Meaning of Measurement Error in the Case of DNA Profiling," *Social Studies of Science* 30(6): 803–45.

Duster, T. (1990) *Backdoor to Eugenics* (New York: Routledge).

Duster, T. (2001) "The Sociology of Science and the Revolution in Molecular Biology," in J. R. Blau (ed), *The Blackwell Companion to Sociology* (London and New York: Blackwell): 213–26.

Etzkowitz, H. & A. Webster (1995) "Science as Intellectual Property," in S. Jasanoff, G. E. Markle, J. C. Petersen, & T. Pinch (eds), *Handbook of Science and Technology Studies* (Thousand Oaks, CA: Sage): 480–505.

Finkler, K. (2000) *Experiencing the New Genetics: Family and Kinship on the Medical Frontier* (Philadelphia: University of Pennsylvania Press).

Finkler, K., C. Skrzynia, & J. P. Evans (2003) "The New Genetics and Its Consequences for Family, Kinship, Medicine and Medical Genetics," *Social Science and Medicine* 57(3): 403–12.

Fortun, Michael (1997) "Projecting Speed Genomics," in M. Fortun & E. Mendelsohn (eds), *The Practices of Human Genetics: Sociology of the Sciences Yearbook,* vol. XXI (Netherlands: Kluwer): 25–48.

Fortun, Michael (2001) "Mediated Speculations in the Genomics Futures Markets," *New Genetics and Society* 20: 139–56.

Fortun, Michael (2005) "For an Ethics of Promising or: A Few Kind Words About James Watson," *New Genetics and Society* 24(2): 157–73.

Fujimura, Joan (1987) "Constructing 'Do-able' Problems in Cancer Research," *Social Studies of Science* 17: 257–93.

Fujimura, Joan (1988) "The Molecular Biological Bandwagon in Cancer Research," *Social Problems* 35(3): 261–83.

Fujimura, Joan (1996) *Crafting Science: A Sociohistory of the Quest for the Genetics of Cancer* (Cambridge, MA: Harvard University Press).

Fujimura, Joan (1999) "The Practices of Producing Meaning in Bioinformatics," in M. Fortun & E. Mendelsohn (eds), *The Practices of Human Genetics: Sociology of the Sciences Yearbook*, vol. XXI (Netherlands: Kluwer): 49–87.

Garland, A. (1997) "Modern Biological Determinism: The Violence Initiative, the Human Genome Project, and the New Eugenics," in M. Fortun & E. Mendelson (eds), *The Practices of Human Genetics: Sociology of the Sciences Yearbook*, vol. XXI (Netherlands: Kluwer): 1–23.

Gaudillière, J.-P. (1998) "How Weak Bonds Stick: Genetic Diagnosis Between the Laboratory and the Clinic," in P. Glasner & H. Rothman (eds), *Genetic Imaginations: Ethical, Legal and Social Issues in Human Genome Research* (Aldershot, U.K.: Ashgate): 21–40.

Gibbon, Sarah (2002) "Family Trees in Clinical Cancer Genetics: Re-examining Geneticization," *Science as Culture* 11(4): 429–57.

Glasner, Peter (2002) "Beyond the Genome: Reconstituting the New Genetics," *New Genetics and Society* 21(3): 267–77.

Glasner, Peter & Harry Rothman (2004) "From Commodification to Commercialisation," in *Splicing Life? The New Genetics and Society* (Aldershot, U.K.: Ashgate).

Gottweis, H. (1995) "German Politics of Genetic-Engineering and Its Deconstruction," *Social Studies of Science* 25(2): 195–235.

Gottweis, H. (1998) *Governing Molecules: The Discursive Politics of Genetic Engineering in Europe and in the United States* (Cambridge, MA: MIT Press).

Gottweiss, H. (2005) "Emerging Forms of Governance in Genomics and Post-genomics: Structures, Trends, Perspectives," in R. Bunton & A. Petersen (eds), *Genetic Governance: Health, Risk and Ethics in the Biotech Age* (London: Routledge): 189–208.

Groenewegen, P. & P. Wouters (2004) "Genomics, ICT and the Formation of R&D Networks," *New Genetics and Society* 23(2): 167–85.

Halfon, S. (1998) "Collecting, Testing and Convincing: Forensic DNA Experts in the Courts," *Social Studies of Science* 28(5/6) (Oct.–Dec.): 801–28.

Hall, E. (2004) "Spaces and Networks of Genetic Knowledge Making: The 'Geneticisation' of Heart Disease," *Health and Place* 10(4): 311–18.

Harvey, M., A. McMeekin, & I. Miles (2002) "Genomics and Social Science: Issues and Priorities," *Foresight* 4(4): 13–28.

Heath, D. (1998) "Locating Genetic Knowledge: Picturing Marfan Syndrome and Its Traveling Constituencies," *Science, Technology & Human Values* 23(1): 71–97.

Hedgecoe, Adam (2001) "Schizophrenia and the Narrative of Enlightened Geneticization," *Social Studies of Science* 31(6): 875–911.

Hedgecoe, Adam (2003a) "Expansion and Uncertainty: Cystic Fibrosis: Classification and Genetics," *Sociology of Health and Illness* 25(1): 50–70.

Hedgecoe, Adam (2003b) "Terminology and the Construction of Scientific Disciplines: The Case of Pharmacogenomics," *Science, Technology & Human Values* 28(4): 513–37.

Hedgecoe, Adam (2004) *The Politics of Personalised Medicine: Pharmacogenetics in the Clinic* (Cambridge, U.K.: Cambridge University Press).

Hedgecoe, Adam & Paul Martin (2003) "The Drugs Don't Work: Expectations and the Shaping of Pharmacogenetics," *Social Studies of Science* 33(3): 327–64.

Hieter, P. & M. Boguski (1997) "Functional Genomics: It's All How You Read It," *Science* 278: 601–2.

Hilgartner, Stephan (1995) "Biomolecular Databases: New Communication Regimes for Biology?" *Science Communication* 17(2): 240–63.

Hilgartner, Stephan (1997) "Access to Data and Intellectual Property: Scientific Exchange in Genome Research," in National Academy of Sciences, *Intellectual Property and Research Tools in Molecular Biology: Report of a Workshop* (Washington, D.C.: National Academy Press): 28–39.

Hilgartner, Stephan (1998) "Data Access Policy in Genome Research," in A. Thackray (ed), *Private Science* (Philadelphia: University of Pennsylvania Press): 202–18.

Hilgartner, Stephan (2004) "Making Maps and Making Social Order: Governing American Genomics Centers, 1988–1993," in J.-P. Gaudillière & H.-J. Rheinberger (eds), *From Molecular Genetics to Genomics: The Mapping Cultures of Twentieth-Century Genetics* (London and New York: Routledge): 113–27.

Hine, C. (1995) "Information Technology as an Instrument of Genetics," *Genetic Engineer and Biotechnologist* 15(2–3): 113–24.

Irwin, A. (2001) "Constructing the Scientific Citizen: Science and Democracy in the Biosciences," *Public Understanding of Science* 10: 1–18.

Jasanoff, Sheila (1998) "The Eye of Everyman: Witnessing DNA in the Simpson Trial," *Social Studies of Science* 28(5/6) (Oct.–Dec.): 713–40.

Jasanoff, Sheila (2005) *Designs on Nature: Science and Democracy in Europe and the United States* (Princeton, NJ, and Oxford: Princeton University Press).

Jones, Mavis & Brian Salter (2003) "The Governance of Human Genetics: Policy Discourse and Constructions of Public Trust," *New Genetics and Society* 22(1): 21–41.

Jordan, Kathleen & Michael Lynch (1998) "The Dissemination, Standardization and Routinization of a Molecular Biological Technique," *Social Studies of Science* 28 (5/6) (Oct.–Dec.): 773–800.

Kaufman, Alain (2004) "Mapping the Human Genome at Généthon Laboratory: The French Muscular Dystrophy Association and the Politics of the Gene," in J.-P. Gaudillière & H.-J. Rheinberger (eds), *From Molecular Genetics to Genomics: The Mapping Cultures of Twentieth-Century Genetics* (London and New York: Routledge): 129–57.

Kay, Lily E. (1993) *The Molecular Vision of Life: Caltech, the Rockefeller Foundation and the Rise of the New Biology* (Oxford: Oxford University Press).

Kay, Lily E. (1995) "Who Wrote the Book of Life? Information and the Transformation of Molecular Biology, 1945–1955," *Science in Context* 8: 609–34.

Kay, Lily E. (2000) *Who Wrote the Book of Life? A History of the Genetic Code* (Stanford, CA: Stanford University Press).

Keating, P. & A. Cambrosio (2004) "Signs, Markers, Profiles, and Signatures: Clinical Haematology Meets the New Genetics (1980–2000)" *New Genetics and Society* 23(1): 15–45.

Keating, P., C. Limoges, & A. Cambrosio (1999) "The Automatic Laboratory: The Generation and Replication of Work in Molecular Genetics," in M. Fortun & E. Mendelsohn (eds), *The Practices of Human Genetics: Sociology of the Sciences Yearbook,* vol. XXI (Netherlands: Kluwer): 125–42.

Keller, Evelyn Fox (1995) Refiguring Life: Metaphors of Twentieth-century Biology (New York: Columbia University Press).

Keller, Evelyn Fox (2001) *The Century of the Gene* (Cambridge, MA: Harvard University Press).

Kerr, Anne (2000) "Reconstructuring Genetic Disease: The Clinical Continuum Between Cystic Fibrosis and Male Infertility," *Social Studies of Science* 30: 847–94.

Kerr, Anne (2003a) "Governing Genetics: Reifying Choice and Progress," *New Genetics and Society* 22: 111–26.

Kerr, Anne (2003b) "Genetics and Citizenship," *Society* 40(6): 44–50.

Kerr, Anne (2003c) "Rights and Responsibilities in the New Genetics Era," *Critical Social Policy* 23(2): 208–26.

Kerr, Anne (2004) *Genetics and Society: A Sociology of Disease* (London: Routledge).

Kerr, Anne (2005) "Understanding Genetic Disease in a Socio-historical Context: A Case Study of Cystic Fibrosis," *Sociology of Health and Illness* 27(7): 873–96.

Kerr, Anne & S. Cunningham-Burley (2000) "On Ambivalence and Risk: Reflexive Modernity and the New Human Genetics," *Sociology* 34(2): 283–304.

Kerr, Anne, S. Cunningham-Burley, & A. Amos (1997) "The New Genetics: Professionals' Discursive Boundaries," *Sociological Review* 45(2): 279–303.

Kerr, Anne, S. Cunningham-Burley, & A. Amos (1998a) "Eugenics and the New Genetics in Britain: Examining Contemporary Professionals' Accounts," *Science, Technology & Human Values* 23(2): 175–98.

Kerr, Anne, S. Cunningham-Burley, & A. Amos (1998b) "Drawing the Line: An Analysis of Lay People's Discussions About the New Genetics," *Public Understanding of Science* 7(2): 113–33.

Kerr, Anne, S. Cunningham-Burley, & A. Amos (1998c) "The New Genetics and Health: Mobilizing Lay Expertise," *Public Understanding of Science* 7(1): 41–60.

Koch, L. & D. Stemerding (1994) "The Sociology of Entrenchment: A Cystic Fibrosis Test for Everyone?" *Social Science and Medicine* 39(9): 1211–20.

Lakoff, A. (2005) "Diagnostic Liquidity: Mental Illness and the Global Trade in DNA," *Theory and Society* 34(1): 63–92.

Lippman, Abby (1994) "The Genetic Construction of Prenatal Testing: Choice, Consent or Conformity for Women?" in K. H. Rothenberg & E. J. Thomson (eds), *Women and Prenatal Testing: Facing the Challenges of Genetic Testing* (Miami: Ohio State University Press): 9–34.

Lippman, Abby (1998) "The Politics of Health: Geneticization Versus Health Promotion," in S. Sherwin (ed), *The Politics of Women's Health: Exploring Agency and Autonomy* (Philadelphia: Temple University Press): 64–82.

Loeppky, Roddy (2005) Encoding Capital: The Political Economy of the Human Genome Project (New York: Routledge Press).

Lynch, Michael (1998) "The Discursive Production of Uncertainty: The O. J. Simpson 'Dream Team' and the Sociology of Knowledge Machine," *Social Studies of Science* 28(5–6): 829–68.

Lynch, Michael (2002) "Protocols, Practices, and the Reproduction of Technique in Molecular Biology," *British Journal of Sociology* 53(2): 203–20.

MacKenzie, A. (2003) "Bringing Sequences to Life: How Bioinformatics Corporealizes Sequence Data," *New Genetics and Society* 22(33): 315–32.

Martin, Paul (1995) "The American Gene Therapy Industry and the Social Shaping of a New Technology," *Genetic Engineer and Biotechnologist* 15: 155–67.

Martin, Paul (1999) "Genes as Drugs: The Social Shaping of Gene Therapy and the Reconstruction of Genetic Disease," *Sociology of Health and Illness* 21: 517–38.

Martin, Paul (2001) "Genetic Governance: The Risks, Oversight and Regulation of Genetic Databases in the U.K.," *New Genetics and Society* 20(2): 157–83.

Martin, Paul & Rob Frost (2003) "Regulating the Commercial Development of Genetic Testing in the U.K.: Problems, Possibilities and Policy," *Critical Social Policy* 23: 186–207.

McCann-Mortimer, P., M. Augoustinos, & A. Lecouteur (2004) " 'Race' and the Human Genome Project: Constructions of Scientific Legitimacy," *Discourse and Society* 15(4): 409–32.

Michael, Mike (2000) "Futures of the Present: From Performativity to Prehension," in N. Brown, B. Rappert, & A. Webster (eds), *Contested Futures: A Sociology of Prospective Techno-science* (Aldershot, U.K.: Ashgate).

Nash, C. (2004) "Genetic Kinship," *Cultural Studies* 18(1): 1–33.

Nelis, A. (2000) "Genetics and Uncertainty," in N. Brown, B. Rappert, & A. Webster (eds), *Contested Futures: A Sociology of Prospective Techno-science* (Aldershot, U.K.: Ashgate): 209–28.

Nelkin, D. (1994) "Promotional Metaphors and Their Popular Appeal," *Public Understanding of Science* 3: 25–31.

Nelkin, D., & S. Lindee (1995) *The DNA Mystique: The Gene as a Cultural Icon* (New York: W. H. Freeman).

Nelkin, D., & S. Lindee (1999) "Good Genes and Bad Genes: DNA in Popular Culture in the Practices of Producing Meaning in Bioinformatics," in M. Fortun & E. Mendelsohn (eds), *The Practices of Human Genetics: Sociology of the Sciences Yearbook,* vol. XXI (Netherlands: Kluwer): 155–67.

Nelkin, D. & L. Tancredi ([1980] 1994) *Dangerous Diagnostics: The Social Power of Biological Information* (Chicago: University of Chicago Press).

Nightingale, Paul & Paul Martin (2004) "The Myth of the Biotech Revolution," *Trends in Biotechnology* 22(11): 564–69.

Novas, Carlos & Nikolas Rose (2000) "Genetic Risk and the Birth of the Somatic Individual," *Economy and Society* 29(4): 485–513.

Nukaga, Y. (2002) "Between Tradition and Innovation in New Genetics: The Continuity of Medical Pedigrees and the Development of Combination Work in the Case of Huntington's Disease," *New Genetics and Society* 21(1): 39–64.

Owen-Smith, J., & W. W. Powell (2001) "Careers and Contradictions: Faculty Responses to the Transformation of Knowledge and Its Uses in the Life Sciences," *Research in the Sociology of Work* 10: 109–40.

Packer, K. & A. Webster (1996) "Patenting Culture in Science: Reinventing the Scientific Wheel of Credibility," *Science, Technology & Human Values* 21(4): 427–53.

Parthasarathy, S. (2003) "Knowledge Is Power: Genetic Testing for Breast Cancer and Patient Activism in the United States and Britain," in N. Oudshoorn & T. Pinch (eds), *How Users Matter: The Co-construction of Users and Technologies* (Cambridge, MA: MIT Press): 133–50.

Parthasarathy, S. (2004) "Regulating Risk: Defining Genetic Privacy in the United States and Britain," *Science, Technology & Human Values* 29(3): 332–52.

Parthasarathy, S. (2005) "Architectures of Genetics Medicine: Comparing Genetic Testing for Breast Cancer in the USA and the U.K.," *Social Studies of Science* 35(1): 5–40.

Petersen, A. (2002) "The New Genetic Citizens," in A. Petersen & R. Bunton (eds), *The New Genetics and the Public's Health* (London: Routledge): 180–207.

Petersen, A. (2005) "Securing Our Genetic Health: Engendering Trust in U.K. Biobank," *Sociology of Health and Illness* 27(2): 271–92.

Polzer, J., S. L. Mercer, & V. Goel (2002) "Blood is Thicker Than Water: Genetic Testing as Citizenship Through Familial Obligation and the Management of Risk," *Critical Public Health* 12(2): 153–68.

Purdue, D. (1999) "Experiments in the Governance of Biotechnology: A Case Study of the U.K. National Consensus Conference," *New Genetics and Society* 18(1): 79–99.

Rabinow, Paul (1996a) "Artificiality and Enlightenment: from Sociobiology to Biosociality," in P. Rabinow, *Essays on the Anthropology of Reason* (Princeton, NJ: Princeton University Press): 91–111.

Rabinow, Paul (1996b) *Making PCR: A Story of Biotechnology* (Chicago: University of Chicago Press).

Rabinow, Paul (1999) *French DNA: Trouble in Purgatory* (Chicago and London: University of Chicago Press).

Rabinow, Paul & Talia Dan-Cohen (2005) *A Machine to Make a Future: Biotech Chronicles* (Princeton, NJ, and Oxford: Princeton University Press).

Rapp, R. (1998) "Refusing Prenatal Diagnosis: The Multiple Meanings of Biotechnology in a Multicultural World," *Science, Technology & Human Values* 23(1): 45–70.

Rapp, R. (2000) *Testing Women, Testing the Foetus: The Social Impact of Amniocentesis in America* (New York: Routledge).

Reardon, Jenny (2001) "The Human Genome Diversity Project: A Case Study in Coproduction," *Social Studies of Science* 31(3): 357–88.

Reardon, Jenny (2004) *Race to the Finish: Identity and Governance in an Age of Genomics* (Princeton, NJ: Princeton University Press).

Robins, R. (2001) "Overburdening Risk: Policy Frameworks and the Public Uptake of Gene Technology," *Public Understanding of Science* 10: 19–36.

Rose, Nikolas (2001) "The Politics of Life Itself," *Theory, Culture and Society* 18(6): 1–30.

Rose, Nikolas & Carlos Novas (2004) "Biological Citizenship," in A. Ong & S. Collier (eds), *Global Assemblages: Technology, Politics, and Ethics as Anthropological Problems* (Malden, MA: Blackwell).

Salter, B. & M. Jones (2002) "Regulating Human Genetics: The Changing Politics of Biotechnology Governance in the European Union," *Health, Risk and Society* 4(3): 325–40.

Smart, Andrew (2003) "Reporting the Dawn of the Post-genomic Era: Who Wants to Live Forever?" *Sociology of Health and Illness* 25(1): 24–49.

Stockdale, A. (1999) "Waiting for the Cure: Mapping the Social Relations of Human Gene Therapy Research," *Sociology of Health and Illness* 21(5): 579–96.

Sunder Rajan, Kaushik (2003) "Genomic Capital: Public Cultures and Market Logics of Corporate Biotechnology," *Science as Culture* 12(1): 87–121.

Sunder Rajan, Kaushik (2005) "Subjects of Speculation: Emergent Life Sciences and Market Logics in the United States and India," *American Anthropologist* 107(1): 19–30.

Taussig, K. S., R. Rapp, & D. Heath (2003) "Flexible Eugenics: Technologies of the Self in the Age of Genetics," in A. H. Goodman, D. Heath, & M. S. Lindee (eds), *Genetic Nature/Culture: Anthropology and Science Beyond the Two-culture Divide* (Berkeley: University of California Press): 58–76.

Turney, J. (1998) *Frankenstein's Footsteps: Science, Genetics and Popular Culture* (London: Yale University Press).

Turney, J. & J. Turner (2000) "Predictive Medicine, Genetics and Schizophrenia," *New Genetics and Society* 19(1): 5–22.

Tutton, R. (2004) "'They Want to Know Where They Came From': Population Genetics, Identity, and Family Genealogy," *New Genetics and Society* 23(1): 105–20.

Tutton, R., A. Kerr, & S. Cunningham-Burley (2005) "Myriad Stories: Constructing Expertise and Citizenship in Discussions About the New Genetics," in M. Leach, I. Scoones, & B. Wynne (eds), *Science and Citizens: Globalization and the Challenge of Engagement* (London: Zed Press): 101–12.

Van Lente, H. (1993) *Promising Technology: The Dynamics of Expectations in Technological Developments*, Ph.D. diss., Twente University, Enschede, Netherlands.

Ventura Santos, R. & M. Chor Maio (2004) "Race, Genomics, Identities and Politics in Contemporary Brazil," *Critique of Anthropology* 24(4): 347–78.

Waldby, C. (2001) "Code Unknown: Histories of the Gene," *Social Studies of Science* 31(5): 779–91.

Webster, Andrew (2005) "Social Science and a Post-genomic Future: Alternative Readings of Genomic Agency," *New Genetics and Society* 24(2): 227–38.

Willis, E. (1998) "The 'New' Genetics and the Sociology of Medical Technology," *Journal of Sociology* 34: 170–83.

Wright, S. (1994) *Molecular Politics: Developing American and British Regulatory Policy for Genetic Engineering, 1972–1982* (Chicago: University of Chicago Press).

Wright, S. (1996) "Molecular Politics in a Global Economy," *Politics and the Life Sciences* 15: 249–63.

Wright, S. & D. A. Wallace (2000) "Varieties of Secrets and Secret Varieties: The Case of Biotechnology," *Politics and the Life Sciences* 19: 45–57.

Wyatt, S. (2000) "Talking About the Future: Metaphors of the Internet," in N. Brown, B. Rappert, & A. Webster (eds), *Contested Futures: A Sociology of Prospective Techno-science* (Aldershot, U.K.: Ashgate).

Yoxen, E. J. (1982) "Constructing Genetic Disease," in P. Wright & A. Treacher (eds), *The Problem of Medical Knowledge: Examining the Social Construction of Medicine* (Edinburgh: Edinburgh University Press): 144–61.

Yoxen, E. (1983) *The Gene Business: Who Should Control Biotechnology?* (London: Pan Books).

33 Emerging Medical Technologies

Linda F. Hogle

At a recent meeting on nanotechnology, a speaker described the following scenario. A person opens a pill bottle to take a daily dose of medication. In doing so, biosensors on the container transmit information about the person's biochemical status to the primary physician, and an inventory of remaining medication is reported to suppliers. Health status and other information is then relayed back to the person's Black-Berry,[1] stimulating him or her to follow recommendations to purchase goods, change daily patterns, or if nothing more, be aware of his or her body's condition on a daily (or more frequent) basis.

Discussions of medical technologies are often freighted with such fantastical future scenarios, but one need not go that far to see how intimately connected biomedicine is with other domains of life and labor. In fact, it is rare to pick up a newspaper, listen to workplace conversations, or watch entertainment without some reference to medical technology in one of its myriad forms. Medical technologies permeate all aspects of human experience from birth to death, whether one is healthy, disabled, or ill. In addition to diagnosing disorder and replacing bodily function, medical technologies can compile and disseminate information about bodies, monitor physical and mental states, ameliorate or create new forms of suffering, or make people "better than well." Technological systems and the information they provide also affect family and work life, regulate individuals and societies using medically derived norms, and participate in the selection and application of resources to certain groups (and not others). The medical shaping of social identity is thus a significant aspect of medical devices, diagnostic tools, and data dissemination that deserves analysis.

The scenario is a good tool for considering what comprises medical technologies and how tightly connected they have become with other aspects of daily life, commerce, and governance. Medical technologies can be defined as the various devices, instruments, and therapies used for diagnostic, therapeutic, rehabilitative, preventive, or experimental purposes as well as the practices and procedures associated with them. Yet there are conceptualizations of users, of the nature of illness and susceptibility, and of the relations among technologies and the body that animate emerging technologies and create certain kinds of connections in interaction with institutional and technical means. What sort of medical technological system was this scientist imagining? What contributions by clinical practitioners, political authorities, insurers, population health planners, or industrial developers and suppliers of goods and services might lead to this particular assembly of medical-biological, communications, and engineering technologies, and what new knowledge and entities might emerge as a result?

The diversity and extension of technologies into many domains presents a challenge to those who would analyze the field as a set of techniques, knowledge forms, and practices. While it is impossible to cover all technologies, uses, historical precedents, and contemporary dilemmas, this chapter uses representative work from the social and historical study of medicine to illustrate key themes and approaches to studying medical technologies.

The chapter is organized into three parts. The first deals with the centrality of technologies in diagnosis, that is, the determination of the nature and cause of disease. Diagnostic and research data from instruments are essential to such determinations but are in constant interaction with systems of expertise, theories, and the institutions in which they exist. Rather than passively supplying information, technologies may change what constitutes evidence of both the presence of disorder and of the utility of certain therapeutic approaches. Medical technologies, in conjunction with concepts of disease, can categorize individuals into culturally constructed states of normality or pathology and have become a central part of decision-making about managing health problems in certain ways, including prognosis and decisions about which therapies to use. Diagnoses can determine treatments (how and where people will or will not be treated) and prognosis (probabilities and what is to be done). For these reasons, STS researchers have become interested in new forms of subjectivity as technologies affect peoples' lives and work in tangible ways.

The drive toward ever more specific connections of causal mechanisms to illness stimulates a desire for more evidence about which interventions work and under what conditions. The second section deals with testing and evaluating emerging technologies, as this is the phase that links the analysis of diagnosis to therapy. Testing produces various forms of knowledge. Greater volumes of data and specific kinds of proofs are demanded in order to make the link between mechanism, disease, and therapy and to reduce the variability of practices and products thought to create inefficiencies. New products must also be tested to pass regulatory oversight and financial review, as the state, private payers, and other authorities have a stake in decisions about availability and costs of health services. The kinds of evidence sought (predictive, classificatory, economic) are looped back to pragmatic problems of testing, because the definitions, protocol design, and interpretations of results may frame medical problems in particular ways. At the same time, products are reconfigured through early interactions with potential users and those who have something at stake in the introduction of new technologies or in preventing their use.

The final section deals with technological modifications to the body, including therapeutic, aesthetic, and life-extending ones. While medicine has been thought to be about repair, restoration, and the alleviation of suffering, other goals (such as longer life, the elimination of traits perceived to be disabling to individuals and society, the expression of individuality and for some a search for perfection) are increasingly involved, in some cases aligning technologies with identity politics. To see how humans and technologies constitute each other, a number of works in STS explore the expectations, categorizations, hopes, and desires embedded in such emerging technologies and the ways they are deployed.

Selected examples of STS work illustrate various approaches to these themes, and a discussion of recent innovations, in particular, regenerative medicine, will illustrate emerging forms of technological systems that will have broad implications for biological and social life in contemporary global economies. Other chapters in this volume deal with specific technologies including organ transplantation and genetic tests (chapter 34), genetic and reproductive technologies (chapter 32), imaging (chapter 13), and pharmaceuticals (chapter 29), and these topics are only touched upon here.[2]

WAYS OF KNOWING: DIAGNOSIS, DISEASE CLASSIFICATION, AND TECHNOLOGIES

In a seminal paper on what he calls the "tyranny of diagnosis," Charles Rosenberg draws attention to the pivotal role of diagnosis and the ways it has been reconfigured as medicine becomes more technical, specialized, and bureaucratized. He argues that agreed-upon disease categories based on assumptions of ontologically real and specific disease entities have become the core organizing principle in medicine (Rosenberg, 2002). The codification of concepts into bureaucratic systems then becomes the way to control costs, manage deviance, and legitimate certain sick roles (but not others). Ultimately, the resulting taken-for-granted categorizations of patients and disorders structure clinical and patient practices. Integral parts of the way knowledge is produced and standardized are the various instruments, techniques, information, and communication systems collectively called medical technologies.

Apparatuses can be used to extract information with which to establish a body's biological and social status, monitor it over time and circumstances, and report the findings to various types of experts across widespread networks. From this information, large databases can be created with which to define health and illness, reformulate categories of normal and abnormal, make judgments about individuals and populations, provide predictors of risk, and then plan future services and technologies. In this way, assumptions about deservedness, capability, and behaviors are built in to both the technologies and interpretation of data they produce.

Diagnosis, broadly understood, has been studied in STS work by a variety of approaches. The following discussion groups these into historical studies of specific technologies; constructivist, actor-network, and assemblage analyses; studies of classification and standardization processes; and emerging forms of subjectivity.

Technology Histories
Historians have shown how medical technologies often emerge from their original use as research tools and how the development of diagnostic instruments was connected

with theories about disease and bodily function (Marks, 1993). In an era of interest in the mechanical properties of the body, for example, the thermometer was developed to measure temperature changes and the sphygmomanometer to measure the pressure of blood flow to test the heart's pumping efficiency (Porter, 2001). Yet developments of instruments in turn profoundly affected theories of the body and disease. Most notably, microscopy changed what was presumed to be true about cells and their structure. As optical and histological techniques improved, the enhanced ability to observe tissues linked knowledge about anatomy and physiology (Davis, 1981).

Marks (1993) advocates the study of medical technologies by looking at the "life histories" of medical machines. This approach enables understanding the particular skills and techniques that develop around particular instruments. Tracing the role of patients and various users in the design and deployment of specific technologies reveals the multiple origin stories that bear on a technology's biography. However approached, historical studies are critical to understanding the interplay between instruments, theories of disease, and biological and social responses.

Technology, Organization, and the Medical-Industrial Complex

Some authors have extended the study of particular instruments to make visible the ways that work and medical work spaces are affected. For example, equipment such as diagnostic imaging requires specific skills, leading to the development of new professional groups, and costly, large-scale equipment often necessitates architectural changes and clinical facilities with the capacity and expertise to handle it. Sophisticated diagnostics may then be bundled with related services at centralized, often urban locales (Barley, 1988; Blume, 1992; Howell, 1995). Others expand analyses to include the broader informational, organizational, economic, and political systems in which technologies exist, trying to capture power relations in terms of how technologies will be used and by whom. Stanley Reiser in particular drew attention to the situation of technology within more general problems related to the medical-industrial complex. His landmark book, *Medicine and the Reign of Technology*, was a significant contribution during a time of alarming health care cost increases (Reiser, 1978).

Another key work in this vein was Nelkin and Tancredi's *Dangerous Diagnostics* (1989). Writing about the explosion of diagnostic tests, the authors exposed the connection of tests to reimbursement patterns from insurance plans. Payers, interested in limiting costs, may either create an imperative to use tests (as in the screening of potential policy holders for costly diseases or when a high reimbursement rate creates financial incentives for physicians to test many patients) or may restrict access to costly tests (sophisticated studies may be ordered when inexpensive lab tests are inconclusive, but may or may not aid in diagnosis). Insurers desire diagnostic data to estimate life spans and employers to estimate productivity and limit liability.

In a remarkable example of the linkage of state economic interests, the medical industry, and the diagnosis of disease, Plough (1986) demonstrated how concepts of cost-benefit and clinical efficiency were built into medical definitions in the newly created disease category of end-stage renal disease (ESRD). Essentially, the high costs of chronic illness through the mid-twentieth century became the lens through which the complex physiology of kidney and other organ failure were viewed. Ultimately, treatment options were narrowed to dialysis rather than other possible therapeutic options, in large part because of intensive lobbying by manufacturers of the new technology. Plough's work exemplifies a shift to understanding technologies as being constituted by interactions among various elements at differing levels, rather than as having a unidirectional impact "upon" society.

Social Constructions, Material Practices, and Assemblages

The move in STS more generally toward social construction of technology examined the social nature of the way truth claims are made and facts are stabilized. This led a number of researchers to revisit the content of specific artifacts, rather than their use alone. Other constructivists took more of a systems approach, examining artifacts within their institutional environments, which helped link close-in studies of specific technologies to more macro-level views (Bijker et al., 1987). Edward Yoxen's (1990) study of ultrasound's development into a key diagnostic tool is an example. The move from a nonmedical domain (the military) to medicine, and its ultimate use for diagnosing problems in fluid-filled areas of the body, required consensus among diverse groups of clinical medical, engineering, and physics professionals and negotiation across professional, technical, and institutional domains about appropriate applications. Also, the images were difficult to interpret for clinical users accustomed to chemical or radiological data. Perceptual blocks from some potential users could be ameliorated by making the images simpler and easier to read, but this was possible only at the expense of technical complexity. Image data were thus produced not simply as a matter of theoretical science or accurate reproductions of bodily interiors but as a compromise and a result of a series of tradeoffs between reliability and ease of interpretation necessary to make the technology usable in the clinic.

An extension of social studies was actor-network theory, which took seriously both human and nonhuman actors as having a form of agency. Technologies are not passive in this view; rather, they actively intervene in the situations in which they are put to use. Annemarie Mol (2000) illustrates by showing that self-measurement devices for glucose do more than allow for the measurement of preexisting facts. Instead, they alter the value of the facts by changing the target of treatment (more frequent measurements report glucose levels on a different, higher curve than the previous normative ideal). This in turn ratchets down the level of blood glucose deemed to be acceptable. The device made to detect abnormal blood sugar alters what counts as abnormal, Mol argues, creating a type of nonhuman agency.

Social constructivist perspectives are often criticized as placing too much emphasis on social determinants, with insufficient consideration of possible agendas built into technology design and deployment or of the kinds of knowledge being produced. Actor-network studies are criticized because they tend to focus on managers and elite experts in technological domains, with insufficient attention to those who may be less visible but yet are affected by the technology. Using a "social worlds" approach,

Clarke and Montini (1993) point out that there may be actors downstream who may not be directly involved in networks of innovation but are certainly implicated in assumptions and decisions being made on their behalf (see the following section). Another approach to analyzing medical technologies is to consider how the material practices of doing research and clinical work constitute medical knowledge. That is, the cell culture techniques, methods of quantifying and visualizing biological phenomena, and other routine activities in the lab may, for example, affect the way disease models are formulated or how life forms get defined. Similarly, practices involved in categorizing pathologies, handling data, establishing testing or treatment protocols, and determining where patients will be treated (and by whom) are all linked to assumptions about health, illness, and appropriate care (Casper & Berg, 1995; Pickering, 1992).

Observing material practices shows how tools may be made to be the "right tools for the job" (Clarke & Fujimura, 1998). The process of "making it right" may occur even after the technology has been introduced into routine use, as in the case of the Pap smear (Casper & Clarke, 1998). A number of tinkering strategies, including changing definitions and techniques, were required by pathologists, clinicians, public health officials, and others before the technique became accepted as a diagnostic screening tool for cancer. The coordinating and negotiating activities that take place across disciplines and domains have become a key to understanding innovation and knowledge production. In her work on cancer researchers, Fujimura (1987) argued that the work of articulation and alignment in order to gain agreement and stabilize facts is what makes problems "doable."

Alignment of interests, theories, and techniques may affect acceptance, rejection, or routinization of a technology, but so may political exigencies, cultural values, or ethical concerns. Legal concerns, values about life extension, or political issues related to the termination of particular lives may become the dominant factor in the determinations of dead, dying, or salvageable life, trumping network alignments or even evidence of a technology's efficacy (Kaufman & Morgan, 2005; Timmermans, 2002). A number of recent works reflect on cultural influences on technologies, healing traditions in various cultures, and power relationships in the clinic and lab that are important contributions to STS literature on technology (Brown & Webster, 2004; Lock et al., 2000).

Another important aspect of studying diagnosis is the way facts are stabilized. Cambrosio and Keating, among others, demonstrated the subtle ways that medical knowledge is constituted through nomenclature, tacit knowledge, and procedural rituals (1992). For knowledge to be durable, data must be made to be intelligible. Otherwise it has little clinical utility. Test results must also be able to be compared across patients and conditions. Yet protocols to collect and interpret information are based on criteria that are often arbitrary and site-specific and may be limited by capabilities of local expertise. Nevertheless, data have to be intelligible to have clinical utility. Burri and Dumit (chapter 13 in this volume) describe difficulties of interpreting data in imaging technologies, which are particularly problematic. Visual records produced by computerized tomography, ultrasound, PET scanners, and magnetic resonance imaging are not photographic captures of reality but mathematically constructed representations of structures or metabolic functions. Image interpretation requires considerable skill and agreement on what the images really show as well as cross-referencing to other ways of mapping anatomy (see also Cartwright, 1995; Dumit, 2004; Prasad, 2005).

Computerized medical decision-making tools were meant to streamline decision-making at the bedside and increase objectivity by comparing patient information to reference databases and standardized care plans. Although these tools operate on supposedly stabilized facts about diagnoses, other social assumptions about patients and their disorders get built into the systems, as demonstrated ethnographically by Berg (1997) and Forsythe (1996). Although information systems were developed as data interpretation tools to aid in classifying ailments and rationalizing variant and costly practices, they function in multiple roles, including reordering work patterns in the clinic, changing the content of bedside work, and in some cases, reifying power disparities between patients and caregivers.

Blending some ideas from constructivist and network perspectives, a number of researchers view technologies as an assemblage of machines, knowledges, practices, people, histories, and futures. This framing enables a different understanding of the embeddedness and potential power of medicine in our everyday lives. The innovation of the polymerase chain reaction (PCR) for example, illustrates how a concept (the manipulation of genetic material) led to a technique (the ability to identify and amplify DNA), which itself was transformed into a form of knowledge production that has profoundly influenced cultural change in science and in popular understandings of biological life (Rabinow, 1996). Analyzing such transformations sheds light on the emerging forces that animate predictions such as those which opened this chapter.

Keating and Cambrosio successfully illustrate key points about heterogeneity of practices and settings, coordination, and standardization in their extensive study of practices in immunology laboratories. Their recent work is concerned less with laboratory-level phenomena and the production of local knowledge than with inter-laboratory traffic, with attention to the configuration of instruments, people, methods, concepts, and substances that traverse domains of biology and medicine, science and technology, and disciplines within biomedical sciences (Keating & Cambrosio, 2003). They argue that the existence of such networks is necessary for the establishment of classifications from which diagnoses and prognoses are made. The authors call such networks "biomedical platforms." Platforms are more than passive infrastructural or coordinating activities, however. They generate new kinds of biomedical entities that sometimes slip between clinically or laboratory-based definitions of pathology and make networks possible. In this way, the authors distinguish platforms from social or technical networks (theory-methods packages or actor networks).

Using the example of leukemias and lymphomas, diseases that target the immune system, the authors observed local patterns of interpretation that emerged when new techniques and types of expertise were grafted onto existing practices and

organization of work. For example, in the United States it is visually oriented pathologists who are in charge of the labs, whereas in France it is medical biologists, accustomed
to mathematically derived measures. This made a difference in the scoring of cell
markers and, hence, which markers were seen to be clinically relevant. In turn, this had
an effect on attempts to create classification systems with which to diagnose, categorize,
and give prognoses for diseases. But classifications change with new data collected from
additional patients, and they do more than simply order information. Classifications
themselves, then, are tools leading to new knowledge about disease entities.

Classification and Standardization

Classifying patients and diseases involves processes of standardization, which are also
critical for making protocols and instruments work across locales. The less visible work
of standards setting is where cultural forms, power relations, and gate-keeping are
established in ways that not only enable work to proceed across incommensurate
models and data sets but also legitimate particular ways of thinking about disease
(Bowker & Star, 1999).

Standardization activities were central in the transformation of healing practices
into scientific, technological medicine. By the mid-nineteenth century, efforts had
been made to increase the reliability of clinical judgments that previously had been
made by observation of bodily signs and by the physician's senses of touch, smell,
and sight. Newly introduced instruments provided quantifiable measurements of
bodily function, visualizations of bodily interiors, and graphic representations of relationships over time and across subjects.

The quantification of information from and about patients' bodies was meant to
provide an objective snapshot of bodily conditions but also served to create indicators of pathological mechanisms that were thought to be linked to identifiable disease
entities. Whereas diseases had earlier been seen as idiosyncratic with multiple possible causes, concepts of disease categories could now be understood apart from particular bodies and circumstances (Rosenberg, 2002). Furthermore, data from instruments
could be more easily aggregated in ways that could also be used to govern populations. Foucault's (1974) notion of biopower has been influential in this regard. By the
nineteenth century, statistics and other administrative means were employed to survey
and analyze populations and plan state programs for health and welfare. As life itself
became an object of political scrutiny and intervention, both individual bodies and
populations could be subjugated through techniques that included the constant monitoring, testing, and improving of the self (Foucault, 1978; see also Rabinow, 1992;
Turner, 1996).

On the one hand, the increasing specificity of diagnosis matched by ever more targeted tests (whether or not interventions are available) appears to make medicine more
oriented to individuals, while on the other hand, informational technologies enable
data to become more abstracted at the level of populations. Modern biomedicine seeks
to see, chemically analyze, or otherwise detect changes in individuals' bodies down
to the genetic and molecular levels, and considerable investments have been made in

making or adapting tools to do so. At the same time, the data are pooled both to make claims about causal links and to generate standardized, rationalized care plans applicable to large groups.

The effort to standardize clinical practice guidelines involves increased scientific review of new and old therapies to produce comparable, quantifiable proofs of efficacy. This concept, known as evidence-based medicine and public health, has been a powerful trend in health policy, influencing trials of new therapies, payment patterns, and clinical decision-making. Although the intent is to promote best practices for making decisions about patients, current models and proofs often do not take into account the many political, cultural, and behavioral realities that affect interactions among patients, physicians, the health care system, and the environment. At the same time, the way evidence about bodily conditions and medical therapies is produced says much about the mutual penetration of research, industry, the clinic, and the state.

Techniques of biopower can be seen today in the connections between formal medical classification systems and the state system. In their study of the International Classification of Disease (ICD) system, Bowker and Star (1999) outline the links between medical and other welfare systems in which the state has a central role. The authors suggest that an elaborate information system that collects data on many aspects of human life on an ongoing basis and can be mined for a variety of purposes is essential to the state's interest in the health and well-being of citizens, which are also concerns for the good of the state. The result can be improved quality of clinical decision-making, cost savings, and healthier citizens, but it also means increased surveillance and the potential for discrimination for those in- or out-of-category.

Still, there are tensions between attempts to standardize, normalize, and unify bodies and technological practices and the diversity that bodies display under varying conditions, as the set of studies by Berg and Mol (1998) illustrates. The authors argue that diseases and the technologies used to diagnose and ameliorate them are not a single thing to be understood, but rather they become different kinds of objects through material and social practices. Such studies focus on the stories that are told about medical-scientific objects in diverse environments to show how norms get established.

Subjectivity, Identity, and Emerging Medical Technologies

Using diagnostic technologies to name and classify diseases not only provides a means for generalizing across populations, time, and locales but also provides a rationale for justifying giving or withholding treatments and labeling individuals and groups as being ill, aberrant, or "at risk." Diagnostic technologies and classifications thus alter human experiences and subjectivity. Along with theories about the body and its well-being, technologies can serve to sort individuals into groups and reorder social relationships on the basis of classifications. One example comes from Biehl, Coutinho, and Outeiro's study of HIV/AIDS testing in Brazil (2001). Counterintuitively, the people who most requested testing (and repeated testing) were those who were seronegative. The authors argue that testing capitalized on anxiety in target healthy populations, where individuals reported AIDS-like symptoms that often were linked to worries about other social issues such as sexuality and oppressive gender roles. This created a ready market for tests, which people used as way of formulating sexual orientation and identities.

Diagnostic tests can provide evidence of existing or potential pathology, but when data are used to create categories based on probability and risk rather than "normal" or "ill," new forms of subjectivity are created. Genetic tests are increasingly being used to create these new categories. When applied to other nonmedical institutional settings, such categories have far-reaching implications for governance and for individual lives. Susceptibility to substance abuse or chemical sensitivity, potential psychiatric disorders, probability of manifesting a genetic disorder, or even being in the state of carrying a gene can have serious consequences in terms of workplace discrimination or the courts (Dumit, 2000; Rapp, 1999).

Paul Rabinow (1992) argues that the ability to know and administer information about individuals' genetic makeups through genetic testing and the development of genomic biology will have the effect of reordering social relations and the way societies think about therapies. Administrative management of individuals and populations, particularly in the form of risk calculations, will be as critical as or more critical than direct interventions, creating a new form of what he calls "biosociality"—new subjectivities based on medical-administrative categories rather than traditional social relations.

Sorting people into categories of normality on the basis of interpretive data is problematic, as the previous discussion of imaging suggested. "Normal" on radiological scans may mean a "typical" looking or even an ideal type, but it may not necessarily mean "healthy" (Dumit, 2004). Visual images can project other forms of subjectivity, as when images of a developing fetus became cultural icons as much as pieces of medical data to identify abnormalities. The images are thought to create emotional bonding to the fetus and to establish a separate identity for the fetus as a person (Petchesky, 1987; Hartouni, 1997). The additional use of images helped solidify ultrasound's position as an unquestioned clinical tool, even though experience has shown that its use makes little difference in pregnancy outcomes.

TRIALS OF EMERGING TECHNOLOGIES

Trials for drugs, devices, and therapies become an extension of diagnosis through the way knowledge is produced in both local and global contexts. The experiment-therapy continuum is most visible during this part of emerging technology development. During such trials, the less visible participants may appear and controversies might arise.

Oudshoorn chose the clinical testing phase of a new drug as a way to study the way technologies and potential users co-construct each other (2003). The innovation of contraceptives for men (the "male pill") involved the destabilization of dominant cultural narratives about concepts of masculinity. The negotiations of meanings in the testing phase revealed that in order to be culturally as well as technically feasible, an identity of test subjects and potential users as caring, responsible men had to be constructed. Importantly, by viewing a technology's networks from the consumer's point of view and challenging assumptions of who "the user" is, she was able to see the role of subjects' partners in such constructions as well as how various users' perspectives affected product design and potential acceptance.

Clarke and Montini (1993) explored the various interpretations of controversial, early-stage technologies in the context of contentious social issues with their study of the abortifacient RU486. Using a social worlds approach enabled the authors to identify not only the human experts active in the field (reproductive scientists, FDA personnel, politicians, physicians, lobbyists) and the nonhuman actors, but those actors who may be less visible in early development, yet implicated by virtue of the fact that actions taken in the arena will have consequences for them. In this way, rather than viewing the events as a simple story about the relations of domination in reproductive politics, they demonstrated the competing claims and representations of the drug as a "second-generation" birth control pill, a means of mobilizing feminists, a chemical dangerous for women, or a safe way to have an abortion.

With increased involvement of the state in the organization of health care and the need to provide objective measures to replace individual skill and judgment, the randomized clinical trial became the "gold standard" by which more subjective evaluations of new diagnostics and therapies were evaluated scientifically (Marks, 1997; Timmermans & Berg, 2003).[3]

With so many new therapies to evaluate, and with complex innovations involving hybrid components (biological, chemical, computing), there is a demand for expertise in protocol design. Contract research organizations (CROs), generally for-profit groups outside the usual clinical review processes, became an institutional innovation that linked technology companies with health care providers, disease concepts, and human subjects on a global scale. Product manufacturers sponsoring the trials, however, may be concerned with how data from trials can be formulated in a way that serves not only regulatory purposes (proof of safety and efficacy) but also marketing purposes, according to Petryna (2005), who studied global recruitment of human subjects for such trials.

Petryna's main points have to do with the mutability of international guidelines for protecting human subjects in experimental research for the sake of expediency and to pave the way for an emerging human subjects research "industry."[4] Of interest to the analysis of emerging technologies, however, is the observation that North American and Western European populations are too contaminated by technologies to test new ones; that is, the pool of potential subjects is shrinking because too many are already taking drugs or therapies that might interfere with the new therapy being tested. Dumit (2003), in his work on the ubiquity of pharmaceuticals, would argue that this says a great deal about the necessity of chemical technologies to sustain life, at least in North American society.

Since CROs must recruit large numbers of subjects, they do so in countries where medical technologies are less available. The requirement of testing experimental therapies in wealthy countries thus becomes a form of health care delivery in states that cannot afford their own drug and device technologies.[5] A result is what Petryna calls ethical variability, that is, compromises on the type and degree of oversight required based on assumptions about the potential users and their greater willingness to participate within political and economic contexts.

As a technology moves from the status of experimental device to routine therapy, meanings may change. Barbara Koenig (1988) argued that many technologies are quickly taken up because of a "technological imperative" whereby physicians strive to use the latest equipment, even in the absence of much evidence of effectiveness. Illustrating with the history of the adoption of therapeutic plasma exchange for autoimmune diseases, she explored the social processes that led from an infatuation with the new to the somewhat chaotic, learn-as-you-go environment of introduction, to the change in roles and rituals that indicated its acceptance as a routine therapy. The people involved—including nurses, physicians, and patients but also manufacturers' representatives—adapt social roles, supplies, and procedures to make the technology work in a more routinized way. In the process, the adoption of the new tool becomes a moral imperative as well, as the failure to provide a new therapy may come to be seen as unethical.

The examination of knowledge production through clinical trials, instrument design, and data interpretation demonstrates several things: evidence is malleable and takes multiple forms to do what we ask it to do. It may prove a benefit of technologies compared to cost or effectiveness (however that is defined), determine fair allocation, and define users and agenda-setters. Yet the literature on the production of knowledge through medical technology trials reveals a gap in STS studies, that is, the differences in cultural and scientific authority and how local biologies do or do not get incorporated into medical technological systems (Lock et al., 2000).

Diagnosis may have been a revolutionary way of understanding disease through the twentieth century, but the concept itself may now be changing. What is the meaning of diagnosis when many of the emerging technologies meant to detect disorder are being used to test *other technologies*, rather than the person, as in the case of personalized genomics used to determine whether a drug is effective or safe for a specific person, or when brain imaging is used to calibrate other tests? What about imaging used to detect which parts of the brain are aroused in "rational choice" games so that the information can be used for marketing? Home diagnostic tests—often performed without clinical expertise or intervention—are already being used for everything from justifying lifestyle choices to tracing ancestral roots. Individuals in their role as consumers can buy test kits or visit walk-in screening centers without an order from their physician. The media (including the Internet), advertising, and novel, emerging forms of commerce should be further analyzed in relation to these new roles. In addition, disease concepts are not so stable as bureaucratic systems presume. As anyone familiar with STS perspectives knows, definitions made for medical, legal, or social purposes change over time and in varying contexts. Categories built on definitions of "gender identity" or "chemical sensitivity" may not be settled by simply acquiring more bio-logical or other data (Dumit, 2000).

At the very least, ways of analyzing the technologies involved may need to be revisited. Most of the emerging technologies are hybrids of computational, com-munications, mechanical, chemical-pharmaceutical, and biological elements. Each component brings in various kinds of expertise, participants, and ways of defining the medical problem to be solved. Many are being used for multiple purposes simultane-ously, such as a therapy for a disorder being used to enhance performance in a healthy person, which positions them differently both socially and technologically in each setting. In the next section, I recount additional forms of emerging technologies that challenge our current ways of understanding medical technologies' role in contem-porary life.

TECHNOLOGICAL MODIFICATIONS OF THE BODY

By the mid-1980s, interest in medical and social practices affecting life and health turned toward a focus on the body. While some writers suggested that the body is being displaced by information, imaging, and other representational technologies (Hayles, 1999; Martin, 1992; Waldby, 2000), others were drawn to the materiality of the body, asking what it means to be "biological" and what the implications of medical technologies are for how people live, work, conduct research, or receive health care (Franklin & Lock, 2003). Issues of identity and subjectivity were central to all these explorations, as emerging medical sciences and technologies disrupted received notions about cultural meanings of the body and its social relations.

Yet it is alterations to the body, including prosthetics and surgical or pharmaceuti-cal interventions, that most visibly transform identities and the way we interact in the world. There have long been concerns about the boundaries of the human. The distinctions between living and dead, persons or things, the natural and the cultural or technological, human and other species have been questioned throughout history, as evidenced in religious concerns, literature, and public controversies in reac-tion to innovations in science and medicine, particularly experiments with extending or altering life forms. Concerns about boundaries have become blurred in new ways with the advent of artificial organs, recombinant gene techniques, and other biotech-nologies that combine biological, mechanical, and sometimes other species compo-nents. A number of works in the 1990s challenged romanticized ideals of pure or natural categories, including older concepts of individuals as bounded, autonomous subjects. Instead, these works used the image of the cyborg or hybrid to explore emerg-ing relationships among information and machine technologies, humans, and animals. Through research on the transplantation of human or animal tissue, artifi-cially assisted reproduction, artificial life forms, bioinformatics, and other boundary-crossing technologies, they questioned taken-for-granted categories of nature and culture (Brown & Michaels, 2001; Gray, 2001; Hayles, 1999; Latour, 1993; Strathern, 1992).

Donna Haraway (1991, 1997) in particular led new thinking about human and non-human relationships by drawing attention to the way that lives and identities are fashioned through boundary crossings. Boundaries, often assumed to be sacred and static, are easily transgressed by emerging biological technologies that can create artificial cells and chromosomes, life forms made from minimal genomes. Symbiotic relations are being established between machines and humans at the intimate, subcellular level with nanotechnologies, computer-mediated representations, patented life forms, and the marriage of information technologies with biology and medicine. Haraway's subjects are the knowledge-power practices that become inscribed in particular material ways and their consequences for a politics of the body in contemporary technoscience. The work of Haraway, Hayles (1999), and others has stimulated discussion and critique of both seductive and ominous notions of what it might mean to be "trans" or "post" human. Beyond STS communities, however, the moral quality of such debates has resulted in backlash from neoconservatives in political and public life (Fukiyama, 2002).

To avoid getting stuck in moral judgments about such fantastical possibilities, it may be more useful to trace the history of techniques for augmenting human function, asking the question of what particular kinds of subjects are being created by the way augmentations are conceptualized and executed.

Prosthetics, Bionics, and "Being Fit"

Humans have always fashioned replacements for failed or lost body parts and tissue. The array of contemporary parts replacements is stunning. It includes, to name a few, artificial hearts, artificial respiration machines, artificial retinas and cochlear implants to restore partial sensory function, and myoelectric limbs that can automatically adjust to environments and sense user movement intention.

A central question in the design and deployment of augmentations is what it means to have an able body and how appearance and function affect identity. For example, as several medical historians have suggested, the sight of disfigured bodies can create pubic anxieties about vulnerability and the ability of citizens to take care of themselves. There may be social pressure to "fill in" missing parts or to provide prosthetics to serve as implements with which to perform certain work tasks. In this sense, prosthetics function on several levels: to demonstrate progress in medical technology and the ability to fix anything, to assuage national conscience about bodies mutilated in war or industrial trauma, and to enable individuals to function socially as well as physically (Ott et al., 2002; Thomson, 1997).

Lupton and Seymour (2000), in a study of disabled individuals' relations to their assistive technologies, argue that technologies have both the potential to enhance self-hood at the same time that they exacerbate the meanings of disability. Assistive technologies such as communication boards and wheelchairs may augment function and provide independence and control, but they may also suggest dependence and difference. The theme of fitness was somewhat differently pursued by Emily Martin, who writes about the alignment of national and corporate competitiveness goals and social expectations to have flexible, immunologically adaptable and physically strong bodies (1994).

There is a complicated set of issues about the normalization of nonstandard bodies that has not yet been fully explored in STS studies. In particular, processes of defining who is disabled, who may receive what kind of replacement device or therapy, and how this relates to national policy and views of technology use need further explication (L. Davis, 1995; Kohrman, 2003).[6] Blume's (1997) remarkable, detailed study of the introduction of cochlear implants to amplify sound showed how the deaf resisted the designation of "disabled," claiming that theirs is a community identity based on a variance from other parts of the population and should be left alone. The deaf felt the use of implants was an attempt to use medical technology to discriminate against them and normalize or change their identity.[7]

If individuals experience their body as expression of identity, then differing kinds of prosthetics (for mobility, visible normality, or sex change or cosmetic purposes) may bring private bodies and public identities into alignment. "What, after all, did the natural body mean if an engineered body approximated one's sense of self far more persuasively than the body one was born into?" (Serlin, 2002: 3).

Steven Kurzman is among the few social scientists who examine how individuals and their bodies learn to "perform" with a prosthesis in his work on amputees (Kurzman, 2002). Somewhere between the patient's descriptions of discomfort and mobility needs and the practitioner's analysis of gait and the biomechanics of the adapted device, tacit knowledge and a mutual reading of body signs work together to align the body with its new body part. In this example, technologies (the prosthetic, measuring, and testing devices and the medical record, among other things) are one facet of the articulation of need, solution, and identity as a person-with-artificial-parts.

Steven Kurzman's observations of artificial leg design in the United States, rural India, and Cambodia illustrate the work of incorporating local body culture into body modifications (2003). The Jaipur limb uses locally available materials rather than high-tech Western imports, with designs that suit local geography and needs, such as the ability to walk on different kinds of terrains, with or without shoes, or to squat or sit low, rather than in chairs. Significantly, the design principles were meant to be a part of an overall community-based rehabilitation effort, using local skills and knowledge. In this way, these prosthetics became enabling not only to the wearer but also in the way they were situated in transnational production systems and networks of power.

The replacement and exchange of body parts is troubling for some analysts. In their pioneering analyses of organ substitution dating from the 1950s, Renée Fox and Judith Swazey expressed their fundamental concern about the social obligations involved in exchanging human body parts as valuable resources (1974, 1992). In their interpretation, the enterprise of "remaking people" through kidney dialysis and transplantation was based on assumptions that the body is mechanistic and fragmentary. Such a reductionist view of the body shifts the focus of medical practitioners and product developers to replacing worn-out parts rather than dealing with the root causes of suffering, dying processes, and the proper provision of care. At the same time, the concern over supply and demand issues arising from the view of organ failure as an allocation problem contributed to the transformation of health care into a consumption activity (see chapter 34 in this volume for a more extended discussion of organ transplantation and its researchers).

Their continued concern about a "spare parts" approach to medical therapy was evident in later work on the development of an artificial heart and, more recently, analysis of the clinical trials for the AbioCor totally implantable artificial heart (Fox & Swazey, 1992, 2004). By describing the suffering of the trial participants (most of whom died within a short time) and the ambiguity created when some bodily functions are replaced even while others are failing (the recipients had to be in end-stage heart failure to be eligible for the trial, and the mechanical heart could have outlasted the patient), Fox and Swazey raised the question of what "success" means in the attempt to create synthetic substitutes for vital organs. They also observed what they felt was an explicit deployment of American cultural symbols to describe both the trial and its participants. Experimental subjects were described as pious and hard-working community members who became pioneers in the frontier of medicine, and as such, American heroes. For the authors, the recruitment of subjects as cultural symbols raises questions about what other social purposes are being pursued by experimental medicine, in particular, selling the need to use medical technologies to demonstrate national progress and superiority.

Enhancement Technologies

With the technological capability and social acceptability of modifying and augmenting the body, there has been rapid growth in the kinds of procedures and assists made available. Most of these have been based on repair and restoration of function, but increasingly, techniques are being employed to improve mental and physical traits beyond what is considered to be normal or necessary for life. These so-called "enhancement technologies" raise unexamined questions about cultural notions of deficiency and ability in contemporary societies, as well as what constitutes "therapy" and adequate care (L. Davis, 1995; Hogle, 2005; Sinding, 2004).

The use of medical interventions to improve human performance begs the question, what are the proper goals of medicine and is it ever possible to distinguish them from other social goals? Foucault (1978) argued that modern states regulate their subjects not through repressive means, but rather through social institutions such as biomedicine. Medicine can define health as a certain type of fitness toward the goal of producing good citizens and culling out deficient ones. Large social projects such as eugenics and policies that selectively favor or disfavor certain states of being are examples. Biopower relates to governments' concerns with fostering the life of the population, which is achieved through disciplines (regulation) of the body. Individuals are encouraged to participate in the constant monitoring, testing, and improving of the self (Foucault, 1978; Turner, 1996). Viewed through the neoliberal lens, responsibility for betterment has shifted from the state to individuals themselves, who are expected to strive toward goals of ever higher physical and mental functioning—a sort of therapeutic culture of the self (Rose, 2001; Rose & Novas, 2005). The goals are often defined by commercial interests in selling improvement products as much as state interests in regulating bodies (Bordo, 1998; Dumit, 2003; Featherstone, 1991). Rather than operating at the level of improving societies then, the body becomes the object of improvement work.

Some explain enhancements more simply as being consistent with a culture that pursues perfection and desires to avoid the difficult problems of everyday life, or as a symptom of less tolerance for variation and flaws (Fukiyama, 2002; President's Council on Bioethics, 2003). Others suggest that enhancements are liberatory, that they are essentially a type of self-chosen evolution (Bostrom, 2003). Caplan and Elliott (2004) argue that attempts to achieve better performance should be allowed on the basis that science should be provided free rein for discovery. Such moral judgments about the rightness of enhancements, and about who should receive them and under which circumstances, are linked with ways of ordering and valuing individuals in various societies (Parens, 1998; Elliott, 2003). For example, when aging is viewed as an economic and social "problem" rather than a normal life process, there are several possible responses: invest social resources in preventive care, innovate technologies to ameliorate suffering from degenerative disorders associated with aging, or develop techniques to extend lives or reverse the aging process. The aged may feel stigmatized, less able to compete for jobs, or in other ways subjects in need of technological intervention and may either seek a more youthful appearance through cosmetic surgery or seek rejuvenation through other anti-aging enhancements (Post & Binstock, 2004).

The confluence of technology, identity, and consumerism has been most thoroughly analyzed in relation to individuals' decisions to alter their bodies for cosmetic purposes. The history of aesthetic surgery illustrates the intertwined influences of the demand for technical expertise in reconstructive surgery to treat injuries from war and the increase in injuries with the industrial revolution, trends in postwar medical practice, and cultural concerns about appearances during a time when increasing immigration to the United States drew attention to ethnic differences (Gilman, 1999; Haiken, 1997). Yet the number and kinds of techniques to improve appearance have proliferated in many countries,[8] becoming one of the most profitable forms of medical technology available. In addition to procedures to make individuals more youthful or beautiful (according to changing cultural notions of what that means), new procedures such as collagen injections into foot pads, toe shortening, and navel repositioning are performed to accommodate changing fashion trends. Hand rejuvenation and chin implants, popular among business and sales professionals, suggest a desire to change appearance for competitive reasons (American Society of Plastic Surgeons, 2004). Susan Bordo (1998) and Anne Balsamo (1992) argue that gender is the filter through which cultural norms of beauty and treatment of bodies are interpreted and practices of power are played out and that such procedures are ultimately harmful to women because negative messages are reinforced for other women. Kathy Davis's (1995) ethnography of plastic surgery patients, however, suggests that women are not passive victims; rather, they become active agents in changing their position in society by changing their appearance.

Arguments about autonomy and the choice to reconstruct bodies and selves are at the heart of many enhancement technologies. Carl Elliott (2003) argues that the rapid rise of psychopharmaceuticals is as much due to an individual's search for authenticity in modernity and the commercially stimulated obsession with personal identity as it is a treatment for disorders. Some of his patients, for example, claim to feel "more like themselves" on the drugs than off.

Decisions about how such enhancements are viewed are consequential, since they may determine how technologies may be used, who is allowed access, and who will pay. Breast augmentation, for example, is considered to be a cosmetic procedure and would normally not be covered in many insurance systems, but the case of reconstruction after breast removal raises the question of whether the procedure is critical enough to a woman's identity and well-being to warrant calling it a therapy. Such policy dilemmas make moral judgments about the social importance of breasts more visible. Complicating policy and practice decisions further is the growing practice of using therapies for nonintended ("off-label") uses to enhance performance, as in the case of drugs for Alzheimer's disease and narcolepsy being used to improve memory or cognitive performance in normal individuals, or of erythropoietin or gene therapy for sports performance (Behar, 2004; Hall, 2003a).

Neural enhancements become particularly troublesome, as they may involve changes in memory, cognition, and behavior. The introduction of prototypes for a prosthetic hippocampus and a microchip for memory processing (to increase, pattern, or erase memory) and drugs to alter brain chemistry will require new analyses of subjectivity and personhood in its most intimate relation to technology (Gray, 2001; Farah & Wolpe, 2004; Hall, 2003a; Healy, 2004; Rose 2003).

Further analysis is needed to explore the way enhancement technologies entail various life strategies and define multiple subjectivities. Scrutiny of the kinds of decisions being made about the appropriateness of using biology to solve social problems has been central in studies of biopolitical projects such as past eugenics programs, contemporary genetic testing, public health screening for risk predisposition, the human genome project, and others (Rapp et al., 2001). Yet enhancements demonstrate that there are niches where medical problem-solving around specific social issues of fairness and physical and social advantage may link up with concepts of bodily ability in more complex ways. Future studies should include critique of the persistent lack of historical and social contexts in much of bioethics and health policy literature that has framed debates up to this point (Hogle, 2005).

The human-technology interface, as demonstrated in the use of prosthetics, implants, and enhancements, can transform identities and disrupt received notions of what it means to be human. The regulation of newborns, the elderly, the brain-injured, or others' bodies through technological assistance and social and biological orders is similarly destabilizing (Lock, 2002; chapter 34 in this volume; Kaufman & Morgan, 2005; Timmermans, 2002). The ambiguous states that result often call for further legal-medical management to resolve the dissonance. In some cases, as with widely publicized cases of persistent vegetative state and controversial new techniques that manipulate life forms in unexpected ways, the state may become more centrally involved. In the following section, I elaborate the example of regenerative medicine to illustrate the multiple mediations that medical innovations can make.

Regenerative Medicine

Like previous benchmark medical technologies, regenerative medicine transforms human identity and kinship; the relations of physicians, researchers, and patients; and social and economic forms of exchange. Regenerative medicine thus becomes a model case for studying long-standing conversations in the social study of science and technology about the production of knowledge and reframings of life, work, and social relationships. Technoscientific production based on bodily materials also has significant implications for transnational and local governance and for the role of democratic participation in science and medicine.

The neologism regenerative medicine (RM) stands for the set of sciences and technologies involved in the collective project meant to coax the body to repair itself and potentially to extend the lifespan.[9] Common to definitions and descriptions are the intertwined notions of production (of therapies, capital goods, and solutions to problems of health and aging) and the promise of the controlled ability to design, proliferate, and dispense new forms of living tissue.

A dominant theme in the RM narrative has to do with the naturalness of the sources for healing (as compared to mechanical or chemical fixes); however, the need to control processes requires technological assistance. The ability to redirect certain cells to become other kinds of cells, to make them perform functions alien to them, to get them not only to proliferate but also to form three-dimensional structures and to stay where clinicians want them to be involves remaking life sciences and engineering technologies in ways that disturb cultural ideas about the relations of bodies to bodily constituents and what constitutes life. Two RM researchers explicitly compare the techniques to other kinds of engineering design challenges: "Coaxing cells to form tissues in a reliable manner is the quintessential engineering design problem that must be accomplished under the classical engineering constraints of reliability, cost, governmental regulation and societal acceptance" (Griffith & Naughton, 2002: 1010). Technological assistance in creating life forms has been elaborated in work on reproductive medicine, cloning, and preimplantation genetic diagnosis techniques (Cussins, 1998; Franklin, 2001, 2005; Franklin & Roberts, 2006; Strathern, 1992; Franklin & Ragoné, 1998).

STS researchers have recently developed a more nuanced view of the complex relationships of technical tools, economies, and exchange systems in the production of human tissues, pushing analyses of body commodification beyond limited discussions of dehumanization, reduction, and deconstruction of the body into parts, and commercialization processes that have characterized many studies of technologies and the body. This view also differs from previous analyses of the discursive nature of the body as information without materiality. Haraway (1997) uses the term "corporeal fetishism" to describe how the new techniques are not just a way of using and exchanging living things, but also involve appropriating nature so as to change the relations of human and nonhuman things and create new forms of material-semiotic-work objects.

Waldby uses the term "biovalue" to describe the way biological technologies increase both use value and exchange value of tissue that would otherwise have little use. By this she means "the yield of vitality produced by the biotechnical reformulation of living processes" (Waldby, 2002: 310). Waldby argues that older forms of gift exchange give way to a complex new set of relationships among donors, recipients, caregivers, and institutions as a result. Franklin and Lock (2003) note that the very meanings of terms such as production, capital, and value are changing as life itself is discursively and materially converted into forms that can enter domains of explicitly commercial exchange. The ability not only to reuse human parts but also to redesign and repli-cate them on demand opens up possibilities for multiple platforms for either thera-peutic (tissue repair) or industrial (drug discovery, diagnostics, bioweaponry) use on a global scale. After Charis Cussins, they argue that older forms of production as a basis for capital accumulation are being displaced in biotechnology by reproduction. As they put it, bodies that reproduce, particularly by proliferating replicas of their tissue, are regulated under different systems and ways of thinking than bodies that labor (Franklin & Lock, 2003: 7). Central to these shifts is the concept of promissory capital, that is, capital raised for speculative ventures based on promised future returns (Franklin, 2005; Hogle, 2003a; see also chapter 34 in this volume for a discussion of body commodification).

There may be multiple economies, however, based on where RM is allowed to proceed and in what form, across which national borders its products may be traded, and whether they are developed as "off the shelf" products readily available for poten-tial users or matched specifically to individuals using a patient's own cells. Such economies develop in interactions with other systems, including health programs that prioritize individualized medicine, cultural and historical situatedness of reproductive medicine practices and uses of embryos, global trade systems, and forms of governance over areas of science, health, and ethics (Faulkner & Kent, 2001; Gottweis, 1998; Prainsack, 2004).

Viewed from another perspective, the new cellular techniques hail a new era in the material culture of experimental systems. Examining the history of cell culture, Landecker (2002, 2007) shows how existing techniques were recombined to create a new experimental object—cell cultures that were observable outside the body—which in turn became a new concept: life *in vitro*. In Landecker's analysis, the shock was not that material could exist outside the body (that is, the problem of conceptualizing the relationship of parts-to-wholes). Rather, it represented a shift from thinking about bio-logical processes as internal and unobservable to those that could be external, visible, and malleable. The conceptual shift to thinking of tissue as living, not in the process of dying, was essential to efforts to perfuse organs and tissues outside the body during the same time period. The spatial and temporal reorganization involved in making cells more amenable to experimentation and preservation thus had the effect of transforming biological theories and practices.

Similarly, in contemporary cloning and stem cell work, de-differentiation and reprogramming cells to perform non-native functions involves temporal functions that move processes forward and backward in time and spatial functions of attracting and activating proteins in non-native patterns. The ability to arrest and pace the development and proliferation of cells according to chosen specifications introduces a new temporality that not only enables biological phenomena thought to be impossible, but allows procedures to be stabilized and standardized (Sunder Rajan, 2003; Waldby, 2002).

Still, cells must have legitimacy as a tool. The "rightness" of the tool is not only the outcome of the articulation and alignment of skills, institutions, theories, equipment, and funding but also of the political environment in which they exist (Clarke & Fujimura, 1998). European researchers have taken the lead in STS work on emerging forms of governance over regenerative medicine (Gottweis, 1998, 2002; Kent & Faulkner, 2002; Salter, forthcoming; Webster, 2005). Brian Salter and colleagues have analyzed economic indicators and potential markets for RM products, tracing funding flows from venture and other forms of capital in a way that reveals both the articulation points with national economic projects and private commercial interests, locates the gaps and volatility in market support, and attempts to create international partnerships to move state and scientific interests forward (2005). Their comparative European survey indicated the need to acknowledge that powerful cultural and religious values affect scientific issues. Salter concluded that the traditional technocratic means for deciding on science policy issues on the basis of scientific authority has given way to a different kind of cultural authority necessary for the research to proceed. In both the EU funding initiatives and the United Nations debates about global bans on cloning techniques, there was a process of negotiation and "cultural trading" of values in order to reach compromises that would enable transnational research to proceed. At the same time, the compromises are unstable as narratives shift, new techniques are promoted by particular groups, and new participants who may not be involved in attempts at political compromise become involved. The entry of Korea, China, India, and other non-Northern sites is perhaps most threatening to countries that have dominated the production of knowledge and products in biomedical science.

Adult and embryonic stem cell research raises questions about the moral status of the embryo, which is not covered in detail here.[10] Rooted in the literature on controversy, a few content analyses are appearing that trace the way narratives about new RM techniques shape and are shaped by ethical debates and political battles (Leach, 1999; Hornig Priest, 2001). Popular accounts of cloning have also been written by science journalists and researchers (Hall, 2003b; Kolata, 1998; Wilmut et al., 2000). As with many new medical technologies, exaggerated claims of potency, endless flexibility, and potential profitability on the one hand, and the consequences of making immoral choices on the other, play an essential role in communication about RM.

Franklin (2005: 59) beautifully describes the symbiotic relationship between communication vehicles, investors, consumers, and various public audiences. She uses an analogy to the cellular "feeder layer" required to make stem cells thrive in culture: "A direct media feeder system links developments in stem cell research to the possibility of treatment for severe, disabling and often fatal conditions—binding stem cell technology securely into a rhetorical fabric of hope, health and an improved future through increasing biological control." Considerable effort has been devoted to representing stances on RM research as being aligned with dominant political, religious, or scientific points of view, making RM a dramatic example of the high stakes medical science can have in politics. Alternative naming schemes, theories about transdifferentiation or fusion, and alternative techniques for acquiring stem cells recently proposed by politicians, not scientists, are attempts to enable research and commerce to proceed without losing support from political constituencies. Such work on the part of scientists, politicians, and public interest groups may exemplify a new form of ethics in which responses to public accountability are built in to emerging technologies. Ultimately, acellular biomaterials may win out as a way to avoid controversy altogether.

Although some observers marvel that a "science matter" has become so caught up in public politics, historical studies of embryology and developmental biology show how past scientific debates continue to inform current debates (Hopwood, 2000; Maienschein, 2003). The kinds of evidence and reasoning applied to identifying the beginnings, growth, and endings of life established the context for contemporary theories and research practice but with somewhat different objectives in mind.[11] Instruments and techniques of dealing with embryos—specimen collections, microscopes, and microtomes for dissection—became critical in transforming the field into contemporary developmental biology. Such tools were used by early embryologists to bring gestational development into the realm of biomedicine, making this effort a part of the project of disciplining and regulating interpretations of embryos (Morgan, 1999).[12] Contemporary interpretations continue to be highly contested but only partly because of new techniques. Debates about the status of embryos are clearly inflected by early negotiations and representations.

Engineering approaches to controlling life are today also dominating emerging medical technologies, through new techniques in synthetic biology and systems biology. The shift from gene to three-dimensional system that took place in the late twentieth century is moving to the subcellular level in the twenty-first, as nanobiotechnologies are further transforming fundamental understandings of how life works.

Nanobiotechnologies (NBTs) operate at the most fundamental molecular building-block levels (Toumey, 2004). One nanometer (nm) is one millionth of one millimeter, and most NBTs function in the range of 0.1 to 100 nm, manipulating individual atoms and molecules. When biological processes are used to create new devices and materials, as in the self-assembly of rods and wires for biosensors, or when metal atoms can be attached to proteins in the body to form indwelling diagnostic systems, old questions about what is the proper relationship of human bodies to technologies are resurrected.

Biomedical applications for NBTs include both diagnostic and therapeutic uses (biochemistry monitors, drug delivery systems, heat-emitting molecules that home in on tumor cells rather than large areas). Additionally, they enable ways to observe phenomena at the atomic level (quantum dots to observe DNA movement; microfluidics and gating systems to observe chemotaxis, which is the ability of cells to attract or be repelled by stimuli). In addition to questions about risk, the idea of indwelling sensors and signaling systems raise concerns about security, privacy, and external control.

Self-assembled DNA, artificial cells, and minimal genes are inherently ambiguous. Directed by humans but fabricated in the body, how are we to think of such entities for legal-regulatory purposes or for the meaning of being human, or for that matter, for being biological? How may such intimately indwelling devices change cultural concepts of the "normal" body, and how do we understand changes to the human technology interface? Who are the "users," and who has intention and agency? Importantly, to whom does it matter and for what purposes? These technologies have enormous moral, legal, scientific, and commercial implications, and the public and political debates about them may affect the field.

THE FUTURE OF STS STUDIES OF MEDICAL TECHNOLOGIES

Bionics, enhancement technologies, regenerative medicine, and nanobiotechnologies represent the kinds of emerging technologies Strathern (1992) described in her commentaries on technoscience in late modernity. As she sees it, through hyperquantified, cultural techniques of design and control, Nature can be assisted in ways that modify traits to suit essentialized notions of what natural entities should be. Yet efforts to optimize corporeal existence through medical technologies are not possible without a circuit of enterprise, biology, medicine, and culture. Residing in increasingly complex relations among these domains, the new technologies "include emergent life forms that refigure traditional understandings of economy governance and biology" (Franklin, 2005: 60).

Biological enterprises such as stem cell and genome projects involve global cooperation but also competition based on national agendas and desires to capture biocapital. In her writings on bioinformatics and genomic science, Fujimura (2003) suggests that Japanese scientists may manipulate culture as another tool to produce or reorder worlds in alignment with both national economic priorities and ideals of Japanese society. Like Fujimura, Rabinow and Dan-Cohen (2005; Rabinow, 1996) are concerned with the way entrepreneurial biomedical science makes futures of a certain sort through the introduction of enabling technologies and social systems. The challenge for social scientists is how to study such far-reaching, globe-spanning phenomena.

Increasingly, there is crossover to other domains not previously thought of as having medical social relations. The defense industry and governmental agencies have long researched medical interventions and enhancements to alter soldiers' performance and survival (Gray, 2001; Hoag, 2003; Talbot, 2002). New product marketing and sales techniques utilize brain scanning to study brand loyalty (Huang, 2005). The primary market for cellular technologies will most likely be as drug discovery–enabling tools (Hogle, 2005) rather than therapies. There is a need to understand how such technologies have medical and nonmedical significance, how assumptions made in one domain traverse territories, and what the implications are for the systems in which therapy, rehabilitation, cure, and normalization may develop new meanings.

The extent to which medical technologies become a part of informal economies is receiving insufficient attention in STS. Most analyses of medical technologies assume the boundaries of state-sanctioned exchange systems and regulatory schemes and thus analyze only the visible forms they take in hospitals, clinics, and labs. People in cash-poor countries know how to acquire and exchange goods, ranging from common or exotic pharmaceuticals to laboratory equipment, outside recognized, formal channels and across national borders. Such "shadow economies," as Nordstrom (2004) calls them, are not confined to illegal activities; rather, they often cross legal, illegal, and quasi-legal divides and involve extensive networks of people and resources. Street vendors, middlemen, and the wealthy form stable, normalized, international networks that link flows of cash to and from Western and Eastern manufacturers, INGO development programs, and informal economies. Military medical supplies in war zones and nonmonitored goods from development agencies that end up in shadow economies may not only provide an affordable means of health care for many more people than their original targets, but the profit from their sale is returned to both local and transnational economies. The use of immunosuppressive drugs has the unintended effect of creating brokerage systems for organs.

In a more formalized way, many countries are benefiting from the limitations caused by the high cost of high-tech medical care in industrialized nations by setting up clinics, surgery services, and experimental therapy shops specifically geared toward health "tourists." Thailand, India, Brazil, and former Soviet states now attract patients seeking less expensive knee replacements, cosmetic surgery, cancer treatments, and experimental cellular therapies from physicians who likely were trained in the United States or United Kingdom. Infertility treatments and organ transplants have become sought-after and profitable services, despite some limited regulation on nonresident access to such therapies. Unproven therapies with unknown risks performed by unlicensed practitioners will continue to attract global trade, but this newer phenomenon is a twist on the pattern of desperate patients traveling to get treatments that are not approved in their home countries. More studies need to be made of both advanced technologies and their fate in poor countries as well as innovations arising out of less-studied countries (Prasad, 2005).

Transplant patients trade their immunosuppression drugs in parking lots to fellow patients who have hit the limit of what insurers will pay for their medications. Older

Americans get their prescriptions filled in Canada. High school and college students sell their Ritalin to peers for better test performance in a competitive environment. Athletes buy blood to boost their oxygen-use capacity, brokers sell ova on world markets, and patients travel to China to get stem cell therapies for their degenerative disorders. Medical technologies thus are implicated in creating new forms of economic possibilities and political power that are inseparable from mainstream economies and product planning.

With such dramatic effects on life, labor, and governance, it is little wonder that the study of medical technologies has been an area of intense focus in studies of science, technology, and society. The techniques, forms of knowledge, and practices emerging in the twenty-first century promise to be fruitful ground for future studies of medical technology and will contribute to understandings of science, technology, and society more generally.

Notes

A portion of the material in this chapter is based on work supported by the National Science Foundation under grant #0539130. Any opinions expressed in this material are those of the author and do not necessarily reflect the views of the National Science Foundation.

1. Hand-held personal communication device.

2. It is impossible to cover all technologies and themes exhaustively. The reader is referred to review papers as well as studies of medical technologies that illustrate the interactions between technologies, health care delivery systems, professional practices, and roles and organizations.

3. Randomized, large-scale clinical trials produce volumes of data, but unless carefully designed, may not produce evidence that can guide meaningful action. There are ethical concerns as well: testing of technology on humans means some must do without known therapies. Controlled trials often require a "wash-out" period when a subject must be on no medications that may conflict with the test, raising an ethical concern about denial of the standard-of-care.

4. Richard Rettig (2000) has also flagged the increase in for-profit clinical research, and he raises concerns about conflicting interests and types of evidence produced when the conditions of testing may not be neutral.

5. A different situation exists in societies where participation in and access to clinical and scientific information may enable consumer activists to claim a role in setting the conditions and interpretations of clinical trials. Such was the case in Epstein's story of AIDS activists, which argues that changes in lay authority and expertise are affecting the way technologies are defined and tested (1996). Löwy, however, argues that the significance of political and economic considerations varies depending on the social position of participants and the symbolic interpretations of diseases such as AIDS, as compared with cancer (2000).

6. There is an extensive and rich literature in disability studies; however, few of these sources deal specifically with the social study of assistive technologies, and only a few with cultural variations in meanings of disabled bodies (Kohrman, 2003; Kurzman, 2003).

7. The ethical questions quickly became more complicated when developers decided that it was young children who should be implanted while, it was thought, they still had a chance to develop "normal" speech and social interaction. This raises an important question that is relevant for a number of surgical or chemical interventions that would likely be performed in children, such as sex

assignment or the use of growth hormone for short stature. Who should make the decisions about interventions with body modifications, under what conditions, and what are the cultural implications?

8. In the United States alone, there were 8.3 million procedures in 2003, and of these, 37 percent of patients had already had at least one procedure performed (ASPS, 2004).

9. Tissue engineering combines life sciences and engineering to develop biological substitutes and to stimulate tissue generation in the body. Products are often hybrids of cells, biological materials such as growth factors, and biomaterials (scaffolds, computer chips).

Stem cells are early-stage, undifferentiated cells with the ability to both self-replenish and generate cells that can become several types of tissue (pluripotency). For example, stem cells derived from adult tissue can develop into mesenchymal cells (an intermediate, progenitor type of cell), which in turn can produce bone, tendon, and ligament. Embryonic stem cells (hESCs) are derived from a 4- to 6-day-old embryo (called a blastocyst) and are considered to be totipotent, since they can generate all types of tissue.

Nuclear transfer or cloning is a technique used to create a blastocyst from which hESCs could be removed. The nucleus is removed from a donor egg, then genetic material from another individual is inserted into the egg and the entity is allowed to mature to the blastocyst stage. Stem cells removed from the blastocyst would be immunologically and genetically similar to the genetic donor, so if they are used to repair tissue or treat a disease in that donor, there would be no need for immunosuppression (as in organ transplantation or artificial organs). This is referred to as therapeutic cloning. If, on the other hand, the blastocyst were implanted into a woman's uterus with the intention of producing a living child, rather than having its stem cells removed, it would be called reproductive cloning, which is currently prohibited in most countries. Variations on nuclear transfer techniques are prohibited in some countries and states within the United States. See Hall (2003b) for a detailed history of the early participants in cloning.

10. The questions of whether embryos should have special protection from destruction and what sorts of research should be allowed are covered in Holland et al. (2001; see also Mulkay, 1994). Attempts to delineate moral and legal permissibility of how to treat blastocysts have been mired in competing ways to define what constitutes a human life—stage, number of cells, biomarkers, ability to implant in a uterus, or the act of conception itself (McGee & Caplan, 1999). The search for a bright line with which to define and bound personhood in early-stage life reveals it to be a Maginot line, dependent on local politics, historically situated religious and cultural influences, and economically driven national policies.

11. Embryos have long been invoked as political actors, particularly as visualization techniques and displayed embryo specimen collections made embryos more visible both to embryologists and some members of the public (Morgan, 1999). Hartouni (1997), Rapp (1999), and others have observed that in contemporary embryo politics, the availability of images of the developing fetus (from ultrasound imaging, amniocentesis reports, and other representations that make fetuses less abstract and more recognizable) resulted in personification in ways that enhanced antiabortion activists' campaigns. Printed displays of ultrasound images rapidly became a normalized part of the pregnancy experience in popular culture. More recently, obstetricians have been supplying to couples photos of the blastocysts created during *in vitro* fertilization procedures, visualizing "life" at earlier stages (about 8–16 cells). Yet Morgan argues that in the early twentieth century, images themselves did not result in personification. Rather, embryos at that time were more connected with problems of race, evolution, and the relations of human and nonhuman species than with abortion issues (Morgan, 1999). The term embryo had little relevance to early embryologists because they were less concerned with naming stages of the entity than with the processes of morphogenesis and identifying commonalities or distinctions that might give clues to species descent and developmental divergence (Maienschein, 2003).

12. Historical studies are also enlightening with regard to contemporary thinking about generating life through experimental means. Experiments on regeneration by Roux and Drietsch at the end of the

nineteenth century (on frogs and sea urchins, respectively) were concerned with questions of regulation of generative processes, that is, whether development could be initiated from environmental factors or was driven from within the cells (Maienschein, 2003; Pauly, 1987). In a detailed review of early experiments on parthenogenesis, Pauly shows how Jacques Loeb attempted to find proof that all life can be engineered. Loeb believed there were mechanical explanations for all biological phenomena, including behaviors.

References

American Society of Plastic Surgeons (ASPS) (2004) "Procedural Statistics Trends 1992–2004." Available at: http://www.plasticsurgery.org.

Balsamo, Anne (1992) "On the Cutting Edge: Cosmetic Surgery and the Technological Production of the Gendered Body," *Camera Obscura* 28: 207–39.

Barley, Stephen (1988) "The Social Construction of a Machine: Ritual, Superstition, Magical Thinking, and Other Pragmatic Responses to Running a CT Scanner," in M. Lock & D. Gordon (eds), *Biomedicine Examined* (Boston: Kluwer): 497–540.

Behar, Michael (2004) "Will Genetics Destroy Sports? A New Age of Biotechnology Promises Bigger, Faster, Better Bodies," *Discover* 25(7): 40–45.

Berg, Marc (1997) *Rationalizing Medical Work: Decision Support Techniques and Medical Problems* (Cambridge, MA: MIT Press).

Berg, Marc & Annemarie Mol (eds) (1998) *Differences in Medicine: Unraveling Practices, Techniques, and Bodies* (Durham, NC: Duke University Press).

Biehl, João, Denise Coutinho, & Ana Luzia Outeiro (2001) "Technology and Affect: HIV/AIDS Testing in Brazil," *Culture, Medicine, and Psychiatry* 25: 87–129.

Bijker, Wiebe, Thomas Hughes, & Trevor Pinch (eds) (1987) *The Social Construction of Technological Systems* (Cambridge, MA: MIT Press).

Blume, Stuart (1992) *Insight and Industry: On the Dynamics of Technological Change in Medicine* (Cambridge, MA: MIT Press).

Blume, Stuart (1997) "The Rhetoric and Counter-Rhetoric of a 'Bionic' Technology," *Science, Technology & Human Values* 22(1): 31–56.

Bordo, Susan (1998) "Braveheart, Babe, and the Contemporary Body," in E. Parens (ed), *Enhancing Human Traits: Ethical and Social Implications* (Washington, DC: Georgetown University Press): 189–220.

Bostrom, Nick (2003) "Human Genetic Enhancements: A Transhumanist Perspective," *Journal of Values Inquiry* 37(4): 493–506.

Bowker, Geoffrey & Leigh Star (1999) *Sorting Things Out: Classification and Its Consequences* (Cambridge, MA: MIT Press).

Brown, Nik & Mike Michaels (2001) "Switching Between Science and Culture in Transpecies Transplantation," *Science, Technology, & Human Values* 26(1): 3–23.

Brown, Nik & Andrew Webster (2004) *New Medical Technologies and Society: Reordering Life* (Malden, MA: Polity Press).

Cambrosio, Alberto & Peter Keating (1992) "A Matter of FACS: Constituting Novel Entities in Immunology," *Medical Anthropology Quarterly* 6(4): 362–84.

Caplan, Arthur & Carl Elliott (2004) "Is It Ethical to Use Enhancement Technologies to Make Us Better Than Well? *Public Library of Science Medicine* 1(3): 172–75. Available at: http://www.plosmedicine.org.

Cartwright, Lisa (1995) *Screening the Body: Tracing Medicine's Visual Culture* (Minneapolis: University of Minnesota Press).

Casper, Monica & Marc Berg (1995) "Constructivist Perspectives on Medical Work: Medical Practices and Science and Technology Studies," *Science, Technology, & Human Values* 20(4): 395–407.

Casper, Monica & Adele Clarke (1998) "Making the Pap Smear into the 'Right Tool' for the Job: Cervical Cancer Screening in the USA, Circa 1940–95," *Social Studies of Science* 28(2): 255–90.

Clarke, Adele & Joan Fujimura (1998) "What Tools? Which Jobs? Why Right?" in A. Clarke & J. Fujimura (eds), *The Right Tools for the Job* (Princeton, NJ: Princeton University Press): 3–44.

Clarke, Adele & Theresa Montini (1993) "The Many Faces of RU486: Tales of Situated Knowledges and Technological Contestations," *Science, Technology, & Human Values* 19(1): 42–78.

Cussins, Charis (1998) "Producing Reproduction: Techniques of Normalization and Naturalization in Infertility Clinics," in A. Franklin & H. Ragoné (eds), *Reproducing Reproduction: Kinship, Power, and Technological Innovation* (Philadelphia: University of Pennsylvania Press): 66–101.

Davis, Audrey (1981) *Medicine and Its Technology: An Introduction to the History of Medical Instrumentation* (Westport, CT: Greenwood Press).

Davis, Kathy (1995) *Reshaping the Female Body: The Dilemma of Cosmetic Surgery* (New York: Routledge).

Davis, Lennard (1995) *Enforcing Normalcy: Disability, Deafness and the Body* (New York: Verso).

Dumit, Joseph (2000) "When Explanations Rest: 'Good-Enough' Brain Science and the New Sociomedical Disorders," in M. Lock, A. Young, & A. Cambrosio (eds), *Living and Working with the New Biomedical Technologies: Intersections of Inquiry* (Cambridge: Cambridge University Press): 209–32.

Dumit, Joseph (2003) "Liminality, Ritual, and the Brain Image: Pharmaceutical Reflections," plenary talk at the ASA Decentennial, Manchester, U.K., July 14–18.

Dumit, Joseph (2004) *Picturing Personhood* (Princeton, NJ: Princeton University Press).

Elliott, Carl (2003) *Better Than Well: American Medicine Meets the American Dream* (New York: Norton).

Epstein, Stephen (1996) *Impure Science: AIDS, Activism, and the Politics of Knowledge* (Berkeley: University of California Press).

Farah, M. & Paul Wolpe (2004) "Monitoring and Manipulating Brain Function: New Neuroscience Technologies and Their Ethical Implications," *Hastings Center Report* 34(3): 35–45.

Faulkner, Alex & Julie Kent (2001) "Innovation and Regulation in Human Implant Technologies: Developing Comparative Approaches," *Social Science and Medicine* 53: 895–913.

Featherstone, Mike (1991) "The Body in Consumer Culture," in M. Featherstone, M. Hepworth, & B. Turner (eds), *The Body: Social Process and Cultural Theory* (London: Sage): 170–96.

Forsythe, Diana (1996) "New Bottles, Old Wine: Hidden Cultural Assumptions in a Computerized Explanation System for Migraine Sufferers," *Medical Anthropology Quarterly* 10(4): 551–74.

Foucault, Michel (1974) *The Birth of the Clinic: An Archaeology of Medical Perception* (New York: Random House).

Foucault, Michel (1978) *The History of Sexuality,* vol. 1: *An Introduction* (New York: Pantheon).

Fox, Renée & Judith Swazey (1974) *The Courage to Fail: A Social View of Organ Transplants and Dialysis* (Chicago: University of Chicago Press).

Fox, Renée & Judith Swazey (1992) *Spare Parts: Organ Replacement in American Society* (New York: Oxford University Press).

Fox, Renée & Judith Swazey (2004) "'He Knows That Machine Is His Mortality': Old and New Social and Cultural Patterns in the Clinical Trial of the AbioCor Artificial Heart," *Perspectives in Biology and Medicine* 47(1): 74–99.

Franklin, Sarah (2001) "Culturing Biology: Cell Lines for the Second Millennium," *Health* 5(3): 335–54.

Franklin, Sarah (2005) "Stem Cells R Us: Emergent Life Forms and the Global Biological," in A. Ong & S. Collier (eds), *Global Assemblages: Technology, Politics, and Ethics as Anthropological Problems* (Malden, MA: Blackwell): 59–78.

Franklin, Sarah & Margaret Lock (eds) (2003) *Remaking Life and Death: Toward an Anthropology of the Biosciences* (Santa Fe, NM: School of American Research).

Franklin, Sarah & Helena Ragoné (1998) *Reproducing Reproduction: Kinship, Power and Technological Innovation* (Philadelphia: University of Pennsylvania Press).

Franklin, Sarah & Celia Roberts (2006) *Born and Made: An Ethnography of Preimplantation Genetic Diagnosis* (Princeton, NJ: Princeton University Press).

Fujimura, Joan (1987) "Constructing Do-able Problems in Cancer Research: Articulating Alignment," *Social Studies of Science* 17: 257–93.

Fujimura, Joan (2003) "Future Imaginaries: Genome Scientists as Sociocultural Entrepreneurs," in A. Goodman, D. Heath, & S. Lindee (eds), *Genetic Nature/Culture: Anthropology and Science Beyond the Two-Culture Divide* (Berkeley: University of California Press): 176–99.

Fukiyama, Frances (2002) *Our Posthuman Future: Consequences of the Biotechnology Revolution* (New York: Farrar, Straus, and Giroux).

Gilman, Sander (1999) *Making the Body Beautiful: A Cultural History of Aesthetic Surgery* (Princeton, NJ: Princeton University Press).

Gottweis, Herbert (1998) *Governing Molecules: The Discursive Politics of Genetic Engineering in Europe and the United States* (Cambridge, MA: MIT Press).

Gottweis, Herbert (2002) "Stem Cell Policies in the United States and Germany: Between Bioethics and Regulation," *Policy Studies Journal* 30: 444–69.

Gray, Chris H. (2001) *Cyborg Citizen: Politics in the Posthuman Age* (New York: Routledge).

Griffith, Linda & Gail Naughton (2002) "Tissue Engineering: Current Challenges and Expanding Opportunities," *Science* 295(5557): 1009–14.

Haiken, Elizabeth (1997) *Venus Envy: A History of Cosmetic Surgery* (Baltimore, MD: John Hopkins University Press).

Hall, Stephen (2003a) "The Quest for a Smart Pill," *Scientific American* 289(3): 54–65.

Hall, Stephen (2003b) *Merchants of Immortality: Chasing the Dream of Human Life Extension* (New York: Houghton Mifflin).

Haraway, Donna (1991) *Simians, Cyborgs and Women: The Reinvention of Nature* (New York: Routledge).

Haraway, Donna (1997) *Modest Witness@Second Millenium—FemaleMan Meets OncoMouse: Feminism and Technoscience* (New York: Routledge).

Hartouni, Valerie (1997) *Cultural Conceptions: On Reproductive Technologies and the Remaking of Life* (Minneapolis: University of Minnesota Press).

Hayles, N. Katherine (1999) *How We Became Posthuman: Virtual Bodies in Cybernetics, Literature, and Information* (Chicago: University of Chicago Press).

Healy, Melissa (2004) "Botox for the Brain," *Los Angeles Times*, December 20, p. F1.

Hoag, Hannah (2003) "Remote Control: Could Wiring Up Soldiers' Brains to the Fighting Machines They Control Be the Future Face of Warfare?" *Nature* 423: 796–97.

Hogle, Linda F. (2003a) "Life/Time Warranty: Rechargeable Cells and Extendable Lives," in S. Franklin & M. Lock (eds), *Remaking Life and Death: Toward an Anthropology of the Biosciences* (Santa Fe, NM: School of American Research): 61–96.

Hogle, Linda F. (2003b) "The Anthropology of Bioengineering," *Anthropology Newsletter* 44(4).

Hogle, Linda F. (2005) "Enhancement Technologies and the Body," *Annual Review of Anthropology* 34: 695–716.

Holland, Suzanne, Karen Lebacqz, & Laurie Zoloth (eds) (2001) *The Human Embryonic Stem Cell Debate: Science, Ethics, and Public Policy* (Cambridge, MA: MIT Press).

Hopwood, Nick (2000) "Producing Development: The Anatomy of Human Embryos and the Norms of Wilhelm His," *Bulletin of the History of Medicine* 74(1): 29–79.

Hornig Priest, Susanna (2001) "Cloning: A Study in News Production," *Public Understanding of Science* 10: 59–69.

Howell, Joel (1995) *Technology in the Hospital: Transforming Patient Care in the Early Twentieth Century* (Baltimore, MD: Johns Hopkins University Press).

Huang, Gregory (2005) "The Economics of Brains," *Technology Review* (May): 74–76.

Kaufman, Sharon & Lynn Morgan (2005) "The Anthropology of the Beginnings and Ends of Life," *Annual Review of Anthropology* 34: 337–62.

Keating, Peter & Alberto Cambrosio (2003) *Biomedical Platforms: Realigning the Normal and the Pathological in Late-Twentieth-Century Medicine* (Cambridge MA: MIT Press).

Kent, Julie & Alex Faulkner (2002) "Regulating Human Implant Technologies in Europe: Understanding the New Regulatory Era in Medical Device Regulation," *Health, Risk and Society* 4(2): 189–209.

Koenig, Barbara (1988) "The Technological Imperative in Medical Practice: The Social Creation of a 'Routine' Treatment," in M. Lock & D. Gordon (eds), *Biomedicine Examined* (Boston: Kluwer): 465–96.

Kohrman, Matthew (2003) "Why Am I Not Disabled? Making State Subjects, Making Statistics in Post-Mao China," *Medical Anthropology Quarterly* 17(1): 5–24.

Kolata, Gina (1998) *Clone: The Road to Dolly and the Path Ahead* (New York: William Morrow).

Kurzman, Stephen (2002) "There Is No Language for This: Communication and Alignment in Contemporary Prosthetics," in K. Ott, D. Serlin, & S. Mihm (eds), *Artificial Parts, Practical Lives: Modern Histories of Prosthetics* (New York: New York University Press): 227–48.

Kurzman, Stephen (2003) *Performing Able-Bodiedness: Amputees and Prosthetics in America*, Ph.D. diss., University of California, Santa Cruz.

Landecker, Hannah (2007) *Culturing Life: How Cells Become Technologies* (Cambridge, MA: Harvard University Press).

Landecker, Hannah (2002) "New Times for Biology: Nerve Culture and the Advent of Cellular Life in Vitro," *Studies in History and Philosophy of Biological and Biomedical Sciences* 33: 567–623.

Latour, Bruno (1993) *We Have Never Been Modern* (Cambridge, MA: Harvard University Press).

Leach, J. (1999) "Cloning, Controversy, and Communicaton," in E. Scanlon, R. Hill, & K. Junker (eds), *Communicating Science: Professional Contexts* (New York: Routledge): 218–30.

Lock, Margaret (2002) *Twice Dead: Organ Transplants and the Reinvention of Death* (Berkeley: University of California Press).

Lock, Margaret, Alan Young, & Alberto Cambrosio (eds) (2000) *Living and Working with the New Medical Technologies: Intersections of Inquiry* (New York: Cambridge University Press).

Löwy, Ilana (2000) "Trustworthy Knowledge and Desperate Patients: Clinical Tests for New Drugs from Cancer to AIDS," in M. Lock, A. Young, & A. Cambrosio (eds), *Living and Working with the New Medical Technologies* (New York: Cambridge University Press): 49–81.

Lupton, Deborah & Wendy Seymour (2000) "Technology, Selfhood, and Physical Disability," *Social Science and Medicine* 50(12):1851–62.

Maienschein, Jane (2003) *Whose View of Life? Embryos, Cloning and Stem Cells* (Cambridge, MA: Harvard University Press).

Marks, Harry (1993) "Medical Technologies: Social Contexts and Consequences," in W. Bynum & R. Porter (eds), *Companion Encyclopedia of the History of Medicine*, vol. 1 (London: Routledge): 1592–618.

Marks, Harry (1997) *The Progress of Experiment: Science and the Therapeutic Reform in the United States, 1900–1990* (New York: Cambridge University Press).

Martin, Emily (1992) "The End of the Body?" *American Ethnologist* 19(1): 121–40.

Martin, Emily (1994) *Flexible Bodies: Tracking Immunity in American Culture: From the Days of Polio to the Age of AIDS* (Boston: Beacon Press).

McGee, Glenn & Arthur Caplan (1999) "What's in the Dish? Ethical Issues in Stem Cell Research," *Hastings Center Report* 29(2): 109–36.

Mol, Annemarie (2000) "What Diagnostic Devices Do: The Case of Blood Sugar Measurement," *Theoretical Medicine and Bioethics* 21: 9–22.

Mol, Annemarie (2002) *The Body Multiple: Ontology in Medical Practice* (Durham, NC: Duke University Press).

Morgan, Lynn (1999) "Materializing the Fetal Body: Or What Are Those Corpses Doing in Biology's Basement?" in L. Morgan & M. Michaels (eds), *Fetal Subjects, Feminist Positions* (Philadelphia: University of Pennsylvania Press): 43–60.

Mulkay, Michael (1994) "The Triumph of the Pre-embryo: Interpretations of the Human Embryo in Parliamentary Debate Over Embryo Research," *Social Studies of Science* 24: 611–39.

Nelkin, Dorothy & Laurence Tancredi (1989) *Dangerous Diagnostics: The Social Power of Biological Information* (New York: Basic Books).

Nordstrom, Carolyn (2004) *Shadows of War: Violence, Power, and International Profiteering in the Twenty-first Century* (Berkeley: University of California Press).

Ott, K., D. Serlin, & S. Mihm (2002) *Artificial Parts, Practical Lives: Modern Histories of Prosthetics* (New York: New York University Press).

Oudshoorn, Nelly (2003) *The Male Pill: A Biography of a Technology in the Making* (Chapel Hill, NC: Duke University Press).

Parens, Erik (1998) *Enhancing Human Traits: Ethical and Social Issues* (Washington, DC: Georgetown University Press).

Pauly, Philip (1987) *Controlling Life: Jacques Loeb and the Engineering Ideal in Biology* (New York: Oxford University Press).

Petchesky, Rosalind (1987) "Fetal Images: The Power of Visual Culture in the Politics of Reproduction," *Feminist Studies* 13(2): 263–92.

Petryna, Adriana (2005) "Ethical Variability: Drug Development and Globalizing Clinical Trials," *American Ethnologist* 32(2): 183–97.

Pickering, Andrew (1992) "From Science as Knowledge to Science as Practice," in A. Pickering (ed), *Science as Practice and Culture* (Chicago: University of Chicago Press): 1–28.

Plough, Alonzo (1986) *Borrowed Time: Artificial Organs and the Politics of Extending Lives* (Philadephia: Temple University Press).

Porter, Roy (2001) "Medical Science," in R. Porter (ed), *Cambridge Illustrated History of Medicine* (Cambridge: Cambridge University Press): 154–242.

Post, Stephen & Robert Binstock (2004) *Fountain of Youth: Cultural, Scientific and Ethical Perspectives on a Biomedical Goal* (New York: Oxford University Press).

Prainsack, Barbara (2004) "Negotiating Life: The Biopolitics of Embryonic Stem Cell Research and Human Cloning in Israel," paper presented to the EASST conference, Paris, August 27.

Prasad, Amit (2005) "Scientific Culture in the 'Other' Theater of 'Modern Science': An Analysis of the Culture of Magnetic Resonance Imaging Research in India," *Social Studies of Science* 35(3): 463–89.

President's Council on Bioethics (2003) "Beyond Therapy: Biotechnology and the Pursuit of Happiness." Available at: http://www.bioethics.gov/reports/beyondtherapy.

Rabinow, P. (1992) "Artificiality and Enlightenment: From Sociobiology to Biosociality," in F. Delaporte (ed), *Incorporations* (New York: Zone Books): 234–51.

Rabinow, Paul (1996) *Making PCR: A Story of Biotechnology* (Chicago: University of Chicago Press).

Rabinow, Paul & Talia Dan-Cohen (2005) *A Machine to Make a Future: Biotech Chronicles* (Princeton, NJ: Princeton University Press).

Rapp, Rayna (1999) *Testing Women, Testing the Fetus: The Social Impact of Amniocentesis in America* (New York: Routledge).

Rapp, Rayna, Deborah Heath, & Karen-Sue Taussig (2001) "Genealogical Disease: Where Hereditary Abnormality, Biomedical Explanation and Family Responsibility Meet," in S. Franklin & S. McKinnon (eds), *Relative Values: Reconfiguring Kinship Studies* (Durham, NC: Duke University Press): 384–412.

Reiser, Stanley (1978) *Medicine and the Reign of Technology* (New York: Cambridge University Press).

Rettig, Richard (2000) "The Industrialization of Clinical Research," *Health Affairs* 19(2): 129–46.

Rose, Nikolas (2001) "The Politics of Life Itself," *Theory, Culture, Society* 18(6): 1–30.

Rose, Nikolas (2003) "Neurochemical Selves," *Society* 41(1): 46–59.

Rose, Nikolas & Carlos Novas (2005) "Biological Citizenship," in A. Ong & S. Collier (eds), *Global Assemblages: Technology, Politics, and Ethics as Anthropological Problems* (Malden, MA: Blackwell): 439–63.

Rosenberg, Charles (2002) "The Tyranny of Diagnosis: Specific Entities and Individual Experience," *Milbank Quarterly* 80(2): 237–60.

Salter, Brian (forthcoming) "The Global Politics of Human Embryonic Stem Cell Science," *Global Governance*.

Serlin, David (2002) *Replaceable You: Engineering the Body in Postwar America* (Chicago: University of Chicago Press).

Sinding, Christiane (2004) "The Power of Norms," in J. H.Warner & F. Huisman (eds), *Locating Medical History: The Stories and Their Meanings* (Baltimore, MD: Johns Hopkins University Press): 262–84.

Strathern, Marilyn (1992) *Reproducing the Future: Anthropology, Kinship and the New Reproductive Technologies* (New York: Routledge).

Sunder Rajan, Kaushik (2003) "Genomic Capital: Public Culture and Market Logics of Corporate Biotechnology," *Science as Culture* 12(1): 87–121.

Talbot, David (2002) "Super Soldiers," *Technology Review* 105(8): 44–50.

Thomson, Rosalind Garland (1997) *Extraordinary Bodies: Figuring Physical Disability in American Culture and Literature* (New York: Columbia University Press).

Timmermans, Stefan (2002) *Sudden Death and the Myth of CPR* (Philadelphia: Temple University Press).

Timmermans, Stefan & Marc Berg (2003) *The Gold Standard: The Challenge of Evidence-Based Medicine and Standardization in Health Care* (Philadelphia: Temple University Press).

Toumey, Chris (2004) "Cyborgs in Nanotech: Hopes and Fears for the Human," paper presented at "Imaging and Imagining the Nanoscale," Columbia, SC.

Turner, Bryan (1996) *The Body and Society*, 2nd ed (Thousand Oaks, CA: Sage).

Waldby, Catherine (2000) *The Visible Human Project: Informatic Bodies and Posthuman Medicine* (New York: Routledge).

Waldby, Catherine (2002) "Stem Cells, Tissue Cultures and the Production of Biovalue," *Health: An Interdisciplinary Journal for the Social Study of Health, Illness and Medicine* 6(3): 305–23.

Webster, Andrew (2005) "Introduction: International Comparison of Health Technologies," paper presented at the Workshop on International Health Technology Assessment, Rome, June 20–22.

Wilmut, Ian, Keith Campbell, & Colin Tudge (2000) *The Second Creation: Dolly and the Age of Biological Control* (New York: Farrar, Straus, and Giroux).

Yoxen, E. (1990) "Seeing with Sound: A Study of the Development of Medical Images," in W. Bijker, T. Hughes, & T. Pinch (eds), *The Social Construction of Technological Systems: New Directions in the Sociology and History of Technology* (Cambridge, MA: MIT Press): 281–303.

34 Biomedical Technologies, Cultural Horizons, and Contested Boundaries

Margaret Lock

[I]nside and outside, as experienced by the imagination, can no longer be taken in their simple reciprocity; consequently, by omitting geometrical references . . . by choosing more concrete, more phenomenologically exact inceptions, we shall come to realize that the dialectics of inside and outside multiply with countless diversified nuances.[1]

With increasing facility, practitioners of biomedical technologies are able to manipulate the human body, not merely in their effort to represent and cure pathology but with intent to enhance what "nature" has endowed. In carrying out these activities, a straightforward reciprocity between inside and out can no longer be sustained. The body of modernity with its stable boundaries is increasingly under threat owing to an incremental circulation of body organs, tissues, and molecules. The new technologies have extensive ethical, social, and political repercussions. In this essay I review practices associated with two particular biomedical technologies: organ procurement and transplantation, and genetic testing for single gene disorders and complex diseases. Although both technologies are designed primarily to accomplish medical ends— in the one case to save lives and in the other to assess future risk of disease—their effects extend beyond the medical sphere, creating new forms of biosociality and subjectivity.

I will emphasize here the rupture and reformulation of material, social, and national boundary demarcations associated with these particular technologies. In the case of organ transplants, the biologies of self and other are hybridized, frequently resulting in temporary identity confusion or more permanent identity transformation. Genetic test results have the potential to create molecular genealogies that vie with kinship and ethnic ties based on a shared, real, or presumed heredity. These boundary reformulations challenge normative expectations about embodiment, identity, and relationships of individuals to familial and other social groups, as well as conventional cultural horizons and even global politics. Professional and public discussions about social and ethical matters affect the further development of the technology in question and the guidelines or limits put on its implementation.

From a review of research about different geographical locations, this essay will expose cultural assumptions and the local moral, individual, and social consequences associated with the implementation of these technologies. Needless to say, extensive social repercussions are by no means limited to the two technologies under consideration (see, e.g., Hogle, chapter 33 in this volume). Moreover, organ transplants and genetic testing have little in common as far as actual material practices are concerned and the people who become most intimately involved with these technologies do not resemble each other closely. Those who undergo transplants are very ill; those who opt for genetic testing hope to learn something about future risk of illness. But once the performance of these technologies is situated in a broader context, their commonalities immediately become apparent.

First, both organ transplantation and molecular genetics and genomics belong to the era of biomedicalization (Clarke et al., 2003; see also Condit, 1999). These technologies can only come to fruition and be put into practice where the apparatus of technomedicine is present. Both rely heavily on a complex infrastructure involving advanced computer technologies, local and global networking, and the storage and transfer of data and samples. Second, both technologies are based on the knowledge of molecular biology that started to accumulate from the first half of the twentieth century—immunology and tissue matching in the case of transplants, and DNA typing in the case of genetic testing and screening.

Third, the question of ownership of the human body and body parts comes to the fore in connection with both technologies. Do individuals have property rights in their own body organs, tissues, and genetic material? Do their families have rights with respect to deceased bodies and body parts? Should communal ownership of genetic material be recognized under certain circumstances—among self-identified indigenous peoples, for example? Or does genetic material belong to humanity at large? Fourth, applications of these technologies inevitably bring human relatedness to the fore and raise questions about hybridity, body boundaries, embodiment, and identity. A fifth feature common to the use of both these technologies involves decisions as to which lives count as worthwhile and should be "saved" and which can be justifiably "sacrificed." Related to this is a sixth point: the practice of organ transplants and the global collection of DNA material as well as genetic testing are technologies that have the potential to exacerbate and, indeed, magnify already existing inequities.

My approach is grounded in recognition of a co-construction among those individuals who create and apply these particular technologies, the technological artifacts in question, and the individuals on whom the technologies are practiced. Such an approach avoids both technological reductionism and essentialization of users (Oudshoorn & Pinch, 2003: 3; see also Oudshoorn & Pinch, chapter 22 in this volume). Adding to the complexity, biomedical knowledge and associated technological practices depend on the agency of the material body; here too co-construction is evident.

My examination of the "social life" of transplanted organs and DNA materials (Kopytoff, 1986) focuses on individuals directly affected by the practices of these specific technologies. I do not deal in depth with the "performance" of solid organs or DNA samples in preparation sites and operating rooms, genomics centers, and genetics laboratories (see, e.g., Hogle, 1999: 140–85). Instead, I interrogate embodiment, relatedness, and identity and consider the ownership and commodification of the body alongside broader social and political aspects of both these technologies.

CULTURAL HORIZONS

A decade ago, Andrew Feenberg argued that the "legitimating effectiveness of technology depends on unconsciousness of the cultural-political horizon under which it is designed" (1995: 12). He suggested that a "recontextualizing critique of technology can uncover that horizon, demystify the illusion of technical necessity, and expose the relativity of the prevailing technical choices" (1995: 12). Today, with the systematic introduction of institutional review boards, regulatory committees, Royal commissions, and the like, it has become *de rigueur* to monitor the bioethical issues associated with any given technology, but these activities usually leave the cultural-political horizon unexamined. Setting out to consider only the ethical consequences that follow from the implementation of biotechnologies is insufficient for a project of "recontextualization," as Feenberg understands it.

Research in the social sciences, science studies, and feminist studies has shown repeatedly that the presence of the social precedes and is embedded within any kind of technoscientific project, initiative, or discovery (Franklin, 2003; Grove-White, 2006; Haraway, 1991, 1997; Lock, 2002b, 2003; Strathern, 2005; Wright, 2006). Historical and comparative research into the development and implementation of biomedical technologies provides an excellent opportunity to glimpse cultural-political horizons at work (Adams, 2002; Cohen, 1998; Hogle, 1999; Latour, 1988; Lock, 1993, 2002a; Rapp, 1999; Shapin & Schaffer, 1985). Such research encourages reflexivity and the denaturalization of all normative practices, including bioethical reviews.

Before turning to the specific technologies under consideration, I will briefly revisit the concept of culture. Once the significant cross-cultural differences in the application of biomedical technologies are recognized, it is all too easy to account for this by means of cultural relativism while at the same time assuming that practices in the West are in effect devoid of culture and better accounted for in terms of politics.

In common with many anthropologists today, Marilyn Strathern is adamant that if the problematic concept of culture is to be used at all, then it must be applied ubiquitously, to all societies. She argues that the concept of culture "draws attention to the way things are formulated and conceptualized as a matter of practice or technique. People's values are based in their ideas about the world; conversely ideas shape how people think and react." She adds, "ideas always work in the context of other ideas, and contexts form semantic (cultural) domains that separate ideas as much as they connect them" (1997: 42). However, culture is neither static nor totalizing. Values are never distributed equally across populations and are inevitably implicated in relationships of power and the maintenance of inequalities. Moreover, values are perennially open to dispute. Arjun Appadurai (1990: 5) argues that a major problem today is "the tension between cultural homogenization and cultural heterogenization." He points out that by homogenization is usually meant "Americanization" and/or "commoditization." A second process, that of indigenization, often goes unnoticed even though it transforms newly diffused ideas, knowledge, behaviors, technologies, and material goods to "fit" with the cultural horizons of their new localities. It has been shown repeatedly that artifacts, including biomedical technologies, can be introduced successfully into new cultural settings without a simultaneous adoption of the use originally associated with them (Lock & Kaufert, 1998; van der Geest & Whyte, 1988). New meanings and social relations coalesce around such transported artifacts. This is not an argument for the autonomy of artifacts (or for that matter for the autonomy of culture) but rather for their inherent heterogeneity as social objects. Alternatively, some artifacts and technologies, notably when they threaten entrenched values, are actively rejected, fail to take root, or are severely restricted.

At the national level, the idea of culture is often self-consciously appealed to in order to reaffirm a shared tradition. This reinvented history is frequently imagined as uncorrupted by either colonial forces or modern influences. Thus, mytho-history is invoked to create an idealized past out of which culture can be turned into an "exclusionary teleology" (Daniel, 1991: 8) sometimes having profound effects on the application of technology (Hogle, 1999; Lock, 2002a).

Marking out differences among peoples is not the only exclusionary use made of the concept of culture. Culture is also exclusionary when conceptualized in opposition to nature, as the "natural" order, not created by human endeavor but by a higher power or by the forces of evolution. This margin, where culture is perceived to encroach on the natural world, can become a site for the emergence of disputative moralizing discourses laying bare the limits of what is believed tolerable within a given community (Brodwin, 2000).

USERS AS CONSUMERS AND CITIZENS

In situations of extreme deprivation, or where violence is the norm, technologies can be used to victimize people—forced sterilization being a case in point. However, research in gender studies has shown that recipients of technological interventions are rarely passive participants (see, e.g., Casper & Clarke, 1998; Cowan, 1987; Cussins, 1998; Ginsburg & Rapp, 1995; Lock & Kaufert, 1998; Trescott, 1979; and others). Furthermore, Casper and Clarke have highlighted diversity among users, insisting on their status as implicated actors (1998; see also Lie & Sørensen, 1996).

Rose and Blume (2003) take issue with the way in which users of technology are so often conceptualized as consumers. They point out that in the "developed" world, users of biomedical technologies are first and foremost citizens who are participants in government-run health care systems (with the notable exception of the United States). Governments and associated advisory bodies, including relevant professional organizations and NGOs, set up guidelines and regulations about who can have access to specific technologies and under what circumstances. Restrictions apply for example to the use of certain technologies (e.g., gene therapy and xenotransplants) in experimental settings. Procurement of organs and selection of recipients are also carefully controlled by national or regional organizations. The state or organizations delegated with the appropriate power thus "configure" eligibility and access to biomedical technologies. In other instances, claims of equal access for everyone are belied by the limited number of publicly run facilities. As a result, reproductive technologies and genetic testing are also available through private clinics or genomic companies on a fee-for-service basis. Advertising aimed directly at consumers plays a large role in promoting such practices. Discussing vaccines, Rose and Blume (2003: 108) argue that while "individuals are always and necessarily implicated" as both citizens and consumers, this relationship must be established separately for each particular biomedical technology. One cannot be a consumer of organs on the free market without breaking the law in most countries. The majority of genetic tests are not available to users simply by virtue of their being citizens, because health care services do not support such expenditures. It is not difficult, however, to become a consumer of genetic tests for a price—this is regarded as a matter of individual choice.

AGENCY, IDENTITY, AND TRANSFORMED SELVES

Charis Cussins (1998: 168) has pointed out that a considerable amount of research has shown the "dependence of science and technology on social, individual, and political factors" but that what she terms the "other direction"—the dependence of selves on technology—has received relatively little attention. In her study of infertility clinics, Cussins challenges the ideas of agency or of selves as preexisting categories. She argues instead for an "ontological choreography" in which the subjects on whom infertility technologies are enacted experience a temporary fragmentation of self into body functions and parts. However, this bodily objectification neither eliminates agency entirely nor permanently transforms the self. This process has similarities to the experience of individuals who undergo organ transplantation and of many who are tested for specific genes. The theme of fragmentation and transformation of self is extended below to take into account events outside the clinic and to include the social repercussions of these technologies on familial and communal life.

Even though human biologicals are singularized and objectified in research and clinical settings, symbolic, emotional, and anticipatory meanings are inevitably associated with their uses. Consequently, biotechnological practices can bring about radical transformations in self, identity, and human relationships. In the case of organ transplants, individual modes-of-being in the world in particular may be profoundly altered.

COMMODIFICATION OF HUMAN BIOLOGICALS

The transformation of biologicals into artifacts, including the commodification of human body parts as therapeutic tools, has a long history that raises social and moral issues. Human corpses, and body parts procured from the living and the dead, have had value as trophies of war, religious relics, anatomical specimens, and therapeutic materials and medicinals for many hundreds of years. In Europe, the commodification of human bodies for medical purposes was more often than not associated with violence. Vivisection of humans and animals by Herophilus in fourth century BCE Alexandria earned him a lasting reputation as the "father of scientific anatomy" (Potter, 1976). Anatomists performed public dissections of the corpses of criminals and vagrants in church precincts in thirteenth century Italy (Park, 1994), and such dissections continued until the early nineteenth century in civic anatomy theaters built in many parts of Europe, ensuring that the bodies of individuals on the margins of society accrued enormous medical value. Richardson argues that in Europe, by the seventeenth century, the human corpse was bought and sold like any other commodity (Richardson, 1988; see also Linebaugh, 1975). The Anatomy Act, designed to prohibit the sale of dead bodies in the United Kingdom, was signed in 1831 and remains the foundation for modern law in that country. However, workhouses and other institutions that housed the poor, including hospitals, were defined as "lawfully in possession of the dead." These institutions could legally confiscate bodies when no claimant came forward or when no money was available to pay for a funeral (Richardson, 1996: 73; Lacqueur, 1983). In the interests of medicine, then, the poor were effectively defined as socially dead, their commodified bodies not due the respect given to the rest of society. Only at the end of the nineteenth century did public outcry against such practices bring them to a halt.

Only after medical men succeeded in conceptualizing corpses as biological objects, as wholly part of nature, and therefore without cultural baggage, did it become relatively easy to strip *all* bodies—not just those of the "socially dead"—of social, moral, and religious worth. Commodification for the benefit of scientific advancement then became both legal and laudable (Mantel, 1998). However, as Richardson (1988) has shown, for families, the bodies of deceased relatives were not so easily divested of social meanings.

Clearly, the commodification of human biologicals is not simply a feature of the globalized economy of modernity. However, because the necessary technologies of procurement and preservation were lacking, human tissues and organs (except for blood) could not be routinely incorporated into systems of exchange prior to the second half of the twentieth century. Certain of these technological limitations remain: it is not yet possible, for example, routinely to store unfertilized human eggs, and although techniques have improved for preservation of solid organs outside the human body, they remain of use for transplantation only for a limited time.

Recent advances in biomedical technologies, and the consequent proliferation of machine/human hybrids make it impossible to sustain the fiction of a radical dichotomy between the human and material worlds (a fiction clearly identified long ago by Marx). While the dichotomy, with its accompanying objectification, served to justify commodification, the present blurring of the "natural" boundaries of self and other inevitably raises moral issues.

Legitimization of biomedical technologies is accompanied by rhetoric about their value: the assumption that they contribute to scientific progress and fulfill human "needs." As Strathern (1992) has noted, however, opposition that accompanies the introduction of so many of the new biomedical technologies makes it clear that they are frequently assumed to be a threat to moral order. Legal regulations and guidelines for professional conduct in research laboratories and the clinic are meant to justify such practices and damp down anxiety. On the other hand, citizens of democratic societies often assume that they have a "right" to the full range of biomedical technologies, and some individuals, concerned more with personal "empowerment" and their own health than with civic virtue, lobby for unhampered access to technologies such as genetic testing, genetic engineering, reproductive technologies, organ transplants, and so on. These contradictions tend to be only provisionally resolved because, as the technologies are modified, the practices they make possible inevitably change.

Another problematic issue associated with human biologicals, as Linebaugh (1975) intimates, is that of individual ownership of physical bodies and body parts. For example, thirty years ago a new death—brain death—was created in order that organs could be procured from patients whose brains have irreversibly lost all integrated function but whose hearts and lungs continue to "live" with the assistance of artificial ventilators. Although both a good number of involved families and some health care professionals who work in intensive care units remain ambivalent about the status of brain-dead patients, this new definition of death permitted the majority of those countries with the necessary technologies to conceptualize and legally establish these "living cadavers" as dead enough to become organ donors (Lock, 2002a,b). Even so, a strong resistance remains to the sale of solid human organs procured from cadavers; organs must be gifted.

Property rights in almost all European countries and in North America are invested in *living* individuals, following John Locke, who argued forcefully that every man is the "proprietor of his own person" (see de Witte and ten Have, 1997, on the body as property). Capitalizing on the dead raises a dilemma for property law. Moreover, the vision of families selling parts of their relatives even as they die appears ghoulish to most of us. As noted above, in all countries where organ transplants are routinized, an elaborate network has been set up to monitor procurement and distribution of organs, not only for quality but also for fair play. These networks depend on organs that are freely donated by family members without expectation of any recompense other than gratitude. Even in those European countries where the State has the legal authority to take organs regardless of the wishes of deceased individuals and their families, organs are not taken without family cooperation, although clearly the expectation is that families should be willing to give. Similarly, solid organs procured from living donors must also be "gifted" (in contrast to renewable human biologicals or those believed to be surplus to requirements, e.g., blood, sperm, and eggs). As has been clearly demonstrated in countries where organ sales routinely take place (Cohen, 2002; Scheper-Hughes, 2002), a market model makes individuals vulnerable to exploitation. This is the case, even when, as in Iran, the government has attempted to set up a regulated market (Zargooshi, 2001). Lesley Sharp (2006) argues that the activities of the transplant world are responsible for the current situation, in which the human body may be worth more than $230,000 on the open market. Not surprisingly, the value of specific body parts depends on geographical location. A kidney can be purchased for about $250 to $30,000 depending on where it is sold (Scheper-Hughes, 2003; Zargooshi, 2001).

When it comes to genetic material, the assumption is that individuals relinquish their rights of ownership the moment these materials are removed from their body. For example, drug companies in a search of unusual DNA sequences in human populations send out forays to isolated regions to undertake "gene prospecting." The stakes are high because the hope (unfulfilled thus far) is to produce new vaccines and medications from what are assumed to be rare DNA samples. The greatest furor in connection with this "biopiracy" has to do with patenting of DNA sequences procured in this way. Unless negotiations are carefully carried out ahead of time, individuals lose all control over the uses to which their body materials are put and are excluded from any resultant profits (Everett, 2003; Lock, 2002a).

The technological processing of DNA into immortalized cell lines results in a hybrid simultaneously naturally and culturally produced. This hybrid status permits patent claims on DNA sequences because cell lines can be classed as inventions (Strathern, 1996). Moreover, some are convinced that if human organs and tissue are taken with informed consent, then the donor relinquishes all rights of control—the individual no longer has an "interest" in his or her body parts. The fact that the material is procured from the human body means, however, that the door remains ajar for dissent. The case of John Moore's spleen is perhaps the best-known example of contestation about ownership of body parts potentially valuable to bioscientists (Boyle, 1996). In the case of organ donation, except when donor and recipient are close relatives, donors may not "direct" who will be the recipient of their organ; their rights to do so are forfeited at the time of donation.

It is worth noting that much of what goes on in connection with the practice of biomedical technologies takes place away from the public domain—in laboratories, clinics, committee meetings, and computer networking centers. When commenting on the application of technologies, social scientists frequently highlight the "unintended consequences" resulting from the introduction of new technologies (Winner, 1986). These possible outcomes are often dismissed by involved scientists, expert committees, and governments as irrational fears on the part of the public (Grove-White, 2006). Following adoption of the Nuremburg Code after World War II, regulation and institutionalized checks of the management of human subjects and care of patients have been applied with increasing vigor in Europe, North America, and Australasia. In these countries, at least, it is now difficult to "externalize" or dismiss as irrelevant unintended consequences for patients by appealing to the greater good. Similar care is not extended to organ donors because they are not recognized by the transplant world as patients (though gradually it is becoming clear that the health of kidney donors may at times be put in jeopardy [Crowley-Matoka & Switzer, 2005]). In the so-called developing world, fewer "restrictions" are in place. Drug trials that would not be acceptable to review boards in North America and Europe are common, as is the dumping of unwanted (Petryna, 2006), out-of-date medications and the procurement of organs without appropriate clinical care (Scheper-Hughes, 2003). In these countries, negative consequences, unintended or otherwise, are common. Their citizens find themselves in a position similar to that of disenfranchised individuals who in early modern Europe were classed as "beyond the pale" and whose bodies were made into targets for medical purposes.

A SHORTAGE OF ORGANS

The idea of organ transplants, a pervasive fantasy from mythological times, became a viable possibility only after Alexis Carrol, the 1913 Nobel Prize winner for medicine, showed that cells and tissues could not only be kept in suspended animation but could be made to function and reproduce independently of the donor body (Hendrick, 1913). Despite a considerable amount of experimentation, attempts at allografts (transplants from one individual to another) or xenografts (cross-species transplants) were failures until the 1950s. The first successful kidney transplant was carried out between identical twins in 1954. Extensive animal experimentation and transplantation between identical twins (isografts) increased knowledge about the immune system and led to the recognition of the importance of long-term immunosuppression of the recipient's body. However, throughout the 1960s immunosuppressant use had mixed success, and only after the development of cyclosporine in 1978 was organ transplant technology widely routinized. Throughout this time many philosophical and ethical issues were also raised about transplantation. In both Paris and Boston, where transplant technology was most strongly promoted, it was argued that organ transplants "transcended the laws of nature" (Kuss, cited in Tilney, 2003: 48) and desecrated the human body.

In Europe and North America, the practice of transplant technology is entirely dependent on voluntary donations. Human organs have from the outset been thought of as scarce commodities. A metaphor of a "shortage" of organs, firmly embedded in transplant discourse, is so powerful that it affects both the market value of human body parts and the globalization of the enterprise. Indeed, the claim today is often made that there is a *growing* shortage of organs on the assumption that donation rates have fallen. There is no doubt that waiting lists for organs, especially kidneys, are long and growing; however, several obvious reasons account for this state of affairs. First, there are fewer car accidents today than 20 years ago. Second, trauma units are more effective at preventing patients with traumatized brains from becoming brain-dead. Third, populations in technologically advanced societies are aging rapidly. These changes mean that the potential donor pool has decreased considerably over the past two decades.

On the recipient side of the equation, the demand for organs has increased because the population is aging and, owing to complications associated with increasing rates of diabetes and hepatitis C among younger people, there are more cases of end-stage kidney and liver disease. These diseases are intimately associated with poverty, alienation, and social inequality and are first and foremost public health problems. Finding sufficient organs to deal with a burgeoning problem of this magnitude is neither appropriate nor feasible. The perception of an increasing shortage of organs is exacerbated by an exponential increase in the number of patients deemed eligible to become recipients. As a result of changing public expectations and decisions made by committees constituted by transplant communities, transplants are to be available for tiny infants, individuals over 80, and patients with co-morbidities. Furthermore, second or third transplants are routinely carried out when earlier ones fail. In other words, the transplant world has broadened its sights and increased the "need" at a time when there are fewer potential donors. This discrepancy goes virtually unnoticed in official discourse.

Given the assumption of a shortage, the frequent discussions on how to increase the supply of organs tend to focus on inducing families to cooperate more willingly with donation of the organs of their brain-dead relatives. In North America, for example, it is estimated that fewer than 50 percent of possible organ procurements from brain-dead patients whose donor cards are signed are actually accomplished because families do not agree to donation (Siminoff & Chillag, 1999). Even though their consent is not legally required, organ procurement facilitators will not override family sentiment. Joralemon (1995) regards this as evidence of a "cultural rejection" of the transplant enterprise. One result of the growing "need" for organs is that, whereas until recently organs were primarily procured from brain-dead donors, in the United States, as of 2001, more than 50 percent of organs are acquired from living donors.

Four assumptions run through the debates about an organ shortage. First, organs go to waste if not donated, and every citizen should be willing to contribute to their use in the transplant enterprise. Second, organs are regarded as simply mechanical entities void of any symbolic or affective meaning or value. Third, diagnosis of brain death is seen as straightforward and acceptable as human death by everyone involved. Moreover, families should be willing to interrupt the grieving process for up to 24 hours while organs are procured. Finally, it is assumed that donation, being eminently worthwhile, is likely to assist families in the mourning process.

THE SOCIAL LIFE OF HUMAN ORGANS

Before organs can be removed from donors and prepared as living substitutes, they must be tacitly recognized as fungible, and cadaver donors must be designated as dead. Agreement that the body will not be violated through organ removal is easier when organs are seen as objects. However, because organs procured for transplant must remain biologically alive, even the involved physicians cannot fully reduce them to mere things (Lock, 2002b)—organs retain a hybrid status.

Mixed metaphors associated with human organs encourage confusion about their worth. The language of medicine insists that human body parts are material entities, devoid entirely of identity, whether located in donors or recipients. However, to promote donation, organs are endowed with a life force that can be gifted, and donor families are not discouraged from the belief that their relatives "live on" in the bodies of recipients or even that they are "reborn" (Sharp, in press). Organ donation is commonly understood as creating meaning out of senseless, accidental deaths—a technological path to transcendence.

Research has also shown that owing to the enforced anonymity that surrounds donated organs, large numbers of recipients experience a frustrated sense of obligation about the need to repay the family of the donor for the extraordinary act of benevolence that has brought them back from the brink of death (Fox & Swazey, 1978, 1992; Simmons et al., 1987; Sharp, 1995). The "tyranny of the gift" is well documented in the transplant world (Fox, 1978: 1168), but people also desire to know something about the donor because donated organs very often represent more than mere biological body parts. They are experienced by recipients as personified with an agency that manifests itself in some surprising ways and profoundly influences the recipient's sense of self.

A good number of organ recipients worry about the gender, ethnicity, skin color, personality, and social status of their donors, and many experience a radically changed mode of being-in-the-world thanks to the power and vitality diffusing from the organ they have received. Sharp (1995) points out that receiving an organ is a personally transformative experience, influencing recipients' assessment of their social worth. She argues that this transformation takes place both subjectively, when a recipient's sense of self is extended to include qualities attributed to the donor, and through interactions with family, communities, and the medical profession. Sharp notes, as does Hogle (1995, 1999), that the language used in connection with organ procurement depersonalizes bodies and body parts but that many recipients re-personalize organs through narratives about their rebirth. The organ takes on a biography of its own, independent of the person in whom it resides (see also Crowley-Matoka, 2001; Lock, 2002b).

Fetishism is doubly at work: the fetishism of objectification, postulated by Marx, and the fetishism in which gifts (including human body parts), having entered a system of exchange, remain infused with a personal essence, as described by Marcel Mauss. Contradictions are rife: if recipients attribute animistic qualities to this "life-saving" transplanted organ, they are severely reprimanded (Sharp, 1995). As Nicholas Thomas (1991) suggests with respect to commodified objects in general, human organs are "promiscuous"—at once things-in-themselves and diffused with a life force and an agency that is manifestly social. Thomas's description applies equally to genetic material, though its promiscuity is performed in remarkably different ways depending on geographical context and local cultural and political horizons.

TWICE DEAD, TWICE BORN

The linked networks of organ donation, procurement, and transplantation can be blocked or facilitated in a variety of ways in different locations. Difficulties are often not a result of a lack of technical expertise but of cultural and political considerations. For example, several countries, among them Sweden, Germany, Denmark, Japan, and Israel, have, since the late 1960s, conducted protracted public discussions as to whether or not brain death—a condition that cannot exist unless body function is sustained by means of a ventilator—can be legally recognized as the end of human life (Rix, 1999; Schöne-Seifert, 1999). In Japan, the debate among lawyers, medical professionals, intellectuals, and members of the public in professional forums and the media has far outstripped discussions about abortion or any other bioethical matter. Results of the numerous national opinion polls make the lack of consensus clear. The 1997 Japanese law places medical interests second to family concerns and recognizes brain death as the end of human life only when the diagnosed patient and his or her family have given prior notice of a willingness to donate organs. As of 2005, organs had been procured from fewer than forty brain-dead donors.

Numerous factors have contributed to this impasse including a lack of trust in the medical profession, a conservative legal profession, extensive media criticism of hospital practices, and the mobilization of citizen groups to block the formal recognition of brain death as the end of human life. Equally important are culturally informed practices in connection with death, notably, the centrality of the family in making end-of-life decisions (Long, 2003), a reluctance on the part of many people to permit commodification of dead bodies, a strong resistance to "gifting" body parts, and fears that a brain-dead body is murdered when taken off the ventilator. Religious organizations have not been outspoken in these debates (Hardacre, 1994); rather, reservations arise from what many Japanese assume to be rational, common sense responses to an extraordinary technology that threatens moral order. Of course, by no means do all Japanese respond in the same way (Lock, 2002a,b).

Hogle (1999) shows how disputes in Germany about the commodification of human body parts and their use as therapeutic tools are powerfully influenced by the history of National Socialism experimentation and its practices of eugenics. Reluctance to cooperate with the transplant enterprise is also rooted in medieval beliefs about the diffusion of the essence of life throughout the entire human body. Although the ideas of "solidarity" (a powerful metaphor from the former East Germany) and Christian "charity" are both used to encourage organ donation, making organ donation into a social good in multicultural Germany remains fraught with difficulties (Hogle, 1999: 192).

In Mexico, as in Japan, virtually all organ transplants are "living related" donations between close relatives. A common nationalist sentiment, shared by many political leaders, is that procurement of organs from brain-dead bodies is an inhumane activity in which only a country such as the United States could participate. On the basis of extensive fieldwork, Crowley-Matoka (in press) argues that in Mexico, the family, as the core of social and moral life, is regarded as both a "national" and a "natural" resource for organs. Above all, it is mothers who are expected to donate, partly because of their prime role as nurturers, and partly because their bodies are seen as more expendable than those of working men. Donation patterns "fit" with the brutal reality of an impoverished life and the accepted division of labor in Mexican households. Among recipients Crowley-Matoka finds evidence of concerns among men about sexual potency, of being like a gelding or a half-woman.

Lesley Sharp (2006), in her ethnographic account of donation, procurement, and transplantation of organs in the United States, shows clearly that the origins of organs for transplant are deliberately dehumanized and sanitized. Nevertheless, numerous donor kin cannot accept the biomedical trajectory of a technologically diagnosable material death. In the course of many years of studying the U.S. transplant enterprise, Sharp has observed how social relationships between donors and recipients have been transformed. Initially, such relations were based entirely on imagination, owing to enforced anonymity; more recently, they became something that can be celebrated in the public sphere. It is common today to build edifices as donor memorials and to hold public gatherings in which donors and recipients come together to celebrate donor's lives. The *leitmotiv* of such gatherings is one of loss and redemption and of birth and rebirth. Speakers are organ recipients who often know exactly who was their donor (Sharp, 1995; see also Lock, 2002b). Metaphors derived from Christianity are drawn on liberally, and testimonies are delivered in a manner similar to that used in Pentecostal churches, although organ recipients are not necessarily believers.

The problems associated with body commodification and biomedical technologies become overtly political in countries where an enormous disparity exists between rich and poor. Cohen (2002), Das (2000), and Scheper-Hughes (1998) have shown how the disenfranchised are particularly vulnerable to exploitation. By tracing complex networks of activities associated with organ procurement and their transplantation involving organ brokers, unscrupulous doctors, and at times the unwilling participation of live kidney donors, these researchers make it clear that societal inequities are reproduced and even magnified through the practices of transplant technology. On the basis of research in India, Das (2000) is critical of both contract law and globally applied bioethics grounded in the language of rights. She argues that such language masks the politics of violence and suffering involved in organ procurement where gross inequalities are present in social life and where bribery and corruption are not uncommon. On the other hand, Crowley-Matoka (2001) shows that in Mexico poor people often become organ recipients and that economic assistance from Mexicans living in America, and at times their organ donations, enrich the bonds between immigrants and their relatives at home.

Because organ recipients, wherever they reside, are, in effect, permanently invalided, many organ recipients and some donors can no longer provide for their families; some experience discrimination when looking for jobs or when dealing with insurance companies; those who are single may have trouble finding partners; and for women, the risk of child-bearing is increased. However, among those recipients who have few problems with organ rejection, many feel young again or even reborn.

The impact of transplant technologies on the everyday lives of people directly involved with this enterprise is context dependent. Such technologies force reconsideration of unexamined assumptions about the basic social contract, of what counts as self and other, and of the accepted boundaries between nature and culture. The discussion provoked by this technology can thus act as a touchstone for political debates about nationalism, modernization, progress, equity, whose lives are valuable and whose can be sacrificed, what counts as death, and more generally about the commodification of the human body and the possibility for creating new social relationships as a result of the breaching of body boundaries involved in transplantation.

LAISSEZ-FAIRE EUGENICS

From its outset in the 1960s, the institutionalization of genetic screening of specific high-risk populations, followed two decades later by the implementation of genetic testing of pregnant women, was made difficult by the historical links between the world of clinical genetics and that of eugenics (Duster, 1990). The guiding principle of the eugenics movement of the first half of the twentieth century was a belief that the elimination of "poor" genes could be justified for the good of society at large (Kevles, 1985; Kitcher, 1996).The only method available to dispose of such genes was to enact policies by means of which the reproductive lives of individuals designated as genetically unworthy and as a burden to society were managed by medical and governmental representatives.

Today, a different rhetoric informs interventions that may result in the termination of pregnancy. Individual choice is presented as dominant, and the role of government is rendered invisible. Decisions about termination of a pregnancy on the basis of genetic test results inevitably involve moral choices, not simply about the act of abortion *per se* but also about what counts as normal and abnormal. The grounds on which such decisions are made, however, are relatively rarely explicitly examined (Duster, 1990; Lock, 2002c), and such practices have been characterized by some as a "neo-eugenics" (Kitcher, 1996).

Over 20 years ago, the historian of science Edward Yoxen (1982) pointed out that although the role of genetics in disease etiology was recognized throughout the twentieth century, it was only after the advent of molecular genetics that the notion of "genetic disease" came to dominate this discourse, often obscuring the role of other contributory factors. Keller (1992) argues that this conceptual shift made the Human Genome Project both reasonable and desirable for scientists. The objective of mapping the human genome was to create a baseline norm, which in fact would not correspond to the genome of any single individual. In theory, everyone is deviant (Lewontin, 1992). Moreover, many involved scientists believed that it would soon be possible to "guarantee all human beings an individual and natural right, the right to health" (Keller, 1992: 295). A 1988 report published by the U.S. Office of Technology Assessment argued that genetic information will be used "to ensure that . . . each individual has at least a modicum of normal genes," justified by the belief that "individuals have a paramount right to be born with a normal, adequate hereditary endowment." The planned use of genetic information in this way is described in the report as a "eugenics of normalcy" (Office of Technology Assessment, 1988; cited by Keller, 1992).

Although documents such as these mention the improvement of the quality of the human gene pool, they do not focus on social policy or the good of the species. A belief in individual choice is dominant, and it is assumed that genetic information is indispensable to realize the individual's inalienable right to health.

Virtually everyone agrees that the eugenics of the first part of the twentieth century was grounded in invalid science, and its practices are roundly criticized. However, the social cost of treating and caring for "defective" children is still made explicit when justifying the implementation of screening programs. For example, the State of California introduced maternal serum α-fetoprotein screening for all pregnant women in the early 1990s in the hope of reducing the number of infants born with neural tube defects and thereby saving costs (Caplan, 1993). In 1990 the guidelines of the International Huntington Association advocated refusing to test women who were unwilling to provide assurance that they would terminate their pregnancy if the Huntington gene were found. As Paul and Spencer (1995: 304) point out, "Those who made this recommendation certainly did not think they were promoting eugenics. Assuming that eugenics is dead is one way to dispose of deep social, political and ethical questions. But it may not be the best one." Similarly, Ginsburg and Rapp (1995) argue that biological and social reproduction are inevitably bound up with the production of culture. Rapp's (1999) ethnographic study of amniocentesis, a technology used primarily to detect Down syndrome and single-gene disorders, shows how, despite a policy of nondirective counseling, American genetic counselors tailor the way in which they convey test results to their clients' ethnicity. Often inadvertently, these counselors encourage the persistence of "stratified reproduction" in which "some categories of people are empowered to nurture and reproduce, while others are disempowered" (Ginsburg & Rapp, 1995: 3). Rapp's ethnography also makes clear that, when confronted with this type of testing, many women, especially those who are neither white nor middle class, become noncooperative and frequently reinterpret or resist the risk information they are given.

Ambivalence and resistance are common responses to genetic testing in general: it is estimated that only between 15 and 20 percent of people considered at risk for adult-onset genetic disease have made use of testing (Quaid & Morris, 1993; Beeson & Doksum, 2001), and pregnant women have actively refused to be tested (Rapp, 1998) or ignore test results (Hill, 1994; Rapp, 1999).

Even though extreme caution would seem in order, we forge ahead rapidly with the routinization of genetic testing and screening on the assumption that people will be able to make rational choices about abortion and about suitable, genetically compatible marriage partners and thus avoid bringing diseased children into the world (Beeson & Doksum, 2001). There is no doubt that some programs—notably screening for sickle cell trait in the United States and elsewhere (Duster, 1990)—are associated with a long history of racism and discrimination. In contrast, screening for thalassemia and Tay-Sachs disease has brought enormous relief to certain families (Angastiniotis et al., 1986; Kuliev, 1986; Mitchell et al., 1996), and the Cuban government reports success with a screening program for sickle cell disease (Granda et al., 1991).

Program success is measured in terms of reduction in the incidence of the disease in question. This is usually achieved by means of genetic testing of teenagers deemed to be at risk who are at liberty to draw on these results later when making decisions about marriage, reproduction, and, when necessary, abortion. In Montréal, more than 25 years of screening for thalassemia and Tay-Sachs disease have led to an almost 100 percent reduction in incidence. The majority of involved families state that without such a program they would not have had children at all for fear of the disease and that they are now at ease in the knowledge that their offspring will be spared great suffering (Mitchell et al., 1996).

Willis (1998) points out that abortion politics and vocal "right to life" campaigners might affect the implementation and spread of screening technologies. Disability rights activists are also critical of testing because "a single trait stands in for the whole (potential) person. Knowledge of the trait is enough to warrant the abortion of an otherwise wanted fetus" (Parens & Asch, 1999: S2).

GENETICIZATION, GENETIC RESPONSIBILITY, AND GENETIC CITIZENSHIP

In 1992, Abby Lippman coined the term geneticization to capture an ever-growing tendency to distinguish people on the basis of their genetic makeup. She was concerned above all with possible reinforcement of racism, inequalities, and discrimination of various kinds as a result of a renewed conflation of social realities and biological difference (Lippman, 1998: 64).

More recently, Adam Hedgecoe (2001) used a concept of "enlightened geneticization" to show how current scientific discourse about schizophrenia prioritizes genetic explanations and subtly diverts attention away from nongenetic factors even while paying lip service to the contribution of environmental and other nongenetic factors to disease causation (see also Spallone, 1998). Although Hedgecoe agrees with Lippman that genetic determinism is at work, he points out that geneticization, and medicalization more generally (see Lock & Kaufert, 1998; Lock, Lloyd, & Prest, 2006) also have some positive effects. For example, medical recognition of a given condition as a disease reduces social stigma and allocation of responsibility to the individual and family (McGuffin et al., 2001). Moreover, many families appear to take comfort in being told that a disabling condition is the result of faulty genetics and therefore, by implication, has nothing to do with moral shortcomings (Turney & Turner, 2000).

Social scientists have also studied responses of individuals and families directly affected by the new technologies of genetic testing and screening. Rayna Rapp and colleagues document how networks of families claim "genetic citizenship" and increasingly coalesce around lethal and highly disabling single-gene diseases that afflict their children. Such groups provide mutual social support and lobby the U.S. Congress for improved research funding; similar activities happen in other countries (Callon & Rabeharisoa, 2004). These activists are painfully aware that only rarely will drug companies invest in research for the rare diseases that affect their families (Rapp, 2003).

Edward Yoxen (1982) suggested long ago that our new abilities to detect "presymptomatically ill" individuals would ensure that virtually all of us would shortly become subject to increased medical surveillance. More recently, Paul Rabinow (1996) created the concept of biosociality to describe the constitution of new group identities based on shared alleles. Nikolas Rose (1993) has outlined the emergence of a new form of governance in which individuals are expected to exhibit prudence with respect to embodied risk, and Novas and Rose (2000) have examined what it means to be designated as "genetically at risk."

KINSHIP AND EMBODIED RISK

The introduction of molecular genetics to clinics and public health screening programs has profoundly affected individual behavior and family dynamics (Kerr et al., 1998; Michie et al., 1995; Hallowell, 1999; Konrad, 2005). At the same time, individuals interpret available knowledge about molecular genetics to "fit" their preconceived ideas about family risk for specific diseases. People also frequently resist using the results of genetic testing alone to account for the illnesses that "run" in their families (Condit, 1999; Lock, Freeman, Sharples, & Lloyd, 2006). When genetic information is actively incorporated into accounts about illness causation, such information supplements previously held notions of kinship, heredity, and health. Cox and McKellin (1999: 140) have demonstrated that lived experience of genetic risk and lay understandings of heredity conflict with theories of Mendelian genetics because "theories of Mendelian inheritance frame risk in static, objective terms. They abstract risk from the messiness of human contingency and biography." Kerr and colleagues (1998) write that lay persons are their own authority when it comes to appreciating and understanding how genetics may shape their lives.

These findings suggest that the new forms of community that Rabinow has envisioned under the rubric of "biosociality" are by no means self-evident. The technologies of genetic testing and screening have the power to reveal embodied risk, but to date the majority of people refuse such information, choosing not to divine the future. Many individuals are sensitive to the way in which knowledge about DNA inevitably transcends body boundaries and has immediate significance for families and, at times, communities; they worry about the effects of testing on the family as a whole (Gibbon, 2002). This type of information challenges the foundations of contemporary bioethics grounded in the idea of the autonomy of individuals (Hayes, 1992). Perhaps of more importance, it has the potential to cause both ruptures and alliances among kin, creating "unnatural" molecular boundaries of inclusion and exclusion (Gibbon, 2002).

In her study of Huntington's disease, Monica Konrad (2005) has recently examined how diagnostic tools in clinical genetics are creating "pre-symptomatic persons as new social identities." Konrad's ethnography shows how knowledge gained from genetic testing is situated among "moral systems of foreknowledge" held by involved families. She argues that "culture" is put to work in dealing with the paradox of testing for genes that provide a prognosis about a disease for which there is no cure. Konrad points out, as do I and my colleagues in our research on Alzheimer's disease (Lock, Freeman, & Lloyd, 2006), that families draw on the concept of "blended inheritance" (Richards, 1996) to prophesy who among them will become ill. While genetic testing enhances predictability, it is far from obvious how people will respond to test results, negative or positive. In the case of Huntington's disease, for example, certain people whose test results were negative have committed suicide, ostensibly because of guilt at having escaped the family disease (Almqvist et al., 1999, 2003; Quaid & Wesson, 1995).

Genetic test results are rarely associated with certainty. Even the autosomal dominant gene associated with Huntington's disease is not 100 percent penetrant so that not quite everyone with the gene will get the disease (McNeil et al., 1997). Moreover, as with many similar diseases, the age of onset is variable and cannot be predicted with accuracy. After the Huntington gene was mapped, many of the risk estimates formerly given to people were found to be inaccurate, with disturbing social repercussions in several cases (Almqvist et al., 1997).

Women in particular come to think of themselves as responsible for circumventing their family risk by undergoing genetic testing and planning reproduction accordingly. Genetic discourse constructs women as the "bearers of 'nature's defects'" (Steinberg, 1996). Referring to genetic testing for breast and ovarian cancer, Hallowell (1999) suggests that by constructing genetic risk and risk management as moral issues, women relinquish their right not to know about their own genetic risk. Kenen (1999) also argues that genetic testing has the potential to fundamentally alter the way in which we think of ourselves in relation to others, making visible an "interdependent self."

Genetic testing for complex disease is becoming increasingly common, notably in the private sector where people who pay to be tested are usually simply informed about the presence or absence of a particular polymorphism in their genome and given little or no information about the statistical probabilities for disease incidence associated with it. Furthermore, developments in postgenomic science, notably in connection with epigenetics, are making it abundantly clear that not only is scientific knowledge about the genomics of complex disease in a primitive state, but inevitably calculations of risk are exceedingly problematic (Lock, 2005). Susceptibility genes for complex disease are for the most part neither necessary nor sufficient to cause the disease in question. One study strongly suggests that first-degree relatives of Alzheimer patients who are tested for the susceptibility gene most commonly associated with this disease do not undergo any fundamental reconceptualization of embodiment or subjectivity as a result of testing. There are at least four likely explanations for this situation: the disease has late onset; relative risk estimates given to those individuals believed to be at the highest risk (based on sex, genotype, and number of affected family members) are just over 50 percent by age 85; many people living in Alzheimer families already understand themselves as profoundly affected by this disease independently of genotyping, and such families almost without exception believe that nongenetic factors may be amenable to modification, whereas genes are not (Lock,

Freeman, & Lloyd, 2006). Further research will show whether social scientists have inadvertently participated in a genetic hype that assumes significant changes in embodiment based on knowledge about a person's genotype. In this respect, the impact of being an organ recipient may well be significantly different from that associated with genetic testing and screening.

CONCLUSIONS

It is clear that much remains to be learned. For example, we have next to no data on subjective transformations experienced by those transplant recipients who buy organs illegally. Nor do data exist about how identity is affected over the long term in families in which one or more individuals has undergone genetic testing. What findings we have about non-Mendelian, complex diseases suggest that disembodied, abstracted knowledge about molecularized identities does not appear to have the leverage to replace identification based on "blood" and heredity, although information about genes associated with breast cancer may be an exception.

Space has not permitted more than minimal elaboration on the ethnographic findings described throughout this chapter. The ethnographic approach allows us to avoid both reductionism and essentialism and provides a powerful tool for examining how biomedical technologies are co-constructed with the material. Ethnography also lays bare cultural and sociopolitical constraints on practices associated with the use of these technologies and shows that limiting attention to individuals—whether they are characterized as users, consumers, or responsible citizens—privileges the bounded, autonomous body of modernity. The practices of transplant technology and of genetic testing and screening make it abundantly clear that researchers have no choice but to recognize the ubiquitous presence of hybrid, postmodern bodies, fluid subjectivities, and shifting human collectivities, which in turn are associated with the potential for new forms of embodiment and identity. Documenting the profound social effects of these technologies anchors a reflexive, critical analysis of biomedical technologies in action, both globally and locally.

Note

1. Gaston Bachelard (1964) *The Poetics of Space* (New York, Orion Press): 216.

References

Adams, V. (2002) "Establishing Proof: Translating 'Science' and the State in Tibetan Medicine," in M. Nichter & M. Lock (eds), *New Horizons in Medical Anthropology: Essays in Honour of Charles Leslie* (Reading, U.K.: Harwood): 200–20.

Almqvist, Elisabeth, Shelin Adam, Maurice Bloch, Anne Fuller, Philip Welch, Debbie Eisenberg, Don Whelan, David Macgregor, Wendy Meschino, & Michael R. Hayden (1997) "Risk Reversals in Predictive Testing of Huntington Disease," *American Journal of Human Genetics* 61: 945–52.

Almqvist, Elisabeth, Maurice Bloch, Ryan Brinkman, David Craufurd, & Michael R. Hayden (1999) "A Worldwide Assessment of the Frequency of Suicide, Suicide Attempts, or Psychiatric Hospitalization after Predictive Testing for Huntington Disease," *American Journal of Human Genetics* 64: 1293–304.

Almqvist, E. W., R. R. Brinkman, S. Wiggins, M. R. Hayden, & The Canadian Collaborative Study of Predictive Testing (2003) "Psychological Consequences and Predictors of Adverse Events in the First 5 Years After Predictive Testing for Huntington's Disease," *Clinical Genetics* 64: 300–309.

Angastiniotis, M., S. Kyriakidou, & M. Hadjiminas (1986) "How Thalassaemia Was Controlled in Cyprus," *World Health Forum* 7: 291–97.

Appadurai, Arjun (1990) "Disjuncture and Difference in the Global Cultural Economy," *Public Culture* 2: 1–24.

Beeson, Diane & Teresa Doksum (2001) "Family Values and Resistance to Genetics Testing," in B. Hoffmaster (ed), *Bioethics in Social Context* (Philadelphia: Temple University Press): 153–79.

Boyle, James (1996) *Shamans, Software, and Spleens: Law and the Construction of the Information Society* (Cambridge, MA: Harvard University Press).

Brodwin, Paul E. (ed) (2000) *Biotechnology and Culture: Bodies, Anxieties, Ethics* (Bloomington: Indiana University Press).

Callon, M. & V. Rabeharisoa (2004) "Gino's Lesson on Humanity: Genetics, Mutual Entanglements and the Sociologist's Role," *Economy and Society* 33: 1–27.

Caplan, Arthur L. (1993) "Neutrality Is Not Morality: The Ethics of Genetic Counseling," in D. M. Bartels, B. S. LeRoy, & A. L. Caplan (eds), *Prescribing Our Future: Ethical Challenges in Genetic Counseling* (Hawthorne, NY: Aldine de Gruyter): 149–65.

Casper, M. & A. Clarke (1998) "Making the Pap Smear into the Right Tool for the Job: Cervical Cancer Screening in the United States, c. 1940–1995" *Social Studies of Science* 28(2/3): 255–90.

Clarke, Adele E, Janet K. Shim, Laura Mamo, Jennifer Ruth Foskett, & Jennifer Fishman (2003) "Biomedicalization: Technoscientific Transformations of Health, Illness and U.S. Biomedicine," *American Sociological Review* 68: 161–94.

Cohen, Lawrence (1998) *No Aging in India: Alzheimer's, the Bad Family, and Other Modern Things* (Berkeley: University of California Press).

Cohen, Lawrence (2002) "The Other Kidney: Biopolitics Beyond Recognition," in N. Scheper-Hughes & L. Wacquant (eds), *Commodifying Bodies* (London: Sage): 9–30.

Condit, Celeste M. (1999) "How the Public Understands Genetics: Non-Deterministic and Non-Discriminatory Interpretations of the 'Blueprint' Metaphor," *Public Understanding of Science* 8: 169–80.

Cowan, R. S. (1987) "The Consumption Junction: A Proposal for Research Strategies in the Sociology of Technology," in W. Bijker (ed), *The Social Construction of Technological Systems* (Cambridge, MA: MIT Press).

Cox, S. & W. McKellin (1999) "'There's This Thing in Our Family': Predictive Testing and the Construction of Risk for Huntington Disease," in P. Conrad & J. Gabe (eds), *Sociological Perspectives on the New Genetics* (London: Blackwell): 121–48.

Crowley-Matoka, Megan (2001) "Modern Bodies, Miraculous and Flawed: Imaginings of Self and State in Mexican Organ Transplantation," Ph.D. diss., University of California, Los Angeles.

Crowley-Matoka, Megan (in press) *Producing Transplanted Bodies: Life, Death and Value in Mexican Organ Transplantation* (Durham, NC: Duke University Press).

Crowley-Matoka, Megan and Galen Switzer (2005) "Nondirected Living Donation: A Preliminary Survey of Current Trends and Practices" *Transplantation* 79(5): 515–19.

Cussins, Charis M. (1998) "Ontological Choreography: Agency for Women Patients in an Infertility Clinic," in M. Berg & A. Mol (eds), *Differences in Medicine: Unraveling Practices, Techniques, and Bodies* (Durham, NC: Duke University Press): 166–201.

Daniel, Valentine (1991) *Is There a Counterpoint to Culture?* Wertheim Lecture, Center for Asian Studies, Amsterdam.

Das, V. (2000) "The Practice of Organ Transplants: Networks, Documents, Translations," in M. Lock, A. Young, & A. Cambrosio (eds), *Living and Working with the New Medical Technologies: Intersections of Inquiry* (Cambridge: Cambridge University Press): 263–87.

de Witte, Joke I. & Henk ten Have (1997) "Ownership of Genetic Material and Information," *Social Science and Medicine* 45: 51–60.

Duster, Troy (1990) *Back Door to Eugenics* (New York: Routledge).

Everett, Margaret (2003) "The Social Life of Genes: Privacy, Property and the New Genetics," *Social Science in Medicine* 56: 53–65.

Feenberg, Andrew (1995) "Subversive Rationalization: Technology, Power and Democracy," in A. Feenberg & A. Hannay (eds), *Technology and the Politics of Knowledge* (Bloomington: Indiana University Press): 3–22.

Fox, Renée (1978) "Organ Transplantation: Sociocultural Aspects," in W. T. Reich (ed), *Encyclopedia of Bioethics* (New York: Free Press): 1166–69.

Fox, Renée & Judith P. Swazey (1978) *The Courage to Fail: A Social View of Organ Transplants and Dialysis* (Chicago: University of Chicago Press).

Fox, Renée & Judith P. Swazey (1992) *Spare Parts: Organ Replacement in American Society* (Oxford: Oxford University Press).

Franklin, Sarah (2003) "Ethical Biocapital: New Strategies of Cell Culture," in S. Franklin & M. Lock (eds), *Remaking Life and Death: Toward an Anthropology of the Biosciences* (Santa Fe, NM: School of American Research Press).

Friedlaender, Michael M. (2002) "The Right to Sell or Buy a Kidney: Are We Failing Our Patients?" *Lancet* 359: 971–73.

Gibbon, Sahra (2002) "Re-examining Geneticization: Family Trees in Breast Cancer Genetics," *Science as Culture* 11: 429–57.

Ginsburg, Faye & Rayna Rapp (eds) (1995) *Conceiving the New World Order: The Global Politics of Reproduction* (Berkeley: University of California Press).

Granda, H., S. Gispert, A. Dorticos, M. Martin, Y. Cuadras, M. Calvo, G. Martinez, M. A. Zayas, J. A. Oliva, & L. Heredero (1991) "Cuban Programme for Prevention of Sickle Cell Disease," *Lancet* 337: 152–53.

Grove-White, Robin (2006) "Britain's Genetically Modified Crop Controversies: The Agricultural Environment Biotechnology Commission and Its Negotiation of 'Uncertainty,'" *Community Genetics* 9: 170–77.

Hallowell, Nina (1999) "Doing the Right Thing: Genetic Risk and Responsibility," *Sociology of Health and Illness* 5: 597–621.

Haraway, Donna (1991) "A Cyborg Manifesto: Science, Technology, and Socialist-Feminism in the Late Twentieth Century," in *Simians, Cyborgs, and Women: The Reinvention of Nature* (New York: Routledge): 149–81.

Haraway, Donna (1997) *Modest_Witness@Second_Millennium: FemaleMan_Meets_Oncomouse* (New York: Routledge).

Hardacre, Helen (1994) "The Response of Buddhism and Shinto to the Issue of Brain Death and Organ Transplants," *Cambridge Quarterly of Healthcare Ethics* 3: 585–601.

Hayes, C. (1992) "Genetic Testing for Huntington's Disease: A Family Issue," *New England Journal of Medicine* 327: 1449–51.

Hedgecoe, Adam (2001) "Schizophrenia and the Narrative of Enlightened Geneticization," *Social Studies of Science* 31: 875–911.

Hendrick, Burton J. (1913) "On the Trail of Immortality," *McClure's* 40: 304–17.

Hill, Shirley (1994) *Managing Sickle Cell Disease in Low-Income Families* (Philadelphia: Temple University Press).

Hogle, Linda (1995) "Standardization Across Non-Standard Domains: The Case of Organ Procurement," *Science, Technology & Human Values* 20: 482–500.

Hogle, Linda (1999) *Recovering the Nation's Body: Cultural Memory, Medicine and the Politics of Redemption* (New Brunswick, NJ: Rutgers University Press).

Joralemon, Donald (1995) "Organ Wars: The Battle for Body Parts," *Medical Anthropology Quarterly* 9(3): 335–56.

Keller, Evelyn Fox (1992) "Nature, Nurture, and the Human Genome Project," in D. J. Kevles & L. Hood (eds), *The Code of Codes: Scientific and Social Issues in the Human Genome Project* (Cambridge, MA: Harvard University Press): 281–89.

Kerr, A., S. Cunningham-Burley, & A. Amos (1998) "The New Human Genetics and Health: Mobilizing Lay Expertise," *Public Understanding of Science* 7: 41–60.

Kevles, Daniel J. (1985) *In the Name of Eugenics: Genetics and the Uses of Human Heredity* (Cambridge, MA: Harvard University Press).

Kitcher, Philip (1996) *The Lives to Come: The Genetics Revolution and Human Possibilities* (New York: Simon and Schuster).

Konrad, Monica (2005) *Narrating the New Predictive Genetics: Ethics, Ethnography and Science* (Cambridge: Cambridge University Press).

Kopytoff, Igor (1986) "The Cultural Biography of Things: Commoditization as Process," in A. Appadurai (ed), *The Social Life of Things: Commodities in Cultural Perspective* (Cambridge: Cambridge University Press): 64–91.

Kuliev, A. M. (1986) "Thalassaemia Can Be Prevented," *World Health Forum* 7: 286–90.

Lacqueur, Thomas (1983) "Bodies, Death and Pauper Funerals," *Representations* 1: 109–30.

Latour, Bruno (1988) *The Pasteurization of France* (Cambridge, MA: Harvard University Press).

Lewontin, Richard C. (1992) "The Dream of the Human Genome," in H. C. Plotkin (ed), *New York Review of Books* (New York: Wiley): 31–40.

Lie, M. & K. H. Sorensen (eds) (1996) *Making Technology Our Own? Domesticating Technology into Every-day Life* (Oslo, Norway: Scandinavian University Press).

Linebaugh, Peter (1975) "The Tyburn Riot: Against the Surgeons," in D. Hay, P. Linebaugh, J. Rule, E. P. T. Thompson, & C. Winslow (eds), *Albion's Fatal Tree: Crime and Society in Eighteenth-Century England* (London: Allen Lane): 65–117.

Lippman, Abby (1998) "The Politics of Health: Geneticization Versus Health Promotion," in S. Sherwin (ed), *The Politics of Women's Health: Exploring Agency and Autonomy* (Philadelphia: Temple University Press).

Lock, Margaret (1993) *Encounters with Aging: Mythologies of Menopause in Japan and North America* (Berkeley: University of California Press).

Lock, Margaret (2002a) "Alienation of Body Tissue and the Biopolitics of Immortalized Cell Lines," in N. Scheper-Hughes & L. Waquant (eds), *Commodifying Bodies* (London: Sage): 63–92.

Lock, Margaret (2002b) *Twice Dead: Organ Transplants and the Reinvention of Death* (Berkeley: University of California Press).

Lock, Margaret (2002c) "Utopias of Health, Eugenics, and Germline Engineering," in M. Nichter & M. Lock (eds), *New Horizons in Medical Anthropology* (London: Routledge): 239–66.

Lock, Margaret (2003) "On Making up the Good-as-Dead in a Utilitarian World," in S. Franklin & M. Lock (eds), *Remaking Life and Death: Toward an Anthropology of the Biosciences* (Santa Fe, NM: School of American Research Press).

Lock, Margaret (2005) "Eclipse of the Gene and the Return of Divination," *Current Anthropology* 46: S47–S70.

Lock, Margaret & Patricia Kaufert (eds) (1998) *Pragmatic Women and Body Politics* (Cambridge: Cambridge University Press).

Lock, Margaret, Stephanie Lloyd, & Janalyn Prest (2006) "Genetic Susceptibility and Alzheimer's Disease: The 'Penetrance' and Uptake of Genetic Knowledge," in A. Leibing & L. Cohen (eds), *Thinking About Dementia: Culture, Loss, and the Anthropology of Senility* (New Brunswick, NJ: Rutgers University Press): 123–54.

Lock, Margaret, Julia Freeman, Rosemary Sharples, & Stephanie Lloyd (2006) "When It Runs in the Family: Putting Susceptibility Genes into Perspective," *Public Understanding of Science* 15(3): 277–300.

Long, Susan O. (2003) "Reflections on Becoming a Cucumber: Culture, Nature, and the Good Death in Japan and the United States," *Journal of Japanese Studies* 29(1): 33–68.

Mantel, Hilary (1998) *The Giant, O'Brien* (Toronto: Doubleday).

McGuffin, P., B. Riley, & R. Plomin (2001) "Toward Behavioral Genomics," *Science* 291(5507): 1242–49.

McNeil, S. M., A. Novelletto, J. Srinidhi, G. Barnes, I. Kornbluth, M. R. Altherr, J. J. Wasmuth, J. F. Gusella, M. E. MacDonald, & R. H. Myers (1997) "Reduced Penetrance of the Huntington's Disease Mutation," *Human Molecular Genetics* 6: 775–79.

Michie, S., H. Drake, M. Bobrow, & T. Marteau (1995) "A Comparison of Public and Professionals' Attitudes Towards Genetic Developments," *Public Understanding of Science* 4: 243–53.

Mitchell, John J., Annie Capua, Carol Clow, & Charles R. Scriver (1996) "Twenty-Year Outcome Analysis of Genetic Screening Programs for Tay-Sachs and β-Thalassemia Disease Carriers in High Schools," *American Journal of Human Genetics* 59: 793–98.

Novas, Carlos & Nikolas Rose (2000) "Genetic Risk and the Birth of the Somatic Individual," *Economy and Society* 29(4): 485–513.

Office of Technology Assessment (1988) *Mapping Our Genes* (Washington, DC: Government Printing Office).

Oudshoorn, Nelly & Trevor Pinch (eds) (2003) *How Users Matter: The Co-Construction of Users and Technologies* (Cambridge, MA: MIT Press).

Parens, Eric & Adrienne Asch (1999) "The Disability Rights Critique of Prenatal Genetic Testing: Reflections and Recommendations," *Hastings Centre Report* 29(5): S1–S22.

Park, Katherine (1994) "The Criminal and the Saintly Body," *Renaissance Quarterly* 47: 1–33.

Paul, Diane B. & Hamish G. Spencer (1995) "The Hidden Science of Eugenics," *Nature* 374: 302–4.

Petryna, Adriana (2006) "Globalizing Human Subjects Research" in A. Petryna, A. Lakoff, & A. Kleinman (eds), *Global Pharmaceuticals: Ethics, Markets, Practices* (Durham and London: Duke University Press): 33–60.

Potter, Paul (1976) "Herophilus of Chalcedon: An Assessment of His Place in the History of Anatomy," *Bulletin of the History of Medicine* 50: 45–60.

Quaid, K. A. & M. Morris (1993) "Reluctance to Undergo Predictive Testing: The Case of Huntington Disease," *American Journal of Medical Genetics* 45: 41–45.

Quaid, Kimberly A. & Melissa K. Wesson (1995) "Exploration of the Effects of Predictive Testing for Huntington Disease on Intimate Relationships," *American Journal of Medical Genetics* 57: 46–51.

Rabinow, Paul (1996) *Essays on the Anthropology of Reason* (Princeton, NJ: Princeton University Press).

Rapp, Rayna (1998) "Refusing Prenatal Diagnosis: The Meanings of Bioscience in a Multicultural World," *Science, Technology & Human Values* 23(1): 45–71.

Rapp, Rayna (1999) *Testing Women, Testing the Fetus: The Social Impact of Amniocentesis in America* (New York: Routledge).

Rapp, Rayna (2003) "Cell Life and Death, Child Life and Death: Genomic Horizons, Genetic Diseases, Family Stories," in S. Franklin & M. Lock (eds), *Remaking Life and Death: Toward an Anthropology of the Biosciences* (Santa Fe, NM: School of American Research Press).

Richards, Martin (1996) "Lay and Professional Knowledge of Genetics and Inheritance," *Public Understanding of Science* 5: 217–30.

Richardson, Ruth (1988) *Death, Dissection, and the Destitute* (London: Routledge).

Richardson, Ruth (1996) "Fearful Symmetry: Corpses for Anatomy, Organs for Transplantation," in R. C. Fox, L. J. O'Connell, & S. J. Youngner (eds), *Organ Transplantation: Meaning and Realities* (Madison: University of Wisconsin Press): 66–100.

Rix, Bo Andreassen (1999) "Brain Death, Ethics, and Politics in Denmark," in S. J. Youngner, R. M. Arnold, & R. Shapiro (eds), *The Definition of Death: Contemporary Controversies* (Baltimore, MD: Johns Hopkins University Press): 227–38.

Rose, Dale & Stuart Blume (2003) "Citizens as Users of Technology: An Exploratory Study of Vaccines and Vaccination," in N. Oudshoorn & T. Pinch (eds), *How Users Matter: The Co-Construction of Users and Technologies* (Cambridge, MA: MIT Press).

Rose, Nikolas (1993) "Government, Authority and Expertise in Advanced Liberalism," *Economy and Society* 22(3): 283–99.

Scheper-Hughes, Nancy (1998) "Truth and Rumor on the Organ Trail," *Natural History* 107(8): 48–56.

Scheper-Hughes, Nancy (2002) "Bodies for Sale: Whole or in Parts," in N. Scheper-Hughes & L. Wacquant (eds), *Commodifying Bodies* (London: Sage): 31–62.

Scheper-Hughes, Nancy (2003) "Rotten Trade: Millennial Capitalism, Human Values and Global Justice in Organs Trafficking," *Journal of Human Rights* 2: 197–226.

Schöne-Seifert, Bettina (1999) "Defining Death in Germany: Brain Death and Its Discontents," in R. M. Arnold, R. Shapiro, & S. J. Youngner (eds), *The Definition of Death: Contemporary Controversies* (Baltimore, MD: Johns Hopkins University Press): 257–71.

Shapin, S. & S. Schaffer (1985) *Leviathan and the Air-Pump: Hobbes, Boyle and the Experimental Life* (Princeton, NJ: Princeton University Press).

Sharp, Lesley A. (1995) "Organ Transplantation as a Transformative Experience: Anthropological Insights into the Restructuring of the Self," *Medical Anthropology Quarterly* 9(3): 357–89.

Sharp, Lesley A. (2006) *Strange Harvest: Organ Transplants, Denatured Bodies, and the Transformed Self* (Berkeley: University of California Press).

Siminoff, Laura A. & Kata Chillag (1999) "The Fallacy of the 'Gift of Life,'" *Hastings Center Report* 29(6): 34–41.

Simmons, Roberta G., Susan K. Marine, Robert Simmons, Susan D. K. Marine, Richard L. Simmons (1987) *Gift of Life: The Effect of Organ Transplantation on Individual, Family, and Societal Dynamics* (New Brunswick, NJ: Transaction Books).

Spallone, Pat (1998) "The New Biology of Violence: New Geneticisms for Old?" *Body and Society* 4: 47–65.

Steinberg, D. L. (1996) "Languages of Risk: Genetic Encryptions of the Female Body," *Women: A Cultural Review* 7: 259–70.

Stenberg, Avraham (1996) "Ethical Issues in Nephrology: Jewish Perspectives," *Nephrology, Dialysis, Transplant* 11: 961–63.

Strathern, Marilyn (1992) *Reproducing the Future: Anthropology, Kinship, and the New Reproductive Technologies* (New York: Routledge).

Strathern, Marilyn (1996) "Cutting the Network," *Journal of the Royal Anthropological Institute* 2(3): 517–35.

Strathern, Marilyn (2005) "Robust Knowledge and Fragile Futures," in A. Ong & S. J. Collier (eds), *Global Assemblages: Technology, Politics and Ethics as Anthropological Problems* (Malden, MA: Blackwell): 464–81.

Thomas, Nicholas (1991) *Entangled Objects: Exchange, Material Culture, and Colonialism in the Pacific* (Cambridge, MA: Harvard University Press).

Tilney, Nicholas L. (2003) *Transplant: From Myth to Reality* (New Haven, CT: Yale University Press).

Trescott, M. M. (ed) (1979) *Dynamos and Virgins Revisited: Women and Technological Change in History* (Lanham, MD: Scarecrow Press).

Turney, John & Jill Turner (2000) "Predictive Medicine, Genetics and Schizophrenia," *New Genetics and Society* 19(1): 5–22.

Van der Geest, Sjaak & Susan Reynolds Whyte (eds) (1988) *The Context of Medicines in Developing Countries: Studies in Pharmaceutical Anthropology* (Dordrecht, Netherlands: Kluwer).

Willis, Evan (1998) "Public Health, Private Genes: The Social Context of Genetic Biotechnologies," *Critical Public Health* 8(2): 131–39.

Winner, Langdon (1986) *The Whale and the Reactor: A Search for Limits in an Age of High Technology* (Chicago: University of Chicago Press).

Wright, Susan (2006) "Reflections on the Disciplinary Gulf Between the Natural and the Social Sciences," *Community Genetics* 9(3): 161–69.

Yoxen, E. (1982) "Constructing Genetic Diseases," in P. Wright & A. Treacher (eds), *The Problem of Medical Knowledge: Examining the Social Construction of Medicine* (Edinburgh: University of Edinburgh): 144–61.

Zargooshi, Javaad (2001) "Iranian Kidney Donors: Motivations and Relations with Recipients," *Journal of Urology* 165: 386–93.

35 STS and Social Studies of Finance

Alex Preda

Over the past two decades, custom-tailored technologies and theoretical models have become ubiquitous features of financial markets. Contemporary markets mean screens displaying an uninterrupted flow of prices in public places, financial products designed with the help of complex mathematical models, software programs for the instant display and analysis of financial data, and much more. Against the background of a global expansion, this massive presence, together with the growing dependence of financial transactions on both technology and formal modeling, raises the question of the impact of science and technology on a fundamental institution of modern societies. The relevance of this question can be better understood if we take into account the historical dimension of the processes through which science and technology have penetrated financial transactions. Historians of economics and sociologists alike have recently acknowledged that this impact should be measured in centuries rather than decades (e.g., Sullivan & Weithers, 1991; Harrison, 1997; Jovanovic & Le Gall, 2001). How do they contribute, then, to the preeminent position occupied by financial institutions in developed societies? To what extent is finance shaped by science and technology?

Since the mid-1990s, scholars from STS have become increasingly aware of these questions. Working initially independently of each other, several scholars started research projects on the role of science and technology in financial markets. The output of these projects has materialized in books, journal articles, Ph.D. dissertations, conferences, informal exchange networks, coordinated projects, as well as national associations (e.g., the *Association d'études sociales de la finance* in France). Research hosted at several universities in Western Europe and North America has grown at a steady pace, attracting doctoral students, research funding, together with the interest of academic publishers, and cross-fertilizing academic fields such as behavioral finance, economic sociology, economic anthropology, international political economy, and geography.

One question arising here is that of the background against which the interest of STS scholars was directed toward finance. Several developments frame this moment, independently of particular interests and motivations. (1) After the fall of the Iron Curtain and toward the mid-1990s, the acceleration of global financial expansion highlighted the central position occupied by technology and by formal models of finance. (2) More or less celebratory media representations of the wave of financial expansion contrasted with several severe crises toward the end of the 1990s, crises in which formal models played an important role (e.g., the Long-Term Capital Management crisis of 1998). These events triggered renewed discussions about the capacity of financial markets to replace social policies and raised issues of trust, legitimacy, and market constitution, directly involving both technology and financial theories. (3) Since the mid-1980s, criticism of the central assumptions of neoclassical economics had increased its pace in economic sociology as well as in the history of economics. Insights and theoretical approaches developed in science and technology studies had been fruitfully transferred to the history of economics, especially in the work of Philip Mirowski (1989). Additional research in the history of financial economics (e.g., Mehrling, 2005; Bernstein, 1996) also highlighted the conceptual links between physics (especially thermodynamics) and financial theory.

Against this background, a transfer of research topics, concepts, and approaches from STS to the study of financial markets took place, to the effect that social studies of finance (SSF) emerged as a new field of inquiry. Yet, SSF (which comprises different emerging paradigms) cannot be seen as a mere extension or as an application of science and technology studies to finance. First, there has been cross-fertilization with other disciplinary fields, most notably perhaps with economic sociology. Second, SSF did not simply take over already existing STS concepts but modified and enriched them, developing its own research agenda. In the following, I discuss some of the most important conceptual and topical links between STS and the social studies of finance, thus exploring the SSF research agenda. In the first step of the argument, I show how various SSF approaches conceptualize the relationship between knowledge and financial action, analogous to the STS conceptualization of the link between scientific knowledge and practical action. In a second step, I examine how SSF approaches the demarcation problem with regard to financial economics and to markets. I argue that the social studies of finance take over, reformulate, and expand the demarcation problem examined in science and technology studies. In the third step, I discuss the concept of agency developed in SSF and show its similarities and differences with concepts of agency present in science and technology studies as well as in economic theory. The conclusion reviews the research agenda of the social studies of finance and discusses potential cross-fertilization with the STS agenda.

FINANCIAL INFORMATION AND PRICE AS EPISTEMIC THEMES

Information has become a crucial concept of economic theory in the 1970s as a result (and continuation) of efforts started during World War II in operations research (e.g., Klein, 2001: 131; Mirowski, 2002: 60), efforts aiming at optimizing action outcomes based on random, incomplete data (e.g., tracking airplanes with guns and message encryption). This required mathematical tools for transforming randomness into determined patterns, tools that were combined with the notion (formulated by

Friedrich von Hayek and the Austrian School of economics in the 1930s) that markets can be seen as gigantic distributors of information, similar to a telephone switchboard (Mirowski, 2002: 37). This fusion between a view of allocation processes as determined by information on the one hand, and the formal processing of random signals in order to identify determined patterns on the other, led to conceptualizing information as additive signals, independently of the cognitive properties of the receiver. The effect was to separate information from cognition; while the former was treated as a sort of telephone signal, triggering a reaction from the receiver, cognition was deemed to be irrelevant. Noise was equated with uncertainty (Knight, [1921]1985) and seen as a blurring of determined (or meaningful) patterns, analogously to an encryption machine that scrambles the message by inserting (apparently) random signals.

This concept of information as signals, which has proved influential in economic sociology too (e.g., White, 2002: 100–101) is being contested by the game-theoretical notion of information as choice of actions relative to signals under a fixed decision rule (Mirowski, 2002: 380). This introduces the idea of rational expectations on the part of economic actors (Sent, 1998: 22); expectations contain deterministic patterns that filter the random signals. This second notion of information maintains the distinction to cognition, seen not as entirely irrelevant but as statistical inference.

According to Mirowski (2002: 389; 2006), there is a third concept of information as symbolic computation, coming from artificial intelligence, which has proved less influential than the other two. Relevant in this context is the fact that "information," as it is used in neoclassical economic theory, is seen analogously to phone signals. Uncertainty (or noise) is understood as random signals, with no underlying meaningful pattern, while cognition is taken either as irrelevant or as reducible to statistical inferences.

Financial markets can be seen thus as information processors, sending out price signals (Paul, 1993: 1475) on the basis of which actors make their choices according to (rational) decision rules. In this process, actors reciprocally anticipate their respective expectations and incorporate them into signals. In turn, these anticipations are accompanied by dispersion and volatility, understood as a measure of ignorance and uncertainty in the marketplace (e.g., Stigler, 1961: 214). Along with price observation (Biais, 1993: 157), networks of relationships (e.g., Baker, 1984; Abolafia, 1996) and specialization (Stigler, 1961: 220) contribute to reducing noise.

Price signals are regarded as fully reflecting all the information available to market actors (Stigler, 1961). This is also a key assumption of the efficient market hypothesis (EMH). The presence of a large number of actors in the market, acting independently of each other, handling all the relevant information they can get, is a fundamental condition for market efficiency and liquidity (Fama, 1970, 1991; Jensen, 1978). These participants "compete freely and equally for the stocks, causing, because of such competition and the full information available to the participants, full reflection of the worth of stocks in their prevailing prices" (Woelfel, 1994: 328).

EMH is related to the random walk hypothesis (RWH), which can be followed back to Louis Bachelier's treatment of stock price movements as a Brownian motion

([1900]1964) and to Jules Regnault, a mid-nineteenth-century French broker (Jovanovic & Le Gall, 2001). Prices are conceived of as similar to gas molecules, moving independently of each other, with future movements being independent of past movements. This tenet grounds models for computing the probability of future price movements, such as the Black-Merton-Scholes formula (Mehrling, 2005; MacKenzie, 2006). The EMH tenet was contested early by Benoit Mandelbrot, who noticed that price fluctuations are inconsistent with a Gaussian distribution of securities prices (they generate "fat tails") and that prices are scale-invariant (Mirowski, 2004: 235, 239: Mehrling, 2005: 97–98).

The assumption of market efficiency presupposes that at any given time economic agents can distinguish between (meaningful) signals and noise, between the relevant and the irrelevant, *without* recourse to issues of cognition. Several epistemological problems arise here. (1) The distinction between prices and price data: prices as signals cannot be separated from price data, which are not neutral with respect to production and recording processes, as well as to their material support. Recording data implies the use of technology; therefore, the question arises about how price recording technologies shape price data and financial transactions with them (see the Social and Cultural Boundaries of Financial Economics section). (2) The generation and recording of data are not independent of formal and informal theoretical assumptions about veridicality, consistency, homogeneity, reproducibility, comparability, and memorization, assumptions that are incorporated into recording procedures and technologies and reflected in analysis and interpretation. How are these assumptions produced, and which social forces are involved in this process? (3) The use of price data by financial actors implies observation, monitoring, and representation. These processes, in their turn, require interpretation (provided by financial theories), skills, and tacit knowledge.

Seen in this perspective, price data neither appear as given, natural, or determined by the inherent rationality of financial actors, nor do they appear as analogous to phone signals that trigger the recipients' reactions. Rather, these data appear as praxeological structures (Lynch, 1993: 261), that is, as routine, accountable sequences of social action. In this perspective, information, the key concept of financial economics (Shleifer, 2000: 1–3), is not treated as the natural starting point of investigation but as a practical problem for financial actors. When using price data, academic economists share a set of epistemic assumptions with nonacademic financial actors: assumptions about veridicality, consistency, homogeneity, reproducibility, and so on. The scientific work of financial modeling or experimenting does not appear as embedded in a type of understanding or rationality radically different from (and superior to) the lay one. At the same time, since theoretical models are used in financial transactions, they have not only a representational but also an instrumental quality. How do they affect, then, the very assumptions they rely on? A first task on the research agenda is, therefore, to investigate these price-related epistemic themes.

I begin with price observation: what does it mean to observe securities prices as objective and given? Karin Knorr Cetina and Urs Bruegger have studied how dispersed traders observe prices in trading rooms with the help of computer screens. They argue that price observation is above all a collective work (2002: 923–24) of reciprocal coordination, which takes place over considerable geographical distances and does not require spatial co-presence. What it requires is temporal co-presence: the observation of the same price data at the same moment in time. Temporal co-presence, in its turn, is achieved in a form of interaction which Knorr Cetina and Bruegger call face-to-screen (2002: 940), in opposition to Erving Goffman's face-to-face situation (1982): personal interaction mediated and determined by the flow of prices on the computer screen. Reciprocal coordination determines that price data can be accounted for as objective and reproducible while being continuously generated in conversational interactions. Whereas in the scientific laboratory spatial coordination (Gieryn, 2002: 128) plays an important role in the observation of scientific objects, in the trading room it is temporal coordination that appears as crucial.

The laboratory appears as an "'enhanced' environment that 'improves upon' the natural order as experienced in everyday life in relation to the social order" (Knorr Cetina, 1995: 145). The trading room, by contrast, does not work as a system that modifies and integrates an external (natural) order into the social order. Rather, the trading room constitutes a reflexive system of data observation and projection (Knorr Cetina 2005: 40) that brackets out the outside world: the price data it operates with are generated in the system's own conversational interactions. In the process of reciprocal coordination, however, the data become objectified and treated as external with respect to the system's operations. A key role in this process is played by the computer screen, on which financial actors project the outcomes of their interactions (i.e., the price data). At the same time, similar to the scientific lab, trading rooms constitute heterogeneous frameworks of distributed cognition (Beunza & Stark, 2004: 92), where instruments and actors with different properties and skills, respectively, produce and categorize the objects (i.e., financial products) of action.

This raises the question of the role played by price-recording and -displaying technologies with respect to epistemic themes such as veridicality and homogeneity. Veridicality of price data implies that participants ascribe them a referential quality while investing them with trust at the same time. Homogeneity implies that price data are accessible in the same form to every participant (i.e., standardized), a requirement derived from the condition of actors' mutual coordination based on data observation. The relationship between trust, standardization, and technology has been a central STS issue during the last two decades (e.g., MacKenzie & Wajcman, 1985; MacKenzie, 2001a; Porter, 1995): technology disentangles data from the particular skills of individual persons and invests it with abstract authority. Trust is displaced from personal relationships and individual reputations and put on a mix of abstract competences and iterable rules, incorporated in technology. With respect to price data, historical studies of competing price-recording technologies show how their introduction to financial markets in the late 1860s changed the veridicality of price data (Preda, 2003). While one technology (the pantelegraph) attempted to confer veridicality on price data by reproducing the signature of transaction partners, its competition (the stock ticker) disentangled price data from individuals and tied them to each other. Data thus appeared as self-sufficient, abstract representations of a flow of transactions. Their veridicality was grounded in the technology's set of simple, iterable rules, which could reproduce these data across various contexts.

Standardization of financial information involves calculative agencies (Callon, 1998: 6–12; 1999: 183)—that is, procedures and techniques through which the "economic" is disentangled from the "social." These procedures, provided by theoretical models, are instruments through which a certain type of economic rationality is enacted. In a study of standardized cotton prices in world markets, Koray Çalişkan (forthcoming) investigates the social processes through which different stages of standardization are attained. These stages, which Çalişkan, following Callon, calls "prosthetic prices," involve (1) the reciprocal fine-tuning of the traders' pricing models and expectations, (2) the projection of future prices based on commonly acknowledged calculations, and (3) the narrative framing of pricing formulas.

A complementary aspect of standardization is how price data—made abstract and taken out of the concrete contexts of their generation—are used by financial actors to calculate and thus construct paths of collective action. A central dimension of financial calculation is that discursive sense-making procedures frame the data and make it accountable—that is, practically intelligible—to financial actors. Several case studies have examined the practices of accountants, who are confronted with the task of meeting formal rationality criteria when dealing with financial information. These studies show that accountants do not treat financial data as abstract, disembedded, and universal but rather as depending on local procedures through which they are made practically intelligible; these include negotiation, storytelling, and tinkering, among others (e.g., Kalthoff, 2004: 168; 2005). Since the accountants' criteria of formal rationality depend on the generation of intelligible data, and the latter depend on local sense-making procedures, it follows that in practice there can be no clear-cut distinction between formal, abstract rationality, on the one hand, and practical intelligibility, on the other. Several authors have stressed the need for studies of "ethnoaccountancy" (e.g., Heatherly et al., forthcoming; Vollmer, 2003), which should focus on the practical methods through which financial data are generated and invested with formal qualities. Examples are profit and costs as historical categories of financial knowledge, local methods of accounting for financial data, and practical rules for the classification of these data.

Observation, representation, and calculation of financial data are approached as epistemic themes, in a manner that is both directly and indirectly influenced by science and technology studies. One of the contributions of SSF is to show that price data—regarded as unproblematic both by financial economics and by economic sociology—are constituted in a web of interactions involving both human actors and technological artifacts. While economic sociology has focused mainly on the study of social-structural embeddedness of economic transactions, social studies of finance show that information is the outcome of complex, multilayered interaction processes and indistinguishable from cognition. At the same time, rationality criteria do not merely build a normative horizon for financial action but are actually generated and used as practical tools in the actors' transactions. This link between local practices and theoretical horizons questions the relationship between financial theory—understood both as prescription and as representation—and practical action. I turn now to this aspect.

SOCIAL AND CULTURAL BOUNDARIES OF FINANCIAL ECONOMICS

As an established academic discipline, financial economics claims to build a theoretical horizon for concrete actors and practices by enunciating the ideal conditions of rationality under which efficient action becomes possible. As shown in the previous section, a cornerstone of financial economics is the EMH, with the assumption that all action-relevant information quickly becomes fully incorporated into securities prices, and therefore actors can make transaction-relevant decisions based on data about price variations. This incorporation mechanism is public; sufficiently large numbers of actors have access to data about price variations so that no single person or group can consistently control transactions. The probability of gaps between future and actual prices can be computed according to a formal model and tested against empirical data. In this account, the EMH, which has known several varieties, can be seen as a deductive theoretical model of price behavior.

At this point, several questions arise: (1) about financial theory as the product of a historical development and about the social and cultural factors playing a role here, (2) about how the boundaries of this model were drawn, and (3) about the relationship between the theoretical model and the empirical data against which it is tested. The historiography of economics has presented modern financial theory as the result of a straightforward development beginning with Louis Bachelier (and Jules Regnault earlier) and continuing in the 1960s and the 1970s with the work of Eugene Fama and Paul Samuelson, among others (e.g., Dimson & Moussavian, 1998: 93). Yet, a more illuminating approach would be to follow the history of financial theory not as a string of disembodied, asocial thoughts but as a series of social and cultural processes through which its language, concepts, and objects of investigation take shape. Starting from this premise, Alex Preda (2004a) has investigated the nineteenth-century prehistory of financial theory and shown how a vernacular "science of financial investments" reconfigured investor behavior as rational, grounded in attention and observation, while linking the concept of price to those of news and information. This "science" disentangled financial securities from gambling and prepared the field for a formal treatment of price movements. At the same time, brokers like Jules Regnault applied physical principles to the study of price variations (Jovanovic & Le Gall, 2001). Formal models like Bachelier's shifted from investor to price behavior, represented in an algebraic not a geometrical fashion. We are confronted here with the emergence of several cultural boundaries (between rational and nonrational behavior, gambling and investing, human actors and prices) that lay the ground for the formal theory of efficient markets.

Although the prehistory of financial theory traced these cultural and conceptual boundaries, the theory's growth into a full-blown deductive, formal model took place between the 1950s and the early 1970s. The more general intellectual background of this process was a sustained program of economic research into information and optimization algorithms, initiated during World War II at several U.S. research institutes. Whereas neoclassical economics operated until then with a concept of utility modeled on classical mechanics' notion of energy, this research program had at its core the concept of information, understood as patterns of signals similar to phone codes (Mirowski, 2002: 7, 21). The growth of financial theory into the dominant academic model, however, required further boundary work, concerning (1) theorists and practitioners of formal pricing models and (2) financial theorists and the nonfinancial economists in the academic world. The setting in which this second boundary was traced was provided by U.S. business schools, which underwent a rapid "academicization" in the 1960s, providing a home for financial economics, which otherwise was sometimes marginalized in the more established economics departments.

As to the first boundary—although in the beginning practitioners were hostile to pricing models and to the general assumptions of the EMH, some of them enrolled this theoretical apparatus as a handy tool in their controversies and feuds with other practitioners (Mehrling, 2005; MacKenzie, 2006). A central case studied by Donald MacKenzie and Yuval Millo (2003) is that of the option pricing formula developed by Fischer Black, Myron Scholes, and Robert C. Merton in the early 1970s. In the early stages of its use, empirical data did not fit the predictions of the Black-Scholes-Merton formula. Yet, traders on the Chicago Board Options Exchange (CBOE) used it because of its cognitive simplicity, academic reputation, and free availability. The Black-Scholes-Merton pricing formula offered traders a tool for coordinating their actions and a guide to trading and hedging. The use of the Black-Scholes-Merton formula, together with innovations in financial products, led to an increasing fit between empirical data and theoretical predictions and thus ultimately to the academic and practical success of this model.

The establishment of financial economics as a successful academic discipline and, with it, of the EMH as a dominant theoretical model was the outcome of complex social processes that traced the boundaries of finance as a domain of legitimate theoretical conceptualization and empirical investigation. This was accompanied by jurisdictional claims of practitioners, conflicts of interest among academic and nonacademic groups, and a reconceptualization of market exchanges as optimization algorithms. The boundaries between academic financial theory and practice, between academic and other forms of expertise, appear as porous and shifting; vernacular concepts of price as information have played a role in preparing the conceptual foundations of financial theory, while the interests, practices, and institutions of nonacademic groups have contributed in an essential fashion to the overall success of formal pricing models.

Although a central tenet of EMH is that securities prices move in a random fashion and cannot be predicted, technical analysis (or chartism) maintains that prices move according to predictable patterns. In spite of this inconsistency with (and of attacks from) academic theory, chartism has been successful with financial practitioners for a century. How can a vernacular form of expertise coexist with an established academic theory asserting the opposite? How can it maintain success with practitioners over long periods of time? The investigation of these issues, pertaining to studies of demarcation and expertise (Evans, 2005; Collins & Evans, 2003), recently has been started (e.g., Preda, 2004b). At the same time, the impact of financial theory on markets, together with the prominent role played by technology, raises the issue of agency: how are the structures of financial action changed by formal pricing models, by price-recording and data-processing technologies? How is the organization of markets affected by them?

IMPACT OF THEORETICAL MODELS AND TECHNOLOGY: AGENCY IN FINANCIAL MARKETS

The "technologization" of stock exchanges started in the late 1860s with the stock ticker, followed by cinema screens in the 1920s, teletypewriters in the 1930s, and computers in the early 1960s. In the 1950s, the New York Stock Exchange (NYSE) drafted plans for computer recording of trading data, and in 1962 it formulated the aim of developing a "complete data processing system" that "will mechanize virtually all present manual operations in the Exchange's stock ticker and quotation services" (NYSE, 1963: 48–49). In 1963, a special study of the Securities Exchange Commission (SEC) recommended to the U.S. Congress the automation of financial markets.

In foreign exchange markets, Reuters introduced the first monitor screen and keyboard in 1967 and the Monitor Dealing Service (a system of computerized transactions) in 1970. In the early 1980s, the PC won over proprietary systems in brokerage offices, a process that facilitated the automation of major financial exchanges such as Euronext (formed in 2000 by the merger of the Paris, Brussels, and Amsterdam stock exchanges). The coexistence of automated and nonautomated financial exchanges has highlighted technology-induced differences in price and volatility patterns (Franke & Hess, 2000: 472), raising the question of the role of technology in the constitution of securities prices. In the late 1990s, the first electronic exchange networks (ECNs) were approved by the SEC as platforms for financial transactions. In 2006, ECNs like Archipelago merged with the NYSE.

Neoclassic economic theory, for its part, has conceived agents as isolated individuals, endowed with calculative capacities, desires, and preferences, which remain unaffected by their relationship with other human beings or with artifacts (Davis, 2003: 167). Combined with the prevailing notion of information as signal, this has led to conceiving economic agents as atomistic calculators who process external signals and take decisions (Mirowski, 2002: 389). Nevertheless, studies of market microstructure question these agential assumptions (e.g., O'Hara, 1995: 5, 11).

One of the lasting theoretical and empirical contributions of social studies of science has been to stress the irreducibility of agency to human intentionality or will and to show that scientific theories and technological artifacts shape future paths of collective action. At least two concepts mark the STS contribution: (1) theoretical (or disciplinary) agency, concerning the ways in which conceptual artifacts (like scientific models or mathematical formalisms) change cultural and social structures (e.g., Pickering, 1995: 145), and (2) sociotechnical agency, concerned with the role of material arrangements and of technological artifacts (e.g., Bijker, 1995: 192, 262; Bijker et al., 1987). The STS conceptualization of agency differs from technological determinism in that technology (1) is not seen as preconfiguring paths of action, (2) implies not only constraints but also social resistance, and (3) is not seen as distinct from but as a form of social action. Consequently, the computerization of financial exchanges is not seen as inevitable but as the result of specific social interests, conflicts, and group mobilization.

Studies of theoretical and sociotechnical agency have investigated (1) how the production of formal models and technologies shape future paths of action (the producer side) and (2) how the use of theories and technological artifacts affect collective action and transform communities (the user side). It has been argued that user groups act as market intermediaries, thereby playing a special role with respect to social diffusion and agency (Pinch, 2003). With respect to the field of finance, it becomes relevant to examine how theoretical models and technologies are produced and adopted in financial markets and how their use affects financial transactions and changes the markets' organizational patterns.

Financial models (like the Black-Scholes-Merton formula) do not merely formulate a set of rules that, when applied, will ensure that these transactions meet efficiency and rationality criteria. If we take these models as normative, we risk a determinist position, according to which financial agents simply follow theoretical prescriptions. If we accept the representational character of formal models, we take financial transactions as an isolated asocial domain of investigation and assume a naturalist stance (MacKenzie, 2001b).

To avoid these conceptual difficulties while preserving a notion of theoretical agency, Michel Callon (1998) has suggested the concept of *performativity*. According to Callon, economic theory shapes the way in which transactions are conducted and markets are organized; it has a performative character. A program of research on performativity should involve an investigation of the social forces, groups, interests, and mechanisms through which successful theoretical intervention is performed. An example in this respect (Callon, 1998) is the reshaping of an agricultural produce market by economic consultants, a reshaping that enacts a normative model of rationality. This enactment, however, is not automatic but involves conflicts between interest groups, persuasion, and the mobilization of organizational structures and artifacts.

Theoretical agency (performativity) consists of two opposite yet closely intertwined processes. The first is the demarcation of the boundaries between the economic and the social, or disentanglement (Callon, 1999: 186)—a process through which ethical and social aspects are redefined as outside the sphere of transactions. The second process is the social entanglement between producer and user groups, through which they reciprocally tune their interests and enroll heterogeneous resources to realize these interests. In the language of actor-network theory, performativity then implies the creation of a heterogeneous network that defines its interests and mobilizes adequate resources while tracing conceptual and cultural boundaries in such a manner that the outcome of this process (e.g., empirical data, results) appears to reinforce the resources (e.g., confirm the abstract model). However, since the outcome of boundary-marking (data) is neither independent of the resources used (model) nor interest-neutral, it follows that model and data circularly reinforce each other, in a way similar to the bond existing between theory producers and users.

Theoretical agency (or performativity) combines then the normative aspect of economic theories with the reflexive character of economic knowledge: normative models of economic processes are developed by academic researchers and at the same time monitored by market actors, who adopt and adapt these models to their own interests, practices, and situations.

Empirical studies such as MacKenzie and Millo's (2003) historical analysis of the Black-Scholes-Merton option pricing formula have highlighted how traders on the Chicago Board Options Exchange (the user community) imposed Black-Scholes prices on those who believed them to be too low. In using the formula as a tool in hedging and trading, traders started pushing down options prices; in doing so, they generated prices that fitted those predicted by the theoretical model. This, in turn, acted as an empirical confirmation of the theoretical model. The use of the formula by options traders, together with the introduction of new financial products, narrowed the gap between theoretical predictions and actual prices. At the same time, the use of the option pricing formula changed the organization of derivative markets, their legal definition, and the structure of financial products. On these grounds, MacKenzie (2004; 2006: 17) distinguishes *generic performativity, Barnesian performativity,* and *counter-performativity.* While generic performativity designates the use of the model as a tool by practitioners, Barnesian performativity means that users will generate such data as to confirm the model's predictions, without the data being directly derived from the model. Counter-performativity, by contrast, designates the situation where the use of a theoretical model engenders counter-productive imitation: price data generated by imitative trades no longer match the model's predictions (e.g., "fat tail" distributions).

Other studies, such as Philip Mirowski and Edward Nik-Khah's (2007) investigation of wavelength auctions, have argued that the boundary between the economic and the social is never perfect, since group interests and structures play a dominant role. Moreover, this boundary is marked by conflicts of interests between competing user and producer groups, who form alliances. Mirowski and Nik-Khah show how in the auctions of mobile phone wavelengths competing alliances between phone companies (user groups) and experimental economists (theory producers) were formed, with objectives and agendas that fused theoretical and political aspects. They also argue that the complexity and heterogeneity of financial expertise (ranging from academics to securities analysts, accountants, and merger lawyers), together with the prominent role of group interests, require a more nuanced approach to theoretical agency than that provided by the concept of performativity. The overall argument resonates with the requirement for a more intense analysis of various, even contradictory forms of financial expertise needed for a better understanding of how boundaries are produced and maintained in finance (see also Miller, 2002).

The second aspect of agency is related to the massive reliance of global financial markets on technological systems for data processing and transactions. Enmeshed with this aspect are issues such as (1) the social forces that advance the technologization of financial transactions, both in a historical and in a contemporary perspective; (2) the assumptions that underlie the design of trading programs; (3) the effects of technology on the organization of financial exchanges and the perception of financial data; and (4) the ties between technology and forms of financial expertise like securities analysis.

With respect to the first issue, recent historical studies have shown that, when the first price-recording technology (the stock ticker) was introduced on the NYSE, user and producer groups (stock brokers and telegraph companies, respectively) formed alliances to promote their monopoly and control price data. This technology displaced bodily recording techniques, standardized data, and disentangled authority and credibility from individual actors (Preda, 2006). In her investigation of the Chicago Board of Trade, Caitlin Zaloom (2003) has confronted the question of the CBOT's bitter resistance to automation, in contrast to the latter's enthusiastic adoption by the Paris Bourse in the late 1980s, as studied by Fabian Muniesa (2000, unpublished). Zaloom's argument is that trading technologies are multilayered and embedded in local settings, being represented not only by software programs but also by the body techniques and spatial arrangements traders use to communicate and gather relevant information. Lack of trading automation does not mean the absence of any technique; traders rely on a set of distributed, heterogeneous techniques for solving informational problems. The body techniques employed by traders have developed into change-resistant routines, intertwined with networks of personal relationships and with a social hierarchy on the trading floor. This constellation of specific routines, spatial arrangements, and social relationships is perceived by participants as proprietary and as inaccessible to outsiders. Automation is resisted by traders and perceived as a menace to their privileges, to the existing networks of relationships, and, above all perhaps, to established ways of gathering and processing information. Instead of being perceived as reducing informational uncertainties, automated trading is seen as increasing social uncertainties. Resisting it does not mean that the CBOT traders resist any kind of technology. On the contrary: they mobilize the existing techniques as a unique resource in fighting off attempts to change the ways in which they gather and process information.

In contrast to the CBOT's resistance to computerized trading, Fabian Muniesa shows how automation was successfully introduced to the Paris Bourse. In the 1980s, the problem of the Paris Bourse was to attract customers by offering distinct features that other stock exchanges did not possess. This competitive pressure had been heightened by the relatively marginal position of the Bourse with respect to other major exchanges (London and New York), by the deregulation of the London Stock Exchange in 1986, and by the latter's subsequent technological upgrading. The management of the Paris Bourse adopted (and adapted) a system of computerized trading (CATS, or computer-assisted trading system) introduced on the Toronto Stock Exchange in 1975 with limited success. Yet, in modifying the CATS system (which became CAC, or *Cotation assistée en continu*), the Paris Bourse was confronted with the problem of the assumptions (among others, about equilibrium and fairness) that should underlie the trading algorithms. These assumptions determine the design of the trading algorithm software and, consequently, the processes through which securities prices are formed (or what market participants call "price discovery").

From the beginning, it becomes clear that pricing is not a natural process involving the identification (or discovery) of an already existing "ideal" or "objective" price. Rather, pricing appears as a complex social process of negotiation involving interest groups, software, economic theories, and computer networks, among others. The absence of human intermediaries (i.e., brokers) does not imply the absence of any negotiation process but rather displacement and distribution among heterogeneous actors. While working in the tradition of the actor-network theory characteristic of the Paris school of STS, Muniesa highlights both producer- and user-related aspects of sociotechnical agency. On the producer side, he shows that the successful introduction of automated trading on the Paris Bourse was brought about by an alliance of managers, brokers, and software engineers who reciprocally tuned their positions and adapted existing technologies to local constellations of interests. An outcome of this reciprocal tuning was the presentation of trading software as embodying "a vision of the market" (Muniesa 2000: 303)—that is, a "perfect" theory of market equilibrium that was not imported as a given into this alliance but produced by it. The agential character of formal equilibrium models has less to do with their normative character than with their role as a resource in such a heterogeneous alliance.

On the user side, the trading software enables participating actors to compute prices, which they afterward project as "true," "real," and "discovered." This mode of computing differs from previous ones (which used statistical means) and implies standardization, without being reducible to it. As performed by the software, the calculation of prices is standardized and displayed to actors from a central source. Traders appear as anonymous participants in transactions, known only to a central, data-providing authority (the computer). Yet, exactly because participation in trading is anonymous and routed via a technological authority, actors need to reciprocally coordinate their expectations by inferring personal or categorical identities from the computerized display of price data. Coordination of expectations, in turn, allows traders to project future courses of actors and to construct the market as a collective movement of human and nonhuman agents, a movement that grounds evaluations of market fairness and justice. Personal agency and technical agency combine to configure the market as an entity *sui generis*, with a life of its own.

FINANCIAL MODELS, TECHNOLOGY, AND RISK

The starting point of my argument (presented in the Financial Information and Price as Epistemic Themes section) has been the centrality of the concept of information in financial economics. Acknowledging this position means investigating the epistemic premises of this concept, its cultural trajectory in the history of economics, as well as its links with technology. A significant link is that between information and risk: a standard argument of financial economics (taken over by economic sociology as well) is that economic actors gather and distribute information to process uncertainties into risks (e.g., Stinchcombe, 1990: 5), thereby enabling economic decisions. Yet, if information cannot be separated from (tacit and explicit) forms of knowledge and expertise, depending on heterogeneous constellations of human actors and artifacts, it follows that the said forms of knowledge, together with group relations and concrete technologies, will have an impact on how financial risks are produced and managed. Since financial risk constitutes a major problem in a global world (as repeatedly illustrated by the crises of the late 1980s and 1990s), investigation of this area offers a potential for practical contributions as well.

On a first, micro-interaction level, financial risk appears as a discursive device that, combined with body technique and with price-recording technologies, is employed in managing the "trading self" (Zaloom, 2004: 379). While more general economic discourse ascribes a negative connotation to risk, the practice of financial actors is to approach it as something that is not entirely manageable through calculations and formulas but requires narrative framings and classifications (see also Mars, unpublished; Kalthoff, 2005).

On a different, organizational level, financial risk is made sense of with the help of technologies like software programs and formal models, which saw a rapid, worldwide expansion in the 1990s. Tracing the sources of this expansion, Michael Power (2004) argues that technologies such as enterprise risk management (ERM) originated in a cultural shift that put emphasis on shareholder value and on increased performances of company stock prices in the market. ERM was implemented in banks all over the world to control financial exposure and to prevent overengagement in financial trades. Yet, since such technologies are based on algorithms that automatically overrule human actors' decisions, a reciprocal tuning of traders and software is no longer possible. The introduction of standardized risk measurement technologies, managed from outside the trading floor, blocks out the local skills and personal knowledge of human actors, which play an important role in avoiding financial loss. Risk-measurement technologies are not instruments that measure an external given reality ("risk") but tools of financial action (Holzer & Millo, 2004: 16). These models change the very phenomena they are supposed to represent; consequently, their use does not automatically diminish financial risks and volatility (see also MacKenzie, 2005: 78). While traders use models to calculate option prices and exposures, they also observe and imitate each other, to the effect that "superportfolios" emerge. In situations of financial instability, the use of the same pricing formulas in the same way, with the same trades, can have destructive effects.

CONCLUSION

I have argued that a distinctive feature of social studies of finance is the investigation of scientific models, technology, and forms of expert knowledge in financial institutions. Is SSF then to be regarded as a subfield of STS? Are financial institutions complex enough to support an emerging discipline over longer periods of time? What would the SSF research program look like?

Undoubtedly, the majority of SSF studies has been done by academics trained in the sociology of science and technology, or who had an established reputation in STS. Many of them continue to conduct parallel research projects in both fields. The major themes of investigation—such as observation, representation, boundary marking, agency, and risk, to name but a few—had already been successfully investigated with respect to science and technology. Yet, in spite of the clear affinities and influences, SSF does not appear as a mere subdomain of STS. There are several reasons: the first is that SSF combines epistemic topics with the study of problems relevant in areas like economic sociology and behavioral finance, bringing a genuine contribution to the study of financial institutions. One of these problems is the pricing mechanism: while financial economics has noticed the impact of technology, it has been the role of SSF studies to show how price data, theoretical assumptions, trading software, and computer networks influence the constitution of securities prices. Another genuine contribution is related to the analysis of information as the cornerstone of financial markets. While financial economics and economic sociology have understood information as signal processing and treated it as a black box, SSF has highlighted the social and institutional origins of this concept as well as the epistemic and cultural assumptions on which financial information is constituted.

A second reason for the growing disciplinary autonomy of SSF is that it has made conceptual contributions, acknowledged as such, in disciplines such as sociology, behavioral finance, and the history of economics, an example being the concept of performativity, which can be seen as an extension and modification of the notion of agency developed in the sociology of science and technology. Another example is the concept of markets as a reflexive system, built on an analogy with the concept of laboratory. This indicates growing disciplinary autonomy, without affecting the ties between STS and SSF. Owing to the close personal and intellectual ties between these fields, I expect them to stay in a lively dialogue.

A further question with respect to the possibility of disciplinary autonomy is whether the field of inquiry is deep enough to support continuous SSF research in the long run. I can confidently venture the following: the research done since the mid-1990s is a mere scratch on the surface of the field. There is a wealth of uninvestigated or under-investigated topics, both historical and contemporary. A short list would include the social and epistemic history of competing price-recording technologies, the development of trading software and the interface between the software industry and financial markets, trading robots, the social history of financial information as a commodity, the emergence of epistemic intermediaries like financial analysts, the growing role of financial expertise, the relationship between formal financial models and vernacular economics, the relationship between academic theories and nonacademic ones, and vernacular forms of financial knowledge and theories. The field shows enough depth and relevance to support research in the long run.

While there is neither a formal research program, comparable, for instance, with the strong program in the sociology of scientific knowledge (but see Preda, 2001), nor a single school (comparable to the Edinburgh, Paris, or Bath/Cardiff schools in STS), this can be seen rather as an advantage, since it allows the inclusion of various research interests and approaches. Nevertheless, the possibility cannot be excluded that formal research programs will emerge and that we will witness more internal differentiation after the initial growth period. Already several distinct approaches are configuring: one centered on the concept of performativity and influenced by (but not limited to) the actor-network theory perspective and another one grounded in the tradition of laboratory studies and centered on field work in the trading room. I expect that further empirical studies and theoretical contribution will deepen the differentiation process. In any case, the prominence of financial institutions in our world, together with the growing role of financial theories, expertise, and technologies, make this one of the most exciting developments to have emerged from STS.

References

Abolafia, Mitchel (1996) *Making Markets: Opportunism and Restraint on Wall Street* (Cambridge, MA: Harvard University Press).

Bachelier, Louis ([1900]1964) "Theory of Speculation," in P. H. Cootner (ed), *The Random Character of Stock Market Prices* (Cambridge, MA: MIT Press): 17–78.

Baker, Wayne (1984) "The Social Structure of a National Securities Market," *American Journal of Sociology* 89: 775–811.

Bernstein, Peter L. (1996) *Against the Gods: The Remarkable Story of Risk* (New York: Wiley).

Beunza, Daniel & David Stark (2004) "How to Recognize Opportunities: Heterarchical Search in a Trading Room," in K. Knorr Cetina & A. Preda (eds), *The Sociology of Financial Markets* (Oxford: Oxford University Press): 84–101.

Biais, Bruno (1993) "Price Formation and Equilibrium Liquidity in Fragmented and Centralized Markets," *Journal of Finance* 48(1): 157–185.

Bijker, Wiebe E. (1995) *Of Bicycles, Bakelites, and Bulbs: Toward a Theory of Sociotechnical Change* (Cambridge, MA: MIT Press).

Bijker, Wiebe E., Thomas P. Hughes, & Trevor Pinch (1987) *The Social Construction of Technological Systems: New Directions in the Sociology and History of Technology* (Cambridge, MA: MIT Press).

Çalişkan, Koray (forthcoming) "Markets' Multiple Boundaries: Price Rehearsal and Trading Performance in Cotton Trading at Izmir Mercantile Exchange," in M. Callon, Y. Millo, & F. Muniesa (eds), *Market Devices: Sociological Review Monograph Series* (Oxford: Blackwell).

Callon, Michel (1998) "Introduction," in M. Callon (ed), *The Laws of the Markets* (Oxford: Blackwell): 1–57.

Callon, Michel (1999) "Actor-Network Theory: The Market Test," in J. Law & J. Hassard (eds), *Actor-Network Theory and After* (Oxford: Blackwell): 181–95.

Collins, Harry M. & Robert Evans (2003) "The Third Wave of Science Studies: Studies of Expertise and Experience," *Social Studies of Science* 32(2): 235–96.

Davis, John B. (2003) *The Theory of the Individual in Economics: Identity and Value* (London: Routledge).

Dimson, Elroy & Massoud Moussavian (1998) "A Brief History of Market Efficiency," *European Financial Management* 4(1): 91–103.

Evans, Robert (2005) "Demarcation Socialized: Constructing Boundaries and Recognizing Difference," *Science, Technology & Human Values* 30(1): 3–16.

Fama, Eugene (1970) "Efficient Capital Markets: A Review of Theory and Empirical Work," *Journal of Finance* 25: 383–417.

Fama, Eugene (1991) "Efficient Capital Markets: II," *Journal of Finance* 46: 1575–617.

Franke, Günter & Dieter Hess (2000) "Information Diffusion in Electronic and Floor Trading," *Journal of Empirical Finance* 7: 455–78.

Gieryn, Thomas (2002) "Three Truth-Spots," *Journal of History of the Behavioral Sciences* 38(2): 113–32.

Goffman, Erving (1982) *Interaction Ritual: Essays on Face-to-Face Behavior* (New York: Pantheon).

Harrison, Paul (1997) "A History of an Intellectual Arbitrage: The Evolution of Financial Economics," in J. B. Davis (ed), *New Economics and Its History*, annual supplement to *History of Political Economy* 29(suppl.): 172–87.

Heatherly, David, David Leung, & Donald MacKenzie (forthcoming) "The Finitist Accountant: Classifications, Rules, and the Construction of Profits," in T. Pinch & R. Swedberg (eds), *Living in a Material World: On Technology, Economy, and Society* (Cambridge, MA: MIT Press).

Holzer, Boris & Yuval Millo (2004) "From Risks to Second-Order Dangers in Financial Markets: Unintended Consequences of Risk Management Systems," Discussion Paper 29 (London: CARR/LSE).

Jensen, Michael (1978) "Some Anomalous Evidence Regarding Market Efficiency," *Journal of Economic Literature* 6: 95–101.

Jovanovic, Franck & Philippe Le Gall (2001) "Does God Practice a Random Walk? The 'Financial Physics' of a Nineteenth-Century Forerunner, Jules Regnault," *European Journal of the History of Economic Thought* 8(3): 332–62.

Kalthoff, Herbert (2004) "Financial Practices and Economic Theory: Outline of a Sociology of Economic Knowledge," *Zeitschrift für Soziologie* 33(2): 154–75.

Kalthoff, Herbert (2005) "Practices of Calculation: Economic Representation and Risk Management," *Theory, Culture and Society* 22(2): 69–97.

Klein, Judy L. (2001) "Reflections from the Age of Economic Measurement," in J. L. Klein & M. S. Morgan (eds), *The Age of Economic Measurement*, annual supplement to *History of Political Economy* 33(suppl.): 111–36.

Knight, Frank ([1921]1985) *Risk, Uncertainty, and Profit* (Chicago: University of Chicago Press).

Knorr Cetina, Karin (1995) "Laboratory Studies: The Cultural Approach to the Study of Science," in S. Jasanoff, G. E. Markle, J. C. Petersen, & T. Pinch (eds), *Handbook of Science and Technology Studies* (Thousand Oaks, CA: Sage): 140–66.

Knorr Cetina, Karin (2005) "How Are Global Markets Global? The Architecture of a Flow World," in K. Knorr Cetina & A. Preda (eds), *The Sociology of Financial Markets* (Oxford: Oxford University Press): 38–61.

Knorr Cetina, Karin & Urs Bruegger (2002) "Global Microstructures: The Virtual Societies of Financial Markets," *American Journal of Sociology* 107(4): 905–50.

Lynch, Michael (1993) *Scientific Practice and Ordinary Action: Ethnomethodology and Social Studies of Science* (Cambridge: Cambridge University Press).

MacKenzie, Donald (2001a) *Mechanizing Proof: Computing, Risk, and Trust* (Cambridge, MA: MIT Press).

MacKenzie, Donald (2001b) "Physics and Finance: S-Terms and Modern Finance as a Topic for Science Studies," *Science, Technology & Human Values* 26: 115–44.

MacKenzie, Donald (2004) "Is Economics Performative? Option Theory and the Construction of Derivatives Markets," paper presented at the Harvard-MIT Economic Sociology Seminar, November 16.

MacKenzie, Donald (2005) "How a Superportfolio Emerges: Long-Term Capital Management and the Sociology of Arbitrage," in K. Knorr Cetina & A. Preda (eds), *The Sociology of Financial Markets* (Oxford: Oxford University Press): 62–83.

MacKenzie, Donald (2006) *An Engine, Not a Camera: Finance Theory and the Making of Markets* (Cambridge, MA: MIT Press).

MacKenzie, Donald & Yuval Millo (2003) "Constructing a Market, Performing a Theory: The Historical Sociology of a Financial Derivatives Exchange," *American Journal of Sociology* 109: 107–45.

MacKenzie, Donald & Judy Wajcman (eds) (1985) *The Social Shaping of Technology: How the Refrigerator Got Its Hum* (Philadelphia: Open University Press).

Mars, Frank (unpublished) *Wir sind alle Seher: Die Praxis der Aktienanalyse*, Ph.D. diss., Bielefeld, Germany.

Mehrling, Perry (2005) *Fischer Black and the Revolutionary Idea of Finance* (Hoboken, NJ: Wiley).

Miller, Daniel (2002) "Turning Callon the Right Way Up," *Economy and Society* 31(2): 218–33.

Mirowski, Philip (1989) *More Heat Than Light: Economics as Social Physics, Physics as Nature's Economics* (Cambridge: Cambridge University Press).

Mirowski, Philip (2002) *Machine Dreams: Economics Becomes a Cyborg Science* (Cambridge: Cambridge University Press).

Mirowski, Philip (2004) *The Effortless Economy of Science?* (Durham, NC: Duke University Press).

Mirowski, Philip (2006) "Twelve Theses on the History of Demand Theory in America," in W. Hands & P. Mirowski (eds), *Agreement of Demand*, supplement to vol. 38 of *History of Political Economy*: 343–79.

Mirowski, Philip & Edward Nik-Khah (2007) "Markets Made Flesh: Performativity, and a Problem in Science Studies, Augmented with Consideration of the FCC Auctions," in D. MacKenzie, F. Muniesa, & L. Siu (eds), *Do Economists Make Markets? On the Performativity of Economics* (Princeton, NJ: Princeton University Press): 190–224.

Muniesa, Fabian (2000) "Performing Prices: The Case of Price Discovery Automation in the Financial Markets," in H. Kalthoff, R. Rottenburg, & H.-J. Wagener (eds), Facts and Figures: Economic Representations and Practices (Marburg, Germany: Metropolis): 289–312.

Muniesa, Fabian (unpublished) "Des marchés comme algorithms: Sociologie de la cotation électronique à la Bourse de Paris," Ph.D. diss., Ecole des Mines, Paris.

NYSE (1963) "The Stock Market Under Stress: The Events of May 28, 29, and 31, 1962: A Research Report by the New York Stock Exchange" (New York: New York Stock Exchange).

O'Hara, Maureen (1995) Market Microstructure Theory (Oxford: Blackwell).

Paul, Jonathan M. (1993) "Crowding Out and the Informativeness of Securities Prices," Journal of Finance 48(4): 1475–96.

Pickering, Andrew (1995) The Mangle of Practice: Time, Agency, and Science (Chicago: University of Chicago Press).

Pinch, Trevor (2003) "Giving Birth to New Users: How the Minimoog Was Sold to Rock and Roll," in N. Oudshoorn & T. Pinch (eds), How Users Matter: The Co-construction of Users and Technologies (Cambridge, MA: MIT Press): 247–70.

Porter, Theodore M. (1995) Trust in Numbers: The Pursuit of Objectivity in Science and Public Life (Princeton, NJ: Princeton University Press).

Power, Michael (2004) "Enterprise Risk Management and the Organization of Uncertainty in Financial Institutions," in K. Knorr Cetina & A. Preda (eds), The Sociology of Financial Markets (Oxford: Oxford University Press): 250–68.

Preda, Alex (2001) "Sense and Sensibility: Or, How Should Social Studies of Finance Be(have)? A Manifesto," Economic Sociology: European Electronic Newsletter 2(2): 15–18.

Preda, Alex (2003) "Les hommes de la Bourse et leurs instruments merveilleux: Technologies de transmission des cours et origins de l'organisation des marches modernes," Réseaux 21(122): 137–66.

Preda, Alex (2004a) "Informative Prices, Rational Investors: The Emergence of the Random Walk Hypothesis and the Nineteenth-Century 'Science of Financial Investments,'" History of Political Economy 36(2): 351–86.

Preda, Alex (2004b) "Epistemic Performativity: The Case of Financial Chartism," paper presented at the workshop Performativities of Economics, École des Mines, Paris.

Preda, Alex (2006) "Socio-technical Agency in Financial Markets," Social Studies of Science 36(5): 753–82.

Sent, Esther-Mirjam (1998) The Evolving Rationality of Rational Expectations: An Assesment of Thomas Sargent's Achievements (Cambridge: Cambridge University Press).

Shleifer, Andrei (2000) Inefficient Markets: An Introduction to Behavioral Finance (Oxford: Oxford University Press).

Stigler, George (1961) "The Economics of Information," Journal of Political Economy 69(3): 213–25.

Stinchcombe, Arthur L. (1990) Information and Organizations (Berkeley: University of California Press).

Sullivan, Edward J. & Timothy M. Weithers (1991) "Louis Bachelier: The Father of Modern Option Pricing Theory," Journal of Economic Education 22(2): 165–71.

Vollmer, Hendrik (2003) "Bookkeeping, Accounting, Calculative Practice: The Sociological Suspense of Calculation," Critical Perspectives on Accounting 3: 353–81.

White, Harrison (2002) *Markets from Networks: Socioeconomic Models of Production* (Princeton, NJ: Princeton University Press).

Woelfel, Charles (1994) *Encyclopedia of Banking and Finance* (Chicago: Irwin).

Zaloom, Caitlin (2003) "Ambiguous Numbers: Trading Technologies and Interpretation in Financial Markets," *American Ethnologist* 30(2): 258–72.

Zaloom, Caitlin (2004) "The Productive Life of Risk," *Cultural Anthropology* 19(3): 365–91.

36 Nature and the Environment in Science and Technology Studies

Steven Yearley

KNOWING NATURE

In the decade since the first STS handbook to include a chapter on the environment (Yearley, 1995), the significance of environmental topics to the science and technology studies community has grown with startling rapidity. In part this is because there has been an increasing number of detailed studies on topics such as environmental controversies (Carolan & Bell, 2004; Krimsky, 2000), the relationship between research and environmental policy (Bocking, 2004; Sundqvist et al., 2002), environmental modeling (Shackley, 1997a; Sismondo, 1999), ecosystem management practices (Helford, 1999), citizen participation in environmental understanding and decision-making (Bush et al., 2001; Petts, 2001; Yearley et al., 2001), the shaping of environmental research (Jamison, 2001; Zehr, 2004), and the development of innovative institutions for the production of certified environmental knowledge (most famously the Intergovernmental Panel on Climate Change, discussed below). STS authors have also contributed to theoretical and conceptual analyses of environmental themes and of ideas about environmentalism (e.g., Latour's [2004] on political ecology and Yearley's [1996, 2005a: 41–53] on the globalization of environmentalism). These two considerations alone would merit a fresh discussion, but such discussion is now pressingly needed for two additional reasons.

First, it has become clear that the earlier framing of this issue as "STS studies of the environment and environmental science" is too narrow. It is now evident that the environment is critical to STS, not just as one more site to study but because studying it affords key insight into the status of "the natural" in advanced modernity. At the simplest level, scientific knowledge is indispensable to contemporary environmental policies because science offers to tell us how nature is. Plants and animals, let alone the climate, cannot speak for themselves; ecologists, oceanographers and meteorologists have become their proxies. This idea is institutionalized in such things as "environmental impact assessments" in which professional advisers are employed to figure out the impact of a new development (like a freeway or harbor) on the surrounding environment. But such practices inevitably construct "nature" as a baseline condition at the same time as they disclose the presumed impacts of the new development. Even on a small scale, such construction is far from straightforward. At the planetary level—in a dynamic ecosystem where even the heat radiated from the sun is believed to vary and where the climate has undergone large fluctuations within recorded history—one cannot build the idea of humanly induced climate change without constructing what the "natural" climate would, counterfactually, have been. In a sense, the larger the environmental impact, the more counterfactual must the natural baseline be. Though this is not how he meant it, McKibben (1989; see also Yearley, 2005b) implicitly recognized this point in his celebrated announcement of the "end of nature." For McKibben, humanly caused global climate change meant that one can no longer find a purely natural environment anywhere on Earth. For the STS scholar, a question of at least equal interest is how the natural is constructed in the very course of advancing such claims.[1]

Commonsensically, for most environmental issues the "natural" condition is fit, healthy, and desirable. Evolution by natural selection ensures that nature is finely tuned. But this comforting observation rapidly runs into problems. For one thing, the contemporary countryside—perhaps most acutely in Europe—is more or less wholly unnatural. It is a managed landscape, run for the cultivation of plants and animals that otherwise would never have existed in such profusion. In Britain even the officially designated "Areas of Outstanding Natural Beauty" (AONBs) are, with unremarked irony, thoroughly unnatural. Worse still for the commonsense, benign view of nature, many things that are taken as "bads" and routinely combated are also natural: diseases, pests, and earthquakes. Nature is thus not unproblematically good nor desirable. Accordingly, STS work around environmental topics has had the opportunity, perhaps not taken as forcefully as we might wish, to face up to "nature" and "the natural." But the environment is not the sole arena for contests over nature; parallel disputes are under way in relation to the new biology and genetic engineering. In a shorthand way, in the case of the environment the problem is that humans are conducting an unheeding experiment on external nature while, in the case of our species' biological nature, humans are wrestling with how to regulate increasing control over our species being. Natural variation was typically assumed to govern human life and reproduction, but once such matters are understood as being under conscious human control, the notions of luck, fortune, and fairness that more or less worked for centuries can no longer function in the same ways. Accordingly, I need to devote a little time to both these realms of "nature."

The second additional reason why a review is required is a substantive one. Though there has been STS investigation of a very wide range of issues within nature and the environment, it is clear that a great deal of recent work has clustered around three substantive topics: humanly induced climate change, genetically modified crops and foodstuffs, and genomics and human reproduction. All three focus attention on "the natural" although only the first two would typically be classed as environmental issues. All three are important for any STS conception of nature and environment, however, not only because of the amount of attention paid to them in the field but additionally because they are at the frontier of STS engagement with policy, social theory, and social change.

It is also worth pointing out that nature and environment cannot be discussed exclusively in terms of STS publications. In part that is because some highly influential authors (McKibben, 1989; Fukuyama, 2002; Beck, 1992, 1995) come almost wholly from outside STS. But it is also because STS ideas influence the work of many authors who see themselves more as environmental sociologists (McCright & Dunlap, 2000, 2003), geographers (Castree & Braun, 2001; Demeritt, 2002), or policy analysts (Hajer, 1995). It is also because STS authors have drawn on work from other traditions, for example, the literature on globalization or anthropological work on kinship and natural relations (Strathern, 1992; J. Edwards, 2000).

In brief, in this chapter I argue that the conceptual key to recent STS work on the environment is about the matter of *knowing nature*. Science and technology are valuable for environmental management precisely because they offer authoritative, far-ranging and powerful ways of comprehending the natural world. The distinctive contribution of STS research is to see that the very business of "knowing nature" shapes the knowledge that results; this decisively influences how effective or not such knowledge is in other public contexts.

GLOBAL WARMING AND HUMANLY INDUCED CLIMATE CHANGE

At first sight, the issue of climate change resembles numerous other environmental controversies that STS scholars have studied. A claim about a putative environmental problem is raised by scientists and taken up and amplified by the media and environmental groups; in time, a policy response follows. As is well known, meteorologists—already aware that the climate had undergone numerous dramatic fluctuations in the past—began in the second half of the twentieth century to offer ideas and advice about the possibility of climate changes affecting our civilization in the longer term (Boehmer-Christiansen, 1994a; P. Edwards, 2001; Jäger & O'Riordan, 1996; Miller & Edwards, 2001; Kim, 2005). Though skeptics like to point out that initial warnings also included the possibility that we might be heading out of an interglacial warm period into the cold, as early as the 1950s there was a focus on atmospheric warming (P. Edwards, 2000). As such climate research was refined, largely thanks to the growth in computer power in the 1970s and 1980s, the majority opinion endorsed the earlier suggestion that enhanced warming driven by the build-up of atmospheric carbon dioxide was the likely problem. Environmental groups are reported to have been initially wary of this claim (F. Pearce, 1991: 284) since it seemed such a long shot and with such high stakes. With acid rain on the agenda and many governments active in denying scientific claims about this effect, it seemed hubristic to warn that emissions might be sending the whole climate out of control.

Worse still, at a time when environmentalists were looking for concrete successes, the issue seemed almost designed to provoke and sustain controversy. The records of past temperatures and particularly of past atmospheric compositions were often not good, and there was the danger that rising trends in urban air-temperature measurements were simply an artifact; cities had simply become warmer as they grew in size. The heat radiating from the sun is known to fluctuate, so there was no guarantee that any warming was a terrestrial phenomenon due to "pollution" or other human activities. Others doubted that additional carbon dioxide releases would lead to a build-up of the gas in the atmosphere, since the great majority of carbon is in soils, trees, and oceans, so sea creatures and plants might simply sequester more carbon. Even if the scientific community was correct about the build-up of carbon dioxide in the atmosphere, it was fiendishly difficult to work out what the implications would be.

Hart and Victor track the interaction between climate science and U.S. climate policy from the 1950s up to the mid-1970s by which time greenhouse emissions had "been positioned as an issue of pollution" (1993: 668); the climate, "scientific leaders discovered, could be portrayed as a natural resource that needed to be defended from the onslaught of industrialism" (1993: 667). Subsequently, according to Bodansky (1994: 48), the topic's rise to policy prominence was assisted by other considerations. There was, for example, the announcement of the discovery of the "ozone hole" in 1987; this lent credibility to the idea that the atmosphere was vulnerable to environmental degradation and that humans could unwittingly cause harm at a global level. Also important was the coincidence in 1988 between Senate hearings into the issue and a very hot and dry summer in the United States. In his election campaign, George Herbert Walker Bush even spoke of combating the greenhouse effect with the "White House effect," denying that politicians were powerless to act in the face of this newly identified threat. Still, most politicians responded to the warnings in the 1980s with a call for more research. Although environmental campaigners countered that there was no need for more research before taking measures to increase energy efficiency and use more renewables, most spokespersons concurred with the view that further knowledge would be important, particularly if some warming had already been set in train by emissions to date. One significant outcome of this support for research was the setting up in 1988 of a new form of scientific organization, the Intergovernmental Panel on Climate Change (IPCC), under the aegis of the World Meteorological Organization and the United Nations Environment Program (Agrawala, 1998a,b). The aim of the IPCC was to collect together the leading figures in all aspects of climate change with a view to establishing in an authoritative way the nature and scale of the problem. This initiative was highly important and a novel phenomenon as far as the STS community was concerned: "While by no means the first to involve scientists in an advisory role at the international level, the IPCC process has been the most extensive and influential effort so far" (Boehmer-Christiansen, 1994b: 195; see also Shackley, 1997b; Miller, 2001b).

STS interest matched the wealth and diversity of issues available here. The first issue to attract attention was the novel conjunction between this form of scientific organization and its dependence on super-fast computing facilities required to do the climate modeling; this dependence ensured, for most of the period, that key work could only be done at a handful of centers worldwide. In a series of papers, Shackley and Wynne (1995, 1996) examined how modeled knowledge was produced, made credible, and rendered serviceable for the policy community (see also Shackley et al., 1998, 1999). Thus, writing with two Dutch colleagues (van der Sluijs et al., 1998), they investigated the strikingly consistent nature of estimates of climate sensitivity over a series of models and policy reviews. Their puzzle was that "[t]he estimated range of the climate sensitivity to CO_2-doubling of 1.5°C–4.5°C has remained remarkably stable over two decades, despite the huge growth of climate science" (1998: 315). Their interpretation was that factors within the sociology of this community tended to make changes in the policy prescriptions much less likely than continuity. In any case, the estimate was broad enough to admit of numerous different interpretations with little friction among the scientific contributors, even if the estimate tacitly excluded more catastrophic scenarios. Sociological factors specific to this community seemed to influence the knowledge it produced. Lahsen carried out ethnographic work on the climate modeling community, examining how the models (known as GCMs, or general circulation models) gained credibility (2005b; see also Sundberg, 2005: 166–84). By their nature, such models cannot be tested against the future. Nor can they really be adequately tested against data about past climates, since they are constructed precisely in the light of information about the past (P. Edwards, 2000: 232). Accordingly, the models are inevitably to some extent conjectural, and one form of test consists of running them against each other; Lahsen investigates the way the unreality and circularity of these procedures is managed by practicing modelers. Modeling remains very time-consuming and expensive: "Despite vast increases in computer power, full runs of today's state-of-the-art GCMs still require hundreds of supercomputer hours, since modelers add complexity to the models even more rapidly than computers improve" (P. Edwards, 2000: 232). Given that the climate science community is not homogeneous, Shackley (2001) argues for the existence of contrasting "epistemic lifestyles" within the modeling community. Some modelers are concerned with developing the most comprehensive model they can, arguing that this is a necessary route to meaningful climate prediction. Others are concerned to establish as quickly as possible models capable of addressing long-term trends so that projections can be made and fed into the policy process (see also Sundberg, 2005: 136–37). The latter group tends to be dominated by thermodynamicists, who argue that the climate system can be treated as a black box exchanging energy with the rest of the universe. Shackley goes on to point out that the existence of these differences interacts with the research-funding system (see P. Edwards, 1996; see also Bloomfield, 1986). In the United States there are many centers with different disciplinary focuses—they give conflicting advice roughly along disciplinary lines. In the United Kingdom, where there is only one center, scientists are forced to be more cooperative and consensus-oriented.

Other STS work focused on the shaping of the negotiations within the IPCC. Given the huge scale of the IPCC and its novelty both as an institution and in terms of the phenomena it was trying to assess, a key issue was how it would reach judgments distilled from all the detail. One specific, if not typical, case was the question of the economic valuation of lives threatened by climatic changes. In terms of policy responses, there appear to be two broad possibilities: either we try to limit the build-up of greenhouse gases (by reducing emissions or boosting sequestration and so on), or we take steps to adapt to a changed climate by building better sea defenses, relocating housing, increasing provision for cooling buildings and associated measures. To work out a reasonable balance somewhere between "all abatement" or "all adaptation," one needs to know the relative pros and cons. Both strategies had costs and benefits, and economists working on the 1995 assessment argued that the various policy paths could not be evaluated without a worldwide analysis of these advantages and costs. After such an analysis had been completed, the equations could then be solved to get a mix of policies that provided the greatest net benefit at the lowest cost (Fankhauser, 1995). In short, they wanted to work out both the economic costs associated with greenhouse-gas abatements and those associated with people becoming victims of the adaptation route. Among other things, this entailed putting a price on the typical life income of people from the various countries, and it turned out, for example, that each South Asian (many of whom are likely to suffer from sea-level rises) was calculated to "cost" their country much less than each Westerner whose income might be lost. The economists argued that they were not evaluating the worth of people's lives, only putting a price on the forgone earnings of typical individuals, but the procedure appeared to value the life of a South Asian at about one fifteenth the worth of a Northern citizen. The valuations were critical, since the relative cheapness of South Asians meant that the "rational" global policy orientation was for relatively little abatement (since abatement was costly, as it tended to impact high-earning Northerners) and a good deal of adaptation (mostly in the developing world); the adaptation appeared relatively inexpensive because it tended to impact people with low incomes. This line of reasoning, though retained in chapter 6 of volume III of the 1995 Assessment Report (D. Pearce et al., 1996), was widely criticized among NGOs (notably the Global Commons Institute, which was founded precisely around this issue). In the end, the economistic argument was largely disavowed in the summary for policy makers with which the volume began. In the section on the social costs of humanly caused climate change, the summary asserted that:

The literature on the subject of this section is controversial . . . There is no consensus about how to value statistical lives or how to aggregate statistical lives across countries. Monetary valuation should not obscure the human consequences of anthropogenic climate change damages, because the value of life has meaning beyond monetary considerations (Bruce et al., 1996: 9–10).

While this revealed deep philosophical divisions over the very conceptualization of the scientific climate change issues (O'Riordan & Jordan, 1999), this kind of approach was less in evidence in later assessments. This prompted economics-enthusiast Lomborg (2001: 301) to lament, "it is regrettable that [such economic issues are] not rationally assessed in the latest [i.e. subsequent] report."

Though this point is made particularly prominent by the use of an example from economics, it highlights a more general issue. The IPCC has to arrive at summary judgments, and these judgments (again as van der Sluijs et al.'s 1998 study indicates; see also van der Sluijs, 1997) are not narrowly determined by the vast array of scientific results in the reports. It is clear that sociological and social psychological considerations factors enter into the formulation of these judgments (for the related case of the UNFCCC [see below], see Miller, 2001a). Moreover, the IPCC reports are characterized by a further level of judgment, since each report volume is introduced with a summary for policy-makers (e.g., Bruce et al., 1996) that has to be approved in detail by the countries' representatives; it is "thus an *intergovernmentally negotiated* text" as the Preface makes clear (1996: x, emphasis added).

A third leading interest of STS scholars has been the relationship between the IPCC—indeed, the whole climate-change regulation community—and its critics (Lahsen, 2005a). Critics were quick to point to the supposed vested interests of this community. Its access to money depends on the severity of the potential harms that it warns about; hence—or so it was argued—it inevitably has a structural temptation to exaggerate harms. This highlights one of the outstanding feature of the IPCC: though there have been other mass scientific projects (including the Human Genome project [see chapter 32 in this *Handbook*]), the IPCC is unusual in that the science with which it had to deal was more controversial and more complex than the obvious comparators (Nolin, 1999). Admittedly, the human genome was enormously complicated, and there were sharply diverging views on how the sequencing should be done, but there was a high level of agreement within the profession about what the answer should look like and no organized lobby denying its basic premises. By contrast, the IPCC was trying to offer policy-relevant analyses that many other policy advisers, including some respected scientists, were explicitly trying to junk. As it was working in such a multidisciplinary area, the IPCC attempted to extend its network widely enough to include all the relevant scientific authorities. But this meant that the IPCC ran into problems with peer reviewing and perceived impartiality; there were virtually no "peers" who were not already within the IPCC (for an analysis of the accusations that could be leveled, see P. Edwards and Schneider, 2001). In line with the classic script of "science for policy," the IPCC legitimated itself in terms of the scientific objectivity and impartiality of its members. But critics were able to point out that the scientific careers of the whole climate change "orthodoxy" depended on the correctness of the underlying assumptions. Worse, the IPCC itself selected who was in the club of the qualified experts and thus threatened to be a self-perpetuating community with a vested interest in continuing to find evidence for the importance of the phenomenon to which its members' careers were shackled (see Boehmer-Christiansen, 1994b: 198). When just one chapter in the 2001 Third Assessment Report has ten lead authors and over 140 contributing authors,[2] then it is clear that this departs from the standard notion of scientific knowledge production. This was of course on top of all the well-recognized problems of science for policy, which Weinberg (1972: 209) had referred to as "trans-science" and which Collingridge and Reeve (1986) came to describe in their "over-critical" model of science advising (on this issue in relation to climate change, see Yearley, 2005c: 160–73). And it was on top of the peculiar difficulty of trying to model future climates in a system of unknown (though enormous) complexity.

The range of critics has been enormous. At one end there have been scholars and moderate critics who have concerns that the IPCC procedure tends to marginalize dissenting voices and that particular policy proposals (such as the Kyoto Protocol) are maybe not so wise or so cost-effective as proponents suggest (e.g., Boehmer-Christiansen, 2003; Boehmer-Christiansen & Kellow, 2002). There are also many consultants backed by the fossil-fuel industry who are employed to throw doubt on claims about climate change (see Freudenburg, 2000 for a discussion of the social construction of "non-problems"); these claims-makers have entered into alliance with right-leaning politicians and commentators to combat particular regulatory moves (McCright & Dunlap, 2000, 2003). Informal networks, often Web-based, have been set up to allow "climate-change skeptics" to exchange information, and they have welcomed all manner of contributors, whether direct enemies of the Kyoto Protocol or more distant allies such as opponents of wind farms (Haggett & Toke, 2006) or anti-nuclear conspiracy theorists. Gifted cultural players including Rush Limbaugh and Michael Crichton have waded into this controversy, with Crichton's 2004 novel *State of Fear* having a "technical appendix" on the errors in climate science. At the same time, mainstream environmental NGOs have tended to argue simply that one should take the scientists' word for the reality of climate change, a strategy about which they have clearly been less enthusiastic in other cases (Yearley, 1993: 68–69, 1992).

There is a second major way in which the IPCC was distinctive: its commitment to include economic, social scientific, and policy aspects of the issues. Correspondingly, other STS work has focused on the role of the social sciences in analyzing climate change and, to some extent, on the IPCC's own social science. Though, according to the self-understanding of the IPCC, these disciplines could not have the precision and exactitude to which the physical sciences aspired, it was clear that global climate change could not be studied in the absence of societal analyses for two reasons. On the one hand, the things that worry us about climate change are chiefly the implications for people, commerce, cities, and to some extent wildlife. The actual impacts that will arise clearly depend on how people respond. Without expert advice on these policy matters, there could be no sensible modeling of the "output" side of the climatologists' work. On the other hand, possible policy responses to climate change (if it is happening) again depend on people's willingness to accept the policy prescriptions—to forgo air travel or to put up with climate risks and so on. The IPCC handled this issue by dividing its procedures into three parallel tracks dealing with the physical sciences, the socioeconomic impacts and possible policy responses. In a four-volume work, edited by Rayner and Malone (1998), STS and social science scholars were invited to turn the question around and to focus, so to speak, on the climate impacts of global human change. This innovative enterprise was clearly aimed to mirror the IPCC's work and to highlight the disciplinary orientations overlooked by the IPCC. Alterations in greenhouse gas concentrations are largely due to emissions from people and from their activities, and thus the rate of such atmospheric change depends on the speed and nature of economic growth, the size of future populations, the technologies chosen by people, the cultures of consumption and leisure they develop, and so on. The institutional assumption of the IPCC is that the only relevant social science is economics; many of the contributors to Rayner and Malone's volumes focus on the role of culture, often from the standpoint of Mary Douglas's cultural theory (Douglas et al., 1998).

Social science engagement with climate issues has also taken the form of studies of public participation in policy responses to global warming. If scientific understanding about environmental issues is uncertain, as it admittedly is with significant aspects of climate change, then—so the argument goes—policy decisions cannot simply be led by expert advice. Rather, decisions will inevitably be matters of political judgment, and in democratic societies such decisions should be democratic and transparent. In the set of studies summarized in Kasemir et al. (2003), participatory techniques are proposed as one powerful means for democratizing the handling of such topics. This work was primarily based on a large-scale European project known as Ulysses (for *Urban lifestyles, sustainability, and integrated environmental assessment*). This project was based on seven European cities (Athens, Barcelona, Frankfurt, Manchester, Stockholm, Venice, and Zurich), and much of its innovative character derived from its use of extensive focus-group type workshops to get citizens to reflect on the ways in which urban lifestyles could change to address climate change and sustainable living. These group meetings commonly acquainted participants with computer-based models of such issues as greenhouse gas emissions so that citizens could use the models to investigate the likely consequences of their proposed lifestyle changes (Guimarães Pereira et al., 1999). The chief drawback of this study was that the models participants employed had been devised for the purposes of the research and were not used by local governments or environmental authorities, so that the study's practical payoff was necessarily limited (contrast the modeling study reported in Yearley et al., 2003; see also Yearley, 1999, 2006).

Finally, STS scholars have been interested in the scientific community's—and specifically the IPCC's—role in the wider policy process (see also Skodvin, 2000; Demeritt, 2001). In the late 1980s and early 1990s it seemed that getting the science right would be enormously important to the policy process, as was commonly thought to have happened in the ozone case (Benedick, 1991; Christie, 2000; Grundmann, 1998, 2006). STS attention typically focused on how the IPCC and others generated this knowledge. But from the outset, climate scientists advised that states needed to act rapidly if greenhouse gas concentrations were to be regulated; pressure grew for the introduction of some form of international treaty, and in 1990 the United Nations took the initiative in setting up an intergovernmental negotiating committee (INC) for a Framework Convention on Climate Change (FCCC) (Bodansky, 1994: 60).[3] The FCCC, set up in 1992, eventually gave rise to the Kyoto Protocol of 1997, which set out a process for introducing a binding treaty committing participating nations to greenhouse gas emission targets. The irony of this development was that the negotiating apparatus was in place even before the IPCC had finished its second assessment report, and details of the science began to matter less once the horse-trading started. Commentators quickly noted tensions between the IPCC and the UNFCCC. The UNFCCC had different institutional sponsors, and it effectively took the lead in greenhouse politics, threatening to make the results of the more sophisticated GCMs redundant (Miller, 2001a). In the scramble to get states to sign up to the Kyoto Protocol, political and policy considerations won out over scientific ones. The reasons that Canada ratified and Australia did not have more to do with economics, local political campaigning, and the perceived suitability of the international convention to local needs than to the relative credibility of the science in Canada and Australia (Padolsky, 2006; see also Victor, 2001). The relationship of the IPCC to the policy process remains complex; much interest focuses on the potential for sudden climate impacts (known as "tipping points") and the kick-starting of runaway positive feedback loops.[4]

In conclusion, climate change has proved to be a major area for STS research in the last decade because of the complexity of the relationship between knowledge and policy advising. Even so, it would only count as a highly complex case study were it not for two novel factors critical to STS audiences. There is first the way in which the IPCC process runs science-advising up against the very limits of legitimation through peer review. Second, there is the fact that, through its own deployment of social science, the IPCC inevitably raises a "reflexive" question about the role of the social sciences (of course including STS; see Jasanoff & Wynne, 1998). Additionally, by suggesting the capacity for human influence on the environment at a global level—a level so pervasive that anthropogenic emissions may change sea level radically and increase the severity of storms and hurricanes—it indicates how precarious the naturalness of nature may be (on the ideological aspects of nature, see Sunderlin, 2003, and in a different sense, Douglas, 1982 on natural symbols).

GENETIC MODIFICATION AND GM PLANTS AND FOODS

The second issue to dominate environmental policy and debate, and also to dominate STS inquiries, in the last decade has been genetically modified (GM) crops and food. Despite the intensity of recent public controversies over genetic foodstuffs, work on genetic engineering has a three-decade-long history, and STS studies, particularly of laboratory safety and the regulation of genetically modified entities, commenced early (Bennett et al., 1986; see also Jasanoff, 2005: 42–63). Through the 1980s companies and university scientists were working on developing products, and this area was routinely identified in science policy analyses as one of the potential hotspots for economic growth in the recession-affected economies of the late 1980s and early 1990s. Environmental campaigners too were preparing their positions. But the period of quiet preparation came to an end in the 1990s when products started to come to market.

The principal issue was testing. Here was a new product, whether GM crop, animal, or bacterium, that needed to be assessed. Of course, all major industrialized countries had some established procedures for the safety testing of new foodstuffs. But the leading question was how novel were GM products taken to be. For some, the potential for the GM entity to reproduce itself or to cross with living relatives in unpredictable ways suggested that this was an unprecedented form of innovation that needed unprecedented forms of caution and regulatory care. On the other hand, industry representatives and many scientists and commentators claimed that it was far from unprecedented. People had been introducing agricultural innovations for millennia by crossing animals, allowing "sports" to flourish, and so on. Modern (though conventional) plant breeding already used extraordinary chemical and physical procedures to stimulate mutations that might be beneficial. On this view, regulatory agencies were already well prepared for handling innovations in living reproductive entities. As Jasanoff points out (2005: 49), the ground for the regulatory battle was prepared to a large degree in the United States, where the courts had endorsed the regulators' decision that it was products (particular foods or seeds, and so on) and not processes (the business of genetic modification) that should be at the heart of the test (see also Kloppenburg, 2004: 132–40).

GM crops were first certified in the United States, where they passed tests set by the Department of Agriculture, the Food and Drug Administration, and the EPA. Though an early product, the Flavr Savr [sic] tomato found little acceptance on the market, success came with GM corn (maize), soya, various beets, and canola (rape). Essentially, GM versions of these crops offered two sorts of putative benefits: either the crops had a genetic resistance to a pest or they had a tolerance to a particular proprietary weed-killer. The potential advantages of the former are rather evident (even if there is a worry about pests acquiring resistance); the supposed benefits of the latter are more roundabout.[5] The idea is that weed-killer can be used at a later stage in the growing season, since the crops are immune. Weeds can be killed off effectively with minimal spraying. Companies also benefit, of course, since farmers are obliged to buy the weed-killers that match the seeds, and this may even extend the period of market protection beyond the expiry of patents.

European companies were not far behind their U.S. counterparts in bringing these products to market, but European customers were far less accepting of the technology than those in North America (Levidow, 1999). STS responses to this issue have taken two main points of focus. There has been some interest in trying to explain the different responses as between European and North American polities. A great deal of attention also has been paid to the range of regulatory logics available.

To begin with the latter, it is clear that European regulators have tended to be more precautionary about this technology than have U.S. officials. But examination of the precautionary principle in practice indicates that the principle itself does not tell the regulator how precautionary to be (Levidow, 2001; see also Marris et al., 2005). Arguments about the regulatory standard have simply switched to arguments about the meaning of precaution (see also Dratwa, 2002). Discordant interpretations of precautionarity have taken a more precise form in disputes over the standard known as "substantial equivalence." As Millstone, Bruner, and Mayer pointed out in a contribution to *Nature* (1999), some starting assumptions are required to decide how to test the safety of GM food. Precisely because GM crops are—by definition—different from existing crops at the molecular level, it needs to be decided at what level to begin to test for any differences that might give cause for concern or even rule out the new crop technology. According to Millstone et al. (1999, 525; emphasis added):

The biotechnology companies wanted government regulators to help persuade consumers that their products were safe, yet they also wanted the regulatory hurdles to be set as low as possible. Governments wanted an approach to the regulation of GM foods that could be agreed internationally, and that would not inhibit the development of their domestic biotechnology companies. The FAO/WHO [UN Food and Agriculture Organization/World Health Organization] committee recommended, therefore, that GM foods should be treated by analogy with their non-GM antecedents, and evaluated primarily by comparing their *compositional data* with those from their natural antecedents, so that they could be presumed to be similarly acceptable. Only if there were glaring and important compositional differences might it be appropriate to require further tests, to be decided on a case-by-case basis.

Regulators and industry agreed on a criterion of substantial equivalence as the means for implementing such comparisons.

By this standard, if GM foods are compositionally equivalent to existing foodstuffs, they are taken to be substantially equivalent in regard to consumer safety. Thus, GM soya beans have been accepted for consumption after they passed tests focusing on a "restricted set of compositional variables" (1999: 526). However, as Millstone et al. argue, with just as much justification, regulators could have chosen to view GM foodstuffs as novel chemical compounds coming into people's diets. Before new food additives and other such innovative ingredients are accepted, they are subjected to extensive toxicological testing. These test results are then used conservatively to set limits for "acceptable daily intake" (ADI) levels. Of course, with GM staples (grains and so on), the small amounts that would be able to cross the ADI threshold would be commercially insufficient. However, safety concerns would be strongly addressed. These authors' point is not so much that GM foods should be treated as food additives or pharmaceuticals but that the decision to introduce the substantial equivalence criterion is not itself based on scientific research. That decision is the basis on which subsequent research is done (see Jasanoff, 1990, for the same "logic"). For proponents of the technology, substantial equivalence is a straightforward and common-sensical standard. But the standard conceals possible debate about what the relevant criteria for sameness are. As Millstone, Bruner and Meyer point out, for other purposes the GM seed companies are keen to stress the distinctiveness of their products. GM material can only be patented because it is demonstrably novel. How then can one be sure that it is novel enough to merit patent protection but not so novel that differences beyond the level of substantial equivalence may not turn out to matter a decade or two into the future?

This issue was also at the heart of the United Kingdom's widely publicized "Pusztai affair." Arpad Pusztai worked at a largely government-funded research establishment near Aberdeen in Scotland and was part of a team examining ways of testing the food safety of GM crops. He and others were concerned that compositionally similar foodstuffs might not have the same nutritional or food-safety implications. The experiments for which he became notorious were conducted on rats; he fed lab rodents on three kinds of potatoes: non-GM potatoes, non-GM potatoes with a lectin from snowdrops added, and potatoes genetically modified to express the snowdrop lectin. Lectins are a family of proteins some of which are of interest for their possible insecticidal value; it is also known that some lectins (e.g., those in red kidney beans) can cause digestive problems when eaten. His results suggested that the rats fared worse on the GM lectin-producing potatoes than on either of the other samples, possibly implying that it was not the lectins that were causing the problem but some aspect of the business of genetic modification itself.

As Eriksson (2004) has detailed, this controversy unraveled in a surprising way. Pusztai announced his results in a reputable British television program apparently intending not to argue against GM *per se* but to assert that more sophisticated forms of testing would be needed to address safety concerns fully—exactly the kinds of testing which he and colleagues might have been able to perform. But the headline message that came over was that GM foods might cause health problems when eaten. In a muddled and confusing way, Pusztai's conclusions came to be criticized by his own institute and he was ushered into retirement. The ensuing controversy and hasty exercise in news management signally failed to concentrate on his findings and the details of his experimental design. Instead, people lined up around the conduct of the controversy itself, either championing Pusztai as a whistle-blowing researcher who was unjustly disciplined by his bosses for publicizing inconvenient findings or dismissing him as a sloppy scientist who rushed into the public gaze with results that were unchecked and unrefereed. On the face of it, it is a curious sociological phenomenon that such important studies have barely been repeated, even if the Pusztai affair lives on within the wider policy debate. Eriksson's study shows how the controversy, by focusing not on the experiment but on disputes over Pusztai's status as an expert in this field, came to take on such an attenuated form. She also explores the manifold ways in which expertise in such a multi-disciplinary area is constructed and contested.

STS interest in the GM case has also been drawn to errors by the manufacturers and suppliers. No matter how emphatic the assurances have been that the new plant technologies are well tested and under control, there have been a series of problems with, for example, corn (maize) approved only for animal rations ending up in human foodstuffs or with traits engineered into plants arising in wild relatives. The analytical interest of this has been developed by STS scholars who have noted how these difficulties continue to throw up problems of what is to count as a reasonable test in such open-ended and far from comprehensively understood contexts. Moreover, such difficulties pose interesting challenges for one popular strategy for managing the horrors of GM organisms (GMOs): the idea that there should be labeling and strict traceability, even if the use of labels has been contentious (as discussed below). But of course the ideas of labeling and traceability rely on the adequacy of routine methods for identifying, tracing, and containing the technology, and all these points have been disputed (Klintman, 2002; Lezaun, 2006). Moreover, as Lezaun (2004) has neatly illustrated in another case, the uncontrollability of the technology also poses problems for the companies in their attempts to manage the intellectual property rights in their technology. In one Canadian case, a farmer was accused by Monsanto of using GM seeds on his farm illicitly, that is, without buying the seeds and paying the fee. The farmer sought to turn the tables by arguing that his farm had been contaminated by Monsanto's invasive product; he tried to blame the company's product for violating his rights. Lezaun highlights the complexity of this case, since part of what the court was invited to figure out was how much Monsanto should be responsible for the behavior of its products which, according to the farmer at least, had invaded his property. Seen another way however, the plant was simply following the dictates of its nature in its reproductive behavior. This case demonstrates how delicate were the borderlines around "nature" that needed to be drawn in this case: there was first the question of which was a natural and which was a synthetic plant, and then there was a question about whether the "invasive" character of plant behavior should be regarded as humanly controlled or as natural to the plant.

The second major theme to have attracted the attention of STS scholars has been the question of the precise reasons for public resistance and consumer anxiety.[6] Actors within the controversy have clearly faced the same question, but they have tended to account in asymmetrical ways. Proponents of the technology tend to blame public anxieties on scare tactics and protectionism, while opponents see corporate greed combated by the perspicacity of the public. STS authors have taken a more symmetrical approach pointing principally to three factors. First, Europeans were being offered this new food technology in the wake of the BSE, or "mad cow," debacle. The changes in the food-processing procedures which are now thought to have created the conditions for the release and spread of the mad-cow prions had been pronounced safe by the same regulatory authorities. Particularly in Britain, the government initially insisted that the best scientific advice was that there was no danger to humans from the affected beef; subsequently in 1996 they announced a sudden change of mind. Thus, the idea of GMOs being considered safe by the regulatory authorities and by governmental advisers could easily be shrugged off and viewed with distrust. Events such as the Pusztai affair were drawn on to intensify this feeling that the scientific establishment was not to be trusted. In the absence of persuasive and comprehensive assurance, there was also a question about how ordinary citizens made dietary decisions and carried on their lives (these issues have stirred wider sociological interest too [see, e.g., Tulloch & Lupton, 2003]). In the United Kingdom, for example, the reputable supermarkets moved to institutionalize the reassurance that governmental agencies failed to provide (Yearley, 2005a: 171–4). Jasanoff had remarked on a similar response in the case of BSE (1997).

Another explanatory factor resulted from the fact that, as noted above, the European landscape is decisively shaped by centuries of agricultural practice. The natural heritage and farming are inseparable. Thus, there was concern from environmentalists, and even from official nature-protection bodies and from countryside groups, about the effect of this new technology on wildlife. Particularly in France, this was allied to the third explanatory issue: a desire to protect traditional rural lifestyles in the face of the perceived threats of globalization and economic liberalization. These new technologies were viewed as further evidence of U.S. attempts to penetrate and reshape the European agricultural market.

This last point came to be reflected in the unfolding trade conflict over GM foods and seeds. U.S. companies have urged that European resistance to GM imports should be combated by appeals to the WTO. A formal complaint was lodged in 2003, with the United States hoping to use the WTO to force open European markets to U.S. farmers and U.S. seed companies. The U.S. argument is that there is no scientific evidence of harm arising from GM food and crops, since these products have all passed proper regulatory hurdles in the U.S. system and the corresponding regulations inside the EU also. Furthermore, on this view, any future strategy of labeling of GM produce in the European market (the procedure favored in many European Union member states as a possible compromise) is discriminatory and an unfair trading practice, since it draws consumers' attention to an aspect of the product that has no relation to its safety (see Klintman, 2002). The label "warns" the customer of the GM content but, if that content is not dangerous, then all the label will do is penalize U.S. and other GM-using suppliers. Accordingly, the WTO should expect to outlaw this labeling practice as an unjustified impediment to trade. European consumer advocates argue, by contrast, that the U.S. testing has not been precautionary enough and that properly scientific tests would require more time and more diverse examinations than have been applied in routine U.S. trials and in their European Union counterparts.

The distinctive difficulty in this case is that, by and large, the official expert scientific communities on opposing sides take diametrically opposing views (though on trans-Atlantic expert contacts, see Murphy & Levidow, 2006). In the United States, the conceptualization of the issue is primarily this: all products have potential associated risks and the art of the policy maker is to ensure that an adequate assessment of risk and of benefits is made. European analysts are more inclined to argue that the risk framework itself leaves something to be desired, since the calculation of risk necessarily implies that risks can be quantified and agreed on. In the case of GM crops, so the argument goes, there is as yet no way of establishing the full range of possible risks, so no "scientific" risk assessment can be completed.

Within their separate jurisdictions, each of these opposing views can be sensibly and more or less consistently maintained. However, the differing views appear to be tantamount to incommensurable paradigms for assessing the safety and suitability of GM crops. There is no higher level of scientific rationality or expertise to which appeal can be made to say which approach is correct, and of course the WTO does not have its own corps of "super scientists" to resolve such issues. However, observers of the WTO fear that its dispute settlement procedures, although supposedly neutral and merely concerned with legal and administrative matters, tacitly favor the U.S. paradigm, since the WTO's approach to safety standards emphasizes the role of scientific proofs of safety and, in past rulings, "scientific" has commonly been equated with U.S.-style risk assessment (Busch et al., 2004, Winickoff et al., 2005). This case may thus not only affect policy toward GMOs but also set a highly significant precedent for how disputed scientific views are handled before the WTO (see also Murphy et al., 2006).

Although the GM case is currently mainly a bone of contention between wealthy Northern countries, other nations have gotten caught up in the struggle. Facing food shortages caused by drought, Zambia in 2002 was offered food aid by the United States, which just happened to consist of genetically engineered cereals.[7] It seemed clear to many that the U.S. was using this case as a Trojan horse to encourage the uptake of GM foods. On the other hand, Zambians realized that there was a danger that their future exports to the European Union would be threatened if they lost their GM-free status. The government thus equivocated over accepting the food aid.

The GM case has also been of great interest to the STS community because of the willingness of governments—particularly in Europe but also, for example, in New Zealand—to initiate various public consultations over the introduction of the technology (Walls et al., 2005). STS work on this topic has taken various forms. As noted by Evans and Collins in chapter 25 in this *Handbook*, there is a general interest in the rationales for and appropriateness of public consultation over technical issues. But these general issues aside, it is reasonably clear that in this case a key consideration in Europe and the Antipodes was the explicit attempt to combat public disquiet by being seen apparently to listen to the public. Hansen (2005) conducted a comparative analysis of consultation exercises in Denmark, Germany, and Britain, while Horlick-Jones et al. (2004; 2007) carried out an external assessment of the British "GM Nation" exercise that ran alongside the exercise itself (see also Pidgeon et al., 2005; Rowe et al., 2005). They pointed out that it was difficult to get an "authentic" consultation, since participants were self-selecting and the process was shaped by its instigators not by the participants. On a smaller scale, Harvey (2005) undertook a participant observation study of a subset of the GM Nation groups. He not only confirmed that participants seemed unclear about their relation to policy-making over GMOs, he also focused on participants' experience of the exercise and on the kinds of topics participants opted to talk about. He noted that the discussion often turned on scientific information. Given a more or less free rein, he observed, participants chose to argue about such issues as safe planting distances, the impact of GM crops on beneficial insects, and so on, and were thus caught in endless and frustrating debates that they were typically unable to resolve in the context of the meetings.

This led Harvey (2005) to take a different view about public responses from that advanced by Wynne (2001). Wynne argued that, in general, citizens had been alienated by the structuring of the policy debate. The debate was being pursued on a basis that was not of the public's choosing and was, in many senses, foreign to their preferred ways of conceptualizing the issues. Even the ethical aspects of the debate—on which the public were supposedly to have some sort of privileged say—had been transformed into the expert meta-ethical discourses of deontological and utilitarian reasoning. Harvey observed that, rather than favoring ethical and broadly political considerations as Wynne's contention would seem to imply, participants focused on "factual" types of claims even if their concerns were not ones that scientists were able to answer authoritatively. The policy debate thus tells us a good deal about GMOs but also much about the detailed nature of public conceptualization of and engagement with such policy issues.

In the United Kingdom, alongside the GM Nation exercise, the government also held a series of farm-scale trials—conducted on volunteered farmland—to test the environmental and practical implications of the new technology. Initially, these trials were treated by anti-GM campaigners as bogus; they were thought to have been set up in such a way as to more or less guarantee success. Accordingly, Greenpeace and other groups attempted to disrupt the trials (Yearley, 2005a: 173–74). But by 2005, when the trial results finally came to be announced, it was reported that GM cultivation had demonstrated a negative effect on wildlife in some cases. It was not that the GM material itself was harmful but that the weed-killing was so effective that wildlife was deprived of seeds and other food. A final irony is that recent STS work has pointed out that the farmers who ran the trials were not given a voice in interpreting the results; on the basis of interviews with farmers, the suggestion is that "participatory" methods involving the farmers' verdicts would likely have been more favorable to GM than the experts' assessments alone, since farmers viewed the management of the trial fields as "unrealistically" inflexible (Oreszczyn & Lane, 2005).

Media coverage of the GM case has also engaged STS scholars. Though there have been studies of media treatment of other environmental issues including climate change (see Carvalho & Burgess, 2005; Mazur & Lee, 1993; Zehr, 2000), the apparent unease of European and some North American consumers with GMOs has prompted a particular emphasis on the role of the media in framing public disquiet in this case. This work has been summarized by Priest, who asserts that "the scientific mainstream remains concerned that the news media have overemphasized and inappropriately legitimized opposition points of view. This does not seem to be borne out by available evidence regarding news coverage of biotech in the U.S. or Europe" (Priest & Ten Eyck, 2004: 178; see also Priest, 2001). Large-scale studies have tracked media—especially press—treatments of the GM case across Europe and North America, providing a key resource for STS analyses (see the collection by Gaskell & Bauer, 2001).

In sum, the case of GMOs is of wide interest to STS researchers because of the light it throws on comparative safety assessment and the interpretation of scientific evidence and precaution in international perspective; it has been the paramount exercise in "technology assessment" of the last decade. But, as with climate change, it also indicates vividly how contemporary societies have responded to the challenge of knowing nature in their handling and regulation of innovative forms of life.

CONSTRUCTING AND PERFORMING NATURE

As anthropological as well as STS authors have observed (e.g., Ingold, 1993), conceptualizations of "nature" and "the natural" sit uneasily in mainstream Western culture. Humans are somehow both in and above nature. Environmental discourses have tended to treat nature as apart from culture, a move that has been echoed—in reverse—in the social sciences. But such boundary work stimulates a feeling of disquiet. It is clear that humans, as animals and as residents of the biosphere, are continuous with nature; equally, human individuals are composed through and through of nature. In contemporary Western cultures there is no tenable basis for segregating the human from the natural. Recent developments in the way that the human genome is understood, and in what Rose refers to as the process of becoming "somatic individuals" (Novas & Rose, 2000), have offered to begin to transform this conceptual terrain further. Within the last decade, authors have again begun to write about human nature (see such different examples as Fukuyama, 2002; Habermas, 2003). Rabinow, in the term "biosociality" (1996), has insightfully commented on the way that cultural transformations are accompanied by changes in the way that cultures draw the nature/culture divide. The connection between STS and this literature is reviewed further in chapter 32 in this volume.

In contrast to conceptual concerns around humans' place in nature, a key site for STS investigations of "the natural" has emerged from various attempts in different countries at nature reconstruction, also known as restoration ecology. In other words, in many industrialized areas where natural habitat has been destroyed or degraded there have been attempts to reinstate what has been lost. Such efforts explicitly ask what the natural state was so that nature can be mimicked or reproduced; in such enterprises, human cultures take it on themselves to produce nature. As with both the climate change and GM cases, there are contrasts between the North American and European approaches. North America (and Australia) was understood as a wilderness, a natural place to which Europeans came late. The typical ambition is to restore habitats to the way they would have been in pre-Columbian times. European history lacks this pivotal moment, and thus the reference point is harder to identify.

The detailed workings of nature reconstruction are often complex as Helford's insightful case study from Northern Illinois shows (1999): attempts to restore the ecology of an oakwood led to the renegotiation of the character of the habitat classification itself. Restoration practitioners argued for the "naturalness" of an oak savanna, whereas academic ecologists had regarded this simply as a transitional zone between oakwoods and prairie, something that was precisely not a habitat in its own right and thus not a worthy goal for restoration activities. Helford charts how the natural character of the oakwood savanna was constructed in the very business of establishing the nature reserve. Within ecology there had already been an overthrow of traditional ideas of a stable "climax" vegetation. It came to be accepted that forests had all along been subject to repeated calamities with no stable end-state. Elsewhere, especially in Australia and in California, the role of fire in forest life had come under scrutiny, and in the Australian case this related specifically to the "naturalness" and desirability of Aboriginal practices of interaction with forests through fire (see Verran, 2002).

In the European context, there has been a good deal of technical work on how to reconstruct vanished habitats: for example, peat bogs in The Netherlands have been a particular focus and STS scholars have become involved in discussions about methodology and principles (Wackers et al., 1997). Scotland too, with large areas of sparsely populated countryside, has been ripe for restoration. Samuel (2001) has given a clear overview of a dispute concerning the management of the west coast island of Rum. The state nature conservation agency aimed to recreate the island's ecology as it would have been before human intervention. However, others used archaeological evidence to claim that deer-hunting humans had been part of the ecology for thousands of years. In the local political context, this was typically viewed as a contrast between an English/southern view of Scotland as wild and unpeopled, and a Highland/local view of an age-old local culture. The "correct" reconstruction of nature was inevitably viewed through the lens of political commitments and sensibilities even if—in the end—all parties agreed that the original condition could not literally be reconstructed, since climate and even the genetic characteristics of the plant life would have changed over the intervening years.

Similar considerations have been raised in the surprising context of weather modification technologies. Through nearly the whole of the twentieth century some enterprising individuals and organizations offered various technologies for weather modification, primarily to do with encouraging or forestalling rain and principally targeted at agricultural communities. Given the unpredictability of the weather, it is always difficult to assess whether these interventions work. For example, one cannot make rain when there are no clouds, but when there are clouds, it is hard to know whether the intervention has caused the rain or not. In separate studies, Matthewman (2000) and Turner (2004) examined the social dynamics of these interventions, dynamics in which the naturalness of erratic weather commonly plays a strong discursive role (see also Kwa, 2001; and on modeling in meteorology more generally, see Jankovic, 2004). Thus, Turner notes that some agricultural communities, disturbed that rain-making for other clients was causing "their" rain to fall on others' fields, campaigned against weather modification by setting up a society for "natural weather." Such studies demonstrate both the difficulty of specifying "the natural" in unambiguous terms and people's continuing propensity to find naturalness in surprising places.

CONCLUSION: NATURE, THE ENVIRONMENT, AND STS

I conclude from this review that "the environment and nature" has been one of the leading research themes in STS in the last decade. Empirical analyses have covered a broad front although the energies of STS scholars have been particularly focused on the two areas of climate change and GM crops. Though these two topics have merited attention because of their policy and political implications, they also demonstrate clearly the sense in which STS work on the environment is fundamentally about the issue of how to know nature authoritatively. I have argued that the distinctive contribution of STS research is to see that the very activity of "knowing nature" shapes the knowledge that results: the way that the IPCC is configured shapes the projections for global temperature change that are produced; the way that risk assessments of

GMOs are organized impacts the estimations of risk. For this reason, I conclude that STS has become the paramount disciplinary basis for understanding "knowing nature" in conditions of advanced modernity.

Notes

I should like to express my thanks to Miriam Padolsky, Joseph Murphy, Eugénia Rodrigues, Mike Lynch, and two anonymous reviewers for their detailed and very helpful comments on this chapter.

1. This is also increasingly a question for cultural geographers, such as Braun and Castree (1998); see also Franklin (2002), Castree and Braun (2001), and Yearley (2005d).

2. My example is chapter 2, "Observed Climate Variability and Change."

3. Loosely defined, a framework convention is an undertaking to set up a forum committed to certain objectives within which particular binding agreements will subsequently be developed in the form of protocols. Thus, the FCCC contained no specific greenhouse gas abatement undertakings, only an agreement to develop and possibly engage in such arrangements in future years; it did, however, set out certain procedural matters concerning decision-making and so on. Many international agreements take this form.

4. The contribution by the U.K. government's leading scientific adviser, Sir David King, indicates one kind of role open to scientific advisers. In January 2004, he gave his judgment that climate change posed a greater threat than terrorism ("U.S. Climate Policy Bigger Threat to World than Terrorism" was the headline in the U.K. newspaper *The Independent* [January 9, 2004]).

5. Such resistance can arise without genetic modification (e.g., "naturally" pesticide-resistant crop strains are in use in Australia) though this is held to be usually a multigene characteristic and thus possibly not identical to GM pesticide resistance.

6. Some analysts sought to identify actors' positions not to explain them but to try to get them to agree on "least worst" ways forward: see Stirling and Mayer (1999) and, for comment, Yearley (2001).

7. see http://news.bbc.co.uk/1/hi/world/africa/2371675.stm.

References

Agrawala, Shardul (1998a) "Context and Early Origins of the Intergovernmental Panel on Climate Change," *Climatic Change* 39: 605–20.

Agrawala, Shardul (1998b) "Structural and Process History of the Intergovernmental Panel on Climate Change," *Climatic Change* 39: 621–42.

Beck, Ulrich (1992) *Risk Society: Towards a New Modernity* (London: Sage).

Beck, Ulrich (1995) *Ecological Politics in an Age of Risk* (Cambridge: Polity).

Benedick, Richard E. (1991) *Ozone Diplomacy: New Directions in Safeguarding the Planet* (Cambridge, MA: Harvard University Press).

Bennett, David, Peter Glasner, & David Travis (1986) *The Politics of Uncertainty: Regulating Recombinant DNA Research in Britain* (London: Routledge & Kegan Paul).

Bloomfield, Brian P. (1986) *Modelling the World: The Social Constructions of Systems Analysts* (Oxford: Blackwell).

Bocking, Stephen (2004) *Nature's Experts: Science, Politics and the Environment* (New Brunswick, NJ: Rutgers University Press).

Bodansky, Daniel (1994) "Prologue to the Climate Change Convention," in Irving M. Mintzer & J. Amber Leonard (eds), *Negotiating Climate Change: The Inside Story of the Rio Convention* (Cambridge: Cambridge University Press): 45–74.

Boehmer-Christiansen, Sonja (1994a) "Global Climate Protection Policy: The Limits of Scientific Advice," part 1, *Global Environmental Change* 4: 140–59.

Boehmer-Christiansen, Sonja (1994b) "Global Climate Protection Policy: The Limits of Scientific Advice," part 2, *Global Environmental Change* 4: 185–200.

Boehmer-Christiansen, Sonja (2003) "Science, Equity, and the War Against Carbon," *Science, Technology & Human Values* 28: 69–92.

Boehmer-Christiansen, Sonja & Aynsley J. Kellow (2002) *International Environmental Policy: Interests and the Failure of the Kyoto Process* (Cheltenham: Edward Elgar).

Braun, Bruce & Noel Castree (eds) (1998) *Remaking Reality: Nature at the Millennium* (London: Routledge).

Bruce, James P., Hoesung Lee, & Erik F. Haites (1996) "Summary for Policymakers," in J. P. Bruce, H. Lee & E. F. Haites (eds), *Climate Change 1995: Economic and Social Dimensions of Climate Change* (Cambridge: Cambridge University Press): 5–16.

Busch, Lawrence, Robin Grove-White, Sheila Jasanoff, David Winickoff, & Brian Wynne (2004) Amicus Curiae Brief Submitted to the Dispute Settlement Panel of the World Trade Organization in the Case of "EC: Measures Affecting the Approval and Marketing of Biotech Products," April 30, 2004.

Bush, Judith, Suzanne Moffatt, & Christine E. Dunn (2001) "Keeping the Public Informed? Public Negotiation of Air Quality Information," *Public Understanding of Science* 10: 213–29.

Carolan, Michael & Michael M. Bell (2004) "No Fence Can Stop It: Debating Dioxin Drift from a Small U.S. Town to Arctic Canada," in Neil Harrison & Gary Bryner (eds), *Science and Politics in the International Environment* (Boulder, CO: Rowman & Littlefield): 385–422.

Carvalho, Anabela & Jacquie Burgess (2005) "Cultural Circuits of Climate Change in U.K. Broadsheet Newspapers, 1985–2003," *Risk Analysis* 25(6): 1457–69.

Castree, Noel & Bruce Braun (eds) (2001) *Social Nature: Theory, Practice and Politics* (Oxford: Blackwell).

Christie, Maureen (2000) *The Ozone Layer: A Philosophy of Science Approach* (Cambridge: Cambridge University Press).

Collingridge, David & Colin Reeve (1986) *Science Speaks to Power: The Role of Experts in Policymaking* (New York: St Martin's Press).

Demeritt, David (2001) "The Construction of Global Warming and the Politics of Science," *Annals of the Association of American Geographers* 91: 307–37.

Demeritt, David (2002) "What Is the 'Social Construction of Nature'? A Typology and Sympathetic Critique," *Progress in Human Geography* 26: 766–89.

Douglas, Mary (1982) *Natural Symbols: Explorations in Cosmology* (New York: Pantheon Books).

Douglas, Mary, Des Gasper, Steven Ney, & Michael Thompson (1998) "Human Needs and Wants," in Steve Rayner & Elizabeth L. Malone (eds), *Human Choice and Climate Change*, vol. 1 (Columbus, OH: Battelle Press): 195–263.

Dratwa, Jim (2002) "Taking Risks with the Precautionary Principle: Food (and the Environment) for Thought at the European Commission," *Journal of Environmental Policy and Planning* 4: 197–213.

Edwards, Jeanette (2000) *Born and Bred: Idioms of Kinship and New Reproductive Technologies in England* (Oxford: Oxford University Press).

Edwards, Paul N. (1996) "Global Comprehensive Models in Politics and Policymaking," *Climatic Change* 32: 149–61.

Edwards, Paul N. (2000) "The World in a Machine: Origins and Impacts of Early Computerized Global Systems Models," in Agatha C. Hughes & Thomas P. Hughes (eds), *Systems, Experts, and Computers: The Systems Approach in Management and Engineering, World War II and After* (Cambridge, MA: MIT Press): 221–54.

Edwards, Paul N. (2001) "Representing the Global Atmosphere: Computer Models, Data, and Knowledge About Climate Change," in Clark A. Miller & Paul N. Edwards (eds), *Changing the Atmosphere: Expert Knowledge and Environmental Governance* (Cambridge, MA: MIT Press): 31–65.

Edwards, Paul N. & Stephen H. Schneider (2001) "Self-Governance and Peer Review in Science-for-Policy: The Case of the IPCC Second Assessment Report," in Clark A. Miller & Paul N. Edwards (eds), *Changing the Atmosphere: Expert Knowledge and Environmental Governance* (Cambridge, MA: MIT Press): 219–46.

Eriksson, Lena (2004) *From Persona to Person: The Unfolding of an (Un)Scientific Controversy*, Ph.D. diss., Cardiff University, Wales.

Fankhauser, Samuel (1995) *Valuing Climate Change: The Economics of the Greenhouse* (London: Earthscan).

Franklin, Adrian (2002) *Nature and Social Theory* (London: Sage).

Freudenburg, William R. (2000) "Social Constructions and Social Constrictions: Toward Analyzing the Social Construction of 'the Naturalized' as Well as 'the Natural,'" in Gerd Spaargaren, Arthur P. J. Mol and Frederick H. Buttel (eds), *Environment and Global Modernity* (London: Sage): 103–19.

Fukuyama, Francis (2002) *Our Post-Human Future: Consequences of the Biotechnology Revolution* (London: Profile).

Gaskell, George & Martin W. Bauer (eds) (2001) *Biotechnology, 1996–2000* (London: Science Museum).

Grundmann, Reiner (1998) "The Strange Success of the Montreal Protocol: Why Reductionist Accounts Fail," *International Environmental Affairs* 10: 197–220.

Grundmann, Reiner (2006) "Ozone and Climate: Scientific Consensus and Leadership," *Science, Technology & Human Values* 31(1): 73–101.

Guimarães Pereira, Ângela, Clair Gough, & Bruna De Marchi (1999) "Computers, Citizens and Climate Change: The Art of Communicating Technical Issues," *International Journal of Environment and Pollution* 11: 266–89.

Habermas, Jürgen (2003) *The Future of Human Nature* (Cambridge: Polity).

Haggett, Claire & David Toke (2006) "Crossing the Great Divide: Using Multi-method Analysis to Understand Opposition to Windfarms," *Public Administration* 84(1): 103–20.

Hajer, Maarten (1995) *The Politics of Environmental Discourse: Ecological Modernization and the Policy Process* (Oxford: Clarendon Press).

Hansen, Janus (2005) *Framing the Public: Three Case Studies in Public Participation in the Governance of Agricultural Biotechnology*, Ph.D. diss., European University Institute, Florence, Italy.

Hart, David M. & David G. Victor (1993) "Scientific Elites and the Making of U.S. Policy for Climate Change Research," *Social Studies of Science* 23: 643–80.

Harvey, Matthew (2005) *Citizens, Experts and Technoscience: A Case Study of 'GM Nation? The Public Debate,'* Ph.D. diss., Cardiff University, Wales.

Helford, Reid M. (1999) "Rediscovering the Resettlement Landscape: Making the Oak Savanna Ecosystem 'Real,'" *Science, Technology & Human Values* 24(1): 55–79.

Horlick-Jones, Tom, John Walls, Gene Rowe, N. F. Pidgeon, Wouter Poortinga, & Tim O'Riordan (2004) "A Deliberative Future? An Independent Evaluation of the *GM Nation?* Public Debate About the Possible Commercialisation of Transgenic Crops in the U.K., 2003," Understanding Risk Working Paper 04-02 (Norwich, U.K.: Centre for Environmental Risk): 1–182.

Horlick-Jones, Tom, John Walls, Gene Rowe, N. F. Pidgeon, Wouter Poortinga, Graham Murdoch, & Tim O'Riordan (2007) *The GM Debate: Risk, Politics and Public Engagement* (London: Routledge).

Ingold, Tim (1993) "Globes and Spheres: The Topology of Environmentalism," in Kay Milton (ed), *Environmentalism: The View from Anthropology* (London: Routledge): 31–42.

Jäger, Jill & Tim O'Riordan (1996) "The History of Climate Change Science and Politics," in Tim O'Riordan & Jill Jäger (eds), *Politics of Climate Change: A European Perspective* (London: Routledge): 1–31.

Jamison, Andrew S. (2001) *The Making of Green Knowledge* (Cambridge: Cambridge University Press).

Jankovic, Vladimir (2004) "Mesoscale Weather Prediction from Belgrade to Washington, 1970–2000," *Social Studies of Science* 34(1): 45–75.

Jasanoff, Sheila (1990) *The Fifth Branch: Science Advisers as Policymakers* (Cambridge, MA: Harvard University Press).

Jasanoff, Sheila (1997) "Civilization and Madness: The Great BSE Scare of 1996," *Public Understanding of Science* 6: 221–32.

Jasanoff, Sheila (2005) *Designs on Nature* (Princeton, NJ: Princeton University Press).

Jasanoff, Sheila & Brian Wynne (1998) "Science and Decisionmaking," in Steve Rayner & Elizabeth L. Malone (eds), *Human Choice and Climate Change*, vol. 1 (Columbus, OH: Battelle Press): 1–87.

Kasemir, Bernd, Jill Jäger, Carlo C. Jaeger, & Matthew T. Gardner (eds) (2003) *Public Participation in Sustainability Science: A Handbook* (Cambridge: Cambridge University Press).

Kim, Sang-Hyun (2005) *Making the Science of Global Warming: A Social History of Climate Science in Britain*, Ph.D. diss., University of Edinburgh.

Klintman, Mikael (2002) "The Genetically Modified Food Labeling Controversy," *Social Studies of Science* 32(1): 71–92.

Kloppenburg, Jack Ralph (2004) *First the Seed: The Political Economy of Plant Biotechnology*, 2nd ed. (Madison: University of Wisconsin Press).

Krimsky, Sheldon (2000) *Hormonal Chaos: The Scientific and Social Origins of the Environmental Endocrine Hypothesis* (Baltimore, MD: Johns Hopkins University Press).

Kwa, Chunglin (2001) "The Rise and Fall of Weather Modification: Changes in American Attitudes Toward Technology, Nature and Society," in Clark A. Miller & Paul N. Edwards (eds), *Changing the Atmosphere: Expert Knowledge and Environmental Governance* (Cambridge, MA: MIT Press): 135–65.

Lahsen, Myanna (2005a) "Technocracy, Democracy and U.S. Climate Politics: The Need for Demarcations," *Science, Technology & Human Values* 30(1): 137–69.

Lahsen, Myanna (2005b) "Seductive Simulations: Uncertainty Distribution Around Climate Models," *Social Studies of Science* 35(6): 895–922.

Latour, Bruno (2004) *Politics of Nature: How to Bring the Sciences into Democracy* (Cambridge, MA: Harvard University Press).

Lezaun, Javier (2004) "Pollution and the Use of Patents: A Reading of *Monsanto v. Schmeiser*," in Nico Stehr (ed), *Biotechnology: Between Commerce and Civil Society* (New Brunswick, NJ: Transaction): 135–58.

Lezaun, Javier (2006) "Creating a New Object of Government: Making Genetically Modified Organisms Traceable," *Social Studies of Science* 36(4): 499–531.

Levidow, Les (1999) "Britain's Biotechnology Controversy: Elusive Science, Contested Expertise," *New Genetics and Society* 18: 47–64.

Levidow, Les (2001) "Precautionary Uncertainty: Regulating GM Crops in Europe," *Social Studies of Science* 31: 845–78.

Lomborg, Bjørn (2001) *The Skeptical Environmentalist: Measuring the Real State of the World* (Cambridge: Cambridge University Press).

Marris, Claire, Pierre-Benoit Joly, Stéphanie Ronda, & Christophe Bonneuil (2005) "How the French GM Controversy Led to the Reciprocal Emancipation of Scientific Expertise and Policy Making," *Science and Public Policy* 32(4): 301–8.

Matthewman, Steve (2000) "Reach for the Skies: Towards a Sociology of the Weather," *New Zealand Sociology* 15(2): 205–25.

Mazur, Allan & Jinling Lee (1993) "Sounding the Global Alarm: Environmental Issues in the U.S. National News," *Social Studies of Science* 23(4): 681–720.

McCright, Aaron M. & Riley E. Dunlap (2000) "Challenging Global Warming as a Social Problem: An Analysis of the Conservative Movement's Counter-claims," *Social Problems* 47: 499–522.

McCright, Aaron M. & Riley E. Dunlap (2003) "Defeating Kyoto: The Conservative Movement's Impact on U.S. Climate Change Policy," *Social Problems* 50: 348–373.

McKibben, Bill (1989) *The End of Nature* (New York: Random House).

Miller, Clark A. (2001a) "Challenges in the Application of Science to Global Affairs: Contingency, Trust and Moral Order," in Clark A. Miller & Paul N. Edwards (eds), *Changing the Atmosphere: Expert Knowledge and Environmental Governance* (Cambridge, MA: MIT Press): 247–85.

Miller, Clark A. (2001b) "Hybrid Management: Boundary Organizations, Science Policy, and Environmental Governance in the Climate Regime," *Science, Technology & Human Values* 26(4): 478–500.

Miller, Clark A. & Paul N. Edwards (eds) (2001) *Changing the Atmosphere: Expert Knowledge and Environmental Governance* (Cambridge, MA: MIT Press).

Millstone, Erik, Eric Brunner, & Sue Mayer (1999) "Beyond 'Substantial Equivalence,'" *Nature* 401: 525–26.

Murphy, Joseph and Les Levidow (2006) *Governing the Transatlantic Conflict over Agricultural Biotechnology* (London: Routledge).

Murphy, Joseph, Les Levidow, & Susan Carr (2006) "Regulatory Standards for Environmental Risks: Understanding the U.S.-European Union Conflict over Genetically Modified Crops," *Social Studies of Science* 36(1): 133–60.

Nolin, Jan (1999) "Global Policy and National Research: The International Shaping of Climate Research in Four European Union Countries," *Minerva* 37(2): 125–40.

Novas, Carlos & Nikolas Rose (2000) "Genetic Risk and the Birth of the Somatic Individual," *Economy and Society* 29: 485–513.

Oreszczyn, S. & A. Lane (2005) "Farmers' Understandings of Genetically Modified Crops Within Local Communities," ESRC Science in Society Programme. Available at: http://www.sci-soc.net/SciSoc/Projects/Genomics/Famers+understandings+of+gentically+modified+crops.htm.

O'Riordan, Tim & Andrew Jordan (1999) "Institutions, Climate Change and Cultural Theory: Towards a Common Analytical Framework," *Global Environmental Change* 9(2): 81–94.

Padolsky, Miriam (2006) *Bringing Climate Change Down to Earth: Science and Participation in Canadian and Australian Climate Change Campaigns.* Ph.D. Dissertation, Department of Sociology and Science Studies Program, University of California, San Diego.

Pearce, David W., et al. (1996) "The Social Costs of Climate Change: Greenhouse Damage and the Benefits of Control," in James P. Bruce, Hoesung Lee, & Erik F. Haites (eds), *Climate Change 1995: Economic and Social Dimensions of Climate Change* (Cambridge: Cambridge University Press): 179–224.

Pearce, Fred (1991) *Green Warriors: The People and the Politics Behind the Environmental Revolution* (London: Bodley Head).

Petts, Judith (2001) "Evaluating the Effectiveness of Deliberative Processes: Waste Management Case Studies," *Journal of Environmental Planning and Management* 44(2): 207–22.

Pidgeon, N. F., W. Poortinga, Gene Rowe, Tom Horlick-Jones, John Walls, & Tim O'Riordan (2005) "Using Surveys in Public Participation Processes for Risk Decision-Making: The Case of the 2003 British *GM Nation?* Public Debate," *Risk Analysis* 25(2): 467–79.

Pimentel, David S. & Peter H. Raven (2000) "Commentary: Bt Corn Pollen Impacts on Non-target Lepidoptera: Assessment of Effects in Nature," *Proceedings of the National Academy of Sciences of the United States of America* 97(July 18): 8198–99.

Priest, Susanna Hornig (2001) *A Grain of Truth: The Media, the Public, and Biotechnology* (Lanham, MD: Rowman & Littlefield).

Priest, Susanna Hornig & Toby Ten Eyck (2004) "Peril or Promise: News Media Framing of the Biotechnology Debate in Europe and the U.S.," in Nico Stehr (ed), *Biotechnology: Between Commerce and Civil Society* (New Brunswick, NJ: Transaction): 175–86.

Rabinow, Paul (1996) *Essays on the Anthropology of Reason* (Princeton, NJ: Princeton University Press).

Rayner, Steve & Elizabeth L. Malone (eds) (1998) *Human Choice and Climate Change*, vols. 1–4 (Columbus, OH: Battelle Press).

Rowe, Gene, Tom Horlick-Jones, John Walls, & N. F. Pidgeon (2005) "Difficulties in Evaluating Public Engagement Initiatives: Reflections on an Evaluation of the U.K. GM Nation? Public Debate about Transgenic Crops," *Public Understanding of Science* 14(4): 331–52.

Samuel, Andrew (2001) "Rum: Nature and Community in Harmony?," *ECOS: A Review of Conservation* 22(1): 36–45.

Shackley, Simon (1997a) "Trust in Models? The Mediating and Transformative Role of Computer Models in Environmental Discourse," in Michael Redclift & Graham Woodgate (eds), *The International Handbook of Environmental Sociology* (Cheltenham: Edward Elgar): 237–60.

Shackley, Simon (1997b) "The Intergovernmental Panel on Climate Change: Consensual Knowledge and Global Politics," *Global Environmental Change* 7(1): 77–9.

Shackley, Simon (2001) "Epistemic Lifestyles in Climate Change Modeling," in Clark A. Miller & Paul N. Edwards (eds), *Changing the Atmosphere: Expert Knowledge and Environmental Governance* (Cambridge, MA: MIT Press): 107–33.

Shackley, Simon & Brian Wynne (1995) "Global Climate Change: The Mutual Construction of an Emergent Science-Policy Domain," *Science and Public Policy* 22(4): 218–30.

Shackley, Simon & Brian Wynne (1996) "Representing Uncertainty in Global Climate Change Science Policy: Boundary-Ordering Devices and Authority," *Science, Technology & Human Values* 21(3): 275–302.

Shackley, Simon, J. Risbey, & M. Kandlikar (1998) "Science and the Contested Problem of Climate Change: A Tale of Two Models," *Energy and Environment* 8: 112–34.

Shackley, Simon, J. Risbey, P. Stone, & Brian Wynne (1999) "Adjusting to Policy Expectations in Climate Change Science: An Interdisciplinary Study of Flux Adjustments in Coupled Atmosphere Ocean General Circulation Models," *Climatic Change* 43: 413–54.

Sismondo, Sergio (1999) "Models, Simulations, and Their Objects," *Science in Context* 12: 247–60.

Skodvin, Tora (2000) "The Intergovernmental Panel on Climate Change," in Steinar Andresen, Tora Skodvin, Arild Underdal, & Jørgen Wettestad (eds), *Science and Politics in International Environmental Regimes* (Manchester, U.K.: Manchester University Press): 146–81.

Stirling, Andy & Sue Mayer (1999) *Re-Thinking Risk: A Pilot Multi-Criteria Mapping of a Genetically Modified Crop in Agricultural Systems in the U.K.* (Brighton, Sussex: Science Policy Research Unit).

Strathern, Marilyn (1992) *After Nature: English Kinship in the Late Twentieth Century* (Cambridge: Cambridge University Press).

Sundberg, Mikaela (2005) *Making Meteorology: Social Relations and Scientific Practice* (Stockholm: Stockholms Universitet).

Sunderlin, William D. (2003) *Ideology, Social Theory and the Environment* (Lanham, MD: Rowman & Littlefield).

Sundqvist, Göran, Martin Letell, & Rolf Lidskog (2002) "Science and Policy in Air Pollution Abatement Strategies," *Environmental Science and Policy* 5(2): 147–56.

Tulloch, John & Deborah Lupton (2003) *Risk and Everyday Life* (London: Sage).

Turner, Roger (2004) "Weather Modification: Trust, Science, and Civic Epistemology," paper presented at 4S-EASST conference, Paris, September.

van der Sluijs, Jeroen (1997) *Anchoring Amid Uncertainty: On the Management of Uncertainties in Risk Assessment of Anthropogenic Climate Change* (Utrecht, Netherlands: Universiteit Utrecht).

van der Sluijs, Jeroen, Josée van Eijndhoven, Simon Shackley, & Brian Wynne (1998) "Anchoring Devices in Science for Policy: The Case of Consensus Around Climate Sensitivity," *Social Studies of Science* 28(2): 291–323.

Verran, Helen (2002) "A Postcolonial Moment in Science Studies: Alternative Firing Regimes of Environmental Scientists and Aboriginal Landowners," *Social Studies of Science* 32: 729–62.

Victor, David G. (2001) *The Collapse of the Kyoto Protocol and the Struggle to Slow Global Warming* (Princeton, NJ: Princeton University Press).

Wackers, G., T. van Hoorn, & W. E. Bijker (1997) "Het Natuurontwikkelingsdebat: Dilemma's van een Open Leerproces," in N. E. van der Poll & A. Glasmeier (eds), *Natuurontwikkeling: Waarom en Hoe? Verslag van een Debat* (Den Haag: Rathenau Instituut): 41–58.

Walls, John, Tee Rogers-Hayden, Alison Mohr, & Tim O'Riordan (2005) "Seeking Citizens' Views on Genetically Modified Crops: Experiences from the United Kingdom, Australia, and New Zealand," *Environment* 47(7): 22–36.

Weinberg, Alvin M. (1972) "Science and Trans-Science," *Minerva* 10: 209–22.

Winickoff, David, Sheila Jasanoff, Lawrence Busch, Robin Grove-White, & Brian Wynne (2005) "Adjudicating the GM Food Wars: Science, Risk and Democracy in World Trade Law," *Yale Journal of International Law* 30: 81–123.

Wynne, Brian (2001) "Creating Public Alienation: Expert Cultures of Risk and Ethics on GMOs," *Science as Culture* 10: 445–81.

Yearley, Steven (1992) *The Green Case: A Sociology of Environmental Arguments, Issues and Politics* (London: Routledge).

Yearley, Steven (1993) "Standing in for Nature: The Practicalities of Environmental Organisations' Use of Science," in Kay Milton (ed), *Environmentalism: The View from Anthropology* (London: Routledge): 59–72.

Yearley, Steven (1995) "The Environmental Challenge to Science Studies," in Sheila Jasanoff, Gerald E. Markle, James C. Petersen, & Trevor Pinch (eds), *Handbook of Science and Technology Studies* (London: Sage): 457–79.

Yearley, Steven (1996) *Sociology, Environmentalism, Globalization* (London: Sage).

Yearley, Steven (1999) "Computer Models and the Public's Understanding of Science: A Case-Study Analysis," *Social Studies of Science* 29: 845–66.

Yearley, Steven (2001) "Mapping and Interpreting Societal Responses to Genetically Modified Food and Plants: Essay Review," *Social Studies of Science* 31: 151–60.

Yearley, Steven (2005a) *Cultures of Environmentalism: Empirical Studies in Environmental Sociology* (Basingstoke, U.K.: Palgrave Macmillan).

Yearley, Steven (2005b) "The 'End' or the 'Humanization' of Nature?" *Organization and Environment* 18: 198–201.

Yearley, Steven (2005c) *Making Sense of Science: Science Studies and Social Theory* (London: Sage).

Yearley, Steven (2005d) "The Wrong End of Nature," *Studies in History and Philosophy of Science* 36: 827–34.

Yearley, Steven (2006) "Bridging the Science-Policy Divide in Urban Air-Quality Management: Evaluating Ways To Make Models More Robust Through Public Engagement," *Environment and Planning C* 24: 701–14.

Yearley, Steven, John Forrester, & Peter Bailey (2001) "Participation and Expert Knowledge: A Case Study Analysis of Scientific Models and Their Publics," in Matthijs Hisschemöller, Rob Hoppe, William N. Dunn, & Jerry R. Ravetz (eds), *Policy Studies Review Annual* 12: *Knowledge, Power and Participation in Environmental Policy Analysis* (New Brunswick, NJ: Transaction Publishers): 349–67.

Yearley, Steven, Steve Cinderby, John Forrester, Peter Bailey, & Paul Rosen (2003) "Participatory Modelling and the Local Governance of the Politics of U.K. Air Pollution: A Three-City Case Study," *Environmental Values* 12: 247–62.

Zehr, Stephen C. (2000) "Public Representations of Scientific Uncertainty About Global Climate Change," *Public Understanding of Science* 9: 85–103.

Zehr, Stephen C. (2004) "Method, Scale and Socio-Technical Networks: Problems of Standardization in Acid Rain, Ozone Depletion and Global Warming Research," *Science Studies* 7: 47–58.

37 Bridging STS and Communication Studies: Scholarship on Media and Information Technologies

Pablo Boczkowski and Leah A. Lievrouw

By any measure, media and information technologies—sociotechnical systems that support and facilitate mediated cultural expression, interpersonal interaction, and the production and circulation of information goods and services—are the backbone of social, economic, and cultural life in many societies today. They are important in themselves as cultural and technical artifacts, and they are embedded in almost every other type of specialized technological system, including those used in finance, manufacturing, extractive industries, transportation, utilities, education, health care, defense, and law enforcement. Indeed, it is difficult to identify any aspect of contemporary life that is not affected in some way by the development and use of media and information technologies.

In light of their ubiquity and societal reach, as well as how rapidly the systems themselves have changed over the past three decades, we might expect that studies of this class of technologies would have been central in the research agendas of communication studies, on the one hand, and science and technology studies, on the other. Both disciplines would seem to have an obvious interest in them. However, in each case the story has been more complicated.

The social, psychological and cultural effects of mediated messages and content have been analyzed in communication studies since the field's founding. Interest in the role of technology in such effects rose in parallel with the growing popularity of television between the 1960s and 1980s (Meyrowitz, 1985; McLuhan, 1964; Postman, 1985; Williams, 1975). However, these debates were largely confined to specialized domains of inquiry within mass communication and cultural studies. Only in the 1970s and 1980s, as networked computing and telecommunications technologies diffused rapidly in corporate, entertainment, and academic settings and converged with and challenged the conventional boundaries among "mass media," interpersonal communication, and organizational communication, did the study of these technologies expand into an intellectual space that linked diverse domains of inquiry and become a major topic of interest in its own right in communication studies (Parker, 1970; Pool, 1977, 1983; Rice et al., 1984; Rogers, 1986; Williams et al., 1988).

The centrality of media and information technologies as objects of inquiry has taken even longer to emerge in STS, a field that has tended to focus on complex technologies with sophisticated engineering knowledge and materials. Certainly, scholars in this field had produced a handful of important studies of media and information technologies by the early 1990s, including examinations of the telephone and videotex as "large technical systems" (Mayntz & Schneider, 1988; Galambos, 1988; Schneider et al., 1991); cultural histories of radio, telephony, and electric media (Douglas, 1987; Fischer, 1992; Martin, 1991) and social studies of computing (Forsythe, 1993; Kling & Iacono, 1989; Star, 1995; Suchman, 1987; Turkle, 1984; Woolgar, 1991). However, media and information technologies have become a major research focus in STS only in the decade or so since the introduction of the World Wide Web, when "the Internet" reached the desktops of scholars, artists, and critics throughout the academy and popular culture and triggered their intellectual curiosity—not only about this technology but also about earlier and contemporary ones.

Today, the study of media and information technologies is a major pursuit in communication studies and STS alike, as a rising tide of related books, articles, conference panels and presentations, and academic tracks in both fields attests. In our view, this shift is partly due to several important intellectual bridges between the two disciplines that have developed around their shared interests. These bridges have energized dialogue between the fields and fostered innovative scholarship. For STS, communication studies has provided an extensive body of social science research and critical inquiry that documents the relationships among mediated content, individual behavior, social structures and processes, and cultural forms, practices, and meanings. For communication studies, STS has provided a sophisticated conceptual language and grounded methods for articulating and studying the distinctive sociotechnical character of media and information technologies themselves as culturally and socially situated artifacts and systems.

Despite their significance, however, these intellectual bridges have not been explicitly articulated in the literature of either field. Therefore, in this chapter we focus on three conceptual bridges that have been especially fruitful in both fields—and that, taken together, map a significant portion of scholarship on media and information technologies at the intersection of STS and communication studies.[1]

- Prevailing notions about *causality* in technology-society relationships
- The *process* of technology development
- The social *consequences* of technological change

In both fields, these bridges have been framed and explored mainly as binaries, with a tension between rival assumptions or approaches. Questions about causality have been framed as a debate between determination and contingency. Questions about technology development have been framed in terms of opposing production and consumption processes. And questions regarding the consequences of media and information technologies have been framed around discontinuous versus continuous modes of social change, of disruptive "revolution" versus incremental "evolution."

The value of binary approaches is that either element of a duality can be foregrounded and contrasted against the other. However, in this chapter we contend that these three dualities can be better understood as dialectic relationships. Each half of the duality presumes, critiques, and builds on the other. By focusing on the complementary dynamics of these relationships, we hope to provide a nuanced and comprehensive account of scholarship on media and information technologies at the intersection of STS and communication studies.

In what follows we examine the three bridges and the conceptual dualities underlying each one. This approach does not exhaust all relevant issues in scholarship about media and information technologies, which encompass an enormous range of theoretical and empirical approaches across numerous disciplines (see Lievrouw & Livingstone, 2006a). Nonetheless, we selectively review research on media and information technologies that sheds light on, first, the mutual intellectual influences between STS and communication studies with regard to this class of technologies over the last few decades, and, second, how the conceptual linkages have shaped the current "territory" of understanding about media and information technologies in society. We begin by defining key terms and concepts and in subsequent sections move to discussions of causality, process, and consequences. We conclude with a summary of the media and information technologies research landscape framed by the three bridges and consider the implications of that landscape for continued intellectual dialogue between the two fields.

MEDIA AND INFORMATION TECHNOLOGIES: EVOLVING DEFINITIONS

How are we to characterize media and information technologies? What distinguishes these technologies as a class? We have chosen the broad label "media and information technologies," as opposed to more familiar terms like "information and communication technologies," "new media," or "IT," for several reasons. Before addressing the terminology, however, we wish to review the different, but related, approaches to defining these technologies that have been taken within communication studies[2] and STS, respectively.

An important tradition of inquiry in communication studies has tended to view technologies according to their technical features, particularly those that support or extend human sensory perception and communicative action across time and space. From the uses of symbols, language, and writing to express and shape thought and experience (Goody, 1981; Ong, 1982), to the cultural fixity and standardization suggested by mechanically printed texts (Eisenstein, 1979), to the "extension" of sounds and images via photography, motion pictures, sound recording, and electronic media (Williams, 1981), to the "separation of communication from transportation" achieved by the telegraph (Carey, 1989: 203), the significance of media technologies within this line of communication scholarship has often hinged on their role as "extensions of man" (McLuhan, 1964).

For example, in his classic analyses of ancient civilizations, Harold Innis (1972) argues that social and political systems evolve differently according to whether they depend on "time-biased" media (i.e., durable, immobile, and difficult to change, such as stone) or "space-biased" media (more ephemeral, portable, and easy to revise, such as parchment or paper). Later, Innis's colleague Marshall McLuhan (1964) classified media technologies into the more abstract categories of "hot" and "cool." Hot media, such as print and radio, he said, elicit intense psychological involvement from the audience, while cool media, such as television, provoke psychological detachment and distance.

Another significant tradition of inquiry within communication studies has taken a behaviorally oriented approach to highlight the complexity of contemporary media technologies and their reliance on computing and telecommunications. Wilbur Schramm (1977), for example, classifies media technologies according to their correspondence to human sensory perception: motion versus still images, sound versus silent, text versus picture, one-way/simplex versus two-way/duplex transmission. But he also brings in their institutional context by contrasting inexpensive, local, small-scale "little media," such as newsletters, print shops, or local radio stations, with "big media" having extensive, expensive, complex infrastructures and organizational arrangements, such as telephone systems, national broadcast networks, or communications satellites. In contrast to "mass media," Rice and his associates (1984: 35) define "new media" as "those communication technologies . . . that allow or facilitate interactivity among users or between users and information" owing to the two-way transmission capabilities of their telecommunications- and computer-based infrastructures. Ithiel Pool (1990: 19) includes "about 25 main devices" that incorporate computing and/or telecommunications technologies in a list of "new" communications media.

Despite the differences between them, both approaches share a persistent focus on technical features and capabilities and an enduring concern, particularly in the United States, with the social and psychological "effects" of media technologies and content on individuals and audiences. Effects researchers continue to explore the nature and extent of media effects and to inform the management and regulation of media channels and content.

Definitions of media and information technologies in STS, on the other hand, have tended to focus more on issues of meaning, practice, and the connection of particular technological systems to a broader "landscape" of artifacts rather than technical features alone. A fundamental tenet of STS is that the material aspect of technology must be situated and studied within its various social, temporal, political, economic, and cultural contexts. The critique of technological determinism that catalyzed so much historical and sociological research in the 1980s, both within and outside STS, was partly based on the idea that the technical attributes of technologies matter less than how they are actually used, given the meanings that people attribute to them. For example, Suchman (1987) showed that human-machine interaction, even in situations where technically skilled individuals operate complex computerized devices such as photocopiers, depends on locally contingent attributions of meaning rather than disembodied, decontextualized rules. Similarly, Kling and Iacono (1987) demonstrated that organizational constraints and culture, and institutional forms, do more to shape computer-based information systems than do data structures, software, or hardware architectures *per se*.

Studies of the origins of radio (Douglas, 1987), telephony (Fischer, 1992; Galambos, 1988), sound technologies (Pinch & Trocco, 2002; Thompson, 2002), videotex (Schneider et al., 1991), and the development of computing and the Internet (Abbate, 1999; Edwards, 1996) have helped establish a broad view of what counts as media and information technologies among STS scholars. Print and broadcasting, computing and telecommunications, "old" and "new" media technologies alike fall within the purview of relevant scholarship in STS. By taking a long-term historical view, and by underscoring issues of meaning and practice, STS has illuminated crucial connections between particular technological systems and the broader world of artifacts and culture.

Interestingly, the historically grounded, meaning- and practice-based scholarship typical within STS resonates with views of media technology commonly held among communication scholars working in the British and European "media studies" tradition. The critical, cultural perspective of this tradition contrasts with the mainly American, "administrative" focus on effects and regulation (Lazarsfeld, 1941). Instead, it emphasizes the cycle of "production–circulation–reception of cultural products" or "media commodities" such as films, television programs, popular music, and fashion (O'Sullivan et al., 2003: 15; see also Williams, 1981). It tends to view media technologies, including newer systems such as mobile telephony and the Internet, as "texts" subject to cultural analysis and critique. They are at once the products and the tools of a cultural and economic system whose aim is the reproduction of social, political, and economic domination, order, and privilege. In different hands, media technologies can also serve the interests of resistance, emancipation, and equity.

For example, in his historical and institutional analysis of television in the United Kingdom and the United States, Raymond Williams (1975) navigates between the material nature of television technologies and programming, and their social and cultural meanings. He warns against both technological determinism and what he calls "symptomatic technology" (1975: 13), that is, technology as an entirely socially determined "symptom" of the culture that produces it. He argues that while certain technologies may evolve into "new social forms" (1975: 18–19), the path of evolution depends on the actors and interests involved and will produce unpredictable or unintended consequences. Although Williams is primarily concerned with television content, his analysis is nonetheless consistent with what many scholars in STS today would call a "mutual shaping" perspective on technology and society, the interplay of materiality and action.

Since the 1980s and 1990s, many of the views about media and information technologies advanced within STS and media studies have been more broadly adopted among communication researchers dissatisfied with the implicit technological determinism of media effects research and the language of "impacts" of new technologies *on* society, behavior, and culture. Coincident with a broader shift within the field in the 1980s, away from the administrative perspective and toward a contextual perspective that stressed local practices, everyday life, subjectivity, interaction, and meaning (Gerbner, 1983), many communication scholars have turned to concepts drawn from STS, such as interpretive flexibility, social shaping, and social construction of technology, in their theorizing and analyses of newer media and information technologies.[3] Today, the deterministic language of "effects" and "impacts" has largely been supplanted in communication technology research by more relational, subjective, and meaning-driven frameworks and concepts. The rejection of technological determinism, and the acceptance of a relatively strong form of social constructionism, has become the prevailing perspective in new media studies in Europe, North America, and elsewhere. This development can be counted as one of the most important cross-disciplinary influences of STS on the field (Lievrouw & Livingstone, 2006b; see also chapter 7 in this volume).

Why "Media and Information Technologies"? Notes on Terminology

As stated previously, we have deliberately chosen the term "media and information technologies," rather than other commonly used labels, to describe the broad class of sociotechnical systems that are studied in both STS and communication studies. In contrast to these other terms, the phrase "media and information technologies" foregrounds four distinctive facets of these systems: their broad historical scope, their infrastructural dimension, their fundamental materiality, and the distinctive interplay of this materiality with symbolic content and meaning.

First, "media and information technologies" is meant to suggest a sense of historical inclusiveness and scope. Consistent with the strong historical, meaning- and practice-oriented approach to technology within STS, these technologies include older craft, mechanical, and electric technologies, such as printing, typewriters, telegraphy, and broadcasting, as well as newer systems such as the Internet, mobile telephony, satellite systems, and search engines. In contrast, terms such as "new media," "information and communication technologies" (ICT), and "information technology" (IT) have been commonly used to privilege computing and telecommunications technologies relative to other types of artifacts.

Second, taking a cue from Star and Bowker's (2006) concept of infrastructure (see also Bowker & Star, 1999), the term media and information technologies is used to suggest that particular artifacts should be conceptually situated within a broader landscape of related, and often unnoticed or invisible, material things, such as filing cabinets, magnetic tape and optical disks, telephone poles, library shelves, or wireless bandwidth, for example. That is, even when the object of study is a novel technology, it should always be seen in its relationships to an installed base of related things. Terms like new media, ICT, and IT, on the other hand, often emphasize the novelty and uniqueness of particular devices and obscure their relationships to the broader world of other artifacts on which they depend for their very functioning.

Third, and related to the point about infrastructure, media and information technologies are fundamentally material. That is, people engage with them in space and time, as embodied, situated beings, as they do with other artifacts. Even supposedly "virtual" media systems and "friction-free" cyberspace are in essence complex configurations of "hard" physical components, from cables to code.

Fourth, drawing from the work of Silverstone and his collaborators (Silverstone & Haddon, 1996; Silverstone & Hirsch, 1992; Silverstone et al., 1992), we want to emphasize the centrality of content and its constitutive articulation with materiality. Media and information technologies are not only artifacts in the material sense but also the means for creating, circulating, and appropriating meaning. Whether they mediate entertainment, arts, interaction, organizing, or data, in no other class of technologies—such as bicycles, missiles, bridges, and electrical grids—are material form and symbolic configurations so intimately tied and mutually constructed. We might say that media and information technologies are at once cultural material and material culture. That is, on the one hand, they are cultural products in themselves, in which constellations of textual, aural, and visual symbols play a central role. On the other hand, they are a key part of the material culture of mediated communication, in which ensembles of technologies acquire a prominence much higher than in unmediated communication. This distinctive quality is to a large extent what has made them so compelling to STS and communication scholars alike.

In a definition that draws from STS and communication research, Lievrouw and Livingstone (2006b) argue that media and information technologies comprise the material systems themselves and their social contexts, including the *artifacts* or *devices* used to mediate, communicate, or convey information; the *activities* and *practices* in which people engage to communicate or share information; and the *social arrangements* or *organizational forms* that develop around the devices and practices. In light of the preceding discussion, we would refine the definition of media and information technologies to highlight the interplay of symbolic content and meaning with the artifacts, practices, and social arrangements that are associated with them. We return to this point in the conclusion of this chapter.

THREE BRIDGES

As we noted at the start of this chapter, over the last few decades the study of media and information technologies, whether in communication or STS, has centered on certain fundamental questions or issues that have ordinarily been framed as binary oppositions between two competing concepts, with a camp of advocates on each side. In our view, three important issues in particular have served as "bridges" between the two fields: causality in technology-society relationships, the technology development process, and the social consequences of technological change. In this section we examine each bridge and the opposing concepts involved in them, illustrating the discussion with relevant examples from the literature in both fields.

Causality

Scholarship about media and information technologies has raised important questions about causality in the relationship between technology and society. Research in STS and communication studies has often espoused different perspectives on this issue, partly as a result of their different intellectual traditions and orientations. On the one hand, given its history of behavioral and cultural theorizing, communication research has tended to see technology as a factor that can generate, or help generate, distinctive social effects, rather than as an object of inquiry worthy of social explanation in itself. On the other hand, STS technology research—with its grounding in contextualist history and constructivist sociology of technology—has often made the social factors that shape the development and, to a lesser extent, the use of technology the central focus of inquiry and has been hesitant to say much about technology's large-scale societal effects.

These different notions of causality, and their associated conceptual and methodological preferences, can be appreciated by contrasting two highly regarded studies of print technology: Eisenstein's *Printing Press as an Agent of Change* (1979) and Johns's *Nature of the Book: Print and Knowledge in the Making* (1998)—as well as the debate between the two authors published in a recent issue of *The American Historical Review* (Eisenstein, 2002a,b; Johns, 2002).[4]

The Printing Press as an Agent of Change has been enormously influential in communication technology scholarship and many other fields. It argues that the advent of the printing press led to the emergence of a "print culture" that reflected the distinctive attributes of the press as a technological system, as contrasted with scribal manuscript production. In turn, this culture ushered in a series of revolutionary transformations that altered almost every aspect of "Western civilization." In Eisenstein's view, a crucial attribute of print is "typographical fixity," that is, a printed text's content and format is preserved in print and thus becomes independent from its use. Prior to mechanical printing, "information had to be conveyed by drifting texts and vanishing manuscripts" (1979: 114). According to Eisenstein (1979: 113),

> The great tomes, charts, and maps that are now seen as "milestones" [of the "varied intellectual 'revolutions' of early-modern times"] might have proved insubstantial had not the preservative powers of print also been called into play. Typographical fixity is a basic prerequisite for the rapid advancement of learning. It helps to explain much else that seems to distinguish the history of the past five centuries from that of all prior eras.

To Eisenstein, "the implications of typographical fixity . . . involve the whole modern 'knowledge industry' . . . [as well as] issues that are . . . geopolitical" (1979: 116–17), from the "linguistic map of Europe" (1979: 117)—"a 'mother's tongue' learned 'naturally' at home would be reinforced by inculcation of a homogenized print-made language mastered . . . when learning to read" (1979: 118)—to its legal infrastructure— "laws pertaining to licensing and privileges . . . have yet to be examined as by-products of typographical fixity" (1979: 120).

Johns's (1998) *Nature of the Book* opposes critical aspects of *Printing Press as an Agent of Change* and Eisenstein's theoretical and methodological approach. According to Johns, in Eisenstein's account "printing itself stands outside history" (1998: 19). Therefore, "its 'culture' ... is deemed to exist inasmuch as printed texts *possess* some key characteristic ... The origins of this property are not analyzed" (1998: 19). To resolve what he considers to be the limitations of this approach, Johns (1998: 19–20) proposes that

We may consider fixity not as an *inherent* quality, but as a *transitive* one ... We may adopt the principle that fixity exists only inasmuch as it is recognized and acted upon by people—and not otherwise. The consequence of this change in perspective is that print culture itself is immediately laid open to analysis. It becomes a *result* of manifold representations, practices, and conflicts, rather than just the monolithic *cause* with which we are often presented. In contrast to talk of a "print logic" imposed on humanity, this approach allows us to recover the construction of different print cultures, in particular, historical circumstances.

The differences between Johns's and Eisenstein's notions of causality are intertwined with epistemic choices that guide the process of inquiry. For example, in his debate with Eisenstein in the *The American Historical Review*, Johns (2002) notes, "Where Eisenstein asks what print culture itself is, I ask how printing's historic role came to be shaped. Where she ascribes power to a culture, I assign it to communities of people. Most generally, where she is interested in qualities, I want to know about processes."

A revealing aspect of Johns's representation of their respective epistemic choices, to some extent echoed by Eisenstein (2002b) in her rebuttal of Johns's comments, is that he frames their choices in oppositional terms. This use of oppositional terms has been a persistent feature of discussions about causality in both communication studies and STS, principally as the debate between societal versus technological determinism.[5] Yet, although it may be rhetorically advantageous to cast one's arguments against a perceived polar opposite, this strategy can also limit the understanding of phenomena that may exhibit evolving combinations of the features that are portrayed as mutually exclusive.

To overcome this shortcoming, Lievrouw (2002: 192) has proposed to recast this type of opposition as "a dynamic relationship between determination and contingency." In her framework, "determination and contingency are interdependent and iterative, and ... this relationship can be seen at key junctures or 'moments' in ... media development and use" (2002: 183). When causality is considered in this way, different factors may determine or be contingent at different points in time as media and information technologies develop. This approach thus casts a broader conceptual net that captures both the social shaping of technology development and use, and the emergence of broad, persistent societal effects.

Such a causal framework aligns with a conceptual move within STS toward understanding technology as an object of inquiry, in terms of an ensemble of social and material elements in which dynamic combinations of determination and contingency generate different sociomaterial configurations (Bijker, 1995a; Callon, Law, & Rip, 1986; Jasanoff, 2004; Latour, 1996; Pickering, 1995). In a recent application of this view to the study of media and information technologies, Boczkowski (2004: 11) used the following lens to look at the development of online newspapers:

Media innovation unfolds through the interrelated mutations in technology, in communication, and in organization. I make sense of any of these three elements in the context of its links to the others, much like a triangle in which the function and meaning of any one side can be understood only in connection to the other two.

While sharing this basic stance regarding causality, and of technologies as sociomaterial ensembles, different scholars have underscored different dimensions in the relationships between determination and contingency. Three of these dimensions—discourse, practice, and pragmatics—demonstrate the value of taking a more encompassing and complex perspective on causality that at the same time allows for different conceptual foci.

Edwards's (1996) study of the interpenetration of politics, technology, and popular culture in America during the Cold War furnishes a powerful illustration of an analysis that highlights the discursive dimension.[6] According to Edwards (1996: 120), this period was marked by a "closed-world discourse" in which computerized technologies were at once symbol, tool, embodiment, and conduit and always deeply integrated with military procedures, cultural life, and subjective experiences.

The Cold War can be best understood in terms of *discourses* that connect technology, strategy, and culture: it was quite literally fought inside a quintessentially semiotic space, existing in models, language, iconography, and metaphor, embodied in technologies that lent to these semiotic dimensions their heavy inertial mass. In turn, this technological embodiment allowed closed-world discourse to ramify, proliferate, and entwine new strands.

Edwards uses the notion of discourse neither to highlight computerized technologies' discursive "impact" on society nor the discursive "choices" made by groups of powerful actors to shape these technologies, but he "views technology as one focus of a *social process* in which impacts, choices, experiences, metaphors, and environments all play a part" (1996: 41). This social process is a quintessentially dynamic one that unfolds over time and in which different material and nonmaterial elements shift from more determined to more contingent, and vice versa.

The role of practice is illuminated in a study of the production and consumption of sound reproduction technologies by Sterne (2003), in which he examines, among other issues, practice under the label of "audile technique."[7] By choosing the term "technique" rather than "practice" to make sense of actions related to the manipulation of sound reproduction technologies, the author blends the material and nonmaterial. In his analysis, the emergence of a set of audile techniques is contingent on constellations of bodily, cultural, material, and economic factors. But once stabilized as part of people's sociomaterial repertoire, techniques can play a determining role in the emergence of novel technologies and their associated sensations, symbols, and markets. Thus, in opposition to the argument that media and information technologies cause or constitute an extension of human senses and sensorial practices, as argued by McLuhan (1964), Ong (1982), and Stone (1991), among others, Sterne (2003: 92) shows that

All the *technologies* of listening that I discuss emerge out of *techniques* of listening. Many authors have conceptualized media and communication technologies as prosthetic sense. If media do, indeed, extend our senses, they do so as crystallized versions and elaborations of people's prior practices—or techniques—of using their senses.

Finally, in their study of classification systems and standards embodied in infrastructures, Bowker and Star (1999) propose a turn toward pragmatism to account for the development and use of information and media technologies. Following the lead of W. I. and Dorothy Thomas ([1917]1970), Bowker and Star (1999: 289) invite scholars to focus on the "definition of a situation," because "that definition . . . is what people will shape their behavior toward." Their approach to causality turns consequences from determined to determining and remains open about the social and material factors that affect the emergence of consequences:

[This approach] makes no comment on where the definition of the situation may come from—human or nonhuman, structure or process, group or individual. It powerfully draws attention to the fact that the materiality of anything . . . is drawn from the consequences of its situation. (Bowker & Star, 1999: 289–90)

To summarize, scholarship on media and information technologies at the intersection of STS and communication studies has historically enacted a treatment of causality that focused on the agency of either technological or societal factors. An alternative treatment has more recently gained currency by characterizing technology as sociomaterial configurations in which the different elements exhibit different degrees of determination and contingency at different moments in the unfolding of their relationship.

Process

Production and consumption form one of the major conceptual pairs in social and cultural theorizing, including work in STS and communication studies. As with notions of causality, general theorizing in both fields has espoused different orientations toward the relationships between production and consumption in the process of technology development.

On the one hand, because most of the initial technology scholarship in STS centered on articulating alternatives to technological determinism, studies during this period tended to focus more on the production of new artifacts and less on their consumption. As Bijker (2001: 15524) put it in a review of the social construction of technology model, until the mid-1990s, "the issue of technology's impact on society . . . had been bracketed for the sake of fighting technological determinism."

On the other hand, technology research in communication studies has centered on either production dynamics, often with a political economy focus (Gandy, 1993; Mosco, 1989; Robins & Webster, 1999; Schiller, 1999), or on the consumption side (Meyrowitz, 1985; Katz & Rice, 2002; Reeves & Nass, 1996; Walther, 1996), but less on the connection between the spheres of production and consumption. For instance, the diffusion of innovation framework, very popular in communication studies' technology research, commonly begins the process of inquiry once artifacts have been developed. As Rogers (1995: 159) wrote in a review of this framework, "past diffusion researchers usually began with the first adopters of an innovation . . . [and did not address] events and decisions occurring previous to this point."

Building on these traditions of inquiry, but also extending them, the thrust of scholarship on media and information technologies at the intersection of STS and communication studies has been to interrogate the links between production and consumption, developing concepts that shed light on the different processes that connect these two spheres.

STS researchers began to open the "black box" of production in ways that shed light on consumption by the early 1990s. For example, Woolgar (1991) showed that the process of software production "configures the user;" that is, it embeds the producer's vision of consumers and consumption practices in the design of the technology and thus influences technological adoption. Drawing from this notion as well as from Akrich's (1992, 1995) related idea of "inscription,"[8] a growing line of research bridging STS and communication studies has argued that in the technology development process, technical choices are made, artifacts are symbolically framed, and regulatory environments are fostered in ways that have consequences for consumption. Two recent studies of media and information technologies illustrate this approach at two extremes of social experience: the personal, small-scale realm of the body, and the impersonal, large-scale domain of the market.

In his account of Douglas Engelbart's role in the development of computer interface technologies such as the mouse, Bardini (2000) shows that Engelbart and his collaborators incorporated their ideas about users' bodies into their technical design choices, which subsequently influenced consumption. "Engelbart wasn't interested in just building the personal computer. He was interested in building the person who could use the computer to manage increasing complexity efficiently" (Bardini, 2000: 55). Engelbart and his colleagues thought that interface alternatives that took greater advantage of bodily capabilities had better chances of succeeding, that is, of "augmenting" users' cognition. This notion guided the design of tools such as the mouse, which complemented the movement of the hand and the dynamics of hand-eye coordination:

The user's hands and eyes were limited input and output devices in the human-computer interface. In developing the mouse and the chord keyset in the early 1960s, Engelbart and his group at [the Stanford Research Institute] made a quantum leap in human-computer interaction: the introduction of the body as whole as a set of connected, basic sensory-motor capabilities. (Bardini, 2000: 102)

The market is another important dimension for exploring the relationships between production and consumption. The commercial success of new artifacts depends not only on their technical functionality but also on their appropriation by users. Instead of seeing markets as asocial entities that obey only economic laws of supply and demand, scholars looking at the commercial fate of media and information technologies have focused on how market-making affects production and consumption simultaneously, and on the social construction of goods and their cultures of consumption (Douglas, 1987; Millard, 1995; Smulyan, 1994; Yates, 2005). For instance, in their history of electronic music synthesizer technologies, Pinch and Trocco (2002) examined the practices involved in the creation and growth of markets for musical instruments. They found that selling strategies affected both production and consumption of different kinds of synthesizers and proposed that salespeople "are a crucial link between the worlds of production and consumption. Whether through their interactions with users or by moving from use to sales, salespeople tie the world of use to the world of design and manufacture" (Pinch & Trocco, 2002: 313).

Parallel to opening the black box of production, scholarship on media and information technologies has also aimed to unpack consumption practices in ways that illuminate their links to production dynamics.[9] This effort partly originated in analyses of these technologies that account for the agency of users in both historical (Douglas, 1987; Fischer, 1992; Martin, 1991; Marvin, 1988) and contemporary settings (Ang, 1991, 1996; Lull, 1990; Morley, 1992; Silverstone, 1994).[10] This line of research has made substantive progress toward a better conceptual understanding of this agency particularly on three fronts: the domestication of new artifacts, the role of users as agents of technological change, and the resistance to new technologies.

Combining a focus on meaning informed by audience research and an approach to materiality inspired by social constructionist technology scholarship, Silverstone and Hirsch (1992) argue that when users bring new artifacts into the familiar household setting, they "domesticate" them by investing them with meaning and situating them within a material environment, both of which are locally contingent. In other words, in the process of domestication "new technologies . . . are brought (or not) under control by and on behalf of domestic users. In their ownership and in their appropriation into the culture of family or household and into the routines of everyday life, they are at the same time, cultivated. They become familiar, but they also develop and change" (Silverstone & Haddon, 1996: 60). Domestication unfolds in four stages—appropriation, objectification, incorporation, and conversion—in which new communication opportunities are opened up for both actors and artifacts (Aune, 1996; Laegran, 2003; Silverstone & Haddon, 1996).

Whereas the notion of domestication underscores the interpretive agency of users, research on the role of users as agents of technological change examines situations in which unanticipated user practices trigger material transformations of artifacts, and the mechanisms by which makers incorporate such changes into subsequent versions of their design (Boczkowski, 1999; Feenberg, 1992; Fischer, 1992; Orlikowski et al., 1995; Suchman, 2000).[11] For instance, Douglas (1987: 301–2) has shown that users of early radio broadcasting equipment were instrumental in turning what was initially a point-to-point communication system into a mass communication medium:

The amateurs and their converts had constructed the beginnings of a broadcasting network and audience. They had embedded radio in a set of practices and meanings vastly different from those dominating the offices at RCA. Consequently, the radio trust had to reorient its manufacturing priorities, its corporate strategies, indeed, its entire way of thinking about the technology under its control.

A third stream of work that highlights user agency examines resistance to new technologies, particularly the intentional opposition to technological change and its implications for production dynamics (Bauer, 1997; Kline, 2000, 2003; Wyatt et al., 2002). In his study of the introduction of the telephone in rural America in the early parts of the twentieth century, Kline (2000) has documented that established traditions of country life such as eavesdropping and visiting informed the ways that people in rural areas used the telephone: they listened to others' conversations and participated in multiple-party calls via party lines. Telephone companies tried to discourage these practices, but users actively resisted their attempts: "recognizing the difficulty of exerting social discipline over thousands of far-flung, rather independent-minded consumers ... commercial firms redesigned the telephone network to fit the social practices of this 'class' of customer" (2000: 48). Thus, Kline argues, "producers, rather than consumers, adapted the new technology to fit the social patterns of daily life" (2000: 48).

To sum up, the treatment of the technology development process in scholarship on media and information technologies has challenged stark distinctions between the spheres of production and consumption as well as built theoretical resources to illuminate the various forms and mechanisms that connect these two spheres.

Consequences

Debates have also ensued in both communication studies and STS about the social consequences of media and information technologies. Although historians have noted that utopian and dystopian claims have been made about virtually every new communication device or information service to come along (Lubar, 1993), as Marvin (1988) points out, predictions about technologies are not always borne out by their actual consequences. In STS and communication studies, two main views of the consequences of media and information technologies have emerged.

On the one hand, the technologies are thought to be "revolutionary," that is, they are a challenge to, and a radical departure from, existing media and information systems and impose new practices and institutional arrangements. Eisenstein's work, discussed above, takes this strong revolutionary view regarding the advent of the printing press. In the case of newer technologies, advocates of the revolutionary perspective contend that, because the technologies are designed, built, organized, distributed, and used differently from conventional mass media and information systems, they have the potential to overturn the social relations, work patterns, cultural practices, and economic and political orders created and fostered by industrial-era communication and information technologies (Beniger, 1986; Castells, 2001; Harvey, 1989; Pool, 1983; Zuboff, 1988). This position has been characterized as the "discontinuity" perspective (Schement & Curtis, 1995; Schement & Lievrouw, 1987; Shields & Samarajiva, 1993; Webster, 2002).

After the Second World War, the discontinuity perspective was fostered by inventors, engineers, designers, and planners involved in the defense projects, academic labs, and industries where many of the technologies were first developed (Light, 2003). They foresaw the integration of broadcasting and print with computer- and telecommunications-based systems that would provide interactive services and information delivery on demand. The dramatic growth of new computing and media technologies in this period prompted a number of prominent intellectuals and social scientists to look for corresponding changes in Western society and culture (e.g., Drucker, 1968; McLuhan, 1964; Mumford, 1963). Some asked whether these new technologies might be driving a transition as important as that from agricultural to industrial society in Europe and the United States in the eighteenth and nineteenth centuries, ushering in a late-twentieth-century "post-industrial" or "information" society (Bell, 1973; Machlup, 1962; Porat & Rubin, 1977; see also Schement & Lievrouw, 1987). Some speculated that a "communications revolution" might well be at hand (Gordon, 1977; Williams, 1983; see also Cairncross, 2001).

The opposing continuity view rejects the revolutionary rhetoric and asserts that the social consequences of technological change tend to be more gradual and incremental because they are necessarily situated within the context of established technologies, practices, and institutions. Partly in relation to its historical and ethnographic grounding and its focus on practice and meaning, STS scholarship has generally adopted the continuity view. Johns, for example, takes this more gradualist approach to the consequences of the printing press in his account of "print and knowledge in the making," discussed above.

Within communication studies, the continuity perspective was first articulated in the 1970s and 1980s by scholars trained in political economy and critical theory. In their view newer media and information technologies, like earlier mass media systems, are conceived, organized, and operated according to the logic of mass production, capitalism, commodification, and market economics. They reinforce inequitable systems of social and economic organization and control and help extend those systems into domains that were formerly resistant to rationalization and the industrial model of production (e.g., education, health care, law, and cultural production). According to this view, even if information rather than physical goods is the new commodity, the commodity system itself still rules, and its negative consequences persist (Garnham, 1990; Mosco, 1996; Robins & Webster, 1999; Schiller, 1981; Slack & Fejes, 1987; Traber, 1986).

By the early 1990s, the continuity and discontinuity perspectives had come to an impasse, despite attempts to negotiate a middle view (Schement & Curtis, 1995; Schement & Lievrouw, 1987) or to identify a range of views on media and information technologies and social change (Shields & Samarajiva, 1993). Influenced by the political economy of media, the critical/cultural turn noted above, and the critique of technological determinism advanced by STS, younger researchers in both communication and STS have increasingly tended to reject the revolutionary "new technologies, new society" discourse of information society research and have focused on the micro-scale, everyday, social and cultural contexts, uses, and meanings of newer com-munication technologies. Continuity has become the predominant perspective in social-scientific studies of media and information technologies and social change since the 1990s (Lievrouw & Livingstone, 2006b).

The discontinuity view was not dead, however. Artists, creative writers, historians, and critics who encountered networked computing for the first time in the early 1990s were well aware of the dangers of technological determinism; nonetheless, many of them used the novel technical features of these technologies as a point of departure for conceptualizing new kinds of digital media products (Bolter & Grusin, 1999; Hayles, 1999; Manovich, 2001; Murray, 1998; Poster, 1990; Stone, 1995). This schol-arship presents a different stance on the continuity-discontinuity issue by balancing claims about the perceived newness of novel digital artifacts with an understanding of their links to previously developed media and information technologies and the symbolic and social processes associated with them.

As media and information technologies have become commonplace over the last decade, some scholars in both STS and communication studies have begun to con-sider the consequences of new technologies as infrastructures, that is, as they become embedded in an existing technological base, transparent, and visible only when they break down (Star & Bowker, 2006; Star & Ruhleder, 1996). As Edwards (2003: 185) puts it, "the most salient characteristic of technology in the modern—industrial and postin-dustrial—world is the degree to which technology is *not* salient for most people, most of the time." For example, although the gradual integration of media and informa-tion technologies into existing systems and practices has made them more usable, convenient, and reliable, it has also created vast new possibilities for undetected surveillance and invasions of privacy (Agre & Rotenberg, 1997). It also has generated tools that allow individuals to resist such intrusions (Brook & Boal, 1995; Phillips, 2004).

The increasingly routine quality of media and information technologies has also been characterized as "banalization" (Lievrouw, 2004). For instance, contributors to a recent special issue of *New Media & Society* suggest that the late-twentieth-century information technology "revolution" is over, supplanted by incremental improve-ments in stability, security, reliability, ubiquity, and ease of use. The current sense is one of "slouching toward the ordinary" (Herring, 2004: 26), of "new and improved without the new" (Lunenfeld, 2004: 65). Stephen Graham (2004), a critic of the dis-course of technological discontinuity, revolution, and "transcendence," finds that rou-tinization largely confirms the continuity perspective. Calabrese (2004) argues that the reassertion of a familiar, mass-media "pipeline" style of sales and distribution online by traditional media and content industries has produced new media genres that look much like the old.

Whether, and to what extent, media and information technologies have become "banal" remains an open question. What is certain, however, is that as they have become more pervasive, familiar, and integrated into everyday practices and larger social, cultural, and institutional arrangements and structures, it is no longer possible to view the consequences of media and information technologies as a matter of *either* continuity *or* discontinuity. Recent studies at the intersection of communication studies and STS have adopted a view of social change that encompasses both the continuous and the discontinuous, the evolutionary and the revolutionary qualities and characteristics of media and information technologies and their effects (Boczkowski, 2004; Thompson, 2002; Turner, 2005).

CONCLUDING REMARKS: IMPLICATIONS AND DIRECTIONS FOR NEW RESEARCH

In the preceding sections we have proposed that the study of media and information technologies, especially in communication research and STS over the last twenty years, can be mapped around three main conceptual bridges: *causality*, comprising a tension between determination and contingency; *process*, conceived as multiple relationships between production and consumption; and *consequences*, contrasting continuity and discontinuity views of social change. These three concepts have often been represented in terms of opposing binaries; however, we have argued that they are better viewed as mutually determining, dialectic pairs in which each half of the pair assumes and builds on the other.

The map presented here is descriptive in that it organizes two broad, disparate bodies of work in terms of their common concerns, problematics, and mutual intellectual influence. But maps are not only descriptive tools; they also have a performative function. They help people navigate territories, locate landmarks in space, arrive at known destinations, discover previously unknown places, and make new connections between old and new locations. Like a map, the framework proposed here provides a tool for navigating the "problem space" of the social study of media and information technologies, both within and beyond communication studies and STS. It may also suggest new connections among the different disciplinary and intellectual traditions engaged in the study of these systems.

These connections have become essential as media and information technologies have proliferated and become more ubiquitous, and as mediation has become a central feature of social life over the last century. The technologies have been incorporated into a vast range of artifacts, practices, and social arrangements, including many that lie outside of what have been traditionally seen as "media" or "information technologies," such as finance, transportation, and health care. Recent empirical research at the intersection of STS and communication studies has demonstrated the growing ubiquity and centrality of mediation over time and in a variety of social and cultural contexts (Bowker & Star, 1999; Boczkowski, 2004; Downey, 2002; Light, 2003; Sterne, 2003; Thompson, 2002; Turner, 2005). In parallel, this proliferation and ubiquity may recently have helped rekindle interest in media and information technologies in fields where the topic has long been considered peripheral, such as economics (Hamilton, 2004), anthropology (Ginsburg et al., 2002), and sociology (Starr, 2004).

Taking advantage of the pervasiveness of media and information technologies today and of the dramatic rise of interest in them and their social/cultural contexts and implications, and building on the conceptual framework advanced here, we would like to suggest three possible avenues for continuing scholarship at the intersection of STS and communication studies. Consistent with our framework, they broadly concern the relationship between technology and society, technology development processes, and the consequences of sociotechnical change.

First, with regard to the causal relation between technology and society, and the tension between determination and contingency, given the growing turn to "mutual shaping" or "co-production" approaches, future work might address the particular conditions that may tilt the balance toward determination or contingency, or the specific mechanisms and processes that "harden" sociotechnical configurations under certain conditions or make them more malleable in other conditions. Scholarship that takes a historical or comparative perspective could be especially useful in both cases. For example, future studies might take as their point of departure a still-emerging body of research that takes an environmental perspective, analyzing technological systems, social structures and relations, and action together. These studies often seek to identify factors that can make such environments more determined, or "closed," on the one hand, or more contingent or open on the other (Davenport, 1997; Lievrouw, 2002; Nardi & O'Day, 1999; Verhulst, 2005).

Second, regarding the roles of production and consumption in the technology development process, two complementary directions for further work might contrast cases in which the boundary between production and consumption blurs or even disappears with those where production and consumption are so clearly segregated that they have minimal influence on each other. For instance, in the domain of so-called "citizen journalism," the success of South Korea's *OhMyNews*, which thousands of citizens-turned-journalists have transformed into a popular and politically influential online news site, might be compared with the failure of the *Los Angeles Times's* attempt to utilize WIKI TOOLS to make its editorials user-driven. The forum was shut down days after being launched because editors felt that some postings had become too aggressive. The first case demonstrates that people's engagement with media and information technologies is not easily reduced to the roles of producers *or* consumers,[12] while the second case shows that the production-consumption divide is still an important dynamic in many media and information contexts. Perhaps casting these as a dynamic of integration and separation could shed additional light on production and consumption as heuristic constructs.

Third, regarding the consequences of sociotechnical change, the increased sense of ordinariness and banality of media and information technologies could open the way for future work that might reconcile or at least recast the relationships between observed continuities and observed discontinuities, whether at the micro-scale of everyday life, practice, particular inventions, and meanings or at the macro-level of large-scale social relations and change.[13] Continuities and discontinuities are both observable across many levels of analysis, yet few theorists have attempted to integrate or frame them relative to each other.

We must add one critical point about all three suggested avenues for study: they must also account for the tightly interwoven relationship between the material and the symbolic, which, as we noted earlier, distinguishes media and information technologies from other types of sociotechnical infrastructures. Although it is tempting to classify and analyze these two dimensions of media and information technologies as distinct phenomena, they are in fact inextricably bound together. Future studies must confront the ways that meaning and forms of content contribute to influence material alternatives, and by the same token, how the physical materiality, durability, and format of specific technological devices and systems help shape content and meaning. This fundamental dialectic is at the heart of the interplay of determination and contingency, production and consumption, and continuity and discontinuity.

To conclude, we have proposed that concerns with causality, process, and consequences have delineated the domain of media and information technologies across STS and communication studies alike. Our aim has been to propose a broad framework for articulating shared concepts, problems, and interests in this rapidly growing area of study. Causality, process, and consequences, regardless of the particular contexts, settings, or applications in question, are fundamental concerns in the understanding of these and other technologies. Building on and transcending the binaries that have characterized research and scholarship to date may also help build dialogue and collaboration across these two traditions of inquiry and institutional boundaries.

Notes

We would like to thank our chapter's editor, Judy Wajcman, and four anonymous reviewers for their most helpful comments. We are also grateful for the valuable suggestions made by Jen Light, Doug Thomas, and session participants at the 2005 annual conference of the Society for Social Studies of Science, where an earlier version of this chapter was presented. In addition, Boczkowski would like to acknowledge the feedback received from the students—Max Dawson, Bernie Geoghegan, Divya Kumar, Dan Li, Limin Liang, Bhuvana Murthy, Ben Shields, and Gina Walejko—who took a quarter-long seminar on the ideas presented in this chapter at Northwestern University in fall 2005. Finally, we dedicate this essay to the memory of Roger Silverstone, who pioneered the dialogue between Communication Studies and Science and Technology Studies.

1. These bridges also correspond to fundamental issues in social, cultural, and historical studies of all technologies.

2. At several points in this chapter, we make a distinction between two schools of thought or traditions of inquiry within communication studies. On the one hand is a broadly behaviorist, medium-oriented, social science–based tradition that has tended to focus on the social and psychological effects of media and applied research regarding media professions and industries. The other tradition draws more from critical/cultural theory and political economy and tends to focus on issues of economic inequities and power, institutional structures, and cultural domination/hegemony. We have attempted to show how both traditions have played a role in the linkages between communication studies and STS. We thank an anonymous reviewer for reminding us that the first tradition, historically located in North America and East Asia, is often viewed critically by adherents of the second tradition, which is historically associated with the British/Birmingham school of media studies and is the predominant perspective in the United Kingdom and parts of Europe and Latin America.

3. In organizational communication research, where a substantial body of administrative research already existed regarding the implementation and management of ICTs in the workplace, the move to the contextual perspective, and the influence of concepts from STS, was particularly significant (see, e.g., Fulk, 1993; Jackson, 1996; Jackson et al., 2002; Orlikowski & Gash, 1994).

4. In addition to illustrating two different treatments of causality in technology-society relationships, these two books are also examples of two ways of conceptualizing technology as an object of inquiry, both discussed in the introductory section of this chapter. Einsenstein's book, influenced by the work of medium theorists like Innis and McLuhan, is inscribed within the tradition of scholarship that has characterized technology in terms of its technical features. Johns's book, drawing from constructivist scholars like Shapin and MacKenzie, is part of a mode of inquiry that has tended to stress issues of meaning, practice, and broader cultural connections of technological systems.

5. For an extended treatment of this matter, see chapter 7 in this volume. For additional discussions about this matter in general, see Bijker (1995b), Brey (2003), MacKenzie (1984), Staudenmaier (1989), and Williams and Edge (1996). For discussions focused on media and information technologies, see Dutton (2005), Edwards (1995), Kling (1994), Pfaffenberger (1988), Slack and Wise (2002), and Winner (1986).

6. It is important to note that Edwards's treatment of the notion of discourse draws partly from Foucaultian theory, which emphasizes the ties between symbolism and materiality in discursive configurations. We include Edwards's work as a powerful illustration of the discursive dimension precisely because his multilayered attention to symbolism, from micro-level metaphoric language to macro-level constructions of popular culture, is not in opposition to materiality but inextricably tied to it. For additional treatments on discursive aspects of media and information technologies, see, for instance, Bazerman (1999), Carey (1989), Gillespie (2006), and Wyatt (2000).

7. For a broader discussion on the "turn to practice" in social and cultural theory, see Schatzki et al. (2001). For additional treatments on practice issues in the study of media and information technologies, see, for instance, Boczkowski and Orlikowski (2004), Foot et al. (2005), Heath and Luff (2000), and Orlikowski (2000).

8. According to Akrich (1992: 208), producers "define actors with specific tastes, competences, motives, aspirations, political prejudices, and the rest, and they assume that morality, technology, science, and economy will evolve in particular ways. A large part of the work of innovators is that of *'inscribing'* this vision of—or prediction about—the world in the technical content of the new object."

9. Mackay et al. (2000: 737) have argued that this move has been part of a larger shift in social and cultural theorizing: "the turn to 'the user' is a feature of broader discourses, including that of the social sciences, not just the sociology of technology." For more on this matter in STS, see Oudshoorn and Pinch (2003) and chapter 22 in this volume.

10. Another early example of this line of work is Rice and Rogers's notion of "reinvention" in the diffusion of innovations, defined as "the degree to which an innovation is changed by the adopter in the process of adoption and implementation after its original development" (1980: 500–501). Subsequent research on reinvention added significant empirical detail, but provided not so much conceptual elaboration about the dynamics of user agency.

11. "Users" need not be individuals: in her study of the co-evolution of users and technologies in the life insurance industry, Yates (2005) has shown the value of focusing on a previously overlooked level of analysis, that of the collective—as opposed to individual—user. According to the author, "although individual agents clearly played critical roles, they could not act alone but had to mobilize those above and below them in the company hierarchy, as well as their peers, to acquire and apply such technology . . . This firm and industry focus illuminates a level thus far studied on the producer side but rarely on the user side" (2005: 259).

12. In communication studies, a reassessment of the notion of "audience," which equates engagement with media and information technologies with consumption, has been under way for over a decade (Abercrombie & Longhurst, 1998; Ang, 1991; Gray, 1999; Livingstone, 2004). Interactivity, another fruitful window into the production-consumption relationship, has been a locus of STS scholarship since the pioneering work of Suchman (1987). In communication studies, interactivity and related concepts, such as telepresence and propinquity, have been investigated since the 1970s (see Rafaeli, 1988; McMillan, 2006).

13. This is not a technology research issue that is new in either communication studies or STS, as evidenced in both early scholarship such as Marvin (1988) and recent scholarship such as Boczkowski (2004) and Yates (2005). But more remains to be done in specifying the more general mechanisms whereby discontinuity arises from continuity.

References

Abbate, J. (1999) *Inventing the Internet* (Cambridge, MA: MIT Press).

Abercrombie, N. & B. Longhurst (1998) *Audiences: A Sociological Theory of Performance and Imagination* (Thousand Oaks, CA: Sage).

Agre, P. E. & M. Rotenberg (eds) (1997) *Technology and Privacy: The New Landscape* (Cambridge, MA: MIT Press).

Akrich, M. (1992) "The De-Scription of Technical Objects," in W. Bijker & J. Law (eds), *Shaping Technology/Building Society: Studies in Sociotechnical Change* (Cambridge: MIT Press): 205–24.

Akrich, M. (1995) "User Representations: Practices, Methods and Sociology," in A. Rip, T. Misa, & J. Schot (eds), *Managing Technology in Society* (London: Pinter): 167–84.

Ang, I. (1991) *Desperately Seeking the Audience* (New York: Routledge).

Ang, I. (1996) *Living Room Wars: Rethinking Media Audiences for a Postmodern World* (London: Routledge).

Aune, M. (1996) "The Computer in Everyday Life: Patterns of Domestication of a New Technology," in M. Lie & K. Sorensen (eds), *Making Technology Our Own? Domesticating Technology into Everyday Life* (Stockholm: Scandinavian University Press): 91–120.

Bardini, T. (2000) *Bootstrapping: Douglas Engelbart, Coevolution, and the Origins of Personal Computing* (Stanford, CA: Stanford University Press).

Bauer, M. (1997) "Resistance to New Technology and Its Effects on Nuclear Power, Information Technology and Biotechnology," in M. Bauer (ed), *Resistance to New Technology: Nuclear Power, Information Technology and Biotechnology* (Cambridge: Cambridge University Press): 1–41.

Bazerman, C. (1999) *The Languages of Edison's Light* (Cambridge, MA: MIT Press).

Bell, D. (1973) *The Coming of Post-Industrial Society: A Venture in Social Forecasting* (New York: Basic Books).

Beniger, J. (1986) *The Control Revolution: Technological and Economic Origins of the Information Society* (Cambridge, MA: Harvard University Press).

Bijker, W. (1995a) *Of Bicycles, Bakelites, and Bulbs: Toward a Theory of Sociotechnical Change* (Cambridge, MA: MIT Press).

Bijker, W. (1995b) "Sociohistorical Technology Studies," in S. Jasanoff, G. Markle, J. Petersen, & T. Pinch (eds), *Handbook of Science and Technology Studies* (Thousand Oaks, CA: Sage): 229–56.

Bijker, W. (2001) "Social Construction of Technology," in N. J. Smelser & P. B. Baltes (eds), *International Encyclopedia of the Social and Behavioral Sciences,* vol. 23 (Oxford: Elsevier): 15522–27.

Bijker, W. E. & J. Law (eds) (1992) *Shaping Technology/Building Society: Studies in Sociotechnical Change* (Cambridge, MA: MIT Press).

Bijker, W. E., T. P. Hughes, & T. Pinch (eds) (1987) *The Social Construction of Technological Systems: New Directions in the Sociology and History of Technology* (Cambridge, MA: MIT Press).

Boczkowski, P. (1999) "Mutual Shaping of Users and Technologies in a National Virtual Community," *Journal of Communication* 49: 86–108.

Boczkowski, P. (2004) *Digitizing the News: Innovation in Online Newspapers* (Cambridge, MA: MIT Press).

Boczkowski, P. & W. Orlikowski (2004) "Organizational Discourse and New Media: A Practice Perspective," in D. Grant, C. Hardy, C. Oswick, N. Philips, & L. Putnam (eds), *The Handbook of Organizational Discourse* (London: Sage): 359–77.

Bolter, J. D. & R. Grusin (1999) *Remediation: Understanding New Media* (Cambridge, MA: MIT Press).

Bowker, G. & S. Star (1999) *Sorting Things Out: Classification and Its Consequences* (Cambridge, MA: MIT Press).

Brey, P. (2003) "Theorizing Modernity and Technology," in T. Misa, P. Brey, & A. Feenberg (eds), *Modernity and Technology* (Cambridge, MA: MIT Press): 33–71.

Brook, J. & I. A. Boal (eds) (1995) *Resisting the Virtual Life: The Culture and Politics of Information* (San Francisco: City Lights).

Cairncross, F. (2001) *The Death of Distance: How the Communications Revolution Is Changing Our Lives* (Boston: Harvard Business School Press).

Calabrese, A. (2004) "Stealth Regulation: Moral Meltdown and Political Radicalism at the Federal Communications Commission," *New Media and Society* 6(1): 106–13.

Callon, M., J. Law, & A. Rip (eds) (1986) *Mapping the Dynamics of Science and Technology: Sociology of Science in the Real World* (Basingstoke, U.K.: Macmillan).

Carey, J. (1989) *Communication as Culture: Essays on Media and Society* (Boston: Unwin Hyman).

Castells, M. (2001) *The Internet Galaxy: Reflections on the Internet, Business and Society* (Oxford: Oxford University Press).

Davenport, T. H. (1997) *Information Ecology: Mastering the Information and Knowledge Environment* (New York: Oxford University Press).

Douglas, S. (1987) *Inventing American Broadcasting, 1899–1922* (Baltimore, MD: Johns Hopkins University Press).

Downey, G. (2002) *Telegraph Messenger Boys: Labor, Technology and Geography, 1850–1950* (New York: Routledge).

Drucker, P. (1969) *The Age of Discontinuity: Guidelines to Our Changing Society* (New York: Harper & Row).

Dutton, W. (2005) "Continuity or Transformation? Social and Technical Perspectives on Information and Communication Technologies," in W. Dutton, B. Kahin, R. O'Callaghan, & A. Wyckoff (eds), *Transforming Enterprise: The Economic and Social Implications of Information Technology* (Cambridge, MA: MIT Press): 13–24.

Edwards, P. (1995) "From 'Impact' to Social Process: Computers in Society and Culture," in S. Jasanoff, G. Markle, J. Petersen, & T. Pinch (eds), *Handbook of Science and Technology Studies* (Thousand Oaks, CA: Sage): 257–85.

Edwards, P. (1996) *The Closed World: Computers and the Politics of Discourse in Cold War America* (Cambridge, MA: MIT Press).

Edwards, P. (2003) "Infrastructure and Modernity: Force, Time, and Social Organization in the History of Sociotechnical Systems," in T. Misa, P. Brey, & A. Feenberg (eds), *Modernity and Technology* (Cambridge, MA: MIT Press): 185–225.

Eisenstein, E. (1979) *The Printing Press as an Agent of Change: Communications and Cultural Transformations in Early Modern Europe* (Cambridge: Cambridge University Press).

Eisenstein, E. (2002a) "An Unacknowledged Revolution Revisited," *American Historical Review* 107(1): 87–105.

Eisenstein, E. (2002b) "Reply," *American Historical Review* 107(1): 126.

Ettema, J. & T. Glasser (1998) *Custodians of Conscience: Investigative Journalism and Public Virtue* (New York: Columbia University Press).

Feenberg, A. (1992) "From Information to Communication: The French Experience with Videotex," in M. Lea (ed), *Contexts of Computer-Mediated Communication* (London: Harvester-Wheatsheaf): 168–87.

Fischer, C. S. (1992) *America Calling: A Social History of the Telephone to 1940* (Berkeley and Los Angeles: University of California Press).

Foot, K., B. Warnick, & S. Schneider (2005) "Web-Based Memorializing After September 11: Toward a Conceptual Framework," *Journal of Computer-Mediated Communication* 11(1). Available at: http://jcmc.indiana.edu/vol11/issue1/foot.html.

Forsythe, D. (1993) "Engineering Knowledge: The Construction of Knowledge in Artificial Intelligence," *Social Studies of Science* 23: 445–47.

Fulk, J. (1993) "Social Construction of Communication Technology," *Academy of Management Journal* 36: 921–50.

Galambos, L. (1988) "Looking for the Boundaries of Technological Determinism: A Brief History of the U.S. Telephone System," in R. Mayntz & T. P. Hughes (eds), *The Development of Large Technical Systems* (Frankfurt and Boulder, CO: Campus and Westview): 135–53.

Gandy, O. (1993) *The Panoptic Sort: A Political Economy of Personal Information* (Boulder, CO: Westview).

Garnham, N. (1990) *Capitalism and Communication: Global Culture and the Economics of Information* (London: Sage).

Gerbner, G. (1983) *Ferment in the Field.* Special issue of *Journal of Communication* 33(3) Summer.

Gillespie, T. (2006) "Engineering a Principle: 'End-to-End' in the Design of the Internet," *Social Studies of Science* 36(3): 427–57.

Ginsburg, F., L. Abu-Lughod, & B. Larkin (eds) (2002) *Media Worlds: Anthropology on New Terrain* (Berkeley and Los Angeles: University of California Press).

Goody, J. (1981) "Alphabets and Writing," in R. Williams (ed), *Contact: Human Communication and Its History* (New York: Thames and Hudson): 105–26.

Gordon, G. N. (1977) *The Communications Revolution: A History of Mass Media in the United States* (New York: Hastings House).

Graham, S. (2004) "Beyond the 'Dazzling Light': From Dreams of Transcendence to the 'Remediation' of Urban Life: A Research Manifesto," *New Media and Society* 6(1): 16–25.

Gray, A. (1999) "Audience and Reception Research in Retrospect: The Trouble with Audiences," in P. Alasuutari (ed), *Rethinking the Media Audience: The New Agenda* (Thousand Oaks, CA: Sage): 22–37.

Hamilton, J. (2004) *All the News That's Fit to Sell: How the Market Transforms Information into News* (Princeton, NJ: Princeton University Press).

Harvey, D. (1989) *The Condition of Postmodernity: An Enquiry into the Origins of Cultural Change* (Cambridge, MA: Blackwell).

Hayles, K. (1999) *How We Became Posthuman: Virtual Bodies in Cybernetics, Literature, and Informatics* (Chicago: University of Chicago Press).

Heath, C. & P. Luff (2000) *Technology in Action* (Cambridge: Cambridge University Press).

Herring, S. C. (2004) "Slouching Toward the Ordinary: Current Trends in Computer-Mediated Communication," *New Media and Society* 6(1): 26–36.

Innis, H. A. (1972) *Empire and Communications* (Toronto: University of Toronto Press).

Jackson, M. (1996) "The Meaning of 'Communication Technology': The Technology-Context Scheme," in B. Burleson (ed), *Communication Yearbook* 19 (Thousand Oaks, CA: Sage): 229–67.

Jackson, M., M. S. Poole, & T. Kuhn (2002) "The Social Construction of Technology in Studies of the Workplace," in L. Lievrouw & S. Livingstone (eds), *The Handbook of New Media* (London: Sage): 236–53.

Jasanoff, S. (ed) (2004) *States of Knowledge: The Co-Production of Science and Social Order* (London: Routledge).

Johns, A. (1998) *The Nature of the Book: Print and Knowledge in the Making* (Chicago: University of Chicago Press).

Johns, A. (2002) "How to Acknowledge a Revolution," *American Historical Review* 107(1): 106–25.

Katz, J. & R. Rice (2002) *Social Consequences of Internet Use: Access, Involvement and Interaction* (Cambridge, MA: MIT Press).

Kline, R. (2000) *Consumers in the Country: Technology and Social Change in Rural America* (Baltimore, MD: Johns Hopkins University Press).

Kline, R. (2003) "Resisting Consumer Technology in Rural America: The Telephone and Electrification," in N. Oudshoorn & T. Pinch (eds), *How Users Matter: The Co-Construction of Users and Technology* (Cambridge, MA: MIT Press): 51–66.

Kling, R. (1994) "Reading 'All About' Computerization: How Genre Conventions Shape Nonfiction Social Analysis," *Information Society* 10: 147–72.

Kling, R. & S. Iacono (1989) "The Institutional Character of Computerized Information Systems," *Office: Technology and People* 5(1): 7–28.

Laegran, A. (2003) "Escape Vehicles? The Internet and the Automobile in a Local-Global Intersection," in N. Oudshoorn & T. Pinch (eds), *How Users Matter: The Co-Construction of Users and Technology* (Cambridge, MA: MIT Press): 81–100.

Latour, B. (1996) *Aramis: Or the Love of Technology* (Cambridge, MA: Harvard University Press).

Lazarsfeld, P. F. (1941) "Remarks on Administrative and Critical Communications Research," *Studies in Philosophy and Social Science* 9: 3–16.

Lievrouw, L. A. (2002) "Determination and Contingency in New Media Development: Diffusion of Innovations and Social Shaping of Technology Perspectives," in L. A. Lievrouw & S. Livingstone (eds), *The Handbook of New Media: Social Shaping and Consequences of ICTs* (London: Sage): 183–99.

Lievrouw, L. A. (2004) "What's Changed About New Media?" Introduction to the fifth anniversary issue of *New Media and Society* 6(1) (February): 9–15.

Lievrouw, L. A. & S. Livingstone (eds) (2006a) *The Handbook of New Media,* updated student ed. (London: Sage).

Lievrouw, L. A. & S. Livingstone (2006b) "Introduction to the Updated Student Edition," in L. A. Lievrouw & S. Livingstone (eds), *The Handbook of New Media,* updated student ed. (London: Sage): 1–14.

Light, J. S. (2003) *From Warfare to Welfare: Defense Intellectuals and Urban Problems in Cold War America* (Baltimore, MD: Johns Hopkins University Press).

Livingstone, S. (2004) "The Challenge of Changing Audiences: Or, What Is the Researcher To Do in the Age of the Internet?" *European Journal of Communication* 19: 75–86.

Lubar, S. (1993) *Infoculture: The Smithsonian Book of Information Age Inventions* (Washington, DC: Smithsonian).

Lull, J. (1990) *Inside Family Viewing: Ethnographic Research on Television's Audience* (London: Routledge).

Lunenfeld, P. (2004) "Media Design: New and Improved Without the New," *New Media and Society* 6(1): 65–70.

Machlup, F. (1962) *The Production and Distribution of Knowledge in the United States* (Princeton, NJ: Princeton University Press).

Mackay, H., C. Carne, P. Beynon-Davies, & D. Tudhope (2000) "Reconfiguring the User: Using Rapid Application Development," *Social Studies of Science* 30(5): 737–57.

MacKenzie, D. (1984) "Marx and the Machine," *Technology and Culture* 25: 473–502.

Manovich, L. (2001) *The Language of New Media* (Cambridge, MA: MIT Press).

Martin, M. (1991) *"Hello Central?" Gender, Technology and Culture in the Formation of Telephone Systems* (Montreal: McGill-Queen's University Press).

Marvin, C. (1988) *When Old Technologies Were New: Thinking About Electric Communication in the Late Nineteenth Century* (New York and Oxford: Oxford University Press).

Mayntz, R. & V. Schneider (1988) "The Dynamics of System Development in a Comparative Perspective: Interactive Videotex in Germany, France, and Britain," in R. Mayntz & T. P. Hughes (eds), *The Development of Large Technical Systems* (Boulder, CO: Westview Press): 263–98.

McLuhan, M. (1964) *Understanding Media: The Extensions of Man* (New York: McGraw-Hill).

McMillan, S. (2006) "Exploring Models of Interactivity from Multiple Research Traditions: Users, Documents and Systems," in L. A. Lievrouw & S. Livingstone (eds), *The Handbook of New Media,* updated student ed. (London: Sage): 205–29.

Meyrowitz, J. (1985) *No Sense of Place: The Impact of Electronic Media on Social Behavior* (New York: Oxford University Press).

Millard, A. (1995) *America on Record: A History of Recorded Sound* (Cambridge: Cambridge University Press).

Morley, D. (1992) *Television, Audiences, and Cultural Studies* (London: Routledge).

Mosco, V. (1989) *The Pay-Per Society: Computers and Communication in the Information Age: Essays in Critical Theory and Public Policy* (Norwood, NJ: Ablex).

Mosco, V. (1996) *The Political Economy of Communication: Rethinking and Renewal* (London: Sage).

Mumford, L. (1963) *Technics and Civilization* (New York: Harcourt, Brace & World).

Murray, J. (1998) *Hamlet on the Holodeck: The Future of Narrative in Cyberspace* (Cambridge, MA: MIT Press).

Nardi, B. & V. L. O'Day (1999) *Information Ecologies: Using Technologies with Heart* (Cambridge, MA: MIT Press).

Ong, W. (1982) *Orality and Literacy: The Technologizing of the Word* (London: Methuen).

Orlikowski, W. J. (2000) "Using Technology and Constituting Structures: A Practice Lens for Studying Technology in Organizations," *Organization Science* 11: 404–28.

Orlikowski, W. & D. Gash (1994) "Technological Frames: Making Sense of Information Technology in Organizations," *ACM Transactions on Information Systems* 12: 174–207.

Orlikowski, W., J. Yates, K. Okamura, & M. Fujimoto (1995) "Shaping Electronic Communication: The Metastructuring of Technology in the Context of Use," *Organization Science* 6: 423–44.

Oudshoorn, N. & T. Pinch (2003) "Introduction: How Users and Non-Users Matter," in N. Oudshoorn & T. Pinch (eds), *How Users Matter: The Co-Construction of Users and Technology* (Cambridge, MA: MIT Press): 1–25.

Parker, E. (1970) "The New Communication Media," in C. S. Wallia (ed), *Toward Century 21: Technology, Society and Human Values* (New York: Basic Books): 97–106.

Pfaffenberger, B. (1988) "The Social Meaning of the Personal Computer: Or, Why the Personal Computer Revolution Was No Revolution," *Anthropological Quarterly* 61: 39–47.

Phillips, D. J. (2004) "Privacy Policy and PETs: The Influence of Policy Regimes on the Development and Social Implications of Privacy Enhancing Technologies," *New Media and Society* 6(6): 691–706.

Pickering, A. (1995) *The Mangle of Practice: Time, Agency, and Science* (Chicago: University of Chicago Press).

Pinch, T. & F. Trocco (2002) *Analog Days: The Invention and Impact of the Moog Synthesizer* (Cambridge, MA: Harvard University Press).

Pool, I. de S. (ed) (1977) *The Social Impact of the Telephone* (Cambridge, MA: MIT Press).

Pool, I. de S. (1983) *Technologies of Freedom* (Cambridge, MA: Belknap/Harvard University Press).

Pool, I. de S. & E. M. Noam (eds) (1990) *Technologies of Boundaries: On Telecommunications in a Global Age* (Cambridge, MA: Harvard University Press).

Porat, M. U. & M. R. Rubin (1977) *The Information Economy*, OT Special Publication 77–12, 9 vols. (Washington, DC: U.S. Department of Commerce, Office of Telecommunications).

Poster, M. (1990) *The Mode of Information: Poststructuralism and Social Context* (Chicago: University of Chicago Press).

Postman, N. (1985) *Amusing Ourselves to Death: Public Discourse in the Age of Show Business* (New York: Viking-Penguin).

Rafaeli, S. (1988) "Interactivity: From New Media to Communication," in R. P. Hawkins, J. M. Wiemann, & S. Pingree (eds), *Advancing Communication Science: Merging Mass and Interpersonal Processes* (Newbury Park, CA: Sage): 110–34.

Reeves, B. & C. Nass (1996) *The Media Equation: How People Treat Computers, Television and New Media Like Real People and Places* (Stanford, CA: CSLI Publications; Cambridge: Cambridge University Press).

Rice, R. E. & Associates (eds) (1984) *The New Media: Communication, Research and Technology* (Beverly Hills, CA: Sage).

Rice, R. E. & E. M. Rogers (1980) "Reinvention in the Innovation Process," *Knowledge: Creation, Diffusion, Utilization* 1(4): 499–514.

Robins, K. & E. Webster (1999) *Times of the Technoculture: From the Information Society to the Virtual Life* (London: Routledge).

Rogers, E. M. (1986) *Communication Technology: The New Media in Society* (New York: Free Press).

Rogers, E. (1995) *Diffusion of Innovations,* 4th ed. (New York: Free Press).

Schatzki T. R., K. Knorr Cetina, & E. von Savigny (eds) (2001) *The Practice Turn in Contemporary Theory* (London: Routledge).

Schement, J. R. & T. Curtis (1995) *Tendencies and Tensions of the Information Age: The Production and Distribution of Information in the United States* (New Brunswick, NJ: Transaction).

Schement, J. R. & L. A. Lievrouw (1987) *Competing Visions, Complex Realities: Social Aspects of the Information Society* (Norwood, NJ: Ablex).

Schiller, D. (1999) *Digital Capitalism: Networking the Global Market System* (Cambridge, MA: MIT Press).

Schiller, H. I. (1981) *Who Knows: Information in the Age of the Fortune 500* (Norwood, NJ: Ablex).

Schneider, V., T. Charon, J. M. Graham, I. Miles, & T. Vedel (1991) "The Dynamics of Videotex Development in Britain, France, and Germany: A Cross-National Comparison," *European Journal of Communication* 6: 187–212.

Schramm, W. (1977) *Big Media, Little Media: Tools and Technologies for Instruction* (Beverly Hills, CA: Sage).

Shields, P. & R. Samarajiva (1993) "Competing Frameworks for Research on Information-Communication Technologies and Society: Toward a Synthesis," in S. A. Deetz (ed), *Communication Yearbook 16* (Newbury Park, CA: Sage): 349–80.

Silverstone, R. (1994) *Television and Everyday Life* (London: Routledge).

Silverstone, R. & L. Haddon (1996) "Design and Domestication of Information and Communication Technologies: Technical Change and Everyday Life," in R. Silverstone & L. Haddon (eds), *Communication by Design: The Politics of Information and Communication Technologies* (New York: Oxford University Press): 44–74.

Silverstone, R. & E. Hirsch (1992) *Consuming Technologies: Media and Information in Domestic Spaces* (London and New York: Routledge).

Silverstone, R., E. Hirsch, & D. Morley (1992) "Information and Communication Technologies and the Moral Economy of the Household," in R. Silverstone & E. Hirsch (eds), *Consuming Technologies: Media and Information in Domestic Spaces* (London: Routledge): 15–31.

Slack, J. D. & F. Fejes (eds) (1987) *The Ideology of the Information Age* (Norwood, NJ: Ablex).

Slack, J. D. & J. M. Wise (2002) "Cultural Studies and Technology," in L. A. Lievrouw & S. Livingstone (eds), *The Handbook of New Media: Social Shaping and Consequences of ICTs* (London: Sage): 485–501.

Smulyan, S. (1994) *Selling Radio: The Commercialization of American Broadcasting, 1920–1934* (Washington, DC: Smithsonian Institution Press).

Star, S. L. (ed) (1995) *The Cultures of Computing* (Oxford: Blackwell).

Star, S. L. & G. Bowker (2006) "How to Infrastructure," in L. A. Lievrouw & S. Livingstone (eds), *The Handbook of New Media,* updated student ed. (London: Sage): 230–45.

Star, S. L. & K. Ruhleder (1996) "Steps Toward an Ecology of Infrastructure: Design and Access for Large Information Spaces," *Information Systems Research* 7: 111–34.

Starr, P. (2004) *The Creation of the Media: Political Origins of Modern Communications* (New York: Basic Books).

Staudenmaier, J. (1989) *Technology's Storytellers: Reweaving the Human Fabric* (Cambridge, MA: MIT Press).

Sterne, J. (2003) *The Audible Past: Cultural Origins of Sound Reproduction* (Durham, NC: Duke University Press).

Stone, A. R. (1991) "Will the Real Body Please Stand Up? Boundary Stories About Virtual Cultures," in M. Benedikt (ed), *Cyberspace: First Steps* (Cambridge, MA: MIT Press): 81–118.

Stone, A. R. (1995) *The War of Desire and Technology at the Close of the Mechanical Age* (Cambridge, MA: MIT Press).

Suchman, L. (1987) *Plans and Situated Actions: The Problem of Human-Machine Communication* (Cambridge: Cambridge University Press).

Suchman, L. (2000) *Working Relations of Technology Production and Use,* paper presented at the Heterarchies Seminar, Columbia University, New York, February.

Thomas, W. & D. Thomas ([1917]1970) "Situations Defined as Real Are Real in Their Consequences," in G. Stone & H. Farberman (eds), *Social Psychology Through Symbolic Interaction* (Waltham, MA: Xerox): 54–155.

Thompson, E. (2002) *The Soundscape of Modernity: Architectural Acoustics and the Culture of Listening in America, 1900–1933* (Cambridge, MA: MIT Press).

Traber, M. (ed) (1986) *The Myth of the Information Revolution: Social and Ethical Implications of Communication Technology* (London: Sage).

Turkle, S. (1984) *The Second Self: Computers and the Human Spirit* (New York: Simon and Schuster).

Turner, F. (2005) "Where the Counterculture Met the New Economy: Revisiting the WELL and the Origins of Virtual Community," *Technology and Culture* 46(3): 485–512.

Verhulst, S. (2005) *Analysis into the Social Implication of Mediation by Emerging Technologies,* position paper for the MIT-OII Joint Workshop, "New Approaches to Research on the Social Implications of Emerging Technologies" (Oxford: Oxford Internet Institute, Oxford University), April 15–16. Available at: http://www.oii.ox.ac.uk.

Walther, J. B. (1996) "Computer-Mediated Communication: Impersonal, Interpersonal, and Hyperpersonal Interaction," *Communication Research* 23: 3–43.

Webster, R. (2002) *Theories of the Information Society,* 2nd ed. (London: Routledge).

Williams, F. (1983) *The Communications Revolution* (rev. ed.) (New York: New American Library).

Williams, F., R. E. Rice, & E. M. Rogers (1988) *Research Methods and the New Media* (New York: Free Press).

Williams, R. (1975) *Television: Technology and Cultural Form* (New York: Schocken Books).

Williams, R. (ed) (1981) *Contact: Human Communication and Its History* (New York: Thames and Hudson).

Williams, R. & D. Edge (1996) "The Social Shaping of Technology," *Research Policy* 25: 865–99.

Winner, L. (1986) "Mythinformation," in L. Winner (ed), *The Whale and the Reactor: A Search for Limits in an Age of High Technology* (Chicago: University of Chicago Press): 98–117.

Woolgar, S. (1991) "Configuring the User: The Case of Usability Trials," in J. Law (ed), *A Sociology of Monsters: Essays on Power, Technology and Domination* (London: Routledge): 57–99.

Wyatt, S. (2000) "Talking About the Future: Metaphors of the Internet," in N. Brown, B. Rappert, & A. Webster (eds), *Contested Futures: A Sociology of Prospective Techno-Science* (Aldershot: Ashgate Press): 109–26.

Wyatt, S., G. Thomas, & T. Terranova (2002) "They Came, They Surfed, They Went Back to the Beach: Conceptualising Use and Non-Use of the Internet," in S. Woolgar (ed), *Virtual Society? Technology, Cyberpole, Reality* (Oxford: Oxford University Press): 23–40.

Yates, J. (2005) *Structuring the Information Age: Life Insurance and Technology in the Twentieth Century* (Baltimore, MD: Johns Hopkins University Press).

Zuboff, S. (1988) *In the Age of the Smart Machine: The Future of Work and Power* (New York: Basic Books).

38 Anticipatory Governance of Nanotechnology: Foresight, Engagement, and Integration

Daniel Barben, Erik Fisher, Cynthia Selin, and David H. Guston

I. INTRODUCTION

The widespread understanding that nanotechnology constitutes an emerging set of science-based technologies with the collective capacity to remake social, economic, and technological landscapes (e.g., Crow & Sarewitz, 2001) has, in itself, generated tangible outcomes. In the first years of the new millennium, governments around the world created national nanotechnology programs that spent billions of dollars (Roco, 2003), reconfigured institutional arrangements, and constructed new sites for research and development (R&D). Large transnational corporations have similarly made significant investments in R&D at the nanoscale, and venture capitalists have funded start-up companies—often launched by university researchers—specializing in a broad array of nanotechnologies (Lux Research, 2006). Many of these actors present nanotechnology as an enabling platform for other transformative innovations that will become even more powerful through its "convergence" with biotechnology, information technology, and cognitive science. The magnitude and speed of such transformations demand critical reflection on the role of technology in society and the composition of desirable futures. The presumed nascent state of nanotechnology suggests that critical reflection along with other forms of response may actually contribute to such outcomes. Nanotechnology thus affords crucial opportunities for researchers in science and technology studies (STS) to participate in the construction of safe, civil, and equitable nanotechnological developments.

The future prospects for nanotechnology, or nanoscale science and engineering (NSE), are fundamentally uncertain. In its novelty, complexity, uncertainty, and publicity, nanotechnology represents "postnormal science" (Funtowicz & Ravetz, 1993). It thus occasions new approaches to the conduct of research evaluation and assessment that require the engagement of a variety of potential users and stakeholders in the production of knowledge (Gibbons et al., 1994), as well as new organizations that span the boundary between knowledge production and public action (Guston, 2000). Not only is it unclear which scientific and technological potentials out of the many that theoretically exist might actually come to pass, but the shape and desirability of eventual sociotechnical outcomes may in part depend on the work of these new interactions and approaches. Indeed, nanotechnology can also be thought of as a metaphor for even more inchoate potential futures of other new technologies, the history of technological emergence, and the role of technoscience in destabilizing social systems—for better and for ill.

The case of nanotechnology thus has broader applicability, for such fundamental uncertainties pose challenges for science and technology decision making in public and private sectors, as well as for STS scholarship. The challenges for STS include the continued consideration of the place of its scholarship, especially when—as explored below—it is invited by policy makers and others to have a role in the pursuit and development of science and technology. Accepting this invitation, as this chapter suggests, may mean not only attending to areas of research that are not fully developed, but also attempting to create a different scope, scale, and organization of STS research.

Notably, a great deal of the study of the societal aspects of nanotechnology is bound up in the rhetoric of novelty. With this in mind, this chapter provides a brief overview of how prominent actors define nanotechnology and frame some of the societal issues associated with it. Set within this disputed context of the novelty of NSE itself and its attendant societal issues, the chapter then surveys a unique set of policies that has emerged across several countries. Generally, these policies do not presume the automatic provision of social goods from NSE research. Instead, policy mandates call for nanoscale R&D to be situated within broader social processes. Next, the chapter considers some of the unique interactions that, in part inspired by these policies, have emerged among STS researchers and policy makers, scientists, and the public by reviewing and analyzing some key features of foresight, engagement, and integration that mark these efforts. Finally, the chapter emphasizes the novelty of the scope, scale, reach, and context of much of this STS research. Specifically, the authors believe that the main contribution of this largely unprecedented multipronged, large-scale STS approach to nanotechnology is the creation of a broad capacity for "anticipatory governance" (Guston & Sarewitz, 2002).

II. DEFINING NANOTECHNOLOGY AND ITS ISSUES

No definition can encompass the complex research and policy realm that nanotechnology signifies (Woodhouse, 2004). Nevertheless, a variety of scientific and bureaucratic interests seek a concrete definition. In the United States, the National Nanotechnology Initiative (NNI) has tinkered with its original definition, most recently defining nanotechnology broadly as "the understanding and control of matter at dimensions of roughly 1 to 100 nanometers, where unique phenomena enable novel applications" (NNI, 2007). The nongovernmental standard-setting body, ASTM International, similarly defines nanotechnology as "a wide range of technologies that measure, manipulate, or incorporate materials and/or features with at least one dimension between approximately 1 and 100 nanometers (nm). Such applications exploit the properties, distinct from bulk/macroscopic systems, of nanoscale components" (Active Standard E2456-06).

Such definitions fall under the conception of "mainstream nanotechnology" (Keiper, 2003), which is largely an immediate extension of chemistry and materials science that originally might not have attracted much political attention or funding, and clearly exclude "molecular nanotechnology" (Drexler, 2004), which focuses on longer-term, directed self-assembly techniques that critics characterize as science fiction but which lent a great deal of verve to early nanotechnology promotions. The NNI situated nanotechnology between mainstream and molecular conceptions so that investment, which had in part been conceived as a response from the physical sciences to the exploding biomedical research funding of the 1990s, included biology. And like genetic engineering before it, nanotechnology under these definitions blurs boundaries not only among technical disciplines but also between science and engineering and between research and manufacturing—thus building in the promise of economic payoffs from research at the onset. The bridging of disciplines as well as the hyperbolic promises to society mark nanotechnology as the "new frontier."

However sufficient broad definitions might be for promoting research programs, they are hard for social scientists to operationalize. Bibliometric research has struggled to define nanotechnology in order to track its intellectual and geographic dynamics. Such work (e.g., Porter, Youtie, & Shapira, 2006) has identified four broad and overlapping areas of inquiry—nanodevices and electronics, nanostructure chemistry and nanomaterials, nanomedicine and nanobiology, and metrology and nanoprocesses. This categorization nearly replicates a taxonomy derived by the Royal Society and Royal Academy of Engineering (2004). The definition of nanotechnology is furthermore expected to change over time. For instance, prominent nanotechnology "roadmaps" predict an evolution from nanomaterials to passive nanosystems to active nanosystems (Roco & Renn, 2006). It is thus more accurate to talk of a plurality of nanotechnologies, even while acknowledging the prominence and persistence of the abstract singular term resulting from a combination of advances in instruments and research communities (Mody, 2006) and political agendas and alliances (McCray, 2005).

Frank and brazen optimism on behalf of nanotechnology—even the government sponsors who eschew the molecular nanotechnology vision hail it as "the next industrial revolution"—contrasts with equally compelling arguments about its unintended consequences (Sarewitz & Woodhouse, 2003), giving rise to an urgency to address issues of equity, ethics, and engagement. However, the almost protean form of nanotechnologies conspires with broad time horizons to further complicate the recognition and critique of related cultural, ethical, legal, educational, economic, and environmental (henceforth "societal") issues. While issues need not be new to warrant consideration, a particular search for novelty has accompanied the societal debate: What is new about nanotechnologies that leads to pressing societal issues?

As implied in the definitions quoted above, the standard technical explanation for novelty stresses the properties of matter that manifest at the nanoscale. Thus, although nanotechnologies reinforce the continuing miniaturization that leads to the potential unobtrusiveness, embeddedness, and ubiquity of microtechnologies and nanotechnologies, there are also new electrical, optical, magnetic, and mechanical properties derived from surface-to-volume ratios, quantum mechanics, and other rules that apply to small sizes, numbers, or aggregates of particles. This uniqueness means, for example, that some nanoparticles are able to permeate boundaries previously seen as impervious, e.g., the blood-brain barrier. Much of the publicity accorded to nanotechnology has thus been due to a lively discourse on risk assessment that has focused on the toxicological profiles of a range of engineered nanoparticles (e.g., carbon, silver, titanium dioxide) that may not match that of their larger counterparts.

A number of observers have catalogued societal issues that emerging nanotechnologies may raise. The early treatment by Roco and Bainbridge (2001), for example, includes "implications" of economic, political, educational, medical, environmental, and national security import, as well as potential consequences for privacy and global equity (the "nanodivide") and a sea change in what it means to be human through the possibilities of nano-enabled enhancements. Moore (2002) divides the "implications" of nanotechnology into three categories: social, including environmental, health, economic, and educational; ethical, including academic-industry relations, abuse of technology, social divides, and concepts of life; and legal, including concepts of property, intellectual property, privacy, and regulation.

As Lewenstein (2005b) argues, such lists—while thoughtful and relatively complete—frame nanotechnologies in a determinist fashion as things that have "implications" for society but are not themselves influenced by society. Similarly, Baird and Vogt (2004) reframe most of these issues in terms of "interactions," and they add to their list what they call "hypertechnology"—the too-fast pace of innovation. Grunwald (2005) recapitulates many of these issues as well, arguing however that they are not novel enough to warrant the name "nanoethics," which now appears in the title of a journal launched by Springer in 2006 and in an entry in Macmillan's *Encyclopedia of Science, Technology, and Ethics* (Berne 2006a).

While the novelty of the societal issues surrounding nanotechnologies may not be as obvious as the novelty of some nanoscale properties, nanotechnologies clearly have inspired a great deal of attention. The next section picks up on the theme of novelty regarding the role of STS in the development of nanotechnologies, as national governments have summoned social scientists to participate in their initiatives.

III. THE POLICY MANDATE

Since the late 1990s, public and private sector decision makers have promoted NSE as a linchpin for creating economic wealth and solving a vast number of societal problems. Correspondingly, governments around the world have invested heavily in NSE, attempting to create internationally competitive national infrastructures of NSE R&D by tying together the "triple helix of industry, government and academia" (Etzkowitz & Leydesdorff, 2000).

The emphasis on economic advantage and the transformative capacities of nanotechnologies helped catalyze the rapid growth of NSE R&D and commercialization programs, but it also took shape against cautionary discursive backgrounds developed by such prominent individuals as Bill Joy (2000) and Charles, the Prince of Wales (2004), as well as activist groups such as Greenpeace (Arnall, 2003) and the ETC Group (2003). Months after the inauguration of the NNI, Joy presented a catastrophic vision of self-replicating "nanobots" and considered "relinquishment" as a strategy for avoiding this disastrous fate (Joy, 2001). Less spectacular than Joy's "grey goo" scenario, biotechnology also began to be associated with nanotechnology, in particular the widespread experience of skepticism, criticism, and antagonism in the fields of agricultural and food biotechnology and embryonic stem cell research. The ETC Group (formerly, Rural Advancement Foundation International, or RAFI), which forged coalitions between activists in the global North and South to work against agricultural biotechnology and related intellectual property rights, has repeatedly called for a moratorium on particular forms of NSE R&D because of environmental health and safety concerns.

Sensitive to these activist responses, policy makers appear to have been infected with "nanophobia-phobia" (Rip, 2006) from dystopian doomsday scenarios (Bennett & Sarewitz, 2006) and genetically modified foods in Europe (NRC, 2002). They have responded by sponsoring a more proactive approach to societal issues that emphasizes not only the study of ethical, legal, and social issues but the integration of social science research and public interventions into the R&D process (Fisher & Mahajan, 2006a). Distinct from policies promoting biotechnology research, nanotechnology policy does not approach R&D as if it would automatically produce the most desirable outcomes. Instead, policy makers now endorse a conception of R&D that requires the integration of broader societal considerations in order to serve the public good and support decision making.

Under the language of "responsible innovation," government institutions in the United States and European Union, among others, have thus proposed integrating social science research into NSE programs at an early stage (Commission of the European Communities, 2004; NSTC, 2004). In an effort to advance socially desirable outcomes for NSE, policies have prescribed broader guidelines for integrating societal concerns and perspectives, thus inviting STS research to play a formative role in the sociotechnical context of developing nanotechnologies.

The move is particularly compelling in the case of the United States because it occurs in a political context that, since the closing of the congressional Office of Technology Assessment, has paid little attention to technology assessment. Several European nations and EU institutions have also become much more receptive to public engagement in the aftermath of large-scale technoscience controversies, including HIV-tainted blood, "mad cow" disease, and GMOs. Before the U.S. Congress passed the Twenty-First Century Nanotechnology Research and Development Act in 2003 (Public Law 108–153), STS scholars Langdon Winner and Davis Baird testified to Congress about the integration of STS research with NSE. Winner (2003) recommended "open deliberations about technological choices" that would occur at early, premarket stages, yet disparaged the idea of creating a field of "nanoethics" based on the model of bioethics. In order to avoid a "drift toward moral and political triviality" on the part of social and ethical researchers he suggested engaging broader publics, from ordinary citizens to laboratory researchers. Likewise, Baird (2003) proposed instituting the collaboration of ethics researchers with nanotechnology researchers in the laboratory. The criticality of early intervention drew from decades of research into the generation and shaping of technologies (e.g., Collingridge, 1980; Dierkes & Hoffmann, 1992; Sørensen & Williams, 2002).

The resulting legislation went "beyond assessment" (Fisher, 2005) and differed from earlier efforts at institutionalizing reflexivity, such as the Human Genome Project's Ethical, Legal, and Social Implications (ELSI) program. Significantly, the legislation—and other policies like it around the world—explicitly invoked the notion of "integrating" societal research and public inputs into NSE R&D and policy. It also implied that such efforts should influence NSE (House Committee on Science, 2003), presenting practical challenges for STS researchers and creating a new, more active role for social science.

Other nations and political entities have supported similar attempts at fostering collaborations among scientists and engineers, social scientists, and the interested public. The European Union (Commission of the European Communities, 2004), the Netherlands (De Witte & Schuddeboom, 2006), the regional government of Flanders, Belgium (Flemish Institute for Science and Technology, 2006), and Brazil and Colombia (Foladori, 2006) have all not only instituted social science research on nanotechnologies, but notably link that research in an integrated fashion to decision making.

The envisioned collaborations across academic cultures suggest pressure to contribute to the social shaping of nanotechnologies in two respects: (1) Social scientists are expected to provide NSE researchers with contextual awareness of the interdependencies among science, technology, and society, thus allowing broader social perspectives to have greater influence on the design and conduct of R&D and its outcomes. (2) Social scientists are expected to learn details of nanotechnologies and the conditions of their emergence, thus allowing them to better elaborate assessments of societal impacts and interact with publics accordingly. The rationales underlying these two motivations—the quality of nanotechnological development and the enrollment of social scientists—point in different directions, suggesting tensions between the diverging expectations. New collaborations between natural and social scientists will thus be an increasingly important activity and site of inquiry.

IV. FORESIGHT, ENGAGEMENT, AND INTEGRATION

Whether summoned and enabled by the policy initiatives described above, local public groups, or individual research laboratories, STS researchers are "being invited in" (Rip, 2006) to engage with NSE in multiple modes and a variety of settings. Together, such endeavors face at least three general challenges: the anticipation and assessment of nanotechnologies that are in the process of emerging; the engagement of publics that are mostly still latent; and the integration of broader considerations into R&D contexts that have been largely self-governing. This section surveys some of the STS research inspired by such considerations, while pointing to some of the challenges—both analytic and practical—to STS and its researchers.

Foresight

Although by one count there were in early 2007 more than 350 NSE products in commerce (WWIC, 2007), these products alone or in collection offer nothing like the societal transformation promised for nanotechnologies. The emergent quality of nanotechnologies means that many discussions are about potential—often bordering on hype (Berube, 2006)—and therefore many social science interventions are analytically attuned to the future.

The future is diversely manifest as scenarios of use, broader comprehensive visions, sociotechnical scenarios, metaphorical-symbolic expectations, and expectations of technoeconomic potentials (Borup & Konrad, 2004). Prominent expectations about nanotechnologies run in two directions: toward an elixir for postindustrial ills through seamless interactions with nature, instantaneous and nonpolluting production, and unprecedented wealth and health (Drexler, 1986; Anton, Silberglitt, & Schneider, 2001; Wood, Jones, & Geldart, 2003) and toward an Armageddon wrought by self-replicating nanobots (Joy, 2000) or, more soberly, environmental hazards, unintended consequences (Tenner, 2001), shifts in privacy and security (MacDonald, 2004), and greater economic inequalities (Meridian Institute, 2005). The act of attaching oneself to the short or the long term, to the mundane or the exotic visions, is often an act of affiliation with "serious" science or with science fiction (Selin, 2007). As elixir or armageddon, the futures of nanotechnologies have become a focus of the popular press, government programs, and industry analyses.

STS investigations in foresight, each with a different theoretical and empirical approach, have focused sociological interest on expectations (Selin, 2007; Van Lente, 1993; Brown & Michael, 2003), visions (Grunwald, 2004), or "guiding visions" (Meyer & Kuusi, 2004), future imaginaries (Fujimura, 2003), and emerging irreversibilities (van Merkerk & Rip, 2005). Expectations research often employs actor-network theory (ANT), while Rip's scenario work draws from co-evolutionary theory (Rip, 2005). Lösch's (2006) investigations into nanotechnology's futuristic visions argue for discourse theory (e.g., Luhmann, 1995) to crystallize the distributed nature of "the future" as a means of communication. There are also investigations drawing on literary theory and the role of science fiction in the development of nanotechnologies (Milburn, 2004) and the moral vision of its practitioners (Berne, 2006b). Each of these perspectives provides its own prescription for what to do analytically with the future (e.g., trace agency, identify communicative pathways, employ a cultural critique).

There are several distinct approaches to anticipating the longer-term implications of nanotechnologies: forecasting, public deliberation, scenario development, foresight, and vision assessment. Forecasting can be set apart from these other approaches in its orientation toward accurate predictions and allegiance to technological determinism.

However, the methods of forecasting and predictive modeling figure prominently in roadmapping exercises and also address powerful industrial and governmental actors' need for limiting uncertainty (Bunger, forthcoming). The other approaches share a more pluralistic epistemology that suggests multiple futures and intrinsic uncertainty, due at least to the heterogeneous production of technology and society.

Public deliberation exercises often treat the future as a linguistic effect, that is, talk about the future. In 2005, the EU launched a 6th Framework project called Nanologue (2007) in order to "establish a common understanding . . . and to facilitate a Europe-wide dialogue among science, business and civil society about its benefits and potential impacts." After a mapping and polling exercise, the study created, also through participatory methods, three scenarios which then were circulated in order to help structure the debate about responsible innovation. The Center for Nanotechnology in Society at Arizona State University (CNS-ASU) also uses scenarios to help frame debates about the societal implications of new technologies. Different from the Nanologue scenarios, the CNS-ASU scenarios are co-constructed in a large-scale, virtual format through multiple wiki sites. These scenarios serve as inputs for public engagement as well as for social scientific analysis.

While scenarios are often synonymous with foresight, foresight includes such diverse methodologies as life cycle assessment, Delphi studies, cross-impact assessment, future-oriented bibliometrics, and novel ways of performing technology assessment. These sorts of interventions are usually strongly linked with technological innovation and seek to integrate reflection with everyday decision making. Foresight thus aims to enrich futures-in-the-making by encouraging and developing reflexivity in the system.

Building reflexivity in innovation systems highlights a key feature of nanotechnology foresight: the connection with decision making and governance. Sorting through certainties and uncertainties and determining viable options need not be idle speculations, but can be a means toward prudent action. The Danish government, for example, supported a Green Technology Foresight project (Joergensen et al., 2006) in order to support its priority setting. The project was an unparalleled effort to interview and engage a diverse selection of actors working in NSE. The United Kingdom Economic and Social Research Council commissioned the James Martin Institute for Science and Civilization to create scenarios about converging technologies which describe alternative trajectories for the development of nanotechnology and are intended to inform ESRC's research strategy. The Woodrow Wilson International Center also has a foresight and governance project that focuses on the emergence of nanotechnologies by using scenarios, public deliberation, and risk analysis with a particular eye to effecting policy.

These projects are novel in their focus on early intervention, their use of methodologies that have a nuanced relation to futures, and their attempts to allow NSE researchers to characterize the outcomes of their knowledge production. These interventions are thus unique experiments in handling the demands of postnormal science by seeking to build reflexivity through foresight.

Engagement

NSE has only recently become known to wider constituencies as a new interdisciplinary and cross-sectoral field. However, social scientists who have specialized in the
analysis of the Public Understanding of Science and Technology (PUST)—a field that
has developed in the past four decades in the context of contested technologies, starting with nuclear power—have already brought to bear on NSE-related issues the vast
array of research instruments on the public perception and acceptance of S&T. Even
so, this research can only portray publics who have a vague idea of nanotechnology
(Bainbridge, 2002). Thus the finding that the general public is largely in favor of nanotechnology does not necessarily carry much insight, and it is likely to change with
further development of nanotechnology or with social events (Currall et al., 2006).
The same may be true for the correlation between public perception of risks and trust
in regulatory systems (Cobb & Macoubrie, 2004).

As described above, the policy mandates for public involvement in nanotechnology
go beyond opinion polls to more substantive engagement that is consonant with the
shift in some of the literature from public understanding of to public engagement in
S&T (Lewenstein, 2005a). Thus, new roles for social scientists have been created that
extend beyond the supposedly independent and external analysis of public perceptions and understandings to new kinds of engagement with publics.

Over the last two decades, science museums have become more prominent intermediary actors in communicating S&T issues to the public. The Science Museum of
London, for example, has gained an exemplary prominence in combining its traditional role of exhibiting vast collections of items with a new role of sponsoring and
conducting PUST studies, which include experiments with public participation (e.g.,
Durant, 1992; Durant, Bauer, & Gaskell, 1998). With the advent of NSE, science
museums have become part of significant efforts to educate and engage the public.
The U.S. National Science Foundation has committed 20 million dollars over five years
to science museums under the auspices of the Nanoscale Informal Science Education
Network (NISE Net), which brings together museum professionals, researchers, and
informal science educators to inform and engage the public about NSE through traditional museum exhibits and less traditional public forums and Internet venues.

NSE has also been the site of more direct forms of public participation and engagement. Nanojury UK, a consensus conference or citizens' panel held in the United
Kingdom in 2005, demonstrates a commitment to upstream engagement in nanotechnology, where "upstream" means involving the public in detailed activities at a
time when they have very little substantive knowledge of the issues (Rogers-Hayden
& Pidgeon, 2006).

In France, public debates have been organized by NGOs and in some cases sponsored by local officials facing anti-nanotechnology activism. For instance, Entreprises
Pour l'Environnement (Companies for the Environment) sponsored a so-called
"citizen consultation" in October 2006.

In the United States, consensus conferences focusing on nanotechnologies have
been held in university communities in Wisconsin (Powell & Kleinman, forthcoming)

and North Carolina (Hamlett & Cobb, 2006), and the CNS-ASU is conducting an integrated set of six consensus conferences in a National Citizens' Technology Forum. The Center for Nanotechnology in Society at the University of California, Santa Barbara (CNS-UCSB) is conducting participatory exercises, as is the University of South Carolina, and several nano-in-society groups have collaborated with NISE Net in hosting public forums. Despite the mandate in U.S. nanotechnology law for public engagement, social science reflection on approaches to and experiences with public engagement is more advanced in Europe (Joss & Durant, 1995; Abels & Bora, 2004), where such activities have been part of the toolkit of parliamentary technology assessment and have been continually pioneered, particularly in the context of biotechnology (e.g., the large-scale GM Nation exercise in the UK [Steering Board, 2003]).

Integration

The anticipatory and engagement exercises described above are meant to be taken up into ongoing sociotechnical processes to shape their eventual outcomes. While numerous sites of science and technology governance allow for "sociotechnical integration" to be observed, facilitated, or affected (Fisher, Mahajan, & Mitcham, 2006), there has been gathering interest in "revisiting" (Doubleday, forthcoming) one of the classic sites of STS scholarship—the laboratory. Here, at the myth-laden headwaters of scientific knowledge, traditional laboratory studies mingle with more interactive approaches and collaborations, as the considerable but often unacknowledged role of laboratory researchers in implementing and influencing research policies has been cast as an intricate part of the networks of agency that shape NSE, its technological trajectories, and sociotechnical outcomes (Macnaughten, Kearnes, & Wynne, 2005).

As noted, the call for social and natural scientists to work "together in dialog" (Baird, 2003) is unique neither to STS nor to nanotechnologies. More novel is the provision of resources by governments to the task—and the opportunities that have in several cases emerged only as a result of invitations extended by laboratory directors to social scientists and humanists (e.g., Giles, 2003). In accordance with emerging opportunities, several research, education, and engagement programs have sought to encourage "prospective and current nanotechnology researchers to engage—in a thoughtful and critical manner—with [societal] issues as an integral part of their research endeavors" (Sweeney, 2006: 442). The nature of these programs has varied, and some of them overlap with programs of public engagement, foresight, and imagination and of identifying and analyzing ethical and societal issues. What stands out as characterizing many of these efforts is the interest in increasing the reflexivity of the actors and social processes that comprise the objects of study.

Alongside the ethnographic studies of NSE laboratories that have begun to emerge (Glimell, 2003; Kearnes, Macnaghten, & Wilsdon, 2006), several university-based integration-oriented laboratory research projects have also been undertaken (NSTC, 2004). By and large, such "new ethnographies" (Guston & Weil, 2006) seek to "develop the capacity of nanoscientists to reflect on the wider societal dimensions of their work"

(Doubleday, 2005). An implicit and in some cases explicit focus on changes in laboratory practices resulting from the presence and interactions with social researchers can be seen in these projects. One study documents concrete changes in NSE research practices as a result of an iterative protocol for the "modulation" of research decisions (Fisher & Mahajan, 2006b). Another describes the construction of a "trading zone" at the outset of NSE research that informed the eventual project selection (Gorman, Groves, & Catalano, 2004). Attempts to integrate social and humanistic considerations into laboratory and other technoscientific decision processes thus push empirical science studies in new directions. The act of emphasizing the reflexive elements of participant-observation in laboratory studies is a move toward "ethnographic intervention": the integration of social research into technoscientific research by means of collaboratively developed feedback mechanisms that stimulate a more self-critical approach to knowledge generation (Fisher, forthcoming).

Integration projects also include private sector partnerships with nongovernmental organizations (Demos, 2007; Krupp & Holliday, 2005). Together, laboratory integration projects exhibit three, somewhat overlapping trends: efforts to address environmental health and safety considerations (Krupp & Holliday, 2005); efforts aimed at long-term reflective capacity building, such as creating "citizen scientists" (Kearnes, Macnaughten, & Wilsdon, 2006) or occasioning ethical reflection (Berne, 2006b); and efforts that are able to shape the course of R&D work with respect to broader societal considerations (Fisher & Mahajan, 2006b; Gorman, Groves, & Catalano, 2004). The latter trend simultaneously suggests new capacities on the part of STS researchers to influence sociotechnical processes, and challenges to understand the limits of such budding capacity.

V. AN EMERGING PROGRAM

In light of the policy mandates discussed in section III, the STS research and engagement activities described in section IV can be conceived in terms of an emerging yet coherent program that represents a potentially significant development for STS. Such a program is developed at the interface of and in close interaction with key social processes that underlie research conduct, policy making, public education, and the collective anticipation of nanotechnologies. In this way, such a program suggests an evolution in the capacity of STS researchers and institutions to act across a broad front of networks and systems. The fact that this development has largely coincided with the rise of nanotechnology as a cultural and political construct raises opportunities and challenges, as well as ironies, for the STS community. In this section, we describe characteristics that are visible within many smaller- and larger-scale STS research and engagement activities. We then characterize the emerging program as one of building capacity for anticipatory governance. Finally, we consider several questions, motivations, and criticisms that an STS program of this sort will be likely to face in the future as it co-evolves with other new, emerging, and converging technologies.

"Ensemble-ization"

To characterize these developments in STS occasioned by its engagement with nano-technologies, we employ the concept of a "research ensemble," a term that Hackett and his colleagues (2004) use in the context of large-scale fusion research. According to Hackett et al. (2004: 748), a research ensemble denotes an arrangement of "materials, methods, instruments, established practices [. . .] ideas, and enabling theories;" such ties are co-produced by researchers and policy makers to connect a research group to others both within its own field and beyond, and "influences the group's performance and the work of its members." An ensemble so defined helps stage the work that can be accomplished "through interactions with other groups and with policy makers" (ibid.). We choose this concept—as opposed to others, such as systems, networks, boundary organizations, configurations, and the like—because of its concrete focus on the interactions between the work of research groups and the wider social and policy processes that can influence this work.

While we cannot completely theorize the process of "ensemble-ization" here, we maintain that STS engagement with nanotechnologies reveals something of a trend toward it in two central respects: the first has to do with relations among the components of the STS research ensemble, and the second concerns the relation of the ensemble to its objects of study. In the first case, the plurality of methodologies and actors in various large-scale STS entities represent research ensembles at a scale of coordination, collaboration, and focus hitherto not found in STS. The pragmatic mobilization of multiple research technologies—foresight, engagement, and integration—around the single problem of the societal aspects of nanotechnologies creates a tightly arranged, resource-endowed entity that requires coordination, application, and management. In this first sense of ensemble-ization, several large-scale STS entities focused on nanotechnology have begun to surface since the year 2003. Each includes activities focused on anticipation and foresight, public engagement, and sociotechnical integration. This multi-method, mission-driven, action-oriented research characterizes a potentially new form of STS research.

Principal examples include the U.S. Centers for Nanotechnology in Society at UCSB and ASU and the NanoSoc program in Flanders, Belgium. Each is closely related to formal government science policy, and each includes a coordinated set of anticipatory, engagement, and integration activities. Others, such as the Dutch NanoNed research consortium, are part and parcel of government-funded science programs, even if not stemming directly from parliamentary decree. Still others, such as the network of STS scholars and activists in the United Kingdom that centers largely around Lancaster University and the nongovernmental organization Demos, situate their work in the context of statements by policy makers. This group has developed the notion of "upstream public engagement," used alternative future scenarios with publics, and studied future imaginaries in laboratory settings.

Such research ensembles not only represent the larger-scale coupling and coordination of STS researchers and methodologies, but they also embody an increased ability to act. They are evolving with respect to their origin and goals, as is particularly evident with respect to the development of several entities out of or in parallel with policy mandates. Moreover, as shown earlier, both larger and smaller projects seek to facilitate and even participate in the framing and co-construction of dialogues, agendas, expectations, and—notably—decisions pertaining to nanotechnological development trajectories. Thus, STS engagement with nanotechnologies exhibits a second trend toward ensemble-ization insofar as STS research ensembles seek to interact with some of the existing ensembles of science, technology, and policy making that have hitherto been isolated from broader societal influences.

For instance, the upstream engagement activities in the United Kingdom that are focused on the nanoscale are intended to "shape the trajectory of technological development" (Wilsdon, 2005). Similarly, the NanoNed consortium includes a component of constructive technology assessment which, like upstream engagement, has long sought to introduce a more extensive and nuanced array of participants in order to "influence design and technical change" (Schot, 2005). The "real-time technology assessment" at the core of the CNS-ASU ensemble is a coordinated collection of approaches meant "to build into the R&D enterprise itself a reflexive capacity that [. . .] allows modulation of innovation paths and outcomes in response to ongoing analysis and discourse" (Guston & Sarewitz, 2002).

Thus, in the facilitation of interactions, whether among various publics or between STS researchers and various publics, these STS ensembles are aligned with the notion of constructing and shaping decision processes, research practices, levels of public trust, and the transparency of policy processes. Research ensembles help specify linkages among research groups that in turn affect the performance and work of such groups, thus embodying forms of mediation between science and society. As such, they not only can map the "connection between policy and knowledge production" (Hackett et al., 2004: 751), but their alteration and expansion—through STS interventions—can thus shape these very connections.

Anticipatory Governance

As we have suggested, the futuristic discourse of nanotechnologies, as well as their fundamental technical and social uncertainties, requires the cultivation of a societal capacity for foresight, by which we mean not only formal methodologies but also more generalized abilities to bridge the cognitive gap between present and future. Whether through foresight, public engagement exercises, or ethnographic intervention, visions and their assessment have played a prominent role in both representations of and STS research on nanotechnologies. The forward-looking, engagement-oriented, and results-seeking characteristics of this STS research distinguish it from prior work in PUST, ELSI, and observational laboratory studies. The growing capacity to act that the ensemble-ization of STS, both in relation to itself and to its objects of study, builds what we elaborate here as "anticipatory governance" (Guston & Sarewitz, 2002).

Anticipatory governance implies that effective action is based on more than sound analytical capacities and relevant empirical knowledge: It also emerges out of a distributed collection of social and epistemological capacities, including collective self-criticism, imagination, and the disposition to learn from trial and error. For, although action and outcomes are emergent qualities of human choice and behavior, they rarely, if ever, proceed from certainty or prediction, and neither are they based on the simple intentions of individual actors or policies. Rather, as the concept of "anticipation" is meant to indicate, the co-evolution of science and society is distinct from the notion of predictive certainty. In addition, the anticipatory approach is distinct from the more reactionary and retrospective activities that follow the production of knowledge-based innovations—rather than emerge with them. Anticipation implies an awareness of the co-production of sociotechnical knowledge and the importance of richly imagining sociotechnical alternatives that might inspire its use.

In parallel, the notion of "governance" commonly refers to a move away from a top-down government approach to an approach where management by people and institutions becomes possible without detailed and compartmentalized regulation from the top (Lyall & Tait, 2005: 3). The activities implicated by the concept of governance are diverse, ranging from the technological determinism latent in the idea of nanotechnology as the "next industrial revolution" (NSTC & IWGN, 2000) to the radical expression of technological choice in calls for a moratorium. But between adapting to a coming revolution and halting development exists an array of governing options: licensing, civil liability, insurance, indemnification, testing, regulation, restrictions on age or other criteria (rather than on ability to pay), labeling, modulation of designs and research practices, and so on. Some options, like labeling and life cycle analysis, complement private sector governance by providing more complete information necessary for market efficiency. Some, like civil liability and indemnification, distort markets for important reasons of justice or critical technology development. Anticipatory governance seeks to lay the intellectual foundation for (any of) these approaches early enough for them to be effective.

Beyond the role of STS ensembles described above, we can cite two additional but still nascent examples of anticipatory governance: On the macro level, "acceptance politics" (Barben, 2006) denotes the political strategies and practices concerned with influencing the public acceptance of controversial phenomena like nanotechnologies and thus the choice of governance mechanisms. For example, many actors involved in NSE perceive biotechnology and particularly genetically modified organisms as the strategic background against which to shape public acceptance or rejection (e.g., Mehta, 2004; David & Thompson, forthcoming). On the micro level, "midstream modulation" (Fisher, Mahajan, & Mitcham, 2006) refers to the demonstrated phenomenon of a nanoscale engineering research group to adjust its own practices according to broader "upstream" and "downstream" societal contexts, principally as a result of observing decision processes and imagining additional technical alternatives.

Anticipatory governance comprises the ability of a variety of lay and expert stakeholders, both individually and through an array of feedback mechanisms, to collectively imagine, critique, and thereby shape the issues presented by emerging technologies before they become reified in particular ways. Anticipatory governance evokes a distributed capacity for learning and interaction stimulated into present action by reflection on imagined present and future sociotechnical outcomes. STS researchers, projects, and subfields are being tethered together and linked to the contexts they seek to study with the aim of incrementally building the capacity to more broadly anticipate and participate in shaping things to come.

Opportunities, Challenges, and Ironies

Insofar as the policy mandates implicitly rely on STS tenets and expertise, they present a clear opportunity for the STS community to reconceive if not reinvent forms of foresight, engagement, and integration (Macnaughten, Kearnes, & Wynne, 2005). At the same time, the opportunity challenges the community, raising questions of its growing capability to participate more directly and intentionally in shaping sociotechnical change—as well as dilemmas about how far to go in seeking to influence change and pitfalls of ill-conceived approaches to anticipation (Williams, 2006).

Each arena we have examined—foresight, engagement, and integration—sets particular obstacles for researchers. In following the future-oriented discourse of NSE, for example, there is a risk of avoiding or downplaying the present by centering debate in the future. That is, many of the societal issues posed by nanotechnologies, including questions of equality, privacy, and human enhancement, can be meaningfully framed in the present as well. The choice of concentrating on future scenarios rather than on current practices bears a similar ethical burden as the choice to invest resources on "transformative" research rather than to address current ills. Moreover, talk about the future, whether connected to analytical projects, participatory experiments, or scenario-building collaborations with NSE researchers demands that STS researchers be involved explicitly in the construction of possible futures. Because anticipation is performative, there is no sidestepping this responsibility (as opposed to, say, Gieryn's [1995] prescription regarding boundary work that good constructivists watch it rather than do it).

With regard to engagement exercises, the concept of acceptance politics raises the specter of the cooptation of STS research for the purpose of legitimating nanotechnologies and pacifying publics. In conducting participatory investigations into the future of NSE, STS researchers must create constructive links with relevant stakeholders, thus raising the question that the researchers must answer: Who are the carriers of legitimate or authorized visions of nanotechnological futures?

Finally, integration demands a sophisticated balancing of scholarly objectives, the politics of the laboratory, and the prospects for progressive alteration of the research agenda and its anticipated outcomes. The responsibilities of the participant-observer, whether "lab-based sociologist" or "embedded humanist," are likely to be different when the context of the research is a laboratory setting within a larger shared community, university, political system, and culture, as opposed to a geographically and culturally distant setting. Further, episodes that are not necessarily part of the envisioned "sociotechnical integration"—for example, the mistreatment of animal or human research subjects, research misconduct, intellectual property disputes, and the like—may surprise the participant-observer, creating conflicts of commitment even within a framework that incorporates a concept of the public good. In seeking to influence policy and decision making, even in as innocuous a setting as the laboratory, STS researchers subject themselves to being influenced in a heightened way.

How can STS scholarship respond to the rather generous invitations from policy makers to partake in creating the future of nanotechnologies while both retaining its critical perspective and avoiding falling prey to Winner's critique of academic distance? To what end does STS participate in the normatively charged contexts it seeks to describe, and at what cost to its academic integrity and credibility?

Such questions have provoked periodic self-critical reflections (e.g., Fuller, 2005) and injunctions (Jasanoff, 1999). They are reminiscent of the concern, voiced in Winner's congressional testimony, that previous ELSI research may have been co-opted by its patrons. Importantly, the question of acting to what end presents both normative and pragmatic challenges. The challenges of understanding what "socially desirable" goals are and assessing whether present arrangements are likely to produce desired results surface long-standing debates in the STS community about the role of researchers in influencing their objects of study. These concerns have also been expressed in language of the "entanglement" of social and humanist researchers with nanotechnology programs. Having been invited to consider nanotechnology, they lend weight and credibility to an otherwise "malleable, plastic, and elusive" notion that embodies a particular set of agendas: "if [nanotechnology] has social impact, it must be real" (Nordmann, 2006).

Ironically, as STS becomes better endowed with resources, more highly coordinated, and more entangled within innovation systems, it becomes more like its objects of study. In developing ensembles with the ability to anticipate, engage, and integrate, STS researchers become more visible and significant participants in their own right, and—perhaps for the first time—instruments of governance themselves.

Note

Authors listed alphabetically except the final author. This material is based on work supported by the National Science Foundation under cooperative agreement #0531194. Any opinions, findings, and conclusions or recommendations expressed in this material are those of the authors and do not necessarily reflect the views of the National Science Foundation.

References

Abels, G. & A. Bora (2004) *Demokratische Technikbewertung* (Bielefeld: transcript Verlag).

Anton, P. S., R. Silberglitt, & J. Schneider (2001) *The Global Technology Revolution: Bio/Nano/Materials Trends and Their Synergies with Information Technologies by 2015* (Santa Monica, CA: Rand).

Arnall, Alexander Huw (2003) *Future Technologies, Today's Choices: Nanotechnology, Artificial Intelligence and Robotics: A Technical, Political, and Institutional Map of Emerging Technologies* (London: Greenpeace Environmental Trust).

Bainbridge, William Sims (2002) "Public Attitudes Toward Nanotechnology," *Journal of Nanoparticle Research* 4: 561–70.

Baird, D. (2003) Testimony to the Senate Committee on Commerce, Science, and Transportation, May 1. Available at: commerce.senate.gov/hearings/testimony.cfm?id=745&wit_id=2012.

Baird, Davis & Tom Vogt (2004) "Societal and Ethical Interactions with Nanotechnology (SEIN): An Introduction," *Nanotechnology Law and Business Journal* 1(4): 391–96.

Barben, D. (2006) "Visions of Nanotechnology in a Divided World: The Acceptance Politics of a Future Key Technology." Paper presented at the Conference of the European Association for the Study of Science and Technology (EASST), Université de Lausanne: Panel "Nanosciences and Nanotechnologies: Visions Shaping a New World: Implications for Technology Assessment, Communication, and Regulation," August 23–26.

Bennett, I. & D. Sarewitz (2006) "Too Little, Too Late? Research Policies on the Societal Implications of Nanotechnology in the United States," *Science as Culture* 15(4): 309–26.

Berne, R. (2006a) "Nano-Ethics," in C. Mitcham (ed), *Encyclopedia of Science, Technology, and Ethics* (New York: MacMillan Reference USA): 1259–62.

Berne, R. (2006b) *Nanotalk: Conversations with Scientists and Engineers About Ethics, Meaning, and Belief in the Development of Nanotechnology* (Mahwah, NJ: Lawrence Erlbaum).

Berube, D. (2006) *Nano-Hype: The Truth Behind the Nanotechnology Buzz* (Amherst, NY: Prometheus Books).

Borup, M. & K. Konrad (2004) "Expectations in Nanotechnology and in Energy: Foresight in the Sea of Expectations." Background Paper, Research Workshop on Expectations in Science and Technology, Risoe National Laboratory, Roskilde, Denmark, April 29–30.

Brown, N. & M. Michael (2003) "A Sociology of Expectations: Retrospecting Prospects and Prospecting Retrospects," *Technology Analysis and Strategic Management* 15(1): 3–18.

Bunger, M. (forthcoming) "Forecasting Science-Based Innovations," in J. Wetmore, E. Fisher, & C. Selin (eds), *Yearbook of Nanotechnology in Society. Volume 1:* Cobb, M. D. & J. Macoabrie (2004) "Public Perceptions about Nanotechnology: Risks, Benefits, and Trust," *Journal of Nanoparticle Research* 6: 395–405.

Collingridge, D. (1980) *The Social Control of Technology* (London: Frances Pinter).

Commission of the European Communities (2004) *Towards a European Strategy for Nanotechnology* (No. COM [2004]) (Brussels: Commission of the European Communities).

Crow, Michael M. & Daniel Sarewitz (2001) "Nanotechnology and Societal Transformation," in A. H. Teich, S. D. Nelson, C. McEnaney, & S. J. Lita (eds), *AAAS Science and Technology Policy Yearbook* (New York: American Association for the Advancement of Science).

Currall, S. C., E. B. King, N. King, N. Lane, J. Madera, & S. Turner (2006) "What Drives Public Acceptance of Nanotechnology?" *Nature Nanotechnology* 1 (December): 153–55.

David, Kenneth & Paul B. Thompson (eds) (forthcoming) *What Can Nano Learn from Bio? Lessons for Nanoscience from the Debate over Agricultural Biotechnology and GMOs* (New York: Academic Press).

De Witte, Pieter & Paul Schuddeboom (2006) *NanoNed Annual Report 2005* (Utrecht, the Netherlands: NanoNed). Available at: www.nanoned.nl.

Demos (2007) "The Nanodialogues." Available at: http://83.223.102.49/projects/thenanodialogues/overview.

Dierkes, M. & U. Hoffmann (eds) (1992) *New Technology at the Outset: Social Forces in the Shaping of Technological Innovations* (Frankfurt/New York: Campus/Westview).

Doubleday, R. (2005) "Opening up the Research Agenda: Views from a Lab-Based Sociologist," Paper presented at *Research Training in Nanosciences and Nanotechnologies: Current Status and Future Needs,* European Commission Workshop, Brussels, Belgium, April 14–15.

Doubleday, R. (forthcoming) "The Laboratory Revisited: Academic Science and the Responsible Development of Nanotechnology," in T. Rogers-Hayden (ed), *Engaging with Nanotechnologies: Engaging Differently?*

Drexler, E. K. (1986) *Engines of Creation: The Coming Era of Nanotechnology* (New York: Doubleday).

Drexler, E. K. (2004) "Nanotechnology: From Feynman to Funding," *Bulletin of Science, Technology & Society* 24(1): 21–27.

Durant, John (1992) *Biotechnology in Public: A Review of Recent Research* (London: Science Museum for the European Federation of Biotechnology).

Durant, John, Martin Bauer, & George Gaskell (1998) *Biotechnology in the Public Sphere: A European Sourcebook* (London: Science Museum).

ETC Group (2003) *The Big Down: From Genomes to Atoms* (Winnipeg: Action Group on Erosion, Technology, and Concentration).

Etzkowitz, Henry & Loet Leydesdorff (2000) "The Dynamics of Innovation: From National Systems and 'Mode 2' to a Triple Helix of University–Industry–Government Relations," *Research Policy* 29(2): 109–23.

Fisher, E. (2005) "Lessons Learned from the ELSI Program: Planning Societal Implications Research for the National Nanotechnology Program," *Technology in Society* 27: 321–28.

Fisher, E. (forthcoming) "Ethnographic Intervention: Exploring the Negotiability of Laboratory Research," *Nanoethics*.

Fisher, E. & R. L. Mahajan (2006a) "Contradictory Intent? U.S. Federal Legislation on Integrating Societal Concerns into Nanotechnology Research and Development," *Science and Public Policy* 33(1): 5–16.

Fisher, E. & R. L. Mahajan (2006b) "Midstream Modulation of Nanotechnology Research in an Academic Laboratory." Paper presented at the American Society of Mechanical Engineers International Mechanical Engineering Congress and Exposition, Chicago, November 5–10.

Fisher, E., R. Mahajan, & C. Mitcham (2006) "Midstream Modulation: Governance from Within," *Bulletin of Science, Technology & Society* (26)6: 485–96.

Flemish Institute for Science and Technology (FIST) (2006) *NanoSoc: Nanotechnologies for Tomorrow's Society*. Project Description.

Foladori, G. (2006) "Nanotechnology in Latin America at the Crossroads," *Nanotechnology Law & Business* 3(2): 205–16.

Fujimura, J. (2003) "Future Imaginaries: Genome Scientists as Socio-Cultural Entrepreneurs," in A. Goodman, D. Heath, & S. Lindee (eds), *Genetic Nature/Culture: Anthropology and Science Beyond the Two-Culture Divide* (Berkeley: University of California Press): 176–99.

Fuller, S. (2005) "Is STS Truly Revolutionary or Merely Revolting?" *Science Studies* 18(1): 75–83.

Funtowicz, Silvio O. & Jerome R. Ravetz (1993) "The Emergence of Post-Normal Science," in René von Schomberg (ed), *Science, Politics and Morality: Scientific Uncertainty and Decision Making* (Dordrecht, Boston, London: Kluwer): 85–123.

Gibbons, Michael, Camille Limoges, Helga Nowotny, Simon Schwartzman, Peter Scott, & Martin Trow (1994) *The New Production of Knowledge: The Dynamics of Science and Research in Contemporary Societies* (London: Sage).

Gieryn, T. F. (1995) "Boundaries of Science," in S. Jasanoff, G. E. Markle, J. C. Petersen, & T. Pinch (eds), *Handbook of Science and Technology Studies* (Thousand Oaks, CA: Sage): 393–443.

Giles, J. (2003) "What Is There to Fear from Something so Small?" *Nature* 426 (December 18/25): 750.

Glimell, H. (2003) "Challenging Limits: Excerpts from an Emerging Ethnography of Nano Physicists," in Hans Fogelberg & Hans Glimell (eds), *Bringing Visibility to the Invisible: Towards a Social Understanding of Nanotechnology*, STS Research Reports No. 6 (Goteborg, Sweden: Goteborg Universitet).

Gorman, M. E., J. F. Groves, & R. K. Catalano (2004) "Societal Dimensions of Nanotechnology," *IEEE Technology and Society* 23(4): 55–62.

Grunwald, A. (2004) "Vision Assessment as a New Element of the FTA Toolbox," in F. Scapolo & E. Cahill (eds), *New Horizons and Challenges for Future-Oriented Technology Analysis*. Proceedings of the EU-US Scientific Seminar: New Technology Foresight, Forecasting, and Assessment Methods, Seville, Spain: 53–67.

Grunwald, Armin (2005) "Nanotechnology: A New Field of Ethical Inquiry?" *Science and Engineering Ethics* 11(2): 187–201.

Guston, D. (2000) *Between Politics and Science* (New York: Cambridge University Press).

Guston, D. H. & D. Sarewitz (2002) "Real-Time Technology Assessment," *Technology in Society* 24(1–2): 93–109.

Guston, D. & V. Weil (2006) "New Ethnographies of Nanotechnology," *Society for the Social Studies of Science*, Vancouver, British Columbia, Canada, November 1–5.

Hackett, E., D. Conz, J. Parker, J. Bashford, & S. DeLay (2004) "Tokamaks and Turbulence: Research Ensembles, Policy, and Technoscientific Work," *Research Policy* 33(5): 747–67.

Hamlett, Patrick W. & Michael D. Cobb (2006) "Potential Solutions to Public Deliberation Problems: Structured Deliberations and Polarization Cascades," *Policy Studies Journal* 34(4): 629–48.

House Committee on Science and Technology (2003) Report 108-089, United States House of Representatives, 108th Congress, 1st Session (Washington, DC: U.S. Government Printing Office): 1–24.

Jasanoff, S. (1999) "STS and Public Policy: Getting Beyond Deconstruction," *Science Technology & Society* 4: 59–72.

Joergensen, M. S. (2006) *Green Technology Foresight about Environmentally Friendly Products and Materials: The Challenges from Nanotechnology, Biotechnology, and ICT* (Copenhagen: Danish Environmental Protection Agency).

Joss, S. & J. Durant (eds) (1995) *Public Participation in Science: The Role of Consensus Conferences in Europe* (London: Science Museum).

Joy, Bill (2000) "Why the Future Doesn't Need Us," *Wired* 8(04) April. www.wired.com/wired/archive/8.04/joy.html/

Kearnes, M., P. Macnaghten, & J. Wilsdon (2006) *Governing at the Nanoscale: People, Policies, and Emerging Technologies* (London: Demos).

Keiper, A. (2003) "The Nanotechnology Revolution," *The New Atlantis: A Journal of Technology and Society* 1(2): 17–34.

Krupp, F. & C. Holliday (2005) "Let's Get Nanotech Right," *The Wall Street Journal*, June 15.

Lewenstein, B. V. (2005a) "Nanotechnology and the Public," *Science Communication* 27(2): 169–74.

Lewenstein, B. V. (2005b) "What Counts as a 'Social and Ethical Issue' in Nanotechnology?" *HYLE: International Journal for the Philosophy of Chemistry* 11(1): 5–18.

Lösch, A. (2006) "Means of Communicating Innovations: A Case Study for the Analysis and Assessment of Nanotechnology's Futuristic Visions," *Science, Technology and Innovation Studies* 2: 103–25.

Luhmann, N. (1995) *Social Systems* (Stanford, CA: Stanford University Press) (German original 1984: Suhrkamp).

Lux Research (2006) *The Nanotech Report: Investment Overview and Market Research for Nanotechnology*, 4th ed. (New York: Lux Research).

Lyall, C. & J. Tait (eds) (2005) *New Modes of Governance: Developing an Integrated Policy Approach to Science, Technology, Risk, and the Environment* (Aldershot, U.K.: Ashgate).

MacDonald, C. (2004) "Nanotechnology, Privacy and Shifting Social Conventions," *Health Law Review* 12(3): 37–40.

Macnaughten, P., M. Kearnes, & B. Wynne (2005) "Nanotechnology, Governance, and Public Deliberation: What Role for the Social Sciences?" *Science Communication* 27(2): 268–91.

McCray, W. P. (2005) "Will Small Be Beautiful? Making Policies for our Nanotech Future," *History and Technology* 21(2): 177–203.

Mehta, M. D. (2004) "From Biotechnology to Nanotechnology: What Can We Learn from Earlier Technologies?" *Bulletin of Science, Technology & Society* 24(1): 34–39.

Meridian Institute (2005) *Nanotechnology and the Poor: Opportunities and Risks*. Available at: http://www.meridian-nano.org/NanoandPoor-NoGraphics.pdf.

Meyer, M. & O. Kuusi (2004) "Nanotechnology: Generalizations in an Interdisciplinary Field of Science and Technology," *HYLE: International Journal for Philosophy and Chemistry* 10(2): 153–68.

Milburn, C. (2004) "Nanotechnology in the Age of Posthuman Engineering: Science Fiction as Science," in N. K. Hayles (ed), *Nanoculture: Implications of the New Technoscience* (Bristol: Intellect Books): 109–29.

Mody, C. (2006) "Corporations, Universities, and Instrumental Communities: Commercializing Probe Microscopy, 1981–1996," *Technology and Culture* 1: 56–80.

Moore, Fiona N. (2002) "Implications of Nanotechnology Applications: Using Genetics as a Lesson," *Health Law Review* 10(3): 9–15.

Nanologue (2007) "Europe-wide Dialogue on the Ethical, Social, and Legal Impacts of Nanotechnology." Available at: www http://www.nanologue.net/.

National Nanotechnology Initiative (2007) "What Is Nanotechnology?" Available at: http://www.nano.gov/html/facts/whatIsNano.html.

National Research Council (NRC), Division of Engineering and Physical Sciences, Committee for the Review of the National Nanotechnology Initiative (2002) *Small Wonders, Endless Frontier: A Review of the National Nanotechnology Initiative* (Washington, DC: National Academies Press).

National Science and Technology Council (NSTC) (2004) "National Nanotechnology Initiative: Strategic Plan" (Washington, DC: National Science and Technology Council, Committee on Technology, Subcommittee on Nanoscale Science, Engineering, and Technology).

National Science and Technology Council (NSTC), Committee on Technology & Interagency Working Group on Nanoscience, Engineering, and Technology (IWGN) (2000) "National Nanotechnology Initiative: Leading to the Next Industrial Revolution," Supplement to President's FY 2001 Budget (Washington, DC: NSTC).

Nordmann, A. (2006) "Entanglement and Disentanglement in the Nanoworld," Presentation at *TA NanoNed Day*, Utrecht, The Netherlands, July 14.

Porter, Alan, Jan Youtie, Philip Shapira (2006) "Refining Search Terms for Nanotechnology." Paper prepared for presentation at the National Science Foundation, Arlington, VA, August 24.

Powell, M. & D. Kleinman (forthcoming) "Building Citizen Capacities for Participation in Technoscientific Decision Making: The Democratic Virtues of the Consensus Conference Model," *Public Understanding of Science*.

Prince of Wales (2004) "Menace in the Minutiae," *The Independent on Sunday,* July 11.

Rip, A. (2005) A Sociology and Political Economy of Scientific-Technological Expectations Presentation for UNICES Seminar, Utrecht, the Netherlands, September 26.

Rip, A. (2006) "Folk Theories of Nanotechnologists," *Science as Culture* 15(4): 349–66.

Roco, Mihail C. (2003) "Broader Societal Issues of Nanotechnology," *Journal of Nanoparticle Research* 5(3–4) (August): 181–89.

Roco, M. C. & W. S. Bainbridge (eds) (2001) *Societal Implications of Nanoscience and Nanotechnology* (Dordrecht, the Netherlands, and Boston: Kluwer).

Roco, M. C. & O. Renn (2006) "Nanotechnology and the Need for Risk Governance," *Journal of Nanoparticle Research* 8(2): 153–91.

Rogers-Hayden, T. & N. Pidgeon (2006) "Reflecting upon the UK's Citizens' Jury on Nanotechnologies: Nano Jury UK," *Nanotechnology Law and Business* 2(3): 167–78.

Royal Society & Royal Academy of Engineering (2004) *Nanoscience and Nanotechnologies: Opportunities and Uncertainties* (London: Royal Society & Royal Academy of Engineering).

Sarewitz, D. & E. Woodhouse (2003) "Small Is Powerful," in A. Lightman, D. Sarewitz, & C. Desser (eds), *Living with the Genie: Essays on Technology and the Quest for Human Mastery* (Washington, DC: Island): 63–83.

Schot, J. (2005) "The Idea of Constructive Technology Assessment," in C. Mitcham (ed), *Encyclopedia of Science, Technology, and Ethics* (Detroit: Macmillan Reference USA).

Selin, C. (2007) "Expectations and the Emergence of Nanotechnology," *Science, Technology and Human Values* 32(2): 196–220.

Sørensen, K. H. & R. Williams (eds) (2002) *Shaping Technology, Guiding Policy: Concepts, Spaces and Tools* (Cheltenham and Northampton, MA: Edward Elgar).

Steering Board (2003) *GM Nation? The Findings of the Public Debate: Final Report* (London: Steering Board of the Public Debate on GM [Genetic Modification] and GM Crops).

Sweeney, A. E. (2006) "Social and Ethical Dimensions of Nanoscale Science and Engineering Research," *Science and Engineering Ethics* 12(3): 435–64.

Tenner, E. (2001) "Unintended Consequences and Nanotechnology," in M. C. Roco & W. S. Bainbridge (eds) *Social Implications of Nanoscience and Nanotechnology* (Arlington, VA: National Science Foundation): 241–45.

Van Lente, H. (1993) *Promising Technology: The Dynamics of Expectations in Technological Developments.* PhD Thesis (Twente, the Netherlands: Universitet Twente).

Van Merkerk, R. O. & H. van Lente (2005) "Tracing Emerging Irreversibilities in Emerging Technologies: The Case of Nanotubes," *Technological Forecasting & Social Change* 72: 1094–1111.

Williams, R. (2006) "Compressed Foresight and Narrative Bias: Pitfalls in Assessing High-Technology Futures," *Science as Culture* 15(4): 327–48.

Wilsdon, J. (2005) "Paddling Upstream: New Currents in European Technology Assessment," in M. Rodemeyer, D. Sarewitz, & J. Wilsdon (eds), *The Future of Technology Assessment* (Washington, DC: Woodrow Wilson International Center for Scholars): 22–29.

Wilsdon, J. & R. Willis (2004) "See-Through Science: Why Public Engagement Needs to Move Upstream" (Demos).

Winner, Langdon (2003) Testimony to the Committee on Science of the U.S. House of Representatives on the Societal Implications of Nanotechnology, April 9. Available at: gop.science.house.gov/hearings/full03/apr09/winner.htm.

Wood, S., R. Jones & A. Geldart (2003) *The Social and Economic Challenges of Nanotechnology* (Swindon: Economic and Social Research Council).

Woodhouse, E. J. (2004) "Nanotechnology Controversies," *Technology and Society Magazine, IEEE* 23(4): 6–8.

Woodrow Wilson International Center (WWIC) for Scholars (2007) "A Consumer Inventory of Nanotechnology Products." Available at: www.nanotechproject.org.

Contributors

EDITORS

Olga Amsterdamska teaches social studies of science and medicine in the Department of Sociology and Anthropology at the University of Amsterdam. Her research focuses on the development of the biomedical sciences, the history of epidemiology, and the interactions between the laboratory, the clinic, and public health in twentieth-century medicine. She is currently investigating the history of research on autism. She has published *Schools of Thought in Linguistics* (1987) and a number of articles on the history of the medical and biological sciences and on some theoretical problems in the social studies of science. She is the former editor of *Science, Technology, and Human Values*. Her e-mail address is O.Amsterdamska@uva.nl.

Edward J. Hackett is a professor in the School of Human Evolution and Social Change at Arizona State University and Director of the Division of Social and Economic Sciences at the National Science Foundation. He studies the social organization and dynamics of scientific research, including patterns of collaboration, scientific worklife and careers, peer review, and the responsible conduct of research. He also writes about environmental justice and other topics in environmental social science. With Diana Rhoten, he is currently studying the formation of interdisciplinary scientists and the process of integrative research. With Daryl E. Chubin, he is the coauthor of *Peerless Science: Peer Review and U.S. Science Policy* (1990). His e-mail address is ehackett@asu.edu.

Michael Lynch is a professor in the Department of Science and Technology Studies at Cornell University. His research is on practical actions and social interactions in laboratories, criminal courts, and other institutional settings. He is the author of *Art and Artifact in Laboratory Science* (1985), *Scientific Practice and Ordinary Action: Ethnomethodology and Social Studies of Science* (1993), and (with Simon Cole, Ruth McNally and Kathleen Jordan) *Truth Machine: The Contentious History of DNA Profiling* (forthcoming, 2008). He is also the editor of *Social Studies of Science* and President of the Society for Social Studies of Science (2007–2009). His e-mail address is MEL27@cornell.edu.

Judy Wajcman is a professor in the Research School of Social Sciences at the Australian National University. She is a research associate at the Oxford Internet Institute and a visiting professor at London Business School. Her current research explores the impact of information and communication technologies such as the mobile phone on time poverty and work-life balance. She is the author of *Feminism Confronts Technology* (1991), *Managing Like a Man: Women and Men in Corporate Management* (1998), *The Social Shaping of Technology*, Second Edition (with Donald MacKenzie, 1999), *TechnoFeminism* (2004), and *The Politics of Working Life* (with Paul Edwards, 2005). Her e-mail address is judy.wajcman@anu.edu.au.

AUTHORS

Vincanne Adams is Professor of Medical Anthropology in the Department of Anthropology, History and Social Medicine at the University of California, San Francisco. She has published extensively on translating science in Asian regions, particularly in the context of development, medicine, and women's health. Her books include *Tigers of the Snow and Other Virtual Sherpas: An Ethnography of Himalayan Encounters* (1980), *Doctors for Democracy: Health Professionals in the Nepal Revolution* (1998), and (with Stacy Leigh Pigg) *Sex and Development: Science, Sexuality, and Morality in Global Perspective* (2005). Her e-mail address is AdamsV@dahsm.ucsf.edu.

Warwick Anderson is Research Professor in the History Department and the Centre for Values, Ethics and the Law in Medicine at the University of Sydney. Until 2007 he was Robert Turell Professor of Medical History and Population Health, Professor of the History of Science, and Chair of the Department of Medical History and Bioethics at the University of Wisconsin-Madison. With Gabrielle Hecht he edited a special issue of *Social Studies of Science* in 2002 on "Postcolonial Technoscience." He is the author of *The Cultivation of Whiteness: Science, Health and Racial Destiny in Australia* (2003), *Colonial Pathologies: American Tropical Medicine, Race, and Hygiene in the Philippines* (2006), and *The Collectors of Lost Souls: Kuru and the Creation of Value in Science* (forthcoming, 2008). He was recently awarded the Frederick Burkhardt Fellowship (2005–2006) and a Guggenheim fellowship (2007–2008). His e-mail address is whanderson@med.wisc.edu.

Brian Balmer is a reader in Science Policy Studies in the Department of Science and Technology Studies, University College London. He studies the nature of scientific expertise, and the role of experts in science policy formation, especially in the history of chemical and biological warfare. He also does research on the role of volunteers in biomedical research, and the history of the "brain drain" debate in the United Kingdom. He is the author of *Britain and Biological Warfare: Expert Advice and Science Policy, 1935–65* (2001). His e-mail address is b.balmer@ucl.ac.uk.

Daniel Barben is an associate research professor with the Consortium for Science, Policy and Outcomes at Arizona State University. His research interests are in comparative social studies of emerging technologies. He is the author of *Theorietechnik und Politik bei Niklas Luhmann. Grenzen einer universalen Theorie der modernen Gesellschaft* (1996) and *Politische Ökonomie der Biotechnologie. Innovation und gesellschaftlicher Wandel im internationalen Vergleich* (2007) and coeditor of *Biotechnologie–Globalisierung–Demokratie: Politische Gestaltung transnationaler Technologieentwicklung* (2000, with Gabriele Abels). His e-mail address is daniel.barben@asu.edu.

Anne Beaulieu is Deputy Programme Leader of the Virtual Knowledge Studio. Her work focuses on the interaction between new technologies and scientific research practices. She is currently investigating the development of digital infrastructure in women's studies as well as the dynamics of online infrastructure in other scientific fields. She has written about methodological issues regarding laboratory studies and online research; the intellectual agenda of current ethnographic research on the Internet; biomedical digital imaging and databasing technologies, including an ethnographic study of brain imaging; and, more recently, ethnographic research on the Internet. Her e-mail address is anne.beaulieu@vks.knaw.nl.

Wiebe E. Bijker is Professor of Technology and Society at the University of Maastricht, The Netherlands. He was trained as an engineer in physics (Technical University of Delft), studied philosophy (University of Groningen), and holds a Ph.D. in the sociology and history of technology (University of Twente). Bijker is past President of 4S and was a member of the executive council of SHOT. He is also founding coeditor of the MIT Press series *Inside Technology*. Bijker's current research interests relate to science and technology for development and to issues of democracy and vulnerability in technological cultures. His e-mail address is W.Bijker@TSS.unimaas.nl.

Pablo J. Boczkowski is an associate professor in the Department of Communication Studies and in the Program in Media, Technology and Society at Northwestern University. His research examines the transformation of the social and political institutions of print culture in the digital age. He is the author of *Digitizing the News: Innovation in Online Newspapers* (2004) and is currently writing a book on the relationships between materiality and mimicry in the new media environment. His e-mail address is pjb9@northwestern.edu.

Steve Breyman currently serves as Executive Director of Citizens' Environmental Coalition while on sabbatical in 2007–2008. He is also Associate Professor of Science and Technology Studies at Rensselaer Polytechnic Institute in Troy, New York, where he directed the Department's Graduate Program and served as Director of Rensselaer's Ecological Economics, Values and Policy Program. His latest book is *Why Movements Matter: U.S. Arms Control Policy and the West German Peace Movement* (2001). He is currently writing about the U.S. anti-incinerator movement, chemical security in the United States, green chemistry, and chemical policy reform. He is active in struggles for peace, environmental integrity, and justice in the United States and abroad. His e-mail address is breyms@rpi.edu.

Massimiano Bucchi is a professor of Sociology of Science at the University of Trento, Italy. His research addresses the interaction among experts, citizens, and policy makers and the role of the public sphere in scientific debates. He is the author of *Science and the Media: Alternative Routes in Science Communication* (1998) and *Science in Society: An Introduction to Social Studies of Science* (2004) and editor of *Science Communication Handbook* (with Brian Trench, forthcoming). His e-mail address is massimiano.bucchi@soc.unitn.it.

Regula Valérie Burri is an associate research fellow at Collegium Helveticum, Swiss Federal Institute of Technology (ETH) and University of Zurich. Her research interests include the social, cultural, and political implications of scientific knowledge and technological risks, especially in the fields of biomedicine and biosciences, nanotechnologies, and emerging technologies. She is the author of *Doing Images: Zur Praxis medizinischer Bilder* (forthcoming) and a coeditor of *Biomedicine as Culture: Instrumental Practices, Technoscientific Knowledge, and New Modes of Life* (with Joseph Dumit, 2007). Her e-mail address is burri@collegium.ethz.ch.

Michel Callon is Professor of Sociology at the Ecole des Mines de Paris and member of the Centre de Sociologie de l'Innovation. He is a former President of the Society for Social Studies of Science. One of the originators of Actor-network theory (ANT), he has recently applied that approach to study economic life (notably economic markets). His publications include *The Laws of the Markets* (1998), and "Some Elements of a Sociology of Translation: Domestication of the Scallops and the Fishermen of St. Brieuc Bay" (1986). His e-mail address is michel.callon@ensmp.fr.

Nancy Campbell is an associate professor in the Department of Science and Technology Studies at Rensselaer Polytechnic Institute in Troy, New York. She is a historian who studies interactions between science and policy, using basic and clinical research on drug addiction and drug policy as a case study. She also writes about feminist STS, surveillance studies, and science and social inequality. She is the author of *Discovering Addiction: The Science and Politics of Substance Abuse Research* (2007) and *Using Women: Gender, Drug Policy, and Social Justice* (2000). Her e-mail address is campbell@rpi.edu.

Adele E. Clarke is a professor of Sociology and History of Health Sciences at the University of California, San Francisco. Her books include *Disciplining Reproduction: American Life Scientists and* the "*Problem of Se*" (1998) and *Situational Analysis Grounded Theory After the Postmodern Turn* (2005). She coedited *The Right Tools for the Job: At Work in Twentieth Century Life Sciences* (1992; in French, 1996), *Revisioning Women, Health and Healing: Cultural, Feminist and Technoscience Perspectives* (1999), and *Biomedicalization: Technoscience and Transformations of Health and Illness in the U.S.* (forthcoming, 2008). Her e-mail address is Adele.Clarke@ucsf.edu.

Harry Collins is Distinguished Research Professor and Director of the Centre for the Study of Knowledge, Expertise and Science (KES) at Cardiff University. He was awarded the JD Bernal Prize in 1997. His thirteen books cover the sociology of scientific knowledge and the relationship between humans and machines and include the widely read *Golem* series on science, technology, and medicine. His current research covers the sociology of gravitational wave detection (www.cf.ac.uk/socsi/gravwave) and the nature of expertise (www.cf.ac.uk/socsi/expertise). Collins and Evans's: *Rethinking Expertise* was published in 2007, and further volumes on expertise and tacit knowledge are in press or in preparation. His e-mail address is collinsh@Cardiff.ac.uk.

Susan E. Cozzens is Professor of Public Policy, Director of the Technology Policy and Assessment Center (TPAC), and Associate Dean for Research and Faculty Development in the Ivan Allen College, Georgia Institute of Technology. She studies science and technology policies in developing countries and science, technology, and inequalities. Current projects include studies of sectoral innovation in water supply and sanitation and the distributional consequences of emerging technologies. She is formerly Director of the Office of Policy Support at the National Science Foundation. Her e-mail address is scozzens@gatech.edu.

Jennifer L. Croissant is Associate Professor in the Department of Women's Studies at the University of Arizona. She is a coauthor with Sal Restivo and Wenda Bauchspies of *Science, Technology, and Society: A Sociological Approach* (2005) and coeditor with Sal Restivo of *Degrees of Compromise: Industrial Interests and Academic Values* (2001). She has several collaborations in the works with Laurel Smith-Doerr on analysis of gender, science, and institutional change, as well as forthcoming works on vanilla production and pain measurement. Her e-mail address is jlc@email.arizona.edu.

Park Doing is Visiting Assistant Professor in the Department of Science and Technology Studies and the Bovay Program for History and Ethics in Engineering at Cornell University. He studies technical practice, decision-making, and the dynamics of expertise in scientific and engineering settings. He also writes about technical ethics as it pertains to social and environmental justice. His study of the power shift from physics to biology at a modern synchrotron laboratory, *Velvet Revolution at the Synchrotron: Physics, Biology, and Scientific Change* is forthcoming (2008). His e-mail address is PAD9@cornell.edu.

Joseph Dumit is Director of Science and Technology Studies and Associate Professor in the Department of Anthropology at the University of California, Davis. His research interests are in the anthropology of science, technology, and medicine, focusing on brains, drugs, movements and rationalities. He is the author of *Picturing Personhood* (2004) and the coeditor of *Cyborgs & Citadels* (with Gary L. Downey, 1997), *Cyborg Babies* (with Robbie Davis-Floyd, 1998), and *Biomedicine as Culture* (with Regula Valérie Burri, 2007). His e-mail address is dumit@ucdavis.edu.

Steven Epstein is Professor of Sociology at the University of California, San Diego, and Director of UCSD's Science Studies Program. His areas of interest lie at the intersection of studies of biomedicine, social movements, sexuality, gender, and race. His book *Impure Science: AIDS,*

Activism, and the Politics of Knowledge (1996), a study of the politicized production of knowledge in the AIDS epidemic, examines the roles of laypeople and activists in transforming medical research. Recently he published *Inclusion: The Politics of Difference in Medical Research* (2007), which charts the rise and assesses the consequences of new ways of managing difference (especially gender and race) within biomedical research in the United States. His e-mail address is sepstein@ucsd.edu.

Henry Etzkowitz holds the Chair in Management of Innovation, Creativity and Enterprise at Newcastle University Business School, where he directs the Triple Helix and Women in Science and Technology (WIST) research groups. He is also Visiting Professor in the Department of Technology and Society, School of Engineering and Applied Sciences, Stony Brook University. He is the author of *Triple Helix: University, Industry Government Interactions* (forthcoming) and *MIT and the Rise of Entrepreneurial Science* (2002) and coauthor of *Athena Unbound: The Advancement of Women in Science and Technology* (2000). He publishes regularly in such journals as *Research Policy, Science and Public Policy, R&D Management, Technology Analysis and Strategic Management,* and *Minerva.* His e-mail address is henry.etzkowitz@newcastle.ac.uk.

Robert Evans is a senior lecturer in sociology at the Cardiff School of Social Sciences. His research has applied STS to topics ranging from macroeconomic forecasting to stem cell research and participatory research methods. His current research interests focus on the nature of expertise and the role of qualitative research in public participation. He is the author of *Macroeconomic Forecasting: A Sociological Appraisal* (1999) and *Rethinking Expertise* (with Harry Collins, 2007). His e-mail address is EvansRJ1@Cardiff.ac.uk.

Erik Fisher is an assistant research professor in the Center for Nanotechnology in Society and the Center for Single Molecule Biophysics at Arizona State University. His research is on the governance of emerging technologies, sociotechnical integration at macro (policy discourse) and micro (laboratory practice) levels, and the philosophy of science policy. He has published on nanotechnology policy, self-critical science, and the midstream modulation of technological development. He is currently working, with Carl Mitcham, on a history of ethics policies and, with Arie Rip, on a philosophy of laboratory modulation. His e-mail address is fishere@cires.colorado.edu.

Jenny Fry is a lecturer in the Department of Information Science, Loughborough University. Her research is concerned with the disciplinary shaping of networked digital resources and scholarly communication on the Internet. Her publications in this area include "Scholarly Research and Information Practices: A Domain Analytic Approach," in *Information Processing and Management* (2006). She also writes about legal and ethical issues relating to the Internet and is the author of the editorial "Google's Privacy Responsibilities at Home and Abroad," in *Journal of Librarianship and Information Science* (2006). Her e-mail address is j.fry@lboro.ac.uk.

Stefan Fuchs is head of the Regional Research Network at the Institute for Employment Research (IAB) of the German Federal Employment Agency. His research is on the careers of women in science, and he is the author of *Gender, Science, and Scientific Organisations in Germany* (with Jutta Allmendinger and Janina von Stebut, 2001). With Sandra Hanson and Ivy Kennelly he is currently studying perceptions of fairness among scientists (*Journal of Women and Minorities in Science,* forthcoming). He also works on the careers of women in innovation and technology transfer in Europe. His e-mail address is stefan.fuchs@iab.de.

Sonia Gatchair is a doctoral student in the School of Public Policy at Georgia Institute of Technology. Her research interests include the use of technology and innovation strategies for economic development, labor market issues, and policy evaluation. She is currently examining the distributional consequences of high technology economic development policies in the United States, with an emphasis on racial and ethnic inequalities. She formerly worked in the public sector in Jamaica in science and technology policy and management. Her e-mail address is sgatchair@gatech.edu.

Ronald N. Giere is Professor of Philosophy Emeritus as well as a member and former Director of the Center for Philosophy of Science at the University of Minnesota. He is the author of *Understanding Scientific Reasoning* (5th ed., 2006), *Explaining Science: A Cognitive Approach* (1988), *Science Without Laws* (1999), *Scientific Perspectivism* (2006), and editor of *Cognitive Models of Science* (1992) and *Origins of Logical Empiricism* (1996). Professor Giere is a past president of the Philosophy of Science Association and a member of the editorial board of the journal *Philosophy of Science*. His current research focuses on agent-based accounts of models and scientific representation and on connections between naturalism and secularism. His e-mail address is giere@umn.edu and his homepage is: http://www.tc.umn.edu/~giere/.

Thomas F. Gieryn is Rudy Professor and Chair in the Department of Sociology at Indiana University, where he is also Adjunct Professor in the Department of History and Philosophy of Science. His research centers on how processes of "boundary-work" sustain or challenge the cultural authority of science in society. Gieryn's book *Cultural Boundaries of Science: Credibility On the Line* (1999) won the Merton Best Book Award from the Section on Science, Knowledge and Technology of the American Sociological Association. He is now at work on a new book provisionally titled *Truth-Spots: How Place Legitimates Claims and Substantiates Belief.* His e-mail address is gieryn@indiana.edu.

Namrata Gupta is a freelance researcher in Sociology of Science, with a focus on Gender and Academic Science and on Industrial Sociology. She is the coauthor of *Women Academic Scientists in India* (with Arun Sharma, 2002), *Industrial Workers and the Formation of "Working Class Consciousness" in India* (with Raka Sharan, 2004), *Triple Burden on Women in Science: A Cross-Cultural Analysis* (with Carol Kemelgor, Stefan Fuchs, and Henry Etzkowitz, 2005), and *Women Academic Scientists in India, Indian Women in Doctoral Education in Science and Engineering* (forthcoming, 2007). Her e-mail address is namrata432@rediffmail.com.

David H. Guston is Professor of Political Science at Arizona State University and Associate Director of its Consortium for Science, Policy, and Outcomes. He is also Principal Investigator and Director of the NSEC/Center for Nanotechnology in Society at ASU. His book *Between Politics and Science* (2000) won the 2002 Don K. Price award from the American Political Science Association. Professor Guston is the North American editor of the peer-reviewed journal *Science and Public Policy*, and he is a fellow of the American Association for the Advancement of Science. His e-mail address is david.guston@asu.edu.

Adam M. Hedgecoe is a senior lecturer in the Department of Sociology at the University of Sussex. His research focuses on the impact of genetic testing on medical practice, as well as wider debates about the new genetics. He also has a research interest in the sociology of bioethics. He is the author of *The Politics of Personalised Medicine: Pharmacogenetics in the Clinic* (2004) and is currently working on a four-country comparative ethnography of research ethics committees. His e-mail address is a.m.hedgecoe@sussex.ac.uk.

Iina Hellsten is a researcher at the Royal Dutch Academy of Arts and Sciences (KNAW). Her research is concerned with the politics of metaphors and hype in science communication on the Web. Her publications include "Public Communication of Genetics and Genomics: The Politics and Ethics of Metaphorical Framing," in the *Handbook of Public Communication of Science & Technology* (with Brigitte Nerlich, forthcoming, 2008). Her work also has been also published in *Science Communication, New Genetics and Society, Journal of Computer-Mediated Communication,* and *New Media & Society,* and other journals. Her e-mail address is iina.hellsten@vks.knaw.nl.

Christopher R. Henke is Assistant Professor in the Department of Sociology and Anthropology at Colgate University. His research explores the line between social order and social change in the historical and contemporary practice of science. More specific topics include agricultural science, expertise and environmental conflicts, and built environments and the role of "place" in science. He is currently conducting research on the debates surrounding the use of transgenic crops and is the author of *Cultivating Science, Harvesting Power: Science and Industry in California Agriculture* (forthcoming). His e-mail address is chenke@colgate.edu.

David Hess is a professor in the Science and Technology Studies Department at Rensselaer Polytechnic Institute. He studies the political, cultural, scientific, design, and technological aspects of social change movements (including religious, health, environmental, agro-food, and local/green business movements). His work has been published in over a dozen books and edited volumes and 40 peer-reviewed articles. His most recent book is *Alternative Pathways in Science and Industry: Activism, Innovation, and the Environment in an Era of Globalization* (2007). His e-mail address is hessd@rpi.edu.

Linda F. Hogle is Associate Professor of medical social sciences in the Department of Medical History and Bioethics at the University of Wisconsin-Madison, and Director of the Holtz Center for Science and Technology Studies. Her current research focuses on ways that interdisciplinary science transforms concepts and practices of engineering, medicine, and human biology. She is the author of *Recovering the Nation's Body: Cultural Memory, Medicine and the Politics of Redemption* (1999) and is currently writing a book tentatively titled *Regenerative Medicine and the Morphogenesis of Biology, Engineering and the Body.* Her e-mail address is lfhogle@wisc.edu.

Alan Irwin is a professor at Copenhagen Business School. He studies issues of scientific governance and science-public relations. He has an abiding interest in the relationship between technical expertise and democracy, especially as this is enacted within particular contexts and cultures. Professor Irwin is the author of *Risk and the Control of Technology* (1985), *Citizen Science* (1995), *Sociology and the Environment* (2001), and (with Mike Michael) *Science, Social Theory and Public Knowledge* (2003). With Brian Wynne, he was the editor of *Misunderstanding Science?* (1996). His e-mail address is ai.research@cbs.dk.

Sheila Jasanoff is Pforzheimer Professor of Science and Technology Studies at Harvard University's John F. Kennedy School of Government, where she directs the Program on Science, Technology and Society. Her research centers on the role of science and technology in democratic governance, with a particular focus on the production and use of knowledge and public reason in legal and political decision-making. Her books include *The Fifth Branch: Science Advisers as Policymakers* (1990), *Science at the Bar: Law Science and Technology in America* (1995), and *Designs on Nature: Science and Democracy in Europe and the United States* (2005). Her e-mail address is sheila_jasanoff@harvard.edu.

Deborah G. Johnson is the Anne Shirley Carter Olsson Professor of Applied Ethics and Chair of the Department of Science, Technology, and Society in the School of Engineering and Applied Sciences of the University of Virginia. Professor Johnson's work explores the relationship between technology and ethics, especially information technology. In pursuing this territory, she has published on a range of topics including privacy, computers as surrogate agents, the relationship between gender and technology, and engineering ethics. She recently completed an edited volume entitled *Technology & Society: Engineering our SocioTechnical Future* (with Jameson Wetmore, forthcoming, 2008). Her e-mail address is dgj7p@virginia.edu.

David Kaiser is an associate professor in the Program in Science, Technology, and Society at the Massachusetts Institute of Technology and a lecturer in MIT's Department of Physics. His historical research centers on the development of physics during the twentieth century, focusing in particular on pedagogical changes and their effects on the discipline. He is the author of *Drawing Theories Apart: The Dispersion of Feynman Diagrams in Postwar Physics* (2005) and editor of *Pedagogy and the Practice of Science: Historical and Contemporary Perspectives* (2005). His e-mail address is dikaiser@mit.edu.

William Keith is a professor in the Department of Communication at the University of Wisconsin-Milwaukee. He has written about conceptual and historical dimensions of public and scientific argumentation. He edited (with Alan Gross) *Rhetorical Hermeneutics, Invention and Discovery in the Age of Science* (1998). His current research explores public deliberation and civic communication. His latest book is *Democracy as Discussion: Civic Education and the American Forum Movement* (2007). His e-mail address is wmkeith@uwm.edu.

Carol Kemelgor is a psychotherapist and psychoanalyst in private practice in North Salem, New York. She is a member of the Psychoanalytic Association of the Westchester Center for the Study of Psychoanalysis and Psychotherapy. Her research on women in science and interpersonal relations in academic research laboratories has been illuminated by contemporary psychoanalytic theory. She is a coauthor of *Athena Unbound: The Advancement of Women in Science and Technology* (with Henry Etzkowitz and Brian Uzzi, 2000) and has been published in *Science, Minerva,* and *Science and Public Policy,* among others. Her e-mail address is ckemelgor@msn.com.

Kyung-Sup Kim is a practicing attorney in Seoul, South Korea, and formerly a postdoctoral fellow in the Technology Policy and Assessment Center at the Georgia Institute of Technology. His area of research interest is intellectual property law. He has a master's degree in Public Policy from the University of Chicago and a law degree from the University of Texas at Austin. His e-mail address is kskriskim@yahoo.com.

Andrew Lakoff is Associate Professor in the Department of Sociology and the Program in Science Studies at the University of California, San Diego. His research focuses on the interaction of the human and life sciences with broader political and social transformations. He is the author of *Pharmaceutical Reason: Knowledge and Value in Global Psychiatry* (2005) and coeditor (with Adriana Petryna and Arthur Kleinman) of *Global Pharmaceuticals: Ethics, Markets, Practices* (2006). His current research concerns the knowledge practices of security experts in the United States. His e-mail address is alakoff@ucsd.edu.

Bruno Latour is Professor and Dean for Research at Sciences Po, Paris. He has written many books and articles in the field of science, technology, and society. He has recently presented two international exhibitions, and the latest one, with Peter Weibel, *Making Things Public* is also presented in an MIT Press catalog by that title (2005), that bears on the issues of the representation of science, art, and politics. He is also interested in social theory and has recently published a book on actor-network theory called *Reassembling the Social* (2005). He can be contacted at: http://www.bruno-latour.fr.

Leah A. Lievrouw is a professor in the Department of Information Studies at UCLA. Her research and writing focus on the relationship between media and information technologies and social change, particularly the role of technologies in social differentiation and oppositional social and cultural movements, and intellectual freedom in pervasively mediated social settings. She is the author of over 40 journal and proceedings articles, book chapters, and other works related to new media and information society issues. Her current projects include *Media and Meaning: Understanding Communication, Technology, and Society* (forthcoming) and *Understanding Alternative and Activist New Media* (forthcoming). Her e-mail address is llievrou@ucla.edu.

Margaret Lock, an anthropologist at McGill University, is the Marjorie Bronfman Professor in Social Studies in Medicine, Emerita. She is a Fellow of the Royal Society of Canada and an Officier de L'Ordre national du Québec. She has authored or coedited 15 books and written over 180 articles. Her monographs, *Encounters with Aging: Mythologies of Menopause in Japan and North America* and *Twice Dead: Organ Transplants and the Reinvention of Death,* have won numerous awards. Professor Lock's current research is concerned with post-genomic knowledge and its circulation among basic scientists, clinicians, families, and society at large, with an emphasis on the genetics of Alzheimer's disease. Her e-mail address is margaret.lock@mcgill.ca.

Brian Martin is a professor in the School of Social Sciences, Media and Communication at the University of Wollongong. He studies dissent, nonviolence, scientific controversies, democracy, information issues, and strategies for social movements. His latest book is *Justice Ignited: The Dynamics of Backfire* (2007). His e-mail address is bmartin@uow.edu.au.

Paul Martin is Reader in Science and Technology Studies and Deputy Director of the Institute for Science and Society (formerly IGBiS) at the University of Nottingham. His research covers innovation in the biotechnology industry, the social and ethical issues raised by genetics, and the regulation of new medical technologies. He has a strong interest in the sociology of expectations and has recently published two major reports on the development of genomics: *Realising the Promise of Genomic Medicine* (Royal Pharmaceutical Society) and *False Positive? Prospects for the Clinical and Commercial Development of Pharmacogenetics* (University of Nottingham). His e-mail address is Paul.Martin@nottingham.ac.uk.

Philip Mirowski is Carl Koch Chair of Economics and the History and Philosophy of Science, and Fellow of the Reilly Center, University of Notre Dame. He is the author of *Machine Dreams* (2002), *The Effortless Economy of Science?* (2004), *More Heat Than Light* (1989), and the forthcoming *ScienceMart™: A Primer on the New Economics of Science*. His book *Effortless Economy* was awarded the Ludwik Fleck Prize from the Society for the Social Studies of Science in 2006. His e-mail address is mirowski.1@nd.edu.

Cyrus C. M. Mody is an assistant professor in the Department of History at Rice University. His work focuses on the history of (and occasionally policy for) nanotechnology and its antecedent fields, including scanning probe microscopy, molecular electronics, and microfabrication. He has published articles on historical and commercial aspects of nanotechnology in *Social Studies of Science, Technology and Culture* and other journals. His e-mail address is Cyrus.Mody@rice.edu.

Federico Neresini is Professor of Sociology of Science at the University of Padova, Italy. His main research interests are in the area of the sociology of science, in particular the public understanding of science. Recently he has focused on biotechnology issues, with specific attention to cloning and in vitro fertilization. His publications include the book *Sociologia della Salute* (*Sociology of Health: A Handbook*) (edited with M. Bucchi, 2001). His e-mail address is federico.neresini@unipd.it.

Gonzalo Ordóñez is a doctoral student in the Joint Doctoral Program in Public Policy, Georgia Institute of Technology and Georgia State University, and a faculty member at the Universidad

Externado in Bogotá, Colombia. His research interests focus on science and technology policy in developing countries, and his current research is on the impact of international collaboration on building research capacity, using Colombia as a case study. He is a former director of the Colombian Observatory of Science and Technology and staff member at COLCIENCIAS. His e-mail address is gonzalo.ordonez@pubpolicy.gatech.edu.

Nelly Oudshoorn is a professor at the Department of Science, Technology, Health, and Policy Studies at the University of Twente in The Netherlands. Her research interests and publications include the co-construction of users and technologies. She is the author of *Beyond the Natural Body: An Archeology of Sex Hormones* (1994), *The Male Pill: A Biography of a Technology in the Making* (2003), and coeditor of *Bodies of Technologies: Women's Involvement with Reproductive Medicine* (with Ann R. Saetnan and Marta S. M. Kirejczyk, 2000), and *How Users Matter: The Co-Construction of Users and Technologies* (with Trevor Pinch, 2003). Her e-mail address is n.e.j.oudshoorn@utwente.nl.

Trevor Pinch is Professor and Chair of Science and Technology Studies at Cornell University. He has published fourteen books and numerous articles on aspects of the sociology of science and technology. His studies include quantum physics, solar neutrinos, parapsychology, health economics, the bicycle, the car, and the electronic music synthesizer. His most recent books are *How Users Matter* (edited with Nelly Oudshoorn, 2003), *Analog Days: The Invention and Impact of the Moog Synthesizer* (with Frank Trocco, 2002), and *Dr. Golem: How to Think About Medicine* (with Harry Collins, 2005). *Analog Days* won the 2003 Book of the Year award from *Foreword Magazine*. His current research includes a study of the online music community ACIDplanet.com. His e-mail address is tjp2@cornell.edu.

Alex Preda is a reader in Sociology in the School of Social and Political Studies at the University of Edinburgh. His present research is on social interactions in electronic, anonymous financial markets, and on the technological design supporting such transactions. He is author of *AIDS, Rhetoric, and Medical Knowledge* (2005) and coeditor (with Karin Knorr Cetina) of *The Sociology of Financial Markets* (2005). His e-mail address is a.preda@ed.ac.uk.

Marina Ranga is a lecturer in Innovation Management in the Newcastle University Business School, UK. Her research is in policy analysis of national and regional innovation systems and strategic governance of R&D, Triple Helix interactions and impact of entrepreneurial activities on the performance of university research groups, and gender in science, technology transfer and incubation activities. She currently works with the European Commission and several UN agencies on topics related to these areas. Together with Henry Etzkowitz, she examined research policies in nine non-European countries in relation to the EU's 3% Action Plan and preconditions for implementing Triple Helix–based innovation strategies in transition countries. Her e-mail address is L.M.Ranga@ncl.ac.uk.

Brian Rappert is Associate Professor of Science, Technology and Public Affairs in the Department of Sociology and Philosophy at the University of Exeter (UK). His long-term interest has been the examination of how choices can be, and are, made about the adoption and regulation of security-related technologies, particularly in conditions of uncertainty and disagreement. His recent books include *Controlling the Weapons of War: Politics, Persuasion, and the Prohibition of Inhumanity* (2006), *Biotechnology, Security and the Search for Limits: An Inquiry into Research and Methods* (2007), *Technology & Security* (2007) and *A Web of Prevention* (2007). His e-mail address is b.rappert@ex.ac.uk.

Matt Ratto is a member of the Humlab and a visiting researcher in the History of Ideas Department at Umea University, Sweden. He is also a senior research fellow at the Center for

Science Technology and Society at Santa Clara University and a research affiliate with the Meta-media Collaboratory at Stanford University. His research examines how information technologies are involved in the processes by which cultures, societies, and institutions simultaneously make sense of the world and work to construct and organize it. His e-mail address is matt.ratto@gmail.com.

William Rehg is Associate Professor of Philosophy at Saint Louis University. His chief areas of interest include science and technology studies, moral-political theory, and argumentation theory. He is the author of *Insight and Solidarity: The Discourse Ethics of Jürgen Habermas* (1994), translator of Habermas's *Between Facts and Norms* (1996), and coeditor (with James Bohman) of *Deliberative Democracy* (1997) and *Pluralism and the Pragmatic Turn* (2001). His current writing projects focus on the topics of scientific argumentation and social solidarity. His e-mail address is rehgsp@slu.edu.

Cynthia Selin is an assistant research professor at the Center for Nanotechnology in Society at Arizona State University and at the Consortium for Science, Policy and Outcomes. Her research explores three interwoven areas: foresight methodologies, the sociology of expectations, and the societal aspects of nanotechnology. Her research seeks to understand the development of new technologies and investigate the methods used to grasp their emergence. She co-edited (with Jameson Wetmore and Erik Fisher) the first volume of the *Yearbook of Nanotechnology in Society*, entitled *Excavating Futures of Nanotechnology* (forthcoming, 2008). Her e-mail address is cynthia.selin@asu.edu.

Esther-Mirjam Sent is Professor of Economic Theory and Policy at the University of Nijmegen in the Netherlands. Her book *The Evolving Rationality of Rational Expectations: An Assessment of Thomas Sargent's Achievements* (1998) was awarded the 1999 Gunnar Myrdal Prize of the European Association for Evolutionary Political Economy. She is the editor (with Philip Mirowski) of *Science Bought and Sold: Essays in the Economics of Science* (2002). She is also the coeditor of the *Journal of Institutional Economics*. Her research interests include the history and philosophy of economics as well as the economics of science. Her e-mail address is e.m.sent@fm.ru.nl.

Steven Shapin is Franklin L. Ford Professor of the History of Science at Harvard University. He works at the intersection of the history and sociology of science. Some of his books include *A Social History of Truth* (1994) and *The Scientific Revolution* (1996). *Science as a Vocation: Scientific Authority and Personal Virtue in Late Modernity* will be published in 2008. His e-mail address is shapin@fas.harvard.edu.

Andrea Scharnhorst is a senior research fellow at the Virtual Knowledge Studio for the Humanities and Social Sciences at the Royal Netherlands Academy of Arts and Sciences in Amsterdam. Her expertise includes physics, scientometrics, evolutionary economics and STS. She has published widely on the application of self-organization and evolution models to the development of science, and on the transfer of concepts and methods from natural to social sciences. Her e-mail address is andrea.scharnhorst@vks.knaw.nl.

Sergio Sismondo teaches in the Departments of Philosophy and Sociology at Queen's University, Canada. He writes on general issues in STS and is the author of *An Introduction to Science and Technology Studies* (2004). He also is doing research on the political economy of pharmaceutical knowledge and has written articles on the pharmaceutical industry's work to publish clinical trial results to best advantage. A third direction of his research concerns deflationary philosophy of science, exploring the natures of truth embodied in ordinary scientific work. His e-mail address is sismondo@queensu.ca.

Laurel Smith-Doerr is a faculty member in the Department of Sociology at Boston University and Program Director in the Science Studies Program at the National Science Foundation. She studies tensions in the institutionalization of science, particularly the life sciences, including networks in the biotechnology industry, commercialization in the university, gendering in hierarchies, and scientists' resistance to ethics education. She is the author of *Women's Work: Gender Equality v. Hierarchy in the Life Sciences* (2004). Her e-mail address is Ldoerr@bu.edu.

Miriam Solomon is a professor in the Philosophy Department at Temple University. Her research is in the areas of social epistemology of science, gender and science, and epistemology of medicine. She is the author of *Social Empiricism* (2001). She is currently writing a book on new paradigms in medical epistemology. Her e-mail address is msolomon@temple.edu.

Susan Leigh Star is Professor at the Center for Science, Technology and Society at Santa Clara University. Her research is on heterogeneous knowledges in computer science, information systems, natural history, feminism, and brain research. Her books include *Regions of the Mind: Brain Research and the Quest for Scientific Certainty* (1989), *Sorting Things Out: Classification and Its Consequences* (with Geoffrey Bowker, 2000), and *Boundary Objects and the Poetics of Infrastructure* (in preparation). She is the current president of the Society for the Social Studies of Science (2005–07) and coeditor of *Science, Technology and Human Values*. Her e-mail address is slstar@scu.edu.

John Stone is a senior lecturer in the Department of War Studies at King's College London. He studies the theory and history of military strategy and is the author of *The Tank Debate: Armour and the Anglo-American Military Tradition* (2000). He is presently writing a book on the political and technological dimensions of strategy. His e-mail address is john.stone@kcl.ac.uk.

Lucy Suchman is a professor of Anthropology of Science and Technology in the Department of Sociology at Lancaster University and the co-director of Lancaster's Centre for Science Studies. She spent 20 years as a researcher at Xerox's Palo Alto Research Center (PARC) before taking up her present position. Her research investigates how human-machine relations are enacted in practice and with what theoretical, practical and political consequences. She is currently engaged in a critical study of "innovation" as a rhetorical invocation and as lived material practice, based on her years at Xerox PARC. She is the author of *Plans and Situated Actions* (1987), which has been revised and reissued under the title, *Human-Machine Reconfigurations* (2007). Her e-mail address is l.suchman@lancaster.ac.uk.

Anupit Supnithadnaporn is a doctoral student at the School of Public Policy, Georgia Institute of Technology. Her specialty is science and technology policy, including the issue of technology diffusion, the effect of foreign direct investment in developing countries on innovative capacity and knowledge spillover. Her minor is environmental policy, particularly air pollution and economic modeling. Before starting her Ph.D. study, she worked at the central government planning agency in Thailand as a planning and policy analyst. Her e-mail address is gtg065t@mail.gatech.edu.

Charles Thorpe is an associate professor in the Department of Sociology and a member of the Science Studies Program at the University of California, San Diego. He has previously taught at Cardiff University and University College London. He is interested in social and political theory, the nature of scientific power and authority, and problems of technology and the university in advanced capitalism. He is the author of *Oppenheimer: The Tragic Intellect* (2006). His e-mail address is c.thorpe@ucl.ac.uk.

Stephen Turner is Graduate Research Professor of Philosophy at the University of South Florida. He is the author of *The Social Theory of Practices* (1994), *Brains/Practices/Relativism* (2002), and

Liberal Democracy 3.0.: Civil Society in an Age of Experts (2003), as well as many articles on practice theory. His writings on science studies range from discussions of philosophical issues of explanation to studies of government geology funding and the Columbia Shuttle catastrophe. He has been an NEH Fellow, a Simon Honorary Visiting Professor at Manchester University, and a fellow of the Swedish Collegium for Advanced Studies in the Social Sciences. His e-mail address is turner@shell.cas.usf.edu.

Katie Vann is a social theorist and organizational ethnographer working with the Virtual Knowledge Studio and with the Center for Science, Technology and Society, Santa Clara. She researches the cultural and institutional technologies of neoliberal sociality and political-economic order, particularly in the context of knowledge work, and, more recently, water resources management. Some essays on social studies of labor appear in the journals *Ephemera: Theory and Politics in Organization* and *Social Epistemology*. An essay on labor in cyberinfrastructure projects appears in *New Infrastructures for Knowledge Production* (2006), and an essay on strategic labor politics appears in *Deleuzian Intersections in Science, Technology and Anthropology* (2007). Her e-mail address is katievann@hotmail.com.

Jameson M. Wetmore is an assistant professor at the Consortium for Science, Policy & Outcomes; the School of Human Evolution and Social Change; and the Center for Nanotechnology in Society at Arizona State University. His work examines the role of values and responsibility in the creation, maintenance, and dissolution of sociotechnical systems through topics such as automobile safety, nanotechnology research, and old order Amish communities. With Deborah Johnson he coedited a book entitled *Technology and Society: Building Our Sociotechnical Future*, to be published by MIT Press in 2008. He is Secretary of the Liberal Education Division of the American Society for Engineering Education. His e-mail address is jameson.wetmore@asu.edu.

Paul Wouters is Programme Leader of the Virtual Knowledge Studio (VKS) and of the Erasmus Virtual Knowledge Studio KNAW, a cooperative centre of the VKS and the Erasmus University Rotterdam. He has published on the history of information science and technology in research, in particular the history of citation analysis and scientometrics. He has coedited books about data sharing and the interaction between technology and society. He also published as a science journalist in a variety of newspapers and Web media. His present interests include emerging scholarly practices, especially in the humanities and social sciences. His e-mail address is paul.wouters@vks.knaw.nl.

Sally Wyatt is a senior research fellow with the Virtual Knowledge Studio for the Humanities and Social Sciences, Royal Netherlands Academy of Arts and Sciences. Her research focuses on the relationship between technological and social change, particularly on issues of social exclusion and inequality. She has been a president of the European Association for the Study of Science and Technology (2000–2004). Professor Wyatt has edited (with Flis Henwood, Nod Miller and Peter Senker) *Technology and In/equality: Questioning the Information Society* (2000). With Andrew Webster, she is a series editor of the series entitled *Health, Technology & Society*, published by Palgrave Macmillan. Her e-mail address is sally.wyatt@vks.knaw.nl.

Steven Yearley is Professor of the Sociology of Scientific Knowledge at the University of Edinburgh (Scotland) and also Director of the ESRC Genomics Policy and Research Forum. He has a long-standing interest in environmental sociology and science studies and has devised techniques for public engagement in environmental knowledge-making. Recently he has authored *Making Sense of Science* (2005) and *Cultures of Environmentalism* (2005) and (with colleague Steve Bruce) co-wrote *The SAGE Dictionary of Sociology* (2006). His e-mail address is steve.yearley@ed.ac.uk.

Name Index

Subject Index